Aquatic and Wetland Plants of Northeastern North America

Publication of this book was made possible, in part, through support from the following:

The Hanes Fund
Northeast Aquatic Plant Management Society (NEAPMS)
Northeast Chapter of Lake Management Society (NEC NALMS)
North America Lake Management Society (NALMS)
New York State Federation of Lake Associations
Pennsylvania Native Plant Society
Anonymous in Memoriam
Anonymous in Memoriam
Figure Foundation

Garrett E. Crow and C. Barre Hellquist

Aquatic and Wetland Plants of Northeastern North America

Second Edition

The University
of Wisconsin Press

The University of Wisconsin Press
728 State Street, Suite 443
Madison, Wisconsin 53706
uwpress.wisc.edu

Gray's Inn House, 127 Clerkenwell Road
London EC1R 5DB, United Kingdom
eurospanbookstore.com

Printed in the United States of America
This book may be available in a digital edition.

Library of Congress Cataloging-in-Publication Data
Names: Crow, Garrett E., author. | Hellquist, C. B., author.
Title: Aquatic and wetland plants of northeastern North America / Garrett E. Crow and C. Barre Hellquist.
Description: Second edition. | Madison, Wisconsin : The University of Wisconsin Press, 2023. | Includes bibliographical references and index.
Identifiers: LCCN 2022048799 | ISBN 9780299343002 (hardcover)
Subjects: LCSH: Aquatic plants—Northeastern States—Identification. | Wetland plants—Northeastern States—Identification. | Aquatic plants—Atlantic Provinces—Identification. | Wetland plants—Atlantic Provinces—Identification.
Classification: LCC QK117 .C84 2023 | DDC 581.7/607307—dc23/eng/20230418
LC record available at https://lccn.loc.gov/2022048799

In the first edition of this book both Barre and I dedicated our efforts in gratitude to our families, who grew up traipsing along with us on extended collecting field trips. This time it seems more appropriate for me to give a special dedication to the late Dr. Harold Z. Snyder, my professor and mentor at Taylor University who, as a founder and teacher at the Au Sable Biological Field Station in 1963 (later the Au Sable Institute), opened my eyes to the world of aquatic biology. It was Doc Snyder who introduced me to N. C. Fassett's *A Manual of Aquatic Plants*—the forerunner of this book. So, in a sense, this effort by Barre and me had its genesis while I was yet an undergraduate student at Taylor University. I would also like to dedicate this reference work to Au Sable Institute, which through the years has trained innumerable students in field biology and allowed me to serve as a professor at the Institute for several summers. It is the hands-on experiences in the field that made all the difference for me, and I see that as an outcome for all the field station students I have had the privilege to teach.

Garrett E. Crow

This second edition is dedicated to the two botanists responsible for my passion for aquatic plants, Dr. Albion Hodgdon and Dr. Eugene Ogden. While a student in the Department of Botany at the University of New Hampshire, I took an aquatic plant course with Dr. Hodgdon using Fassett's *A Manual of Aquatic Plants* as our text. Dr. Hodgdon's enthusiasm for aquatic botany captured my attention, and I asked Dr. Hodgdon if I could study the *Potamogeton* of New England as a Ph.D. student. Sadly, Dr. Hodgdon was in a traffic accident the next summer and died shortly after. In the meantime, I contacted Dr. Eugene Ogden, then the State Botanist for New York State, a leading expert in freshwater botany who specialized in the Potamogetonaceae. Dr. Ogden joined my graduate committee and generously shared his expertise with me. After I received my degree, my friendship with Dr. Ogden and his family continued until his death in his nineties. Dr. Ogden introduced me to an unusual *Potamogeton* in New York that I later named *Potamogeton* ×*ogdenii*, a rare hybrid between *P. hillii* and *P. zosteriformis*. I always felt privileged to have been mentored by these two professors, both former students of the late Merritt Lyndon Fernald of Harvard University. I also dedicate this to my wife, Marion, who died in August 2021. She constantly supported my work on the first edition of this book and encouraged and supported all my fieldwork over the years.

C. Barre Hellquist

Contents

Preface

Dr. Norman C. Fassett's *A Manual of Aquatic Plants* first appeared in 1940 and quickly became a classic in its field. Its success was, in large part, due to Dr. Fassett's aim to provide a manual which would make possible the identification of an aquatic plant in sterile as well as reproductive conditions. He made an all-out effort to construct his keys as simply as possible, coupled with flowering or fruiting characters where essential for proper identification. He copiously cross-referenced characters in his keys to labeled illustrations to facilitate the identification process further.

Shortly before his death in 1954, Dr. Fassett invited Dr. Eugene Ogden to prepare a revised edition. This effort, which appeared in 1957, resulted in reprinting the 1940 edition along with the addition of a 22-page appendix aimed at bringing the nomenclature of the original text up to date. It also provided a much-improved supplementary key to the genus *Potamogeton*, of which he was a leading expert. Both Drs. Fassett and Ogden readily acknowledged that the addition of an appendix, though extremely useful, fell far short of the need for a complete revision. Unhappily, the appendix went largely ignored by users, thus Fassett's *A Manual of Aquatic Plants* remained functionally the 1940 edition. Yet, this was the book we, as young aquatic botanists, "grew up" using and appreciating.

In setting out to write *Aquatic and Wetland Plants of Northeastern North America* (Crow and Hellquist, 2000), we addressed the badly needed updating while attempting to retain the features which made Fassett's work a classic. The region covered by our first edition expanded that of Fassett's and remains the same in this edition: southeastern Manitoba and Minnesota to Missouri, eastward to Newfoundland and Virginia (see Map 1). Fassett's treatment of 752 taxa (plus 95 named forms) within 61 families was greatly expanded in our first edition to include 1,186 taxa within 109 families. In this revised edition of *Aquatic and Wetland Plants of Northeastern North America* we have consulted the voluminous literature over the last twenty-plus years that has seen much input from molecular/DNA studies and phylogenetic updates realigning families and genera, bringing this work to our current understanding of the taxonomy of plants occupying aquatic and wetland habitats—covering 1,223 species (1,267 taxa) within 112 families.

Acknowledgments

Illustrations have come from a variety of sources: these sources are identified in parentheses in the captions, either by author or by an acronym we've assigned. A large number of illustrations have been used from Fassett's *A Manual of Aquatic Plants* (F), the forerunner of our first edition of *Aquatic and Wetland Plants of Northeastern North America*. Many drawings were taken from our Aquatic Vascular Plants of New England series, published as bulletins of the New Hampshire Agricultural Experiment Station (NHAES): those from Part 1 were drawn by C. Barre Hellquist, and those from Parts 2–8 were prepared by Ms. Pamela Bruns Brayton. Ms. Bruns Brayton (PB) also prepared additional illustrations specifically for our first edition of this reference. Ms. Tess Feltes (TF) prepared illustrations of *Eriophorum* specifically for our first edition; we have also utilized some of Ms. Feltes' illustrations which appeared in Crow's (1982) *New England's Rare, Threatened and Endangered Plants* (Crow). A few drawings were prepared specifically for this second edition by C. Barre Hellquist (CBH).

We gratefully acknowledge permission to use illustrations from the following publications: Helen I. Ashton (1977), *Aquatic Plants of Australia* (APOA); Ernest O. Beal (1977), *A Manual of Marsh and Aquatic Vascular Plants of North Carolina* (Beal); E. O. Beal and J. W. Thieret (1986), *Aquatic and Wetland Plants of Kentucky* (B&T); E. Lucy Braun (1961), *The Woody Plants of Ohio* (OHIO); E. Lucy Braun (1967), *The Vascular Flora of Ohio*, Vol. 1, *Monocotyledonae* (Braun); N. L. Britton and A. Brown (1896–1898), *An Illustrated Flora of the Northern United States, Canada, and the British Possessions* (B&B); L. G. Chafin, J. C. Putnam Hancock, H. O. Nourse, and C. Nourse (2007), *Field Guide to the Rare Plants of Georgia* (J. C. Putnam Hancock); Agnes Chase (1964), *First Book of Grasses* (Chase); Donovan S. Correll and Helen B. Correll (1972), *Aquatic and Wetland Plants of Southwestern United* States (C&C); I. D. Cowie, P. S. Short, and M. Osterkamp Madsen (2000), *Floodplain Flora: A Flora of the Coastal Floodplains of the Northern Territory, Australia* (FNTA); G. W. Douglas, D. V. Meidinger, and J. Pojar (1999), *Illustrated Flora of British Columbia. Volume 3: Dicotyledons (Diapensianceae through Onagraceae)* (Ill. Fl. B.C.); W. H. Fitch and G. W. Smith (1880), *Illustrations of the British flora: a series of wood engravings, with dissections of British Plants* (Fitch and Smith, 1880); Henry A. Gleason (1952), *The New Britton and Brown Illustrated Flora of the Northeastern United States and Adjacent Canada* (Gleason); J. G. Gmelin (1769), *Flora Sibirica: sive historia plantarum Siberiae*, Vol. 4 (Flora Sibirica); Robert K. Godfrey (1988), *Trees, Shrubs, and Woody Vines of Northern Florida and Adjacent Georgia and Alabama* (G); Robert K. Godfrey and Jean W. Wooten (1979, 1981), *Aquatic and Wetland Plants of Southeastern United States*, Vols. 1, 2 (G&W); A. Haines (2011), *Flora Novae Angliae: A manual for the identification of higher vascular plants of New England* (NPT); Frederick J. Hermann (1970), *Manual of the Carices of the Rocky Mountains and Colorado Basin* (RMC); A. S. Hitchcock (1950), *Manual of Grasses of the United States*, 2nd ed., revised by Agnes Chase (Hitchcock); C. L. Hitchcock, A. Cronquist, M. Ownbey, and J. W. Thompson (1955–1969), *Vascular Plants of the Pacific Northwest* (HCOT); Herman Kurz and Robert K. Godfrey (1962), *Trees of Northern Florida* (K&G); C. L. Lundell (1961–1969), *Flora of Texas* (Lundell); K. K. Mackenzie (1940), *North American Caricaea* (Mackenzie); Herbert L. Mason (1957), *A Flora of the Marshes of California* (Mason); G. C. Oeder (1767), *Flora Danica* (Oeder, in *Flora Danica*); C. D. Preston and J. M. Coft (1997), *Aquatic Plants in Britain and Ireland* (P&C); Clyde F. Reed (1970), *Selected Weeds of the United States* (Reed); O. W. Robuck (1985), *The Common Plants of Muskegs of Southeast Alaska* (USFS); G. A. Ryan (1978), *Native Trees and Shrubs of Newfoundland and Labrador* (Ryan); C. S. Sargent (1890–1902), *The Silva of North America* (Sargent); C. S. Sargent (1905), *Manual of the Trees of North America* (Sargent, 1905); P. H. Strausbaugh and E. L. Core (1952–1964), *Flora of West Virginia* (WVA); L. A. Viereck and E. L. Little, Jr. (1972), *Alaska Trees and Shrubs* (USDA); Zhengyi Wu (2022), *Flora of China Illustrations*, Vol. 8, Missouri Botanical Garden (Fl. China).

Permission graciously provided by the New York State Museum is acknowledged for use of illustrations published in several of their scientific bulletins (NYS Museum): Richard S. Mitchell and E. O. Beal (1979), *Magnoliaceae through Ceratophyllaceae of New York State*; Richard S. Mitchell and J. Kenneth Dean (1978), *Polygonaceae (Buckwheat Family) of New York State*; Richard S. Mitchell and J. Kenneth Dean (1982), *Ranunculaceae (Crowfoot Family) of New York State*; Richard S. Mitchell and Charles J. Sheviak (1982), *Rare Plants of New York State*; Charles J. Sheviak (1982), *Biosystematic Study of the Spiranthes cernua Complex*; and Eugene C. Ogden (1981), *Field Guide to Northeastern Ferns*.

We extend our thanks to the following persons and journals for permission to use illustrations from their publications (indicated by author/date): I. J. Bassett, C. W. Crompton, J. McNeill, and P. M. Taschereau (*Agriculture Canada Monographs*); O. Ceska, A. Ceska, and P. D. Warrington (*Brittonia*); A. J. Delahoussaye and J. W. Thieret (*Sida*); M. L. Fernald (*Rhodora*); H. Gil, K. Lee, Y. Ha, C. Jang, and D. K. Kim (*Korean Journal Plant Taxonomy*); L. K. Henry, W. E. Buker, and D. L. Pearth (*Castanea*); R. Kral (*Sida, J. Arnold Arboretum, Annals Missouri Bot. Garden*); C. T. Mason and H. H. Iltis (*Transactions of the Wisconsin Academy of Sciences, Arts, and Letters*); R. V. Lansdown (*Novon*); M. C. Pace and K. M. Cameron (*Systematic Botany*); P. J. Laureto and J. S. Pringle (*The Michigan Botanist*); N. H. Russell (*Sida*); J. G. Smith (Annual Report of Missouri Botanical Garden, published in Fassett's 1940 edition), and H. K. Svenson (*Rhodora*). The Field Museum of Natural History, Chicago, kindly permitted us to use the plates from E. E. Sherff (*The Genus* Bidens, Nat. Hist., Bot. Ser., 1937) (FIELD MUS).

Numerous taxonomic treatments have now been published in the volumes of *Flora North America North of Mexico*, most of which are available on the internet (http://beta.floranorthamerica.org/Main_Page). The Flora of North America Association has graciously given permission to utilize many illustrations published in various volumes. These are designated in captions as (FNA), artist and volume number. We herein indicate our thanks to the following artists who have prepared the illustrations used in FNA publications: Barbara Alongi (*Elatine triandra*, V. 12, p. 351; *E. rubella*, V. 12, p. 351). Tanya Harvey (*Utricularia vulgaris* subsp. *macrorhiza* bladder, V. 18, in press). Linny Heagy (*Drosera intermedia*, V. 6, p. 417). Amanda Humphrey (*Isolepis pseudosetacea*, V. 23, p. 139). John Myers (*Isoetes valida*, as *I. engelmannii*, V. 2, p. 71; *Isoetes virginica*, as *I. piedmontana*, V. 2, p. 71). Susan A. Reznicek (*Carex*, all in V. 23: *C. gynocrates*, p. 296; *C. recta*, p. 382, *C. muskingumensis*, p. 364; *C. albolutescens*, p. 364; *C. longii*, p. 367; *C. festucacea*, p. 370; *C. tribuloides* var. *tribuloides*, p. 364; *C. heleonastes*, p. 314; *C. seorsa*, p. 328; *C. scabrata*, p. 487; *C. pellita*, p. 496; *C. lonchocarpa*, p. 516; *C. prasina*, p. 464; *C. magellanica* subsp. *irrigua*, p. 418; *C. retrorsa*, p. 505; *C. grayi*, p. 513; *C. intumescens*, p. 513; *C. gigantea*, p. 513; *Schoenoplectus smithii* var. *smithii*, p. 55; *Schoenoplectus purshianus* var. *purshianus*, p. 55). Linda A. Vorobik (*Muhlenbergia glomerata*, V. 25, p. 155, copyright Utah State University, permission granted). Yevonn Wilson-Ramsey (*Utricularia ochroleuca*, V. 18, in press; *U. vulgaris* subsp. *macrorhiza*, V. 18, in press; *Glossostigma cleistanthum*, V. 17, online). Elizabeth Zimmerman (*Eleocharis mamillata*, V. 23, p. 73).

Websites have become wonderful resources for obtaining botanical information and images since our first edition appeared. The UF/IFAS Center for Aquatic and Invasives Plants at the University of Florida (UF/IFAS) granted permission for use of aquatic plant drawings from their database available on their website (https://plants.ifas.ufl.edu/); the Native Plant Trust (NPT) provides numerous drawings and photographs on their website, including illustrations from Haines' (2011) *Flora Novae Angliae* (available at: https://gobotany.nativeplanttrust.org/); The Michigan Flora Online database (https://michiganflora.net/) provides basic information from *Field Manual of Michigan Flora* (Voss and Reznicek, 2012) in a searchable/browsable form, with keys, maps, and numerous photographs. The Jepson e-Flora (Jepson Flora) kindly provided the illustration of *Elatine brachysperma* (https://ucjeps.berkeley.edu/eflora/). The PLANTS Database website hosted by the USDA Natural Resources Conservation Service (https://plants.sc.egov.usda.gov/home) has a wealth of information, including distribution synonyms, maps, illustrations—especially valuable online source for the public domain drawings from Britton and Brown (1896–1898)—as well as other illustrations, cited as (USDA-NRCS).

Numerous botanists graciously gave valuable assistance in reviewing various sections of the manuscript for the first edition. The keys were rigorously field tested by students taking Aquatic Plant classes from C. B. Hellquist (University of Michigan Biological Station), G. E. Crow (University of New Hampshire), and R. L. Stuckey (Ohio State University). We have also benefited from comments from many users of the first edition. For this second edition Donald H. Les contributed to the taxonomic treatment of the Hydrocharitaceae, as well as advised us on a number of aquatic monocot groups for which he conducted DNA studies. Peter Zika thoroughly reviewed the Juncaceae and provided suggestions for improving the keys to *Juncus*.

Introduction

The aim of this work is to aid in the identification of vascular plants which are native or have become naturalized and are now growing wild, occupying aquatic and wetland habitats of the Northeast. As a taxonomic/floristic work, it is designed to be of value to biologists, students of biology, conservationists, environmental consultants, personnel of local, state, and federal agencies, and any individuals with general ecological interests. Recognizing that the prospective users will vary greatly as to experience in plant identification, we have attempted to utilize less technical language wherever possible. Nevertheless, botanical terminology is a necessity, thus a glossary of plant terms as well as a glossary of habitat terms are included for clarification. Illustrations are also provided to aid the users; the 633 plates include figures of 1,107 taxa, with 87 percent of the 1,267 taxa covered in the book fully or partly illustrated. To facilitate the identification process further, references to the figures are frequently included directly in the keys to direct the user to drawings that will help in understanding the particular term or condition.

The geographical range covered by this work includes the region from Newfoundland west along the 50th parallel to southeastern Manitoba and Minnesota, south to Virginia and Missouri. Thus, along the southern edge, our range abuts the range covered by Drs. Robert K. Godfrey and Jean W. Wooten in their 2-volume manual, *Aquatic and Wetland Plants of the Southeastern United States.*

The coverage in this reference includes plants of wetland habitats as well as sites that might be considered "truly" aquatic habitats. Although we have included plants of bogs, fens, and salt marsh habitats, we have not made a specific effort to include plants which may occur in low woods along rivers and streams subject to frequent flooding, in woodland springs, or along vernal woodland pools or seasonal alpine brooks. Admittedly it is sometimes difficult to determine the outer boundary of a wetland habitat, therefore the decision to include or exclude certain species has been necessarily somewhat subjective. Our tendency has been toward inclusion rather than exclusion. A number of plants which are only occasionally found in wet sites may be mentioned in the text but omitted from the keys.

The keys in this revised edition treat a total of 1,223 species (1,267 taxa with subspecies/varieties included) representing 325 genera in 112 families of vascular plants. The families presented follow the phylogenetic classification system of the Pteridophyte Phylogeny Group (2016) for Pteridophytes, the phylogenetic system of Lu et al. (2014) for Gymnosperms (see also Gymnosperm Database, https://www.conifers.org/zz/gymnosperms.php), and for Angiosperms, the classification system of the Angiosperm Phylogeny Group (APG IV, 2016; see also Angiosperm Phylogeny Website. Version 14, July 2017 [and more or less continuously updated since], http://www.mobot.org/MOBOT/research/APweb/). Genera and species are then arranged in the order in which they key out. Infraspecific taxa considered important from an aquatic standpoint have been included in the keys. Because of the considerable variation in vegetative morphology exhibited by many aquatic plants, formal recognition of the nomenclatorial rank *forma* has not been applied here. All too often a single plant can change morphologically in response to habitat conditions, changing from one morphological form to another as the season progresses and/or water levels change in aquatic and wetland habitats. This morphological variability has, however, been taken into account in constructing the keys; thus, some species may key out in more than one place in the key.

The generic descriptions are intended to be diagnostic, providing additional information which may not be utilized in the construction of the keys but may be useful in the identification process. We have tried to include as many illustrations as possible to aid also in the process of identifying specimens, and we have especially attempted to provide figures that show the important key features. Because of spatial constraints, we have elected to include only selected family descriptions; they are readily available in numerous works.

We have attempted to utilize vegetative features and good "field characters" as much as possible to facilitate the identification of plants in vegetative condition; however, it must be recognized that for many species a reliable determination of the identity of a plant to the level of genus or species will require flowering and/or fruiting material. This is especially true of numerous families, especially the Cyperaceae, Juncaceae, Poaceae, Typhaceae, Orchidaceae, Asteraceae, Onagraceae, Haloragaceae, and many more. In general, it is much more difficult to identify aquatic and wetland plants early in the growing season. Ideally, in floristic inventory work one should be able to make observations and collect specimens over the entire growing season. Thus, one can observe the relationship of growth forms in aquatics such as *Sclerolepis uniflora* whereby the plants remain vegetative in the submersed aquatic form, but as the water level drops, the stems of the aquatic growth form become stranded along a newly exposed shore and then give rise to upright, seemingly unrelated fertile shoots. While the aquatic form of this species will key out in Key 3 as well as Key 7 of the General Keys, the fertile plant keys out in Key 8; keying out this species in the family key of the Asteraceae requires fertile material.

The geographical distributions and habitat information presented for the taxa in this work have been drawn widely from the published literature, such as regional floras and manuals, monographs and revisions, floristic papers, and from numerous taxonomic treatments prepared for the multivolume *Flora of North America North of Mexico*, from online e-floras and the USDA PLANTS website and online herbarium specimens that may document locations not cited in the literature. Likewise, we draw heavily from our own field experience. Generally, ranges are written from east to west, then southward to give an overview of the distribution in North America. For instance, a northern species may be encountered from Newfoundland westward to Alaska and southward to New England, northern New Jersey, Pennsylvania, northern Ohio, northern Indiana, northern Illinois, Minnesota, Saskatchewan, Montana, Idaho, and Washington. Sometimes a northern species may also extend southward along the Appalachians as far as North Carolina, or in the Rocky Mountains southward to Colorado and in the Cascades southward to northern California. A species which chiefly occurs on the Atlantic and Gulf coastal plain may range from southeastern Massachusetts on Cape Cod southward to Long Island, New York, southern New Jersey, Delaware, Maryland, Virginia, the Carolinas, Georgia, and Florida, then westward along the Gulf Coast to Alabama, Mississippi, Louisiana, and eastern Texas, and not infrequently extend northward from the Gulf along the Mississippi embayment as far as southwestern Illinois and southeastern Missouri, with some disjunct populations even as far north as west Michigan. Some coastal plain species also occur as disjunct populations in southern parts of Nova Scotia, as relics of exposed coastal plain migrations northward during the last Ice Age. Occurrence in the West Indies, Central America, and South America is noted. Taxa introduced into our flora from the Old World are so noted. However, we have not attempted to include adventive introductions of our own native species in the Old World.

As a guide to literature pertaining to aquatic and wetland taxa which might be helpful to the user of the manual, selected references have been provided. Full citations are found in the References section. In bringing nomenclature up to date we have attempted to provide a partial listing of some of the more significant synonyms to aid the user in comparing taxa treated here with those in other botanical references. Users desiring a more comprehensive compilation of synonyms we refer to the following valuable websites: Tropicos hosted by the Missouri Botanical Garden (http://legacy.tropicos.org/Home.aspx), the International Plant Name Index (IPNI) (https://www.ipni.org, a collaboration of Royal Botanic Gardens, Kew, Harvard University Herbaria, Australian National Herbarium), Plants of the World Online (Royal Botanic Gardens Kew), and the Database of Vascular Plants of Canada (VASCAN, https://data.canadensys.net/vascan/search). Abbreviations of authors of scientific names follow those of Tropicos and IPNI.

In 1988 the U.S. Fish and Wildlife Service (Reed, 1988) prepared a national list of plant species that occur in wetlands. Sometimes inventory work may involve an analysis of habitat by using wetness index values; these values are assigned based on the updated 2012 and 2016 National Wetland Plant List (Lichvar et al., 2012; U.S. Army Corps of Engineers, 2016, http://wetland-plants.usace.army.mil/nwpl_static/v33/home/home.html). These wetland indicators can vary from region to region; thus, we have elected not to repeat this information. But it can be readily accessed from the PLANTS database (USDA, NRCS website, http://plants.usda.gov), with information provided species by species for those considered obligate wetland species (OBL), facultative wetland species (FACW), or facultative species (FAC, equally likely to occur in wet or dry sites), but not for upland plant species.

Abbreviations

adv.	adventive
Ala.	Alabama
Alask.	Alaska
Alta.	Alberta
Am.	America, American
Ariz.	Arizona
Ark.	Arkansas
B.C.	British Columbia
Ber.	Bermuda
c.	central
C.Am.	Central America
Calif.	California
Can.	Canada
C.B.I.	Cape Breton Island, Nova Scotia
Colo.	Colorado
Conn.	Connecticut
D.C.	District of Columbia
Del.	Delaware
Delmarva Pen.	Delaware-Maryland-Virginia Peninsula
e.	east, eastern, eastward
Eur.	Europe
f.	forma
Fla.	Florida
Ga.	Georgia
Green.	Greenland
Guat.	Guatemala
I.	Island, Islands
Ida.	Idaho
Ill.	Illinois
Ind.	Indiana
intro.	introduced
Kans.	Kansas
Ky.	Kentucky
L.	Lake
La.	Louisiana
Lab.	Labrador
L.I., N.Y.	Long Island, New York
Man.	Manitoba
M.I.	Magdalen Islands, Quebec
Mass.	Massachusetts
Md.	Maryland
Me.	Maine
Mex.	Mexico
Mich.	Michigan
Minn.	Minnesota
Miss.	Mississippi
Mo.	Missouri
Mont.	Montana
mts.	mountains
n.	north, northern, northward
natzd.	naturalized
N.Am.	North America
N.B.	New Brunswick
N.C.	North Carolina
N.D.	North Dakota
N.E.	New England
ne.	northeast, northeastern
Neb.	Nebraska
Nev.	Nevada
Nfld.	Newfoundland
N.H.	New Hampshire
N.J.	New Jersey
N.M.	New Mexico
N.S.	Nova Scotia
nw.	northwest, northwestern
N.W.T.	Northwest Territories
N.Y.	New York
Okla.	Oklahoma
Ont.	Ontario
Oreg.	Oregon
Pa.	Pennsylvania
P.E.I.	Prince Edward Island
Pen.	Peninsula
Que.	Quebec
R.	River
R.I.	Rhode Island
s.	south, southern, southward
S.Am.	South America
St. P. et Miq.	St. Pierre et Miquelon Islands
Sask.	Saskatchewan
S.C.	South Carolina
sc.	southcentral
S.D.	South Dakota
subsp.	subspecies
sw.	southwest, southwestern
Tenn.	Tennessee
Tex.	Texas
Trop. Am.	Tropical America
Va.	Virginia
var.	variety
Vt.	Vermont
w.	west, western, westward
Wash.	Washington
W.I.	West Indies
W.Va.	West Virginia
Wisc.	Wisconsin
Wyo.	Wyoming
Yuk.	Yukon Territory

Map 1. Geographical area covered

Nuisance Aquatic Plants of the Northeast

Aquatic plants are becoming an increasingly common nuisance in lakes, ponds, and streams of the northeastern United States and southeastern Canada. Areas in the northern portion of this manual's range tend to be relatively free of some of the more severe weed problems affecting more southern areas. Yet, fishermen, boaters, swimmers, water-skiers, and lake-front property owners find aquatic weeds to be locally problematic when they increase at a great rate. In some cases, weeds eventually cover large areas of open water and hinder its recreational use. It should be noted that many plants that have the potential to be invasive are, unfortunately, sold for water gardens and aquarium use. A number of states now have laws to prevent sales of many aggressive aquatic plants within their states as well a prohibition of transporting invasive aquatics across their state borders. A good source for information is https://nationalplantboard.org/laws-and-regulations/state-regulated-noxious-weeds/.

Once aquatic weeds become well established, various costly methods may be employed in an attempt to eradicate or control the plants. Some of these methods include: 1) spraying with chemical herbicides, 2) mechanical harvesting of dense beds of plants, 3) dredging to deepen waters and installing benthic mats to inhibit aquatic plant growth, and 4) introducing a biological control organism, such as various natural predators from where the plants are native. Some of the natural predators used for biological control include parasitic fungi, aquatic insects that feed on plant tissues, nematodes that infect the root systems, and the introduction of sterile triploid amur (grass carp) to feed on aquatics. Control methods typically require application by licensed professionals and/or state permits.

Boaters and fishermen who move boats from one body of water to another are often vectors for the introduction of invasive plants into new sites, and weedy species most frequently appear for the first time in lakes near public access boat launch sites. Therefore, when a boat is removed from a lake, it is critical for all weeds to be safely discarded from inside the boat as well as cleaned from the propeller before moving the boat to another location; many states levy heavy fines for non-compliance.

Aquatic plants can easily become established in a region because most of them reproduce both vegetatively and by seed (Philbrick and Les, 1996). For some species every fragment can develop roots and form a new plant, thus rapidly building up a population. Many produce winter buds (turions) or tubers that serve as a means for vegetative propagation, dispersal, or overwintering without even forming seeds. A newly introduced invasive aquatic plant can produce numerous turions late in the season, setting the stage for greatly expanded populations as they emerge the following spring.

Although some of our native aquatic species can become problematic weeds, many of our nuisance aquatics have been introduced from foreign countries over time and have become naturalized, permanently established as components of the North American flora, and these often out-compete the indigenous species (Stuckey, 1993). In studying aquatic and wetland habitats for various management policies and strategies, correct identification of aquatic plants is very important. Although control of weeds is highly desirable, misidentifications of natives can lead to possible inadvertent eradication of important or rare indigenous species (Hellquist, 1993) and thus have a negative impact on the biodiversity of aquatic habitats of a region which has a much greater level of aquatic species richness than previously known (Crow, 1993).

In the Northeast the genus causing the greatest concern is *Myriophyllum*, the Water-milfoils. Four species are of particular concern: *M. spicatum*, *M. heterophyllum*, *M. quitense*, and *M. aquaticum*.

The dreaded European introduction, *Myriophyllum spicatum* (Eurasian Water-milfoil), has now spread throughout the United States. Although it has only become a species of concern since about 1970, it now is frequently *the* dominant weedy species in highly alkaline waters of our region. It has replaced much of our native Northern Water-milfoil, *Myriophyllum sibiricum* (*M. exalbescens*). Although typically a hard water species, it has also shown up in the acid and brackish waters of eastern New England, causing concern regarding its adaptability to a broader range of water alkalinity. It now occurs in acid waters with *M. heterophyllum*.

Myriophyllum heterophyllum (Variable Water-milfoil), a species native mainly in southern and midwestern states, is particularly problematic. This introduced species has become established in many of the acid lakes and ponds of New England where it is not native, especially in those of eastern Massachusetts, central New Hampshire, and southern and central Maine, as well as New Brunswick, Canada. This species typically occurs in waters of higher alkalinity in other portions of its native range, such as in Michigan and Oklahoma. In these areas where it is native it is often not weedy. *Myriophyllum heterophyllum* can form extensive stands, with individual plants becoming extremely robust. A study by Moody and Les (2002) indicated that both native (non-aggressive) and non-native (invasive) populations occur in New England. Les and Mehrhoff (1999) speculated that highly invasive populations in the Northeast resulted from populations spreading northward from a southerly native range; this was confirmed by Thrum et al. (2011) based on molecular genetics, indicating multiple introductions from native populations both from the southern and midwestern portions of the native range. The picture is further complicated by the presence in New England of hybrid populations of *M. heterophyllum* × *M. pinnatum*, which always exhibit invasiveness (Moody and Les, 2002).

In more southern areas of the U.S. *Myriophyllum aquaticum* (Parrot Feather) may be problematic. It is a South American species that can be obtained from suppliers of water garden plants (even on the internet) and is often used as an attractive shallow-water species for aquatic gardens. It readily escapes from cultivation and becomes weedy; for this reason, it is banned for sale in several states. The species often takes on a hardy terrestrial form when water levels become low, thus making it less susceptible to control by carrying out late season reservoir drawdowns. Presently it occurs as far north as Long Island, New York, Connecticut, southern Maine, Ohio, and Missouri, and might be expected to spread further northward, perhaps as a consequence of climate change, and especially to northeast coastal regions, where the maritime climate is less harsh in the winter.

Myriophyllum quitense (Andean Water-milfoil), a native of high elevations of the Andes Mountains from Tierra del Fuego northward to Venezuela, with disjunct populations in northwestern North America and in the Canadian Maritimes, is now considered to be native rather than introduced (Ceska et al., 1986; Moody and Les, 2010; Ritter and Crow, 1998). McAlpine et al. (2007) discuss the occurrence of *M. quitense* in southern New Brunswick and Prince Edward Island and note that it has become well established in the southern portion of the St. John River estuarine system which forms a border with the United States along much of the river upstream. It's just a matter of time until it is found on the U.S. side, so it would be prudent to look for it in eastern Maine. In the East *M. quitense* vegetatively most resembles *M. sibricum* but is easily identified when fertile due to the large bracts formed in the inflorescence.

The family Hydrocharitaceae has several species of concern. These include *Egeria densa*, *Hydrilla verticillata*, *Hydrocharis morsus-ranae*, *Najas minor*, and, most recently, *Stratiotes aloides*.

Egeria densa (Waterweed) is a tropical and subtropical plant which has now become established at a few sites in the Northeast. This species is known to be a problem in the states of Alabama, Georgia, Florida, Louisiana, North Carolina, South Carolina, and Oklahoma, and populations occur as far north as coastal northeastern Massachusetts and southern New Hampshire west to Ohio, Illinois, and Missouri. This aquatic plant is commonly sold in the aquarium trade and by biological supply companies and is often sold under the name Elodea; thus, it can be confused with our native *Elodea canadensis*. Another species, *Egeria najas*, is commonly marketed in the aquarium trade as Anacharis (an old generic name for *Elodea*). Vegetatively it often has the appearance of *Hydrilla* due to the leaf serrations. This species does not appear to have become established and naturalized.

Hydrilla verticillata (Hydrilla) is a common aquatic plant throughout Eurasia, Africa, and Australia, occurring as far north as central Siberia (personal observation by the authors). This species was introduced into the southern United States, southern California, and the American tropics. Hydrilla is considered by many wetland biologists as potentially the most dreaded species of aquatic weed in warm temperate areas, such as the southeastern United States. Presently it occurs in scattered sites as far north as Connecticut, Massachusetts, southern Maine, central New York, Indiana, and Iowa (Les et al., 1997). This species is similar to our native *Elodea* and is occasionally sold as *Elodea*. Every effort must be made to prevent the introduction and spread of this species into the northeast region. If Hydrilla were to become established, it has the potential to be one of the most difficult species to eradicate. And while seeds seldom mature in North America, plants are able to persist through winter by an abundance of tuber-like turions the plants produce.

Hydrocharis morsus-ranae (European Frogbit) was introduced into North America in the early 1930s near Ottawa, Ontario, Canada (Catling and Dore, 1982; Roberts, Stuckey, and Mitchell, 1981). Since then it has become locally established and is a nuisance species in southeastern Canada and northern New York, Lake Champlain, Vermont, central Maine, and southern Michigan. Frogbit is aggressive in its vegetative reproduction and overwinters by forming an abundance of winter buds. It has been theorized by some that the advent and spread of Frogbit in the Northeast probably arose from its introduction by sportsmen's clubs to provide food and cover for waterfowl. However, as with many aquatic plants, it may have inadvertently been introduced by people disposing of plants from summer water gardens into a local body of water.

Najas minor (Minor Naiad) was discovered in the Hudson River, New York, in 1934 (Merilainen, 1968). Since its discovery, this species has spread quickly westward, occurring widely throughout much of our range with the greatest concentrations in the Midwest. Only recently has it started moving eastward. It has now been found in alkaline waters in western Vermont, Massachusetts, and in southeastern

New Hampshire (Padgett and Crow, 1993a). Since it is a species of alkaline waters, and most of New England's waters are acidic, further spreading in the region is likely to be slow. Because this species is an annual, reproducing abundantly by seed, infestations often appear to develop later in the summer. During late summer it also fragments readily and may spread from one body of water to another. *Najas guadalupensis*, a native Naiad species in much of the United States, has also been spreading eastward and acting as an invasive/adventive plant in western New England, and is now found in all six New England states.

Stratiotes aloides (Water Soldier or Water Aloe), a hardy species from northern Europe, was discovered in 2008 in southern Ontario, occurring in the Trent Severin Waterway (near Trent, Ont.) and Black River (near Sutton, Ont.). It has the potential of spreading, especially by hitchhiking on motorboat propellers, and can readily spread by stolons locally and by the production of winter buds (Snyder, 2016). Now that it has become established in the Northeast, further spread of this species has the potential to cause a serious aquatic weed problem.

Cabomba caroliniana (Fanwort) is one of the most difficult weeds to eradicate in the Northeast. This species is native to the southern United States. However, it is a very popular aquarium plant and can be readily purchased at water garden stores and online, thus has considerable potential to escape and become established in more northern areas. *Cabomba* is hardy and extremely aggressive. Once established in a pond it is almost impossible to eradicate. It is resistant to various herbicides and is not a favored food of the amur (grass carp), a fish species often used to control a wide variety of aquatic weeds. This species is very common in acid-water lakes of eastern New England (Hellquist and Crow, 1986). This species is prohibited from import into Canada. Another species, *Cabomba furcata* (Forked Fanwort), native to Central and South America, is now found in southern Florida. This purple-flowered Fanwort is available for sale on the internet and may have the potential to become established further north.

Many species of *Potamogeton* (Pondweed) are native to northeastern North America. However, the non-native, and unwelcome, *Potamogeton crispus* (Curly-leaved Pondweed) has occurred as an aquatic weed in the United States for many years. Documentation in eastern New England goes back more than 170 years. Presently it is found in waters that are highly polluted or are extremely alkaline (Hellquist, 1980). *Potamogeton crispus* is found throughout the United States and southern Canada, even extending into Central America (Crow, 2003b). This species has an interesting life cycle, because, as an annual, it reaches its maximum development rather early in June in the Northeast. At this time fruits are produced, although fruiting is sparse. More importantly, by early July turions (hibernacula or winter buds) have formed, and they begin to dislodge from the plants as the main portion of the plant dies off. Surprisingly, by mid-August the turions start to develop and small plants appear, rooted in the substrate. These plants remain small, and the species overwinters in this form. In the following spring the maximum growth occurs, leading to the robust plants seen in June.

Trapa natans (Water-chestnut), native to Eurasia, is a rooted annual aquatic. Upon reaching the surface, plants rapidly develop floating leaves and spread out over a water surface by producing many floating rosettes; these can choke out large areas of open water. This species has been established in the Northeast for many years. Presently it is known in the Champlain Valley of Vermont and New York, the Mohawk and Hudson River systems of New York, throughout Massachusetts, the Connecticut River in Connecticut, and the Chesapeake Bay region (Trudeau, 1982). In 1998 it was documented for the first time in the Nashua River in southern New Hampshire (Crow specimen, University of New Hampshire Herbarium). The species was already known from the Nashua River within Massachusetts, but since the river flows north into New Hampshire, it was just a matter of time before populations would become established in the neighboring state. This species can occupy very large areas, but it is one that can be effectively controlled by persistent mechanical harvesting. Since the plant is an annual, the floating rosettes should be harvested early in the summer *before* the plants produce their large caltrop nuts. This species can rival the tropical *Eichhornia crassipes* (Water-hyacinth) in its ability to choke out a water surface.

One species of aquatic fern, *Marsilea quadrifolia* (European Water-clover), occasionally becomes aggressive. There is some question as to whether this Eurasian species might also be native in some areas of North America. It is extremely hardy when compared with the many tropical species, such as *Marsilea mutica* from Australia, that may be purchased through aquatic plant suppliers. Care should be taken not to introduce any species of *Marsilea* into our wetlands. Subtropical species planted in a shallow pond are known to grow during the summer at a phenomenal rate—up to a foot a week. *Marsilea quadrifolia* grows at a much slower rate, but still can become weedy.

Nymphoides peltata (Yellow Floating-heart) is another aggressive Eurasian aquatic that has not yet become abundantly established in the northeast region, but it has the potential to become a real nuisance. The earliest known herbarium specimen for this species was collected in 1882 from eastern Massachusetts (Stuckey, 1974). This species has the potential to colonize an area quickly. The plants produce many blossoms, which, in turn, produce numerous seeds. Plants also spread rapidly by producing fast-growing shallow rhizomes. Although typically a plant of warmer regions, it is hardy even in the Northeast. More concern should be paid to this species.

Callitriche stagnalis (Pond Water-starwort) is a Eurasian species found scattered throughout the Northeast. It has been established in Ossipee Lake, New Hampshire, for over 50 years, but has never become a nuisance. However, in a neighboring lake it recently has become very aggressive (Hellquist, pers. observ., 2020).

A common weedy marsh plant throughout the United States is *Lythrum salicaria* (Purple Loosestrife). This tall aggressive Eurasian marsh plant was reported adventive as early as 1824, growing in meadows in New England and Canada (Stuckey, 1980a). This species very likely escaped from flower gardens, because it has been grown for many years as an attractive ornamental. Purple Loosestrife remains available for sale on the internet, in spite of its aggressiveness. However, states have vigorously opposed planting this species. It was once believed that it was "safe" to plant the sterile hybrid, but even these hybrids have the ability to spread vigorously from rhizomes. Throughout the Northeast, Purple Loosestrife has colonized hundreds to thousands of acres of land, especially seasonally wet sites that tend to dry by late summer. The attractive magenta-purple flower produces a spectacular display during midsummer. Of greatest concern is the tendency of this species to crowd out and displace native vegetation.

Butomus umbellatus (Flowering-rush) has become well established along the St. Lawrence River drainage and the shores of Lake Champlain. It is also well established along marshes of the southern side of Lake Erie in Ohio and in Michigan. At these locations it has become a major component of the emerged shoreline and marsh vegetation. It has very attractive pink flowers and perhaps might be considered desirable, until one discovers it is an adventive plant which can become aggressive. Since about 1970 this species has continued spreading westward and may now be found at widely scattered locations across the northern United States and southern Canada. The plants of the St. Lawrence River region are more like those of Asia, which were treated by Russian botanists as a segregate species, *B. junceus* Turcz. (Fedchenko, 1934; Czerepanov, 1981). Whereas other populations appear to have been introduced from Europe, Gaskin et al. (2021) found that plants of the western North American invasion to be genetically distinct from those in eastern North America, with triploids dominant in the West, whereas diploids dominate in the East; they were also able to match several samples from western North American invasives to a single genotype from the Netherlands as well as a match from Hungary. It is noteworthy that western triploids, sexually sterile, reproduce primarily by rhizomes and fragmentation, whereas diploids tend to reproduce by seeds and bulbils that form in the inflorescence, as well as rhizomes (Lui et al., 2005; Thompson and Eckert, 2004).

Probably one of the most invasive wetland plants in the Northeast is *Phragmites australis* subsp. *australis* (Common Reed), a plant with a nearly worldwide distribution. This large grass (often up to 4 meters tall) is extremely aggressive and readily colonizes almost any area where drainage has been impeded. Roadside ditches are common places to observe this species. Probably the largest population of this species occurs in the northern New Jersey meadowlands just west of New York City. Eradication of this species should take place when new sites are discovered, because it will quickly out-compete the present vegetation. On the other hand, we have a native member of the genus, *Phragmites americana* (American Reed), which is not at all aggressive, and in some states is considered in need of conservation.

Although many of our aquatic weeds are aliens, there is a potential for many native aquatic plants in the Northeast to become aggressive if introduced into new area—beyond their natural range. They may also become more weedy if a local aquatic habitat becomes greatly disturbed. Such plants include Waterweed, *Elodea canadensis* and *E. nuttallii*, and various species of *Potamogeton*, for example, *Potamogeton illinoensis*, *P. natans*, and the various species of the narrow-leaved Potamogetons. Likewise, the White Waterlilies (*Nymphaea odorata* subsp. *odorata* and subsp. *tuberosa*) and the Yellow Waterlilies (*Nuphar*, especially *N. advena* and *N. variegata*) can all spread and become weedy. Pickerelweed (*Pontederia cordata*) and Wild-rice (*Zizania* spp.) often form near monocultures in shallow water. *Equisetum fluviatile* (Water Horsetail) readily invades and colonizes, often dominating, shallow waters. *Ceratophyllum demersum* (Coontail), a free-floating submersed species, also often chokes out the middle and upper layers of a body of water, greatly reducing light needed by submersed vegetation.

The various native cattails (*Typha*) can also become invasive in a wetland or along the margins of lakes and ponds. *Typha latifolia* and the hybrid *T.* ×*glauca* (the hybrid between native *T. latifolia* and non-native *T. angustifolia*) easily form large clones along the shore of waterways. They are often the first aquatic plant to invade newly created bodies of water. A potential problem on the increase is that water-garden nurseries throughout the country are offering exotic species of cattails for sale. These various species are all hardy and may lead to weed problems in the future. They include *Typha minima*, *T. laxmannii*, and *T. gracilis*. Care should be exercised not to introduce any wetland plants to a new body of water.

Care should also be taken in planting the showy Lotus species (*Nelumbo*), both our native Yellow Water Lotus (*Nelumbo lutea*) and the East Asian Indian Lotus (*N. nucifera*). There are large aggressive populations of both species in scattered localities in the East. They often are planted for horticultural purposes because of the large, showy colorful blossoms.

Three tropical and subtropical weedy species are Alligator Weed (*Alternanthera philoxeroides*), Water Lettuce (*Pistia stratiotes*), and Water-hyacinth (*Eichhornia crassipes*)

(Aurand, 1982). These species are presently reported in our range from southeastern Virginia, where they are most likely short-lived. The two latter species are readily available for purchase from garden centers promoting aquatic gardens. Although they are very problematic in the South, these should not become a problem in the Northeast.

There are various other exotic species that have essentially become naturalized in our flora and the potential to become a nuisance in some localities. Some of these are Moneywort (*Lysimachia nummularia*), Water Cress (*Nasturtium officinale* and *N. microphyllum*), Great Hairy Willowherb (*Epilobium hirsutum*), and Forget-me-not (*Myosotis scorpioides*).

A major source of introductions of non-native wetland plants occurs in wetland mitigation projects. According to the federal EPA, "Under Section 404 of the Clean Water Act, wetlands may be legally destroyed, but their loss must be compensated for by the restoration, creation, or enhancement of other wetlands. This strategy should result in 'no net loss' of wetlands" (https://www.epa.gov/wetlands/wetlands-restoration-definitions-and-distinctions). One major problem is that often the plants purchased for newly constructed wetlands are obtained from nurseries in the western or Plains states. This creates a philosophical problem of introducing new genetic strains, even of native plants, as well as introducing western species into the East (Padgett and Crow, 1993a, 1993b, 1994a, 1994b).

Presently, almost any species of aquatic plant can be purchased from some source in the United States. It is a mistake for anyone to purchase exotic plants and then plant or discard them into the wild where they may easily spread. Information on federal and state listed invasive species can be searched for on the USDA PLANTS database (at https://plants.usda.gov/home). The International Waterlily and Water Gardening Society website also provides valuable information (https://iwgs.org/invasive-species/regulated-and-prohibited-aquatic-plants-usa/). The USGS Nonindigenous Aquatic Species website provides species lists of non-native aquatic plants (https://nas.er.usgs.gov/taxgroup/Plants/default.aspx).

General Keys to Families

1. Plants woody, trees, shrubs, or subshrubs.. . . . . Key 1, xxiv
1. Plants herbaceous.
  2. Plants free-floating on the surface (fig. 2) (in *Lemna trisulca* floating just below the surface, lacking distinct stems and leaves, fig. 58f,g). . . . . Key 2, xxvii
  2. Plants submersed, emersed, terrestrial, or with floating leaves, usually rooted (fragmented specimens often free-floating).
    3. Stems and leaves, or their divisions, limp, thread-like (fig. 3), or ribbon-like (fig. 5), more than 10 times as long as wide (figs. 3, 4, 5, 6). . . . . Key 3, xxix
    3. Stems and leaves various, but not limp and thread-like or ribbon-like.
      4. Leaves compound, cut into several flat leaflets (fig. 7). . . . . Key 4, xxxv
      4. Leaves simple (sometimes pinnate-pinnatifid, fig. 37b); or stem naked or appearing naked (figs. 10a, 13).
        5. Leaves deeply lobed (figs. 8b,d,e,g), or blade with basal lobes (fig. 9e–k), or leaves peltate, petiole attached near center of blade (fig. 9a).
          6. Leaves deeply lobed, or basal lobes extending laterally or forward (fig. 9g). . . . . Key 5, xxxvii
          6. Leaves with basal lobes extending below junction of blade and petiole (fig. 9e–k), or leaves peltate (fig. 9a). . . . . Key 6, xxxix
        5. Leaves unlobed to shallowly lobed (leaf bases sometimes cordate, but not with distinct basal lobes), never peltate; or stems naked or appearing naked.
          7. Stems conspicuously jointed, with small, tooth-like leaves united in a sheath at each joint (figs. 22, 23, 24). . . . . Equisetaceae, 7
          7. Stems not jointed or having tooth-like leaves united in a sheath.
            8. Leaves basal or cauline, more than 10 times as long as wide; or stems naked or appearing naked, with leaves reduced to bladeless sheaths. . . . . Key 7, xlii
            8. Leaves basal or cauline, less than 10 times as long as wide. . . . . Key 8, l

Key 1 Plants woody, trees, shrubs, or subshrubs

1. Leaves scale-like (fig. 1e) or subulate.

2. Leaves scale-like, opposite; leafy twigs somewhat to strongly flattened (fig. 1e)…… Cupressaceae, 30

2. Leaves subulate, alternate; leafy twigs not flattened (*Taxodium ascendens*)…… Cupressaceae, 30

1. Leaves needle-like (fig. 1b,c) or flat and narrow to broad (fig. 1a,f,h).

3. Leaves needle-like.

4. Trees; branches bearing cones, with seeds borne on cone scales, true flowers not produced.

5. Leaves usually in clusters on spur shoots (fig. 1b), scattered on new growth of stems (*Larix*)…… Pinaceae, 30

5. Leaves scattered on the twigs (fig. 1c).

6. Leaves flattened, flexible, 2-ranked (fig. 1c); stems somewhat smooth; cone scales peltate (*Taxodium distichum*)…… Cupressaceae, 30

6. Leaves mostly 4-sided, stiff, spirally arranged; stems roughened with persistent peg-like petioles; cone scales attached at base (*Picea*)…… Pinaceae, 30

4. Shrubs, low, spreading, much branched; branches bearing true flowers, with seeds enclosed in black, fleshy fruits (*Empetrum*)…… Ericaceae, 617

3. Leaves flat and broad, or if narrow, then not needle-like.

7. Leaves and buds fragrant, aromatic when crushed.

8. Leaves and young stems resin-dotted; flowers borne in catkins on last year's stems; perianth absent, bracts subtending flowers brown (*Myrica*)…… Myricaceae, 517

8. Leaves and young stems lacking resin dots; flowers borne in clusters from buds on last year's stems; perianth yellow (*Lindera*)…… Lauraceae, 50

7. Leaves and buds not aromatic.

9. Leaves pinnately compound (fig. 1f).

10. Leaves opposite (fig. 632e).

11. Stem pith large; lenticels large, conspicuously raised; flowers bisexual, with white petals; fruit a berry-like drupe (*Sambucus*)…… Viburnaceae, 816

11. Stem pith small; lenticels small, not conspicuously raised; flowers unisexual, lacking petals; fruit a samara.

12. Leaflets mostly 3–5, lobed or coarsely serrate; twigs glaucous (*Acer negundo*)…… Sapindaceae, 549

12. Leaflets mostly 5–11, entire or serrate; twigs not glaucous (*Fraxinus*)…… Oleaceae, 652

10. Leaves alternate (fig. 390a,c).

13. Stipules present, adnate to lower portion of petiole (figs. 7g, 390c); flowers pink or yellow. solitary; non-poisonous.

14. Stems prickly; flowers pink; fruit a cluster of achenes enclosed in a red hip (*Rosa*)…… Rosaceae, 493

14. Stems not prickly; flowers yellow; fruit a cluster of achenes, but not enclosed in a hip (*Dasiphora fruticosa*)…… Rosaceae, 493

13. Stipules absent; flowers cream-colored, in somewhat pendulous axillary panicles (fig. 432g); poisonous, causing severe dermatitis (*Toxicodendron*)…… Anacardiaceae, 549

9. Leaves simple.

15. Leaves opposite or whorled.

16. Leaves and young stems succulent; plants chiefly of coastal salt marshes.

17. Leaf margins toothed; head of flowers consisting of only disk flowers (*Iva*)…… Asteraceae, 745

17. Leaf margins entire; head of flowers with both disk and ray flowers (*Borrichia*)…… Asteraceae, 745

16. Leaves and young stems not succulent; plants of freshwater sites.

18. Leaves lobed (*Acer*)…… Sapindaceae, 549

18. Leaves not lobed.

19. Leaves with translucent dots (visible when held to light) (*Hypericum*)…… Hypericaceae, 460

19. Leaves lacking translucent dots.

20. Leaves with lateral veins curving somewhat parallel to margin, running toward and ending near apex (fig. 472a,d) (*Cornus*)…… Cornaceae, 602

20. Leaves with lateral veins running toward and ending at or near margin.

21. Leaves evergreen, somewhat leathery. narrowly elliptic; flowers with stamens bent over and anthers held in pouches in corolla; fruit a capsule (*Kalmia*)…… Ericaceae, 617

21. Leaves deciduous, not leathery, shape variable, but not narrowly elliptic; flowers with stamens not bent and anthers not held in pockets; fruit a drupe or a nutlet.

22. Plants flowering before leaves emerge; fruits below leaves on stems of previous year's growth; flowers lacking corolla; leaves long-tapering at both ends (*Forestiera*)…… Oleaceae, 652

22. Plants flowering after leaves emerge; fruits on stems of new growth; flowers with corolla; leaves various, but not long-tapering at both ends.

23. Leaves often in whorls of 3 or 4 (some leaves opposite in *Cephalanthus*, even on same branch); flowers and fruits in dense globose heads or in axillary clusters; fruit dry, a nutlet or capsule.

24. Stems ascending, distinctly woody, lacking spongy tissue; flowers and fruits in globose heads; flowers white (*Cephalanthus*)…… Rubiaceae, 629

24. Stems strongly arching, frequently touching the water herbaceous on upper portion, but woody toward base, lower portion of stem surrounded by spongy tissue; flowers and fruits in axillary clusters; flowers magenta (*Decodon*)........ .... Lythraceae, 525

23. Leaves strictly opposite; flowers and fruits in axillary pairs or in much-branched cymes; fruit succulent, a berry or drupe (*Lonicera*)...................... Caprifoliaceae, 812

15. Leaves alternate.

25. Plants viny, often climbing, becoming woody toward base, herbaceous upward to tip (*Solanum dulcamara*). .... Solanaceae, 645

25. Plants trees, shrubs, or subshrubs.

26. Leaves lobed.

27. Leaf margins entire, blades with distinct midvein and branching pinnate veins; late summer twigs with clustered bud at tip; fruit an acorn (*Quercus*, fig. 399). .... Fagaceae, 509

27. Leaf margins toothed; blades with primary veins 3, palmate; late summer twigs lacking clustered buds; fruit a cluster of 3–5 follicles or a globose head of achenes.

28. Trees; leaf blades coarsely, irregularly toothed; axillary bud completely hidden beneath base of petiole; flowers unisexual; flowers and fruits borne in pendent globose heads; fruit achenes; bark of older trunks sloughing in thin plates, giving mottled appearance (*Platanus*, fig. 353)...... Platanaceae, 443

28. Shrubs; leaf blades finely, regularly toothed; axillary bud not hidden; flowers bisexual; flowers and fruits borne in a corymb; fruit follicles; bark of older stems peeling in strips (*Physocarpus*, fig. 388e). .... Rosaceae, 493

26. Leaves not lobed.

29. Leaf margins entire or very minutely serrulate or crenulate so as to appear entire.

30. Trees; stem pith diaphragmed (with transverse septae) in longitudinal section.

31. Leaves leathery, with stipules, somewhat evergreen (especially southward); flowers very large, showy, petals 6–9; fruit an aggregate of follicles (*Magnolia*)......................... Magnoliaceae, 50

31. Leaves not leathery, lacking stipules, deciduous; flowers small, petals 5 or often absent; fruit a drupe (*Nyssa*)... Nyssaceae, 602

30. Trees, shrubs, or subshrubs; stem pith continuous (not diaphragmed) in longitudinal section.

32. Leaves mucronate at apex (margins sometimes very minutely toothed, yet appearing entire), glabrous; corolla of separate petals; fruit a red drupe, rarely yellow, with 4 or 5 seed-like nutlets (*Ilex mucronata*). .... Aquifoliaceae, 733

32. Leaves acute to obtuse or rounded at apex, but not mucronate, either glabrous or pubescent; corolla united (if petals separate, then densely tomentose beneath, *Rhododendron groenlandicum*); fruit a capsule, a red, blue to black, many-seeded berry or a black drupe with 10 seed-like nutlets. .... Ericaceae, 617

29. Leaf margins serrate, dentate, or crenate.

33. Leaves at least 3 times as long as wide; lateral buds covered by a single bud scale (*Salix*). .... Salicaceae, 476

33. Leaves less than twice as long as wide; lateral buds covered by 2 or more bud scales.

34. Flowers greatly reduced, borne in catkins (fig. 1i,j).

35. Plants monoecious; fruit a nutlet; pistillate catkins persisting; trees or shrubs...... Betulaceae, 509

35. Plants dioecious; fruit a capsule; pistillate catkins falling off early; trees (*Populus*)........ .... Salicaceae, 476

34. Flowers not reduced, borne variously but not in catkins.

36. Leaf bases rounded, usually oblique; twigs usually zig-zag.

37. Leaf margins double-serrate (with some teeth lacking a vein running to margin); fruit a bur-like nut or a samara; trunk scaly (becoming shreddy in *Planera*), but not warty (*Planera*, *Ulmus*). .... Ulmaceae, 502

37. Leaf margins serrate, but not double-serrate; fruit a drupe; trunk typically warty (*Celtis occidentalis*). .... Cannabaceae, 505

36. Leaf bases tapering, acute, or if occasionally rounded, then not oblique; twigs not zig-zag.

38. Flowers with hypanthium present (free or adnate to ovary); stamens 15 or more; fruit a cluster of follicles (*Spiraea*) or a pome (*Aronia*)............................. Rosaceae, 493

38. Flowers lacking hypanthium; stamens 10 or fewer; fruit a capsule, berry, or drupe.

39. Twigs with dense stellate pubescence; fruit a 3-valved capsule (*Clethra*)........ .... Clethraceae, 617

39. Twigs glabrous or pubescent, but hairs not stellate; fruit a 5-valved capsule, berry, or drupe.

40. Corolla united, urceolate; fruit a capsule or a many-seeded berry (*Eubotrys*, *Lyonia*, *Vaccinium*)... Ericaceae, 617

40. Corolla only slightly united at base (petals appearing separate) or absent; fruit a 3–10-seeded drupe.

41. Corolla present; drupes with 5–10 seeds (stones), reddish, or if black, then leaves evergreen and leathery (*Ilex*).................. Aquifoliaceae, 733

41. Corolla absent; drupes with 3 seeds (stones), black (*Rhamnus*). .... Rhamnaceae, 502

Fig. 1. a. *Nyssa sylvatica*, portion of branch (F); b. *Larix laricina*, portion of branch (Sargent); c. *Taxodium distichum*, branchlet (deciduous as a unit) (F); d. *Cephalanthus*, opposite leaves (whorled not shown) (F); e. *Thuja occidentalis*, portion of branch (Sargent); f. *Fraxinus*, compound leaf (F); g. *Forestiera*, opposite leaves (F); h. *Alnus incana*, leaf (F); i. *Alnus*, staminate catkins (F); j. *Alnus*, pistillate catkins (F).

Key 2 Plants free-floating on or just below the surface

1. Plants with whorls of inflated leaf bases.
   2. Whorled leaf bases floating on the surface, subtending flower scapes (fig. 2i); carnivorous traps consisting of bladders occurring on portion of plant extending submersed below floating rosette (*Utricularia inflata*, *U. radiata*). Lentibulariaceae, 683
   2. Whorled leaf bases along the stem, submersed just below the surface, carnivorous traps consisting of a terminal halfmoon-shaped snap-trap at tip of inflated portion, subtended by 5–6 bristles (fig. 446a,b), bladders absent (*Aldrovanda*). . . . . . . Droseraceae, 568
1. Plants lacking whorls of inflated leaf bases (petioles appearing inflated in *Eichhornia*, fig. 2b, and *Trapa*, fig. 2c); carnivorous traps absent.
   3. Plants dichotomously forked or lobed.
      4. Plants with overlapping leaves, upper surface velvety (fig. 28a,f) (*Azolla*). . . . . .Salvinianceae, 13
      4. Plants lacking leaves, thalloid, upper surface glabrous (*Riccia*, fig. 2g, *Ricciocarpus*, fig. 2h, non-vascular plants, a bryophyte family of liverworts).. . . . . Phylum Marchantiophyta, Ricciaceae—not covered in book.
   3. Plants not dichotomously forked.
      5. Plants not distinctly differentiated into stems and leaves, small to minute (figs. 2d,e,f,k, 58f,g) (*Lemna*).. . . . . Araceae, 53
      5. Plants differentiated into stems and leaves, small or larger.
         6. Plants small, not forming rosettes; leaves lacking petioles, 0.5–15 mm long; plants reproducing by spores borne in sporocarps.
            7. Leaves appearing opposite (actually whorled, 2 leaves floating, 1 submersed), not overlapping, upper surface covered with erect forked hairs (fig. 31g) (*Salvinia*). . . . . .Salviniaceae, 13
            7. Leaves alternate, overlapping, upper surface velvety (fig. 28a) (*Azolla*). . . . . .Salviniaceae, 13
         6. Plants larger, forming rosettes; leaves petiolate (fig. 2a,b,c,j), blades more than 20 mm long; plants reproducing by flowers and seeds (detached plants of *Nymphoides*, fig. 2a, and *Trapa*, fig. 2c, normally rooted, might key out here).
            8. Leaves with petioles appearing greatly inflated (fig. 2b), blades not spongy; flowers showy, bisexual (*Eichhornia*). . . . . Pontederiaceae, 413
            8. Leaves with petioles not inflated, but blades spongy (fig. 2j); flowers not showy, unisexual (*Limnobium*, *Hydrocharis*). . . . . Hydrocharitaceae, 62

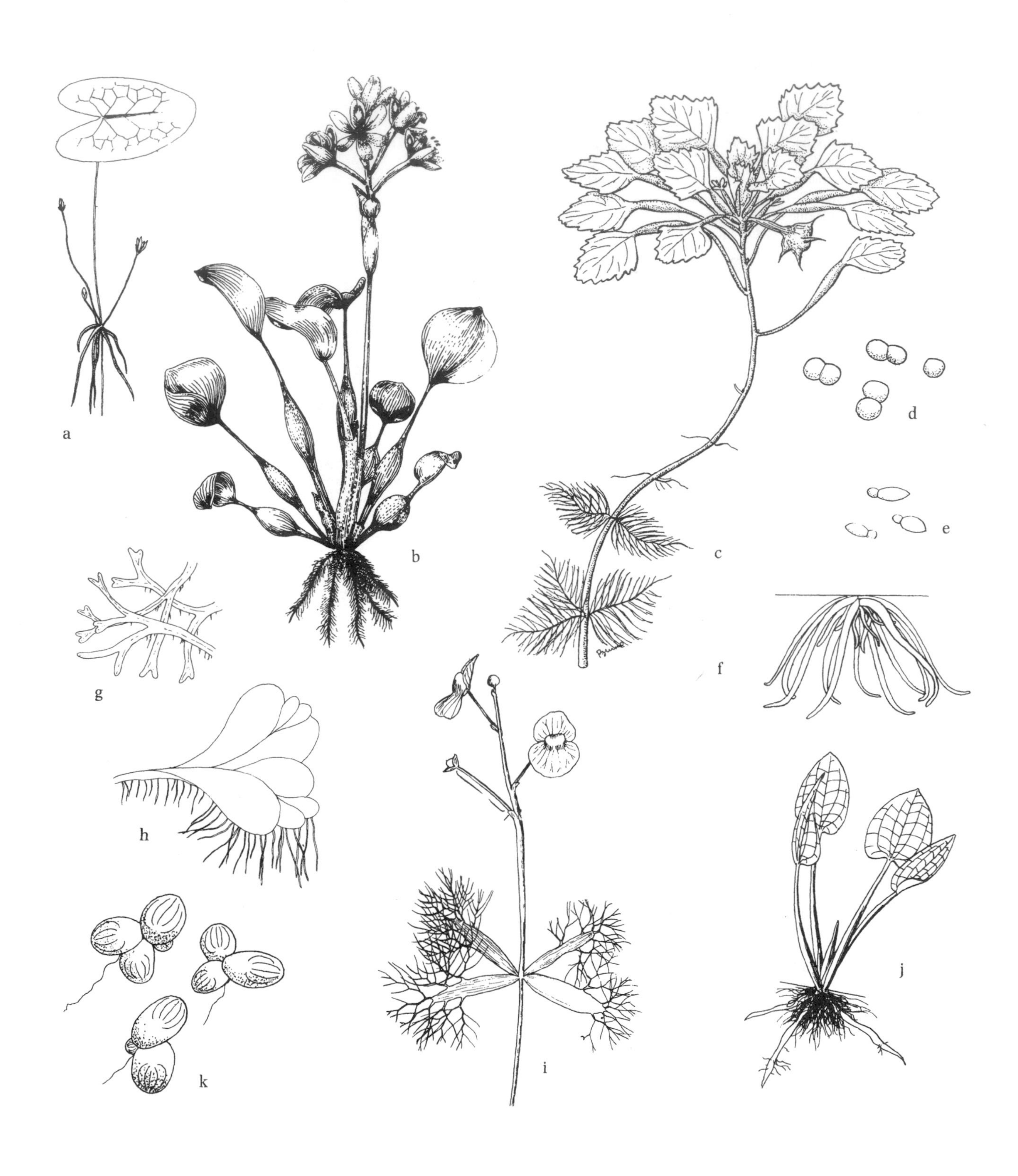

Fig. 2. a. *Nymphoides*, upper portion of plant (F); b. *Eichhornia crassipes*, habit (Beal); c. *Trapa*, upper portion of plant (NHAES); d, e. *Wolffia*, habit (F); f. *Wolffiella*, habit (F); g. *Riccia*, habit (F); h. *Ricciocarpus*, habit (F); i. *Utricularia radiata*, upper portion of plant (F); j. *Limnobium*, floating plant (F); k. *Lemna*, habit (F).

Key 3 Stems and leaves, or their divisions, limp, thread-like or ribbon-like, more than 10 times as long as wide

1. Leaves (or leaf-like branches) scattered along stem (fig. 3a,c).
    2. Leaves flat, ribbon-like, undivided.
        3. Leaves opposite (fig. 16b), whorled (fig. 16c), or subopposite, with fascicles, thus appearing whorled.
            4. Leaves whorled, or subopposite, with fascicles (appearing whorled).
                5. Leaves distinctly whorled; leaves not sheathing.
                    6. Leaves 3–6 per whorl, longer leaves usually 1–2.5 cm long; stems solid; flowers (on emersed plants) in terminal heads (fig. 606a) (*Sclerolepis*). . . . . Asteraceae, 745
                    6. Leaves 6–12 per whorl, longer leaves usually 2–4 cm long; stems hollow; flowers (on emersed plants) in axillary whorls (fig. 509c) (*Hippuris*). . . . . Plantaginaceae, 655
                5. Leaves subopposite, with fascicles, appearing whorled; leaves with basal sheaths (*Juncus repens*, sterile submersed form). . . . . Juncaceae, 200
            4. Leaves opposite.
                7. Stem with 2 narrow wings running down from each leaf base (fig. 6h) (*Didiplis*). . . . . Lythraceae, 525
                7. Stem lacking narrow wings from leaf bases.
                    8. Leaf margins spinulose to conspicuously toothed, or if margins entire, at least leaf sheath spinulose (*Najas*). . . . . Hydrocharitaceae, 62
                    8. Leaf margins entire.
                        9. Submersed leaves emarginate at apex; leaves toward surface more or less obovate, weakly 3-nerved, often forming floating rosettes; fruit heart-shaped, with two 1-seeded segments, smooth (*Callitriche*). . . . . Plantaginaceae, 655
                        9. Submersed leaves acute at apex; all leaves similar; fruit more or less ellipsoidal, 1-seeded, toothed on one side (*Zannichellia*). . . . . Potamogetonaceae, 108
        3. Leaves alternate.
            10. Leaf blades somewhat filiform or round in cross-section; stipule adnate to leaf base for at least 10 mm or more (up to its entire length), forming a sheath around stem.
                11. Stipule adnate its entire length (tapering, fig. 4j, or rounded at summit, figs. 96g, 97), appearing inflated; flowers usually 2, borne on a slender peduncle; fruits long-stalked, in umbel-like clusters (*Ruppia*). . . . . Ruppiaceae, 108
                11. Stipule partly adnate with a free ligule-like tip (fig. 4h); flowers several to many, in a spike borne on a stout peduncle; fruits in sessile to subsessile clusters (*Stuckenia*). . . . . Potamogetonaceae, 108
            10. Leaf blades flattened; stipules absent, or if present, adnate to leaf base for less than 10 mm, forming a short sheath around stem, or free.
                12. Leaves with a distinct midvein; flowers common; perianth parts 4 or absent.
                    13. Flowers and fruits in a cylindrical spike, lacking a spathe (although sometimes sheathed by stipule); plants of fresh, brackish, or weakly saline waters (*Potamogeton*). . . . . Potamogetonaceae, 108
                    13. Flowers and fruits in a flattened spike, sheathed by a long, leaf-like spathe (fig. 98b); plants of coastal saline waters (*Zostera*). . . . . Zosteraceae, 108
                12. Leaves lacking a midvein (fig. 6e); flowers uncommon; perianth parts 6, yellow (inconspicuous cleistogamous flowers often on submersed stems) (*Heteranthera dubia*). . . . . Pontederiaceae, 413
    2. Leaves or their divisions thread-like (if segments flattened, then not ribbon-like).
        14. Plants of swift-moving water, especially river rapids, clinging to rocks by holdfast-like structures, often appearing moss-like or alga-like (*Podostemum*). . . . . Podostemaceae, 464
        14. Plants of ponds, lakes, and streams, anchored by roots or root-like structures.
            15. Leaves simple.
                16. Leaves opposite (fig. 4b,c), whorled, or with clusters of smaller leaves in axils (fig. 6f), thus appearing whorled.
                    17. Leaves distinctly whorled (3–6 leaves per node), lacking leaf sheaths at base; plants remaining vegetative while submersed, becoming fertile when stranded on shore, producing an erect terrestrial growth form bearing a terminal capitate inflorescence (*Sclerolepis*). . . . . Asteraceae, 745
                    17. Leaves opposite, or opposite with axillary clusters of smaller leaves giving superficial whorled appearance, sheathing at base; plants typically becoming fertile while completely submersed, producing axillary flowers and fruits.
                        18. Leaf margins spinulose to conspicuously toothed, or if margins entire, then at least leaf sheath spinulose; fruits sessile, entire (*Najas*). . . . . Hydrocharitaceae, 62
                        18. Leaf margins entire; fruits stalked, usually dentate on one side (*Zannichellia*). . . . . Potamogetonaceae, 108
                16. Leaves alternate.
                    19. Stems arising singly.
                        20. Stipule free (figs. 4a, 6b,c,e) or partly adnate to leaf (fig. 4h), with a free ligule-like tip; flowers several to many, in a spike borne on a thick peduncle; fruits in dense sessile to subsessile clusters (spike). . . . . Potamogetonaceae, 108
                        20. Stipule adnate its entire length (rounded at summit) (fig. 4j); flowers usually 2, borne on a slender peduncle; fruits long-stalked, in loose umbel-like clusters. . . . . Ruppiaceae, 108
                    19. Stems arising in clumps (sterile, submersed forms of *Juncus pelocarpus*, *J. subtilis*). . . . . Juncaceae, 200
            15. Leaves compound (fig. 4d,e,f,i,k), or leaves dissected resembling compound leaves (fig. 4g).
                21. Leaves with 1 central axis (fig. 4i,k).

22. Each leaflet cut into numerous thread-like divisions (fig. 4k) (*Rorippa aquatica*). . . . . .Brassicaceae, 554
22. Each leaflet consisting of a single thread-like segment (fig. 4i).
23. Flowers with petals absent (or caducous), axillary or in spike-like terminal inflorescences subtended by bracts.. . . . . Haloragaceae, 446
23. Flowers with petals present (white or white to pinkish) borne in terminal verticillate inflorescences (fig. 475b), or crowded in verticils at upper portion of floating whorl of inflated branches (fig. 475a) (*Hottonia*). . . . . Primulaceae, 604
21. Leaves or leaf-like branches repeatedly divided/forked lacking 1 central axis (fig. 4d,e,f,g).
24. Leaves or leaf-like branches alternate (fig. 4e,g).
25. Plants bearing small bladders (fig. 4g) (*Utricularia*). . . . . Lentibulariaceae, 683
25. Plants lacking bladders (fig. 4e)
26. Leaves with sheathing leaf base; flowers not borne in a head.. . . . . Ranunculaceae, 425
26. Leaves lacking a sheathing leaf base; flowers borne in few-flowered heads subtended by involucral bracts, the heads numerous on paniculate inflorescences (fig. 614, *Eupatorium capillifolium*). . . . . Asteraceae, 745
24. Leaves or leaf-like branches opposite (fig. 4d,f) or whorled (fig. 3b).
27. Leaves or leaf-like branches whorled; plants lacking roots, free-floating, or sometimes anchored by modified branches (rhizoids).
28. Plants bearing small bladders at tips of leaves (leaf-like branches); segments not spinulose; leaf whorls scattered somewhat evenly along stem (*Utricularia purpurea*). . . . . Lentibulariaceae, 683
28. Plants lacking bladders; leaf segments usually spinulose (fig. 3b); leaves especially densely compacted near tip of stem.. . . . . Ceratophyllaceae, 425
27. Leaves opposite (in *Bidens beckii* sometimes appearing whorled, fig. 4d); plants anchored by true roots.
29. Leaves stalked, extending from opposite sides of stem (fig. 4f) (*Cabomba*). . . . . Cabombaceae, 38
29. Leaves not stalked, much divided immediately at base, extending in all directions around stem (fig. 4d) (sometimes appearing whorled) (*Bidens beckii*). . . . . Asteraceae, 745

1. Leaves basal (fig. 3d) or leaves and leaf-like stems arising in clusters along a rhizome (fig. 3g).
30. Leaves or stems thread-like or quill-like.
31. Plants with a bulbous base (figs. 3e, 12e), rhizomes absent. . . . . Isoetaceae, 5
31. Plants lacking a bulbous base, spreading by rhizomes.
32. Leaves uncoiling at tips as plants mature; plants reproducing by spores borne in sporocarps on short stalks from rhizomes (fig. 31e) (*Pilularia*). . . . . Marsileaceae, 17
32. Leaves or stems not coiled prior to maturation; plants reproducing by flowers and seeds.
33. Rhizome scarcely thicker than the thread-like stems (fig. 10c), often forming dense mats (vegetative submersed forms of *Eleocharis acicularis, E. radicans*). . . . . Cyperaceae, 220
33. Rhizome thicker than the thread-like green leaves or stems (fig. 3g) (thicker erect stems usually fertile).
34. Rhizomes 2 mm or more in diameter (*Juncus militaris*). . . . . Juncaceae, 200
34. Rhizomes approximately 1 mm in diameter.
35. Thread-like structures leaves (fig. 3g); fertile stems somewhat round in cross-section; inflorescence appearing somewhat lateral, with a single, short involucral bract (fig. 261b); tubers borne on slender stolons (*Schoenoplectus subterminalis*). . . . . Cyperaceae, 220
35. Thread-like structures stems (with short bladeless sheaths at base); fertile stems triangular in cross-section; inflorescence terminal (fig. 266f); tubers absent (*Eleocharis robbinsii*). . . . . Cyperaceae, 220
30. Leaves flat, ribbon-like (fig. 5).
36. Leaves with ligule at junction of sheath and blade (fig. 5a) (sometimes submersed and sterile. with floating leaves, especially *Zizania* and *Glyceria*). . . . . Poaceae, 366
36. Leaves lacking ligule.
37. Rhizome short, thick (fig. 11a) (submersed form of *Butomus*). . . . . Butomaceae, 76
37. Rhizome slender or absent.
38. Leaves with a distinct midvein (fig. 5b,c,d).
39. Leaves with a broad central lacunae band (fig. 5b) (*Vallisneria*).. . . . . Hydrocharitaceae, 62
39. Leaves lacking a distinct central lacunae band (fig. 5c,d) (*Sagittaria*). . . . . Alismataceae, 76
38. Leaves with a weak midvein or none (fig. 5e,f).
40. Leaves with a weak midvein (fig. 5e); plants vegetative (submersed, juvenile form of *Pontederia cordata*). . . . . Pontederiaceae, 413
40. Leaves lacking any midvein (figs. 5f, 6a); plants vegetative or fertile, monoecious; flowers and fruits in globose heads (fig. 151a,c) (*Sparganium*). . . . . Typhaceae, 174

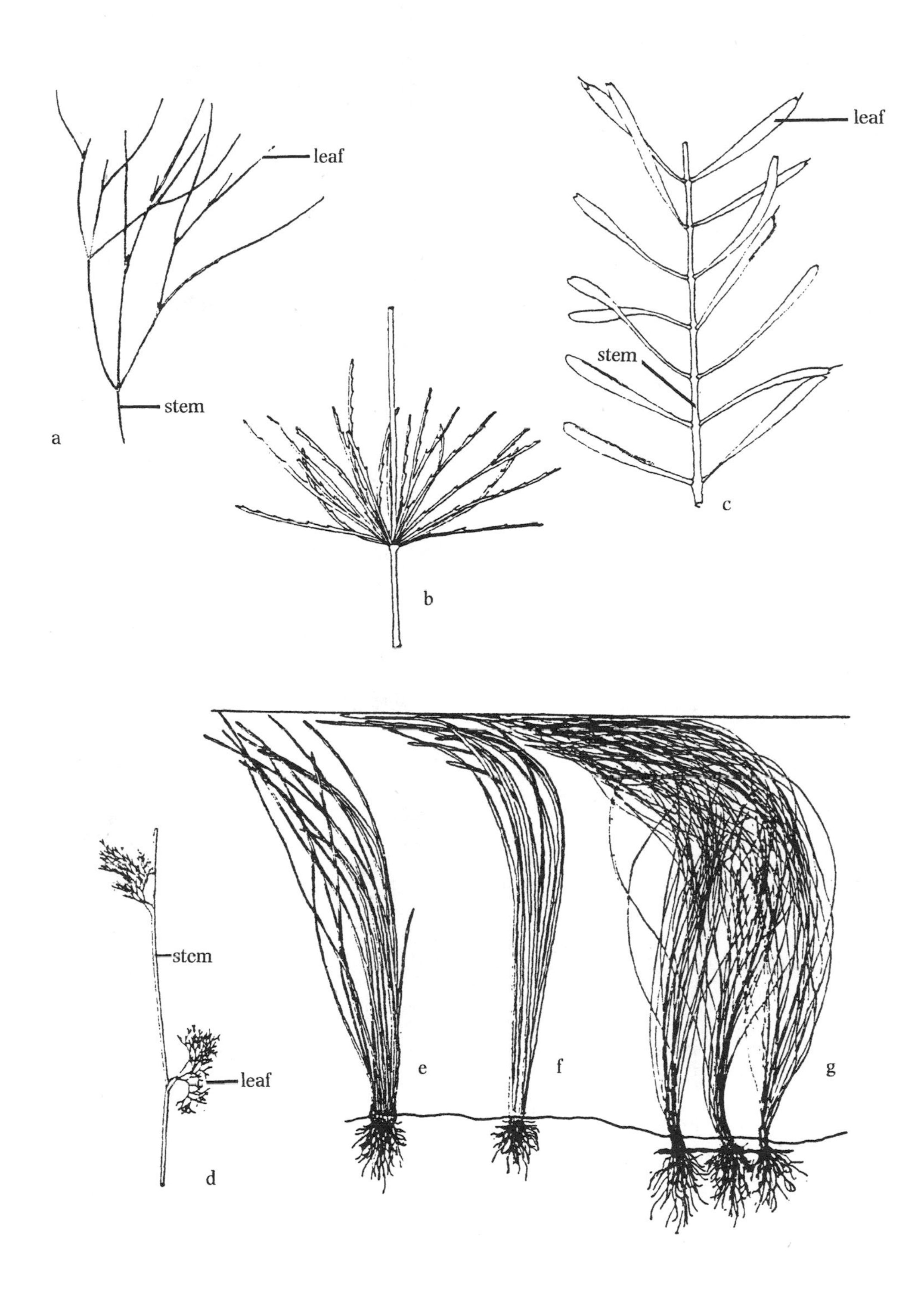

Fig. 3. a. Scattered simple thread-like leaves on a branched stem, generalized; b. *Ceratophyllum*, leaf; c. flat, opposite, ribbon-like leaves on stem, generalized; d. *Ranunculus longirostris*; e. *Isoetes*, generalized; e. ribbon-like leaves arising from one point, generalized; g. thread-like leaves or stems arising in clusters along a rhizome (F).

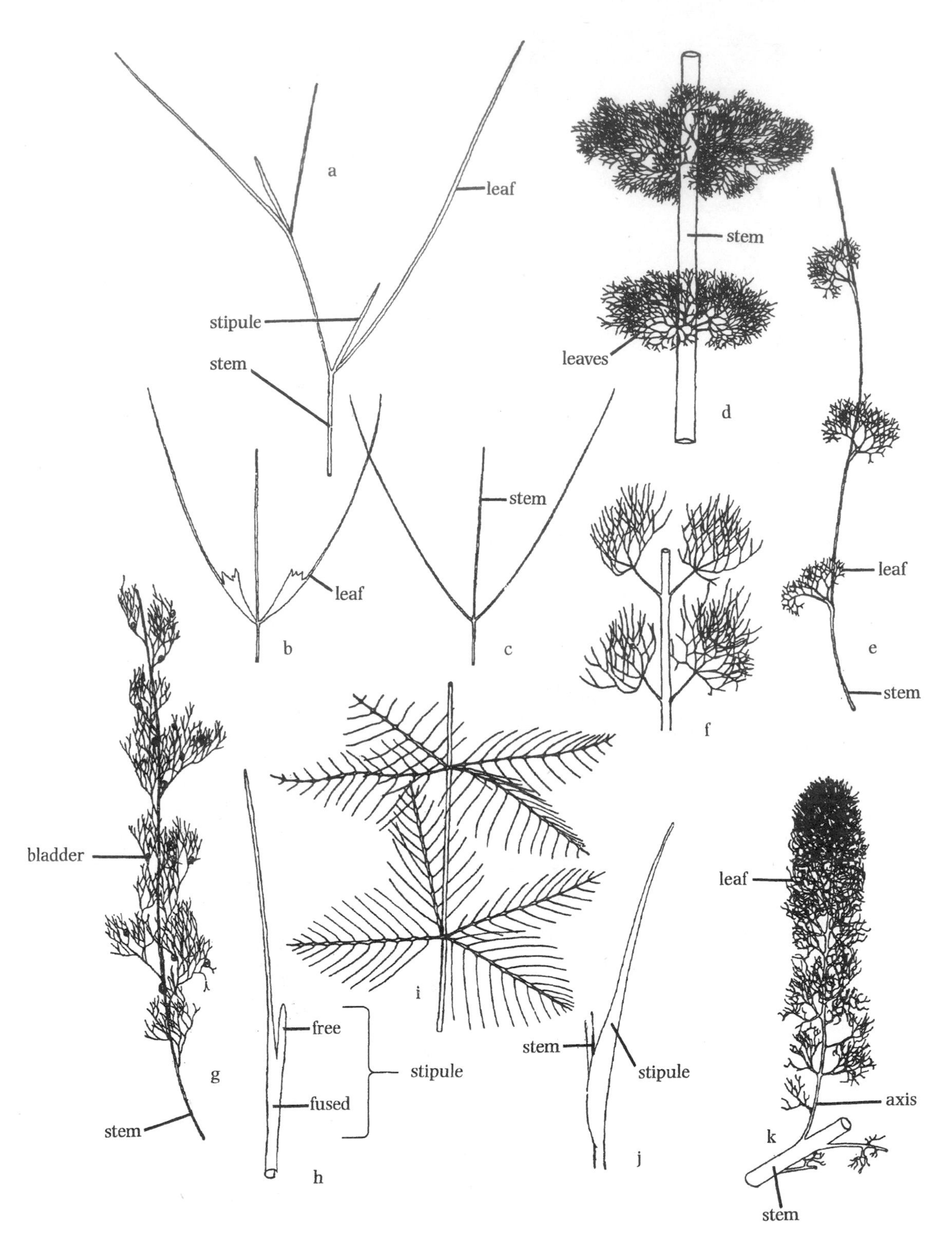

Fig. 4. a. *Potamogeton*, diagrammatic; b. *Najas*, sheathing leaf bases generalized; c. *Zannichellia*, slender leaf bases, generalized; d. *Bidens beckii*, section of stem with leaves opposite, but appearing whorled; e. *Ranunculus*, dissected leaves with sheathing leaf bases, generalized; f. *Cabomba*, leaves dissected; g. *Utricularia*, leaves dissected, bearing small bladder traps, generalized; h. *Stuckenia pectinata*, leaf, showing stipule with lower portion adnate (fused) to leaf, upper portion free, diagrammatic; i. *Myriophyllum*, showing leaves pinnately dissected, whorled, generalized; j. *Ruppia*, leaf, generalized; k. *Rorippa aquatica*, section of stem with leaf (F).

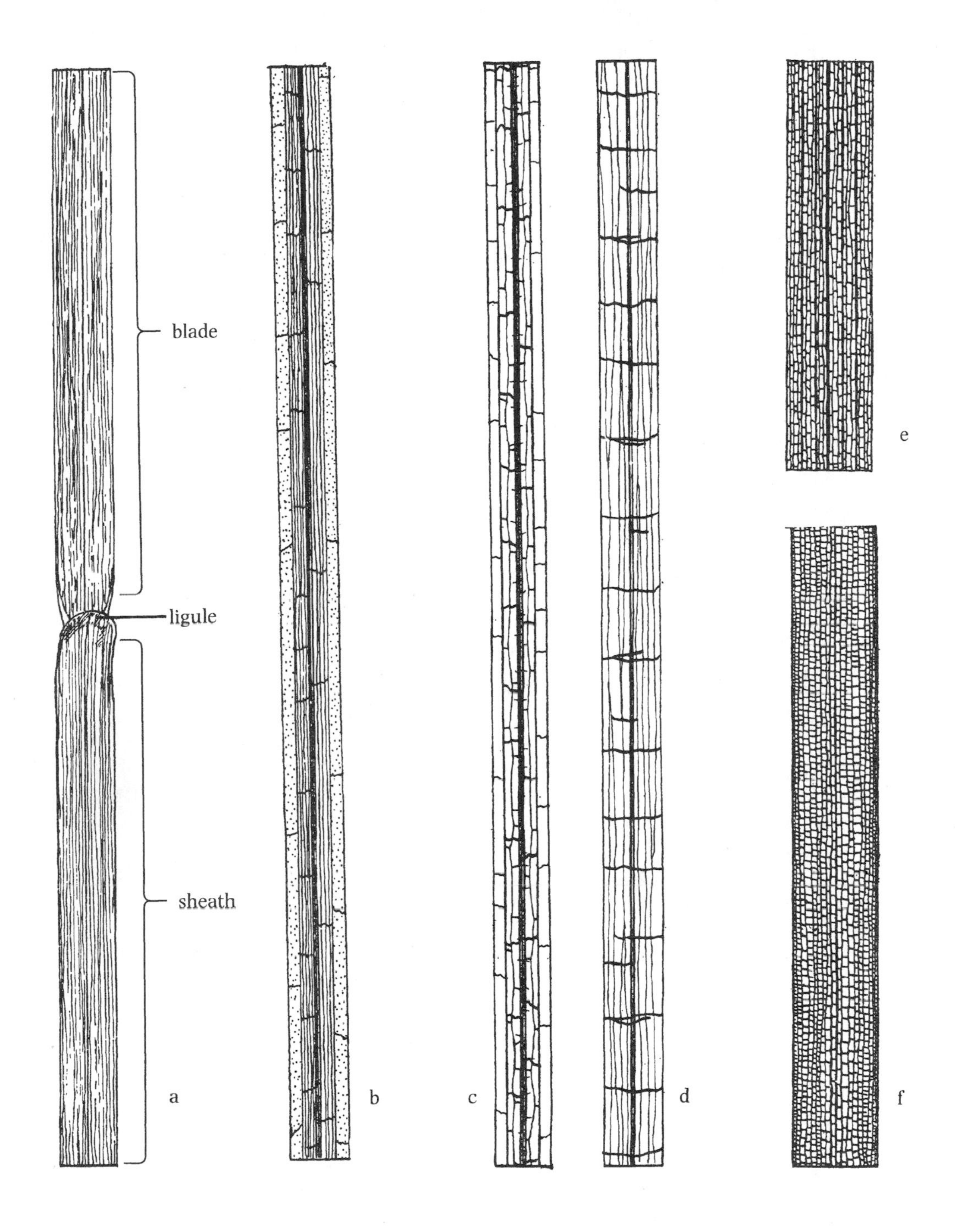

Fig. 5. Sections of leaves, about natural size: a. Poaceae (grasses), generalized; b. *Vallisneria*; c. *Sagittaria*; d. *Sagittaria*; e. *Pontederia cordata*, submersed form; f. *Sparganium*, generalized (F).

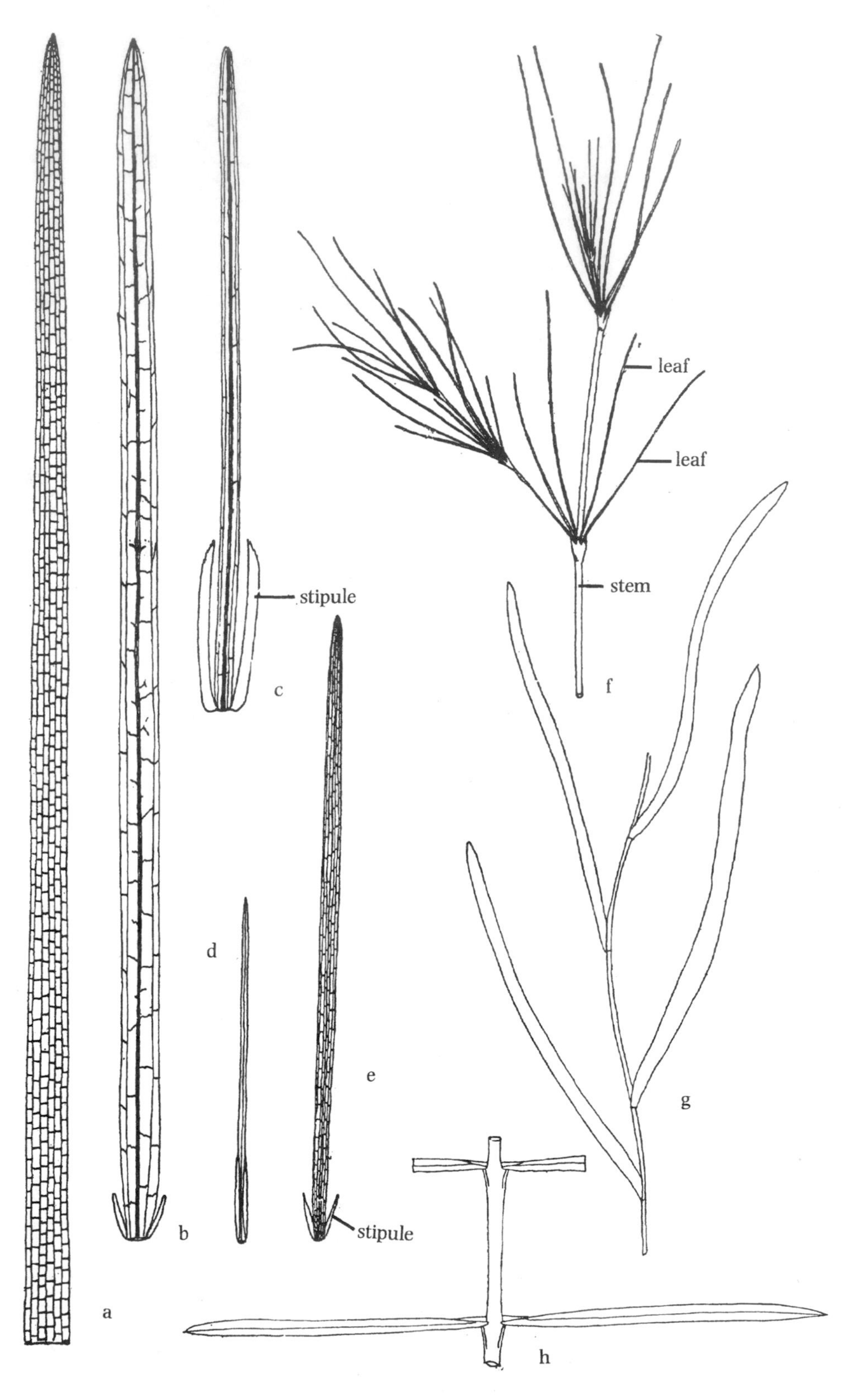

Fig. 6. a. *Sparganium*, leaf, generalized; b–d. *Potamogeton*, leaves, each with 2 stipules at base; e. *Heteranthera dubia*, leaf with 2 stipules at base; f. *Najas*, axillary leaf fascicles; g. portion of plant with alternately arranged ribbon-like leaves, diagrammatic; h. *Didiplis diandra*, aquatic form, section of stem with opposite leaves, showing 2 narrow wings running down from each leaf base (F).

Key 4 Leaves compound, or divided into several leaflets

1. Plants distinctly fern-like, reproducing by spores (spores never in sporocarps) (figs. 8a, 22).
    2. Sterile and fertile fronds similar, sporangia borne on lower surface of frond.
        3. Fronds from upright rhizomes (appearing clumped) (*Dryopteris*)....Dryopteridaceae, 27
        3. Fronds from creeping rhizomes.
            4. Sori round, sparsely or densely spaced along center of pinnules (fig. 35d); rhizome 1–2 mm in diameter; stipes black, lacking scales (*Thelypteris*)....Thelypteridaceae, 22
            4. Sori elongate, appearing continuous along midrib of pinnae and pinnules (fig. 8a); rhizome 6–10 mm in diameter; stipes purplish-brown, with scales, deciduous at base (*Woodwardia virginica*)....Blechnaceae, 22
    2. Sterile and fertile fronds different (fig. 26b,c) (*Osmundastrum*), or fronds divided into sterile and fertile portions (fig. 25) (*Osmunda*). ....Osmundaceae, 13
1. Plants not fern-like, reproducing by flowers and seeds, or by spores borne in sporocarps.
    5. Leaves compound, with 4 leaflets, cross-shaped (fig. 31a,b) (similar to 4-leaved clover); plants reproducing by spores borne in sporocarps (*Marsilea*)....Marsileaceae, 17
    5. Leaves various, but not with 4 leaflets; plants reproducing by flowers and seeds.
        6. Plants not rooted, usually free-floating below surface in tangled mats; plants thalloid, consisting of branched, long-stalked fronds, giving compound appearance (fig. 58f,g) (*Lemna trisulca*)....Araceae, 53
        6. Plants rooted, not free-floating; plants with distinct stems and leaves.
            7. Divisions of leaves fine, less than 1 mm wide (fig. 7a,c).
                8. Flowers with petals absent (or caducous), axillary or in spike-like terminal inflorescences subtended by bracts. ....Haloragaceae, 446
                8. Flowers with petals present (white or white to pinkish) borne in terminal verticillate inflorescences (fig. 475b) or crowded in verticils at upper portion of floating whorl of inflated branches (figs. 7a, 475a) (*Hottonia*)....Primulaceae, 604
            7. Divisions of leaves or leaflets broader, greater than 1 mm wide (fig. 7d–i), or if less than 1 mm, then more than once compound (fig. 622b).
                9. Leaves opposite, or both whorled and alternate.
                    10. Leaves opposite (*Bidens*)....Asteraceae, 745
                    10. Leaves of upper stems whorled, lower leaves alternate (*Hottonia palustris*)....Primulaceae, 604
                9. Leaves alternate.
                    11. Leaves with stipules (figs. 392f, 393b), leaflets toothed; sepals, petals, and stamens borne on a shallow hypanthium (*Geum, Potentilla, Sanguisorba*)....Rosaceae, 493
                    11. Leaves lacking stipules, or if present (Fabaceae), then leaflets entire; sepals, petals, and stamens not borne on a hypanthium.
                        12. Base of petiole expanded into a sheath (fig. 7d,e).
                            13. Stem completely encircled by sheathing base of petiole (figs. 617a, 624a)....Apiaceae, 795
                            13. Stem not encircled by base of petiole (or if sheathing, then not completely encircling stem (*Menyanthes*).
                                14. Leaves pinnately compound (fig. 474b).
                                    15. Leaflets usually 15 or more; corolla blue; stamens 5 (*Polemonium*)....Polemoniaceae, 604
                                    15. Leaflets usually 5; corolla white; stamens 3 (*Valeriana*)....Caprifoliaceae, 812
                                14. Leaves palmately compound (figs. 351d, 574a).
                                    16. Corolla white, united, lobes conspicuously fringed; flowers bisexual, with 1 pistil and 5 stamens (*Menyanthes*)....Menyanthaceae, 739
                                    16. Corolla yellow, not united, petals never fringed, or corolla absent; flowers bisexual or unisexual, with several to many pistils and/or stamens....Ranunculaceae, 425
                        12. Base of petiole not expanded into a sheath.
                            17. Flowers borne in few-flowered heads, subtended by involucral bracts (*Eupatorium capillifolium*). ....Asteraceae, 745
                            17. Flowers borne in few-to many-flowered racemes.
                                18. Flowers radially symmetrical, 4-merous, cross-shaped; fruit a silique; terminal leaflet usually the largest (fig. 7h)....Brassicaceae, 554
                                18. Flowers bilaterally symmetrical, 5-merous; fruit a legume (fig. 387d) or loment (fig. 387e); terminal leaflet about the same size as lateral leaflets....Fabaceae, 490

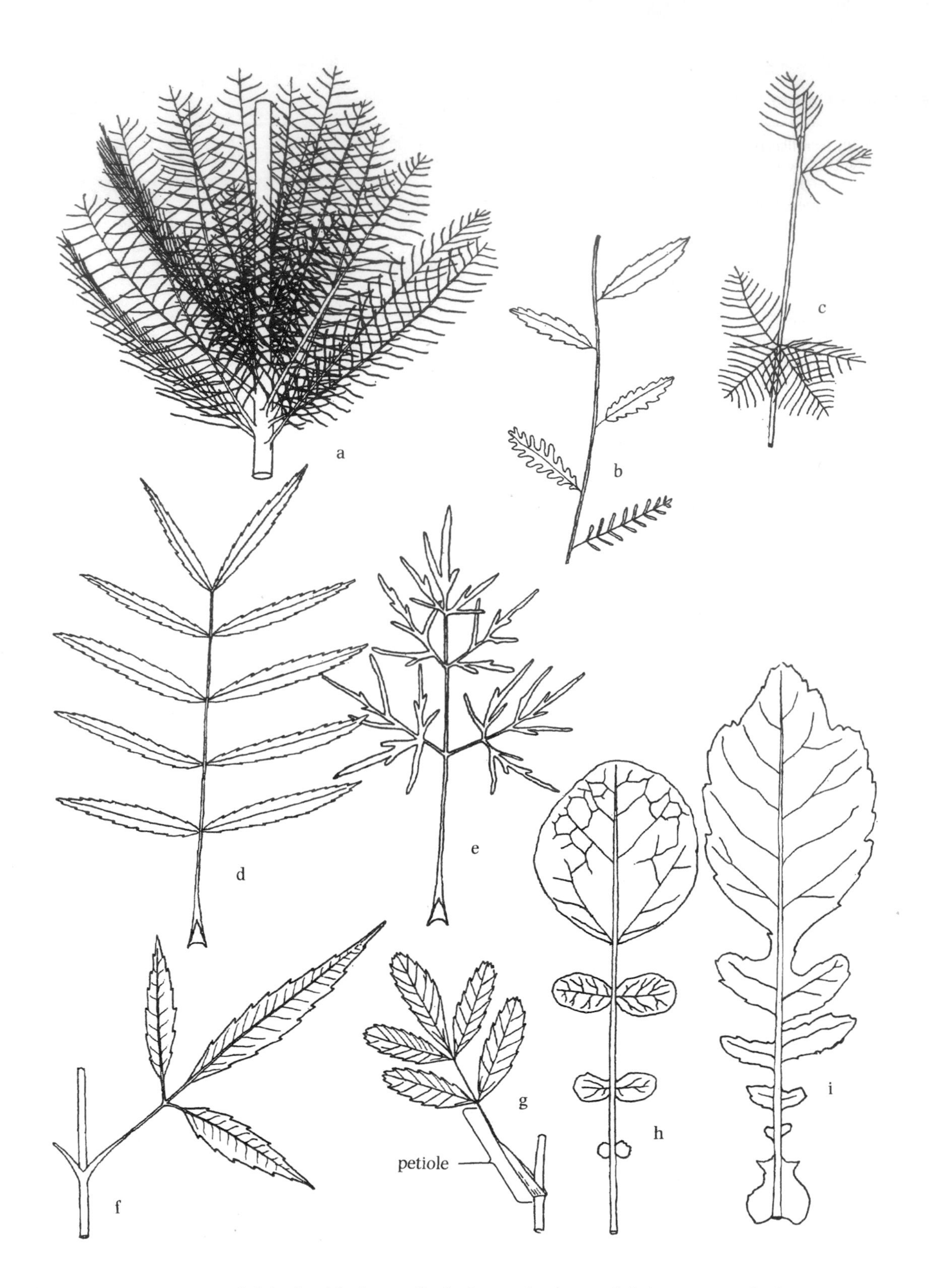

Fig. 7. a. *Hottonia*, crowded feather-like leaves, finely dissected submersed; b. *Proserpinaca*, leaf variation on a single plant; c. *Myriophyllum*, generalized, with 2 leaves alternately arranged (*above*) and 4 whorled leaves at a single node (*below*); d. *Sium*, emersed compound leaf; e. *Sium*, highly divided submersed leaf; f. *Bidens frondosa*, tripartite leaf; g. *Comarum palustre*, leaf with stipule adnate to petiole; h. *Nasturtium*, leaf; i. *Rorippa palustris*, compound leaf, with terminal leaflet largest (F).

Key 5 Leaves simple, with blades deeply lobed, or with basal lobes extending laterally or forward

1. Plants distinctly fern-like, reproducing by spores.
   2. Fertile fronds brown (green while developing), spike-like, appearing woody, segments bead-like, nearly enclosing sporangia (figs. 8i, 32c,e) (*Onoclea*). . . . . Onocleaceae, 22
   2. Fertile fronds green, herbaceous, segments narrow, linear, with sporangia in linear sori along midveins on lower surface (fig. 8a) (*Woodwardia areolata*). . . . . Blechnaceae, 22
1. Plants not fern-like, reproducing by flowers and seeds.
   3. Leaves somewhat succulent, surface mealy or scurfy; plants of salt marshes (*Atriplex*). . . . . Amaranthaceae, 589
   3. Leaves not succulent, surface glabrous or pubescent, but not mealy or scurfy; plants of freshwater sites.
      4. Leaves opposite.
         5. Stems square in cross-section.
            6. Flowers strongly 2-lipped, borne in axillary clusters; lower leaves not hastate, lobes extending forward (fig. 247c) (*Lycopus americanus*). . . . . Lamiaceae, 702
            6. Flowers weakly 2-lipped (fig. 544i), borne in terminal spikes; lower leaves hastate (*Verbena hastata*). . . . . Verbenaceae, 696
         5. Stems round in cross-section.
            7. Leaf margins with lobes sharply serrate or coarsely toothed (figs. 8h, 582f,g,h,i); flowers borne in dense heads, subtended by involucral bracts (simulating a single flower); fruit a cluster (head) of awned achenes (fig. 582a,d,e), barbs of individual achenes readily attaching to clothes (*Bidens*). . . . . Asteraceae, 745
            7. Leaf margins with lobes entire or with a few coarse, blunt teeth; flowers axillary, solitary or paired; fruit a capsule (*Leucospora multifida*). . . . . Plantaginaceae, 655
      4. Leaves alternate.
         8. Leaves palmately veined (fig. 8e).
            9. Petioles expanded at base to form a sheath around stem (fig. 347a); fruit a globose cluster of achenes or follicles (*Ranunculus, Aconitum*). . . . . Ranunculaceae, 425
            9. Petioles not sheathing; fruit a capsule or schizocarp.
               10. Petiole attached near center of blade (fig. 8f); stamens 5, filaments separate; ovary inferior; fruit a schizocarp splitting into 2 segments (*Hydrocotyle ranunculoides*). . . . . Araliaceae, 795
               10. Petiole attached at base of blade (fig. 9g); stamens numerous, filaments fused (monadelphous), enclosing style; ovary superior; fruit a capsule, or a schizocarp (splitting into numerous segments). . . . . Malvaceae, 549
         8. Leaves pinnately veined.
            11. Flowers in heads (*Cotula*). . . . . Asteraceae, 745
            11. Flowers various, but never in heads.
               12. Plants viny, herbaceous above, but becoming woody at base; corolla united, 5-lobed, deep purple; fruit fleshy, a red berry (*Solanum dulcamara*). . . . . Solanaceae, 645
               12. Plants not viny, herbaceous; corolla absent, or of 4 separate petals, white, yellow, or pinkish to purplish (but not deep purple); fruit dry, a silique (fig. 437d), or a nutlet.
                   13. Leaves pinnatifid, sometimes pectinate on lower portion of stem (fig. 8j); corolla absent; fruit a 3-sided nutlet (*Proserpinaca*, amphibious plant with leaves varying in terrestrial growth form). . . . . Haloragaceae, 446
                   13. Leaves variously lobed (fig. 8c,e,g), but never pectinate; corolla present, petals 4, cross-shaped; fruit a silique. . . . . Brassicaceae, 554

Fig. 8. a. *Woodwardia virginica*, portion of fertile frond with elongate sori (NYS Museum); b, c. *Rorippa aquatica*, deeply lobed submersed leaves (F); d. *Ranunculus*, leaf with sheathing leaf base, generalized (F); e. *Ranunculus sceleratus*; lobed leaf (F); f. *Hydrocotyle*, leaf blade peltate (petiole attached in center), generalized (F); g. *Rorippa palustris*, basal leaves (F); h. *Bidens connata*, tripartite compound leaf with coarsely toothed margins (F); i. *Onoclea*, leaf and fertile fronds (NYS Museum); j. *Proserpinaca palustris*, habit showing progression from base of plant to top of development of pectinate to pinnatifid to coarsely toothed leaves as water level drops (F).

Key 6 Leaves with basal lobes extending below junction of blade and petiole, or leaves peltate

1. Leaves peltate (fig. 9a).
  2. Leaves large, more than 15 cm wide; flowers large, showy, yellow or pink; pistils embedded in a broad, flat-topped receptacle (*Nelumbo*)........ Nelumbonaceae, 443
  2. Leaves smaller, not more than 8 cm wide; flowers small, red-purple or white; pistils not embedded in a broad, flat-topped receptacle.
    3. Leaves elliptic, submersed parts covered with a thick mucilaginous coating; petals red-purple; carpels 4–10, separate (*Brasenia*)........ Cabombaceae, 38
    3. Leaves circular, lacking mucilaginous coating; petals white; carpels 2, united (*Hydrocotyle*)........ Araliaceae, 795
1. Leaves with basal lobes extending below junction of blade and petiole (fig. 9c,e,f,h,i,j,k).
  4. Plant a vine, climbing, twining, and sprawling; flowers borne in a head. subtended by involucral bracts (simulating a single flower) (*Mikania*)........ Asteraceae, 745
  4. Plants not a vine (sometimes creeping and rooting at the nodes); flowers borne variously, but not in a head.
    5. Leaf margins toothed to crenulate.
      6. Flowers bilaterally symmetrical, lower petal spurred (*Viola*)........ Violaceae, 473
      6. Flowers radially symmetrical, petals not spurred.
        7. Stamens numerous; pistils simple, numerous, separate; ovary superior; fruit a cluster of follicles (*Caltha natans*)........ Ranunculaceae, 425
        7. Stamens 5; pistils compound, carpels 2, united; ovary inferior; fruit a schizocarp, splitting into 2 segments (fig. 616d,f).
          8. Leaves circular, peltate, or deeply notched (appearing peltate) (*Hydrocotyle*, fig. 616)........ Araliaceae, 795
          8. Leaves ovate to oblong, cordate to truncate at leaf base (*Centella*, fig. 618d,e,f)........ Apiaceae, 795
    5. Leaf margins entire.
      9. Plants grass-like (fig. 95a,e); leaves very narrow, somewhat round in cross-section (slightly succulent) (*Triglochin*)........ Juncaginaceae, 101
      9. Plants not grass-like; leaves broad, flat, or if narrow, not round in cross-section.
        10. Stems bearing recurved barbs (fig. 449a,e); stipular sheath present (fig. 79e), surrounding stem at nodes (*Persicaria arifolium*, *P. sagittatum*)........ Polygonaceae, 572
        10. Stems lacking barbs; stipular sheath absent.
          11. Leaf bases sagittate or hastate; leaves emersed.
            12. Leaf blades with both a single prominent vein extending into each basal lobe (fig. 55b) and a distinct marginal vein (fig. 55b); flowers borne in a spadix, surrounded by a green spathe (*Peltandra*)........ Araceae, 53
            12. Leaf blades lacking both a single prominent vein extending into each basal lobe and a distinct marginal vein; flowers borne in a spike or in whorls of 3 in a raceme.
              13. Basal lobes rounded (fig. 9c); petals violet-blue; fruit a utricle, crested with toothed ridges (*Pontederia*)........ Pontederiaceae, 413
              13. Basal lobes pointed (fig. 9j); petals white; fruit a globose cluster of achenes (*Sagittaria*)........ Alismataceae, 76
          11. Leaf bases various, cordate to reniform or having a deep sinus between lobes (if sinus so wide such that base appears sagittate, then leaf floating); leaves submersed, floating, or emersed.
            14. Plant with strong skunk-like odor when crushed; spathe mottled, purplish-green and/or reddish-brown, emerging early Spring, before leaves (*Symplocarpus*)........ Araceae, 53
            14. Plant lacking a strong odor; spathe, if present, white or greenish-white.
              15. Leaves floating, plants rooted in water.
                16. Leaves net-veined, veins branching and running toward margin; sepals 4–9; petals numerous.
                  17. Leaves with a distinct midvein; flowers solitary; petals numerous, separate; stamens numerous........ Nymphaeaceae, 38
                  17. Leaves with midvein as prominent as other veins; flowers borne in umbels (usually 1 open at a time); petals 5, united; stamens 5 (*Nymphoides*)........ Menyanthaceae, 739
                16. Leaves parallel-veined, veins converging toward apex (fig. 9d); sepals 3; petals 0–3 (*Hydrocharis*, *Limnobium*)........ Hydrocharitaceae, 62
              15. Leaves emersed or plants rooted in mud near edge of water.
                18. Leaves usually broadly rounded at apex (in *Limnobium* sometimes slightly pointed at apex).
                  19. Leaves large, usually more than 12 cm wide, with a distinct midvein and pinnate lateral veins (fig. 50a); flowers yellow (*Nuphar*)........ Nymphaeaceae, 38
                  19. Leaves smaller, usually less than 6 cm wide, distinct midvein lacking, veins parallel, running from the base, converging toward apex (fig. 9d); flowers white to yellowish-green or pale blue.
                    20. Leaf blades nearly circular to slightly longer than wide (fig. 9d); flowers unisexual; fruit berry-like (*Hydrocharis*, *Limnobium*)........ Hydrocharitaceae, 62
                    20. Leaf blades usually reniform, wider than long (fig. 338c); flowers bisexual; fruit a capsule (*Heteranthera reniformis*)........ Pontederiaceae, 413
                18. Leaves usually tapering to acute or obtuse apex, or abruptly pointed at apex.
                  21. Flowers borne in a spadix, subtended by a conspicuous white spathe (fig. 400b); leaf blade broadly cordate, with a distinct marginal vein, abruptly pointed at apex (fig. 56a) (*Calla*)........ Araceae, 53
                  21. Flowers borne in a spike or in whorls in a raceme, not subtended by a white spathe; leaf blade ovate to lanceolate to linear-lanceolate, lacking a marginal vein, tapering to acute or obtuse apex.

22. Petals violet-blue, in a dense, erect spike; fruit a utricle, crested with toothed ridges (*Pontederia*). . . . . Pontederiaceae, 413
22. Petals white, in a dense nodding spike-like raceme, or in whorls on a raceme; fruit a globose cluster of 3–5 nutlets, or numerous achenes.
    23. Leaves net-veined, with a midvein; flowers in a dense spike-like raceme, nodding at tip (fig. 53a); fruit a globose cluster of 3–5 nutlets (Saururus). . . . . Saururaceae, 50
    23. Leaves parallel-veined, lacking a midvein; flowers in whorls in a raceme; fruit a globose cluster of numerous achenes (Echinodorus cordifolius).. . . . . Alismataceae, 76

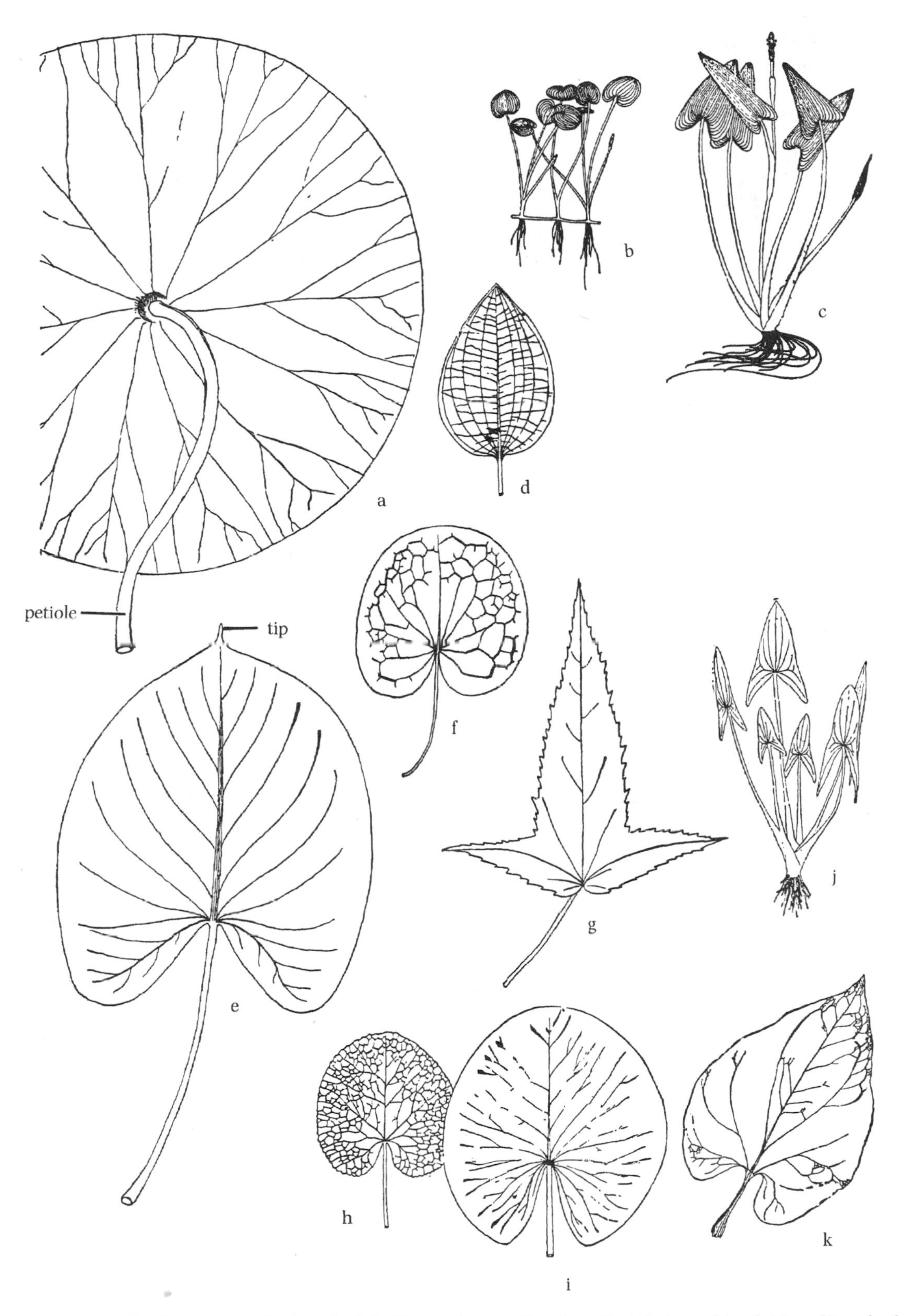

Fig. 9. a. *Nelumbo*, portion of peltate leaf; b. *Heteranthera reniformis*; c. *Pontederia cordata*; d. *Limnobium*, leaf; e. *Calla*, leaf; f. *Nymphoides*, leaf; g. *Hibiscus laevis*, hastate leaf; h. *Caltha*, leaf; i. *Nuphar*, leaf with basal lobes; j. *Sagittaria latifolia* with arrow-shaped (sagittate) leaves; k. *Saururus cernuus*, leaf (F).

Key 7 Leaves basal or cauline, more than 10 times as long as wide; or stems naked or appearing naked, with leaves reduced to bladeless sheaths

1. Underground parts (root-like) in substrate with tiny bladders; upright stems leafless, with yellow or purple spurred flowers (fig. 541a) (*Utricularia resupinata, U. cornuta, U. juncea, U. subulata*)... Lentibulariaceae, 683
1. Underground parts in substrate lacking bladders; upright stems naked or with leaves, flowers various.
   2. Leaves greatly reduced, appearing as scales (figs. 464b,d, 494a,f) or bumps (fig. 10f along stems, stems thus appearing naked.
      3. Plants distinctly succulent; plants of saline sites. usually coastal (*Salicornia*)................................ Amaranthaceae, 589
      3. Plants not succulent; plants of non-saline sites.
         4. Stems several, arising along a creeping rhizome; leaves rounded at apex, appearing as minute bumps (figs. 10f, 355) (*Myriophyllum tenellum*).. Haloragaceae, 446
         4. Stems single; leaves acute at apex.
            5. Flowers 3-merous, perianth tube 3-angled or 3-winged; capsule 3-winged (fig. 122i) (*Burmannia*)........ Burmanniaceae, 140
            5. Flowers 4-merous. perianth tube round; capsule round in cross-section (*Bartonia*)........................ Gentianaceae, 635
   2. Leaves not reduced to scales or bumps along stems (although sometimes reduced to bladeless sheaths).
      6. Leaves basal (fig. 11a,e,g); or stems naked (or appearing naked), all arising from one point (figs. 10a, 13b), or arising scattered along a creeping rhizome (fig. 12a,b).
         7. Plants covered with conspicuous, sticky glandular hairs (fig. 447c) (*Drosera filiformis*)......................... Droseraceae, 568
         7. Plants lacking glandular hairs.
            8. Leaves or ascending naked stems usually 15 cm or more long.
               9. Erect structures (usually leaves) flat.
                  10. Base of plant triangular in cross-section (sterile plants)... Cyperaceae, 220
                  10. Base of plant not triangular in cross-section.
                      11. Leaf with a ligule at junction of blade and sheath (fig. 11b); stems hollow.......................... Poaceae, 366
                      11. Leaf without a ligule, sheath at base of blade present or absent; stems, if present, not hollow.
                          12. Flowers bilaterally symmetrical, with 1 larger petal forming a distinct lip or labellum (figs. 124b, 126a); stamens fused to style forming a specialized structure, the column, the pollen aggregated in dense waxy mass (pollinium).. Orchidaceae, 140
                          12. Flowers radially symmetrical, without a lip; stamens and pistils appearing normal, not specialized.
                              13. Leaf with a keeled center (fig. 11f,h,i), or a thick, raised midrib.
                                  14. Leaf base, when crushed, with an aromatic odor; flowers and fruits borne in a spadix, appearing lateral on the leaf-like axis (fig. 54a,b) (*Acorus*). .... Acoraceae, 53
                                  14. Leaf base, when crushed, lacking an aromatic odor; flowers and fruits borne in globose heads (fig. 151) (*Sparganium*).. Typhaceae, 174
                              13. Leaf lacking a keeled center or a raised midrib.
                                  15. Flowers and fruits borne in a dense cigar-like cylindrical spike; junction of sheath and blade abrupt, summit of sheath truncate (fig. 11c) (*Typha domingensis*, sheath tapering to blade, but plants usually 2–3 m tall, fig. 11d)... Typhaceae, 174
                                  15. Flowers not borne in a dense, cylindrical spike; junction of sheath and blade transitional, sheath gradually tapering to blade, or lacking a distinct blade.
                                      16. Perianth wooly.. Haemodoraceae, 413
                                      16. Perianth glabrous.
                                          17. Leaves 5 mm or more wide.
                                              18. Plants with a conspicuous rhizome; flowers blue or yellow, more than 6 cm wide (*Iris*). .... Iridaceae, 167
                                              18. Plants tufted, lacking a rhizome (fig. 11e); flowers white or yellow, 1 cm or less wide.
                                                  19. Leaf margins with sharp teeth, pointing toward apex (*Stratiotes*)........ .... Hydrocharitaceae, 62
                                                  19. Leaf margins entire.
                                                      20. Flowers white, borne in branched or unbranched verticils; fruit a cluster of many achenes, in a flat-topped ring (*Alisma*) or globose head (*Baldellia*)... Alismataceae, 76
                                                      20. Flowers yellow, borne in a globose head-like spike (one flower opening at a time) (fig. 161, 164); fruit a capsule (*Xyris*).............. Xyridaceae, 189
                                          17. Leaves less than 5 mm wide.
                                              21. Leaves arising from a rhizome.
                                                  22. Flowers borne in an umbel.
                                                      23. Flowers pink, carpels 6, separate; stalk of inflorescence tall, up to 1.5 m; rhizome short, thick (fig. 11a) (*Butomus*). .... Butomaceae, 76
                                                      23. Flowers white, carpels 2, united; stalk of inflorescence short (near the ground), 3–10 mm; rhizome elongate, slender (*Lilaeopsis*)........ Apiaceae, 795
                                                  22. Flowers borne in a raceme or spike.......................... Asparagaceae, 174
                                              21. Leaves arising in a dense tuft, lacking a rhizome (fig. 11e). base sometimes bulbous (*Xyris*).. Xyridaceae, 189
               9. Erect structures (leaves or stems) round to oval, U-shaped, square, or triangular in cross-section.

24. Ascending structures consisting of leaves.
 25. Leaves triangular in cross-section; flowers and fruits borne in globose heads (fig. 151) (*Sparganium*). .... Typhaceae, 174
 25. Leaves round to oval or U-shaped in cross-section, or tubular (fig. 480b); flowers solitary or borne in a spike, or if flowers not produced, then plants reproducing by spores in sporangia embedded in leaf bases.
  26. Plants with a hard, bulbous base (fig. 12e); plants reproducing by spores borne in sporangia embedded in leaf bases (fig. 20b) (*Isoetes*).......... Isoetaceae, 5
  26. Plants lacking a hard, bulbous base; plants reproducing by flowers and seeds.
   27. Leaves tubular, open at upper end (water accumulating in tubular portion), not succulent (fig. 480b) (*Sarracenia*).......... Sarraceniaceae, 617
   27. Leaves not tubular, somewhat succulent.
    28. Leaves sheathing at base (fig. 95b); pistils 3–6 (*Triglochin*). .... Juncaginaceae, 101
    28. Leaves clasping at very base, but not sheathing; pistils 1 (*Plantago*).......... Plantaginaceae, 655
24. Ascending structures consisting of stems (usually with basal, bladeless sheaths).
 29. Plants vegetative.
  30. Stems 1 m or more tall (Scirpus).......... Cyperaceae, 220
  30. Stems less than 1 m tall.
   31. Stems 3-angled or 4–8-angled in cross-section.
    32. Stems 3-angled (*Scirpus, Bulboschoenus*).......... Cyperaceae, 220
    32. Stems 4–8-angled (*Eleocharis elliptica, E. quadrangulata, E. tenuis*). .... Cyperaceae, 220
   31. Stems nearly round in cross-section.
    33. Sheaths open, edges often overlapping, but not united to form a tube around stem (fig. 174d) (*Juncus*).......... Juncaceae, 200
    33. Sheaths closed, forming a tube around stem (figs. 173d, 259a).
     34. Stems single or few, arising at intervals along a rhizome (figs 10c, 12b, 272b) (*Eleocharis*). .... Cyperaceae, 220
     34. Stems tufted, lacking a creeping rhizome (*Eleocharis, Fimbristylis Isolepis, Lipocarpha, Schoenoplectiella*; these genera are difficult to distinguish in vegetative condition). .... Cyperaceae, 220
 29. Plants fertile, with flowers/fruits, borne at or near tips of stems (figs. 10a,g, 13b,d,f).
  35. Flowers reduced, hidden by overlapping scales (fig. 276b); perianth absent (fig. 276f,g) or reduced to bristles (figs. 276c,e, 263d,h); fruit an achene.
   36. Inflorescence (spikelet) terminal, with no involucral bract extending beyond (fig. 13a,i).
    37. Achene with a distinctive tubercle formed by the persistent style base (fig. 268) (*Eleocharis*).. .... Cyperaceae, 220
    37. Achene with persisting style base, but lacking a distinct tubercle (involucral bract so short as to appear absent; *Trichophorum, Blysmus*). .... Cyperaceae, 220
   36. Inflorescence appearing somewhat lateral, with a single involucral bract extending beyond (fig. 13d,f) (*Schoenoplectus*).......... Cyperaceae, 220
  35. Flowers small, but readily visible; perianth parts 6, in 2 whorls, scale-like, brownish (fig. 13e,g); fruit a capsule (*Juncus*).......... Juncaceae, 200
8. Leaves or ascending naked stems less than 15 cm long.
 38. Leaves paired, directly arising from slender creeping stem (fig. 564), rooting at the nodes (*Glossostigma*)........ .... Phrymaceae, 718
 38. Leaves not paired, arising directly from slender creeping stem, or if arising from creeping stem, then leaves or stems in clusters (fig. 18e).
  39. Flowers irregular, with 1 large petal forming a distinct lip (figs. 125a, 128b,c,d); stamens fused to style forming a specialized structure, the column, the pollen aggregated in dense waxy mass (pollinium)......... .... Orchidaceae, 140
  39. Flowers regular or irregular; stamens and pistils appearing normal, not specialized; or plants without flowers, reproducing by spores.
   40. Leaves hollow in cross-section, appearing like 2 united tubes (fig. 12d) (*Lobelia dortmanna*)............ .... Campanulaceae, 733
   40. Leaves not hollow in cross-section, not appearing as 2 united tubes.
    41. Plants with a bulbous, hard base (figs. 12e, 162a).
     42. Leaves flat, 2-ranked; plants reproducing by flowers and seeds (*Xyris torta*). .... Xyridaceae, 189
     42. Leaves round in cross-section, not 2-ranked (fig. 12e); plants reproducing by spores borne in sporangia embedded in individual leaf bases (fig. 20b) (*Isoetes*).......... Isoetaceae, 5
    41. Plants lacking a bulbous, hard base.
     43. Leaves 10 mm or more wide at base.
      44. Leaf margins with sharp teeth, pointing toward apex; flowers borne singly on a scape (*Stratiotes*).......... Hydrocharitaceae, 62
      44. Leaf margins entire; flowers borne in branched or unbranched verticils; fruit a cluster of many achenes.......... Alismataceae, 76
     43. Leaves less than 10 mm wide at base.

45. Tufts of leaves connected by slender rhizomes or runners (figs. 10c, 12f).
46. Erect green leaves or stems of essentially the same width throughout, or somewhat expanded into a narrow blade toward apex (fig. 18e).
47. Tufts of leaves connected by green runners (fig. 18e); petals yellow (*Ranunculus reptans*).. Ranunculaceae, 425
47. Tufts of leaves, solitary leaves, or stems connected by rhizomes (often white); petals white to purplish, or absent.
48. Leaves usually borne singly along a rhizome (fig. 618c) (*Lilaeopsis*)........ .... Apiaceae, 795
48. Leaves or stems in clumps along a rhizome.
49. Erect green structures (leaves or stems) pointed at apex; flowers borne in spikes, racemes, or whorled clusters.
50. Erect green structures stems; flowers in terminal spikes, petals lacking (*Eleocharis*)................................... Cyperaceae, 220
50. Erect green structures leaves; flowers in racemes or whorled clusters, petals white.
51. Roots septate (fig. 78a, 81a,b); pistils numerous; fruit a globose head of achenes (fig. 81d) (*Sagittaria*). .... Alismataceae, 76
51. Roots not septate; pistils 1; fruit a capsule (*Tofieldia*). .... Tofieldiaceae, 61
49. Erect green structures (leaves) rounded at apex; flowers solitary, on recurved stalks (*Limosella*). .... Scrophulariaceae, 678
46. Erect green leaves or stems widest at base, tapering toward apex.
52. Leaves hollow with septa (seen as minute cross-lines when held to light; hand lens needed); flowers in spikes or clusters, bisexual.
53. Sheaths open, edges overlapping but not united to form a tube around stem; flowers visible, with perianth of 3 scale-like sepals and 3 scale-like petals (fig. 13g); fruit a capsule (*Juncus*). .... Juncaceae, 200
53. Sheaths closed, forming a tube around stem; flowers hidden by overlapping scales, with perianth absent or reduced to slender bristles, petals lacking; fruit an achene (*Eleocharis*, especially *E. acicularis*). .... Cyperaceae, 220
52. Leaves not hollow, lacking septa; flowers solitary, unisexual (*Littorella*)........ .... Plantaginaceae, 655
45. Tufts of leaves not connected by slender rhizomes.
54. Leaves dimorphic; vegetative leaves short, curly, grass-like; fertile leaves erect, pinnatifid, appearing 1-sided (fig. 27); plants reproducing by spores (*Schizaea pusilla*). .... Schizaeaceae, 13
54. Leaves all similar; plants reproducing by flowers and seeds.
55. Leaves slightly to decidedly succulent.
56. Leaves sheathing at base; pistils 3–6 (*Triglochin*). .... Juncaginaceae, 101
56. Leaves clasping at very base, but not sheathing; pistils 1 (*Plantago maritima*).. .... Plantaginaceae, 655
55. Leaves not succulent.
57. Leaves widest at base and tapering toward apex.
58. Roots septate (fig. 12i) (*Eriocaulon*)......................... Eriocaulaceae, 193
58. Roots not septate.
59. Base of plant with persistent old leaves; flowers numerous, borne in a dense, button-like head (*Lachnocaulon*). .... Eriocaulaceae, 193
59. Base of plant lacking persistent old leaves; flowers few, borne in a raceme (fig. 12h) (*Subularia*).......................... Brassicaceae, 554
57. Leaves widest near or above middle.
60. Leaves broadly rounded at apex (fig. 12j); flowers solitary; pistils and stamens numerous (*Myosurus*).......................... Ranunculaceae, 425
60. Leaves pointed (sometimes slightly twisted) at apex (fig. 11e); flowers borne in a capitate spike (fig. 11e); pistils 1, stamens 3 (*Xyris*). .... Xyridaceae, 189
6. Leaves scattered along stem; stems never appearing naked.
61. Leaves whorled, (2)3–12 per node.
62. Leaves (2)3 or 4 per whorl; stems not hollow (small lacunar canals present).
63. Leaves (2)3 per whorl, not tipped by a tiny callus (*Elodea*). .... Hydrocharitaceae, 62
63. Leaves (3)4–6 per whorl, tipped by a tiny callus (*Sclerolepis*). .... Asteraceae, 745
62. Leaves 6–12 per whorl; stems hollow (*Hippuris*).. Plantaginaceae, 655
61. Leaves alternate or opposite.
64. Flowers greatly reduced, hidden in axils of overlapping scales, forming spikes or spikelets; perianth absent or reduced to bristles.

65. Stems solid (except *Dulichium*, hollow including nodes; *Scirpus cyperinus*, sometimes with pith breaking down), triangular in cross-section, angles sometimes rounded, but pith angular; leaves 3-ranked; sheaths closed; flowers/fruits each subtended by a single bract (fig. 189).. Cyperaceae, 220
65. Stems hollow, solid at nodes, round in cross-section; leaves 2-ranked; sheaths open (except *Glyceria*); flowers/fruits of the spikelets each enclosed by 2 bracts (lemma and palea), each spikelet subtended by 2 bracts (1st and 2nd glumes) (fig. 301). .... Poaceae, 366
64. Flowers not reduced (although sometimes small), readily visible, solitary, or forming various inflorescences; perianth present, or if absent, then surrounded by perianth-like spathe.
66. Leaves not sheathing at base.
67. Flowers in dense heads (often simulating a single flower), subtended by involucral bracts (*Symphyotrichum, Euthamia, Solidago*). .... Asteraceae, 745
67. Flowers borne variously, not in heads, lacking involucral bracts.
68. Leaves and stems scabrous (fig. 13h) (*Campanula aparinoides*). .... Campanulaceae, 733
68. Leaves and stems not scabrous.
69. Leaves alternate.
70. Flowers bilaterally symmetrical, blue or pale-blue (*Lobelia*).................... Campanulaceae, 733
70. Flowers radially symmetrical, white, yellow, or green.
71. Leaves succulent; flowers 5-merous; ovary superior.
72. Flowers white with yellow eye; inflorescence a stalked spike, terminal spikes often paired, axillary spikes single (*Heliotropium curassavicum*)................ Boraginaceae, 652
72. Flowers green (petals absent); inflorescence a sessile compact cluster of mostly 3 flowers, axillary (*Suaeda*)...................................... Amaranthaceae, 589
71. Leaves not succulent; flowers 4-merous; ovary inferior (fig. 411c,d,e) (*Ludwigia linearis*). .... Onagraceae, 517
69. Leaves opposite.
73. Corolla of 4 separate petals, borne at summit of floral tube; stamens separate from petals; ovary inferior (*Epilobium palustre*).. Onagraceae, 517
73. Corolla united, lobes 4 or 5, floral tube absent; stamens inserted on corolla tube; ovary superior.
74. Petals deep blue or purple; corolla lobes fringed or toothed at apex (*Gentianopsis*)...... .... Gentianaceae, 635
74. Petals blue, pink, lavender, or white; corolla lobes not fringed or toothed at apex.
75. Corolla lobes 4, blue, slightly asymmetrical, corolla tube inconspicuous; stamens 2 (*Veronica*)... Plantaginaceae, 655
75. Corolla lobes 5, bilaterally symmetrical, corolla tube well-developed; stamens 4 or 5 (including staminode).
76. Flowers pink to lavender (rarely white); upper corolla lip 3-lobed, lower lip 2-lobed; stamens 4, lacking a staminode (*Agalinis*).............. Orobanchaceae, 728
76. Flowers white, greenish-yellow, or rose; upper corolla lip 2-lobed, hood-like, lower lip 3-lobed; stamens 5, 4 fertile with a single (non-fertile) staminode (*Chelone glabra*).. Plantaginaceae, 655
66. Leaves sheathing at base.
77. Flowers in dense heads.
78. Flower/fruit heads flat-topped (*Cotula*).. Asteraceae, 745
78. Flower/fruit heads globose.
79. Flower/fruit heads axillary or supra-axillary and terminal; leaves keeled (*Sparganium*)......... .... Typhaceae, 174
79. Flower/fruit heads terminal; leaves not keeled (*Eryngium aquaticum*)..................... Apiaceae, 795
77. Flowers borne variously, not in heads.
80. Leaf margins spinulose to conspicuously toothed, or if margin entire, at least leaf sheath at base spinulose (*Najas*).. Hydrocharitaceae, 62
80. Leaf margins entire.
81. Flowers bilaterally symmetrical, with 1 larger petal forming a distinct lip (fig. 125b); stamens fused to style forming a specialized structure, the column, the pollen aggregated in dense waxy mass (pollinium). .... Orchidaceae, 140
81. Flowers radially symmetrical, lacking a lip; stamens and pistils appearing normal not specialized.
82. Leaves acute to blunt at apex, lacking a terminal pore; perianth yellow; fruit a capsule (*Heteranthera dubia*)... Pontederiaceae, 413
82. Leaves blunt at apex, with a terminal pore (having the appearance of a fingernail, fig. 93f); perianth greenish; fruit an inflated follicle (fig. 93e) (*Scheuchzeria*). .... Scheuchzeriaceae, 101

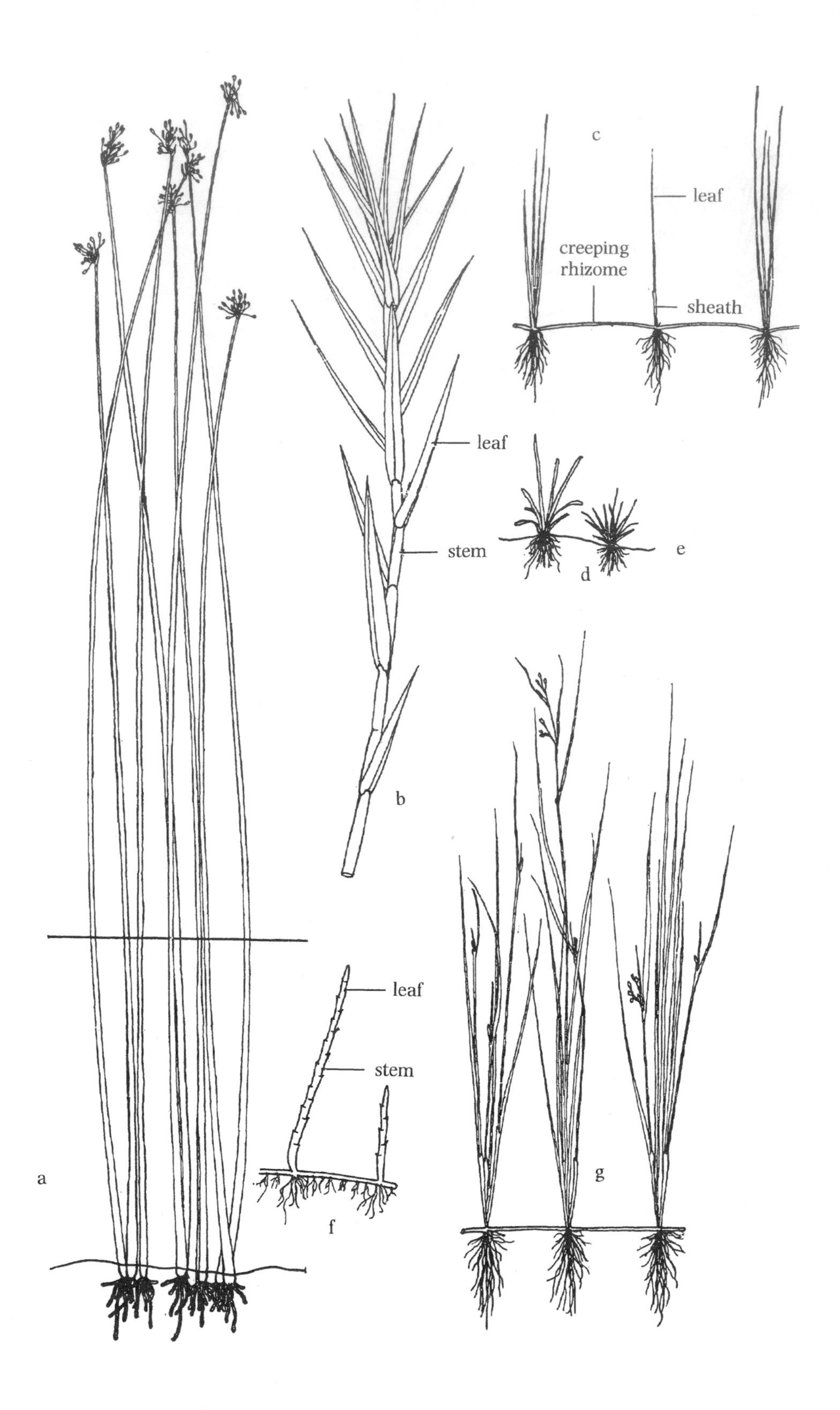

Fig. 10. a. *Schoenoplectus acutus* or related bulrushes; b. *Dulichium*, portion of stem with leaves; c. *Eleocharis acicularis* or *E. radicans*, generalized; d, e. rosette plants (such as *Sagittaria*, *Eriocaulon* spreading by creeping rhizomes), generalized; f. *Myriophyllum tenellum* stems with leaves appearing as bumps; g. Juncaceae (rushes) generalized (F).

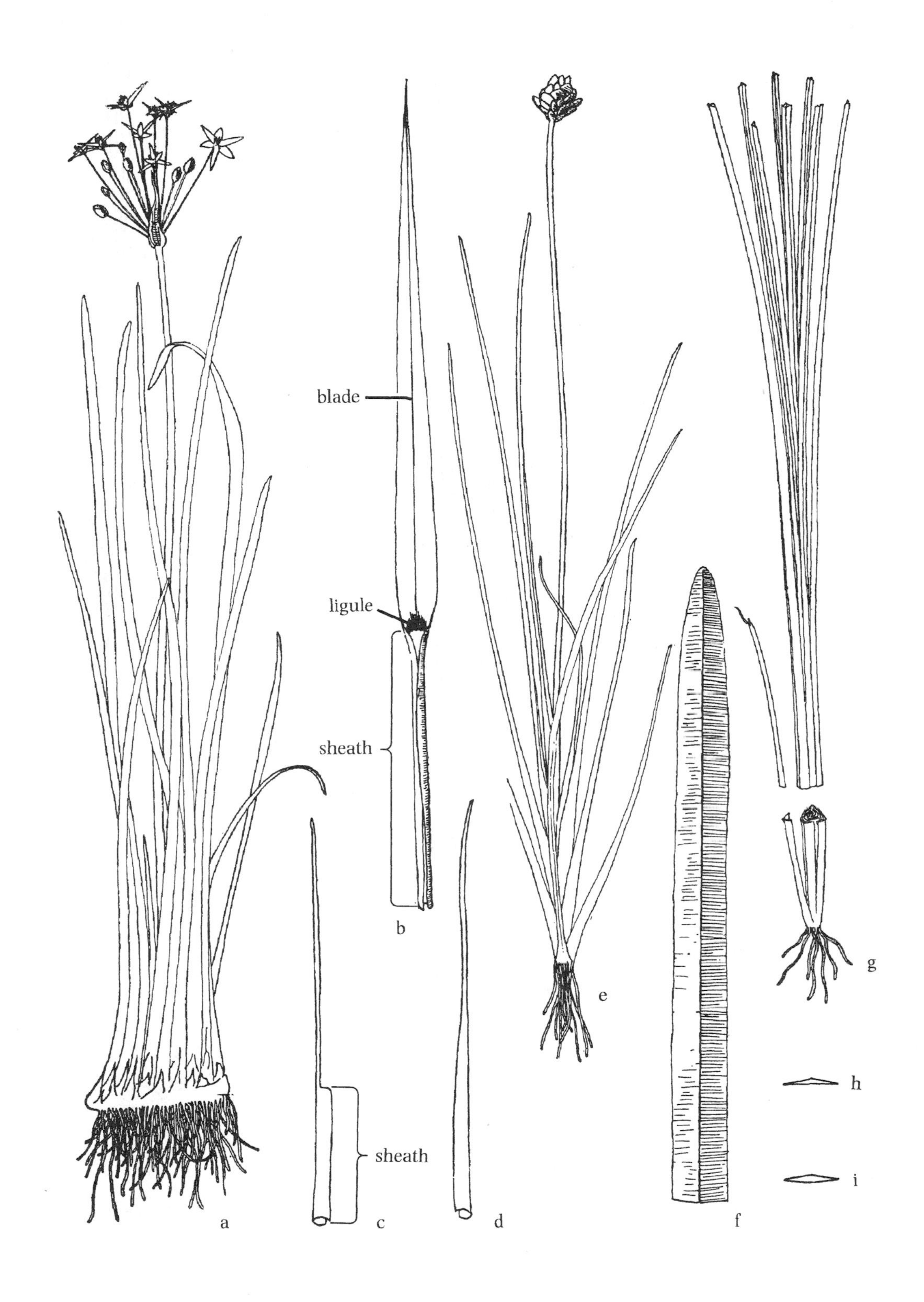

Fig. 11. a. *Butomus* with compact rhizome; b. Poaceae (grasses), leaf, generalized; c. *Typha*, leaf, with auricle at summit of sheath, diagrammatic; d. leaf with tapered sheath, diagrammatic; e. *Xyris*, generalized; f. *Sparganium*, portion of leaf, generalized; g. Cyperaceae (sedges), triangular cross-section, plant generalized; h, i. *Sparganium* or *Acorus*, leaf sections, diagrammatic (F).

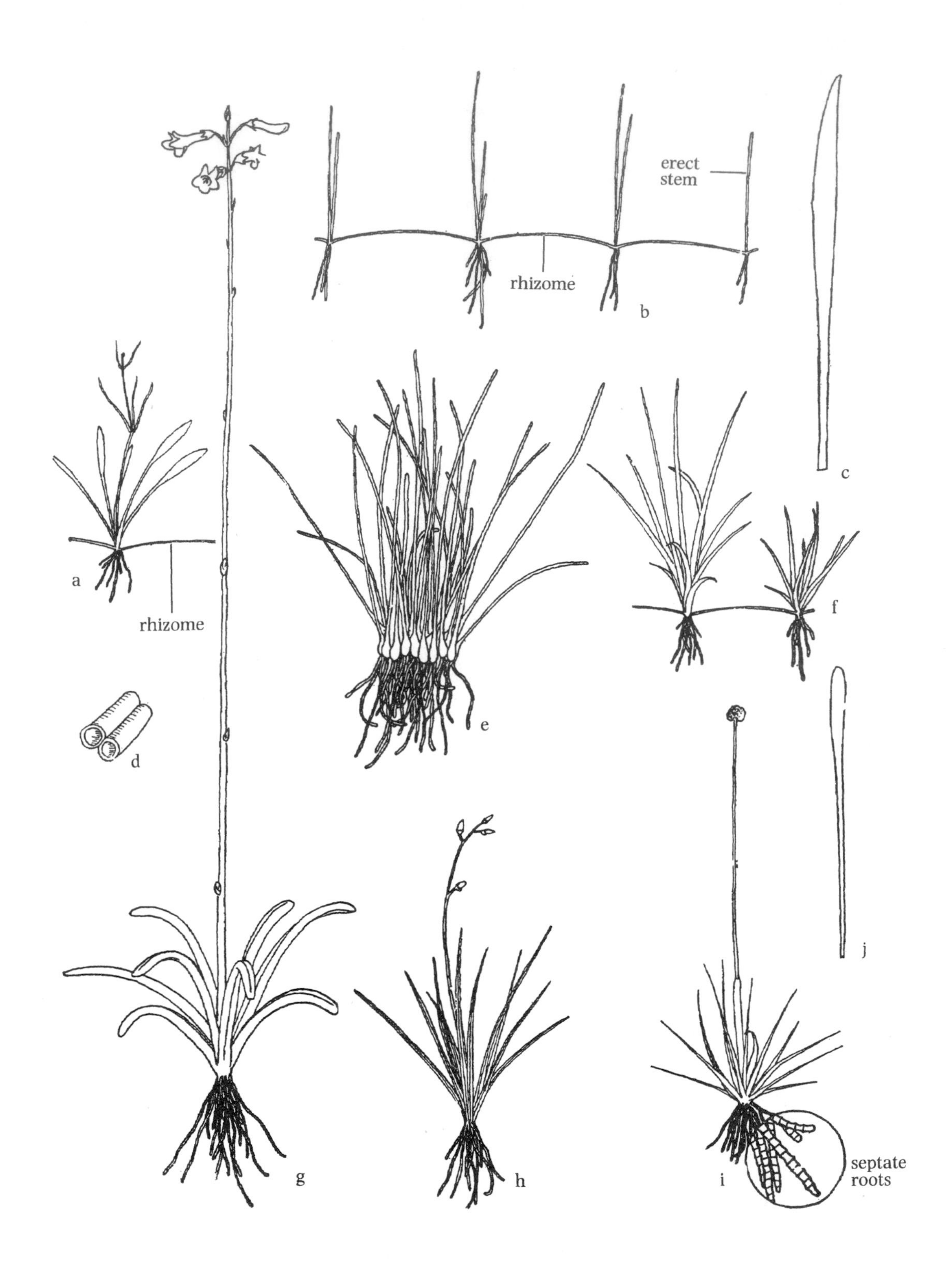

Fig. 12. a. *Ranunculus reptans* var. *ovalis*; b. *Eleocharis acicularis*; c. *Xyris*, leaf, generalized; d. *Lobelia dortmanna*, leaf appearing as 2 hollow tubes in cross-section, diagrammatic; e. *Isoetes*, habit, generalized; f. *Juncus pelocarpus*, submersed form; g. *Lobelia dortmanna*, habit (the basal leaves submerged, the flower on emergent portion of stem); h. *Subularia*, habit (usually submersed); i. *Eriocaulon*, habit, enlargement showing septate roots; j. *Myosurus*, leaf, generalized (F).

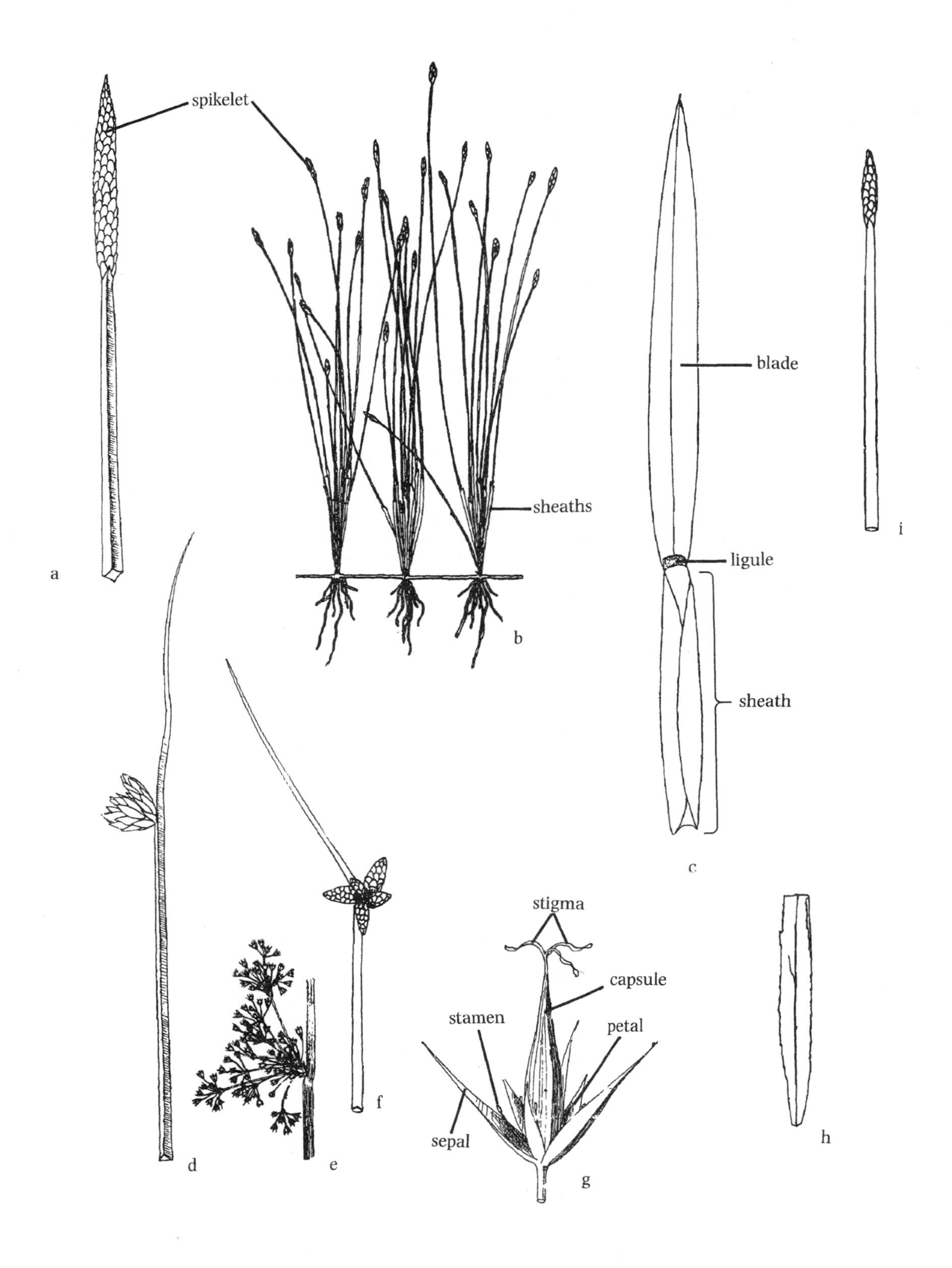

Fig. 13. a. *Eleocharis quadrangulata*, showing inflorescence and square stem cross-section; b. *Eleocharis*, generalized; c. Poaceae (grasses), leaf, generalized; d. *Schoenoplectus pungens* or related species, showing inflorescence and triangular stem cross-section; e. *Juncus effusus*, inflorescence; f. *Schoenplectiella purshiana* or related species, triangular stem section with inflorescence and ascending involucral bract; g. *Juncus*, flower, generalized; h. *Campanula aparinoides*, basal portion of leaf; i. *Eleocharis palustris* or related species, showing bladeless stem with terminal spike (F).

Key 8 Leaves basal or cauline, less than 10 times as long as wide

1. Leaves many-ranked (appearing spiraled), linear-subulate; plants reproducing by spores, sporangia borne in a strobilus (fig. 19). . . .. Lycopodiaceae, 3
1. Leaves alternate, opposite, or whorled; plants reproducing by flowers and seeds.
  2. Flowers with stamens fused to style forming a specialized structure, the column, the pollen aggregated in dense waxy mass (pollinium), the column (orchids unlikely to be collected in vegetative condition); flowers bilaterally symmetrical, with 1 larger petal forming a distinct lip (figs. 127f,g,h, 141d,g)........ Orchidaceae, 140
  2. Flowers with stamens and pistils appearing normal, not specialized; flowers radially or bilaterally symmetrical (lip present in *Pinguicula* and *Thalia*).
    3. Leaves basal; flowering scape naked or bearing a small leaf or stem so short that leaves appear to arise mostly from one point.
      4. Leaves covered with sticky glands.
        5. Leaves petiolate, blades spatulate, oval, or linear, glands distinctly stalked (fig. 447f); corolla white, of separate petals, lacking a spur; flowers numerous, borne in a 1-sided inflorescence (scorpioid cyme) (*Drosera*)........ Droseraceae, 568
        5. Leaves lacking a petiole, blade broad, spatulate to elliptic, glands appearing as dots (hand lens), not stalked; corolla lavender, united, with a long spur; flowers solitary borne, on a scape (*Pinguicula*)........ Lentibulariaceae, 683
      4. Leaves lacking sticky glands.
        6. Leaves tubular, open at upper end (water accumulating in tubular portion) (fig. 480a,b,f,g) (*Sarracenia*)........ Sarraceniaceae, 617
        6. Leaves flat.
          7. Leaves with 3 or more veins running more or less parallel from base toward apex (fig. 14e).
            8. Leaves with stipules at base of petiole (fig. 18c) (*Potamogeton*; short-stemmed forms of normally long-stemmed plants which have been stranded in the mud, especially common in *P. gramineus*, fig. 117c, *P. natans*, *P. pulcher*, and *P. polygonifolius*)........ Potamogetonaceae, 108
            8. Leaves lacking stipules.
              9. Plants usually more than 5 cm tall (few plants of *Parnassia parviflora* shorter).
                10. Inflorescence an elongate golden-yellow spadix (fig. 54d) (*Orontium*)........ Araceae, 53
                10. Inflorescence various or flowers solitary.
                  11. Flowers 4- or 5-merous.
                    12. Flowers solitary; scape with a single leaf near middle (*Parnassia*)........ Celastraceae, 456
                    12. Flowers in an inflorescence; scape lacking any leaves.
                      13. Petals lavender, 5-merous, borne in a panicle (*Limonium*)........ Plumbaginaceae, 589
                      13. Petals whitish (scarious), 4-merous, borne in a spike (*Plantago cordata*)........ Plantaginaceae, 655
                  11. Flowers 3-merous.
                    14. Perianth pink, in a compact inflorescence; leaves evergreen (*Helonias*)........ Melanthiaceae, 140
                    14. Perianth white, solitary or in an open inflorescence; leaves not evergreen.
                      15. Flowers in an inflorescence, spathes absent; leaf blades much longer than wide........ Alismataceae, 76
                      15. Flowers solitary, arising from a spathe; leaf blades nearly round to slightly longer than wide (fig. 60) (*Hydrocharis*, *Limnobium*)........ Hydrocharitaceae, 62
              9. Plants 5 cm or less tall.
                16. Rosettes of leaves connected by runners as thick as the petioles (fig. 18e) (*Ranunculus reptans* var. *ovalis*, common)........ Ranunculaceae, 425
                16. Rosettes connected by white thread-like runners (*Limosella aquatica*, rare)........ Scrophulariaceae, 678
          7. Leaves with 1 major vein running from base toward apex, usually with lateral veins branching from it (fig. 14d,g).
            17. Leaves with a strong skunk-like odor when crushed; flowers borne in a globose spadix, subtended and surrounded by a spathe, flowering early Spring before leaves emerge (fig. 57c) (*Symplocarpus*)........ Araceae, 53
            17. Leaves lacking a skunk-like odor; flowers solitary, or inflorescence various, not borne in a spadix.
              18. Leaves floating, margins coarsely toothed. petiole inflated; fruit a 4-spined nut (fig. 426c) (*Trapa*)........ Lythraceae, 525
              18. Leaves emersed, margins entire to finely toothed or crenate; fruit a capsule, achene, or utricle.
                19. Plants large, 1–3 m tall; blades 30–50 cm long (*Thalia*)........ Marantaceae, 421
                19. Plants smaller, less than 1 m tall; blades 20 cm or less long.
                  20. Flowers solitary, bilaterally symmetrical, spurred (*Viola*)........ Violaceae, 473
                  20. Flowers more than 1, borne in an inflorescence, radially symmetrical, not spurred (*Micranthes*)........ Saxifragaceae, 452
    3. Leaves cauline or cauline and basal; flowers not borne on a scape.
      21. Plants (at least leaves) distinctly succulent; plants mostly of coastal or of inland saline or alkaline sites.
        22. Cauline leaves opposite (fig. 15a).
          23. Leaves extremely minute, scale-like, stems appearing leafless (fig. 464) (*Salicornia*)........ Amaranthaceae, 589
          23. Leaves larger, not scale-like.
            24. Leaves sessile; flowers variously colored, but not green.
              25. Leaves with dots (hand lens needed); flowers weakly bilaterally symmetrical (*Bacopa*)........ Plantaginaceae, 655
              25. Leaves lacking dots; flowers radially symmetrical.

26. Leaf bases connate (fig. 16d); perianth parts separate; pistils 3 or 4; fruit a follicle (*Crassula*). . . . . Crassulaceae, 446
26. Leaf bases not connate; perianth parts united; pistils 1; fruit a capsule (*Lysimachia maritima*). . . . . Primulaceae, 604
24. Leaves petiolate (often some leaves hastate); flowers greenish (*Atriplex*). Amaranthaceae, 589
22. Cauline leaves alternate (fig. 15c), at least upper leaves.
27. Cauline leaves distinctly petiolate.
28. Leaves with distinct fetid or camphoric odor; flowers lavender, purple, or pink, borne in terminal heads (*Pluchea odorata* var. *succulenta*). . . . . Asteraceae, 745
28. Leaves lacking fetid or camphoric odor (often some leaves hastate); flowers greenish, borne in axillary clusters (*Atriplex*). . . . . Amaranthaceae, 589
27. Cauline leaves sessile.
29. Leaves small, linear, nearly round in cross-section; flowers greenish (*Suaeda*). . . . . Amaranthaceae, 589
29. Leaves larger, broad and flat; flowers yellow, white, or bluish.
30. Flowers yellow, borne in a head (*Solidago sempervirens*). . . . . Asteraceae, 745
30. Flowers white or bluish, borne in a scorpioid cyme (*Heliotropium*). . . . . Boraginaceae, 652
21. Plants not succulent (or only slightly succulent); plants generally of non-saline sites.
31. Cauline leaves opposite (fig. 16e) or whorled (fig. 16a).
32. Plants submersed.
33. Leaves expanded at base (fig. 4b), sheathing stem (*Najas*). . . . . Hydrocharitaceae, 62
33. Leaves not sheathing.
34. Leaves opposite (fig. 16e); seed coats pitted (*Elatine*). . . . . Elatinaceae, 469
34. Leaves whorled, (2)3–12 per node; seed coats not pitted.
35. Leaves (2)3–8 per whorl; stems not hollow (small lacunar canals present).
36. Leaves not tipped by a tiny callus; flowers solitary, axillary, rising to and floating at water surface (*Egeria, Elodea, Hydrilla*). . . . . Hydrocharitaceae, 62
36. Leaves tipped by a tiny callus; flowers usually absent on submersed plants (flowers typically appearing on stranded plants, aggregated in a solitary, terminal head) (*Sclerolepis*). . . . . Asteraceae, 745
35. Leaves 6–12 per whorl; stems hollow (*Hippuris*). . . . . Plantaginaceae, 655
32. Plants emersed, or stranded, or with leaves floating at surface.
37. Leaf margins toothed or distinctly wavy.
38. Stems square in cross-section.
39. Flowers radially symmetrical, 4-merous, petals separate; hypanthium urn-shaped (fig. 428b,e) (*Rhexia*). . . . . Melastomataceae, 545
39. Flowers bilaterally symmetrical, weakly to strongly 2-lipped, 5-merous, petals united; hypanthium absent.
40. Leaves scabrous; flowers weakly 2-lipped (*Verbena*). . . . . Verbenaceae, 696
40. Leaves glabrous or pubescent, but not scabrous; flowers strongly 2-lipped, or if weakly irregular, then with a minty odor.
41. Flowers 1 per leaf axil; ovary not lobed; fruit a capsule.
42. Calyx tube 5-angled; calyx lobes shorter than or equaling calyx tube; stamens 4, all fertile (*Mimulus, Erythranthe*). . . . . Phrymaceae, 718
42. Calyx tube not angular; calyx lobes longer than calyx tube; stamens 4, 2 fertile and 2 sterile (*Lindernia*). . . . . Linderniaceae, 678
41. Flowers in axillary whorls or terminal spikes; ovary 4-lobed; fruit 4 nutlets. . . . . . Lamiaceae, 702
38. Stems round in cross-section.
43. Leaves connate at base, encircling stem (fig. 15j) (rarely separate in *Eupatorium perfoliatum*). . . . . Asteraceae, 745
43. Leaves free at base (fig. 15a).
44. Leaves whorled (fig. 15b) (*Eupatorium, Eutrochium, Fleischmannia*). . . . . Asteraceae, 745
44. Leaves opposite, or lower leaves opposite and upper alternate.
45. Flowers borne in a head, yellow (*Bidens*). . . . . Asteraceae, 745
45. Flowers not borne in a head, colors various, usually not yellow.
46. Plants prostrate, creeping to weakly ascending at tip, or floating.
47. Petals united; stamens 2, inserted on petals; fruit a capsule.
48. Calyx lobes 4.
49. Flowers tiny, less than 2 mm, 1 per axil; petals readily falling off (*Micranthemum*). . . . . Linderniaceae, 678
49. Flowers greater than 2 mm, borne in axillary racemes; petals persisting (*Veronica*). . . . . Plantaginaceae, 655
48. Calyx lobes 5 (*Bacopa*). . . . . Plantaginaceae, 655
47. Petals separate; stamens 4–10, not inserted on petals; fruit a 2-lobed capsule or a schizocarp splitting into 2 segments.

50. Flowers blue, borne in heads; leaves petiolate, alternate or in clusters; plants from Mo. and Va. southward (*Eryngium prostratum*). . . . . Apiaceae, 795
50. Flowers greenish-yellow to greenish-red, solitary; leaves usually sessile, upper leaves alternate; plants mostly northern (*Chrysosplenium americanum*). . . . . Saxifragaceae, 452
46. Plants upright.
51. Flowers unisexual, greenish, borne in small clusters on spikes (fig. 397a) or panicles (fig. 398a); petals absent. . . . . Urticaceae, 505
51. Flowers bisexual, variously colored, not green, borne variously, but not in clusters; petals present.
52. Flowers 4-merous, petals separate; ovary inferior (*Epilobium*). . . . . Onagraceae, 517
52. Flowers 5-merous, petals united to form a tube; ovary superior.
53. Flowers radially symmetrical or nearly so.
54. Flowers borne in 1-sided spike-like cymes; stamens 5 (*Cynoctonum*). . . . . Loganiaceae, 635
54. Flowers solitary, in leaf axils; stamens 4, (2 fertile, 2 sterile) (*Lindernia*). . . . . Linderniaceae, 678
53. Flowers bilaterally symmetrical, solitary or borne in various inflorescences, but not 1-sided cymes.
55. Calyx tube 5-angled; calyx lobes shorter than or equaling calyx tube (*Mimulus, Erythranthe*). . . . . Phrymaceae, 718
55. Calyx tube not angular; calyx lobes longer than calyx tube, or sepals separate (*Bacopa, Chelone, Gratiola, Mercardonia*). . . . . Plantaginaceae, 655
37. Leaf margins entire (occasionally margin scabrous, but not toothed, e.g. *Diodia, Galium*).
56. Leaves with numerous translucent dots (visible with a hand lens. especially when held to light). . . . . Hypericaceae, 460
56. Leaves lacking translucent dots or, if dots present, then brown or black, but not translucent.
57. Petiole and/or leaf base fringed (ciliate) (fig. 15i) (*Lysimachia*). . . . . Primulaceae, 604
57. Petiole and/or leaf base not fringed, or leaves sessile.
58. Leaves with dots (hand lens needed).
59. Flowers radially symmetrical: stamens 5 (*Lysimachia*). . . . . Primulaceae, 604
59. Flowers strongly to weakly bilaterally symmetrical: stamens 4 or fewer.
60. Flowers solitary; stamens 4; leaves 6 cm or less long (*Bacopa, Gratiola, Mecardonia*). . . . . Plantaginaceae, 655
60. Flowers borne in a short to elongate spike; stamens 2; leaves 3–20 cm long (*Justicia*). . . . . Acanthaceae, 683
58. Leaves lacking dots.
61. Plants with milky sap (*Asclepias*). . . . . Apocynaceae, 645
61. Plants with clear sap.
62. Leaves petiolate.
63. Stem weak and thread-like, supported by the water, sometimes plants growing stranded on mud; leaves rounded at apex (fig. 15f) (*Callitriche palustris, C. heterophylla*). . . . . Plantaginaceae, 655
63. Stem thick, usually rigid enough to support itself; leaves acute at apex.
64. Leaf blade about twice as long as wide; petals yellow; ovary inferior (*Ludwigia*). . . . . Onagraceae, 517
64. Leaf blade much more than twice as long as wide (fig. 15h); petals white to pinkish. or magenta; ovary superior (*Decodon, Rotala*). . . . . Lythraceae, 525
62. Leaves sessile (tapering to base in some).
65. Leaves in whorls.
66. Distance between leaf whorls less than leaf length; leaf margins not scabrous (fig. 509, 606).
67. Flowers axillary, in upper leaves; stems hollow; leaves not tipped by a tiny callus (*Hippuris*). . . . . Plantaginaceae, 655
67. Flowers terminal, in heads; stems solid; leaves tipped by a tiny callus (*Sclerolepis*). . . . . Asteraceae, 745
66. Distance between leaf whorls greater than leaf length; leaf margins scabrous (fig. 491) (*Galium*). . . . . Rubiaceae, 629
65. Leaves opposite.
68. Leaves with tuft of cottony hairs within leaf base; leaf apex usually with tiny spine (*Alternanthera*). . . . . Amaranthaceae, 589
68. Leaves lacking tuft of hairs within leaf base; leaf apex lacking spine.
69. Leaf margins scabrous (*Diodia*). . . . . Rubiaceae, 629

69. Leaf margins not scabrous.
70. Stems with ridges running down from leaf bases (figs. 16h,i, 18b).
71. Leaves truncate at base (fig. 16i). . . . . .Lythraceae, 525
71. Leaves tapered at base.
72. Stamens numerous; leaves with conspicuous veins branching from midrib (*Hypericum, Triadenum*). . . . . Hypericaceae, 460
72. Stamens 5 or fewer; leaves with obscure veins branching from midrib (*Didiplis, Rotala*).. . . . . Lythraceae, 525
70. Stems lacking ridges running down from leaf bases.
73. Leaf bases connate, forming a boat-shaped cup around stem (fig. 16d) (*Crassula*). . . . . .Crassulaceae, 446
73. Leaf bases, if touching, not fused.
74. Leaf apices acute.
75. Flowers bilaterally symmetrical (*Lindernia*). . . . . Linderniaceae, 678
75. Flowers radially symmetrical. . . . . .Gentianaceae, 635
74. Leaf apices blunt, rounded, or notched.
76. Flowers conspicuous, terminal, petals 4–8 mm long, 4-merous; fruit an elongate capsule (*Epilobium palustre*). . . . . Onagraceae, 517
76. Flowers inconspicuous, axillary, petals less than 4 mm long, 2-, 3-, or 5-merous, or perianth absent; fruit subglobose.
77. Plants with rosette of floating leaves accompanied with linear submersed leaves (slightly notched at apex; fruit a schizocarp, splitting into four 1-seeded segments (fig. 514d) (*Callitriche heterophylla, C. stagnalis, C. palustris*). . . . . Plantaginaceae, 655
77. Plants lacking rosette of floating leaves.
78. Leaves with slight notch at apex; fruit a schizocarp, splitting into four 1-seeded segments (fig. 254) (*Callitriche hermaphroditica, C. stenoptera*). . . . . .Plantaginaceae, 655
78. Leaves not notched at apex; fruit a capsule.
79. Flowers with a floral tube, perianth and stamens inserted at top of tube; seeds not pitted (*Lythrum portula*). . . . . .Lythraceae, 525
79. Flowers lacking a floral tube; seeds having pitted seed coats (*Elatine*, figs. 16f, 95b,c,h,i) (*Bergia*, fig. 374). . . . . .Elatinaceae, 469
31. Cauline leaves alternate or single.
80. Leaves minute, scale-like (figs. 18a, 123a,e).
81. Leaves rounded at apex, borne along simple stems arising from a rhizome (fig. 18a); flowers axillary, borne along stem toward tip (fig. 355d) (*Myriophyllum tenellum*). . . . . Haloragaceae, 446
81. Leaves acute at apex, borne along a solitary stem (fig. 123a,e), rhizome absent; flowers terminal, single or in a small cluster. . . . . .Burmanniaceae, 140
80. Leaves much larger.
82. Leaf margins toothed or irregularly lobed.
83. Leaves sessile (tapering to base in some).
84. Plants submersed; leaves with stipules at base, wavy-margined (*Potamogeton crispus*).. . . . . .Potamogetonaceae, 108
84. Plants emersed to terrestrial: leaves lacking stipules, not wavy-margined.
85. Flowers borne in heads.. . . . . Asteraceae, 745
85. Flowers not borne in heads.
86. Stems square in cross-section; leaves scabrous (*Phyla*). . . . . Verbenaceae, 696
86. Stems round in cross-section; leaves not scabrous.
87. Flowers bilaterally symmetrical, 5-merous; each leaf vein ending in a white gland at margin (fig. 17f) (*Lobelia*). . . . . .Campanulaceae, 733
87. Flowers radially symmetrical, 4-merous; leaf veins not ending in a white gland at margin.

88. Flowers with elongate hypanthium forming a floral tube at the summit of the ovary; stamens 8; ovary inferior (*Epilobium*)........................ Onagraceae, 517
88. Flowers lacking hypanthium; stamens 6 (4 long, 2 short); ovary superior. ... .Brassicaceae, 554
83. Leaves petiolate.
89. Leaves velvety on lower surface (hairs star-shaped) (*Hibiscus*).......................... Malvaceae, 549
89. Leaves not velvety on lower surface.
90. Leaf blade about as long as wide.
91. Leaves coarsely serrate (fig. 14d); petiole inflated (leaves floating) (fig. 17c); flowers solitary (*Trapa*)... Lythraceae, 525
91. Leaves wavy-margined, but not serrate; petiole not inflated: flowers in a raceme...... ....Brassicaceae, 554
90. Leaf blade much longer than wide.
92. Base of petiole sheathing, partly or completely encircling stem.
86. Base of petiole forming a tubular sheath surrounding stem (ochrea) (fig. 17d); flowers inconspicuous; perianth reddish-green (*Rumex*). .... Polygonaceae, 572
86. Base of petiole sheathing stem, but not forming a tube; flowers conspicuous, perianth yellow, or white to pinkish (*Caltha*)......................... Ranunculaceae, 425
92. Base of petiole not forming a sheath.
93. Leaves all similar, not dimorphic.
94. Leaf margin evenly and uniformly serrate, never lobed.
95. Base of petiole fringed with white hairs; ovary inferior (*Campanulastrum americana*)..Campanulaceae, 733
95. Base of petiole lacking fringe of hairs (fig. 17a); ovary superior.
96. Plants with rhizomes, perennial; petals tiny, greenish, not spurred (*Penthorum sedoides*).................................. Penthoraceae, 446
96. Plants lacking rhizomes, annual; petals larger, yellow or orange, spurred (fig. 473) (*Impatiens*). ....Balsaminaceae, 604
94. Leaf margin irregularly serrate, becoming lobed near base (fig. 17e) (*Rorippa palustris*)...Brassicaceae, 554
93. Leaves of two types, pinnatifid to pectinate on lower portion of stem, serrate on upper portion (especially as stem become emergent) (figs. 8j, 17b) (upper leaves often becoming pinnatifid to pectinate late in season on stranded plants) (*Proserpinaca*). .... Haloragaceae, 446
82. Leaf margins entire (fig. 18c).
97. Veins several to many, running from base of leaf to apex without branching (fig. 14c), midrib lacking.
98. Leaf blades plicate (*Veratrum*). .... Melanthiaceae, 140
98. Leaf blades flat.
99. Plants submersed, leaves submersed (figs. 102a, 105a,b, 110, 119) or both floating and submersed (figs. 18g, 102c, 104, 117a,b); stipules present, free from the petioles (fig. 105c); leaves not sheathing at base; flowers not showy, greenish, 4-merous (*Potamogeton*)....... .... Potamogetonaceae, 108
99. Plants emersed or terrestrial; stipules absent; leaves sheathing at base; flowers showy, white, blue to purplish-blue, 3-merous.
100. Leaf base cordate to truncate, petiolate; perianth united to form a tube; stamens borne on perianth tube, 6, or if 3, then anthers dimorphic, 1 larger, oblong to sagittate, 2 smaller, ovate (*Heteranthera, Pontederia*)..................... Pontederiaceae, 413
100. Leaf base narrowly tapering to a sessile or subpetiolar sheath; perianth of separate parts; stamens not borne on perianth, 3, anthers all alike (*Maianthemum*)........... .... Asparagaceae, 174
97. Veins branching from midrib, or only midrib conspicuous (lateral veins obscure).
101. Stem encircled by stipular sheath (ochrea) at petiole base (fig. 18i).................. Polygonaceae, 572
101. Stem lacking stipular sheath.
102. Stem square or 3-angled, or appearing nearly square to roundish, with ridges running down from leaf bases (*Lythrum*, fig. 18b).
103. Flowers in heads, subtended by involucral bracts (*Oclemena nemoralis*)...... Asteraceae, 745
103. Flowers solitary or in axillary clusters, or in an inflorescence, but not forming a head subtended by involucral bracts.
104. Flowers with calyx and corolla adnate, forming a floral tube, petals usually 6(4–7), separate, arising from summit of floral tube; petals purple, magenta. or pale lilac to pink or white (if white, then flowers axillary) (*Lythrum*)...... ....Lythraceae, 525
104. Flowers with calyx and corolla separate from each other, petals 5, united, arising from base or summit of ovary; petals light blue to white (if white, then flowers terminal).

105. Inflorescence a scorpoid cyme; corolla salverform, light blue with a yellow center; ovary superior, 4-lobed (*Myosotis*). . . . . Boraginaceae, 652
105. Inflorescence a spike-like raceme or flowers solitary, terminal; corolla bell-shaped, light blue to white, lacking a yellow center; ovary inferior, not lobed (*Campanula*). . . . . Campanulaceae, 733

102. Stem not square or angled, lacking ridges running down from leaf bases.
106. Leaves with 3 conspicuous veins running from near base toward apex (fig. 18h); flowers with numerous pistils (*Ranunculus*). . . . . Ranunculaceae, 425
106. Leaves with midrib and branching lateral veins; flowers with 1 pistil (often deeply 4-lobed in Boraginaceae).
107. Plants pubescent.
108. Petals light blue with yellow center, borne in a terminal, scorpioid cyme; axillary spines absent (*Myosotis*). . . . . Boraginaceae, 652
108. Petals deep blue, borne in axillary or terminal cymes; axillary spines usually present (fig. 502a) (*Hydrolea*). . . . . Hydroleaceae, 645
107. Plants glabrous.
109. Leaf blades usually much less than 10 mm long, and nearly as wide (*Nasturtium officinale*, a dwarfed growth form). . . . . Brassicaceae, 554
109. Leaf blades usually 10 mm or more long, much longer than wide.
110. Cauline leaves usually less than twice as long as wide; ovary half-inferior (*Samolus*). . . . . Primulaceae, 604
110. Cauline leaves more than twice as long as wide (fig. 18f); ovary superior or inferior.
111. Leaves usually 5 mm or less wide, with whitish gland-like callus along margin at terminus of each lateral vein (fig. 17f) (*Lobelia*). . . . . Campanulaceae, 733
111. Leaves mostly 5 mm or more wide, lacking whitish gland-like calluses along margin.
112. Flowers in heads, subtended by involucral bracts (figs. 584a, 585c, 607). . . . . Asteraceae, 745
112. Flowers solitary or in spikes, not subtended by involucral bracts.
113. Flowers yellow, 4-merous, solitary in leaf axils (fig. 15g) (*Ludwigia*). . . . . Onagraceae, 517
113. Flowers green, 5-merous, borne in spikes (*Amaranthus*). . . . . Amaranthaceae, 683

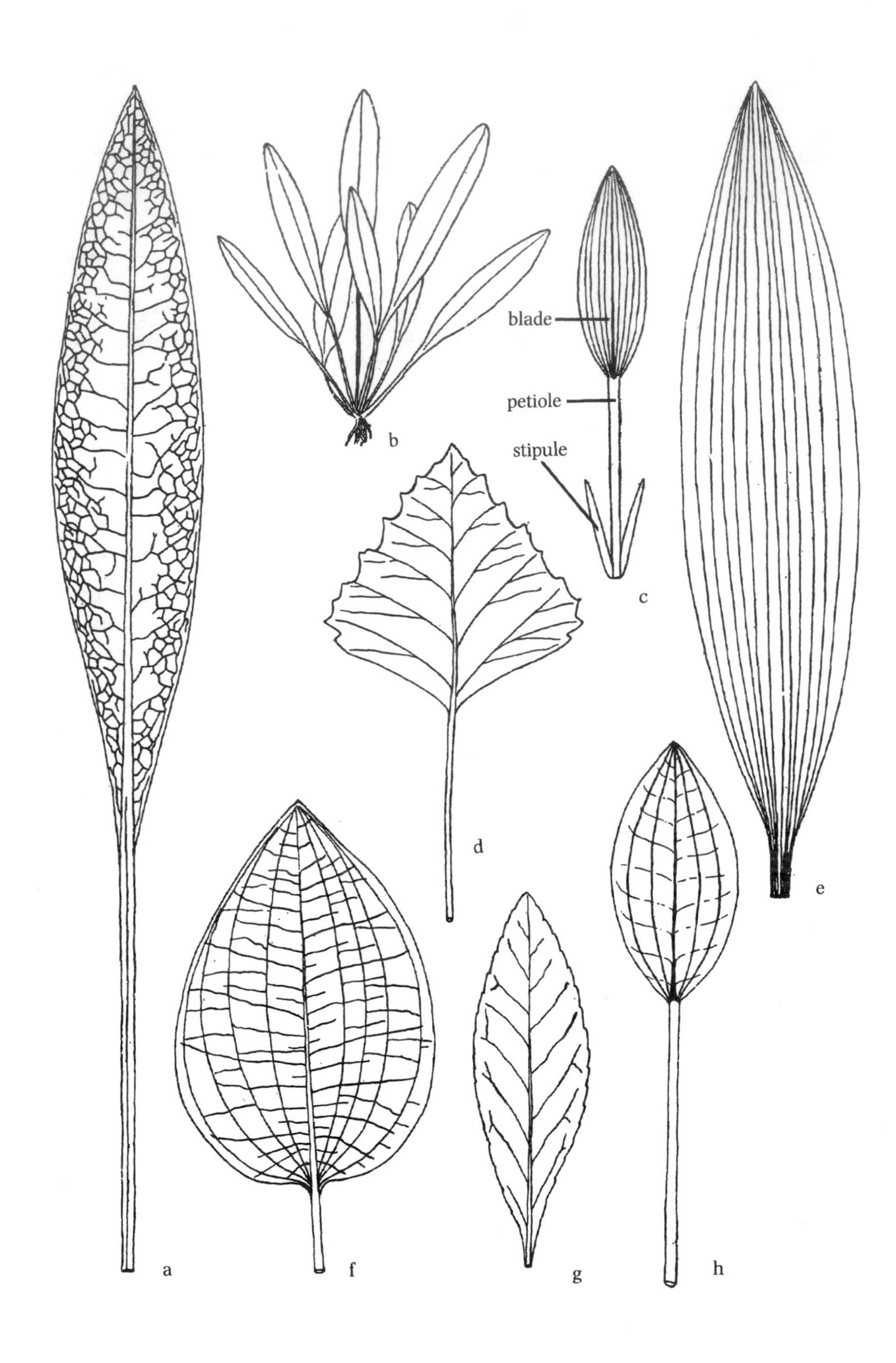

Fig. 14. Leaf venation patterns: a. *Rumex*, leaf; b. *Rumex*, basal leaves with strong midvein, arising from one point; c. *Potamogeton*, terrestrial form, showing parallel venation with 2 stipules at base of petiole; d. *Trapa*, leaf; e. *Orontium*, leaf; f. *Limnobium*, leaf (often spongy viewed from under side); g. *Viola lanceolata*, lowermost basal leaf showing typical venation pattern; h. Alismataceae (*Alisma*, water-plantains), leaf, generalized, showing strong parallel venation pattern (F).

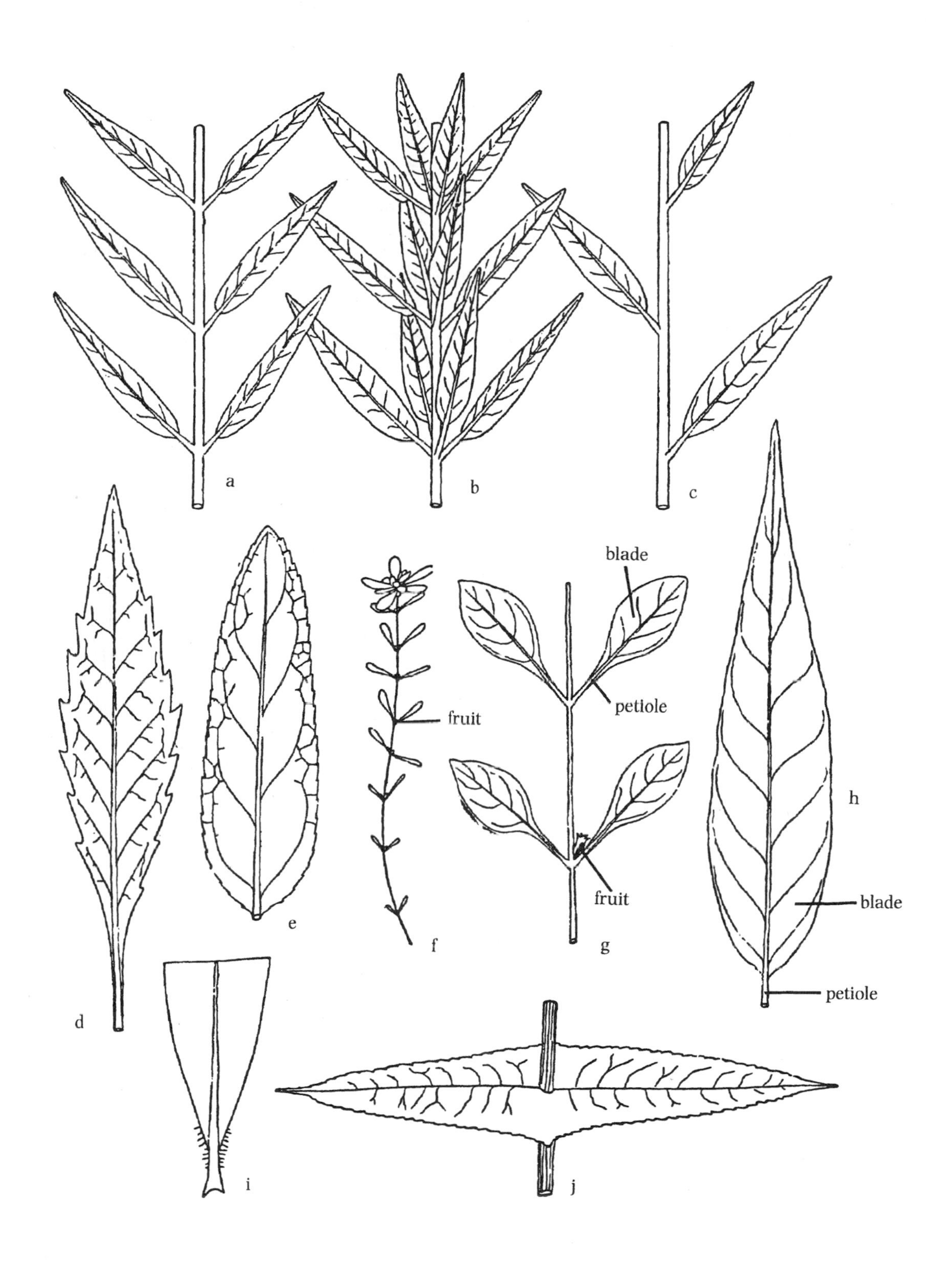

Fig. 15. a. Opposite leaves, diagrammatic; b. whorled leaves, diagrammatic; c. alternate leaves, diagrammatic; d. *Bidens*, leaf; e. *Veronica*, leaf, diagrammatic; f. *Callitriche*, showing floating rosette of uppermost leaves and fruits in axils of opposite, submersed leaves; g. *Ludwigia palustris*, showing an axillary fruit; h. *Decodon*, leaf; i. *Lysimachia*, lower portion of leaf showing ciliate petiole and leaf base, generalized; j. *Eupatorium perfoliatum*, pair of leaves, bases fused (connate) (F).

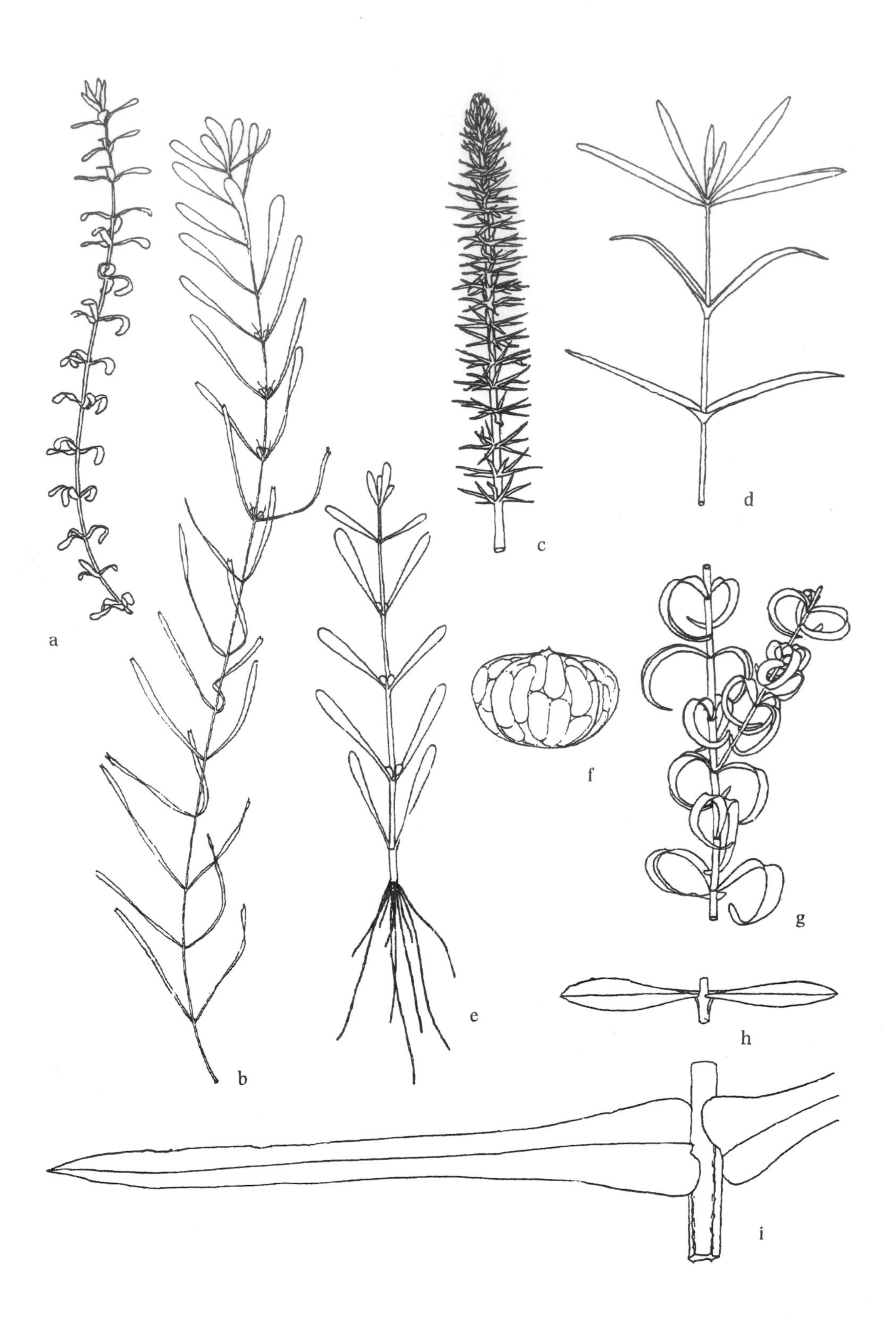

Fig. 16. a. *Elodea*, showing whorled leaves; b. *Callitriche*, generalized, with some axillary fruits on submersed stems; c. *Hippuris*, newly emersed upper portion; d. *Crassula aquatica*, upper portion showing opposite leaves with connate bases forming boat-shaped cup around stem; e. *Elatine triandra*, submersed form; f. *Elatine*, capsule, generalized; g. *Elodea*, portion with leaves opposite; h. *Didiplis diandra*, terrestrial form; i. *Ammannia coccinea*, leaf and leaf base, showing truncate leaf base and ridge running down 4-angled stem from leaf base (F).

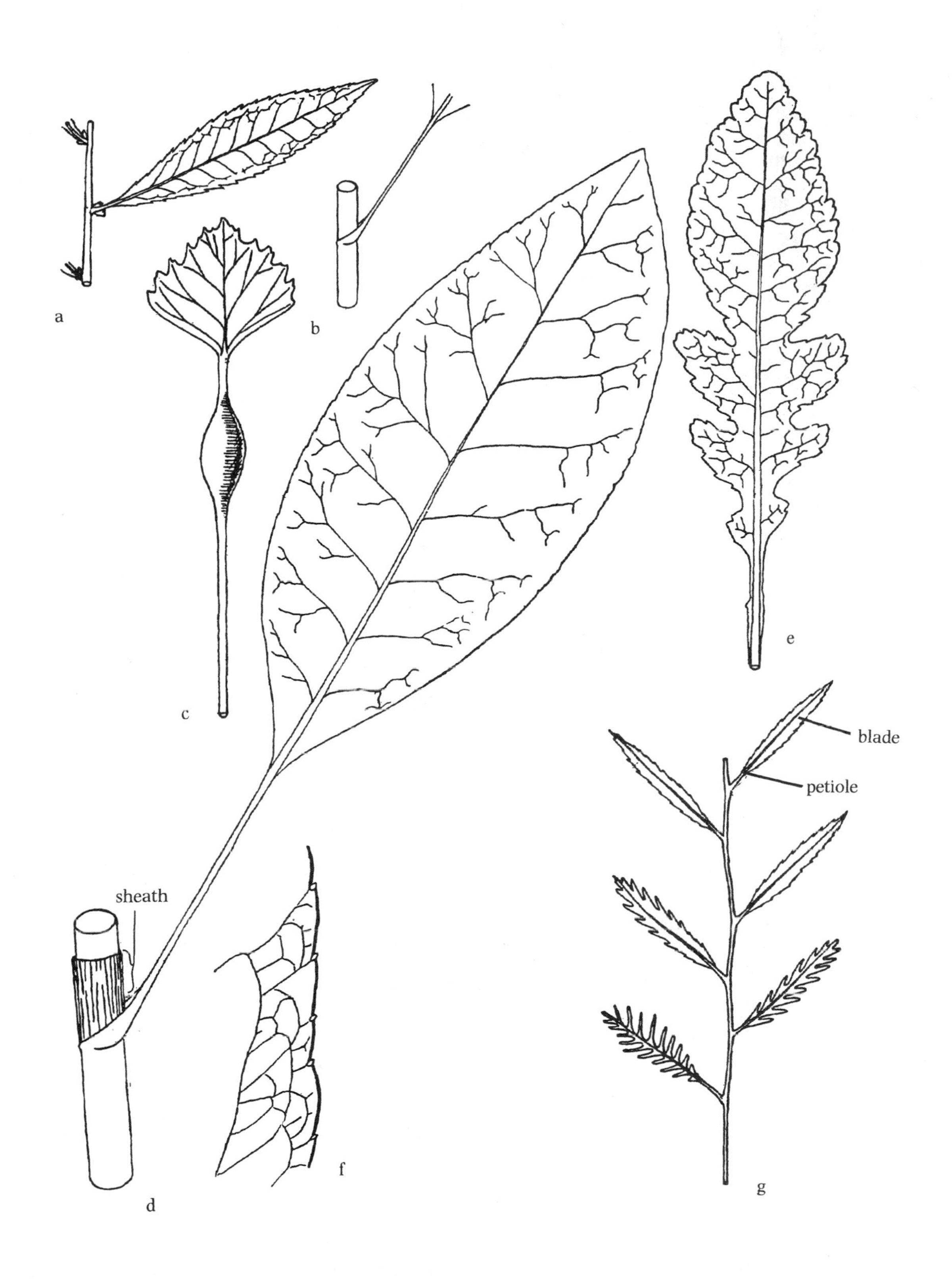

Fig. 17. a. *Penthorum*, leaf with pair of stipules at base; b. petiole with clasping leaf base; c. *Trapa*, leaf with petiole inflated with spongy tissue; d. *Rumex*, section of stem with leaf, showing stipular sheath (ochrea); e. *Rorippa palustris*, leaf; f. *Lobelia cardinalis*, portion of leaf showing gland-like callus at vein tips along margin; g. *Proserpinaca palustris*, showing leaf variation on a single plant (F).

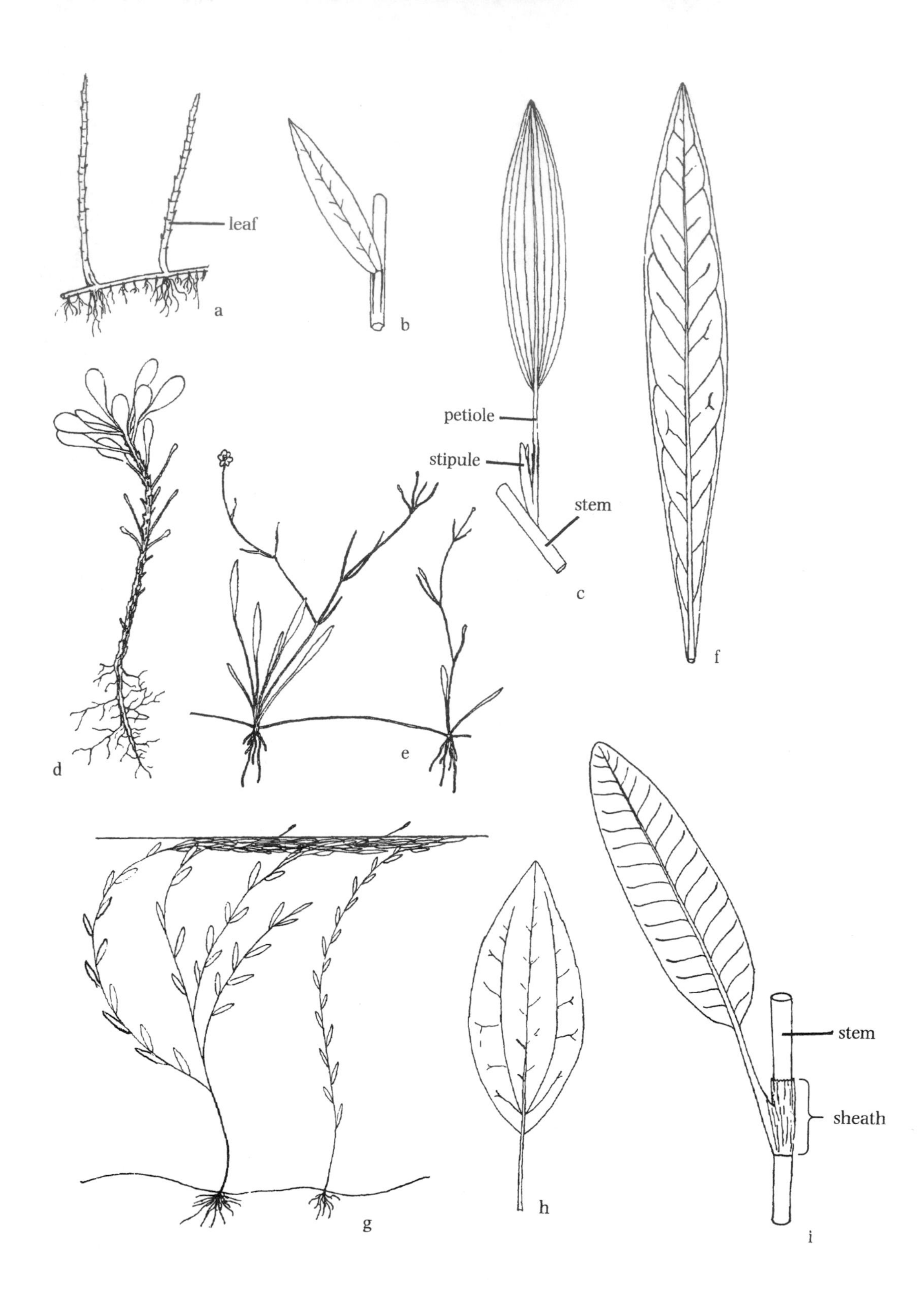

Fig. 18. a. *Myriophyllum tenellum*, stems with leaves appearing as bumps; b. *Lythrum alatum*, section of stem with leaf, showing ridges running down stem from leaf base; c. *Potamogeton*, leaf with pair of stipules at base, generalized; d. *Nasturtium officinale*, a small submersed growth form; e. *Ranunculus reptans*, terrestrial form; f. *Ludwigia polycarpa*, leaf; g. *Potamogeton*, generalized, stems reaching and partly floating on surface; h. *Ranunculus laxicaulis*, leaf; i. *Persicaria*, section of stem with leaf, showing stipular sheath (ocrea), generalized (F).

Aquatic
and Wetland Plants
of Northeastern
North America

Pteridophytes (Ferns and Fern Allies)

REFERENCES: Kramer, 1993; Lellinger, 1985; Mickel, 1979; Palmer, 2018; Pteridophyte Phylogeny Group, 2016; Schneider et al., 2009; Smith et al., 2006; Tryon and Tryon, 1982.

Lycopodiaceae / Club-moss Family

REFERENCES: Pteridophyte Phylogeny Group, 2016; Wagner and Beitel, 1992, 1993.

---

1. Sporophylls of strobilus yellowish, wider and shorter than erect stem leaves; leaves of erect stems few, scattered, scale-like; leaves on horizontal stems dimorphic, lateral leaves larger than the median leaves. . . . . 1. *Pseudolycopodiella*
1. Sporophylls of strobilus greenish, similar to erect stem leaves; leaves of erect stems numerous, crowded, unmodified; leaves on horizontal stems all similar. . . . . 2. *Lycopodiella*

---

1. *Pseudolycopodiella* (Slender Bog Club-moss)

Perennials, stems creeping an erect; leaves of stem small, 1-nerved, sparse, scale-like; sporangia borne in strobili, in axils of small, bracteate sporophylls; strobilus not conspicuously differentiated from erect stem; homosporous.

1. *P. caroliniana* (L.) Holub Slender Bog Club-moss Fig. 19
Wet pine barrens, damp peats, ditches, and wet sands. Coastal plain, local, c. Mass. and L.I., N.Y., s. to s. N.J., Md. and Va.; more common, N.C. s. to Fla., w. to c. Tex. and Mex.; W.I., C.Am. and S.Am. (*Lycopodium carolinianum* L.; *Lycopodiella caroliniana* (L.) Pichi Sermolli)

2. *Lycopodiella* (Bog Clubmoss)

Perennials, stems creeping and erect; leaves of erect stem small, 1-nerved, numerous; sporangia borne in strobili, in axils of sporophylls; strobilus conspicuously differentiated from erect stem; sporophylls longer than erect stem leaves; homosporous.

---

1. Erect fertile stems 3–7(10) cm long; strobili with sporophylls spreading; leaves of horizontal stems up to 6 mm long. . . . . 1. *L. inundata*
1. Erect fertile stems (4)8–35(45) cm long; strobili with sporophylls appressed; leaves of horizontal stems 4–13 mm long.
   2. Leaves of horizontal stems typically with toothed margins; horizontal stems strongly arching, 2–4 mm wide; erect stems 5–15 mm in diameter (including leaves); strobili 10–20(25) mm in diameter, sporophylls wide-spreading; horizontal stems arching, rooting where touching ground. . . . . 2. *L. alopecuroides*
   2. Leaves of horizontal stems typically entire, or with sparse scattered marginal teeth; erect stems 2–9 mm in diameter (including leaves); strobili 3–4(8) mm diameter, sporophylls appressed, incurved; horizontal stems prostrate (flat on ground, not arching) rooting along entire length.
      3. Erect fertile stems 15–40 cm; horizontal stem leaves with scattered marginal teeth. . . . . 3. *L. appressa*
      3. Erect fertile stems 9–15 cm; horizontal stem leaves lacking marginal teeth.
         4. Horizontal stems 1–1.7(2) mm diameter, leaves typically 4–6 mm long, subappressed (arising at an acute angle); fertile shoots 9–13 cm with short strobilus, the strobili 2–4 cm long. . . . . 4. *L. subappressa*
         4. Horizontal stems 1.8–2.2(3) mm diameter; leaves typically 6–13 mm long, spreading (often perpendicular to stem); fertile shoots, 13–17 cm with longer strobilus, the strobili 5–8 cm long. . . . . 5. *L. margueritae*

---

1. *L. inundata* (L.) Holub Northern Bog Club-moss Fig. 19
Damp sands, peats, swamps, lakeshores, bogs, and gravel pits. Lab. and Nfld. w. to Ont. and Minn., s. to Va., W.Va., and Wisc.; Sask. w. to s. Alask., s. to w. Mont. and Wash. (*Lycopodium inundatum* L.)

2. *L. alopecuroides* (L.) Cranfill Foxtail Bog Club-moss Fig. 19
Bogs, sandy peats, savannas, wet barrens, ditches and gravel pits. Coastal plain, Nantucket I., Mass., R.I., and L.I., N.Y., s. to Fla., w. to La. and e. Tex.; w. Ky. (*Lycopodium alopecuroides* L.; *L. inundatum* var. *alopecuroides* (L.) Tuck.)

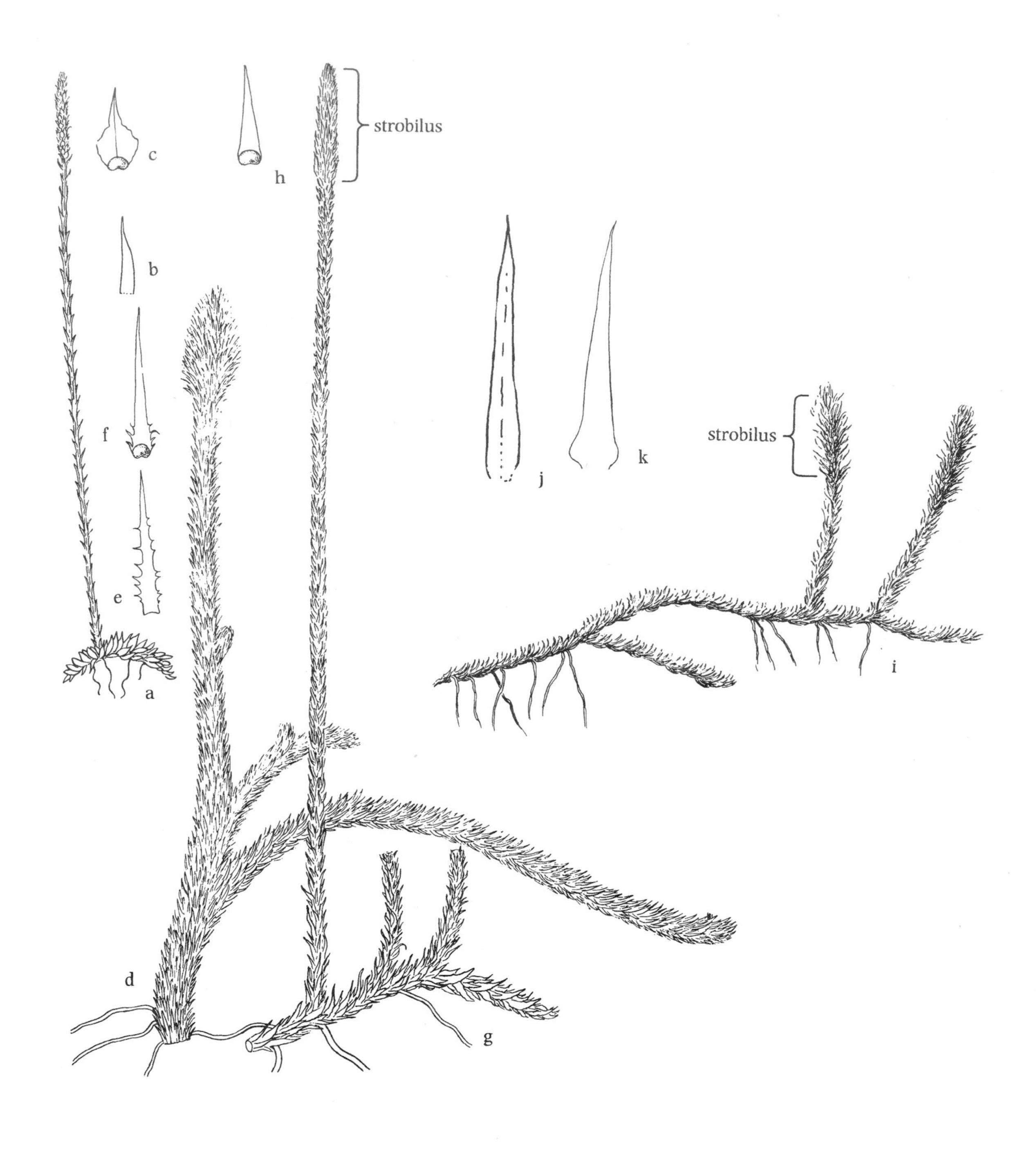

Fig. 19. *Pseudolycopodiella caroliniana*: a. habit; b. leaf; c. sporophyll (Lundell).
Lycopodiella alopecuroides: d. habit; e. leaf; f. sporophyll (Lundell).
Lycopodiella appressa: g. habit; h. sporophyll (Lundell).
Lycopodiella inundata: i. habit; j. leaf; k. sporophyll (HCOT).

3. *L. appressa* (Chapm.) Cranfill Appressed Bog Club-moss
Fig. 19
Wet sandy and peaty shores, bogs, savannas, ditches, and gravel pits. N.S., and N.E. s. to Del.; coastal plain and piedmont, Va. s. to Fla., w. to Mo., e. Kan., e. Okla., and e. Tex. (*Lycopodium appressum* (Chapm.) Lloyd & Underw.; *L. inundatum* var. *appressum* Chapm.; *L. inundatum* var. *bigelovii* Tuck.)

4. *L. subappressa* J. G. Bruce Northern Appressed Clubmoss
Wet, acidic ditches and sandy borrow pits. Endemic to sw. Mich.; a sexually reproducing polyploid (Bruce et al., 1991).

5. *L. margueritae* J. G. Bruce Northern Prostrate Club-moss
Wet, acidic ditches and sandy borrow pits. Endemic to Mich.; a sexually reproducing polyploid (Bruce et al., 1991).

Isoetaceae / Quillwort Family

1. *Isoetes* (Quillwort)

Perennials, arising from a short corm-like rootstock; leaves simple, linear, spirally arranged, expanded at base; sporangia borne singly in leaf bases, at least partly covered by the velum, containing either megaspores or microspores; megaspores white, globose, ca. 0.5 mm in diameter with an equatorial ridge and three radial ridges converging at the spore's proximal pole; microspores gray or brown in mass, dust-like.

The species of quillworts look very much alike, the plants always consisting of a number of linear leaves from a somewhat swollen base. Surface ornamentation and size of mature, dry megaspores are critical for identification. A hand lens is essential for resolving megaspore ornamentation, and a compound microscope fitted with an ocular micrometer is necessary for determining spore diameter. Twenty megaspores should be measured to determine average spore size.

References: Britton and Goltz, 1990; Brunton and Britton, 1991, 1996; Hickey, 1986; Kott, 1981; Kott and Britton, 1980, 1983, 1985; Mickel, 1979; Pfeiffer, 1922; Reed, 1965; Taylor et al., 1993.

1. Megaspores averaging 500 μm or less in diameter.
  2. Megaspore surface with spines (fig. 20c). . . . . 1. *I. echinospora*
  2. Megaspore surface with ridges (fig. 20d) or wrinkled/tuberculate (fig. 20j).
    3. Megaspore surface with high thin ridges.
      4. Ridges forming complete reticulum with even crests (fig. 20d). . . . . 2. *I. engelmannii*
      4. Ridges forming broken reticulum with uneven crests. . . . . 3. *I. valida*
    3. Megaspore surface obscurely to distinctly wrinkled/tuberculate (fig. 20j).
      5. Megaspores distinctly tuberculate to wrinkled. . . . . 4. *I. virginica*
      5. Megaspores obscurely wrinkled or tuberculate, the girdle obscure.
        6. Velum incompletely covering sporangium; megaspore girdle prominent, wide; habitat tidal rivers of e. Va. . . . . . 5. *I. mattaponica*
        6. Velum completely covering sporangium; megaspore girdle obscure; habitat of temporary pools, ditches and rivulets, somewhat widespread, or of deep, cold lakes of N.S., N.B., Me.
          7. Leaf bases typically lustrous black; plants of temporary pools, ditches and rivulets, more widespread in distribution. . . . . 6. *I. melanopoda*
          7. Leaf bases pale; plants of deep, cold lakes of N.S., N.B., Me. . . . . 7. *I. prototypus*
1. Megaspores averaging more than 500 μm in diameter.
  8. Megaspores averaging more than 600 μm in diameter. . . . . 8. *I. lacustris*
  8. Megaspores averaging less than 600 μm in diameter.
    9. Megaspore surface with reticulate (fig. 20h) or isolated, high, thin ridges (fig. 20e).
      10. Leaves olive-green, stout, stiff, recurved at tip; megaspore surface with reticulate ridges (fig. 20h) and densely papillate girdle along distal side of equatorial ridge. . . . . 9. *I. tuckermanii*
      10. Leaves bright green, often fine, lax, twisted; megaspore surface with mostly isolated ridges and without densely papillate girdle along distal side of equatorial ridge (fig. 20c). . . . . 10. *I. riparia*
    9. Megaspore surface with reticulate bold wrinkles or low, round ridges. . . . . 11. *I. acadiensis*

1. *I. echinospora* Durieu Spiny-spored Quillwort
Fig. 20
Shallow water and wet shores. Green. and Lab, w. to Alask., s. to Nfld., N.E., N.J., n. Pa., n. Ohio, Mich., Wisc., Minn., c. Sask., w. Mont., Ida., n. Colo., n. Utah and Calif. (*I. echinospora* var. *muricata* (Durieu) Engelm.; *I. echinospora* var. *braunii* (Durieu) Engelm.; *I. muricata* Durieu; *I. braunii* Durieu)

2. *I. engelmannii* A. Braun Engelmann's Quillwort
Fig. 20
Ponds, wet shores, and muddy ditches. S. N.H. and Vt. w. to N.Y., Ohio, s. Ont., sw. Mich., s. Ill. and e. Mo., s. to n. Fla. and Ala. (*I. engelmannii* var. *canadensis* Engelm.)

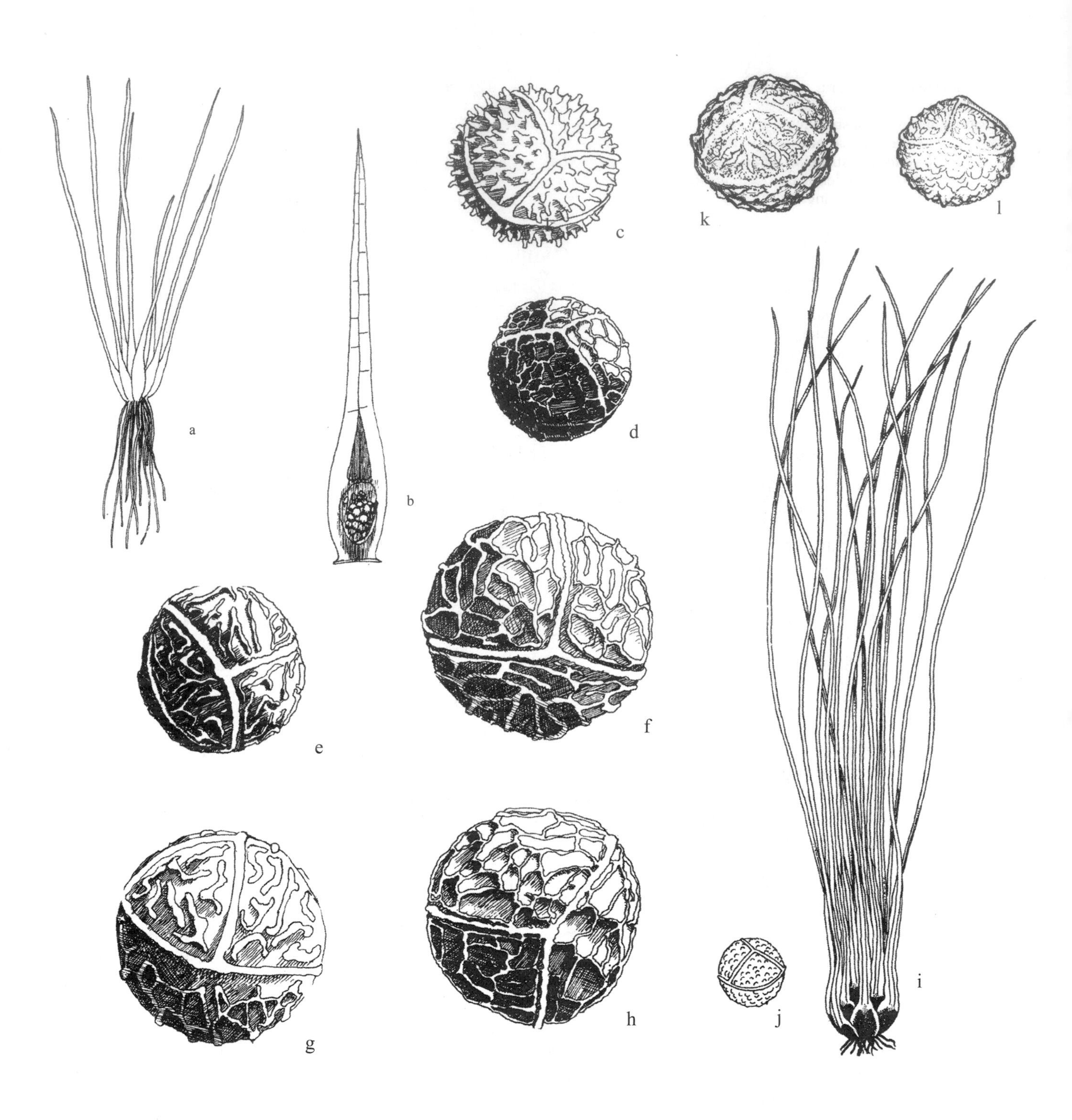

Fig. 20. *Isoetes echinospora*: a. habit; b. inner surface leaf; c. megaspore (F).
Isoetes engelmannii: d. megaspore (F).
Isoetes riparia: e. megaspore (F).
Isoetes lacustris: f, g. megaspore (F).
Isoetes tuckermanii: h. megaspore (F).
Isoetes melanopoda: i. habit; j. megaspore (B&T).
Isoetes valida: k. megaspores (FNA).
Isoetes virginica: l. megaspore (FNA).

3. *I. valida* (Engelm.) Clute Quillwort Fig. 20
Shallow water of upland lakes, bogs, peaty woodland seepages and pools, and shaded stream banks. Del., s. Pa., and W.Va. s. to Va., N.C., e. Tenn., n. Ga. and Ala. Brunton and Britton (1996) noted that *I. valida* constitutes an earlier valid name for the Appalachian/Piedmont endemic previously known as *I. caroliniana*. (*I. engelmannii* var. *valida* Engelm.)

4. *I. virginiana* N. Pfeiff Virginia Quillwort Fig. 20
Mud flats and in temporary depression pools on and around granitic outcrops. Va., N.C., w. S.C., Ga. and e. Ala.

5. *I. mattaponica* Musselman & W. C. Taylor
An aquatic amphibious plant of freshwater tidal marshes of Chesapeake Bay. Known only from the n. Chickahominy, Pamunkey, and Mattaponi Rivers of e. Va.

6. *I. melanopoda* Gay & Durieu Black-footed Quillwort Fig. 20
Vernal pools, rivulets, ditches, temporary shallow ponds and seasonally wet pockets. Local, N.J., Va., N.C. and S.C.; w. Ind., w. to S.D., s. to nw. Ga., Tenn., w. Miss. and e. Tex., Utah.

7. *I. prototypus* D. M. Britton Big Quillwort
Deep water, cold lakes. N.B., N.S. and e. Me.

8. *I. lacustris* L. Lake Quillwort Fig. 20
Ponds, lakes, and streams. S. Greenl., Lab., and Nfld. w. to Ont., Man., and Sask., s. to R.I., N.Y., Mich. and Wisc.; n. Va., se. Tenn. According to Taylor et al. (1993) *I. macrospora* is conspecific with the European *I. lacustris*. (*I. hieroglyphica* A. A. Eaton; *I. macrospora* Durieu)

9. *I. tuckermanii* A. Braun Fig. 20
Estuaries and shallow water of lakes and ponds. Nfld. and N.S. w. to se. Que., s. to N.E. and N.J.; se. Ont.

10. *I. riparia* Engelm. ex A. Braun Riverbank Quillwort Fig. 20
Shallow water, lakes, ponds, rivers, estuaries and tidal shores. S. Que. and s. Ont. s. to N.E., N.J., e. Pa., Del., Md., Va. and n. W.Va.; N.C., S.C. and Fla.

11. *I. acadiensis* Kott Acadian Quillwort
Ponds, lakes and rivers. Nfld., N.S., and N.B. s. to Me., N.H., e. Mass. and N.Y.

Selaginellaceae / Spike-moss Family

1. *Selaginella* (Selaginella, Spike-moss)

References: Alston, 1955; Jermy, 1990; Valdespino, 1993.

---

1. Strobili flattened, sporophylls appressed; leaf margins serrate or minutely toothed, but not spiny; axillary rhizophores present.
   2. Leaves with numerous minute teeth, apex of median leaves long-attenuate, with midvein reaching apex; fertile stems spreading, only slightly elevated above surface, strobili 1–4 cm long. . . . . 1. *S. eclipes*
   2. Leaves serrate, apex acute, or attenuate, midrib not extending to apex; strobili 1–2 cm long. . . . . 2. *S. apoda*
1. Strobili cylindric, sporophylls spreading; leaf margins short-spiny; rhizophores absent. . . . . 3. *S. selaginoides*

---

1. *S. eclipes* W. R. Buck Selaginella
Open, often wet calcareous sites, in fens, wet meadows, wet sandy, rocky, or marly lakeshores and river margins. Ont. and Que., w. to Wisc. and Iowa, s. to N.Y., Mich., Ind., Ill., Ark. and Okla. Sometimes treated as a subspecies of *S. apoda* (Buck, 1977).

2. *S. apoda* (L.) C. Morren Meadow Spikemoss Fig. 21
Basic to acid soils of swamps, wet meadows, marshes, open woods and stream banks. Me. w. to Ohio, Ind., Ill., Mo. and Okla., s. to Fla. and Tex.; Mex.

3. *S. selaginoides* (L.) P. Beauv. ex Schrank & Mart. Northern Spikemoss Fig. 21
Sunny to partial shade, neutral to calcareous substrates, mossy stream banks, bogs, fens, and lakeshores, mossy hummocks of White Cedar swamps; often hidden among mosses and denser vegetation. Lab., Que. and Ont. w. to B.C., Yuk. and Alask., s. to Me., n. Mich., Wisc., Minn., mts. to N.M. and Nev.; circumboreal. In the field it can be confused with *Lycopodiella inundata*; (*Lycopodium selaginoides* L.; *Selaginella spinosa* P. Beauv.)

Equisetaceae / Horsetail Family

1. *Equisetum* (Horsetail, Scouring Rush)

Perennials, arising from rhizomes; stems erect or reclining, hollow, jointed; leaves small, tooth-like, whorled, forming a sheath at the joints; sporangia in terminal cone-like strobili; homosporous.

References: Hauke, 1963, 1965, 1978, 1993.

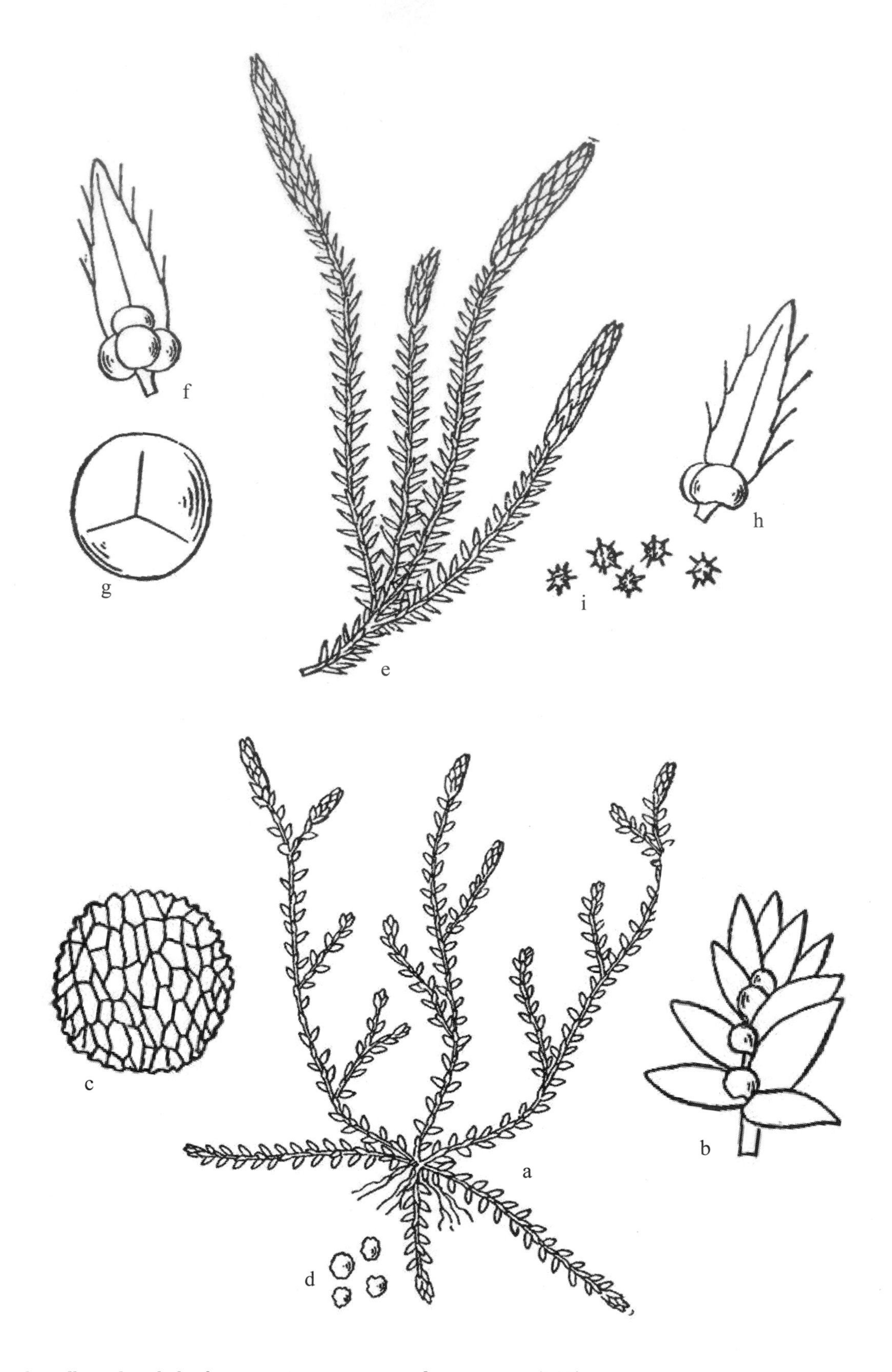

Fig. 21. *Selaginella apoda*: a. habit; b. sporangia; c. megaspore; d. microspores (B&B).
Selaginella selaginoides: e. habit with strobili at plant tips; f. megasporangium; g. megaspore; h. microsporangium; i. microspores (B&B).

1. Stems branched.
  2. Side cavities of stem approximately same diameter as central cavity (fig. 22c); central cavity 0.15–0.3 times the stem diameter. .... 1. *E. palustre*
  2. Side cavities of stem much smaller in diameter than central cavity or absent (fig. 24f); central cavity diameter greater than 0.3 times the stem diameter.
    3. Sheath teeth reddish-brown, papery; branches arching, recurved. .... 2. *E. sylvaticum*
    3. Sheath teeth dark, firm, branches spreading to ascending.
      4. Central cavity diameter less than 0.6 times the stem diameter, branches in regular whorls along stem, ascending. .... 3. *E. arvense*
      4. Central cavity diameter 0.9 times the stem diameter (pressing very flat); branches irregular, from middle nodes, spreading. .... 4. *E. fluviatile*
1. Stems unbranched.
  5. Plants dimorphic (fig. 23b,e); fertile stems brown, emerging before green vegetative stems (withering in early spring as branched sterile stems emerge). .... 3. *E. arvense*
  5. Plants not dimorphic; fertile and vegetative stems both green (not withering).
    6. Plants evergreen; strobilus with a firm, sharp point at apex.
      7. Sheath teeth persistent, with dark centers and white margins; central cavity diameter smaller, usually 0.3 times the stem diameter or less. .... 5. *E. variegatum*
      7. Sheath teeth early deciduous, leaving a white sheath with a dark ring at its summit and its base; central cavity diameter very large, usually 0.75 times the stem diameter or more (pressing very flat). .... 6. *E. hyemale*
    6. Plants not evergreen, dying back at end of season; strobilus lacking s sharp point at apex. .... 7. *E. laevigatum*

1. *E. palustre* L. Marsh Horsetail, Meadow Horsetail Fig. 22
Marshes, meadows, wet shores and wet woods. Lab. w. to Alask., s. to Nfld., n. N.E., n. Pa., Mich., Minn., N.D., Mont. and Oreg.; c. Calif.

2. *E. sylvaticum* (L.) Wood Horsetail
Cool, wet woods, woodland streamlets, wet banks, swamps, wet meadows and wet prairies. Greenl. and Lab. w. to Alask., s. to Va., Ohio, Mich., Wisc., Iowa, S.D. and Wash.

3. *E. arvense* L. Common Horsetail, Field Horsetail, Bottle Brush Fig. 23
Meadows, sterile embankments, damp woods, thickets, and occasionally in wet sands or marly areas. Green. w. to Alask., s. to N.C., n. Ala., n. Ark., n. Tex., N.M. and Calif. A sterile hybrid between *E. arvense* and *E. fluviatile* has been called *E.* ×*litorale* Kühlew. ex Rupr. (fig. 32e) and is often found when the ranges of the parents overlap. Hauke (1965) notes that *E.* ×*litorale* is morphologically intermediate in several characters; its branches are about as long as or shorter than *E. arvense*, but they tend to spread, in contrast with the branches of *E. arvense*, which are long and ascending. Although the hybrid plants are sterile, they may produce strobili with poorly developed spores.

4. *E. fluviatile* L. Water Horsetail Pipes Fig. 24
Shallow water, wet shores, swales and meadows. Lab. w. to Alask., s. to Va., Ohio, Ill., Neb., e. Wyo. and Oreg. Although *E. fluviatile* is initially unbranched, relatively short, spreading branches typically develop with maturity; an unbranched, somewhat robust growth form is occasionally found. According to Hauke (1965) in wet habitats the hybrid, *E.* ×*litorale* (fig. 32), is morphologically closer to *E. fluviatile*, but in drier sites it becomes progressively more like *E. arvense*.

5. *E. variegatum* Schleich. ex Weber & C. Mohr Variegated Horsetail Fig. 22
Calcareous meadows, marly sands, bogs and shores. Green. and Lab. w. to Alask., s. to Nfld., n. and w. N.E., N.Y., n. Ind., Minn., w. Neb., Colo., Utah and ne. Oreg. A circumboreal species. A sterile hybrid between *E. hyemale* and *E. variegatum* is known as *E.* ×*mackayi* (Newman) Brichan; it's likely to key to *E. variegatum*, but is more robust, with a wider stem diameter up to 3.6 mm.

6. *E. hyemale* L. Common Scouring Rush Fig. 24
Sandy shores, river margins, ditches, wet depressions in wood, and moist banks. Nfld. and Que. w. to B.C., s. to Fla., Tex., Calif. and Mex.; Guat. Our taxon is subsp. *affine* (Engelm.) Calder & Taylor. *Equisetum* ×*ferrissii* Clute is a frequent sterile hybrid of *E. hyemale* and *E. laevigatum* occurring in the same habitats as its parents—sandy shores, beaches, dunes, borrow pits, fields, ditches, roadsides and railway embankments and persisting (like *E. hyemale*) in forests). Such specimens would key to *E. hyemale*, but the hybrid can be recognized having an upper dark band on the sheath, but lacking a dark band at the sheath base.

7. *E. laevigatum* A. Braun Smooth Scouring Rush Fig. 24
Sandy or clayey banks, marshes, prairies, beaches and ditches. Que. and Ont. w. to s. Man. and s. B.C., s. to Ohio, Ind., Mo., Tex., n. Mex. and Calif.

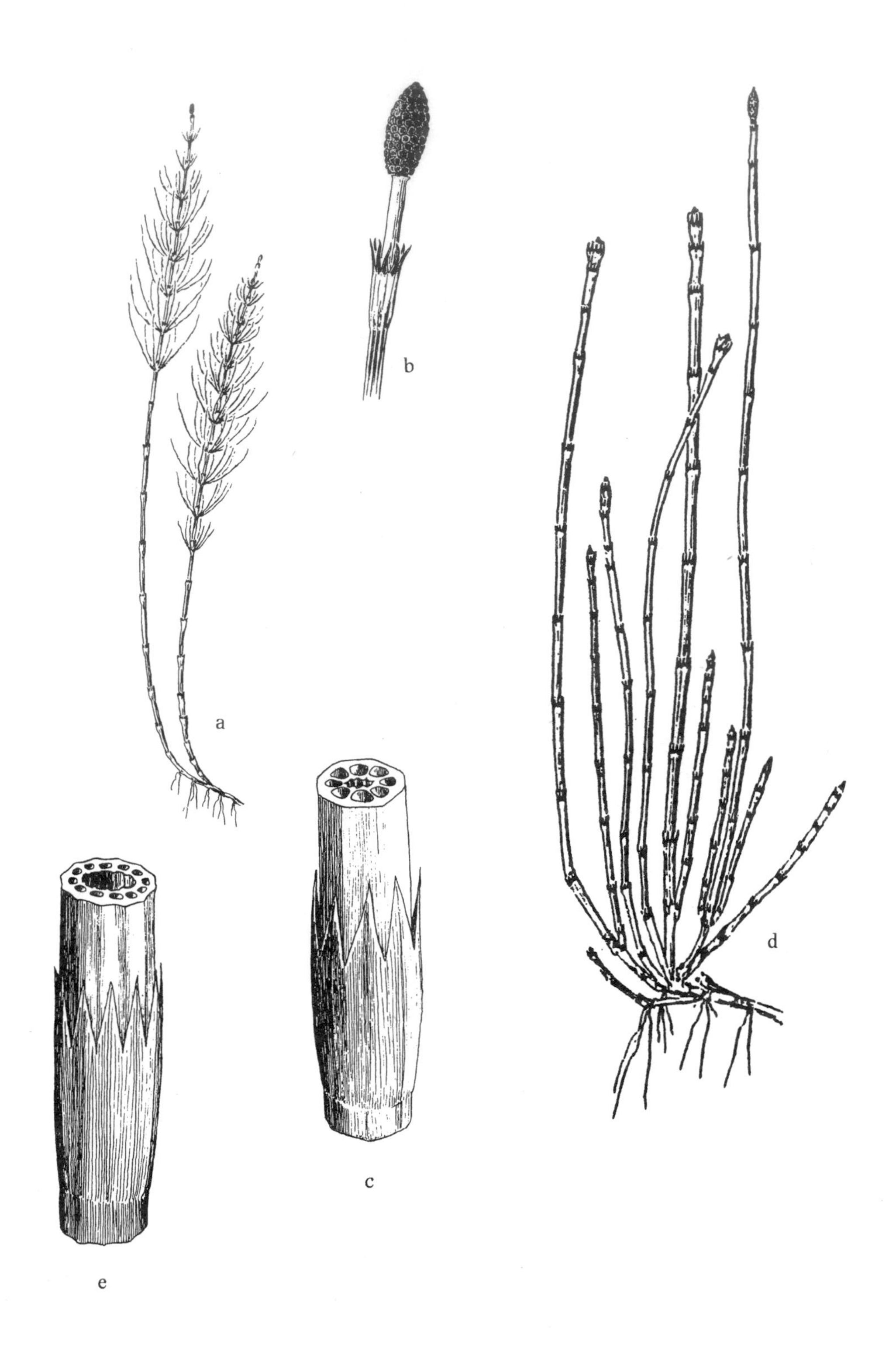

Fig. 22. *Equisetum palustre*: a. habit (HCOT); b. upper portion of plant with strobilus (HCOT); c. sheath and stem, cross-section (F).
Equisetum variegatum: d. habit (HCOT).
Equisetum ×litorale: e. sheath and stem, cross-section (F).

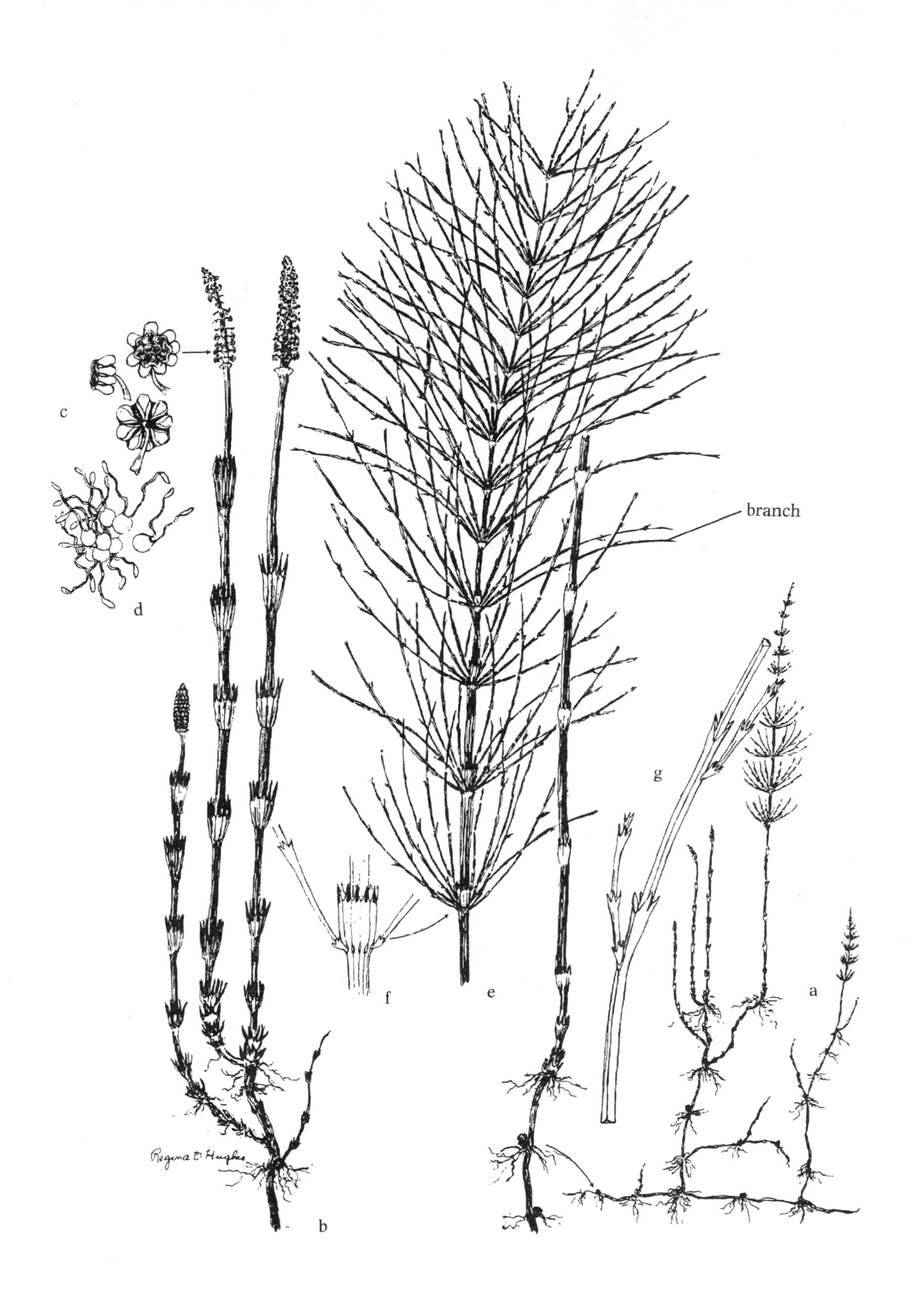

Fig. 23. *Equisetum arvense*: a. habit with fertile and young vegetative stems; b. fertile stems; c. sporangiophores; d. spores with elators; e. vegetative stem; f. sheath; g. enlarged vegetative branch (Reed).

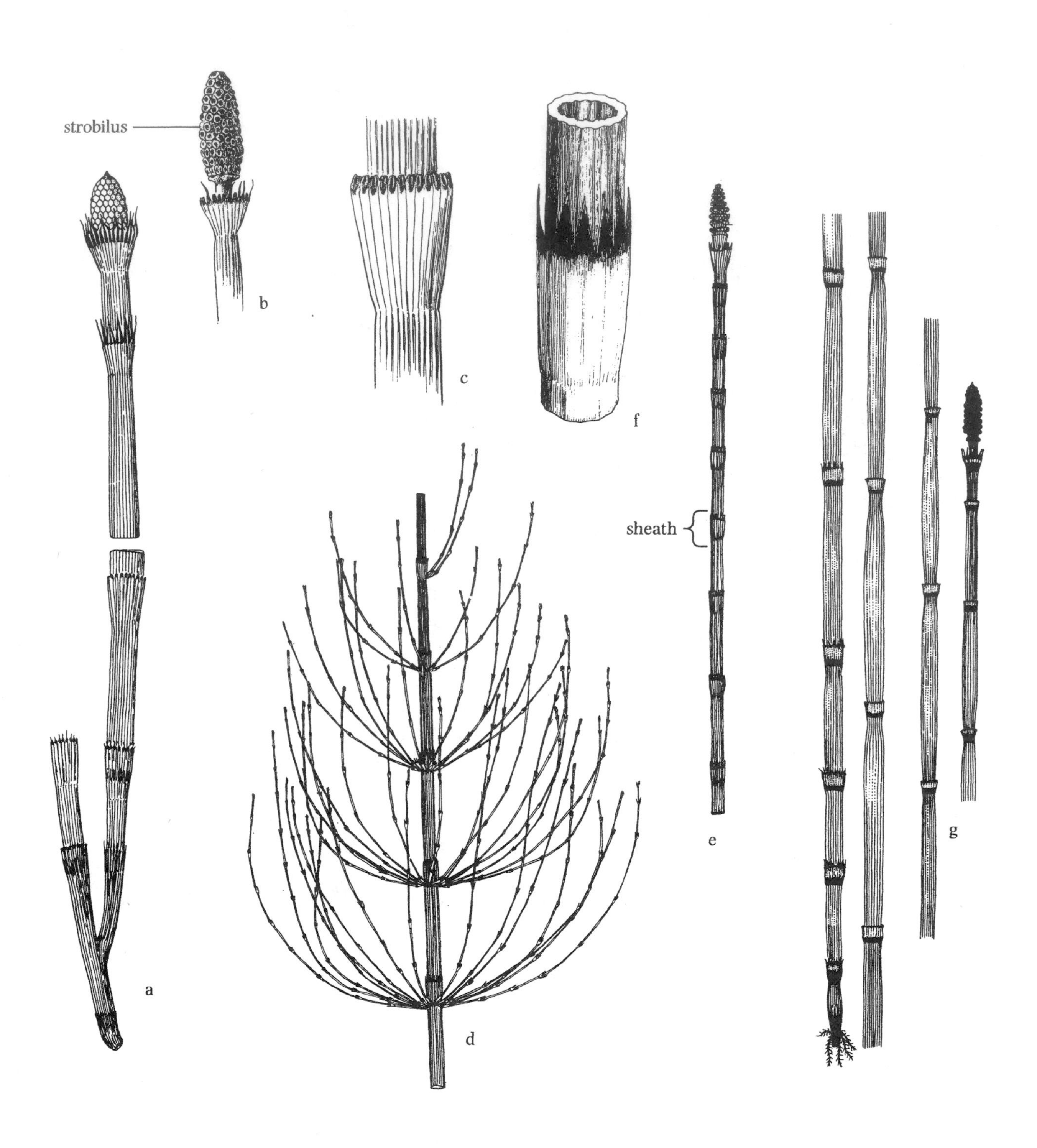

Fig. 24. *Equisetum laevigatum*: a. habit (Lundell); b. strobilus (HCOT); c. sheath (HCOT).
Equisetum fluviatile: d. portion of branched form; e. upper portion of unbranched form; f. sheath and stem, cross-section (F).
Equisetum hyemale: g. habit (B&T).

Osmundaceae / Flowering Fern Family

Perennials, arising from short, thick, rhizomes; fronds clumped coarse, compound; fertile fronds with sporangia borne on modified portions (fertile pinnae) of leaves, or separate dimorphic fronds, homosporous.

REFERENCE: Whetstone and Atkinson, 1993.

1. Fronds bipinnate (fig. 25b); fertile and vegetative fronds similar except for fertile pinnae in upper portion of frond. . . . . 1. *Osmunda*
1. Fronds pinnate-pinnatifid (fig. 26b); fertile and vegetative fronds dissimilar (dimorphic, fertile frond cinnamon-colored, completely spore producing, then withering). . . . . .2. *Osmundastrum*

1. *Osmunda* (Flowering Fern)

1. *O. regalis* L. Royal Fern Fig. 25
Low woods, thickets, swales, swamps and meadows. Nfld. w. to Ont. and Minn., s. to Fla., Miss., Tex. and Mex.; Trop. Am. The North American taxon is var. *spectabilis* (Willd.) A. Gray.

2. *Osmundastrum* (Cinnamon Fern)

1. *O. cinnamomeum* (L.) C. Presl Cinnamon Fern Fig. 26
Swamps, low woods, thickets and stream banks. Lab. and Nfld. w. to Ont. and Minn., s. to Fla., Tex. and N.M.; Mex.; Trop. Am. (*Osmunda cinnamomea* L.)

Schizaeaceae / Curly-grass Family

1. *Schizaea* (Curly-grass)

Perennials, tiny, arising from short, creeping rhizomes; fronds dimorphic; sterile fronds short, curly, and grass-like; fertile fronds erect, folded fist-like or pectinate; homosporous.

REFERENCES: LeBlond and Weakley, 2002; Stolze, 1987; Wagner, 1993.

1. *S. pusilla* Pursh Curly-grass Fern Fig. 27
Acid, sphagnous bogs, damp peats, wet ledges and shores. Nfld., St. P. et Miq., N.S. and N.B.; margin of interdunal ponds, L.I., N.Y., and wet sands and peats of pine barrens in N.J. and Del.; Atlantic White Cedar Swamp forest, N.C.; Peru. Locally abundant in Nfld. and s. N.J., rare elsewhere.

Salviniaceae / Water-fern Family

1. Leaves whorled, but appearing paired, 2 floating and simple, less than 5 mm long, 1 submersed and highly divided (root-like, fig. 31g). . . . . 1. *Salvinia*
1. Leaves alternate, overlapping in 2 rows, ca. 0.5–1 mm long, all floating (figs. 28, 29). . . . . 2. *Azolla*

1. *Salvinia* (Water Sprangles)

Annuals, free-floating; floating leaves rounded and slightly folded along midrib, upper surface covered with erect, forked hairs; submersed leaves fine, root-like; sporocarps globose; heterosporous.

REFERENCE: Mitchell and Thomas, 1972.

1. *S. minima* Baker Floating Fern, Water Sprangles Fig. 31
Floating in still water of ponds and marshes. Intro. from Trop. Am. This taxon has become locally established in the South and is short-lived elsewhere.

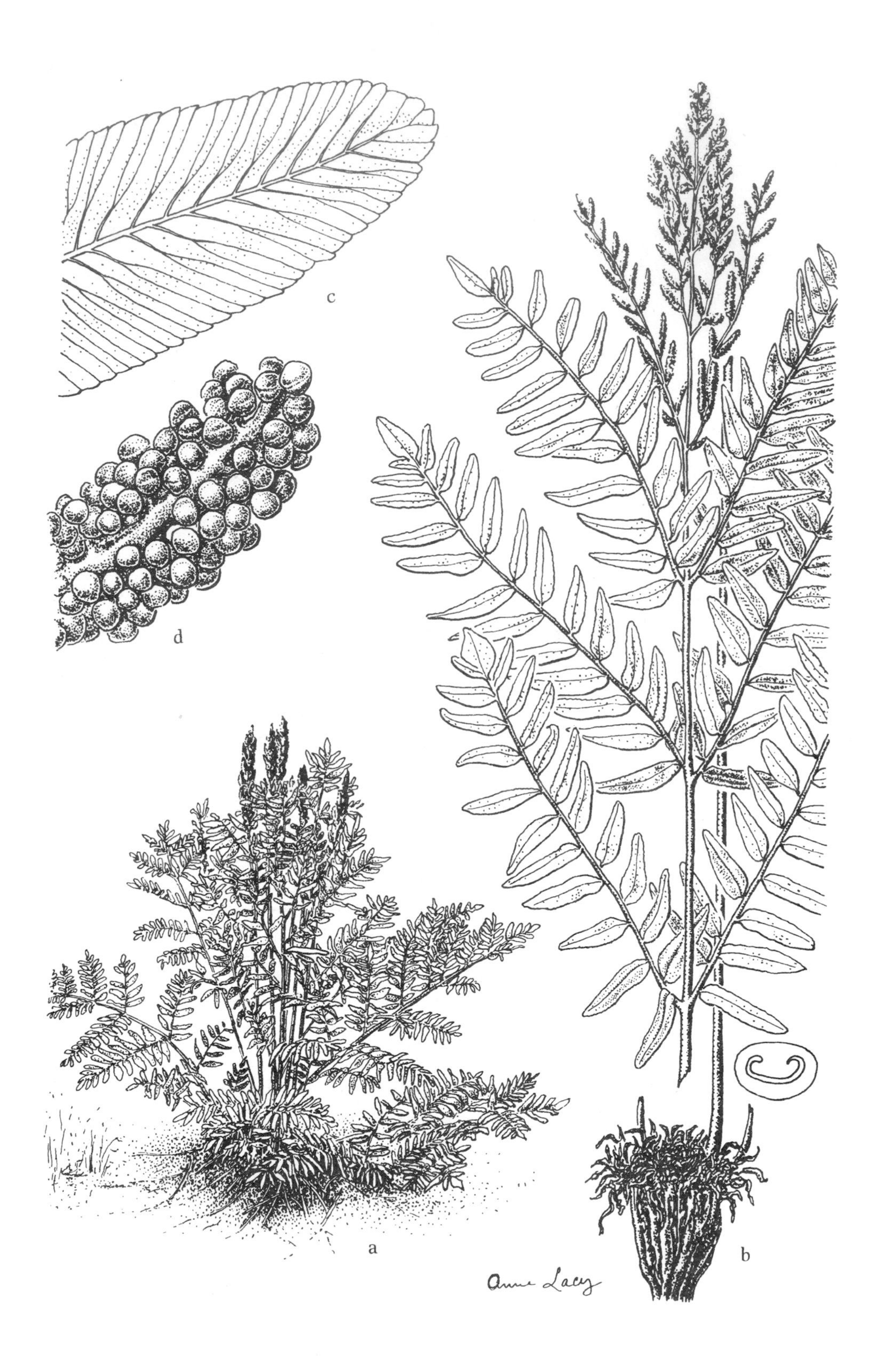

Fig. 25. *Osmunda regalis*: a. habit; b. frond with vegetative portion below, fertile portion above; c. vegetative pinnule; d. fertile pinnule with sporangia (NYS Museum).

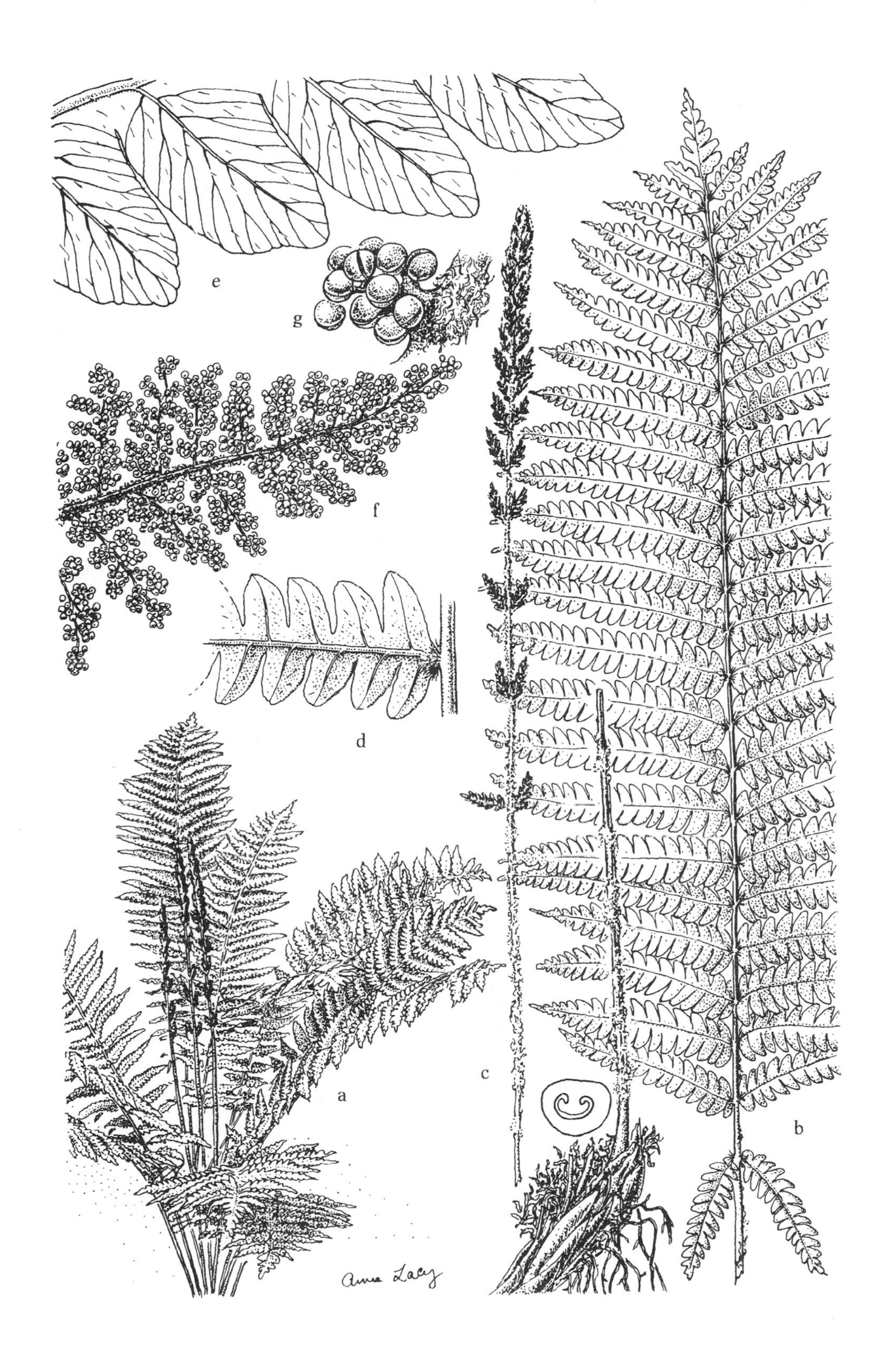

Fig. 26. *Osmundastrum cinnamomum*: a. habit; b. vegetative frond; c. fertile frond; d, e. portion of vegetative frond; f. portion of fertile frond; g. sporangia (NYS Museum).

Fig. 27. *Schizaea pusilla*: a. habit; b. fertile frond (NYS Museum).

2. *Azolla* (Mosquito Fern)

Annuals, free-floating; roots thread-like; stems branched, hair-like; leaves minute, scale-like, in 2 rows; plants green or red, with velvety appearance; sporocarps globose; heterosporous; microsporangia aggregated into 4 massulae, bearing numerous hooked appendages (glochidia). Because plants are often sterile, the following key is based chiefly on vegetative characters, but fertile material is needed to be confident of correct identification (cf. Svenson, 1944).

Azolla is sometimes segregated into a monogeneric family, the Azollaceae, as in the FNA treatment (Lumpkin, 1993).

References: Bates and Browne, 1981; Lumpkin, 1993; Moore, 1969; Svenson, 1944.

---

1. Leaves ca. 0.5 mm long, surface smooth; plants 0.5–1 cm long. .... 1. *A. caroliniana*
1. Leaves ca. 1 mm long, surface papillate; plants 1–6 cm long.
   2. Stems dichotomously branched; plants 1–3 cm long; glochidia on microspore massulae with several septae. .... 2. *A. microphylla*
   2. Stems pinnately branched; plants (2)3–6 cm long; glochidia on microspore massulae with few or no septae. .... 3. *A. filiculoides*

---

1. *A. caroliniana* Willd. Fig. 28
Quiet waters. N.E. and N.Y. w. to s. Ont., Mich., Wisc., and S.D., s. to Fla., Okla., Tex. and Mex.; W.I. and S.Am. Populations are short-lived in the North.

2. *A. microphylla* Kaulf. Fig. 29
Quiet waters. Sw. Wisc., se. Minn., Ill., Iowa, Mo., Neb., and ne. Colo. s. to Ark., Okla., w. Tex., N.M. and Ariz.; B.C. s. Calif. and Mex.; C.Am. and S.Am.

3. *A. filiculoides* Lam. Fig. 29
Quiet waters. Alask. and B.C. s. to Calif. and Mex.; C.Am. and S.Am. Sometimes escaped from horticultural water gardens within our range.

Marsileaceae / Pepperwort Family

---

1. Leaves with distinct blades and petioles, leaflets 4 (fig. 30). .... 1. *Marsilea*
1. Leaves lacking blades, filiform (fig. 31e). .... 2. *Pilularia*

---

1. *Marsilea* (Water-clover)

Perennials, arising from slender rhizomes; leaflets 4, fan-shaped, sporocarps globose, on short peduncles at base of leaves along rhizome; heterosporous.

References: Johnson, 1986, 1988, 1993.

---

1. Sporocarps 1 per peduncle, arising from juncture of petiole and rhizome; leaves pubescent; southern and western species. .... 1. *M. vestita*
1. Sporocarps 2 or 3 per peduncle, arising from near base of petiole; leaves glabrous; introduced in Northeast. .... 2. *M. quadrifolia*

---

1. *M. vestita* Hook. & Grev. Hair Water-clover Fig. 30
Shallow ponds, pools, ditches and damp shores. W. Minn. and nw. Iowa w. to N.D., s. Sask., s. Alta. and s. B.C., s. to Ark., w. Miss., sw. Ala. and w. Fla., Tex., N.M., Ariz., Calif. and n. Mex.; Peru. (*M. mucronata* A. Braun)

2. *M. quadrifolia* L. European Water-clover Fig. 31
Shallow ponds, slow-flowing rivers, and muddy shores. Local, Me. and Mass. w. to Iowa, s. to Md., Ky. and Mo.; intro. from Eur.

2. *Pilularia* (Pillwort)

Perennial, arising from slender, creeping rhizomes; leaves filiform, uncoiling at apices with maturation, reduced to petioles, lacking blades; sporocarps globose, on short peduncles at base of leaves along rhizome; heterosporous.

References: Dennis and Webb, 1981; Johnson, 1993; Key, 1982.

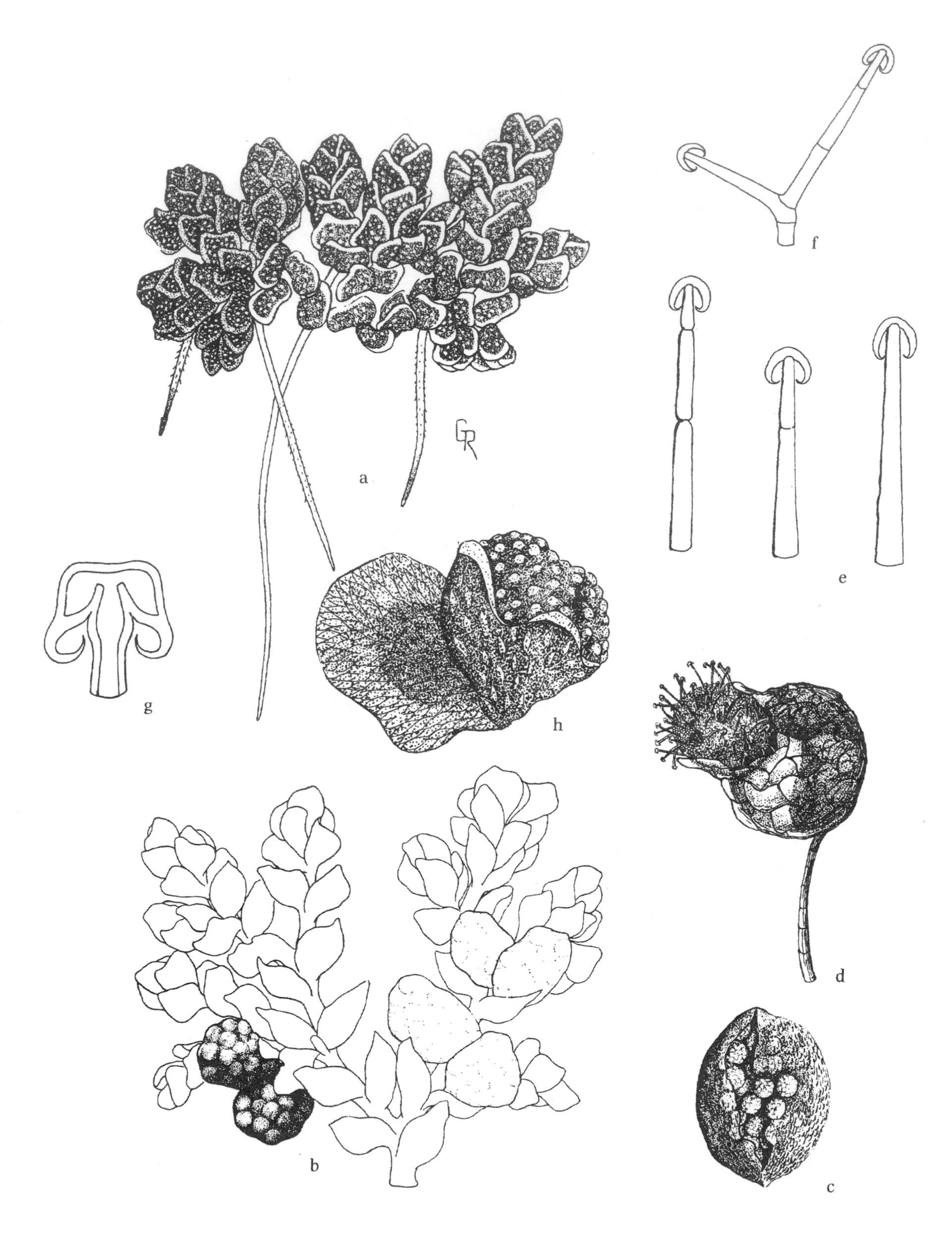

Fig. 28. *Azolla caroliniana*: a. habit; b. lower surface of plant with microsporocarps; c. microsporocarp; d. microsporangium with massula being discharged; e. glochidia types from a single massula; f. single-branched glochidium; g. tip of glochidium highly magnified; h. 2-lobed leaf (Godfrey, Reinert, and Houk, 1961).

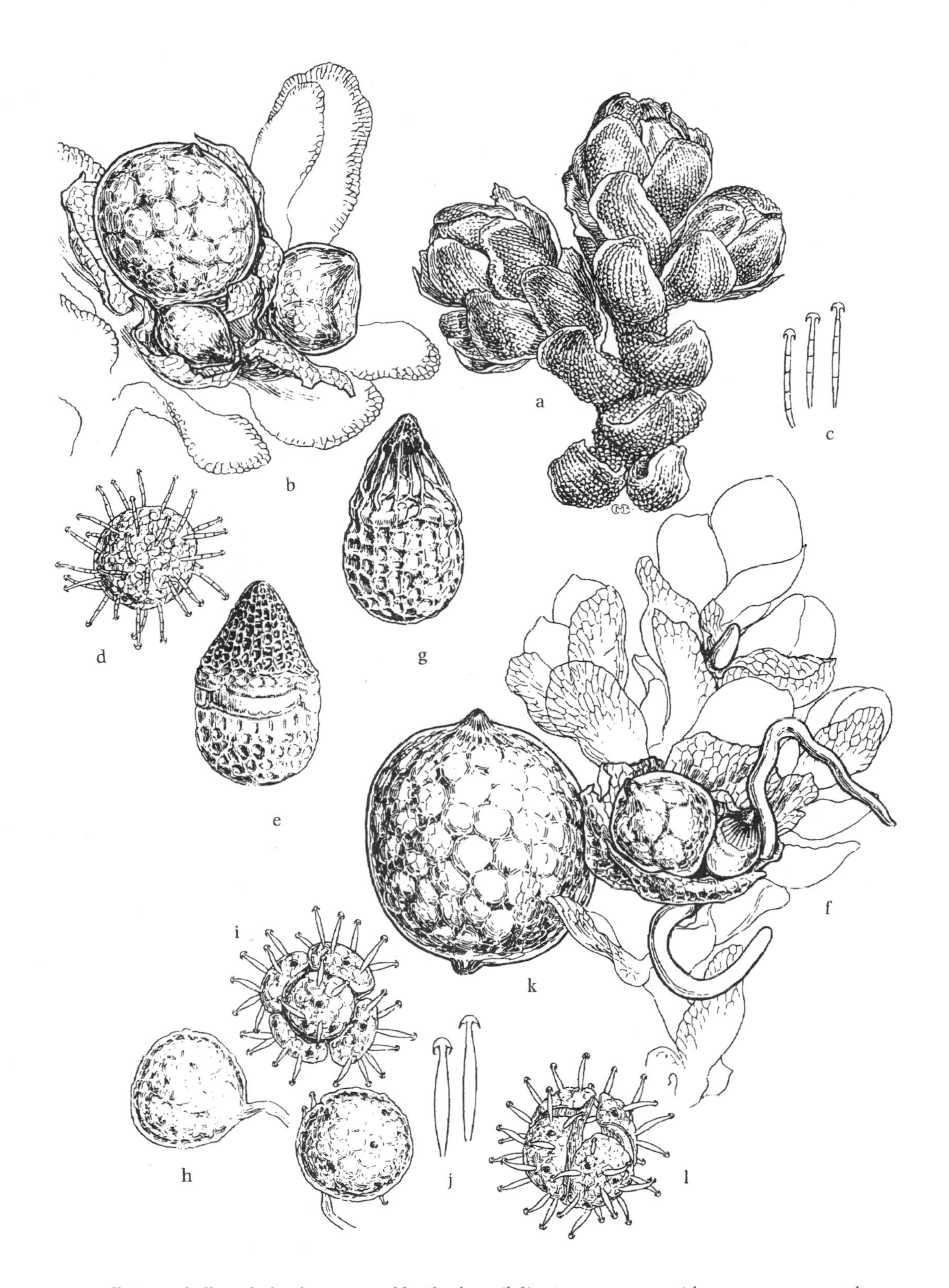

Fig. 29. *Azolla microphylla*: a. habit; b. portion of fertile plant: (*left*) microsporocarp with megasporocarp at base, (*right*) pair of megasporocarps enclosed in 1 indusium (uncommon); c. septate glochidia of microsporic massulae; d. microsporic massula; e. megaspore covered by tip of indusium (Mason).
Azolla filiculoides: f. portion of fertile plant, lower surface, showing roots and small microsporocarp with megasporocarp at base; g. megaspore covered by tip of indusium; h. young, stalked microsporangia, showing a few glochidia of massulae protruding from ruptured wall; i. separating massulae; j. nonseptate glochidia; k. microsporocarp containing large number of microsporangia; l. separating massulae (Mason).

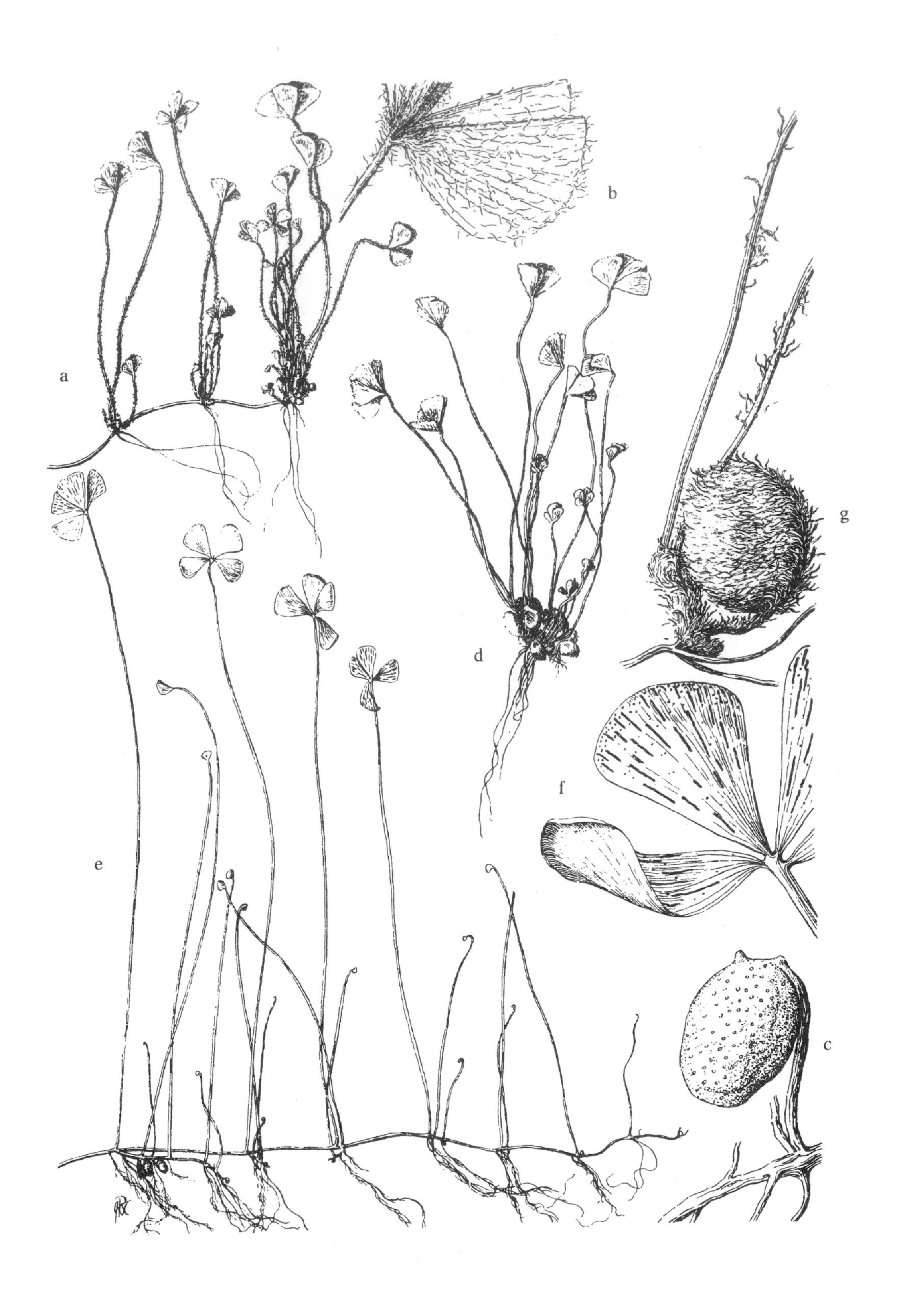

Fig. 30. *Marsilea vestita*: a. habit, aquatic form; b. leaf, aquatic form; c. sporocarp, aquatic form; d, e. habit, terrestrial form; f. leaf, terrestrial form; g. sporocarp, terrestrial form (Mason).

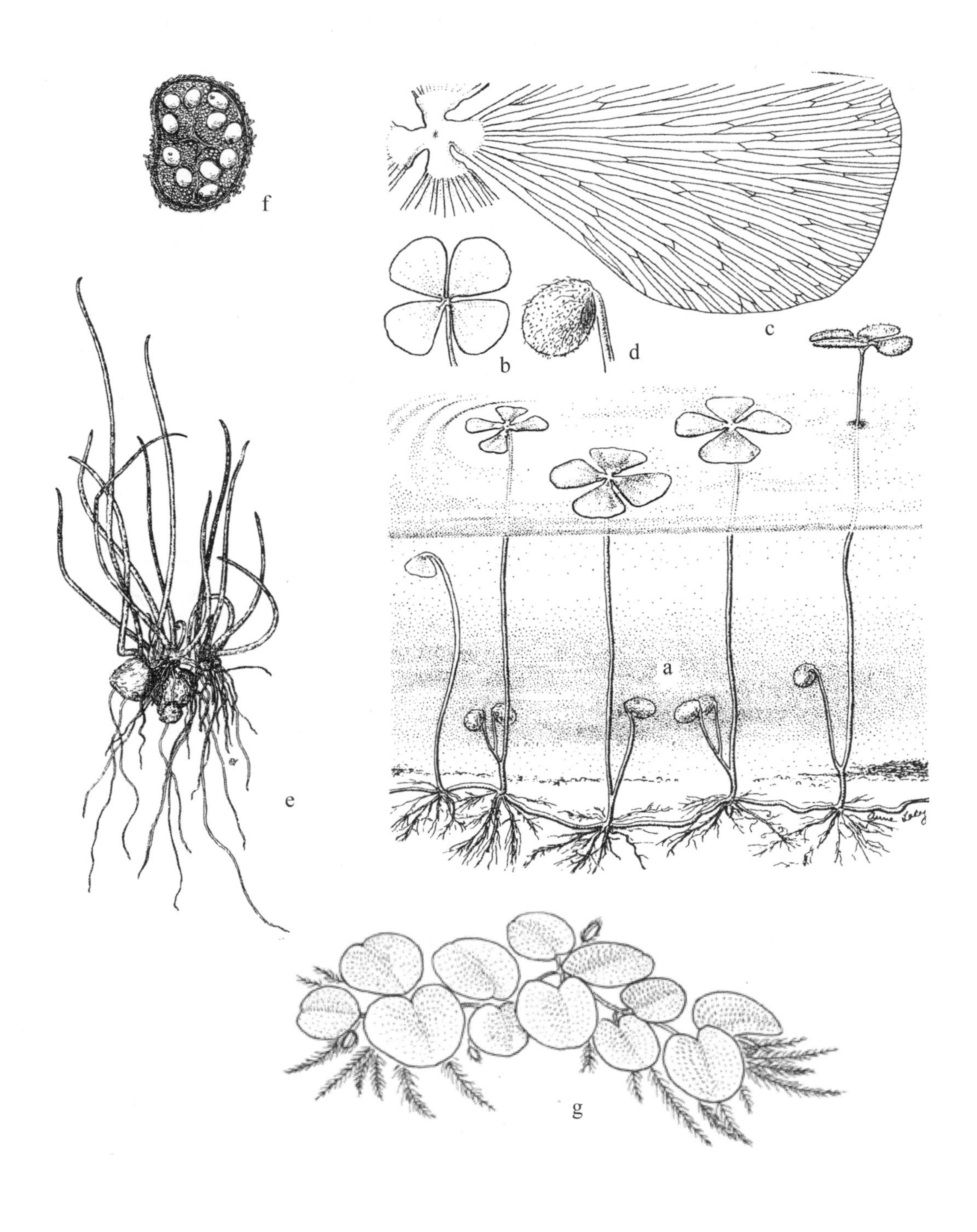

Fig. 31. *Marsilea quadrifolia*: a. habit; b. leaf; c. portion of leaf; d. sporocarp (NYS Museum).
Pilularia americana: e. habit; f. sporocarp, cross-section showing sporangia (Mason).
Salvinia minima: g. habit (UF/IFAS).

1. *P. americana* A. Braun American Pillwort Fig. 31
Shallow temporary pools or ponds, pools on granite flat rocks, clayey depressions on prairies, and mud along margins of lakes or streams. Tenn. and Ga.; se. Mo. and Ark. w. to Neb. and Tex.; Oreg. and Calif.; Baja, Calif.

Onocleaceae / Sensitive Fern Family

1. *Onoclea* (Sensitive Fern)

Perennials, arising from rhizomes; fronds dimorphic; fertile fronds erect, appearing woody, with bead-like segments, brown; sterile fronds deltoid-ovate, deeply pinnatifid; homosporous.

1. *Onoclea sensibilis* L. Sensitive Fern Fig. 32
Low woods, open ground, alluvial plains, swamps and meadows. Nfld. w. to Man., s. to Fla., La. and e. Tex.; Colo.

Blechnaceae / Chain-fern Family

1. *Woodwardia* (Chain-fern)

Perennials, arising from scaly rhizomes; fronds with stipes as long as blades, not jointed; sori borne as a series of clusters along midveins, indusia conspicuous, covering the sporangia, withering at maturity; homosporous.

Reference: Cranfill, 1993.

1. Fronds, both fertile and sterile, similar, blades pinnate-pinnatifid (fig. 33), with petioles dark purple to black basally, straw-colored above, swollen at base. 1. *W. virginica*
1. Fronds strongly dimorphic, fertile fronds long and narrow, with linear unlobed segments (fig. 34), petioles reddish-brown basally, straw-colored above, not swollen at base, sterile fronds short and broad, petioles green. 2. *W. areolata*

1. *W. virginica* (L.) Sm. Virginia Chain-fern Fig. 33
Acid swamps and bogs. P.E.I. and N.S. w. to Ont., Mich. and n. Ind., s. chiefly along the coastal plain to Fla. and e. Tex.; Ber.

2. *W. areolata* (L.) Moore Netted Chain-fern Fig. 34
Acid swamps, peats and wet woods. Coastal plain and piedmont, Sw. N.S. to Fla. w. to Tex.; inland to Ohio, s. Ill., Mo. and e. Okla.; sw. Mich.

Thelypteridaceae / Marsh Fern Family

1. *Thelypteris* (Marsh Fern)

Perennials, arising from slender, blackish rhizomes; fronds bipinnate or pinnate-pinnatifid, segments entire or nearly so; sori on veins; indusia small, sometimes absent, reniform, ciliate or glanduliferous; homosporous.

Reference: Smith, 1993.

1. *T. palustris* Schott Marsh Fern, Meadow Fern, Snuff-box Fern Fig. 35
Marshes, bogs, swamps and wet woods. Nfld. w. to Man., s. to Fla. and e. Tex. The N.Am. taxon is var. *pubescens* (G. Lawson) Fernald. (*Dryopteris thelypteris* (L.) Sw.)

Fig. 32. *Onoclea sensibilis*: a. habit; b. vegetative frond; c. fertile frond; d. portion of vegetative frond; e. portion of fertile frond with bead-like pinnae (NYS Museum).

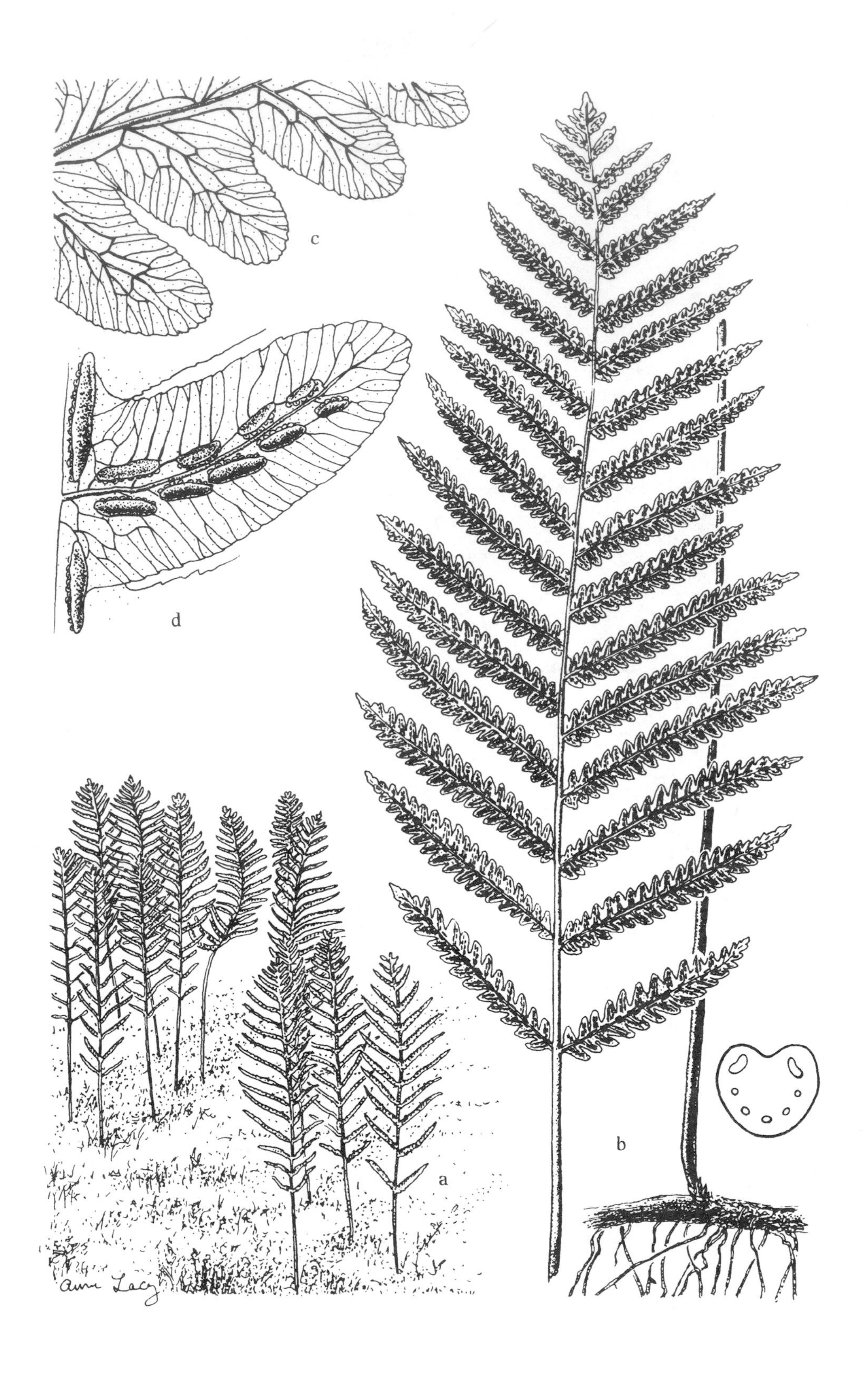

Fig. 33. *Woodwardia virginica*: a. habit; b. frond; c. upper surface of frond; d. lower surface of frond with sori (NYS Museum).

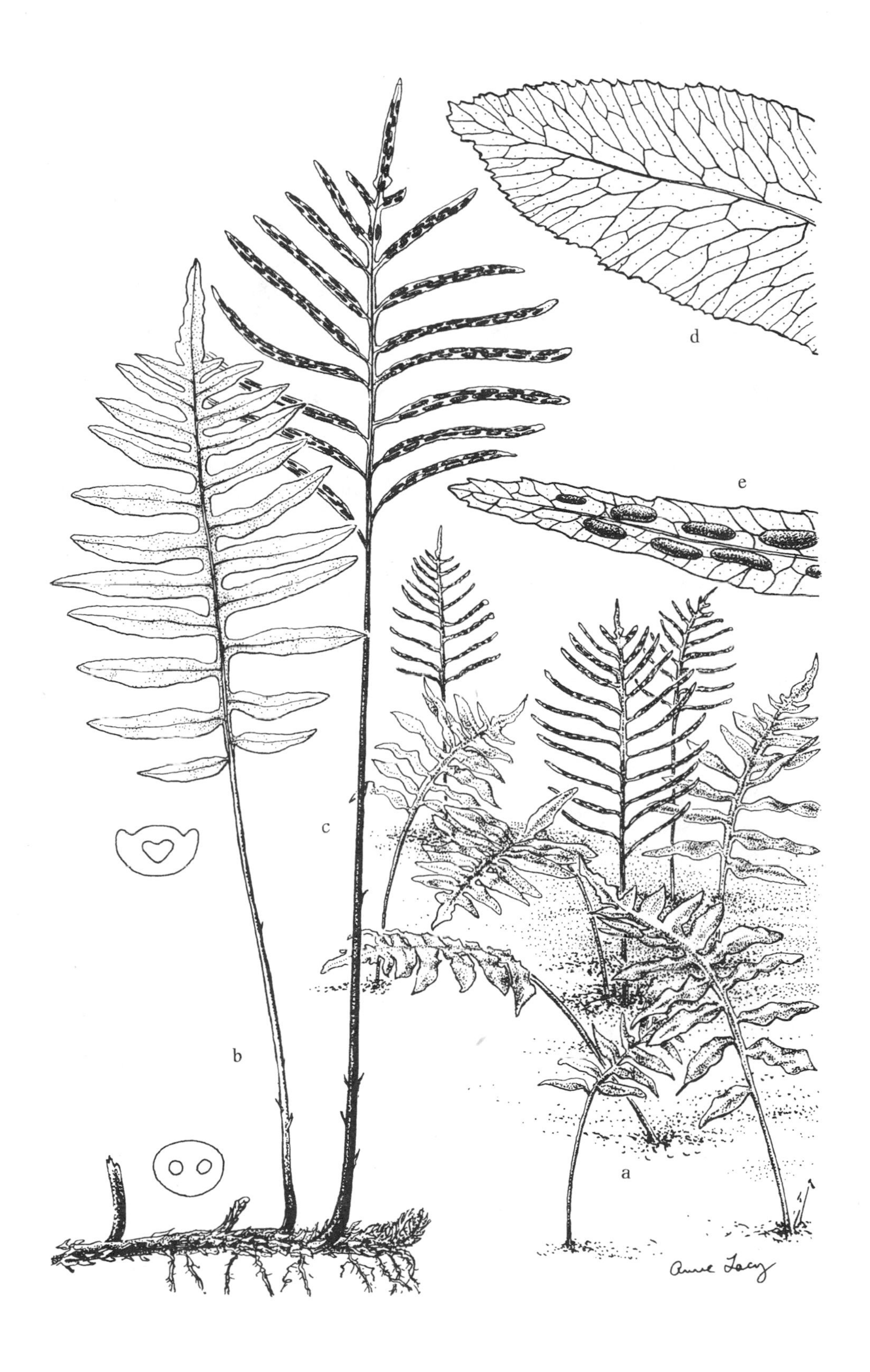

Fig. 34. *Woodwardia areolata*: a. habit; b. vegetative frond; c. fertile frond; d. portion upper surface of vegetative frond; e. portion lower surface of fertile frond with sori (NYS Museum).

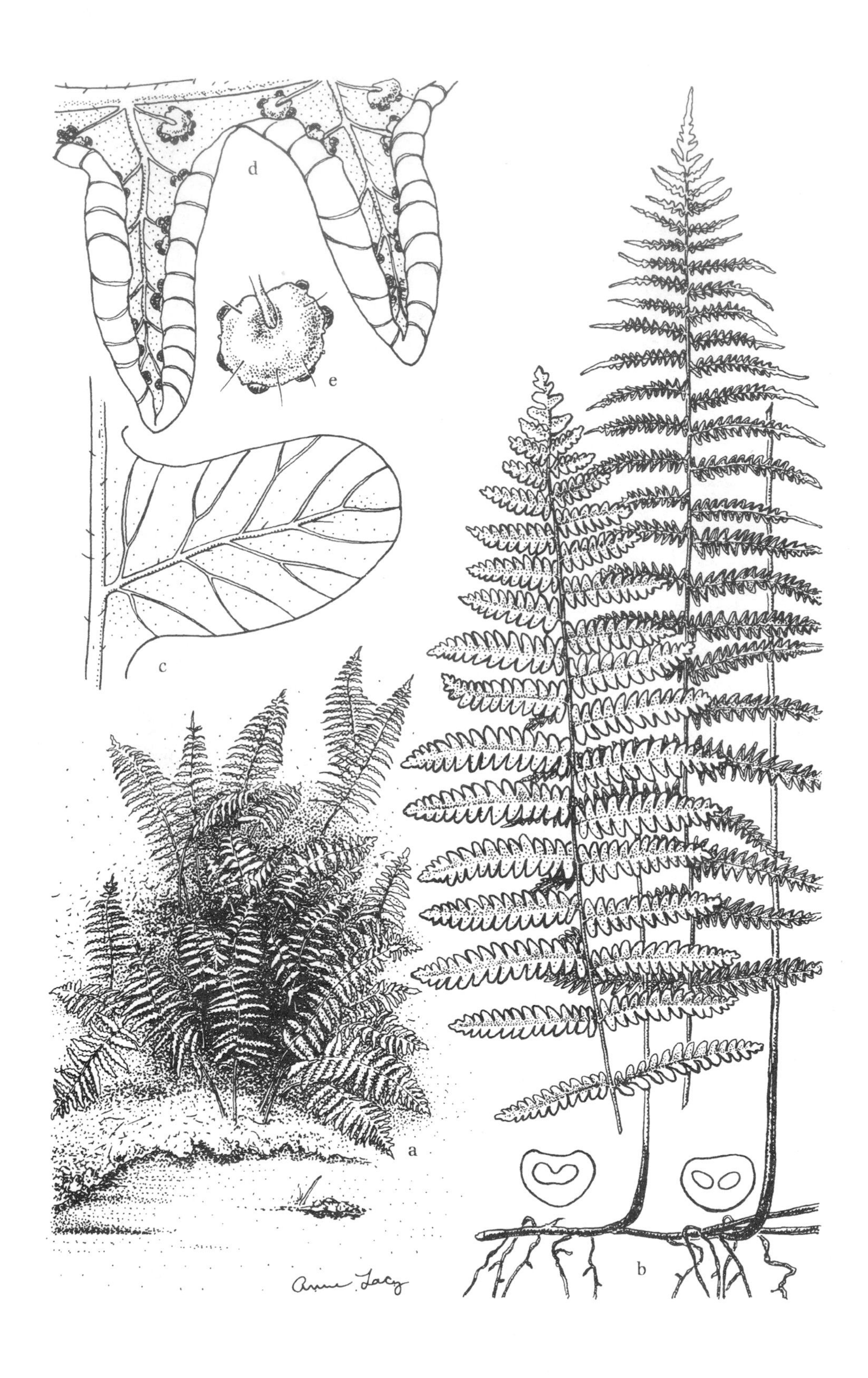

Fig. 35. *Thelypteris palustris*: a, b. habit; c. portion of frond; d. lower surface of frond with sori, partially hidden by revolute margins; e. indusium (NYS Museum).

Dryopteridaceae / Shield Fern Family

1. *Dryopteris* (Shield Fern)

Perennials, arising from rhizomes; fronds pinnate-pinnatifid to tripinnate; evergreen or deciduous, stipes thick, shorter than blades; sori borne on veins; indusia reniform, persistent, glabrous, homosporous.

REFERENCE: Montgomery and Wagner, 1993.

---

1. Fronds lanceolate, slightly narrowed toward base; pinnae 8–14 cm long; fertile and sterile fronds alike. . . . . 1. *D. clintoniana*
1. Fronds narrowly lanceolate, strongly narrowed toward base; pinnae 5–8 cm long; fertile and sterile fronds somewhat dimorphic, the fertile frond more erect, with pinnae oriented in horizontal plane. . . . . 2. *D. cristata*

---

1. *D. clintoniana* (D. C. Eaton) Dowell Clinton's Wood Fern Fig. 36

Swamps and wet woods. N.B. and Que. w. to Mich., s. to N.J., Pa., Ohio and Ind. (*D. cristata* var. *clintoniana* (D. C. Eaton) Underw.)

2. *D. cristata* (L.) A. Gray Crested Shield Fern, Crested Wood Fern Fig. 37

Marshes, bogs, thickets and wet woods. Nfld. w. to s. Alta. and se. B.C., s. to N.C., e. Tenn., c. Ill. and e. Iowa; Ala.; Neb.

Fig. 36. *Dryopteris clintoniana*: a. habit; b. frond; c. portion of frond; d. portion of fertile frond with sori (NYS Museum).

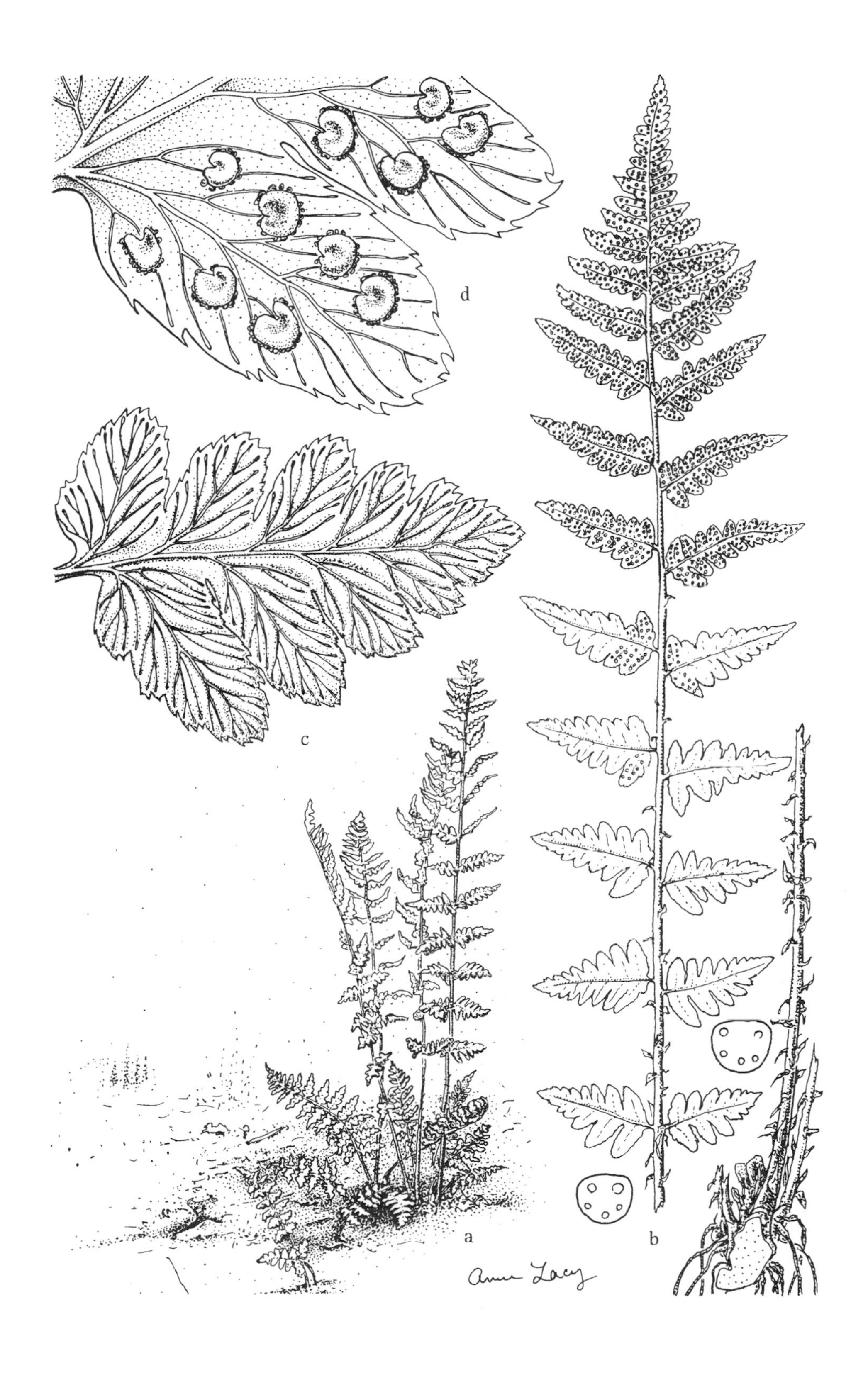

Fig. 37. *Dryopteris cristata*: a. habit; b. frond; c. portion of frond; d. portion of lower surface of fertile frond with sori (NYS Museum).

Gymnosperms

Pinaceae / Pine Family

1. Leaves stiff, angled in cross-section, 4-sided, borne singly along stem; evergreen, remaining thorough winter, petioles persistent on older twigs, stems roughened. . . . . 1. *Picea*
1. leaves soft, flat, clustered on short spurs on older twigs, borne singly on new shoots; deciduous in autumn, petioles not persistent, older twig not roughened. . . . . 2. *Larix*

1. *Picea* (Spruce)

Trees, evergreen; stems roughened with persistent peg-like petioles; leaves needle-like, spirally arranged; ovuliferous cones pendulous, maturing during the first year, scales persistent; seeds winged.

1. *P. mariana* (Mill.) Britton, Sterns & Poggenb. Black Spruce, Bog Spruce Fig. 38

Bogs and cold slopes. Lab. and Nfld. w. to Alask., s. to n. N.J., n. Pa., Mich., Wisc., Minn., Sask. and B.C.

2. *Larix* (Larch)

1. *Larix laricina* (Du Roi) K. Koch Tamarack, Hackamatack, American or Black Larch Fig. 39

Swamps and bogs, occasionally in upland habitats. Lab. w. to Alask., s. to n. N.J., W.Va., n. Ohio, n. Ill., Minn., Man., Sask. and ne. B.C.

Cupressaceae / Cypress Family

1. Leaves scale-like (fig. 40) or subulate.
    2. Leaves scale-like, opposite; leafy twigs somewhat to strongly flattened (figs. 40, 41).
        3. Cones ovoid-oblong; cone scales overlapping, not peltate, tan or brown, not glaucous. . . . . 1. *Thuja*
        3. Cones globose; cone scales peltate, not overlapping, bluish and glaucous. . . . . 2. *Chamaecyparis*
    2. Leaves subulate, alternate; leafy twigs not flattened (*T. ascendens*, fig. 43). . . . . 3. *Taxodium*
1. Leaves needle-like, flattened, flexible, 2-ranked (*T. distichum*, fig. 42b). . . . . 3. *Taxodium*

References: Chaw et al., 1997; Lu et al., 2014; Wang and Ran, 2014; Watson and Eckenwalder, 1993.

1. *Thuja* (Arborvitae, White Cedar)

Trees, evergreen; branchlets flattened, in horizontal sprays; leaves scale-like, often bearing a dorsal gland, opposite, overlapping; ovuliferous cones ovoid-oblong, scales basally attached, overlapping.

1. *T. occidentalis* L. Northern White Cedar, Arborvitae Fig. 40

Calcareous soils of swamps, riverbanks, bogs, gravelly shores and upland sites. N.S., P.E.I., and s. Que. w. to Man., s. to n. and w. N.E., N.Y., mts. to N.C., Tenn., Ohio, n. Ind., n. Ill., Wisc. and Minn.

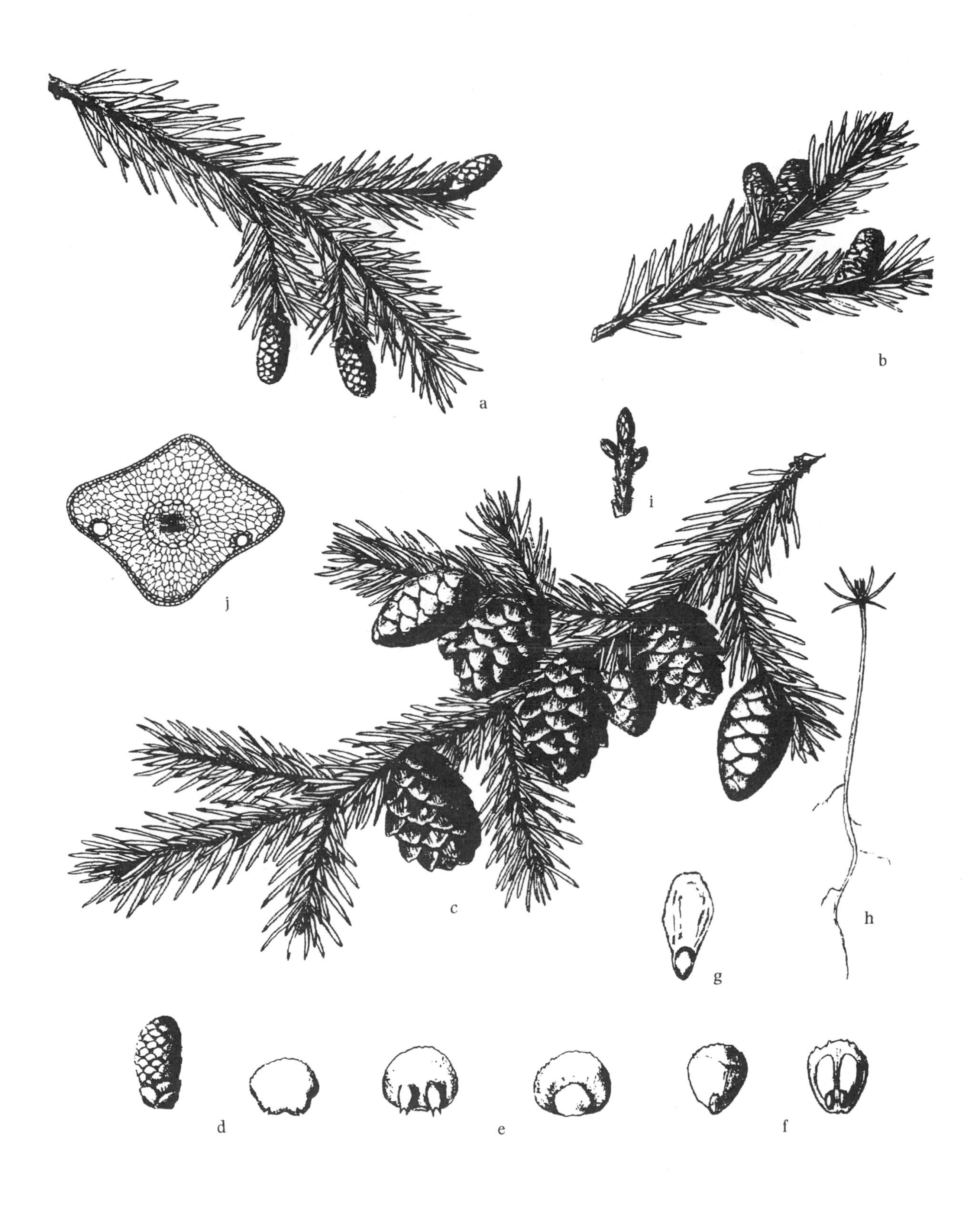

Fig. 38. *Picea mariana*: a. branch with male cones; b. branch with young female cones; c. branch with mature female cones; d. male cone with microsporangia; e. female cone scale, inner side showing 2 ovules, and outer side showing bract; f. mature female cone scales, outer side and inner side with 2 winged seeds; g. seed; h. seedling; i. winter bud; j. leaf, cross-sections (Sargent).

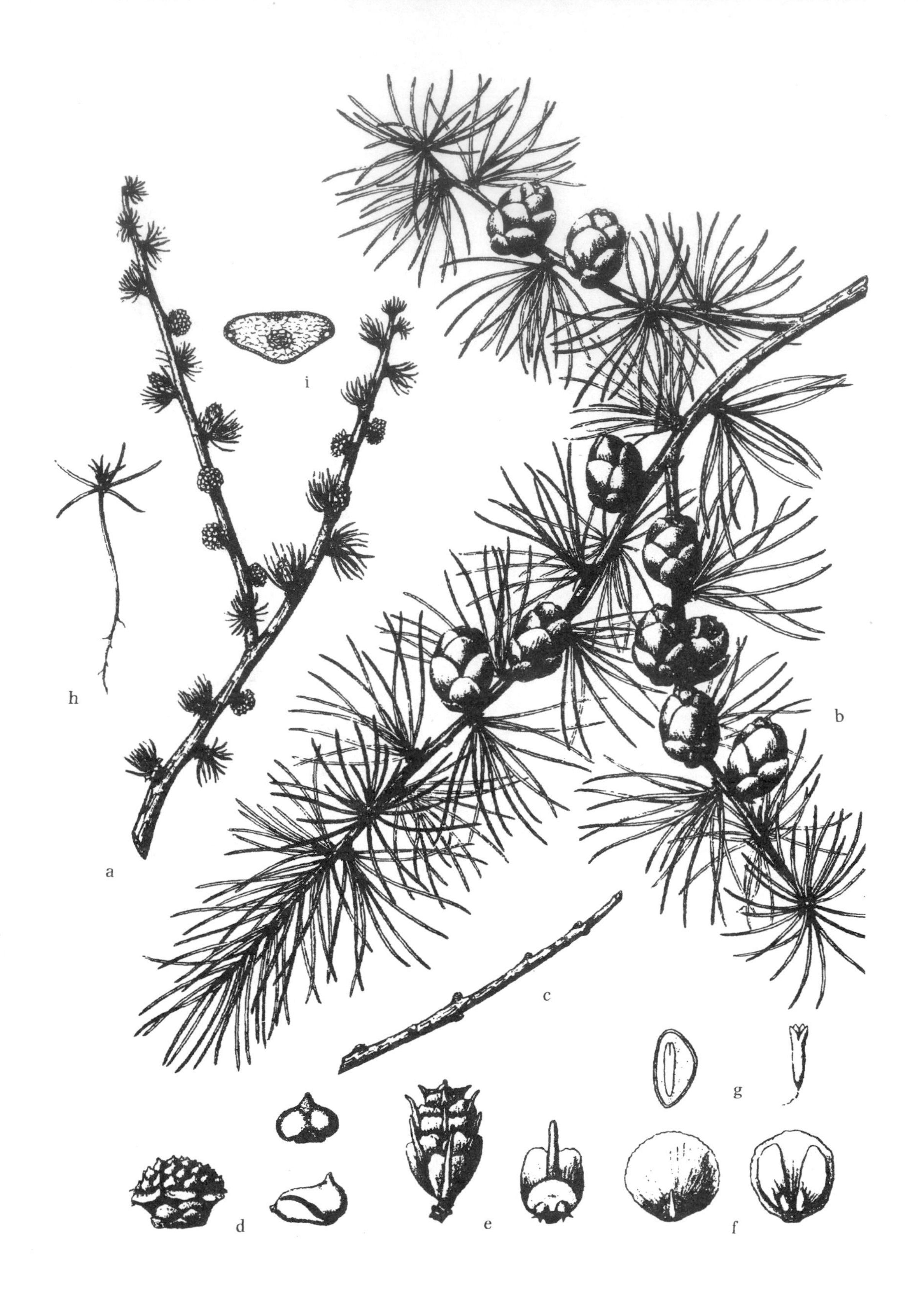

Fig. 39. *Larix laricina*: a. branch with emerging leaves and male cones; b. mature branch with female cones; c. winter twig; d. male cone with microsporangia, two views; e. young female cone with scale showing 2 ovules; f. mature female cone scales, outer side and inner side with 2 winged seeds; g. embryo surrounded by nutritive tissue, and embryo enlarged; h. seedling; i. leaf, cross-section (Sargent).

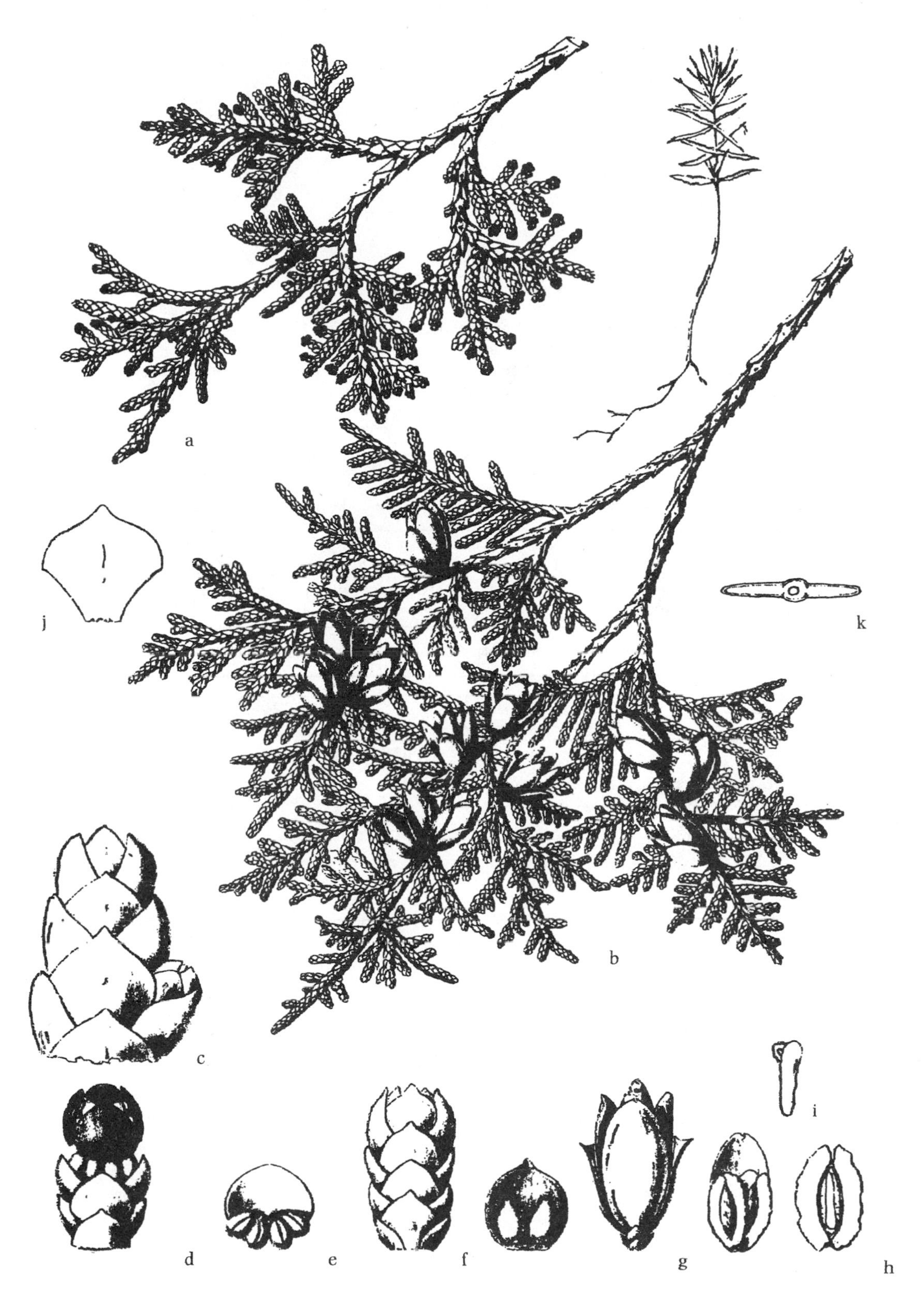

Fig. 40. *Thuja occidentalis*: a. branch with male cones; b. branch with mature female cones; c. end of branchlet; d. branchlet with male cone; e. scale with microsporangia; f. young female cone with scale showing 2 ovules; g. mature female cone and scale with 2 winged seeds; h. seed; i. embryo; j. leaf; k. leaf, cross-section (Sargent).

2. *Chamaecyparis* (Atlantic White Cedar, Cypress)

Trees, evergreen; branchlets flattened, in horizontal sprays; leaves scale-like, often bearing a dorsal gland, opposite, overlapping; ovuliferous cones globose, scales peltate, often glaucous.

1. *C. thyoides* (L.) Britton, Sterns & Poggenb. White Cedar, Atlantic White Cedar Fig. 41

Acid swamps, peats and wet sands. Coastal plain, s. Me. and s. N.H. s. to n. Fla., w. to Miss.

3. *Taxodium* (Bald Cypress)

Trees; stems with axillary buds persistent, branchlets bearing leaves deciduous as a unit in autumn; leaves narrow, linear; microstrobili in drooping terminal panicles; ovuliferous cones subglobose, scales peltate.

Bald cypress trees are typically enlarged at the base and frequently have "knees" from the roots protruding above the water. Formerly treated in a distinct family, the Taxodiaceae has recently been included within the Cupressaceae.

1. Short shoots with leaves 2-ranked, spreading horizontally to pendent; leaves mostly narrowly linear, 5–17 mm long, divergent. .... 1. *T. distichum*
1. Short shoots with leaves mostly ascending vertically; leaves narrowly lanceolate, 3–10 mm, appressed, overlapping. .... 2. *T. ascendens*

1. *T. distichum* (L.) Rich. Bald Cypress, Swamp Cypress Fig. 42

Swamps and ponds. Chiefly coastal plain, N.Y. and Del. s. to Fla., w. to se. Tex. and ne. Mex.; Mississippi embayment n. to s. Ind., s. Ill., se. Mo. and e. Okla. (*T. distichum* var. *nutans* (Aiton) Sweet)

2. *T. ascendens* Brongn. Pond Cypress Fig. 43

Swamps and ponds. Chiefly coastal plain, Del. and Va., s. to Fla., w. to Ala., Miss. and e. La.

Fig. 41. *Chamaecyparis thyoides*: a. branch with male cones; b. branch with mature female cones; c. tip of branchlet; d. male cone; e. scale with microsporangia, two views; f. young female cone with scale showing 2 ovules; g. mature female cone; h. seed; i. embryo surrounded by nutritive tissue, and embryo enlarged; j. seedling; k. leaf; l. branchlet, cross-section (Sargent).

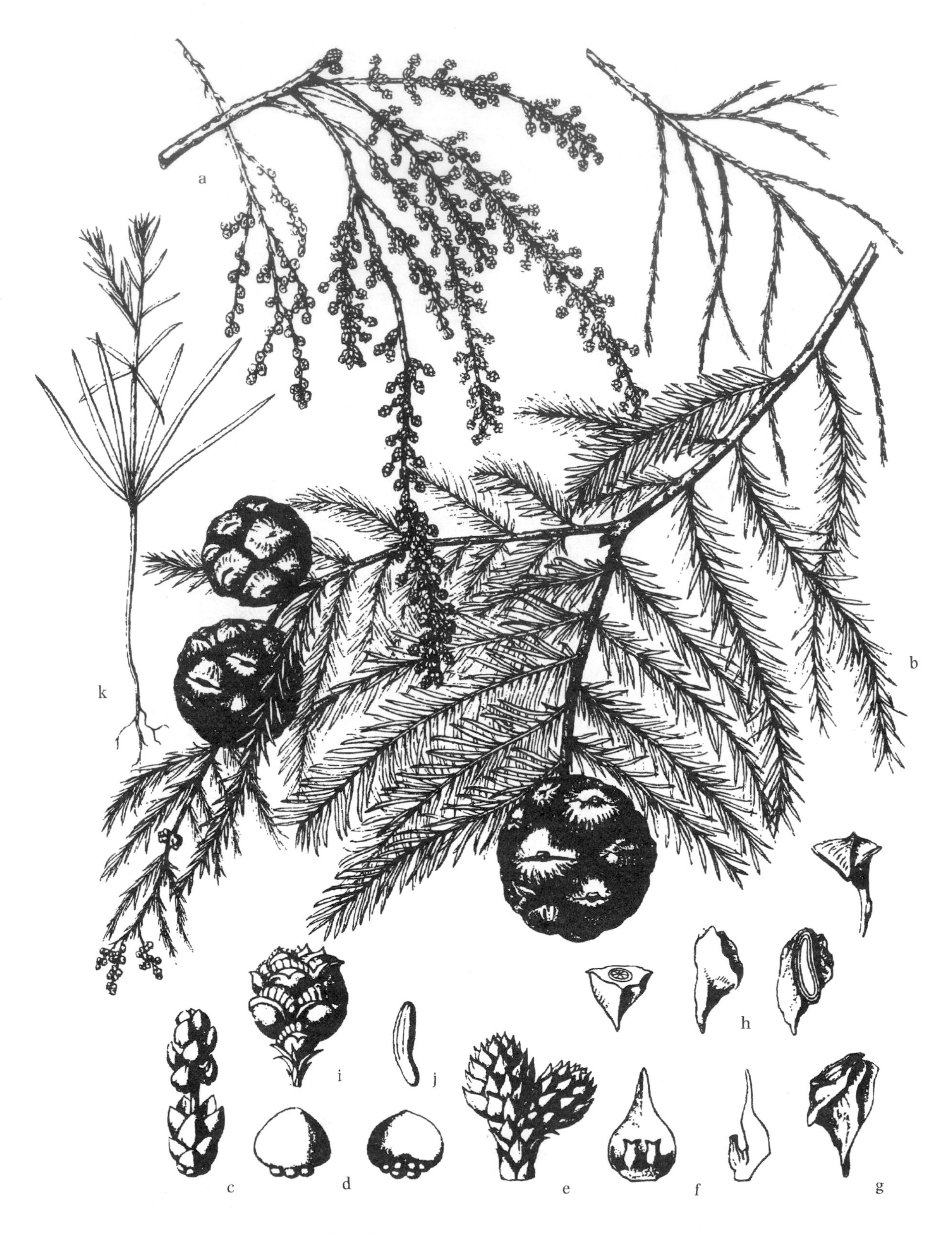

Fig. 42. *Taxodium distichum*: a. branch with numerous male cones and 2 young female cones at tip; b. branch with mature female cones and winter buds; c. branchlet with male cone; d. scale with microsporangia, two views; e. young female cones; f. female cone scale, view of inner surface showing 2 ovules, and side view; g. scale with seeds; h. whole seed with cross-section (*left*) and vertical section (*right*); i. developing female cone; j. embryo; k. seedling (Sargent).

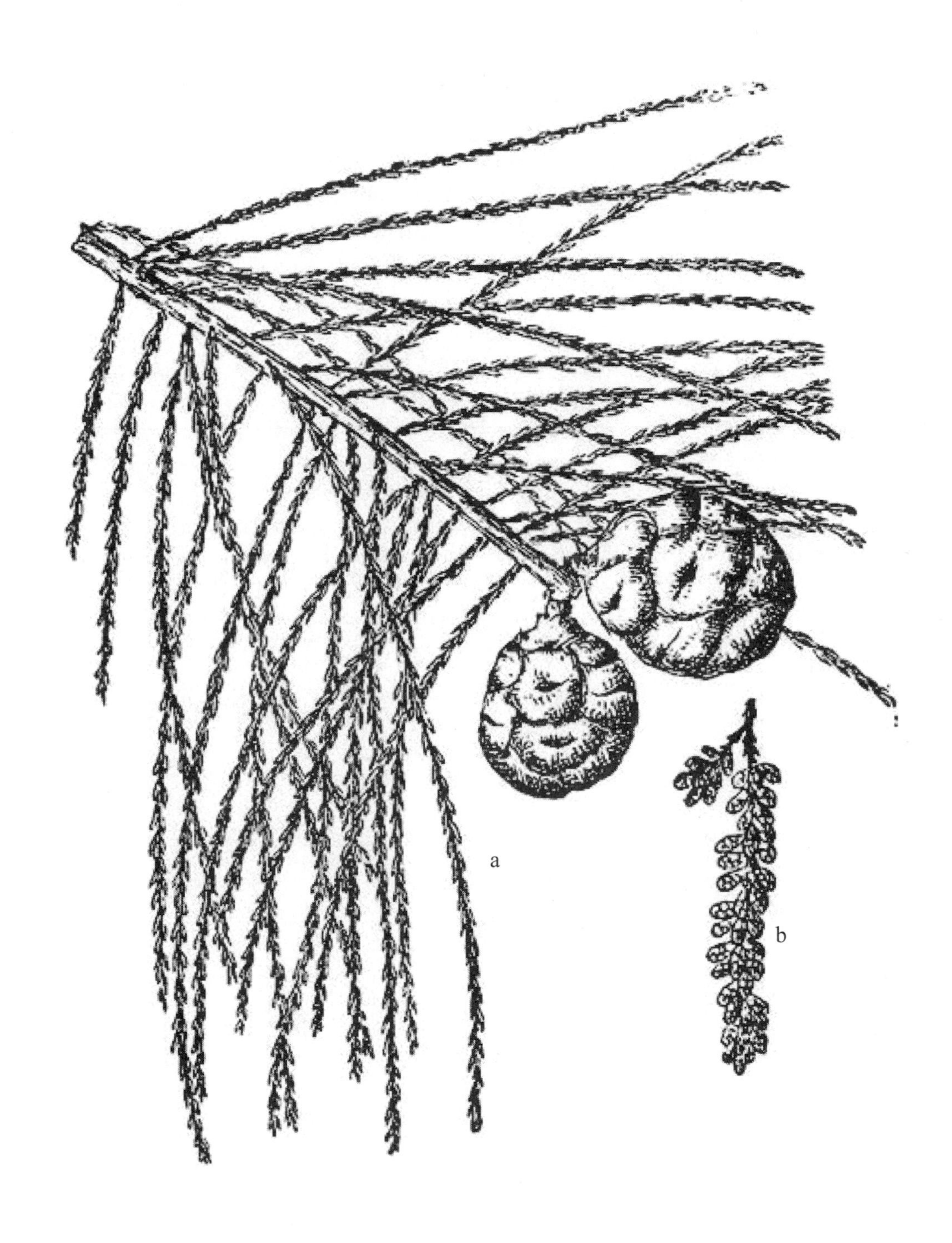

Fig. 43. *Taxodium ascendens*: a. portion of branch with female cones; b. male cones (B&B).

Angiosperms (Flowering Plants)

Primitive Angiosperms

Cabombaceae / Water-shield Family

1. Submersed leaves opposite, dissected into linear segments (fig. 44a); floating leaves small, inconspicuous, oblong to linear-elliptic, peltate, less than 2 cm long, subtending flowers; submersed portions of plants lacking mucilaginous coating; flowers white to pinkish; stamens 3–6. . . . . 1. *Cabomba*
1. Submersed leaves absent, except on young plants, then alternate; floating leaves conspicuous, peltate, elliptic, margins entire, 2–10 cm long (fig. 45a); submersed portions of plants typically with a mucilaginous coating; flowers red-purple; stamens 18–36 or more. . . . . 2. *Brasenia*

1. *Cabomba* (Fanwort)

Perennials, arising from short rhizomes with fibrous roots; submersed leaves opposite, petioled, fan-like with flat, finely dissected divisions; floating leaves small, oblong to linear-elliptic, peltate; flowers usually solitary, borne in axils of floating leaves, perianth white to pinkish, about 12 mm long; stamens 3–6; carpels (2)3(4), separate; fruit leathery, indehiscent, 1–3-seeded.

References: Crow, 2020; Fassett, 1953a; Moseley et al., 1984; Ørgaard, 1991; Schneider and Jeter, 1982; Wiersema, 1997.

1. *C. caroliniana* A. Gray Fanwort Fig. 44
Ponds, lakes, pools in marshes, reservoirs, quiet streams and ditches. S. N.H. and Mass. w. to N.Y., s. Ont., s. Mich., s. Ill., and Mo., s. to Fla., e. Okla. and e. Tex.; s. Wash., nw. Oreg. and n. Calif.; W.I.; native throughout much of southeastern U.S., natzd. N.J., Pa., and Ohio northward. (*C. caroliniana* var. *pulcherrima* R. M. Harper; *C. pulcherrima* (R. M. Harper) Fassett)

2. *Brasenia* (Water-shield)

Perennials, arising from creeping rhizomes, overwintering by turions coated with mucilage; leaves alternate, floating (submersed leaves present only on young plants), elliptic, peltate; submersed parts typically covered with a thick, mucilaginous coating; flowers solitary, axillary; perianth red-purple, 12–20 mm long; carpels 4–10, separate; fruit fusiform, leathery, indehiscent, 1- or 2-seeded.

References: Adams, 1969; Osborn and Schneider, 1988; Wiersema, 1997.

1. *B. schreberi* J. F. Gmel. Water-shield Fig. 45
Ponds, lakes and slow streams. Nfld., P.E.I. and s. Que. w. to s. Ont. and Minn., s. to Fla. and Tex.; s. Alask., s. B.C. and Alta. s. to Oreg. and Calif.; Mex., W.I. and C.Am. (*B. peltata* Pursh; *B. purpurea* (Michx.) Casp.)

Nymphaeaceae / Waterlily Family

References: Li, 1955; Mitchell and Beal, 1979; Wiersema and Hellquist, 1997; Wood, 1959.

1. Flowers white to pinkish, broadly spreading; sepals 4, greenish; petals numerous, conspicuous, elliptic to spatulate to oblanceolate; stigmas radiating from globose ovary summit extending into linear, incurved sterile appendages; leaves circular, lobes with recurved tips (fig. 46a, d); venation essentially palmate (fig. 46a,d). . . . . 1. *Nymphaea*

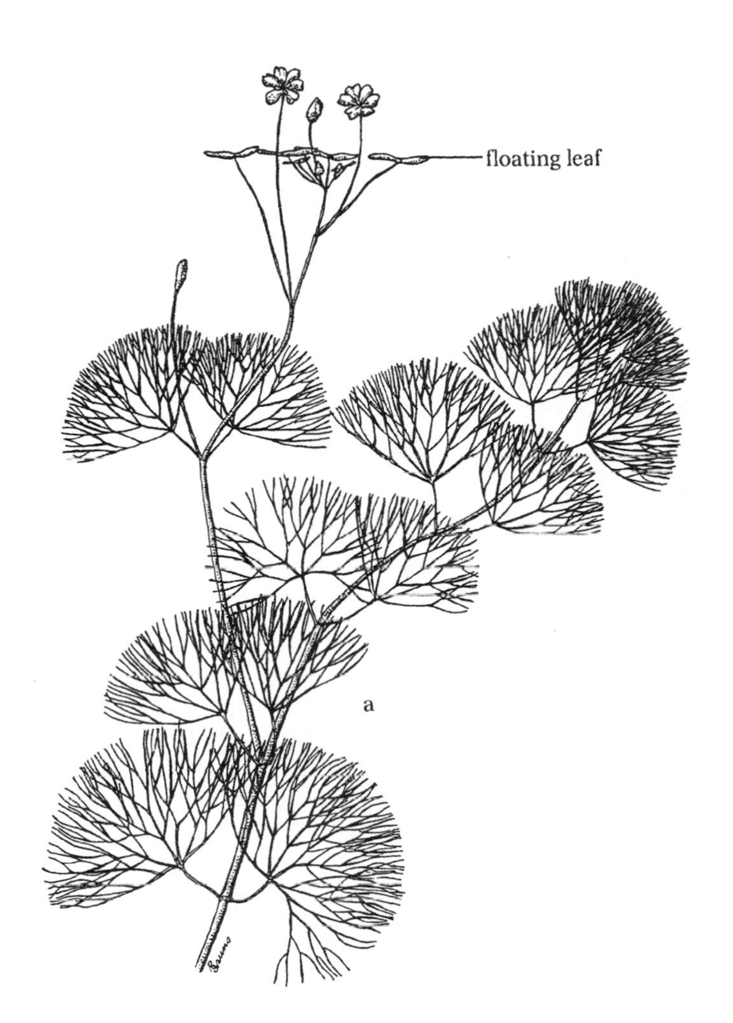

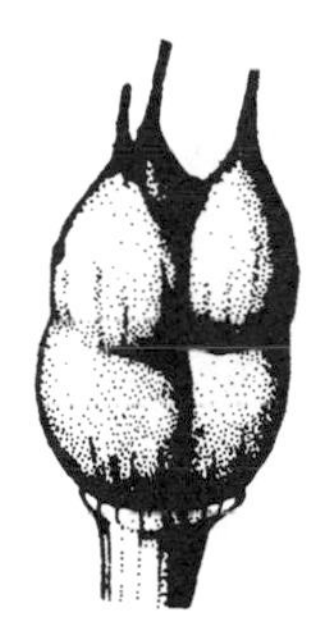

Fig. 44. *Cabomba caroliniana*: a. habit (NHAES); b. fruit (NYS Museum).

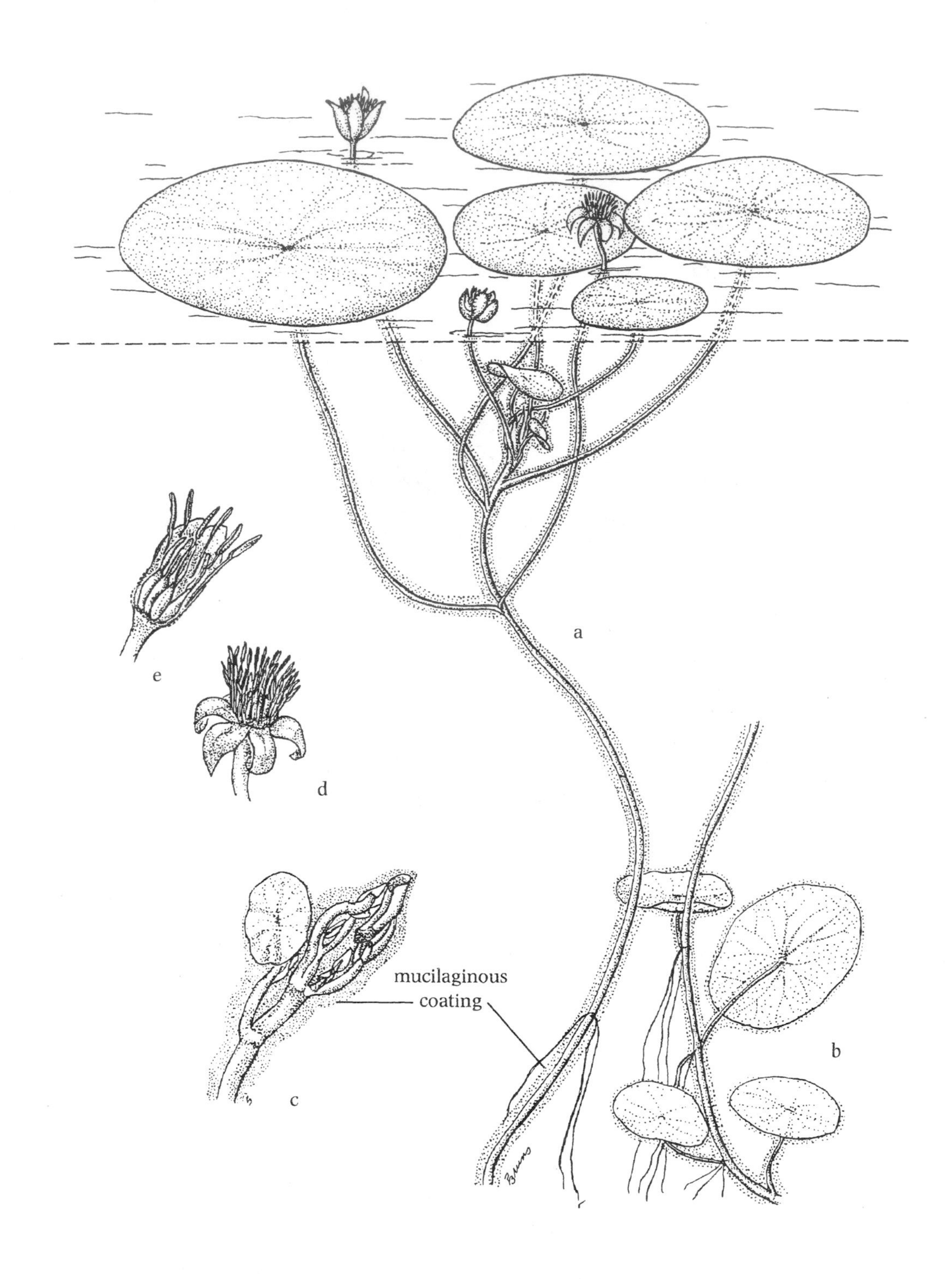

Fig. 45. *Brasenia schreberi*: a. habit; b. submersed leaves of young plant; c. winter bud with thick mucilaginous coating; d. flower; e. section of flower showing separate carpels (NHAES).

1. Flowers yellow; sometimes greenish or reddish, subglobose; sepals 5 or 6(9); petals numerous, obscure, strap-like, usually shorter than, but resembling, the stamens recurved, stigmas sessile, radiating on a disk (fig. 49d), appendages absent, leaves circular to kidney-shaped, ovate to oblong, lobes rounded, venation essentially pinnate (fig. 50a). . . . . 2. *Nuphar*

1. *Nymphaea* (Waterlily, Pondlily)

Perennials, arising from rhizomes; leaves submersed, floating, and/or emersed, deeply cleft, long-petioled; flowers showy; sepals 4, greenish; petals numerous, white to pink, with the inner ones usually transitional to stamens; fruit a leathery berry; seeds arillate, maturing underwater.

REFERENCES: Conard, 1905; Monson, 1960; Schneider and Chaney, 1981; Wiersema, 1988, 1996; Wiersema and Hellquist, 1997; Williams, 1970; Woods et al., 2005.

1. Flowers 6–19 cm wide; petals 17–43; filaments widest below middle; leaves 5–40 cm wide, orbicular to ovate, typically with a narrow or closed sinus; rhizomes horizontal.
   2. Petioles green or purple, not striped, rounded at junction with leaf blade; leaves red to purple (occasionally green) on lower surface; petals elliptic, subacute at tip (fig. 46c); branches of rhizomes not strongly constricted at base; seeds 1.5–2.3 mm long. . . . . 1a. *N. odorata* subsp. *odorata*
   2. Petioles green with brown-purple stripes, slightly flattened and channeled at junction with leaf blade; leaves green to faintly red-purple on lower surface; petals spatulate to oblanceolate, rounded at tip (fig. 46f); branches of rhizomes strongly constricted at base, tuberiferous, readily detached; seeds 2.8–4.4 mm long. . . . . 1b. *N. odorata* subsp. *tuberosa*
1. Flowers 3–7.5 cm wide; petals 8–17; filaments widest above middle; leaves 2–15 cm wide, broadly ovate to obovate, often with a wide sinus; rhizome vertical.
   3. Receptacle on bud and open flower rounded or only slightly 4-angled (fig. 47b,d); carpellary appendages up to 1.5 mm long (fig. 47e). . . . . 2. *N. leibergii*
   3. Receptacle on bud and open flower distinctly 4-angled (fig. 47f,g); carpellary appendages mostly 3 mm or more long (fig. 47h). . . . . 3. *N. tetragona*

1. *N. odorata* Aiton Fragrant Waterlily

1a. *N. odorata* subsp. *odorata* Fig. 46
Ponds, lakes, slow streams and ditches. Nfld. and sw. Que. w. to Ont., Minn., Man. and Sask., s. to Fla., Tex. and Mex.; W.I. and C.Am. Light pink-flowered individuals frequently encountered are treated by some authors as f. *rubra* Guillon. Occasionally, plants with dark pink to light red flowers are found which are likely to be a horticultural variety of the European *Nymphaea alba* var. *rubra* Lönnr., differentiated from *N. odorata* by having 20–24 petals which are rounded at the tips. *Nymphaea odorata* has 17–43 petals which gradually taper to subacute at the tips.

1b. *N. odorata* subsp. *tuberosa* (Paine) Wiersema & Hellq. Fig. 46
Ponds, lakes, slow streams and ditches. E. Me. and sw. Que. w. to n. Ont. and Minn., s. to Pa., Ohio, Ind., Ill., Iowa and Kans. Floating leaves are often very large (up to 40 cm wide), and plants along the shoreline may have leaves elevated above the surface. The best characteristic to separate the two subspecies is the readily detached tubers in subsp. *tuberosa*. Populations east of Lake Champlain in New England are sporadic and show intermediate characteristics which includes the blush-colored underside of leaves rather than green undersides. (*N. tuberosa* Paine)

2. *N. leibergii* Morong Leiberg's Waterlily Fig. 47
Ponds, lakes and slow streams. Que., n. Me. and n. Vt. w. to Isle Royale, Mich., n. Minn., Sask. and Alta., s. to nw. Mont., n. Ida. and Wash. In the United States this rare species is most abundant in northern Maine and northern Minnesota. Populations in Minnesota and Manitoba have been observed with larger floating leaves. Sterile hybrids with *N. odorata* subsp. *odorata* have been observed in northern Maine and northern Vermont. In central Manitoba and Central Saskatchewan, just north of our range, fertile hybrids occur and have been named *Nymphaea loriana* Wiersema, Hellq. & Borsch. (*Nymphaea tetragona* subsp. *leibergii* (Morong) Porsild)

3. *N. tetragona* Georgi Pygmy Waterlily Fig. 47
Ponds, lakes and slow streams. Very rare in our range. Se. Man., n. Sask. w. to n. Alta., B.C. and Alask., s. to nw. Wash. This species occurs widely in northern Eurasia and extends into nw. N.Am. Presently it is known in our range only from se. Man. where the hybrid with *N. odorata* subsp. *tuberosa* also occurs.

Nymphaea mexicana Zucc. has been reported south of our range in northeast N.C.
This yellow-flowered species produces "banana-like" structures at terminal nodes of stolons when overwintering. It hybridizes with *N. odorata* forming *N.* ×*thiona* D. B. Ward. This natural hybrid has been introduced into nc. Ky. Many horticultural hybrids with *N. mexicana* have been developed for the aquatic garden trade and some are relatively hardy in northern states.

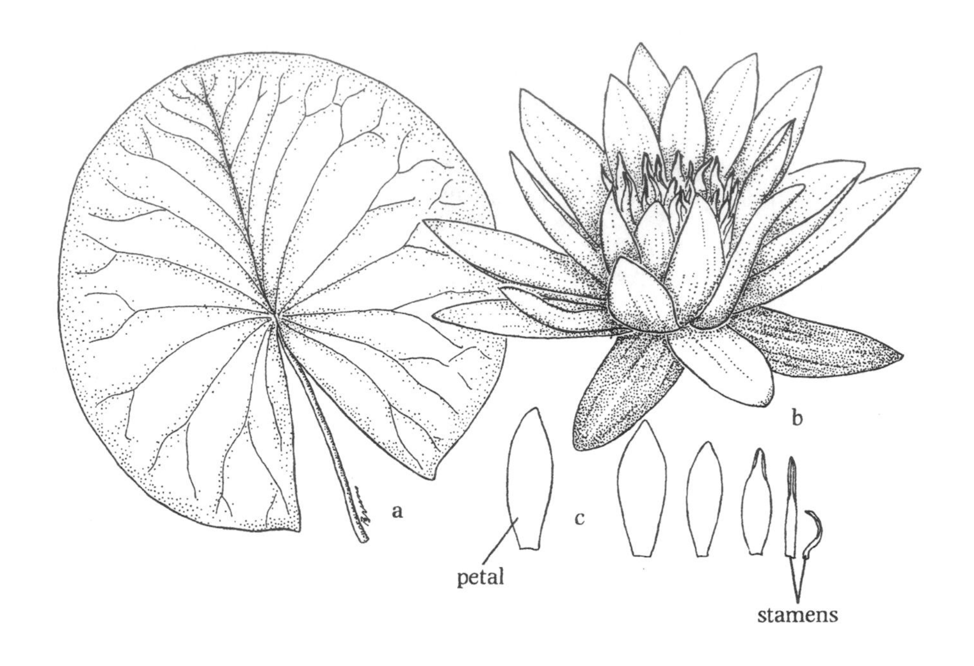

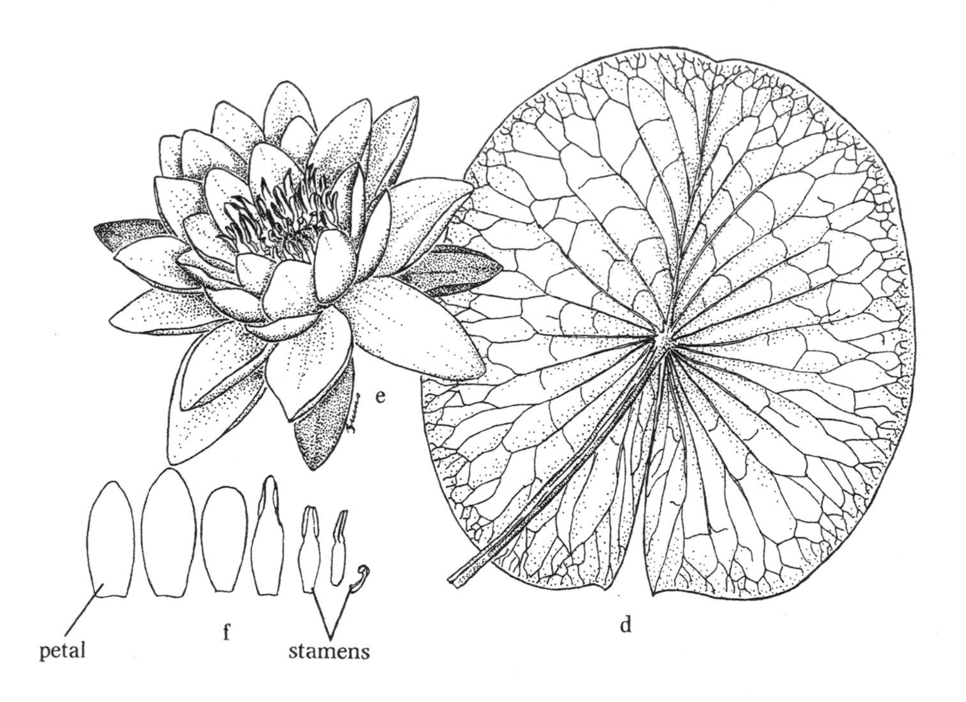

Fig. 46. *Nymphaea odorata* subsp. *odorata*: a. leaf; b. flower; c. series showing transition from petals to stamens (NHAES).
Nymphaea odorata subsp. *tuberosa*: d. leaf; e. flower; f. series showing transition from petals to stamens (NHAES).

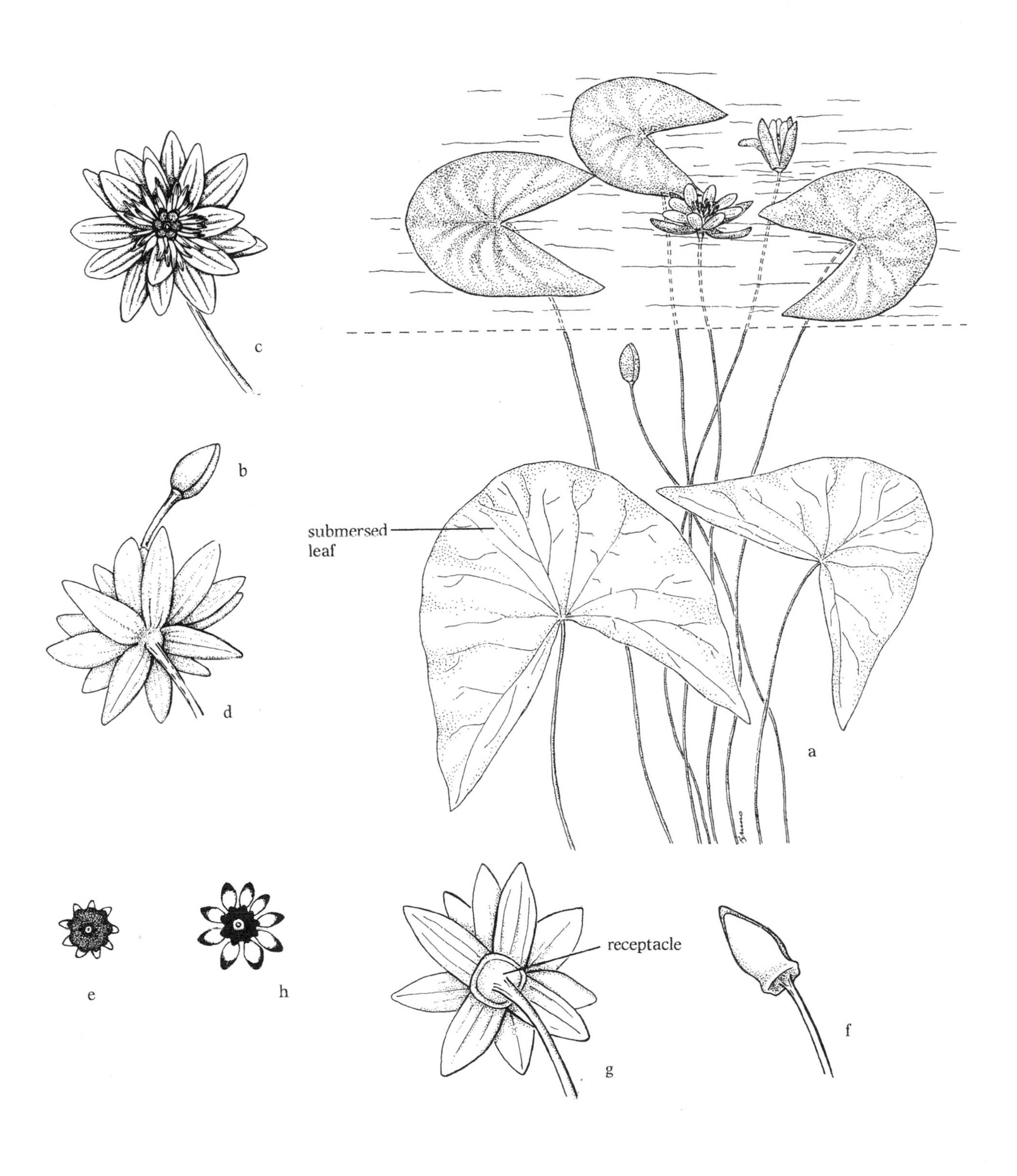

Fig. 47. *Nymphaea leibergii*: a. habit, showing both floating leaves and submersed leaves (NHAES); b. bud showing rounded receptacle; c. upper side of flower; d. lower side of flower showing receptacle; e. carpellary appendages (Wiersema, 1996).
Nymphaea tetragona: f. bud showing 4-angled receptacle; g. lower side of flower showing 4-angled receptacle; h. carpellary appendages (Wiersema, 1996).

2. *Nuphar* (Yellow Waterlily, Spatterdock, Pondlily)

Perennials, arising from large, thick, cylindrical, creeping rhizomes; leaves submersed when young, floating and/or erect at maturity, blades deeply cleft at base, with long petioles; flowers globose, floating or elevated above surface; sepals 5 or 6(9), yellow, sometimes greenish or reddish tinged; petals numerous, yellow, small, strap-like, resembling the stamens; stamens numerous, filaments flat; stigmas sessile, radiating on a disk; fruit an ovoid, leathery berry, with numerous non-arillate seeds per locule, maturing above water surface. Seeds are known to be dispersed by turtles as well as waterfowl (Padgett et al., 2010).

The treatment of *Nuphar* by E. O. Beal (1956), not followed here, recognized only one species, *N. lutea* (L.) Sm., in North America and Europe, with nine subspecies. Recent monographic and molecular studies (Padgett et al., 1998, 1999; Padgett, 2007) strongly indicated that not only should several species be recognized, but that there is a distinction between taxa with Old World affinities and New World taxa. On the basis of this information, it would be inaccurate to apply the name *N. lutea* to any of the New World taxa.

References: Beal, 1956; Beal and Southall, 1977; DePoe and Beal, 1969; Heslop-Harrison, 1955; Padgett, 1997, 2007; Padgett et al., 1998, 1999, 2010, 2021; Schneider and Moore, 1977.

---

1. Leaf blades of mature plants 3.5–20 cm long, 3.5–14.5 cm wide; fruit conspicuously constricted below the stigmatic disk (fig. 48f); stigmatic disk usually red.
   2. Flowers 2 cm or less wide; stigmatic disk with 6–10 deep crenations (fig. 48d); leaf blades 3.5–10 cm long, 3.5–7.5 cm wide; blade notch a third or less the length of midrib. . . . . . 1. *N. microphylla*
   2. Flowers 3 cm or more wide; stigmatic disk with 8–15 shallow crenations (fig. 48f); leaf blades 5–20 cm long, 4.5–14.5 cm wide; blade notch half the length of the midrib. . . . . . 2. *N. ×rubrodisca*
1. Leaf blades of mature plants 7.0–40 cm long, 4.0–25 cm wide; fruit only slightly constricted or not constricted below the stigmatic disk (fig. 49e); stigmatic disk greenish.
   3. Leaves floating; petiole strongly flattened on one side and winged (fig. 49c); inner sepals often maroon or reddish on inner surface. . . . . . 3. *N. variegata*
   3. Leaves emergent or emergent and floating; petiole round in cross-section, not winged (fig. 50c); inner surface of sepals green and/or yellow on inner surface.
      4. Leaf blades less than 3 times as long as wide, basal sinus deeper, 2.5–14.5 cm deep, lobes overlapping to divergent; leaves mostly erect, elevated above the water (fig. 50b); fruit strongly ribbed throughout.
         5. Sepals green to yellow on outer surface; fruit green, up to 5.5 cm long. . . . . . 4a. *N. advena* subsp. *advena*
         5. Sepals red to purple on outer surface; fruit reddish, up to 2.5 cm long. . . . . . 4b. *N. advena* subsp. *ozarkana*
      4. Leaf blades 3–7 times as long as wide, basal sinus shallow, 2–7 cm deep, lobes seldom overlapping; leaves emersed or floating; fruit smooth at basal portion, smooth to ribbed above. . . . . . 5. *N. sagittifolia*

---

1. *N. microphylla* (Pers.) Fernald Small-leaved Pondlily Fig. 48

Ponds, lake margins and slow streams. N.S. and N.B. w. to s. Man., s. to N.E., N.J., e. Pa. (extirpated?), N.Y., n. Mich., n. Wisc. and ne. Minn. This taxon has affinities with the Old World taxa of sect. *Nuphar* and appears to be most closely related to the Eurasian *N. pumila* (Padgett et al., 1998; Padgett, 2007). Beal (1956) had treated these two as conspecific, under the name *N. lutea* subsp. *pumila*, but Padgett (2007) maintains them as distinct species.

2. *N. ×rubrodisca* Morong Red-disk Spatterdock Fig. 48

Ponds, lakes and slow streams. Nfld. w. to e. Ont., s. to N.S., Conn., N.J., e. Pa., N.Y., n. Mich., n. Ill., n. Wisc. and e. Minn. Long referred to as *N. ×rubrodisca*, with its distinctive bright red stigmatic disk on the flower and fruit, this plant was believed to be a fertile hybrid derived from the dwarf yellow waterlily, *N. microphylla*, which has a red stigmatic disk, crossed with *N. variegata*, with a greenish stigmatic disk. Padgett (Padgett, 1997, 2007; Padgett et al., 1998) has obtained strong evidence based on morphologic, pollen fertility, and molecular studies that the plants are, indeed, a result of hybrid crossing between *N. microphylla*, a taxon belonging to an Old World group (sect. *Nuphar*), and *N. variegata*, of New World affinities (sect. *Astylus*). But Padgett regards the hybrid plants as having strongly reduced fertility and believes that scattered populations of the hybrid across the sympatric range of the parents may have been generated from independent hybrid events. (*N. lutea* subsp. *rubrodisca* (Morong) Hellq. & Wiersema)

3. *N. variegata* Engelm. ex Durand Bullhead Pondlily, Yellow Pondlily Fig. 49

Ponds, lakes, slow streams and ditches. Nfld. and Lab. w. to Yuk., s. to Del., ne. Pa., n. Ohio, n. Ind., n. Ill., e. S.D., e. Neb., n.w. Mont. and B.C. (*Nuphar lutea* subsp. *variegata* (Engelm. ex Durand) Beal)

4. *N. advena* (Aiton) Aiton f. Erect Spatterdock Fig. 50

4a. *N. advena* subsp. *advena*

Ponds, lakes, swamps, slow streams and ditches. Se. Me., e. Mass. and sw. Conn. w. to c. N.Y., Pa., s. Ont., s. Mich., s. Wisc., Mo. and se. Neb., s. to Fla., Tex. and e. Mex. Populations previously recognized as *Nuphar ozarkana* Mill. & Standl., from the Ozark region of Missouri and Arkansas, were regarded by Beal (1956) as questionably distinct from typical *N. advena*; however, Padgett (2007) has since recognized these at subspecific rank within *N. advena*. They are distinguished by floating leaves and by red-tinged fruits and inner surfaces of sepals, reminiscent of two northern taxa, *N. microphylla* and *N. variegata*. A third taxon, subsp. *orbiculata*, of se. U.S. is not in our range. (*Nuphar fluviatile* (Harper) Standl.; *N. lutea* subsp. *advena* (Aiton) Kartesz & Gandhi; *Nuphar lutea* subsp. *macrophylla* (Small) Beal)

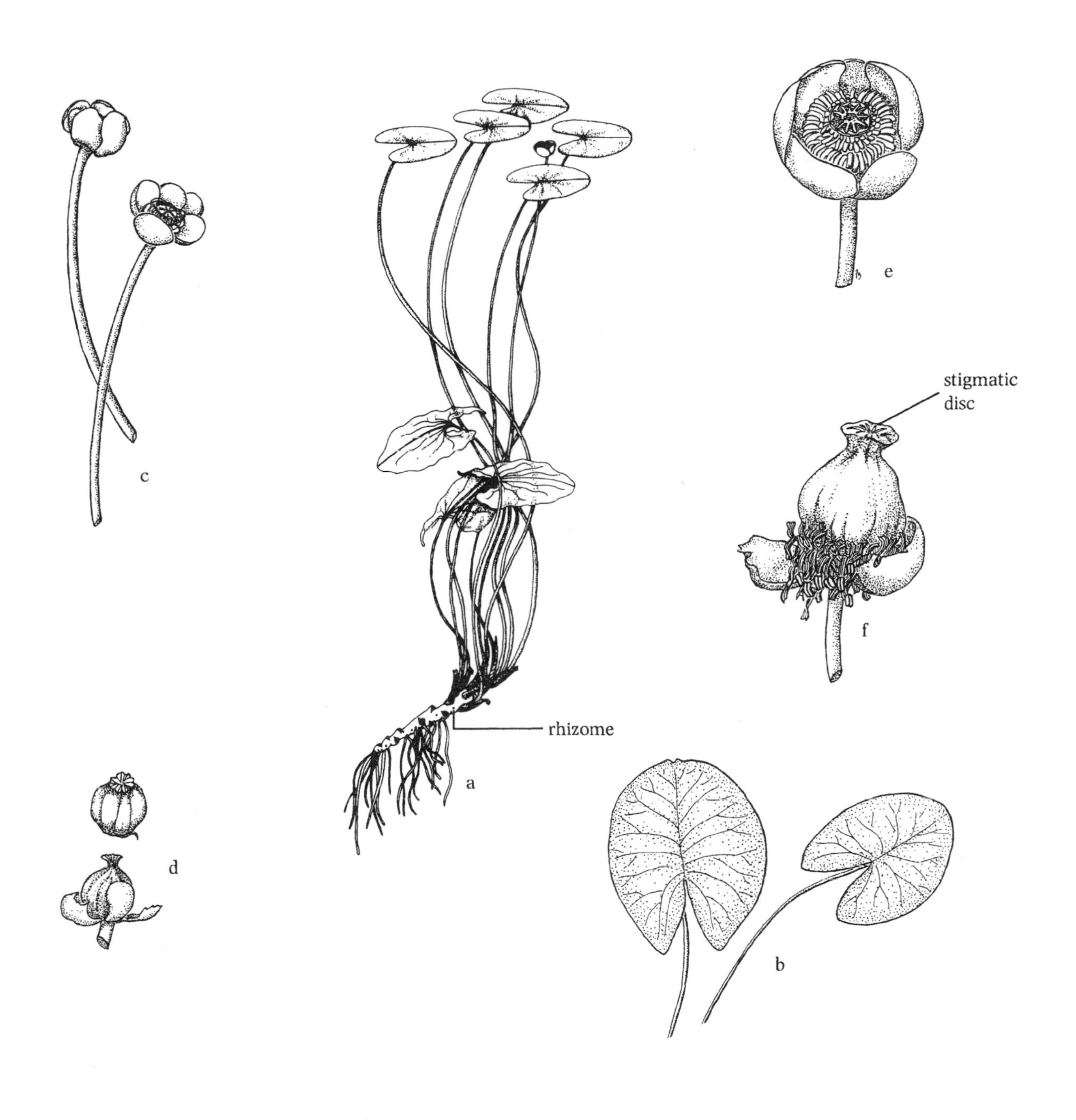

Fig. 48. *Nuphar microphylla*: a. habit (NYS Museum); b. leaves (NHAES); c. flowers (NHAES); d. fruit (NHAES). *Nuphar ×rubrodisca*: e. flower; f. fruit (NHAES).

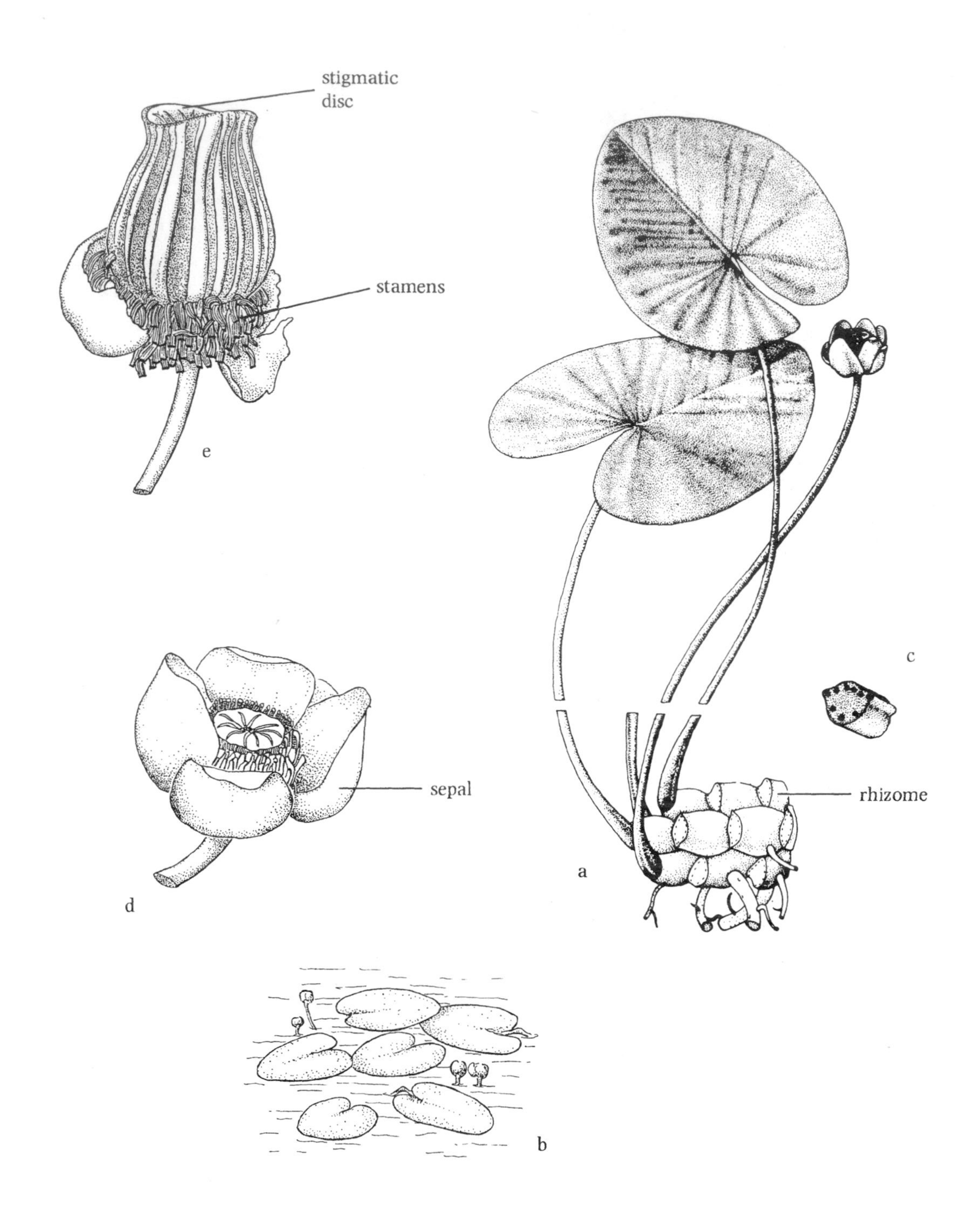

Fig. 49. *Nuphar variegata*: a. habit (NYS Museum); b. habit, showing characteristically floating leaves (NHAES); c. stem, cross-section (NYS Museum); d. flower (NHAES); e. fruit (NHAES).

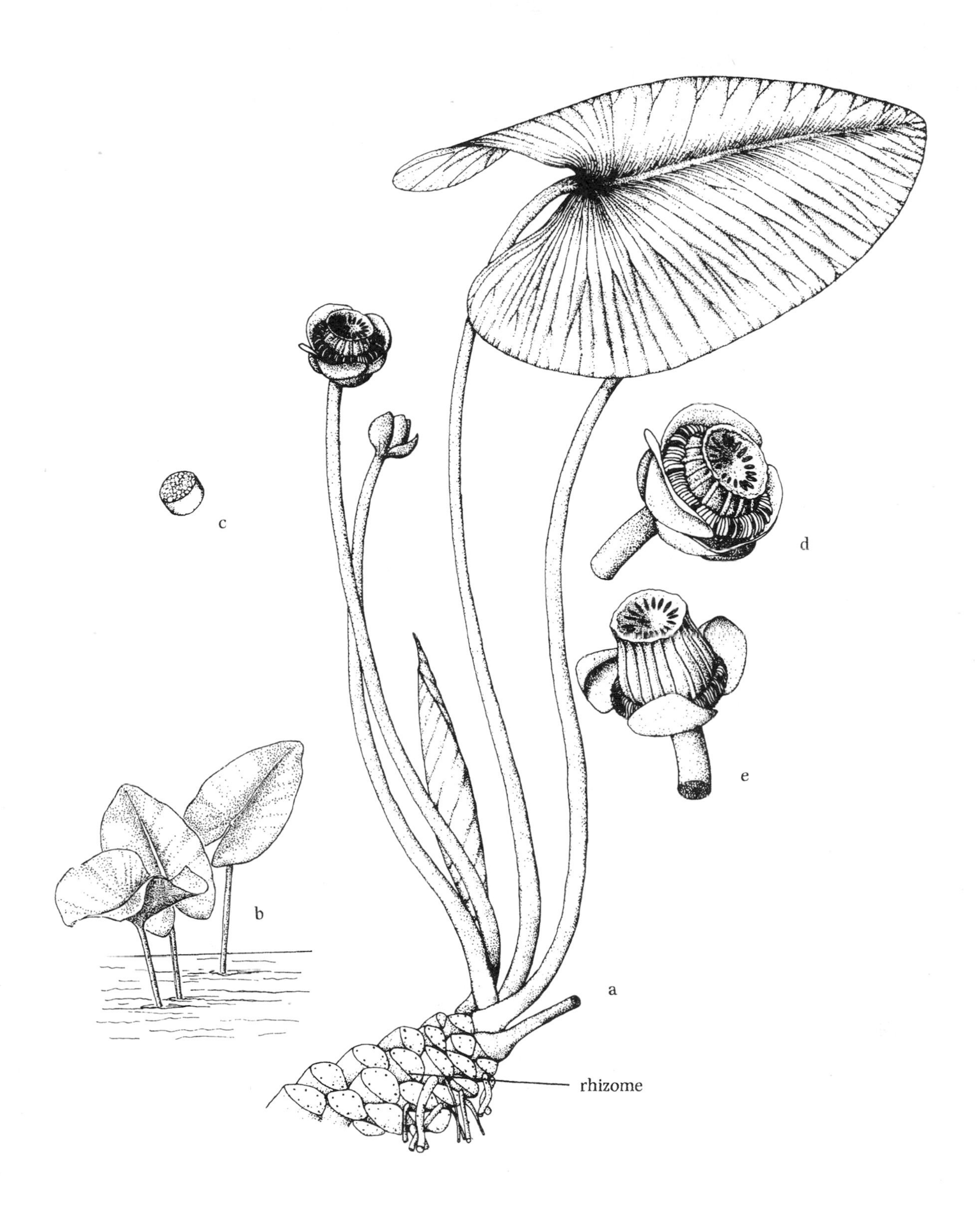

Fig. 50. *Nuphar advena*: a. habit (NYS Museum); b. habit, showing characteristically emersed leaves (NHAES); c. stem, cross-section (NYS Museum); d. flower (NYS Museum); e. fruit (NYS Museum).

4b. *N. advena* subsp. *ozarkana* (G. S. Mill. & Standl.) Padgett Ozark Spatterdock

Ponds, lakes, streams, slow rivers and ditches. S. Mo. and n. Ark. Populations previously recognized as *Nuphar ozarkana* are confined to warm waters of the Ozark Mountains. It is a diminutive plant very similar to subsp. *advena*, but distinguished by red to purple sepals and reddish fruits (Padgett, 2007).

5. *N. sagittifolia* (Walter) Pursh Cape Fear Spatterdock Fig. 51

Freshwater tidal rivers, sloughs, blackwater lakes, ditches, rivers, streams and bayous. Coastal plain and Piedmont; se. Va., N.C. and ne. S.C. (*N. lutea* subsp. *sagittifolia* (Walter) Beal)

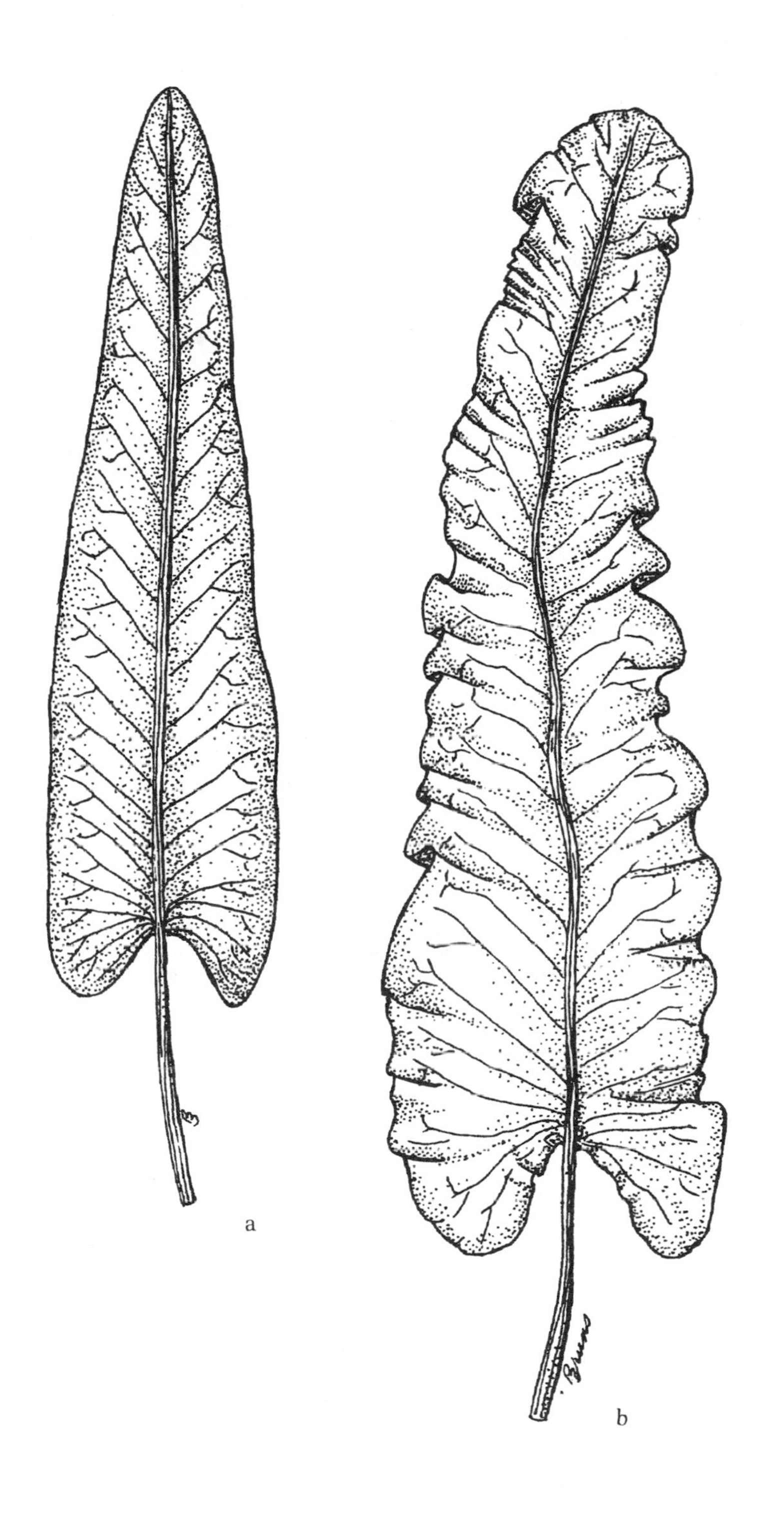

Fig. 51. *Nuphar sagittifolia*: a. floating leaf; b. submersed leaf (PB).

Magnoliids

Magnoliaceae / Magnolia Family

1. *Magnolia* (Magnolia)

Trees or shrubs, evergreen or deciduous; leaves alternate, simple, margins entire; flowers solitary, large, showy, fragrant; sepals 3, small; petals 6–12, large; stamens numerous, spirally arranged; pistils numerous, coherent, on an elongate receptacle; fruit an aggregate of follicles; seeds large, surrounded by a fleshy red to pink aril.

1. *M. virginiana* L. Sweetbay, Swampbay, Laurel
Fig. 52
Wet woods and swamps, wet flatwoods, boggy stream margins and savannas. Chiefly coastal plain, e. Mass., L.I., N.Y. and N.J. s. to Fla., w. to Ark and e. Tex. (*M. virginiana* var. *australis* Sarg.)

Lauraceae / Laurel Family

1. *Lindera* (Spicebush)

Shrubs or sometimes small trees; leaves alternate, simple, aromatic when rushed; flowers yellow, in umbel-like clusters, emerging from lateral buds before the leaves appear; plants polygamodioecious; pistillate flowers with rudimentary stamens; staminate flowers with 9 fertile stamens, anthers dehiscing by flaps; fruit a red drupe, rarely yellow.

1. *L. benzoin* (L.) Blume Spicebush, Benjamin-bush
Fig. 52
Rich moist woodlands, stream margins, floodplain forests and seeps of wooded slopes. S. Me., s. N.H. and Vt. w. to s. Ont., s. Mich., Ill., Mo. and se. Kans., s. to nw. Fla. and Tex. Plants with pubescent leaves and twigs in the southern portion of our range and southward have been treated as var. *pubescens* (Palmer & Steyerm.) Rehder.

Saururaceae / Lizard's-tail Family

1. *Saururus* (Lizard's-tail)

Perennial herbs, arising from rhizomes, simple or freely branching, forming clones; stems 0.15–1.2 m tall, jointed at nodes; leaves alternate, cordate; inflorescence a long, slender, white, spike-like raceme, often drooping near tip; flowers 175–350, small, lacking perianth; stamens usually (3)6(8), filaments whitish; carpels usually 3(4–5), united only at base; fruit a schizocarp, splitting into two dry 1-seeded mericarps, the mericarps indehiscent.

REFERENCES: Buddell and Thieret, 1997; Hall, 1940.

1. *Saururus cernuus* L. Lizard's-tail Fig. 53
Shallow water along stream margins, muddy shore, swamps and wet woods. Mass., R.I., Conn. and sw. Que. w. to s. Mich., Wisc., Ill., Mo. and e. Kans., s. to Fla. and Tex.

Fig. 52. *Magnolia virginiana*: a. branch with flower; b. stem tip with winter bud; c. fruit (NYS Museum). *Lindera benzoin*: d. vegetative branch; e. flowering branch; f. pistillate and staminate flowers; g. fruiting branch (NYS Museum).

Fig. 53. *Saururus cernuus*: a. habit; b. flower; c. fruit; d. seed (G&W).

Monocots

Acoraceae / Sweet-flag Family

1. *Acorus* (Sweet-flag, Calamus)

Emersed perennials, arising from thick creeping rhizomes; aromatic oils, especially in rhizomes; leaves long-linear, equitant, sword-shaped, with a thick raised midrib; inflorescence a long, thick, cylindric spadix, spathe absent; flowers bisexual, tepals 6, greenish; fruit a berry, with seeds 1–6, embedded in mucilage.

REFERENCES: Buell, 1935; Grayum, 1987; Löve and Löve, 1957; Thompson, 2000.

1. Leaves with a single raised midvein; plants producing fruit. . . . . 1. *A. americanus*
1. Leaves with single raised midvein plus 1–5 addition veins; plants not producing mature fruit. . . . . 2. *A. calamus*

1. *A. americanus* (Raf.) Raf. Fig. 54
Wet meadows, marshes, ponds and stream banks. Nfld. w. to Ont., Sask., N.W.T. and Alask., s. to Va., Ill., Iowa, N.D., n. Ida. and n. Wash. This species is often confused with vegetative specimens of *Iris* spp., which have flat leaves, but is easily distinguished by its keeled leaves (raised midvein) and sweet odor of the lower portion of the plant; emergent species of *Sparganium* have keeled leaves but lack the sweet odor. Native populations are diploid and fertile, whereas *A. calamus* in N.Am. are sterile triploids.

2. *A. calamus* L.
Wet meadows, marshes, ponds and stream banks. N.S. and Que. w. to Mich., S.D., and Colo. s. to Ga., Miss. and Tex.; Oreg. and Calif; Eurasia. Plants were originally introduced into North America for medicinal purposes; these are sterile triploids, but readily spread by rhizomes. Diploid and tetraploid populations occur in Asia (Thompson, 2000).

Araceae / Arum Family

REFERENCES: Hellquist, 2013; Hellquist and Crow, 1982; Landolt, 1986, 2000; Landolt and Kandeler, 1987; Thompson, 2000; Wilson, 1960.

1. Plants large, rooted in soil.
    2. Spadix golden-yellow; spathe obscure, forming a sheath at base of long scape (fig. 54d); leaves broadly oblong-elliptic. . . . . 1. *Orontium*
    2. Spadix green or brown; spathe broad, conspicuous; leaves ovate (fig. 57b), cordate (fig. 56a), hastate or sagittate (fig. 55b).
        3. Leaf blades hastate or sagittate (juvenile leaves unlobed), with prominent vein extending into each basal lobe (fig. 55b). . . . . 2. *Peltandra*
        3. Leaf blades broadly rounded to cordate, lacking prominent vein extending into basal lobes (if present).
            4. Spathe white; spadix on a long peduncle (fig. 56), emerging after leaves; plants lacking a strong odor when bruised. . . . . 3. *Calla*
            4. Spathe mottled, purplish-green and/or reddish-brown; spadix nearly sessile (fig. 57a), emerging very early spring, before leaves; plant with a strong skunk-like odor when bruised. . . . . 4. *Symplocarpus*
1. Plants very small, free-floating.
    5. Plants lacking roots.
        6. Fronds (leaves) 1.5 m or less long, globose or ellipsoidal (fig. 58a,b,c). . . . . 5. *Wolffia*
        6. Fronds 6–8 mm long, slender-elongate; sickle-shaped (fig. 58d). . . . . 6. *Wolfiella*
    5. Plants with roots (fig. 58j).
        7. Individual fronds with 1 root, fronds with 1–5(7) veins. . . . . 7. *Lemna*

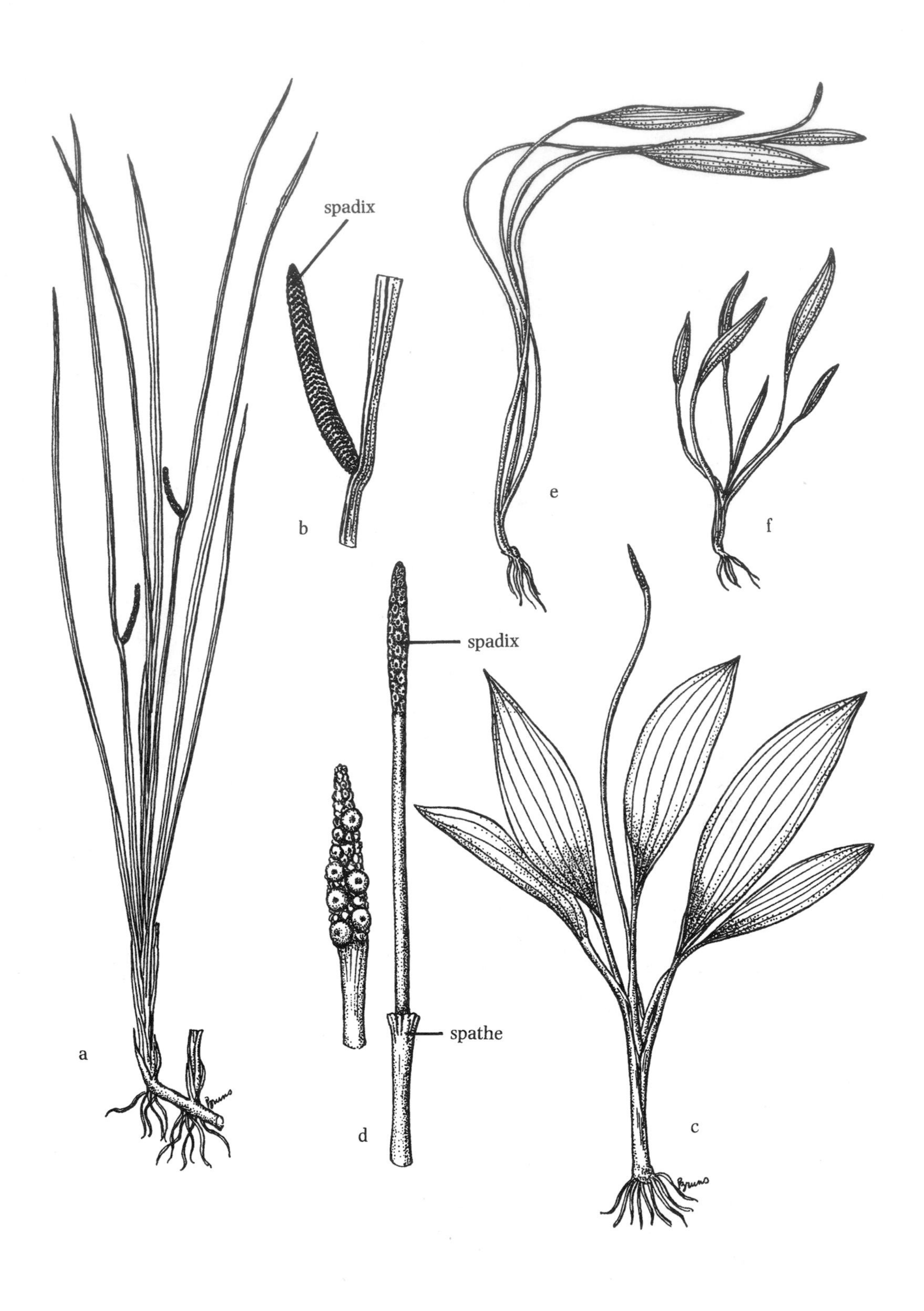

Fig. 54. *Acorus americanus*: a. habit; b. spadix (NHAES).
Orontium aquaticum: c. habit, terrestrial plant; d. spadix with reduced spathe; e. habit, submersed plant; f. submersed juvenile stage (NHAES).

7. Individual fronds with 2 or more roots, fronds with (3)5–21 veins.
   8. Fronds 7–16(21)-veined, roots (5)7–16(21), often with a red-spot on top, often reddish-maroon on lower surface. . . . . . 8. *Spirodela*
   8. Fronds (3)5–7-veined, roots (1)2–5(12), upper surface lacking red-spot, lower surface not red or maroon. . . . . 9. *Landoltia*

---

1. *Orontium* (Golden-club)

Perennial herbs, growing from a deep rhizome; leaves basal, long-petioled, entire, often floating; flowers bisexual, borne in a golden-yellow spadix; spathes reduced, obscure sheath at base of scape; fruit a blue-green or brownish utricle.

Reference: Grear, 1966.

1. *O. aquaticum* L. Fig. 54
Shallow water, swamps and shores of rivers and ponds. Chiefly coastal plain, Mass. and c. N.Y., sw. to Pa., W.Va., Ky. and Tenn., s. to Fla., Miss., La. and Tex.

2. *Peltandra* (Arrow-arum)

Emersed perennials, with thick fibrous or subtuberous roots; leaves basal, long-petioled; blades hastate or sagittate; upper flowers of spadix staminate, lower ones pistillate, fruit an amber or green berry.

Reference: Blake, 1912.

1. *P. virginica* (L.) Schott & Endl. Fig. 55
Bogs, swamps and margins of lakes and ponds, rivers and streams, freshwater tidal rivers, and ditches. C. Me. and s. Que. w. to Ont., Mich., Minn. and Wisc., s. to Fla., se. Okla. and Tex. Vegetative plants are often confused with several species of *Sagittaria* or *Pontederia cordata*, but can easily be differentiated by the conspicuous vein paralleling the leaf margin (fig. 55b), by a single prominent midvein, and a prominent vein extending into each basal lobe. This highly variable species has had many dubious forms named solely on the basis of leaf shape; an odd popula tion northern Vermont had ovate leaves, but could be recognized by its distinct marginal vein.

3. *Calla* (Water-arum, Wild Calla)

Emersed perennials, arising from a long creeping rhizome; leaves cordate, long-petioled; flowers bisexual, borne on a spadix; spathe white, ovate, tapering to a point at apex; fruit an ovoid cluster of red berries.

1. *C. palustris* L. Fig. 56
Wooded swamps, marshes, bogs, and marshy margins of lakes, ponds and rivers. Lab. and Nfld. w. to Alask., s. to n. N.J., w. Md., n. Ind., n. Ill., Wisc., Minn., Sask. and B.C.

4. *Symplocarpus* (Skunk-cabbage)

Emersed perennials, arising from a thick, erect rhizome; leaves ovate to cordate, short-petioled; flowers borne on a spadix, subtended by a mottled purplish-green and/or reddish-brown spathe; ovaries imbedded in spadix, globose or ovoid mass; entire plant producing a foul skunk-like odor when bruised.

1. *Symplocarpus foetidus* (L.) Nutt. Skunk-cabbage Fig. 57
Wet woods, bogs, swamps and stream borders. N.S. and s. Que. w. to s. Ont. and e. Minn., se. Man., s. to n. N.C., ne. Tenn., nc. Ky., Ohio, Ind., c. Ill. and Iowa.

5. *Wolffia* (Water-meal)

Minute free-floating plants, rootless; fronds tiny (up to 1.6 mm), globular, ovoid, or boat-shaped; flowers rarely produced; vegetative reproduction extensive, often covering large areas of surface of ponds.

This genus contains the smallest known flowering plants. The different species are extremely difficult to identify, especially if allowed to dry. Every effort should be made to identify *Wolffia* while material is fresh or preserved in liquid; see technique described by Brunton (2019).

References: Dore, 1957; Hess, 1986.

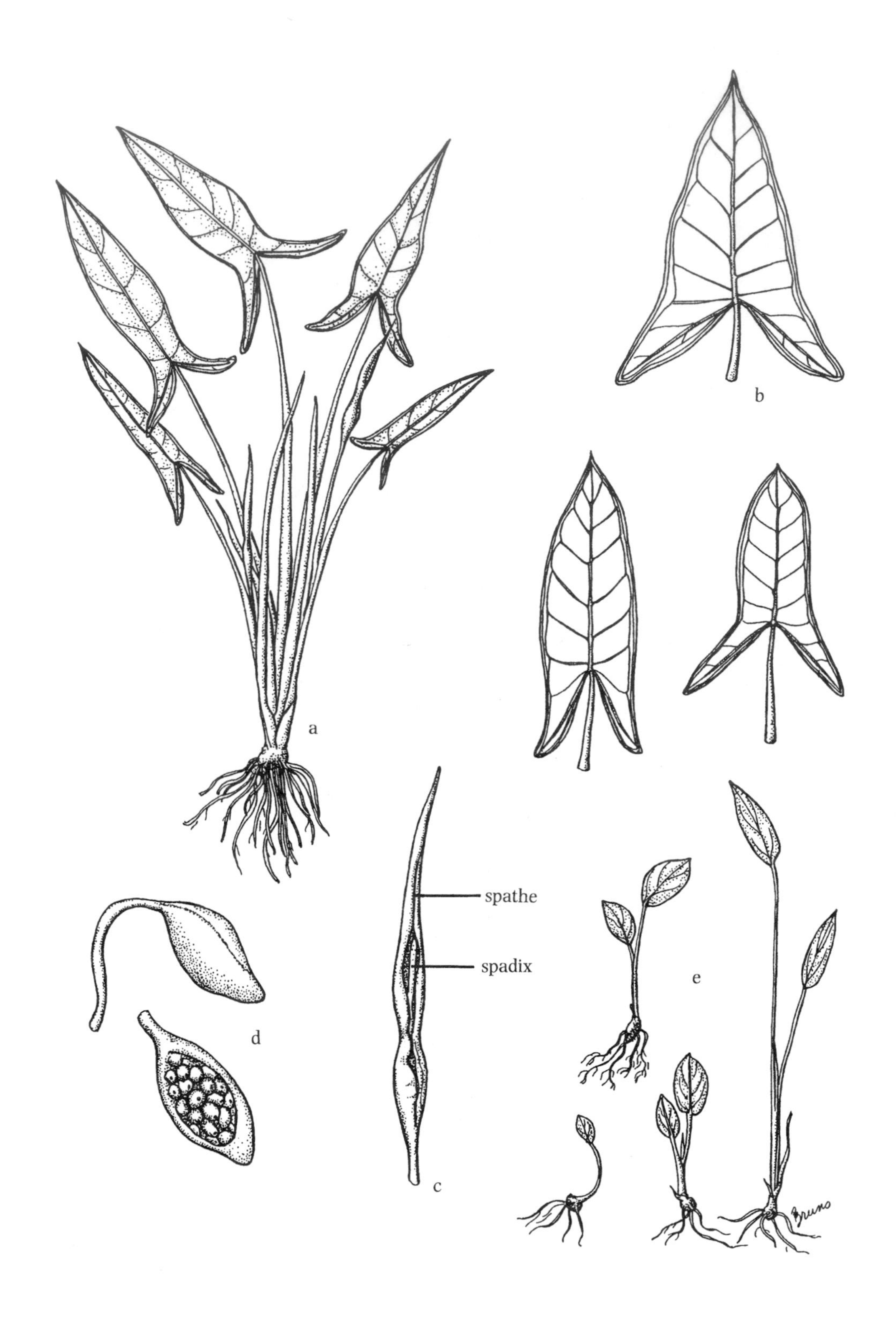

Fig. 55. *Peltandra virginica*: a. habit; b. leaf variations; c. spathe and spadix; d. spadix at fruiting stage and with spathe cut away, below; e. seedlings to juvenile stages (NHAES).

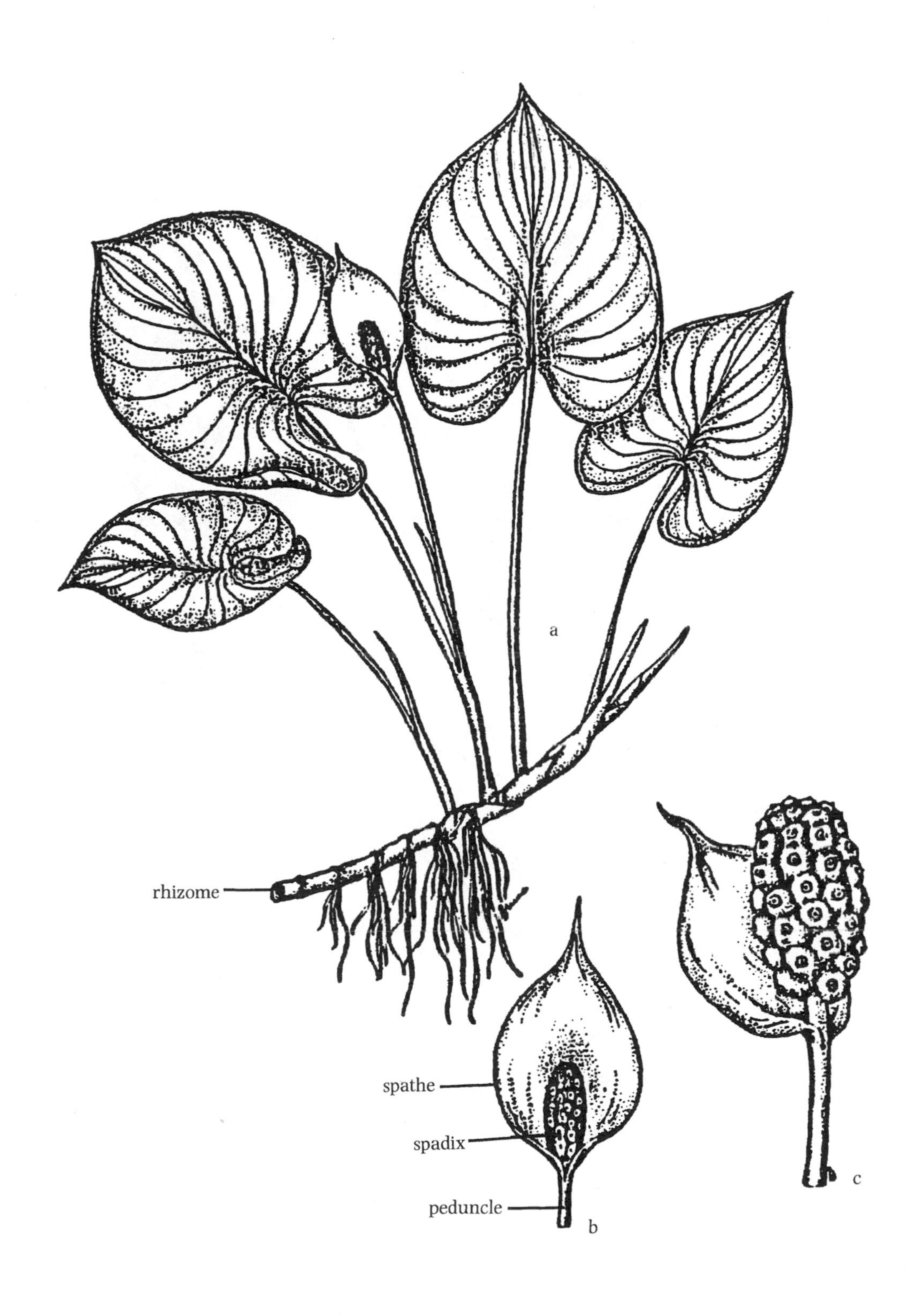

Fig. 56. *Calla palustris*: a. habit; b. spathe and spadix; c. spadix at fruiting stage (NHAES).

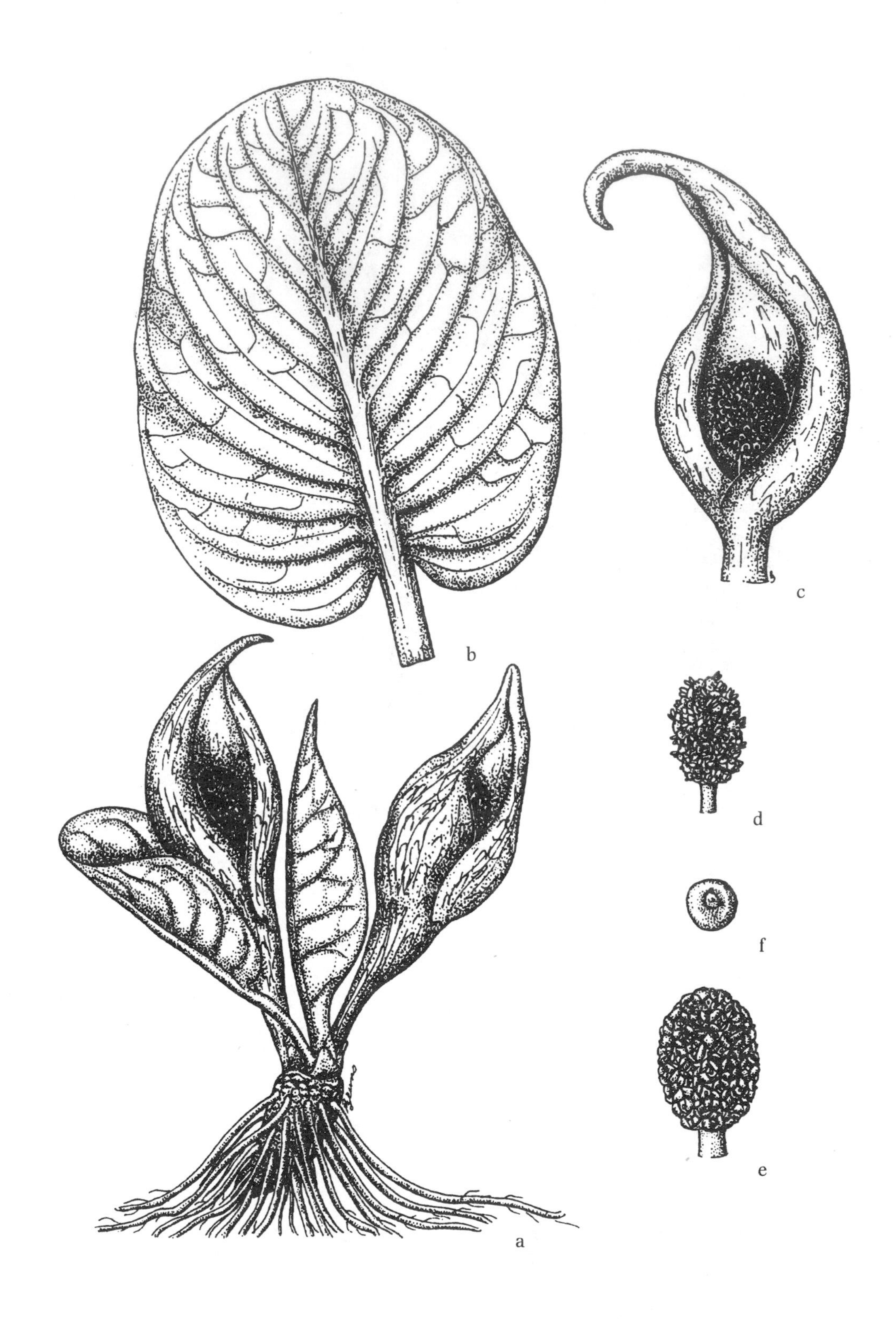

Fig. 57. *Symplocarpus foetidus*: a. habit; b. leaf; c. spathe and spadix; d. spadix of male flowers; e. spadix of female flowers; f. fruit (NHAES).

1. Plants ovoid, rounded at tip; fronds globose (fig. 58a), not dotted; some plants floating below surface under crowded conditions. . . . . . 1. *W. columbiana*
1. Plants longer than broad, pointed at tip; fronds ellipsoidal (fig. 58b,c), dotted; plants floating on surface even when crowded.
   2. Upper surface of fronds flattened (fig. 58b). . . . . . 2. *W. borealis*
   2. Upper surface of fronds raised to a conspicuous papule (fig. 58c). . . . . . 3. *W. brasiliensis*

1. *W. columbiana* Karst. Fig. 58
Floating plants of ditches, pond, lakes and slow-moving rivers. S. Me., c. N.H. and sw. Que. W. to s. Ont., Mich., Wisc., Minn. and Sask., s. to Fla., and Tex.; w. Oreg., Calif. and Mex; C.Am. and n. S.Am. This and other species of *Wolffia* are often associated with stagnant and/or polluted waters and is often found growing mixed with other *Wolffia* spp., especially *W. borealis.*

2. *W. borealis* (Engelm.) Landolt Fig. 58
Floating plants of quiet waters of ditches, ponds, lakes and slow-moving rivers. N.H. and Vt. w. to sw. Que., s. Ont. and s. Minn., s. Sask. and s. B.C., s. to Tenn., Okla., e. Colo., n. Utah, and Calif.; sporatic in western N.Am.

3. *W. brasiliensis* Wedd. Fig. 58
Floating plants of quiet waters of ditches, ponds and streams. Mass. and Conn. w. to w. Mich., Wisc., Mo., and Kans., s. to Fla., Tex. and Mex.; w. Wash., w. Oreg. and nc. Calif. (*W. papulifera* C. H. Thomp.)

6. *Wolfiella* (Mud-midget, Bog-mat)

Tiny plants floating just below water surface; fronds (leaves) reduced, hollow, gradually narrowed from base to apex, 6–8 mm long, asymmetrical, linear-attenuate or falcate or sigmoid, many times longer than wide; fronds occurring singly or cohering at base and radiating in a stellate manner; flowers rare, reproduction primarily vegetative.

1. *W. gladiata* (Hegelm.) Hegelm. Fig. 58
Quiet waters of pond, lakes and ditches. Mass. w. to sw. Mich., Ill., se. Mo. and e. Okla., s. to Fla. and e. Tex. (*W. floridana* (Donn. Sm.) C. H. Thomp.)

7. *Lemna* (Duckweed)

Small, flattened, free-floating plants, frond 1–5-nerved with 1 root per frond; flowers minute, perianth lacking, with 2 staminate flowers (of a single stamen each) and 1 pistillate flower enclose in sac-like spathe; fruit a utricle; vegetative reproduction extensive.

References: Landolt, 1980, 1981, 2000; Reveal, 1990; Urbanska-Worythiewicz, 1975.

1. Plants submersed, free-floating just beneath the surface; primary and secondary fronds abruptly tapered to narrow stalk, 6–15 mm long, with lateral fronds usually remaining attached to the parent plant, often forming large tangled mats) (fig. 58f,g). . . . . . 1. *L. trisulca*
1. Plants free-floating on the surface; fronds elliptic to linear-oblong, 1–6 mm long, forming single plants or clusters of up to 10 fronds.
   2. Fronds 1-veined.
      3. Vein mostly prominent; fronds in clusters of (2)8–10, obovate to elliptic, distinctly asymmetrical at base, 2.5–5 mm long. . . . . . 2. *L. valdiviana*
      3. Vein sometimes indistinct; fronds solitary or in pairs, ovate to broadly elliptic, nearly symmetrical, 1–2.5 mm long. . . . . . 3. *L. minuta*
   2. Fronds 3–5(7)-veined.
      4. Root sheath winged at base, root usually with acute tip; fronds with pointed at tip, with 1 or more distinct papillae near the apex.
         5. Fronds 1–1.7 times as long as broad; root sheath wing 2–3 times as long as wide. . . . . . 4. *L. perpusilla*
         5. Fronds 1–3 times as long as broad; root sheath wing 1–2.5 times as long as wide. . . . . . 5. *L. aequinoctialis*
      4. Root sheath not winged at base, root usually notched at tip; fronds rounded at tip, with or without distinct papillae near the apex.
         6. Fronds not reddish on the lower surface; upper surface lacking a row of distinct papillae. . . . . . 6. *L. minor*
         6. Fronds often reddish on lower surface or both surfaces; upper surface with distinct papillae.
            7. Reddish coloration beginning at point of attachment of root (more intensely on lower surface than upper).
               8. Fronds flat, with line of papillae along midline of upper surface; plants producing olive to brown rootless turions (late in season, sinking to bottom). . . . . . 7. *L. turionifera*

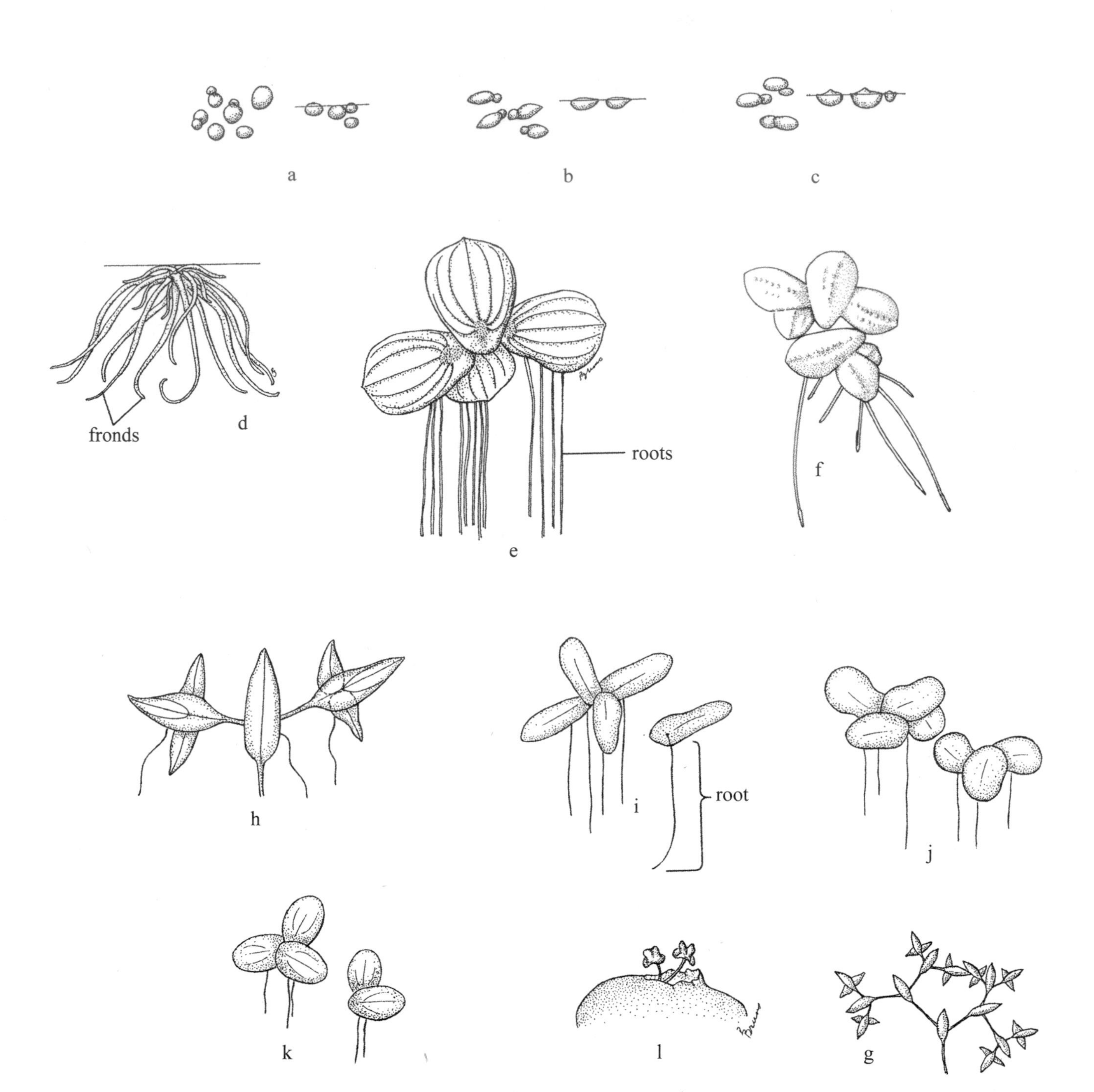

Fig. 58. *Wolffia columbiana*: a. habit and floating view (NHAES).
Wolffia borealis: b. habit and floating view (NHAES).
Wolffia brasiliensis: c. habit and floating view (NHAES).
Wolffiella gladiata: d. floating view (NHAES).
Spirodela polyrhiza: e. habit (NHAES).
Landoltia punctata: f. habit (UF/IFAS).
Lemna trisulca: g. habit; h. habit of parent frond with lateral fronds remaining attached (NHAES).
Lemna valdiviana: i. habit (NHAES).
Lemna perpusilla: j. habit (NHAES).
Lemna minor: k. habit; l. staminate flowers (NHAES).

8. Fronds usually gibbous, with very distinct papillae near apex (with smaller indistinct papillae along midvein) on upper surface; plants not producing turions. . . . . 8. *L. obscura*

7. Reddish coloration often on upper surface beginning at margins and red spots on lower surface beginning near apex. . . . . 9. *L. gibba*

---

1. *L. trisulca* L. Star Duckweed Fig. 58
Quiet waters of ponds, lakes and streams. M.I., Que., P.E.I. and N.S. w. to Alask., s. to Fla., Tex. and Mex; S.Am.

2. *L. valdiviana* Phil. Pale Duckweed Fig. 58
Quiet waters, N.H. and N.Y. w. to Ohio, s. Mich., Ill., Neb., Wyo. and Calif., s. to Fla., Tex. and Mex.; W.I., Ber., C.Am. and S.Am.

3. *L. minuta* Kunth
Quiet waters, widely scattered from s. Ohio w. to s. Ill., Neb., s. Mont. and sw. Wash., s. to Fla. and Tex. (*L. minima* Phil.; *L. miniscula* Herter)

4. *L. perpusilla* Torr. Fig. 58
Quiet waters, s. Que. n. Vt. and Mass. w. to Ohio, Ill., Minn., and Neb. s. to N.C., Tenn., Ark., Tex. and Mex.

5. *L. aequinoctialis* Welw.
Quiet waters, Va. w. to Ky, Ind., Ill., Wisc., and Neb., s. to Fla., Tex., Ariz., and Calif.; Mex., C.Am. and S.Am.; Old World. Landolt (1980, 1981, 2000) distinguished this largely tropical to subtropical species from *L. perpusilla* chiefly by seed characteristics and the presence of a single papule.

6. *L. minor* L. Fig. 58
Quiet, often highly eutrophic or polluted waters. Que. w. to B.C., s. to Fla., La. and Calif.; Mex. and C.Am.; widely in Old World.

7. *L. turionifera* Landolt Red Duckweed
Quiet waters. Widespread in North America, Nfld. and Lab. w. to n. Ont., N.W.T. and Alask, s. to W.Va., Mo., Tex., Calif.; n. Ala; Mex. This species produces rootless turions that sink to bottom for overwintering.

8. *L. obscura* (Austin) Daubs Little Duckweed
Quiet waters. N.Y. w. to Ohio, Ind., Wisc., Minn. and S.D., s. to Fla., Okla. and Tex.; c. Mex.; n. S.Am. (*L. minor* var. *obscura* Austin)

9. *L. gibba* L. Swollen Duckweed
Quiet waters, fresh or brackish. Ind. and Ill.; Ala.; Neb. w. to Wyo., Mont., and Oreg., s. to Okla., Tex., N.M., Ariz., Calif. and Mex.; S.Am., Eur., Cent. Asia, Afr. Common and widespread in western N.Am., especially where winters are mild, but sparse eastward. This species does not produce turions.

8. *Spirodela* (Great Duckweed, Duck-meat)

Small, flattened, free-floating plants, fronds 5–11-veined with 1–18 roots, united in groups; flowers consisting of 1 pistillate and 2 or 3 staminate flowers enclosed by sac-like spathe; vegetative reproduction extensive.

References: Daubs, 1962; Jacobs, 1947; Wohler, Wohler, and Hartman, 1965.

1. *S. polyrhiza* (L.) Schleid. Fig. 58
Standing and slow-moving waters of ditches, ponds, lakes and rivers. N.S., P.E.I., N.E. and se. Que. w. to s. B.C., s. to Fla., Tex., Calif. and Mex.; W.I., C.Am. and n. S.Am.; nearly worldwide.

9. *Landoltia* (Giant Duckweed)

Small free-floating plants, roots (1)2–7(12), surrounded by a tubular sheath, 0.5–3 cm long; fronds obovate to elliptic, with 1–10 leaves fronds united in groups, (3)5–7-veined.

1. *Landoltia punctata* (G. Mey.) Les & D. J. Crawford
Standing and slow-moving waters of ditches, ponds and streams. Chiefly coastal plain, Del., Md. and Va., s. to N.C. Fla., w. to La, and e. Tex.; Mississippi embayment n. to s. Ky., se. Mo. and s. Ill; n. Pa.; Wash., Oreg., Calif., and Ariz.; Trop. Am. Intro. from Old World tropics and to be looked for throughout the Northeast. (*Spirodela punctata* (G. Mey.) C. H. Thomp.)

Tofieldiaceae / False Asphodel Family

Perennial herbs, arising from short rhizomes; leaves, 2-ranked, basal; inflorescence a terminal spike or raceme; flower subtended by an involucre of bracteoles; perianth green or white; ovary superior, 3-locular; fruit a septicidal capsule.

Reference: Packer, 2002.

1. Plants glabrous; fruiting raceme 5–8 mm wide; seeds lacking appendages. . . . . 1. *Tofieldia*
1. Plants glandular or pubescent (fig. 59b); fruiting raceme 10–20 mm wide; seeds with appendages 1 or 2. . . . . 2. *Triantha*

1. *Tofieldia* (False Asphodel)

1. *T. pusilla* (Michx.) Pers. Scotch False Asphodel Fig. 59
Wet rocks and calcareous alpine or arctic marshes. Greenl., Lab. and Nfld. w. to Alask., s. to Gaspé Pen., Que., n. Mich., ne. Minn., nw. Mont. and s. B.C.

2. *Triantha* (False Asphodel)

1. Styles fused at lower portion forming a column up to ⅔ length; capsule as long as or shorter than perianth; seed appendages one, or if 2, then with one usually much shorter than the other, not contorted. . . . . 1. *T. racemosa*
1. Styles separate to base; capsule much longer than perianth; seed appendages 2, equal in length, one contorted. . . . . 2. *T. glutinosa*

1. *T. racemosa* (Walter) Sm. Coastal False Asphodel Fig. 59
Wet sand and peat, boggy areas, pine savannas and seeps. Coastal plain, s. N.J. s. to Fla., w. to se. Tex. (*Tofieldia racemosa* (Walter) Britton, Sterns & Poggenb.)

2. *T. glutinosa* (Michx.) Baker Sticky Tofieldia Fig. 59
Calcareous soils of shores, gravels, marshes, wet meadows and damp ledges. Lab and Nfld. w. to N.W.T. and Alask., s. to Me., w. N.H. e. N.Y., mts. to Va., W.Va., N.C., Tenn. and n. Ga., Ohio, n. Ill., Wisc., Minn., N.D., Sask., Alta. and B.C.; w. Oreg. (*Tofieldia glutinosa* var. *glutinosa* (Michx.) Pers.)

Hydrocharitaceae / Frogbit Family

1. Plants floating on surface (or rooted when stranded) not submersed; leaves basal, with distinct petiole and blade (fig. 60).
    2. Leaves petiolate, blades reniform to ovate to orbicular, leaf margins entire.
        3. Petals absent or, if present, then less than 1.5 times as long as sepals; staminate flowers with anthers ca. 3.5 mm long, elongate, filaments united at least half the length; pistillate flowers with 3–9 styles, deeply 2-fid nearly to base; aerenchyma spaces in central portion of floating leaf, 0.3–1.7 mm in diameter (view from underside, fig. 60b); stipules 1, adnate to petiole base; leaf blade about 2 times as long as basal lobe. . . . . 1. *Limnobium*
        3. Petals present, more than 1.5 times the length of sepals; staminate flowers with anthers ca. 1 mm long, oval, nearly as wide as long, filaments separate nearly to the base; pistillate flowers with 6 styles, shallowly 2-fid, less than half length; aerenchyma spaces in central portion of floating leaf small, 0.22–0.54 mm in diameter (view from underside, fig. 60d); stipules 2, free from petiole; leaf blade about 1.5 times as long as basal lobe. . . . . 2. *Hydrocharis*
    2. Leaves sword-shaped, leaf margins with sharp teeth, pointing toward apex. . . . . 3. *Stratiotes*
1. Plants submersed, rooted, or fragments free-floating beneath water surface; leaves basal or cauline, sessile.
    4. Leaves basal, long, 10–100 cm or more.
        5. Leaves sword-shaped, lacking a lacunae band (submersed thinner, more flaccid than when floating); leaf margins with sharp teeth, pointing toward apex (fig. 61b). . . . . 3. *Stratiotes*
        5. Leaves long, linear and ribbon-like (fig. 62) with a distinct medial lacunae band (fig. 62g); leaf margins entire (or serrate only near apex). . . . . 4. *Vallisneria*
    4. Leaves cauline, short, less than 5 cm, opposite or whorled.
        6. Leaves distinctly whorled, less than 5 times as long as wide.
            7. Leaves 1–4 cm long, 5 in a whorl; flowers with nectaries. . . . . 5. *Egeria*
            7. Leaves 0.5–1.3 cm long, in whorls of 2–7; flowers lacking nectaries.
                8. Leaves in whorls of 3–8; leaf margins conspicuously toothed, midvein of lower surface often with spine-like teeth (fig. 65c,d) (fresh specimens rough to the touch); spathe of staminate flowers spiny. . . . . 6. *Hydrilla*
                8. Leaves in whorls of mostly 3(–7); leaf margins appearing entire to naked eye (minute teeth visible with hand lens), midvein of lower surface lacking spine-like teeth (fresh specimens smooth to the touch); spathe of staminate flowers smooth. . . . . 7. *Elodea*

Fig. 59. *Triantha racemosa*: a. habit; b. flowers; c. fruits (C&C).
Triantha glutinosa: d. habit (USFS).
Tofieldia pusilla: e. habit; f. flower (Gleason).

6. Leaves opposite or pseudo-whorled, very narrow, more than 10 times as long as wide.
   9. Leaf margins entire (*Elodea bifoliata*)........ 7. *Elodea*
   9. Leaf margins denticulate or dentate........ 8. *Najas*

---

1. *Limnobium* (American Frogbit)

Perennial herbs, stoloniferous; leaves basal, floating or emersed, with distinct blade and petiole; plants monoecious; staminate flowers 3–9 per spathe; pistillate flowers solitary in spathe; petals less than 1.5 times the length of sepals; stamens 3–12; fruit an ovoid-ellipsoid, many-seeded berry; seeds subglobose.

References: Catting and Dore, 1982; Cook and Urmi-König, 1983a; Haynes, 2000; Lowden, 1992.

1. *L. spongia* (Bosc) Rich. ex Steud. American Sponge-plant Fig. 60
Shallow water or shores of ponds and streams, ditches and swamps. Chiefly coastal plain; Conn., w. N.Y., s. N.J., Del. and Md. s. to Fla., w. to e. Tex.; Mississippi embayment n. to se. Okla., se. Mo., nw. Tenn., w. Ky., s. Ill. and s. Ind. Our taxon is subsp. *spongia.*

2. *Hydrocharis* (European Frogbit)

Perennial herbs, stoloniferous; leaves basal, free-floating or rooted and emersed, with distinct blade and petiole; plants monoecious; staminate flowers 1–4 per spathe; pistillate flowers solitary in spathe; petals greater than 1.5 times the length of sepals; stamens 9–12; fruit an ellipsoid-globose, many-seeded berry; seeds ellipsoidal.

References: Catling and Dore, 1982; Catling and Porebski, 1995; Cook and Lüönd, 1982a; Dore, 1954, 1968; Haynes, 2000; Roberts et al., 1981.

1. *H. morsus-ranae* L. European Frogbit Fig. 60
Shallow water of marshes, ponds, and slow-moving streams. s. Que., s. Ont., Me., w. Vt., n., w. and se. N.Y., n. N.J. and s. Mich.; w. Wash. This European plant was initially introduced in the St. Lawrence R. system and has spread significantly. It readily overwinters and spreads by turions. The species is similar to *Limnobium spongia*, and the two be can difficult to differentiate in the vegetative state.

3. *Stratiotes* (Water Soldier, Water Aloe)

Perennial herbs, submersed, essentially free-floating, rarely and barely rooted in substrate; stems short, with rosettes of leaves, readily producing stolons with terminal rosettes of leaves (or turions ca. 5 cm long late in growing season); leaves sessile, slightly clasping at base, linear or narrowly triangular (more or less sword-shaped), in spiraling rows forming rosette; submersed leaves thinner, flaccid, emergent leaves somewhat fleshy, rigid; leaf margins spinulose-serrate; plants dioecious, flowering when emergent; flowers showy, sepals green with purplish stripes, petals white, nearly orbicular; staminate plants with 3–6 flowers, 3-merous, stamens up to 40, the innermost 5–15 fertile; pistillate flowers usually solitary (or 2), styles 6, 2-fid to base, staminodes 15–30; fruit a berry-like capsule, with 12–24 seeds enclosed in mucilage. Rarely sets seed.

Stratiotes is presently considered a monotypic genus, with 15 known fossil species.

References: Cook and Urmi-König, 1983b; Snyder et al., 2016.

1. *Stratiotes aloides* L. Fig. 61
Shallow water of marshes and ponds. This Eurasian weed discovered adventive in s. Ont. in 2008 is known presently from the Trent Severin Waterway (near Trent, Ont.) and the Black River (near Sutton, Ont.); it has the potential for spreading by hitchhiking on motorboat propellers. This species readily and effectively spreads by winter buds and stolons, and can produce large populations. Water soldier is used as an ornamental plant in aquatic gardens and is the likely source of introduction.

4. *Vallisneria* (Tape-grass, Wild Celery, Eel-grass)

Perennial herbs, submersed; roots septate; leaves basal, long and ribbon-like; plants dioecious; pistillate flowers solitary, sessile, enclosed in a tubular spathe, reaching surface by peduncle elongation; staminate flowers numerous, enclosed in spathes borne on short peduncles, released and floating to surface; fruit elongate, cylindrical, peduncle recoiling after fertilization, submersing fruit.

References: Fernald, 1918; Haynes, 2000; Lowden, 1982; Marie-Victorin, 1943; Wylie, 1917.

1. *V. americana* Michx. Fig. 62
Ponds, lakes and streams. N.S., N.B., and Que. w. to N.D., Man. and Neb., s. to Fla., Okla. and Tex.; scattered populations, B.C., w. Wash., n. Ida. and Oreg. s. to N.M., Ariz., Nev. and s. Calif. Contrary to the statement by Haynes (2000) that *V. americana* does not possess septate roots, we have, indeed, observed septate roots

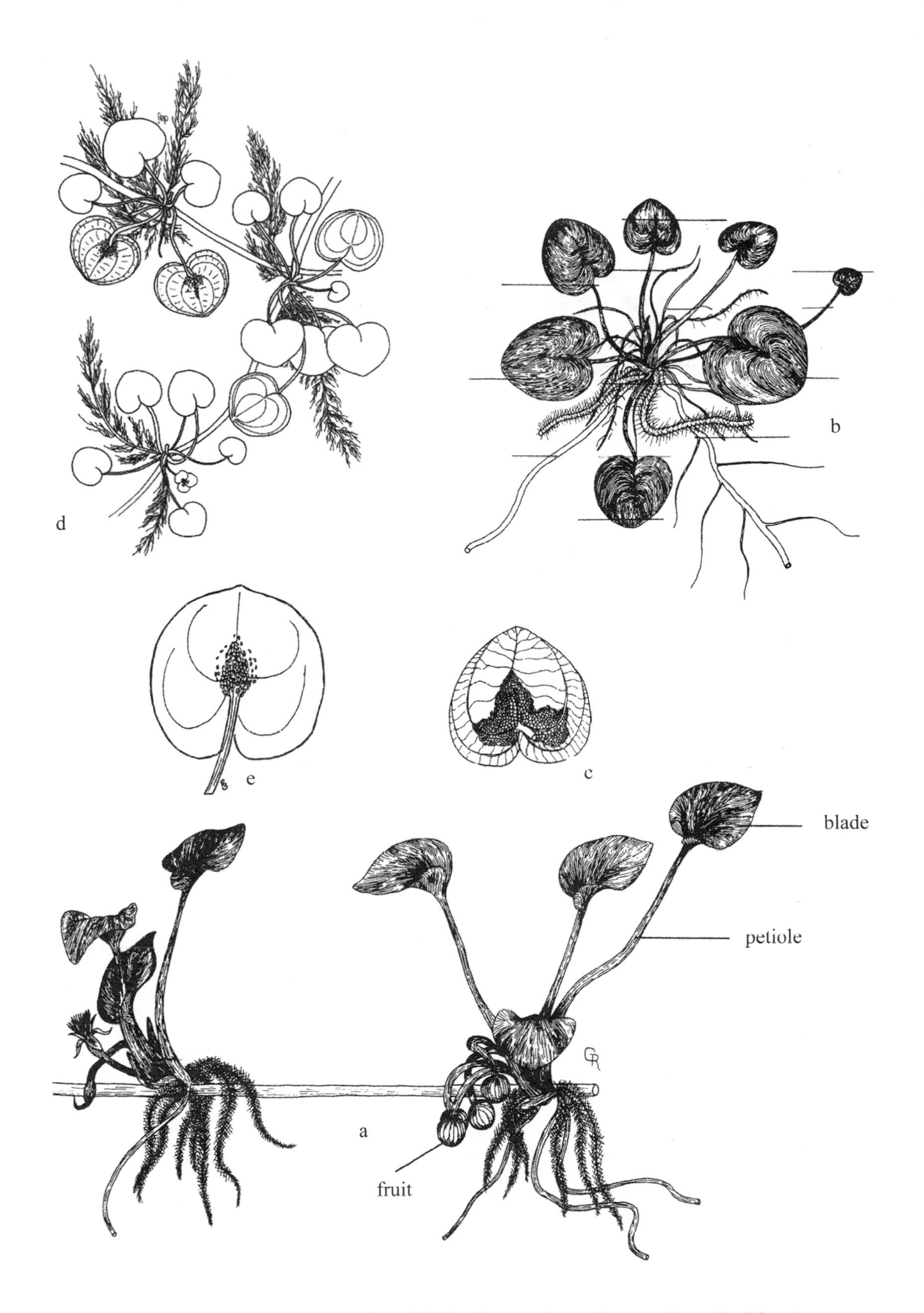

Fig. 60. *Limnobium spongia*: a. habit, floating plant; b. habit, floating plant; c. lower surface of leaf showing aerenchyma tissue (G&W).
Hydrocharis morsus-ranae: d. habit, floating plant (UF/IFAS); e. lower surface of leaf showing aerenchyma tissue in center (PB).

Fig. 61. *Stratiotes aloides*: a. habit; b. leaf; c. male inflorescence; d. female inflorescence (Oeder, in *Flora Danica*, adapted by Snyder et al., 2016).
Elodea bifoliata: e. portion of stem (CBH).

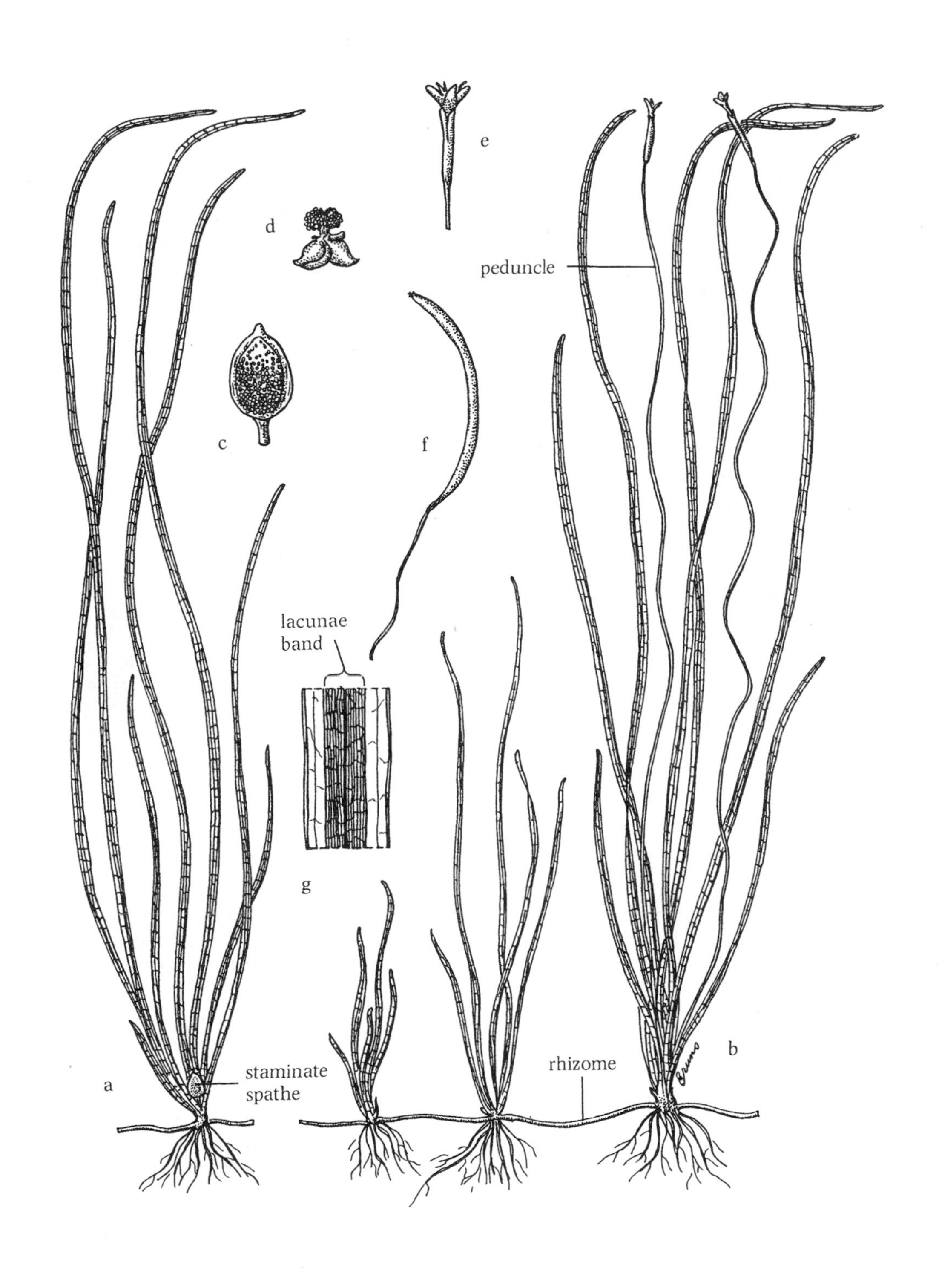

Fig. 62. *Vallisneria americana*: a. habit. staminate plant; b. habit, pistillate plant; c. staminate spathe; d. staminate flower; e. pistillate flower; f. fruit; g. section of leaf showing lacunae band (NHAES).

in populations in New Hampshire (Hellquist, pers. obs.) and in Michigan (Crow, pers. ob.); the character is clear on young white roots, but becomes quite obscured by the buildup of epiphytic algae, giving the roots an orangish-brownish appearance. In a molecular/morphological study of the systematics of *Vallisneria*, Les et al. (2008) recognized several species world-wide, including *V. neotropicalis* Vict., a distinct species very closely related to *V. americana*, occurring in N.Am. only in Florida and the Gulf States; it can be recognized as a generally larger plant than *V. americana*, as well as by its foliage which is suffused with reddish pigment striations, and by an absence of a conspicuous broad lacunae band that is so characteristic of *V. americana*.

5. *Egeria* (Waterweed)

Perennial herbs, submersed; leaves sessile, in whorls of 4(6) along a round stem; plants dioecious; staminate flowers in spathes, borne in upper leaf axils, raised to surface of water on a thread-like hypanthium; pistillate plants presently unknown in the United States.

REFERENCE: Haynes, 2000.

1. *E. densa* Planch. Fig. 63
Ponds, lakes, streams and ditches. N.H., Vt., e. Mass., and R.I. w. to Ill., Neb., Kans., and Okla., s. to Fla. and Tex.; Oreg., Calif. and Ariz. This common aquarium plant has been introduced from South America. It is also widely used in botany classes to observe cytoplasmic streaming. Populations are occasionally established at widely scattered areas, reproducing vegetatively. This species is more common in the southern U.S., often producing fairly large clones. (*Elodea densa* (Planch.) Casp.; *Anacharis densa* (Planch.) Vict.

6. *Hydrilla* (Hydrilla, Water-thyme)

Perennial herbs, submersed; leaves in whorls of (2)4–8; leaf margins sharply serrate, midvein serrate on lower surface of leaf; plants monoecious or dioecious; staminate flowers short-stalked, detaching and floating to surface; pistillate flowers raised to surface on a slender elongated hypanthium; fruit a soft spiny, cylindrical or narrowly conic, few-seeded capsule.

REFERENCES: Cook and Lüönd, 1982b; Netherland, 1997; Yeo et al., 1984.

1. *H. verticillata* (L. f.) Royle Fig. 65
Introduced; ponds, lakes, streams, impoundments, and ditches. Me., Mass., Conn., Md., Del., D.C., and N.C. s. to Fla., w. to Tex.; Iowa (apparently eradicated); w. Wash. and Calif.; Mex., W.I., C.Am. and S.Am. Believed native to Uganda and Tanzania, and occurring widely in Eurasia (as far north as central Siberia) and Australia, this plant has spread widely in many parts of the world; it is considered by many to be one of the most troublesome weeds in the southern United States. Seeds seldom mature in North America, but plants persist through winter by abundant tuber-like turions.

7. *Elodea* (Waterweed)

Perennial herbs, submersed; leaves along stem in whorls of 3(7) (some opposite); plants monoecious, dioecious, or polygamodioecious; flowers with 3 sepals and 3 petals; staminate flowers with stamens monadelphous, flowers rising to surface by elongation at base of hypanthium, or flowers sessile, but breaking loose at maturity, rising to surface where they open and float by divergent sepals; pistillate flowers raised to surface by an elongated hypanthium; fruit leathery, ovoid, few-seeded capsule.

Differentiating our two most common species, *Elodea canadensis* from *E. nuttallii*, by vegetative specimens alone is very unreliable; accurate identifications require flowering/fruiting material. D. Les (pers. comm., 2020) found that fertile populations were needed to corroborate DNA information with reproductive morphology to confirm species; that DNA studies do confirm that there are two genetic lineages, but that hybrids detected by DNA analysis are not uncommon.

REFERENCES: Cook and Urmi-König, 1985; Haynes, 2000; Marie-Victorin, 1931; St. John, 1920, 1965.

1. Staminate spathes 4 mm long or less; styles usually 2 mm or less; anthers 1–1.4 mm long; seeds with long hairs at base; leaves mostly 0.3–1.5 wide, acute at apex, not densely overlapping toward stem tip (fig. 63e). . . . . 1. *E. nuttallii*
1. Staminate spathes 6 mm or more; styles usually more than 2 mm; seeds with or without hairs; leaves mostly 1–5 mm wide, blunt at apex, densely overlapping at stem tip.
   2. Anthers 1.7–3 mm long; peduncles abscising just before or during anthesis; seeds fusiform, 4–6.7 mm long, hairs absent; leaves mostly in whorls of 3, occasionally only 2, opposite on lower stem. . . . . 2. *E. canadensis*
   2. Anthers 3–4.5 mm long; peduncles abscising after anthesis; seeds ellipsoidal, 2.8–3 mm, hairs densely covered in long hairs; leaves in whorled of 3 with some only 2 per node (opposite), especially lower on stem. . . . . 3. *E. bifoliata*

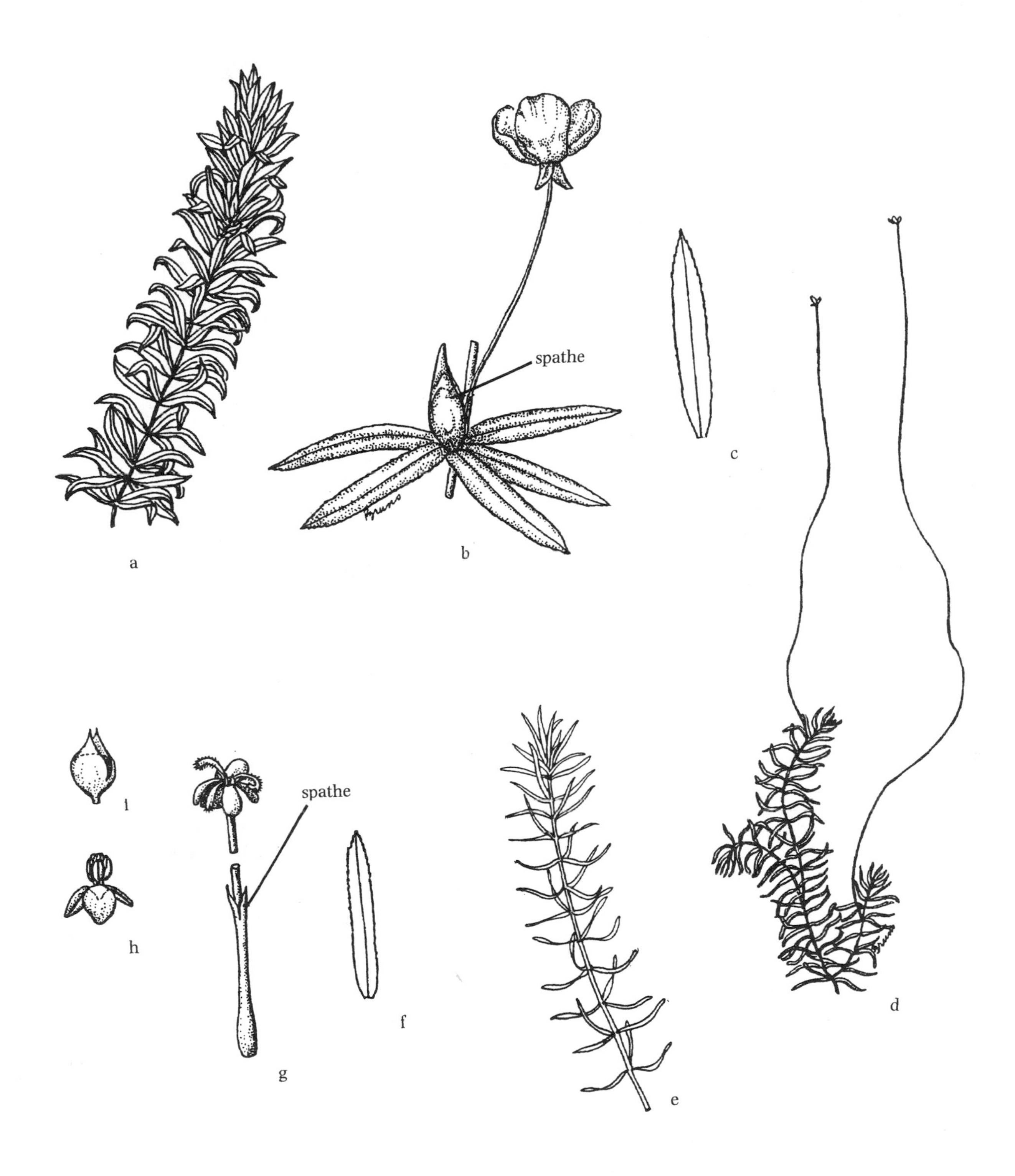

Fig. 63. *Egeria densa*: a. habit; b. portion of plant with staminate flower; c. leaf (NHAES).
Elodea nuttallii: d. habit (NHAES); e. habit (F); f. leaf (NHAES); g. pistillate flower (NHAES); h. staminate flower (NHAES); i. staminate bud within spathe (NHAES).

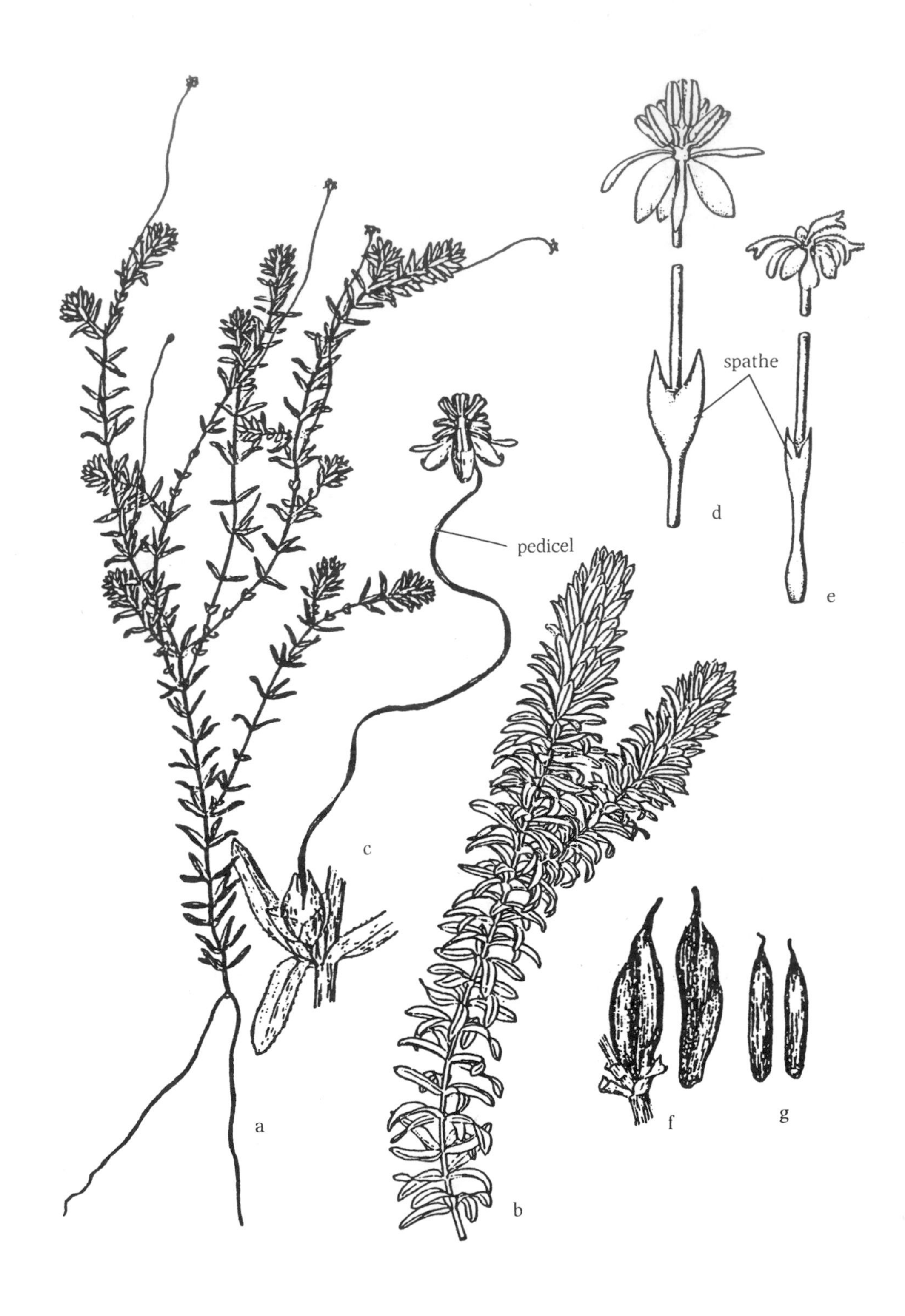

Fig. 64. *Elodea canadensis*: a. habit (Reed); b. tip of branch (F); c. staminate flower (Reed); d. staminate flower (NHAES); e. pistillate flower (NHAES); f. capsules (Reed); g. seeds (Reed).

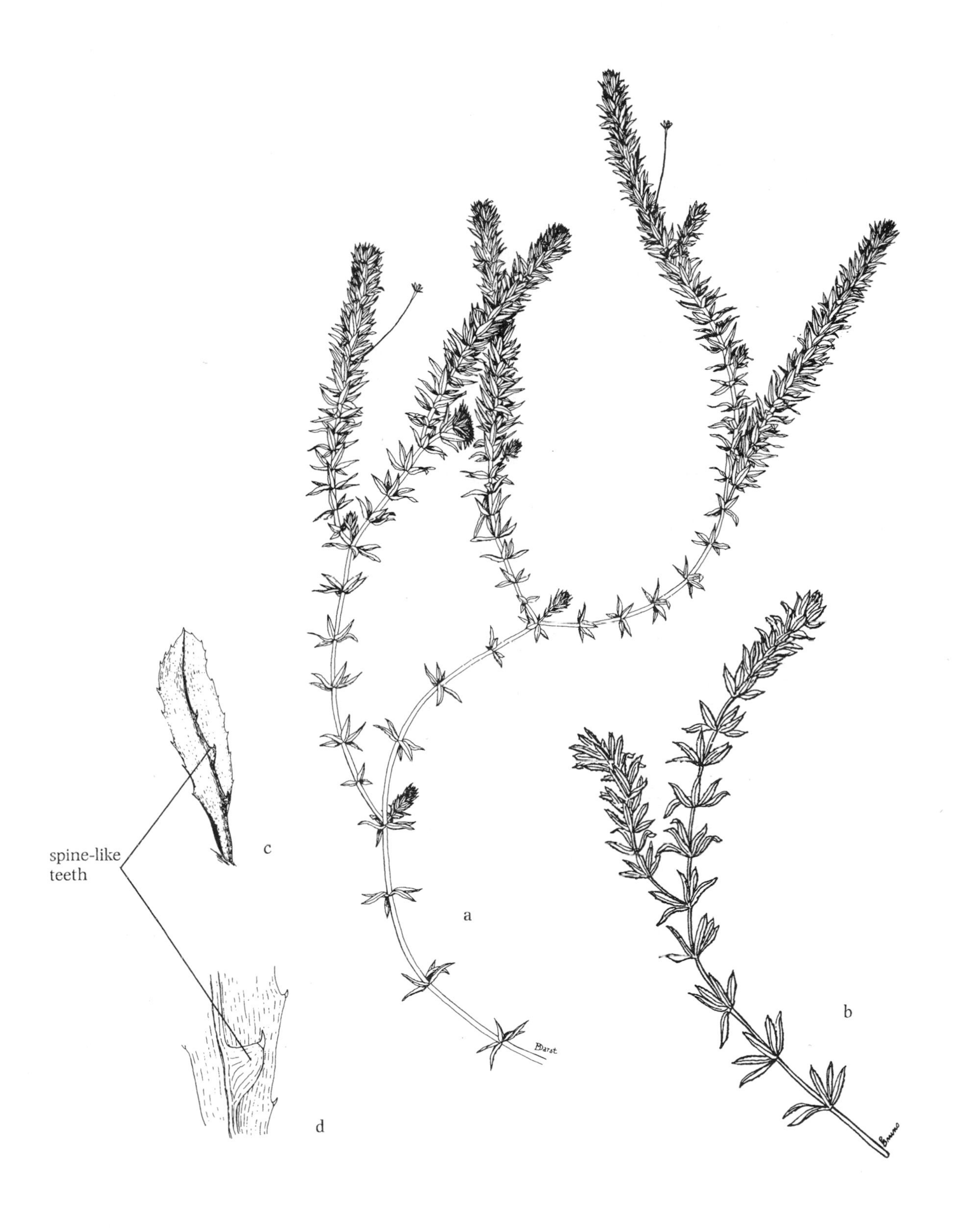

Fig. 65. *Hydrilla verticillata*: a. habit (G&W); b. habit (PB); c. lower surface of leaf (G&W); d. enlargement, portion of lower leaf surface (G&W).

1. *E. nuttallii* (Planch.) St. John Western Waterweed Fig. 63
Acid to moderately alkaline streams, lakes and ponds. N.B., s. Que., and N.E., w. to Ont., Wisc., Minn., Man., Mont., Ida., and B.C., s. to N.C., Ala., Miss., Ark., Okla., N.M., Oreg. and n. Calif. (*E. occidentalis* (Pursh) H. St. John; *Anacharis occidentalis* (Pursh) Vict.)

2. *E. canadensis* Michx. Canada Waterweed Fig. 64
Alkaline to neutral lakes, ponds, and streams. Que. w. to Sask., Mont. and B.C., s. to N.C., Fla., Ala., n. Ark., Okla., N.M., Ariz., and Calif. Staminate plants are rarely encountered; they have leaves which tend to be longer and more linear or lanceolate-oblong than the leaves of pistillate plants. (*Anacharis canadensis* (Michx.) Planch.)

3. *E. bifoliata* H. St. John Two-leaf Waterweed Fig. 61
Neutral to alkaline waters of rivers, lakes and reservoirs. Minn. sporadically w. to s. Man., e. N.D., s. Sask., se. Alta., nc. Mont. and w. B.C., s. to Kans., N.M., Utah, n. Nev., w. Oreg. and n. Calif.; N.Y. The NY site is the Hudson River, where it possibly was introduced through the discharge of ship-ballast water. (*E. longivaginata* H. St. John; *E. nevadensis* (Planch) H. St. John)

E. schweinitzii (Planch.) Casp., a plant very similar to *E. canadensis* but with bisexual flowers, has been described and is known only from the vicinity of Bethlehem, Pa. It has not been collected since 1832 and is presumed extirpated.

8. *Najas* (Naiad, Water-nymph, Bushy Pondweed)

Annuals, submersed; leaves appearing opposite or when fascicled or crowded, appearing whorled, margins dentate, denticulate, or appearing entire; flowers unisexual (plants usually monoecious), axillary, enclosed in perianth-like spathe; fruit indehiscent, with a membranous pericarp, 1-seeded.

Najas has traditionally been treated as a member of a monogeneric family, the Najadaceae. Morphological evidence based on seed coat structure in *Najas* suggested a strong relationship with the Hydrocharitaceae (Shaffer-Fehre, 1991). Various molecular data sets corroborate this viewpoint (Les, 2020; Les et al., 2006). Based on these recent studies the Angiosperm Phylogeny Group IV (2016) has included *Najas* within the Hydrocharitaceae (with Najadaceae relegated to synonymy).

References: Clausen, 1936, 1937; Fernald, 1923; Haynes, 1979, 2000; Haynes and Hellquist, 1996; Les, 2020; Les et al., 2006, 2010; Merilainen, 1968; Stuckey, 1985; Wentz and Stuckey, 1971. Key and partial text contributed by Donald H. Les.

---

1. Leaf margin coarsely dentate and coarsely spiculate (fig. 66b), midribs on lower leaf surface and/or internodes often prickly or spiny (fig. 66a,b); fruit 4–7.5 mm long; plants dioecious. . . . . 1. *N. marina*
1. Leaf margin merely denticulate (figs. 67d, 69c), the denticles very conspicuous (some with expanded bases) or fine and inconspicuous (almost appearing entire), midribs on lower leaf surface and internodes smooth; fruit 2–3.5(4) mm long; plants monecious.
   2. Leaf bases (sheaths) tapered gradually to blade (figs. 67d, 69c).
      3. Seed ("fruit") surface smooth, shiny; leaves tapering to long, slender or broad apex; styles slender.
         4. Seeds slender, more than 3 times as long as wide; leaf apex often broad. . . . . 2. *N. canadensis*
         4. Seeds broader, less than 3 times as long as wide; leaf apex often narrowly tapered. . . . . 3. *N. flexilis*
      3. Seed ("fruit") surface variously areolate or pitted, dull; styles stout. . . . . 4. *N. guadalupensis*
   2. Leaf bases (sheaths) abruptly truncate, auriculate, or lobed at junction with blade (figs. 70b, 71b).
      5. Seed axis often curved, the surface with longitudinal ladder-like (scalariform) rows of striate pits (fig. 70d); leaves often recurved, 0.3–0.5 mm wide, the margins usually conspicuously denticulate (but not deeply dentate, fig. 70a,b); apex of sheath somewhat rounded and very finely toothed (fig. 70b). . . . . 5. *N. minor*
      5. Seed axis symmetrical, the surface with small, randomly scattered areolae; leaves fine, delicate, never recurved, less than 0.3 mm wide; the margins inconspicuously denticulate (fig. 71b); apex of leaf base auriculate and coarsely, irregularly toothed (erose) (fig. 71b). . . . . 6. *N. gracillima*

---

1. *N. marina* L. Spiny Naiad Fig. 66
Salt springs, brackish and highly alkaline waters. C. N.Y. w. to Mich., Minn., se. N.D. and ne. S.D., s. to n. Ohio, n. Ind., and n. Ill.; Fla., w. Okla., and Tex. w. to Utah, Nev. and Calif., s. to Mex.; W.I. and C.Am.

2. *N. flexilis* (Willd.) Rostk. & Schmidt Slender Water Nymph Fig. 67
Acid, alkaline, or brackish waters of streams, ponds, and lakes. Nfld. w. to Minn., N.D., Alta., and B.C., s. to Va., Ohio, Ind., Mo., Neb., and Oreg.

3. *N. canadensis* Michx. Fig. 68
Softwater, alkaline, or brackish waters of streams, ponds, and lakes. Me. w. to Minn., and B.C., s. to Va., Ohio, Ind., Id., Wash. and Oreg. Both *N. canadensis* and *N. flexilis* represent cryptic species, distinct, yet cryptic in that morphologically they are difficult to distinguish, and therefore have been treated as a single species (*N. flexilis*). Clausen (1936) was able to discern a difference in seed morphology in *N. flexilis* leading him to consider *N. flexilis* as an aggregate species and thus described *N. muenscheri* as a new species (unaware of Michaux's earlier publication of *N. canadensis* for this taxon, which has nomenclatural priority). Examining this problem using both seed morphology and DNA analysis, Les et al. (2015) found two genetically different entities, though closely related enough that hybrids between the two cryptic species can be discerned through molecular analysis. (*N. muenscheri* R. T. Clausen; *N. guadalupensis* var. *muenscheri* (R. T. Clausen) Haynes; *N. guadalupensis* subsp. *muenscheri* (R. T. Clausen) Haynes & Hellq.)

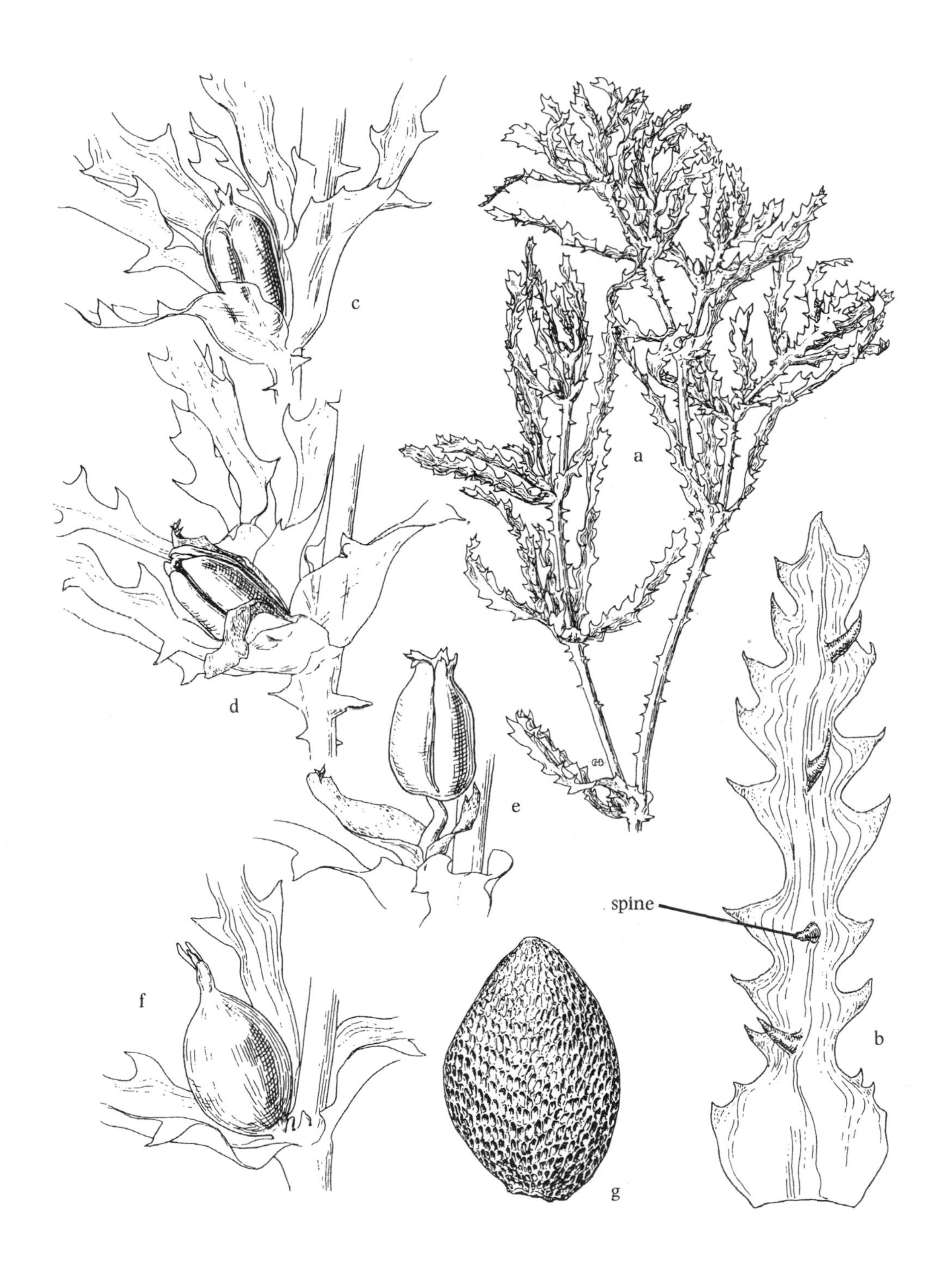

Fig. 66. *Najas marina*: a. upper portion of plant; b. leaf; c–e. development of staminate flower; f. pistillate flower; g. seed (Mason).

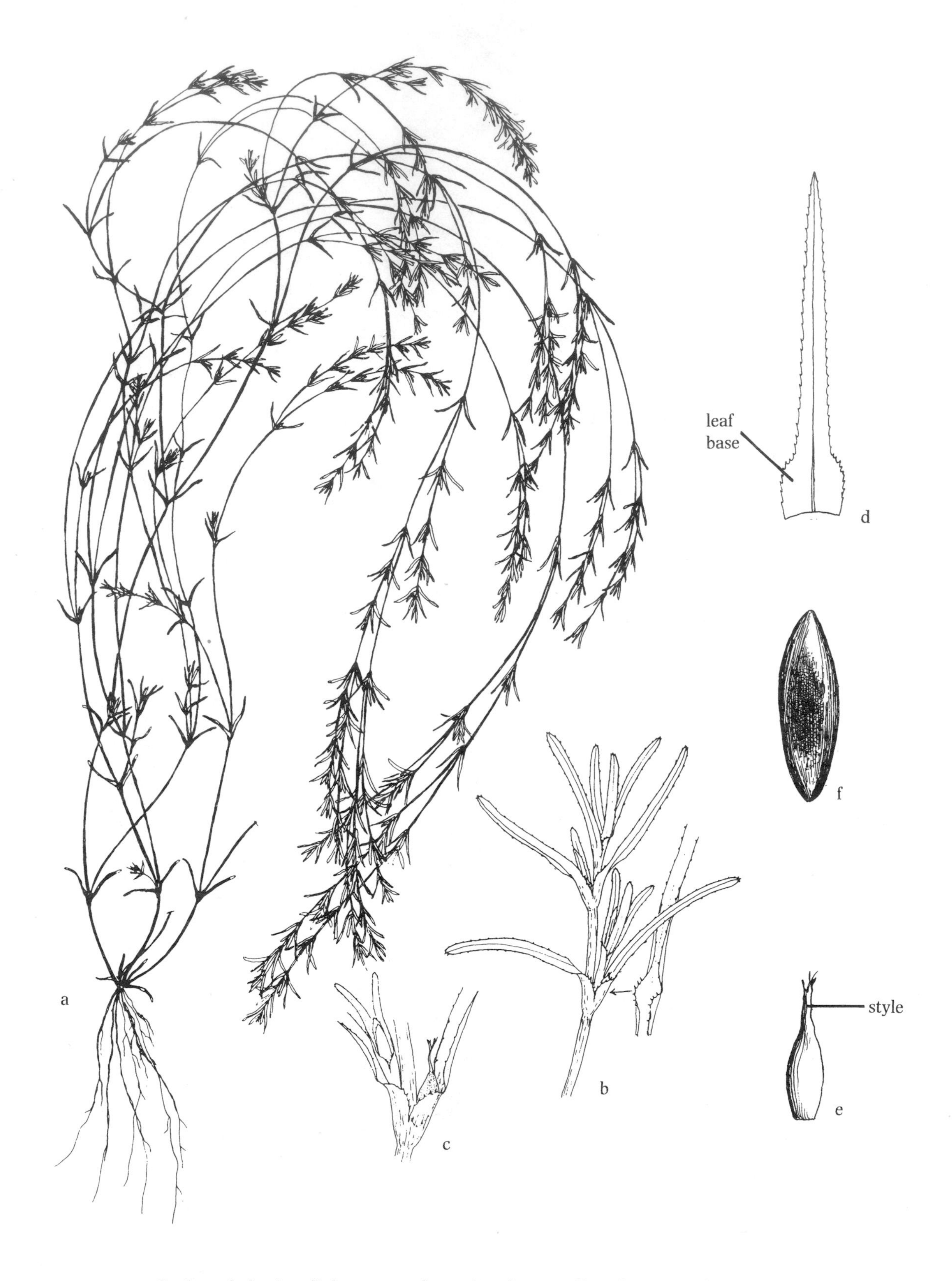

Fig. 67. *Najas flexilis*: a. habit (Reed); b. portion of stem (Reed); c. pistillate flower (Reed); d. leaf (F); e. fruit (F); f. seed (F).

Fig. 68. *Najas canadensis*: a. habit (two views); b. leaf; c. fruit (CBH).

4. *N. guadalupensis* (Spreng.) Magnus
Common Water-nymph Fig. 69
Alkaline lakes and streams, and brackish or saline coastal ponds. S. Me., sw. N.H. Mass., Vt. and sw. Que. w. to Ont., Mich., Minn., s. N.D., S.D., Mont., Ida., Wash. and Oreg., s. to Fla., Tex., Calif. and Mex.; W.I. and C.Am. DNA evidence (Les, 2020; Les et al., 2010) has shown that plants referred to as *N. guadalupensis* subsp. *olivacea* (Rosend. & Butt.) Haynes & Hellq. were actually hybrids of *N. flexilis* × *N. guadalupensis*, thus it is no longer regarded as a distinct taxon.

5. *N. minor* All. Minor Naiad Fig. 70
Alkaline waters of streams, ponds and lakes. E. Me., se. N.H., w. Vt. and Mass. w. to s. Ont., se. Mich., Ill. and e. Mo., s. to ne. Fla., Miss., Ark., La. s. Okla. and e. Tex. Intro. from Eur., spreading rapidly in the U.S.

6. *N. gracillima* (A. Braun ex Engelm.) Magnus
Slender Water-nymph Fig. 71
Acid waters of ponds and lakes. N.S., N.B. and N.E. w. to s. Ont., s. Mich., Wisc. and Minn., s. to N.C., Ala., Miss. and Mo.; n. Calif.

Ottelia alismoides (L.) Pers., Ducklettuce, a federally listed noxious weed, has been recently reported from extreme se. Mo. It occurs as several populations in s. La., with scattered localities in w. Fla., e. Tex. and nc. Calif. (https://www.invasiveplantatlas.org/subject.html?sub=4619, accessed 22 March 2022).

Butomaceae / Flowering-rush Family

1. *Butomus* (Flowering-rush)

Perennials, arising from short fleshy rhizomes; leaves basal, erect or floating; flowers pink, bisexual, numerous, borne in an umbel; perianth of 3 sepals, 3 petals; stamens 9; carpels 6, separate, connate at very base; fruit a whorl of follicles with long beaks.

References: Anderson et al., 1974; Core, 1941; Gaiser, 1949; Gaskin et al., 2021; Haynes, 2000; Lui et al., 2005; Roberts, 1972; Staniforth and Frego, 1980; Stuckey, 1968; Thompson and Eckert, 2004; Zenkert, 1960.

1. *B. umbellatus* L. Fig. 72
Marshes and shores. Initially intro. into St. Lawrence R. system in s. Que., L. Champlain valley of Vt. and N.Y., c. Me., e. Mass., Connecticut River valley in Conn.; also intro. in w. L. Erie, Ohio, Mich. and s. Ont.; nw. Ind. and ne. Ill.; spreading to scattered sites in Wisc., Minn., N.D., S.D., Neb., Mont., Ida., Man. Alta. and B.C.; natzd. from Eur. Anderson et al. (1974) note that the presence of *Butomus umbellatus* s.l. in North America is the result of introductions from two different geographical regions of the Old World: the populations of the Great Lakes and more westerly localities are related to European plants; the plants of the St. Lawrence River region are more like those of Asia, which were treated by Russian botanists as *B. junceus* Turcz. (Fedchenko, 1934; Czerepanov, 1981). Using genetic data Gaskin et al. (2021) found that plants of the western N.Am. invasion genetically distinct from those in eastern N.Am., with triploids dominant in the West whereas diploids dominate in the East; they were also able to match several samples from western N.Am. invasives to a single genotype from the Netherlands as well as a match from Hungary. It is noteworthy that western triploids, sexually sterile, reproduce primarily by rhizomes and fragmentation, whereas diploids tend to reproduce by seeds and inflorescence bulbils as well as rhizomes (Lui et al., 2005; Thompson and Eckert, 2004) (*B. umbellatus* var. *junceus* (Turcz.) Micheli).

Alismataceae/ Water-plantain Family

References: Beal, 1960b; Haynes and Hellquist, 2000; Hellquist and Crow, 1982.

1. Pistils and achenes in a single whorl, a flat-topped ring (fig. 73c): flowers all bisexual, stamens 6. . . . . 1. *Alisma*
1. Pistils and achenes in a dense globose head (fig. 75f, 88d): flowers bisexual or unisexual, stamens mostly 9 to many (6 in *Baldellia*).
    2. All flowers bisexual; achenes plump, conspicuously ribbed, not winged (fig. 75c,g); leaf blade never sagittate; roots not septate.
        3. Stamens 9 to many; styles lateral on pistils/ achenes; flowering stems not producing roots or leaves; plants lacking a strong pungent odor of coriander. . . . . 2. *Echinodorus*
        3. Stamens 6; styles terminal on pistils/achenes (or nearly so); flowering stems often producing roots and tufts of leaves; plants with a strong pungent odor of coriander. . . . . 3. *Baldellia*
    2. Upper flowers of inflorescence unisexual, usually staminate, lower flowers pistillate or sometimes bisexual (petals then bearing a basal spot, *S. calycina*), or plants sometimes dioecious (*S. latifolia*): achenes flattened, winged, not ridged (fig. 88e); leaf blade often sagittate; roots septate (figs. 78a, 82a, 88a). . . . . 4. *Sagittaria*

1. *Alisma* (Water-plantain)

Perennial herbs, arising from corm-like rhizomes; leaves all basal, ribbon-like, ovate, or elliptic; flowers bisexual, borne in a panicle; petals white; fruit achenes, borne in a circle on a flattened receptacle.

References: Bjorkquist, 1968; Hendricks, 1957; Rhoades, 1962.

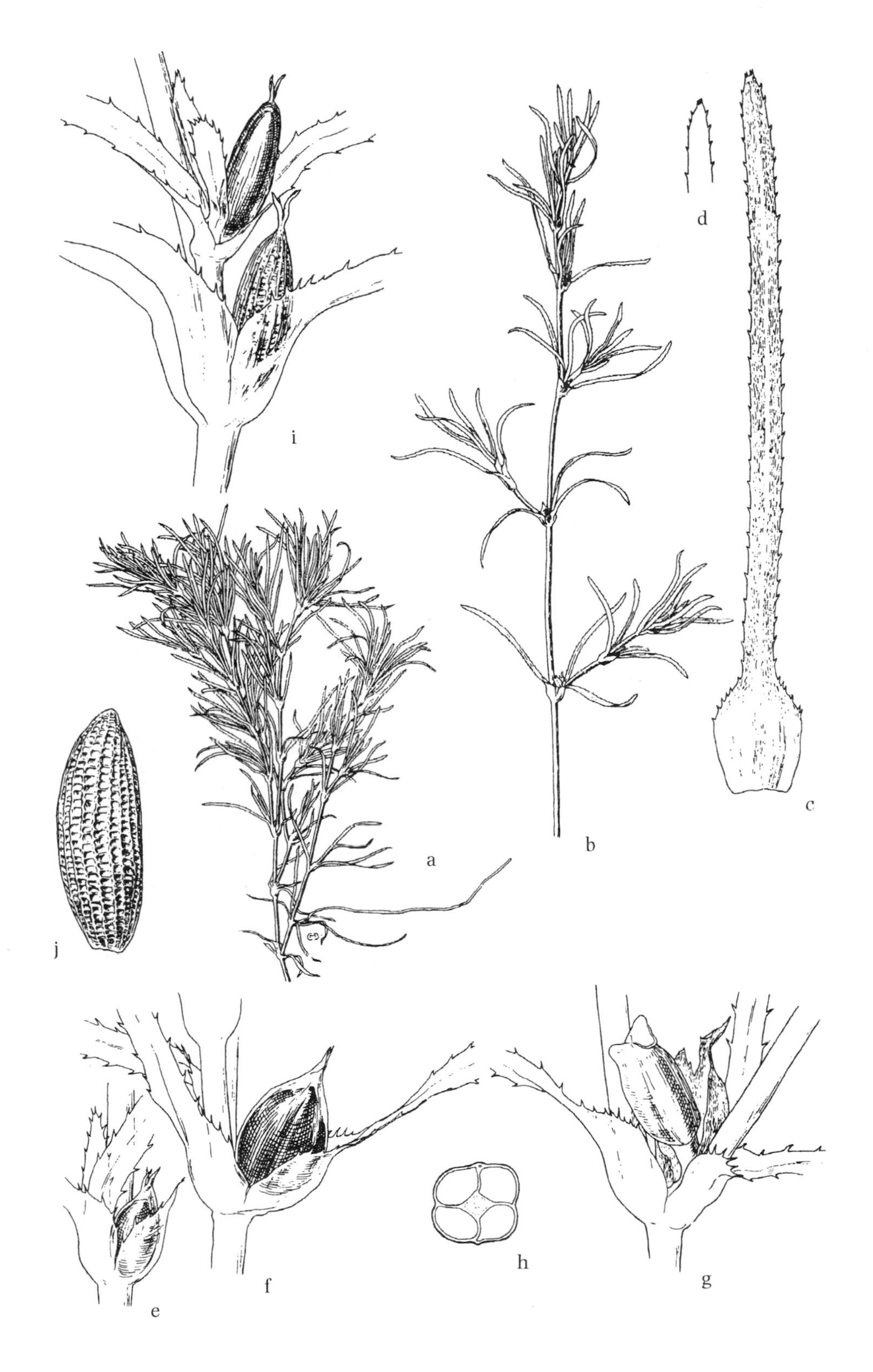

Fig. 69. *Najas guadalupensis*: a, b. upper portion of plant; c. leaf; d. leaf apex; e, f. young staminate flowers; g. mature anther; h. anther, cross-section; i. pistillate flowers; j. seed (Mason).

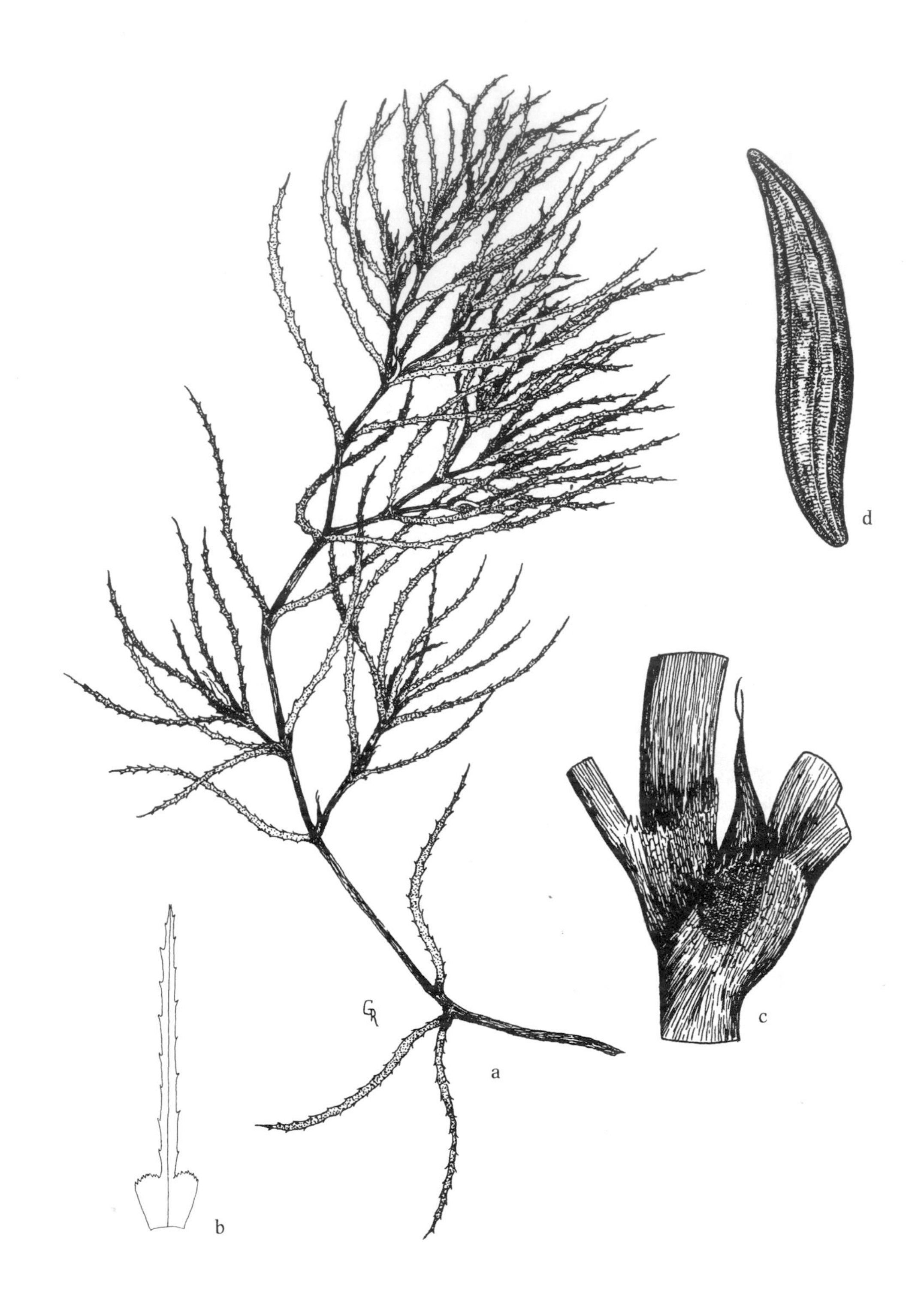

Fig. 70. *Najas minor*: a. habit (G&W); b. leaf (F); c. pistillate flower (G&W); d. seed (G&W).

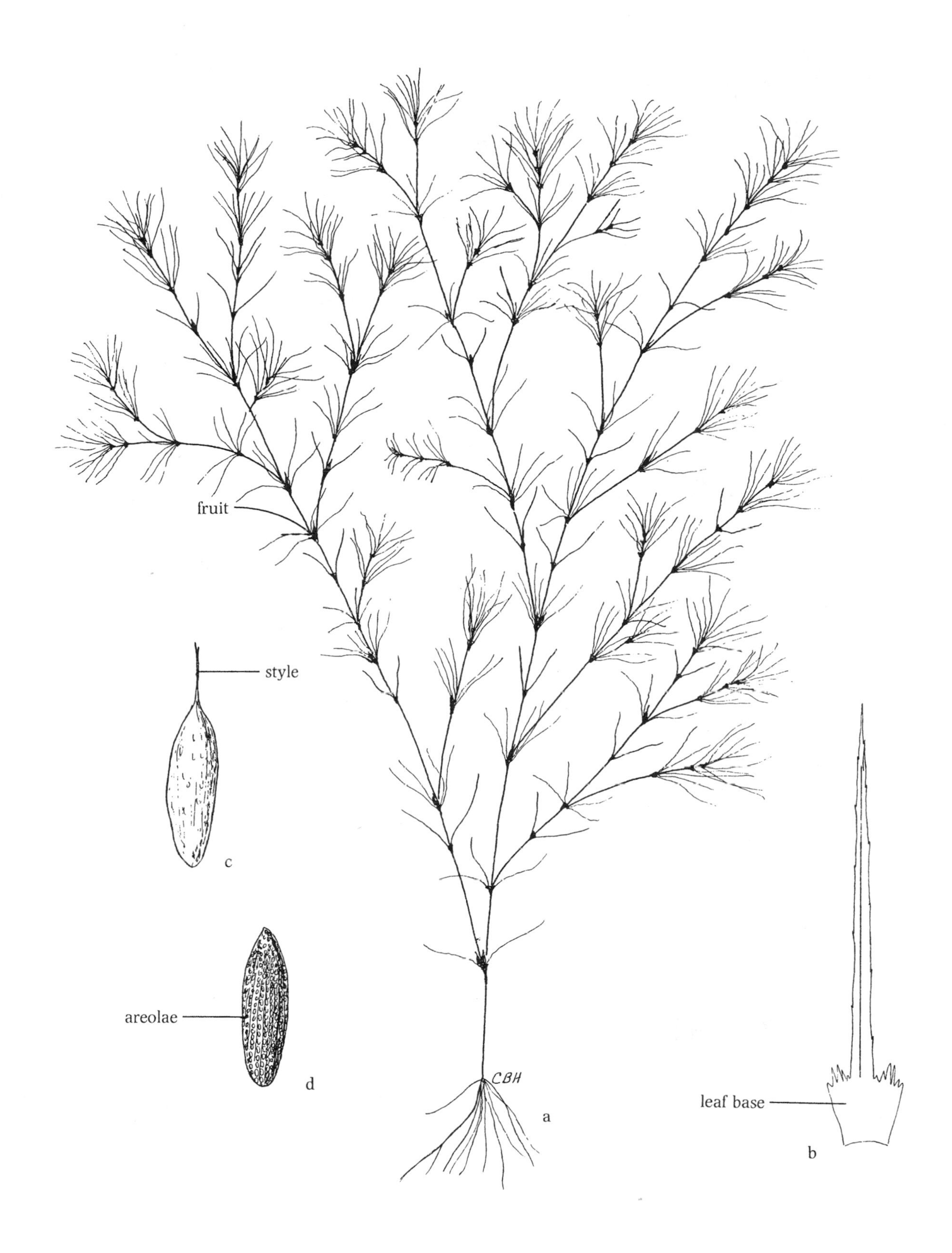

Fig. 71. *Najas gracillima*: a. habit (NHAES); b. leaf (F); c. fruit (NHAES); d. seed (F).

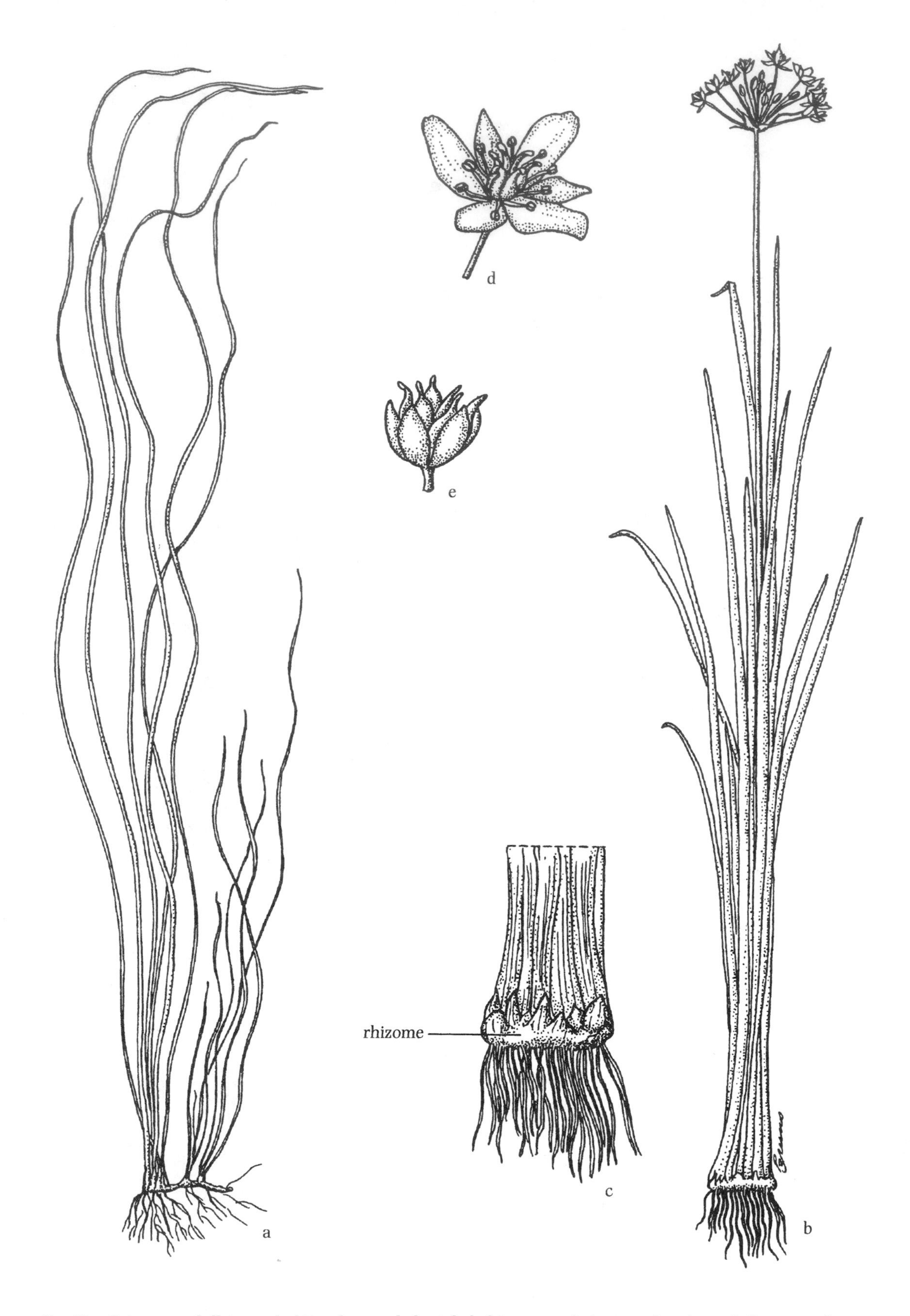

Fig. 72. *Butomus umbellatus*: a. habit, submersed plant; b. habit, emersed plant; c. plant base; d. flower; e. cluster of follicles (NHAES).

1. Leaves submersed and ribbon-like (fig. 73b) or, if emersed, then lanceolate to narrowly elliptic, 0.5–2 cm wide (fig. 73a): achenes with 2 dorsal grooves and a central ridge (fig. 73c). . . . . 1. *A. gramineum*
1. Leaves emersed with ovate to elliptic blades (fig. 74a), up to 15 cm wide, occasionally submersed or floating (Fig. 74b); achenes with single dorsal groove (fig. 74f).
   2. Flowers 7–13 mm wide; sepals at anthesis 3–4 mm long; petals 3.5–6 mm long; fruiting heads 4–7 mm in diameter; achenes 2.2–3 mm long. . . . . 2. *A. triviale*
   2. Flowers 3–3.5 mm wide; sepals at anthesis 2–2.5 mm long; petals 1–2 mm long; fruiting head 3–4 mm in diameter; achenes 1.5–2 mm long. . . . . 3. *A. subcordatum*

1. *A. gramineum* Lej. Narrow-leaved Water Plantain Fig. 73
Calcareous waters and muddy shores. This variable species is sometimes divided into two or three dubious varieties primarily on the basis of vegetative features which appear to be environmentally induced. Local, nw. Vt. and sw. Que. w. to Wisc., Alta. and B.C. s. to Ariz., N.M., Mo. and Va.

2. *A. triviale* Pursh Water-plantain Fig. 74
Shallow ponds, stream margins, marshes and ditches. Nfld. and Que. w. to B.C., s. to N.J., Md., W.Va., Mich., Iowa, n. Okla., N.M., Ariz., Calif. and n. Mex. (*A. plantago-aquatica* var. *americanum* Schult. & Schult. f.)

3. *A. subcordatum* Raf. Water-plantain Fig. 74
Shallow pools, stream margins, marshes and ditches; Me. w. to N.Y., Ont., Minn. and Neb., s. to Calif., Ariz., Tex. and Fla.; Wyo. Both *A. triviale* and *A. subcordatum* have previously been treated as varieties of *A. plantago-aquatica* L. However, Bjorkquist (1968) has demonstrated through chromosome studies and crossing experiments that *A. triviale*, a tetraploid (2n = 28), and *A. subcordatum*, a diploid (2n = 14), have strong sterility barriers between each other and with all other *Alisma* taxa, including *A. plantago-aquatica*. Both of these species are quite variable, and characters sometime appear intermediate between the two taxa. (*A. plantago aquatica* var. *parviflorum* (Pursh) Torr.)

2. *Echinodorus* (Burhead)

Annual or perennial herbs, often spreading by creeping rhizomes; leaves basal; submerse leaves lanceolate-linear, phyllodial or ribbon-like; emersed leaves petiolate with ovate, lanceolate-truncate, to cordate blades; flowers bisexual, borne in umbellate to whorled to paniculate inflorescences; petals white; fruit achenes, in globose heads.

1. Plants small, seldom reaching 15 cm tall; leaves lanceolate-linear (fig. 75a); pistils 15–20; stamens 9. . . . . 1. *E. tenellus*
1. Plants large, more than 15 cm tall; tall; leaves ovate or broadly lanceolate-truncate, occasionally cordate; pistils 45–250; stamens 12–22.
   2. Flowering stems erect, flowers 6–11 mm wide; stamens 9–15; veins of sepals smooth. . . . . 2. *E. berteroi*
   2. Flowering stems decumbent to arching, often rooting at nodes (fig. 75d): flowers 12–20(25) mm wide; stamens 20–22; veins of sepals papillate-ridged.. . . . . 3. *E. cordifolius*

1. *E. tenellus* (Mart. ex Schult. & Schult. f.) Buchenau Dwarf Burhead, Mudbabies Fig. 75
Sandy shores, pond margins, and wet depressions. Local, Mass., Conn., N.Y. and s. N.J. w. to sw. Mich., Ill., Mo. and e. Kans., s. to c. Fla., Ala., e. Tex. and Mex.; W.I., C.Am., and S.Am. (*E. tenellus* var. *parvulus* (Engelm.) Fassett; *E. parvulus* Engelm.; *Helanthium tenellum* (Mart.) Britton)

2. *E. berteroi* (Spreng.) Fassett Upright Burhead Fig. 76
Shallow ponds, muddy shores, marshes and ditches. Local in the East, s. Ohio and Ind., w. to Ill., s. Wisc., Iowa, se. S. D., and e, Neb., s. to Fla., Tex., e. N.M. and Mex.; Utah, Nev., sw. Ariz. and Calif.; W.I. and S.Am. (*E. rostratus* (Nutt.) Engelm.; *E. berteroi* var. *lanceolatus* (Engelm.) Fassett)

3. *E. cordifolius* (L.) Griseb. Creeping Burhead Fig. 75
Swamps, ponds, streams and ditches. Md. and Va.; w. Ky., sw. Ind. and Ill. w. to Iowa, Mo. and Kans., s. to nw. Fla., La., e. Okla. and e. Tex.; W.I. and S.Am. (*E. radicans* (Nutt.) Engelm.)

3. *Baldellia* (Lesser Water-plantain)

Perennial herbs with both a submersed and terrestrial form, rhizomatous; inflorescence arising from a slender rhizome; plants often reproducing vegetatively at nodes forming roots and leaves; leaves having a strong, pungent "coriander" odor; flowering stalk bearing 1–2 umbels with up to 5 long peduncular rays, each with a single white flower; stamens 6; pistils numerous, separate; fruit a globose head of achenes.

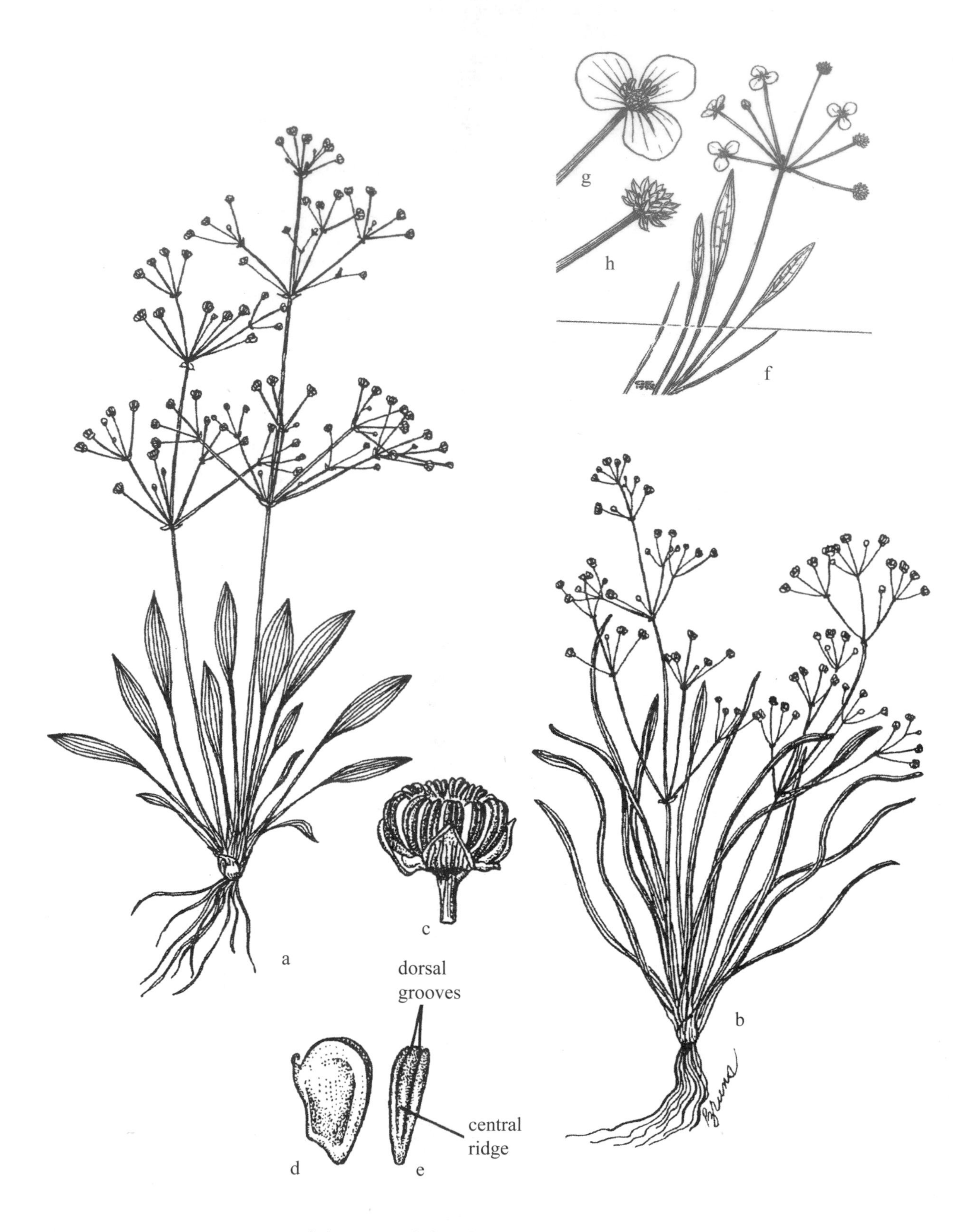

Fig. 73. *Alisma gramineum*: a. habit, emersed plant; b. habit, submersed plant; c. fruiting head; d. achene, side view; e. achene, dorsal view (NHAES).
Baldellia ranunculoides: f. habit; g. flower; h. fruit (G. M. S. Easy in P&C).

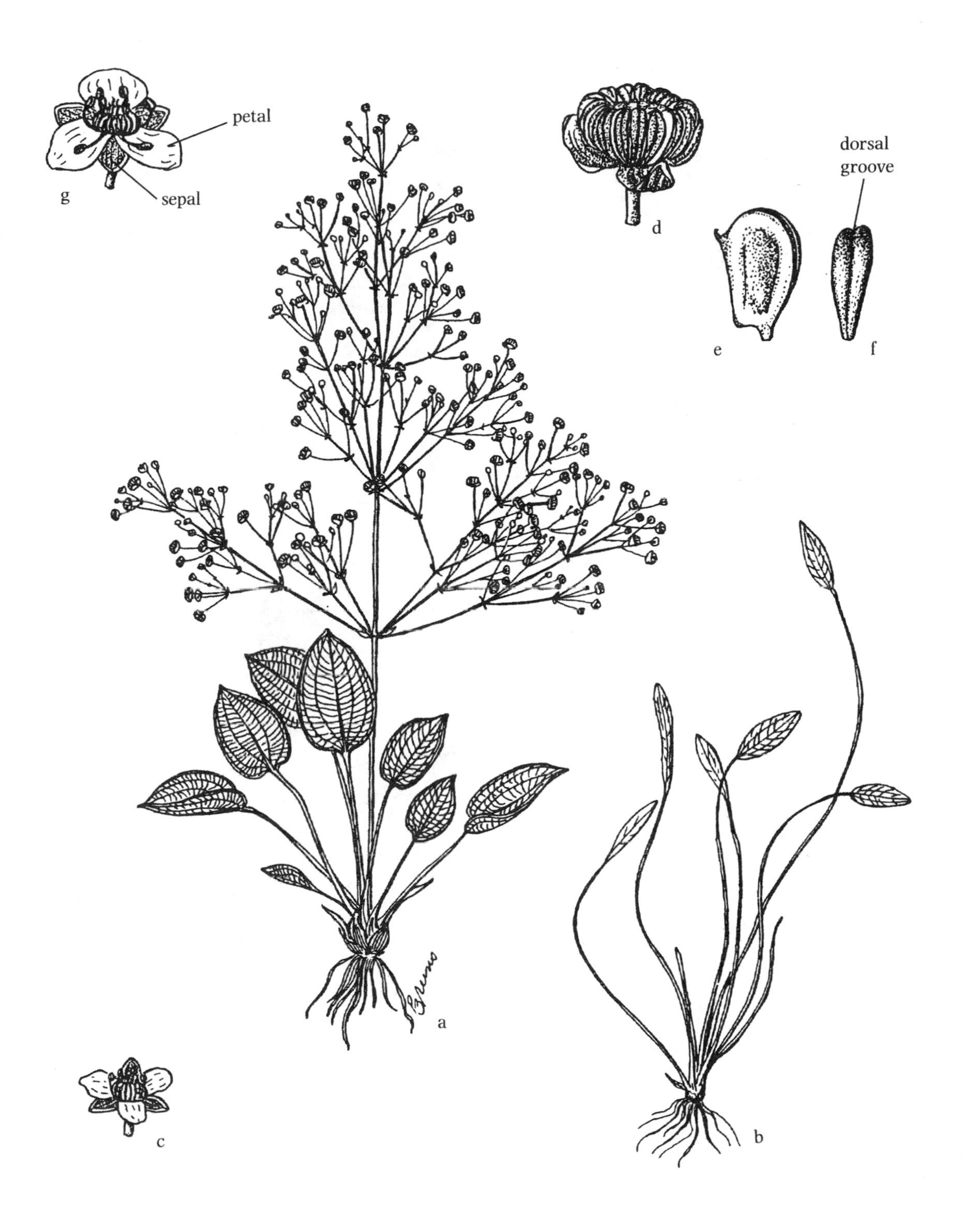

Fig. 74. *Alisma subcordatum*: a. habit, emersed plant; b. habit, submersed plant; c. flower; d. fruiting head; e. achene, side view; f. achene, dorsal view (NHAES).
Alisma triviale: g. flower (NHAES).

Fig. 75. *Echinodorus tenellus*: a. habit; b. nodding flower; c. achene, oblique, side, and dorsal views (NHAES). *Echinodorus cordifolius*: d. habit; e. flower; f. fruiting head; g. achene; h. seed (G&W).

Fig. 76. *Echinodorus berteroi*: a. habit, emersed plant; b. habit, submersed plant; c. portion of inflorescence; d. flower; e. stamen; f. portion of infructescence; g. achene (Mason).

1. *Baldellia ranunculoides* (L.) Parl.
Lesser Water-plantain Fig. 73

Shallow water of shorelines. Intro. from Eur., presently known only from Avalon Pen., e. Nfld. and Puget Sound, Wash. In the sterile state it can resemble the European *Ranunculus repens*, which is also known from the Avalon Pen., Nfld. Kozlowski et al. (2007) provide morphological, ecological and molecular evidence for recognizing two subsp. in Europe, both of which are in need of conservation; the taxon introduced in N.Am. is *B. ranunculoides* subsp. *repens* (Lam.) Á. Löve & D. Löve.

4. *Sagittaria* (Arrowhead, Wapato, Duck-potato)

Perennial or annual herbs, emersed or submersed, spreading by stolons, some perennating by tubers; roots septate, fibrous; leaves extremely variable, blades sagittate or hastate to lanceolate, or phyllodial, ribbon-like to subulate, submersed, floating, or erect; bracts membranous; uppermost flowers of inflorescence usually staminate; lower flowers pistillate or sometimes bisexual (*S. latifolia* typically dioecious); petals white; fruit achenes, laterally flattened, in globose clusters.

Sometimes long ribbon-like leaves of *Sagittaria* are formed in deep water and can be distinguished from those of *Vallisneria* by the venation pattern (fig. 5b,c,d); but both species have septate roots. There are usually several rhizomes, each ending in an edible tuber, hence the name Duck-potato.

REFERENCES: Adams and Godfrey, 1961; Beal, Wooten, and Kaul, 1982; Bogin, 1955; Haynes and Hellquist, 1996; Smith, 1895; Wooten, 1970, 1971, 1973a, 1973b.

---

1. Petals with a green or purple basal spot; sepals of mature flowers appressed (fig. 78d); lower flowers bisexual; stamens 9–15.
   2. Petals with a green basal spot; leaf blades present, not spongy, broadly sagittate or ovate (fig. 77b); plants of freshwater. .... 1a. *S. montevidensis* subsp. *calycina*
   2. Petals with a purple basal spot; leaf blades absent, or if present, spongy, lanceolate to obovate, or rarely sagittate (fig. 78) plants of tidal marshes. .... lb. S. *montevidensis* subsp. *spongiosa*
1. Petals all white; sepals of mature flowers reflexed (fig. 79a) or spreading (fig. 88c,d); lower flowers pistillate, or all flowers unisexual (pistillate or staminate); stamens numerous (more than 10).
   3. Pedicels of pistillate flowers recurved in fruit (fig. 79a); leaf blades unlobed.
      4. Plants emersed, leaves erect with blades, 5–20 cm long, narrow to broadly ovate, occasionally with 1 or 2 basal projections; filaments pubescent. .... 2. *S. platyphylla*
      4. Plants submersed, leaves linear, often recurved, to ribbon-like, bladeless, or sometimes expanding into narrow blades up to 4 cm long; filaments glabrous.
         5. Leaves linear-awl-shaped, often firm, recurved, up to 30 cm long (fig. 80c); plants usually estuarine. .... 3. *S. subulata*
         5. Leaves flexuous, ribbon-like, over 30 cm long (fig. 80a), sometimes developing elliptic, oval or ovate floating leaf blades; plants often in fast-moving freshwater. .... 4. *S. filiformis*
   3. Pedicels of pistillate flowers divergent, ascending in fruit (fig. 81a,b); leaf blades lobed or unlobed.
      6. Stamens with filaments pubescent (figs. 81c, 83c); blades unlobed (rarely lobed in *S. rigida*).
         7. Emergent and submersed leaves round in cross-section, quill-like (fig. 81). .... 5. *S. teres*
         7. Emergent leaves flattened or with petiole and distinct elliptic blade, submersed leaves flattened, not quill-like.
            8. Pistillate flowers and fruiting heads sessile or subsessile (fig. 82e); scape often bent (fig. 82a). .... 6. *S. rigida*
            8. Pistillate flowers and fruiting heads pedicelled (fig. 84a); scape straight, upright.
               9. Stamen filaments linear. .... 7. *S. lancifolia*
               9. Stamen filaments dilated at base.
                  10. Achene 1.5–2.5 mm long; beak of achene less than a fourth as long as width of achene anthers as long as or longer than filaments.
                      11. Phyllodia less than 1 cm wide, linear-lanceolate, acute at apex; pedicels of mature pistillate heads 0.5–3 cm long. .... 8a. *S. graminea* subsp. *graminea*
                      11. Phyllodia 1–2.5 cm wide, linear, rounded at apex; pedicels of mature pistillate heads 2.1–5 cm long. .... 8b. *S. graminea* subsp. *weatherbiana*
                  10. Achene 2.5–3 mm long; beak of achene about half as long as width of achene (fig. 86c); anthers shorter than filaments. .... 9. *S. cristata*
      6. Stamens with filaments glabrous; blades with basal lobes (except rarely lobed in *S. ambigua*).
         12. Leaves linear-lanceolate to elliptic, unlobed (rarely lobed). .... 10. *S. ambigua*
         12. Leaves sagittate (rarely unlobed, then mostly in juvenile plants).
             13. Achenes with beak horizontal (fig. 88e); bracts 1.4 cm or less long; when plants staminate, flowers with 21–40 stamens. .... 11. *S. latifolia*

13. Achenes with beak ascending or erect (fig. 92d); bracts 0.7–4 cm long; when plants staminate (*S. cuneata*), flowers with 10–24 stamens.
  14. Achenes 1.8–2.6 mm long, with tiny erect beak 0.2–0.5 mm long (fig. 90e); submersed plants forming ribbon-like leaves (fig. 90a) and/or floating lanceolate or sagittate leaves (fig. 90c); emersed plants often with recurved petioles. . . . . 12. *S. cuneata*
  14. Achenes 2.1–4 mm long, with erect or oblique beak 0.4–2.1 mm long; plants always emersed, petioles ascending.
    15. Achenes lacking facial resin ducts, facial wings not extending onto beaks; bracts form, papery pointed at tip.
      16. Petiole sharply 5-winged; fruiting heads 1–1.5 cm in diameter; pedicels of pistillate flowers 3–12 mm long. . . . . 13. *S. australis*
      16. Petiole ridged, not winged; fruiting heads (1.2)1.7–2.2 cm in diameter; pedicels of pistillate flowers 10–20 mm long. . . . . 14. *S. brevirostra*
    15. Achenes with facial resin ducts (fig. 92d), facial wings extending onto beaks; bracts thick, herbaceous, blunt at tip. . . . . 15. *S. engelmanniana*

---

1. *S. montevidensis* Cham. & Schlect. Giant Arrowhead

1a. *S. montevidensis* subsp. *calycina* (Engelm.) Bogin Hooded Arrowhead Fig. 77
Swamps ponds, lakes and ditches. N.J., Del. and Md. w. to Ohio, se. Mich., s. Ind., Ill., sw. Wisc., s. Minn., and, S.D. s. to Va., S.C., Tenn., Ala., Miss., La., Tex.; n. Mont, Colo., N.M., Ariz., Oreg., Calif. and Mex. (*Lophotocarpus calycinus* (Engelm.) J. G. Sm.; *S. calycina* Engelm.)

1b. *S. montevidensis* subsp. *spongiosa* (Engelm.) Bogin Spongyleaf Arrowhead Fig. 78
Mud flats of estuaries. Coastal, e. N.B. and e. Que. s. to Va. Although the leaves are typically thick, spongy and phyllodial, occasionally elliptic to narrowly sagittate blades occur. (*Lophotocarpus spongiosus* (Engelm.) J. G. Sm.; *S. calycina* var. *spongiosa* Engelm.; *S. spatulata* (J. G. Sm.) Buchenau)

2. *S. platyphylla* (Engelm.) J. G. Sm. Delta Arrowhead Fig. 79
Shallow water and muddy shores of ponds, streams, ditches and swamps. Va., W.Va., Ohio, and Ky. w. to Ill., Mo. and Kans., s. to n. Fla., La., Ala., Okla., e. Tex. and Mex.; Puget Sound, Wash.; Panama. (*S. graminea* var. *platyphylla* Engelm.; *S. mohrii* J. G. Sm.)

3. *S. subulata* (L.) Buchenau Awl-leaved Arrowhead Fig. 80
Tidal waters and shores of marshes, streams and sloughs, inland in fresh or brackish water. Coastal plain, s. N.E., se. N.Y. and N.J. s. to Fla., w. to Ala. and ne. La.; W.I. and n. S.Am.

4. *S. filiformis* J. G. Sm. Threadleaf Arrowhead Fig. 80
Ponds, lakes and streams. Coastal plain, Me. and Mass. s. to N.J., e. Pa., N.C., Ga. and Fla., w. to Ala. This species is frequently encountered sterile; it occasionally produces flowers, but has not been observed in fruit in the northern portion of its range. (*S. stagnorum* Small; *S. subulata* var. *gracillima* (S. Wats.) J. G. Sm.; *S. subuata* var. *natans* (Michx.) J. G. Sm.)

5. *S. teres* S. Wats. Dwarf Wapato, Slender Arrowhead, Quill-leaved Arrowhead Fig. 81
Locally abundant in sandy and coastal plain freshwater ponds of se. Mass.; rare elsewhere, s. N.H., Mass. s. to R.I., s. N.J. and e. Md.

6. *S. rigida* Pursh Sessile Fruited Arrowhead Fig. 82
Locally abundant along alkaline or tidal shores. Que. and Me. w. to Ont. and Minn., s. to Va., Ky., Tenn., Mo. and Neb.; Ida., Wash. and n. Calif. This species has leaf blades which are quite variable, ranging from slender lanceolate to broadly oval, occasionally becoming somewhat sagittate with the development of 1 or 2 basal lobes. A submersed bladeless form is sometimes encountered. (*S. heterophylla* Pursh; *S. heterophylla* var. *rigida* (Pursh) Engelm.)

7. *S. lancifolia* L. Bull-tongue Arrowhead Fig. 83
Marshy shores, streams, rivers (often deep water), swamps and occasionally tidal waters. Coastal plain Va., s. to Fla., w. to Tex. and Mex.; C.Am. and S.Am. Our plants belong to subsp. *media* (Micheli) Bogin. (*S. falcata* Michx.)

8. *S. graminea* Michx.

8a. *S. graminea* subsp. *graminea* Grassy Arrowhead, Grass-leaved Arrowhead Fig. 84
Freshwater ponds and lakes. S. Lab. and Nfld. w. to Ont. and Minn., s. to Fla., Ill., Mo. and Tex.; Wash.; Cuba. In the Northeast this species produces abundant sterile rosettes; when flowering occurs, fruits are seldom matured. (*S. eatoni* J. G. Sm.)

8b. *S. graminea* subsp. *weatherbiana* (Fernald) Haynes & Hellq. Weatherby's Arrowhead Fig. 85
Shallow water. fresh to slightly saline, wooded swamps, and streams. Coastal plain, Va. s. to. Fla. (*S. weatherbiana* Fernald; *S. graminea* var. *weatherbiana* (Fernald) Bogin)

9. *S. cristata* Engelm. Crested Arrowhead Fig. 86
Streams, lake bottoms and shores. Bruce Pen., Ont., w. to Mich., Wisc., Minn., s. to Mo., Neb. and n. Iowa. Vegetative rosettes, often abundant in 1–2 m of water, are difficult to distinguish from those of *S. graminea*. (*S. graminea* var. *cristata* (Engelm.) Bogin)

10. *S. ambigua* J. G. Sm. Kansas Arrowhead Fig. 87
Pond and lake shores, swamps, shallow water, ditches, and damp areas. Ill., Ind., Mo., Kans. and Okla.

11. *S. latifolia* Willd. Arrowhead, Wapato, Duck-potato Figs. 88, 89
Ponds, lakes, streams, sloughs and marshes. Nfld. and N.B. w. to B.C., s. to n. Fla., Ala., La., Tex., Colo., Calif. and Mex.; C.Am. and S.Am. The erect leaves are extremely variable, ranging from narrow bladeless leaves to narrowly sagittate to broadly sagittate leaves. In the southern portion of the range, plants which are sparsely pubescent, their bracts and calyx often densely pubescent, have been recognized as var. *pubescens* (Muhl.) J. G. Sm. Several varieties and forms have been based on leaf shapes, but such leaf variation may all occur on a single plant. (*S. latifolia* var. *obtusa* (Muhl. ex Willd.)

Fig. 77. *Sagittaria montevidensis* subsp. *calycina*: a. portion of inflorescence; b. leaf; c. flower; d. fruiting heads; e. achene (C&C).

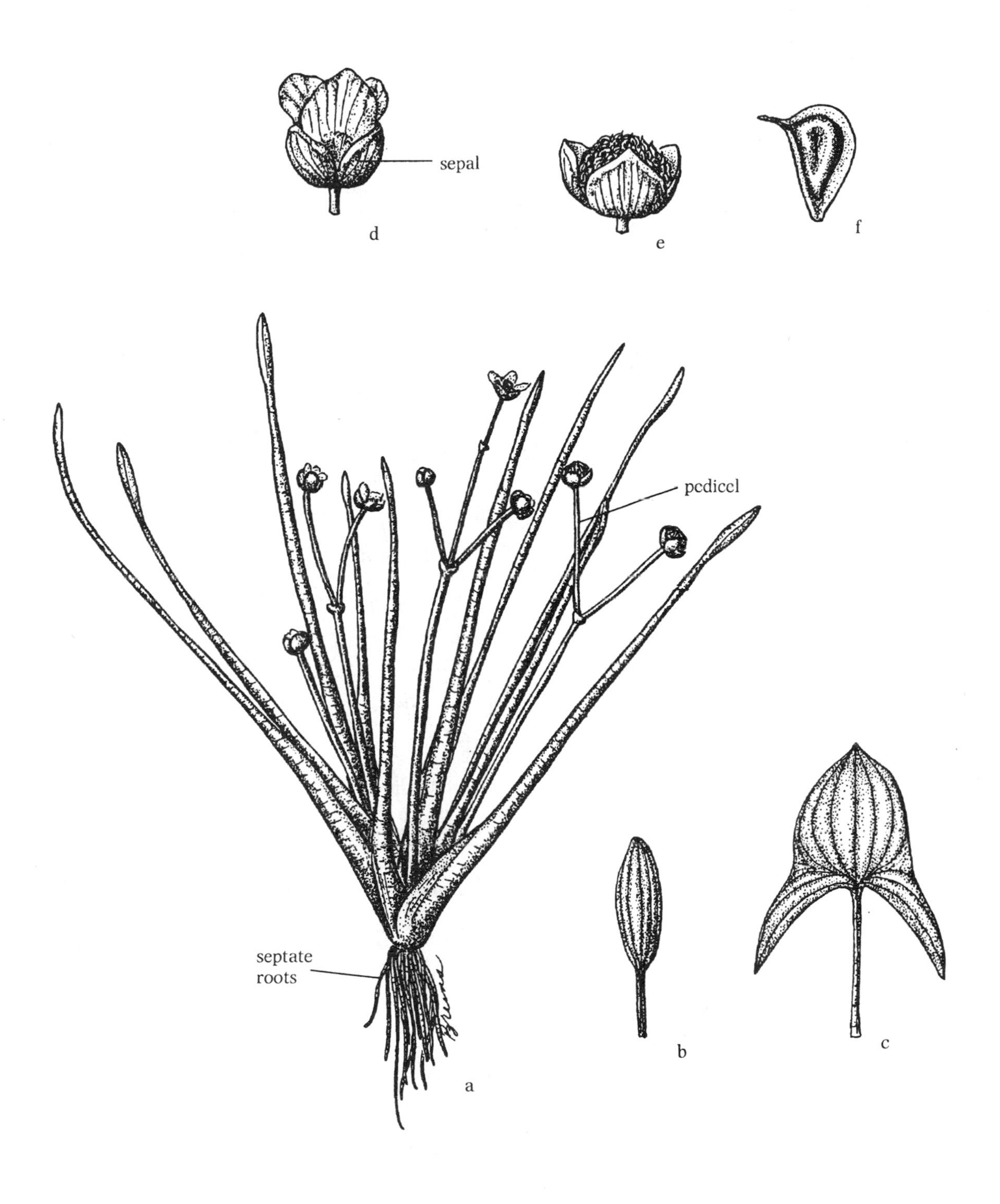

Fig. 78. *Sagittaria montevidensis* subsp. *spongiosa*: a. habit; b, c. leaf blade variations; d. flower; e. fruiting head; f. achene (NHAES).

Fig. 79. *Sagittaria platyphylla*: a. habit; b. achene (PB).

Fig. 80. *Sagittaria filiformis*: a. habit, submersed plant; b. inflorescence (NHAES).
Sagittaria subulata: c. habit, emersed plant; d. habit, submersed plant; e. fruiting head; f. achene (NHAES).

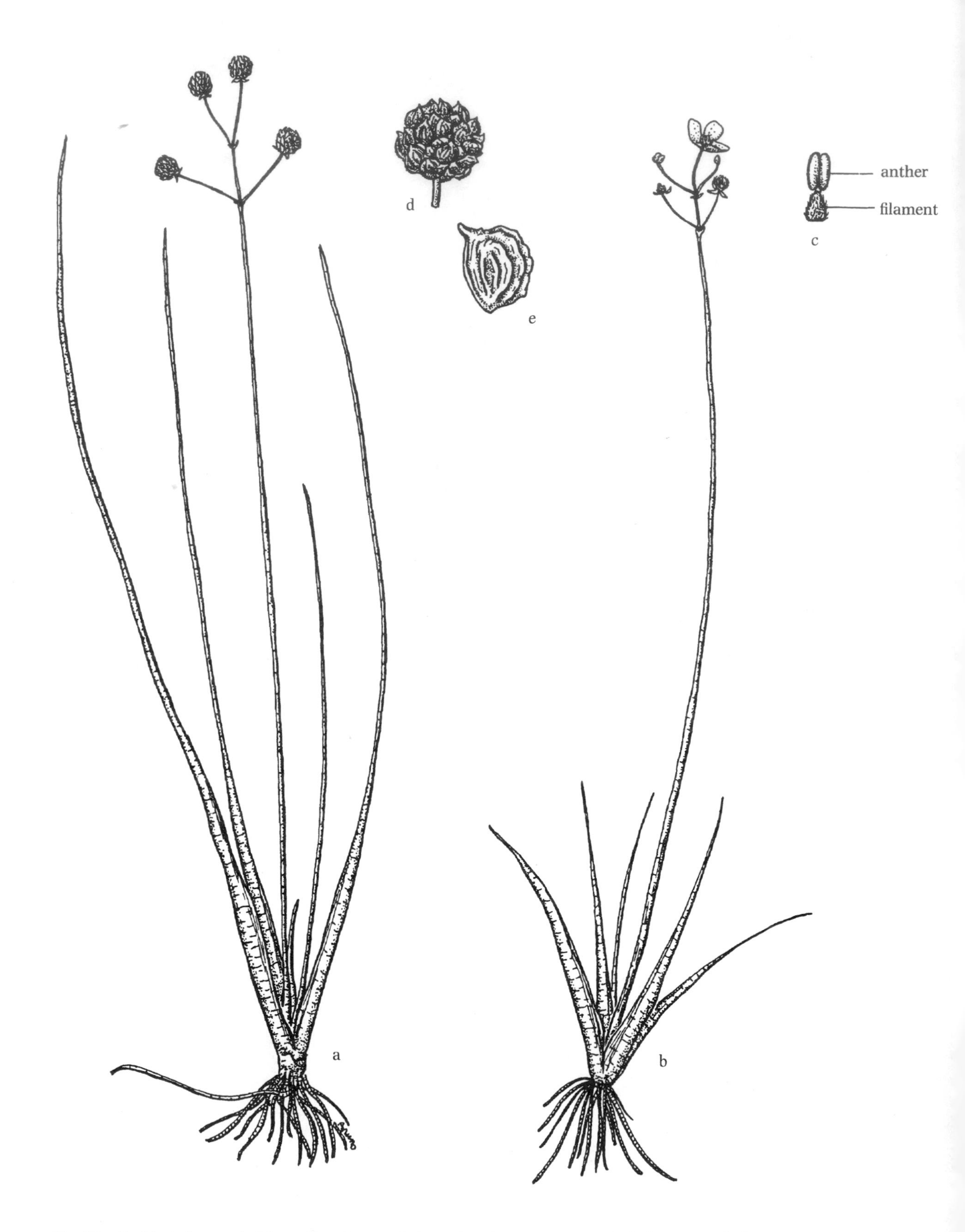

Fig. 81. *Sagittaria teres*: a. habit, partly submersed plant, showing septate roots; b. habit, emersed plant, showing septate roots; c. stamen, showing pubescent filament; d. fruiting head; e. achene (NHAES).

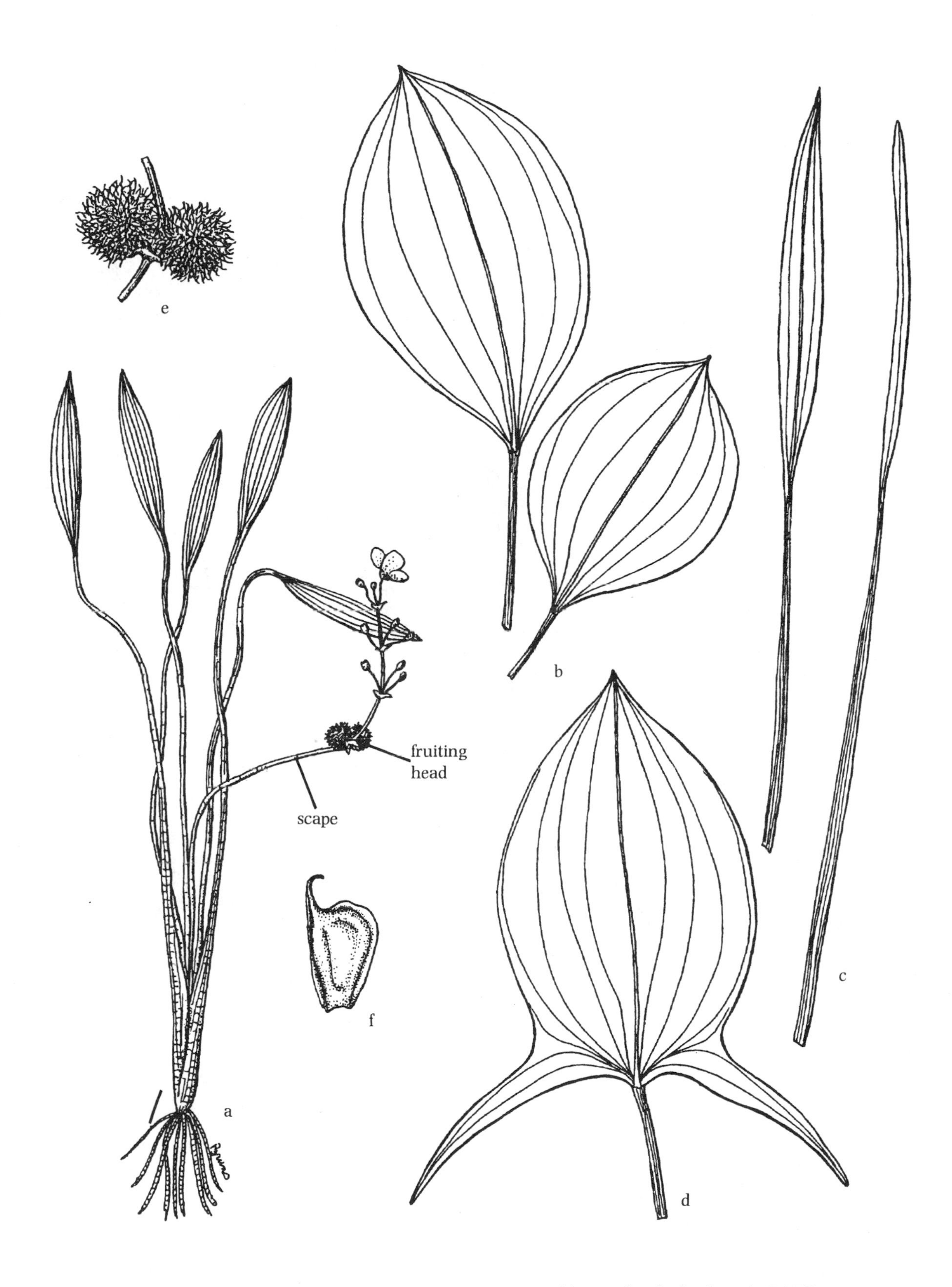

Fig. 82. *Sagittaria rigida*: a. habit; b–d. leaf blade variations; e. paired fruiting heads; f. achene (NHAES).

Fig. 83. *Sagittaria lancifolia*: a. habit (J. G. Smith, 1895); b. flower (C&C); c. stamen (J. G. Smith, 1895); d. achene (J. G. Smith, 1895).

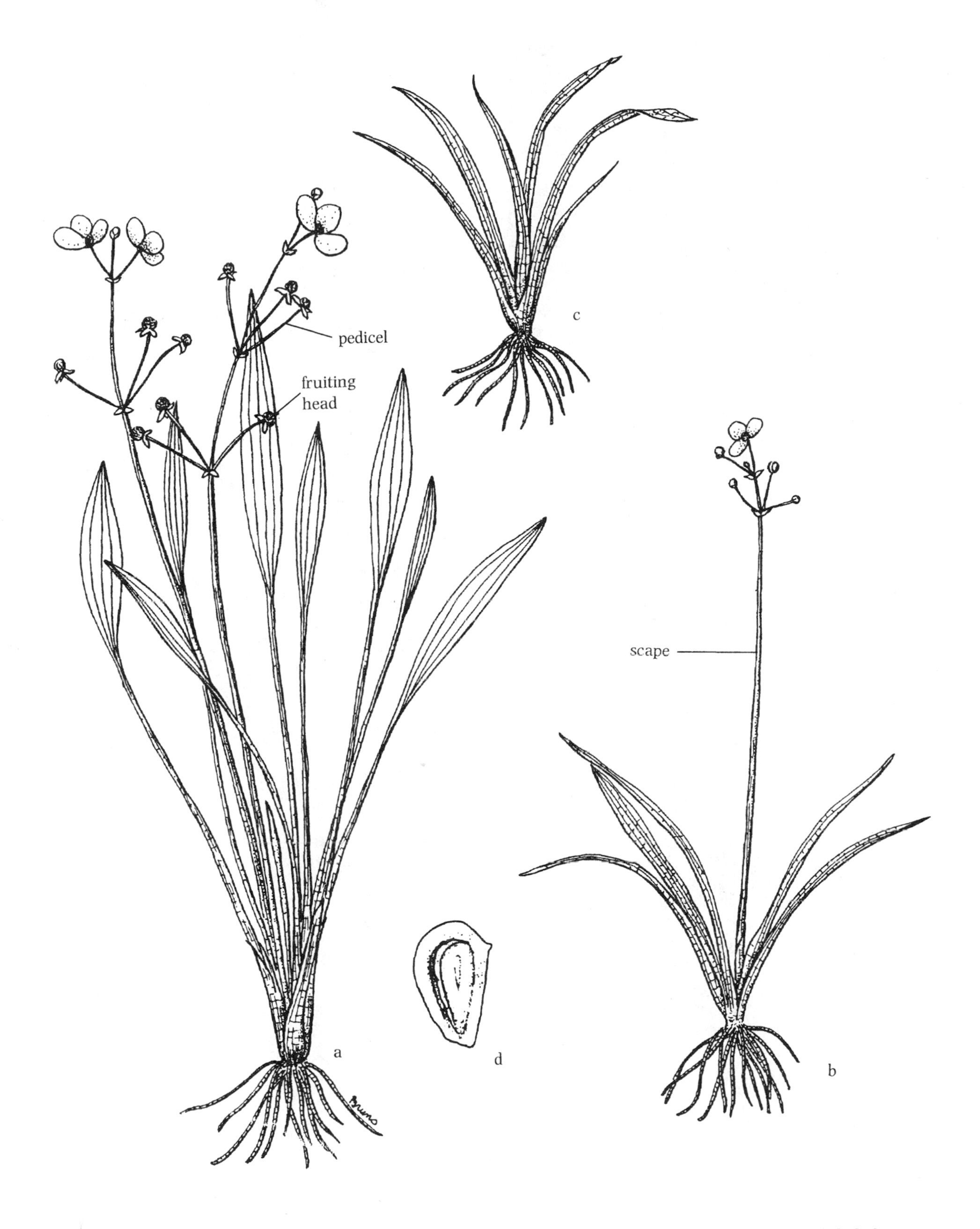

Fig. 84. *Sagittaria graminea* subsp. *graminea*: a. habit, plant with lanceolate, tapering blades (NHAES); b. habit, plant with linear phyllodia (NHAES); c. submersed sterile rosette (NHAES); d. achene (J. G. Smith, 1895).

Fig. 85. *Sagittaria graminea* subsp. *weatherbiana*: a. habit, emersed plant; b. habit, submersed plant; c. achene (PB).

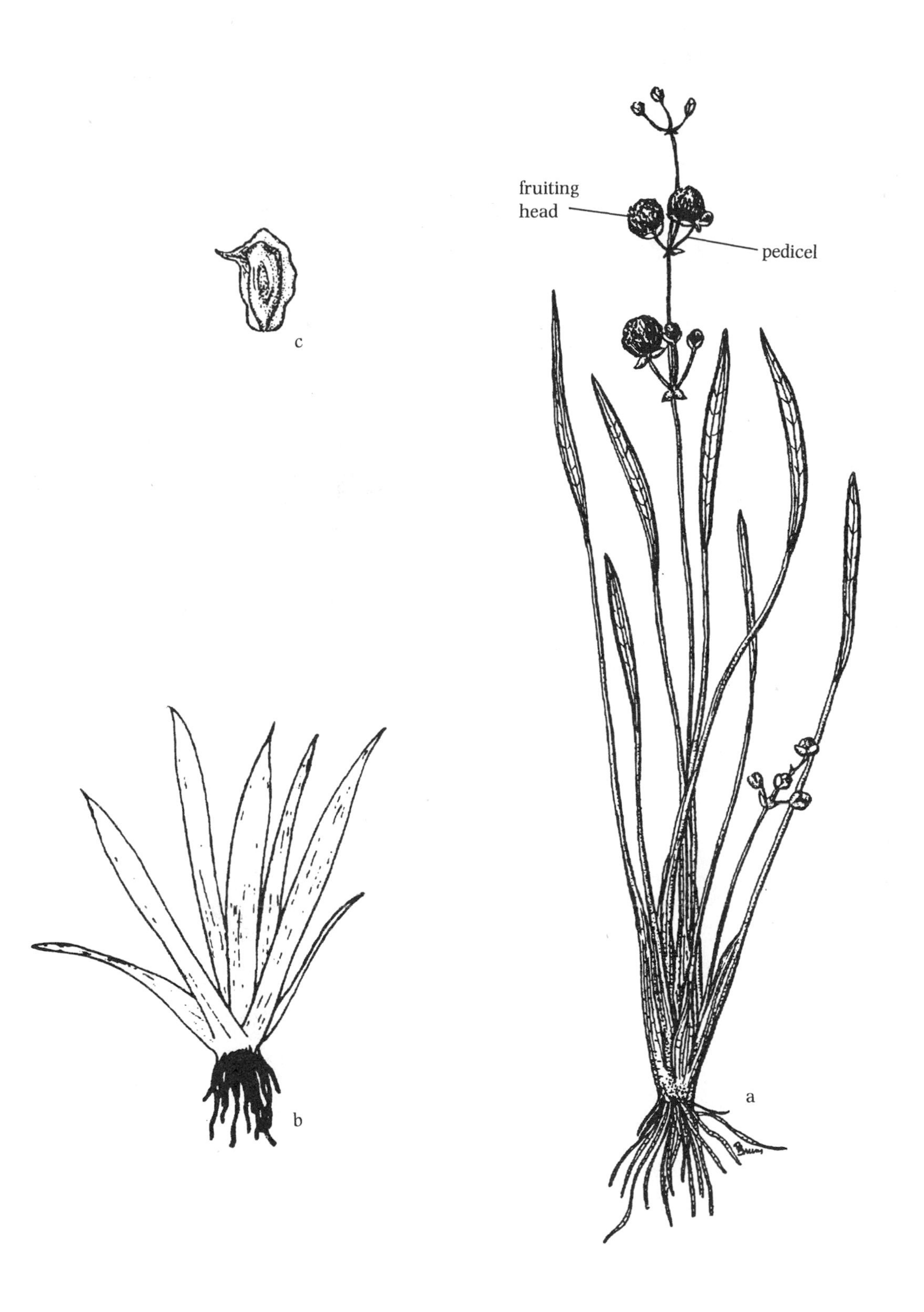

Fig. 86. *Sagittaria cristata*: a. habit, emersed plant (PB); b. habit, submersed sterile rosette (J. G. Smith, 1895); c. achene (PB).

Fig. 87. *Sagittaria ambigua*: a. habit (PB); b. stamen (J. G. Smith, 1895); c. achene (PB).
Sagittaria australis: d. habit; e. achene (Braun).

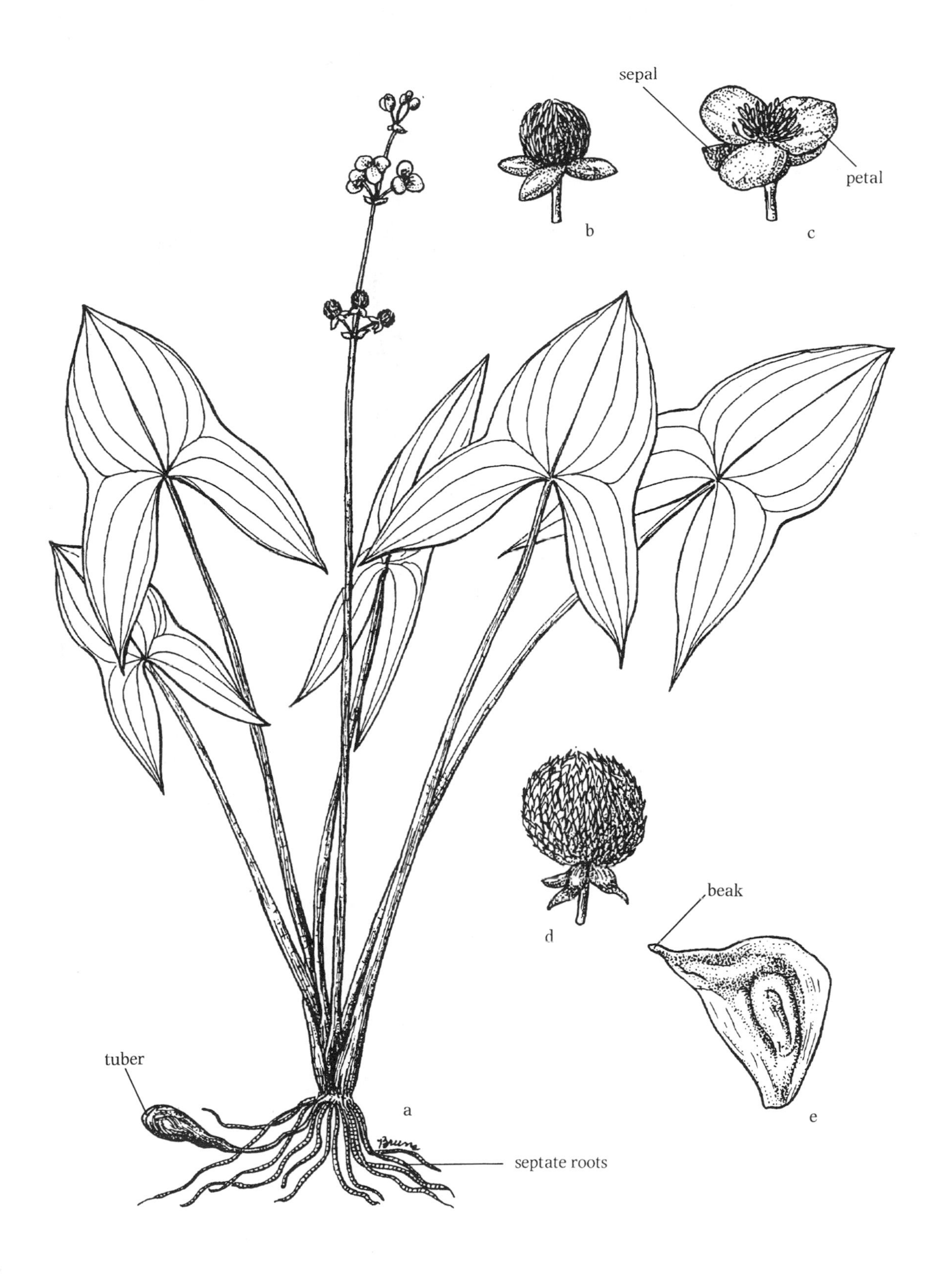

Fig. 88. *Sagittaria latifolia*: a. habit; b. pistillate flower; c. staminate flower; d. fruiting head; e. achene (NHAES).

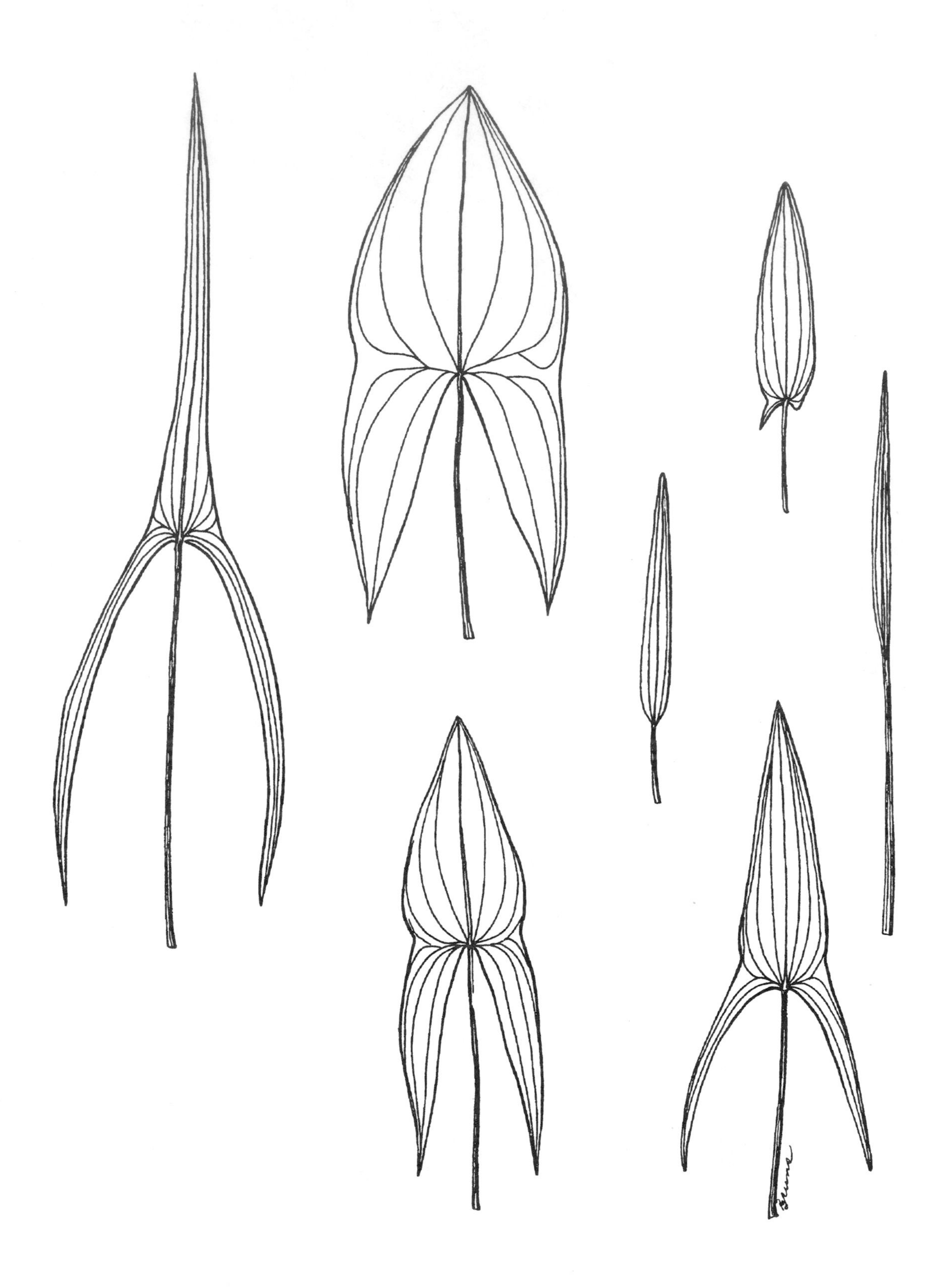

Fig. 89. *Sagittaria latifolia*: leaf variations (NHAES).

12. *S. cuneata* Sheldon Arum-leaved Arrowhead, Northern Arrowhead Fig. 90
Alkaline waters of rivers, ditches, ponds, lakes and streams. Nfld., Lab. and Que., w. to B.C. and Alask., s. to N.E., N.Y., Ohio, Ill., Iowa, Kans., Tex., N.M., Ariz. and Calif. This is a highly variable species. In deep water it may produce vegetative rosettes similar to those of *S. graminea*, but may also produce floating leaves with lanceolate or sagittate blades. Extremely broad, flat, linear leaves are often found in flowing water. When emersed the plant forms sagittate or ovate leaves that usually droop rather than stand erect. (*S. arifolia* Nutt.)

13. *S. australis* (J. G. Sm.) Small Long-beak Arrowhead Fig. 87
Springs, marshes and ponds. N.Y. N.J. and Pa. w. to Ohio, Ind. and s. Ill., s. to Fla., Ala., Miss., Iowa and Mo. (*S. longirostra* var. *australis* J. G. Sm.; *S. engelmaniana* subsp. *longirostra* (Micheli) Bogin)

14. *S. brevirostra* Mackenz. & Bush Short-beak Arrowhead Fig. 91
Sloughs and wet shores. Ohio, s. Ont. and Mich. w. to Wisc., Minn. and e. S.D., s. to Va., Tenn., Ark., c. Okla., Colo. and ne. Tex.: Alta and Calif. (*S. engelmanniana* subsp. *brevirostra* (Mackenz. & Bush) Bogin).

15. *S. engelmanniana* J. G. Sm. Engelmann's Arrowhead Fig. 92
Acid waters of marshes, bogs and sandy ponds. Chiefly coastal plain and piedmont, N.S., e. Mass. s. to Fla., w. to Ala. and Miss.: s. Ont. c. N.Y.

Scheuchzeriaceae / Scheuchzeria Family

1. *Scheuchzeria* (Pod-grass, Rannoch-rush)

Perennial herbs, arising from creeping, jointed rhizomes; leaves linear, grass-like, with terminal pore at apex, broadly sheathing at base; flowers borne in a loose raceme; fruit an inflated follicle.

Reference: Nienaber, 2000.

1. *S. palustris* L. Fig. 93
Sphagnum bogs. Lab. and Nfld. w. to Man., N.W.T. and Alask., s. to N.J., Pa., W.Va., n. Ohio, n. Ill., n. Iowa, N.D., nw. Wyo., w. Mont., Ida. and n. Calif. A circumboreal species, our taxon has long been treated as distinct (subsp. *americana* (Fernald) Hultén); Nienaber (2000), however, noted that variation in the characters of follicle and stigma between Eurasia and N.Am. indicate that no infraspecific taxa warrant recognition. The best diagnostic character for sterile plants is the terminal pore at the apex of the leaf. (*S. palustris* var. *americana* Fernald)

Juncaginaceae / Arrow-grass Family

1. *Triglochin* (Arrow-grass)

Perennial herbs, arising from a short rhizome, some producing stolons; leaves basal, slender, slightly succulent, rush-like; flowers in a spike-like raceme; fruit a cluster of 3–6 follicles separating from a persistent rachis.

References: Crow and Hellquist, 1982; Fernald, 1903; Ford and Ball, 1988; Haynes and Hellquist, 2000; Löve and Lieth, 1961; Löve and Löve, 1958.

1. Fruit cluster wider than long, globose-deltoid (fig. 94f,g). . . . . 1. *T. striata*
1. Fruit cluster longer than wide.
  2. Leaves typically as long as scape (fig. 94a), very slender, curving outward from sheath at 30–40-degree angle; spike 2–7 cm long. . . . . 2. *T. gaspensis*
  2. Leaves typically shorter than scape (fig. 95a,e), slender to somewhat thick, ascending, nearly erect from sheath; spike 6–45 cm long.
    3. Carpels and stigmas 3; mature fruit 1 mm wide; about 5–7 times as long as wide; stolons present, filiform, bearing small bulbs. . . . . 3. *T. palustris*
    3. Carpels and stigmas 6; mature fruit 2–3 mm wide, about twice as long as wide; stolons absent. . . . . 4. *T. maritima*

1. *T. striata* Ruiz & Pav. Fig. 94
Salt marshes and land-shore near the coast. Coastal, Del. and Md. S. to Fla., w. to La.: Oreg. Calif. and Mex.; Cuba, Bahamas, C.Am. and S.Am.

2. *T. gaspensis* Leith & D. Löve Fig. 94
Salt marshes, Nfld. s. to Gaspé Pen., Que., P.E.I., N.S. and Washington Co. Me. In contrast to the clumped habit of *T. maritima*, plants of *T. gaspensis* tend to form lawn-like patches in the salt marsh.

3. *T. palustris* L. Fig. 95
Saline, brackish and calcareous marshes. Greenl. and Lab. w. to Alask., s. to R.I., s. N.Y., nw. Pa., n. Ohio, Ill., Neb., Colo., N.M., Calif. and Mex.; S.Am.

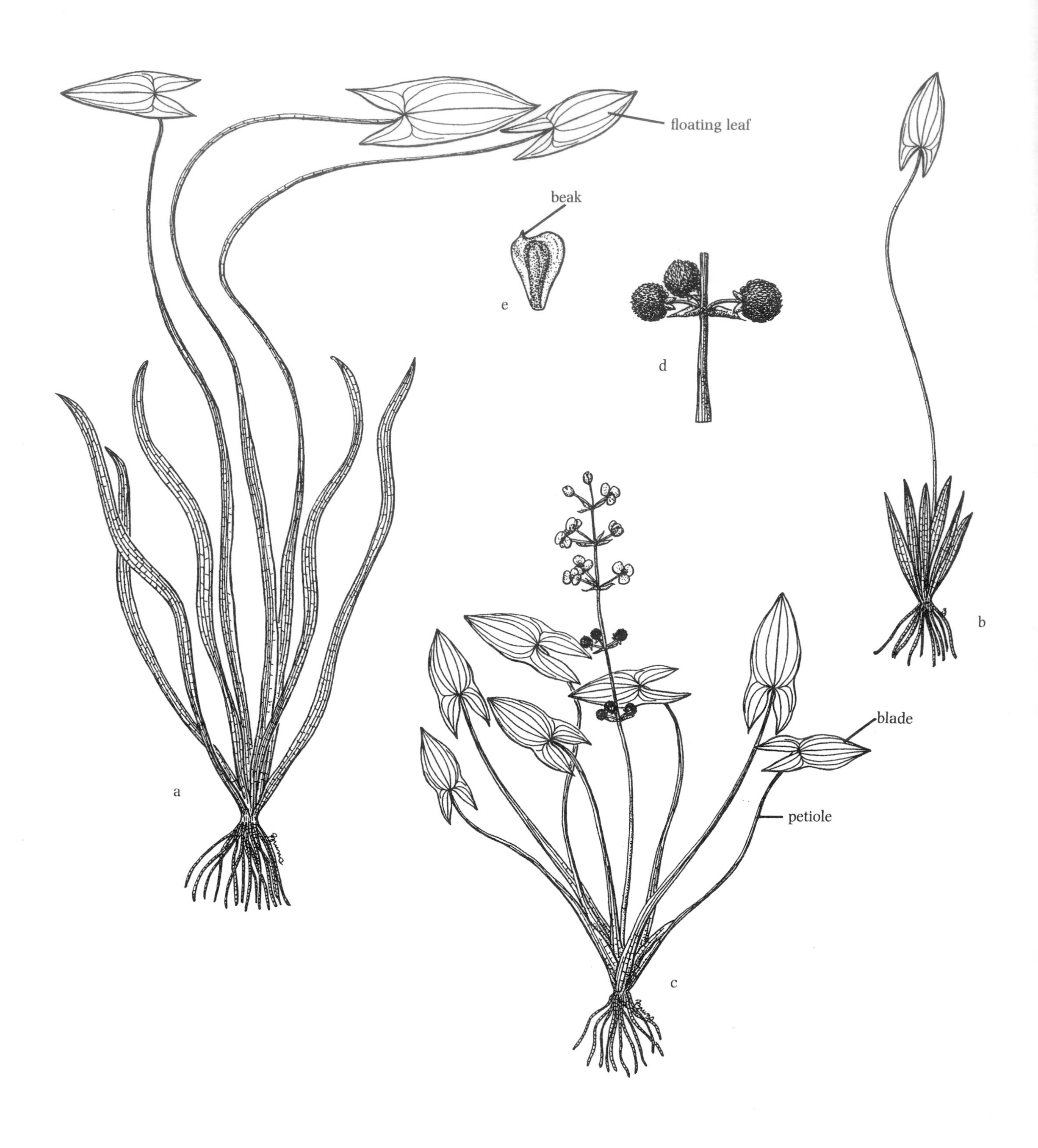

Fig. 90. *Sagittaria cuneata*: a. habit, submersed plant; b. sterile submersed rosette; c. habit, emersed plant; d. fruiting heads; e. achene (NHAES).

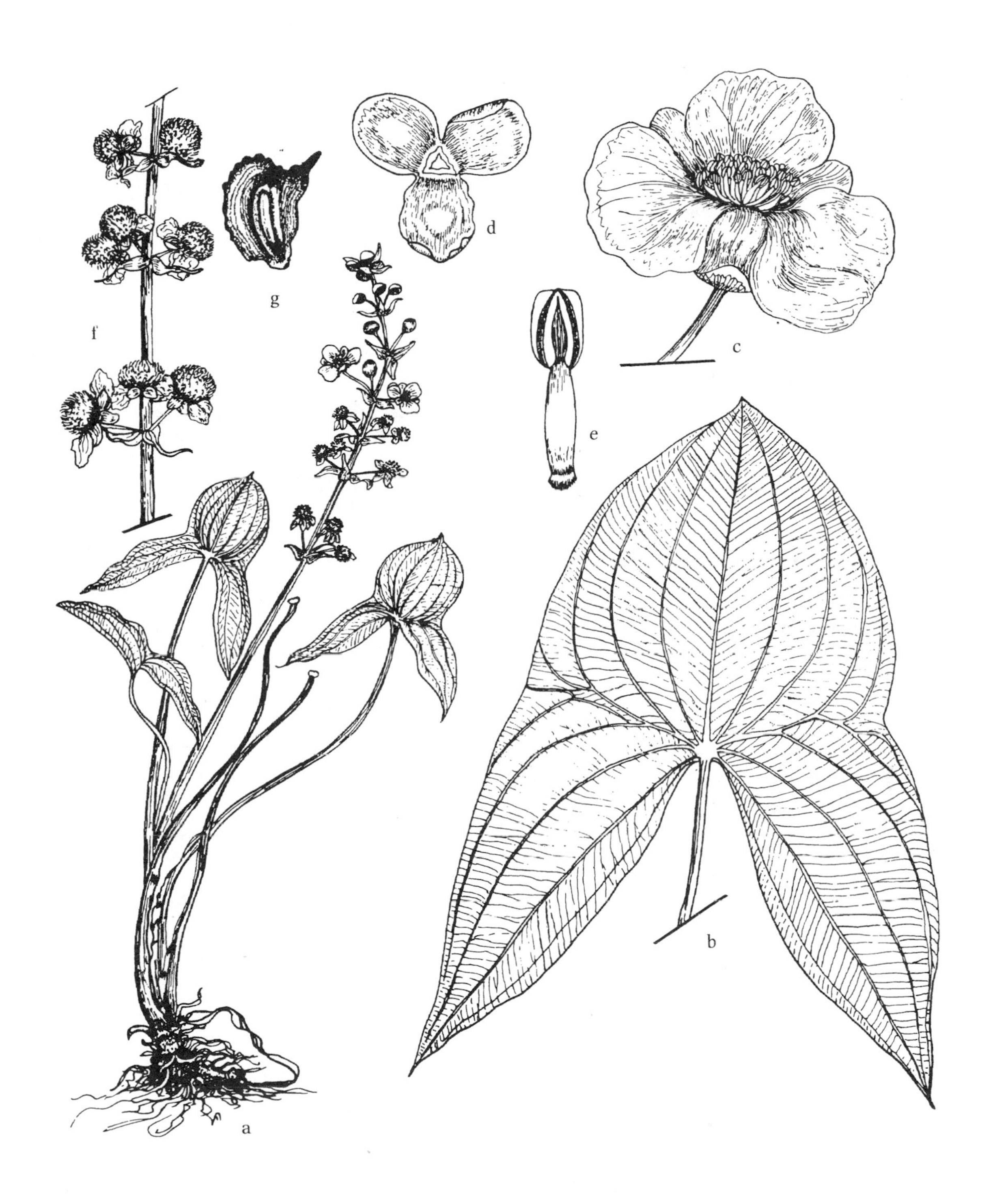

Fig. 91. *Sagittaria brevirostra*: a. habit; b. leaf; c. flower; d. calyx; e. stamen; f. portion of infructescence; g. achene (C&C).

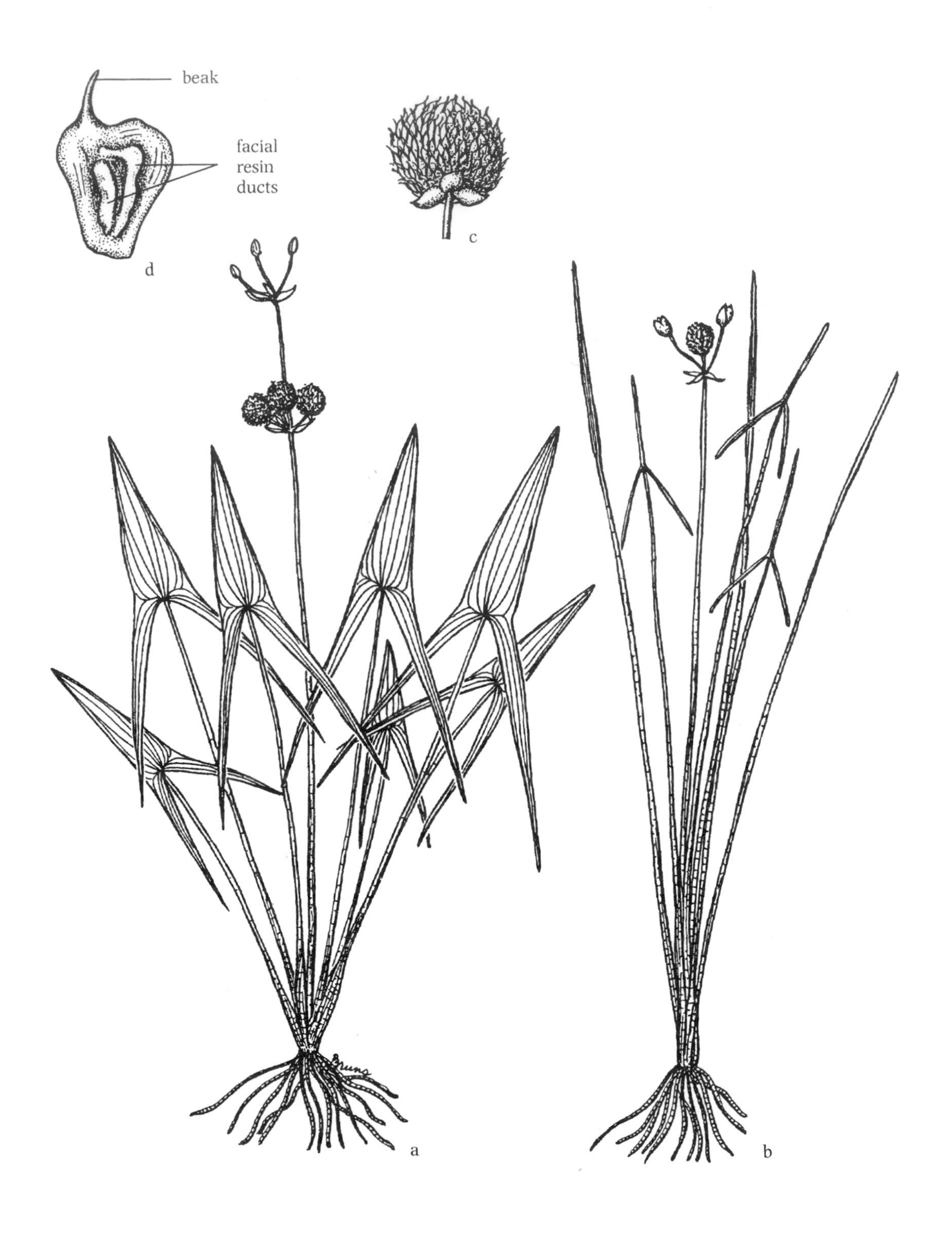

Fig. 92. *Sagittaria engelmanniana*: a, b. habit; c. fruiting head; d. achene (NHAES).

Fig. 93. *Scheuchzeria palustris*: a. habit, flowering plant; b. habit. fruiting plant; c. habit, sterile submersed plant; d. flower; e. fruit; f. leaf apex with terminal pore, with appearance of a fingernail (NHAES).

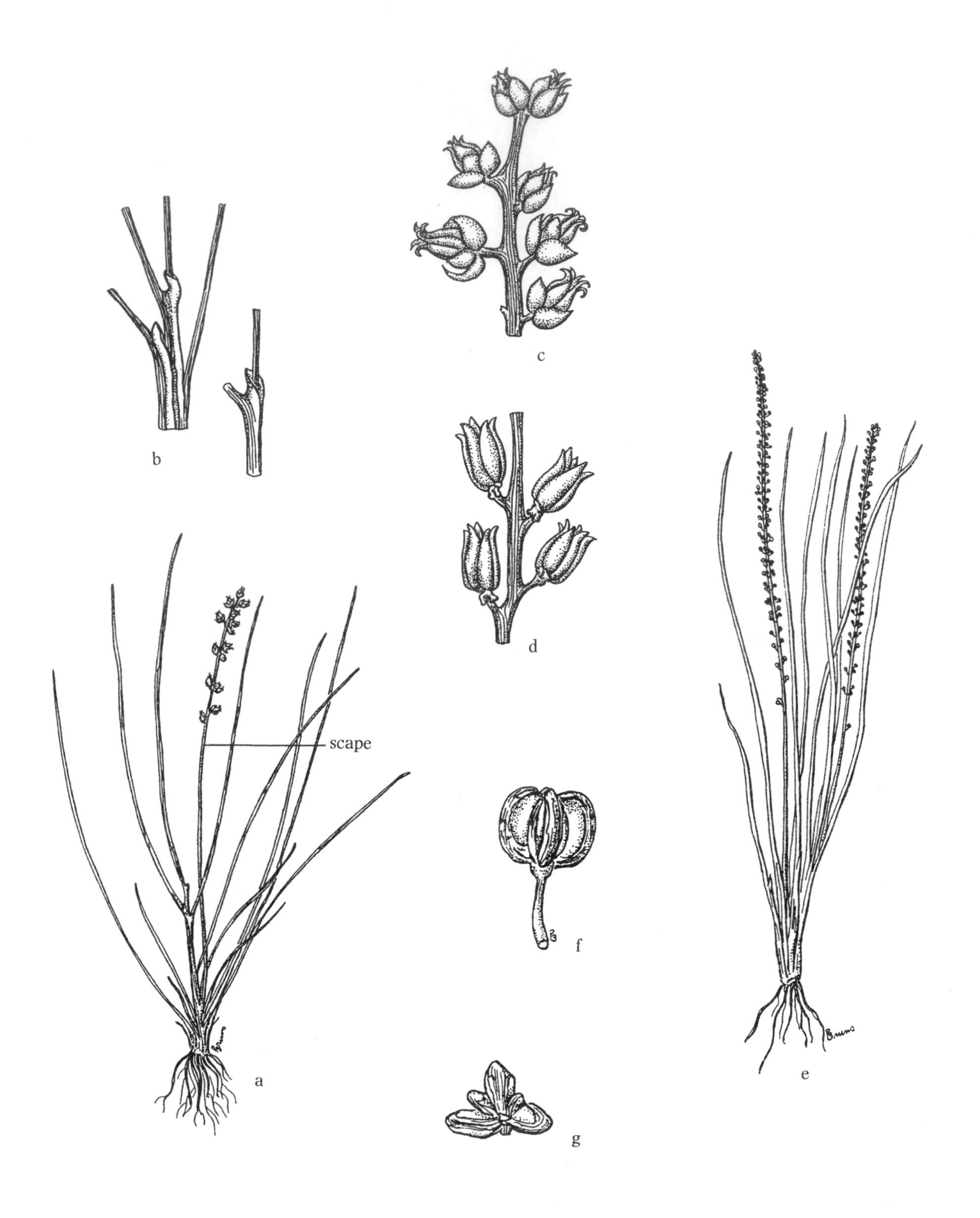

Fig. 94. *Triglochin gaspense*: a. habit; b. leaf sheath; c. flowers; d. fruits (NHAES).
Triglochin striata: e. habit; f. fruit, side view; g. fruit, top view (PB).

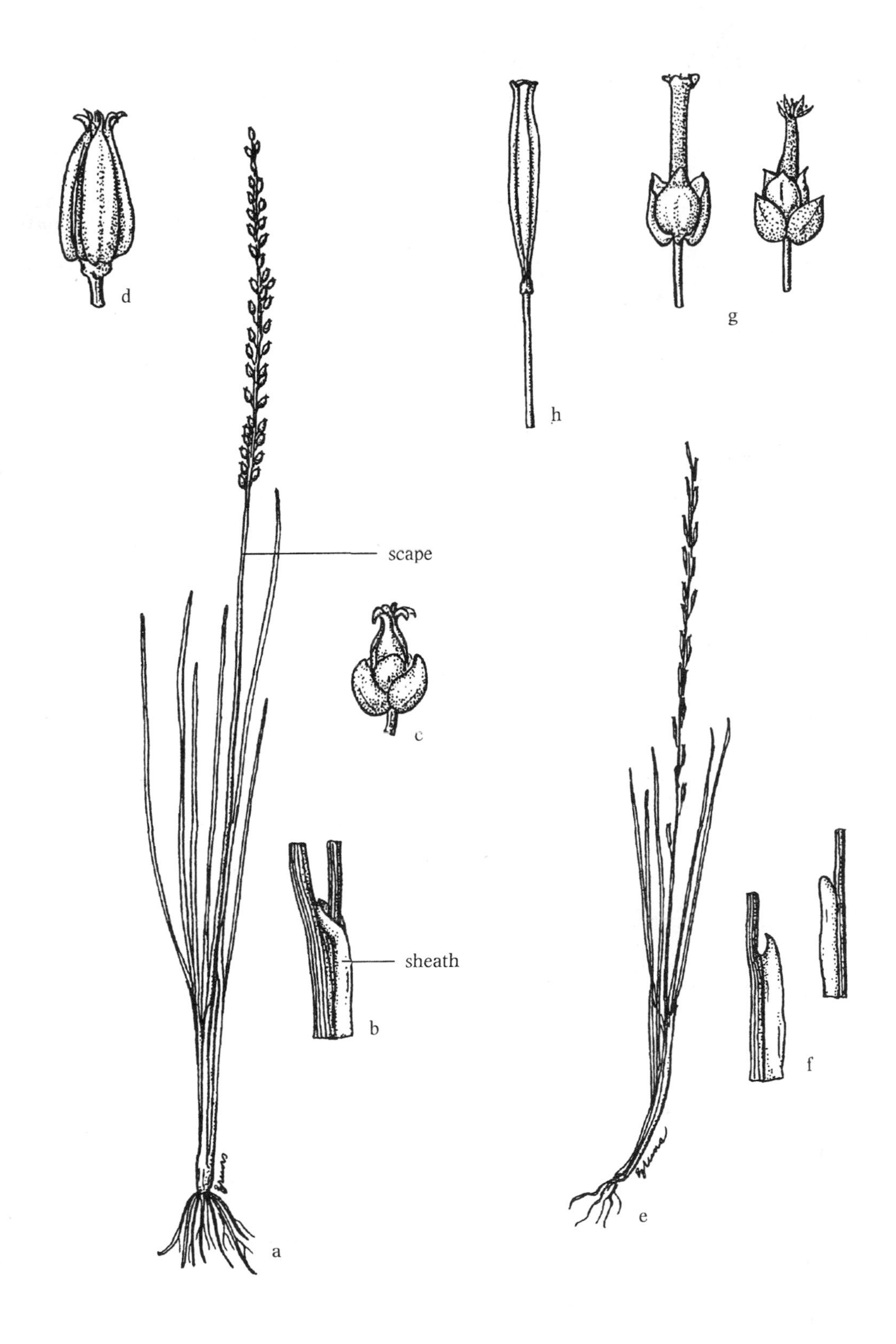

Fig. 95. *Triglochin maritima*: a. habit; b. leaf sheath; c. flower; d. fruit (NHAES).
Triglochin palustris: e. habit; f. leaf sheaths; g. flowers, two stages; h. fruit (NHAES).

4. *T. maritima* L. Fig. 95
Salt marshes; inland in fens and calcareous marshes. Lab. and Nfld. w. to Alask. S. to Del., w. N.Y., nw. Pa., n. Ohio, n. Ill., Neb., Kans., Colo., N.M., Calif and n. Mex.; S.Am. This polymorphic species complex appears to be best treated as a single species (Ford and Ball, 1988) (*T. elatum* Nutt.).

Ruppiaceae / Ditch-grass Family

1. *Ruppia* (Widgeon-grass, Ditch-grass)

Perennial or annual aquatic herbs; leaves alternate, stipules adnate to blades; floating leaves absent; flowers 2, enclosed in leaf sheath at anthesis; ovaries separate, 4, sessile; fruit 4 per flower, on long slender stalk (podogyne, a growth of the base of the gynoecium during fruit maturation), usually below water surface.

Ruppia has been, at times, included within the Potamogetonaceae. However, molecular evidence, however, indicates a very close relationship between *Ruppia* and genera of the Cymodoceaceae, thus leading Les et al. (1997a) to doubt its distinction as the Ruppiaceae, and to place *Ruppia* within the Cymodoceaceae (Les, 2020). APG-IV (2016) acknowledges they are sister groups, but retains the family Ruppiaceae.

References: Fernald and Wiegand, 1914; Haynes, 2000; Les, 2020; Les et al., 1997a; Richardson, 1980, 1983; Thieret, 1971.

1. Podogynes (stalks) in fruit 2–25 mm, coils 0–4, fruit tip more or less obtuse; mainly coastal locations. . . . . 1. *R. maritima*
1. Podogynes (stalks) in fruit 30–300 mm, coils 5–30, fruit tip acute; mainly inland locations (in our range). 2. *R. cirrhosa*

1. *R. maritima* L. Widgeon-grass, Ditch-grass Figs. 96, 97
Brackish and saline waters. Chiefly coastal, Nfld., Lab. and Que. s. to Fla., w. to La. and e. Tex.; coast of James Bay, Sask.; Alask., B.C. and Wash. s. to Calif. and Mex.; W.I., C.Am., S.Am. and Old World. Studies by Richardson (1980, 1983) have revealed considerable variability in *Ruppia*, casting doubt on the distinctiveness of any of the varieties recognized by Fernald and Wiegand (1914). Further, Richardson (1983) demonstrated that in coastal New Hampshire two entities occur which appear to warrant recognition at the species level. One is an annual adapted to estuarine sites with chiefly underwater self-pollination and is diploid. The other is adapted to tidal streams with chiefly surface pollination and is tetraploid. Although the range of variability of these two entities overlaps, the two appear to be reproductively isolated and can be differentiated by a combination of morphological characters.

2. *R. cirrhosa* (Petagna) Grande.
Freshwater lakes, ponds and streams of high concentrations of calcium and sulfur. Ohio, n. Mich. w. to Man., N.W.T., Yuk. and Alask., s. to Ill., Mo., Okla. Tex., N.M., Ariz., Calif. and Mex.; C.Am., S.Am., and Eurasia. Many of the populations have extremely robust plants. (*R. maritima* var. *occidentalis* (S. Watson) Graebn.; *R. cirrhosa* subsp. *occidentalis* (S. Watson) Á. Löve & D. Löve)

Zosteraceae / Eelgrass Family

1. *Zostera* (Eelgrass)

Perennial herbs, submersed; stems flattened, arising from rhizomes; leaves alternate, up to 2 m long; flowers unisexual, borne in rows surrounded by a sheathing leaf-like spathe; fruit a utricle.

References: Ackerman, 1986; Haynes, 2000; Setchell, 1929, 1933.

1. *Z. marina* L. Eelgrass Fig. 98
Common in saline water of estuaries and bays and in sea water along the open coast. Greenl., Lab., Nunavut, and James Bay; Nfld. s. to Fla.; West Coast, Alask., s. to Calif. and Mex.; C.Am.; widespread coastal northern hemisphere.

Potamogetonaeae / Pondweed Family

Perennial aquatic herbs, from rhizomes or winter buds; leaves, alternate or opposite to subopposite with a characteristic midvein often with distinct bands or rows of lacunae; floating leaves present in many species; flowers 4-merous, borne in spikes or axillary clusters (*Zannichellia*), or in some enclosed in leaf sheaths at time of anthesis, submersed or emersed or both; fruit drupe-like, often with a beak and/or keel, long or short-stalked.

This treatment reflects recent evidence from molecular studies (Les et al., 1995). Reevaluation of the genus (Les and Haynes, 1996) recognized the need to treat the taxa of *Potamogeton* subgenus *Coleogeton* as a distinct genus; the generic name *Stuckenia* having been applied to *P. filiformis* and *P. pectinatus* as early as 1912 (Holub, 1997). This has been further confirmed by work on the Asian *Stuckenia* by Kaplan (2008). The Angiosperm Phylogeny Group-VI (2016) recognizes *Zannichellia* within the Potamogetonaceae, rather than a distinct family.

Fig. 96. *Ruppia maritima*: a, b. habit; c. peduncle bearing 2 flowers, each with 2 anthers and 4 pistils; d. 2 flowers after fertilization; e. elongate podogynes with fruits at tips, and recoiled peduncle; f. fruit; g. stipular sheaths; h. habit (Mason).

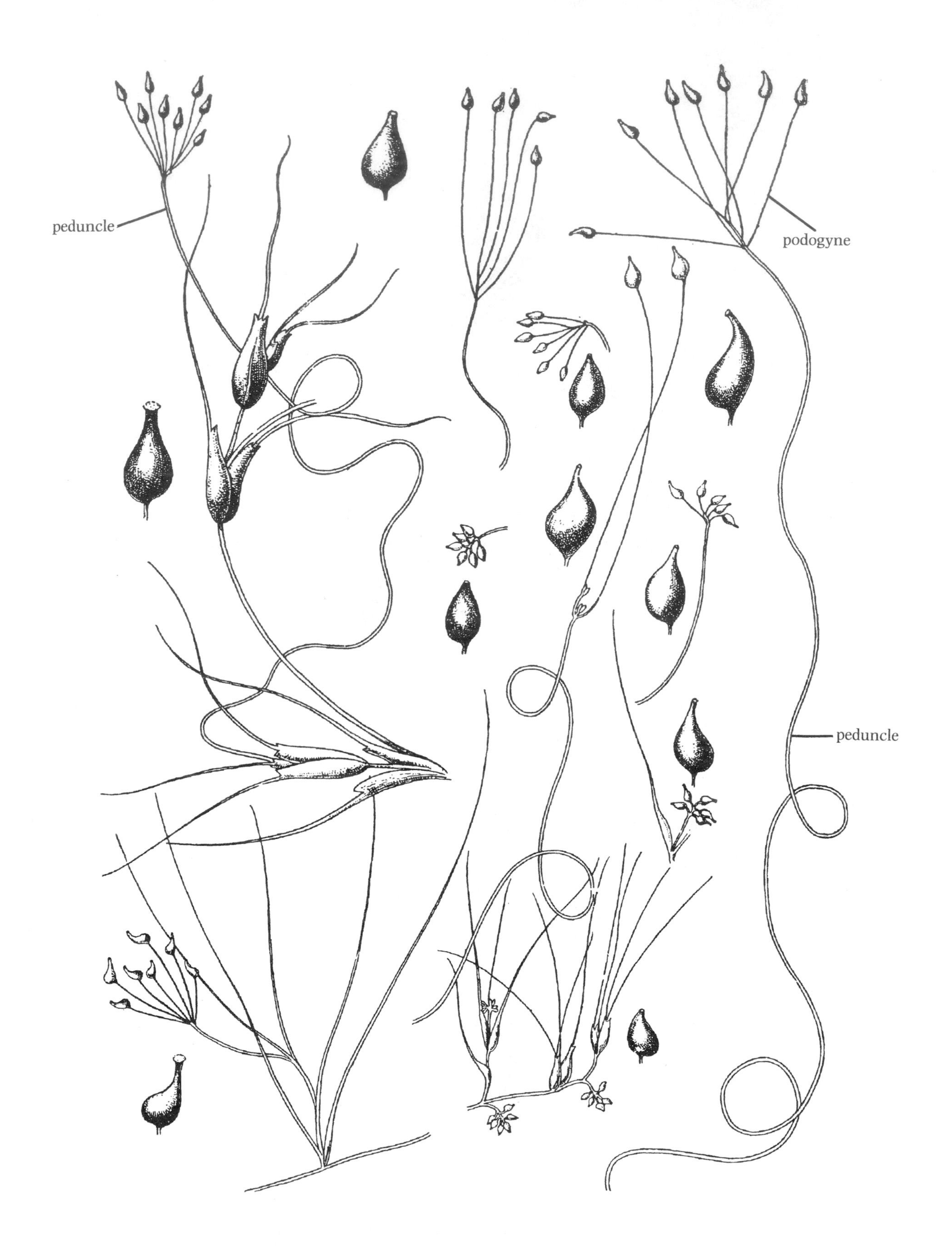

Fig. 97. *Ruppia maritima*: fruit, podogyne, and peduncle variations (Fernald and Wiegand, 1914).

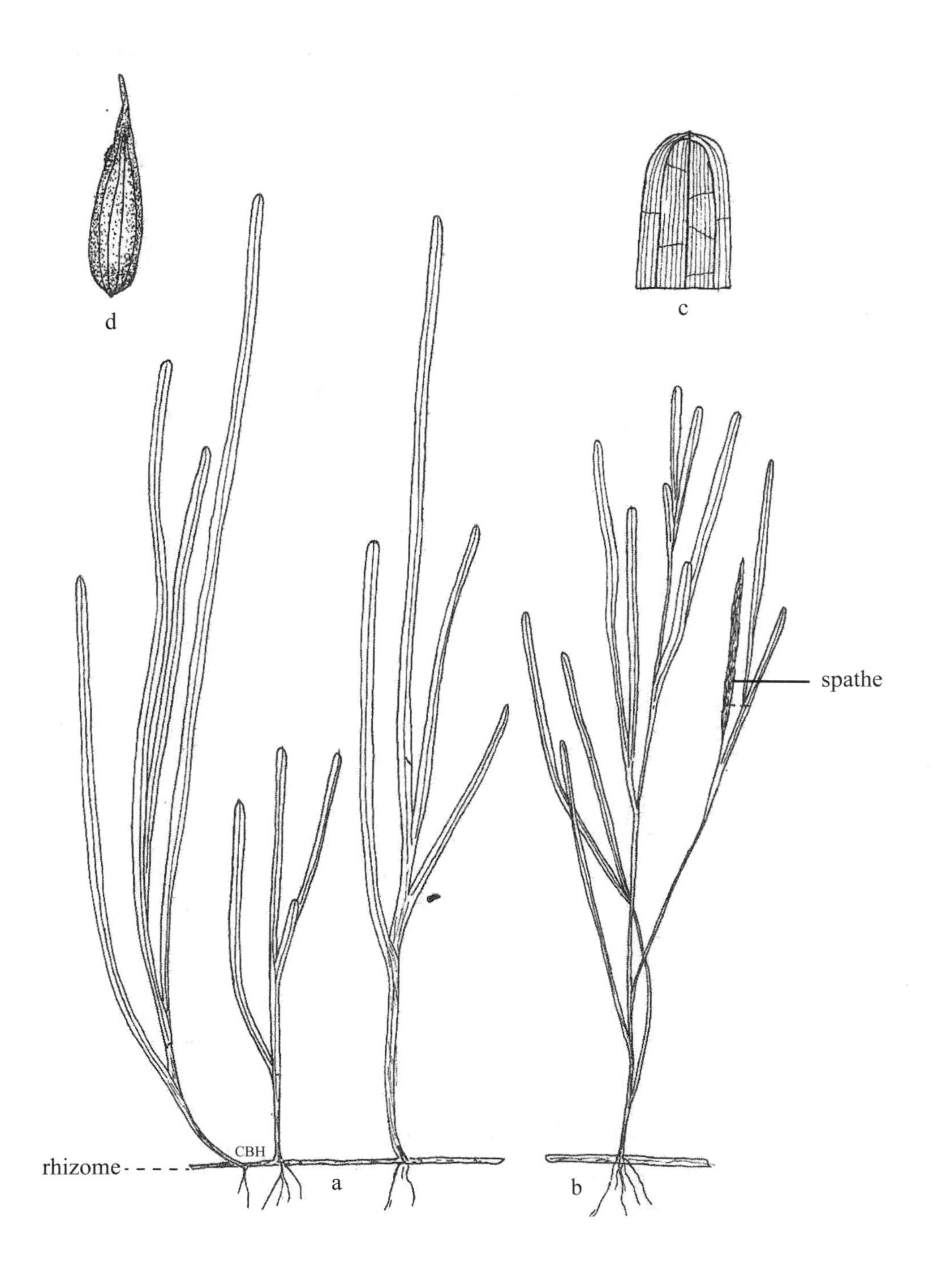

Fig. 98. *Zostera marina*: a. habit; b. habit with spathe; c. leaf tip; d. fruit (NHAES).

References: Catling and Dobson, 1985; Crow, 2003b; Fernald, 1932b; Hagström, 1916; Haynes, 1974, 1978, 1980, 1985; Haynes and Hellquist, 1996, 2000; Haynes and Holm-Nielsen, 1987; Haynes and Williams, 1975; Hellquist, 1980, 1984; Hellquist and Crow, 1980, 1986; Hellquist and Hilton, 1983; Hellquist et al., 1988; Holub, 1997; Hunt and Lutz, 1959; Kaplan, 2002, 2008; Kaplan et al., 2009; Kaplan and Marhold, 2012; Les, 1983; Les et al., 1995; Les et al., 2009; Les and Haynes, 1996; Les and Sheridan, 1990a, 1990b; Moore, 1915; Muenscher, 1936; Ogden, 1943; Philbrick, 1983, 1988; Reznicek and Bobette, 1976; Stuckey, 1978; Wehremeister and Stuckey, 1992; Weigleb and Kaplan, 1998; Yeo, 1965.

---

1. Submersed leaves opposite or subopposite; floating leaves absent; flowers unisexual, in axillary clusters; fruits stalked, dentate or papillate on one side (fig. 99) . . . . 1. *Zannichellia*
1. Submersed leaves alternate; floating leaves present or absent; flowers bisexual, in axillary or terminal spikes; fruits sessile, sometimes with a smooth or undulate lateral keel, but not dentate or papillate.
   2. Submersed leaves with stipular sheaths adnate to leaf base for two-thirds of more of stipule length (fig. 101a,i); submersed leaves opaque, channeled, septate (fig. 100b); peduncles flexible, inflorescence, if reaching water surface, floating on surface. . . . . 2. *Stuckenia*
   2. Submersed leaves with stipulate sheaths free from leaf blade base (fig. 111b), or if adnate for less than half the length of stipule (fig. 103c); submersed leaves translucent, not channeled or septate; peduncles stiff, inflorescence, if reaching surface, emersed above the water. . . . . 3. *Potamogeton*

---

1. *Zannichellia* (Horned Pondweed)

Perennial herbs; submersed; stems arising from rhizomes; leaves linear, opposite to subopposite; fruit in axillary clusters of 2–6 bilaterally symmetrical, usually dentate or papillate on one side, short-stalked.

Reference: Haynes and Holm-Nielsen, 1987.

1. *Z. palustris* L. Horned Pondweed Fig. 99
Saline waters along the coast, alkaline and brackish waters inland. Nfld., e. Que., w. to Alask., s. throughout the U.S. to Mex.; C.Am. and S.Am. (*Z. palustris* var. *major* (Boenn. ex. Rchb.) W. D. J. Koch).

2. *Stuckenia* (Pondweed)

Perennial, from rhizomes; leaves alternate, submersed, filiform to narrowly ribbon-like, with characteristic midvein, channeled, flowers 4-merous, borne in spikes, usually floating on water surface; fruit drupe-like. Hybrids sometimes occur, usually in flowing leaves blunt, obtuse, notched (fig. 369e), or rarely apiculate; stipules connate, fused at base, at least on young plants; fruit beakless or minutely beaked (fig. 369g,h), dark green to brown-colored.

---

1. Leaves acute (fig. 100d), apiculate on young plants; stipules convolute, but open at base; fruit distinctly beaked (fig. 100g), yellow-brown-colored. . . . . 1. *S. pectinata*
1. Leaves blunt, obtuse, notched (fig. 101e), or rarely apiculate; stipules connate, fused at base, at least on young plants; fruit beakless or minutely beaked (fig. 101g,h), dark green to brown-colored.
   2. Ligule, extending beyond stipule sheath distinct, up to 20 mm long (fig. 101i); summits of midstem stipulate sheaths only Slightly inflated, less than twice the width of stem; fruit 2–3 mm long. . . . . 2. *S. filiformis*
   2. Ligule barely formed extending beyond stipule sheath 1 mm or less; summits of midstem stipular sheaths distinctly inflated, twice the diameter of stem; fruit 3–3.5 mm long. . . . . 3. *S. vaginata*

---

1. *S. pectinata* (L.) Börner Sago Pondweed Fig. 100
Calcareous, saline, and alkaline waters. Nfld. w. to Alask., s. to Fla., Calif. and Mex.; W.I., C.Am. and S.Am. Leaves vary in length and thickness, often coarser than illustrated in fig. 368a. This species is easily recognized by the much-branched stems and numerous acute-tipped filiform leaves spreading in fan-like fashion; tubers are frequently produced. Hybrids with *S. filiformis* = *S.* ×*suecica* (K. Richt.) Holub; with *S. vaginata* = *S.* ×*fennica* (Hagstr.) Holeb, often occur in swift-flowing waters. (*Coleogeton pectinatus* (L.) Les & R. R. Haynes; *Potamogeton interruptus* Kit.; *P. pectinatus* L.).

2. *S. filiformis* (Pers.) Börner Fineleaf Pondweed Fig. 101
Cold, usually calcareous or brackish, still or flowing waters. W. Nfld. and N.S. w. to N.W.T. and Alask., s. to Me., N.Y., N.J., Mich., Minn., n. N.D., Colo., N.M. and Calif. Robust plants often sterile and growing in deep cold or flowing waters are treated as *S. filiformis* subsp. *occidentalis* (Robbins) Haynes, Les, and Král. These plants are easily confused with hybrids. Hybrids with *S. pectinata* = *S.* ×*suecica* (K. Richt.) Holeb; with *S. vaginata* = *S.* ×*fennica* Hagstr. (Holub). (*Coleogeton filiformis* subsp. *alpinus* (Blytt) Les & R. R. Haynes; *C. filiformis* subsp. *occidentalis* (Robbins)

Fig. 99. *Zannichellia palustris*: a. habit; b. portion of stem, showing stipular sheaths and axillary flowers; c. single staminate flower and 2 pistillate flowers surrounded by spathe; d. fruit, longitudinal section; e, f. fruit variations; g. fruit, outer coat deteriorated away (Mason).

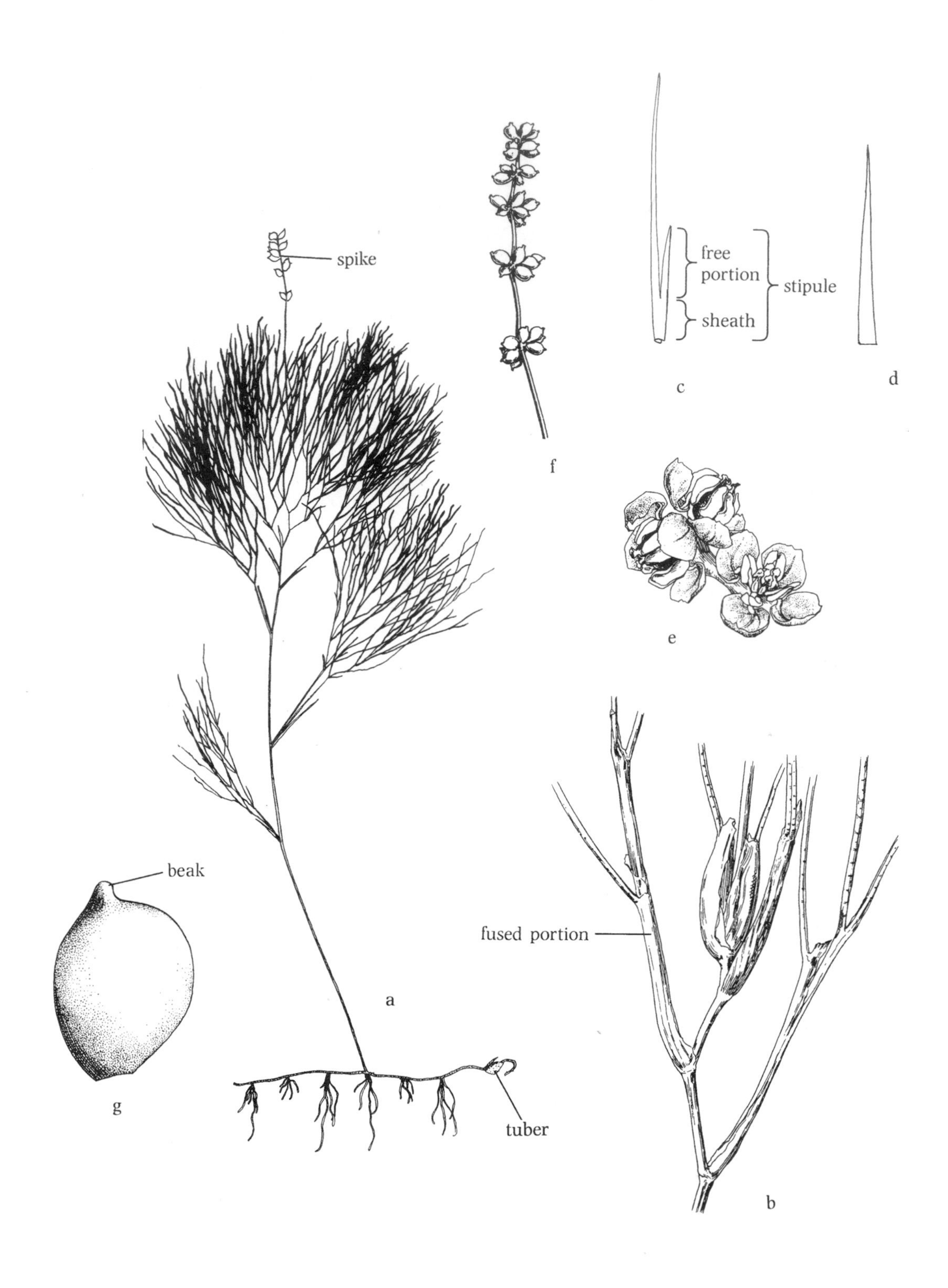

Fig. 100. *Stuckenia pectinata*: a. habit (F); b. stem, showing adnate stipules (Mason); c. leaf with stipule (F); d. leaf apex (F); e. flowers (Mason); f. spike (Mason); g. fruit (F).

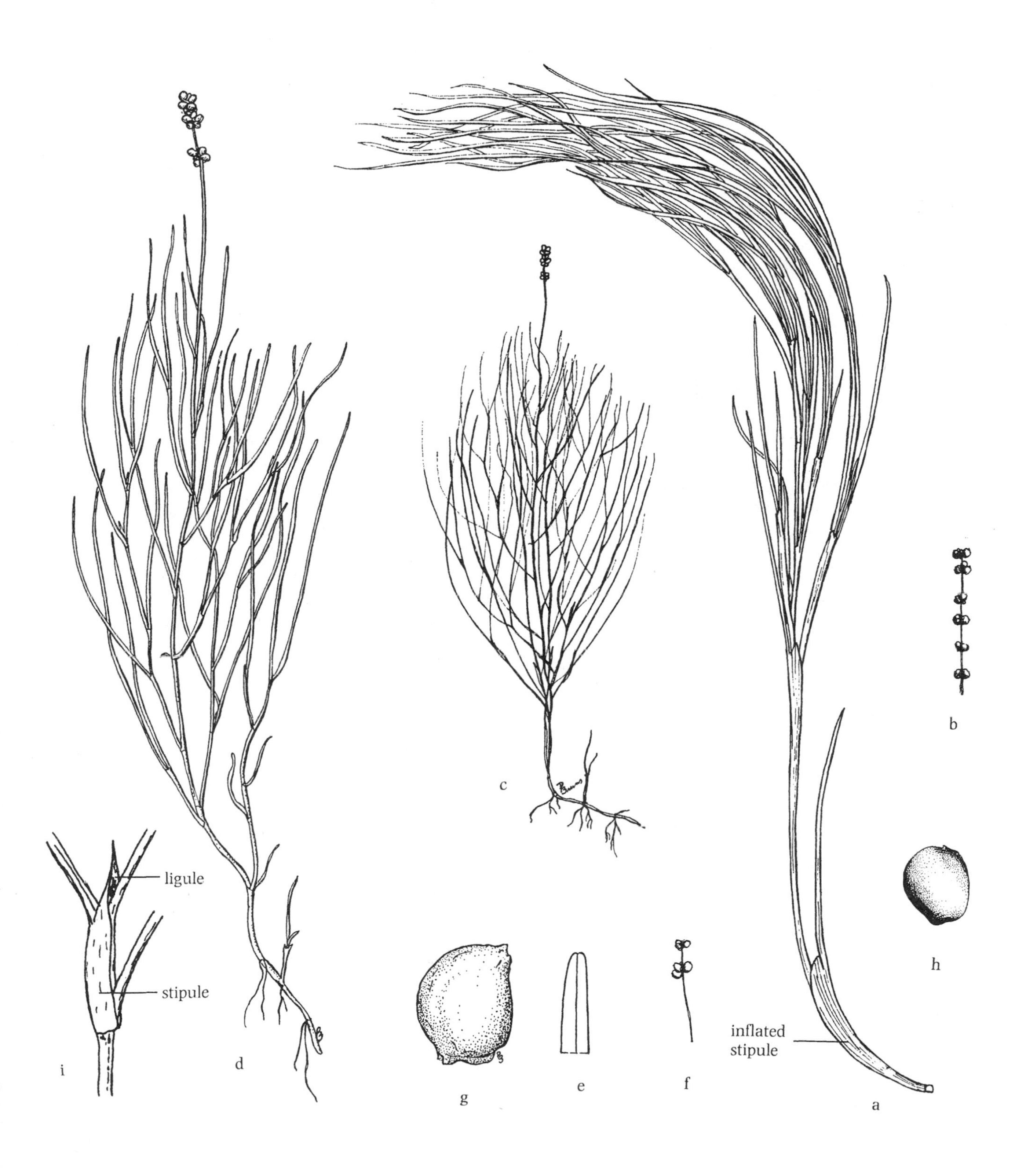

Fig. 101. *Stuckenia vaginata*: a. habit (PB); b. inflorescence (F).
Stuckenia filiformis: c. habit, narrow-leaved form (PB); d. habit, broad-leaved form (PB); e. leaf apex (PB); f. inflorescence (F); g. fruit (PB); h. fruit (F); i. stipule (CBH).

Les & R. R. Haynes; *Potamogeton filiformis* var. *alpinus* (Blytt) Aschers & Graebr.; *P. filiformis* var. *borealis* (Raf.) St. John; *P. filiformis* var. *macounii* Morong; *Stuckenia filiformis* subsp. *alpina* (Blytt) R. R. Haynes, Les, & Král)

3. *S. vaginata* (Turcz.) Holub. Sheathed Pondweed Fig. 101
Calcareous or brackish, usually still waters. N. Ont. w. to Alask., s. to s. Wisc., ne. Minn., n. Colo., Utah, w. Mont., Wyo., Ida. and c. Oreg. Most plants previously thought to belong to *S. vaginata* from Michigan eastward are hybrids. The hybrids usually grow in flowing water while most *S. vaginata* occur in still, deep water. Hybrids with *S. filiformis* = *S.* ×*fennica* (Hagstr.) Holub; with *S. pectinata* = *S.* ×*bottnica* (Hagstr.) Holeb. (*Coleogeton vaginatus* (Turcz.) Les & R. R. Haynes; *Potamogeton vaginatus* Turcz.)

3. *Potamogeton* (Pondweed)

Perennial, or rarely annual, aquatic herbs, from rhizomes or winter buds; leaves alternate, submersed or submersed and floating, variable with a characteristic midvein, sometimes with distinct bands or rows of lacunae; flower 4-merous, borne in spikes, axillary or terminal, submersed or emergent; fruit drupe-like. Hybrids are common, occurring especially in flowing waters.

---

1. Stipular sheaths fused to leaf base (fig. 103c).
 2. Leaf margins slightly serrate to entire, blades stiff, strongly two-ranked (fig. 102a) with basal lobe at junction with the stipule (fig. 102b); plants submersed; floating leaves absent; fruits rarely produced. . . . . 1. *P. robbinsii*
 2. Leaf margins entire, blades lax, not two-ranked and lacking a basal lobe; plants submersed, floating leaves present or absent; fruits commonly produced.
  3. Submersed and floating leaves blunt-tipped; fruit keel blunt (fig. 102g), not well developed; beak absent. . . . . 2. *P. spirillus*
  3. Submersed and floating leaves acute or blunt-tipped; fruit keel well-developed (fig. 103h,i); beak present (fig. 103d) or absent.
   4. Floating leaves 3–7-veined; submersed leaves 0.1–0.6 mm wide, no rows of lacunae; fruits lacking beak. . . . . 3. *P. bicupulatus*
   4. Floating leaves 5–23-veined; submersed leaves 0.9–2.0 mm, 1-many rows of lacunae; fruits with beak.
    5. Fruits 1–1.5 mm long, beak 0.1 mm long (fig. 103d) with distinct coil (fig. 103h), keel pointed (fig. 103h); floating leaves 5–15-veined; all stipules fused at base. . . . . 4. *P. diversifolius*
    5. Fruits 2.5–3 mm long, beak 0.5 mm long, coil lacking, keel acute, lacking points (fig. 104f); floating leaves 9–23-veined; most stipules free, few fused. . . . . 5. *P. tennesseensis* (in part)
1. Stipular sheaths completely free from leaf base (fig. 105c, 112a).
 6. Plants with floating leaves.
  7. Floating leaves 5–9-veined; submersed leaves, thread-like, 1(3)-veined (fig. 111). . . . . 6. *P. vaseyi* (in part)
  7. Floating leaves many-veined (up to 49 veins); submersed leaves rarely thread-like, 1–37-veined.
   8. Submersed leaves linear, flattened (fig. 104a), or phyllodial, appearing as a bladeless petiole (fig. 112a,c).
    9. Submersed leaves flat, linear 1–13-veined with a conspicuous broad lacunae band on either side of mid-vein (fig. 104b); fruits with distinct acute keels (fig. 104c).
     10. Floating leaves acute at tip; submersed leaves 1–3(7)-veined, sometimes only a few of the stipules fused. . . . . 5. *P. tennesseensis* (in part)
     10. Floating leaves rounded at tip; submersed leaves 3–13-veined, stipules free at leaf base. . . . . 7. *P. epihydrus* (in part)
    9. Submersed leaves phylloidal (fig. 112a) (similar to leaf petiole) to slightly linear, 0.8–2 mm wide, 1–5-veined, lacking lacunae band; fruit if produced lacking keels.
     11. Floating leaves 17–35-veined with the petiole at leaf attachment a different color than rest of petiole, cordate at base (not cordate in flowing waters); fruit 3.8–5 mm long with keels obscure (fig. 112b). . . . . 8. *P. natans* (in part)
     11. Floating leaves 7–23-veined with petiole at leaf attachment not or rarely with a different color than rest of petiole, rounded or tapering base; fruit usually 2.5–3.5 mm long, keels prominent (fig. 112d). . . . . 9. *P. oakesianus* (in part)
   8. Submersed leaves broadly linear-oblong to lanceolate to elliptic.
    12. Submersed leaves petiolate (figs. 114c, 115d).
     13. Floating leaves 19–49-veined; submersed leaf margins crisped (fig. 114a,c); stems rusty to black-spotted; Fruits keeled (fig. 114b,e)
      14. Stems black-spotted; submersed leaves, 9–21-veined, 10–25(35) mm wide, slightly arcuate (fig. 114c); fruits 3.1–4.1(4.7) mm. . . . . 10. *P. pulcher* (in part)
      14. Stems often rusty-spotted; submersed leaves 27–49-veined, 25–75 mm wide, distinctly arcuate (fig. 114a); fruits 3.9–5.2(5.7) mm. . . . . 11. *P. amplifolius* (in part)
     13. Floating leaves 11–29-veined; submersed leaf margins entire, rarely crisped; stem lacking spots.

15. Fruits (1.6)2–2.5 mm, lacking well-developed keel (fig. 115c); submersed leaf margin entire; only of e. Nfld., St. P. & Miq., and Sable I., N.S. . . . . 12. *P. polygonifolius* (in part)
15. Fruits 2.5–4.3 mm, with well-developed keel; submersed leaf margins minutely denticulate to entire; widespread in North America.
16. Submersed leaf petioles (7)11–15 cm long, leaves with 2–5 rows of lacunae on either side of mid-vein, margins entire, leaf tips acute to rounded; fruits 2.7–4.3 mm long. . . . . 13. *P. nodosus* (in part)
16. Submersed leaf petioles 0.5–4 cm long, leaves with 1–2 rows of lacunae on either side of mid-vein, margins minutely denticulate, leaf tips acute to awl-shaped (fig. 118a); fruits 2.5–3.6 mm long.. . . . . 14. *P. illinoensis* (in part)
12. Submersed leaves sessile.
17. Stems branched (fig. 117a); submersed, green to brownish, leaf margin denticulate, tip acute to obtuse, stipules acute; fruits laterally compressed, sessile 1.9–2.3 mm long. . . . . 15. *P. gramineus* (in part)
17. Stems unbranched; submersed leaves usually reddish, leaf margin entire, tip blunt, stipules blunt; fruits plump, pedicellate, (2.5)3–3.5 mm long. . . . . 16. *P. alpinus* (in part)
6. Plants lacking floating leaves.
18. Leaf margins distinctly serrate to the unaided eye (fig. 113e). . . . . 17. *P. crispus*
18. Leaf margins entire (or if finely denticulate, visible only under magnification).
19. Leaves linear, ribbon-like, thread-like, flattened or phyllodial.
20. Plants strongly rhizomatous (figs. 105a, 117a).
21. Leaves delicate, thread-like, 0.1–0.5 mm wide; peduncles (3)5–25 cm long (typically 15–25 cm long, fig. 105a).. . . . . 18. *P. confervoides*
21. Leaves flattened, ribbon-like (fig. 104b,e) or appearing as bladeless petioles (fig. 112a), 2–11 mm wide; peduncles 1.5–8 cm long.
22. Submersed leaves flat, linear 1–13-veined with a broad lacuna band on either side of midvein (fig. 104b,e); fruits with distinct pointed keels.
23. Submersed leaves 0.2–1(2) mm wide, 1–3(7)-veined, stipules free or sometimes with few stipules fused to leaf base. . . . . 5. *P. tennesseensis* (in part)
23. Submersed leaves 1–10 mm wide, 3–13-veined, with no stipules fused to leaf blade.. . . . . 7. *P. epihydrus*
22. Submersed leaves phyllodial (appearing as a leaf petiole) to slightly linear, 1–5-veined, lacking lacuna band; fruit, if produced, lacking keels.
24. Submersed leaves 0.7–2.5 mm wide; veins 3–5. . . . . 8. *P. natans* (in part).
24. Submersed leaves 0.3–1 mm wide; veins 3. . . . . 9. *P. oakesianus* (in part)
20. Plants lacking rhizomes or not apparent.
25. Leaves 1(3)-veined.
26. Leaf width 0.1–1 mm.
27. Nodal glands usually absent; stipules 4–12 mm long; peduncles 6–8 m. . . . . 6. *P. vaseyi* (in part)
27. Nodal glands present, stipules 5–20 mm long; peduncles 10–35 mm. . . . . .19. *P. gemmiparus*
26. Leaf width (0.6)1.2–5.0 mm.
25. Leaves 3–15-veined.
28. Nodal glands absent (rarely present in *P. hillii*).
29. Leaf blades with 3–5 veins.
30. Leaves acute at tip, 3(5)-veined; peduncles club-shaped, 0.3–1.1(3.7) cm long, axillary (fig. 106d); fruits 0.4–2.7 mm long, beak 0.2–0.6 mm, 1-keeled (fig. 106g). . . . . .20. *P. foliosus*
30. Leaves, bristle-tipped (fig. 106b), apiculate (rarely blunt-tipped), 3-veined; peduncles slightly club-shaped, 0.6–1.4 cm, long, axillary and terminal; fruits 2.3–4 mm, beak 0.3–0.7 mm, 3–7 keeled. . . . . 21. *P. hillii*
29. Leaf blades with 3 bold veins and up to ± 30 additional faint veins. . . . . 23. *P. zosteriformis* (in part)
28. Nodal glands present (fig. 107f, 109c).
31. Stipules fibrous (fig. 107b) often whitish or brown.
32. Winter buds with leaves and stipules 90° to each other (fig. 107i); leaf apex rounded or apiculate (fig. 107g), blades lacking a bold marginal vein. . . . . 22. *P. friesii*
32. Winter buds with leaves and stipules oriented in same direction; leaf apex acute, bristled or rarely rounded, blades lacking a bold marginal vein.

33. Leaves with 3 bold veins and ± 30 additional faint veins; stem broad 0.6–3.2 mm wide, extremely flattened and stiff; fruit 4–5 mm long. . . . . 23. *P. zosteriformis* (in part)

33. Leaves with 3–5(7)-veins; stems; narrow, less than 1 mm wide, not flattened; flexuous; fruits 1.9–2.1 mm long. . . . . 24. *P. strictifolius*

31. Stipules not fibrous, green, brown, or white.

34. Fruits 2.5–3.6 mm long; winter buds 3.5–7.8 cm × 2.3–5.1 mm; leaves often reddish. . . . 25. *P. obtusifolius*

34. Fruits 1.9–2.8 mm long; winter buds 0.9–3.2 cm × 0.3–1.8 mm; leaves not reddish.

35. Stipules mostly connate (fused surrounding stem); peduncles usually terminal, 1–3 per plant; inflorescence usually of 2–4 distinct, interrupted whorls of flowers/fruits. . . . . 26. *P. pusillus*

35. Stipules convolute (wrapped around stem); peduncles terminal and axillary, more than 3 per plant; inflorescence with flowers/fruits in a crowded spike.. 27. *P. berchtoldii*

19. Leaves broadly linear-oblong, lanceolate, elliptic, or subcircular.

36. Submersed leaves petiolate.

37. Submersed leaf margins crisped (fig. 114c); stems rusty to black-spotted.

38. Stems black-spotted; submersed leaves 9–21-veined, 10–25(35) mm wide, slightly arcuate; fruits 3.1–4.1(4.7) mm with distinct dorsal keel. . . . . 10. *P. pulcher* (in part)

38. Stems often rusty-spotted; submersed leaves 27–49-veined, 25–75 mm wide, distinctly arcuate; fruits 3.9–5.2(–5.7) mm with indistinct keel. . . . . 11. *P. amplifolius* (in part)

37. Submersed leaf margins rarely crisped; stem lacking spots.

39. Fruits (1.6)2–2.5 mm, lacking well-developed keel; submersed leaf margin entire; only of e. Nfld., St. P. & Miq. and Sable I., N.S. . . . . 12. *P. polygonifolius* (in part)

39. Fruits 2.5–4.3 mm, with well-developed keel; submersed leaf margins minutely denticulate to entire; plants widespread in North America.

40. Submersed leaf petioles (7)11–15 cm long, blades with 2–5 rows of lacunae on either side of midvein; leaf tips acute to rounded; fruits 2.7–4.3 mm long. . . . . 13. *P. nodosus* (in part)

40. Submersed leaf petioles 0.5–4 cm long, blades with 1–2 rows of lacunae on either side of on either side of midvein; leaf tips acute to awl-shaped (fig. 118a); fruits 2.5–3.6 mm long. . . . . . 14. *P. illinoensis* (in part)

36. Submersed leaves sessile or sessile and clasping stem.

41. Leaf blades sessile, not clasping.

42. Stems branched; submersed leaf margins mostly denticulate, leaves green-brownish, tip often apiculate; stipules acute.

43. Leaves (1)3–9-veined with 1–2 rows of lacunae on either side of midvein, rarely phyllodial; fruits 1.9–2.3 mm with beak 0.3–0.5 mm long. . . . . 15. *P. gramineus* (in part)

43. Leaves 7–19-veined with 2–5 rows of lacunae on either side of midvein; fruits 2.5–3.6 mm with beak 0.5–0.8 mm long. . . . . 14. *P. illinoensis* (in part)

42. Stems unbranched; submersed leaf margin entire, leaves often reddish, tip not apiculate; stipules blunt. . . . . 16. *P. alpinus* (in part)

41. Leaf blades clasping stem (fig. 118d, 119a,c, 120b).

44. Leaf tip with boat-shaped (curved upward), tip (fig. 118e) splitting when pressed; margins entire; stem "zig-zag" from node to node; fruits 4–5.7 mm long with beak 0.6–1 mm; rhizomes spotted rusty-red. . . . . 28. *P. praelongus*

44. Leaf tip not boat-shaped and splitting when pressed, margins mostly denticulate; stem not "zig-zag"; fruits 1.6–4.2 mm with beak 0.4–0.7 mm, rhizome unspotted.

45. Stipules delicate and deteriorating, soon disappearing; leaves broadly lanceolate, orbiculate, or ovate with 7–33 strong veins; fruits 1.6–3 mm long. . . . . 29. *P. perfoliatus*

45. Stipules firm, disintegrating to coarse, white persistent fibers (fig. 119a); leaves ovate-lanceolate to narrowly lanceolate with 7–33 strong veins; fruits 2.2–4.2 mm long. . . . . 30. *P. richardsonii*

---

1. *P. robbinsii* Oakes Robbins' Pondweed Fig. 102
Ponds, lakes, and slow streams, often deep water. Lab. and Que. w. to Alask. and B.C. s. to N.J., Md., Ind., Ala. Utah, and Calif. This species is typically found vegetative, but is distinctive by its stiff 2-ranked habit; it rarely forms fruit.

2. *P. spirillus* Tuck. Spiral Pondweed Fig. 102
Acid to moderately alkaline waters. Nfld. w. to Man. and S.D., s. to Del., Md., e. Ohio, Iowa and e. Neb.

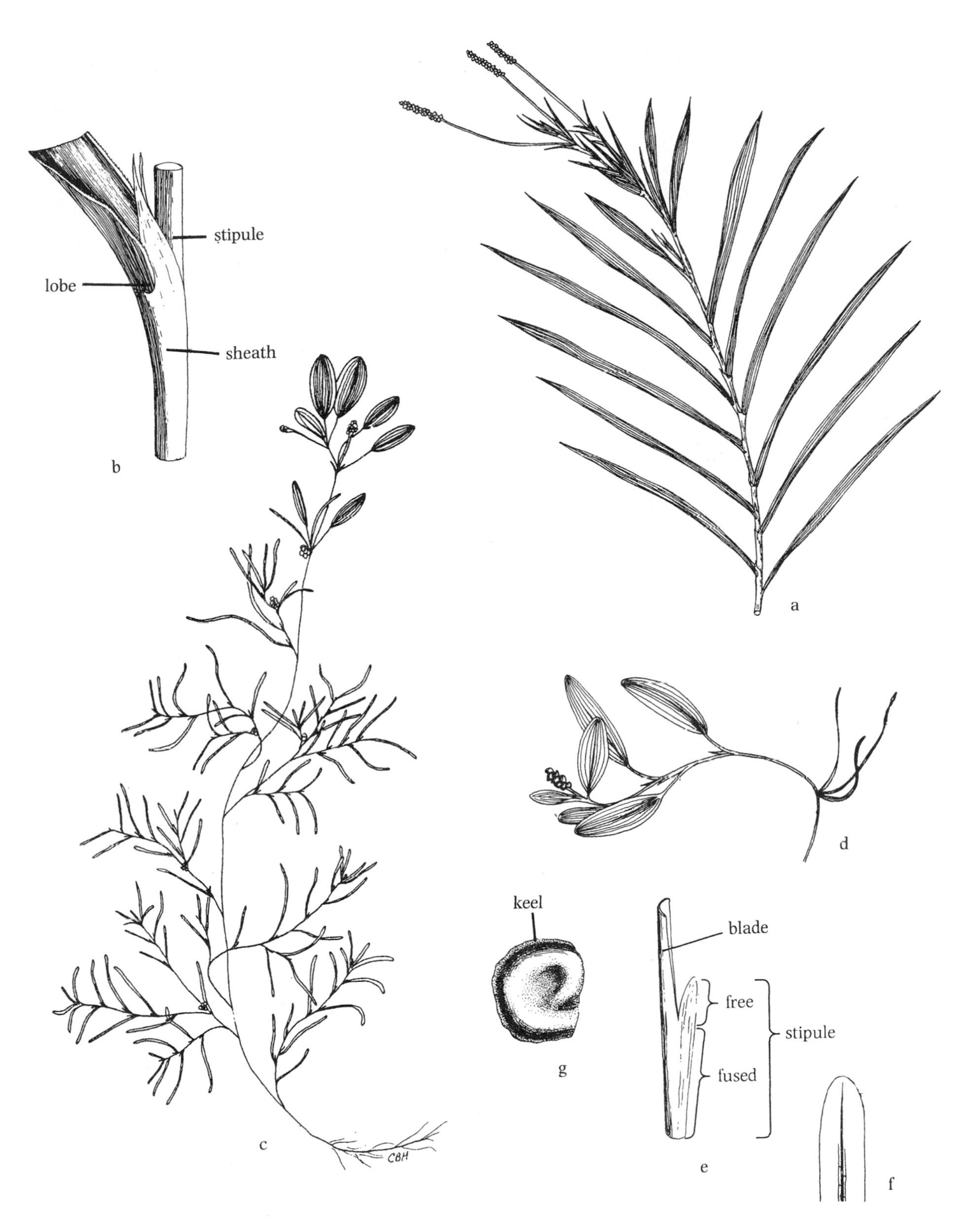

Fig. 102. *Potamogeton robbinsii*: a. upper portion of plant (PB); b. section of stem with leaf base (F). *Potamogeton spirillus*: c. habit (NHAES); d. upper portion of plant (F); e. leaf base with stipule (F); f. leaf apex (F); g. fruit (F).

3. *P. bicupulatus* Fernald Small Pondweed Fig. 103
Acid waters. Me. w. to scattered locations in N.B., Ont., Mich., Wisc., and e. Minn., s. to N.J., N.Y., Pa. and n. Ind.; Va. and Tenn. Hybrids with *P. epihydrus* = *P. ×aemulans* Z. Kaplan, Hellq. & Fehrer. (*P. diversifolius* var. *trichophyllus* Morong)

4. *P. diversifolius* Raf. Water-thread Pondweed Fig. 103
Moderately alkaline waters. S. N.Y. and Pa. w. to n. Ohio, Ill., Minn., Mont., sw. B.C. and Oreg., s. to Fla., Tex., Calif. and Mex.; W.I.

5. *P. tennesseensis* Fernald Tennessee Pondweed
Fig. 104
Streams and rivers. W. Pa. and e. Ohio s. to W.Va., e. Ky., Tenn. and Miss. This taxon is confined mostly to the Cumberland Plateau and can be confused with diminutive plants of *P. epihydrus*.

6. *P. vaseyi* Robbins Vasey's Pondweed Fig. 111
Acid to alkaline waters. N.B., s. Que. and Ont. w. to Wisc. and Minn. s. to Conn., Pa., ne. Ohio, n. Ind. and n. Ill. This species is difficult to locate in late summer because plants with floating leaves typically deteriorate by mid-August. *Potamogeton lateralis* Morong is no longer recognized; it was based on a type specimen which consisted of a mixed collection of *P. vaseyi* and *P. pusillus* (Hellquist et al., 1988).

7. *P. epihydrus* Raf. Leafy Pondweed Fig. 104
Alkaline and acid waters. S. Lab. and Nfld. w. to Minn., Man. and c. B.C., s. to Ga., Ala. and Miss.; Wyo., Ida., B.C. and Wash., s. to Colo., Nev. and n. Calif. Hybrids with *P. bicupulatus* = *P. ×aemulans* Z. Kaplan, Hellq. & Fehrer; with *P. gramineus* = *P. ×concilius* Ar. Benn; with *P. nodosus* = *P. ×subsessillis* Hagstr.; with *P. perfoliatus* = *P. ×versicolor* Z. Kaplan, Hellq., & Fehrer.

8. *P. natans* L. Floating Brownleaf, Floating-leaved Pondweed Fig. 112
Acid to alkaline waters. Greenl. and Nfld. w. to Alask., s. to N.J., Md., n. Ind., Kans., N.M., Ariz. and s. Calif. A hybrid with *P. oakesianus* has been confirmed (Kaplan, pers. comm.)

9. *P. oakesianus* Robbins Oakes' Pondweed Fig. 112
Acid waters, ponds, lakes, bogs and quiet waters. Nfld., w. to Ont., Mich., Wisc. and Minn., s. to N.J., Del., Va. and ne. S.C; Mont. and s. B.C. Hybrids with *P. gramineus* = *P. ×mirabilis* Z. Kaplan, Hellq. & Fehrer with *P. pulcher* = *P. ×floridanus* Small; and with *P. natans* has been confirmed by Kaplan (pers. comm.).

10. *P. pulcher* Tuck. Spotted Pondweed Fig. 114
Acid waters. Chiefly coastal plain, N.S., s. Me. and c. NH s. to Fla., w. to La. and e. Tex.; Mississippi embayment and n. to scattered locations in se. Mo., Ind., Ohio, s. Ont., Mich., Wisc., and e. Minn. Hybrids with *P. amplifolius* = *P. ×rhodensis* Ar. Benn.; with *P. oakesianus* = *P. ×floridanus* Small.

11. *P. amplifolius* Tuck. Large-leaved Pondweed, Bass Weed, Muskie Weed Fig. 114
Acid to alkaline waters, often at great depths. Nfld., N.S. and Gaspé Pen., Que. w. to c. B.C. s. to Va., Ga., Ala., Ark., Okla., Mont., Ida. and Calif. Hybrids with *P. illinoensis* = *P. ×scoliophyllus* Hagstr.; with *P. pulcher* = *P. ×rhodensis* Ar. Benn.

12. *P. polygonifolius* L. Bog Pondweed Fig. 115
Acid waters. Se. Nfld., St. P. & Miq. and Sable I., N.S. DNA analysis by Z. Kaplan (pers. comm.) has shown these native North American populations we previously referred to as *P. oblongus* to be conspecific with the European *P. polygonifolius*. (*P. oblongus* Viviani)

13. *P. nodosus* Poir. Longleaf Pondweed Fig. 115
Fast flowing streams of low alkalinity, or ponds and slow-moving streams of high alkalinity. Que. w. to B.C. s. to Fla., Ala., Tex., Ariz., Calif. and Mex.; W.I., C.Am. and S.Am. Hybrids with *P. gramineus* = *P. ×lanceolatifolius* (Tiselius) C. D. Preston.; with *P. epihydrus* = *P. ×subsessilis* Hagstr.; with *P. illinoensis* = *P.×faxonii*. (*P. americanus* Cham & Schltdl.)

14. *P. illinoensis* Morong Illinois Pondweed Fig. 118
Alkaline waters, Que, e. N.B., Vt. and w. Mass. w. to Man., sw. N.W.T. and B.C., s. to Fla., Tex., s. Calif. and Mex; W.I., C.Am. Plants south of our range typically have longer petioles. Floating leaves are sparingly formed. Hybrids with *P. gramineus* = *P.×deminutus* Hagstr., with *P. nodosus* = *P.×faxonii* Morong; with *P. perfoliatus* = *P.×subdentatus* Hagstr.

15. *P. gramineus* L. Variable Pondweed Fig. 117
Acid to alkaline waters. Greenl., Lab. and Nfld. w. to Alask., s. to N.J., Md., Pa., n. Ohio, n. Ind., n. Iowa, Kans., N.M., Ariz. and Calif. Submersed leaves have been occasionally observed as narrow as 1 mm in width, and even in rare situations phyllodial. Hybrids with *P. epihydrus* = *P. ×concilius* Ar. Benn.; with *P. illinoensis* = *P. ×deminutus* Hagstr.; with *P. natans* = *P. ×sparganiifolius* Laest. ex Fr.; with *P. nodosus* = *P. ×lanceolatifolius* (Tiselius) C. D. Preston; with *P. perfoliatus* = *P. ×nitens* Weber; with *P. richardsonii* = *P. ×hagstroemii* Ar. Benn.; with *P. oakesianus* = *P. ×mirabilis* Z. Kaplan, Hellq. & Fehrer. (*P. gramineus* var. *maximus* Morong ex Ar. Benn; *P. gramineus* var. *myriophyllus* Robbins)

16. *P. alpinus* Balb. Alpine Pondweed Fig. 116
Moderately to strongly alkaline waters. Nfld. w. to Alask., s. to Me., NH, Vt., e. Pa., N.Y., Mich. and n. Minn., s. in mts. to w. S.D., n. Colo., n. Utah and c. Calif. Hybrids with *P. nodosus* = *P. ×subobtusus* Hagstr.; with *P. perfoliatus* = *P. ×prussicus* Hagstr. (*P. alpinus* var. *subellipticus* (Fernald) Ogden; *P. alpinus* var. *tenuifolius* (Raf.) Ogden)

17. *P. crispus* L. Curly-leaved Pondweed Fig. 113
Alkaline and polluted waters. Sw. Que., se. Me., and s. N.H. w. to s. Ont., Mich., Minn., Sask., Alta. and s. B.C., s. to Va., Ga., Ala., La., Tex., Ariz. and Calif. Hybrids with *P. perfoliatus* = *P. ×cooperi* (Fryer) Fryer; *P. praelongus* = *P. ×undulatus* Wolfg.

18. *P. confervoides* Reichenb. Tuckerman's Pondweed Fig. 105
Acid waters, often in bog ponds. Nfld. and Lab. w. to sw. Ont., n. Mich. and n. Wisc., s. to ne. Pa., s. N.J., s. N.C. and n. S.C. (*P. tuckermanii* Robbins)

19. *P. gemmiparus* Robbins Small Pondweed Fig. 111
Acid waters. S. Que., Me, N.H., Mass., R.I. and Conn. (*P. pusillus* L. subsp. *gemmiparus* (Robbins) R. R. Haynes & Hellq.; *P. pusillus* var. *gemmiparus* Robbins; *P. berchtoldii* Fieber subsp. *gemmiparus* (Robbins) Les & Tippery).

20. *P. foliosus* Raf. Leafy Pondweed Fig. 106
Alkaline waters. Nfld., P.E.I. and N.S. w. to c. Alask., s. to w. Fla., La., Tex., Ariz., Calif. and Mex.; W.I. and C.Am. This species rarely produces rhizomes. Our taxon is subsp. *foliosus*; subsp. *fibrillosus* (Fernald) R. R. Haynes & Hellq. is a western taxon. Hybrids with *P. pusillus* = *P. ×turionifer* Hagstr. (*P. foliosus* var. *macellus* Fernald)

Fig. 103. *Potamogeton diversifolius*: a. habit (PB); b. leaf apex (F); c. leaf base with stipule (F); d. fruit (PB). *Potamogeton bicupulatus*: e. habit (PB); f. upper floating leaves (PB); g. leaf apex (F); h. fruit (F); i. fruit (PB).

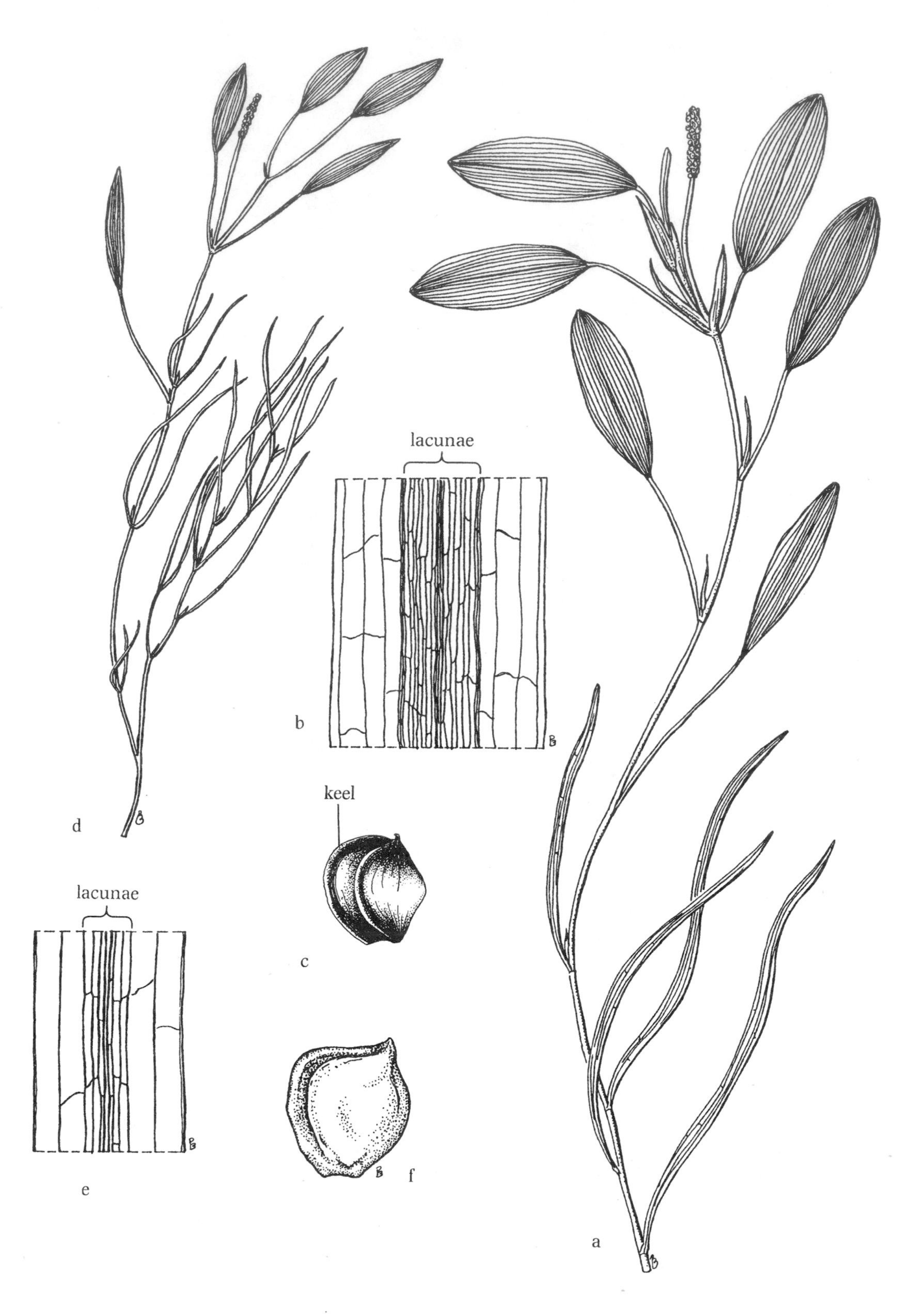

Fig. 104. *Potamogeton epihydrus*: a. upper portion of plant (PB); b. section of leaf showing broad lacunae band (PB); c. fruit (F).
Potamogeton tennesseensis: d. upper portion of plant; e. section of leaf showing lacunae band; fruit (PB).

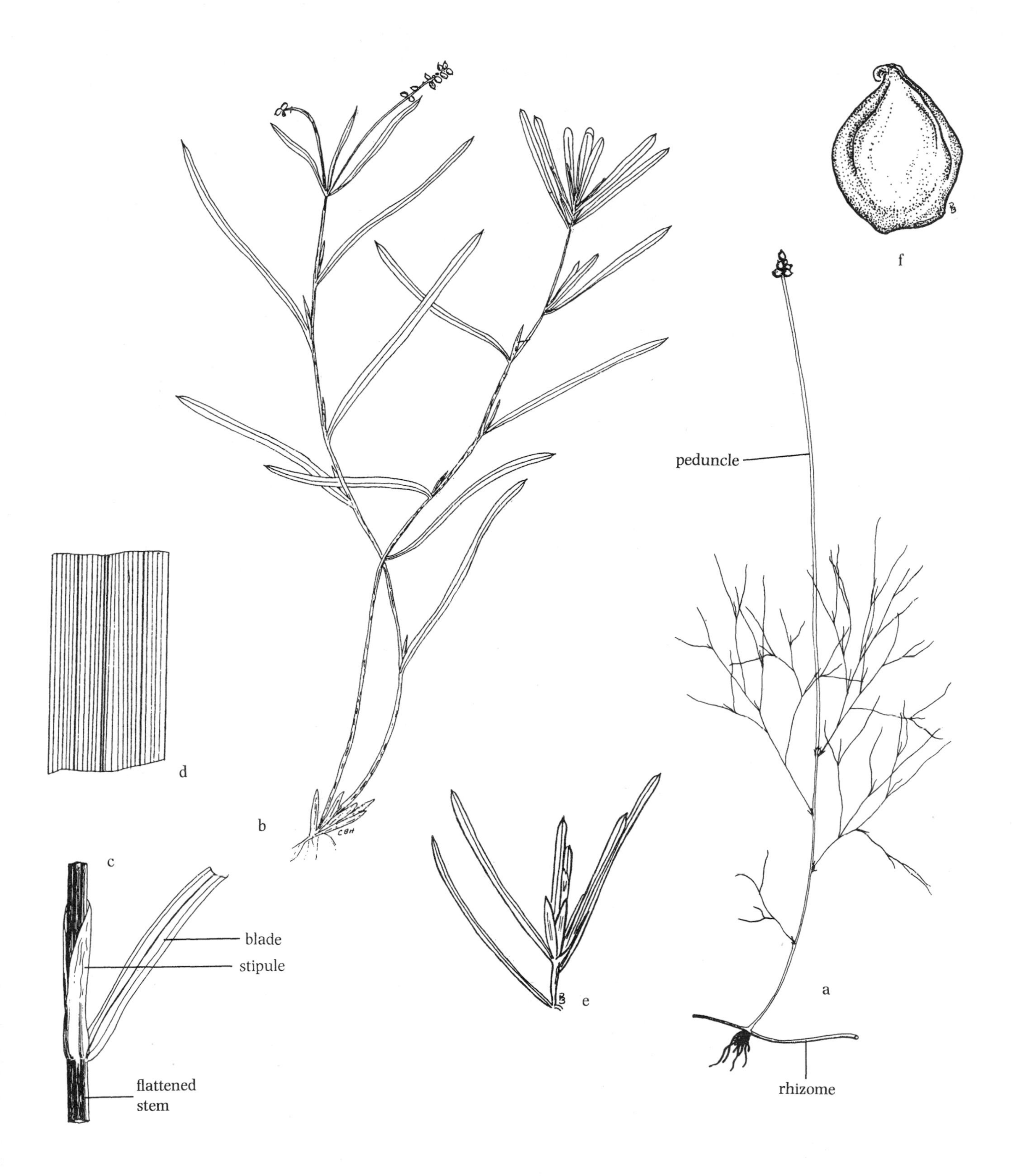

Fig. 105. *Potamogeton confervoides*: a. habit (F).
Potamogeton zosteriformis: b. habit (NHAES); c. section of stem with leaf base (F); d. section of leaf (NHAES); e. winter bud (PB); f. fruit (PB).

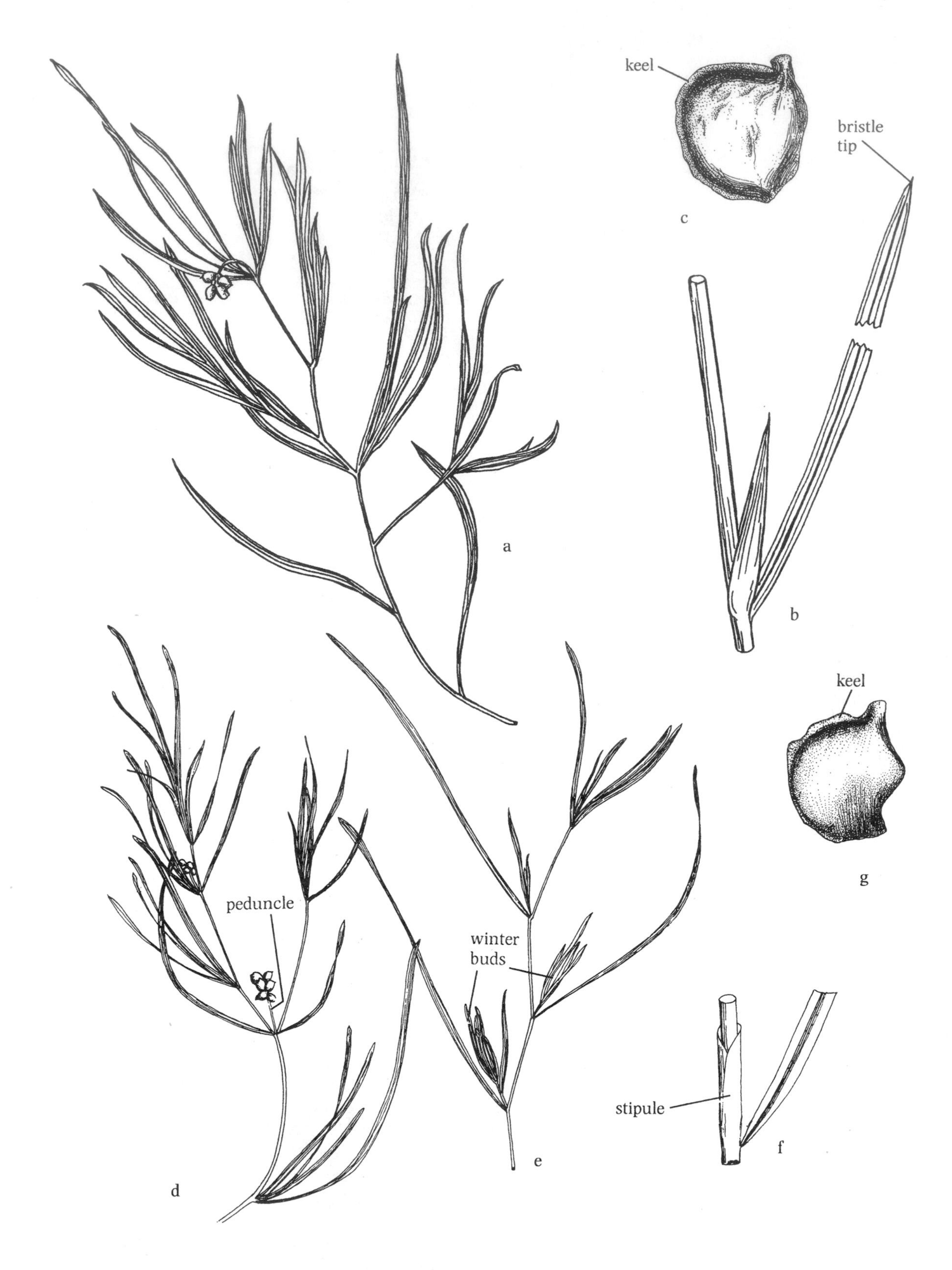

Fig. 106. *Potamogeton hillii*: a. upper portion of plant (PB); b. section of stem with leaf base and stipule (F); c. fruit (F). *Potamogeton foliosus*: d. upper portion of plant; e. habit, showing winter buds; f. section of stem with leaf base; g. fruit (F).

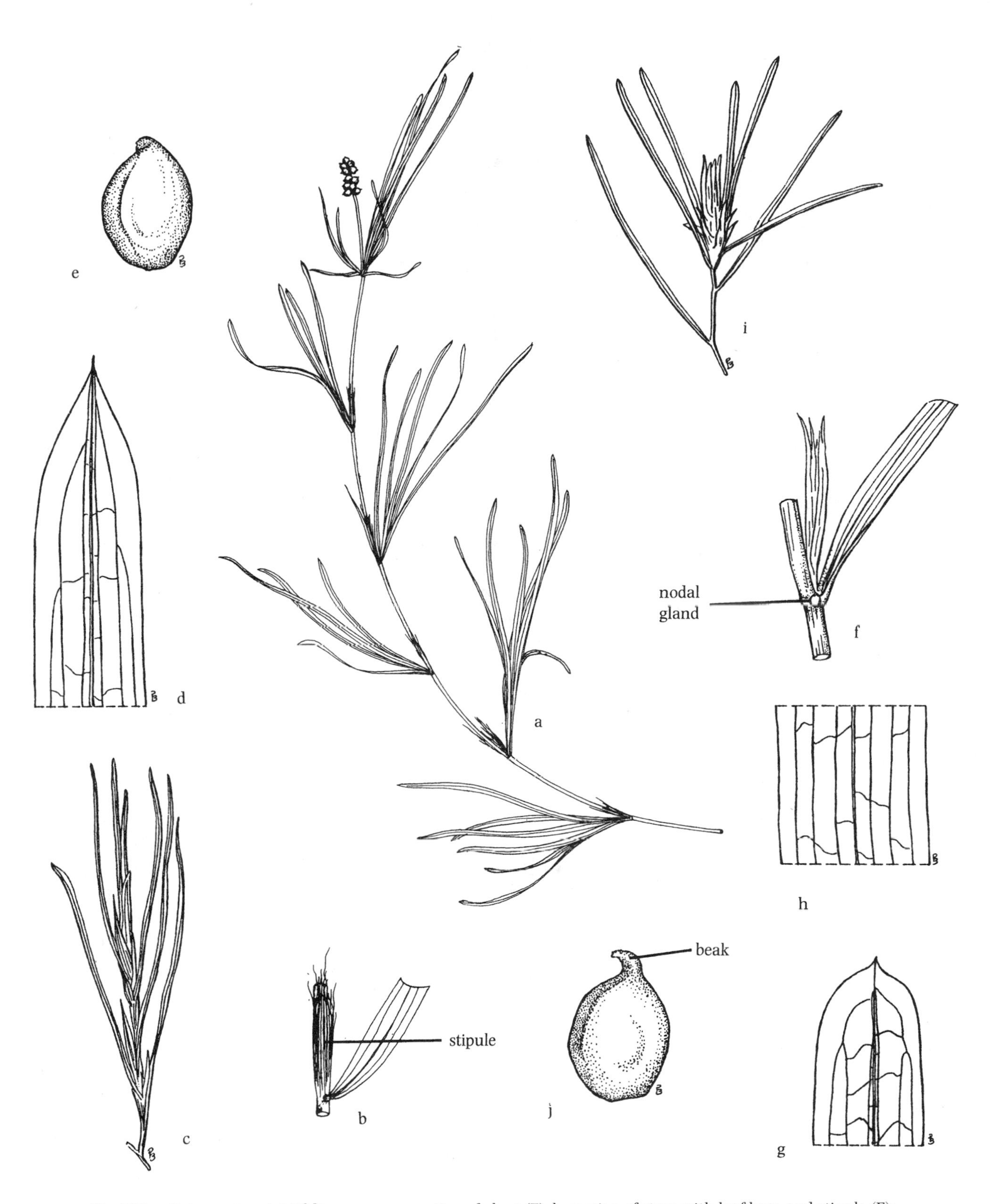

Fig. 107. *Potamogeton strictifolius*: a. upper portion of plant (F); b. section of stem with leaf base and stipule (F); c. winter bud (PB); d. leaf apex (PB); e. fruit (PB).
Potamogeton friesii: f. section of stem with leaf base and fibrous stipule; g. leaf apex; h. midsection of leaf; i. winter bud; j. fruit (PB).

Fig. 108. *Potamogeton ×haynesii*: a. upper portion of plant; b. leaf apex; c. winter bud; d. fruit (PB).
Potamogeton ×ogdenii: e. upper portion of plant; f. leaf apex; g. winter bud; h. fruit (PB).

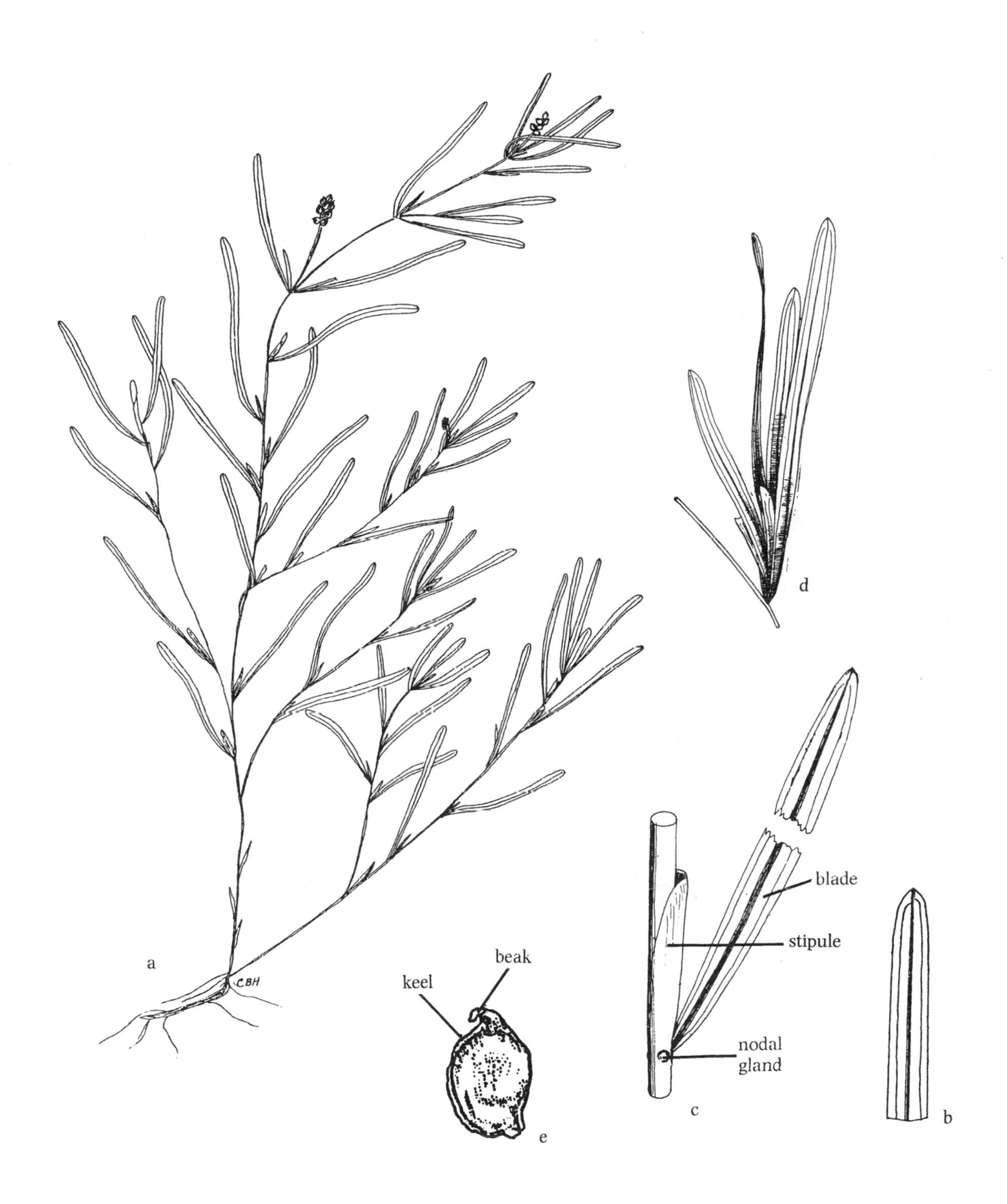

Fig. 109. *Potamogeton obtusifolius*: a. habit (NHAES); b. leaf apex (NHAES); c. section of stem with leaf base and stipule (F); d. winter bud (F); e. fruit (PB).

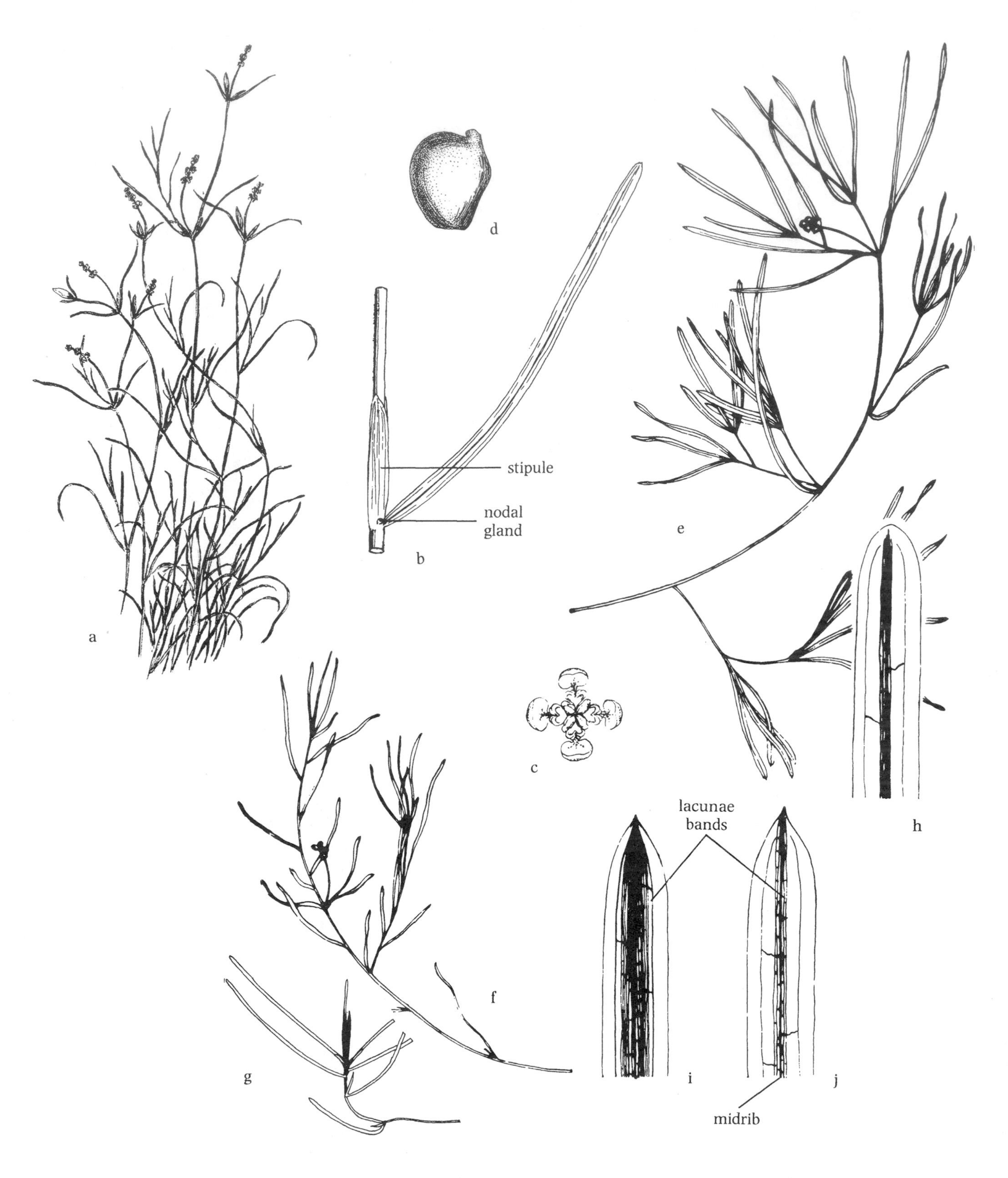

Fig. 110. *Potamogeton pusillus*: a. habit (Reed); b. section of stem with leaf and stipule (F); c. flower (Reed); d. fruit (F). *Potamogeton berchtoldii*: e, f. upper portion of plant; g. winter bud; h–j. leaf apices, showing lacunae band variations (F).

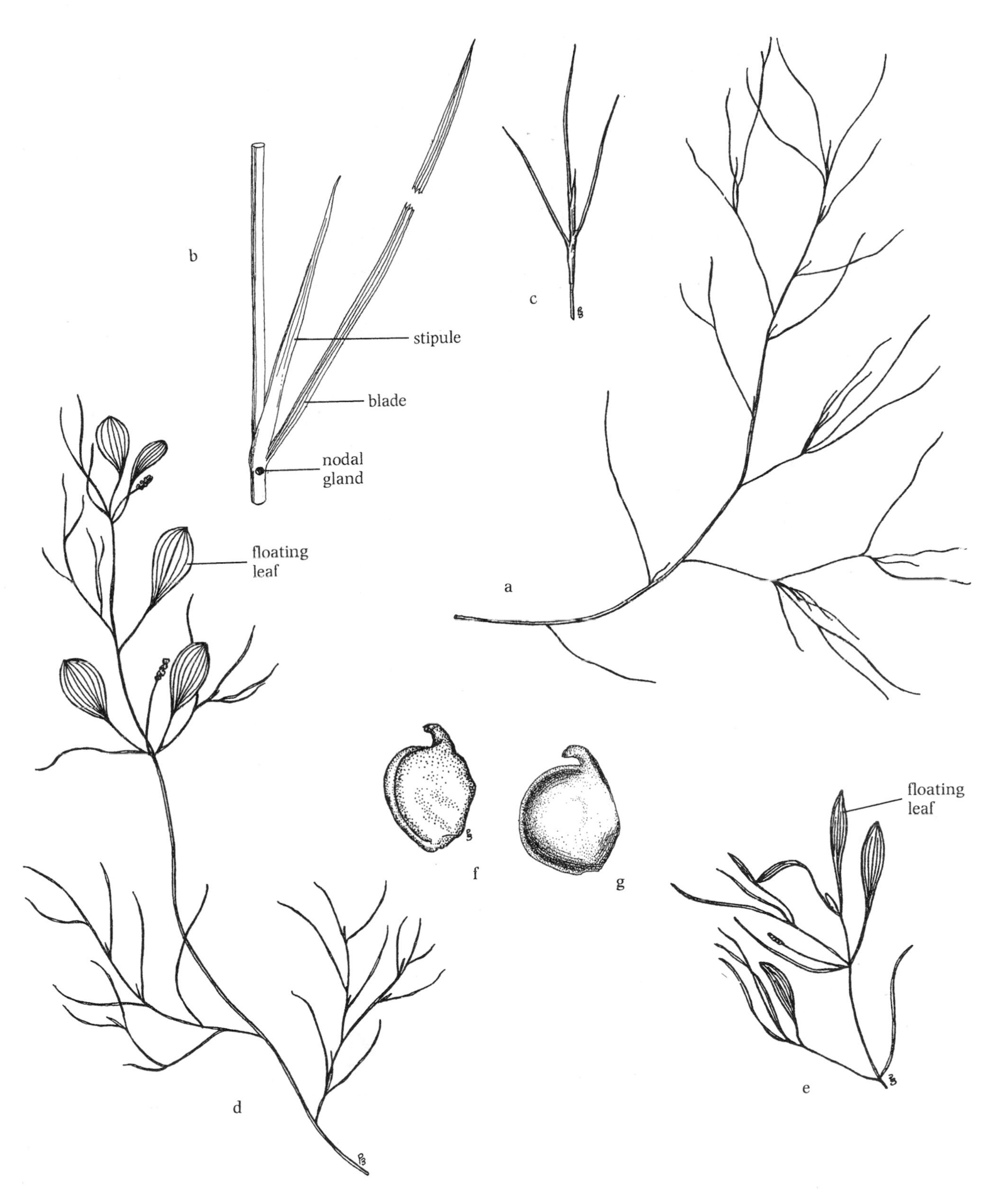

Fig. 111. *Potamogeton gemmiparus*: a. upper portion of plant (F); b. section of stem with leaf base and stipule (F); c. winter bud (PB).
Potamogeton vaseyi: d, e. upper portion of plant (PB); f. fruit (PB); g. fruit (F).

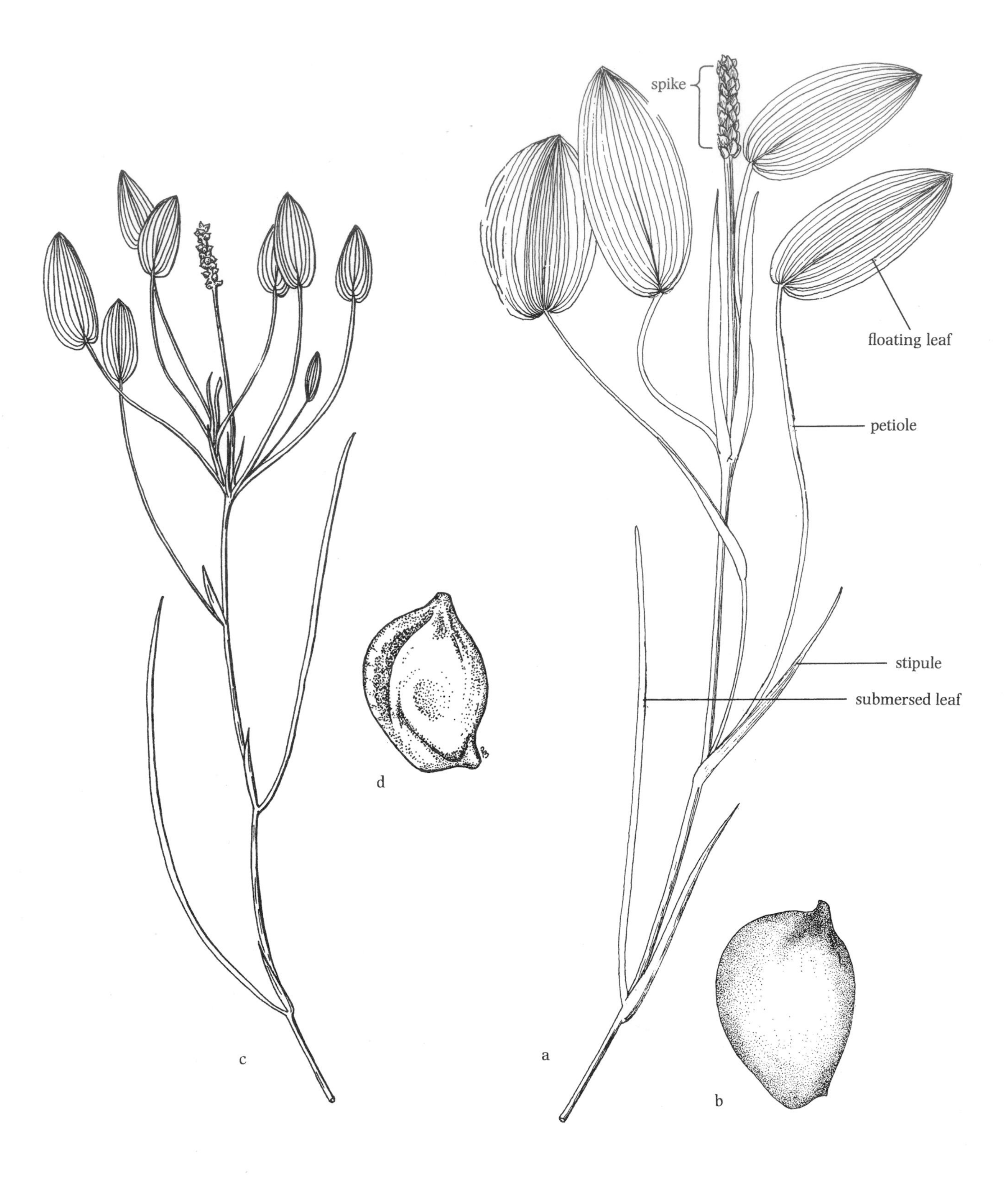

Fig. 112. *Potamogeton natans*: a. upper portion of plant; b. fruit (F).
Potamogeton oakesianus: c. upper portion of plant; d. fruit (PB).

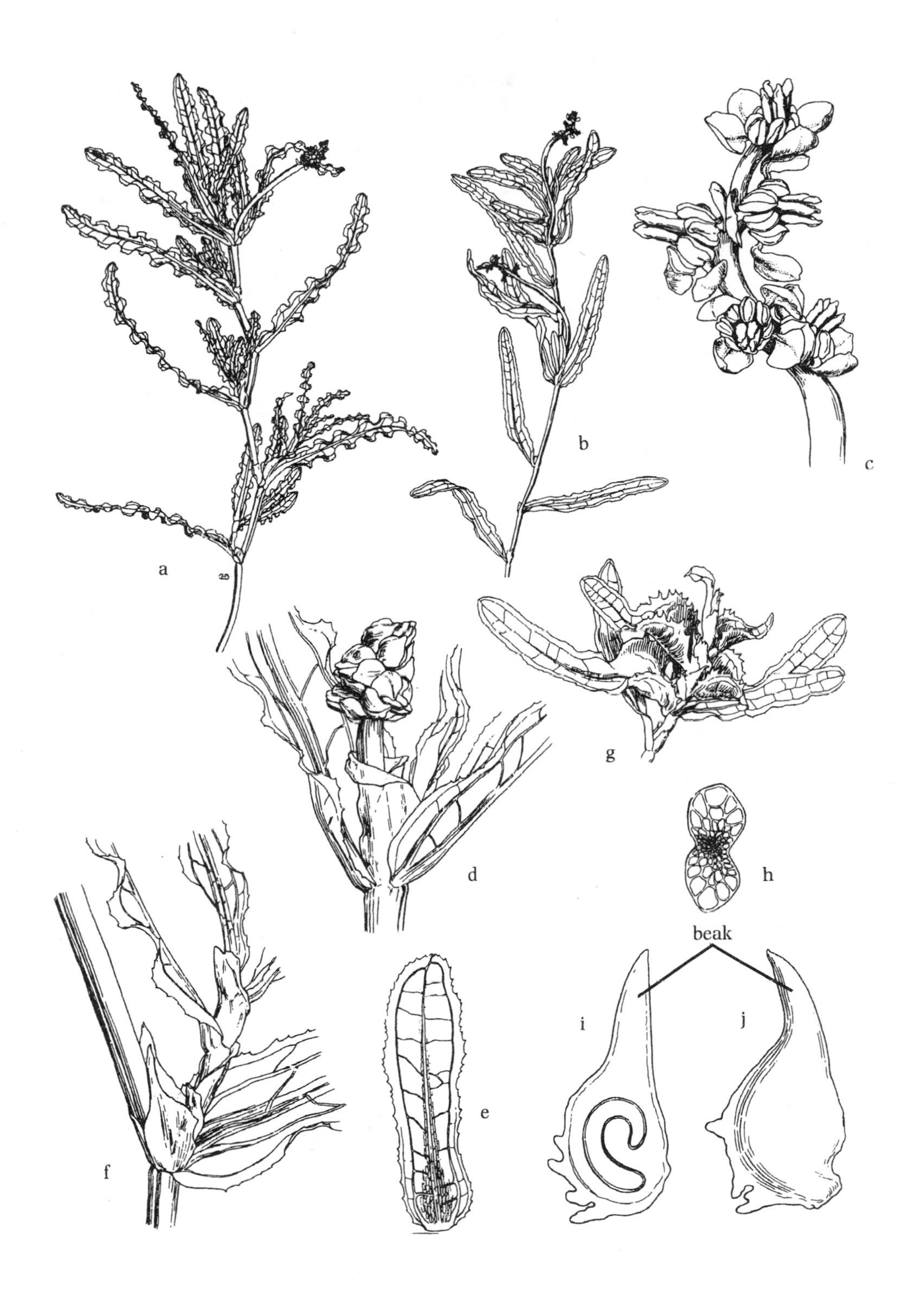

Fig. 113. *Potamogeton crispus*: a, b. upper portion of plant; c. young inflorescence; d. inflorescence; e. leaf; f. section of stem with leaf bases and stipules; g. winter bud; h. stem, cross-section; i, j. fruit (Mason).

Fig. 114. *Potamogeton amplifolius*: a. upper portion of plant (F); b. fruit (PB).
Potamogeton pulcher: c. habit (G&W); d. flower (G&W); e. fruit (PB).

Fig. 115. *Potamogeton polygonifolius*: a. upper portion of submersed plant with floating leaves; b. stranded plant; c. fruit (PB).
Potamogeton nodosus: d. upper portion of plant with floating leaves; e. fruit (PB).

Fig. 116. *Potamogeton alpinus*: a. upper portion of plant, acute-tipped form (F); b. leaf apex, acute form (PB); c. upper portion of plant, blunt-tipped form (F); d. leaf apex, blunt form (PB); e. fruit (PB).

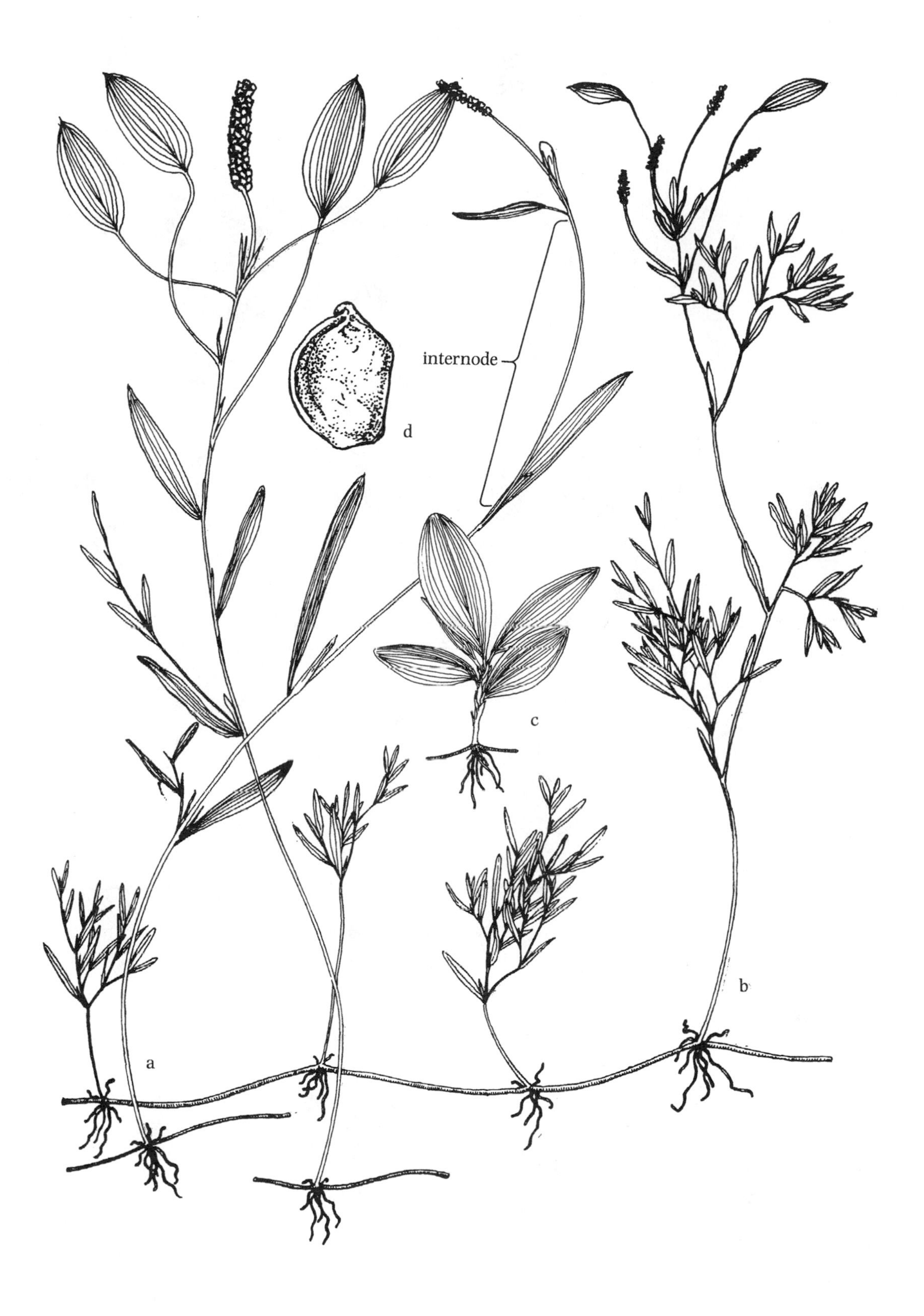

Fig. 117. *Potamogeton gramineus*: a, b. habit, submersed plant (F); c. habit, stranded plant (F); d. fruit (PB).

Fig. 118. *Potamogeton illinoensis*: a. upper portion of plant (PB); b. upper portion of plant with floating leaves (PB); c. fruit (F).
Potamogeton praelongus: d. upper portion of plant (PB); e. boat-shaped apex (F); f. fruit (PB).

21. *P. hillii* Morong Hill's Pondweed Fig. 106
Clear, cold calcareous streams and ditches. Vt., w. Mass. and nw. Conn. w. to s. Ont., n. Mich. and ne. Wisc., s. to N.Y., w. Pa. and e. Ohio; w. Va. The species is uncommon, but most abundant in w. N.E. and n. Mich. Hybrids with *P zosteriformis* are *P. ×ogdenii* Hellq. & Hilton (fig. 376), characterized with leaves basically with 3–9(13) veins and 1–2 rows of lacunae. *Potamogeton ×ogdenii* is known from w. Vt., w. Mass., w. Conn. and e. N.Y. Some hybrids previously believed to be *P. ×ogdenii* have been shown through DNA analysis to be hybrids of *P. berchtoldii × P. zosteriformis* and *P. foliosus × P. zosteriformis* (Z. Kaplan (pers. comm).

22. *P. friesii* Rupr. Fries' Pondweed Fig. 107
Alkaline waters. W. Nfld. w. to c. Alask., s. to Vt., w. MA., w. CT., Pa., w. Neb., n. Utah and Wash.

23. *P. zosteriformis* Fernald Flat-stemmed Pondweed, Eel-grass Pondweed Fig. 105
Alkaline to neutral waters. W. Nfld, and Gaspé Pen., Que. w. to Man., N.W.T. and Alask., s. to n. Va., s. Ohio, Kans., w. Utah, and n. Calif. Hybrids with *P. berchtoldii, P. foliosus, P. hillii*, and *P. strictifolius* have been reported.

24. *P. strictifolius* Ar. Benn. Narrowleaf Pondweed Fig. 107
Alkaline waters. Nw. Nfld., e. Que., n. Me. and Vt. w. to s. Ont., s. Man., s. Alta., Yuk. and s. B.C., s. to nw. Conn., nw. Va., n. Ohio and n. Ind., Wisc., n. Neb., Wyo. and n. Utah. *Potamogeton ×haynesii* Hellq. & G. E. Crow (fig. 108) is a hybrid species derived from *P. strictifolius × P. zosteriformis*; it is characterized by leaves with (5)9–14(18) veins and rows of lacunae typically absent (rarely 2–4). The range is Vt. w. to Ont., Mich., and Minn. (*P. longiligulatus* Fernald)

25. *P. obtusifolius* Mert. & Koch Bluntleaf Pondweed Fig. 109
Moderately alkaline waters. Lab. and w. Nfld. w. to N.W.T., s. Yuk. and c. Alask., s. to Conn., N.J., ne. Pa., N.Y., n. Mich., n. Wisc., nw. Minn., Man., w. Wyo., nw. Mont. and nw. Wash.

26. *P. pusillus* L. Small Pondweed Fig. 110
Alkaline waters. Lab. and w. Nfld. w. to N.W.T., Yuk., and B.C., s. to Fla. Tex., Ariz., Calif., and Mex.; C.Am. Hybrids with *P. foliosus* = *P. ×turionifer* Hagstr.

27. *P. berchtoldii* Fieber Berchtold's Pondweed, Small Pondweed Fig. 110
Mostly acid waters. Nfld. w. to Alask., s. to nw. Fla., La., Wyo., Ariz., n. Nev. and n. Calif.; most abundant in the Northeast. Many specimens of what were previously identified as *P. ×ogdenii* have been shown to be the hybrid of *P. berchtoldii ×P. zosteriformis* through DNA analysis (Kaplan, pers. comm.) Hybrids with *P. perfoliatus* = *P. ×mysticus* Morong. (*P. pusillus* subsp. *tenuissimus* (Mert. & W. D. J. Koch) Haynes & Hellq.; *P. pusillus* var. *tenuissimus* Mert. & W. D. J. Koch.)

28. *P. praelongus* Wulf. White-stemmed Pondweed, Muskie Weed Fig. 118
Deep, moderately alkaline waters. Lab. and Nfld. w. to Alask., s. to N.J., n. Pa., n. Ohio, n. Ind., n. Iowa, Colo., Utah, n. Calif. and Mex. Hybrids with *P. crispus* = *P. ×undulatus* Wolfg.

29. *P. perfoliatus* L. Clasping-leaved Pondweed Figs. 119, 120
Acid, alkaline and saline waters. Greenl. and Nfld. w. to e. Ont., s. to ne. N.C., n. Va., ne. Ohio and Mich.; s. Ala. and s. La.; C.Am. Hybrids with *P. berchtoldii* = *P. ×mysticus* Morong; with *P. crispus* = *P. ×cooperi* (Fryer) Fryer; with *P. epihydrus* = *P. ×versicolor* Z. Kaplan, Hellq. & Fehrer; with *P. gramineus* = *P. ×nitens* Weber; with *P. illinoensis* = *P. ×subdentatus* Hagstr.; with *P. richardsonii* = *P.× absconditus* Z. Kaplan, Hellq. & Fehrer. (*P. perfoliatus* var. *bupleuroides* (Fernald) Farwell)

30. *P. richardsonii* (Ar. Benn.) Rydb. Richardson's Pondweed Fig. 119
Alkaline waters. N.S. and e. Que. w. to N.W.T. and Alask., s. to w. Conn., n. Pa., Ind., n. Iowa, nw. Neb., n. Colo., Utah and n. Calif.; Guat. In very shallow water floating leaves may sometimes develop. Hybrids with *P. gramineus* = *P.×hagstroemii* Ar. Benn.; with *P. perfoliatus* = *P. ×absconditus* Z. Kaplan, Hellq. & Fehrer.

Nartheciaceae / Asphodel Family

1. Inflorescence a glabrous raceme, perianth glabrous; ovary superior; filaments wooly-pubescent; seeds fusiform, 2-tailed (bristle-like). . . . . 1. *Narthecium*
1. Inflorescence a densely white-wooly dichotomous cyme; perianth with crest of long yellow hairs on inner surface; ovary half-inferior; filaments glabrous; seeds ellipsoid, rounded at both ends. . . . . 2. *Lophiola*

1. *Narthecium* (Bog-asphodel)

Perennial herbs, arising from stout rhizomes; leaves basal, erect, narrowly linear; inflorescence a raceme, bracteate; perianth yellow; ovary 3-locular; fruit a capsule; seeds fusiform, 2-tailed.

Reference: Zomlefer, 1997.

1. *N. americanum* Ker Gawl. Fig. 122
Bogs in pine barrens and savannas. Rare, N.J., Del., N.C. and S.C.

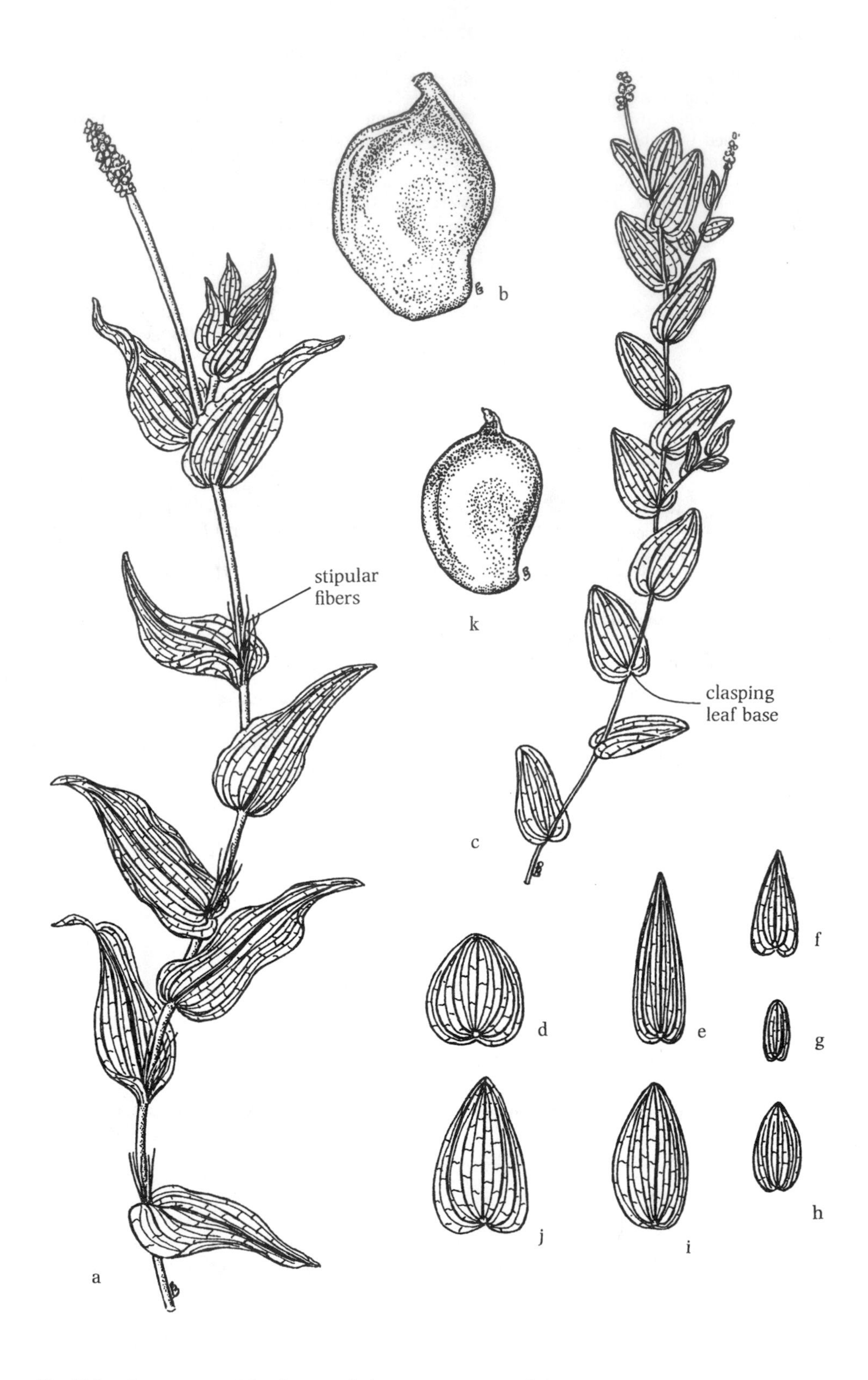

Fig. 119. *Potamogeton richardsonii*: a. habit, upper portion of plant; b. fruit (PB).
Potamogeton perfoliatus: c. upper portion of plant; d–j. leaf variations; k. fruit (PB).

Fig. 120. *Potamogeton perfoliatus*: a. habit; b. branch with inflorescence; c. flower; d. anthers and sepal; e. embryo; f. fruit (G&W).

2. *Lophiola* (Golden-crest)

Perennial herbs, arising from slender to thickened rhizomes; leaves alternate, 2-ranked, linear, basal and cauline; inflorescence a solitary, terminal, dichotomous cyme; perianth yellow, crest of long hairs on inner (adaxial) surface; fruit an ovoid, beaked capsule; seeds ellipsoidal, yellow-white.

REFERENCES: Robertson, 2002; Zomlefer, 1997.

1. *L. aurea* Ker Gawl. Golden-crest Fig. 122
Bogs, savannas, peaty shores and ditches. Coastal plain, w. N.S.; N.J., n. Del.; N.C. s. to n. Fla., w. to Miss. Although *Lophiola* was previously placed with some hesitation in the Haemadoraceae in the FNA treatment (Robertson, 2002), its affinities are better aligned with the Nartheciaceae (Zomlefer, 1997). (*L. americana* (Pursh) Alph. Wood; *L. septentrionalis* Fernald)

Burmanniaceae / Burmannia Family

1. *Burmannia* (Burmannia)

Perennial or annual herbs, small; roots thread-like; leaves alternate, small, bract-like or scale-like; flowers bisexual, solitary, or borne in capitate, racemose, or cymose inflorescences; perianth 6-merous, borne at summit of a floral tube; ovary inferior; fruit a 3-valved capsule, seeds minute.

REFERENCE: Lewis, 2002.

1. *B. biflora* L. Northern Blue-thread; Violet Burmannia Fig. 122
Pond margins, swamps, savannas, southern bogs, pine barrens, ditches and wet woodlands. Coastal plain, Se. Va. (rare, not collected since 1940s) and N.C. s. to Fla., w. to Ala., La. and Tex.: S.Am.

Melanthiaceae / Honey Bush Family

1. *Veratrum* (False Hellebore)

Perennial herbs, large, with coarse fibrous roots; leaves 3-ranked, plicate, broadly oval; inflorescence a large panicle; perianth yellowish-green; ovary 3-lobed; fruit an ovoid capsule.

1. *V. viride* Aiton False Hellebore, Indian Poke Fig. 121
Swamps and wet woods. Que. w. to Minn., s. to N.E., Md., W.Va., mts. to N.C., Ky., Tenn. and Ga. and ne. Ohio.

2. *Helonias* (Swamp Pink)

Perennial herbs, arising from short rhizomes; leaves elongate, basal; scape hollow, with reduced bract-like leaves; inflorescence a spike-like raceme, bracts absent; perianth pink; fruit a 3-lobed loculicidal capsule; seeds 2-tailed.

1. *H. bullata* L. Fig. 121
Swamps and bogs. Coastal plain, se. N.Y. and N.J. s. to e. Va.; mts. Pa. s. to nw. Ga.

Orchidaceae / Orchid Family

The orchids are the largest family of flowering plants, numbering about 26,000 species, and the most advanced family of the Monocots. Because of the beauty of many of the species, members of the Orchid Family are also vulnerable to collection; thus many states have conservation laws protecting their native species of orchids. The family is easily recognized by the following characteristics: flowers bilaterally symmetrical; sepals 3; petals 3, the lateral 2 similar, the third modified into a distinct labellum, typically in a lower position (resupinate) because of 180-degree twisting of the ovary/pedicel; stamens adnate to style, forming column; anthers 1 (2 in *Cypripedium*), with pollen usually densely aggregated into a waxy mass, the pollinium; anther and stigmatic lobes separated by the rostellum and its viscidium (sticky pad detaching with pollinium in pollen transfer); ovary inferior; fruit a capsule, with extremely numerous, tiny, dust-like seeds.

REFERENCES: Brackley, 1985; Case, 1987; Correll, 1950; Dressler, 1981; Fuller, 1933; Henry et al., 1975; Luer, 1975; Morris and Eames, 1929; Romero-González et al., 2002; Sheviak, 1974; Whiting and Catling, 1986.

Fig. 121. *Veratrum viride*: a. habit; b. portion of inflorescence; c. flower (Gleason).
Helonias bullata: d. habit; e. inflorescence; f. flower (J. C. Putnam Hancock).

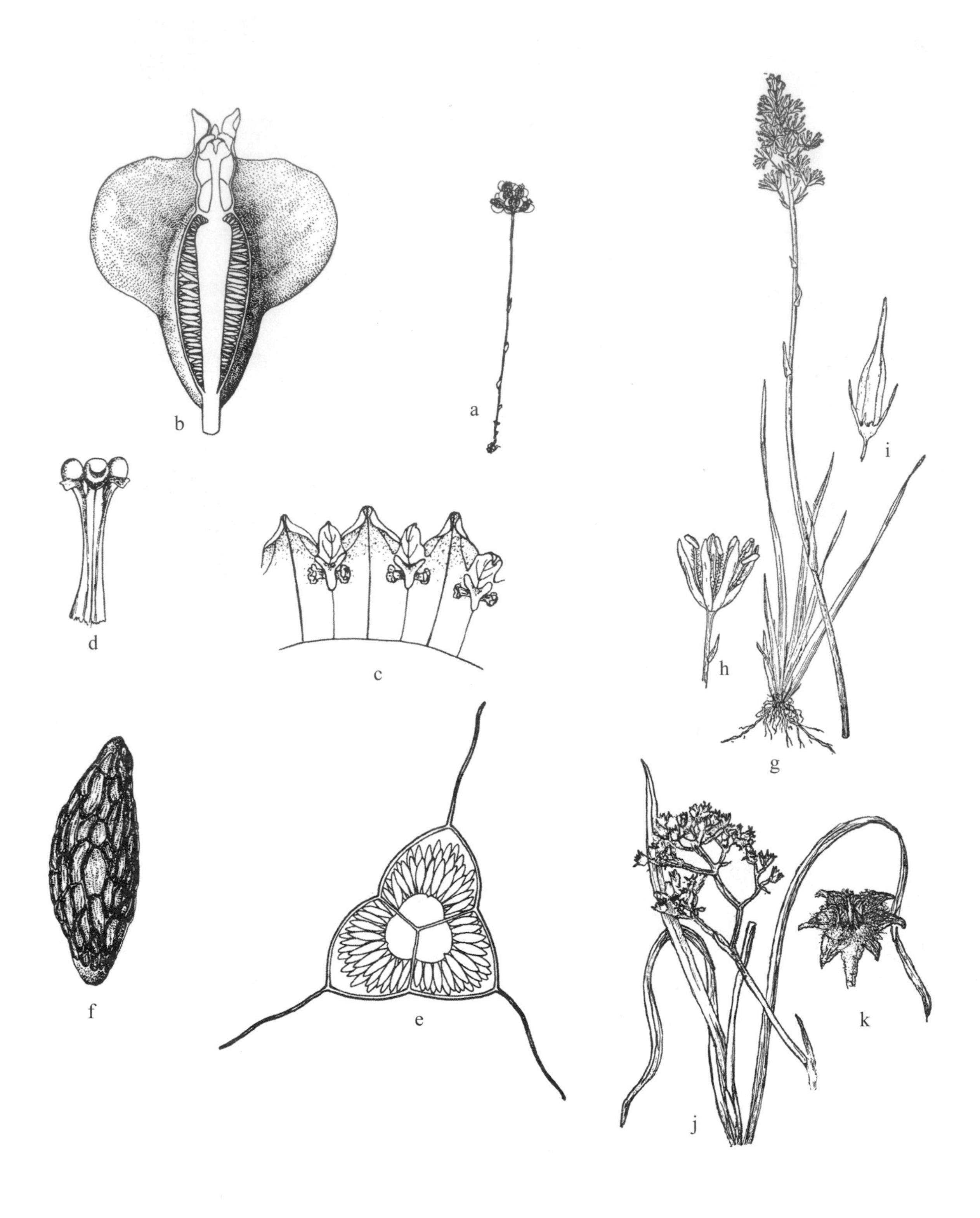

Fig. 122. *Burmannia biflora:* a. habit (C&C); b. flower, with ovary longitudinally sectioned, showing 2 wings (G&W); c. flower, spread open (C&C); d. style and stigmas (C&C); e. ovary, cross-section, showing 3 thin wings (G&W); f. seed (G&W).
Narthecium americanum: g. habit; h. flower; i. capsule (Gleason).
Lophiola americana: j. upper portion of plant; k. flower (Gleason).

1. Labellum of flower pouch-like (fig. 123); leaves plicate; fertile anthers 2 (on lateral sides of column). . . . . 1. *Cypripedium*
1. Labellum of flower various, but not pouch-like; leaves flat, or if ribbed, not plicate; fertile anthers 1(at apex of column).
  2. Leaves whorled at stem tip (fig. 124a); sepals extending far beyond petals (fig. 124a). . . . . 2. *Isotria*
  2. Leaves basal, or basal and cauline, never whorled; sepals about the same length as petals.
    3. Flowers solitary, pink (rarely albino); leaves solitary (rarely 2), or undeveloped at time of flowering.
      4. Leaves linear, basal; labellum 3-lobed, lateral lobes short, middle lobe long, arching downward (fig. 124c). . . . . 3. *Arethusa*
      4. Leaves ovate, borne about midway on stem; labellum unlobed, not arching. . . . . 4. *Pogonia*
    3. Flowers 2 or more, of various colors, inflorescence a raceme or spike; leaves 2 or more.
      5. Leaves cauline.
        6. Leaves several, alternate; flowers spurred (fig. 127c), of various colors. . . . . .5. *Platanthera*
        6. Leaves 2, opposite (fig. 133c); flowers not spurred, green. . . . . 6. *Neottia*
      5. Leaves basal or appearing basal, or absent at time of flowering.
        7. Labellum uppermost, column below it, flowers thus appearing inverted (fig. 136b).. . . . . 7. *Calopogon*
        7. Labellum lowermost, column above it.
          8. Leaves linear to linear-lanceolate; flowers white to ivory, spiraled in a spike (figs. 137a, 140a). . . . . 8. *Spiranthes*
          8. Leaves ovate to elliptic; flowers greenish to yellowish, in a raceme, not spiraled.
            9. Labellum widest at tip (fig. 141g); column elongate (2–3 mm long); leaves 2, stem not swollen at base. . . . . .9. *Liparis*
            9. Labellum tapering to tip (fig. 143b); column very short; leaves solitary or, if 2 or more, then stem conspicuously swollen at base.. . . . . 10. *Malaxis*

1. *Cypripedium* (Lady's-slipper)

Perennial herbs, with coarse, fibrous roots; leaves basal or cauline, plicate, sheathing at base; flowers solitary or few, large, showy; labellum inflated, pouch-like; fertile stamens 2; fruit a capsule.

Reference: Sheviak, 2002.

1. Leaves basal, 2; labellum cleft in front, pink or occasionally white. . . . . 1. *C. acaule*
1. Leaves cauline, 2 or more, labellum not cleft, yellow or white, amid conspicuously marked with purple or pink, if pink, then sepals and petals white.
  2. Labellum broadly rounded at distal end not forming a projection.
    3. Sepals and petals acute to acuminate, petals usually longer than labellum; labellum yellow or white.
      4. Labellum yellow, plants hirsute, the lower 3–6 leaves not crowded (internodes visible).. . . . . 2. *C. parviflorum*
      4. Labellum white; plants sparingly puberulent, the 3 or 4 leaves crowded (internodes concealed). 3. *C. candidum*
    3. Sepals and lateral petals obtuse; lateral petals usually shorter than labellum; labellum pink (or white marked with pink). . . . . .4. *C. reginae*
  2. Labellum prolonged downward in distal end, forming a conic projection. . . . . 5. *C. arietinum*

1. *C. acaule* Aiton Pink Lady's-slipper; Moccasin-flower Fig. 123
Acid soils of dry woodlands, in sphagnum bogs and wet, mossy coniferous woods. Nfld. w. to Alta. and N.W.T., s. to N.C., w. S.C., n. Ga., n. Ala., e. Tenn., n. Ind., n. Ill., Wisc. and Minn.

2. *C. parviflorum* Salisb. Small Yellow Lady's-slipper Fig. 123
Chiefly calcareous swamps, bogs, mossy woods and wet shores. Nfld. and N.S. w. to Ont., Mich., Wisc. and e. Minn., s. to N.J., Pa., mts. to N.C., S.C. and Tenn., Ohio, Ind. and n. Ill. Our taxon is the smaller-flowered var. *makasin* (Farw.) Sheviak, with labellum 15–29 mm, sepals and petals with dark reddish- or purplish-brown and strongly twisted, and lower surface of upper bract nearly glabrous. The typically larger-flowered var. *pubescens* (Willd.) O. W. Knight, with greenish-yellow sepals and petals, slightly twisted, and a larger lower labellum, 2.2–4.8 mm, tends to occur in dryish to moist sites. (*C. calceolus* var. *parviflorum* (Salisb.) Fernald)

3. *C. candidum* Muhl. ex Willd. Small White Lady's-slipper
Fens, swampy meadows, prairies and mossy glades, rare: c. N.Y. w. to s. Ont., s. Mich., Minn., Man. and Sask., s. to n. N.J., e. Pa., Md., Ky., Mo. and Neb.

4. *C. reginae* Walter Showy Lady's-slipper Fig. 123
Chiefly calcareous coniferous swamps, bogs, wet slopes and shores. Nfld. w. to Ont. and e. Sask., s. to n. N.J., mts. to Ga. and Tenn., Ohio, Ill. and Mo.

Fig. 123. *Cypripedium reginae*: a. habit (Henry, Buker, and Pearth, 1975).
Cypripedium acaule: b. habit (WVA).
Cypripedium arietinum: c. habit; d. flower (Crow).
Cypripedium parviflorum: e. habit (WVA).

5. *C. arietinum* R. Br. Ram's-head Lady's-slipper Fig. 123
Coniferous swamps and bogs, especially with White Cedar or Tamarack, and dry, rich, wooded hillsides. N.S. and Que. w. to Man. and Sask., s. to N.E., N.Y., Mich., Wisc. and n. Minn.

2. *Isotria* (Whorled Pogonia)

Perennial herbs, arising from short, erect rhizomes: stem bearing a whorl of leaves at summit; flowers 1, seldom 2; sepals linear, elongate, extending beyond labellum; fruit a capsule.

1. *I. verticillata* (Muhl. ex Willd.) Raf. Whorled Pogonia Fig. 124
Sphagnum bogs and acid woods. Very rare, s. Me. and c. N.H. w. to s. Ont. and Mich., s. to Ga., e. Ind., se. Mo. and e. Tex.

3. *Arethusa* (Arethusa)

Perennial herbs, arising from a bulb; stem bearing a single grass-like leaf; flowers solitary, erect or arching; sepals and petals similar; labellum 3-lobed, middle lobe long, arching; fruit a capsule.

1. *A. bulbosa* L. Arethusa, Dragon's Mouth, Swamp Pink Fig. 124
Sphagnum bogs, peaty meadows and open coniferous swamps. Nfld. w. to Ont. and Minn., becoming more rare southward, s. to N.E., N.J., Del. and Md., mts. to Va., N.C., S.C., n. Ohio, n. Ind. and n. Ill.

4. *Pogonia* (Pogonia)

Perennial herbs, arising from a short rhizome; stems usually bearing a single, ovate leaf; flowers 1 or 2; sepals and petals similar; labellum unlobed, bearded on upper surface; fruit a capsule.

1. *P. ophioglossoides* (L.) Ker Gawl. Rose Pogonia, Snakemouth Orchid Fig. 125
Sphagnum bogs, peaty swales and swamps. Nfld. and Que. w. to Ont. and Minn., s. to N.E., N.J., Va., W.Va., n. Ohio, n. Ind. and n. Ill.; chiefly along the coastal plain and piedmont to Fla., Miss., La. and e. Tex.; mts. of N.C., Ky., Tenn., Mo. and Ark.

5. *Platanthera* (Fringed Orchid)

Perennial herbs, with elongate, tuberous roots; leaves scattered along stem; flowers in racemes, subtended by bracts; labellum various, entire to strongly 3-parted, often toothed to fringed; spur present, conspicuous, often longer than labellum; fruit a capsule.

This temperate group of terrestrial orchids was long treated in the genus *Habenaria* (*sensu lato*) in older literature, but Luer (1975) noted that no true *Habenaria* occur north of the subtropical region of the United States.

References: Schrenk, 1978; Sheviak, 2002.

---

1. Labellum lacerate (fig. 127f) or fringed (fig. 128e).
  2. Labellum deeply 3-parted (fig. 125f).
    3. Flowers purple.
      4. Lobes of labellum clearly fringed.
        5. Flowers usually large, labellum 13–21 mm long, typically deeply fringed; corolla throat not constricted, appearing as 1 opening; anther cells divergent, viscidia oriented to either side of corolla throat . . . . . 1. *P. grandiflora*
        5. Flowers usually smaller, labellum 9–13 mm long, typically shallowly fringed; corolla throat constricted, appearing as 2 openings; anther cells close, viscidia oriented over corolla throat . . . . . 2. *P. psycodes*
      4. Lobes of labellum merely dentate to somewhat dentate-lacerate, but not fringed . . . . . 3. *P. peramoena*
    3. Flowers white to greenish-white or pale yellowish-green.
      6. Flowers white, tinged with cream; petals cuneate . . . . . 4. *P. leucophaea*
      6. Flowers greenish-white or pale yellowish-green: petals linear-oblong to oblong-spatulate . . . . . 5. *P. lacera*
  2. Labellum simple, not 3-parted (figs. 128e, 129c).

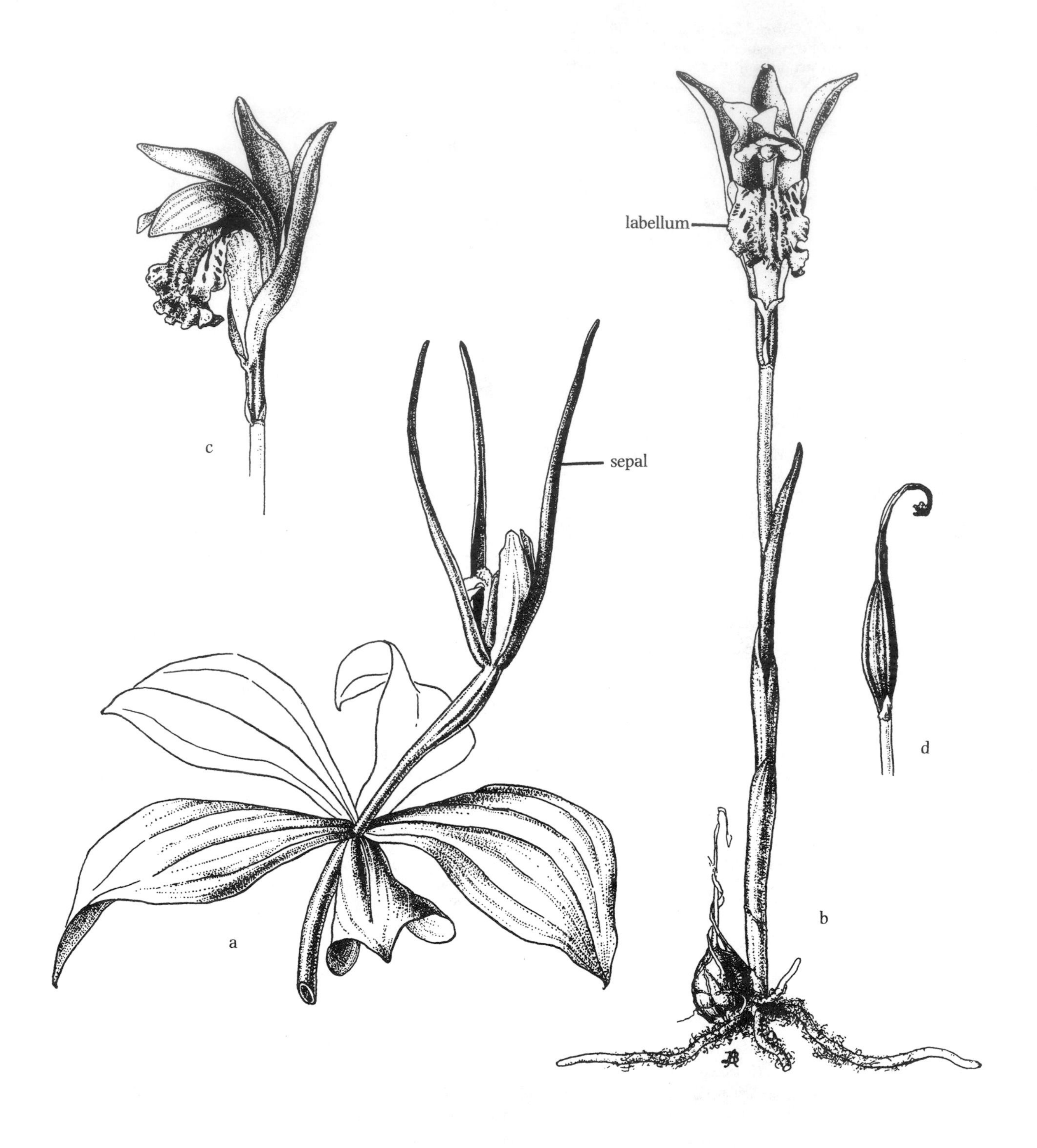

Fig. 124. *Isotria verticillata*: a. upper portion of plant (G&W).
Arethusa bulbosa: b. habit; c. flower; d. capsule (G&W).

Fig. 125. *Pogonia ophioglossoides*: a. habit; b. flower, side view; c. capsule (C&C).
Platanthera psycodes: d. inflorescence (G&W).
Platanthera leucophaea: e. upper portion of plant; f. flower (NYS Museum).

7. Flowers white. . . . . 6. *P. blephariglottis*
7. Flowers orange to orange-yellow.
8. Labellum deeply fringed, 4–8 or 8–19 mm long.
9. Spur longer than ovary and pedicel; body of labellum oblong, 8–19 mm long; spur 20–35 mm long. . . . . 7. *P. ciliaris*
9. Spur shorter than ovary and pedicel; body of labellum ovate, 4–8 mm long; spur 4–10 mm long. . . . . 8. *P. cristata*
8. Labellum erose to somewhat lacerate, not fringed, 3–5 mm. . . . . 9. *P. integra*
1. Labellum entire (fig. 132d) or with 3 apical teeth or shallow lobes (fig. 131e), but not lacerate.
10. Labellum with 3 apical teeth or lobes. . . . . 10. *P. clavellata*
10. Labellum entire.
11. Labellum uppermost, flower appearing inverted (fig. 131b,c); spur usually about 15 mm long (10–16 mm). . . . . 11. *P. nivea*
11. Labellum lowermost; spur 4–10 mm long.
12. Flowers pure white. . . . . 12. *P. dilatata*
12. Flowers whitish-green or yellowish to yellowish-green.
13. Flowers yellowish to yellowish-green, labellum 2.5–6 mm long, spur 2–5 mm long (usually shorter than lip); anther sacs appearing unseparated (less than 0.3 mm), loose pollen masses free from anther sacs. . . . . 13. *P. aquilonis*
13. Flowers whitish-green, labellum 5–12 mm long, spur 4–12 mm long; anther sacs separated by 0.4 mm or more, the pollen masses remaining enclosed in anther sacs. . . . . 14. *P. huronensis*

---

1. *P. grandiflora* (Bigelow) Lindl. Large Purple Fringed Orchid
Swamps, wet meadows, thickets and wet woods. Nfld. w. to s. Ont., Wisc., and e. Minn., s. to n. N.J., mts. to N.C., Tenn. and Ohio. (*Platanthera fimbriata* (Aiton) Lindl.)

2. *P. psycodes* (L.) Lindl. Small Purple Fringed Orchid Fig. 125
Wet meadows, stream margins, margins of bogs, and open wet woods. Nfld. w. to Ont. and e. Minn., s. to N.E., n. N.J., mts. to N.C., Ky., and Tenn., Ohio, n. Ill. and ne. Iowa; disjunct to Mo. and Ark. White color forms are occasionally encountered.

3. *P. peramoena* (A. Gray) A. Gray Purple Fringeless Orchid Fig. 126
Wet flat woods, alluvial forests, stream banks, seeps, marshes, moist prairies, old fields, pastures and ditches. N.J., Del. and Pa. w. to s. Ohio, s. Ind., s. Ill, and se. Mo., s. to Va., w. N.C., w. S.C., Tenn., n. Ga., Miss. and Ark.

4. *P. leucophaea* (Nutt.) Lindl. Prairie Orchid, Prairie White Fringed Orchid Fig. 125
Wet prairies, marly meadows, margins of lakes and fens. Very rare, w. N.Y. (extirpated?) and s. Ont. w. to Mich., s. Wisc., s. Minn., s. Man., and se. N.D., s. to Ohio, Ill., ne. Kans., se. Okla. (extirpated?) and n. La.; disjunct to single stations in n. Me., c. N.B. and Va.

5. *P. lacera* (Michx.) G. Don Ragged Fringed Orchid, Green Fringed Orchid Fig. 127
Marshes, sedge swamps, bogs, wet woods, wet to dry open fields and prairies and ditches. Nfld., N.S., and N.B. w. to Ont. and se. Man., s. to S.C., Ga., Miss., Ark., ne. Tenn. and e. Okla.

6. *P. blephariglottis* (Willd.) Lindl. White Fringed Orchid Fig. 128
Bogs, marshes and wet peaty soils. Nfld. w. to s. Que., s. Ont., and Mich., s. to N.J., Md., Pa. and n. Ohio; c. Ill. Our northeastern variety is var. *blephariglottis*. A taxon with very long spurs, about twice the length of the ovary, has been described as var. *conspicua* (Nash) Luer and occurs on the coastal plain and piedmont from s. N.J. to Fla., w. to Tex.

7. *P. ciliaris* (L.) Lindl. Yellow Fringed Orchid Fig. 128
Bogs, marshes, swamps, wet meadows, peaty or sandy woods and dryish swales. Rare and extirpated in some states, s. N.E. w. to s. Ont., s. Mich., and ne. Ill., s. to Fla., sc. Mo. and e. Tex.

8. *P. cristata* (Michx.) Lindl. Crested Fringed Orchid, Orange-crest Orchid Fig. 129
Meadows, sedge and sphagnum bogs, flatwoods, pine savannas, margins of streams and cypress swamps. Se. Mass., Conn. and se. N.Y. w. to c. Pa., Va., se. Ky., Tenn., and Ark., s. to Fla. and e. Tex.

9. *P. integra* (Nutt.) A. Gray ex L. C. Beck Fig. 130
Wet pine barrens, peaty depressions of pine savannas wet sandy woods. S. N.J., Md. and N.C. w. to Tenn., s. to Fla., Miss., La. and e. Tex.

10. *P. clavellata* (Michx.) Luer Green Woodland Orchid Fig. 131
Wet meadows, shores, stream banks, bogs, mossy woods, thickets and ditches. Nfld. w. to Ont. and Minn., s. to Ga. and La.

11. *P. nivea* (Nutt.) Luer Snowy Orchid Fig. 131
Wet peaty soils, meadows and savannas. Chiefly coastal plain, s. N.J. s. to Fla., w. to e. Tex.

12. *P. dilatata* (Pursh) Lindl. ex Beck White Bog Orchid, Bog Candles Fig. 133
Wet meadows, swales, bogs and mossy swamps. Greenl. and Lab. w. to Alask., s. to n. N.J., nw. Pa., n. Ill., Minn., mts. to N.M., Utah and Calif. Our plants belong to var. *dilatata*.

13. *P. aquilonis* Sheviak Northern Green Orchid Fig. 132
Wet meadows, bogs, swales and peaty thickets. Nfld. and Lab. w. to Que., Ont., N.W.T. and Alask., s. to N.E., N.J., Pa., n. Ind., n. Ill., Wisc., Iowa, Neb., Mont. and Wash. This species has been long confused with the far northern taxon, *P. hyperborea* (L.) Lindl. of Greenland and Iceland (Sheviak, 2002).

14. *P. huronensis* (Nutt.) Lindl. Lake Huron Green Orchid
Wet meadows, marshes, fens stream banks, shores, seeps, roadside ditches and tundra. Lab. and Nfld. w. to Que., Ont., Man., Yuk. and Alask., s. to N.E., N.J., N.Y., Mich., Wisc., S.D., mts. to N.M., Utah, Ida. and Oreg.

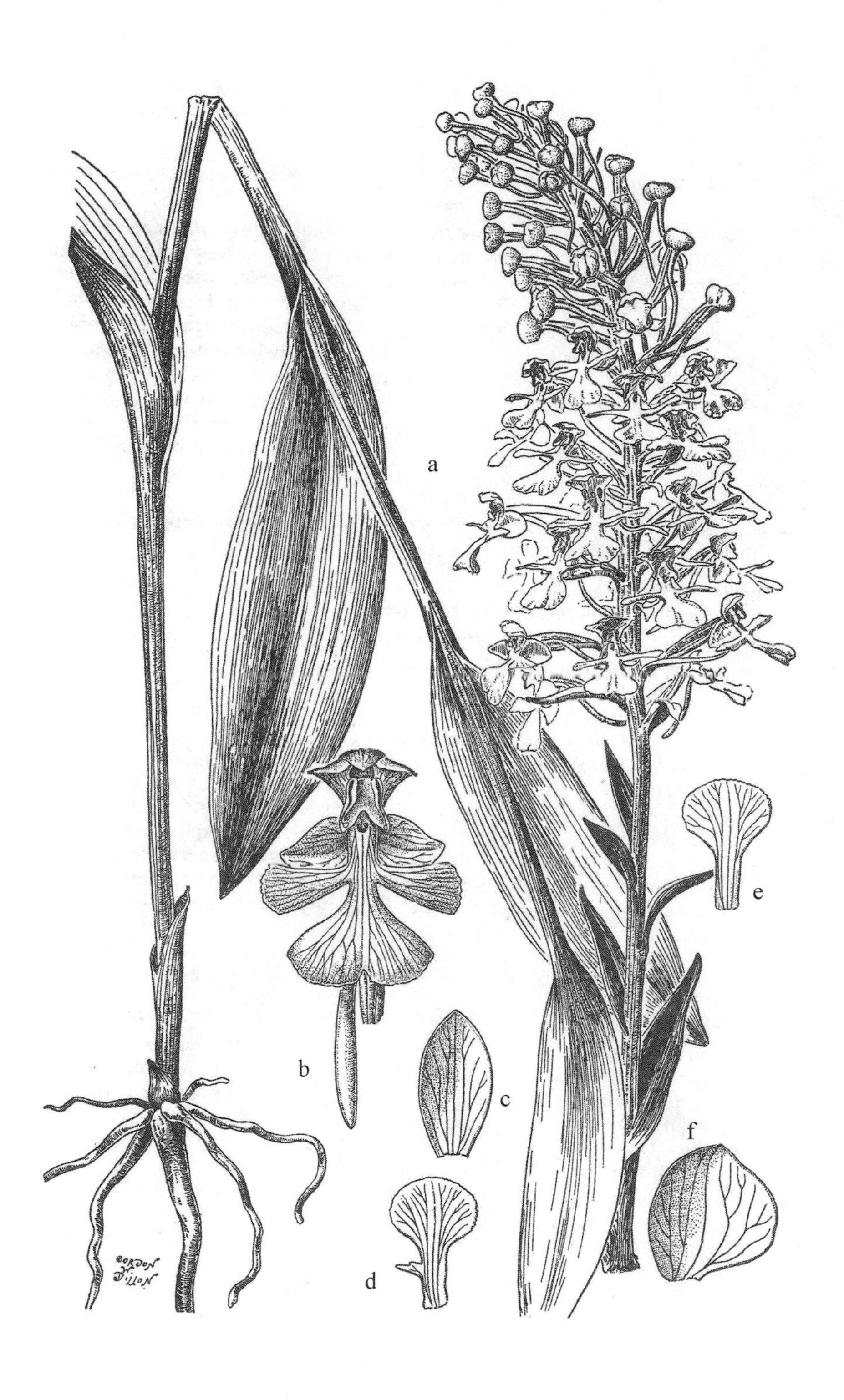

Fig. 126. *Platanthera peramoena*: a. habit; b. flower, front view partly spread open; c. dorsal sepal; d. petal with "spur" on the claw; e. petal "typical"; f. lateral sepal (G&W).

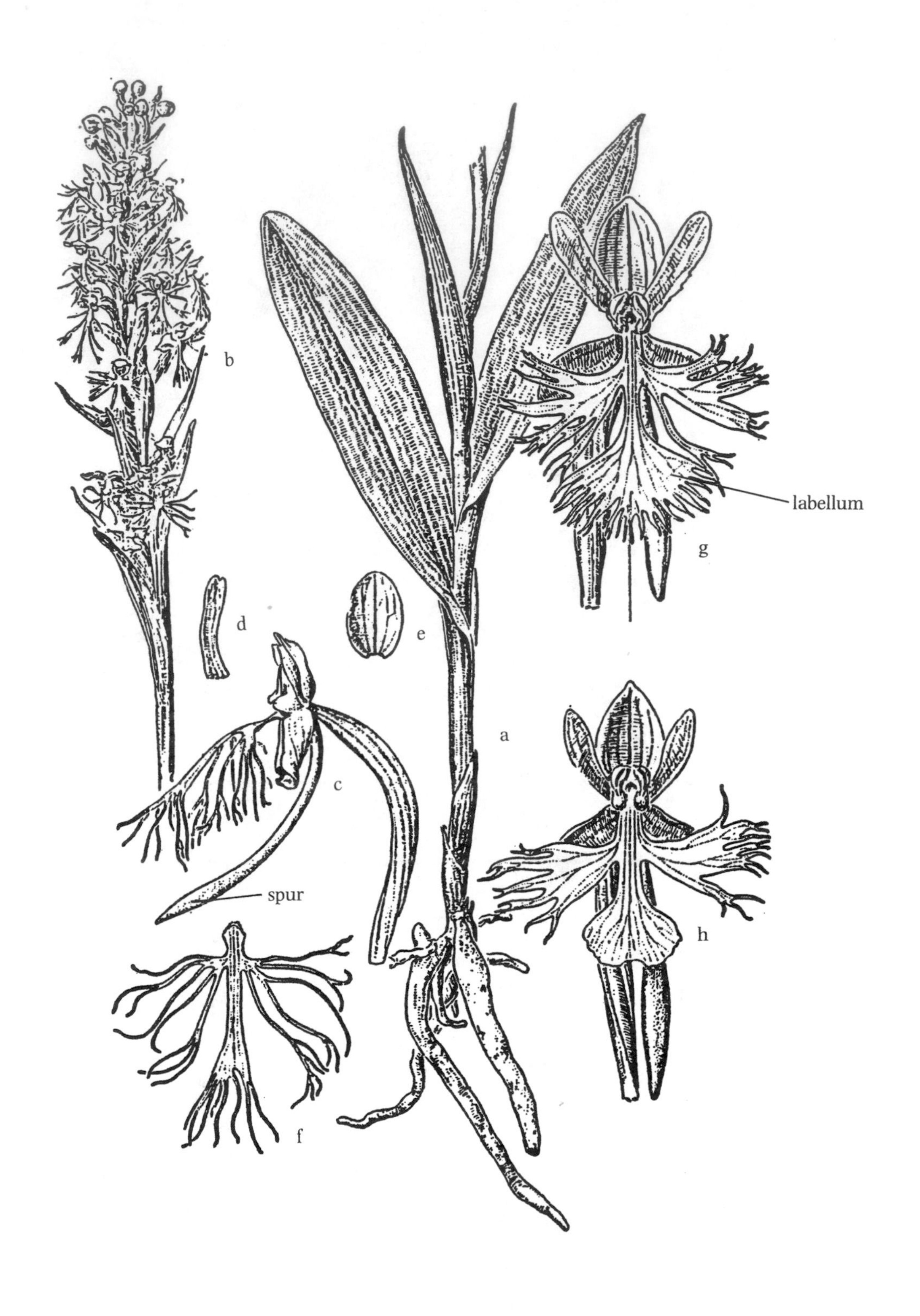

Fig. 127. *Platanthera lacera*: a. lower portion of plant; b. inflorescence; c. flower, side view; d. petal; e. upper sepal; f. labellum; g, h. flowers, hybrids (C&C).

Fig. 128. *Platanthera blephariglottis*: a. habit; b. flower, side view (more southerly variant); c. flower, side view (occasional variant); d. flower, side view (more northerly variant); e. labellum (G&W). *Platanthera ciliaris*: f. inflorescence (C&C).

Fig. 129. *Platanthera cristata*: a. basal portion of plant; b. upper portion of plant; c, d. flower, two views (G&W).

Fig. 130. *Platanthera integra*: a. habit; b. flower, front view spread open; c. flower, side view (G&W).

Fig. 131. *Platanthera nivea*: a. habit; b. flower; c. flower, side view (C&C).
Platanthera clavellata: d. habit; e. flower; f. flower, side view (C&C).

Fig. 132. *Platanthera aquilonis*: a. habit; b. habit of young emerging plant; c. flower, side view; d. flower, front view (HCOT, as *P. hyperborea*).

Fig. 133. *Platanthera dilatata*: a. habit; b. flower (HCOT).
Neottia cordata: c. habit; d. flower; e. flower, side view; f. column, side view (C&C, as *Listera cordata*).

6. *Neottia* (Twayblade)

Perennial herbs, with fibrous roots; stems bearing a single pair of leaves near middle; flowers in racemes; labellum elongate, entire to notched to deeply cleft (ours) at tip; fruit a capsule.

The name *Neottia* has nomenclatural priority over the name *Listera*, when the latter is merged with the European *Neottia* (Pridgeon et al., 2005).

REFERENCES: Magrath and Coleman, 2002; Pridgeon et al., 2005.

---

1. Labellum deeply cleft to half to ⅔ its length into two sharp-pointed narrow apical lobes.
   2. Flowers larger, upper and lateral sepals 2–3 mm; labellum basal auricles/lobes pointed lobes and diverging outward like horns. . . . . 1. *N. cordata*
   2. Flowers smaller, upper and lateral sepals 1.5 mm; labellum basal auricles rounded, projecting backwards. . . . . 2. *N. bifolia*
1. Labellum shallowly cleft, less than half its length, apical lobes rounded.
   3. Pedicels and ovary glabrous; labellum auriculate at base, with rounded lobes, apical lobes lacking tooth in sinus. . . . . 3. *N. auriculata*
   3. Pedicels and ovary nerves finely glandular-puberulent; labellum tapered to narrow base, with rounded auricles on either side above base, apical lobes with small tooth in sinus. . . . . 4. *N. convallarioides*

---

1. *N. cordata* (L.) Rich. Heart-leaved Twayblade Fig. 133
Sphagnum bogs mossy coniferous swamps and alder thickets along rivers. Greenl. and Lab. w. to N.W.T. and Alask., s. to N.E., n. N.J., mts. to Va., N.C., W.Va., ne. Ohio, Ont., n. Mich., n. Wisc., n. Minn., Man., Alta. and B.C. Our taxon is the more diminutive var. *cordata*. (*Listera cordata* (L.) R. Br.)

2. *N. bifolia* (Raf.) Baumbach Southern Twayblade Fig. 134
Open sphagnum bogs, marshes, fern hummocks and rich humus of low moist woods N.S., N.B. and Que. w. to s. Ont., s. to n. Vt. N.Y., Pa., coastal plain to N.J., e. Va., S.C., Fla., Ala., La., and e. Tex.; Mississippi embayment n. to Ark., se. Okla., w. Ky. and Tenn. (*Listera australis* Lindl.)

3. *N. auriculata* (Wiegand) Szlach. Auricled Twayblade Fig. 135
Acidic soils, shrub-carr wetlands, sphagnum bogs, floodplain forest, hardwood and mixed hardwood-coniferous forests. Nfld. and Lab. w. to Man., s. to Me., n. N.H., n. Vt., n. N.Y., n. Mich., n. Wisc. and ne. Minn. (*Listera auriculata* Wiegand)

4. *N. convallarioides* (Sw.) Rich. Broad-leaved Twayblade Fig. 135
Cool coniferous forests, cedar swamps, boggy meadows, mossy/springy sites, wet sandy soil of stream margins, and rich humus of hemlock-hardwood forests. Lab. and Nfld. w. to Ont., n. Mich., n. Wisc. and ne. Minn., s. to n. N.E., and N.Y.; western mts. Alta. and B.C. s. to w. S.D., n. Colo., n. Ariz. and Calif. (*Listera convallarioides* (Sw.) Nutt. ex Elliott)

7. *Calopogon* (Grass-pink)

Perennial herbs, arising from a bulb-like tuber; leaves solitary, basal; flowers pink, in racemes, appearing inverted, labellum uppermost, column below it (flower not resupinate); fruit a capsule.

1. *C. tuberosus* (L.) Britton, Sterns & Poggenb. Grass-pink Fig. 136
Sphagnum bogs, peaty meadows, swamps, marshes and wet shores. Nfld. w. to Ont and sc. Man., s. to Fla. and e. Tex.; W.I.

Calopogon pallidus Chapm., a paler-flowered species of the southern coastal plain, may be encountered in wet acid pine savannas of se. Va. (Goldman et al., 2002).

8. *Spiranthes* (Ladies'-tresses)

Perennial herbs, with a cluster of thick roots; stem bearing leaves at or near base; inflorescence more or less densely spike-like, flowers white to ivory, spiraled; labellum oblong to ovate, margin wavy or recurved, rarely lobed; column short; fruit a capsule.

REFERENCES: Case and Catling, 1983; Catling, 1981, 1982; Pace and Cameron, 2017; Sheviak, 1982; Sheviak and Brown, 2002.

---

1. Flowers of spike tightly spiraled, appearing in 3 or 4 ranks (fig. 137a,d).
   2. Lateral sepals united at base; labellum somewhat fiddle-shaped conspicuously constricted near middle (fig. 137c). . . . . 1. *S. romanzoffiana*
   2. Lateral sepals free at base; labellum ovate to oblong, not constricted, or only slightly so, near middle (fig. 137f).

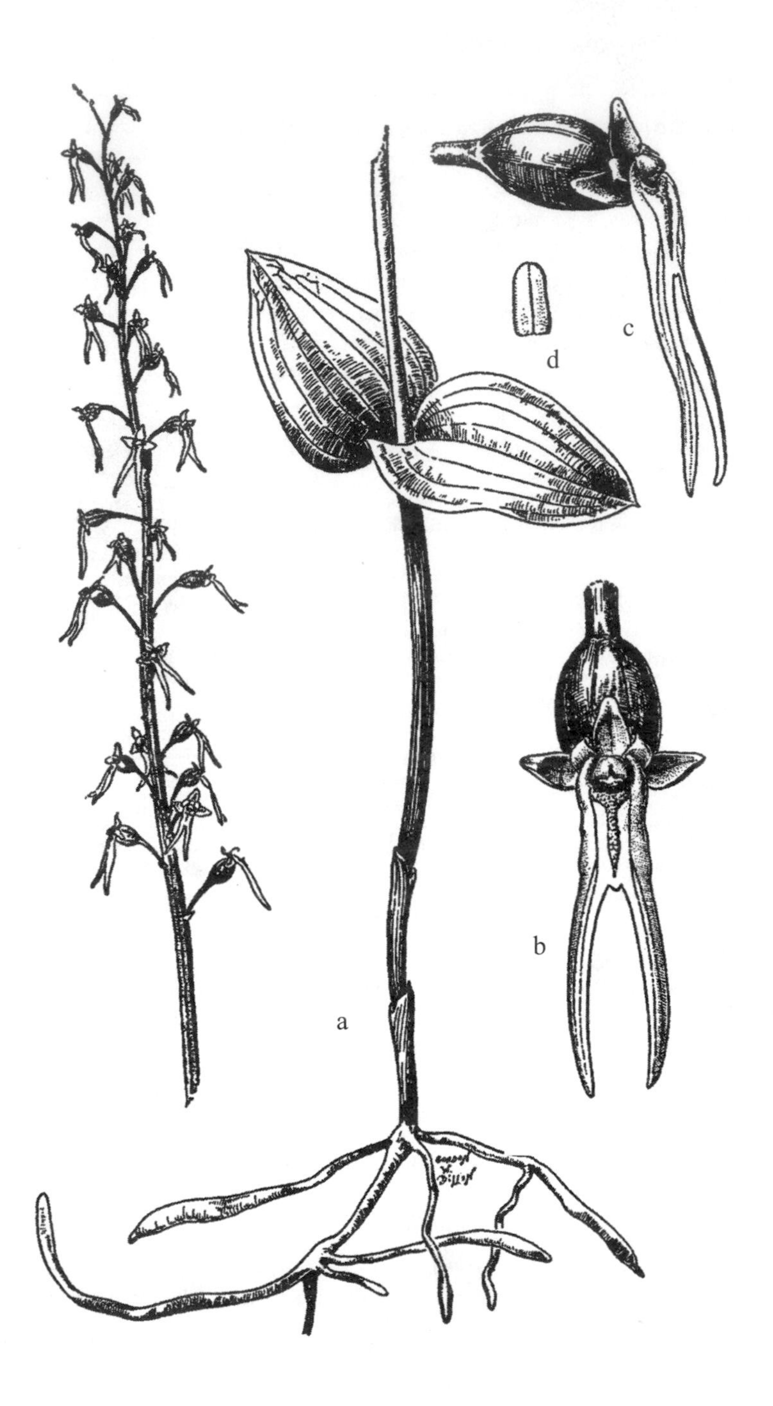

Fig. 134. *Neottia bifolia*: a. habit; b. flower, front view showing deeply cleft labellum; c. flower, side view; d. petal (G&W, as *Listera australis*).

Fig. 135. *Neottia convallarioides*: a. habit; b. flower, front view, spread open, showing barely notched labellum with apical tooth in sinus; c. flower, side view; d. column side view (C&C, as *Listera convallarioides*). *Neottia auriculata*: e. habit; f. flower, side view showing shallowly cleft labellum (B&B).

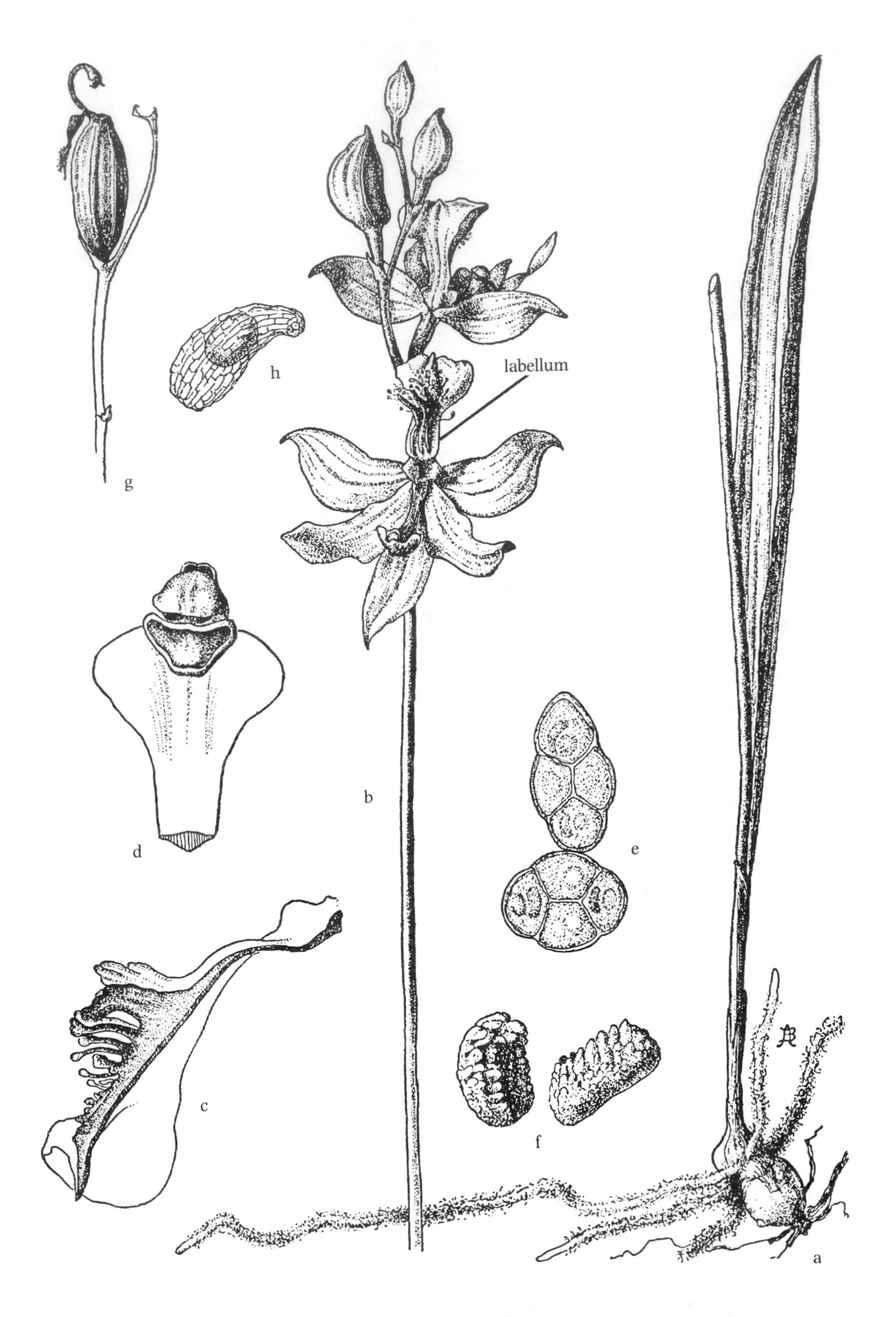

Fig. 136. *Calopogon tuberosus*: a. basal portion of plant; b. inflorescence; c. labellum, longitudinal section, showing papillae; d. portion of column; e. pollen tetrads; f. pollinia; g. capsule; h. seed, highly magnified (C&C).

3. Labellum slightly constricted near middle, rounded to truncate at tip; labellum margin crisped and lacerate; leaves mostly basal, cauline leaves, if present, then on lower stem, with slender blades ascending-spreading, quickly reduced to bracts upward on stem; roots succulent, often somewhat tuberous; plants not stoloniferous.
   4. Flowers frequently nodding; labellum white; lateral sepals lanceolate.
      5. Lateral sepals upward sweeping; labellum apex acute. . . . . 2. *S. cernua*
      5. Lateral sepals downward arching; labellum apex rounded. . . . . 3. *S. arcisepala*
   4. Flowers frequently ascending; labellum yellowish centrally on lower surface (sometimes faintly). . . . . 4. *S. incurva*
3. Labellum not constricted, tapering to obtuse or somewhat acute tip (fig. 140d); labellum margin undulating; leaves basal and cauline, the cauline leaves with blades broad, spreading-recurved on lower portion of stem, well-developed upward on stem to the inflorescence; roots slender; plants stoloniferous, producing colonies of succulent-leaved rosettes. . . . . 5. *S. odorata*

1. Flowers of spike twisted, usually into two 1-ranked groups, secund (fig. 141a).
   6. Labellum widest at base; flowers white to cream, with yellow center; spike pubescent.. . . . . 6. *S. laciniata*
   6. Labellum widest at tip; flowers white, with green veins; spike nearly glabrous or only sparsely pubescent. . . . . 7. *S. praecox*

---

1. *S. romanzoffiana* Cham. Hooded Ladies'-tresses Fig. 137
Meadows, swamps, swales, margins of ponds and alkaline bogs; often calcareous sites. Lab. and Nfld. w. to N.W.T. and Alask., s. to w. Mass., Conn., N.Y., w. Pa., n. Ohio, Mich., n. Ind., Ill., n. Iowa, n. Neb., Mts. of N.M., Ariz. and Calif.

2. *S. cernua* (L.) Rich. Nodding Ladies'-tresses Fig. 137
Wet to dryish meadows and ditches, swales, lakeshores and bogs. P.E.I., N.S., and s. N.B. s. to e. N.E., N.J., se. Pa., Del. and Md.; Va. and W.Va. w. to s. Ohio, Ky, s. Ill., s. Mo. and e. Okla., s. to Ga., w. Fla. Ala., Miss., La. and e Tex. The distribution given by Pace and Cameron (2017) for their revised concept of *S. cernua* is much narrower from that Sheviak and Brown (2002) in FNA. (*S. parksii* Correll)

3. *S. arcisepala* M. C. Pace Fig. 138
Fens, bogs, sphagnous seeps, wet swales, wet roadside ditches. N.B., se. Que., Me., s. to N.H., Vt., w. Mass., e. N.Y., n. N.J., Pa., n. Ohio, ne. Ind. and s. Mich.; Appalachian Highlands s. in w. Md., w. Va., e. W.Va., e. Ky., e. Tenn., and nw. N.C. Occasionally growing interspersed with *S. incurva*. This recently described cryptic species is similar to *S. cernua*; its distinguishing features are its strongly downward-arching lateral sepals, the flowers smaller than those of *S. cernua* and *S. incurva*, and the inflorescence is more open-spiraled.

4. *S. incurva* (Jenn.) M. C. Pace Sphinx Ladies'-tresses Fig. 139
Gaspé Pen., Que. and N.B. w. to s. Ont., Mich., Wisc., Minn., e. S.D. and e. Neb., s. to Me., n. N.H., Vt., n. N.Y., nw. Pa., Ohio, Ind., Ill., n. Mo. and e. Kan.; chiefly north of the Ohio R. valley. This recently described cryptic species is most similar to *S. cernua*, differing by its pale yellow, centrally thickened labellum, more narrowly lanceolate flower parts, ascending and more stellate flowers. Pace and Cameron (2017) have shown it to be of hybrid origin with parental species being *S. cernua* and *S. magnicamporum*. Its incurved callosities at the base of the labellum help distinguish it from *S. magnicamporum* (which has non-incurved, reduced, mounded callosities), with which it is known to frequently grow intermingled.

5. *S. odorata* (Nutt.) Lindl. Fragrant Ladies'-tresses Fig. 140
Shallow water and muddy shores, swamps, marshes and margins of streams. Coastal plain, L.I., N.Y., s. N.J. and Va. s. to Fla., w. to e. Tex.; disjunct in Ky. and Tenn. (*S. cernua* var. *odorata* (Nutt.) Correll)

6. *S. laciniata* (Small) Ames Lace-lip Ladies'-tresses
Wet meadows, marshes, shallow ponds and wet pinelands. Coastal plain, s. N.J., Md. and Va. s. to Fla., w. to La., w. Ark. and e. Tex.

7. *S. praecox* (Walter) S. Watson Giant Ladies'-tresses Fig. 141
Wet meadows, swamps, swales and open wet woods. Coastal plain, N.J., Del., Md. and Va. s. to Fla., w. to La., Ark., se. Okla. and e. Tex.; se. Mo. and w. Ky.

9. *Liparis* (Twayblade)

Perennial herbs, arising from bulbs or tubers; stems with pair of broad, basal leaves; flowers few, in a raceme; labellum entire, widest at tip; column elongate; fruit a capsule.

1. *L. loeselii* (L.) Richard Bog Twayblade, Fen Orchid Fig. 141
Bogs, fens, peaty meadows and mountain ravines. Nfld., N.S. and Gaspé Pen., Que., w. to s. Ont., Man., and B.C., s. to N.E., N.J., Md., to mts. of N.C., Ky. and Tenn., Ohio, Ill., Mo., ne. Kan. and Neb., Mont. and s. Wash.; disjunct to e. Ala.

10. *Malaxis* (Malaxis, Adder's-mouth)

Small perennials, arising from tubers; stems bearing 1 or few broad leaves; flowers small, in dense racemes; petals smaller than sepals, spreading or recurved; labellum auriculate at base, tapering toward tip; column very short; fruit a capsule.

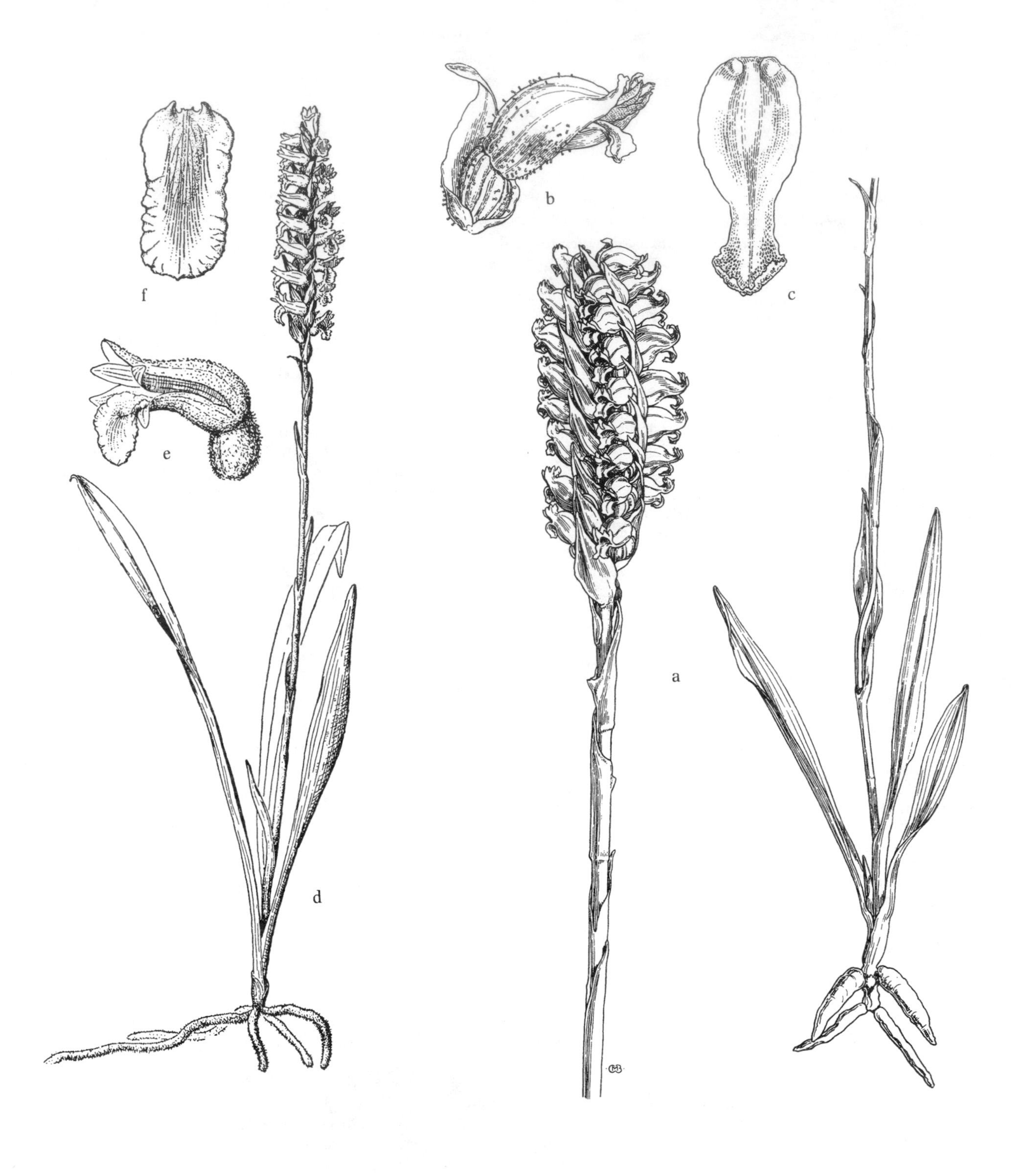

Fig. 137. *Spiranthes romanzoffiana*: a. basal portion of plant and inflorescence; b. flower, side view; c. labellum (Mason).
Spiranthes cernua: d. habit; e. flower, side view; f. labellum (C&C).

Fig. 138. *Spiranthes arcisepala*: a. habit; b. inflorescence; c. flower and floral bract, side view; d–g. flower dissected; d. dorsal sepal; e. dorsal petal; f. lateral sepal; g. labellum; h. flower side view; i. column, dorsal and ventral views; j. column, side view; k. anther; l. 2 pollinia attached to vicidium (Bobbi Angell, in Pace and Cameron, 2017).

Fig. 139. *Spiranthes incurva*: a. habit; b. inflorescence; c. flower and floral bract, side view; d–g. flower dissected; d. dorsal sepal; e. dorsal petal; f. lateral sepal; g. labellum; h. flower, side view; i. column, dorsal and ventral views; j. column side view; k. anther with 2 pollinia (Bobbi Angell, in Pace and Cameron, 2017).

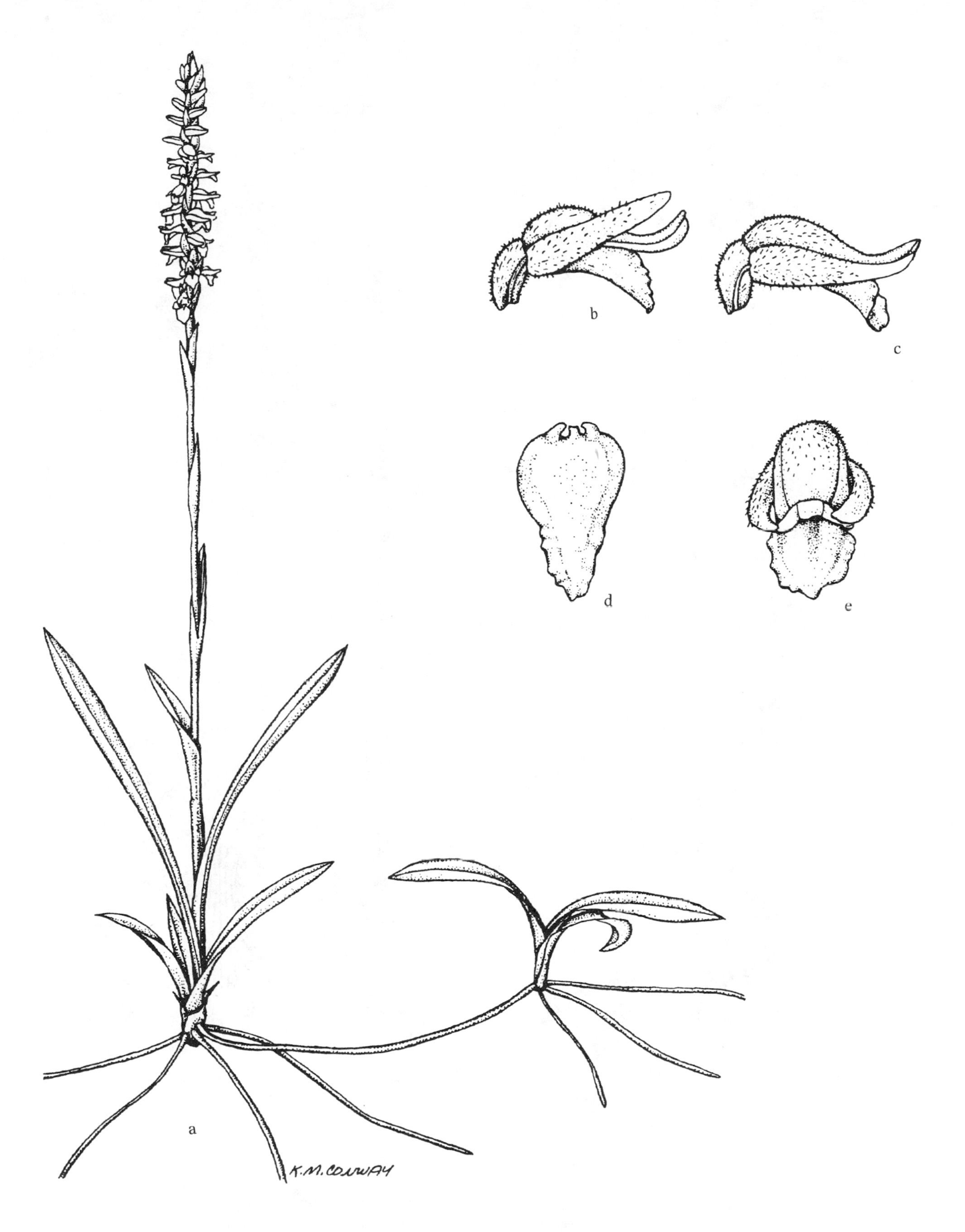

Fig. 140. *Spiranthes odorata*: a. habit; b, c. flowers, side views; d. labellum; e. flower, front view (NYS Museum).

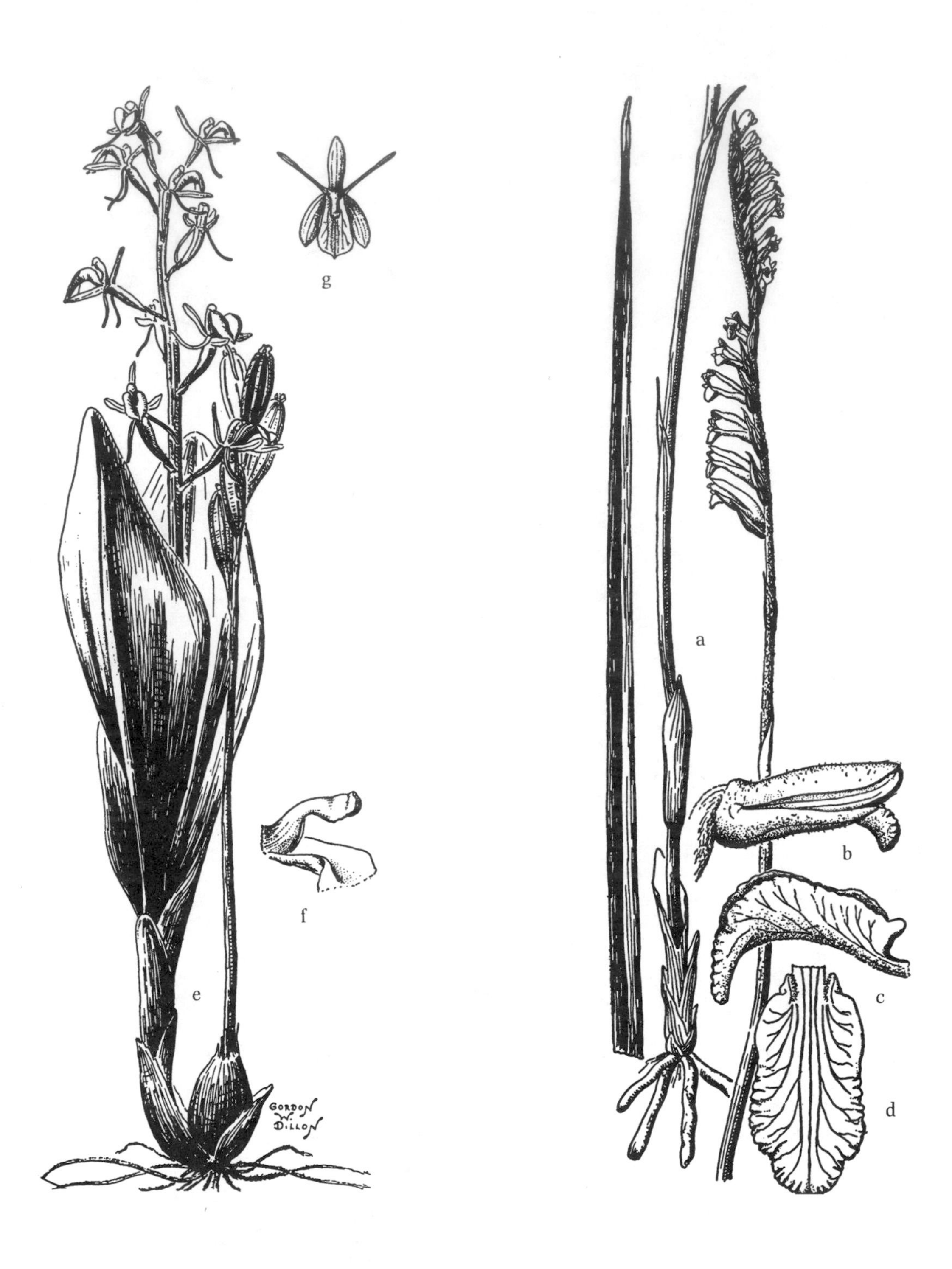

Fig. 141. *Spiranthes praecox*: a. habit; b. flower, side view; c. labellum, side view; d. labellum, front view (C&C). *Liparis loeselii*: e. habit with flowers and with previous year's stem and fruits; f. column and portion of labellum, side view; g. flower, front view (G&W).

1. Leaves solitary (rarely 2); flowers resupinate (labellum oriented upward); stem not swollen at base.
   2. Labellum narrowly pointed at apex .... 1. *M. monophyllos*
   2. Labellum 3-dentate at apex, auriculate at base; pedicels 3.5–10(13) mm long .... 2. *M. unifolia*
1. Leaves 2–5; flowers not resupinate; stem swollen at base, forming a pseudobulb .... 3. *M. paludosa*

1. *M. monophyllos* (L.) Sw. White Malaxis Fig. 143
Damp calcareous gravels, wet meadows, mossy coniferous swamps and thickets, peats, swales and bogs. S. Lab. and Nfld. w. to Ont., Man., Alta., B.C. and s. Alask., s. to N.E., n. N.J., Pa., Mich., ne. Ill. and ne. Minn.; disjunct in c. Colo. and s. Calif. Our taxon is var. *brachypoda* (A. Gray) Morris & Eames. (*M. brachypoda* (A. Gray) Fernald)

2. *M. unifolia* Michx. Green Adder's mouth Fig. 142
Swamps, bogs, thickets, cedar and tamarack swamps, sandy barrens and dry woods. Lab. and Nfld. w. to Man., s. to n. Fla., Miss., e. Kans., e. Okla. and e. Tex.

3. *M. paludosa* (L.) Sw. Bog Malaxis, Bog Adder's-mouth Fig. 143
Sphagnum bogs and swampy woods. Rare, n. Ont. and n. Minn., w. sporadically to N.W.T. and Alask.

Iridaceae / Iris Family

1. *Iris* (Iris)

Perennial herbs, arising from more or less tuberous rhizomes; leaves erect, sword-shaped, linear-lanceolate, equitant; flowers showy, sepals larger than petals; styles petaloid, arching, concealing stamens; ovary inferior; fruit a 3- or 6-angled capsule, seeds in 2 rows per locule.

References. Anderson, 1928, 1936; Cody, 1961; Henderson, 2002.

1. Flowers reddish-brown, coppery, or yellow.
   2. Flowers reddish-brown or coppery, perianth parts notched at tip (fig. 144b,c); capsule 6-angled .... 1. *I. fulva*
   2. Flowers yellow, perianth parts entire at tip; capsule 3-angled .... 2. *I. pseudacorus*
1. Flowers blue to violet-blue.
   3. Ovary and capsule 6-angled; flower stem weak, flexuous, often decumbent .... 3. *I. brevicaulis*
   3. Ovary and capsule 3-angled; flower stem erect.
      4. Leaves 2–5 mm wide; sepals 1.5–2 cm wide, marked with white toward base; ovary and fruit sharply 3-angled .... 4. *I. prismatica*
      4. Leaves 5–50 mm wide; sepals 2–3.5 cm wide, marked with yellow; ovary and fruit bluntly 3-angled (fig. 146f,g).
         5. Petals 2–5 cm long, often shorter than styles; sepals 4–7 cm long, base of blade with minute papillae, enlarged portion with green to yellowish-green spot; leaf bases purple; outermost bracts with margins darkened and somewhat stiff, chartaceous; seeds shiny, finely pebbled, D-shaped in outline .... 5. *I. versicolor*
         5. Petals 3–5 cm, usually longer than styles; sepals 7–10 cm long, base of blade pubescent, enlarged portion with bright yellow spot; leaf bases buff to pale brown; outermost bracts with margins green and soft; seeds dull, irregularly pitted, roundish to irregularly D-shaped in outline .... 6. *I. virginica*

1. *I. fulva* Ker Gawl. Red Iris Fig. 144
Swamps, marshes, wooded swamps and ditches. Ga. w. to La., n. along the Mississippi embayment to se. Mo., sw. Ky. and sw. Ill.

2. *I. pseudacorus* L. Yellow Iris Fig. 145
Marshes, meadows, streams, and lakeshores. Nfld. and N.S. w. to Ont., Minn., and Man., s. to N.C., W.Va., Ohio, n. Ky., and Ill.; se. Tex.; nw. Mont., n. Ida., s. B.C., nw. Wash. and n. Calif.; intro. from Eurasia.

3. *I. brevicaulis* Raf. Zigzag Iris, Short-stem Iris Fig. 147
Swamps, wet woods and bottomlands. S. Ont. and Ohio w. to Ill. and e. Kans., s. to Fla., Ala., La. and e. Tex. (*I. foliosa* Mackenz. & Bush)

4. *I. prismatica* Pursh, Slender Blue Flag
Brackish to saline marshes, freshwater marshes, shores and meadows. Chiefly coastal and piedmont, N.S. and N.E. s. to Del., ne. Md., Va. and S.C.; inland in w. Pa., c. Tenn., w. N.C. and n. Ga. (*I. prismatica* var. *austrina* Fernald)

5. *I. versicolor* L. Blue Flag, Harlequin Blue Flag Fig. 147
Marshes, meadows, stream banks, lake and pond margins and ditches. Lab. and Nfld. w. to Man., s. to Va., n. Ohio, n. Mich., Wisc. and Minn.

6. *I. virginica* L. Southern Blue Flag Figs. 146, 147
Marshes, ditches, wet shores, swamps and shallow water. Sw. Que. w. to s. Mich., Minn., and e. Kans., s. to e. Va., N.C., Fla. and

Fig. 142. *Malaxis uniflora*: a. habit, 4 variations; b. flower subtended by bract, front view; c. labellum, spread out; d. petal; e. lateral sepal; f. dorsal sepal (G&W).

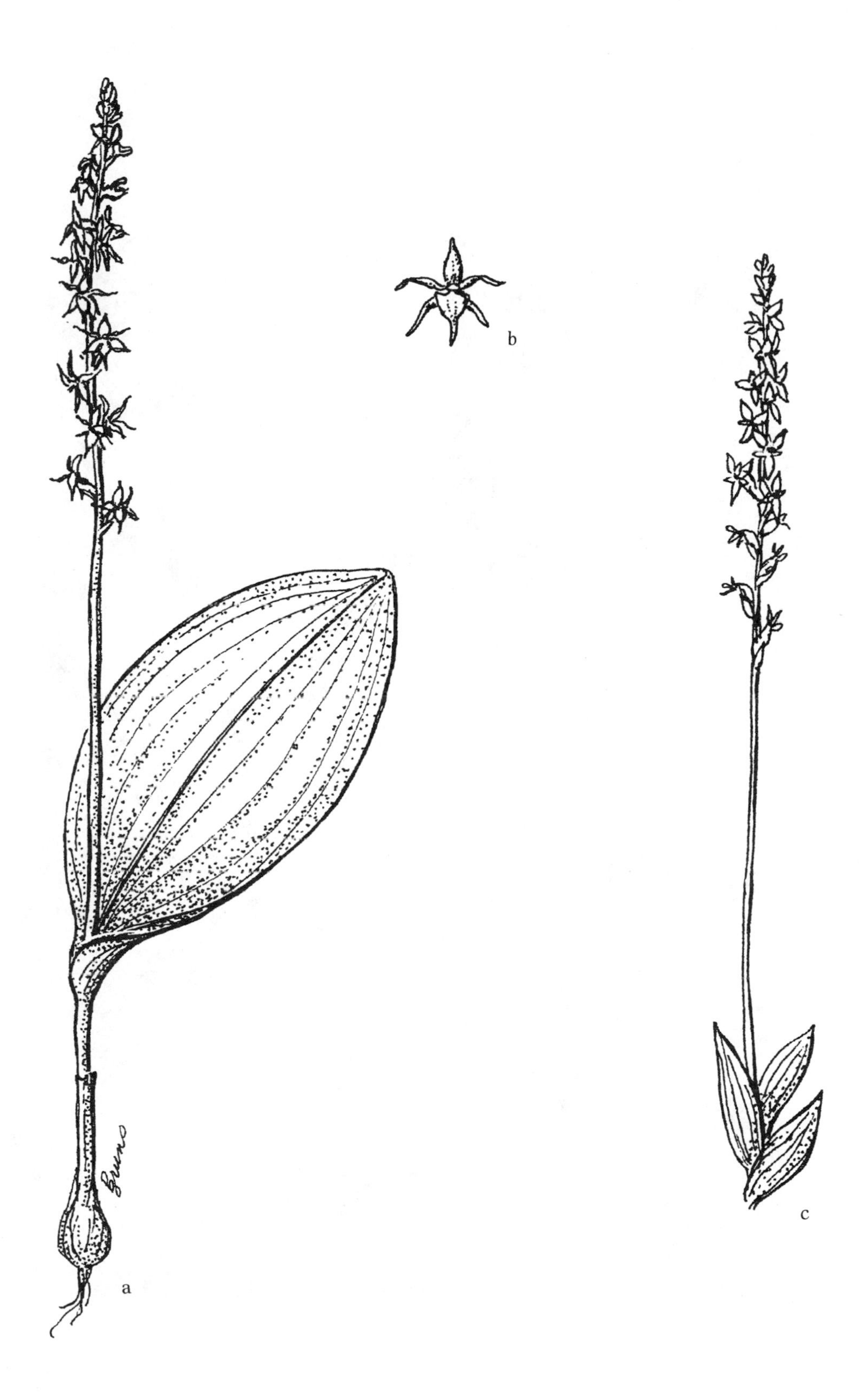

Fig. 143. *Malaxis monophyllos*: a. habit; b. flower (PB).
Malaxis paludosa: c. habit (PB).

Fig. 144. *Iris fulva*: a. habit; b. outer segment, sepal; c. inner segment, petal; d. capsule (C&C).

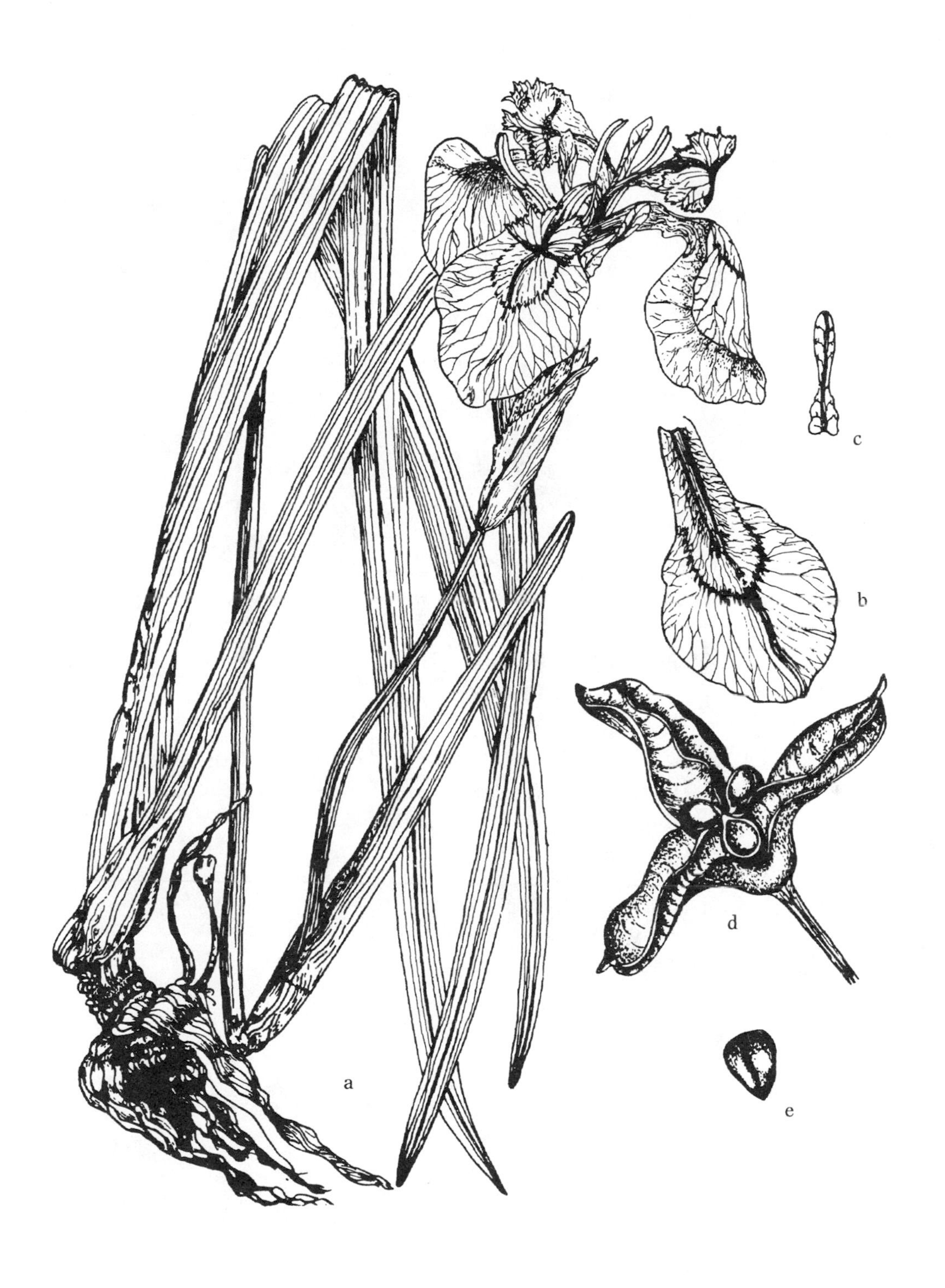

Fig. 145. *Iris pseudacorus*: a. habit; b. sepal (outer segment); c. petal (inner segment); d. capsule, dehisced; e. seed (C&C).

Fig. 146. *Iris virginica*: a. basal portion of plant; b. flowering stem; c. sepal (outer perianth segment); d. petal (inner perianth segment); e. petal; f. fruiting stem; g. capsule, cross-section; h. seeds (G&W).

Fig. 147. *Iris versicolor*: a. upper portion of plant; b. capsule (Braun).
Iris virginica: c. flower; d. upper portion of plant with fruit; e. capsule (Braun).
Iris brevicaulis: f. upper portion of plant (B&T).
Maianthemum trifolium: g. habit (Braun).

e. Tex. Although plants of the southeastern and southcentral U.S. with long-cylindric capsules have been long treated as var. *shrevei* (Small) S. E. Anderson, Henderson (2002) recognized no infraspecific taxa.

Sisyrinchium (Blue-eyed-grass) contains several species occasionally found in damp meadows and ditches. These include *S. atlanticum*, *S. angustifolium*, and *S. montanum*.

Asparagaceae / Asparagus Family

1. *Maianthemum* (False Solomon's-seal)

Perennial herbs, arising from short, thick rhizomes or slender, creeping rhizomes (ours); leaves alternate, sessile or subsessile, narrowing to a subpetiolate sheath; flowers borne in a panicle or raceme (ours); perianth white, 3-merous (ours) or 2-merous; ovary superior, 3-locular; fruit a berry. This is treated by some authors as within the Convallariaceae, as a family distinct from the Asparagaceae.

1. *M. trifolium* (L.) Sloboda False Solomon's-seal Fig. 147
Wet mossy woods, bogs, coniferous swamps and peaty shores. S. Lab. and Nfld. w. to Man., N.W.T. and Alask., s. to N.E., n. N.J., n. Ohio, n. Ill., Minn., Alta. and B.C. (*Smilacina trifolia* (L.) Desf.)

Typhaceae / Cattail Family

1. Flowers and fruits in dense cylindric, spike-like inflorescence, brown to brown-cinnamon; leaves flattened, twisted, erect. . . . . 1. *Typha*
1. Flowers and fruits in globose, bur-like heads, green-brown; leaves keeled and erect or flattened and floating. . . . . 2. *Sparganium*

1. *Typha* (Cattail)

Perennial herbs, ca. 1–3 m tall, arising from creeping rhizomes, colonial; leaves long and narrow, flat on upper surface, slightly rounded on lower surface, erect and slightly twisting, sheathing at base; flowers unisexual, with thousands borne in a dense cylindric terminal spike-like inflorescence (cigar-like), staminate spike above pistillate spike; fruit a minute thin-walled achene, raised on a slender stalk bearing numerous long hairs, style persistent.

Cattails may appear in almost any wet place, often being the first invaders in excavated ponds or ditches. The underground stems spread extensively, so that a large stand of cattails may actually consist of but a few individual plants. A good field character for recognizing *Typha* in the vegetative state is long leaves, flat on top surface, slightly rounded on back, and with a distinctive twisting of the leaves.

References: Crow and Hellquist, 1981; Hotchkiss and Dozier, 1949; Lee, 1975; Smith, 1967, 1986, 1987, 2000; Yeo, 1964.

1. Staminate and pistillate parts of inflorescence usually contiguous (fig. 148a,b) (occasionally separated by a gap up to 5 mm, or even up to 8 mm in some clones), mature fruiting spike at least 2.5 cm in diameter pistillate bractlets absent; stigma lanceolate to ovate-lanceolate, flattened, persistent in fruit; shoots and leaves relatively. . . . . 1. *T. latifolia*
1. Staminate and pistillate parts of inflorescence usually separated by gap of 0.5–8 cm (figs. 149a,c, 150a,b), mature fruiting spike less than 2.5 cm in diameter; pistillate bractlets present, numerous (fig. 149r) (minute in *T.* ×*glauca*); stigmas slender, elongate, often deciduous in fruit; shoots and leaves relatively narrow.
   2. Leaf sheathes with auricles (fig. 149d) (usually disintegrating with age); fruiting spike dark brown; pistillate bractlet tips blunt and dark brown, or narrower than stigmas and very pale.
      3. Dry leaf blades 3–8 mm wide; gap between pistillate and staminate spikes about 1–8 cm; pistillate bractlet tips conspicuous (fig. 149s,t), blunt dark brown; stigmas capillary, usually deciduous in fruit. . . . . 2. *T. angustifolia*
      3. Dry leaf blades about 6–15 mm wide; gap between pistillate and staminate spikes about 0.5–4 cm; pistillate bractlet tips minute (require high magnification), pale; stigmas linear-lanceolate, partly persistent. . . . . 3. *T.* ×*glauca*
   2. Leaf sheaths tapered to blade, lacking auricles; fruiting spike bright cinnamon-brown or orange-brown; pistillate bractlet tips usually pointed, usually paler and wider than stigmas. . . . . 4. *T. domingensis*

1. *T. latifolia* L. Common Cattail Fig. 148
Damp shores, marshes, swamps, marshy river margins, and roadside ditches; often forming extensive stands. Nfld. w. to Alask. s. to Fla., Tex., Calif. and Mex. In this species the staminate and pistillate spikes are typically contiguous, however sometimes plants have a small separation between the staminate and pistillate portions; these have been described as *T. latifolia* f. *ambigua* (Sunder) Kronf. and may be confused with *T. angustifolia* or *T.* ×*glauca*.

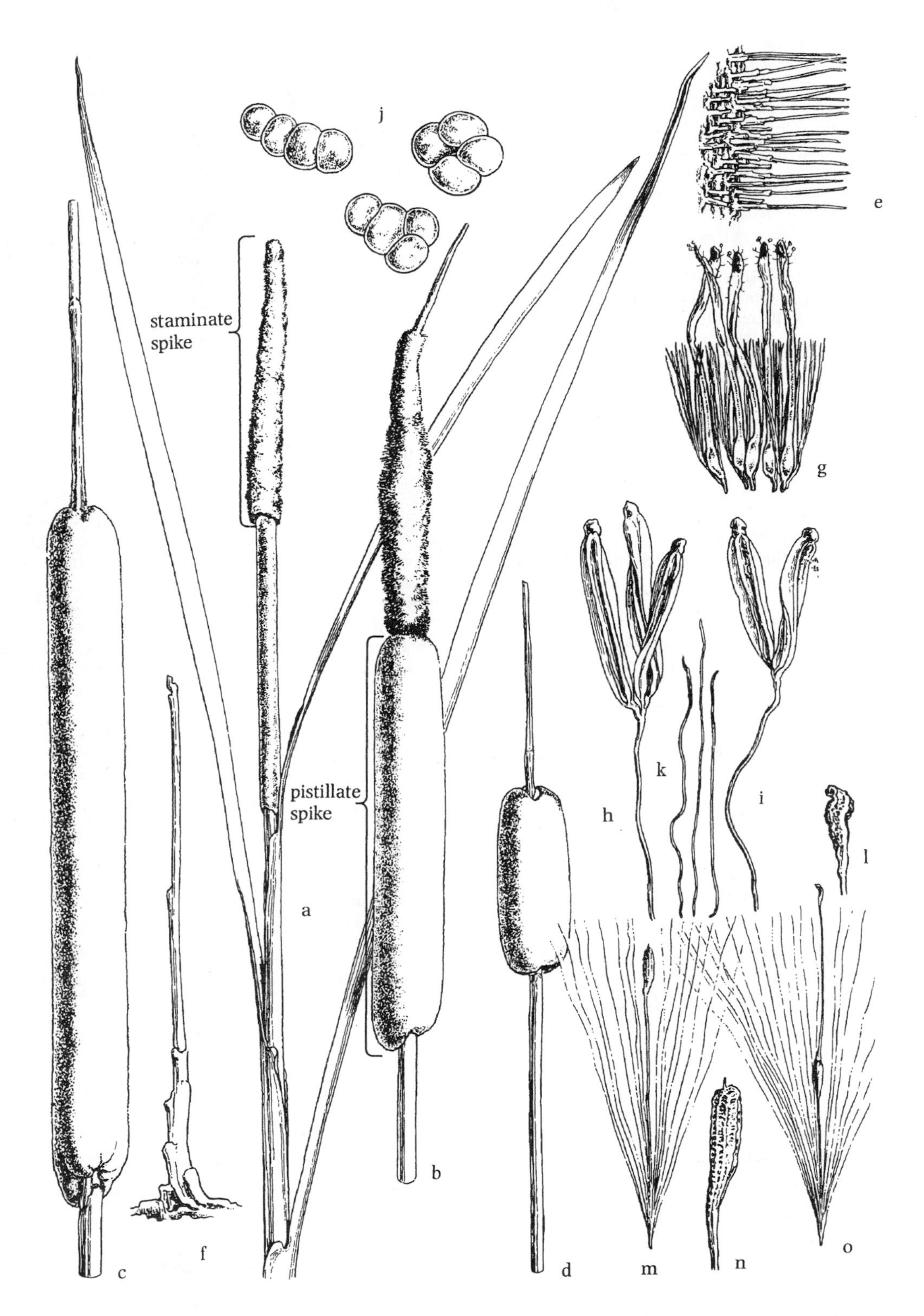

Fig. 148. *Typha latifolia*: a. upper portion of plant with contiguous staminate (*above*) and pistillate (*below*) spike; b. older spike; c, d. pistillate spike variations; e. portion of pistillate spike axis, showing compound pedicels; f. single compound pedicel; g. young pistillate flowers with hair-like bractlets; h, i. anthers on branched filaments; j. pollen tetrad variations; k. staminate bractlets, commonly white-tipped; l. stigma; m. sterile pistillate flower with ellipsoidal aborted ovary; n. aborted ovary; o. fertile pistillate flower (Mason).

2. *T. angustifolia* L. Narrowleaf Cattail
Fig. 149
Coastal salt marshes and inland in nutrient-rich waters. P.E.I. and N.S. w. to s. Que., Ont., s. Sask., Mont. and B.C., s. to S.C., Tenn., Miss., La., Okla., N.M., Oreg., Nev. and Calif. Smith (2000) notes that there is come discrepancy as to whether *T. angustifolia* is native in the U.S. or not, possibly having been introduced from Europe on the Atlantic Coast and spread westward; the USDA PLANTS website (https://plants.sc.egov.usda.gov/core/profile?symbol=TYAN, accessed August 2020) suggests that it is both native and introduced in the U.S., but native in Canada; the Plants of the World Online website (http://plantsoftheworldonline.org/taxon/urn:lsid:ipni.org:names:259280-2, accessed August 2020) indicates that the species is native nearly all around the Northern Hemisphere (introduced in the Russian Far East).

3. *T.* ×*glauca* Godr. Hybrid Cattail
Occasional, often locally abundant, usually with the parental species. This taxon is a hybrid between *T. latifolia* and *T. angustifolia*, and although plants are typically sterile and usually form few seeds, they may persist for many years, spreading widely by rhizomes and sometimes becoming ecologically very important. Plants intermediate between *T. glauca* and either parent, especially *T. angustifolia*, are locally common. Hybrids of *T. domingensis* × *T. latifolia* are very similar and might also be expected on the southeastern and southwestern edges of our range.

4. *T. domingensis* Pers. Southern Cattail Fig. 150.
Coastal marshes, swamps, and pools, and brackish and saline sites. Coastal plain, Del. and e. Md. s. to Fla., w. to La. and Tex.; inland Ill., Neb., sw. Mo. and Kans. w. to Wyo., and Oregon, s. to Okla., N.M., Ariz., and Calif.; Mex.; C.Am., S.Am.

2. *Sparganium* (Bur-reed)

Perennial herbs arising from rhizomes; stems and leaves erect, floating, or submersed; flowers unisexual in globose heads, staminate heads borne above pistillate heads, falling off early after shedding pollen; pistillate heads appearing burr-like; perianth parts of pistillate flowers 3–6, scale-like; fruit a dry, somewhat spongy drupe with 1 or 2 locules, obovoid to fusiform, frequently with a slight median constriction. *Sparganium* is a difficult group, and mature fruit is necessary for the identification of most species, with the exception of *S. eurycarpum* and *S. erectum*, which are the only species with 2 stigmas.

A molecular study by Kim and Choi (2011) of the phylogeny of *Typha*, which included several species of *Sparganium*, supports the recognition of monophyly of the two genera within a single family, the Typhaceae. Systematic studies of *Sparganium* (Sulman et al., 2013) utilizing molecular techniques revealed two lineages (clades) recognizable at the subgeneric level, subgenus *Sparganium* (including *S. eurycarpum* and *S. erectum*), and subgenus *Xanthosparganium* (containing *S. americanum*, *S. androcladum*, *S. angustifolium*, *S. emersum*, *S. fallax*, *S. fluctuans*, *S. glomeratum*, *S. gramineum*, *S. hyperboreum*, *S. japonicum*, *S. natans* and *S. subglobosum*); the study likewise supports placing both *Sparganium* and *Typha* in a single family.

Unusual submersed, thin-leaved plants appearing to be *Sparganium* have been examined using molecular techniques, giving evidence of a intergeneric hybrid between *Sparganium* and *Typha* hybrid (D. Les, pers. comm., 2020); such an intergeneric hybrid would be rather significant and further studies needs to be conducted on this purported hybrid.

References: Beal, 1960a; Cook, 1985; Cook and Nicholls, 1986, 1987; Crow and Hellquist, 1981; Fernald, 1922; Harms, 1973; Kral, 2000; Lakela, 1941; Reveal, 1970; Sulman et al., 2013; Voss, 1966.

---

1. Stigmas 1–2, with 2 on at least some flowers (rarely 3 or 4); fruit ridged, with locules 2.
  2. Pistillate heads in fruit 2–3 cm in diameter, fruit excluding beak 7–9 mm long, 5–7 mm wide; stigmas 2 (rarely 3–4) on (40)60–100% of pistillate flowers; leaf blades 5–7 mm wide. . . . . 1. *S. eurycarpum*
  2. Pistillate heads in fruit 1.5–2.5 m in diameter, fruit excluding beak 5–8 mm long, 3–6 mm wide; stigmas 2 on 5–40% of pistillate flowers; leaf blades 10–28 mm wide. . . . . 2. *S. erectum*
1. Stigmas 1; fruit lacking ridges, with a single locule.
  3. Inflorescence simple (fig. 154).
    4. Staminate heads 1 (figs. 152a, 156a) (if 2, rarely, then staminate heads contiguous with uppermost pistillate head).
      5. Leaves strongly wing-keeled, emergent; fruiting heads (1)1.2–1.6(2) cm in diameter. . . . . 3. *S. glomeratum*
      5. Leaves flat, floating; fruiting heads (0.5)0.7–1.1(1.5) cm in diameter.
        6. Pistillate heads borne directly in axils of leaves or bracts (fig. 152c); fruits with beak 0.5–1.5 mm long. . . . . 4. *S. natans*
        6. Pistillate heads (at least some) borne on peduncles above axils of leaves or bracts (fig. 159d); fruits beakless or with minute beak less than 0.3(0.5) mm long. . . . . 5. *S. hyperboreum*
    4. Staminate heads 2 or more.
      7. Pistillate heads borne directly in axils of leaves or bracts. leaves lax, usually floating, up to 12 dm long, leaves not keeled. . . . . 6. *S. angustifolium*
        8. Fruiting heads 2.5–3.5 cm in diameter; stigmas 2–3.2 mm long; beak 4.5–7 mm long; body of fruit shiny (rarely dull) on upper portion, dull and pitted on lower portion. . . . . 9. *S. androcladum*
        8. Fruiting heads 1.5–2.5 cm in diameter; stigmas 0.8–1.5(3) mm long body of fruit dull, smooth on lower portion. . . . . 10. *S. americanum*
      7. Pistillate heads (at last some) borne above the axils of leaves or bracts (fig. 154).
        9. Fruit beak 1.5–2.2 mm long; stigma 0.6–1 mm long; fruits (fresh) reddish to brownish at base, stems and leaves lax, usually floating, up to 12 dm long, leaves not keeled. . . . . 6. *S. angustifolium*

Fig. 149. *Typha angustifolia*: a–c. upper portion of plant with separate staminate (*above*) and pistillate (*below*) spikes; d. auricled leaf sheath; e. portion of pistillate spike axis, showing compound pedicels; f. single compound pedicel; g. cluster of young anthers surrounded by bractlets, filament not yet elongated; h–j. mature stamens; k–m. staminate bractlets; n. pollen grains; o. sterile pistillate flower with aborted ovary; p. aborted ovary; q. cluster of bractlets, some with enlarged tips; r. swollen tip of pistillate bractlet; s. pistillate bractlets; t, u. fertile pistillate flowers (Mason).

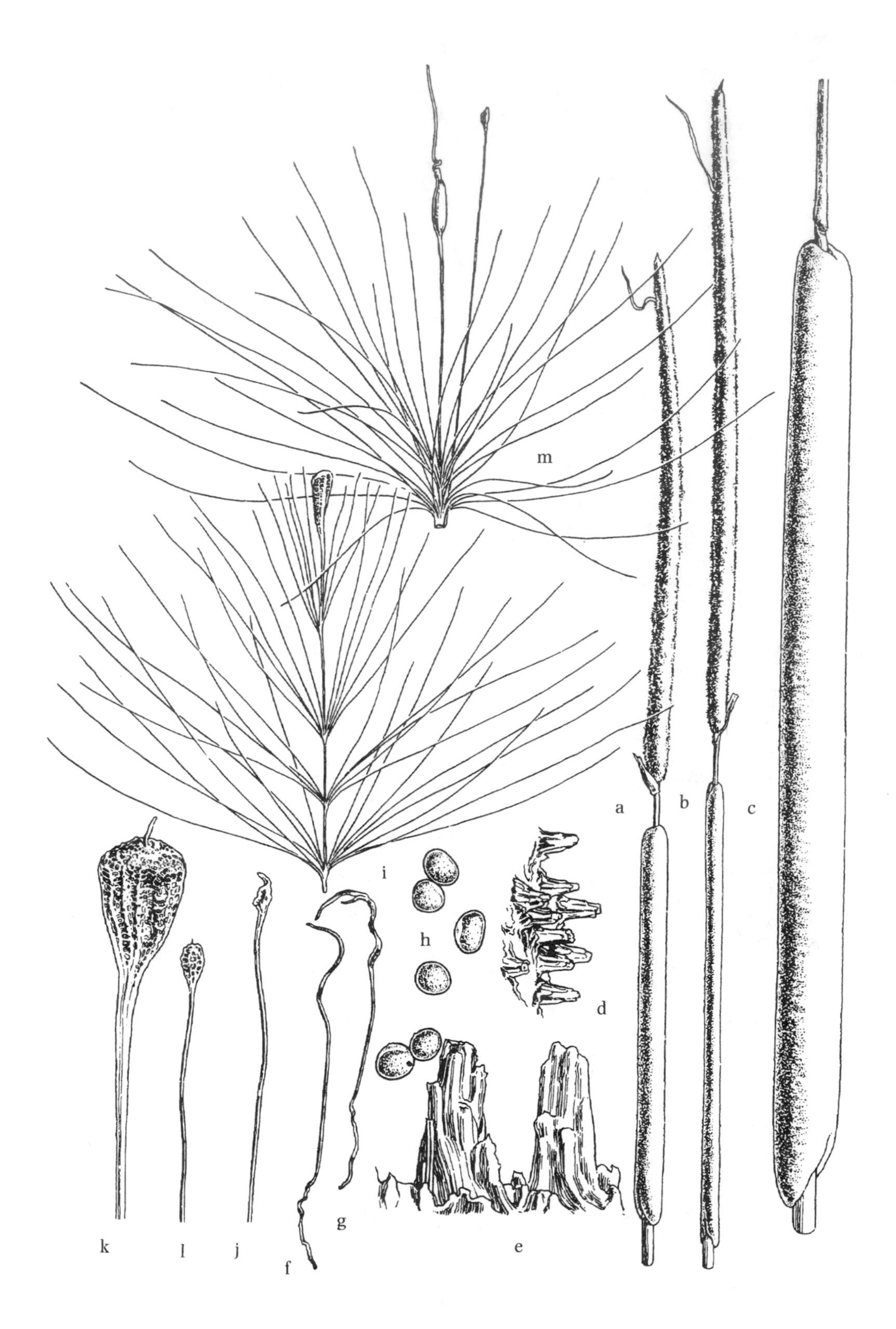

Fig. 150. *Typha domingensis*: a, b. inflorescence, staminate above, pistillate below; c. pistillate spike; d. portion of pistillate spike axis, showing compound pedicels; e. compound pedicels; f, g. bractlets of staminate flowers; h. pollen grains; i. sterile pistillate flower with aborted ovary; j. pistillate bractlet with swollen tip; k, l. aborted ovary; m. fertile pistillate flower (Mason).

9. Fruit beak 2–4 mm long; stigma 0.9–1.2(2) mm long; fruits (fresh) greenish at base; stems and leaves erect, emersed (occasionally floating), up to 8.5 dm long; leaves keeled and triangular in cross-section.. . . . . 7. *S. emersum*

3. Inflorescence branched (figs. 157b, 158b) (*S. fluctuans* occasionally simple, but then fruit beak strongly curved).

10. Plants submersed, with leaves flat, flaccid, upper ones floating; fruit beak strongly curved, sickle-shaped, flattened toward tip (fig. 157c). . . . . 8. *S. fluctuans*

10. Plants emersed with leaves more or less keeled to strongly keeled, ascending to strongly erect; fruit beak not strongly curved, cylindrical to tip.

11. Fruit heads 2.5–3.5 cm in diameter; stigma 2–3 mm long: beak 4.5–7 mm long, body of fruit shiny (rarely dull) on upper portion, dull and pitted on lower portion. . . . . 9. *S. androcladum*

11. Fruit heads 1.5–2.5 cm in diameter; stigma 0.8–1.5(3) mm long; beak 1.5–5 mm long; body of fruit dull, smooth on lower portion. . . . . 10. *S. americanum*

---

1. *S. eurycarpum* Engelm. ex A. Gray Large-fruited Bur-reed Fig. 151

Shallow water. damp shores, swales and marshes. Nfld. and Que. w. to N.W.T. and. B.C., s. to Va., n. Ky., Mo., Okla., N.M., Ariz. and Calif. Our taxon is var. *eurycarpum*; the western taxon, var. *greenei* (Morong) Graebn., ranges coastally from B.C. to California and Baja California, Mex.; both are closely related to the Eurasian species *S. erectum* L.

2. *S. erectum* L. Simple stem Bur-reed Fig. 155

Shallow water and damp soil of marshes, lakes, ponds and river shores. According to the USDA PLANTS website (accessed August 2020) this species is introduced sporadically: Vt., s. N.Y., n. N.J., W.Va. and Wisc. In as much as *S. erectum* appears on the Federal list of Noxious Weeds (Plant Protection Act of 2000) (https://plants.usda.gov/java/noxious?rptType=Federal, accessed August 2020), this species is included here. However, documenting evidence is sparse, casting doubt on its actual presence in N.Am. Localities USDA PLANTS mapped as *S. erectum* subsp. *stoloniferum* (Graebn.) H. Hara in the West, with *S. eurycarpum* var. *greenei* (Morong) Graebn. in synonomy; but molecular evidence of Sulman et al. (2013) supports *S. eurycarpum* (including both var. *greenei* and var. *eurycarpum*) as a distinct species, closely related to the Eurasian *S. erectum*. Likewise, the *Jepson Manual* (Smith, 2012) recognized *S. eurycarpum* var. *greenei* as distinct, while *Flora Nova Angliae* (Haines, 2011) treats it as a synonym of *S. eurycarpum* (negating documentation of "*S. erectum*" in New England. The Plant List website further muddies the water, treating *S. eurycarpum* var. *greenei* as a synonym for *S. stoloniferum* (Buch.-Ham. ex Graebn.) Buch.-Ham. ex Juz., a species of China and Japan (which has sometimes been included in *S. emersum*).

3. *S. glomeratum* (Laest. ex Beurl.) L. Neum. Cluster Bur-reed Fig. 156

Bogs and shallow water. Reported from s. Saguenay Co., Que., Wisc., ne. Minn., Sask., Alta. and B.C.; Cook and Nicholls (1986) confirmed it only from Labrador, but it is now more widely known.

4. *S. natans* L. Northern Bur-reed, Small bur-reed Fig. 152

Lakes, ponds, streams and ditches. Nfld. w. to Alask., s. to n. N.J., n. Pa., Mich., Ind., Ill., Wisc., Minn., Mont., Utah, Colo., Ariz. and Calif. (*S. minimum* (Hartm.) Fries)

5. *S. hyperboreum* Laest. ex Beurl. Fig. 159

Cold ponds, streams and ditches. Arctic regions s. to Nfld., n. N.S., Que., Man. and B.C.

6. *S. angustifolium* Michx. Fig. 153

Lakes, ponds, streams and ditches. Greenl., Lab. and Nfld. w. to Alask., s. to n. N.J., n. Pa., Mich., n. Ill., Minn., Colo., Utah, n. N.M., n. Ariz. and Calif. Vegetative plants are typically narrower-leaved than *S. fluctuans* (*S. emersum* var. *angustifolium* (Michx.) Taylor & MacBryde; *S. multipedunculatum* Rydb.)

7. *S. emersum* Rehmann Fig. 154

Shallow water and damp shores of lakes, ponds, streams, marshes and ditches. Nfld. w. to Ont., Minn., N.D. and s. B.C., s. to N.C., Pa., Ind., Iowa and mts. to Colo. and Calif. Plants with floating leaves are often confused with *S. angustifolium*. The best characters to separate these two species are the fruit color and beak length. Plants with narrow floating leaves (5 mm or less) in the southern portion of our range are usually *S. emersum*. In most older references this taxon was treated as *S. chlorocarpum*, which Cook and Nicholls (1986) regard as conspecific with *S. emersum*. The most common phase of the species in the East is referable to *S. emersum* subsp. *acaule* Cook and Nicholls, which is distinguished by congested pistillate heads (fig. 154b), somewhat smaller fruits, and lower bracts conspicuously longer than the inflorescence. However, we have noted that such plants frequently occur in mixed populations with plants referable to subsp. *emersum* (Crow and Hellquist, 1981). Hence, we are inclined to agree with Voss (1966) in recognizing these plants as merely representing two growth forms (fig. 154). (*S. chlorocarpum* Rydb.; *S. chlorocarpum* var. *acaule* (Beeby) Fernald)

8. *S. fluctuans* (Engelm. ex Morong) B. L. Rob. Fig. 157

Lakes and ponds of low alkalinity. Nfld. w. to s. Que. and n. Alta., and B.C., s. to N.E., n. Pa., n. Mich. and Minn. Vegetative plants may be confused with *S. angustifolium*; the leaf width in *S. fluctuans* ranges from 3–11 mm, but is generally wider than 5 mm, whereas that of *S. angustifolium* ranges from 1.5–5 mm, but is typically less than 5 mm wide.

9. *S. androcladum* (Engelm.) Morong Fig. 158

Damp shores and marshes. Que. w. to Ont. and Minn., s. to s. Va., e. Tenn., Ky., Ill. and Mo. Inflorescences are typically branched, but are occasionally simple; the pistillate heads are rarely supra-axillary.

10. *S. americanum* Nutt. Fig. 159

Shallow water and damp shores. Nfld. w. to Ont., Wisc., Minn., and N.D., s. to Fla., Ala., Okla. and Tex. Populations having floating leaves that otherwise look like *S. americanum* have been analyzed by DNA techniques and shown to be hybrids with *S. emersum* (D. Les, 2020, pers. comm). Beal (1960a) has reported that inflorescence branching and supra-axillary pistillate heads are not infrequent in populations of the coastal plain of the Carolinas.

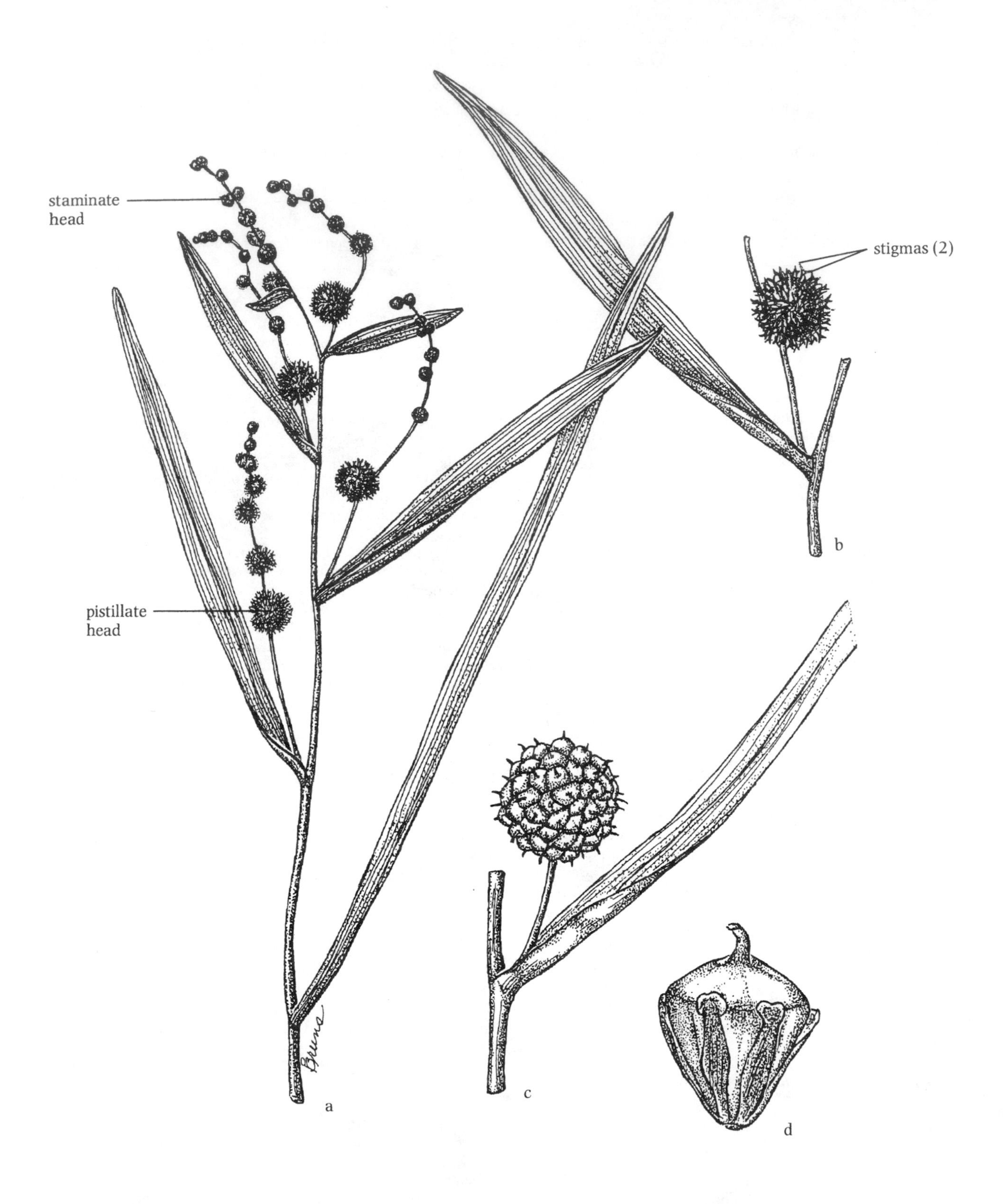

Fig. 151. *Sparganium eurycarpum*: a. upper portion of plant; b. head of flowers, each flower with 2 stigmas; c. fruiting head; d. fruit (NHAES).

Fig. 152. *Sparganium natans*: a. habit, submersed plant; b. habit, emersed plant; c. fruiting head; d. fruit (NHAES).

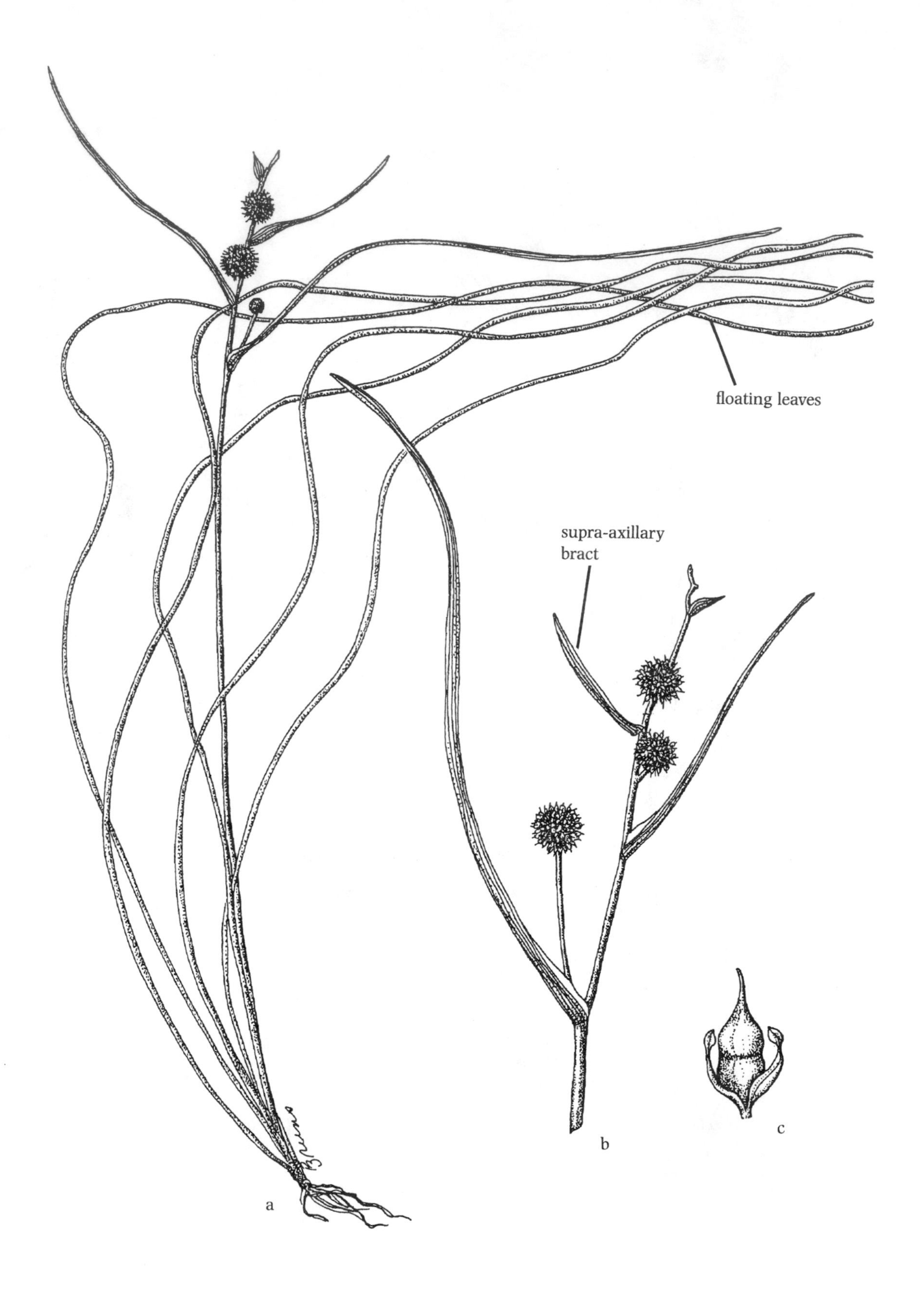

Fig. 153. *Sparganium angustifolium*: a. habit; b. upper portion of plant with fruiting heads; c. fruit (NHAES).

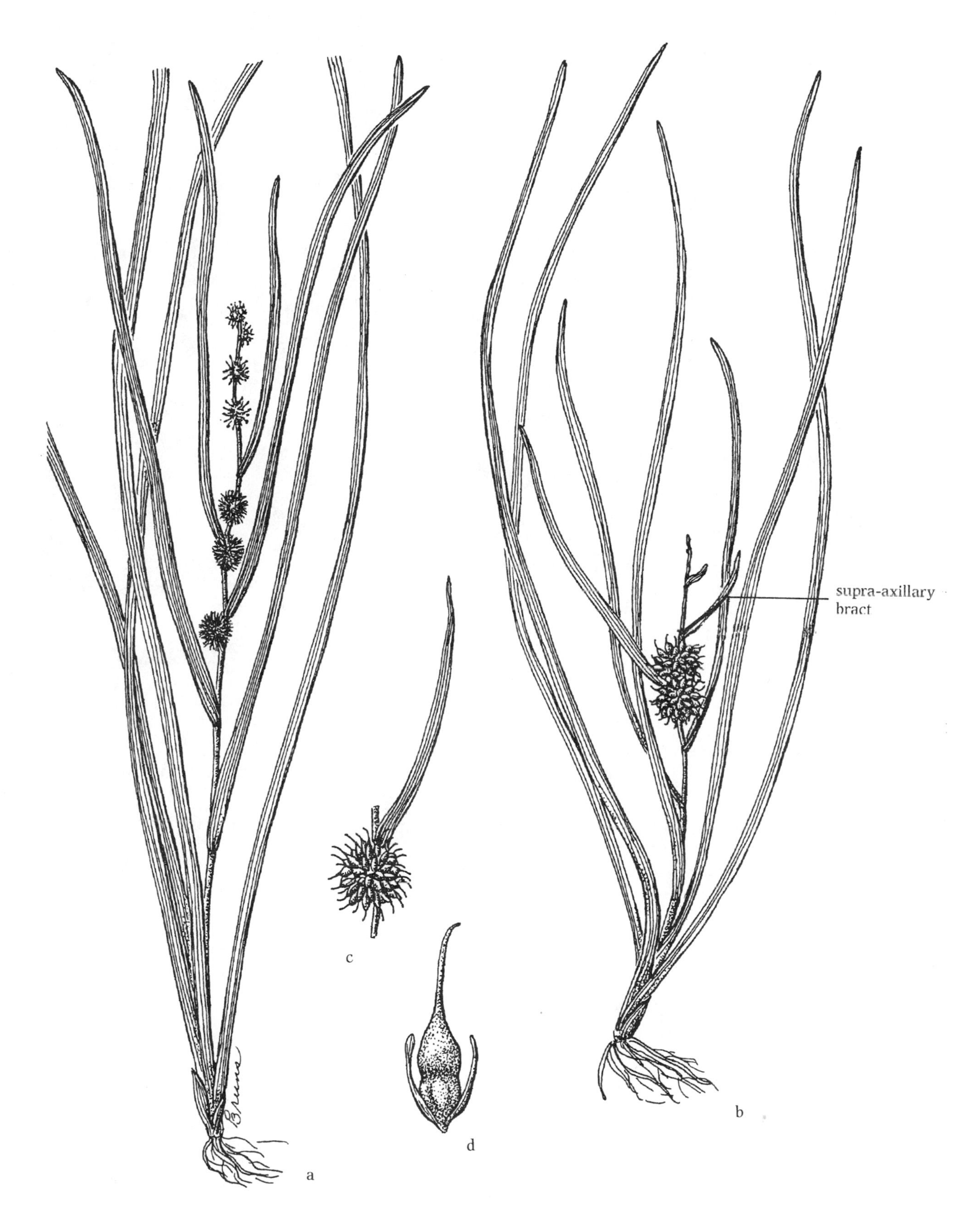

Fig. 154. *Sparganium emersum*: a, b. habit, two growth forms; c. fruiting head; d. fruit (NHAES).

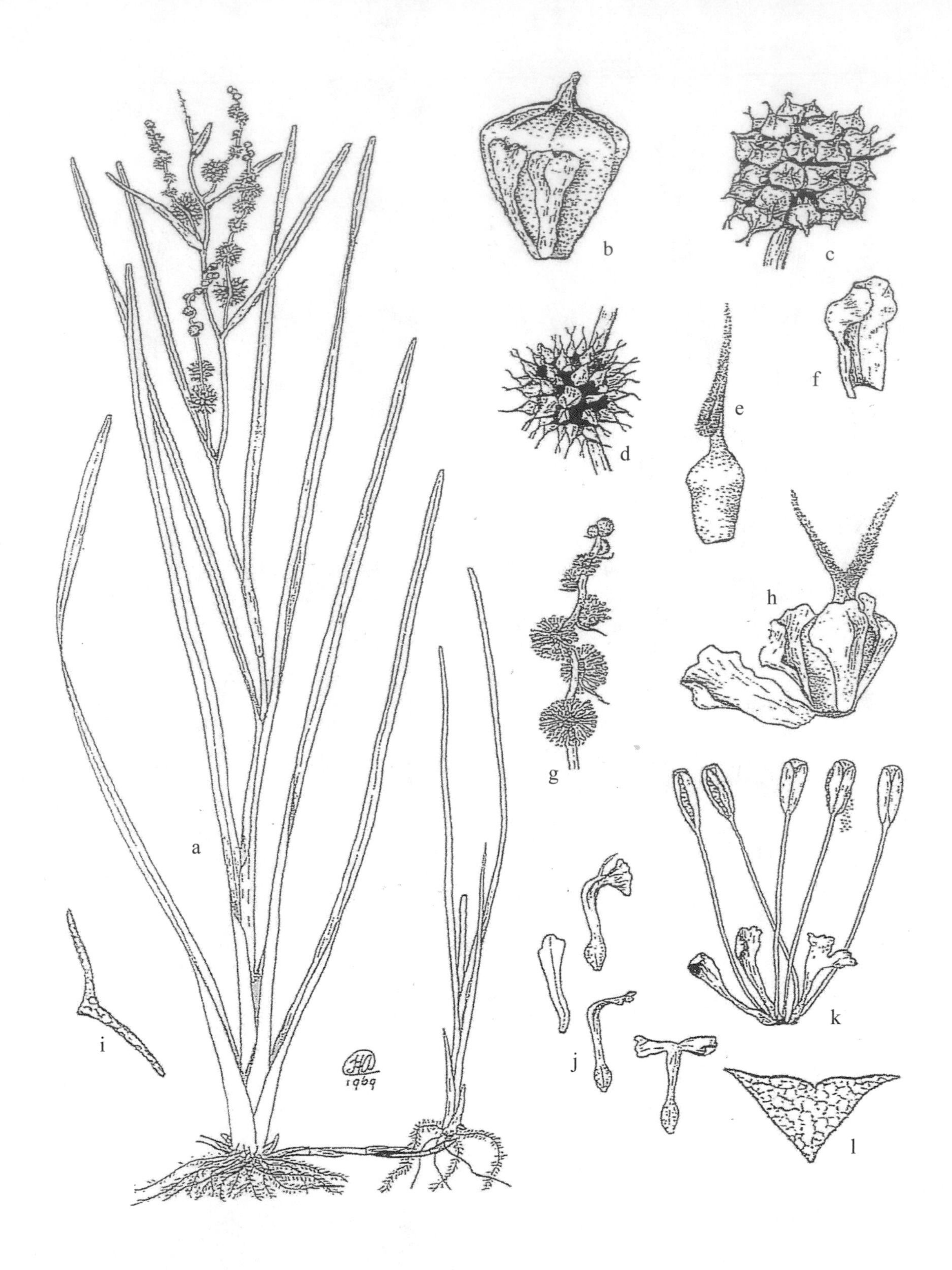

Fig. 155. *Sparganium erectum*: a. habit; b. mature fruit with two perianth segments visible; c. head of mature fruits; d. head of very young fruits, showing some with 1 stigma and some with 2 stigmas; e. pistil (ovary with single style and stigma); f. perianth segment; g. upper (staminate) portion of inflorescence showing lower heads fully at anthesis and unopened heads of male flowers; h. pistil with single style and two stigmas, subtended by perianth segments; i. leaf cross-section just below the apex; j. perianth segments from male flowers; k. male flower, showing stamens subtended by perianth; l. leaf cross-section just above basal sheath showing keeled lower portion (APOA).

Fig. 156. *Sparganium glomeratum*: a. habit; b. upper portion of plant with inflorescence, staminate head terminal, pistillate heads below; c. upper portion with several fruiting burr-like heads; d. cross-sections of distal, middle, and proximal portion of leaf; e. staminate head; f. pistillate head, cross-section (KJPT).

Fig. 157. *Sparganium fluctuans*: a. habit; b. upper portion of plant; c. fruit (NHAES).

Fig. 158. *Sparganium androcladum*: a. habit; b. upper portion of plant; c. fruiting head; d. fruit (NHAES).

Fig. 159. *Sparganium americanum*: a. habit; b. fruiting head; c. fruit (NHAES).
Sparganium hyperboreum: d. upper portion of plant; e. fruit (PB).

Xyridaceae / Yellow-eyed-grass Family

1. *Xyris* (Yellow-eyed-grass)

Emersed perennials; leaves grass-like, basal, 2-ranked (distichous, equitant), blades flat, linear, abruptly or gradually dilated toward sheathing base; inflorescence a compact, head-like spike; flowers bisexual, borne in axils of leathery or chaffy, overlapping bracts, often a single flower open at a time; petals yellow; stamens 3, with 3 petaloid 2-lobed staminodes, each lobe with margin of hairs (appearing fringed); fruit capsular, retained within persisting spike.

References: Kral, 1966b, 1983, 1988, 2000; Malme, 1997.

---

1. Tips of lateral sepals extending beyond tips of subtending bracts (fig. 160b).
   2. Keel of lateral sepals lacerate (fig. 160c). . . . . 1. *X. smalliana*
   2. Keel of lateral sepals ciliate (fig. 160g), fringed (fig. 161c), or entire (fig. 162g).
      3. Plant bulbous and hard at base (deeply set) (fig. 160e). . . . . 2. *X. caroliniana*
      3. Plants soft and flattened at base, not bulbous.
         4. Scapes 80–150 cm long, 1–1.5 mm wide, strongly scabrous, 2-ribbed; sepals conspicuously exerted, keels of lateral sepals, long-fringed with pale hairs, apex yellow-brown. . . . . 3. *X. fimbriata*
         4. Scales 3–45 cm long, 0.5–0.8(1) mm wide, appearing smooth (papillate), 2–4 ribbed; sepals slightly exerted, keels of lateral sepals entire, scarious or lacerate at tip, apex red. . . . . 4. *X. montana*
1. Tips of lateral sepals enclosed by subtending bracts.
   5. Keels of lateral sepals ciliate to ciliolate (fig. 162c) or scabrous.
      6. Plants bulbous and hard at base (fig. 162a); leaf blades twisted; lateral sepals tufted at tip (fig. 162c). . . . . 5. *X. torta*
      6. Plants soft and flattened at base, not bulbous; leaf blades erect or ascending, flat or only slightly twisted; lateral sepals not tufted at tip. . . . . 6. *X. ambigua*
   5. Keels of lateral sepals lacerate (fig. 163g), erose, or dentate.
      7. Plant bases bulbous, lowest outer leaves scale-like (fig. 163a); scapes twisted, often flexuous; upper portion of leaf base conspicuously twisted. . . . . 7. *X. platylepis*
      7. Plant bases not bulbous, lower leaves scale-like; scapes usually not twisted or flexuous; upper portion of leaf base not conspicuously twisted.
         8. Upper scape 4–several-ribbed, nearly round in cross-section. . . . . 8b. *X. difformis* var. *curtissii*
         8. Upper scape 1–2-ribbed, appearing round, or several-ribbed, with 2 ribs wider, giving winged flattened appearance.
            9. Mature fruiting spikes 15 mm long or longer; seed surface mealy 0.8–1.0 mm long or longer. . . . . 9. *X. laxifolia*
            9. Mature fruiting spikes less than 15 mm long (rarely 15 mm); seed surface not mealy, 0.4–0.5 mm long.
               10. Plant bases pinkish or purplish, sometimes greenish, leaves spreading; widespread in eastern U.S. . . . . 8a. *X. difformis* var. *difformis*
               10. Plant bases pale yellow-green to straw-colored; leaves ascending; coastal plain, e. Va. southward. . . . . 10. *X. jupicai*

---

1. *X. smalliana* Nash Small's Yellow-eyed Grass Fig. 160
Acid bogs, swamps, peaty or sandy shallows, lake and pond shore, quiet acidic streams and ditches. Coastal plain, s. Me and se. N.H. s. to s. Ga. and Fla., w. to s. Miss., se La. and e. Tex; C.Am.

2. *X. caroliniana* Walter Carolina Yellow-eyed Grass Fig. 160
Moist sands, peaty pine barrens, coastal dune swales and bogs. Coastal plain, N.J. s. to Fla., w. to s. Ala., La., e. Tex. and Mex.; Cuba, C.Am., and S.Am. This bulbous species is often confused with *X. torta*. *Xyris caroliniana*, a more southerly coastal species, has lateral sepals which are long-fringed toward the tip, whereas *X. torta*, a more ne. N.Am., inland species has a tuft of fringe at the tip.

3. *X. fimbriata* Elliott Fringed Yellow-eyed Grass Fig. 161
Sand, bogs, peat and muck of pineland pools, swamps, slow streams, ditches and pond margins. Coastal plain: N.J., se. Pa. and Del. s. to Fla., w. to Miss., La. and e. Tex.; inland in Tenn. and n. Ala. This is the easiest of all *Xyris* species to identify because of the scabrous scapes and fuzzy appearance of the lateral sepals.

4. *X. montana* Ries Northern Yellow-eyed Grass Fig. 162
Chiefly acidic sites, sphagnum bogs and muskegs, poor fens, peaty sands, seeps and riverbanks. Nfld. N.S. and N.E. w. to s. Que., Ont., n. Mich., Wisc. and ne. Minn. s. locally to n. N.J., and ne. Pa. and w. N.Y.

5. *X. torta* Sm. Slender Yellow-eye Grass Fig. 162.
Acidic sites, sphagnous bogs, acid sandy swamps and swales, pond shore, stream banks, and occasionally on dryish sandy pond shores. C. N.H. and e. Mass. w. to Ont., N.Y., Mich., n. Ind. and e. Minn. s. to Ga., Ala., La. and e. Tex.; Tenn., Mo. and Okla. This is the only bulbous-based species found in the Northeast north of New Jersey.

6. *X. ambigua* Beyr. ex Kunth Coastal Plain Yellow-eyed Grass Fig. 161
Acidic moist or wet sands or sandy peat of bog margins, pond and lake shores, pine flatwoods and ditches. Coastal plain and piedmont, e. Va., s. to Fla., w. to Tenn., Miss., Ark., e. Tex. and Mex.; Cuba and C.Am.

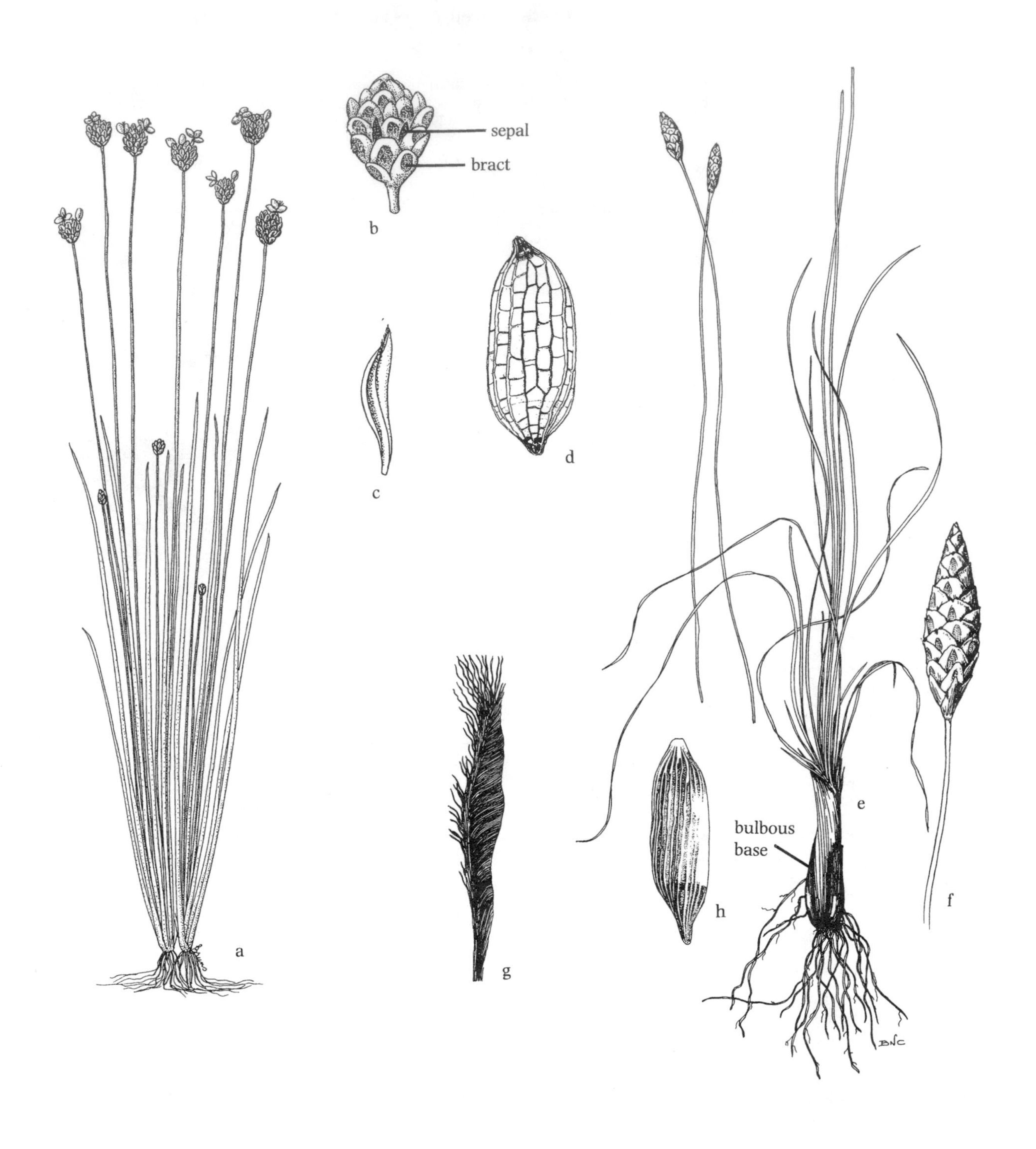

Fig. 160. *Xyris smalliana*: a. habit (NHAES); b. flowering spike (NHAES); c. lateral sepal (NHAES); d. seed (Kral, 1966b).
Xyris caroliniana: e. habit (G&W); f. flowering spike (G&W); g. lateral sepal (Kral, 1966b); h. seed (Kral, 1966b).

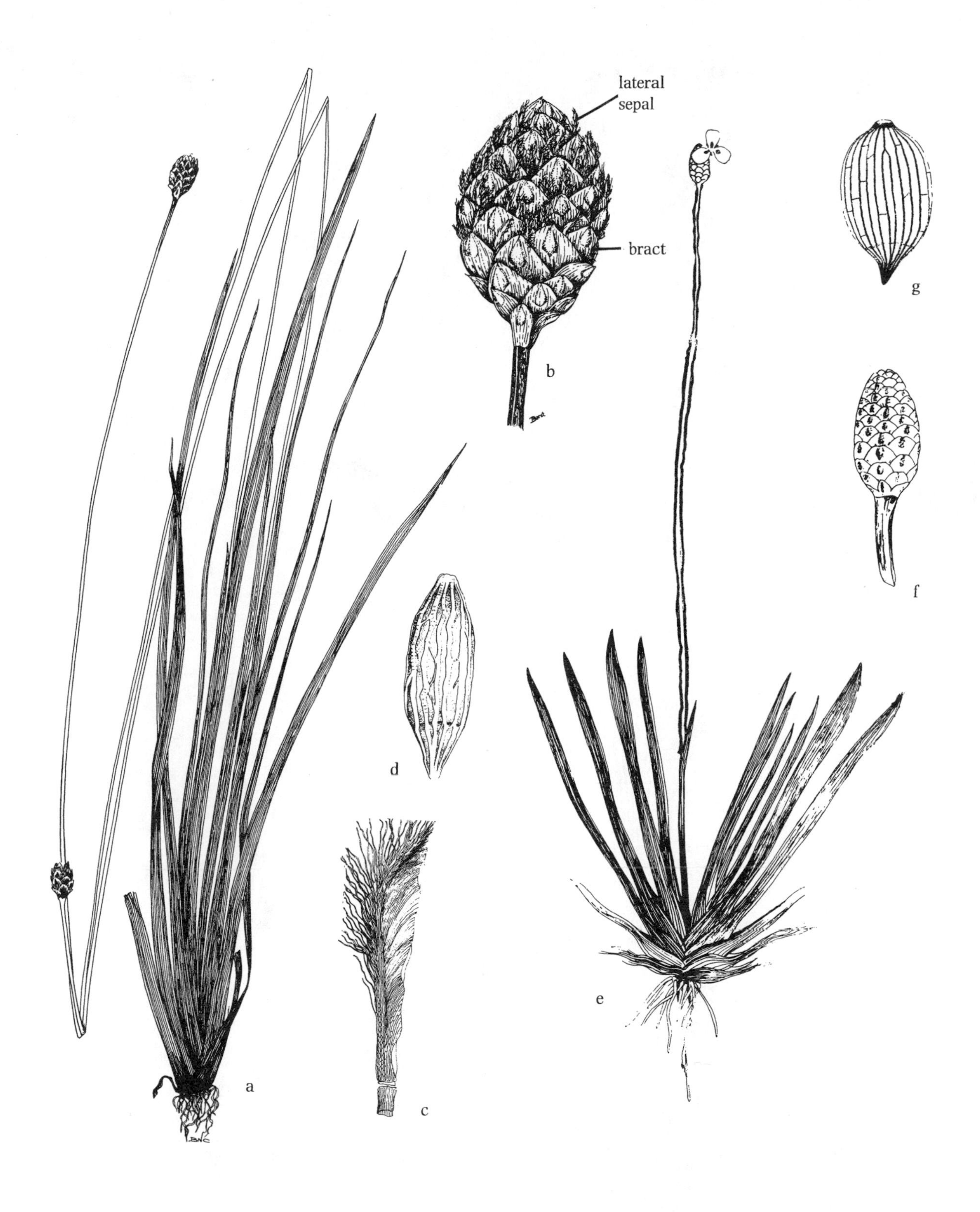

Fig. 161. *Xyris fimbriata*: a. habit (G&W); b. flowering spike (G&W); c. lateral sepal (Kral, 1966b); d. seed (Kral, 1966b).
Xyris ambigua: e. habit; f. flowering spike; g. seed (Kral, 1966b).

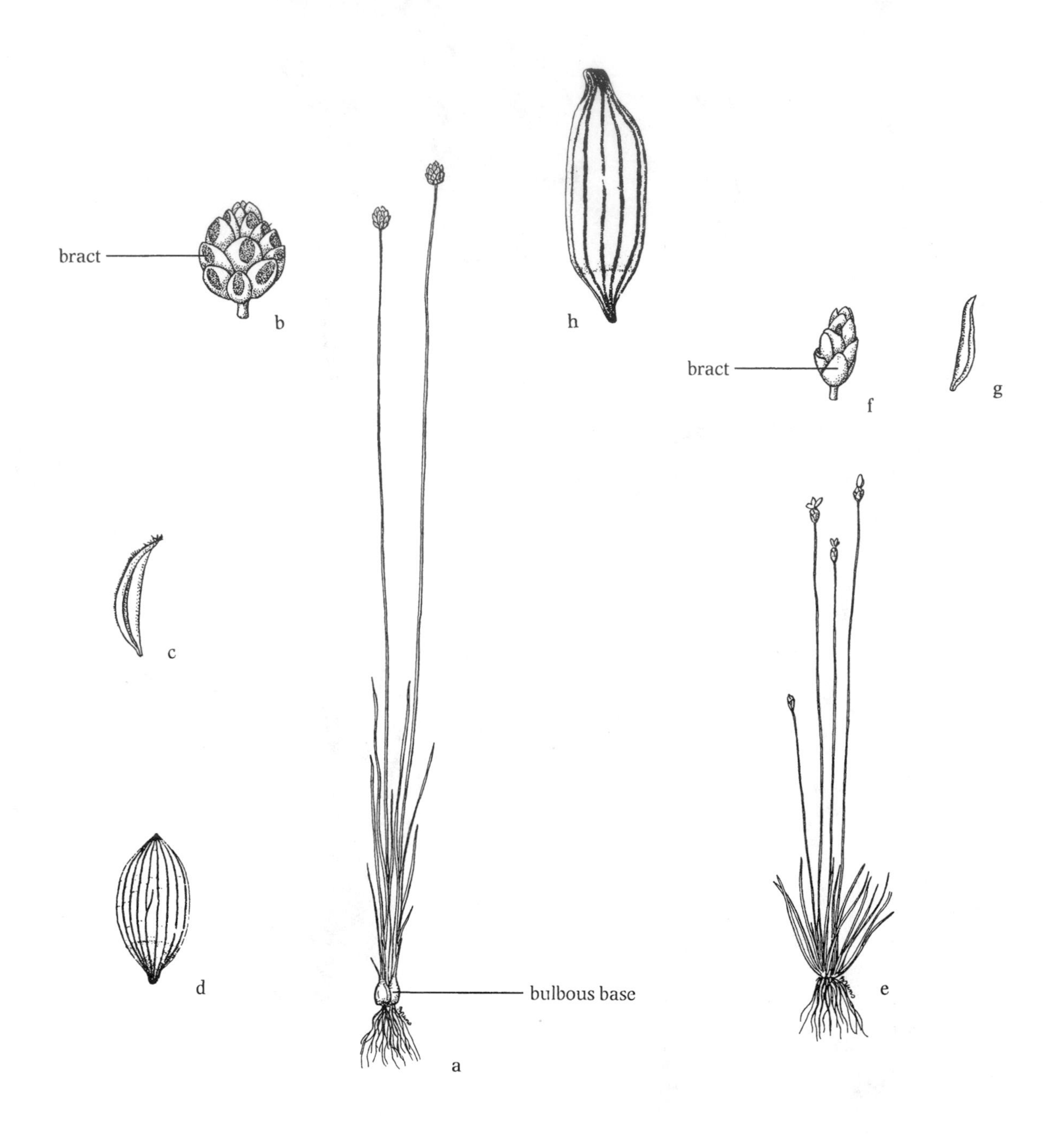

Fig. 162. *Xyris torta*: a. habit (NHAES); b. flowering spike (NHAES); c. lateral sepal (NHAES); d. seed (Kral, 1966b).
Xyris montana: e. habit (NHAES); f. flowering spike (NHAES); g. lateral sepal (NHAES); h. seed (Kral, 1966b).

7. *X. platylepis* Chapm. Tall Yellow-eyed Grass Fig. 163
Wet acidic pine flatwoods, swamps, peats and ditches. Coastal plain, e. Va. s. to Fla., w. to c. and e. Tex. This species is most easily confused with *X. torta* because of its pinkish or red scale-like bulbous leaf bases, but is easily distinguished by the lacerate keels of the sepals.

8. *X. difformis* Chapm.

8a. *X. difformis* var. *difformis* Bog Yellow-eyed Grass Fig. 164
Damp to wet acidic sandy shores of ponds and lakes, sphagnum bogs, poor fens, acidic swamps and in ditches. N.S. and Me. w. to N.Y., Ont., Mich., n. Ind. and Wisc., s. mainly along the coastal plain and piedmont to Fla., La., Ark., s. Okla. and e. Tex. This is the most abundant species in the Northeast.

8b. *X. difformis* var. *curtissii* (Malme) Kral
Curtiss' Yellow-eyed Grass Fig. 164
Wet pine barrens and ditches. Coastal plain, se. Va. s. to Fla., w. to e. Tex. (*X. curtissii* Malme)

9. *X. laxifolia* Mart. Lax-leaf Yellow-eyed Grass Fig. 163
Wet sands, clays and peats of marshes swamps, lake and pond margins and ditches. Coastal plain, se. Va. s. to n. Fla., w. to s. Ark and e. Tex.; Tenn.; C.Am. Our taxon is var. *iridifolia* (Chapm.) Kral. (*X. iridifolia* Chapm.)

10. *X. jupicai* Rich. Richard's Yellow-eyed Grass Fig. 165
Bogs, wet sands, swamps, lakeshores, open woodlands, flatwoods, swales and roadside ditches. Chiefly coastal plain and piedmont, s. N.J. Del, and Md. s. to Fla., w. to Ark., Okla. and e. Tex.; Mex., C.Am. and S.Am.

Eriocaulaceae / Pipewort Family

References: Hellquist and Crow, 1982; Kral, 1966a, 2000; Moldenke, 1937.

1. Scapes glabrous throughout; lacunar tissue in leaves evident without magnification; roots septate (fig. 168a); sepals 2, petals 2; lacunar tissue in leaves visible without magnification.. . . . . 1. *Eriocaulon*
1. Scapes pubescent, at least toward upper portion (fig. 169a); lacunar tissue in leaves not evident without magnification; roots not septate; sepals 3, petals absent; lacunar tissue in leaves not readily visible without magnification. . . . . 2. *Lachnocaulon*

1. *Eriocaulon* (Pipewort)

Submersed or emersed perennials, often tufted, roots conspicuously septate; leaves basal, linear, smooth, loosely cellular; inflorescence a head; plants monoecious or dioecious; flowers subtended by receptacular branchlets; fruit a loculicidal capsule.

1. Heads 7–10 mm in diameter, chalk white; scapes 10–12-ribbed.
   2. Heads soft, easily compressed upon pressing 10–20 mm wide; involucral bracts rounded to obtusely angled at tip, as long as or shorter than flowers, usually dark.. . . . . 1. *E. compressum*
   2. Heads hard, not compressed upon pressing 7–15 mm wide; involucral bracts acute at tip, longer than flowers, pale. . . . . . 2. *E. decangulare*
1. Heads 3–5 mm or less in diameter, dark gray, straw-colored, or dull white; scapes 4–7-ribbed.
   3. Plants of estuaries; scape 4–5-ribbed; mature heads hemispherical, 3–4 mm in diameter (fig. 168f), dull gray or straw-colored; bractlets and perianth parts sparingly clavate-pubescent, some parts smooth; involucral bracts tending to remain ascending on flowering/fruiting heads, concealing bractlets and flowers. . . . . 3. *E. parkeri*
   3. Plants of freshwater; scape (4)5–7-ribbed; mature heads subglobose, 4–5 mm in diameter (fig. 168b), gray, appearing dull white because of perianth parts and bractlets; bractlets distinctly clavate-pubescent; involucral bracts reflexed in flowering/fruiting heads and partly concealed by flowers. . . . . 4. *E. aquaticum*

1. *E. compressum* Lam. Flattened Pipewort Fig. 166
Sandy and peaty shores of shallow pools, depressions, pond and lake shores, and pine barren ponds. Coastal plain and piedmont, s. N.J., Del., e. Md., Va. and N.C. s. to e. Fla., w. to Miss., La. and e. Tex.

2. *E. decangulare* L. Ten-angle pipewort Fig. 167
Sandy and peaty shores of ponds, lakes, pine savannas and ditches, often in water up to 1 m deep. Coastal plain, N.J., N.C. s. to e. Fla., w. to s. Miss., La. and e. Tex.

3. *E. parkeri* B. L. Rob. Estuary Pipewort Fig. 168
Slightly saline waters and freshwater tidal mud flats, along muddy shores of estuaries. Rare, coastal, St. Lawrence R. estuary, Que., N.B., and N.E. s. to N.C. (*E. septangulare* var. *parkeri* (B. L. Rob.) Boivin & Cayouette)

4. *E. aquaticum* (Hill) Druce Seven-angle Pipewort Fig. 168
Acid waters, sandy and peaty shores. Nfld. w. to w. Ont., s. to L.I., N.Y., N.J., Del., mts. of N.C., n. Ohio, n. Ind., Wisc. and Minn. Submersed plants are often noted with scapes of up to 1 m or more in length. (*E. pellucidum* Michx.; *E. septangulare* With.)

Fig. 163. *Xyris platylepis*: a. habit (G&W); b. flowering spike (G&W); c. lateral sepal (Kral, 1966b); d. seed (Kral, 1966b).
Xyris laxifolia var. *iridifolia*: e. habit (G&W); f. flowering spike (G&W); g. lateral sepal (Kral, 1966b); h. seed (Kral, 1966b).

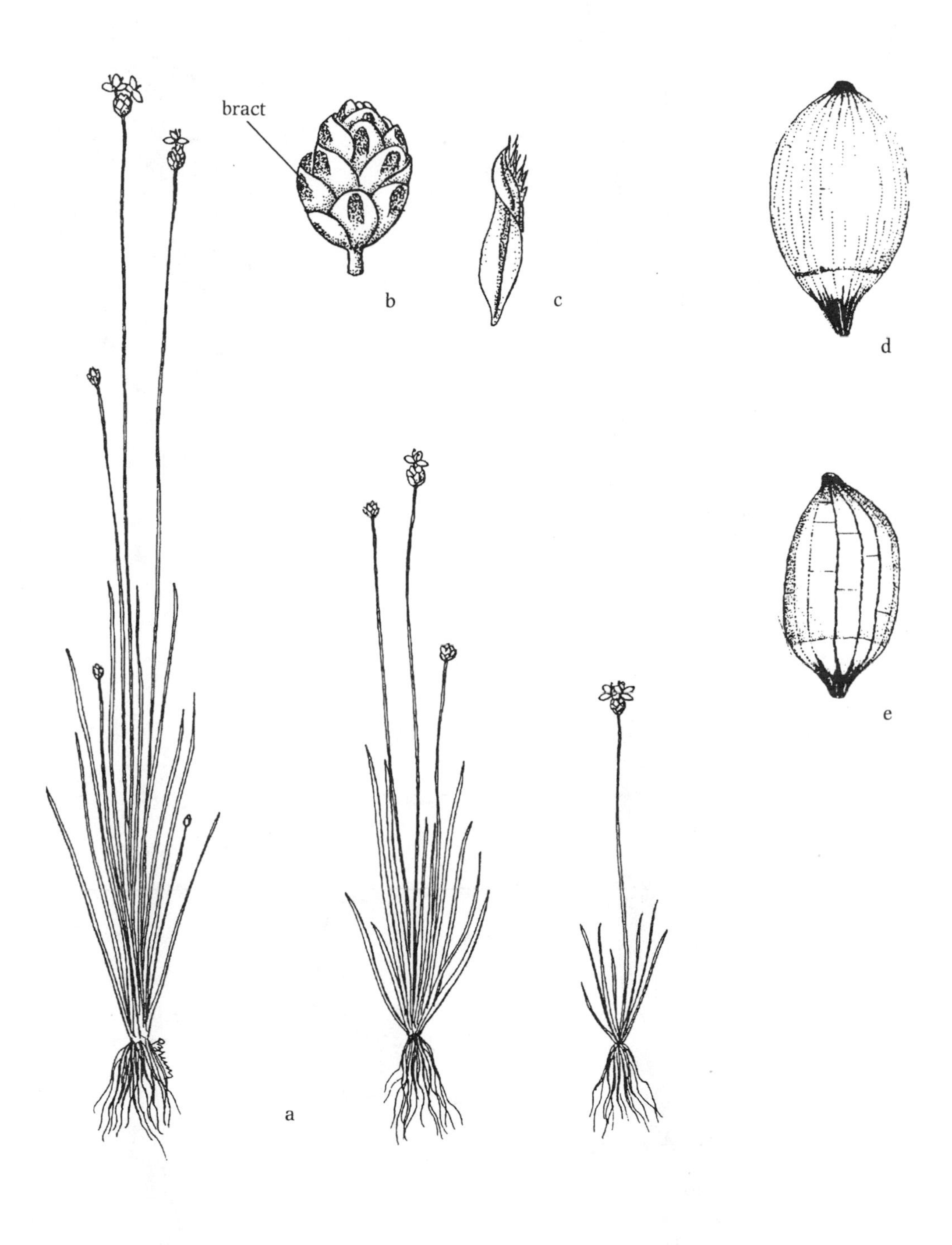

Fig. 164. *Xyris difformis* var. *difformis*: a. habit (NHAES); b. flowering spike (NHAES); c. lateral sepal (NHAES); d. seed (Kral, 1966b).
Xyris difformis var. *curtissii*: e. seed (Kral, 1966b).

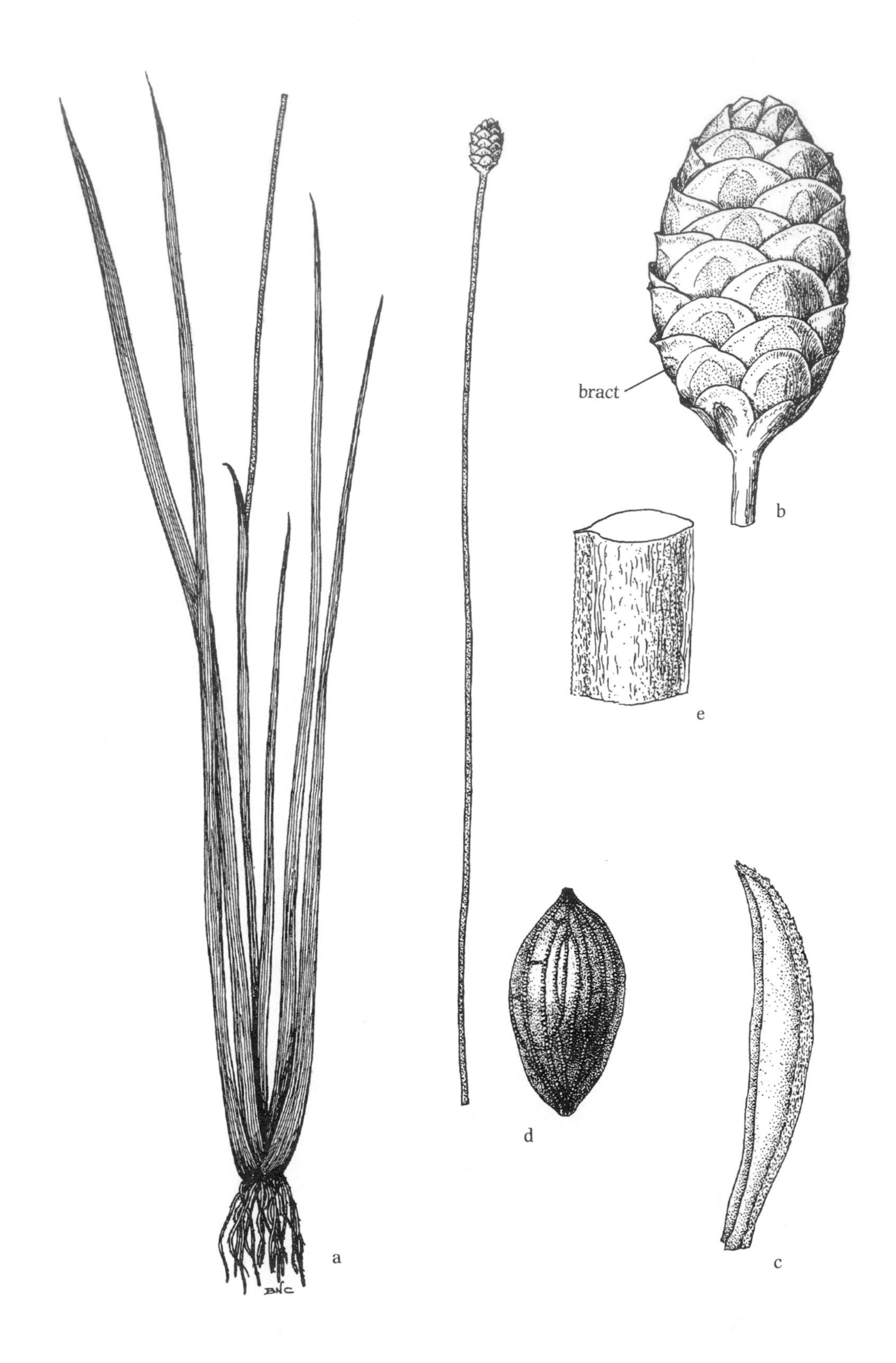

Fig. 165. *Xyris jupicai*: a. habit; b. flowering spike; c. lateral sepal; d. seed; e. scape, cross-section (G&W).

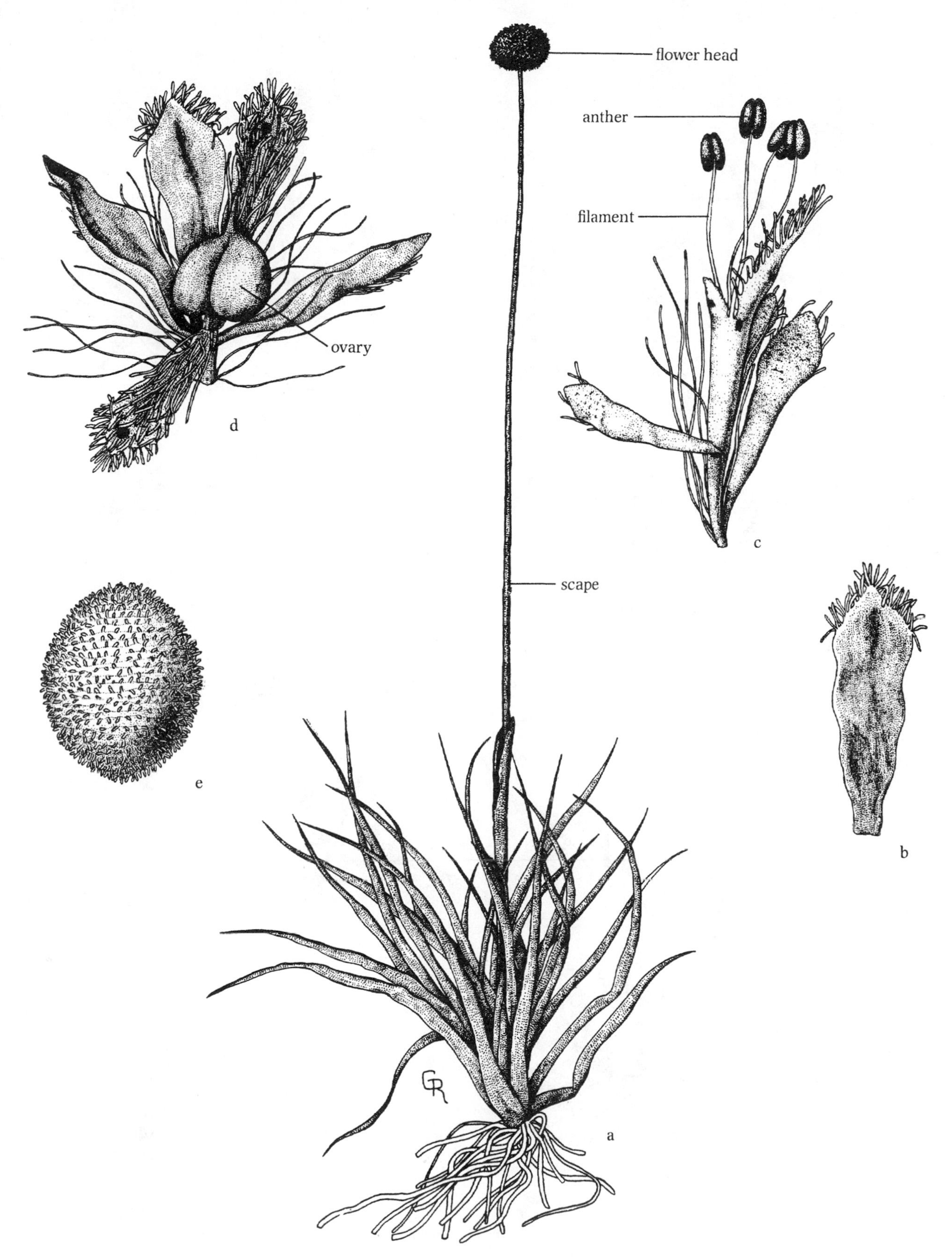

Fig. 166. *Eriocaulon compressum*: a. habit; b. receptacular bract; c. staminate flower; d. pistillate flower; e. seed (G&W).

Fig. 167. *Eriocaulon decangulare*: a. habit; b. staminate flower; c. pistillate flower; d. seed (G&W).

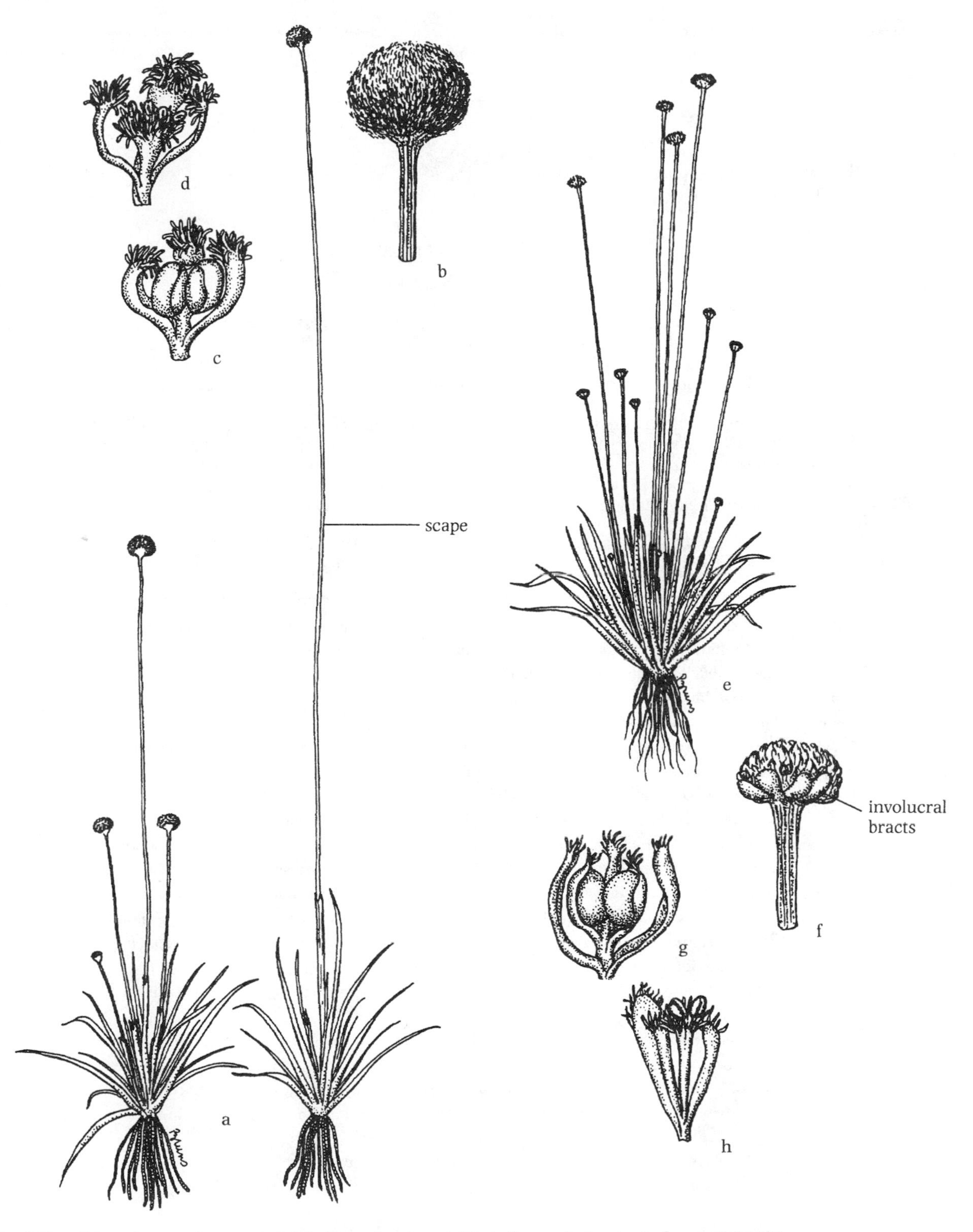

Fig. 168. *Eriocaulon aquaticum*: a. habit; b. flower head; c. pistillate flower; d. staminate flower (NHAES). *Eriocaulon parkeri*: e. habit; f. flower head, subtended by involucral bracts; g. pistillate flower; h. staminate flower (NHAES).

2. *Lachnocaulon* (Hairy Pipewort, Bog-buttons)

Emersed perennials, often tufted; roots branched, fibrous, and slender; leaves linear, spirally arranged appearing basal; inflorescence capitate; plants monoecious; flowers subtended and partly enfolded by 1 or 2 scarious bracts; fruit a 2- or 3-chambered loculicidal capsule.

1. *L. anceps* (Walter) Morong White-head Bog-button
Fig. 169
Moist to dry sands and peats in wet pine barrens and bogs, pond margins, lakeshores, flatwoods and seeps. Coastal plain, N.J., s. to Fla., w. to e. Tex.

Juncaceae / Rush Family

1. *Juncus* (Rush)

Usually perennial herbs, arising from rhizomes, grass-like; stems round in cross-section, pithy or hollow, rarely branching; leaf blades round in cross-section, channeled, flat, or vestigial and awn-like, and in some the leaves bladeless, entirely reduced to sheaths (*J. effusus* complex); flowers small, tepals parts similar (tepals), the outer whorl (sepals) 3, the inner whorl (petals) 3, scale-like; stamens 3 or 6; fruit a 3-valved capsule; seeds numerous, with one end blunt or mucronate, or both ends with distinct tails.

There are numerous species of *Juncus* and they are often distinguished with some difficulty. Fruiting material is usually critical for identification; anthers persist, hidden by tepals in fruit. Rushes are sometimes mistaken for sedges (see fig. 189) or grasses (see fig. 301), but are easily distinguished by flowers with 6 tepals and many-seeded capsules. One of the most common species is *Juncus effusus* (Soft Rush), which may occur in large clumps of several hundred stems and is often abundant in wet meadows, but care must be taken to be sure to distinguish the closely related *J. pylaei*. The true aquatics are *J. militaris*, *J. repens*, *J. subtilis*, and dwarfed forms of *J. marginatus* and *J. pelocarpus*. The others are typically found in marshy wet sites, the plants sometimes occurring in shallow water.

References: Brooks and Clemants, 2000; Brooks and Kuhn, 1986; Clemants, 1990; Cope and Stace, 1978; Eleuterius, 1975; Hämet-Ahti, 1980; Hermann, 1975; Reinking, 1981; Snogerup, 1963, 1980; Snogerup et al., 2002; Stuckey, 1980b, 1981; Zika, 2013, 2015.

---

1. Inflorescence apparently borne laterally, lowest involucral leaf erect, appearing as a continuation of stem (fig. 171a,e).
    2. Leaf sheaths often terminated by cylindrical blade about 1 mm in diameter (fig. 170d). . . . . 1. *J. coriaceus*
    2. Leaf sheaths bladeless (blades vestigial and awn-like).
        3. Individual flowers subtended by 1 bracteole at base of pedicel; flowers/fruits in clusters of 2–6 in branched inflorescences; plants of tidal marshes. . . . . 2. *J. roemerianus*
        3. Individual flowers subtended by 3 bracteoles at base of pedicel (fig. 171f); flowers/fruits 1 per inflorescence axis, or in a single compound cluster; plants of inland freshwater sites, or if coastal, then non-tidal.
            4. Stems borne in large clumps (fig. 171d); stamens 3.
                5. Inflorescences open, bract subtending inflorescence not swollen at base; capsules obtuse at apex.
                    6. Stems with ca. 30–60 fine ridges just below inflorescence, the ridges very closely spaced; the ridges without shiny swollen cells (10× magnification); sheath color pale brown to dark brown or reddish brown at base (never purple-black) the fresh stems shiny and feeling smooth when rolled between your fingers. . . . . 3. *J. effusus*
                    6. Stems with ca. 10–25 ridges just below inflorescence, the ridges separated by about 4–5 times the width of the ridge; the ridges with shiny swollen cells (10× magnification); sheath color purple-black at bases; the fresh stems dull, feeling ridged when rolled. . . . . 4. *J. pylaei*
                5. Inflorescences dense, head-like, bract subtending inflorescence swollen; capsules usually distinctly mucronate at apex. . . . . 5. *J. conglomeratus*
            4. Stems arising singly in rows, or in small clumps along creeping rhizomes (fig. 172a); stamens 6.
                7. Tepals green to brown, without dark stripes; stem with longitudinal ribs.
                    8. Seeds 0.7–1 mm long; basal sheaths red-brown; tepals 1.8–2.7 mm long shorter than capsule. . . . . 6. *J. gymnocarpus*
                    8. Seeds 0.5 mm long; basal sheaths green to pale brown; tepals 2.5–3.5 mm long, equaling or longer than capsule. . . . . 7. *J. filiformis*
                7. Tepals with black or dark red-brown stripes; stem without longitudinal ribs. . . . . 8. *J. balticus*
1. Inflorescence terminating stem, subtended by 1 or 2 ascending or spreading involucral leaves (fig. 173a,f).
    9. Leaves not septate.
        10. Seeds long-tailed at each end (fig. 173h).
            11. Heads with 1–4 flowers: capsule mucronate; seeds 2.5–4 mm long. . . . . 9. *J. stygius*
            11. Heads with 15–50 flowers; capsule blunt (fig. 173g); seeds 1–1.5 mm long. . . . . 10. *J. vaseyi*
        10. Seeds lacking tails (blunt or merely mucronate at apex) (figs. 173c, 174e).

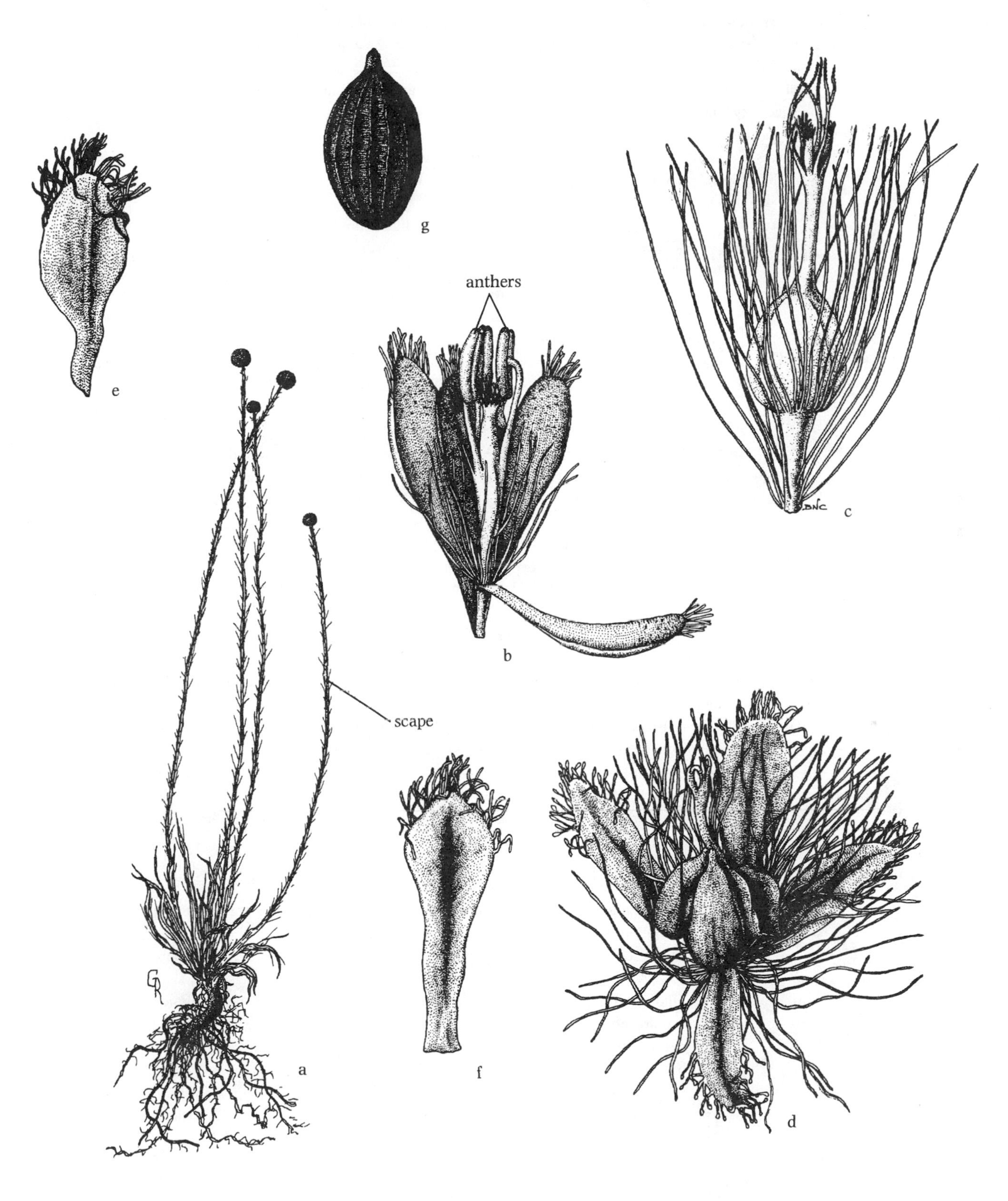

Fig. 169. *Lachnocaulon anceps*: a. habit; b. staminate flower; c. pistillate flower without sepals; d. pistillate flower; e. bract; f. sepal; g. seed (G&W).

12. Leaf blades round in cross-section at least toward apex (sometime channeled basally)
  13. Inflorescence open, greater than a fourth as tall as plant; tepals 2.5 mm long, shorter than capsule; some flowers often replaced by bulblets. . . . . 11. *J. pelocarpus*
  13. Inflorescence contracted, less than a fourth as tall as plant; tepals 3.5–5 mm long, longer than capsule; flowers not replaced by bulblets. . . . . 12. *J. dichotomus*
12. Leaf blades flat and/or involute and channeled to apex.
  14. Flowers in tight heads (figs. 174f, 175e) or congested (sometimes loosely) (fig. 173a); leaves flat.
    15. Leaf blades equitant (strongly laterally compressed in one plane, iris-like), appearing folded at sheath, mostly (1.5)3–6 mm wide. . . . . 13. *J. ensifolius*
    15. Leaf blades not equitant, flat but not folded, mostly less than 3 mm wide.
      16. Stems weakly ascending to prostrate and creeping, or floating, rooting at nodes; tepals parts 5–10 mm long; capsules narrowly ellipsoid (0.8–1.2 mm wide). . . . . 14. *J. repens*
      16. Stems erect; 3–6 mm long; capsules ovoid to oblong, obovoid, or broadly ellipsoid.
        17. Tepals green, lacking scarious margins; capsules broadly ellipsoid, (2.5)2.8–3.5(4.5) mm long; seeds ellipsoid to lunate, 0.3–0.4 mm long. . . . . 12. *J. dichotomus*
        17. Tepals with brown to dark brown conspicuously scarious margins and green mid-stripe; capsules obovoid to oblong, sometimes nearly globose; seeds fusiform or ovoid, 0.4–0.7.
          18. Stamens 6, yellow; tepals 5–7 mm long; stems arising singly, scattered along a rhizome or, if clustered, then lacking a bulb-like base; capsule long-beaked. . . . . 15. *J. longistylis*
          18. Stamens 3, reddish; tepals 2–3.5 mm long; stems 1–3, arising from a bulb-like base; capsule beakless.
            19. Inflorescence with 20–100 heads, each 2–10(12)-flowered; stamens about as long as sepals or longer, persisting with capsules. . . . . 16. *J. biflorus*
            19. Inflorescence with (5)10–25 heads, each 2–5-flowered; stamens shorter than sepals, shriveling at maturity. . . . . 17. *J. marginatus*
  14. Flowers single or in loose clusters; leaves involute and channeled to apex.
    20. Inflorescence a third or more as tall as plant (fig. 175a); plants annual, 4–20(30) cm tall, with soft bases; leaf sheaths not auriculate.
      21. Petals (inner tepals) acute to acuminate, slightly longer than capsule; capsule apex blunt to acute; plants not succulent. . . . . 18. *J. bufonis*
      21. Petals (inner tepals) rounded to mucronate, slightly shorter than capsule; capsule apex blunt to truncate; plants somewhat succulent. . . . . 19. *J. ambiguus*
    20. Inflorescence a fifth or less as tall as plant; plants perennial, typically more than 15 cm tall, with hard bases; leaf sheaths distinctly auriculate.
      22. Plants arising on elongated rhizomes, spreading horizontally, not tufted; leaf sheaths half the length of stem; tepals parts ovate to oblong rounded at apex (fig. 178e).
        23. Capsule shorter than to slightly longer than tepals, ellipsoidal-ovoid or sub-ellipsoidal; anthers 2.5–4 times as long as filaments; style as long as or longer than ovary. . . . . 20. *J. gerardii*
        23. Capsule distinctly exserted beyond tepals (up to 1.5 times as long as tepals); anthers 0.7–2 times as long as filaments; style shorter than ovary. . . . . 21. *J. compressus*
      22. Plants cespitose, tufted and producing clumps with short rhizomes; leaf sheaths short or extending up to a third the length of stem; tepals parts subulate, acute at apex.
        24. Auricles of leaf sheath very thin, conspicuously prolonged, whitish (fig. 179d), lobes longer than 0.5 mm.
          25. Inflorescence typically green; clusters of flowers 1–3(4), longer than the internodes; fruits usually more than 2.5 mm (more than ¾ length of tepals); stem with ridges lacking or only 1 strong ridge per side. . . . . 22. *J. tenuis*
          25. Inflorescence usually orange-red, clusters of flowers 2–5 shorter than the internodes; fruits less than 2.5 mm (shorter than ¾ length of tepals); stem with 2–6 strong ridges per side. . . . . . 23. *J. anthelatus*
        24. Auricles of leaf sheath firm, short, rounded, straw- to copper-colored (fig. 179h), lobes shorter than 0.5 mm long. . . . . 24. *J. dudleyi*

9. Leaves conspicuously septate, with firm internal cross-partitions at regular intervals (typically visible, and can be felt by running finger and thumb up the leaf, but can be further confirmed by longitudinal section).
  26. Seeds with conspicuous white tails, these short or long, up to 2.6 mm long at both ends (figs. 178i, 180h).
    27. Stems and leaves papillose-scabrous; seeds 2–3 mm long (including tails); stamens 6; plants on coastal plain. . . . . 25. *J. caesariensis*

27. Stems and leaves smooth; seeds 0.7–1.8 mm long; stamens usually 3.
  28. Tepals obtuse or subacute, with broad scarious margins, 2–2.6 mm long; seeds ellipsoidal, with short tails (sometimes inconspicuous, about a tenth to a third as long as the body).. . . . . 26. *J. brachycephalus*
  28. Tepals acuminate, with narrow scarious margins, 2–5 mm long; seeds fusiform, with conspicuous tails.
    29. Capsule shorter than to slightly longer than tepals, brown to straw-colored; heads 5–50-flowered.
      30. Mature fruit and tepals 3.8–5 mm long; seeds 1.3–1.8 mm long, with tails two-thirds as long to nearly as long as body length; inflorescence heads often subglobose, 5–50-flowered . . . . 27. *J. canadensis*
      30. Mature fruit and tepals 2–4 mm long; seeds 0.7–0.9(1.2) mm long, with shorter tails, about a third as long as body length; inflorescence heads narrower, hemispherical, 5–20-flowered.. . . . . 28. *J. subcaudatus*
    29. Capsule much longer than tepals, dark brown; heads 2–7-flowered.. . . . . 29. *J. brevicaudatus*
26. Seeds lacking tails (fig. 181d), or tails very short and inconspicuous.
  31. Stems weakly ascending to creeping or floating. . . . . 30. *J. bulbosus*
  31. Stems erect.
    32. Heads globose (fig. 181a), with lower flowers reflexed; capsules lanceolate, tapering to long slender tip (fig. 181c).
      33. Capsules lanceolate, tapering to long slender tip (fig. 181c).
        34. Blades of lower leaves laterally compressed; leaf sheaths approximately same width as blade. . . . . 31. *J. validus*
        34. Blades of lower leaves round to nearly round in cross-section; leaf sheaths distinctly wider than blade.
          35. Uppermost leaf blade (not the bract) as long as or longer than its sheath; sepals hard and firm.
            36. Petals as long as or longer than sepals; sepals 2.5–4 mm long; lobes at summit of leaf sheath 0.5–2.5 mm long; stems 0.7–1.5 mm in diameter near middle.
              37. Lobes at summit of sheath 1–2.5 mm long; heads 2–12(30), each 15–40-flowered, pale green to dull brown; stamens 3.. . . . . 32. *J. scirpoides*
              37. Lobes at summit of sheath 0.5–1 mm long; heads (1)2–15, each 8–20-flowered, reddish brown; stamens 6. . . . . 33. *J. nodosus*
            36. Petals slightly shorter than sepals; sepals 4–5.5 mm long; lobes at summit of leaf sheath 2.5–3.5 mm long; stems 2–4 mm in diameter near middle. . . . . 34. *J. torreyi*
          35. Uppermost leaf blade (not the bract) shorter than its sheath; sepals soft and flexible. . . . . 35. *J. megacephalus*
      33. Capsules obovoid, abruptly contracting to a very short tip (fig. 182h). . . . . 36. *J. brachycarpus*
    32. Heads hemispherical or obconic (fig. 188b), with lower flowers not reflexed.
      38. Stamens 3.
        39. Tepals as long as or slightly longer than capsule.
          40. Inflorescence of (40)50–500 heads; capsule rounded or minutely mucronate; sepals 2.2–2.9 mm long.
            41. Leaves 3.5 mm in diameter, with conspicuous ring-like bands at cross-partitions; roots lacking tubers; petals 1.9–2.3 mm long; capsule 2.3–2.5 mm long. . . . . 37. *J. nodatus*
            41. Leaves 1–3 mm in diameter, with obscure ring-like bands at cross-partitions; some roots bearing tuberous thickenings; petals 2.4–2.8 mm long; capsule 2.4–2.9 mm long. . . . . 38. *J. elliottii*
          40. Inflorescence usually of fewer than 50 heads; capsule acute and apiculate; sepals 3.3–4 mm long. . . . . 39. *J. acuminatus*
        39. Tepals much shorter than capsule.
          42. Tepals of outer whorl shorter than inner tepals; petals (inner whorl) more or less blunt; seeds 0.8–1.2 mm long, tails sometimes inconspicuous (ca. one-tenth body length or less).. . . . . 26. *J. brachycephalus*
          42. Tepals of inner and outer whorls the same length; petals (inner whorl) acute to very sharp-pointed (subulate); seeds less than 0.7 mm long, lacking tails.
            43. Capsule 2.3–3 mm long, lanceolate; tepals lanceolate-acute, two-thirds the length of capsule. . . . . 40. *J. debilis*
            43. Capsule 4–5 mm long, slender, lanceolate-linear; tepals linear-subulate. half the length of capsule. . . . . 41. *J. diffusissimus*
      38. Stamens 6.
        44. Flowers single or in heads of 2 or 3(5), often replaced by bulblets.
          45. Stems lax and creeping, floating, or submersed.
            46. Plants with firm base, not forming mats, if completely submersed, then consisting of tufts of basal leaves, with stems often absent (aquatic form but usually occurs as an erect plant). . . . . 11. *J. pelocarpus*
            46. Plants with soft base, often forming mats, with elongate floating leaves and/or creeping branches. . . . . 41. *J. subtilis*

45. Stems ascending to erect. . . . . 11. *J. pelocarpus*

44. Flowers in heads of (2)3–13(30), never replaced by bulblets.

47. Stems thick, over 50 cm long; fully developed cauline leaf 1, overtopping inflorescence (fig. 184e); rhizomes often giving rise to submersed, thread-like leaves (fig. 184e); flowers 3–4 mm long. . . . . 43. *J. militaris*

47. Stems slender, rarely over 50 cm long; fully developed cauline leaves 2 or more, not overtopping inflorescence; rhizomes not producing submersed thread-like leaves; flowers 2–3 mm long.

48. Branches of inflorescences ascending (fig. 188a); inflorescences usually twice as long as wide; flowers 2–2.5 mm long, petals shorter than or as long as sepals. . . . . 44. *J. alpinoarticulatus*

48. Branches of inflorescence spreading (fig. 188f); inflorescences usually less than twice as long as wide; flowers 2.5–3 mm long, petals as long as or longer than sepals. . . . . 45. *J. articulatus*

---

1. *J. coriaceus* Mack. Leather Rush Fig. 170
Swamps, marshes, and wet woods. S. N.J. w. to Ky., Ark. and Okla., s. to Fla. and e. Tex.

2. *J. roemerianus* Scheele Needle Rush Fig. 171
Saline or brackish marshes. Coastal, Del. and Md. s. to Fla., w. to Tex.

3. *J. effusus* L. Soft Rush, Common Rush Fig. 171
Marshes, swamps, thickets, pond shores, and ditches. Nfld. w. to Ont. and Man., B.C. and Alask., s. throughout the U.S.; C.Am. and S.Am. This is a widely distributed and extremely variable species, with a number of infraspecific taxa described, and while extremes may appear distinct, the presence of intermediates resulted in several being relegated to synonymy. Two subspecies occur within our region. The widely ranging subsp. *solutus* Fernald and Wieg. can be distinguished by having tepals rigid, erect, appressed against capsule; upper sheaths more than 15 cm long, with margins mostly pale and loosely wrapped around stem or flattening and coming free, and open inflorescences. The Eurasian subsp. *effusus*, is amphi-Atlantic in distribution and native in Nfld., N.S., N.B. se. Que., Me. and Mass., but is a rare, introduced plant westward in N.Am.; it is distinguished by having tepals soft, spreading or curving away from capsule; upper sheaths less than 15 cm long, with margins darkened and tightly wrapped around stem, and an inflorescence dense or open (Hämet-Ahti, 1980; Zika, 2013).

Another Old World species that is densely cespitose and an appearance similar to *J. effusus* (but having 6 stamens instead of 3) that has been introduced and is spreading long roadsides and ditches in ne. N.Am. is *J. inflexus* L.; a native of the Caucasus where it occurs in wet soils along streams, ditches and sandy or peaty slopes, this plant is recommended as a cultivar planting for water gardens on pond margins and is sold under the name Hard Rush, Blue Rush or Blue Arrows Rush.

4. *Juncus pylaei* Laharpe
N.E. and N.Y. w. to Mich., Wisc. and Minn., s. to N.J., Va., and S.C., Tenn., and Ill.; introd. in Pacific Northwest, B.C., Mont., Ida., Wash. and Oreg. *Juncus effusus* is closely related to *J. pylaei*; they are geographically sympatric and can occur together in the same wetland. This species, previously regarded as a variety of *J. effusus*, is less robust, but has more strongly ridged and less shiny stems. (*Juncus effusus* var. *pylaei* (Laharpe) Fernald & Wiegand)

5. *Juncus conglomeratus* L. Bunch-flowered Soft Rush
Boggy, peaty and swampy places. Nfld., N.S., s. Que., s. Me, se. N.H., w. Mass. and R.I. w. to N.Y., s. to Pa. and W.Va.; B.C., Wash. and Oreg.; intro. from Europe. This species is very similar to *J. pylaei* and *J. effusus*, having dull ridged stems and red-brown sheaths, but is distinguished by a dense, capitate inflorescence, and the bract subtending the inflorescence with a swollen base. (*J. effusus* var. *conglomeratus* (L.) Engelm.)

6. *J. gymnocarpus* Coville Pennsylvania Rush Fig. 172
Acid, mossy swamps. Local; mts. of Pa., N.C., S.C., and Tenn.; coastal plain of Ga., w. Fla., Ala. and Miss.

7. *J. balticus* Willd. Fig. 172
Brackish and freshwater marshes and shores. Lab. and Nfld. w. to B.C. and Alask., s. to N.E., Pa., n. Ohio, Ill., Mo., Okla., nw. Tex., N.M., Ariz. and Calif.; S.Am. Our taxon belongs to subsp. *littoralis* (Engelm.) Snogerup (Snogerup et al., 2002; Zika, 2013). (*J. arcticus* var. *balticus* (Willd.) Trautv.; *J. arcticus* subsp. *littoralis* (Engelm.) Hultén)

8. *J. filiformis* L. Thread Rush Fig. 170
Wet shores, peat bogs, and ditches. S. Greenl., and Lab. w. to Alask., s. to Mass., N.Y., Md., W.Va., n. Mich., n. Minn., Wyo., s. to Neb., Utah, N.M. and c. Oreg.

9. *J. stygius* L. Moor Rush Fig. 172
Wet sphagnous bogs. S. Lab. and Nfld. w. throughout Can. and Alask., s. to n. Me., N.Y., n. Mich., ne. Wisc., and n. Minn.; Colo.; nw. Pacific Rim. Our taxon is var. *americanus* Buchenau.

10. *J. vaseyi* Engelm. Vasey's Rush Fig. 173
Damp shores, meadows, and thickets. Very rare; Lab. and Que. w. to n. Alta., N.W.T. and B.C., s. to Me., Vt., Mich., Ill., Iowa, Colo., Utah, and Ida. (*J. greenei* var. *vaseyi* (Engelm.) Boivin)

11. *J. pelocarpus* E. Mey. Brownfruit Rush Fig. 174
Sandy and peaty shores, often submersed or floating. S. Lab. and Nfld. w. to Ont. and Minn., s. to N.E., coastal plain to Del., N.C., Ga. s. to Fla.; inland, n. Ind. to Wisc. and Minn.; introd. B.C. s. to Oreg. Flowers are often replaced by bulblets (fig. 174c). Submersed plants (forma *submersus* Fassett) are very similar to the typical floating state of *J. subtilis*. More robust plants ranging southward along the coastal plain, sometimes separated as var. *crassicaudex* Engelm. has been shown to represent clinal variation. (*J. abortivus* Chapm.)

12. *J. dichotomus* Elliott Forked Rush Fig. 173
Damp to wet sands and peats. Coastal plain. Me., and Mass. s. to Fla., w. to Tex., N.M. and Mex.; inland n. to Ky., Ark. and Colo.; introd. in Calif.; Mex., C.Am., W.I. and S.Am.

13. *J. ensifolius* Willd. Swordleaf Rush
Wet ground. Sw. Sask., and s. Alta. w. to B.C. and Alask, s. to N.D., S.D., Colo., Ariz. and Calif. and Utah; disjunct to James Bay, ne. Ont. and nw. Que; adventive in the East, se. N.Y., and n. Wisc.

Fig. 170. *Juncus coriaceus*: a. habit; b. perianth and capsule; c. seed; d. leaf sheath (G&W).
Juncus filiformis: e. habit; f. perianth and capsule; g. seed (HCOT).

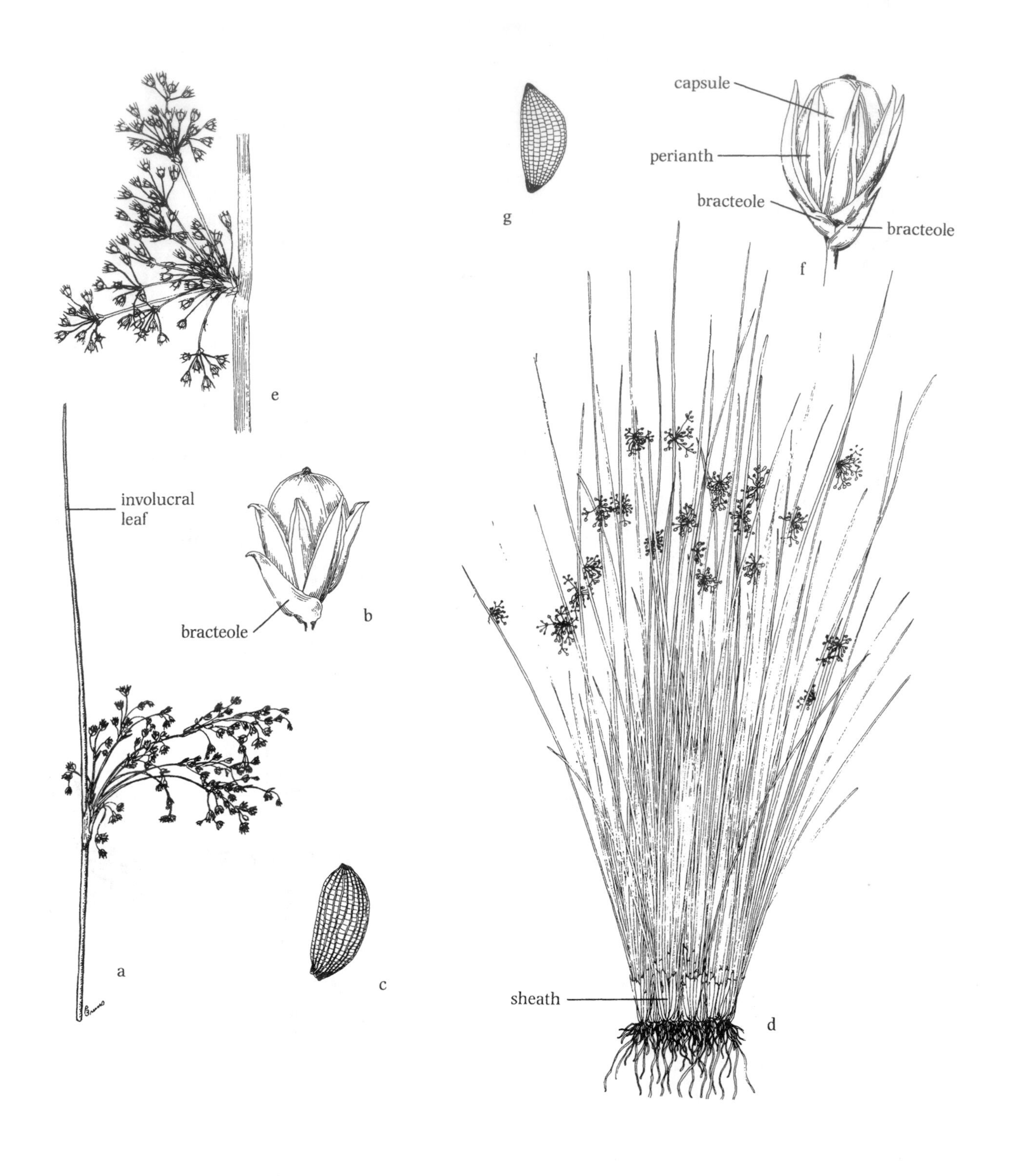

Fig. 171. *Juncus roemerianus*: a. inflorescence (PB); b. perianth and capsule (Beal); c. seed (Beal). *Juncus effusus*: d. habit (F); e. inflorescence (F); f. perianth and capsule (Beal); g. seed (Beal).

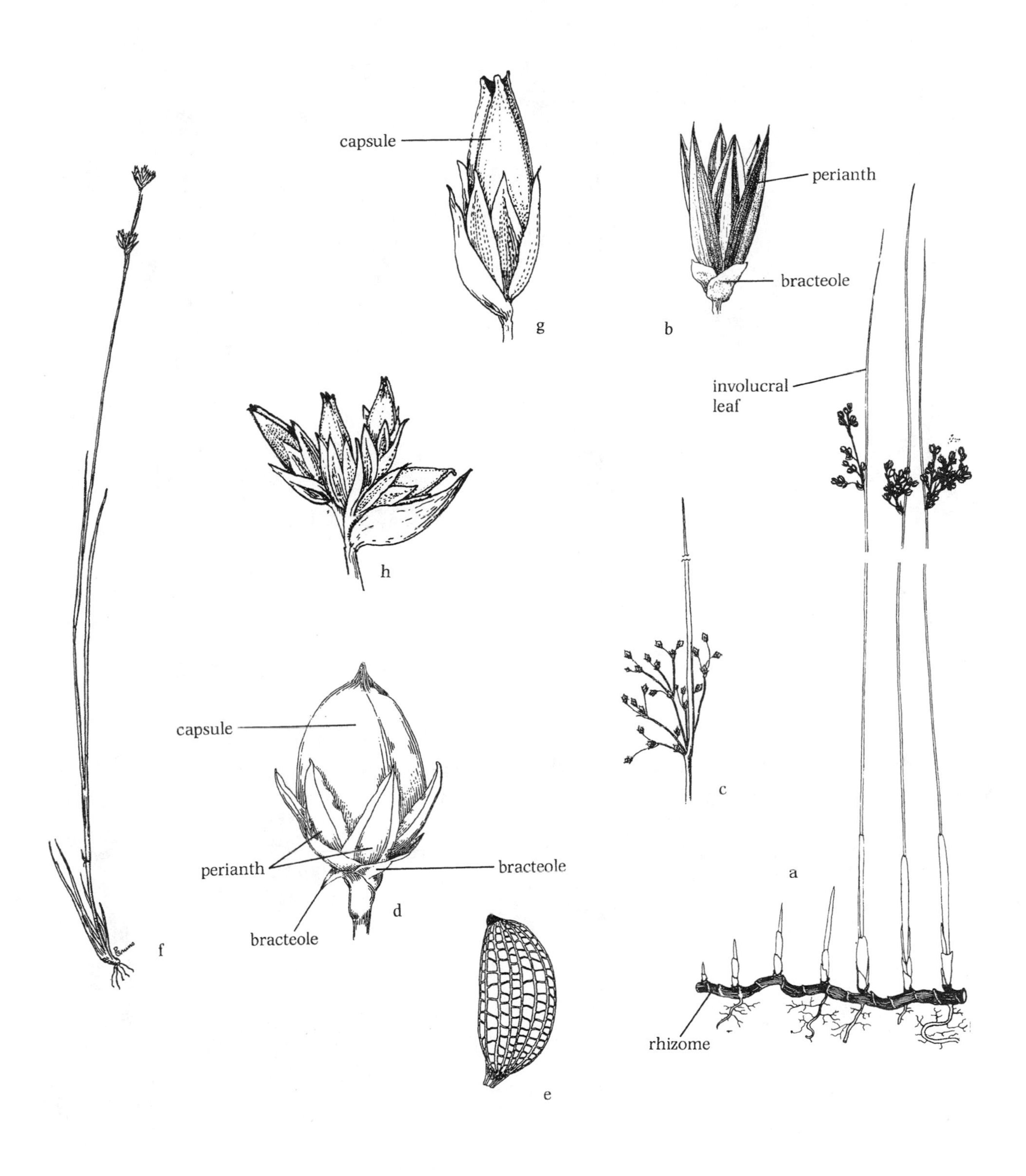

Fig. 172. *Juncus balticus*: a. habit (F); b. perianth (HCOT).
Juncus gymnocarpus: c. inflorescence; d. perianth and capsule; e. seed (Beal).
Juncus stygius: f. habit (PB); g. perianth and capsule (Crow); h. inflorescence (Crow).

Fig. 173. *Juncus dichotomus*: a. habit; b. perianth and capsule; c. seed; d. leaf sheath (G&W).
Juncus vaseyi: e. habit (HCOT); f. inflorescence (PB); g. perianth and capsule (HCOT); h. seed (HCOT); i. leaf sheath (HCOT).

Fig. 174. *Juncus pelocarpus*: a. habit, terrestrial form (F); b. habit, submersed plant (F); c. branch with bulblets (F); d. capsule and perianth (PB); e. seed (PB).
Juncus repens: f. habit, terrestrial form (PB); g. habit, submersed form (PB); h. perianth and capsule (G&W); i. seed (G&W); j. leaf sheath (G&W).

14. *J. repens* Michx. Lesser Creeping Rush Fig. 174
Acid ponds, lakes, depressions, and ditches. Coastal plain, Del. s. to Fla. w. to Tex.; inland n. to Tenn., Ark. and Okla. This species often produces large mats of vegetative plants submersed and floating leaves.

15. *J. longistylis* Torr. Longstem Rush Fig. 175
Damp gravelly and sandy shores, often on stream banks. Man. w. to B.C., s. to Mich., S.D., Neb., N.M., Ariz. and Calif.; locally in Ont., Que., and Nfld.

16. *J. marginatus* Rostk. Grassleaf Rush Fig. 177
Moist sands and peaty soils, marshes, wet clearings, and fields. N.S. and Me. w. to s. Ont., s. Mich., Ill., Mo., Neb. and S.D., s. to Fla., and e. Tex.; Oreg., Calif., Ariz.; S.Am. Another plant belonging to the *J. marginatus* complex, *J. longii* Fernald, is now recognized as a distinct species, and may be encountered in early successional seepages, both temporary and permanent, chiefly on the coastal plain and piedmont of Md., e. Va. and southward (Knapp and Naczi, 2008).

17. *J. biflorus* Elliott Fig. 176
Marshes, low fields, and peaty or sandy bogs. Se. Mass.; se. N.Y. and Pa. w. to Ont., s. Mich., Ill., and e. Kans., s. to Fla., Okla., Tex. and Mex.; Ariz.; C.Am. and S.Am. (*J. marginalis* var. *biflorus* (Elliott) Torr.; *J. odoratus* (Torr.) Steud.) Some authors lump this with *J. marginatus* (Brooks and Clemants, 2000), but in a study of the *J. marginatus* complex Knapp and Naczi (2008) concluded that *J. biflorus*, as well as the mid-Atlantic coastal plain endemic *J. longii*, should be recognized at the species level.

18. *J. bufonius* L. Toad Rush Fig. 175
Damp or dry open areas and muddy shores. Throughout N.Am.; nearly world-wide. Plants with a congested inflorescence often on saline shores, are var. *congestus* Wahlb.

19. *J. ambiguus* Guss. Seaside Rush
Salt marshes and damp, brackish or saline soils. S. Lab. and St. Lawrence Seaway, Que., s. to coastal Mass.; L.I., N.Y.; James Bay; Sask., s. to Neb. and Colo., Calif. and Ariz. (*J. bufonius* var. *halophilus* Buchenau & Fernald)

20. *J. gerardii* Loisel. Black-grass, Saltmeadow Rush Fig. 178
Salt marshes and disturbed saline habitats. Coastal Nfld. and Que. s. to Va.; locally inland w. to N.Y., n. Ky., Ont., s. Man., se. N.D., and Mo., Wyo., Colo. and Utah; sw. B.C. and nw. Wash.

21. *J. compressus* Jacq. Round-fruit Rush
Disturbed saline sites, highway and railroad ditches, and sometimes invading salt marshes. Very local, w. Nfld., C.B.I., St. Lawrence Seaway, Que. w. to Ont., Man., Sask. and B.C., s. to e. Me., c. N.Y., Del., Md. Ohio, Ill., Wisc., nw. Minn., Neb., Colo., nw. Utah, and Oreg.; intro from Eur.

22. *J. tenuis* Willd. Poverty Rush, Path Rush Fig. 179
Damp, wet, or dry open soil of roadsides, thickets, and swamps. Throughout N.Am.; subcosmopolitan. The presence of well-developed auricles is critical to separate *J. tenuis* from *J. dudleyi*. (*J. tenuis* var. *williamsii* Fernald)

23. *J. anthelatus* (Wiegand) R. E. Brooks Large Path Rush
Swamps, wet fields and meadows, margins of wetland, and moist to seasonally dry sandy prairies, borrow pits, paths, and other disturbed habitats. Having the appearance of *J. tenuis*, but a larger plant, often 0.5 m or more tall, with a lax, large inflorescence.

24. *J. dudleyi* Wiegand Dudley's Rush Fig. 179
Damp to dry soil, swamps, and lakeshores, often in calcareous soil. Throughout N.Am. with exception of extreme se. U.S. and n. Mex. (*J. tenuis* var. *dudleyi* (Wiegand) F. J. Herm.)

25. *J. caesariensis* Coville New Jersey Rush Fig. 180
Sphagnous swamps and mossy woods. Very rare; N.S.; s. N.J. s. to Md., se. Va. and N.C.

26. *J. brachycephalus* (Engelm.) Buchenau Smallhead Rush Fig. 181
Calcareous shores, marshes, and meadows. N.S. and N.B. w. to Ont. and N.D. s. to w. N.E., n. N.J., Pa., N.C., Ga., Ala., Tenn. and Ind.; Okla. and Colo.

27. *J. canadensis* J. Gay ex Laharpe Canadian Rush Fig. 178
Marshes, swamps, meadows, peats, and saline and brackish marshes. Nfld. w. to s. Ont., Minn. and Man., s. to s. to Fla., Ala., La., Mo. and Neb.; B.C. s. to w. Wash. and w. Oreg. Similar to *J. brachycephalus* and *J. brevicaudatus* but usually a larger plant; often bulblets replacing flowers in the inflorescence late in season, promoting asexual reproduction.

28. *J. subcaudatus* (Engelm.) Coville & S. F. Blake Woodland Rush Fig. 182
Bogs, mossy woods, and pond margins. Nfld., N.S., Mass., and s. Conn. w. to Pa., W.Va., Ark. and Mo., s. to Ga., Tenn. and Ala.

29. *J. brevicaudatus* (Engelm.) Fernald Narrow-panicle Rush Fig. 182
Muddy, sandy, and wet shores. Lab. and Nfld. w. to n. Alta. and B.C. s. to N.E., e. N.Y., Pa., mts. of N.C. and Tenn., s. Ont., Mich., n. Ill., Minn., N.D., Wyo., Colo., N.M. and Oreg.

30. *J. bulbosus* L. Bulbous Rush Fig. 176
Sandy and peaty shores of ponds and streams, often floating. Nfld., St. P. et Miq., and Sable I., N.S., Mass. and N.Y.; B.C., s. Wash. and Oreg.; intro. from Eur. According to P. Zika (pers. comm., 2020) this species is becoming a pest since it is able to compete with native in peatlands in the Pacific Northwest. This species frequently produces bulblets in the inflorescence like *J. pelocarpus* and *J. articulatus*; its aquatic form is similar to *J. subtilis*.

31. *J. validus* Coville Roundhead Rush Fig. 181
Swales, damp prairies, marshy shores, damp fields, and ditches. Del, Md., se. Va. s. to Fla., w. to Tex.; inland n. to w. Tenn., Mo. and Okla.

32. *J. scirpoides* Lam. Needlepod Rush Fig. 183
Damp soil, pools, wet pinelands, peats, and pond shores. Coastal plain, s. and c. N.Y. s. to Fla., w. to Tex.; n. inland to Ky., n. Ind., Ill., sw. Mich., s. Mo., and Okla.

33. *J. nodosus* L. Needlepod Rush Fig. 183
Moist soil, swamps, and gravels. Nfld. w. to Alask., s. to N.E., n. N.J., w. Va., Ohio, Ill., Iowa, Neb., Tex., N.M., and Nev.

34. *J. torreyi* Coville Torrey's Rush Fig. 180
Moist to wet sandy areas, river and stream banks, and ditches. N.B. Me., sw. Que., Vt., and w. Mass. w. to Sask. and B.C. and

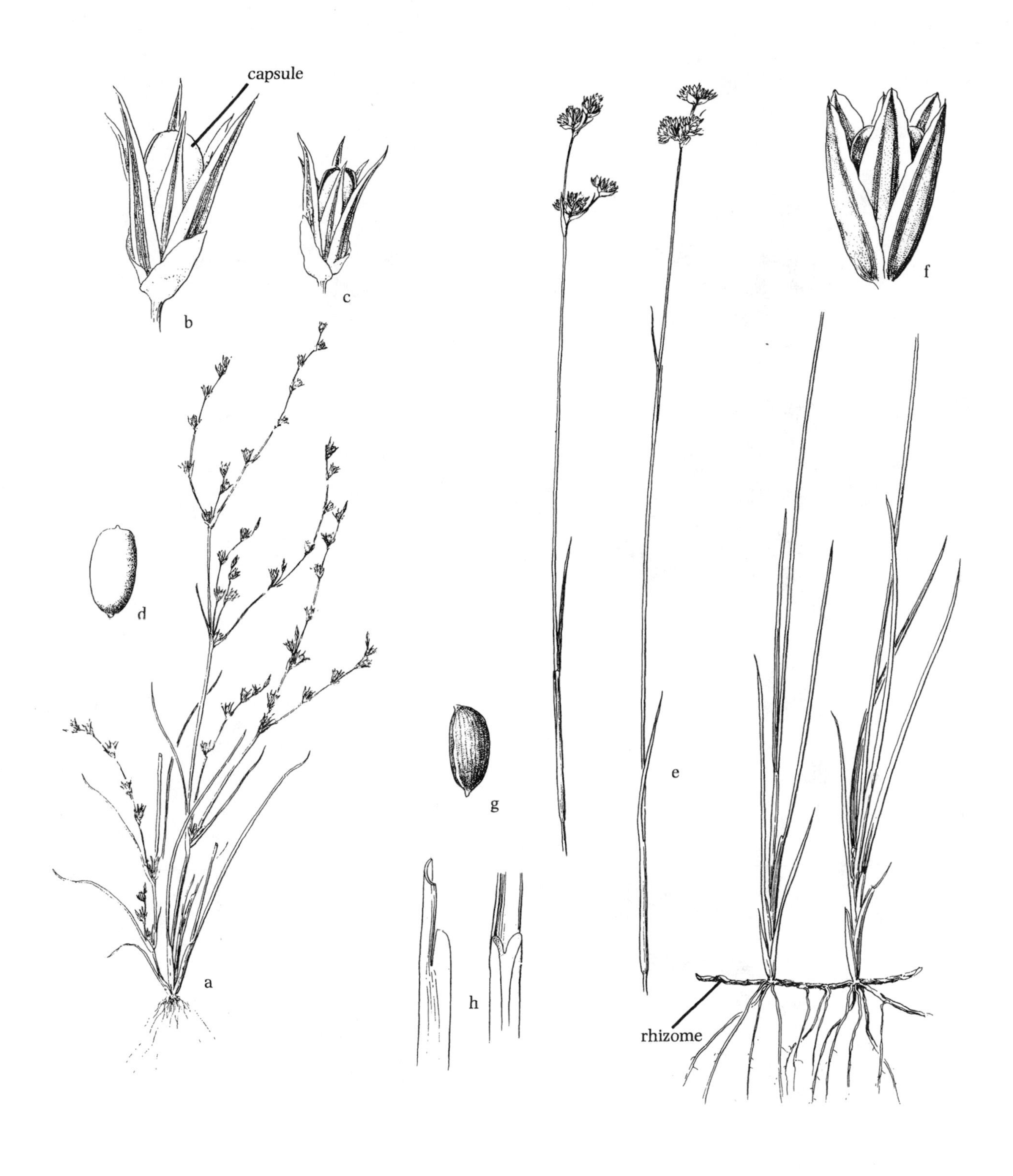

Fig. 175. *Juncus bufonius*: a. habit; b, c. perianth and capsule; d. seed (HCOT).
Juncus longistylis: e. habit; f. perianth and capsule; g. seed; h. leaf sheath, two views (HCOT).

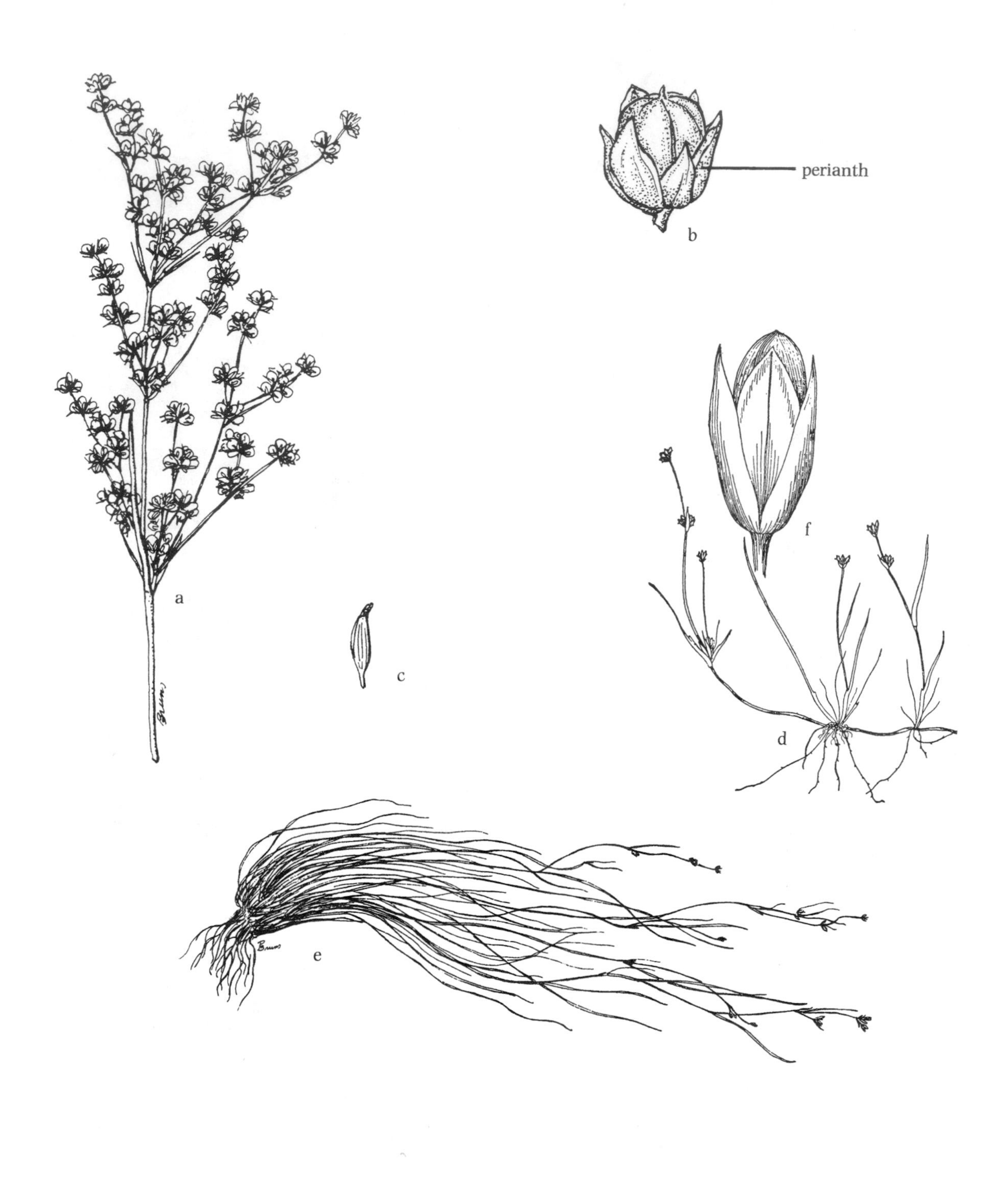

Fig. 176. *Juncus biflorus*: a. inflorescence; b. perianth and capsule; c. seed (PB).
Juncus bulbosus: d. habit (Gleason); e. habit (PB); f. perianth and capsule (Gleason).

Fig. 177. *Juncus marginatus*: a. habit; b. perianth and capsule; c. seed (NYS Museum).

Fig. 178. *Juncus subtilis*: a. habit; b. fascicle of leaves; c. perianth and capsule (Gleason). *Juncus gerardii*: d. habit (F); e. portion of inflorescence (HCOT); f. seed (HCOT). *Juncus canadensis*: g. inflorescence; h. perianth and capsule; i. seed (Beal).

Fig. 179. *Juncus tenuis*: a. inflorescence (HCOT); b. perianth and capsule (G&W); c. seed (G&W); d. leaf sheath with auricles (G&W).
Juncus dudleyi: e. habit; f. perianth and capsule; g. seed; h. leaf sheath, two views of auricles (HCOT).

Fig. 180. *Juncus torreyi*: a. habit with tuberous rhizome (HCOT); b. inflorescence (F); c. perianth and capsule (HCOT); d. seed (HCOT); e. leaf sheath (HCOT).
Juncus caesariensis: f. inflorescence; g. perianth and capsule; h. seed (PB).

Fig. 181. *Juncus validus*: a. habit (G&W); b. inflorescence (PB); c. perianth and capsule (G&W); d. seed (G&W). *Juncus brachycephalus*: e. inflorescence; f. perianth and capsule (PB).

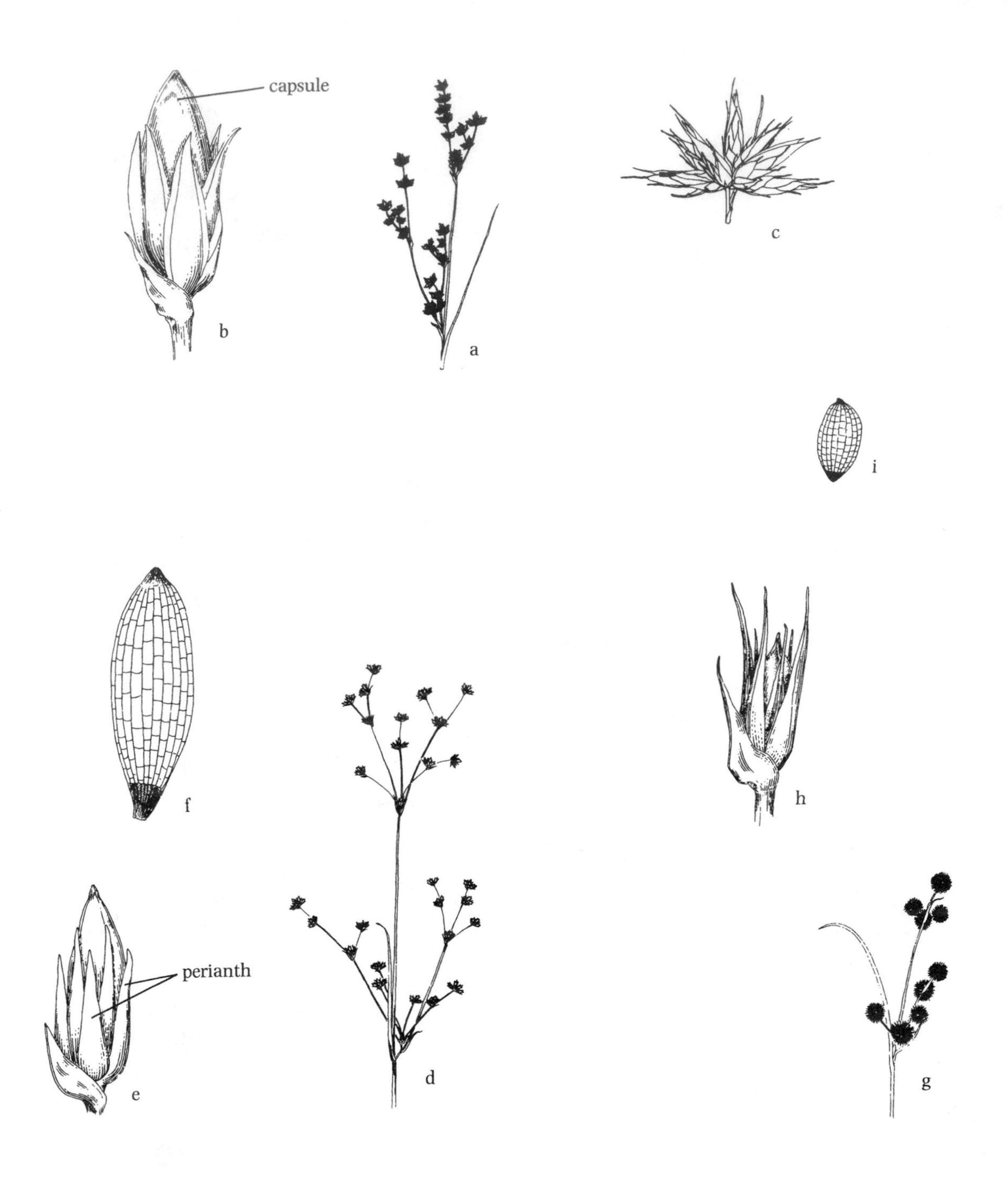

Fig. 182. *Juncus brevicaudatus*: a. inflorescence (F); b. perianth and capsule (Beal); c. inflorescence, modified as insect gall (F).
Juncus subcaudatus: d. inflorescence; e. perianth and capsule; f. seed (Beal).
Juncus brachycarpus: g. inflorescence; h. perianth and capsule; i. seed (Beal).

Fig. 183. *Juncus scirpoides*: a. habit (G&W); b. perianth and capsule (Beal); c. seed (G&W); d. leaf sheath (G&W). *Juncus nodosus*: e. habit; f. flower; g. leaf sheath (C&C).

Wash., s. to Ala., Tex., Calif., and n. Mex. This species is similar to *J. nodosus*, but tends to be a taller, coarser plant, with its heads having a somewhat more bristly appearance.

35. *J. megacephalus* M. A. Curtis Bighead Rush Fig. 184
Freshwater to brackish marshes, peats, sands, marly ponds, and ditches. Coastal plain, Md. and se. Va. s. to Fla., w. to Tex.

36. *J. brachycarpus* Engelm. Whiteroot Rush Fig. 182
Wet sands and peaty soils. Ne. Ohio and. Ont. w. to s. Mich., Ill., Minn., Mo., and Okla., s. to Ga., Fla. and Tex.; locally, e. Mass. and L.I., N.Y.

37. *J. nodatus* Coville Stout Rush Fig. 185
Shallow water, shores, sloughs, peats, and ditches. Ind. w. to Ill., Mo., and Kans., s. to Fla., Ala., Miss., La., and Tex.; Calif.

38. *J. elliottii* Chapm. Elliott's Rush Fig. 185
Wet sands, swales, peats, and pond shores. Coastal plain and piedmont, N.J. and Del. s. to Fla., w. to se. Tex.; inland w. Ky., c. Tenn. and Ark.

39. *J. acuminatus* Michx. Tapertip Rush Fig. 186
Marshes, and shores of ponds, lakes, and streams. N.S. w. to Ont. and Minn., s. to Fla., Tex., and n. Mex.; B.C. s. to Mont., Wash., Calif. and Ariz.

40. *J. debilis* A. Gray Weak Rush Fig. 187
Marshes, shores of acid streams, and wet clearings. Mass., R.I., and Conn. w. to N.Y., Ill., Ky., Tenn., and Mo., s. to. Fla. and e. Tex.

41. *J. diffusissimus* Buckl. Slimpod Rush Fig. 187
Wet clay and sand, marshy shores, ditches, and clearings, often in shallow water. Conn., se. N.Y., Pa. and Md. w. to s. Ohio, s. Ind., s. Ill., Mo. and Kans., s. to c. Fla., La., and e. Tex.; sw. Wash and n. Calif. This species is similar to J. *debilis* but with a more diffuse inflorescence.

42. *J. subtilis* E. Mey. Greater Creeping Rush Fig. 178
Shores of ponds and streams, often floating in currents. Greenl., Nunavut, and Lab. w. to n. Ont., s. to Nfld., N.S., n. Me, St. Lawrence Seaway, and ne. Minn.

43. *J. militaris* J. M. Bigelow Bayonet Rush Fig. 184
Shallow acid waters, sandy gravelly, and peaty shores. Nfld. and N.S. s. to N.E., N.Y., e. Md., and Del.; local in Ont. and Ind. and Mich.

44. *J. alpinoarticulatus* Chaix Northern Green Rush Fig. 188
Wet sandy and gravelly shores, often calcareous. Greenl. and Nfld. w. to Alta., N.W.T., Alask. and B.C., s. to c. Me., w. Vt., N.Y., n. Ohio, n. Ind., Ill., Mo., Neb., Wash., Utah and N.M. A natural hybrid between this species and *J. torreyi* has been described from a local area along Lake Erie in Ohio as *J. ×stuckeyi* Reinking (Reinking, 1981). (*J. alpinus* Vill.)

45. *J. articulatus* L. Jointleaf Rush Fig. 188
Wet sandy and boggy shores, and pond and stream shores. Nfld. w. to B.C. and Alask. s. to N.E., N.J., N.Y., W.Va., Ohio, n. Ind., Minn., Utah, Nev., Oreg., Calif. and N.M. This species is very similar to *J. alpinoarticulatus*, and more than a single character is needed to distinguish them. (*J. articulatus* var. *obtusatus* Engelm.)

Cyperaceae / Sedge Family

The sedges constitute a very large family, superficially resembling the grasses, but they are easily distinguished by several vegetative and reproductive features (fig. 189): stems usually 3-angled in cross-section, solid (except *Dulichium*, some *Carex*, and *Eleocharis equisetoides*, which has septate stems); leaves 3-ranked, sheathing stem with sheath closed; flowers in axils of overlapping scales, each having a single pistil, with 2 or 3 stigmas and 1–3 stamens (unisexual in *Carex* and *Scleria*); perianth absent or reduced to bristles and/or scales; fruit a 2-sided (biconvex) or 3-sided (trigonous) achene. The fruits of sedges are important food for wildfowl.

In this family, divergent opinions exist as to the status of recognized genera. A few genera (i.e. *Dichromena*, *Psilocarya*, and *Hemicarpha*) traditionally recognized by many authors are not recognized here. *Kyllinga* and *Lipocarpha* subsumed into *Cyperus* by some authors, but retained here.

---

1. Achenes enclosed in a perigynium (fig. 190d). . . . . 1. *Carex*
1. Achenes not enclosed in a perigynium.
    2. Flowers unisexual, pistillate spikelets 1-flowered, intermixed with clusters of few-flowered staminate spikelets; achenes white, hard and bony. . . . . 2. *Scleria*
    2. Flowers bisexual (at least some of them); spikelets all alike; achenes various, but not white and bony.
        3. Scales of spikelets distichous (2-ranked, in the same plane).
            4. Stem hollow, round in cross-section (fig. 234a); leaves conspicuously 3-ranked, scattered along stem; inflorescences borne in axils of leaf sheath. . . . . 3. *Dulichium*
            4. Stems solid or nearly so, 3-angled in cross-section, leaves arising from near base of stem.
                5. Perianth bristles present; achenes beaked; inflorescence of 5–8 spikelets. . . . . 4. *Blysmopsis*
                5. Perianth bristles absent; achenes beakless, inflorescence of 1–150 spikelets.
                    6. Inflorescences several to numerous, spicate or digitally arranged; individual spikelets with 5 to many scales. . . . . . 5. *Cyperus*
                    6. Inflorescences with 1–3 dense spikes, globose or lanceolate to ovate-lanceolate; individual spikelets with 1–2(3) scales. . . . . 6. *Kyllinga*
        3. Scales of spikelets spirally overlapping.

Fig. 184. *Juncus megacephalus*: a. habit (G&W); b. perianth and capsule (Beal); c. seed (G&W); d. leaf sheath and leaf section, showing septae (G&W).
Juncus militaris: e. habit (F).

Fig. 185. *Juncus elliottii*: a. habit; b. perianth and capsule; c. seed; d. leaf sheath (G&W). *Juncus nodatus*: e. inflorescence; f. perianth and capsule; g. seed (PB).

Fig. 186. *Juncus acuminatus*: a. lower portion of plant; b. upper portion of plant; c. perianth and capsule; d. seed; e. leaf sheath with auricles (HCOT).

Fig. 187. *Juncus debilis*: a. habit; b. perianth and capsule; c. seed (G&W).
Juncus diffusissimus: d. habit; e. perianth and capsule; f. seed (G&W).

Fig. 188. *Juncus alpinoarticulatus*: a. habit; b. portion of inflorescence; c. perianth and capsule; d. seed; e. leaf sheath (HCOT).
Juncus articulatus: f. habit; g. perianth and capsule; h. seed; i. leaf sheath (HCOT).

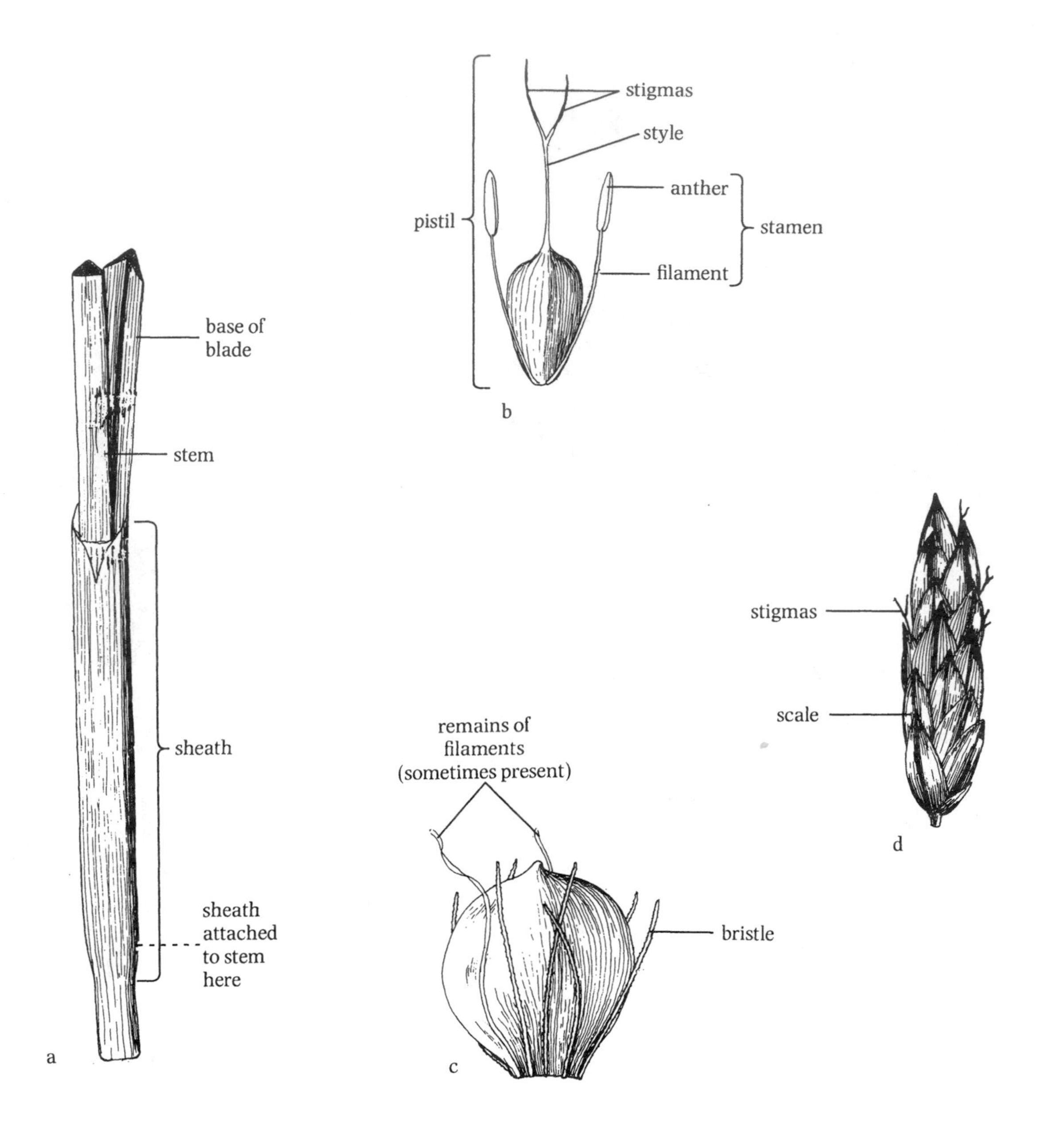

Fig. 189. Structure of Cyperaceae (sedges): a. section of stem with leaf base, generalized; b. flower, generalized; c. achene, generalized; d. spike/spikelet, generalized (F).

7. Perianth of 3 slender bristles, alternating with 3 stalked, dilated scales (paddle-like) (Fig. 244d,h). .... 7. *Fuirena*
7. Perianth of bristles, all the same, or absent.
  8. Perianth bristles (at maturity) greatly exceeding the subtending scales, appearing wooly or cottony.
    9. Bristles bent, curled or tangled, inflorescences of 50 or more spikes.................................. 8. *Scirpus* (in part)
    9. Bristles straight, silky, inflorescences 30 or fewer spikes.
      10. Bristles numerous, 10–25 per flower; leaf blades up to 45 cm long.................................. 9. *Eriophorum*
      10. Bristles 6 (ours); leaf blades up to 1 cm long. .... 10. *Trichophorum*
  8. Perianth bristles absent or short (mostly hidden within subtending scales)
    11. Spikelet solitary.
      12. Spikelet sessile, appearing lateral due to extension of involucral bract (fig. 261b); fertile stems weakly ascending, usually supported by water, often accompanied by long, hair-like leaves; achenes lacking swollen tubercle (*Schoenoplectus subterminalis*). .... 11. *Schoenoplectus* (in part)
      12. Spikelets clearly terminal, lacking an involucral bract; fertile stems erect, leaves consisting of bladeless leaf sheaths, not subtended by hair-like leaves at base (except *Eleocharis robbinsii*); achenes with tubercle formed by swollen, persistent style base (fig. 267). ....12. *Eleocharis*
    11. Spikelets in clusters of 2–3 (ours).
      13. Principal involucral bracts 1, terete, erect, inflorescence appearing lateral.
        14. Plants annuals, cespitose, lacking rhizomes (ours), usually less than 0.8(1) m.
          15. Perianth of bristles shorter than achene; achenes nearly black........................ 13. *Schoenoplectiella*
          15. Perianth lacking; achenes with whitish surface. .... 14. *Isolepis*
        14. Plants perennial, rhizomatous, usually 1 m or more. .... 11. *Schoenoplectus* (in part)
      13. Principal involucral bract flat or folded, inflorescence appearing terminal, even with subtending by involucral bracts.
        16. Spikelets 1–5 fertile florets.
          17. Achenes biconvex or flattened, with distinct tubercle formed by persistent bulbous style base (Fig. 287c,f); styles 2-cleft; stamens 4, perianth consisting of bristles (fig. 285e). .... 15. *Rhynchospora*
          17. Achenes round in cross-section, lacking a tubercle; styles 3-cleft; stamens 2; perianth absent. .... 16. *Cladium*
        16. Spikelets several to many-flowered.
          18. Plants annual, short; perianth bristles absent.
            19. Achene nearly round in cross-section; inflorescence never branched.................... 17. *Lipocarpha*
            19. Achene biconvex or triangular; inflorescence typically branched. .... 18. *Fimbrystylis*
          18. Plants perennial, tall, robust (usually 1 m or more); perianth bristles present.
            20. Stems sharply 3-angled; rhizomes tuberous; spikelets 1–5 cm long; scales pubescent, bifid at summit with awn arising from base of notch; perianth bristles shorter or slightly exceeding the achene. .... 19. *Bolboschoenus*
            20. Stems obtusely 3-angled; rhizomes not tuberous; spikelets 0.2–2.5 cm long; scales glabrous, not bifid, awn if present from scale apex; perianth bristles shorter than or only slightly longer than achene or sometimes absent. .... 8. *Scirpus* (in part)

---

1. *Carex* (Sedge)

Perennial herbs, arising from creeping rhizomes or cespitose; stems usually 3-angled; leaves grass-like, 3-ranked; flowers unisexual, staminate and pistillate flowers in separate spikes (fig. 190i), different parts of same spike, or scattered, mixed in same spike; fruit a 3-angled or biconvex achene enclosed in a perigynium.

This is a large and difficult group; the presence of mature perigynia is essential for the identification of practically all species. To a large extent, the species are based on the nature of the perigynium: it may be beaked or beakless, inflated or closely filled by the achene, or sometimes broadly winged and scale-like.

References: Ball, 2002; Ball and Reznicek, 2002; Bruederle and Fairbrothers, 1986; Crins, 2002; Crins and Ball, 1989a, 1989b; Cochrane, 2002; Cochrane and Naczi, 2002; Mackenzie, 1931–1935, 1940; Mastrogiuseppe et al., 2002; Murray, 2002; Reznicek and Ball, 1974, 1980; Reznicek and Catling, 1986, 2002; Reznicek and Ford, 2002; Rothrock, 2021; Rothrock and Reznicek, 2002; Standley, 1983, 1987a, 2002; Standley et al., 2002; Toivonen, 2002; Voss and Reznicek, 2012; Waterway, 2002.

1. Spikes 1, terminal, entirely staminate or pistillate or mixed (fig. 190b).
  2. Achenes 3-angled; stigmas 3.
    3. Perigynia slender, more than 5 times as long as wide; usually strongly reflexed at maturity. . . . . 1. *C. pauciflora*
    3. Perigynia broader, less than 5 times as long as wide, appressed to ascending. . . . . 2. *C. leptalea*
  2. Achenes biconvex; stigmas 2
    4. Plants densely cespitose, not rhizomatous; perigynia somewhat flattened (convex on one face), strongly, regularly serrulate (teeth minute) on apical portion and beak; anthers 2–3.5 mm long. . . . . 48. *C. exilis*
    4. Plants with 1–3 stems arising from a thread-like rhizome; perigynia plump (convex on both faces), glabrous or only obscurely serrulate.
1. Spikes 2 or more (depauperate plants may have only 1). . . . . 3. *C. gynocrates*
  5. Plants unisexual, inflorescences essentially staminate (occasionally with a few perigynia) or pistillate (rarely with few staminate flowers). . . . . 55. *C. sterilis*
  5. Plants with both staminate and pistillate flowers on same plant (individually spikes often pistillate or staminate or bisexual with staminate flowers at base or at upper portion or even mixed).
    6. Achenes biconvex; stigmas 2 (fig. 191c,g)
      7. Lateral spikes peduncled (fig. 191a), or if sessile, then elongate; terminal spike usually staminate.
        8. Lower pistillate spikes spreading or drooping on slender peduncles (fig. 191a), or if erect, then achene body with constriction (fig. 190m).
          9. Achenes not bent or contorted.
            10. Beak of perigynium usually bent or twisted when dry; perigynia smooth. . . . . 4. *C. torta*
            10. Beak of perigynium straight when dry; perigynia granulose-papillose. . . . . 5. *C. mitchelliana*
          9. Achenes bent or contorted, often with constriction (deeply in some).
            11. Plants stoloniferous, stems 1–few; plants of saline or tidal shores.
              12. Pistillate spikes usually erect (192k). . . . . 6. *C. recta*
              12. Pistillate spikes pendulous (fig. 191e)
                13. Scales of pistillate flowers with awns longer than blades (fig. 191h). . . . . 7. *C. paleacea*
                13. Scales of pistillate flowers lacking awns, or if present (rarely) than shorter than blades. . . . . 8. *C. salina*
            11. Plants tufted; stems several to many; plants of freshwater habitats.
              14. Lower leaf sheaths glabrous; perigynia somewhat inflated, obovoid with abrupt beak.
                15. Awns of lower scales slightly longer than to 4 times the length of perigynia; perigynia 1–2 mm wide, ellipsoidal to obovoid; achenes constricted; pistillate spikes rarely, if ever, staminate on upper portion. . . . . 9a. *C. crinita* var. *crinita*
                15. Awns of lower scales as long to twice as long as perigynia; perigynia 2–3 mm wide, broadly obovoid; achenes not constricted; pistillate spikes often staminate on upper portion. . . . . 9b. *C. crinita* var. *brevicrinis*
              14. Lower leaf sheaths scabrous to hispidulous; perigynia flattened, elliptic, tapering to minute beak. . . . . 10. *C. gynandra*
        8. Lower pistillate spikes erect or ascending (fig. 195k,l), peduncles thicker, or spikes sessile, achene body never constricted.
          16. Perigynia plump, elliptic in cross-section.
            17. Mature perigynia golden-orange (drying brown); terminal spikes usually all staminate; scales whitish to tawny. . . . . 11. *C. aurea*
            17. Mature perigynia whitish; terminal spikes usually staminate only at base; scales brown to purplish. . . . . . 12. *C. garberi*
          16. Perigynia piano-convex or biconvex in cross-section, but not plump.
            18. Stems arising from last year's leaves, seldom bearing bladeless sheaths at base; leaf bases only slightly, if at all, fringed.
              19. Perigynia nerveless (fig. 195c) or obscurely nerved toward base; leaves 2–5 mm wide.
                20. Pistillate scales pale brown with narrow red-brown margins and broad pale midvein. . . . . 13b. *C. aquatilis* var. *substricta*
                20. Pistillate scales red- or purple-brown to purple-black, and narrow pale midvein. . . . . 13a. *C. aquatilis* var. *aquatilis*
              19. Perigynia conspicuously nerved (fig. 196b,d); leaves 1–3 mm wide.
                21. Pistillate scales dark, with a broad central green portion about as wide as dark margins; involucral bracts conspicuously longer than spikes, hyaline portion of bract sheath white. . . . . 14. *C. lenticularis*

21. Pistillate scales dark with a very narrow central green portion much narrower than wide dark margins, often not much wider than midvein; involucral bracts shorter or slightly longer than spikes, hyaline portion of bract sheath purple-brown. . . . . 15. *C. nigra*

18. Stems not arising from last year's leaves, bearing bladeless sheaths at base; leaf bases distinctly fringe.

22. Ligules rounded, about as long as wide or (fig. 197e); pistillate spikes 2–10(15). . . . . 16. *C. emoryi*

22. Ligules acute, longer than wide (fig. 195j); pistillate spikes 2–6(11) cm long. . . . . 17. *C. stricta*

7. Lateral spikes sessile, short; terminal spike usually partly pistillate.

23. Plants with stems usually solitary or few, arising from stolons or offshoots from stolon.

24. Perigynia wingless, lacking sharp margins, not flattened in cross-section.

25. Pistillate spikes, especially lower ones, somewhat separated; terminal spike with flowers pistillate on upper portion. . . . . 45. *C. mackenziei*

25. Pistillate spikes crowded, often appearing as a single head; terminal spike with flowers staminate on upper portion, pistillate on lower. . . . . 18. *C. chordorrhiza*

24. Perigynia winged or with sharply edged margins (fig. 198h), flattened in cross-section. . . . . 19. *C. sartwellii*

23. Plants with stems loosely to densely tufted.

26. Spikes with staminate flowers on upper portion (androgynous).

27. Stems thick, sharply angled, soft, easily compressed.

28. Perigynia 3–5 mm long; beak shorter than or as long as body. . . . . 20. *C. alopecoidea*

28. Perigynia 4.5–9 mm long; beak longer than body.

29. Perigynia 6–9 mm long, base dilated, disk-shaped (fig. 199b); thin portion of leaf sheath strongly purple-dotted, ventral band not transversely wrinkled. . . . . 21. *C. crus-corvi*

29. Perigynia 4–5(6) mm long, base not dilated; thin portion of leaf sheath not purple-dotted, ventral band strongly transversely wrinkled.

30. Leaves not more than 10 mm wide; inflorescence to 10 cm; perigynia 4–5 mm; beak to 2.5 mm. . . . . 22a. *S. stipata* var. *stipata*

30. Leaves 8–15 mm wide, inflorescences to 15 cm; perigynia 5–6 mm beak 3–3.5 mm. . . . . 22b. *S. stipata* var. *maxima*

27. Stems very slender, not sharply angled, firm, not easily compressed.

31. Inflorescences simple, 1–5 perigynia per spikelet; perigynia essentially beakless. . . . . 23. *C. disperma*

31. Inflorescences with spikelets on lateral branches, several to many perigynia per spikelet; perigynia beaked.

32. Perigynia more or less biconvex; pistillate scales acute to slightly cuspidate or mucronate; leaf sheaths not transversely wrinkled.

33. Leaves (2.5)5–8 mm wide; inflorescences 7–18 cm long; perigynia abruptly beaked, tapering basally to short stipe. . . . . 24. *C. decomposita*

33. Leaves 1–3 mm wide; inflorescences 1–6.5(10) cm long; perigynia gradually tapering to beak, rounded to truncate at base.

34. Leaf sheath pale at summit; mature perigynia olive-green to blackish, inner face convex. . . . . 25. *C. diandra*

34. Leaf sheath bronze-colored at summit; mature perigynia light brown to brown, inner face flat. . . . . 26. *C. prairea*

32. Perigynia plano-convex; pistillate scales long-awned; leaf sheaths transversely wrinkled (fig. 201f). . . . . 27. *C. vulpinoidea*

26. Spikes with pistillate flowers on upper portion (gynecandrous).

35. Perigynia with thin-winged margins; achene not filling perigynium to margin; lower portion of perigynium not spongy-thickened; stems hollow.

36. Spikes lanceolate, 12–28 mm long. . . . . 28. *C. muskingumensis*

36. Spikes ovoid to ellipsoid to globose, 5–13(17).

37. Pistillate scales obtuse, acute, or short-acuminate, lacking a tip.

38. Largest perigynia typically more than 2 mm wide (1.5–2.8 mm wide).

39. Perigynia obovate.

40. Styles distinctly bent at base, laterally sinuate; pistillate scales acute at tip. . . . . 29. *C. albolutescens*

40. Styles straight; pistillate scales obtuse at tip. . . . . 30. *C. longii*

39. Perigynia orbicular to elliptic, or diamond shaped (widest at middle with cuneate base).

41. Perigynia diamond shaped, 4–5 mm long; achenes 0.8–1.1 mm wide, elliptic to ovate. . . . . 31. *C. suberecta*

41. Perigynia orbicular to elliptic, 2.5–4.2 mm long; achenes 1–1.3 mm wide, ovate. . . . . 32. *C. festucacea*

38. Largest perigynia not more than 2 mm wide.

42. Perigynia lanceolate to ovate lanceolate or narrowly lanceolate to narrowly ovate.

43. Perigynia lanceolate to ovate lanceolate, 1.1–2 mm wide, 2.9–4.3(3.7) mm long; achene style straight.

44. Perigynia thicker, plano-convex, winged to the base; spike ellipsoid, acute to short attenuate at base (7–16 × 3–9(13) mm long). . . . . 33. *C. scoparia*

44. Perigynia thin, somewhat scale-like, usually not winged to base; spike ovoid-oblong (6–12(16) × 4–8 mm long) to globose, rounded to tapering at base (9–12 × 6–9 mm long).

45. Inflorescences flexuous; spikes mostly separated, wider than long; perigynia ca. 15–40 per spike. . . . . 34. *C. projecta*

45. Inflorescences stiff; spikes densely overlapping, somewhat nodding at summit, longer than wide; perigynia more than 40 per spike.. . . . . . . 35. *C. tribuloides*

43. Perigynia narrowly lanceolate to narrowly ovate, 0.9–1.3 mm wide, 3.4–4.1(4.7) mm long; achene style persistent, sinuate (above base). . . . . .36. *C. crawfordii*

42. Perigynia ovate to elliptic.

46. Spikes usually widely spaced, globose, about as wide as long (4–8 × 4–8 mm); perigynia widely spreading to somewhat recurved; leaf sheath V-shaped at summit, somewhat loose and expanded. . . . . .37. *C. cristatella*

46. Spikes usually overlapping, ovoid-globose, longer than wide (4–10 × 3–7 mm); perigynia ascending to slightly spreading; leaf sheath U-shaped at summit, not expanded. . . . . 38. *C. bebbii*

37. Pistillate scales long-acuminate, often with awned tip.

47. Leaf blades scabrous-margined, 2.5–5.5 mm wide; body of perigynium broadest toward summit (fig. 206b).. . . . . .39. *C. alata*

47. Leaf blades smooth-margined, except at apex, 1–2.5 mm wide; body of perigynium broadest.

48. Spikes rounded at tip; perigynia globose-ovoid, abruptly narrowed to beak; beaks loosely spreading-ascending. . . . . .40. *C. straminea*

48. Spikes tapering to tip; perigynia with nerves on both sides, lanceolate-ovate, tapering gradually to beak; beaks appressed-ascending. . . . . .41. *C. hormathoides*

35. Perigynia not winged; achene filling at least upper portion of perigynium; lower portion of perigynium filled by achene, or in some species, spongy-thickened; stems solid.

49. Margin of perigynium rounded: perigynia fusiform to ovoid, filled by achene, not spongy at base; upper portion of perigynium with very short beak or merely tapering to tip, smooth or scabrous, tip entire to slightly notched.

50. Spikes widely separated, lowermost subtended by long filiform bract, several times as long as spike; perigynia (1)2–3(5) per spikelet. . . . . .42. *C. trisperma*

50. Spikes in close proximity, often crowded, lowermost not subtended by elongate bract; perigynia 3-several per spike, but usually more than 5.

51. Inflorescences short, 0.8–2 cm long; perigynia essentially beakless or to 0.2 mm, tapering to tip, 2.5–3.5 mm long.

52. Perigynium beaks green to pale brown (same color as body); pistillate scales white-hyaline margins with green central portion. . . . . 43. *C. tenuiflora*

52. Perigynium beaks reddish brown with apical hyaline rim (darker than body); pistillate scales reddish brown with pale central portion. . . . . .44. *C. heleonastes*

51. Inflorescences longer, usually longer than 2 cm; perigynia short-beaked, 1.7–3.3 mm long (including beak).

53. Stems smooth throughout; leaves yellowish-green; upper portion of perigynium and beak smooth; plants of saline sites. . . . . 45. *C. mackenziei*

53. Stems scabrous on upper portion; leaves glaucous or deep green; upper portion of perigynium and beak scabrous; plants of freshwater sites.
 54. Leaves glaucous, (1.5)2–4 mm wide; spikes ellipsoidal to subglobose, with 10 to many loosely spreading perigynia.
  55. Culms 15–60 cm; inflorescences 3–5(7) all but proximal spikes approximate or slightly remote. . . . . 46a. *C. canescens* subsp. *canescens*
  55. Culms 30–90 cm; inflorescences 6–12(15) cm, all but uppermost spikes distant, proximal spikes 2–5 cm apart. . . . . 46b. *C. canescens* subsp. *disjuncta*
 54. Leaves green, 1.5–2.5 mm wide; spikes elongate, with (3)5–10 ascending to appressed perigynia. . . . . 47. *C. brunnescens*

49. Margin of perigynium thin-edged; perigynia lanceolate to broadly ovate, only upper portion filled by achene, spongy at base; upper portion of perigynium with prominent, serrulate beak, tip toothed.
 56. Broadest leaves 2.8–5 mm wide.
  57. Perigynia beak smooth-margined, beak 0.2–0.6 mm long.. . . . . 49. *C. seorsa*
  57. Perigynia beak toothed, strongly serrate to at least partially serrulate on margins, beak 0.5–1.35 mm long.
   58. Lower perigynia of spikes usually 1.1–1.7 times as long as wide, usually 2.1–3 mm wide. . . . . 50a. *C. atlantica* subsp. *atlantica*
   58. Lower perigynia of spikes (1.5)1.7–3 times as long as wide, usually 1.2–2 mm wide.
    59. Inflorescences mostly 1.5–3 cm, the lower 2 spikes separated by 1.3–9.5 mm. . . . . 51. *C. wiegandii*
    59. Inflorescences mostly 3–8.5 cm, the lower 2 spikes separated by 10–40 mm. . . . . 52. *C. ruthii*
 56. Broadest leaves 0.8–2.7 mm wide.
  60. Lower perigynia of spikes 2–3 mm wide. . . . . 50a. *C. atlantica* subsp. *atlantica*
  60. Lower perigynia of spikes 0.9–2.2 mm wide.
   61. Lower perigynia usually 1.9–3.8 mm long, 1–2(2.2) times as long as wide; beak usually 0.4–0.9 mm long, usually 0.2–0.5 times as long as body.
    62. Perigynia usually nerveless over area of achene on ventral (adaxial) surface; beak of perigynium distinctly setulose-serrulate. . . . . .53. *C. interior*
    62. Perigynium usually 1–10-nerved over area of achene on both surfaces, beak of perigynium sparsely serrulate, with distinct spaces between teeth.
     63. Broadest leaves 1.6–4(4.5) mm wide; fruiting spikes 18–45 mm long; perigynia 2.3–3.8 mm long. . . . . 50a. *C. atlantica* subsp. *atlantica*
     63. Broadest leaves (0.6)0.8–1.6 mm wide; fruiting spikes 8 20 mm long; perigynia 1.9–3 mm long. . . . . 50b. *C. atlantica* subsp. *capillacea*
   61. Lower perigynia usually 2.1–4.7 mm long, 1.4–3.6 times as long as wide; beak usually 0.9–2 mm long, usually 0.4–0.8 times as long as body.
    64. Plants bisexual, both staminate and pistillate flower present; perigynium lanceolate to ovate, 0.8–2.1 mm wide, beak serrulate, 0.85–2 mm long; achene ovate-lanceolate to rhombic-ovate, 1.3–2.1 mm long. 54. *C. echinata*
    64. Plants unisexual (spikes of pistillate plants rarely with a few staminate flowers; spikes of staminate plants mostly entirely without perigynia); perigynium ovate to deltoid, 1.2–2.2 mm wide, beak setulose-serrate, 0.65–1.6 mm long; achenes ovate to suborbicular, 1–1.7 mm long. . . . . 55. *C. sterilis*

6. Achenes 3-angled (fig. 214d,e); stigmas 3 (fig. 214d).
 65. Perigynia pubescent to scabrous-puberulent.
  66. Perigynia 2.5–4.5(5) mm long; beaks with teeth 0.2–0.7 mm long.
   67. Leaves flat, 2–8(12) mm wide; stems with 3 sharp angles.
    68. Leaves 4–8(12) mm wide; perigynia scabrous to puberulent, beak recurved. . . . . 56. *C. scabrata*
    68. Leaves 2–4.5(6) mm wide; perigynia pubescent, beak straight. . . . . 57. *C. pellita*
   67. Leaves strongly involute, 2 mm or less wide; stems with 3 rounded angles. . . . . .58. *C. lasiocarpa*
  66. Perigynia 6–11 mm long; beaks with teeth 0.8–2.8 mm long. . . . . .60. *C. trichocarpa*
 65. Perigynia glabrous.

69. Pistillate scales with awns nearly as long as or longer than blade.
  70. Beak of perigynium with teeth 1.2–3 mm long.
    71. Leaf blades and sheaths pubescent, especially lower ones; perigynia 7–12 mm long, ascending to somewhat spreading. . . . . 61. *C. atherodes*
    71. Leaf blades and sheaths glabrous; perigynia 5–7 mm long, reflexed. . . . . 62. *C. comosa*
  70. Beak of perigynium with teeth 0.3–1 mm long.
    72. Perigynia reflexed, 3–5 mm long; styles flexuous, but not contorted. . . . . 63. *C. pseudocyperus*
    72. Perigynia ascending to somewhat spreading, 5–8 mm long; styles contorted (fig. 217c).
      73. Staminate scales merely acuminate, lacking awn; stems weakly 3-angled, soft, solitary to few, arising from an elongate rhizome. . . . . 64. *C. schweinitzii*
      73. Staminate scales with rough awn (rarely lacking); stems sharply 3-angled, firm, numerous, forming small to large clumps.
        74. Achenes ellipsoidal, surface papillate; perigynia 7–12-nerved. . . . . 65. *C. lurida*
        74. Achenes cuneate-ovoid, surface smooth; perigynia 13–21-nerved. . . . . 66. *C. hystericina*
69. Pistillate scales blunt to acute, cuspidate, or acuminate, toward spike apex or if awned, then awns shorter than blades (at least most in spike less than half as long as blade)
  75. Perigynia lanceolate, slender, gradually tapering to beak (fig. 219b).
    76. Main leaves 1.6–5 mm wide; bract sheaths concave at apex; anthers 1.1–2.7 mm. . . . . 67. *C. michauxiana*
    76. Main leaves (3.5)5–18 mm wide; bract sheaths elongate, truncate to convex at apex; anthers 3–5 mm.
      77. Larger achenes 3.4–4(4.5) mm, lateral spikes entirely pistillate, widest leaves of vegetative shoots 8–18(21). . . . . 68. *C. folliculata*
      77. Larger achenes 2.3–3.4(3.8) mm.; lateral spikes usually with a conspicuous staminate apex; widest leaves of vegetative shoots (3.5)5–12 mm wide; pistillate scales mostly half to two-thirds as long as perigynia. . . . . 69. *C. lonchocarpa*
  75. Perigynia ovoid or ellipsoidal, abruptly tapering to beak (fig. 220b).
    78. Styles persistent, continuous with achenes; perigynia with conspicuous sharp, stiff teeth at tip (fig. 220b,g).
      79. Perigynia 3–12 mm long; achenes 2–3 mm long; stems not reddish-purple at base (or only slightly suffused with maroon).
        80. Base of plant thick and spongy; leaves with conspicuous cross-markings between veins (fig. 220e).
          81. Staminate spikes 2–4; pistillate spikes 2–5, usually 4–10 cm long.
            82. Pistillate scales, at least some, with distinct scabrous awns; perigynia thick-walled.
              83. Lower leaf sheaths reddish-purplish, becoming fibrillose on inner surface, splitting to form a pinnate network of veins; ligules 12–40 mm long; perianth strongly nerved. . . . . 70. *C. lacustris*
              83. Lower leaf sheaths pinkish to whitish or brownish, seldom fibrillose; ligules 2–10 mm long; perigynia weakly nerved or nerveless. . . . . 71. *C. hyalinolepis*
            82. Pistillate scales acute, but lacking an awn; perigynia thin-walled.
              84. Stems round to obscurely 3-angled, smooth below inflorescence; leaves involute, glaucous, broadest blades 1.5–4(7.2) mm wide. . . . . 72. *C. rostrata*
              84. Stems distinctly 3-angled, scabrous below inflorescence; leaves more or less flat, yellow-green or olive-green; broadest blades 3–11 mm wide.. . . . . 73. *C. utriculata*
          81. Staminate spikes 1 or 2; pistillate spikes 1(2), 2–4 cm long (style articulate, but superficially may appear persistent). . . . . 59. *C. striata*
        80. Base of plant not spongy; leaves with obscure cross-markings between veins, or cross-markings absent.
          85. Achenes asymmetrical, indented on one side; perigynia 5–6.5 mm wide. . . . . 74. *C. tuckermanii*
          85. Achenes symmetrical, not laterally indented; perigynia 1–5 mm wide.
            86. Peduncles of pistillate spikes weak, slender, spikes nodding; perigynia 1–1.5 mm wide. . . . . 75. *C. prasina*
            86. Peduncles or pistillate spikes firm, spikes ascending or spreading, but not distinctly nodding; perigynia 2–5 mm wide.
              87. Leaves involute-filiform, wiry; staminate spikes 1, rarely 2; pistillate scales ovate, blunt. . . . . 76. *C. oligosperma*

87. Leaves flat to W-shaped, not wiry; staminate spikes 2 or more; pistillate scales lanceolate, acute to acuminate.
88. Perigynia firm; beak minutely roughened. . . . . 77. *C. bullata*
88. Perigynia membranous; beak smooth.
89. Perigynia ascending at maturity, (3.6)4–7.5(8.2) mm long, beak 1.1–2.6 mm long, apex abruptly contracted. . . . . 78. *C. vesicaria*
89. Perigynia mostly reflexed at maturity, 6–10 mm long, beak 2.1–4.5 mm. long, apex tapered. . . . . 79. *C. retrorsa*
79. Perigynia 11–20 mm long, with 13–25 strong nerves; achenes 2.2–6 mm long; stems reddish-purple at base.
90. Achenes about as wide or wider than long, somewhat diamond-shaped, with conspicuous angles (fig. 224c).
91. Perigynia ascending to spreading; achenes 3–4.5 mm long, (2.2)2.4–3.4 mm wide, with conspicuous knobbed angles (fig. 224c). . . . . 80. *C. lupuliformis*
91. Perigynia stiffly spreading at right angles to axis; achenes 2.2–2.6 mm long, 2.7–3 mm wide, with angles strongly thickened (not knobbed). . . . . 81. *C. gigantea*
90. Achenes longer than wide, ellipsoid to obovoid or rhomboid, angles rounded.
92. Pistillate spikes globose or nearly so; perigynia ascending and spreading or in all directions.
93. Perigynia radiating out in all directions; achene style nearly straight, withering. . . . . .82. *C. grayi*
93. Perigynia ascending to spreading, with basal-most strongly divergent to reflexed; achene style persisting, contorted. . . . . .83. *C. intumescens*
92. Pistillate spikes cylindrical or short-oblong; perigynia all ascending.
94. Plants loosely clumped, not producing elongate rhizomes; perigynia 11–19 mm long; staminate peduncle 0.5–7 cm long (shorter than or barely exceeding terminal pistillate spike). . . . . 84. *C. lupulina*
94. Plants somewhat colonial, producing long rhizomes; perigynia 10–14 mm long; staminate peduncle 6–18 cm long. . . . . 85. *C. louisianica*
78. Styles deciduous (at least upper portion with stigmas), articulated with achenes; perigynia entire or minutely toothed at tip.
95. Perigynium beakless or with short, obscure beak 0.5 mm or less long.
96. Lower bracts of pistillate spikes sheathless or with a very short sheath.
97. Terminal spikes pistillate on upper portion; perigynia many-nerved. . . . . 86. *C. buxbaumii*
97. Terminal spikes staminate throughout; perigynia few-nerved.
98. Scales as wide or wider than and barely longer than perigynia; pistillate spikes 0.5–2.5 cm long.
99. Stems with 3 sharp angles; leaves involute or flat, not thick or stiff.
100. Leaves involute, 1–2 mm wide; scales ovate to subcircular, as wide as perigynia, persistent. . . . . 87. *C. limosa*
100. Leaves flat, 1–4 mm wide; scales lanceolate, narrower than perigynia, readily deciduous. . . . . 88. *C. magellanica*
99. Stems with 3 rounded angles; leaves flat, thick, and stiff. 89. *C. rariflora*
98. Scales narrower than and much longer than perigynia; pistillate spikes 2–5.5 cm long. . . . . 90. *C. barrattii*
96. Lower bracts of pistillate spikes with well-developed sheath.
101. Leaves whitish, glaucous becoming involute with age; perigynia whitish, glaucous.. . . . . 91. *C. livida*
101. Leaves green to slightly glaucous, flat except at apex of older leaves; perigynia green, not glaucous.. . . . . 92. *C. tetanica*
95. Perigynium with distinct beak 0.5 mm or more long.
102. Perigynia ascending; pistillate spikes 1 or 2; plants stoloniferous, forming large colonies. . . . 59. *C. striata*
102. Perigynia reflexed or spreading; pistillate spikes 2–6; plants forming tufts.
103. Perigynia 2–3.5 mm long, beak a fourth to half as long as body. . . . . .93a. *C. viridula* subsp. *viridula*
103. Perigynia 3–6 mm long, beak half as long as or longer than body.
104. Stems obtusely angled at tip; perigynia obovoid, abruptly beaked. . . . . 93b. *C. viridula* subsp. *brachyrrhyncha*

104. Stems acutely angled at summit; perigynia lanceolate-ovoid to subulate, gradually tapering to beak.
 105. Perigynia 3–4.5(5) mm long, beak smooth. . . . . 94. *C. cryptolepis*
 105. Perigynia 4–6 mm long, beak often serrulate toward tip (fig. 232b). . . . . 95. *C. flava*

1. *C. pauciflora* Lightf. Few-flower Sedge Fig. 190
Acid peat and sphagnum bogs. Lab. and Nfld. w. to s. Alask., s. to N.S., n. and w. N.E., n. Pa., W.Va., n. Ind., Wisc., Minn. and Wash.

2. *C. leptalea* Wahlenb. Bristly-stalked Sedge Fig. 190
Wet open woods and peaty sites, bogs, swamps, swales and clearings. Lab. and Nfld. w. to Alask., s. to Fla., Tex., Colo., N.M. and Calif. Plants with longer (3.5–5 mm), more overlapping perigynia have been segregated as subsp. *harperi* (Fernald) Calder and Taylor.

3. *C. gynocrates* Wormskjöld ex Drejer Northern Bog Sedge Fig. 192
Bogs, peaty shores, and openings in conifer swamps, often on sphagnum hummocks, peaty swales, poor fens, subalpine meadows, tundra outwash gravels and seepages. Greenl., Lab. and Nfld. w. to Nunavut, N.W.T. and Alask., s. to N.S., N.B., Me., N.J., Pa., n. Mich., n. Wisc., n. Minn., N.D., mts. to n. N.M., Utah, n. Nev., ne. Ore., nc. Wash. and B.C. This tiny species is most certainly a much overlooked.

4. *C. torta* Boott Twisted Sedge Fig. 190
Stream margins and swamps. Gaspé Pen., Que., w. to Ont. and Minn., s. to N.E., Del., N.C., Ga., Tenn. and Ark.

5. *C. mitchelliana* M. A. Curtis Mitchell's Sedge Fig. 191
Swales, marshy shores, bogs and wet woods. Coastal plain, Mass. s. to Fla., w. to e. Tex.; inland n. to Ky., Tenn. and n. Ga. (*C. crinita* var. *mitchelliana* (M. A. Curtis) Gleason)

6. *C. recta* Boott Estuary Sedge Fig. 192
Saline and brackish shores, swales, salt marshes and estuaries. Lab. and Nfld. w. to Man., s. to N.H. and Mass. *C. recta* is regarded as a stabilized species of hybrid origin (*C. aquatilis* × *C. paleacea*), occurring within the range of both parents (Ball and Reznicek, 2002). (*C. salina* var. *kattegatensis* (Fries) Almq.)

7. *C. paleacea* Wahlenb. Chaffy Sedge Fig. 191
Saline and brackish marshes and shores. Coastal, Lab. and e. Que. w. to Hudson Bay, Ont. Man. and N.W.T., s. to N.S., Me., N.H. and Mass.

8. *C. salina* Wahlenb. Saltmarsh Sedge Fig. 190
Saline, brackish shores, marshes and swales. Lab., w. Nfld., N.B., Que. and Ont. This species is considered of hybrid origin between *C. paleacea* and *C. subspathacea* (Ball and Reznicek, 2002).

9. *C. crinita* Lam. Fringed Sedge

9a. *C. crinita* var. *crinita* Fig. 193
Swales and damp thickets. Nfld. w. to n. Man., s. to N.S., N.E., Md., n. Ga., Tenn., Mo. and La.

9b. *C. crinita* var. *brevicrinis* Fernald
Wooded swamps and bottomlands. Coastal plain, s. N.E. s. to Fla., w. to e. Tex.; inland n. to Okla., Neb., Ky. and Mo.

10. *C. gynandra* Schwein. Nodding Sedge Figs. 192, 194
Swales and damp thickets. Nfld. w. to Ont., Wisc. and ne. Minn., s. to N.E., Md., N.C., Tenn. and n. Ga. (*C. crinita* var. *gynandra* (Schwein) Schwein. & Torr.)

11. *C. aurea* Nutt. Golden Sedge
Wet meadows, damp shores, bogs, swamps and low woods. Nfld. w. to Alask., s. to N.E., n. Pa., n. Ohio, n. Ind., n. Ill., Minn., Neb., Tex., N.M. and Calif.

12. *C. garberi* Fernald Elk Sedge Fig. 194
Calcareous sands, and marly or gravelly shores. Lab. and Que. w. to Nunavut, N.W.T. and Alask., s. to n. N.E., N.Y., nw. Pa., ne. Ohio, Mich., nw. Ind., ne. Ill., Minn., n. N.D., mts. to Wyo., Utah, Oreg. and Calif.; most abundant along shores of Great Lakes and vicinity.

13. *C. aquatilis* Wahlenb. Water Sedge

13a. *C. aquatilis* var. *aquatilis* Fig. 195
Shallow pools, wet meadows, marshes, shores of ponds, lakes and streams. Lab. and Nfld. w. to Nunavut, N.W.T. and Alask. s. to Me., n. Mich., Minn., N.D., mts. to N.M. Ariz. and Calif. According to Standley et al. (2002), specimens previously identified as *C. aquatilis* var. *altior* should be referred to as *C. emoryi*; the type specimen of *C. aquatilis* var. *altior* is actually an immature specimen of *C. emoryi*, thus the name is placed in synonymy under that species.

13b. *C. aquatilis* var. *substricta* Kük. Fig. 197
Marshes, wet shores and ditches in neutral to calcareous sites. N.S., N.B. and Que. w. to Man. and N.D., s. to Va., Ill., Mo., and Kans.

14. *C. lenticularis* Michx. Lakeshore Sedge Fig. 196
Wet gravelly and sandy shores and meadows. Lab. and Nfld., St. P. et Miq., and Que. w. to Nunavut, N.W.T. and Alask. s. to Mass., e. N.Y., Mich., ne. Minn., Man., Sask., mts. to N.M. and Calif.

15. *C. nigra* (L.) Reichard Smooth Black Sedge Fig. 196
Wet meadows, swales, salt marshes and wet gravels and rocks, especially disturbed sites. Chiefly near the coast; Greenl., se. Lab. and Nfld. w. along St. Lawrence R. to se. Que., s. to N.H. and n. Vt., Mass., R.I., Conn., N.Y.; Mich. and nw. Wisc.; B.C. Standley (1987a) and Standley et al. (2002) noted that *C. nigra* is morphologically variable in North America and that infraspecific taxa should not be recognized. Further, North American populations have not diverged significantly from those in Europe to warrant the recognition of them as a separate taxon. (*C. nigra* var. *strictiformis* (L. H. Bailey) Fernald)

16. *C. emoryi* Dewey Emory's Sedge Fig. 197
Swamps, shores and river margins. N.Y. and N.J. w. to Ind., Wisc., Minn. and Man., s. to Fla., Tex., Wyo. Colo., N.M. and Tex. (*C. aquatilis* var. *altior* (Rydb.) Fernald)

17. *C. stricta* Lam. Upright Sedge Fig. 195
Acid swamps, meadows and low woods. Large tussocks of this species are distinctive in meadows and, at times, stand well above the water in shallow pond margins. Nfld., P.E.I. and N.B. w. to Ont., Man., N.D. and Wyo., s. to N.E., N.C., Tenn., n. Ala., n. Miss. and Tex. (*C. stricta* var. *strictior* (Dewey) Carey)

Fig. 190. *Carex pauciflora*: a. habit; b. inflorescence; c. perigynium; d. achene within perigynium (HCOT).
Carex leptalea: e. habit; f. inflorescence; g. perigynium; h. achene (G&W).
Carex torta: i. inflorescence; j. perigynium (F).
Carex salina: k. inflorescence; l. perigynium; m. achene with constriction; n. pistillate scale (Mackenzie).

Fig. 191. *Carex mitchelliana*: a. inflorescence; b. perigynium; c. achene; d. pistillate scale (Mackenzie).
Carex paleacea: e. inflorescence; f. perigynium; g. achene with constriction; h. pistillate scale (Mackenzie).

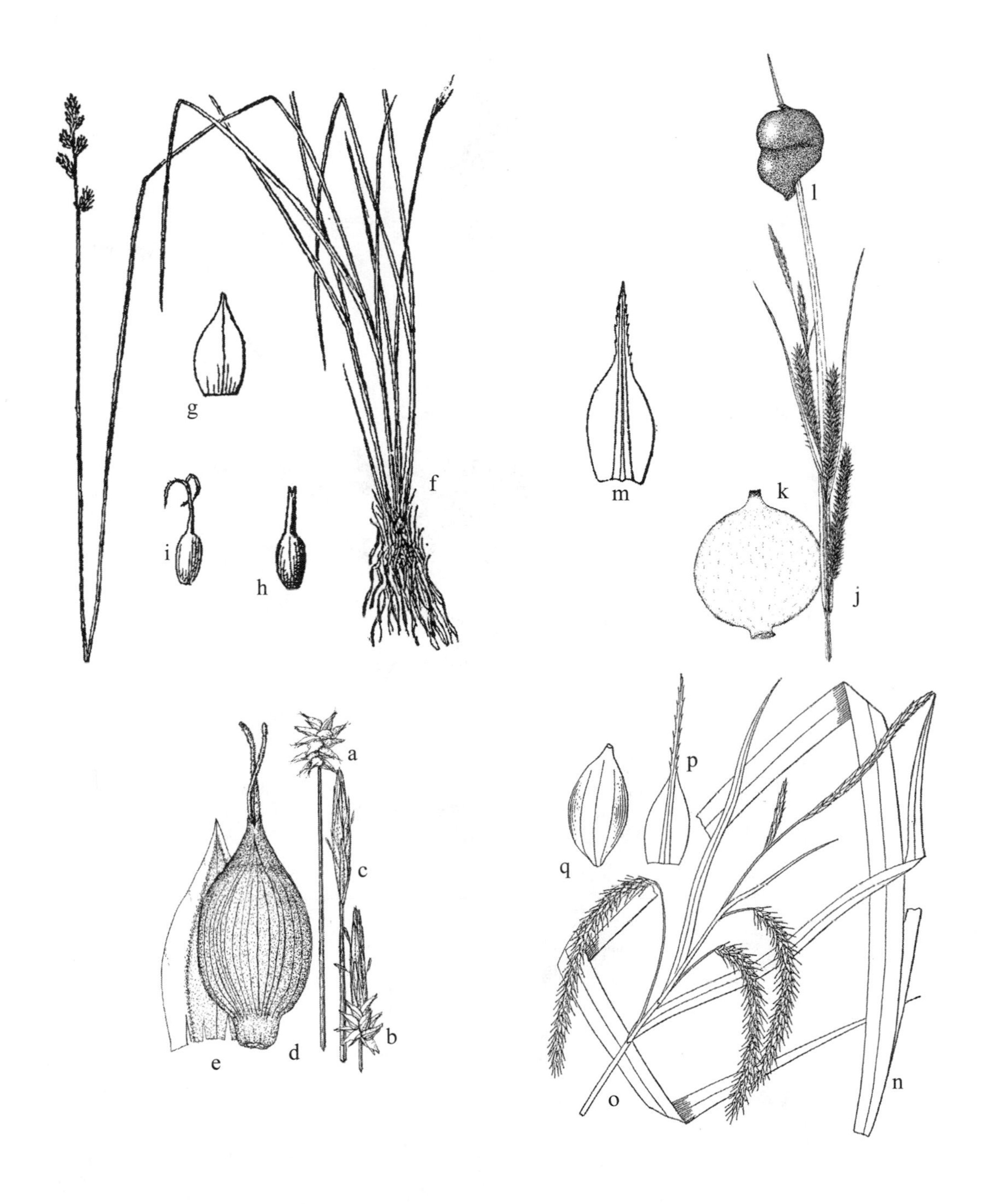

Fig. 192. *Carex gynocrates*: a. pistillate spike; b. androgynous spike; c. staminate spike; d. perigynium; e. scale (FNA).
Carex sterilis: f. habit; g. scale; h. perigynium; i. achene (B&B).
Carex recta: j. upper portion with pistillate spikes (*below*) and staminate spike (*above*) (FNA); k. perigynium; l. achene (FNA); m. scale (B&B).
Carex gynandra: n. portion of leaf; o. inflorescence, pistillate spikes (*below*), staminate spikes (*above*); p. scale with elongated awn; q. achene (B&B).

Fig. 193. *Carex crinita* var. *crinita*: a. inflorescence; b. perigynium; c. achene with constriction; d. pistillate scale (Mackenzie).

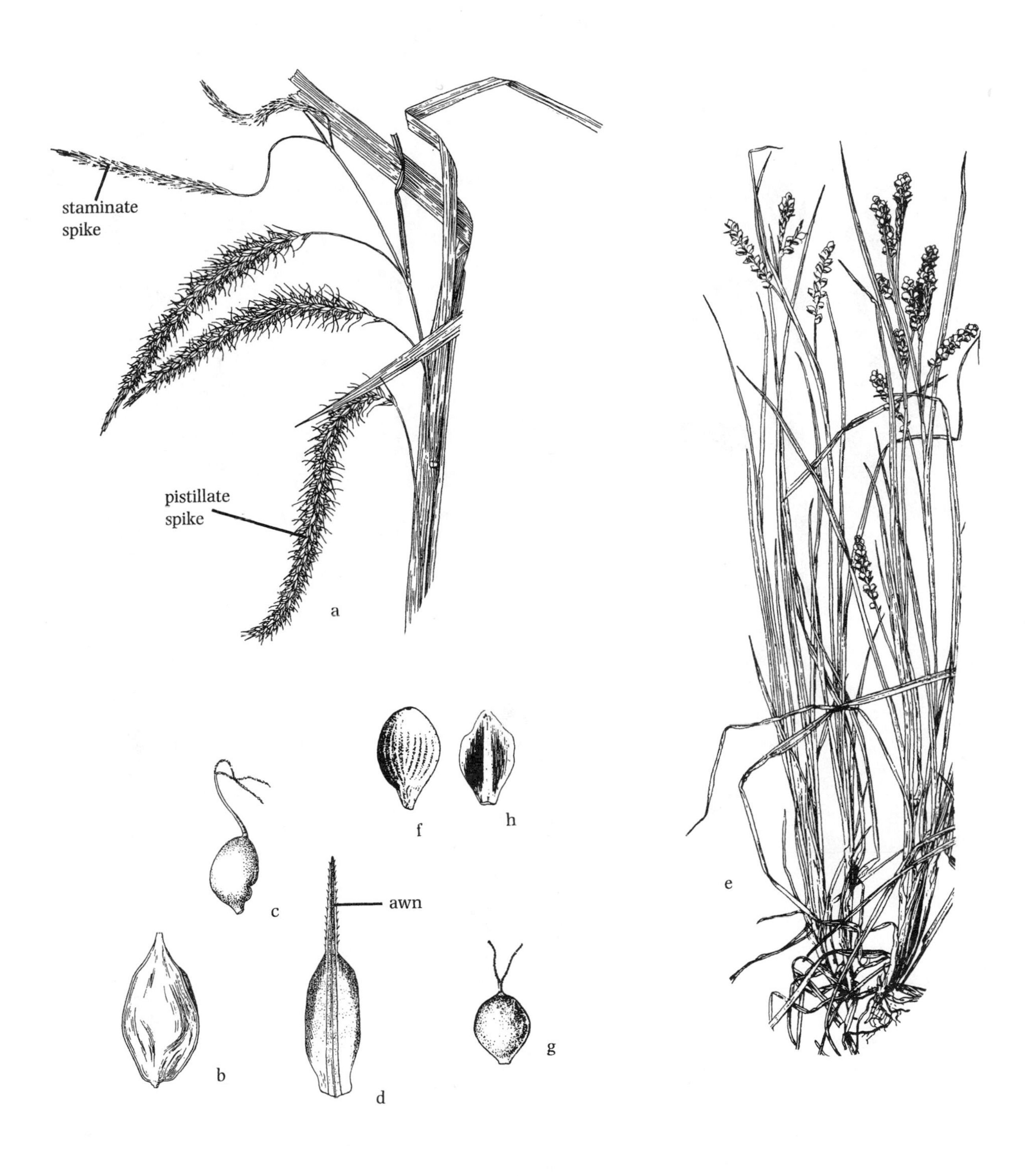

Fig. 194. *Carex gynandra*: a. inflorescence; b. perigynium; c. achene with constriction; d. pistillate scale (Mackenzie).
Carex garberi: e. habit; f. perigynium; g. achene; h. pistillate scale (Mackenzie).

Fig. 195. *Carex stricta*: a. inflorescence (F); b. perigynium; c. achene; d. pistillate scale; e. section of stem with sheath (F); f. inflorescence; g. perigynium; h. achene; i. pistillate scale; j. ligule (Mackenzie). *Carex aquatilis* var. *aquatilis*: k. inflorescence; l. inflorescence; m. perigynium (HCOT).

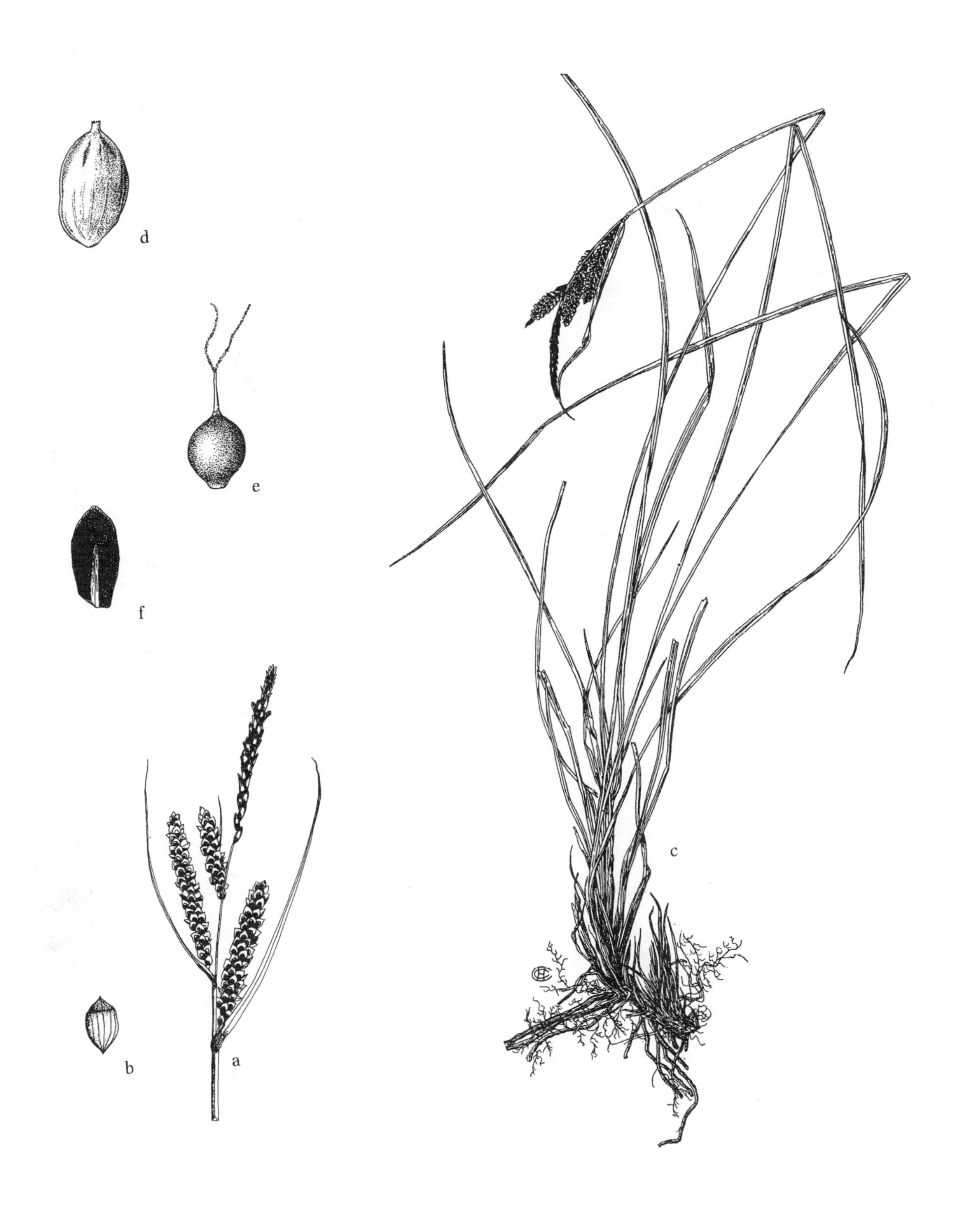

Fig. 196. *Carex lenticularis*: a. inflorescence; b. perigynium (F).
Carex nigra: c. habit; d. perigynium; e. achene; f. pistillate scale (Mackenzie).

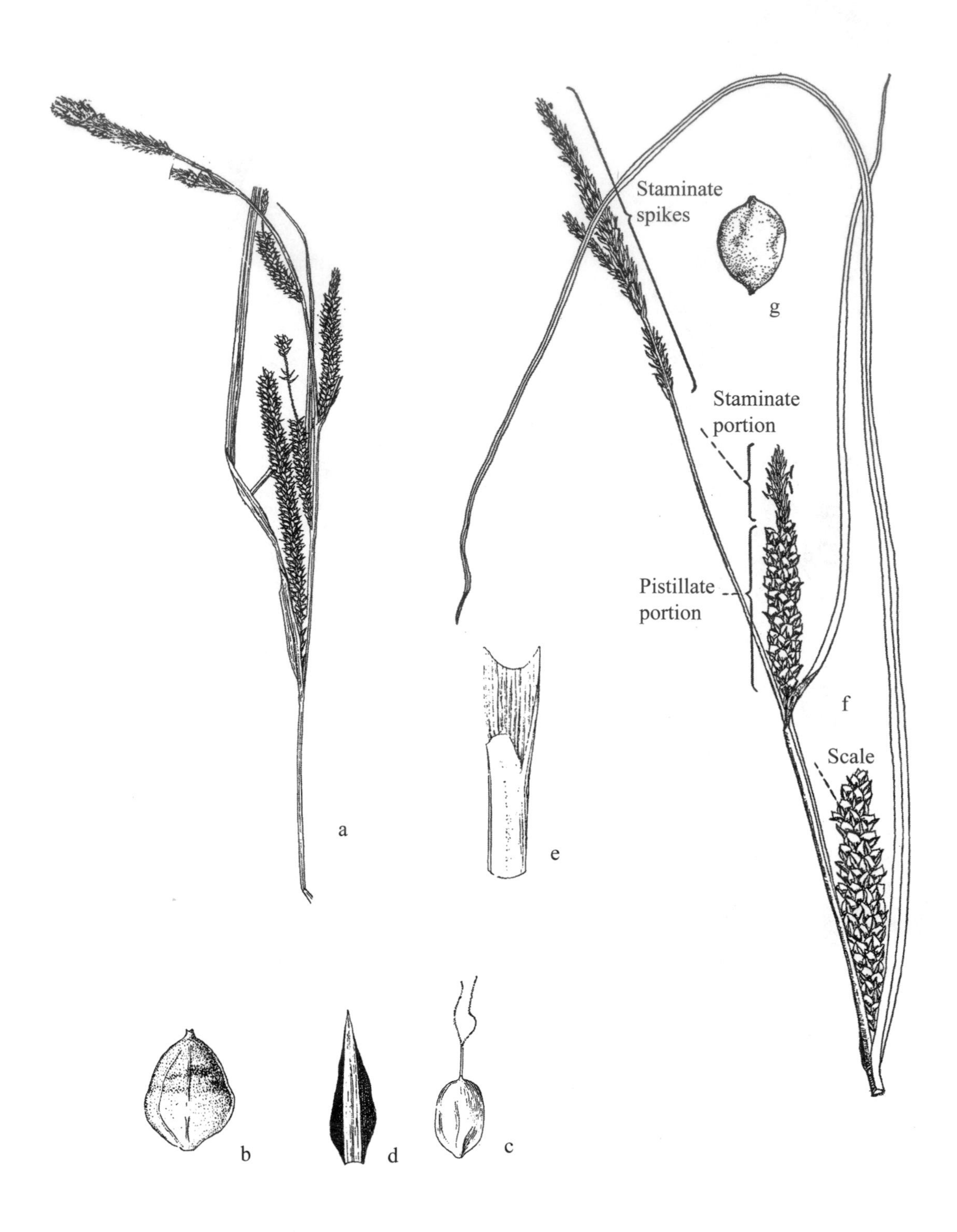

Fig. 197. *Carex emoryi*: a. inflorescence; b. perigynium; c. achene; d. pistillate scale; e. ligule (Mackenzie). *Carex aquatilis* var. *substricta*: f. inflorescence; g. perigynium (F).

18. *C. chordorrhiza* Ehrh. ex L. f. Creeping Sedge Fig. 198
Sphagnum bogs, peaty ground and quagmires. Greenl., S. Baffin I., Que. and Nunavut w. to Alask., s. to Nfld., Me., N.H., sw. Vt., w. Mass., n. Pa., Mich., n. Ind., n. Ill., Iowa, N.D., Mont., n. Ida. and w. Oreg.

19. *C. sartwellii* Dewey Sartwell's Sedge Fig. 198
Calcareous bogs, marshes, swales and lakeshores. Que. w. to James Bay, Man. N.W.T. and Yuk., s. to N.Y., w. Md., Ind., Ill., nw. Mo., Kans., mts. to Colo., Ida. and B.C.

20. *C. alopecoidea* Tuck. Foxtail Sedge Fig. 198
Marshes, meadows and swales. N.S. and s. Que. w. to Man., Sask. and s. to N.E., n. N.J., Md., e. Tenn., Ky., Ill., Mo., S.D. and ne. Wyo.

21. *C. crus-corvi* Shuttlw. ex Kuntze Raven-foot Sedge Fig. 199
Marshes, sloughs, swamps and riverbanks. Chiefly Mississippi embayment; s. Ont. and s. Mich. w. to s. Wisc., s. Minn. and e. Neb., s. to Tenn., Ala. nw. Fla., La., e. Okla. and e. Tex.; fewer populations along Atlantic coastal plain, Va. s. to Ga.

22. *C. stipata* Muhl. ex Willd. Owl-fruit Sedge Fig. 199

22a. *C. stipata* var. *stipata* Owl-fruit Sedge
Wet meadows, marshy shore, swampy woods, floodplains, low woods and damp shaded ground. S. Lab. and Nfld. w. to Man., Alta., B.C. and Alask. panhandle, s. to Ga., Ala., Tenn., Mo., Okla., N.M., Ariz. and Calif.

22b. *C. stipata* var. *maxima* Chapm. ex Boott Stalk-grain Sedge
Seasonally saturated soils in wet meadows, marshes and edges of tidal marshes, cypress swamps. N.J., and Pa., w. to Ohio, Ind., s. Ill. and e. Mo., s. to Fla., Ala., n. Miss. and nw. La.

23. *C. disperma* Softleaf Sedge
Sphagnum bogs, cedar swamps and mossy woods. Greenl., Lab. and Nfld. w. to N.W.T. and Alask., s. to N.E., n. N.J., Pa., n. Ind., ne. Ill., Wisc., Minn., S.D., mts. to N.M., Ariz. and Calif.

24. *C. decomposita* Muhl. Cypress-knee Sedge Fig. 199
Wooded swamps and pond margins, often on rotting logs. N.Y. w. to se. Mich., Ind., s. Ill, and se. Mo., s. to Fla., Miss., La., Okla. and e. Tex.

25. *C. diandra* Schrank Lesser Panicled Sedge Fig. 200
Bogs and peaty swamps, especially calcareous sites. Lab. and Nfld. w. to Yuk. and Alask., s. to N.E., N.J., n. Pa., w. Md., Ohio, n. Ind., n. Ill., Mo., and Neb., mts. to Colo. n. N.M., Utah, Nev. and Calif.

26. *C. prairea* Dewey Prairie Sedge Fig. 200
Calcareous bogs, meadows and wet thickets. Que. w. to Alta. and Yuk., s. to N.E., n. N.J., Va., Ohio, Ind., Ill., Iowa, Neb., Wyo., Mont., Ida. and B.C.

27. *C. vulpinoidea* Michx. Fox Sedge Fig. 201
Wet, low ground, marshes, shores of rivers and streams and near margins of wet wooded areas. Lab and Nfld. w. to s. B.C., s. throughout U.S.

28. *C. muskingumensis* Schwein. Palm Sedge Fig. 202
Hardwood swamps and floodplains, swales, lowland woods and thickets. S. Ont. and Mich. w. to Minn., s. to Tenn., Ark., Kans. and Okla. This is a tall, conspicuous species with large inflorescences and often tall vegetative stems with many tristichous leaves, giving it a superficial miniature palm appearance, hence is often cultivated under the common name Palm Sedge.

29. *C. albolutescens* Schwein. Greenish-white Sedge Fig. 202
Acid swamps, hardwoods swamps, low woods and thickets. N.H. and Mass. w. to s. N.Y., N.J., Pa., n. Ohio, sw. Mich., Ill., s. Mo. and Okla., s. to nw. Fla., La. and e. Tex. Very similar to *C. festucacea*.

30. *C. longii* Mack. Long's Sedge Fig. 202
Borders of marshes, pond margins, moist sandy open ground, savannas, acidic sandy or peaty soils, open woods, thickets and ditches. N.S., Mass. and Vt. w. to N.H., s. Ont., s. Mich., sw. Wisc., Mo. and Okla., s. to Fla., Ala., La. and e. Tex.; sw. Wash., nw. Oreg. and n. Calif.

31. *C. suberecta* (Olney) Britton Prairie Straw Sedge Fig. 200
Calcareous soils of meadows, marshy shores, ditches and prairies. S. Ont. w. to Minn., s. to w. Va., Ohio, Ind., Ill., Mo. and Ark.

32. *C. festucacea* Schkuhr ex Willd. Fescue Sedge Fig. 204
Meadows and fields, swamps margins of wetlands and sedge meadows. S. Me., se. N.H. and Mass. w. to s. N.Y., s. Ont., s. Mich., Wisc., s. Minn., nw. Iowa and Kans., s. to N.C., Ga., nw. Fla., La. and e. Tex.

33. *C. scoparia* Schkuhr ex Willd. Broom Sedge Fig. 203
Open swamps, wet meadows, marshy shores, thickets and moist to dry open sites. Nfld. w. to B.C., s. to S.C., n. Ga., Tenn., Ala., Miss., Ark., La., Okla., mts. to Colo, N.M., Ariz. and n. Calif.

34. *C. projecta* Mack. Necklace Sedge Fig. 204
Swamps, wetlands, moist thickets and depressions in upland forests, wet meadows and ditches and moist northern forests. Lab. and Nfld. w. to Sask., s. to N.E., Va., S.C., n. Ga., n. Tenn., Ill. and Iowa.

35. *C. tribuloides* Wahlenb. Blunt Broom Sedge Fig. 204
Marshes, swales, shores, swamps shrub thickets and ditches. N.S., N.B. and Que. w. to Man. and S.D., s. to Fla. and Tex.; B.C. and n. Oreg. Easily confused with *C. scoparia*; it is critical to collect plants with mature perigynia. Depauperate plants, especially growing in shade are difficult to identify. The widespread var. *tribuloides* is our chief taxon. A second taxon, var. *sangamonensis* Chokey, with smaller perigynia (2.2–2.8(3) times as long as wide), ranges from N.J., Del., and Md. w. to Ky., s. Mich., Ill., Mo. and Kans., s. to Fla. and e. Texas. Voss and Reznicek (2012) indicate that this taxon occurs in southernmost Michigan and is sometimes recognized as a distinct species, but more study is needed.

36. *C. crawfordii* Fernald Crawford's Sedge Fig. 203
Wet meadows, swamps, shores and wet to dry open ground. Lab. and Nfld. w. to N.W.T., and Alask., s. to N.E., n. N.J., N.Y., n. Mich., Ill., e. Mo., Iowa, Man., Sask., Mont., Ida. and Oreg.

37. *C. cristatella* Britton Crested Sedge Fig. 205
Open swamps, marshes, wet meadows, shores, thickets and wet woods. Sw. Que. w. to Sask., s. to N.E., Md., N.C., Ky., Mo. and Kans.

38. *C. bebbii* (L. H. Bailey) Fernald Bebb's Sedge Fig. 205
Wet meadows, swales, and shores, chiefly calcareous soils. Lab. and Nfld. w. to N.W.T. and Alask. s. to N.E., n. N.J., Pa., n. Ohio, n. Ind., Iowa, Neb., mts. to Colo., n. N.M., Ida., ne. Nev., e. Oreg. and Wash.

Fig. 198. *Carex chordorrhiza*: a. upper portion of plant; b. inflorescence; c. perigynium; d. achene; e. pistillate scale (Mackenzie).
Carex sartwellii: f. inflorescence; g. single spike; h. perigynium; i. achene; j. pistillate scale (Mackenzie).
Carex alopecoidea: k. inflorescence; l. single spike; m. perigynium; n. achene; o. pistillate scale; p. ligule (Mackenzie).

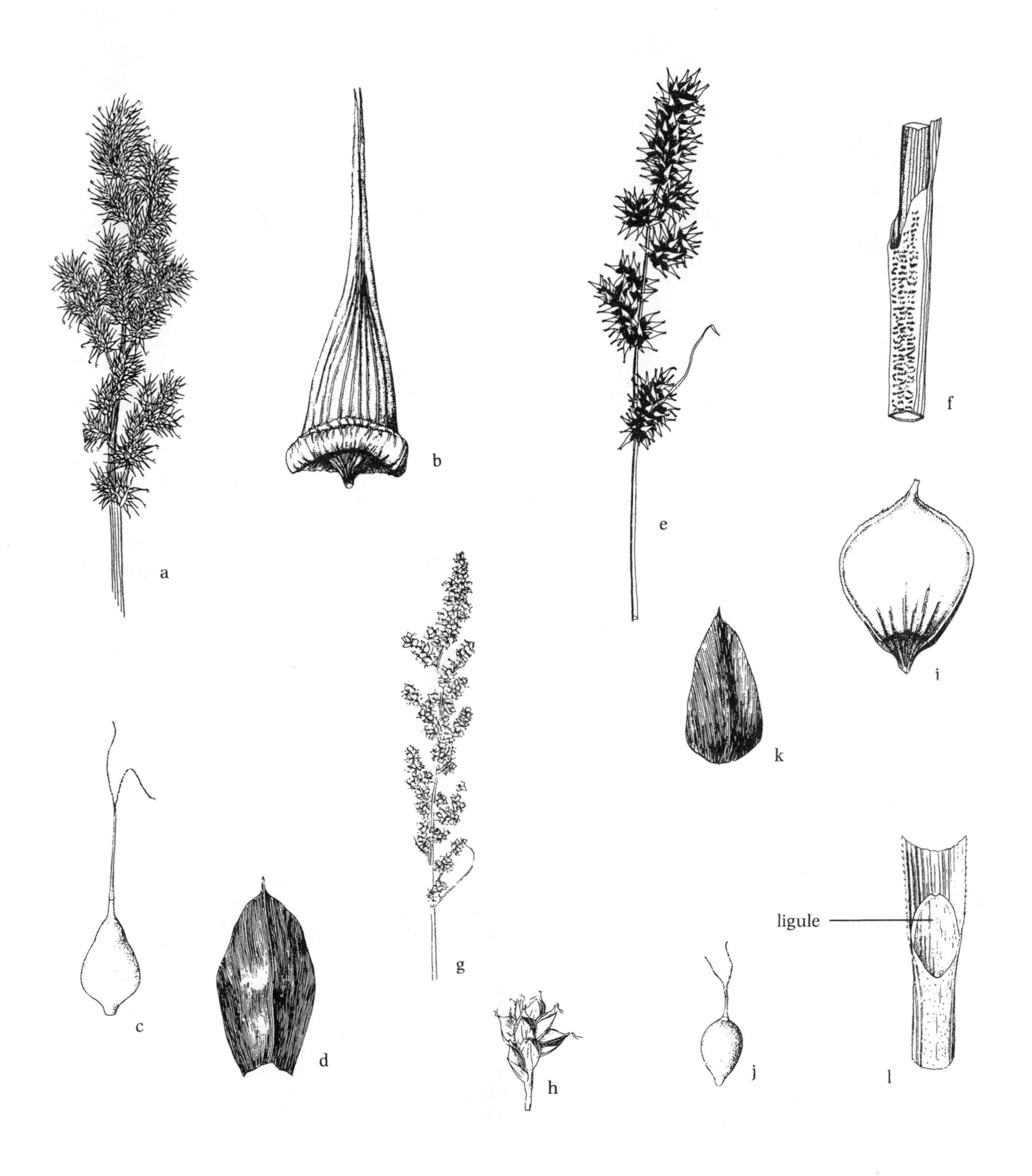

Fig. 199. *Carex crus-corvi*: a. inflorescence (Mackenzie); b. perigynium (G&W); c. achene (Mackenzie); d. pistillate scale (G&W).
Carex stipata: e. inflorescence; f. ligule (F).
Carex decomposita: g. inflorescence (Mackenzie); h. single spike (Mackenzie); i. perigynium (G&W); j. achene (Mackenzie); k. pistillate scale (G&W); l. ligule (Mackenzie).

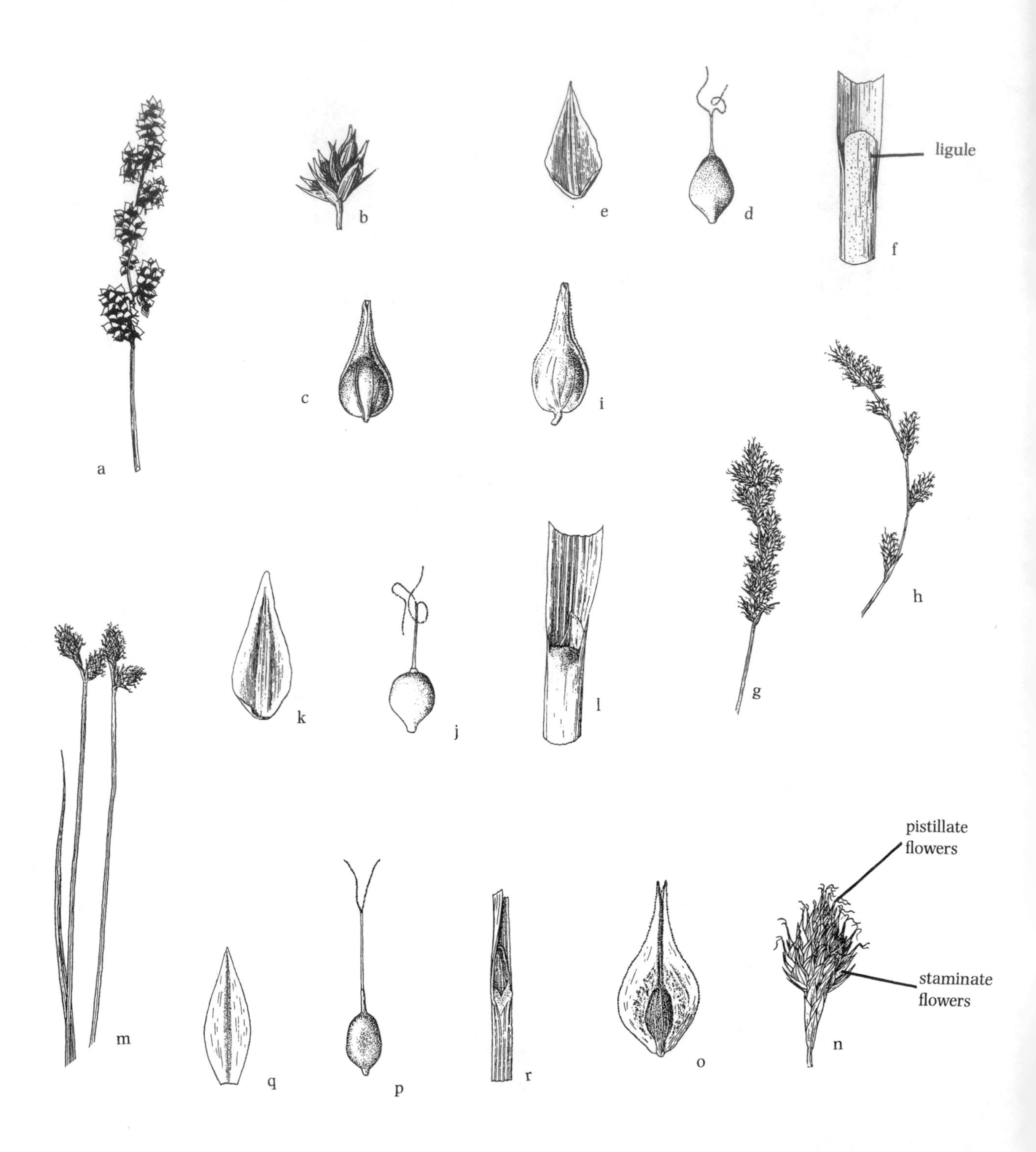

Fig. 200. *Carex diandra*: a. inflorescence (F); b. single spike (Mackenzie); c. perigynium (Mackenzie); d. achene (Mackenzie); e. pistillate scale (Mackenzie); f. ligule (Mackenzie).
Carex prairea: g, h. inflorescences; i. perigynium; j. achene; k. pistillate scale; l. ligule (Mackenzie).
Carex suberecta: m. inflorescence; n. single spike; o. perigynium; p. achene; q. pistillate scale; r. ligule (Mackenzie).

Fig. 201. *Carex vulpinoidea*: a. habit (HCOT); b. inflorescence (G&W); c. perigynium (G&W); d. scale variations (G&W); e. achene (G&W); f. leaf sheath, two views (HCOT).

Fig. 202. *Carex muskingumensis*: a. inflorescence (FNA); b. perigynium (B&B).
Carex albolutescens: c. inflorescence; d. scale; e. perigynium; f. achene with style contorted at base (FNA).
Carex longii: g. inflorescence; h. scale; i. perigynium; j. achene with nearly straight style (FNA).

Fig. 203. *Carex scoparia*: a. inflorescence; b. single spike; c. perigynium; d. achene; e. pistillate scale; f. ligule (Mackenzie).
Carex crawfordii: g. habit (HCOT); h. inflorescence (HCOT); i, j. perigynium, two views (HCOT); k. achene (Mackenzie); l. pistillate scale (Mackenzie); m. ligule (Mackenzie).

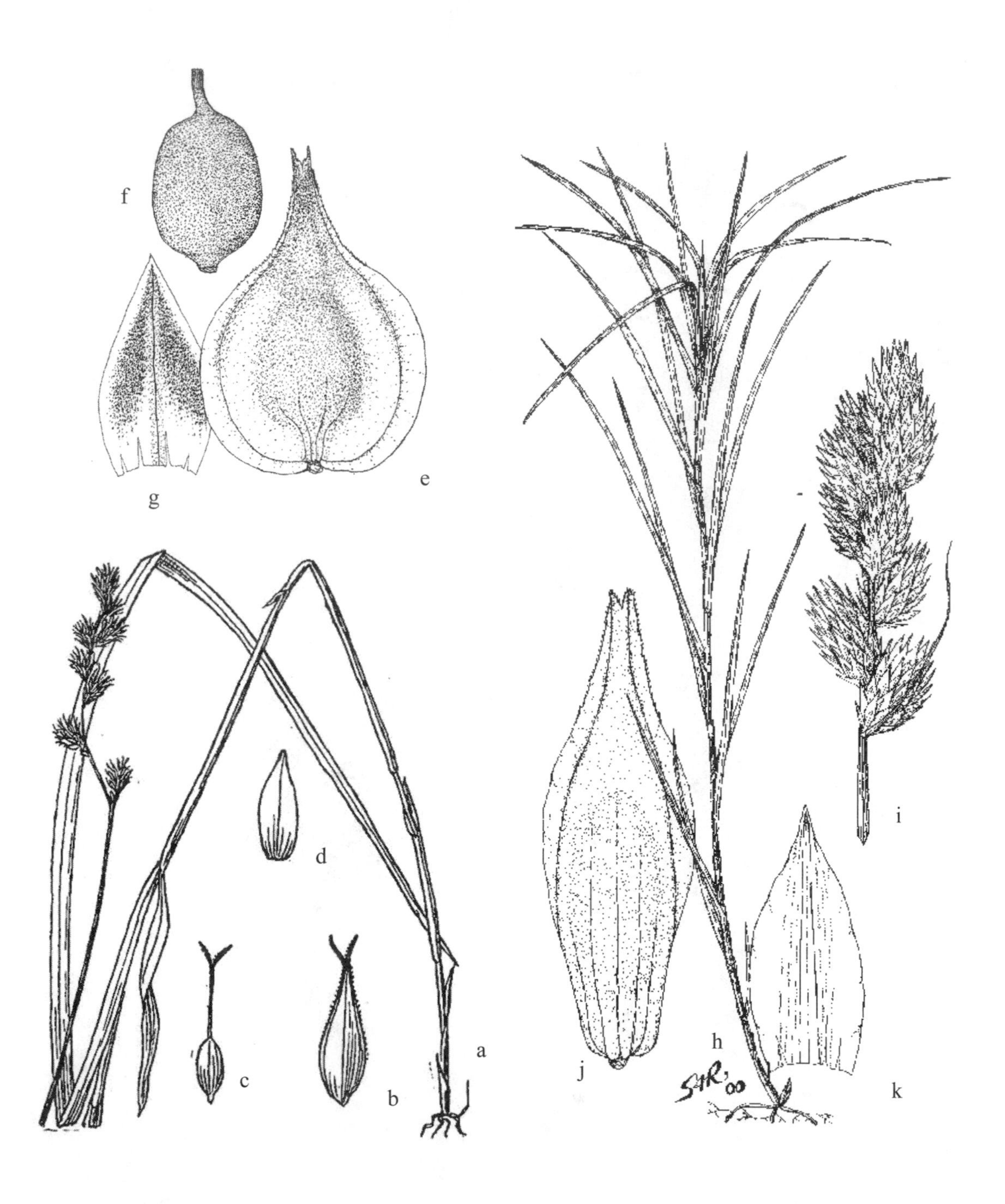

Fig. 204. *Carex projecta*: a. habit; b. perigynium with stigmas exerted; c. achene with style straight, stigmas 2; d. scale (B&B).
Carex festucacea: e. perigynium; f. achene; g. scale (FNA).
Carex tribuloides: h. habit; i. inflorescence; j. perigynium; k. scale (FNA).

Fig. 205. *Carex bebbii*: a. habit (HCOT); b. inflorescence (HCOT); c. perigynium (Mackenzie); d. achene (Mackenzie); e. pistillate scale (Mackenzie); f. ligule (Mackenzie).
Carex cristatella: g, h. inflorescence; i, j. perigynium, two views; k. achene; l. pistillate scale (Mackenzie).

39. *C. alata* Torr. Broad-wing Sedge Fig. 206
Marshes, swamps, bogs, shores, meadows and wet woods. N.H. and Mass. w. to N.Y., nw. Pa., s. Ont., Mich., n. Ind., s. Ill. and s. Mo., s. to Fla., s. Mo. and e. Tex.

40. *C. straminea* Willd. Eastern Straw Sedge Fig. 206
Swamps and marshes. Mass. w. to N.Y., s. Mich. and Wisc., s. to Md., N.C., Ky. and se. Mo. (*C. richii* (Fernald.) Mackenz.)

41. *C. hormathodes* Fernald Marsh Straw Sedge Fig. 206
Saline and freshwater marshes. Coastal, w. Nfld., St. P. et Miq., and e. Que. s. to N.E., se. N.Y., Md., Va. and N.C.

42. *C. trisperma* Dewey Three-seeded Sedge
Coniferous bogs and swamps, mossy woods and boggy hollows. Greenl., Lab. and Nfld. w. to Nunavut, N.W.T. and B.C., s. to N.E., n. N.J., Md., mts. to N.C. and e. Tenn., n. Ohio, n. Ind., ne. Ill., Wisc. and Minn. A closely related smaller plant, *C. billingsii* (O. W. Knight) Kirschb. (sometimes treated as *C. trisperma* var. *billingsii* O. W. Knight) is known throughout the northeastern portion of the *C. trisperma* range; easily overlooked, it has more narrow, filiform-involute leaves 0.3–0.8 mm, occupying hummocks of sphagnum bogs.

43. *C. tenuiflora* Wahlenb. Sparse-flower Sedge
Sphagnum bogs, rich woodland fens, mossy woods, and peaty shores. Baffin I., Lab. and Nfld. w. to N.W.T. and Alask., s. to c. Me., w. Mass., N.Y., Ohio, Mich., Wisc. and Minn., mts. of Colo., Mont. and nc. Wash.

44. *C. heleonastes* L. f. Fig. 211
Bogs, muskegs, mires, wet meadows and lowlands. Lab., and Que. w. to Man., N.W.T., Yuk. and Alask., s. to Nfld., Ont. and n. Mich., Sask., Alta. and B.C.; Eurasia.

45. *C. mackenziei* Wahlenb. Mackenzie's Sedge Fig. 207
Saline shores and marshes. Coastal, Greenl., and Lab. w. to Nunavut, N.W.T. and Alask., s. to N.S., Me., w. Vt., Mass., N.Y. Ohio, Mich., Wisc., n. Minn., Sask., Mont. and n. Wash.

46. *C. canescens* L. Silvery Sedge Fig. 207

46a. *C. canescens* subsp. *canescens*
Swamps, bogs, ponds and stream margins. Greenl., and Lab. w. to Alask., s. to N.E., N.J., Va., Ohio, Ind., Wisc., Minn., N.M., mts. to Wyo., N.M., Ariz. and Calif.

46b. *C. canescens* subsp. *disjuncta* (Fernald) Toivonen
Sphagnum bogs, moist coniferous forests and meadows. Nfld. w. to Ont. and Minn., s. to S.C., W.Va., ne. Ohio, n. Ind., ne. Ill. and Wisc. *C. canescens* subsp. *disjuncta* is a dominating subspecies in the southern portion of the *C. canescens* range.

47. *C. brunnescens* (Pers.) Poir. Brownish Sedge
Low woods, margins of bogs and swamps, and wet turfy hollows and slopes. Greenl., Lab. and Nfld. w. to N.W.T. and Yuk., s. to N.E., N.J., Pa., mts. to N.C., S.C., n. Ga. and e. Tenn., n. Ohio, Mich., n. Ind., ne. Ill., Wisc., and Minn., mts. to w. S.D., Colo., N.M., Nev. and Calif.

48. *C. exilis* Dewey Coastal Sedge Fig. 208
Bogs, fens and open wet sites. Lab. and Nfld. w. to Ont. and n. Minn., s. to N.Y., N.J., Del. and Md.; disjunct in N.C., Ala. and Miss.

49. *C. seorsa* Howe Weak Stellate Sedge Fig. 211
Acid soils, sandy or peaty, hardwood swamps, *Chamaecyparis* swamps, thickets and depressions in forests. Mature specimens critical; immature perigynia may resemble those of *C. brunnescens*, which differs in having the beak at least sparsely serrulate. Se. N.H. and Mass. w. to s. N.Y., s. Ont., Mich. and n. Ind., s. to Va., S.C., Ga., nw. Fla., n. Miss. and Ark. Populations more common along Atlantic coastal plain.

50. *C. atlantica* L. H. Bailey Prickly Bog Sedge Fig. 209

50a. *C. atlantica* subsp. *atlantica* Fig. 209
Bogs and wet, acid thickets and edges of forests. N.S. and N.E w. to s. N.Y., N.J., Pa., n. Ohio, Mich., Ind., s. Ill. and se. Mo., s. to n. Fla., e. Okla. and e. Tex.; most abundant near the coast. (*C. incomperta* E. P. Bicknell; *C. atlantica* var. *incomperta* (E. P. Bicknell) F. J. Herm.)

50b. *C. atlantica* subsp. *capillacea* (L. H. Bailey) Reznicek Fig. 209
Swamps, bogs, wet, acid thickets, floodplains and low woods. N.S., Me. and s. Que. w. to s. Ont., Mich., Ill., and e. Mo., s. to Fla., w. to La. and e. Tex. subsp. *capillacea* and subsp. *atlantica* are quite distinct within our range, but in non-glaciated portions of their ranges intermediates occur (Reznicek and Ball, 1980; Reznicek, 2002). (*C. interior* var. *capillacea* L. H. Bailey; *C. howei* Mackenz.)

51. *C. wiegandii* Mackenz. Wiegand's Sedge Fig. 208
Sphagnum bogs, boggy thickets, lakeshores and wet peaty soils. Rare, Nfld. w. to Ont., s. to Me. and n. Mass., N.Y., Pa. and Mich.

52. *C. ruthii* Mackenz. Ruth's Sedge
Boggy meadows, open or wooded stream banks, springheads and seeps. Appalachian Mts.; w. Va. to w. N.C., w. S.C., e. Tenn. and ne. Ga.

53. *C. interior* L. H. Bailey Inland Sedge Fig. 210
Shores of lakes, ponds, and streams, ditches, wet meadows, fens, wet prairies, and swamps; chiefly calcareous sites. Lab. and Nfld. w. to sw. N.W.T. and s. Alask., s. to N.J., Va., Tenn., Ohio, Mo., Ark., Kan., mts. to Colo., n. Ariz., Nev. and n. Calif.; Mex.

54. *C. echinata* Murray Star Sedge Fig. 210
Wet acid soils, sphagnum bogs, boggy meadows, and sandy lake and river shores. Nfld., and Lab. w. to Ont. and Minn., s. to N.E., N.J., s. in mts. to Pa., Md., Va., W.Va., N.C. and Tenn., n. Ohio, n. Ind., and n. Iowa; N.D. and Sask. w. to B.C., s. in mts. to Wyo., Colo., n. Utah, Nev. and Calif. Our taxon is subsp. *echinata*.

55. *C. sterilis* Willd. Dioecious Sedge Fig. 192
Fens, calcareous sedge meadows, open cedar swamps, calcareous shores and marly flats. Nfld. and Que. w. to N.W.T and Sask., s. to w. N.E., N.J., Va., n. Tenn., n. Ala., se. Mo., Iowa, N.D. and c. Mont. This sedge is unique in having of separate pistillate and staminate plants, dense clumps often being a dominate sedge in fens and calcareous shores (Voss and Reznicek, 2012).

56. *C. scabrata* Schwein. Eastern Rough Sedge Fig. 211
Wet woods, springy thickets, creek borders, ravine bottoms, wet spots in rich deciduous forests, less often in swamps and wet clearings. N.S., N.B. and s. Que. w. to N.Y., Mich. Wisc., s. to N.E., N.J., mts. to Va., N.S., S.C., n. Ga., n. Ala. and e. Tenn., n. Ind.; c. and sw. Mo.

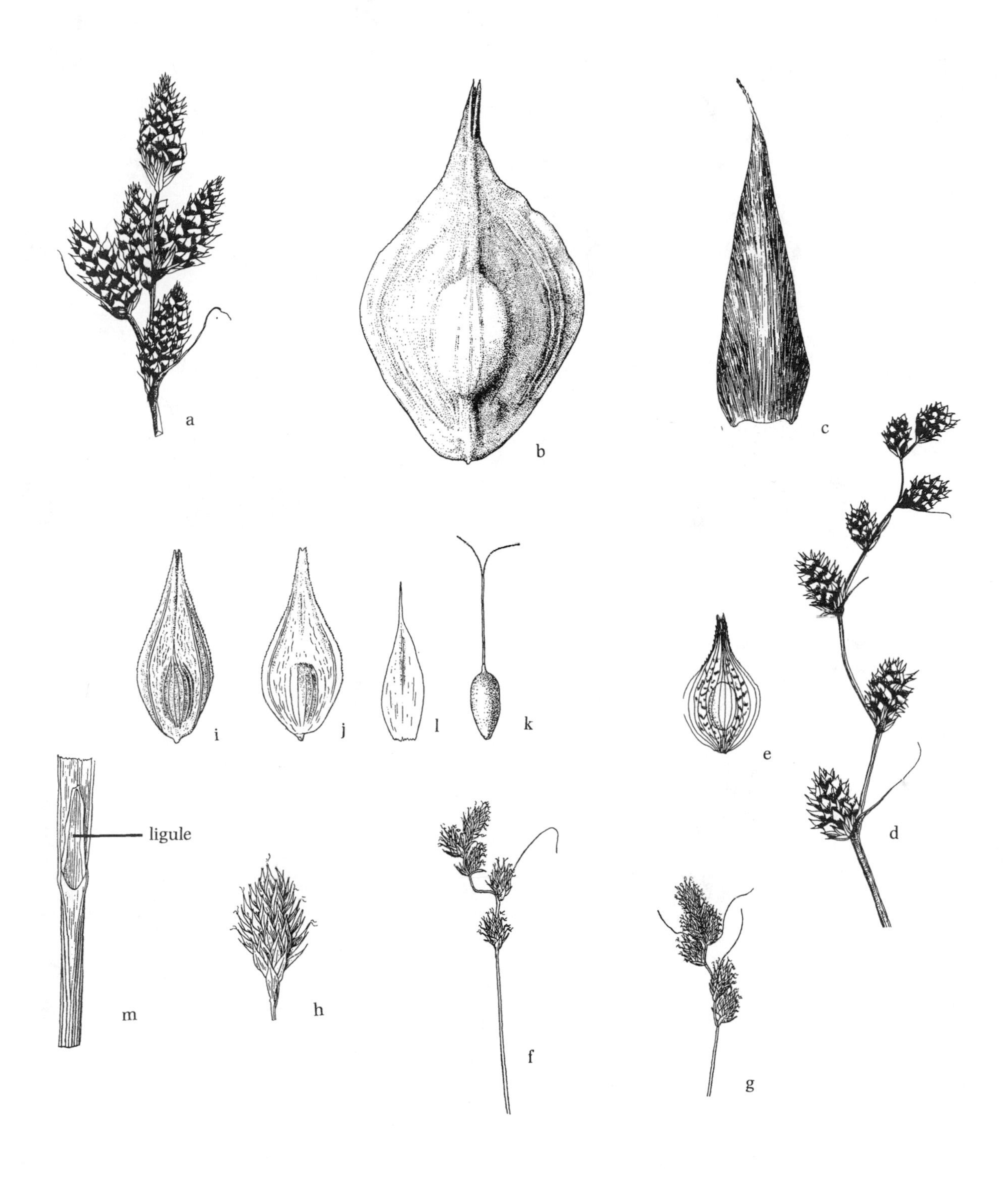

Fig. 206. *Carex alata*: a. inflorescence (F); b. perigynium (G&W); c. pistillate scale (G&W).
Carex straminea: d. inflorescence; e. perigynium (F).
Carex hormathodes: f, g. inflorescences; h. single spike; i, j. perigynium, two views; k. achene; l. pistillate scale; m. ligule (Mackenzie).

Fig. 207. *Carex canescens*: a. habit (HCOT); b. perigynium (Mackenzie); c. achene (Mackenzie); d. pistillate scale (Mackenzie).
Carex mackenziei: e. inflorescence; f. single spike; g. perigynium; h. achene; i. pistillate scale (Mackenzie).

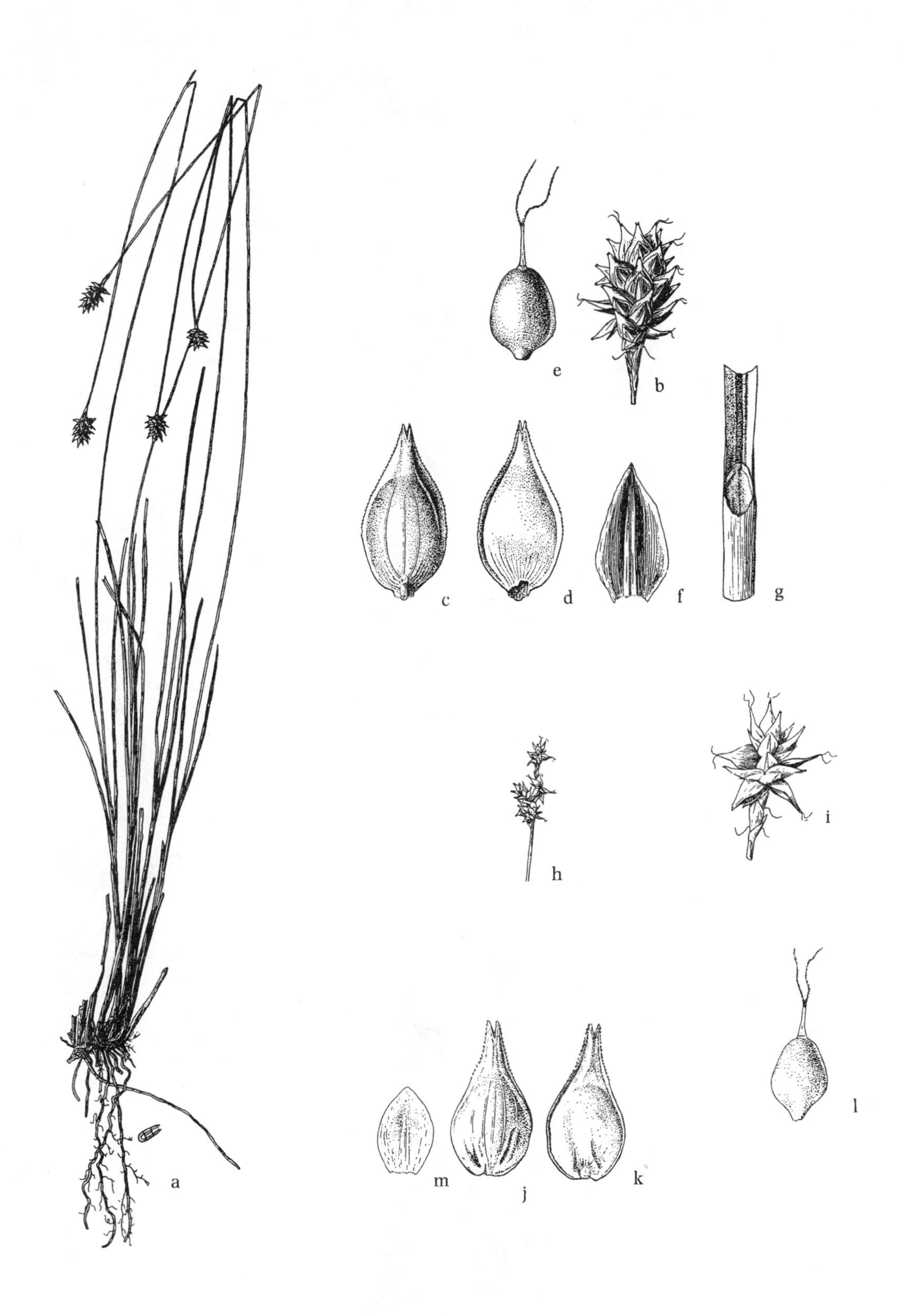

Fig. 208. *Carex exilis*: a. habit; b. single spike; c, d. perigynium, two views; e. achene; f. pistillate scale; g. ligule (Mackenzie).
Carex wiegandii: h. inflorescence; i. single spike; j, k. perigynium, two views; l. achene; m. pistillate scale (Mackenzie).

Fig. 209. *Carex atlantica* subsp. *atlantica*: a. habit (G&W); b. inflorescence (G&W); c, d. perigynium, two views (Mackenzie); e. achene (Mackenzie); f. pistillate scale (Mackenzie); g. ligule (Mackenzie).
Carex atlantica subsp. *capillacea*: h. inflorescence; i. single spike; j, k. perigynium, two views; l. achene; m. pistillate scale; n. ligule (Mackenzie).

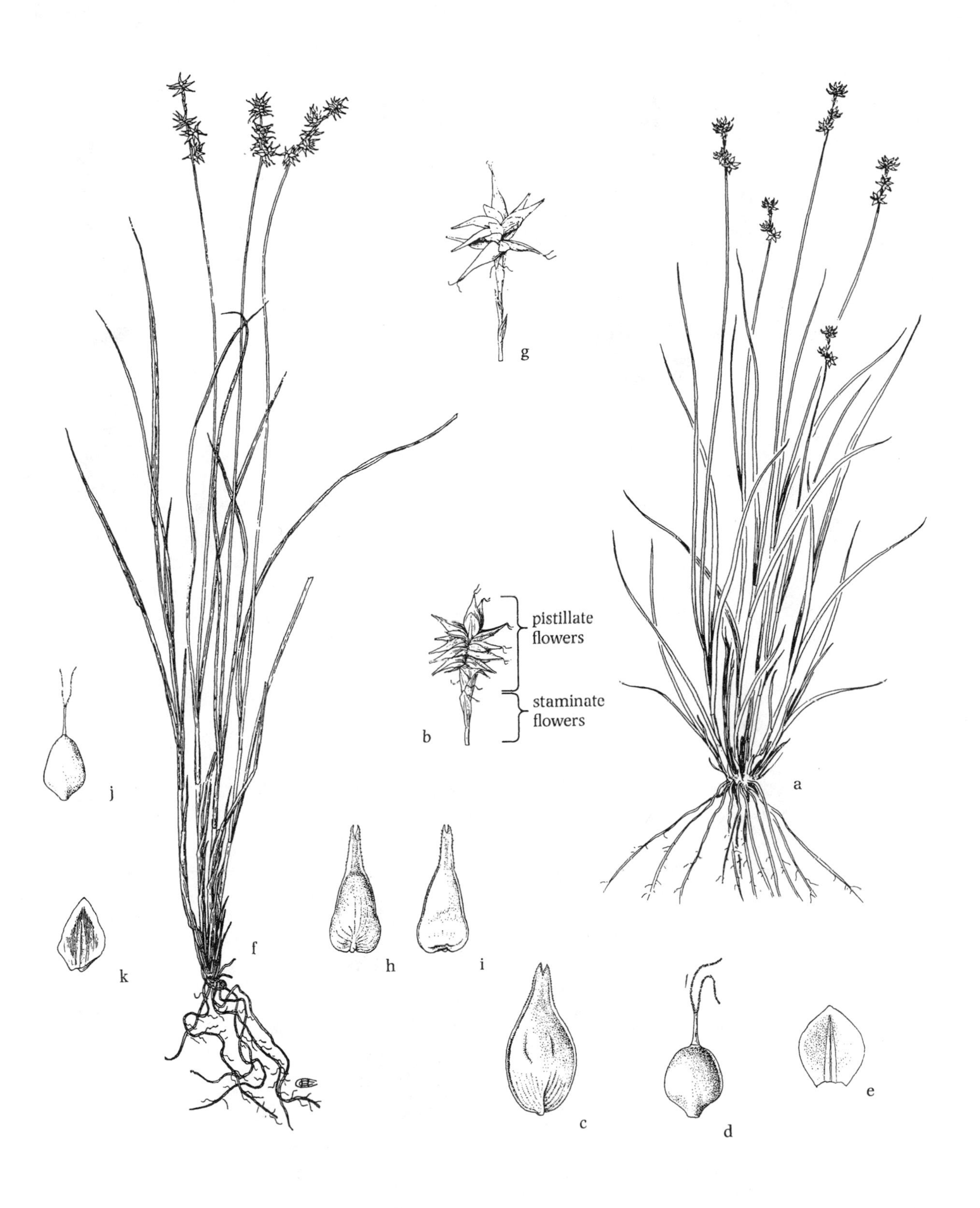

Fig. 210. *Carex interior*: a. habit (HCOT); b. single spike (Mackenzie); c. perigynium (Mackenzie); d. achene (Mackenzie); e. pistillate scale (Mackenzie).
Carex echinata: f. habit; g. single spike; h, i. perigynium, two views; j. achene; k. pistillate scale (Mackenzie).

Fig. 211. *Carex heleonastes*: a. inflorescence; b. perigynium; c. scale (FNA).
Carex serosa: d. inflorescence; e. perigynium (FNA).
Carex scabrata: f. inflorescence, staminate spike (*above*), pistillate spikes (*below*); g. perigynium (FNA).
Carex pellita: h. inflorescence, staminate spikes (*above*), pistillate spikes (*below*); i. perigynium (FNA).
Carex lonchocarpa: j. inflorescence, staminate spike (*above*), pistillate spikes (*below*); k. achene (FNA).
Carex prasina: l. inflorescence, staminate spike (*above*), pistillate spikes (*below*); m, n. perigynium, 2 views (FNA).

57. *C. pellita* Willd. Wooly Sedge Figs. 211, 212
Shores, meadows, swales and riverbanks; often in shallow water. Nfld., Lab and Que. w. to B.C., s. to N.E., Va., Tenn., Ark., Tex., N.M. and s. Calif. This species has been called *C. lanuginosa* Michx., but the type specimen for *C. lanuginosa* turned out to be *Carex lasiocarpa*, therefore the name is placed in synonymy under that taxon (Reznicek and Catling, 2002).

58. *C. lasiocarpa* Ehrh. Wooly-fruit Sedge Fig. 212
Bogs, streams, swales and shallow water. Nfld. w. to B.C., s. to n. N.J., Pa., Va., n. Ohio, n. Ill., ne. Iowa, Man., Sask., Mont., Ida. and Wash.; Calif. The North American plants have been segregated as var. *americana* Fernald (*C. lanuginosa* Michx.)

59. *C. striata* Michx. Walter's Sedge Fig. 213
Acidic soils of pine barren swamps, bogs, boggy depressions and pond margins. Coastal plain, N.H. and se. Mass. s. to L.I., N.Y., N.J., S.C. and n. Fla., w. to Miss. Our taxon, var. *brevis* L. H. Bailey, is distinguished from the southern taxon by having glabrous perigynia. Reznicek and Catling (1986) determined *C. striata* to be the correct name for this plant widely referred to as *C. walteriana* L. H. Bailey.

60. *C. trichocarpa* Muhl. ex Schkuhr Hairy-fruit Sedge Fig. 212
Calcareous meadows, swales, marshes, wet deciduous woods, and riverbanks. Sw. Que., N.H. and Vt. w. to Ont. and Minn., s. to Conn., Del., Va., w. N.C., e. Tenn., W.Va., Ohio, Ind., Ill. and Mo.

61. *C. atherodes* Spreng. Wheat Sedge Fig. 214
Calcareous meadows, shores, marshes and shallow water. S. Me. and s. N.H., w. to Ont., Man., N.W.T. and Alask., s. to w. N.Y., Va., W.Va., Ohio, Ind., Ill., Mo., Kans., Colo., n. N.M., Ariz. and n. Calif.

62. *C. comosa* Boott Longhair Sedge Fig. 215
Swamps, bogs, marshes and shallow water. N.S., N.B. and Me. w. to sw. Que., Ont., Mich., Minn., and S.D., s. to Fla., La. and e. Tex.; Mont., Ida. and Calif.; Mex.

63. *C. pseudocyperus* L. Cypress-like Sedge Fig. 216
Marshes, swamps, bogs and shores; often in shallow water. N.S. and N.B. w. to Man., and Alta., s. to N.E., n. N.J., nw. Pa., Ohio, n. Ind., Wisc., Minn. and ne. N.D.

64. *C. schweinitzii* Dewey ex Schwein. Schweinitz's Sedge Fig. 217
Swamps, meadows. Old springs and streams and low wet woods. W. Vt. and w. Mass w. to s. Ont., Mich. and Wisc., s. to R.I., w. Conn., mts. of Pa., Va., N.C. and Tenn.

65. *C. lurida* Wahlenb. Shallow Sedge Fig. 218
Mostly acid soils; swamps, marshes, wet woods, lakeshores and stream margins. Nfld., Lab, N.S. and s. Que. w. to Ont., Wisc., and Minn., s. to Fla., e. Tex. and e. Mex.

66. *C. hystericina* Willd. Bottlebrush Sedge Fig. 217
Calcareous soils; swamps, swales, wet meadows and shores. Lab., Nfld. and Que. w. to Alta., B.C. and Wash., s. to N.E., Md., mts. to Va., W.Va., n. Ga., Tenn., Ky., Ark., Okla., Tex., mts. to N.M., Ariz. and n. Calif.

67. *C. michauxiana* Boeckl. Michaux's Sedge Fig. 219
Bogs, wet sands and meadows. Nfld. w. to Ont., Man. and Sask., s. to N.S., N.E., n. N.Y., w. Md., Pa., n. Mich., and ne. Minn.

68. *C. folliculata* L. Northern Long Sedge Fig. 219
Swamps, swales, bogs, pond and lake margins, stream banks and wet woods. Nfld. w. to Ont., Mich., and Wisc., s. to N.E., mts. and piedmont to N.C., W.Va., S.C., Ga. and Tenn.; e. Tex. Our taxon is chiefly the northern var. *folliculata*, which passes into the southern var. *australis* L. H. Bailey, a coastal plain taxon ranging from s. N.J. to Fla., w. to La., characterized by having shorter pistillate scales, usually long-acuminate, and seldom awned.

69. *C. lonchocarpa* Willd. ex Spreng. Southern Long Sedge Fig. 211
Sandy peaty or acidic soils of wet savannas, forests seeps, stream banks, and lakeshores. Coastal, N.J., Del., Md. and Va., s. to Fla. w. to Tex.; Tenn. (*C. folliculata* var. *australis* L. H. Bailey)

70. *C. lacustris* Willd. Hairy Sedge Fig. 213
Swamps, ditches, marshes, swales, river and stream margins, Nfld., P.E.I. and Que. w. to Man., Sask. and Alta., s. to N.E., Va., Tenn., Mo. and Kans.; Mont. and Ida. (*C. riparia* var. *lacustris* (Willd.) Kük).

71. *C. hyalinolepis* Steud. Shoreline Sedge Fig. 216
Calcareous or brackish swamps, shores, ditches and wet woods. N.Y. and se. Pa. w. to s. Ont., se. Mich., Ind., Ill., Iowa and e. Neb., s. to Fla. and e. Tex.

72. *C. rostrata* Stokes Beaked Sedge
Swamps, shallow water and wet shores. Greenl. and Lab. w. to N.W.T. and Alask., s. to n. Nfld., n. C.B.I., n. N.B., Mich., n. Ill., n. Wisc. and ne. Minn.; nw. Mont., Ida. and Wash. According to Reznicek and Ford (2002) much of what has been called *C. rostrata* by American authors is referable to *C. utriculata.*

73. *C. utriculata* Boott Northwest Territory Sedge Fig. 220
Swamps, shallow water and wet shores. Baffin I., Nfld. and Lab. w. to Alask., s. to e. Va., N.C., Tenn. Ohio, Ind., Ill., Minn. and Pa., n. W.Va., Ind., Tenn., Iowa and Neb., mts. to Colo., N.M., Ariz., Calif. and n. Baja Calif. (*C. rostrata* var. *utriculata* (Boott) L. H. Bailey)

74. *C. tuckermanii* Dewey Tuckerman's Sedge Fig. 220
Calcareous swamps, meadows, marshes and low woods. N.S., N.B. and Que. w. to Ont. and Minn., s. to N.J., W.Va., Ohio, n. Ind. and ne. Iowa.

75. *C. prasina* Wahlenb. Drooping Sedge Fig. 211
Rich deciduous forests, especially in moist depressions, seeps and along margins of rivers, streams, lakes, swamps, and fens. Me. and s. Que. w. to Ont., Mich. and Wisc., s. to N.J., Va., nw. S.C., n. Ga., n. Ala., n. Miss., se. Mo. and w. Ark.

76. *C. oligosperma* Michx. Few-seed Sedge Fig. 220
Acid swamps, sphagnum bogs and occasionally in shallow water. Lab. and Nfld. w. to Man., N.W.T., Yuk., s. to N.E., Pa., ne. W.Va., nw. Ohio., n. Ind., ne. Ill., Wisc. and ne. Minn.

77. *C. bullata* Schkuhr ex Willd. Button Sedge Fig. 221
Acid swales, meadows and bogs. Chiefly coastal plain and piedmont, N.S. s. to Ga. and Miss.; inland to Ky. and Tenn.

78. *C. vesicaria* L. Blister Sedge Fig. 222
Swamps, marshes and swales. Lab. and Nfld. w. to Ont., Man., s. to N.E., Del., Pa., Va., n. Ky., s. Ind., s. Ill., e. Mo., Wisc., Minn. and N.D.; mts. of Mont. and Wash. s. to Wyo., Utah, Oreg. and Calif.

Fig. 212. *Carex pellita*: a. inflorescence; b. perigynium (F).
Carex lasiocarpa: c. inflorescence; d. perigynium (F).
Carex trichocarpa: e. inflorescence (Braun); f. perigynium (Mackenzie); g. achene (Mackenzie); h. pistillate scale (Mackenzie); i. ligule (Mackenzie).

Fig. 213. *Carex striata*: a. habit (G&W); b. inflorescence (F); c. perigynium (G&W); d. achene (G&W); e. staminate scale (G&W); f. pistillate scale (G&W); g. ligule (G&W).
Carex lacustris: h. inflorescence; i. perigynium (F).

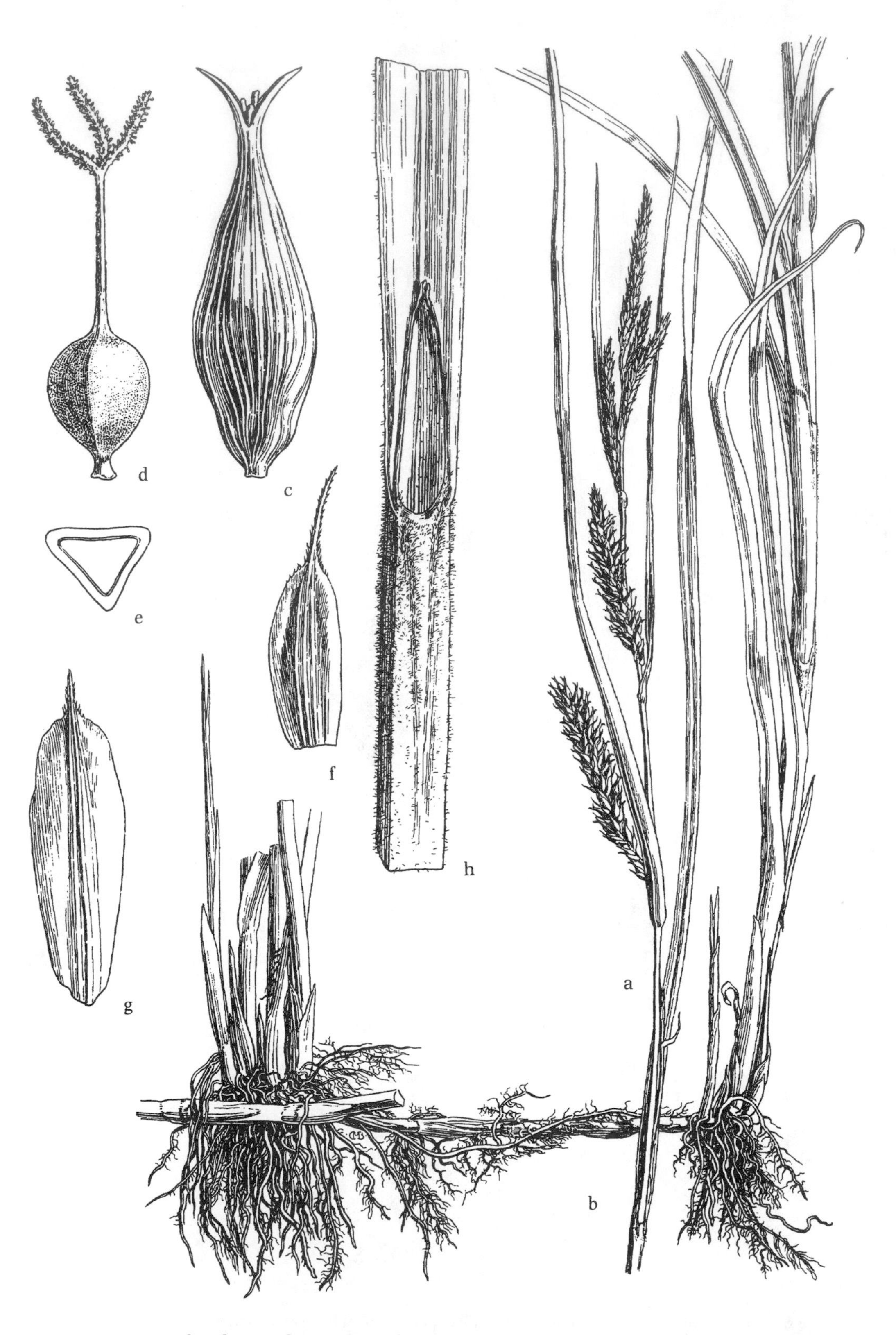

Fig. 214. *Carex atherodes*: a. inflorescence; b. lower portion of plant with thick rhizomes; c. perigynium; d. achene; e. achene, cross-section; f. pistillate scale; g. staminate scale; h. ligule (Mason).

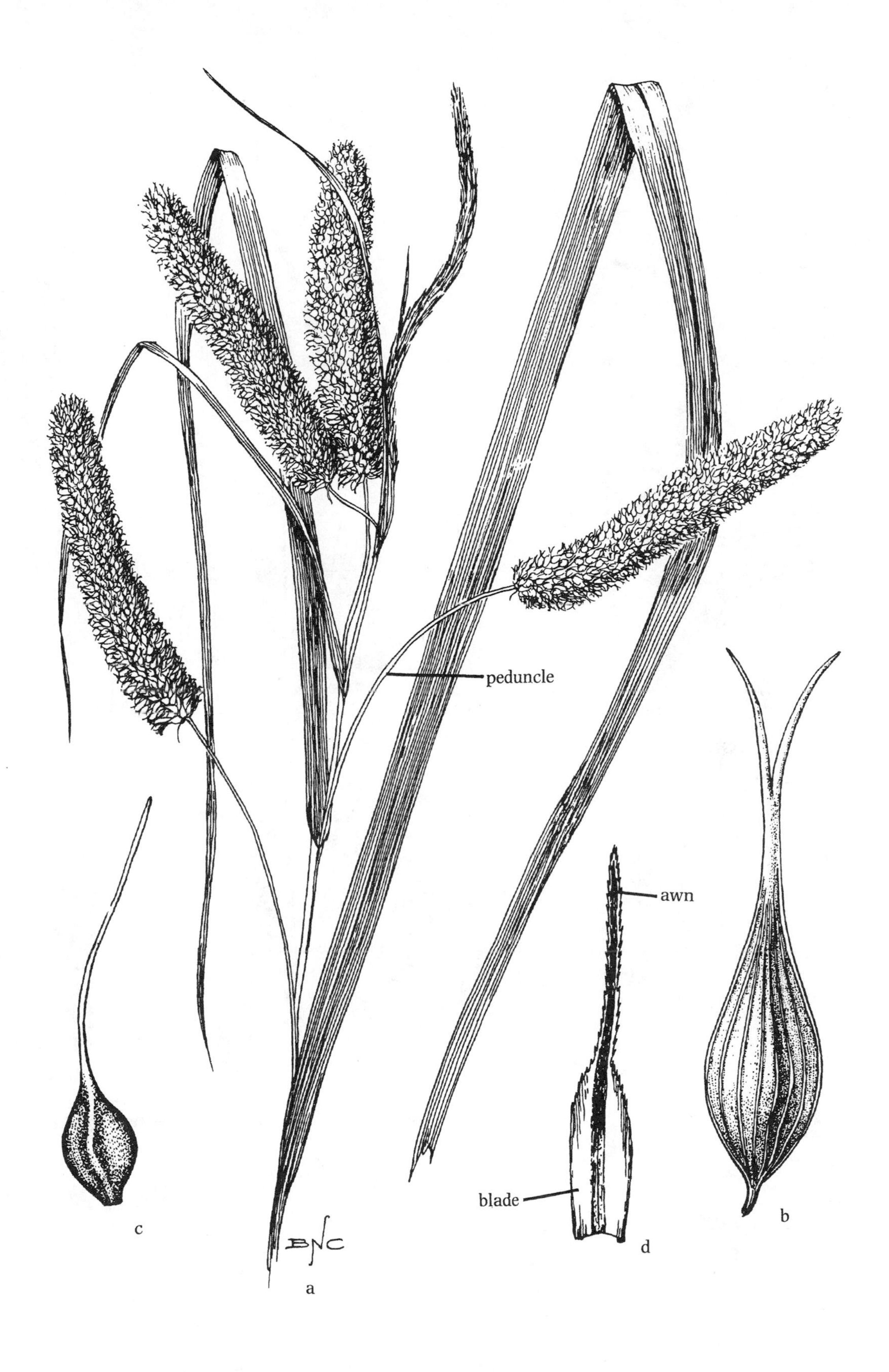

Fig. 215. *Carex comosa*: a. inflorescence; b. perigynium; c. achene; d. pistillate scale (G&W).

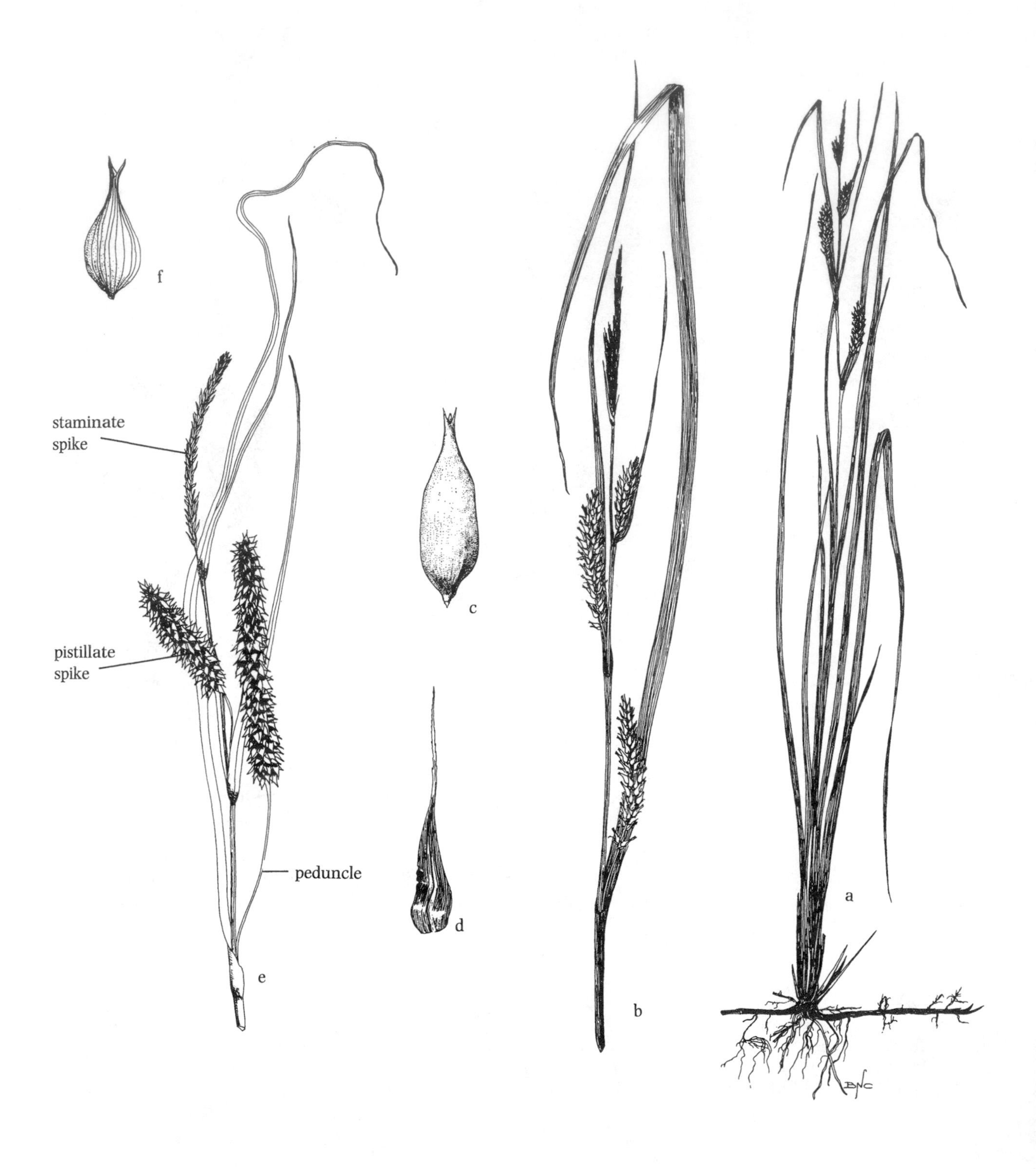

Fig. 216. *Carex hyalinolepis*: a. habit; b. inflorescence; c. perigynium; d. pistillate scale (G&W).
Carex pseudocyperus: e. inflorescence; f. perigynium (F).

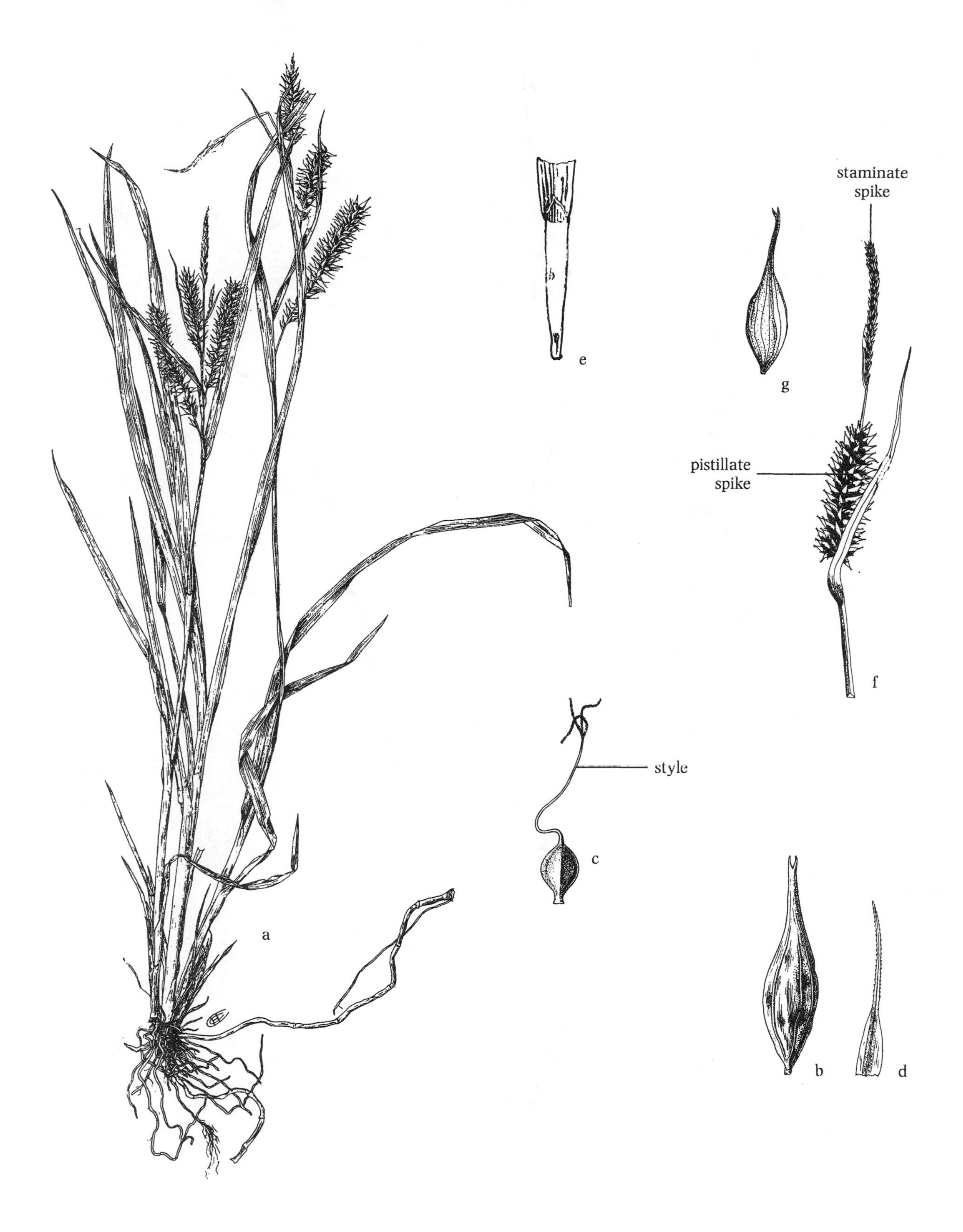

Fig. 217. *Carex schweinitzii*: a. habit; b. perigynium; c. achene; d. pistillate scale; e. ligule (Mackenzie). *Carex hystericina*: f. inflorescence; g. perigynium (F).

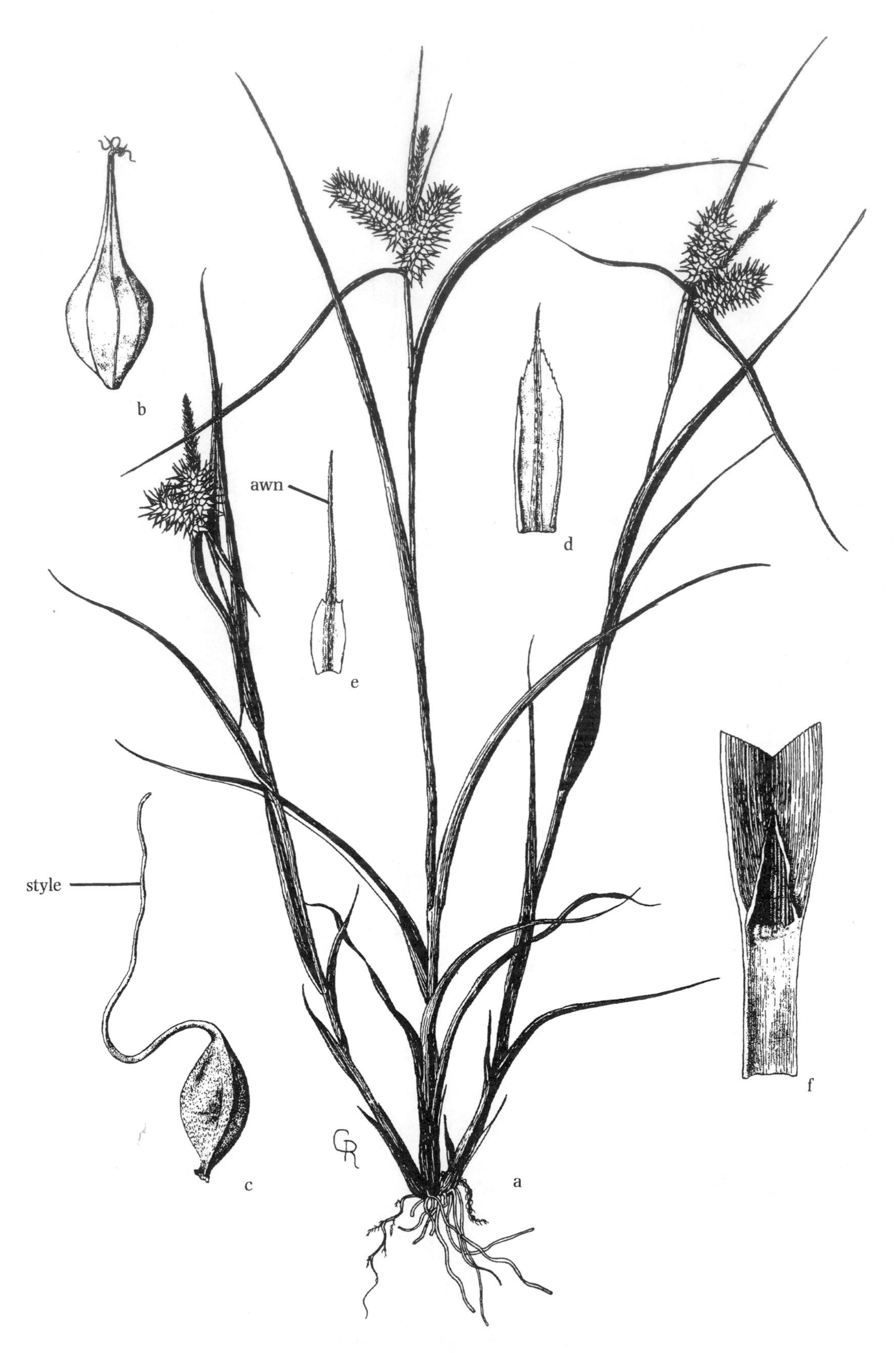

Fig. 218. *Carex lurida*: a. habit; b. perigynium; c. achene; d. staminate scale; e. pistillate scale; f. ligule (G&W).

Fig. 219. *Carex michauxiana*: a. habit; b. perigynium; c. achene; d. pistillate scale; e. ligule; f. section of leaf (Mackenzie).
Carex folliculata: g. inflorescence; h. perigynium; i. pistillate scale (G&W).

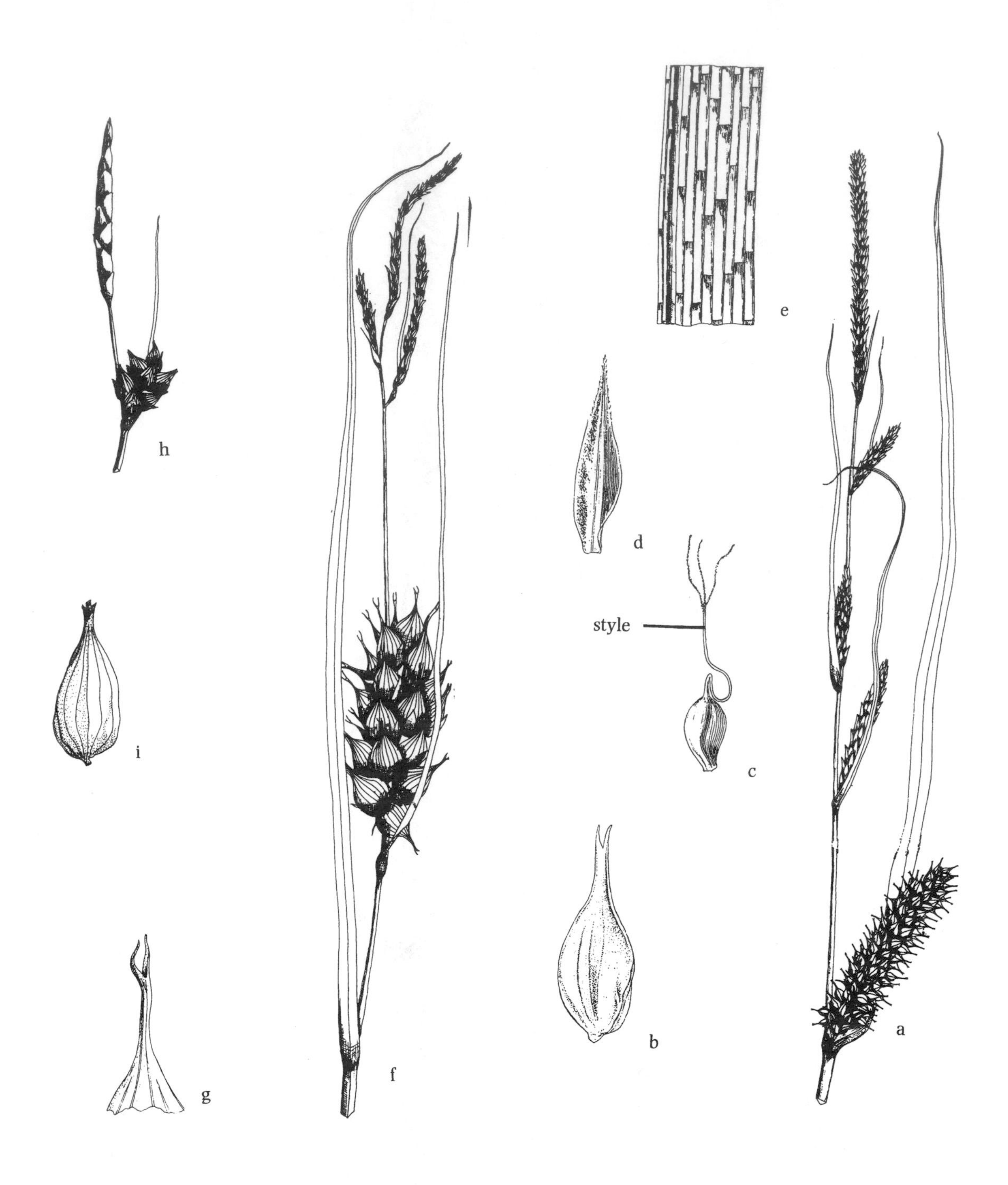

Fig. 220. *Carex utriculata*: a. inflorescence (F); b. perigynium (Mackenzie); c. achene (Mackenzie); d. pistillate scale (Mackenzie); e. section of leaf blade (F).
Carex tuckermanii: f. inflorescence; g. tip of perigynium (F).
Carex oligosperma: h. inflorescence; i. perigynium (F).

Fig. 221. *Carex bullata*: a. habit; b. perigynium; c. achene; d. pistillate scale; e. ligule (Mackenzie).

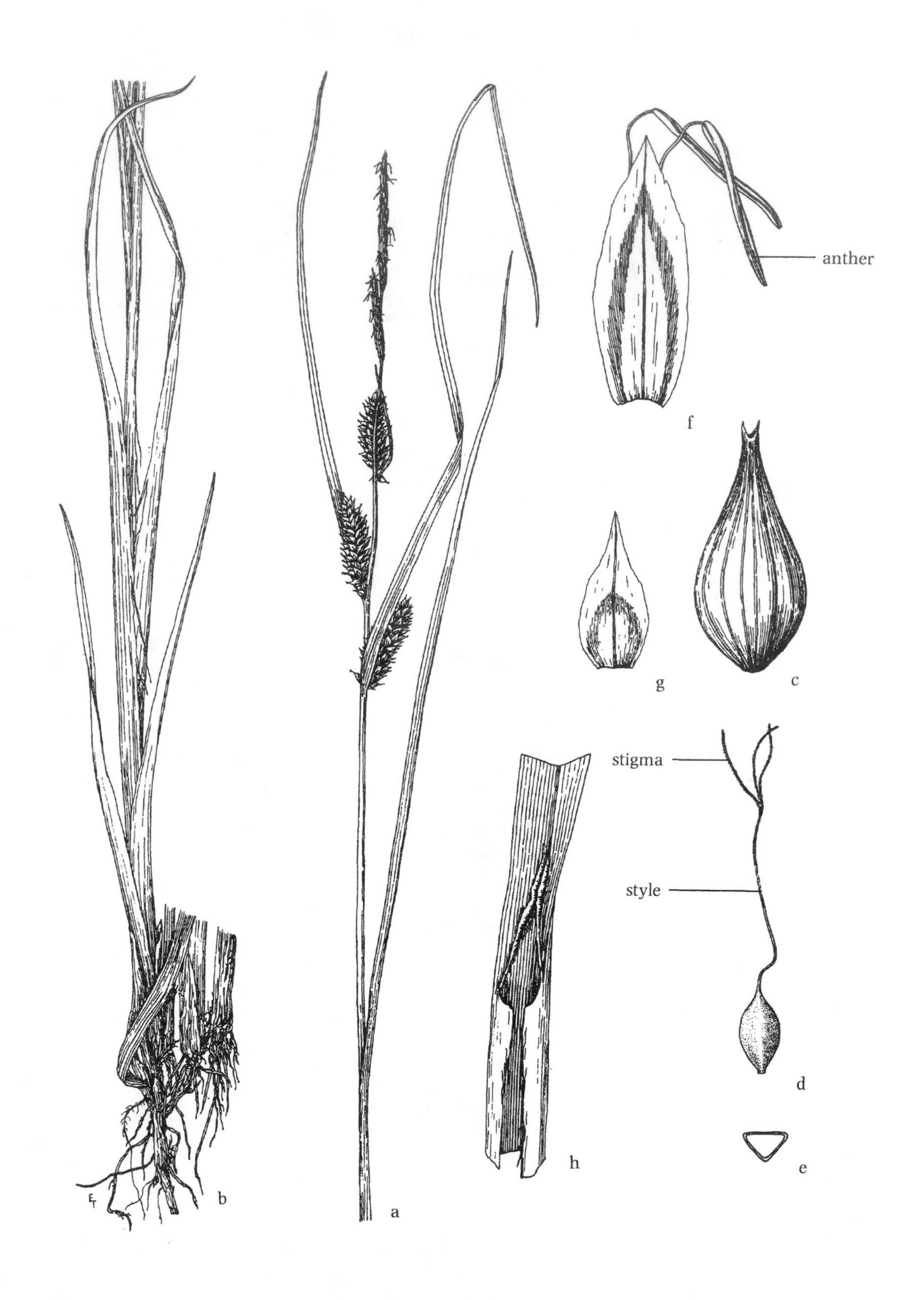

Fig. 222. *Carex vesicaria*: a. inflorescence; b. lower portion of plant; c. perigynium; d. achene; e. achene, cross-section; f. staminate flower and subtending scale; g. pistillate scale; h. ligule (Mason).

79. *C. retrorsa* Schwein. Knot-sheath sedge
Fig. 223
Swamps, marshes, wet thickets, marshy borders of ponds, lakes, streams and rivers, occasional bogs. Nfld. and Que. w. to Man., N.W.T. and B.C., s. to N.E., N.J., Pa., nw. Ohio, Mich., n. Ind., Ill., Iowa, S.D., mts. to Colo., Utah, Nev. and Oreg.

80. *C. lupuliformis* Sartwell ex Dewey False Hop Sedge
Fig. 224
Calcareous swamps, meadows, swales, prairies, and wet wood. Me., w. Vt. and sw. Que., w. to Ont., Mich., Wisc. and Minn., s. to w. N.E., Va., Ky., Ga., Fla., Okla. and e. Tex.

81. *C. gigantea* Rudge Giant Sedge Fig. 225
Open swamps, swamp forests and wet forest opening. Mostly coastal plain and piedmont, Del., Md. and Va. s. to Fla., w. to Ala., La., e. Okla. and e. Tex.; Mississippi embayment n. to Ark., se. Mo., s. Ill., s. Ind. and s. Ohio; Ky. and Tenn.

82. *C. grayi* Carey Gray's Sedge Fig. 223
Floodplains, swamps, along rivers and stream margins and rich deciduous woods. S. Que. and Vt. w. to s. Ont., Mich., Wisc. and se. Minn., s. to Ga., nw. Fla., Ala., Miss., Ark., e. Kans. and se. Okla.

83. *C. intumescens* Rudge Greater Bladder Sedge
Fig. 225
Swamps, margins of wetlands and streams, thickets, depressions in deciduous forests, less often in coniferous forests. Lab. and Nfld. w. to Man., s. to Fla., e. Okla. and e. Tex.; w. S.D. and ne. Wyo.

84. *C. lupulina* Willd. Hop Sedge Fig. 226
Swamps, marshes and wet woods. N.S., N.B. and s. Que. w. to s. Ont., Mich., Wisc., Minn. and Neb., s. to Fla., La., e. Okla. and Tex.

85. *C. louisianica* L. H. Bailey Louisiana Sedge
Swamp forests, floodplains and forest openings. Chiefly coastal plain and piedmont, N.J., Md. and Va., s. to Fla., w. to Miss., La. and e. Tex.; Mississippi embayment north to Ark., se. Mo., s. Ill., s. Ind., Ky. and Tenn.

86. *C. buxbaumii* Wahlenb. Buxbaum's Sedge Fig. 227
Swamps, wet shores and bogs. Greenl., Lab. and Nfld. w. to Alask., s. to N.E., s. Va., N.C., w. S.C. n. Ga., Tenn., Ark., and Kans., mts. s. to Colo. and Utah, Nev. and Calif.

87. *C. limosa* L. Mud Sedge Fig. 226
Marly bogs, sphagnum bogs and marshy pond margins. Nunavut, Lab. and Nfld. w. to Alask., s. to N.E., N.J., Del., Pa., Ohio, n. Ind., n. Ill., ne. Iowa, Neb., mts. to N.M., Utah, Nev. and Calif.

88. *C. magellanica* Lam. Boreal Bog Sedge
Fig. 223, 228
Sphagnum bogs, fens, peaty hollows, coniferous swamps, wet meadows and alpine peats. Greenl., Lab. and Nfld. w. to N.W.T. and Alask., s. to Mass., w. Conn., n. N.J., n. Pa., N.Y., n. Ohio, n. Ind., Wisc., Minn., Colo., N.M., Utah, n. Ida. and Wash. Our taxon is subsp. *irrigua* (Wahlenb.) Hultén. (*C. paupercula* Michx.)

89. *C. rariflora* (Wahlenb.) Sm. Loose-flower Alpine Sedge Fig. 228
Cold bogs, peat barrens and pond margins. Chiefly Arctic; Greenl., and Baffin I. w. to Alask., s. to Nfld., e. Que., N.S., N.B., n. Me., James Bay and n. Man.; Eurasia.

90. *C. barrattii* Schwein. & Torr. Barratt's Sedge
Fig. 228
Peaty swamps, pinelands and wet woods. Coastal plain, Conn., se. N.Y. and s. N.J., s. to N.C and S.C.; mts. of n. Ga., Tenn., and n. Ala.

91. *C. livida* (Wahlenb.) Willd. Livid Sedge Fig. 229
Calcareous meadows and bogs. Lab. and Nfld. w. to Alask., s. to n. Me., N.H., ne. Mass., w. Conn., s. N.J., c. N.Y., ne. Ind., Mich., Minn., mts. of Mont., to Wyo., Colo., Ida., Wash., Oreg. and n. Calif.

92. *C. tetanica* Schkuhr Rigid Sedge Fig. 230
Calcareous bogs, meadows and low woods. N.H., Mass. and Conn. w. to Ont., Man. and Sask., s. to Va., N.C., Ohio, Ky. Ill., e. Mo. and Neb.

93. *C. viridula* Michx. Little Green Sedge

93a. *C. viridula* subsp. *viridula* Fig. 231
Calcareous gravels, shores and springy areas. Greenl., Lab. and Nfld. w. to N.W.T., Alask., s. to N.E., n. N.J., nw. Pa., n. Ohio, n. Ind., n. Ill., Wisc., Minn., S.D., Mont., mts. to N.M., Utah, Nev. and n. Calif. (*C. flava* var. *viridula* (Michx.) L. H. Bailey; *C. oederi* var. *viridula* (Michx.) Kük.)

93b. *C. viridula* subsp. *brachyrrhyncha* (Čelak.) B. Schmid
Fig. 230
Calcareous bogs, fens swamps and gravels. Chiefly coastal. According to Crins (2002) two varieties of this subspecies occur in eastern Canada, based mainly on perigynia, culm and staminate peduncle size to differentiate the taxa: var. *elatior* (Schltdl.) Crins, an introduction from Europe, occurs in Nfld., N.B., N.S. and Que.; var. *saxilittoralis* (Robertson) Crins., a native plant, occurs in similar coastal sites of Lab., Nfld. and N.S. (*C. flava* subsp. *brachyrrhyncha* Čelak.).

94. *C. cryptolepis* Mack. Northeastern Sedge
Fig. 230
Meadows, sandy shore, ditches and swales. Lab. and Nfld. w. to w. Ont. and Sask., s. to N.E., n. N.J., Ohio, Ind., ne. Ill., Wisc. and Minn.

95. *C. flava* L. Yellow Sedge Fig. 232
Wet shore, marshes, swales and meadows; chiefly calcareous sites. Nfld. w. to Man., s. to N.E., n. N.J., Pa., Va., Ohio, Ind., Wisc. and ne. Minn.; Ala., B.C. and coastal Alask. s. to w. Mont. Wyo., Ida. and nw. Wash.

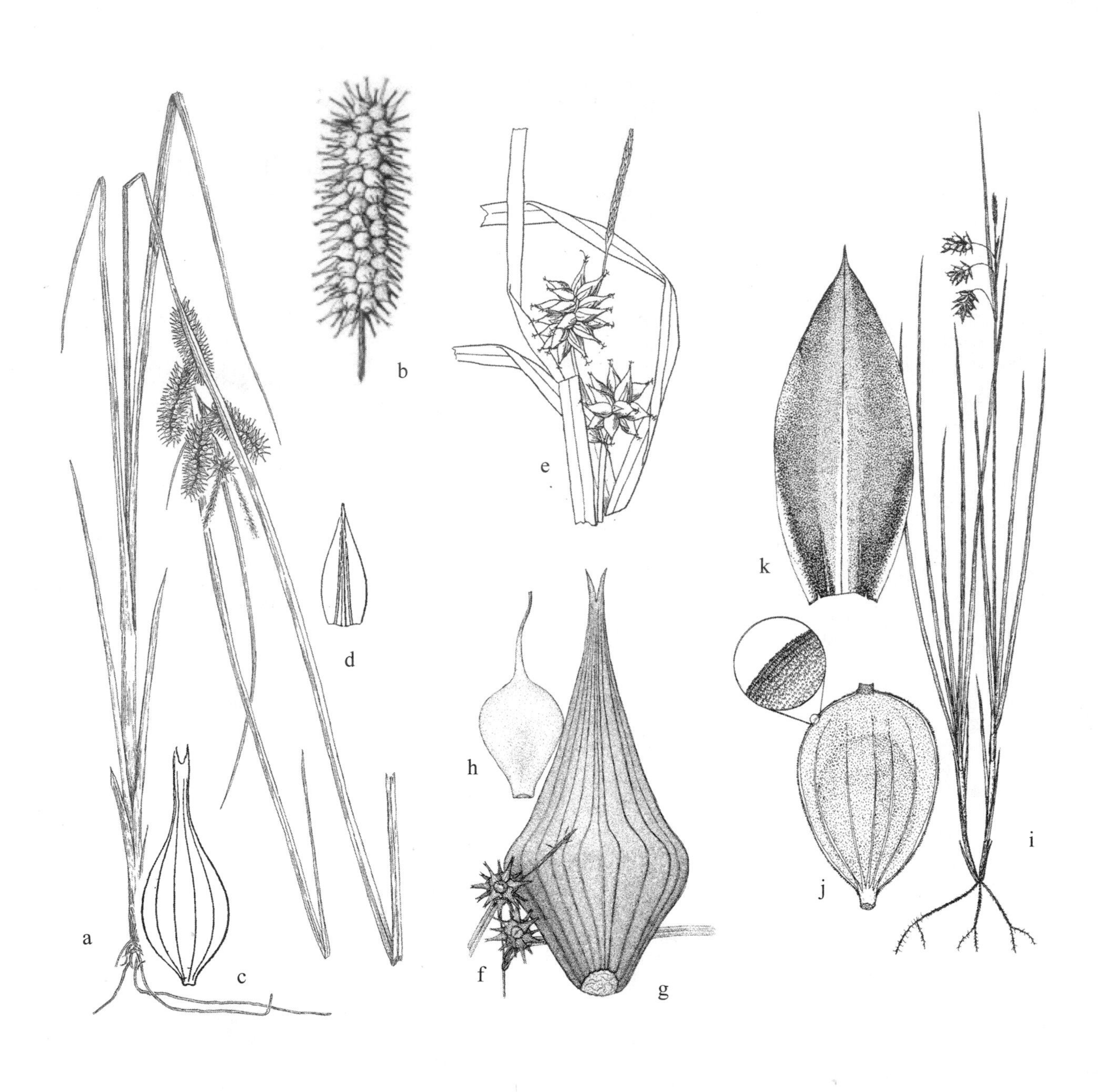

Fig. 223. *Carex retrorsa*: a. habit (RMC); b. pistillate spike (FNA); c. perigynium; d. scale (B&B). *Carex grayi*: e. inflorescence, staminate spike (*above*), pistillate spikes (*below*) (B&B); f. inflorescence, staminate spike (*above*), pistillate spikes (*below*) (FNA); g. perigynium; h. achene with style straight (FNA). *Carex magellanica*: i. habit with nodding pistillate spikes; j. perigynium; k. scale (FNA).

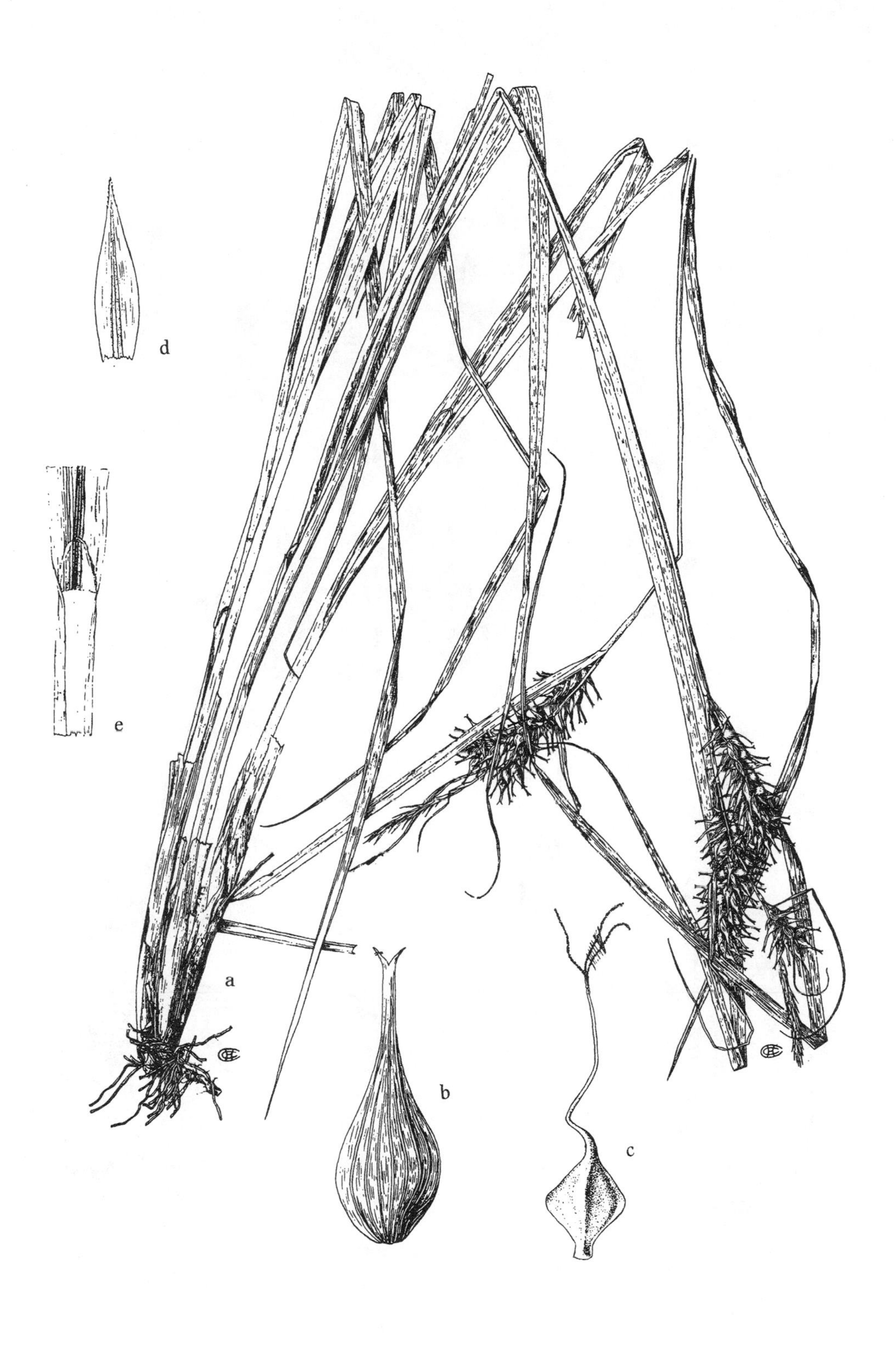

Fig. 224. *Carex lupuliformis*: a. habit; b. perigynium; c. achene with contorted style; d. pistillate scale; e. ligule (Mackenzie).

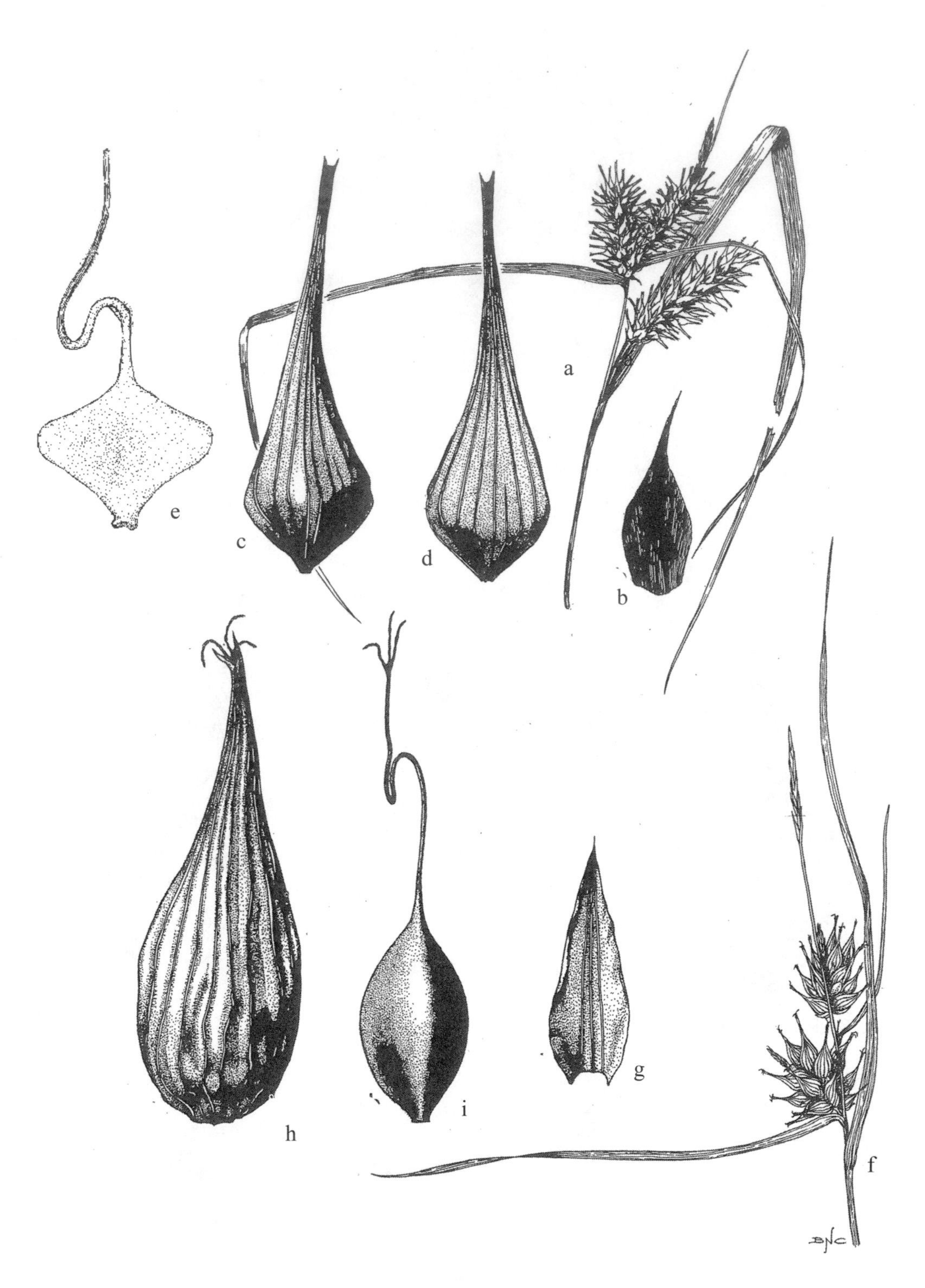

Fig. 225. *Carex gigantea*: a. inflorescence, staminate spike (*above*), pistillate spikes (*below*); b. scale; c. perigynium, dorsal view; d. perigynium, ventral view (C&C); e. achene, somewhat diamond-shaped, with contorted style (FNA).
Carex intumescens: f. inflorescence, staminate (*above*), pistillate spikes (*below*); g. scale; h. perigynium, ventral view; i. achene with contorted style (C&C).

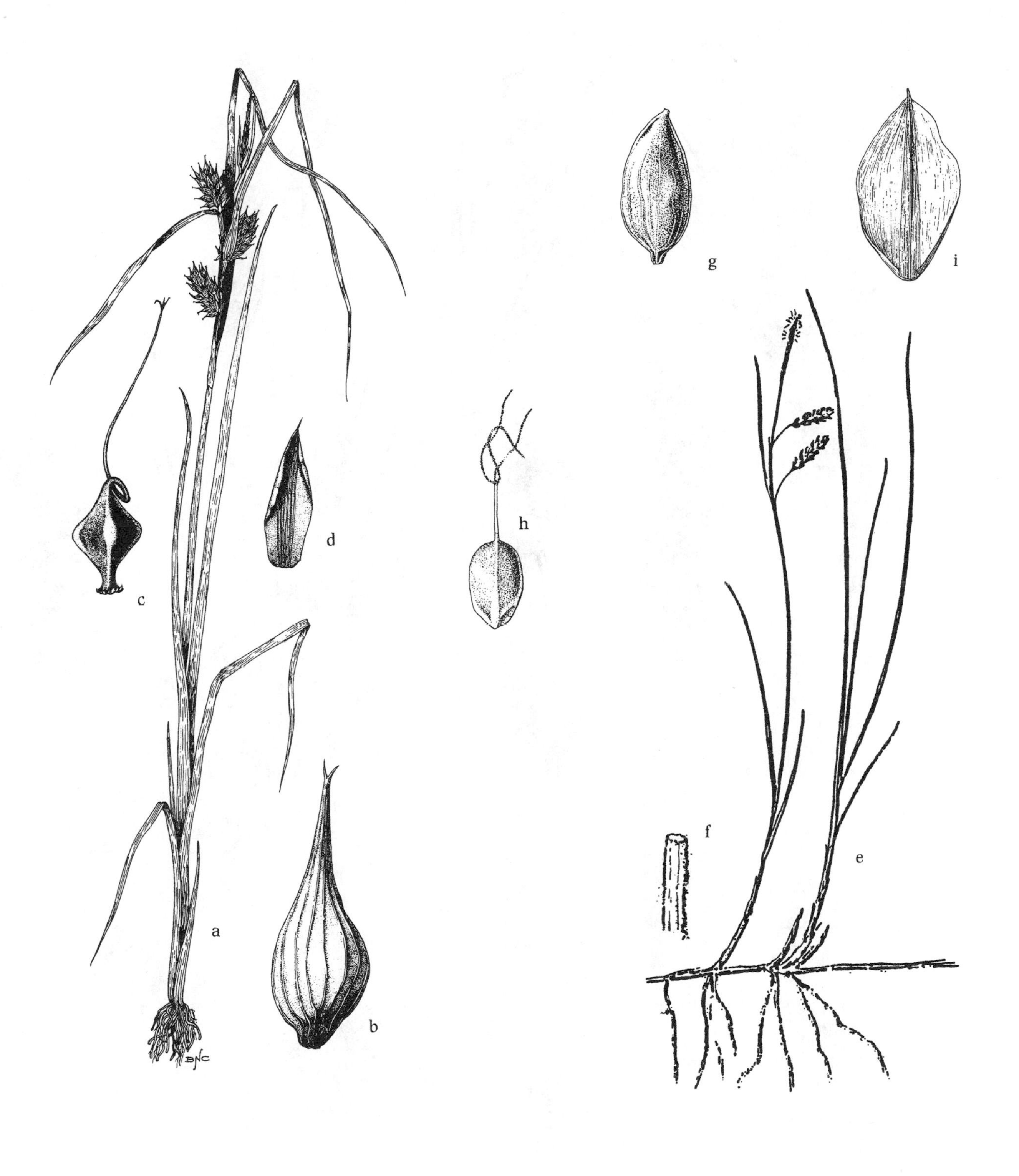

Fig. 226. *Carex lupulina*: a. habit; b. perigynium; c. achene; d. pistillate scale (G&W).
Carex limosa: e. habit (HCOT); f. section of root (HCOT); g. perigynium (Mackenzie); h. achene (Mackenzie); i. pistillate scale (Mackenzie).

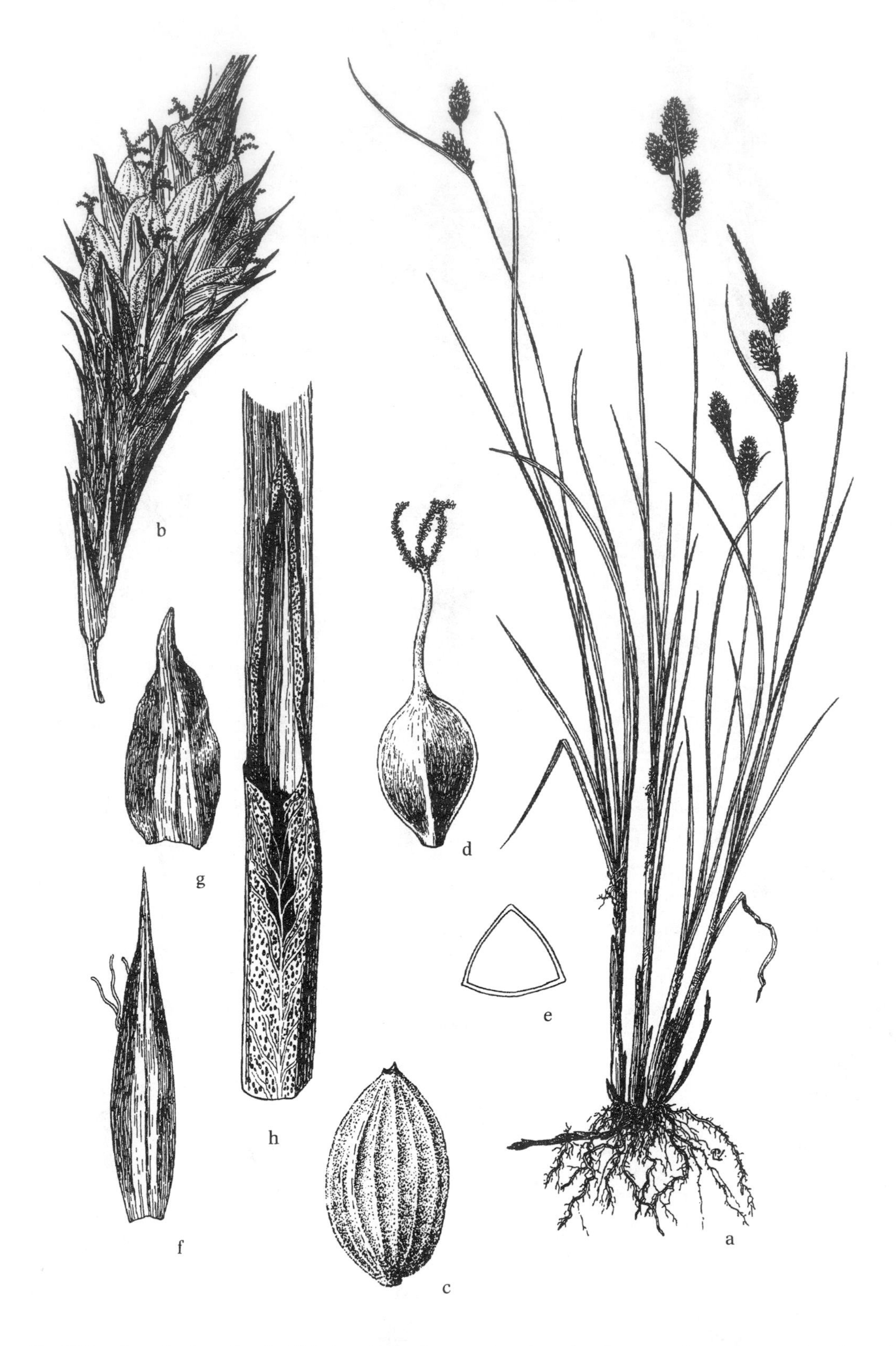

Fig. 227. *Carex buxbaumii*: a. habit; b. terminal spike; c. perigynium; d. achene; e. achene, cross-section; f. staminate scale with filaments; g. pistillate scale; h. ligule (Mason).

Fig. 228. *Carex rariflora*: a. habit; b. perigynium; c. achene; d. pistillate scale (Mackenzie).
Carex barrattii: e. inflorescence; f. perigynium; g. achene; h. pistillate scale (Mackenzie).
Carex magellanica: i. inflorescence; j. perigynium; k. achene; l. pistillate scale (Mackenzie).

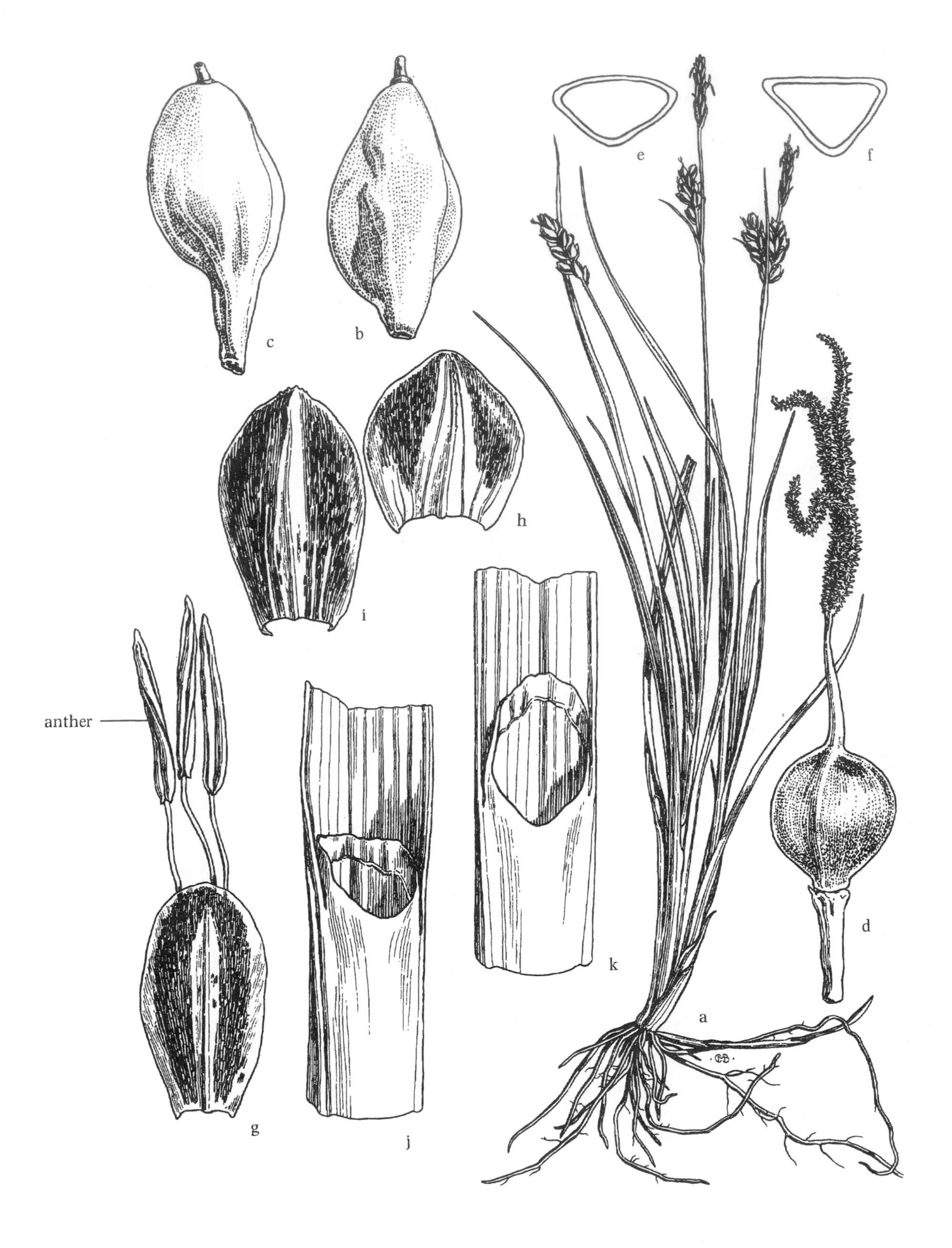

Fig. 229. *Carex livida*: a. habit; b, c. perigynium variations; d. achene; e, f. achenes, cross-sections, showing variations; g. staminate flower with subtending scale; h, i. pistillate scales; j, k. ligules (Mason).

Fig. 230. *Carex tetanica*: a. inflorescence; b. perigynium; c. achene; d. pistillate scale (Mackenzie).
Carex cryptolepis: e. inflorescence; f. perigynium (F).
Carex viridula subsp. *brachyrrhyncha*: g. inflorescence; h. perigynium; i. achene; j. pistillate scale (Mackenzie).

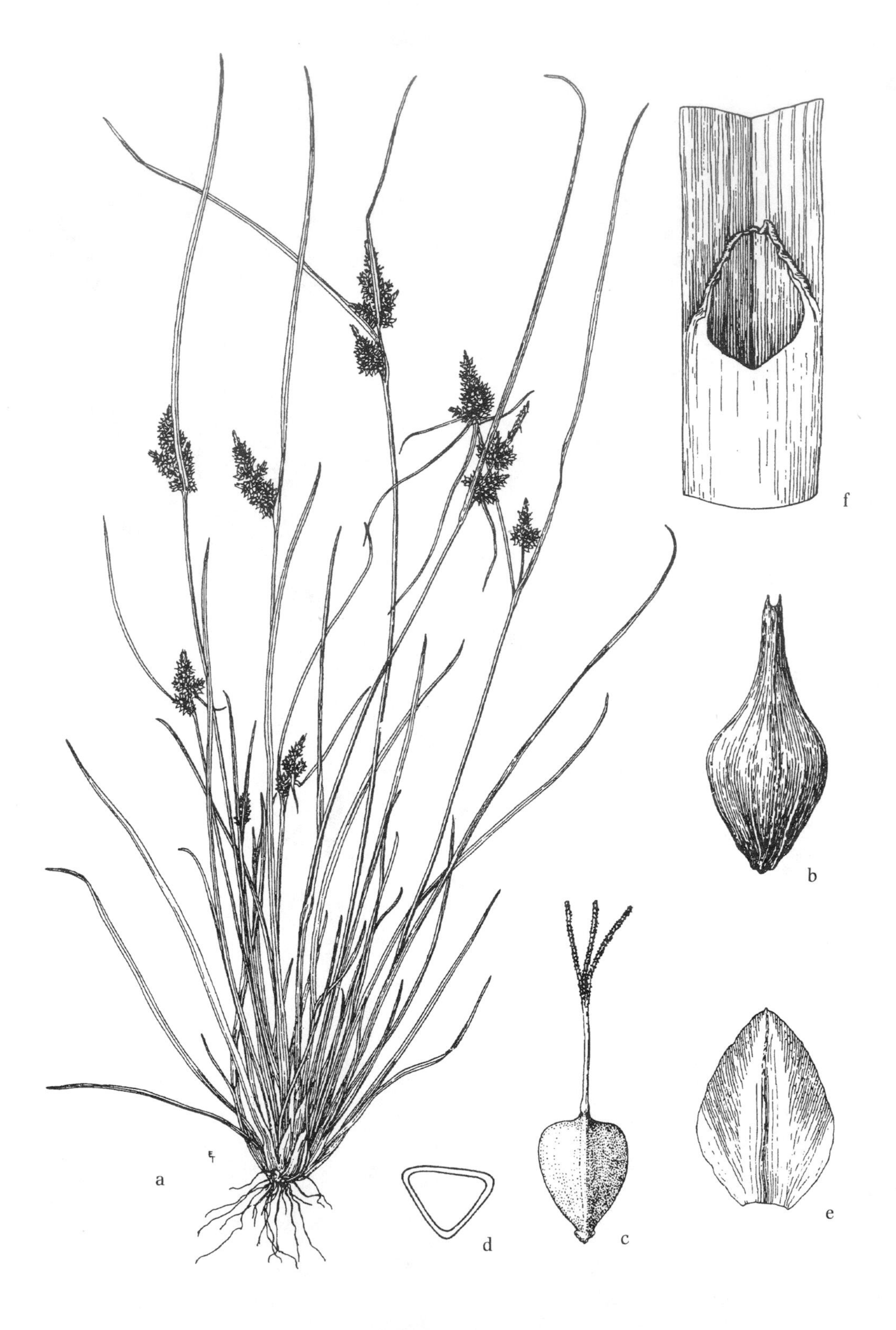

Fig. 231. *Carex viridula* subsp. *viridula*: a. habit; b. perigynium; c. achene; d. trigonous achene, diagram cross-section; e. pistillate scale; f. ligule (Mason).

Fig. 232. *Carex flava*: a. habit; b. perigynium; c. achene; d. pistillate scale; e. section of leaf blade (Mackenzie).

2. *Scleria* (Nut-rush)

Perennial or annual herbs, grass-like; stems sharply 3-angled, leafy; flowers unisexual, pistillate spikelets 1-flowered, intermixed with clusters of few-flowered staminate spikelets; fruit a white stony achene without a tubercle.

REFERENCES: Fairey, 1967; Kessler, 1988; Reznicek et al., 2002.

---

1. Achene smooth, shiny white; swollen foam-like disc (hypogynium) subtending achene 3-angled, white crusty-papillose (fig. 233e,f); leaves (3)5–9 mm wide. . . . . 1. *S. triglomerata*
1. Achene roughened, reticulate or irregularly warty, with low transverse ridges, dull white; disc (hypogynium) absent, or if present, 3-lobed, but not papillose; leaves 0. 5–5 mm wide.
   2. Flower and fruit clusters pedunculate; achene subtended by 3-lobed, bract-like disk (fig. 233c). . . . . 2. *S. reticularis*
   2. Flower and fruit clusters sessile, bracts absent; achene base lacking bract-like disc at base of achene. . . . . 3. *S. verticillata*

---

1. *S. triglomerata* Michx. Whip Nut-rush Fig. 233
Wet to dry sites, borders of marshes, wet peats, thickets, savannas, open woods and prairies. Mass. and Conn. w. to Ont., Wisc., se. Minn., Iowa, and Neb., s. to Fla., La. and e. Tex.

2. *S. reticularis* Michx. Netted Nut-rush Fig. 233
Wet open sands, peaty sands, pond shores, shallow water of cypress ponds, meadows, wet pinelands and savannas. S.N.H. and Mass. w. to Mich. and Wisc. s. to Fla., Mo., Okla., Tex. and Mex.; W.I., C.Am., and S.Am.

3. *S. verticillata* Muhl. ex Willd. Low Nut-rush Fig. 233
Wet sands, sandy peats and marls, swamps, meadows and marshes. Mass. w. to s. Ont., Mich., Wisc., and Minn., s. to Fla., Ala., Ark., Okla. and se. Tex.; Trop. Am.

3. *Dulichium* (Three-way Sedge)

Perennial herbs, arising from rhizomes; stems round in cross-section, hollow, leafy; lower leaves reduced to sheaths; upper leaves with flat blades, conspicuously 3-ranked; inflorescences axillary, spicate; spikelets 3–10 per spike, each with 4–8 scales; scales strictly 2-ranked; flowers bisexual, styles 2-fid, stamens 3; bristles 6–9; achenes biconvex, slender style persistent, linear, but swollen tubercle lacking.

REFERENCE: Mastrogiuseppe, 2002.

1. *D. arundinaceum* (L.) Britton Three-way Sedge Fig. 234
Wet shores, shallow water and bogs. Nfld. and Que. w. to s. B.C., s. to. Fla., e. Tex., Neb., Wyo., Ida. and Calif.

4. *Blysmopsis* (Red Bulrush)

Perennial, from slender rhizomes; culms loosely cespitose; leaves 1–3 basal; inflorescences terminal, spikes solitary, 1–2 cm long, composed of 3–8-flowered spikelets; scales reddish-brown flattened; flowers perfect, stamens 3, styles 2-fid; achenes 4–5 mm long, yellowish-brown, smooth, style base persistent, linear (not swollen).

1. *B. rufa* (Huds.) Oteng-Yeb. Fig. 234
Saline and brackish marshes, rarely in freshwater peaty marshes. Coastal, Nfld., P.E.I. and e. Que, s. to sw. N.S. and sw. N.B.; scattered localities, Hudson Bay; Man., Sask., N.W.T and Alask. (*Scirpus rufus* (Huds.) Schrad.)

5. *Cyperus* (Umbrella-sedge, Galingale, Sweet-rush, Flat-sedge)

Perennial herbs with rhizomes, stolons, or enlarged, hardened bases, or tufted annuals; stems usually 3-angled; leaves chiefly basal; involucral leaves 1 or more, subtending terminal, umbellate inflorescence; spikelets 3-many per spike or head; scales 2-ranked, numerous (up to 76); achene biconvex (style 2-fid) or 3-angled (style 3-fid), style deciduous or if persistent, then without a tubercle.

It is essential to collect the entire plant, preferably in fruiting condition, for accurate identification. Those species with stigmas 2 and biconvex achenes are treated by some authors as a distinct genus, *Pycreus* (Tucker et al., 2002).

REFERENCES: Tucker, 1983, 1985; Tucker et al., 2002.

---

1. Stigmas 2; achenes biconvex.
   2. Scales apex broadly rounded to truncated (with slight acute tip), with conspicuous hyaline margins (white), each scale scarcely overlapping next scale on same side of spikelet. . . . . 1. *C. flavicomus*

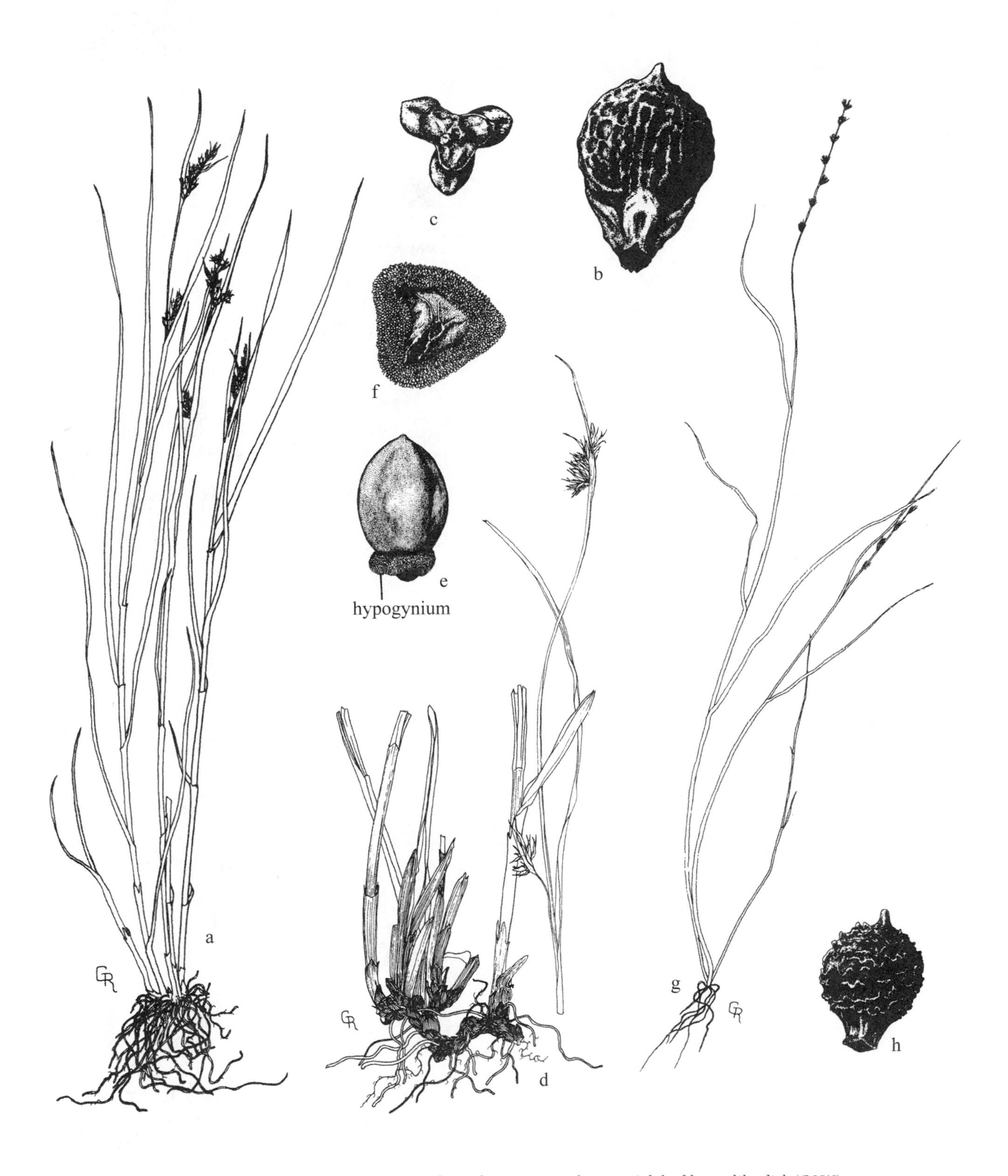

Fig. 233. *Scleria reticularis*: a. habit; b. achene; c. achene, bottom view showing 3-lobed bract-like disk (G&W).
Scleria triglomerata: d. habit; e. achene; f. achene, bottom view (G&W).
Scleria verticillata: g. habit; h. achene (G&W).

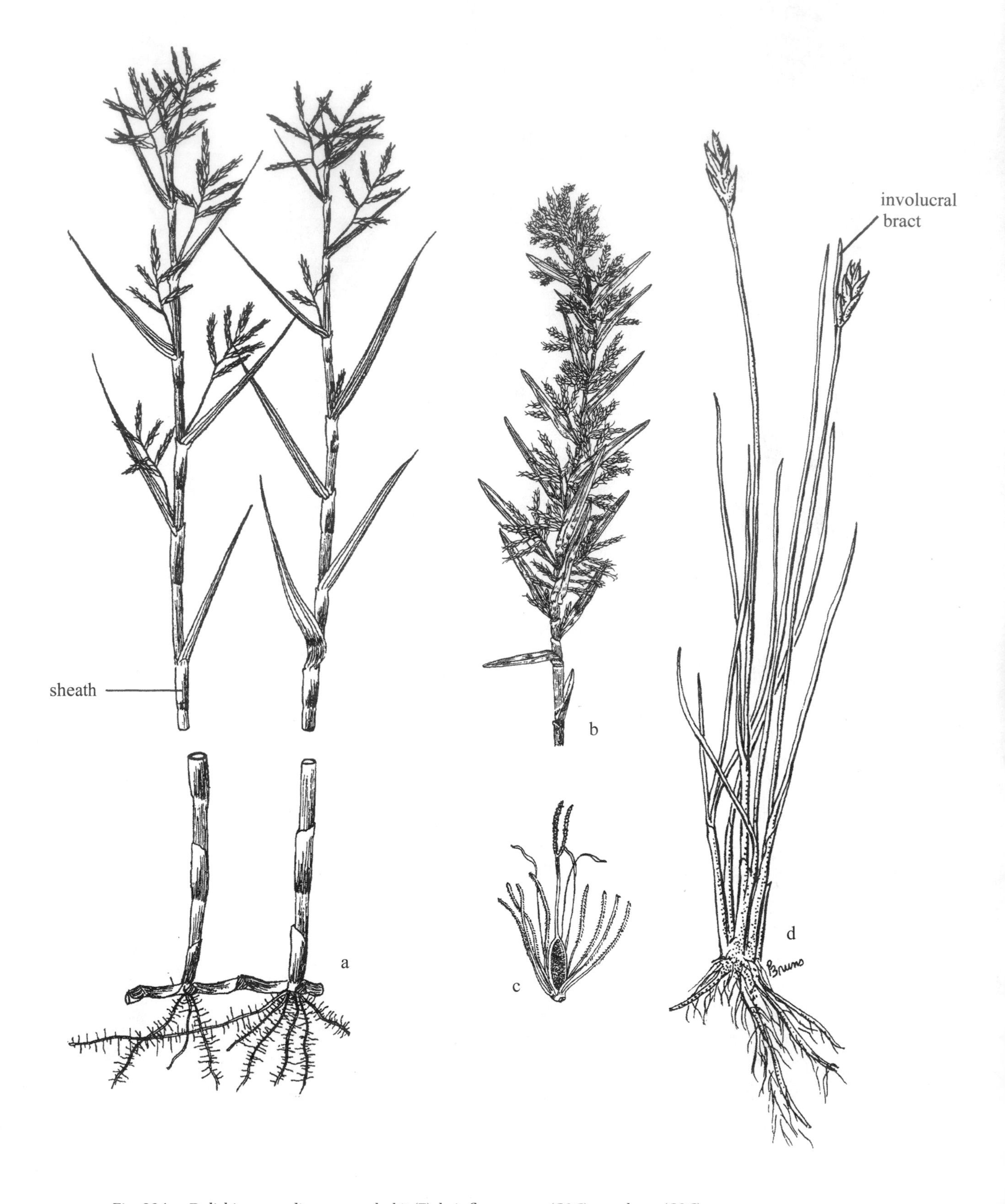

Fig. 234. *Dulichium arundinaceum*: a. habit (F); b. inflorescence (C&C); c. achene (C&C). *Blysmopsis rufa*: d. habit (PB).

2. Scales apex obtuse, acute, or mucronate, with obscure hyaline margins (straw-colored to light brown), each scale overlapping basal quarter to one-third of next scale on same side of spikelet.
 3. Mature achenes black to reddish-brown with transverse lines, appearing wrinkled (fig. 235d)........................ 2. *C. flavescens*
 3. Mature achenes drab or brownish, lacking transverse lines, appearing smooth.
  4. Scales straw-colored (lacking reddish pigment), apex prolonged to a point (fig. 235i); achenes 3 times as long as wide, tip truncate.
   5. Spikelets 2–3 mm wide; scales 2.7–3.6 mm long; achenes 1.2–1.6 mm long..................................3. *C. filicinus*
   5. Spikelets (1)1.2–2 mm wide; scales (1.5)1.8–2.4 mm long; achenes 0.8–1.1(1.2) mm long.................4. *C. polystachyos*
  4. Scales with reddish pigment, apex blunt (fig. 236c); achenes twice as long as wide, tip rounded.
   6. Styles 2-cleft nearly to base (fig. 236f), longer than scales, usually persistent at maturity; scales with marginal band bright red, giving spikelet dark-margined appearance (color especially concentrated at tip)..................... 5. *C. diandrus*
   6. Styles 2-cleft to about middle, hidden by scales, deciduous at maturity; reddish color of scales concentrated near base and toward margins.. 6. *C. bipartitus*
1. Stigmas 3; achenes 3-angled (fig. 237h).
 7. Scale tips divergent (figs. 237d, 238c).
  8. Scales narrowing to elongate or awn-like tip, strongly recurved at apex, strongly (5)7–9(11)-nerved................ 7. *C. squarrosus*
  8. Scales short-tipped, slightly recurved at apex, obscurely 3-nerved...8. *C. acuminatus*
 7. Scale tips straight or slightly incurved.
  9. Lower cauline leaves reduced to bladeless sheaths; achenes distinctly granular to papillose (fig. 238h).................. 9. *C. haspan*
  9. Lower cauline leaves with blades; achenes puncticulate or smooth, but not papillose (except *C. pseudovegetus*).
   10. Scales obovate, nearly as wide as long, appearing bead-like (fig. 239b)... 10. *C. iria*
   10. Scales oblong to lanceolate, clearly longer than wide.
    11. Plants perennial, with rhizomes or stolons or enlarged, bulbous, hard base.
     12. Base of plants enlarged, bulbous, hard (239k); scales and spikelets deciduous at maturity............. 11. *C. strigosus*
     12. Base of plants with slender, scaly rhizomes or stolons, often ending in tubers (fig. 240e); scales and spikelets persistent at maturity.
      13. Scales oriented apically, nearly parallel to axis (fig. 240b); spikelets linear, not densely clustered.
       14. Stems (50)75–110 cm long; scales 3–4 mm long, purple-red or purple-brown............... 12. *C. setigerus*
       14. Stems 10–60(100) cm long; scales 1.8–2.7(3.4) mm long, brown, straw-colored to golden-brown to pale green...13. *C. esculentus*
      13. Scales divergent from axis (fig. 241b); spikelets nearly circular in outline, aggregated in dense clusters (fig. 241a).
       15. Spikelets digitate, radiating from one point on axis; flowers often replaced by bulblets (fig. 240f); plants stoloniferous, tuberiferous.. 14. *C. dentatus*
       15. Spikelets congested, densely clustered; flowers never replaced by bulblets; plants tufted, rhizomes vertical, not stoloniferous..15. *C. pseudovegetus*
    11. Plants annual, with a tuft of roots, lacking rhizomes or enlarged bases.
     16. Scales about 1.3–1.5 mm long; rachilla continuous, not breaking into segments at maturity; wings of rachilla scarious (white), adnate to and extending along margin of scale above..........................16. *C. erythrorhizos*
     16. Scales about 2–3(3.2) mm long; rachilla jointed, breaking into segments at maturity; wings of rachilla firm, not adnate to scales..17. *C. odoratus*

---

1. *C. flavicomus* Michx. White Flat-sedge
Pond shores, swales and ditches. L.I., N.Y., e. Pa., Del, Md. and e. Va. w. to Ky., se Mo., and n. Ark., s. to Ga., n. Fla., La. and Tex.; N.M., s. Ariz., Calif. and Mex.; Trop. Am. (*C. albomarginatus* (Nees) Steud.; *Pycreus albomarginatus* Nees)

2. *C. flavescens* L. Yellow Flat-sedge Fig. 235
Wet sands and peats along pond and lake shores and ditches. se. Mass. and se. N.Y. w. to s. Ont., Mich., Ill., Mo. and Kans., s. to Fla. and Tex.; Calif.; Trop. Am.

3. *C. filicinus* Vahl Fern Flat-sedge
Saline or brackish marshes, rarely freshwater shores. Coastal plain, Me. s. to Ga., w. to La.; s. Ill. (*C. polystachyos* var. *filicinus* (Vahl) C. B. Clark)

4. *C. polystachyos* Rottb. Manyspike Flat-sedge Fig. 235
Damp sands, peats, marshy shores and seasonally wet areas. Chiefly coastal plain and piedmont, Me. and e. Mass. s. to Va., W.Va., S.C. and Fla., w. to e. Tex. and Mex.; inland n. to Okla., Ark., se. Mo., Tenn., and Ky. Trop. Am. In the North this species typically grows in a short, flattened mound. (*Pycreus polystachyos* (Rottb.) P. Beauv.)

5. *C. diandrus* Torr. Low Flat-sedge Fig. 236
Sandy to muddy shores occasionally wet peat. N.S., N.B. and Que. w. to Wisc., Minn. and se. N.D., s. to Va., Tenn., Mo. and Neb.

6. *C. bipartitus* Torr. Slender Flat-sedge Fig. 236
Wet, sandy, gravelly, or muddy soils of marshes, shores of lakes and ponds and ditches. N.B., Que. and N.E. w. to s. Ont., Minn.,

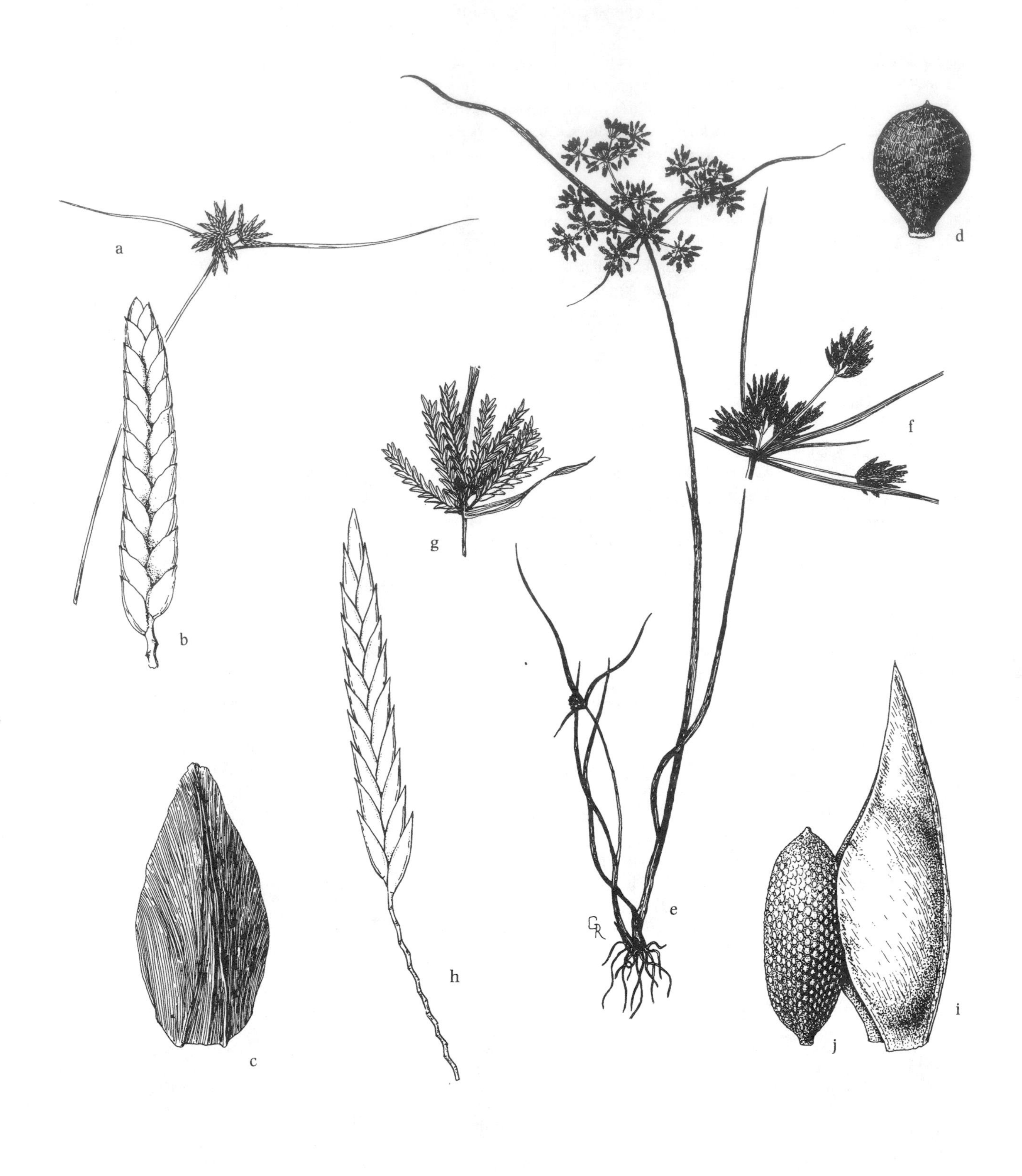

Fig. 235. *Cyperus flavescens*: a. inflorescence; b. spikelet; c. scale; d. achene (G&W).
Cyperus polystachyos: e. habit (G&W); f. inflorescence with strongly ascending glomerules (G&W); g. inflorescence (F); h. spikelet (G&W); i. scale (G&W); j. achene (G&W).

Fig. 236. *Cyperus bipartitus*: a. habit; b. spikelet; c. scale; d. achene (HCOT).
Cyperus diandrus: e. inflorescence; f. achene with 2-cleft style (F).

Mont., Wash. and Oreg., s. to Ga., n. Fla., Miss., ne. Tex., Okla., N.M. and Calif.; S.Am. (*C. rivularis* Kunth; *Pycreus bipartitus* (Torr.) C. B. Clarke)

7. *C. squarrosus* L. Bearded Flat-sedge Fig. 237
Damp sands, silts, riverbanks, shores and wet depressions in rock outcrops. Nfld. and N.B. w. to s. Ont., s. Man., s. Sask., and s. B.C., s. to Fla., Tex., Calif. and Mex.; W.I. and S.Am. (*C. aristatus* Rottb.; *C. inflexus* Muhl.)

8. *C. acuminatus* Torr. & Hook. Tapertip Flat-sedge Fig. 238
Wet limestone soils. N.H.; se. N.Y., se. Pa., and s. Md. w. to Ohio, s. Mich., Ill., Minn., N.D. and B.C. Alta., s. to Ga., Ala., La., Tex., Ariz., Nev., Calif. and ne. Mex.

9. *C. haspan* L. Haspan Flat-sedge Fig. 238
Tidal, fresh to saline waters, marshy shores, stream borders and temporary pools. Coastal plain and piedmont, se. Va. s. to Fla., w. to e. Tex.; inland to s. Ark. and c. Tenn.; Trop. Am.

10. *C. iria* L. Rice-field Flat-sedge Fig. 239
Clearings, ditches, moist fields often in shallow water. Conn and N.Y. w. to Ohio, Ky., s. Mo. and Okla., s. to Va., Fla., e. Tex. and Mex.; inland n. to Ky., Tenn. and Okla.; Calif.; W.I. and C.Am.; intro. from Eurasia.

11. *C. strigosus* L. Straw-colored Flat-sedge Fig. 239
Marshes, meadows, damp clearings and shores. N.B. and N.E. w. to sw. Que., Ont., Mich., Minn., Man., Sask. and S.D., s. to Fla., Tex. and Ariz.; Wash. and Ida. s. to N.M., Ariz. and Calif. (*C. strigosus* var. *robustior* Britton)

12. *C. setigerus* Torr. & Hook. Lean Flat-sedge
Sandy swamps, bottomlands, ditches, and pond and lake shores. Mo., Kans., Okla., Tex. and N.M.

13. *C. esculentus* L. Yellow Nut-grass, Chufa Fig. 240
Waste places, damp sandy soil, fields, riverbanks and marshes; often weedy in cultivated areas. Intro.; N.S. w. to Que; B.C.; native and intro. populations, Me. to N.D., Colo., Utah, Ida. and Wash., s. to Fla., Calif. and Mex.; C.Am. and S.Am.

14. *C. dentatus* Torr. Toothed Flat-sedge Fig. 240
Sandy and gravelly shores of ponds and lakes. N.S. and N.B. and N.E. w. to s. Que. and se. Ont., s. to N.E., Del., Pa., Md., e. Tenn., N.C., S.C., Ga. and s. Ala.; disjunct to nw. Ind.

15. *C. pseudovegetus* Steud. Marsh Flat-sedge Fig. 241
Wet marshy shores, ditches and swales. Se. Mass. (adv.); s. N.J., w. to Ind., Ill., Mo., and se. Kans., s. to Fla. and Tex.

16. *C. erythrorhizos* Muhl. Red-rooted Cyperus Fig. 241
Alluvial or damp soil along marshes, swales and ditches. Se. Me., se. N.H., and Mass. w. to s. Ont., Wisc., N.D. and B.C., s. to Fla., Tex., N.M., Wash. and Calif.

17. *C. odoratus* L. Fragrant Flat-sedge Fig. 242
Marshy shores of rivers, lakes, and ponds, swamp borders, often in brackish or saline areas. Sw. N.H. and Mass. w. to s. Ont. Minn. and Neb., s. to Fla., Okla., Tex., Ariz. and Calif.; Trop. Am. (*C. engelmannii* Steud.; *C. ferruginescens* Boeckl.; *C. ferax* Richard; *C. speciosus* Vahl; *Torulinium odoratum* (L.) S. Hooper)

6. *Kyllinga* (Sedge)

Perennial herbs, usually arising from rhizomes, or tufted annuals; stems 3-angled; leaf blades with scaberulous margins and keel, or sometimes reduced to sheaths; involucral leaves 2–4, subtending dense inflorescence; spikes 1–4, sessile; spikelets 40–150 per spike, each with 2(3) scales; fruit an achene, biconvex, without a tubercle.

References: Delahoussaye and Thieret, 1967; Tucker, 1984, 2002.

1. Spikelets 1.5–2(2.5) mm long; plant tufted, not rhizomatous, annual .... 1. *K. pumila*
1. Spikelets 3.5–4.6 mm long; plant rhizomatous, readily creeping, perennial .... 2. *K. gracillima*

1. *K. pumila* Michx. Low Spike-sedge Fig. 243
Swamps, marshes, ponds, wet clearings and ditches. L.I., N.Y., and N.J. w. to Ind., Mo. and e. Kans., s. to Fla., e. Tex., and Mex.; C.Am., W.I. and S.Am. We have followed Tucker (2002) here, but some authors have accepted the merging of the genus *Kyllinga* into *Cyperus*, following the generic concept of Larridon et al. (2014), in which case the correct name would be *Cyperus hortensis* (Salzm. ex Steud.) Dorr. (*Cyperus tenuifolius* (Steud.) Dandy; *Kyllinga tenuifolia* Steud.)

2. *K. gracillima* Miq. Pasture Spike-sedge Fig. 243
Wet sites, marshy shores and swamp margins. Conn. and N.Y. s. to Ga., w. to Mo., Ark. and Miss. This taxon is similar to *K. brevifolia* Rottb., which is more widespread in the southeastern United States and tropical areas. Both species form large mats from numerous rhizomes. If *Kyllinga* is accepted within the generic concept of *Cyperus*, following Larridon et al. (2004), the correct name would be *Cyperus brevifolioides*. (*Cyperus brevifolioides* Thieret & Delahoussaye; *K. brevifolioides* (Thieret & Delahoussaye) G. C. Tucker)

7. *Fuirena* (Umbrella-grass)

Perennial or annual herbs, grass-like, tufted or rhizomatous; stems leafy; leaves with flat blades, often bearing spreading hairs; spikelets sessile, in tight clusters, subtended by involucral leaf-like bracts; perianth of outer whorl of 3 sharp, slender bristles, alternating with 3 stalked, flat, dilated (blade-like) scales; styles linear, 3-fid, base usually persistent; fruit a 3-angled, stalked, long-beaked achene.

References: Kral, 1978, 2002; Svenson, 1957.

Fig. 237. *Cyperus squarrosus*: a, b. habit; c. inflorescence; d. spikelet; e. scale; f. achene; g. rachis, showing persistent stamens, arrangement of scales and achenes, and 3-cleft styles (Mason).

Fig. 238. *Cyperus acuminatus*: a. habit; b. portion of inflorescence; c. spikelet; d. scale; e. achene (Mason). *Cyperus haspan*: f. habit; g. spikelet; h. achene (G&W).

Fig. 239. *Cyperus iria*: a. inflorescence; b. spikelet; c. scale; d. achene (G&W).
Cyperus strigosus: e. habit (Mason); f. portion of inflorescence (Mason); g. spikelet (Mason); h. portion of winged rachis (Mason); i. scale (Mason); j. achene with style and stamens (Mason); k. bulbous base of plant (F).

Fig. 240. *Cyperus esculentus*: a. inflorescence (G&W); b. spikelet (G&W); c. scale (G&W); d. achene (G&W); e. base of plant (F).
Cyperus dentatus: f. habit, inflorescence proliferous (vegetative bulblets); g. portion of inflorescence, including some bulblets replacing fertile flowers (F).

Fig. 241. *Cyperus pseudovegetus*: a. inflorescence; b. spikelet; c. spike; d. achene (G&W).
Cyperus erythrorhizos: e. habit; f. portion of inflorescence; g. spikelet; h. achene (Mason).

Fig. 242. *Cyperus odoratus*: a. inflorescence; b. spikelet; c. scale; d. achene (G&W).
Fuirena pumila: e. habit (Kral, 1978); f. summit of leaf sheath (Kral, 1978); g. scale (Kral, 1978); h. achene (F).

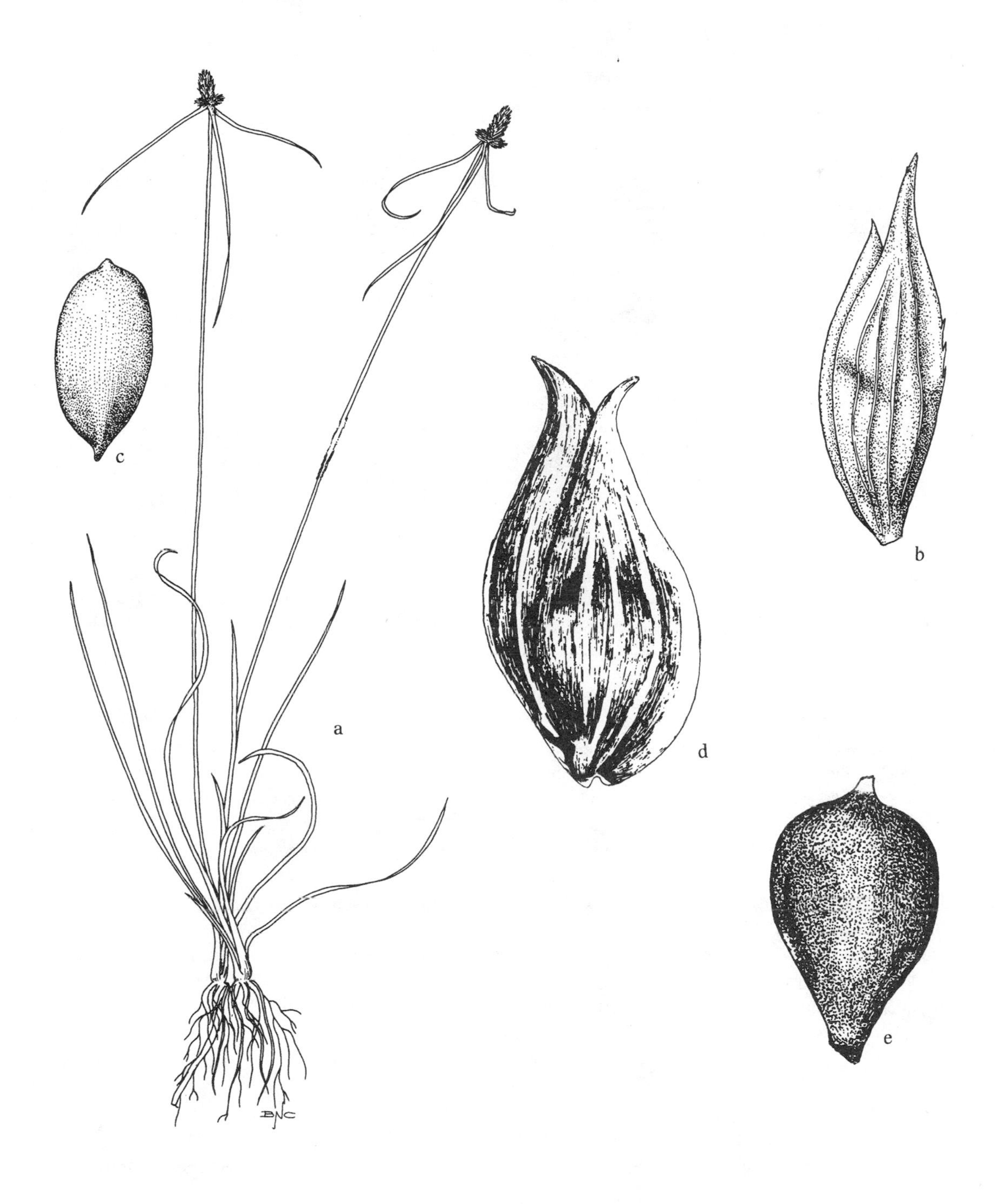

Fig. 243. *Kyllinga pumila*: a. habit; b. scales; c. achene (G&W).
Kyllinga gracillima: d. scales; e. achene (Delahoussaye and Thieret, 1967).

1. Plants annual, tufted, rhizomes absent; anthers 0.5–0.7 mm.
   2. Perianth blades acuminate at apex; e. of Mississippi River. . . . . 1. *F. pumila*
   2. Perianth blades obtuse or blunt and notched; prairie region w. of Mississippi River. . . . . 2. *F. simplex*
1. Plants perennial, stems arising from rhizomes; anthers 1–2 mm.
   3. Perianth bristles short, not extending as far as flat, dilated blades of perianth scales, typically shorter than stipe of achene (fig. 244d); rhizomes lacking corm-like off-shoot buds. . . . . 3. *F. breviseta*
   3. Perianth bristles longer, extending as far as or beyond flat, dilated blades of perianth scales, longer than stipe of achene (fig. 244h); rhizomes producing corm-like off-shoot buds. . . . . 4. *F. squarrosa*

1. *F. pumila* (Torr.) Spreng. Dwarf Umbrella-sedge Fig. 242
Moist to wet sandy or sandy-peaty shores. Chiefly coastal plain, s. N.E., se. N.Y. and s. N.J. s. to Fla., w. to s. Ark. and se. Tex.; disjunct in s. Ont., s. Mich., n. Ind., s. Ill. and sc. Wisc. (*F. squarrosa* var. *pumila* Torr.)

2. *F. simplex* Vahl Western Umbrella-sedge Fig. 245
Swamps, swales, wet peats, seeps and ditches. Sw. Ill., Iowa, nw. Mo., and e. Neb., s. to La., Tex., N.M. and Mex.; W.I. and C.Am. Our taxon is var. *aristulata* (Torr.) Kral.

3. *F. breviseta* (Coville) Coville Saltmarsh Umbrella-sedge Fig. 244
Wet sands, peats and ditches. Coastal plain, se. Va. s. to Fla., w. to se. Tex. (*F. squarrosa* var. *breviseta* Coville)

4. *F. squarrosa* Michx. Hairy Umbrella-sedge Fig. 244
Shores, wet peats, ditches and seeps. Coastal plain and piedmont, L.I., N.Y. and s. N.J. s. to n. Fla., w. to e. Tex.; inland to Ky., Tenn., s. Ark. and se. Okla. (*F. squarrosa* var. *hispida* (Elliott) Chapm.)

8. *Scirpus* (Bulrush)

Perennial, robust, clump-forming herbs, rhizomes non-tuberous; leaves basal and cauline with a prominent midvein, often cross-septate, sheaths closed; stems obtusely 3-angled, tall with terminal more or less open, paniculate inflorescences; inflorescence subtended by one or more bracts; spikelets numerous, small; flowers subtended by scales, bisexual; stamens 3; style 1 or 2–3-fid; perianth mostly 6 bristles; fruit an achene, lenticular, style not persisting.

Reference: Whittemore and Schuyler, 2002.

1. Bristles smooth, strongly contorted or curled, much longer than achene.
   2. Scale of spikelets with midribs conspicuous, green; mature bristles usually hidden by scales; achenes 1.1–1.5 mm long.
      3. Mature stems lax with inflorescences drooping to, or near to, ground; lateral inflorescences 2 or 3, rays with axillary bulblets; bristles usually 2–4 mm long. . . . . 1. *S. lineatus*
      3. Mature stems nearly erect; lateral inflorescences 0 or 1(2), rays lacking axillary bulblets bristles usually longer than 5 mm. . . . . 2. *S. pendulus*
   2. Scales of spikelets with midribs obscure, not green; mature bristles extending beyond scales, appearing wooly; achenes 0.6–1 mm long.
      4. Lateral bracts sticky at base; scales 2.2–3.1 mm long; achenes reddish or reddish-brown; plants with thick elongate rhizomes, not forming tussocks. . . . . 3. *S. longii*
      4. Involucral bracts not sticky at base; scales 1.1–2.1 mm long; achenes whitish; plants with short branching rhizomes, forming tussocks.
         5. Leaves slender, 2–5 mm wide; spikelet scales usually dark blackish-green; achenes maturing in late June and early July. . . . . 4. *S. atrocinctus*
         5. Leaves usually wider, 3–10 mm wide; spikelet scales pale brown, brown, reddish-brown, or occasionally blackish, achenes maturing mid-July to September.
            6. Spikelets usually with 1 central sessile and 2–3 with distinct pedicels; scales pale brown; achenes maturing from mid- to late July. . . . . 5. *S. pedicellatus*
            6. Spikelets in clusters of (2)3–7 or more, mostly sessile (or with poorly developed pedicels); scales brown, reddish-brown, or sometimes blackish; achenes maturing from early August to September. . . . . 6. *S. cyperinus*
1. Bristles with teeth (obscurely so in *S. divaricatus*), usually straight, occasionally with 1 or 2 bends, barely, if at all, longer than achene, or bristles sometimes absent.
   7. Achenes biconvex, styles usually 2-cleft; leaf sheaths red-tinged at base. . . . . 7. *S. microcarpus*
   7. Achenes 3-angled, styles usually 3-cleft; leaf sheaths greenish on lower portion of stem (but with lower sheaths reddish in *S. expansus*).
      8. Spikelets solitary with distinct pedicels; achenes sharply 3-angled, sides concave. . . . . 8. *S. divaricatus*

Fig. 244. *Fuirena breviseta*: a. habit; b. summit of leaf sheath; c. scale; d. perianth and achene (Kral, 1978). *Fuirena squarrosa*: e. habit; f. summit of leaf sheath; g. scale; h. perianth and achene; i. perianth scale variations (Kral, 1978).

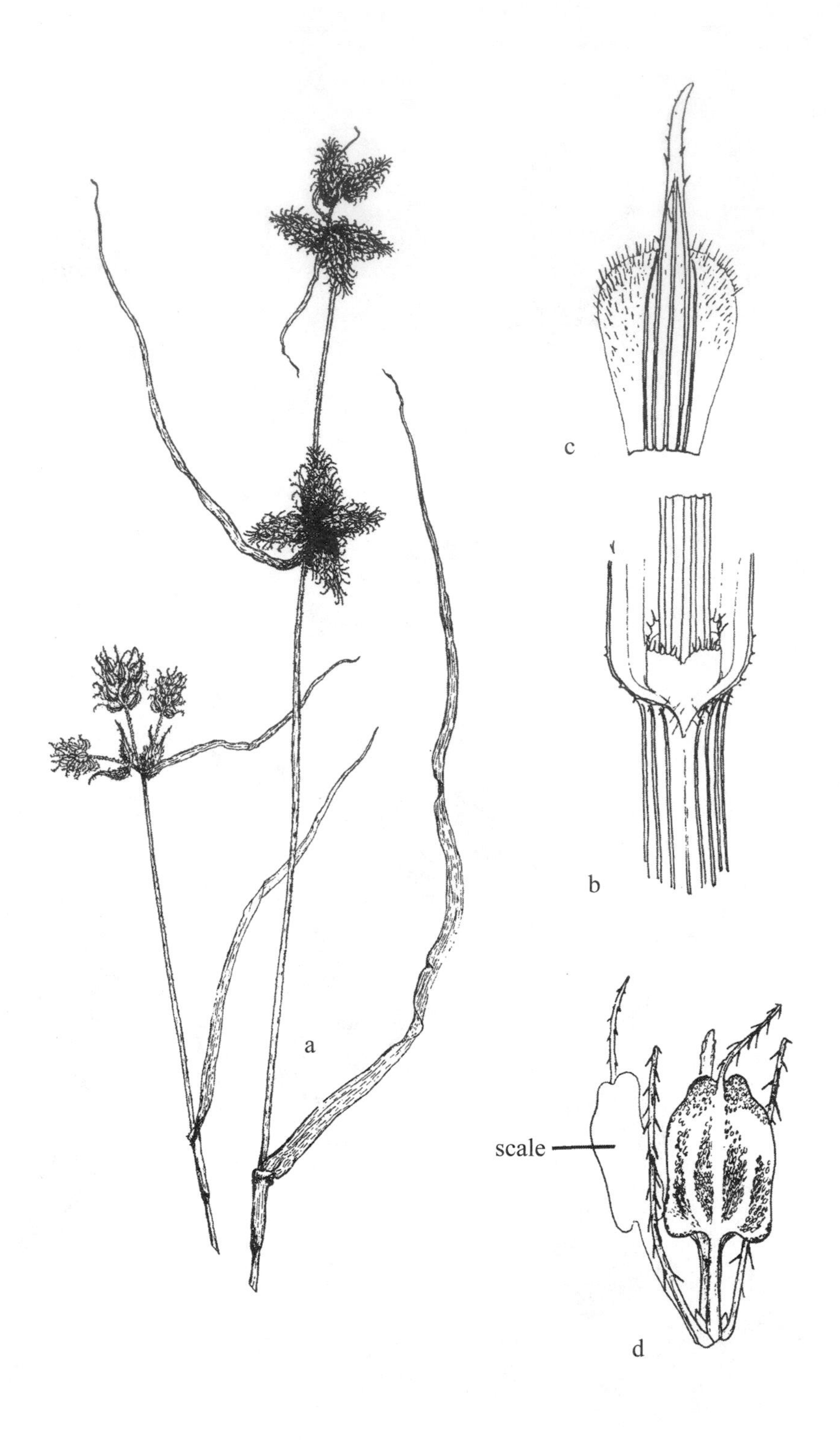

Fig. 245. *Fuirena simplex*: a. upper portion of plant (F); b. summit of leaf sheath (Kral, 1978); c. scale (Kral, 1978); d. perianth scales and bristles (Kral, 1978).

8. Spikelets usually in clusters, appearing sessile (with poorly developed pedicels); achenes not sharply 3-angled, sides convex, flat, or only slightly concave.
 9. Scales reddish-brown, as wide as long (excluding tips); stems 10–20)-leaved. . . . . 9. *S. polyphyllus*
 9. Scales brown or black, usually longer than wide (excluding tips); stems 2–10-leaved.
  10. Bristles with thick-walled, sharp-pointed teeth, along the bristle almost to the base (fig. 250b,f).
   11. Rays of inflorescence ascending to spreading; leaf sheaths on lower portion of stem strongly red-tinged at base; plants with long reddish rhizomes; achenes with bristles readily dropping off; spikelets 1–2.5 mm wide. . . . . 10. *S. expansus*
   11. Rays of inflorescence arching; leaf sheaths on lower portion of stem green at base; plants cespitose, with short brownish rhizomes; achenes with bristles persistent; spikelets 2–3 mm wide. . . . . 11. *S. ancistrochaetus*
  10. Bristles with thin-walled, round-tipped teeth, restricted to upper two-thirds of bristle, or bristles absent.
   12. Bristles absent or 1–3, shorter than achenes. . . . . 12. *S. georgianus*
   12. Bristles 5 or 6, shorter to slightly longer than achenes.
    13. Scale tip awn-like (fig. 250j), about 0.4–0.7 mm long. . . . . 13. *S. pallidus*
    13. Scale tip slightly mucronate, less than 0.4 mm long.
     14. Mature stems lax, with inflorescence drooping to or near to ground, clusters of spikelets usually with fewer than 15; plants of se. Va. coastal plain (within our range). . . . . 14. *S. flaccidifolius*
     14. Mature stems erect, glomerules often with more than 15 spikelets; plants throughout our range.
      15. Lower leaf blades and sheaths usually distinctly nodulose-septate; scales usually brown; bristles straight or curved, following outline of achene, frequently longer than achene. . . . . 15. *S. atrovirens*
      15. Lower leaf blades and sheaths smooth to slightly nodulose-septate; scales usually black; bristles contorted, all shorter than or as long as achene. . . . . 16. *S. hattorianus*

---

1. *S. lineatus* Michx. Fig. 246
Bottomlands, marshes, ditches and wet woods. Coastal plain, se. Va. s. to Fla., w. to La.; se. Mo.

2. *S. pendulus* Muhl. Rufous Wool-sedge Fig. 246
Meadows, marshes stream banks, wet peat and ditches. N.B. and Me. w. to Iowa and e. Colo., s. to Fla., Miss., Okla., Tex. and n. Mex.; Oreg. and Calif.

3. *S. longii* Fernald Long's Wool-sedge Fig. 247
Meadows, swamps and marshes. Coastal, N.S., w. Me, se. N.H., e. Mass., R.I., Conn, and s. N.J. Rare throughout, but locally abundant in s. N.J. Typically encountered sterile, Schuyler (1962a) noted that a large portion of herbarium specimens have charred leaves, suggesting that fire may be an important factor in stimulating flowering.

4. *S. atrocinctus* Fernald Black-girdled Wool-sedge Fig. 248
Meadows, swamps, marshes and ditches often in disturbed areas, Nfld. w. to n. Alta. and B.C., s. to W.Va., Mich., ne. Iowa, S.D. and Wash. *Scirpus atrocinctus* also often grows with and hybridizes with *S. cyperinus* (Schuyler, 1967). According to Schuyler (1961) *Scirpus*×*peckii* Britton—a narrow endemic of N.E., N.Y. and s. Que., characterized by having spikelets borne in clusters on stiff straight inflorescence branches, and the perianth bristles elongate and somewhat bent with a few remote barbules along its length—is sterile and regarded as a species of hybrid origin. Parentage is not clear, but Whittemore and Schuyler (2002) state that the name *Scirpus peckii* has been applied to hybrids between *S. hattorianus* and *S. atrocinctus* or *S. pedicellatus*.

5. *S. pedicellatus* Fernald Stalked Wool-sedge Fig. 247
Marshes, wet lowlands and stream margins. E. Que. w. to ne. Minn., s. to n. N.J., n. Ind., n. Ill. and n. Mo. This species closely resembles *S. atrocinctus* but tends to be more robust, and has pale brown instead of blackish scales. Whittemore and Schuyler (2002) indicate that hybrid swarms sometimes occur as a consequence of hybridization with *S. cyperinus*.

6. *S. cyperinus* (L.) Kunth Common Wool-sedge, Wool-grass Fig. 247
Wet meadows, swamps, marshes and ditches. Nfld. w. to Minn. and Man., s. to Fla. and e. Tex.; B.C. This species is extremely variable, the northern plants tending to have back involucres and bracts, and spikelets with short, ovate, brownish scales (Schuyler, 1967).

7. *S. microcarpus* Presl. Barber-pole Bulrush Fig. 249
Marshes, low tickets, meadows and ditches. S. Lab. and Nfld. w. to Alask., s. to N.E., W.Va., Mich., n. Ill., Iowa, N.M., Ariz. and s. Calif. This species may hybridize with *S. expansus*.

8. *S. divaricatus* Elliott Fig. 249
Bottomlands, woods, swamps, and riverbanks. Se. Va. w. to s. Tenn. and s. Mo. s. to Fla. and La.

9. *S. polyphyllus* Vahl Leafy Bulrush Fig. 249
Low woods, swamps, stream banks and marshes, N.H., Vt., Mass. and N.Y. w. to Ill. s. to Ga. Ala. and Ark.

10. *S. expansus* Fernald Wood Bulrush Fig. 250
Marshes, meadows and low thickets. Me. w. to Ohio and Mich. and Ont., s. to Ga. and Ala.

11. *S. ancistrochaetus* Schuyler Northeastern Bulrush Fig. 250
Pond margins, stream banks, and bogs. Very rare and local, Que., se. Vt., c. Mass., N.Y., Pa., Va. and W.Va.

12. *S. georgianus* Harper Fig. 250
Moist meadows, marshes and ditches, P.E.I. w. to Wisc. and Neb., s. to Ga., Tenn., and e. Tex. This species is rare in the northern part of its range.

13. *S. pallidus* (Britton) Fernald Pale Bulrush Fig. 250
Low ground, marshes, stream margins and ditches. Minn. and Man. w. to Wash., s. to Wisc., Mo., e. Tex., N.M. and Ariz.; e. Pa.

Fig. 246. *Scirpus pendulus*: a. habit (G&W); b. inflorescence (C&C); c. scale (G&W); d. achene (G&W). *Scirpus lineatus*: e. scale; f. achene (G&W).

Fig. 247. *Scirpus longii*: a. inflorescence; b. scale; c. achene (Crow).
Scirpus cyperinus: d. inflorescence (G&W); e. portion of inflorescence (F); f. scale (G&W); g. achene (G&W).
Scirpus pedicellatus: h. portion of inflorescence (F).

Fig. 248. *Scirpus atrocinctus*: a. inflorescence; b. achene; c. scale (PB).
Scirpus hattorianus: d. inflorescence; e. achene; f. scale (PB).

Fig. 249. *Scirpus microcarpus*: a. habit; b. portion of inflorescence; c. scale; d. achene (Mason).
Scirpus divaricatus: e. inflorescence (Beal).
Scirpus polyphyllus: f. achene (F).

Fig. 250. *Scirpus expansus*: a. inflorescence; b. achene (F).
Scirpus ancistrochaetus: c. inflorescence; d. scale; e. achene; f. perianth bristles (Crow).
Scirpus georgianus: g. achene (F).
Scirpus pallidus: h. inflorescence; i. portion of inflorescence; j. scale; k. achene (HCOT).

14. *S. flaccidifolius* (Fernald) Schuyler Fig. 252
Bottomlands, Se. Va. and ne. N.C. (*S. atrovirens* var. *flaccidifolius* Fernald)

15. *S. atrovirens* Willd. Dark-green Bulrush Fig. 251
Meadows, marshes, stream margins, bogs and ditches. Nfld. and s. Que. w. to Minn., and Mont., s. to Ga., Tenn., Mo. and e. Tex. In late summer plants frequently produce leafy tufts in the inflorescence (fig. 251c). This species is closely related to *S. ancistrochaetus*, *S. flaccidifolius*, *S. georgianus*, *S. hattorianus and S. pallidus*, and occasionally hybridizes with them.

16. *S. hattorianus* Makino Fig. 248
Meadows, Marshes and ditches. Nfld. to w. Ont. s. to N.C., Ohio, Ind. and Wisc.

9. *Eriophorum* (Cotton-grass)

Perennial herbs, grass-like, cespitose, or stems arising from creeping rhizomes; flowers bisexual, subtended by membranous scales; perianth of numerous bristles, greatly elongating with maturity; achenes lacking tubercle, subtended by persistent cottony bristles.

References: Ball and Wujek, 2002; Cayouette, 2004; Raymond, 1951; Tucker and Miller, 1990.

---

1. Spikelets solitary (fig. 254d), erect, not subtended by involucral leaves.
  2. Plants with creeping rhizomes, stems chiefly solitary: sterile scales at base of spikelet 7 or fewer.
    3. Perianth bristles bright white.
      4. Spikelets hemispherical, 1–3 mm wide; fertile scales 0.1–0.9 mm long achene narrowly obovoid, 1.7–2.4 mm long, the beak oblique more often than straight. . . . . . 1. *E. scheuchzeri*
      4. Spikelets ellipsoid or obovoid, 2–4 mm wide; fertile scales 0.8–3.2 mm long; achene obovoid or ellipsoid, 2–2.7 mm long, the beak straight more often that oblique. . . . . . 2b. *E. russeolum* subsp. *leiocarpum*
    3. Perianth bristles cinnamon-brown to reddish. . . . . . 2a. *E. russeolum* subsp. *russeolum*
  2. Plants densely cespitose, not spreading by rhizomes; sterile scales at base of spikelet usually 10–15.
    5. Cauline sheath (fig. 254a) with conspicuous white margin (becoming straw-colored on drying); scales pale-margined, with pigmentation strongest near lower middle, divergent or reflexed; achenes 2.5–3.5 mm long.
      6. Perianth bristles reddish or cinnamon-brown. . . . . . 3. *E.* ×*pylaieanum*
      6. Perianth bristles white. . . . . . 4. *E. vaginatum*
    5. Cauline sheath (fig. 255d) lacking white margin; scales with pigmentation stronger toward margin, weaker toward lower middle, ascending; achenes 2–2.3 mm long.
      7. Stems slender, 30–60 cm long; perianth bristles dull white: uppermost cauline sheath located above middle of stem, scarcely inflated (fig. 255d). . . . . . 5. *E. brachyantherum*
      7. Stems thick, 6–20 cm long; perianth bristles bright white; uppermost cauline sheath typically located below middle of stem, conspicuously inflated (fig. 255a). . . . . . 6. *E. callitrix*
1. Spikelets 2-several, drooping or spreading (fig. 257d,g), subtended by 1–several involucral leaves (fig. 257a,d).
  8. Involucral bracts 1, not extending beyond inflorescence (fig. 256a,d), inflorescence appearing lateral; leaves triangular, channeled.
    9. Uppermost cauline leaf with blade shorter than sheath (fig. 256a); scales lead-colored to blackish; achenes 1.5–2 mm long. . . . . . 7. *E. gracile*
    9. Uppermost cauline leaf with blade as long as or longer than sheath (fig. 256d); scales greenish to reddish-brown; achenes 2.5–3 mm long. . . . . . 8. *E. tenellum*
  8. Involucral bracts 2 or more, extending beyond inflorescence (fig. 257a,d), inflorescence appearing terminal; leaves flat (at least below middle).
    10. Spikelets densely crowded, spreading on short peduncles; perianth bristles tawny, at least at base; scales brownish to reddish, the lower with 3–5 nearly equally strong nerves. . . . . . 9. *E virginicum*
    10. Spikelets loosely clustered, nodding on slender, elongate peduncles; perianth bristles white or creamy to buff; scales drab to lead-colored or blackish, with 1 strong nerve.
      11. Scales with strong midrib extending to tip (fig. 257e); upper leaf sheath lacking dark margin; achenes 3–3.4 mm long. . . . . . 10. *E. viridicarinatum*
      11. Scales with strong midrib not extending to tip (fig. 257h); upper leaf sheath with dark margin; achenes 2–2.3(3) mm long. . . . . . 11. *E. angustifolium*

---

1. *E. scheuchzeri* Hoppe White Cotton-grass Fig. 253
Chiefly wet peats of arctic regions, marshy ground, peaty shores of lakes and ponds. Greenl., and Nunavut w. to James Bay, N.W.T. and Alask., s. to Lab., n. Nfld., Sask., mts. of Mont., Wyo., Utah, N.M., B.C. and Wash. *Eriophorum scheuchzeri* is a circumpolar species; the taxon in our range is subsp. *scheuchzeri*; closely related subsp. *arcticum* M. S. Novo. occurs in the Canadian high-Arctic and is circumpolar (Cayouette, 2004).

Fig. 251. *Scirpus atrovirens*: a. basal portion of plant (C&C); b. upper portion of plant (C&C); c. inflorescence with leafy tufts (vegetative propagules) (Braun); d. portion of inflorescence (Braun); e. spikelet (Braun); f. achene (C&C); g. scale (C&C).

Fig. 252. *Scirpus flaccidifolius*: a. inflorescence; b. achene; c. scale (PB).

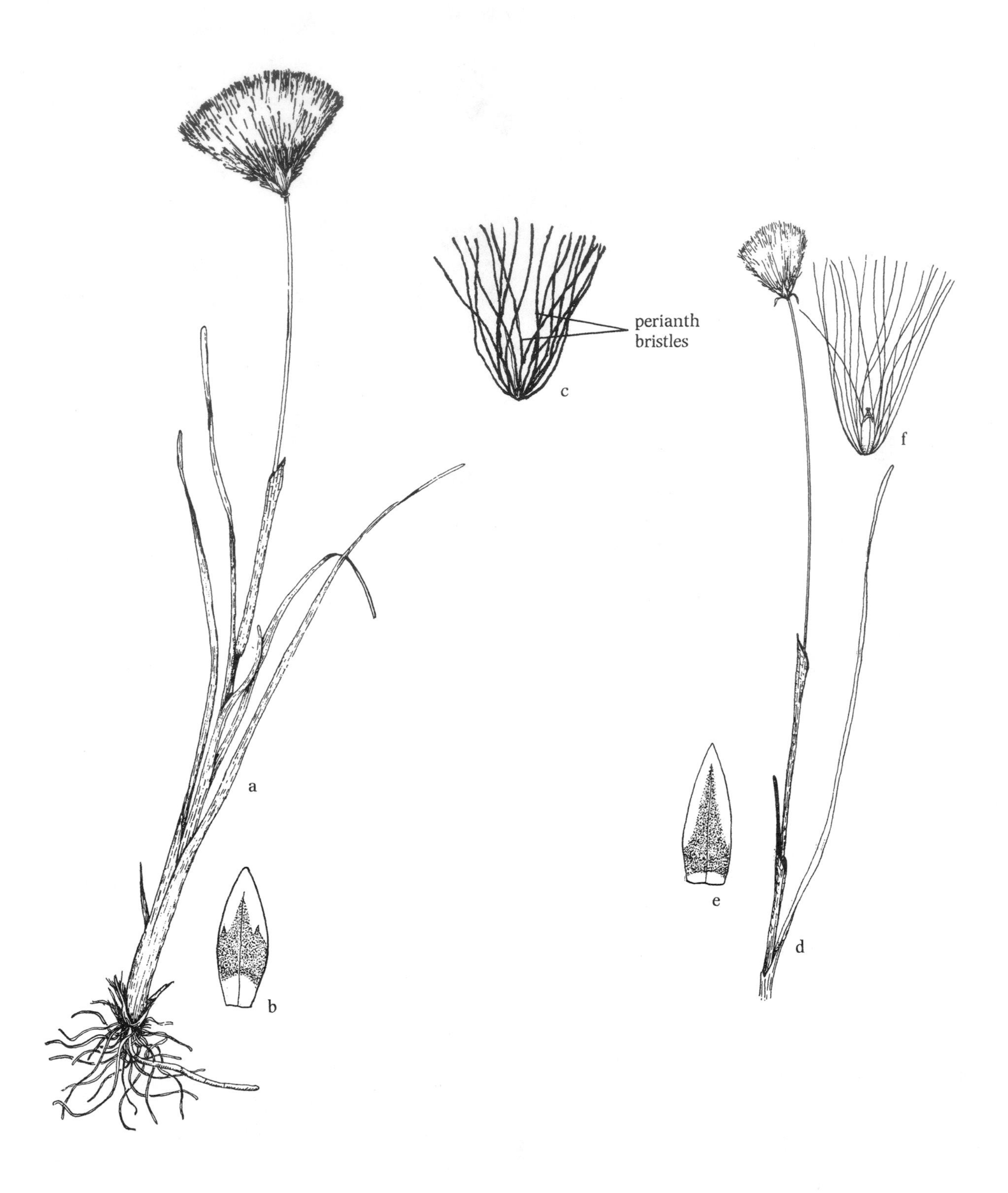

Fig. 253. *Eriophorum russeolum* subsp. *russeolum*: a. habit; b. scale; c. achene (TF).
Eriophorum scheuchzeri: d. habit; e. scale; f. achene (TF).

2. *E. russeolum* Fries Red Cotton-grass

2a. *E. russeolum* subsp. *russeolum* Red Cotton-grass Fig. 253

Edges of ponds, lakeshores, fens and minerotrophic bogs. Nfld. and Lab., N.S., N.B., P.E.I., and Que. w. to James Bay, Ont.; n. Eur. and c. Russia. The more eastern subsp. *russeolum*, an amphi-Atlantic taxon with orange-brown or rufescent spikelet bristles, is easily distinguished from the more western subsp. *leiocarpum*, which has white bristles. A hybrid species, *E. ×medium*, a cross between *E. russeolum* subsp. *russeolum* and *E. scheuchzeri* subsp. *scheuchzeri*, is also an amphi-Atlantic taxon that occurs sporadically where the two parental species occur together (Cayouette, 2004). Although the World Checklist of Cyperaceae (Govaerts and Simpson, 2007) regardes *E. russeolum* as a synonym of *E. chamissonis* C. A. Mey., Cayouette (2004) considers these as distinct species. He further notes that *E. chamissonis* has frequently been mistakenly reported as occurring in eastern North America; it is a species of northwestern North America, from Alaska and B.C.; the Panarctic Flora website is in agreement (www.panarcticflora.org, accessed September 2020).

2b. *E. russeolum* subsp. *leiocarpum* M. S. Novos. White-bristled Russet Cotton-grass

Edges of ponds, lakeshores, fens and minerotrophic bogs. Nfld. and Lab., Que. and Nunavut, w. to N.W.T., Yuk. and Alask., s. to N.S., N.B., n. Mich., n. Wisc., n. Minn., and n. Man. Cayouette (2004) notes that this subspecies is more western and amphi-Beringian, being continuous into central Russia, but becoming somewhat discontinuous further west. (*E. russeolum* var. *albidum* F. Nylander; *E. chamissonis* var. *albidum* (F. Nylander) Fernald; *E. russeolum* subsp. *albidum* (F. Nyl.) Väre)

3. *E. ×pylaieanum* Raymond Cotton-grass Fig. 254

Wet peats, especially disturbed sites. Nfld. and se. Lab. The taxon is a fertile hybrid between *E. russeolum* and *E. vaginatum* subsp. *spissum*, often growing with the latter taxon.

4. *E. vaginatum* L. Tussock Cotton-grass Fig. 254

Wet peats and bogs. Greenl., Baffin I. and Lab. w. to N.W.T. and Alask., s. to N.E., n. N.J., ne. Pa., n. Ind., Wisc., Minn. and Mont. This species flowers in very early spring, and fruits mature in early summer, readily dispersing and leaving empty fruiting stalks. Our taxon is subsp. *spissum* (Fernald) Hultén and had been recognized as *E. spissum* in many earlier manuals. (*E. spissum* Fernald; *E. spissum* var. *erubescens* Fernald)

5. *E. brachyantherum* Trautv. & C. A. Mey. Northland Cotton-grass Fig. 255

Lowland muskeg and calcareous bogs. Nunavut and Lab. w. to N.W.T. and Alask., s. to Nfld., n. Ont., Alta. and n. B.C.; Eurasia. (*E. opacum* (Bjornstr.) Fernald; *E. vaginatum* var. *opacum* Bjornstr.)

6. *E. callitrix* Cham. Arctic Cotton-grass Fig. 255

Calcareous peats. Chiefly arctic; Greenl. and Nunavut w. to Alask., s. to Lab., n. Nfld., James Bay, Alta., mts. of Mont. and Wyo. and B.C.

7. *E. gracile* W. D. J. Koch Slender Cotton-grass Fig. 256

Bogs, wet peat and inundated shores. Nunavut, Lab. and Nfld. w. to Alask., s. to N.E., Del., Pa., Mich., n. Ill., Iowa, Colo., Ida., and e. Calif.

8. *E. tenellum* Nutt. Few-nerved Cotton-grass Fig. 256

Bogs, peaty soils and conifer swamps. Nunavut and Nfld. w. to Ont., s. to N.E., N.J., Pa., Mich., Ill. and Minn. (*E. tenellum* var. *monticola* Fernald)

9. *E. virginicum* L. Tawny Cotton-grass Fig. 257

Bogs and peaty meadows. Se. Lab. and Nfld. w. to Ont., s. to Ga., Tenn., Mich., Iowa and Minn.; B.C.

10. *E. viridicarinatum* (Engelm.) Fernald Thin-leaf Cotton-grass Fig. 257

Bogs, wet peats, peaty meadows and conifer swamps. Lab. and Nfld. w. to N.W.T., and Alask., s. to N.E., N.Y., Ohio, n. Ind., n. Ill., n. Minn., Iowa, N.D., nw. Mont., Colo., Ida., and s. B.C.

11. *E. angustifolium* Honck. Tall Cotton-grass Fig. 257

Bogs, wet peats and boggy shores, Greenl. W. to Alask. s. to n. N.E., N.Y. Ind. n. Ill., Iowa, Colo, N.M. and c. Oreg. A circumpolar species, our taxon is subsp. *angustifolium*. (*E. polystachion* L.)

10. *Trichophorum* (Club-rush)

Perennial, cespitose, or tufted along slender rhizomes; leaves basal or sub-basal; bracts subtending inflorescence 1; inflorescence spikelets solitary; flowers perfect with perianth bristles (ours); stamens 3, styles trifid; achenes plano-convex to trigonous.

Reference: Crins, 2002.

1. Perianth bristles 10–30 mm long, cottony; stems 3-angled; plants in small tufts along creeping rhizomes. . . . . 1. *T. alpinum*
1. Perianth bristles less than 4 mm long, hidden by scales, not cottony; stems nearly round in cross-section; plants in dense tufts, not rhizomatous. . . . . 2. *T. cespitosum*

1. *T. alpinum* (L.) Pers. Alpine Club-sedge, Alpine Bulrush Fig. 258

Bogs, boggy meadows, fens, wet gravels and springy places, especially in calcareous areas. Se. Lab. and Nfld. w. to Alask., s. to N.E., n. N.Y., n. Mich., Minn. and w. Mont. (*Eriophorum hudsonianum* Michx.; *Scirpus hudsonianus* (Michx.) Fernald)

2. *T. cespitosum* L. Deer-hair Sedge Fig. 258

Cold bogs, fens and peats of lowlands and mountain tops. Arctic regions s. to Nfld., N.S., Me., mts. of N.E., disjunct to higher mts. of N.C., Ga. and Tenn., w. to Mich., n. Ill., Minn. and mts. of w. Mont. and higher elevations to Utah and Oreg. (*Scirpus cespitosus* L.; *S. cespitosus* var. *callosus* J. M. Bigelow)

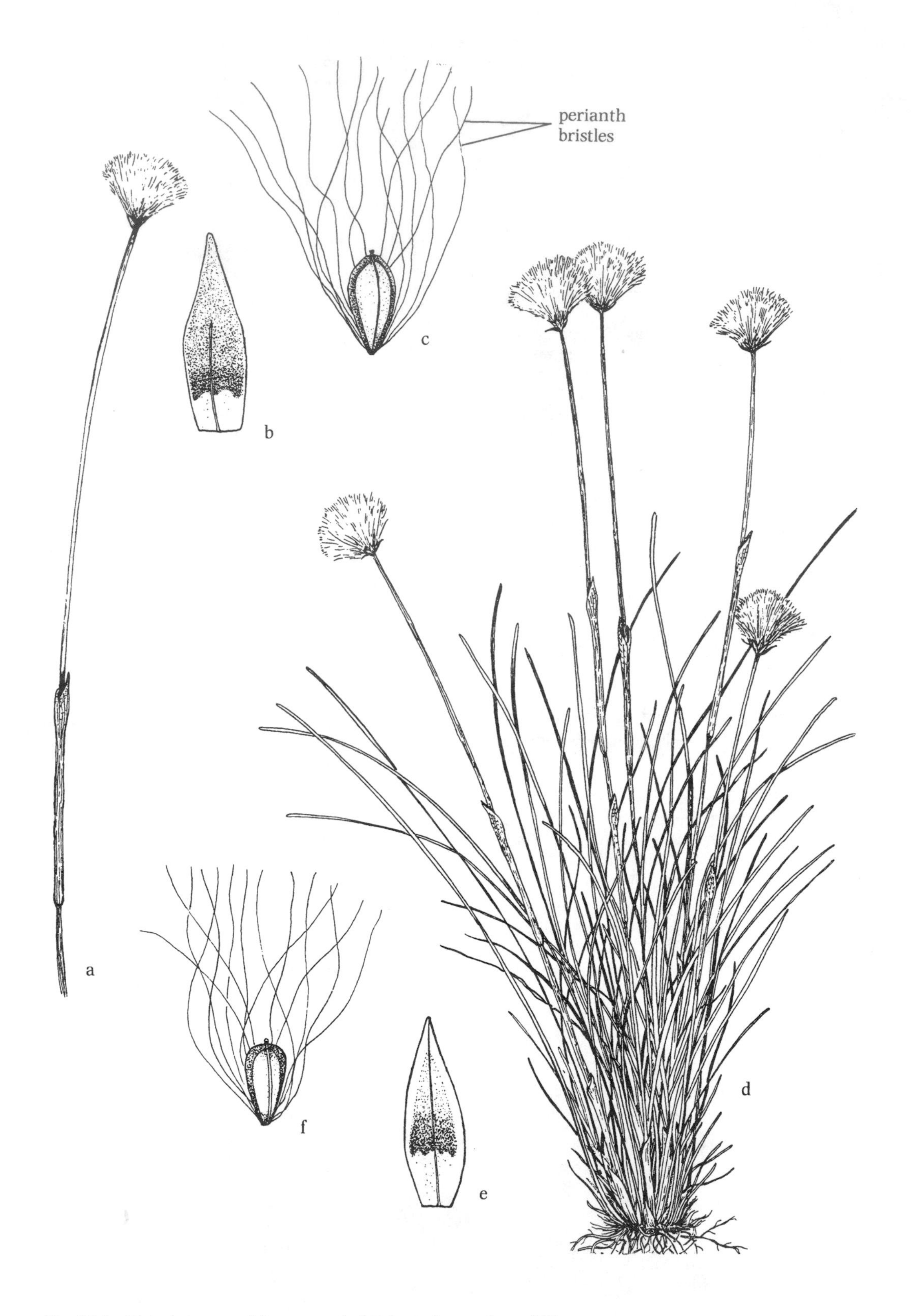

Fig. 254. *Eriophorum* ×*pylaieanum*: a. habit; b. scale; c. achene (TF).
Eriophorum vaginatum: d. habit; e. scale; f. achene (TF).

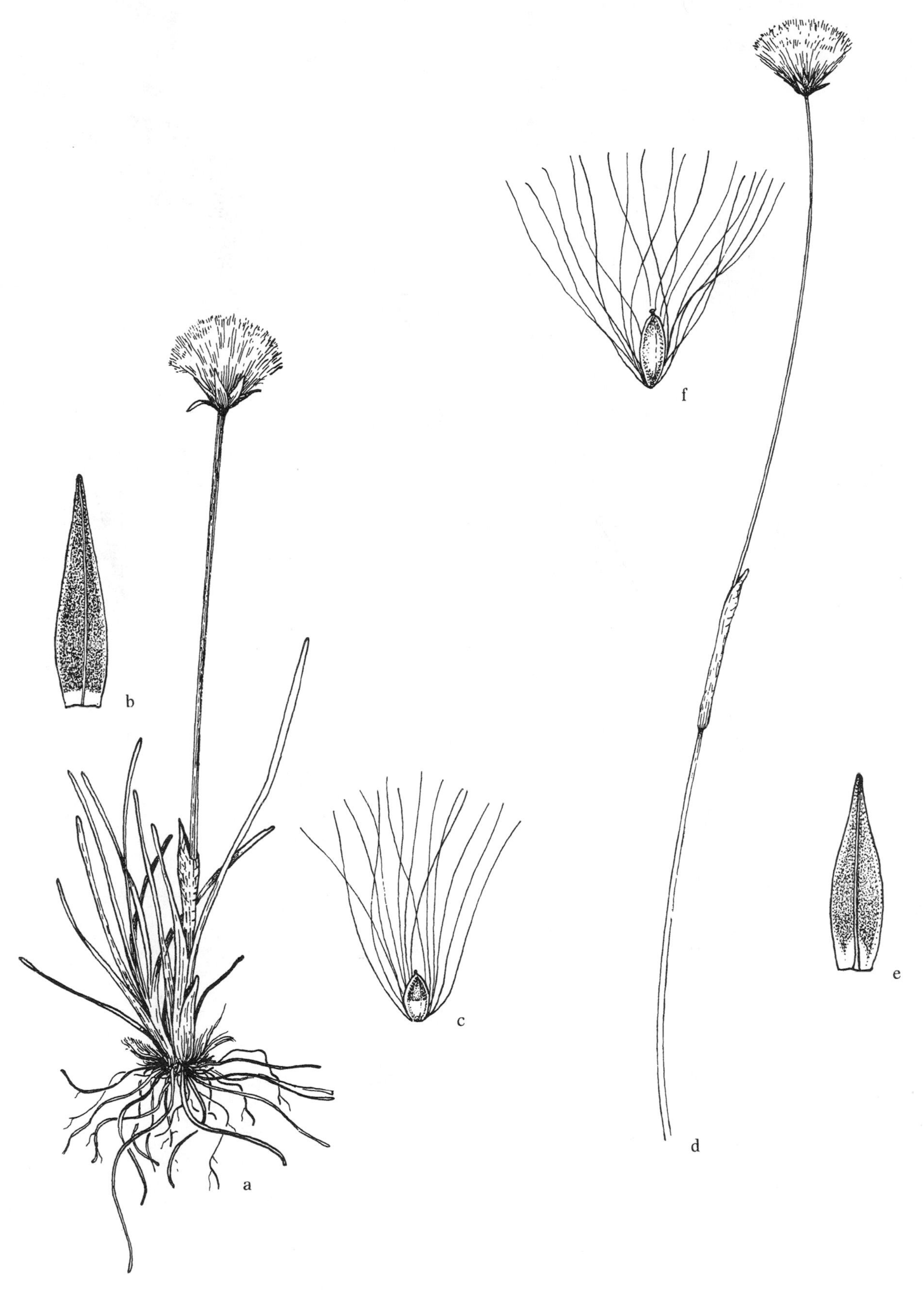

Fig. 255. *Eriophorum callitrix*: a. habit; b. scale; c. achene (TF).
Eriophorum brachyantherum: d. habit; e. scale; f. achene (TF).

Fig. 256. *Eriophorum gracile*: a. habit; b. scale; c. achene (TF).
Eriophorum tenellum: d. habit; e. scale; f. achene (TF).

Fig. 257. *Eriophorum virginicum*: a. habit; b. scale; c. achene (TF).
Eriophorum viridicarinatum: d. habit; e. scale; f. achene (TF).
Eriophorum angustifolium: g. habit; h. scale; i. achene (TF).

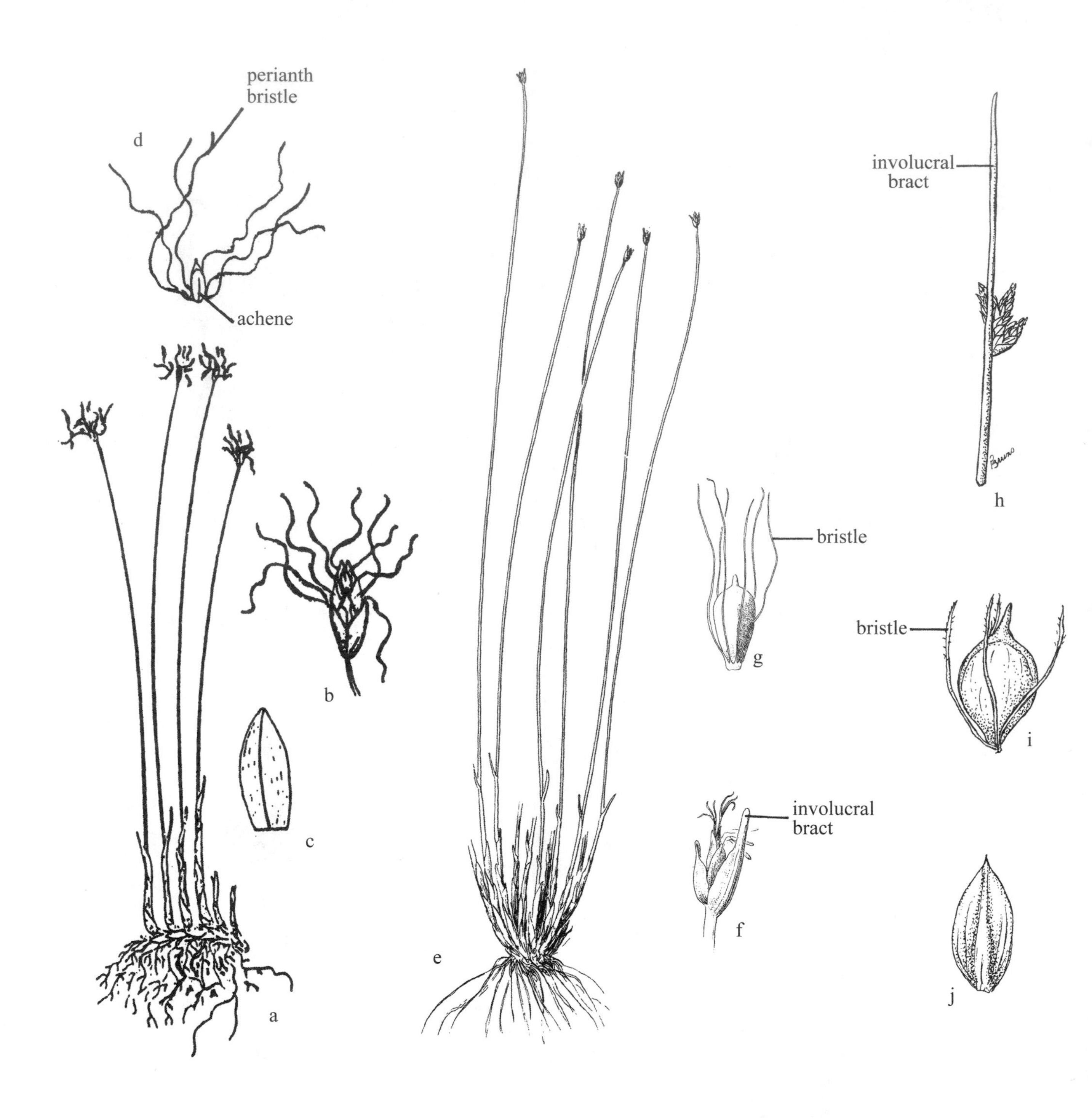

Fig. 258. *Trichophorum alpinum*: a. habit; b. spikelet; c. scale; d. achene (TF). *Trichophorum cespitosum*: e. habit; f. inflorescence; g. achene (HCOT). *Schoenoplectus torreyi*: h. inflorescence; i. achene; j. scale (PB).

11. *Schoenoplectus* (Naked-stem Bulrush)

Annual or perennial (ours); mostly rhizomatous; stems round to strongly 3-angled, often with spongy tissue inside; leaves basal (rarely 1–2 cauline), sheaths tubular, blades well developed or rudimentary; inflorescence appearing lateral on the stem, consisting of solitary spikelets, solitary clusters, to open panicles; involucral bracts 1–5, leaf-like; spikelets 1–100 or more per inflorescence; flowers bisexual; bristles 0–8; achenes plano-convex, biconvex or trigonous, with apical beak, lateral surfaces with transverse wavy ridges.

---

1. Spikelets sessile (fig. 261b).
  2. Fertile stems weakly ascending, usually supported by water, often accompanied by long, hair-like leaves. . . . . 1. *S. subterminalis*
  2. Fertile stems erect or arching; spikelets (1)2 or more, in clusters.
    3. Rhizomes flexible; scales lacking apical notch, margins smooth; styles 3-cleft; achenes 3–4 mm long. .2. *S. torreyi*
    3. Rhizomes firm; scales with obvious apical notch, margins (as least some) erose; styles 2-cleft (rarely 3-cleft); achenes 1.9–3 mm long.
      4. Involucral bracts 1–3.5(5) cm long; scale tips less than 0.5 mm long. . . . . 3. *S. americanus*
      4. Involucral bract 3.5–18 cm long; scale tips 0.5–1 mm long.
        5. Spikelets usually more than 5 per inflorescence; achenes less than 2.5 mm long, 1.6 mm wide; bristles slender, often longer than achenes. . . . . 4. *S. deltarum*
        5. Spikelets fewer than 5 per inflorescence, achenes (2)2.4–3 mm long, greater than 1.5 mm wide, bristles stout, rarely longer than achenes. . . . . .5. *S. pungens*
1. Spikelets usually pedunculate (fig. 262a, 263a), or occasionally short-stalked and appearing subsessile.
  6. Involucral bracts 7–32 cm long (often resembling emerging leaf blade); stems 3-angled, emersed leaves with long triangular, channeled blades; submersed leaves with long ribbon-like blades. . . . . 6. *S. etuberculatus*
  6. Involucral bracts 1–7 cm long; stems round in cross-section; emersed and submersed leaves bladeless or with inconspicuous blades.
    7. Perianth parts 2–4, strap-like fringed with soft, blunt hairs, dark red-brown; spikelet scales smooth or awn sparsely scabrous; leaf blades on upper sheaths absent or up to 20 mm; culms becoming clearly to bluntly trigonous near inflorescence. . . . . .7. *S. californicus*
    7. Perianth parts (4)6(8), bristle-like, 2 usually much shorter, medium brown; spikelet scales scabrous only on the awn; leaf blades on upper sheaths present, usually much shorter than sheath (to longer), 2–220 mm; culms cylindric throughout.
      8. Achenes clearly trigonous, styles 3-fid; perianth bristles 4(5), usually 2 much shorter. . . . . 8. *S. heterochaetus*
      8. Achenes usually plano- or biconvex, rarely trigonous; styles 2-fid (or 2- and 3-fid, rarely all 3-fid); perianth bristles usually 6(4–8), typically nearly equal in length.
        9. Stems soft and easily compressed, pale green when fresh; inflorescence open and lax, spikelets 13–125, usually solitary or in clusters of 2 or 3; achenes usually partly exposed beyond scales; scales bright orange brown with obscure darker spots, awns nearly straight, 0.2–1 mm long. . . . . 9. *S. tabernaemontani*
        9. Stems hard, not easily compressed, dark olive-green when fresh; inflorescence usually stiff, spikelets 3–40, usually solitary or in clusters of 2–7; achenes hidden by scales; scales usually dull grayish-brown, with conspicuous bright brown spots, sometimes dark reddish-brown, awns usually contorted, about 0.5–1.5 mm long. . . . . .10. *S. acutus*

---

1. *S. subterminalis* (Torr.) Soják Water Bulrush Fig. 261
Ponds, bogs, lakes and streams. Nfld., w. to Ont. and Man., s. to N.J., Pa., Mich., and n. Ill.; s. Alask. s. to w. Wyo., nw. Mont., n. Ida. and Calif. This species is often confused with *Eleocharis robbinsii* because both species are often encountered only in the vegetative state, both producing fine hair-like submersed leaves; *E. robbinsii* has more angular flowering stems. (*Scirpus subterminalis* Torr.)

2. *S. torreyi* (Olney) Palla Torrey's Three-square Fig. 258
Sandy and peaty shore of fresh or brackish waters. N.B. w. to Man., s. to L.I., N.Y., Del., Va., n. Ohio, Ill. and Mo. (*Scirpus torreyi* Olney)

3. *S. americanus* (Pers.) Volkart ex Schinz & R. Keller Fig. 259
Saline and brackish marshes. Coastal, N.S. and N.E. s. to Fla., w. to Tex. and Mex.; inland in Mich., Mo., Kans. and Okla.; Pacific Coast, B.C., Wash. and Ida. s. to Calif. Ariz. and N.M.; S.Am. This species is treated as *S. olney* A. Gray in most manuals, but Schuyler (1974b) has shown that name to have been misapplied to this taxon; He further noted that hybrids with *S. pungens* may be abundant in transitional habitats, particularly brackish tidal marshes and shores, and have been called *S. ×contortus* (Eames) T. Koyama.

4. *S. deltarum* (Schuyler) Soják
Alluvial tidal freshwater shores, rarely in saline areas, Fla., Ala. and La.; disjunct to Mo. and Kans. (*Scirpus deltarum* Schuyler)

5. *S. pungens* (Vahl) Palla Three-square Bulrush Fig. 260
Freshwater, brackish and saline shores and marshes. Nfld. w. to s. Ont., Minn., and s. B.C., s. to Fla., Tex., Calif. and Mex.; S.Am. Schuyler (1974a) determined that the name *Scirpus americanus* Pers. had been misapplied to this taxon. (*Scirpus pungens* Vahl)

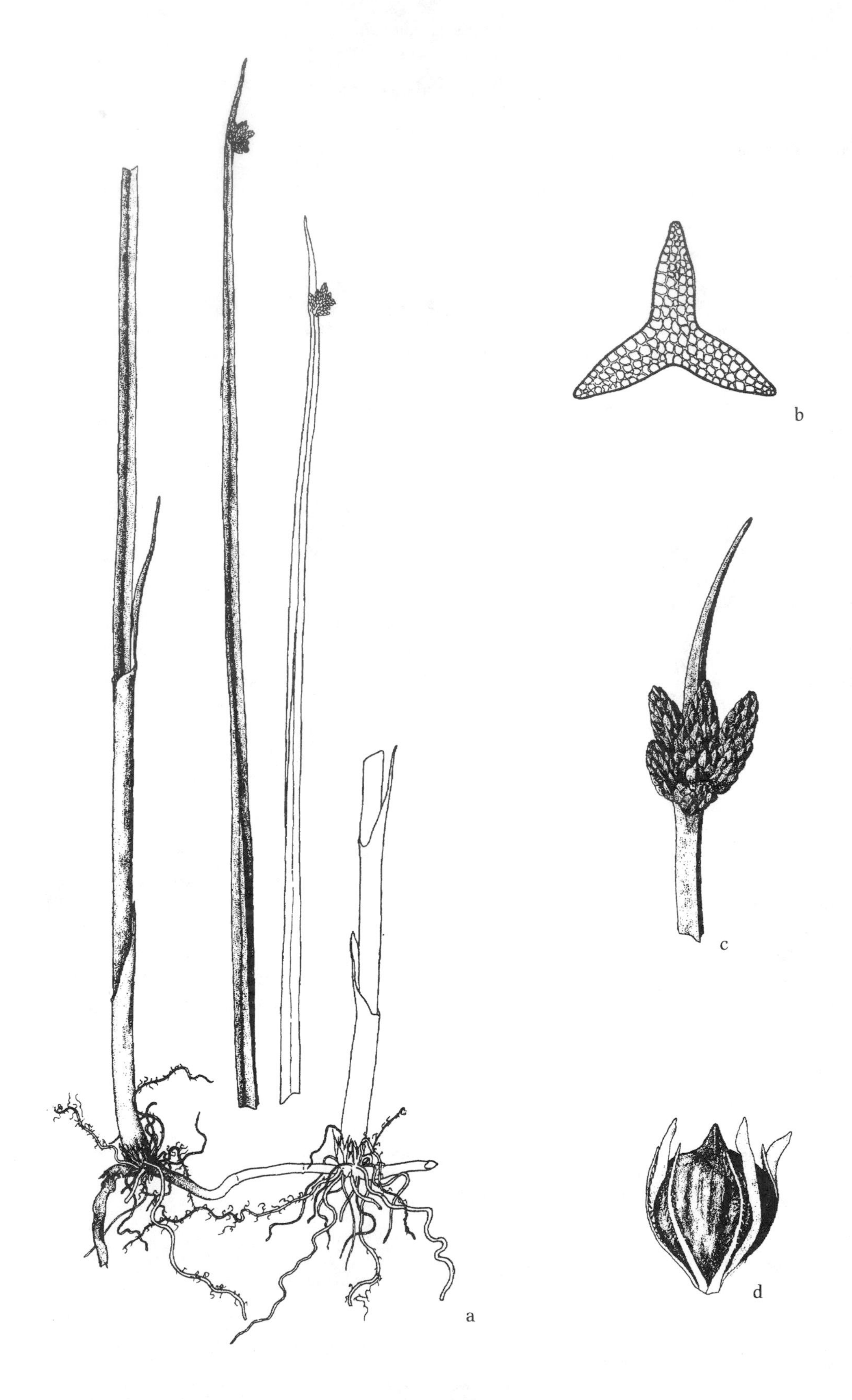

Fig. 259. *Schoenoplectus americanus*: a. habit; b. stem, cross-section; c. inflorescence; d. achene (G&W).

Fig. 260. *Schoenoplectus pungens*: a. habit; b. scale; c. achene; d. triangular stem cross-section, diagramatic (G&W).

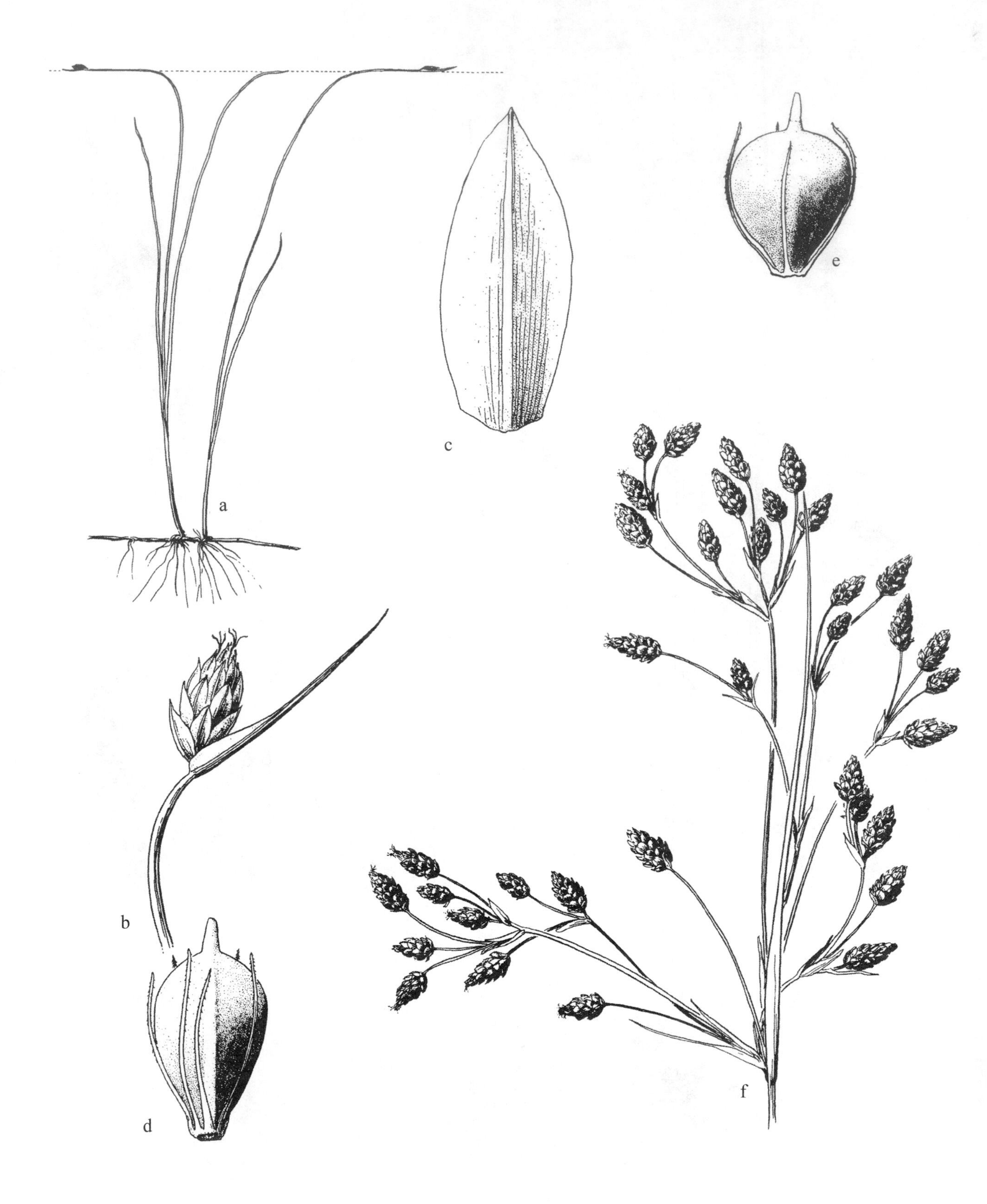

Fig. 261. *Schoenoplectus subterminalis*: a. habit; b. inflorescence; c. scale; d. achene (HCOT).
Schoenoplectus heterochaetus: e. inflorescence; f. achene (HCOT).

6. *S. etuberculatus* (Steud.) Soják Swamp Bulrush Fig. 262
Pond margin, freshwater and brackish marshes. Coastal plain, R.I.; Del. s. to Fla.; w. to e. Tex.; inland n. to s. Mo. (*Scirpus etuberculatus* (Steud.) Kuntze)

7. *S. californicus* (C. A. Mey.) Soják Giant Bulrush Fig. 263
Brackish to freshwater marshes, and shores of ponds and lakes, often emergent. N.C., w. to Mo., s. to Fla., Miss. and Tex.; Calif. and Nev. (*Scirpus californicus* (C. A. Mey.) Steud.)

8. *S. heterochaetus* (Chase) Soják Slender Bulrush Fig. 261
Calcareous shores of ponds, lakes and marshes. Mass., w. Vt., s. to Pa., sw. Ky., Ill., Mo. and Okla.; nw. Ida., Wash., Oreg. and ne. Calif. A similar Eurasian species, *S. lacustris* (L.) Palla, also with 3-cleft styles and 3-angled achenes, but with many spikelets clustered and with 6 bristles has been reported from Camden, New Jersey, and may be sparingly naturalized near the East Coast. (*Scirpus heterochaetus* (Steud.) Kuntze)

9. *S. tabernaemontani* (C. C. Gmel.) Palla Great Bulrush, Softstem Bulrush Fig. 263
Fresh to brackish shores of ponds, lakes, streams and marshes. Nfld. w. to Alask., s. to Fla., Tex., N.M., Calif. and Mex.; S.Am. This species hybridizes frequently with *S. acutus* and rarely with *S. heterochaetus*. Smith (1995) considered *S.* ×*steinmetzii* (Fernald) S. G. Smith to be a hybrid between *S. tabernaemontani* and *S. heterochaetus*, a plant previously known only from a single stream in Maine, but now with new localities recognized from Mass., N.Y., Que., Ont., Wisc., Iowa and Minn.; although it is often sterile, sometimes achenes with fully formed seeds are produced. (*Scirpus tabernaemontani* C. C. Gmel.; *Scirpus validus* Vahl)

10. *S. acutus* (Muhl. ex Bigel.) Á. Löve & D. Löve Hard-stem Bulrush Fig. 264
Freshwater to brackish marshes, pond, and lakes. Nfld. w. to Alask., s. to N.E., W.Va., N.C., Tenn., Mo., Tex., Ariz., Calif., and Mex. This species may hybridize with *S. tabernaemontani* and *S. heterochaetus*. Our taxon is var. *acutus*; however, fig. 264 is drawn from a plant of the western var. *occidentalis* S. Wats.; var. *acutus* typically has scales with highly contorted awns in contrast to the shorter, straighter awns of fig. 264d,e. (*Scirpus acutus* Muhl. ex J. M. Bigelow)

12. *Eleocharis* (Spike-rush, Spike-sedge)

Perennial or annual herbs, tufted or arising from rhizomes or stolons; leaves reduced to bladeless sheaths; spikelet solitary with few to many bisexual flowers, each subtended by a scale; perianth bristles usually 6, occasionally absent; achenes biconvex or 3-angled, with a persistent tubercle. The shape and size of the tubercle are of great taxonomic importance in the genus, and identification is almost impossible without mature achenes.

References: Smith and Gregor, 2014; Smith et al., 2002; Svenson, 1932a.

---

1. Stems nearly as thick as spikelets (fig. 265a, 266b); spikelet scales with 15 or more obscure longitudinal veins, scales persisting following achene maturation.
   2. Stems hollow, septate (transverse septa complete) (fig. 265a). . . . . 1. *E. equisetoides*
   2. Stems spongy, not hollow, not septate or if present, then transverse septa incomplete (*E. robbinsii*).
      3. Stems 1.5–5 mm in diameter, sharply 4-angled (fig. 266db,d). . . . . 2. *E. quadrangulata*
      3. Stems 1–2 mm in diameter, bluntly to sharply 3-angled. . . . . 3. *E. robbinsii*
1. Stems much more slender than spikelets (figs. 268a, 269a); spikelet scales with 1 vein (midrib), scales (and achenes) readily deciduous subsequent to complete achene maturation.
   4. Plants with rhizomes 2 mm or more in diameter (figs. 268a, 269a).
      5. Achenes biconvex; styles 2-cleft.
         6. Basal (proximal) scale clasping ⅔(¾) of stem, with 2 or 3 of subproximal scales empty (sterile) (fig. 267c).
            7. Perianth bristles absent or if present, 4(5), usually shorter than achene. . . . . 4. *E. palustris*
            7. Perianth bristles mostly 5–6(8), usually longer than achene (including tubercle). . . . . 5. *E.mamillata*
         6. Basal (proximal) scale completely encircling stem or clasping ¾ or more (fig. 268e), with subproximal scales empty or with flower.
            8. Basal scale of some or all spikelets completely or clasping ¾ or more of stem; subproximal scale usually empty or with both empty and fertile scales on same plant.. . . . . 6. *E. macrostachya*
            8. Basal scale completely clasping stem (amplexicaulous); subproximal scales fertile.
               9. Spikelets 30–40-flowered; lower and middle scales 1.8–3 mm long; achenes 0.7–1 mm wide. . . . . 7. *E. erythropoda*
               9. Spikelets few-flowered (10–20); lower and middle scales 3–5 mm long; achenes 1–1.8 mm wide.
                  10. Achene tubercles longer than wide, 1.3–1.8 × 1–1.4 mm; achene surface smooth (at 10–20× magnification). . . . . 8. *E. uniglumis*
                  10. Achene tubercles mostly wider than long, 1–1.5 × 0.9–1.25 mm; achene surface finely rugulose (at 10–20× magnification). . . . . 9. *E. ambigens*

Fig. 262. *Schoenoplectus etuberculatus*: a. habit; b. section of submersed leaf; c. section of emersed leaf; d. scale; e. achene (G&W).

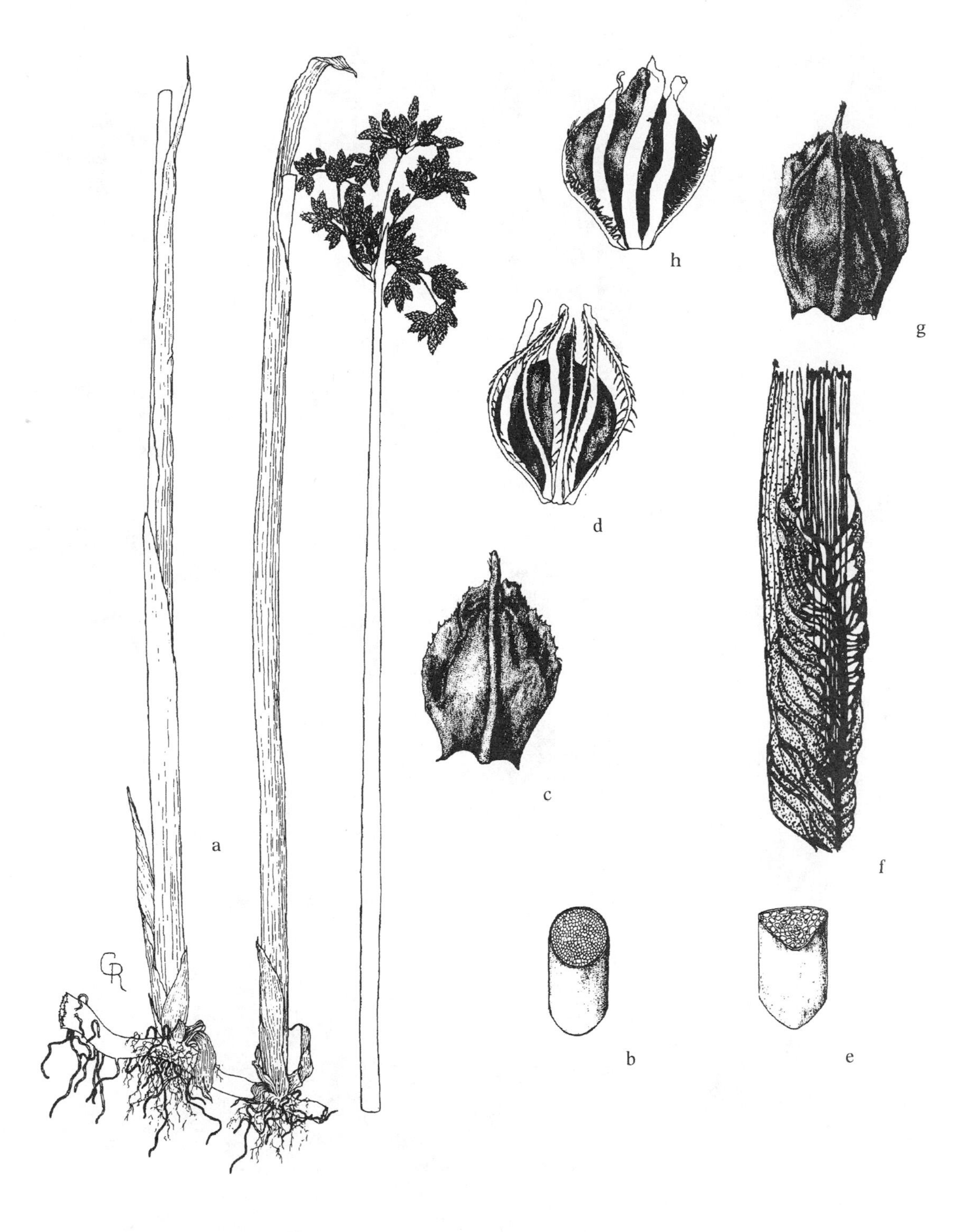

Fig. 263. *Schoenoplectus tabernaemontani*: a. habit; b. upper stem; cross-section; c. scale; d. achene (G&W). *Schoenoplectus californicus*: e. upper stem, cross-section; f. leaf sheath, coarsely pinnate-fimbrillose; g. scale; h. achene (G&W).

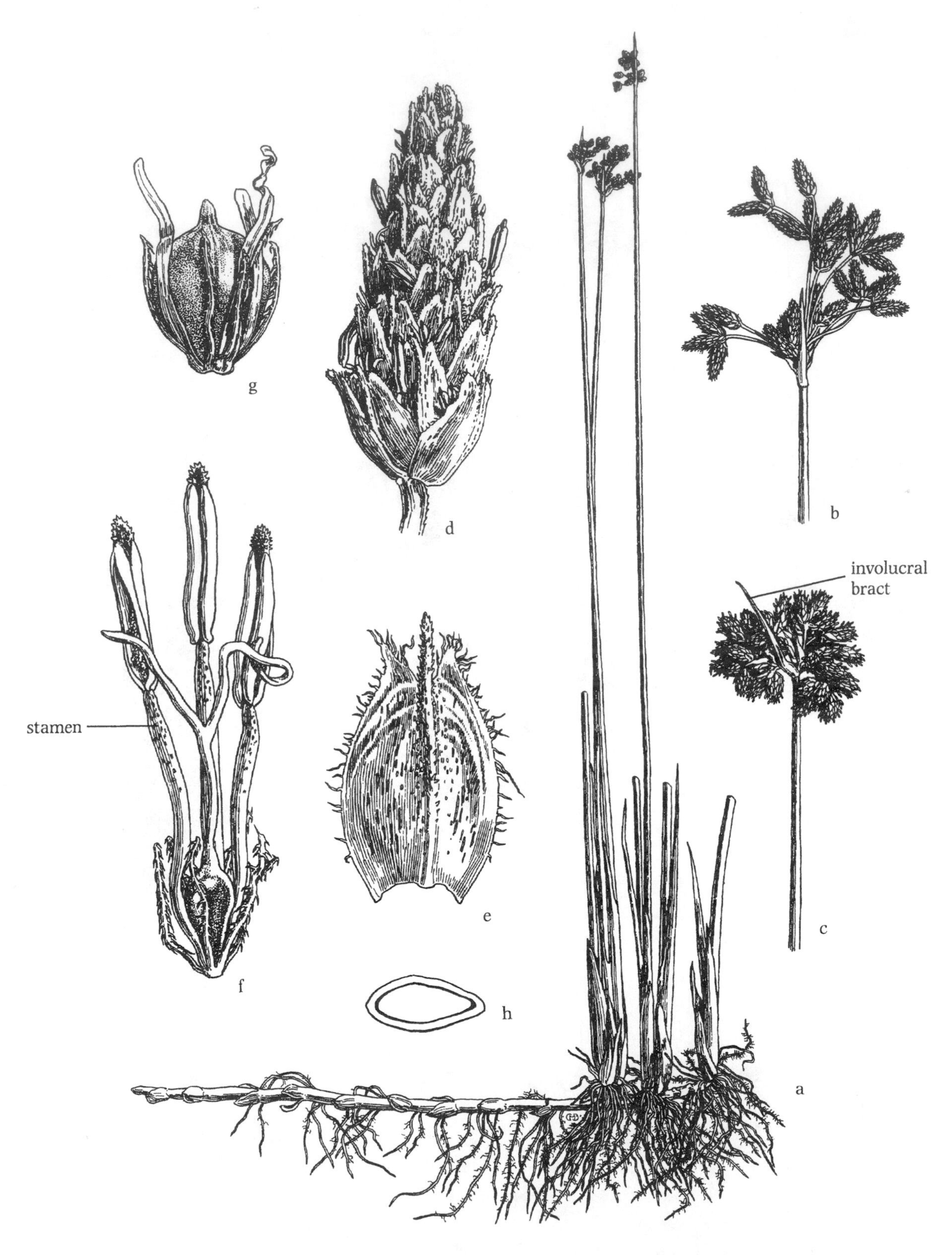

Fig. 264. *Schoenoplectus acutus*: a. habit; b, c. inflorescence variations; d. spikelet; e. scale; f. flower; g. achene with persistent stamen filaments and perianth bristles; h. biconvex achene, diagrammatic cross-section (Mason).

5. Achenes 3-angled (fig. 269c); styles 3-cleft.
 11. Summit of basal sheaths cut at an oblique angle. . . . . 10. *E. fallax*
 11. Summit of basal sheaths cut square (fig. 269d).
  12. Achenes with keel-like angles (fig. 269c). . . . . 11. *E. tricostata*
  12. Achenes lacking keel-like angles.
   13. Stem flattened. . . . . 12. *E. compressa*
   13. Stem angle or wiry.
    14. Tip of upper sheaths whitened, achenes 0.7–1 mm long. . . . . 13. *E. nitida*
    14. Tip of upper sheaths dark-margined; achenes 0.9–1.5 mm long.
     15. Achene surface shallowly pitted; stems 6–8-angled. . . . . 14. *E. elliptica*
     15. Achene surface deeply pitted (fig. 270c,e); stems 4- or 5-angled.
      16. Stems winged; plants 3–9 dm tall. . . . . 15b. *E. tenuis* var. *pseudoptera*
      16. Stems not winged; plants rarely more than 3 dm tall.
       17. Tubercles sharp-pointe (often a fifth as long as achene body) (fig. 270c); stems 4-angled. . . . . 15a. *E. tenuis* var. *tenuis*
       17. Tubercles flattened (fig. 270e); stems 5-angled. . . . . 15c. *E. tenuis* var. *verrucosa*

4. Plants with rhizomes thread-like (fig. 271a,b, 272a) or absent.
 18. Spikelets 3–9 flowered, rarely as many as 15 flowers.
  19. Stems arising from creeping rhizomes (fig. 271b), stems nearly all the same length; achenes grey or brown.
   20. Tubercles distinct from achene body (fig. 271d, f).
    21. Stems spongy; scales light green; stamens 2, anthers 0.3–0.4 mm long. . . . . 16. *E. radicans*
    21. Stems not spongy; scales green with brown marking; stamens 3, anthers 1 mm long. . . . . 17. *E. acicularis*
   20. Tubercles confluent with achene body (fig. 272c,h).
    22. Scales dark-brown, with pale, scarious margins; stems solitary or few; arising from creeping rhizomes; achenes 2–3 mm long. . . . . 18. *E. quinqueflora*
    22. Scales green to yellowish, with dull brown margins; stems many, forming dense mats; achenes 1–1.5 mm long. . . . . 19. *E. parvula*
  19. Stems tufted, rhizomes absent, usually some stems much shorter than others (fig. 273e); achenes olive or yellowish. . . . . 20. *E. intermedia*
 18. Spikelets many-flowered.
  23. Achenes biconvex; styles 2- and 3-cleft.
   24. Summit of basal sheaths white, membranous, loose (fig. 273d).
    25. Achenes 0.8–1 mm long, reddish-brown when young, deep purplish-black when mature; larges scales ca. 2 mm long. . . . . 21a. *E. flavescens* var. *flavescens*
    25. Achenes 1 mm or more long, olive-green; largest scales ca. 2.5 mm long. . . . . 21b. *E. flavescens* var. *olivacea*
   24. Summit of basal sheaths dark-margined, firm, tightly sheathing (fig. 275d).
    26. Plants with rhizomes; tubercles much-narrowed at base (fig. 267). Unusually slender plants of the *Palustres* complex may key out here; return to couplet 6 (*E. palustris*, *E. macrostachya*, *E. erythropoda*, *E. ambigens* and *E. uniglumis*).
    26. Plants lacking rhizomes (fig. 274a); tubercles broader at base (fig. 274c).
     27. Mature achenes black to purplish, 0.7–1 mm long; summit of basal sheaths cut at an angle, triangular tips being longer than diameter of sheaths (fig. 274d,e). . . . . 22. *E. geniculata*
     27. Mature achenes pale brown to deep brown, 1.1–1.5 mm long; summit of basal sheaths cut nearly square, triangular tips being shorter than diameter of sheaths (fig. 275d).
      28. Bristles 2–4 or absent; floral scales colorless to pale brown.
       29. Floral scales rounded, lacking a keel at apex; achenes with perianth bristles 2–4 or absent, tubercles 0.2–0.3 mm long. . . . . 23. *E. aestuum*
       29. Floral scales acute, keeled toward apex; achenes with perianth bristles absent, tubercles 0.1–0.2 mm long. . . . . 24. *E. diandra*
      28. Bristles mostly 6(5–8) or absent; floral scales orange brown to straw-colored.
       30. Tubercle base slightly less than two-thirds as wide as width of achene (fig. 275e); scales purplish-brown. . . . . 25. *E. ovata*
       30. Tubercle base nearly as wide or as wide as the width of the tubercle (fig. 275c,f); scales brown to deep or reddish-brown.

31. Tubercles deltoid, a third to half the length of achene (fig. 275c); spikelets ovoid.........26. *E. obtusa*
31. Tubercles flattened, very short, not more than a fourth the length of the achene (fig. 275f); spikelets usually cylindrical.. 27. *E. engelmanii*
23. Achenes 3-angled; styles 3-cleft.
32. Summit of basal sheaths cut at strongly oblique angle (fig, 276d,i).
33. Achenes with longitudinal ridges; stems flattened.. 28. *E. wolfii*
33. Achenes lacking longitudinal ridges, stems nearly round in cross-section or 4-angled.
34. Tubercles about as wide as long (fig. 276h); stems various.
35. Bristles whitish to light brown extending to or falling short of tubercle of achene (fig. 276c): achenes white to grayish; stems weak, finely capillary; of varying lengths.......................... 29. *E. microcarpa*
35. Bristles reddish, extending beyond tubercle of achene (fig. 276h); achenes olive to olive-brown, maturing to dark brown; stems wiry, erect, all elongate and about equal in length............... 30. *E. albida*
34. Tubercles 3–4 times as long as wide (fig. 273g); stems weak, arching, some short and crowded at base. .... 20. *E. intermedia*
32. Summit of basal sheaths cut nearly square (to slightly oblique (fig. 277d,g).
36. Tubercle nearly or equally as large as achene body (fig. 277c). .... 31. *E. tuberculosa*
36. Tubercle much smaller than achene body.
37. Tubercles much smaller than achene body.
38. Achenes minutely wrinkled, pale yellow to orange; scales 1 mm long, stems capillary. ....13. *E. nitida*
38. Achenes smooth, shiny, black; scales 3–4 mm long; stems flattened, wiry.................. 32. *E. melanocarpa*
37. Tubercles as long as or longer than wide.
39. Achene body smooth; stems thick, often arching and rooting at tips.........................33. *E. rostellata*
39. Achene body reticulate; stems filiform.
40. Achenes 1.3–1.7 mm long, body strongly honeycomb-reticulate; scales 2.5–3 mm long, stems not rooting at tips...34. *E. tortilis*
40. Achenes 0.6–0.8 mm long, body finely reticulate; scales 1.5–2 mm long, stems often arching, rooting at tips...35. *E. vivipara*

---

1. *E. equisetoides* (Elliott) Torr. Jointed Spike-rush; Horsetail Spike-rush; Knotted Spike-rush Fig. 265
Shallow water of marshes, lakes, ponds and ditches. Coastal plain, e. Mass. and Conn. s. to n. Fla., w. to e. Tex.; inland n. to Ont., Mo. and the Great Lakes region.

2. *E. quadrangulata* (Michx.) Roem. & Schult. Square-stem Spike-rush Fig. 266
Shallow water of marshes, lakes and ponds; often tidal along coast. N.H., and e. Mass. w. to Ont., Mich. and Wisc., s. to Fla., Okla., Tex. and Mex.; Oreg. and Calif. (*E. quadrangulata* var. *crassior* Fernald)

3. *E. robbinsii* Oakes Triangle Spike-rush, Robbin's Spike-rush Fig. 266
Shallow water, often submersed, forming large mats in ponds, lakes and marshes. N.S. w. to Ont. and n. Wisc., s. to c. N.Y. and n. Ind.; s. along the coastal plain to n. Fla. and sw. Ala. In the vegetative state this species is easily confused with two other aquatics that produce fine hair-like submersed leaves, *Scirpus subterminalis* and *Juncus militaris.*

4. *E. palustris* (L.) Roem. & Schult. Common Spike-rush Fig. 267
Shallow water and along shores of streams, rivers, lakes and ponds. Greenl. w. to Alask., s. to S.C., Miss., Tex. and Calif. This taxon is a member of an extremely variable, worldwide complex, including *E. macrostachya*, *E. erythropoda*, *E. ambigens* and *E. uniglumis* within our range; several infraspecific taxa are recognized for *E. palustris* in northern Europe, and although at least four variants occur in North America, Smith et al. (2002) formally recognized none. Haines (2011), on the other hand, recognized both subsp. *palustris* and subsp. *vigens* (L. H. Bailey) A. Haines, the latter having a slightly longer achene body and larger stem stomates. (*E. smallii* Britton)

5. *E. mamillata* (H. Lindberg) H. Lindberg Spike-rush Fig. 268
Shallow water and margins of streams, rivers, lakes and ponds. Scattered sites, sw. Que., w. to s. Ont., nw. Mich., nw. Wisc., and ne. Minn. (w. L. Superior); n. Sask. w. to Alta., n. Wash., B.C., N.W.T., Yuk. and Alask; Eur. Easily confused with *E. palustris*, but the tubercles of *E. mamillata* are always sessile, whereas the tubercles of *E. palustris* often have a distinct short neck; the stem of *E. mamillata* tends to be soft, whereas stems are more firm in *E. palustris*; these taxa also differ in stomate morphology (Smith and Gregor, 2014).

6. *E. macrostachya* Britton Pale Spike-rush Fig. 267
Marshes, ditches and shores. Que, w. to Yuk., and Alask, s. to Ill. and Minn., s. to La., Mex. and Calif. (*E. xyridiformis* Fernald & Brackett)

7. *E. erythropoda* Steud. Bald Spike-rush Fig. 268
Shallow water, wet shores and marshes. C.B.I. and Que. w. to N.W.T. and Alask. s. to N.E., Va., Tenn., Ga., Miss., Neb. and Okla. (*E. calva* Torr.; *E. palustris* var. *calva* (Torr.) A. Gray)

8. *E. uniglumis* (Link) Schult. Needle Spike-rush Fig. 269
Wet shores and marshes, salt marshes and saline or brackish shores. Greenl., Nfld. and Lab. w. to N.W.T. and Alask., s. to coastal N.E., L.I., N.Y., ne. N.J., e. Va. and ne. N.C., Ont., Neb.,

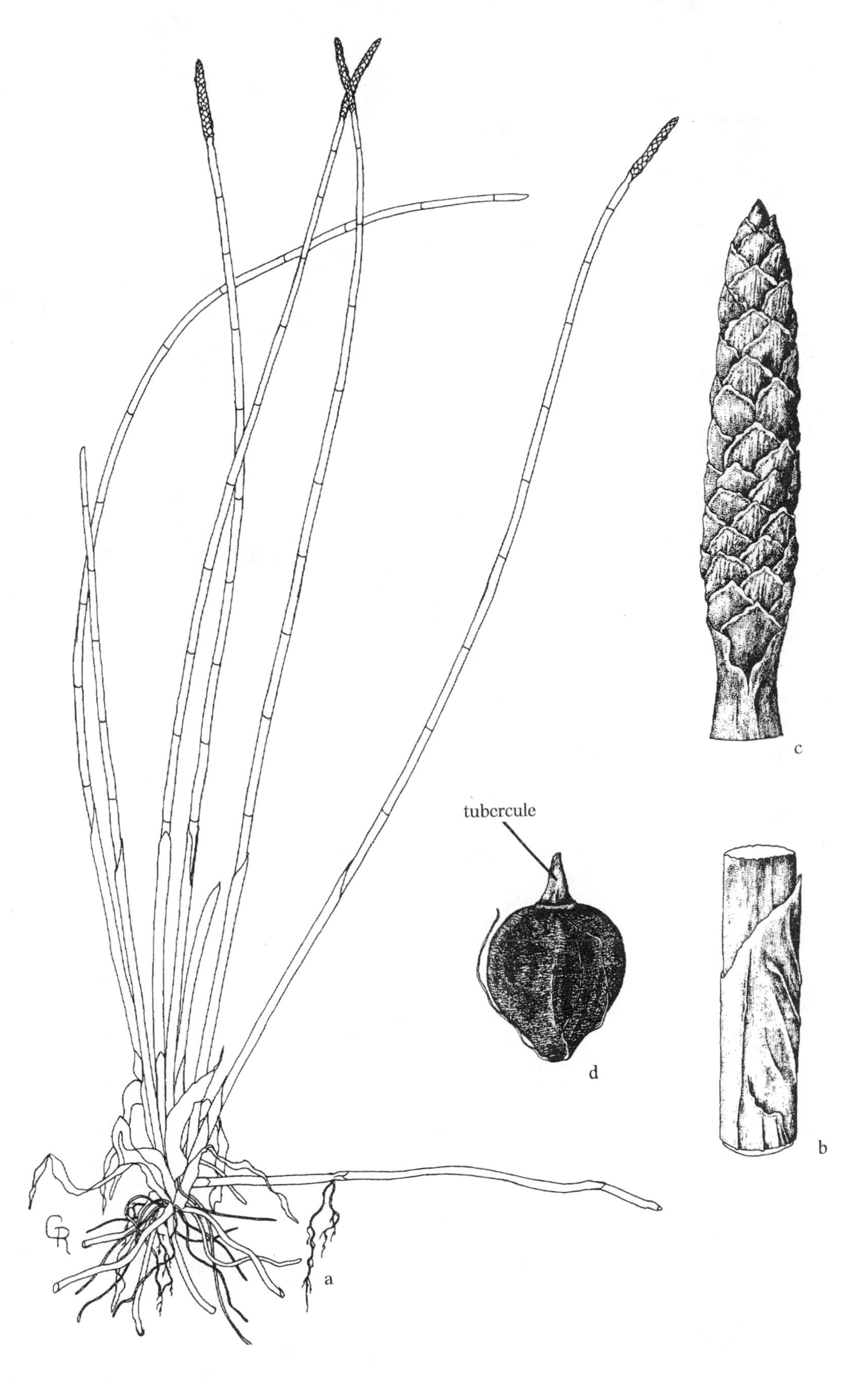

Fig. 265. *Eleocharis equisetoides*: a. habit; b. spikelet; c. achene; d. leaf sheath (G&W).

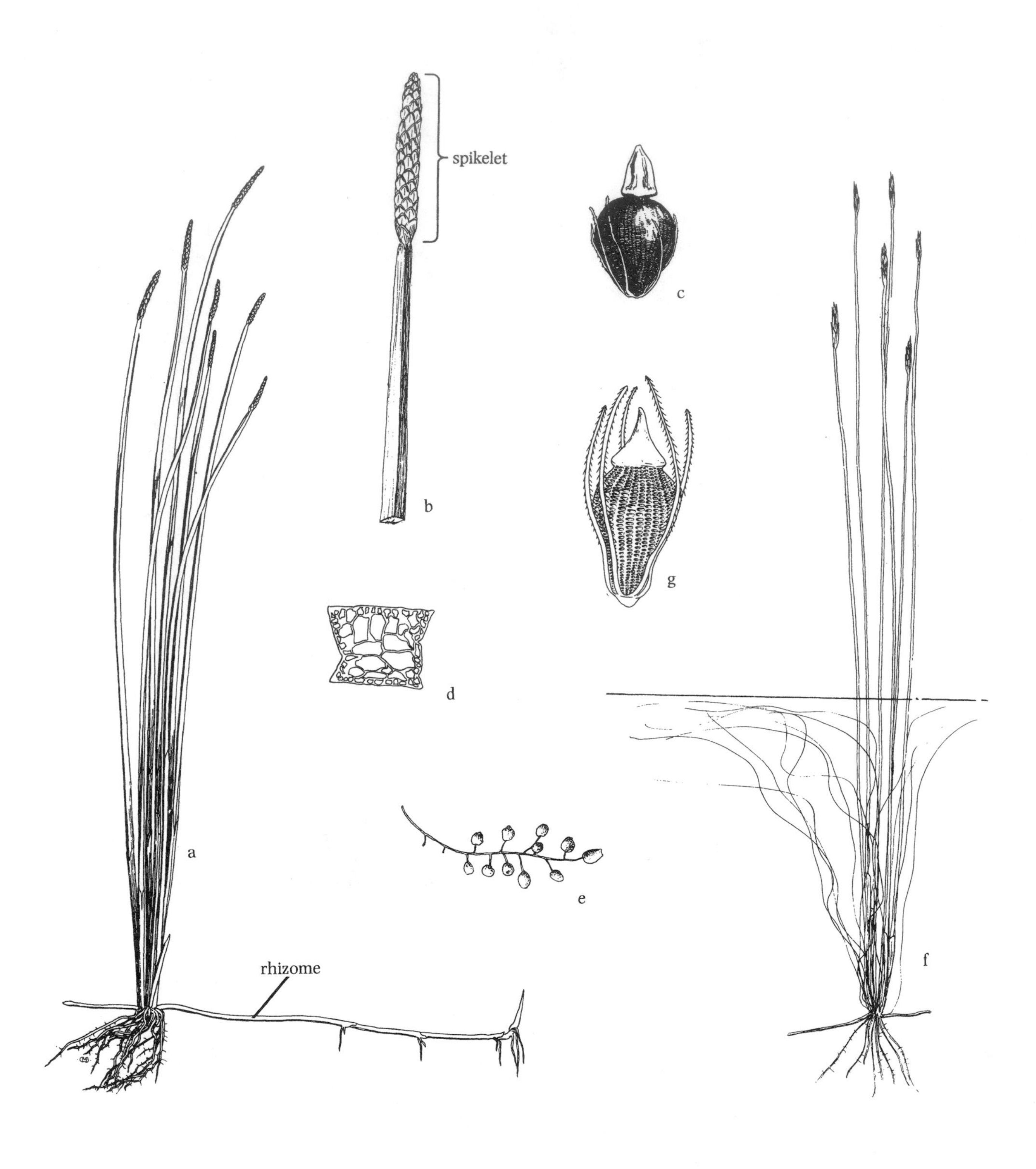

Fig. 266. *Eleocharis quadrangulata*: a. habit (Mason); b. spikelet (F); c. achene (Mason); d. stem, cross-section (Mason); e. tubers (F).
Eleocharis robbinsii: f. habit (F); g. achene (Beal).

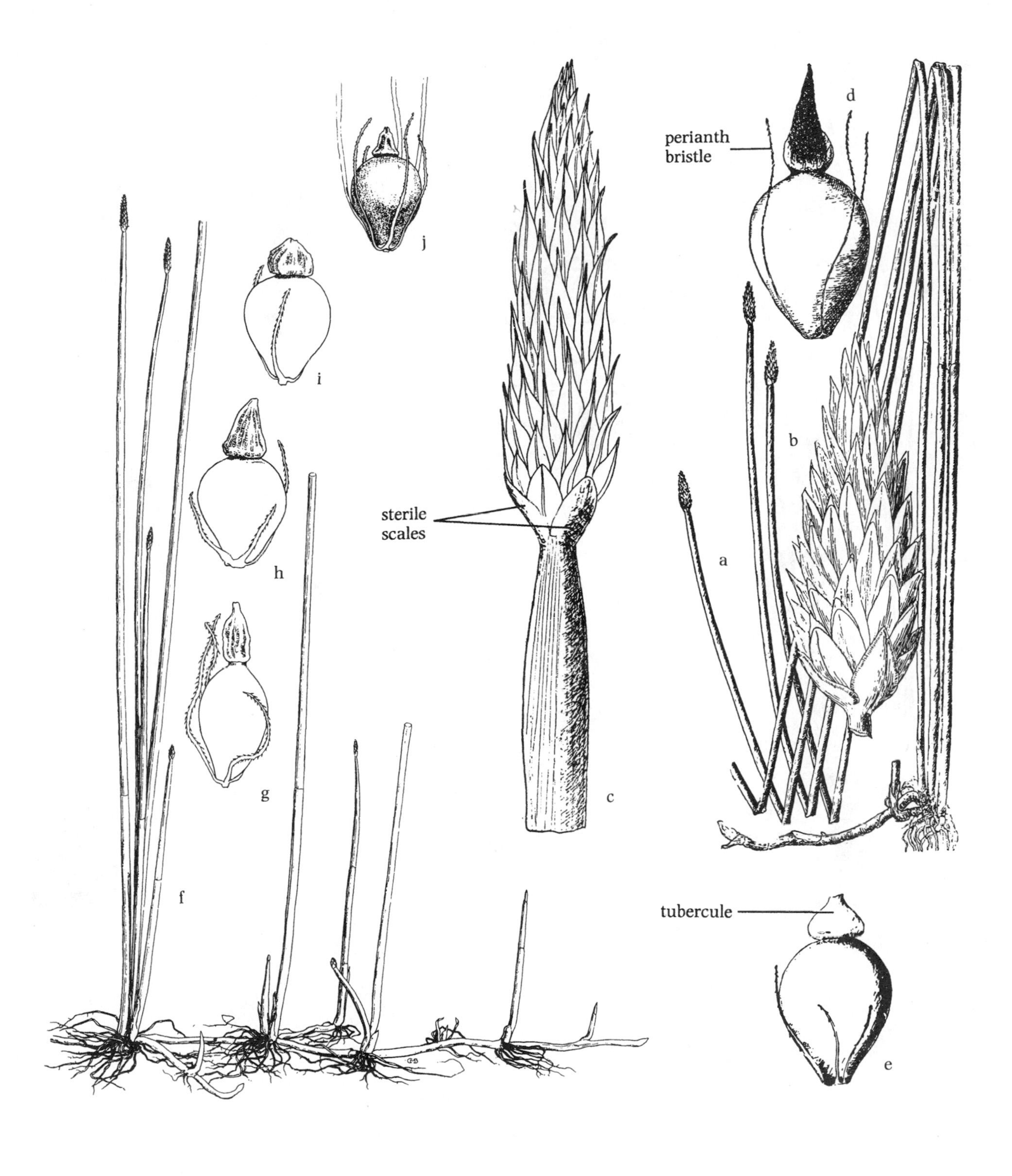

Fig. 267. *Eleocharis palustris*: a. habit; b, c. spikelets; d, e. achene variations (F). *Eleocharis macrostachya*: f. habit; g–j. achene variations (Mason).

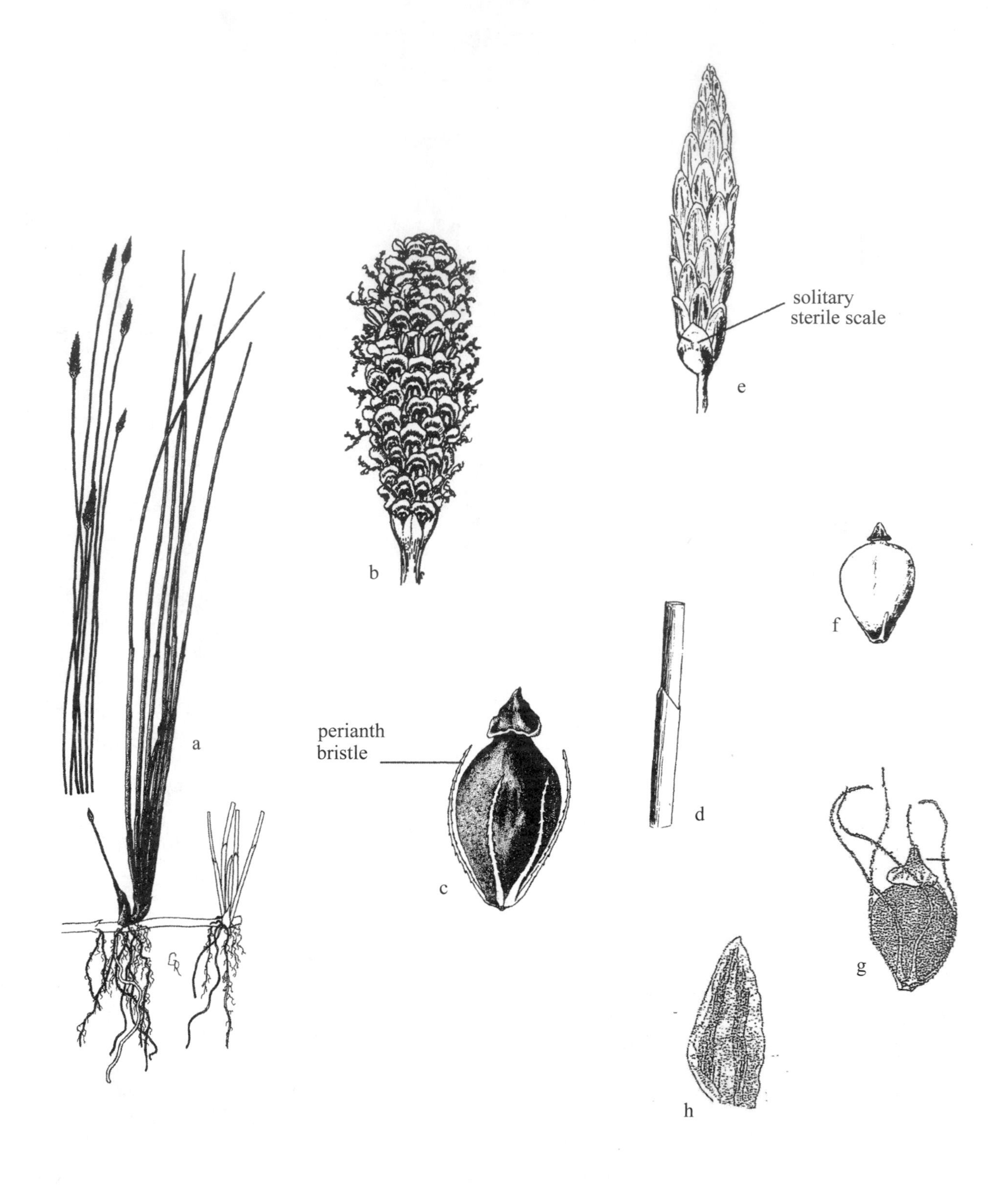

Fig. 268. *Eleocharis ambigens*: a. habit (G&W); b. spikelet (G&W); c. achene (G&W); d. leaf sheath (F).
Eleocharis erythropoda: e. spikelet; f. achene (F).
Eleocharis mamillata: g. habit; h. scale (FNA).

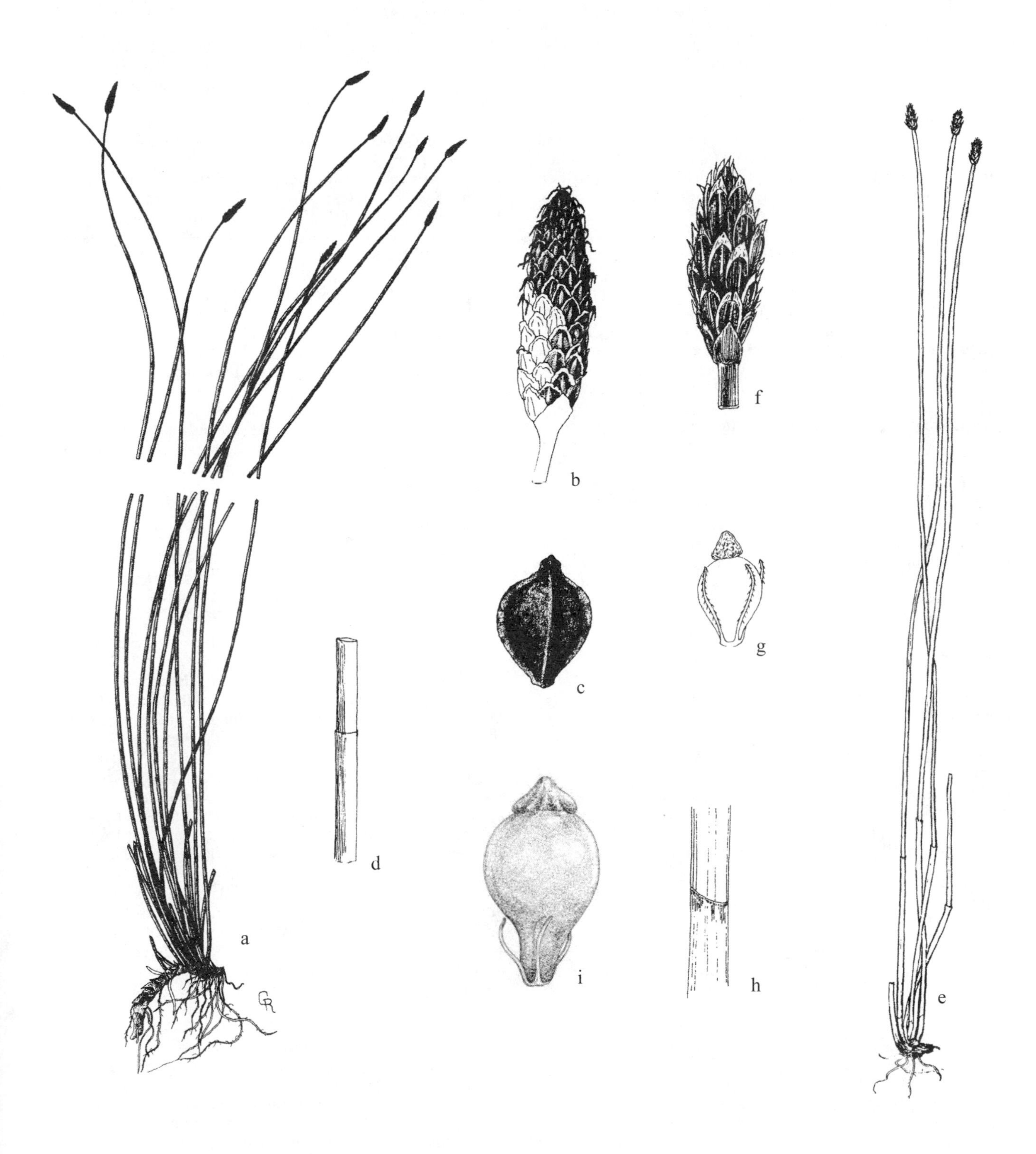

Fig. 269. *Eleocharis tricostata*: a. habit (G&W); b. spikelet (G&W); c. achene (G&W); d. leaf sheath (F).
Eleocharis compressa: e. habit (Svenson, 1932b); f. spikelet (F).
Eleocharis uniglumis: g. achene; h. leaf sheath (Beal).
Eleocharis aestuum: i. achene (Elizabeth Farnsworth, in *Flora Novae Angliae*, NPT).

Colo., Utah, Ida. and Oreg. Plants in saline habitats along coastal areas of Nfld. to Va. have been called *E. halophila*, but the achene characteristics that Fernald (1950) used to differentiate these as two species do not hold true. (*E. halophila* (Fernald & Brackett) Fernald & Brackett)

9. *E. ambigens* Fernald Fig. 268
Freshwater to brackish marshes, pond shores and disturbed places. Coastal plain, Que.; Mass. and Conn. s. to N.J., Va., N.C., and Fla., w. to Miss., La. and e. Tex. *Eleocharis ambigens* has been treated as a synonym of *E. fallax*, but Smith et al. (2002) regard the latter as a distinct species. Furthermore, they regard *E. ambigens* as more closely aligned with *E. uniglumis* and perhaps should be included within that taxon.

10. *E. fallax* Weath. Creeping Spike-rush
Fresh to brackish pond and lake margins and marshes. Endemic to N.S., Mass. and N.J. Smith et al. (2002) noted that the taxonomic status of *E. fallax* is problematic and although it has been often treated as conspecific with *E. ambigens* (differing by having trigonus achenes with 3-fid styles, with tubercles longer than wide, and a rugulose achene surface), they view it best to recognize *E. fallax* as distinct pending further research.

11. *E. tricostata* Torr. Three-angle Spike-rush Fig. 269
Sandy and peaty shores of freshwater to brackish marshes. Coastal plain, se. Mass. s. to N.J., Va. and Fla.; disjunct in w. Mich.

12. *E. compressa* Sulliv. Flatstem Spike-rush Fig. 269
Shallow water, calcareous gravels, sands, and peats. W. Que. w. to Sask., N.W.T. and B.C. s. to N.Y., ne. Va., Ga., n. Ala., Mo., ne, Colo., Tex. and N.M. (*E. elliptica* var. *compressa* (Sulliv.) Drapalik & Mohlenbrock)

13. *E. nitida* Fernald Quill Spike-rush Fig. 270
Damp peaty and sandy sites. Nfld. and Que. w. to Wisc. and ne. Minn., s. to N.S., Me., n. N.H. and Vt.; B.C. and the Aleutian Islands., Alask.

14. *E. elliptica* Kunth Elliptic Spike-rush Fig. 270
Damp marly ground, shores, swales and ditches. Nfld. w. to Alask., Minn., s. to N.E., n. N.J., Ohio, Iowa; infrequent w. to Man., s. Alta., nw. Mont. and sw. B.C. (*E. compressa* var. *atrata* Svenson)

15. *E. tenuis* (Willd.) Schult. Slender Spike-rush

15a. *E. tenuis* var. *tenuis* Slender spike-rush Fig. 270
Wet, damp, and dry sands gravels and peats. N.S. sw. to Ohio, s. to S.C., Ga., Miss. and Tex.

15b. *E. tenuis* var. *pseudoptera* (Weath.) Svenson Fig. 270
Meadows, N.S., Que., L.I., N.Y. and N.J. s. to Va., mts. of N.C. and e. Tenn., Miss. and La. (*E. capitata* var. *pseudoptera* Weath.; *E. elliptica* var. *pseudoptera* (Weath.) L. J. Harms)

15c. *E. tenuis* var. *verrucosa* (Svenson) Svenson) Fig. 270
Moist to wet sandy and peaty soils. Pa., se. Va. and Ky. w. to s. Ind., Ill., S.D., Iowa and Neb. s. to Ark., e. Okla. and e. Tex. (*E. capitata* var. *verrucosa* Svenson; *E. verrucosa* (Svenson) L. J. Harms)

16. *E. radicans* (Poir.) Kunth Rooted Spike-rush Fig. 271
Wet, sandy and mucky shores. Local; se. Va., and Mich.; Okla. and e. Tex. w. to Ariz., Calif. and Mex.; W.I. and S.Am.

17. *E. acicularis* (L.) Roem. & Schult. Needle Spike-rush Fig. 271
Wet sands and mucky shores, often submersed. Greenl. w. to Alask., s. to Fla., Okla., n. Mex. and s. Calif. This species frequently remains vegetative, forming large mats. (*E. acicularis* var. *submersa* (Nilsson) Svenson)

18. *E. quinqueflora* (Hartm) O. Schwarz Few-flowered Spike-rush Fig. 272
Damp, marly shores, ledges and swamps. Greenl. and Nfld. w. to n. Ont., Sask. and B.C., s. to Mass., nw. N.J., N.Y., nw. Pa., Ind., Iowa, Colo., N.M. and Calif. Our plants are separated as subsp. *fernaldii* (Svenson) Hultén from the Eurasian plants. (*E. pauciflora* (Lightf.) Link)

19. *E. parvula* (Roem. & Schult.) Link Dwarf Spike-rush Fig. 272
Wet saline or brackish shores. Nfld. s. along the coast to Fla., w. to Tex. and Mex.; locally inland w. to N.Y., Mich., w. Ont., Minn., Mo., Sask. and B.C., s. along the Pacific Coast from B.C. to Calif.; W.I. and S.Am.

20. *E. intermedia* Schult. Matted Spike-rush Fig. 273
Wet calcareous soils, stream margins, marshy ground and mud flats. E. Que., N.B., Ont., w. to Minn., s. to N.E., n. N.J., Md., W.Va., Tenn., Ill., Iowa and Miss.

21. *E. flavescens* (Poir.) Urban

21a. *E. flavescens* var. *flavescens* Yellow Spike-rush Fig. 273
Wet peats and sands. Coastal plain, N.J. s. to Fla., w. to Ark., e. Tex.; Mont. and Wyo. s. to Calif. and Ariz.; Mex., W.I. and S.Am.

21b. *E. flavescens* var. *olivacea* (Torr.) Gleason Bright Green Spike-rush Fig. 274
Wet sands, peats and mud flats, occasionally in flowing water. Coastal, N.S. s. to Fla.; local inland, Me.; w. to s. Ont., Mich. and Minn., s. to w. Pa. and Ohio, south along coast to Fla. w. to Tex. (*E. olivacea* Torr.)

22. *E. geniculata* (L.) Roem. & Schult. Canada Spike-sedge Fig. 274
Wet, fresh and brackish sands and gravels. B.C., Great Lakes region of s. Ont., Mich. and nw. Ind.; S.C. s. to Fla., w. to Tex.; N.M., Ariz. and Nev., s. Calif.; Trop. Am. (*E. caribaea* (Rottb.) Blake)

23. *E. aestuum* Hines ex A. Haines Tidal Spike-rush Fig. 269
Fresh tidal river shores; rarely inland on muddy shores of lakes and reservoirs. Rare, s. Que., Me., Vt., Conn., N.Y., N.J. and Pa. This recently described species tends to be associated with shores of tidal rivers, especially sand, silt and muddy substrates. Several records of the very similar *E. diandra* from Me. are now referable to *E. aestuum* (Haines, 2011).

24. *E. diandra* C. Wright Wright's Spike-rush
Sandy shores of large lakes and rivers with greatly fluctuating water levels. Endemic, exceedingly rare, N.H., Vt., Mass., Conn., N.Y. and s. Ont. In New England known from Andoscoggin R (N.H.), Connecticut R. (Mass., Conn.), Merrimack R. (Mass.) and L. Champlain (Vt.); recent collections known from Connecticut River in Mass. (1985) and Oneida Lake, N.Y. (1968) (Haines, 2011).

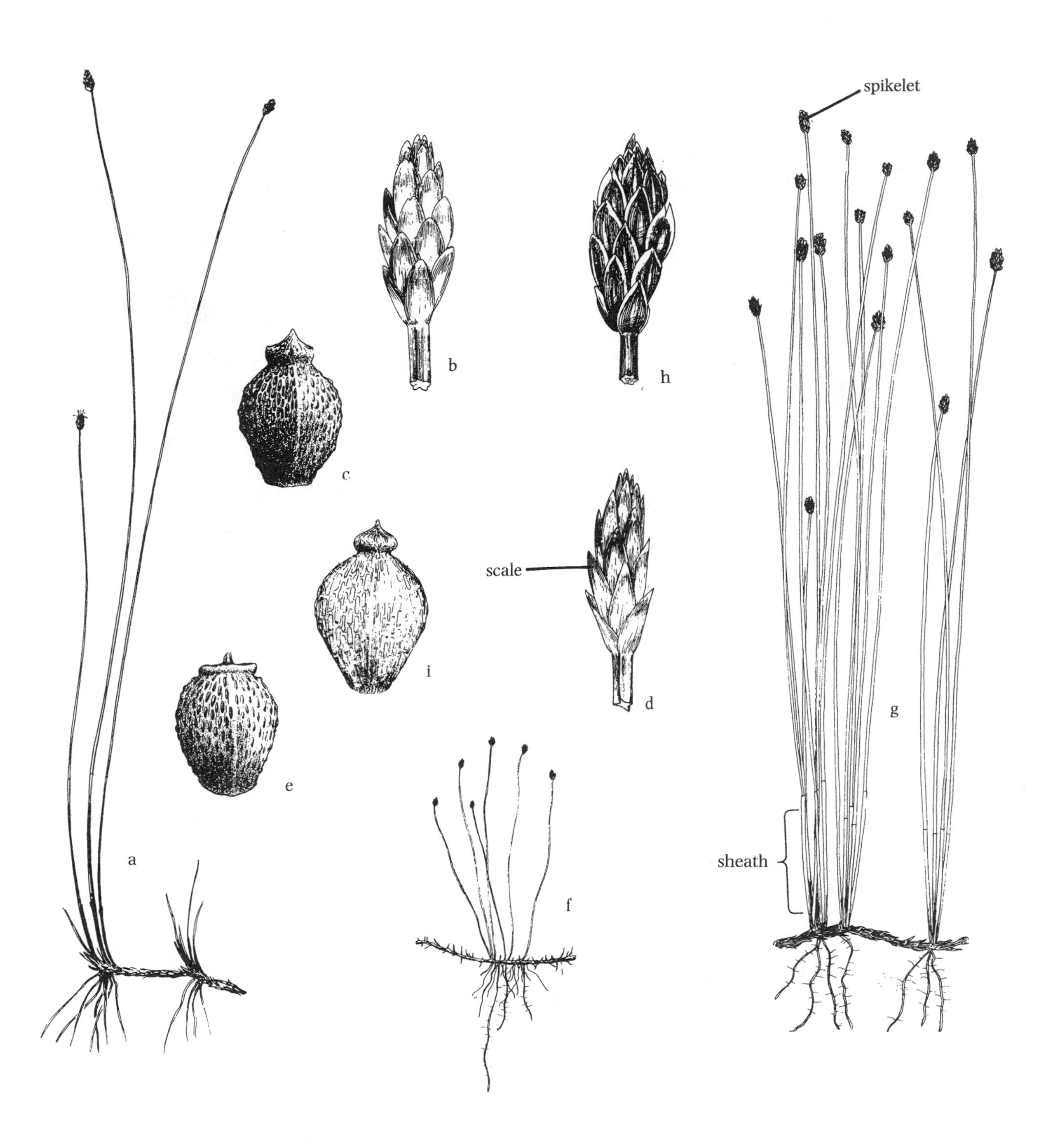

Fig. 270. *Eleocharis tenuis* var. *tenuis*: a. habit (HCOT); b. spikelet (F); c. achene (F).
Eleocharis tenuis var. *pseudoptera*: d. spikelet (F).
Eleocharis tenuis var. *verrucosa*: e. achene (F).
Eleocharis nitida: f. habit (F).
Eleocharis elliptica: g. habit; h. spikelet; i. achene (F).

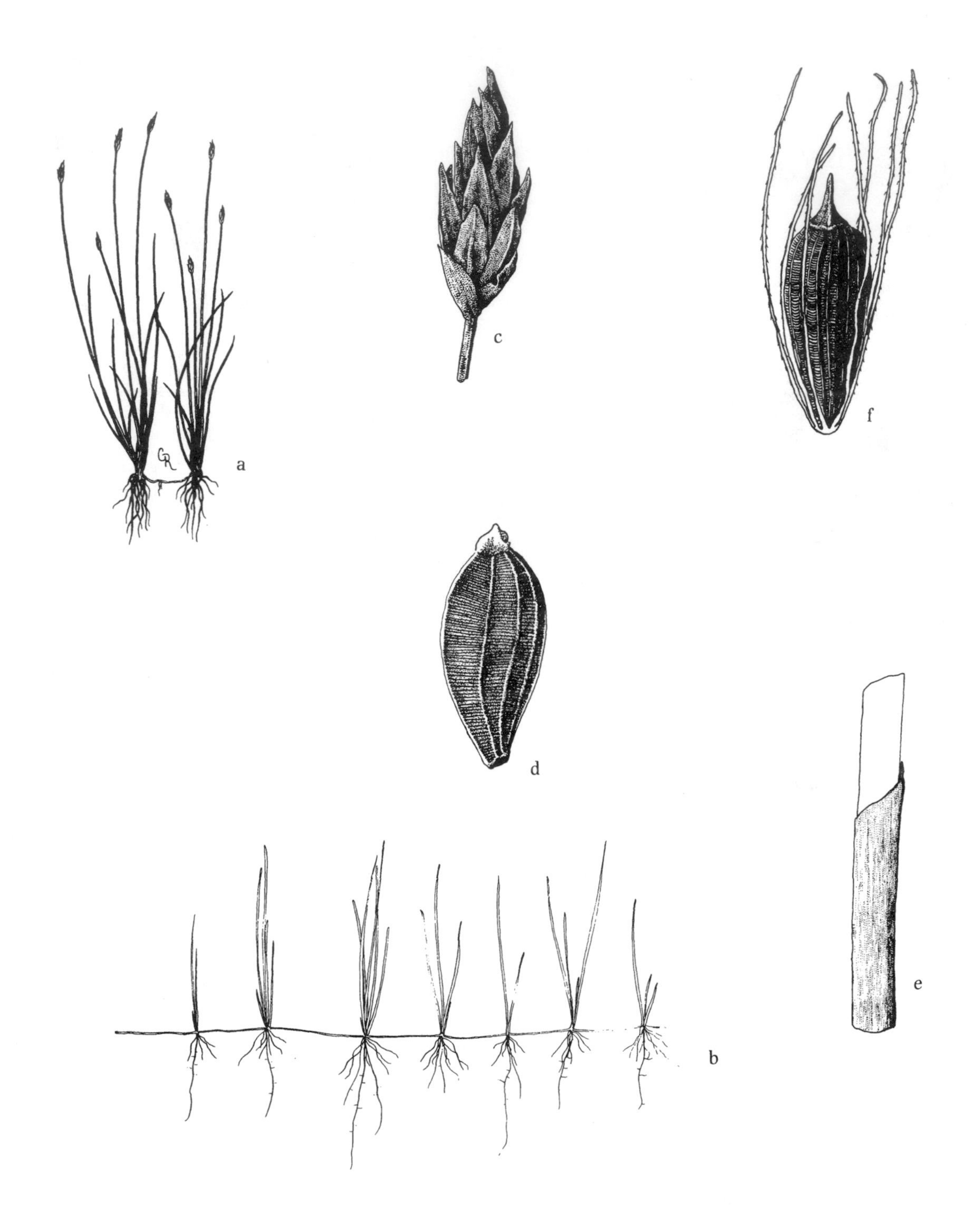

Fig. 271. *Eleocharis acicularis*: a. habit (G&W); b. habit, submersed plant (F); c. spikelet (G&W); d. achene (G&W); e. leaf sheath (G&W).
Eleocharis radicans: f. achene (G&W).

Fig. 272. *Eleocharis quinqueflora*: a. habit; b. spikelet; c. achene (F).
Eleocharis parvula: d. habit (F); e. habit (G&W); f, g. spikelets (G&W); h. achene (G&W); i. leaf sheath (G&W).

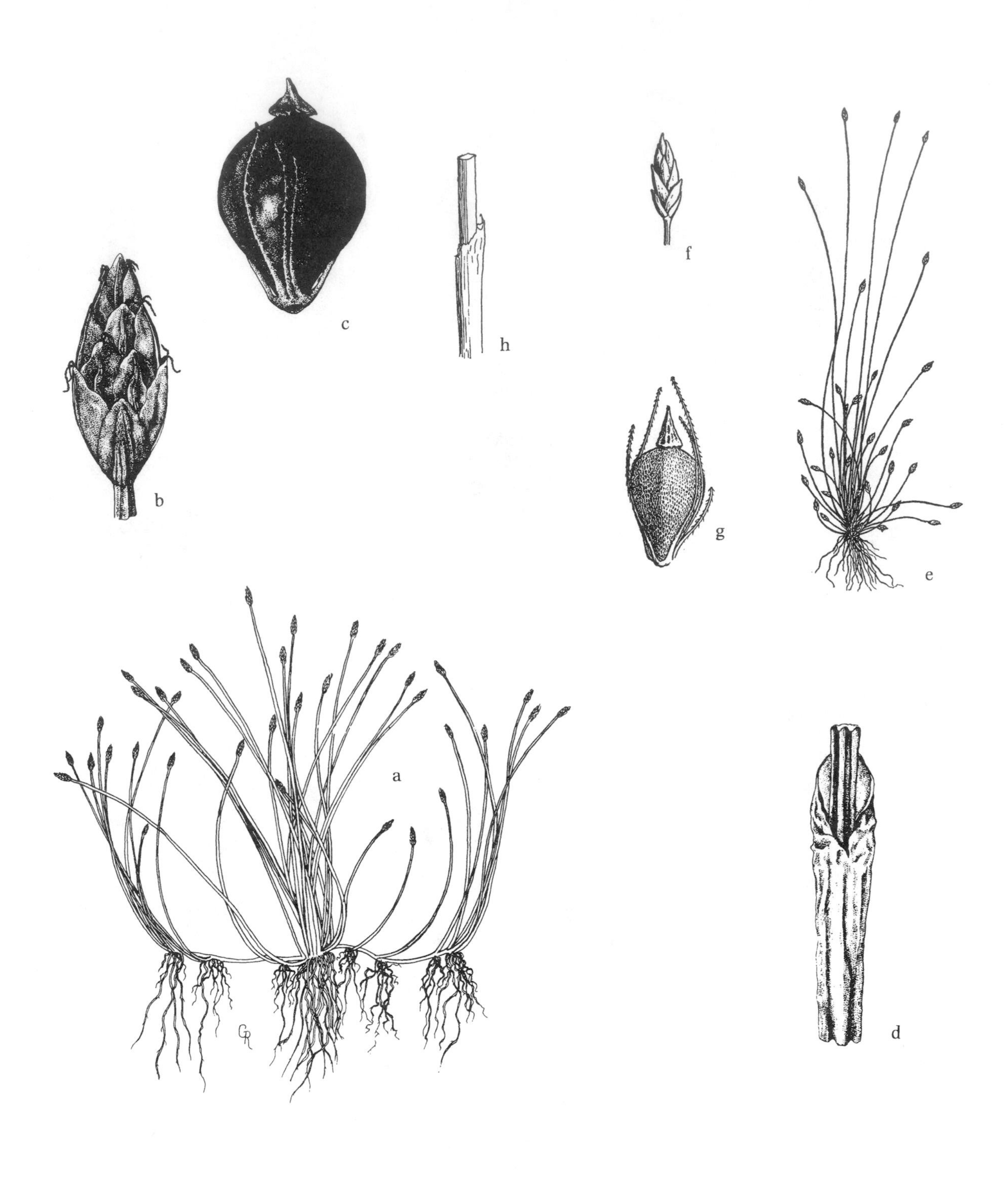

Fig. 273. *Eleocharis flavescens* var. *flavescens*: a. habit; b. spikelet; c. achene; d. leaf sheath (G&W). *Eleocharis intermedia*: e. habit; f. spikelet; g. achene; h. leaf sheath (F).

Fig. 274. *Eleocharis geniculata*: a. habit; b. spikelet; c. achene; d, e. leaf sheath, two views (G&W, as *E. caribaea*). *Eleocharis flavescens* var. *olivacea*: f. habit; g. achene; h. leaf sheath (G&W, as *E. olivacea*).

25. *E. ovata* (Roth) Roem. & Schult. Ovate Spike-rush Fig. 275
Muddy and wet open areas. Local; Nfld. w. to Minn., s. to ne. Mass., Conn., c. N.Y. and n. Ind, Mo. and Okla.; Man., Sask., B.C., Wash. and Oreg.; Ariz. (*E. obtusa* var. *ovata* (Roth) Drapalik & Mohlenbr.)

26. *E. obtusa* (Willd.) Schult. Blunt Spike-rush Fig. 275
Muddy, sandy and peaty shores, often in shallow water, P.E.I. and N.S. w to Alta. and Minn., s. to nw. Fla., Miss., e. Tex., Colo. and N.M.; Ida. and B.C. s. to n. Calif. (*E. obtusa* var. *ellipsoidalis* Fernald; *E. obtusa* var. *jejuna* Fernald; *E. obtusa* var. *peasei* Svenson)

27. *E. engelmannii* Steud. Engelmann's Spike-rush Fig. 275
Wet sand, peat and mud. S. Me. w. to Sask., Alta., B.C. and Wash. s. to e. Va., Ga., Fla., Ark., Tex., Ariz. and s. Calif. (*E. engelmanii* var. *robusta* Fernald; *E. obtusa* var. *engelmannii* (Steud.) Gilly)

28. *E. wolfii* (A. Gray) A. Gray Wolf's Spike-rush Fig. 275
Wet shores and ephemeral prairie ponds and buffalo wallows and river margins. Chiefly Great Plains region, Ohio, Ind. and Ill., w. to Wisc., Minn. and N.D., s. locally to Ga., Tenn., Ala., La., e. Tex., Okla. and ne. Colo.

29. *E. microcarpa* Torr. Small-fruit Spike-rush Fig. 276
Damp sands, swamps and shore, often in shallow water. Coastal plain, se. Mass., e. Conn. and se. N.Y., s. to Fla., w. to La. and Tex.; disjunct to Pa., Tenn., sw. Mich. and nw. Ind. Our taxon is var. *filiculmis* Torr.

30. *E. albida* Torr. White Spike-rush Fig. 276
Brackish marshes and saline shores. Coastal, Del., Md. and e. Va. s. to Fla., w. to La. and e. Tex. and e. Mex.; Ber.

31. *E. tuberculosa* (Michx.) Roem. & Schult. Cone-cup Spike-rush Fig. 277
Wet, sandy and peaty shores, swamps and ditches. Chiefly coastal plain and piedmont, N.S., e. N.E. and se. N.Y. s. to Fla., w. to Ark. and se. Tex.; c. Tenn. and s. Ky.

32. *E. melanocarpa* Torr. Blackfruit Spike-rush Fig. 277
Sandy, peaty and mucky shores. Coastal plain and piedmont, se. Mass., R.I. and s. N.Y. to se. Va. and n. Fla., w. to e. Tex.; disjunct in sw. Mich. and n. Ind.

33. *E. rostellata* Torr. Walking Sedge, Beaked Spike-rush Fig. 278
Saline, brackish and calcareous marshes bogs and swamps. Coastal, N.S. and Me. s. to N.J., Va., N.C., n. Fla. and Ala.; inland w. N.Y. and Ont. w. to Ohio, Mich., n. Ind., n. Ill. and Wisc., Minn., N.D.; chiefly a western species, sw. S.C. and Mont., w. to B.C. and Wash., s. to Kans., Okla., Tex., N.M., Ariz. and Calif.; Mex., Ber., W.I. and S.Am.

34. *E. tortilis* (Link) Schult. Twisted Spike-rush Fig. 279
Springy swamps, wet peat, wet woods and thickets. Coastal Plain, L.I., N.Y., s. N.J., Del. and Md. s. to n. Fla., w. to e. Tex.; sw. Tenn., s. Ark. and e. Okla.

35. *E. vivipara* Link Viviparous Spike-rush; Sprouting Spike-rush Fig. 280
Peaty shores, marshes, pools and ditches. Coastal plain, se. Va. s. to Fla., w. to e. Tex. Most abundant in Florida. (*E. curtissii* Small)

13. *Schoenoplectiella* (Annual Bulrush)

Annuals, occasionally perennial, cespitose with inconspicuous rhizomes; leaves basal, inflorescence terminal appearing lateral, spikelet scales with 3–11 distinct parallel ribs in part of spikelet, apex entire, mostly mucronate or cuspidate to acuminate; perianth of 6 small, barbed bristles or reduced or absent; achenes biconvex, plano-convex or 3-angled, rugulose.

Reference: Smith, 2002.

---

1. Scales not striate; tips awned or cuspidate to acuminate; achenes conspicuously wrinkled transversally (fig. 281f,g).
    2. Style 2-cleft; achenes plano-convex.......... 1. *S. hallii*
    2. Style 3-cleft; achenes 3-angled.......... 2. *S. saximontana*
1. Scales striate; tips blunt or slightly mucronate; achenes smooth of slightly pitted.
    3. Achenes smooth, plano-convex; beak abruptly differentiated from achene body; bristles lacking or, if present, slender.
        4. Perianth bristles lacking or vestigial.......... 3a. *S. smithii* var. *smithii*
        4. Perianth bristles present.
            5. Perianth bristles 4–6, equal to or twice as long as achene.......... 3b. *S. smithii* var. *setosa*
            5. Perianth bristles 1–4, shorter than or equaling achene.......... 3c. *S. smithii* var. *leviseta*
    3. Achenes slightly pitted, biconvex; beak gradually tapering from achene body; bristles stout.
        6. Flowers with perianth bristles.......... 4a. *S. purshiana* var. *purshiana*
        6. Flowers without perianth bristles.......... 4b. *S. purshiana* var. *williamsii*

---

1. *S. hallii* (A. Gray) Lye Hall's Bulrush Fig. 281
Peaty and sandy shore. Rare, in scattered localities, e. Mass., s. Mich., Ill., Wisc., s. Mo, Neb., Kans., Okla., Ky. and Ga. (*Schoenoplectus hallii* (A. Gray) S. G. Smith; *Scirpus hallii* A. Gray)

2. *S. saximontana* (Fernald) Lye Rocky Mountain Bulrush Fig. 281
Shores of ponds, depressions, river bottomlands and ditches. Ohio, Ill. and Mo. w. to S.D., Wyo. Utah, and Calif., s. to Okla. and

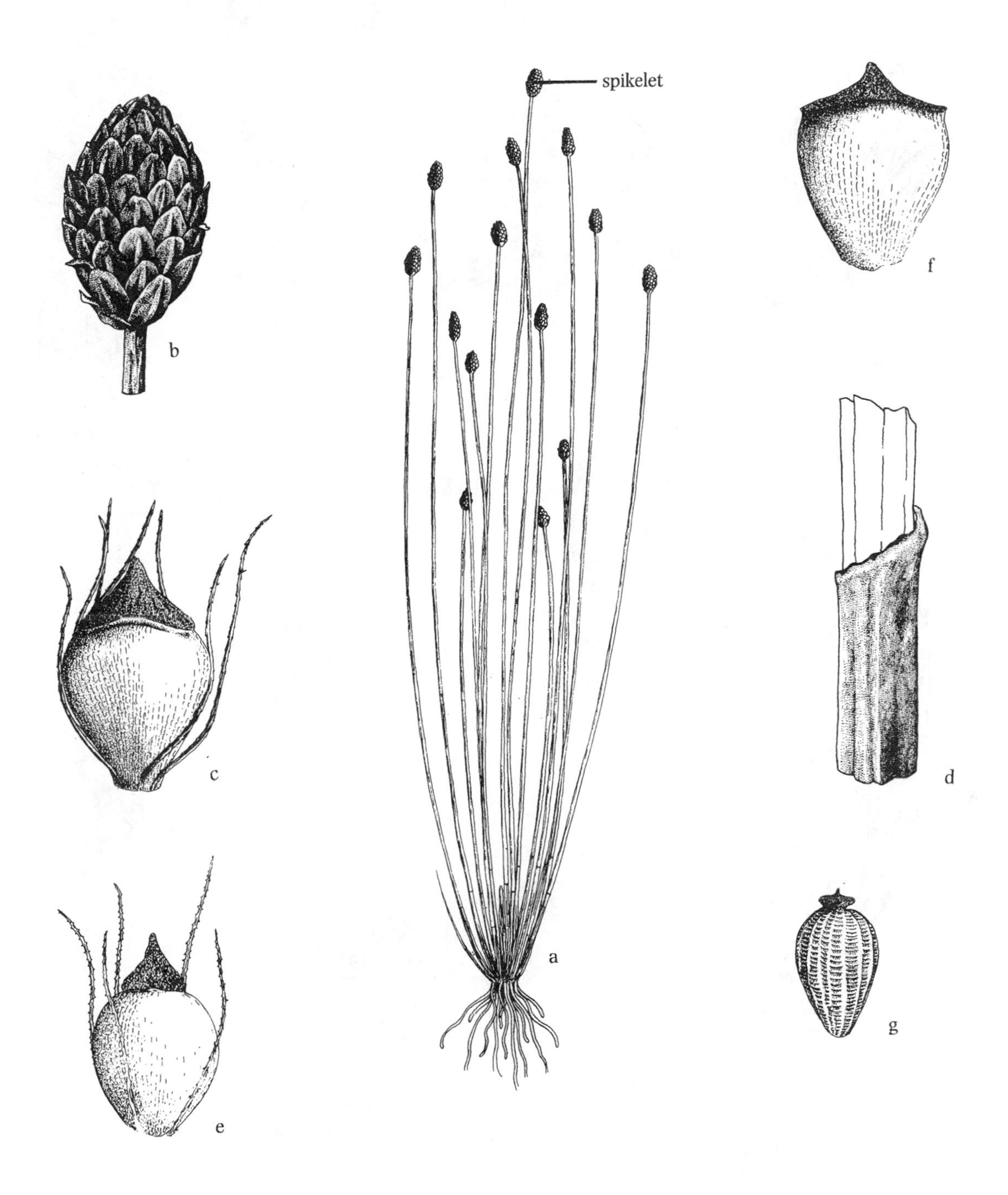

Fig. 275. *Eleocharis obtusa*: a. habit (F); b. spikelet (G&W); c. achene (F); d. leaf sheath (G&W).
Eleocharis ovata: e. achene (F).
Eleocharis engelmannii: f. achene (F).
Eleocharis wolfii: g. achene (F).

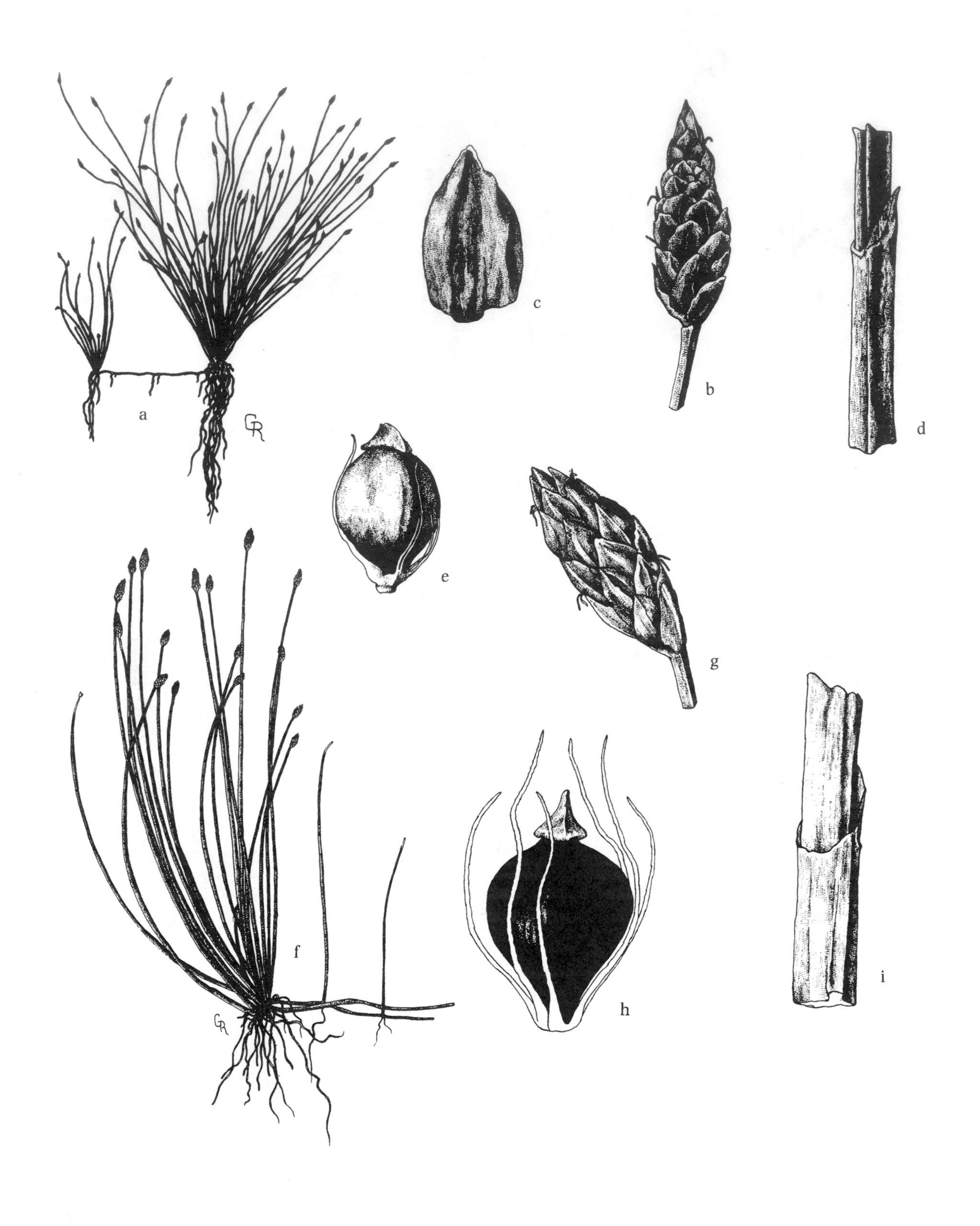

Fig. 276. *Eleocharis microcarpa*: a. habit; b. spikelet; c. scale; d. leaf sheath; e. achene (G&W).
Eleocharis albida: f. habit; g. spikelet; h. achene; i. leaf sheath (G&W).

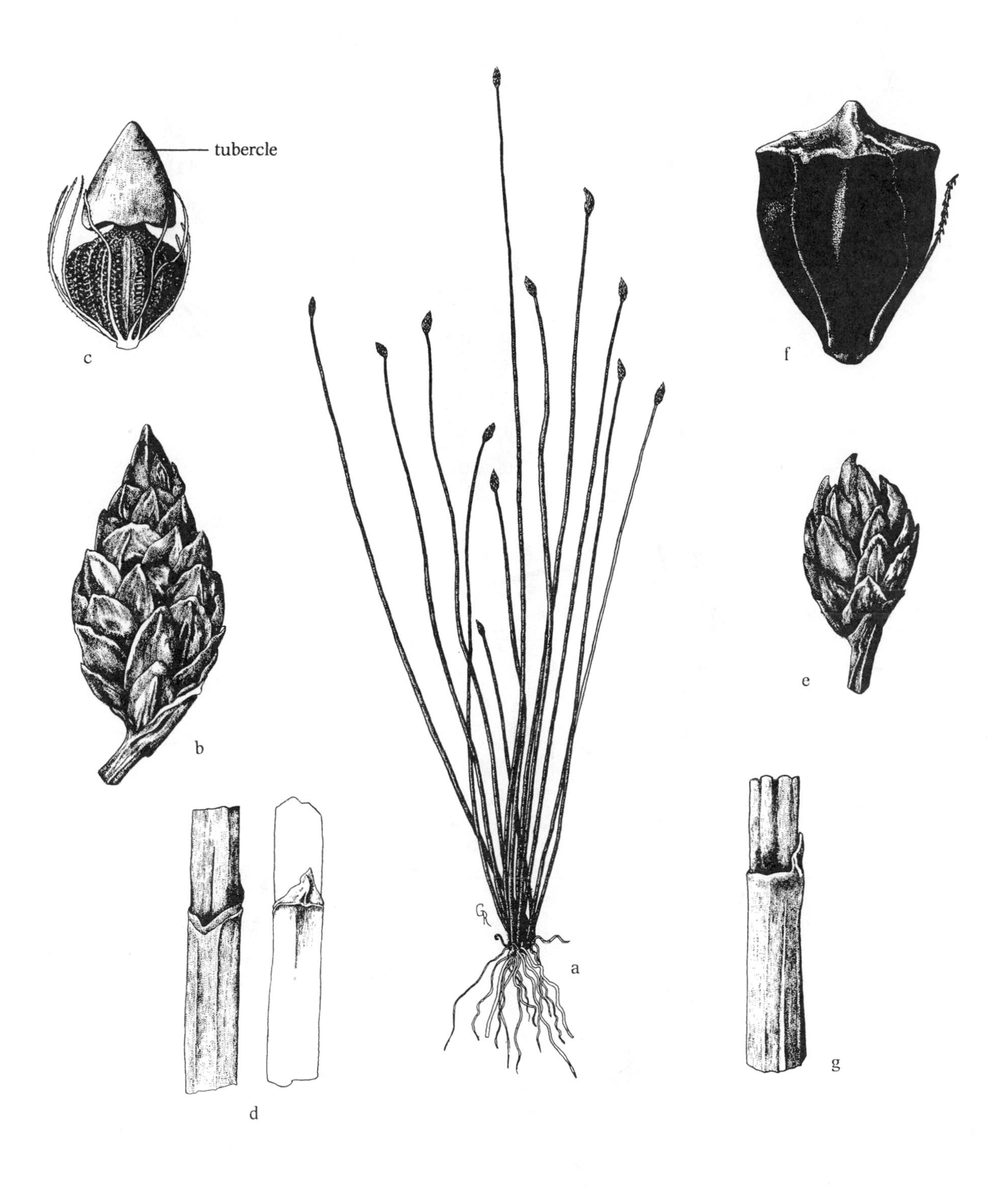

Fig. 277. *Eleocharis tuberculosa*: a. habit; b. spikelet; c. achene; d. leaf sheath (G&W).
Eleocharis melanocarpa: e. spikelet; f. achene; g. leaf sheath (G&W).

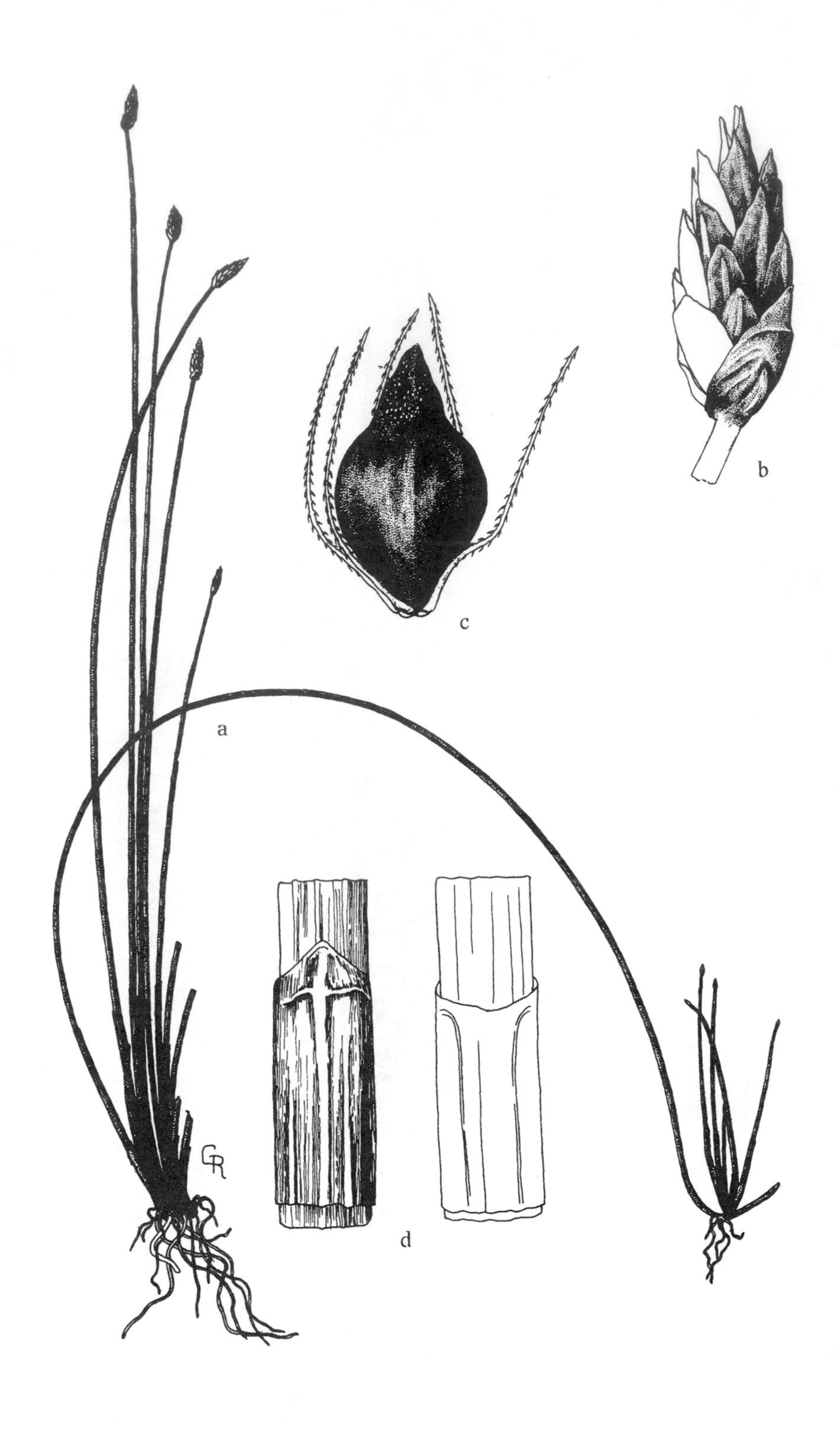

Fig. 278. *Eleocharis rostellata*: a. habit; b. spikelet; c. achene; d. summit of leaf sheath, two views (G&W).

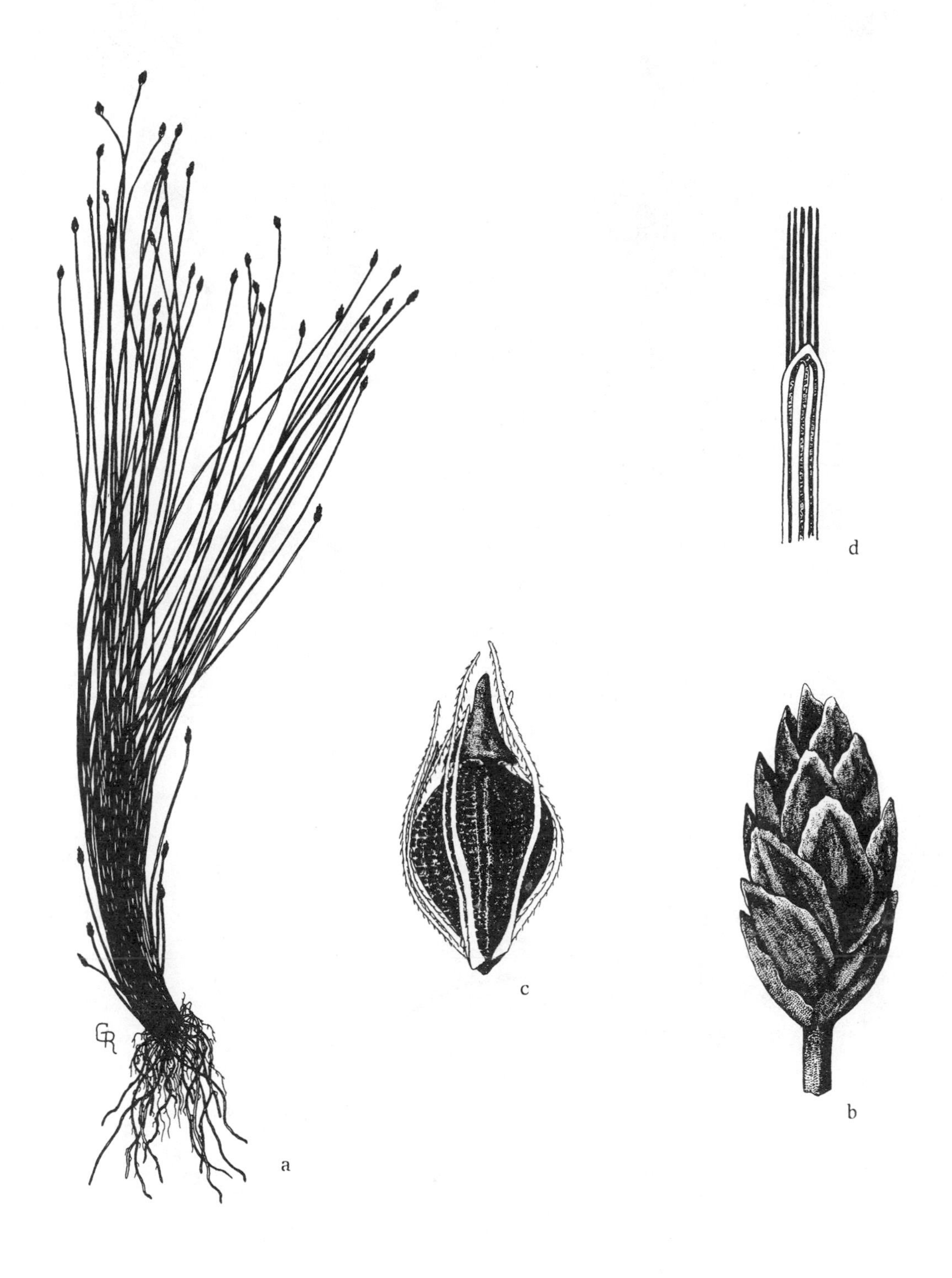

Fig. 279. *Eleocharis tortilis*: a. habit; b. spikelet; c. achene; d. leaf sheath (G&W).

Fig. 280. *Eleocharis vivipara*: a. habit with some spikes proliferated; b. proliferated inflorescence; c. achene; d. summit of leaf sheath (G&W).

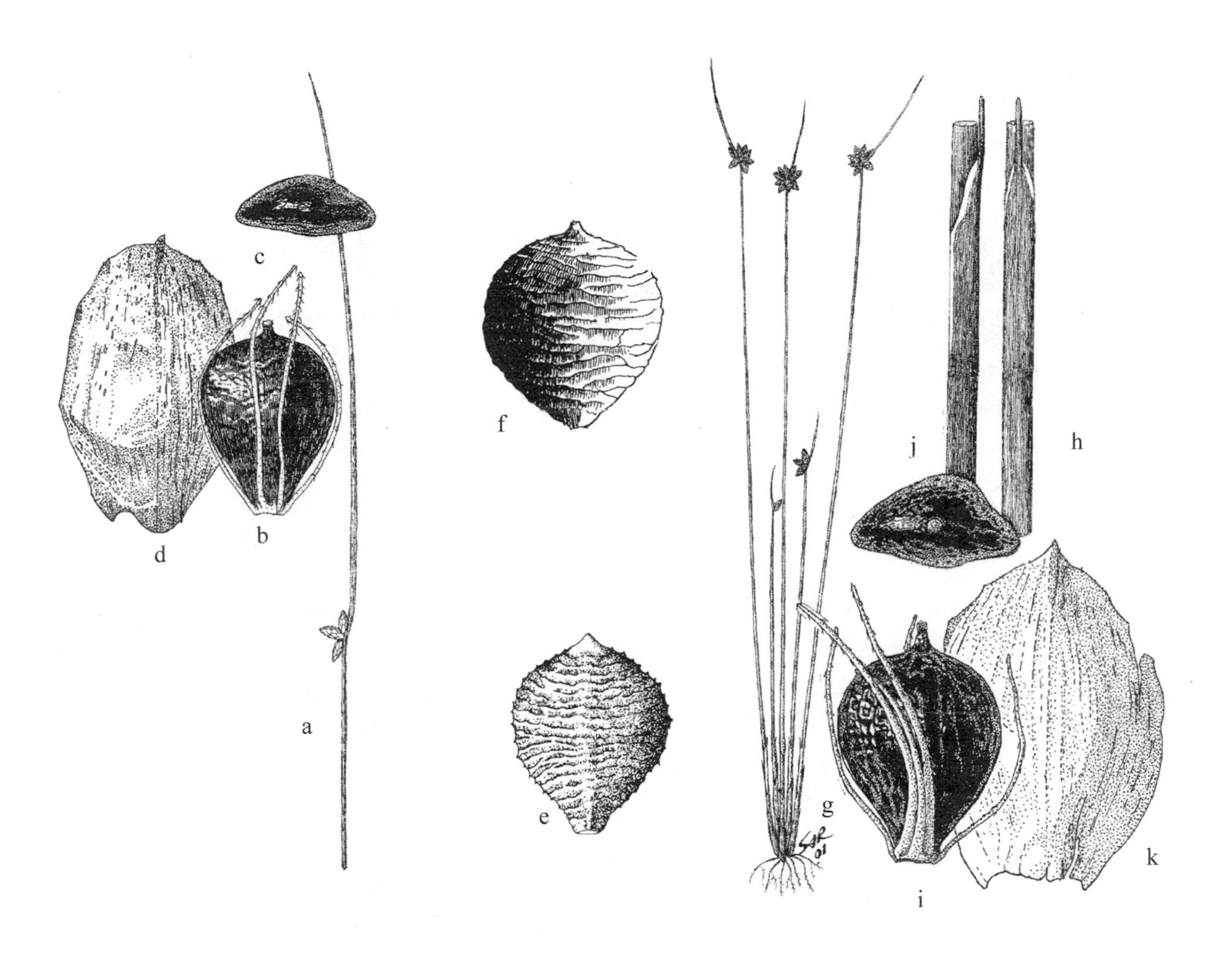

Fig. 281. *Schoenoplectiella smithii* var. *smithii*: a. inflorescence; b. achene; c. achene, cross-section; d. scale (FNA).
Schoenoplectiella hallii: e. achene (F).
Schoenoplectiella saximontana: f. achene (F).
Schoenoplecteilla purshiana var. *purshiana*: g. habit; h. stem sheaths; i. achene; j. achene, cross-section; k. scale (FNA).

Tex.; B.C. (*Schoenoplectus saximontanus* (Fernald) J. Raynal; *Scirpus saximontanus* Fernald)

3. *S. smithii* (A. Gray) Hayasaka Smith's Bulrush Fig. 281

3a. *S. smithii* var. *smithii*
Sandy, peaty, and muddy shores of lakes, ponds and rivers; occasionally tidal. Que. and Ont. w. to Wisc. and Minn., s. to N.E., Del., ne. Va., nw. Pa., Ohio and n. Ind. (*Schoenoplectus smithii* (A. Gray) Soják; *Scirpus smithii* A. Gray)

3b. *S. smithii* var. *setosa* (Fernald) Hayasaka
Sandy and muddy freshwater shores with stable water levels, floating mats and bogs. Que. and Ont., w. to Minn., s. to N.E., and Del., N.C., Ohio, Ind. and Ill. (*Schoenopletus smithii* var. *setosus* (Fernald) S. G. Smith; *Scirpus smithii* var. *setosus* Fernald)

3c. *S. smithii* var. *leviseta* (Fassett) Hayasaka
Coastal, freshwater tidal flats. N.B. and Que., Me. and Va; (*Schoenoplectus smithii* var. *levisetus* (Fernald) S. G. Smith; *Scirpus smithii* var. *levisetus* (A. Gray) Fassett)

4. *S. purshiana* (Fernald) Lye

4a. *S. purshiana* var. *purshiana* Fig. 281
Marshes, bogs, wet peats, acid swamps and sandy shores. Me. w. to s. Mich. and Minn. s. to Ga., Ala. and Miss. *Schoenplectiella purshiana* has long been confused with the closely related *S. hotarui* (Ohwi) Holub, a species of East Asia which does not occur in N.Am.; it was included in the previous edition as *Scirpus hotarui* Ohwi. (*Schoenoplectus purshianus* (Fernald) M. T. Strong; *Scirpus purshianus* Fernald)

4b. *S. purshiana* var. *williamsii* (Fernald) Hayasaka
Freshwater sandy shores often with fluctuating water levels. Mass., Mich., Minn. and Wisc. (*Schoenopectus purshianus* var. *williamsii* (Fernald) S. G. Smith; *Scirpus purshianus* var. *williamsii* Fernald)

14. *Isolepis*

Annuals (ours) or perennial, cespitose or rhizomatous; culms terete; leaves all basal, inflorescences terminal, spikelets 2–3 (ours) (–15); involucral bracts spreading to erect similar to leaf blades; spikelet scales 8–25 each subtending flower; flowers bisexual; perianth absent; stamens persistent; achenes biconvex to trigonous.

REFERENCE: Smith, 2002.

1. Leaf blade longer than sheath, up to 5 cm long, 0.2–0.5 mm wide; scales from middle of spikelet 1.8–2 mm long, with awns 0.2–0.5 mm; achenes 1–1.5 mm long. . . . . . 1. *I. carinata*
1. Leaf blade longer than sheath, up to 3 cm long, 0.2 mm wide; scales from middle of spikelet 1–1.2 mm long, with apex rounded, with mucronate to short abrupt awn 0.1 mm; achenes 0.7–0.9 mm long. . . . . . 2. *I. pseudosetacea*

1. *I. carinata* Hook. & Arn. ex Torr. Keeled Bulrush Fig. 282
Sands, depressions, seeps, shores, temporary wet sites in grasslands and damp woodlands. N.C. and Tenn., w. to sw. Ky., s. Mo. and Okla., s. to Fla., Ala. and Tex; Calif. (*Scirpus koilolepis* (Steud.) Gleason)

2. *I. pseudosetaea* (Daveau) Gandoger Gulf Coast Bulrush Fig. 282
Freshwater, often in temporary wet sites in grassland, open woods, cultivated fields and waste places. Ga. w. to Tex., northward in Ark. and Mo. (*Isolepis molesta* (M. C. Johnston) S. G. Smith; *Scirpus pseudosetacea* Daveau)

15. *Rhynchospora* (Beaksedge, Beak-rush)

Perennial herbs (rarely annual), usually tufted, some producing elongate rhizomes; leaves mostly shorter than stems; stems 3-angled or nearly round in cross-section; flowers in clustered or panicled spikelets, (1)2–4(6) or few to many, lower 2 or 3 spikelet scales empty, upper flower often staminate; perianth consisting of bristles, sometimes absent; fruit a biconvex or flattened achene with a persistent tubercle.

REFERENCES: Gale, 1944; Kral, 2002, Thomas, 1984.

1. Involucral bracts spreading or drooping, basal portions whitish, green toward tips (fig. 283a,b). . . . . . 1. *R. colorata*
1. Involucral bracts ascending, green throughout.
    2. Perianth bristles absent; spikelets few- to many-flowered; annuals.
        3. Tubercles longer than wide, styles persisting (fig. 284f); achene finely roughened or wrinkled. . . . . . 2. *R. scirpoides*
        3. Tubercles much wider than long, styles deciduous (figs. 284d); achene strongly roughened or wrinkled. . . . . . 3. *R. nitens*
    2. Perianth bristles present (fig. 285e,g); spikelets (1)2–4(6)-flowered; perennials.
        4. Tubercles 10 mm or more long (fig. 285g).
            5. Bristles longer than body of achene.

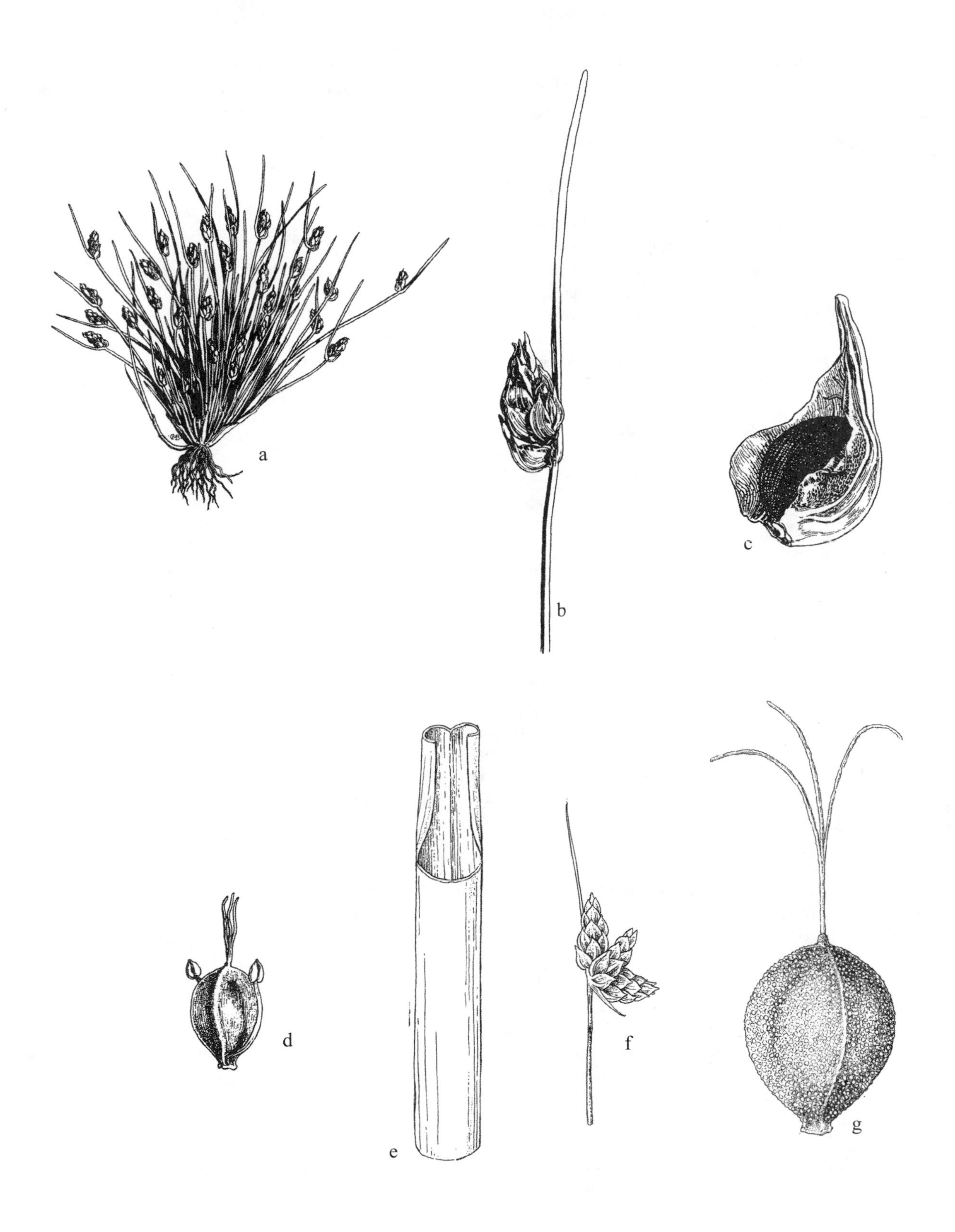

Fig. 282. *Isolepis carinata*: a. habit; b. inflorescence with involucral bract; c. scale and achene; d. achene with stamens persisting (Mason).
Isolepis pseudosetacea: e. stem sheath; f. inflorescence; g. achene (FNA).

6. Spikelets in tight clusters of 10–50 (fig. 285c,f); tubercle (15)18–21 mm long; plants cespitose.............4. *R. macrostachya*
6. Spikelets in open clusters of 1–6 (fig. 285d); tubercle 10–15 mm long; plants stoloniferous.....................5. *R. inundata*
5. Bristles shorter than body of achene........6. *R. corniculata*
4. Tubercles 5 mm or less long.
7. Achenes wrinkled (fig. 286c,g), transversely ridged, rugose, or honeycomb-reticulate.
8. Bristles long-fringed, plumose (especially mid- to base of bristles) (fig. 286)........ 7. *R. oligantha*
8. Bristles merely barbed (fig. 287c).
9. Stems lax and reclining; leaves 1 mm or less wide, involute........8. *R. rariflora*
9. Stems erect or strongly ascending; leaves 1–8 mm wide, flat.
10. Tubercle with a distinct raised rim at base (fig. 287f)........9. *R. harveyi*
10. Tubercle lacking a distinct raised rim at base.
11. Bristles shorter than body of achene.
12. Achene body strongly flattened, lacking a bulge toward tip.
13. Scales awned; achene 1.3–2 mm long, with 10 or more transverse ridges; bristles 6.......10. *R. torreyana*
13. Scales blunt to mucronate; achene 1–1.3 mm long, with 8 or fewer transverse ridges; bristles 0–3. ....11. *R. perplexa*
12. Achene body biconvex, with a bulge toward tip. ....12. *R. globularis*
11. Bristles longer than body of achene.
14. Leaves 1.5–4 mm wide; achene narrowly oblong-ellipsoid, 2–2.5 mm long, body flat...... ....13. *R. inexpansa*
14. Leaves 4–8 mm wide; achene broadly obovate to ovate, 1–1.8 mm long, body biconvex.
15. Achenes 1–1.3 mm long; tubercles smooth; branches of inflorescence spreading, almost at right angles to main rachis. .... 14. *R. miliacea*
15. Achenes 1.4–1.8 mm long; tubercle margins with fine bristles (setaceous) (fig. 290h); branches of inflorescence ascending........ 15. *R. caduca*
7. Achene smooth.
16. Bristles retrorsely barbed.
17. Bristles 8–12, villous at base (fig. 290b); scales of spikelet white to pale brown........16. *R. alba*
17. Bristles 6, generally glabrous at base (fig. 290d); scales of spikelet dark brown to dark reddish-brown.
18. Clusters of spikelets subglobose, the lowest spikelet in each cluster spreading or reflexed (fig. 290c,e).
19. Achenes 2–2.4 mm long, 1.4–1.6 mm wide........17. *R. cephalantha*
19. Achenes 1.4–1.6 mm long, 0.9–1.1 mm wide........ 18. *R. microcephala*
18. Clusters of spikelets ellipsoidal, all spikelets ascending (fig. 289c).
20. Achenes with pale margins, obovoid; leaves flat.
21. Spikelets 1-fruited........19. *R. chalarocephala*
21. Spikelets 2-fruited, rarely 1-fruited, with a sterile floret present.
22. Leaves 3–7 mm wide; spikelets 5–6 mm long; achene body 1. 7–2 mm long with conspicuous central bulge, summit nearly truncate (fig. 289b)........20. *R. glomerata*
22. Leaves 0.5–4 mm wide; spikelets 3–5 mm long; achene body 1.1–1.8 mm long, without central bulge, summit rounded (fig. 289d)........ 21. *R. capitellata*
20. Achenes lacking pale margins, ellipsoidal to narrowly obovoid; leaves with involute margins, nearly filiform........22. *R. capillacea*
16. Bristles antrorsely barbed or smooth, occasionally absent.
23. Base of plants swollen and bulbous; spikelets 1-flowered; bristles absent or 1–3, rudimentary. ....23. *R. pallida*
23. Base of plants not swollen and bulbous; spikelets 2–4(6)-flowered; bristles 5 or 6.
24. Achenes 0.6–1.1 mm wide, distinctly prolonged to a stipe-like base (fig. 292b,d).
25. Plants stoloniferous; achenes 1.3–1.7 mm long; tubercle usually green; spikelets 2- or 3-flowered..... ....24. *R. fusca*
25. Plants cespitose; achenes 1–1.3 mm long; tubercle tan; spikelets 3- or 4(6)-flowered........ 25. *R. filifolia*
24. Achenes 1.1–1.7 mm wide, not stipitate.
26. Tubercles elongate, 1–2.6 mm long, as long as or longer than achene body; bristles as long as or longer than tip of tubercle........26. *R. gracilenta*
26. Tubercles deltoid, 0.4–0.7 mm long, shorter than achene body, bristles typically shorter than achenes; spikelets 2.5–3.2 mm long.

27. Stems and leaves tending to be flexuous and reclining; scales bifid, shorter than achenes; spikelets 2.5–3.2 mm long. .... 27. *R. debilis*

27. Stems and leaves erect, stiff; scales not 2-cleft, longer than achenes; spikelets 3–5 mm long.......... 28. *R. fascicularis*

---

1. *R. colorata* (L.) H. Pfeiffer White-top Sedge, Star-rush White-top Fig. 283
Wet peats, peaty sands, savannas, swales, borders of wet depressions and ditches. Coastal plain, e. Va. s. to Fla., w. to e. Tex. and e. Mex.; W.I. and C.Am. Although the more robust *R. latifolia* (Baldwin) W. W. Thomas has been included by some manuals as occurring in Virginia, both Thomas (1984) and Kral (2002) reported it only as far north as North Carolina. (*Dichromena colorata* (L.) Hitchc.)

2. *R. scirpoides* (Torr.) A. Gray Beak-sedge Fig. 284
Wet, sandy, peaty and mucky shores. Chiefly coastal plain, se. Mass. s. to Fla. panhandle; disjunct to Mich. n. Ind. and c. Wisc. (*Psilocarya scirpoides* Torr.; *P. scirpoides* var. *grimesii* Fernald & Griscom)

3. *R. nitens* (Vahl) A. Gray Short-beak Beak-sedge Fig. 284
Wet sands and peats. Coastal plain, Mass. s. to Fla., w. to Tex.; disjunct to nw. Ind. and sw. Mich. (*Psilocarya nitens* (Vahl) Wood)

4. *R. macrostachya* Torr. ex A. Gray Tall Horned Beak-sedge Fig. 285
Wet sand and peat, along the shores of ponds, lakes, streams and marshes. Coastal plain, s. Me. s. to Fla., w. to Tex.; inland in c. N.Y., c. Ky., sw. Mich., s. Ill., Mo. and Kans. (*R. macrostachya* var. *colpophilia* Fernald & Gale)

5. *R. inundata* (Oakes) Fernald Narrow-fruit Horned Beak-sedge Fig. 285
Peaty and mucky shores of ponds, swamps and ditches. Coastal plain, e. Mass. s. to Fla., w. to Ala., La. and Tex.; rare and local in the northern portion of its range.

6. *R. corniculata* (Lam.) A. Gray Short-beak Horned Beak-sedge Fig. 285
Swamps, ditches, and shores of lakes and streams. Del. s. to Fla., w. to Ky., s. Ill., Mo., La. and Tex.; W.I. (*Rhynchospora corniculata* var. *interior* Fernald)

7. *R. oligantha* A. Gray Featherbristle Beak-sedge Fig. 286
Wet pine barrens, sands, peats, and open savannas, pinelands and southern bogs. Rare, coastal plain s. N.J. s. to n. Fla. w. to e. Tex.

8. *R. rariflora* (Michx.) Elliott Few-flower Beak-sedge Fig. 287
Damps sands and peats of shores, southern bogs, flatwoods, stream banks and savannas. Coastal plain, s. N.J. and Va. s. to Fla. w. to Tex.; inland n. to Tenn., Ark. and Okla.; Mex., W.I. and C.Am.

9. *R. harveyi* Boott Harvey's Beak-sedge Fig. 287
Moist to well-drained sands, sandy clays, peats, open pinelands, savannas, prairies and depressions. Coastal plain, Va. s. to Fla., w. to La., Tex.; inland n. to Tenn., Mo., Kans. and Okla. Our taxon is var. *harveyi*.

10. *R. torreyana* A. Gray Torrey's Beak-sedge Fig. 286
Damp to dry sands and peats of pond margins and pinelands. Coastal plain, se. Mass. s. to Ga. and Miss.

11. *R. perplexa* Britton ex Small Pineland Beak-sedge Fig. 288
Wet sands and peat of pond margins and lakeshores and shallow water, depressions in savanna, and flatwoods. Coastal plain, se. Va. s. to Fla., w. to Tex.; inland n. to c. Tenn.; W.I. (*R. perplexa* var. *virginiana* Fernald)

12. *R. globularis* (Chapm.) Small Globe Beak-sedge Fig. 288
Wet peaty and sandy soils of meadows, savannas, pinelands, depressions and shores. N.J., Md. and se. Pa. s. to Fla., w. to interior N.C., s. Ky., Tenn., Mo., Okla. and Tex.; sporadic in n. Ohio, Ind. and ne. Ill.; n. Calif.; W.I. and C.Am. Our taxon is var. *globularis*. (*R. globularis* var. *recognita* Gale; *R. cymosa* Elliott)

13. *R. inexpansa* (Michx.) Vahl Nodding Beak-sedge Fig. 286
Wet to dry sandy and peaty soils of swamps, ditches and pond margins. Coastal plain, Del., Md. and se. Va. s. to n. Fla., w. to Ark. and e. Tex.; W.I.

14. *R. miliacea* (Lam.) A. Gray Millet Beak-sedge Fig. 289
Stream margins, pine woods, thickets, swamps and ditches. Coastal plain, se. Va. s. to Fla., w. to La. and e. Tex.

15. *R. caduca* Elliott Angle-stem Beak-sedge Fig. 290
Pools, swamps, peat bogs, and borders of wet woods. Coastal plain, se. Va. s. to Fla., w. to e. Tex.; inland n. to n. Ala., nw. Ark. and Okla.

16. *R. alba* (L.) Vahl White Beak-sedge Fig. 290
Bogs, wet peats and sands. Nfld. w. to s. Alask., s. to N.E., Md., e. Va., N.C., n. Ill., Minn., Man., Sask., Alta., B.C., Wash., Ida. and Calif.

17. *R. cephalantha* A. Gray Bunched Beak-sedge Fig. 290
Bogs, savannas, marshy shores of ponds, lakes and streams. Coastal plain, N.Y., N.J. and Va. s. to Fla. w. to La. (*R. cephalantha* var. *pleiocephala* Fernald & Gale)

18. *R. microcephala* Britton Smallhead Beaksedge Fig. 290
Sandy and peaty shores of swamps, savannas and wet pinelands. Coastal plain, N.J. s. to Fla., w. to Miss. and La.; Cuba. (*R. cephalantha* var. *microcephala* (Britton) Kük.)

19. *R. chalarocephala* Fernald & Gale Loosehead Beak-sedge Fig. 291
Pond margins, swamps, savannas, and wet pinelands. Coastal plain, N.Y., N.J. s. to Fla., w. to La. and Tex.

20. *R. glomerata* (L.) Vahl Clustered Beak-sedge Fig. 289
Wet peaty and sandy soil of bogs savannas and meadows. Coastal plain, N.J. s. to Fla., w. to e. Tex.; inland n. to Ill., Ind., Ky., Tenn., Ark., Kans. and Okla.

21. *R. capitellata* (Michx.) Vahl Brownish Beak-sedge Fig. 289
Bogs, sandy shores, low ground, and wet pine savannas. N.S. and N.B. w. to sw. Que., s. Ont., Mich., Wisc., Mo. and Kans., s. to Va., Ga., Miss., and e. Tex.; Oreg. and Calif.

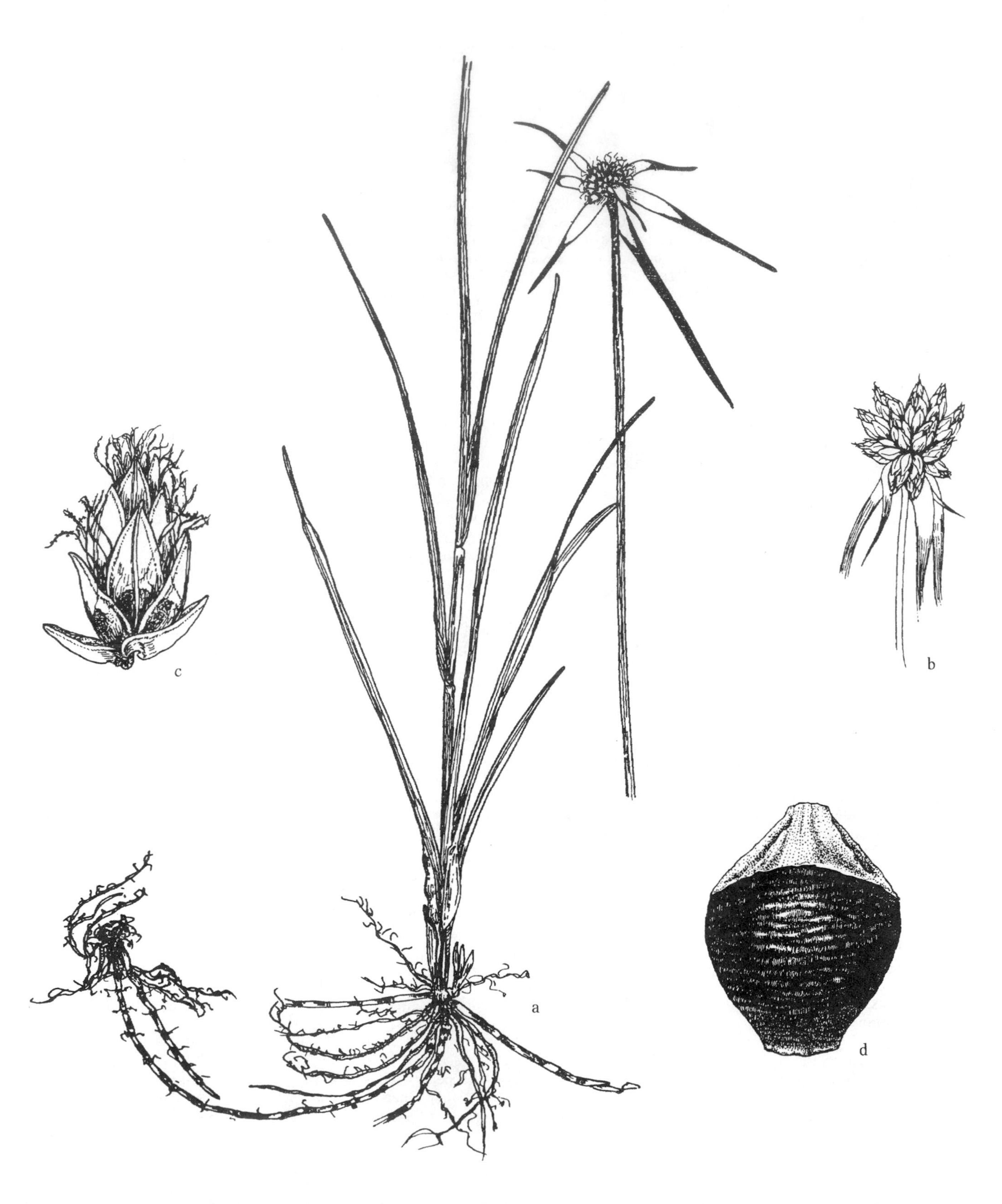

Fig. 283. *Rhynchospora colorata*: a. habit (C&C); b. inflorescence (G&W); c. spikelet (C&C); d. achene (G&W).

Fig. 284. *Rhynchospora nitens*: a. habit; b. sheath; c. spikelet; d. achene (G&W).
Rhynchospora scirpoides: e. habit (F); f. achene (G&W).

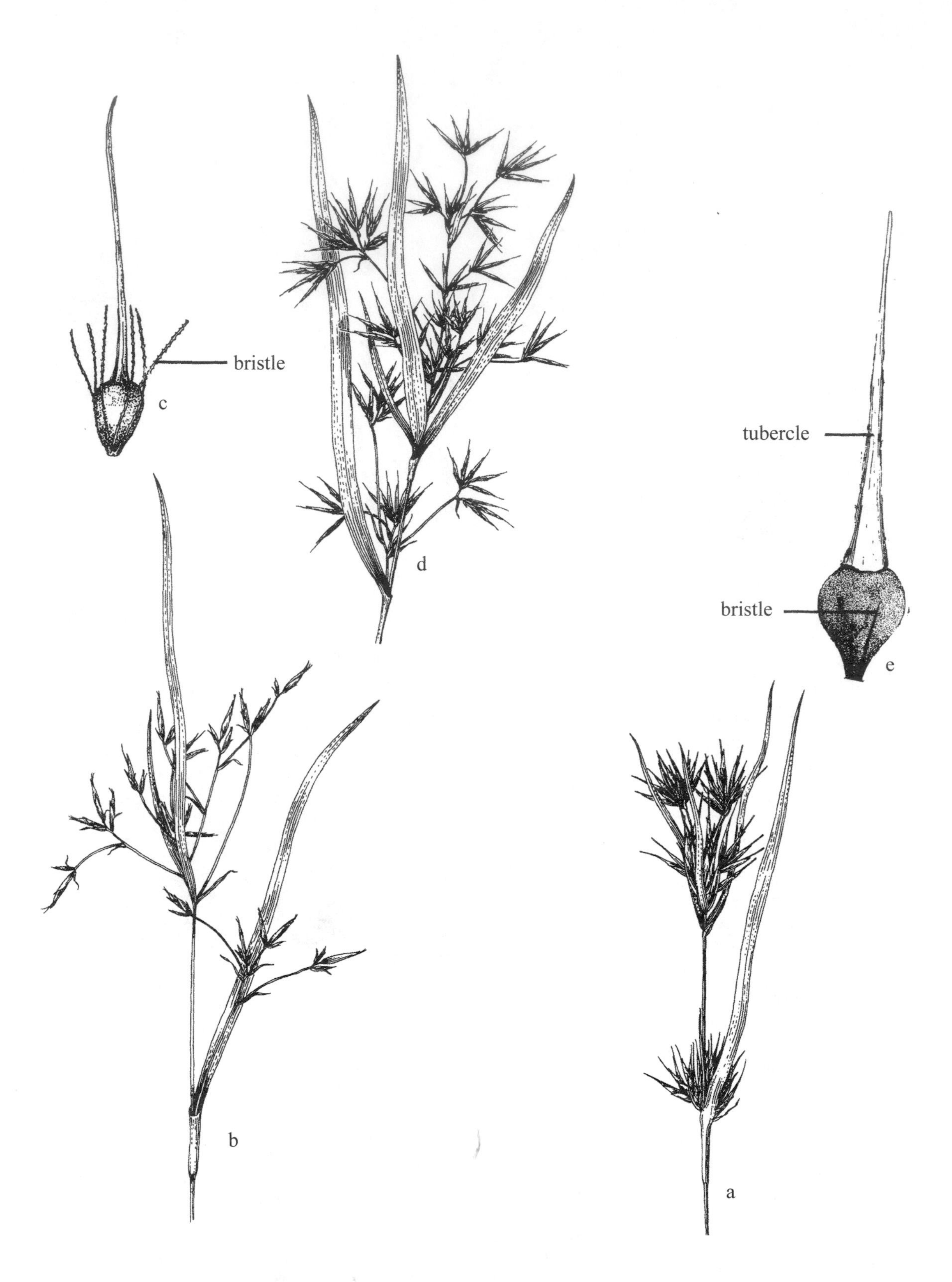

Fig. 285. *Rhynchospora macrostachya*: a. inflorescence (Beal).
Rhynchospora inundata: b. inflorescence (Beal); c. achene (Crow).
Rhynchospora corniculata: d. inflorescence (Beal); e. achene (F).

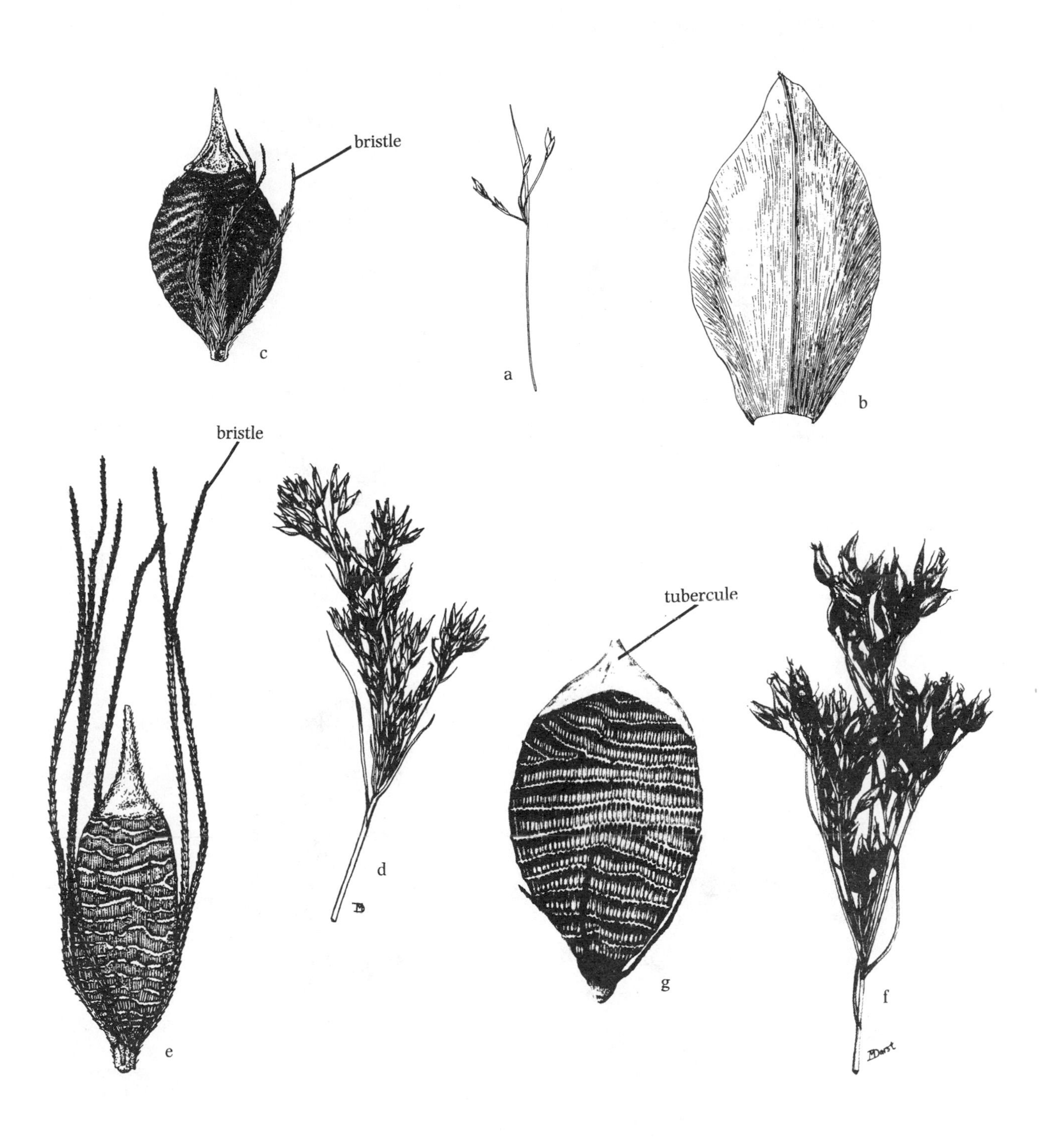

Fig. 286. *Rhynchospora oligantha*: a. inflorescence; b. scale; c. achene (G&W).
Rhynchospora inexpansa: d. inflorescence; e. achene (G&W).
Rhynchospora torreyana: f. inflorescence; g. achene (G&W).

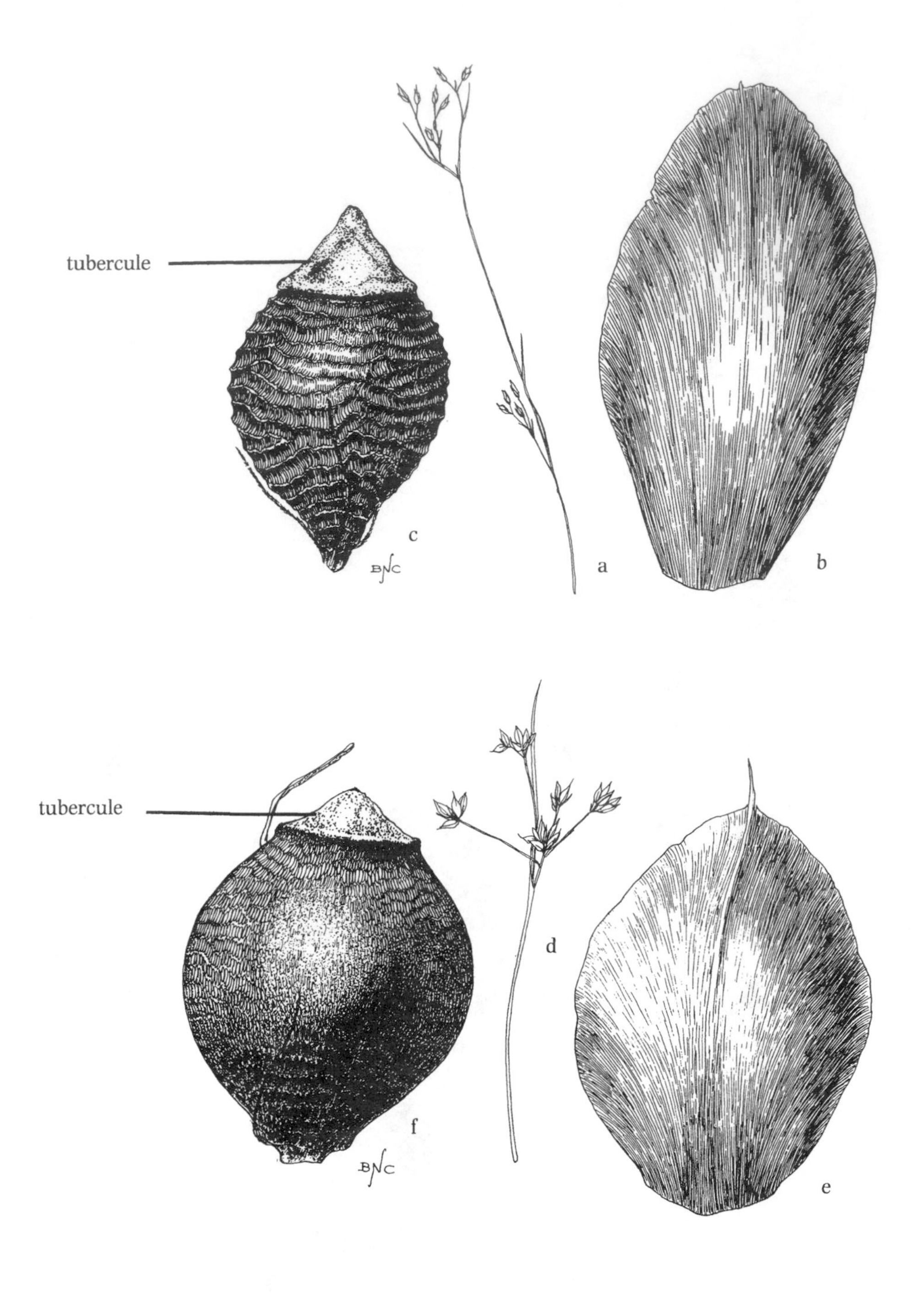

Fig. 287. *Rhynchospora rariflora*: a. inflorescence; b. scale; c. achene (G&W).
Rhynchospora harveyi: d. inflorescence; e. scale; f. achene (G&W).

Fig. 288. *Rhynchospora globularis*: a. inflorescence (F); b. inflorescence (G&W); c. scale (G&W); d. achene (G&W). *Rhynchospora perplexa*: e. inflorescence; f. scale; g. achene (G&W).

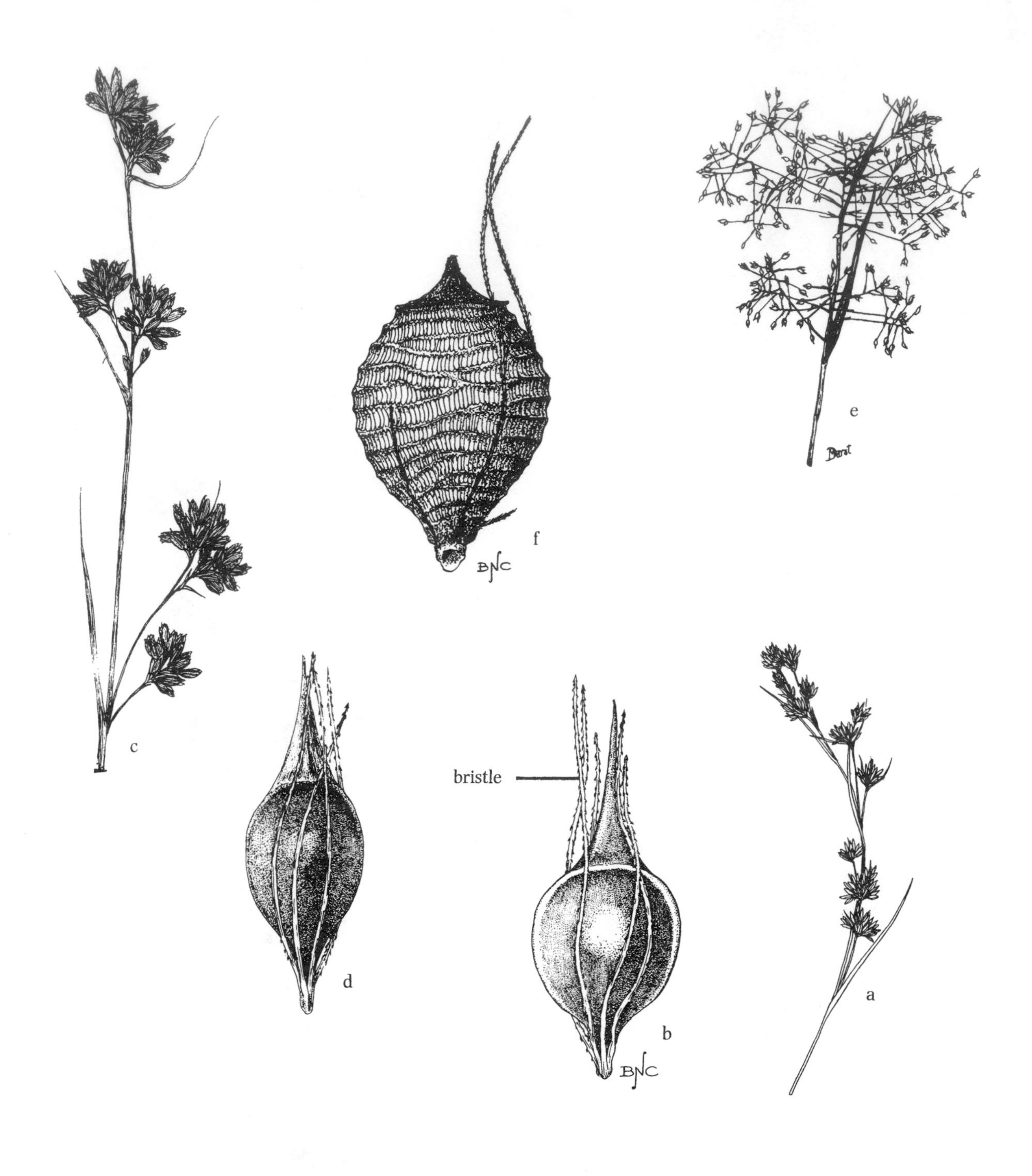

Fig. 289. *Rhynchospora glomerata*: a. inflorescence; b. achene (G&W).
Rhynchospora capitellata: c. inflorescence (F); d. achene (G&W).
Rhynchospora miliacea: e. inflorescence; f. achene (G&W).

Fig. 290. *Rhynchospora alba*: a. habit (HCOT); b. achene (G&W).
Rhynchospora cephalantha: c. inflorescence (F); d. achene (G&W).
Rhynchospora microcephala: e. inflorescence; f. achene (G&W).
Rhynchospora caduca: g. inflorescence; h. achene (G&W).

22. *R. capillacea* Torr. Needle Beak-sedge Fig. 291
Calcareous fens, peaty meadows, marly sands and seeps on limestone ledges. W. Nfld. w. to Sask., and B.C., s. to Me., Conn., n. N.J., se. Pa., sw. Va., ne. Tenn., Ind., Ala., Ark., Okla. and c. Tex.

23. *R. pallida* M. A. Curtis Pale Beak-sedge Fig. 291
Wet peats and wet pine barrens. Coastal plain, L.I., N.Y., s. to N.C. and Ga.

24. *R. fusca* (L.) W. T. Ation Brown Beak-sedge Fig. 292
Bogs, wet peat and sand, wet pine barrens and shores, often marly places. Nfld. w. to Ont., Mich., n. Wisc., ne. Minn. and Sask., s. to L.I., N.Y., Del., Md. and W.Va.

25. *R. filifolia* A. Gray Threadleaf Beak-sedge Fig. 292
Pond margins, swamps and pine savannas. Coastal plain, N.J., Del. and Va. s. to Fla., w. to se. Tex.; Cuba, W.I., C.Am. and S.Am.

26. *R. gracilenta* A. Gray Slender Beak-sedge Fig. 292
Wet pine barrens, peats, swales and sandy depressions. Chiefly coastal plain, N.Y. and N.J. s. to n Fla., w. to e. Tex.; inland n. to Tenn., Ark. and Okla.

27. *R. debilis* Gale Savannah Beak-sedge Fig. 292
Damp sandy and peaty soil of bogs and pine savannas. Coastal plain, se. Va. s. to n. Fla., w. to Ala., La. and Tex.

28. *R. fascicularis* (Michx.) Vahl Fascicled Beak-sedge Fig. 292
Moist to wet pinelands, savannas, pond margins and ditches. Coastal plain, se. Va. s. to Fla., w. to e. Tex. (*R. fascicularis* var. *distans* (Michx.) Kük.)

16. *Cladium* (Twig-rush, Saw-grass)

Perennial herbs, arising from stout rhizomes or stolons; stems leafy; inflorescence much-branched, each branch bearing a cluster of spikelets; spikelets 1- or 2-flowered, perianth absent; fruit an ovoid to globose achene without a tubercle.

Reference: Tucker, 2002.

1. Leaves narrow, 1–3 mm wide, channeled, margins and midrib nearly smooth. . . . . 1. *C. mariscoides*
1. Leaves broad, 5–10 mm wide, flat stiff, margins and midrib saw-toothed. . . . . 2. *C. jamaicense*

1. *C. mariscoides* (Muhl.) Torr. Twig-rush, Smooth Saw-grass Fig. 293
Swamps, sandy, boggy, and marshy shores and swales, often in shallow water, fresh or brackish. Nfld. and N.S. w. to Mich., Minn., and Sask., s. to Va., Ohio, mts. to N.C., S.C., Ga., Ala., Miss. and Tex. (*Mariscus mariscoides* Muhl.)

2. *C. jamaicense* Crantz Saw-grass, Jamaica Swamp-grass Fig. 293
Freshwater swales, marshes, and shores and saline marshes. Coastal, Va. s. to Fla., w. to Ark., Mo., Tex., N.M. and Mex.; W.I. (*Mariscus jamaicensis* (Crantz) Britton; *C. mariscus* subsp. *jamaicense* (Crantz) Kük.)

17. *Lipocarpha* (Half-chaff Sedge)

Annuals, densely tufted; stems 3-angled, leafy at base; leaves narrow, involute; inflorescence a cluster of several sessile, densely flowered spikelets, subtended by 2 long and several short involucral bracts; flowers bisexual, subtended by 1 or 2 small delicate bractlets (easily overlooked), bristles absent; fruit an achene, lacking a tubercle.

References: Frieland, 1941; Tucker, 2002.

1. Leaves reduced to sheaths at base of stem, with very short blades (fig. 294a); achenes 0.4–0.7 mm long, subtended by 1 small bractlet as well as spikelet scale. . . . . 1. *L. micrantha*
1. Leaves with well-developed blades (fig. 295a); achenes 0.7–1 mm long, subtended by 2 small bractlets as well as spikelet scale. . . . . 2. *L. maculata*

1. *L. micrantha* (Vahl) G. C. Tucker Small-flower Half-chaff Sedge Fig. 294
Sandy shores of ponds and streams. Que. and w. Me. w. to Ont., Mich., Wisc., Minn., e. N.D., s to Fla., La., Okla. and Tex.; s. Ida., B.C and Wash., s. to Ida., Calif., Ariz. and N.M.; C.Am. and S.Am. We have followed Tucker (2002), but some authors have accepted the merging of *Lipocharpha* into *Cyperus*, following the generic concept of Bauters et al. (2014), in which case the correct name would be *Cyperus subsquarrosus* (Muhl.) Bauters. (*Scirpus micranthus* Vahl; *Hemicarpha micrantha* (Vahl) Britton)

2. *L. maculata* (Michx.) Torr. American Half-chaff Sedge Fig. 295
Wet sands along pond and stream margins, wet peaty areas, springy places borders of swamps. Coastal plain, N.J., e. Pa., D.C. and Va. s. to Fla., w. to Ala. and Miss.; disjunct to Ill. If one follows the generic concept of Bauters et al. (2014), including *Lipocarpha* within *Cyperus*, the correct name would be *Cyperus neotropicalis* Alain.

Fig. 291. *Rhynchospora capillacea*: a. inflorescence; b. achene (F).
Rhynchospora chalarocephala: c. inflorescence; d. scale; e. achene (G&W).
Rhynchospora pallida: f. habit; g, h. achene variations (NYS Museum).

Fig. 292. *Rhynchospora fusca*: a. inflorescence; b. achene (F).
Rhynchospora filifolia: c. inflorescence; d. achene (G&W).
Rhynchospora gracilenta: e. inflorescence; f. achene (G&W).
Rhynchospora debilis: g. inflorescence; h. achene (G&W).
Rhynchospora fascicularis: i. inflorescence; j. achene (G&W).

Fig. 293. *Cladium jamaicense*: a. inflorescence; b. base of plant; c. section of stem with leaf sheath; d. section of leaf with sharp margins and midrib; e. achene (G&W).
Cladium mariscoides: f. inflorescence; g. achene (F).

Fig. 294. *Lipocarpha micrantha*: a. habit (G&W); b. habit (F); c. inflorescence (G&W); d. scale (G&W); e. achene (G&W).

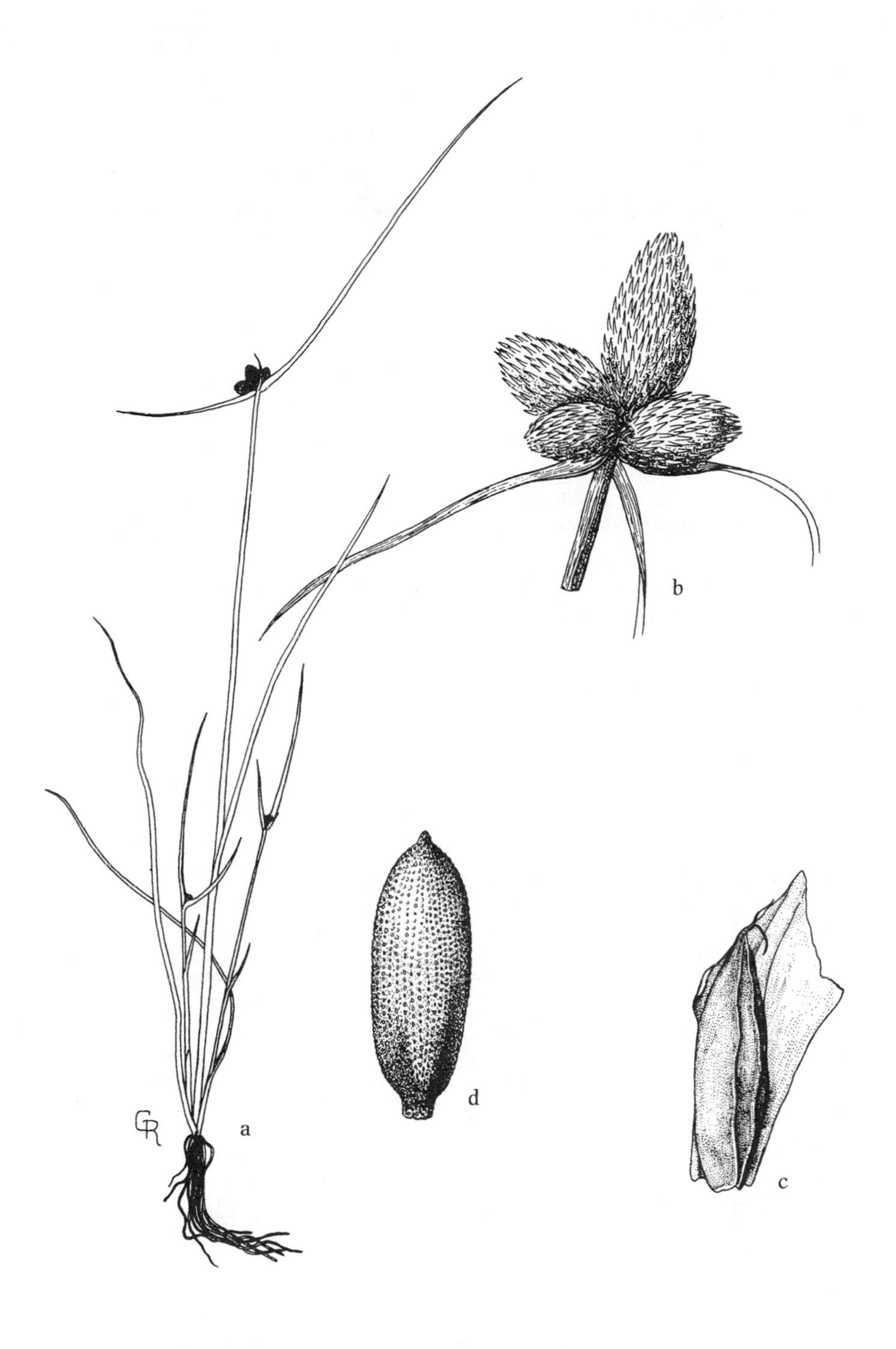

Fig. 295. *Lipocarpha maculata*: a. habit; b. inflorescence; c. scale and bractlets surrounding flower; d. achene (G&W).

18. *Fimbristylis* (Fimbry)

Perennial or annual herbs, grass-like, forming tufts, or culms arising singly from a creeping rhizome; leaves basal; spikelets solitary and terminal on scapes or arranged in umbelliform inflorescences; flowers bisexual, perianth absent; stamens 1–3; styles 2- or 3-cleft, style base flat or swollen, but not persistent at fruit maturity; achenes biconvex or 3-angled, shiny, lacking a tubercle.

Fimbristylis differs from *Rhynchospora* by the style dropping off, thus not forming a tubercle.

References: Kral, 1971, 2002.

---

1. Styles 3-cleft (fig. 296d); achenes 3-angled (fig. 296c) or obovoid and obscurely 3-angled (fig. 297b), but not biconvex.
   2. Spikelets narrowly linear to lanceolate (fig. 296a); achene surface smooth to. .... 1. *F. autumnalis*
   2. Spikelets broadly ovoid to globose (fig. 297a); achene surface reticulate (fig. 297b).......... 2. *F. miliacea*
1. Styles 2-cleft; achenes biconvex.
   3. Spikelets appearing sessile, in dense, capitate clusters (fig. 296g); achenes 0.4–0.5 mm long. .... 3. *F. vahlii*
   3. Spikelets on rays (fig. 296e), rarely congested; achene 1–1.5 or 4–6 mm long.
      4. Plants densely tufted, forming large tussocks, up to 1.5(2) m tall, bases deeply set in substrate.......... 4. *F. castanea*
      4. Plants rhizomatous or tufted, but not forming large tussocks, bases not deeply set in substrate.
         5. Plants perennial; bases hard and bulbous, with rhizomes either slender and somewhat elongate, or short, thick, and knotty or bulbous.
            6. Rhizomes thick, with hard, bulbous, persistent bases, giving knotty appearance; leaf blade lacking ligule.......... 5. *F. puberula*
            6. Rhizomes slender; leaf blade with ligule of short hairs. .... 6. *F. caroliniana*
         5. Plants annual; bases soft, lacking rhizomes.
            7. Leaves linear, 1–2(4) mm wide, ligule of short hairs present; achenes 1–1.2 mm long.......... 7. *F. annua*
            7. Leaves linear-filiform, less than 1 mm wide, ligule absent; achenes 4–6 mm long. .... 8. *F. perpusilla*

---

1. *F. autumnalis* (L.) Roem. & Schult. Slender Fimbry Fig. 296
Moist to wet sands, peats, clays and shores, often in disturbed sites. N.B., N.E., and sw. Que. w. to Ont., Mich., Wisc., Minn., se. S.D. and e. Neb., s. to Fla., e. Tex.; Calif. and Mex.; Trop. Am. (*F. autumnalis* var. *mucronulata* (Michx.) Fernald)

2. *F. miliacea* (L.) Vahl Fimbry Fig. 297
Wet sandy peats, mud flats, shores, savannas and wet ditches and fields. Intro., Coastal plain and piedmont, N.C. s. to Fla., w. to La., e. Tex. and Mex.; inland n. to Ark., Tenn., sw. Ky. and se. Mo; W.I. and C.Am. (*Fimbristylis littoralis* Gaudich.)

3. *F. vahlii* (Lam.) Link Vahl's Fimbry Fig. 296
Damp sands, silts, and clays on margins of ponds, lakes, and streams and disturbed bottomlands. N.J., se. Pa., and N.C. w. to Tenn., sw. Ky., Ill., s. Mo., se. Kans. and Okla., s. to Fla. and Tex.; Calif., Ariz. and Mex.; C.Am.

4. *F. castanea* (Michx.) Vahl Marsh Fimbry Fig. 296
Moist sands, muck, or marl, saline coastal marshes and dune swales. Chiefly coastal plain, se. N.Y., s. N.J. and se. Pa., s. to Fla., w. to e. Tex. and Mex.; W.I.

5. *F. puberula* (Michx.) Vahl Hairy Fimbry Fig. 297
Wet savannas and pinelands, wet peats, meadows and prairies. L.I., N.Y., s. N.J. and se. Pa. w. to sw. Ont., se. Mich., se. Wisc., Mo., Neb., s. to Fla., e. Tex. and se. N.M.; Our taxon is var. *puberula*. (*F. drummondii* (Torr. & Hook.) Boeck.; *F. puberula* var. *drummondii* (Torr. & Hook.) Ward)

6. *F. caroliniana* (Lam.) Fernald Carolina Fimbry Fig. 298
Saline, alkaline, or weakly acid sands or sandy peats of beaches, dune swales, lakeshores, ditches and savannas. Coastal plain, L.I., N.Y., s. N.J. and Del. s. to Fla., w. to e. Tex. and Mex.; Cuba.

7. *F. annua* (All.) Roem. & Schult. Annual Fimbry Fig. 298
Moist open sites, savannas, grasslands and disturbed sites. Coastal plain and piedmont, se. N.Y., N.J. and se. Pa. s. to Fla., w. to Tex., Ariz. and Mex.; n. along the Mississippi embayment to ne. Okla., se. Kans., s. Mo., Tenn., s. Ill. and s. Ind.; W.I. and C.Am. (*F. baldwiniana* (Schult.) Torr.)

8. *F. perpusilla* Harper ex Small & Britton Harper's Fimbry Fig. 298
Alluvial soils. Delmarva Pen.; se. N.C., e. S.C., sw. Ky. and sc. Tenn.

19. *Bolboschoenus* (Bulrush)

Perennial, cespitose or not, rhizomatous; culms strongly trigonous; leaves basal and cauline; blades flat of V-shaped in cross-section, keeled on abaxial surface; inflorescences terminal, sub-umbellate, sub-corymbose, or capitate; involucral bracts 1–5 surpassing inflorescence, spreading or proximal, erect, leaflike; spikelet scales 25 or more, spirally arranged, each subtending the flower, tip notched and awned; flowers bisexual, perianth of 3–6 bristles; achenes biconvex to trigonous, 2.5–5.5 mm long, smooth.

Reference: Smith, 2002.

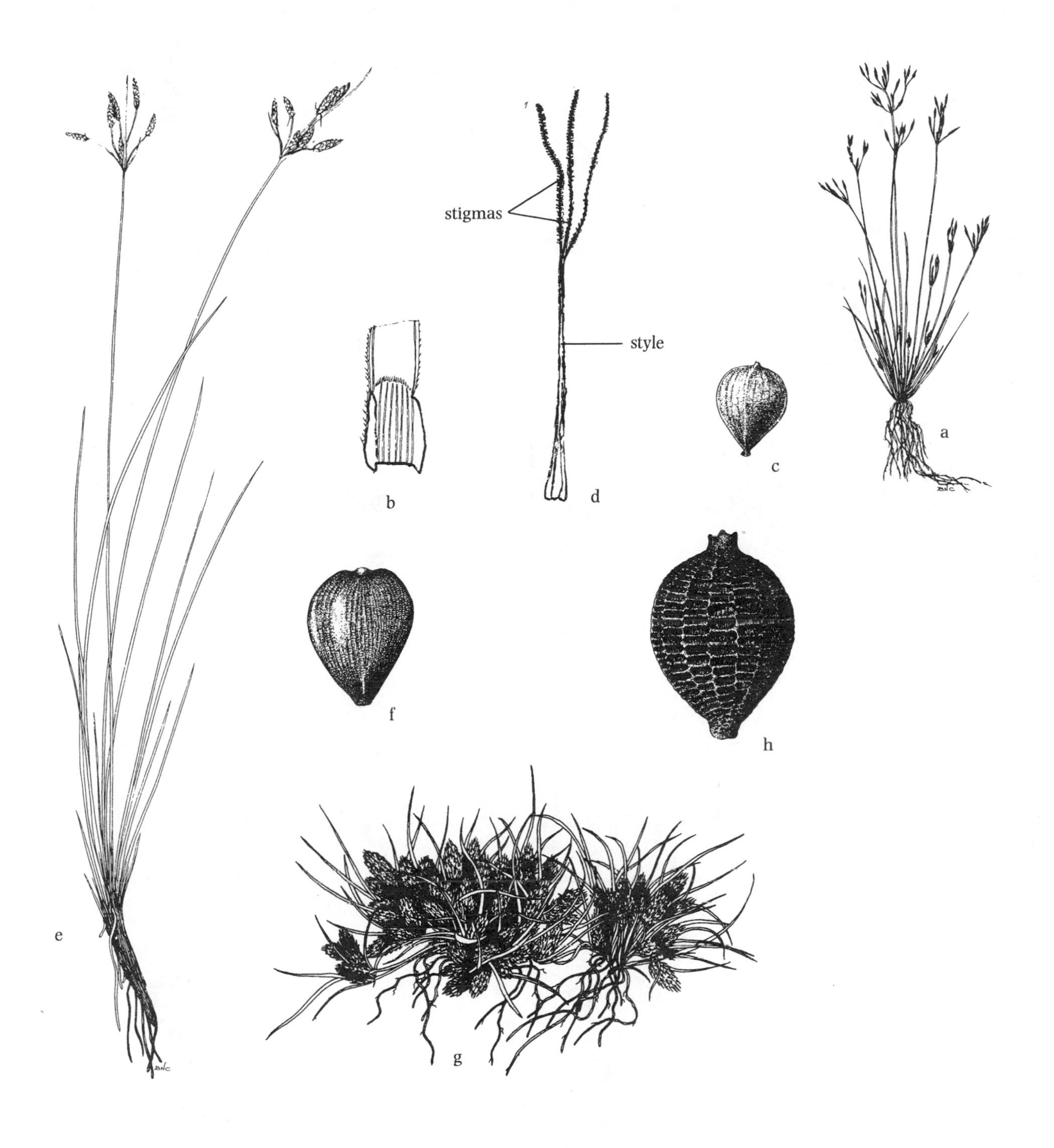

Fig. 296. *Fimbristylis autumnalis*: a. habit (C&C); b. ligule (Kral, 1971); c. achene (C&C); d. stigmas and styles (Kral, 1971).
Fimbristylis castanea: e. habit, portion of plant removed from tussock; f. achene (C&C).
Fimbristylis vahlii: g. habit; h. achene (G&W).

Fig. 297. *Fimbristylis miliacea*: a. habit; b. achene (G&W).
Fimbristylis puberula: c. habit; d. achene (G&W).

Fig. 298. *Fimbristylis caroliniana*: a. habit; b. scale; c. achene; d. section of leaf (Kral, 1971).
Fimbristylis annua: e. habit; f. achene (Kral, 1971).
Fimbristylis perpusilla: g. habit; h. achene (G&W).

1. Inflorescence comparatively open, bristles persisting at base of achene, mostly rounded at summit.
   2. Widest leaf blade 2–6 mm wide; veins diverging proximal to summit of leaf sheath (summit of sheath a membranous, veinless acute triangle); spikelets 3–5 mm wide; achenes 2.5–3.3 mm, beak minute. . . . . 1. *B. glaucus*
   2. Widest leaf blade 7–22 mm wide; veins nearly reaching summit of leaf sheath (summit papery); spikelets 5–10 mm in diameter; achenes 3–5.5 mm, beak distinct, 0.1–0.8 mm.
      3. Achenes acutely 3-angled, plants of freshwater habitats, mostly inland, and freshwater tidal waters. . . . . 2. *B. fluviatilis*
      3. Achenes plano-convex to obtusely 3-angled; plants mostly of mildly saline tidal shores and coastal marshes. . . . . 3. *B. novae-angliae*
1. Inflorescence comparatively compact or congested, bristles usually dropping off base of achene; achenes flattened at summit.
   4. Firm portion of lower leaf sheaths truncate or slightly concave at summit; scales reddish-brown to brown (brown n. of Mass.); awns 0.25 mm wide at base. . . . . 4. *B. robustus*
   4. Firm portion of lower sheaths V-shaped at summit; scales pale grayish-brown to brown; awns 0.5 mm wide at base.
      5. Achenes obtusely compressed trigonous or a few biconvex in same spikelet; styles mostly 3-fid; scales dark to medium brown. . . . . 5a. *B. maritimus* subsp. *maritimus*
      5. Achenes mostly biconvex; styles mostly 2-fid; scales variable, dark to medium brown to stramineous or nearly colorless. . . . . 5b. *B. maritimus* subsp. *paludosus*

1. *B. glaucus* (Lam.) S. G. Smith
Freshwater or brackish shores, marshes and rice-fields. N.H., N.Y.; Ida., Ore. and Calif.; intro. from Eurasia. (*Scirpus glaucus* Lam.)

2. *B. fluviatilis* (Torrey) Soják River Bulrush Fig. 299
Fresh tidal shores along coast, calcareous or alkaline shores and marshes inland. N.B. and Que., w. to B.C., s. to ne. Va., w. Pa., Ohio, Ill., Mo., Kan., N.M., Ariz. and Calif.; Tenn. and Ala. (*Schoenoplectus fluviatilis* (Torrey) A. Gray; *Scirpus fluviatilis* (Torr.) A. Gray)

3. *B. novae-angliae* (Britton) S. G. Smith New England Bulrush
Brackish tidal transition zones in river systems. Coastal, Me. s. to Va.; Ga.; inland in c. N.Y. (*Schoenoplectus novae-angliae* (Britton) M. T. Strong; *Scirpus cylindricus* (Torr.) Britton; *Scirpus novae-angliae* Britton)

4. *B. robustus* (Pursh) Soják Salt Marsh Bulrush Fig. 300
Saline portions of tidal rivers and coastal marshes. Coastal, N.E. and N.Y. s. to Fla., w. to e. Tex.; Calif.; Mex. and S.Am. Schuyler (1974b) notes that plants found from Massachusetts southward usually have ovate spikelets with reddish-brown scales, whereas those northward often have narrow ovate spikelets with brownish scales. Much intergradation occurs between the geological regions. (*Schoenoplectus robustus* (Pursh) M. T. Strong; *Scirpus robustus* Pursh)

5. *B. maritimus* (L.) Palla Bayonnet-grass, Prairie Bulrush

5a. *B. maritimus* subsp. *maritimus*.
Brackish to saline coastal shores and marshes, N.B., P.E I., Que., Me.; intro. from Eur. (*Scirpus fernaldii* E. P. Bicknell; *Scirpus maritima* var. *fernaldii* (E. P. Bicknell) Beetle)

5b. *B. maritimus* subsp. *paludosus* (A. Nelson) T. Koyama. Fig. 300
Brackish to saline coastal and inland shores. P.E.I., N.S. and Que. w. to N.W.T. and Alask. s. to N.J., Mich., Ill., Mo., Tex., N.M., Ariz., Calif. and Mex.; S.Am. Schuyler (1974b) noted that variation within *S. maritimus* subsp. *paludosus* is correlated with the upstream-downstream along rivers or with topographical distribution in marshes. Upstream plants or plants in the upper portion of marshes tend to be taller with open inflorescences and browner scales than plants in more saline sites downstream or in lower portions of marshes. Plants with more open inflorescences occasionally have achenes with prominent dorsal bulges. (*Scirpus paludosus* A. Nelson)

Poaceae (Gramineae) / Grass Family

The grasses belong to the fourth-largest family, yet relatively few of its members occupy aquatic or wetland habitats, unlike the sedges (see Cyperaceae, fig. 189) and the rushes (see Juncaceae), which they superficially resemble. The grasses can be readily distinguished by the following features (fig. 301): stems round in cross-section and hollow, except at nodes; leaves 2-ranked, sheathing stem, with the sheath open and edges often overlapping, and having a ligule at its junction with blade; flowers borne in spikelets, each spikelet with 2 glumes at base and 1–several florets, each floret consisting of a lemma and a palea enclosing the flower with its lodicules (modified petals), a pistil, and 3 or 6 stamens; fruit a grain (caryopsis).

Some grasses, especially *Glyceria* (Manna Grass) and *Zizania* (Wild-rice), are sometimes found submersed and in a vegetative state with floating leaves, ribbon-like and similar to those of *Vallisneria*. These can be recognized as grasses by the presence of a ligule at the junction of the blade and sheath. The presence of a rhizome in *Glyceria* will help to distinguish it from *Zizania*, which being annual, lacks a rhizome and is easily pulled.

References: Barkworth, 2007; Campbell, 1985; Clark and Kellogg, 2007; Dore and McNeill, 1980; Gould, 1975; Gould and Shaw, 1983; Hitchcock, 1950; Pohl, 1980; Soderstrom et al., 1987.

Fig. 299. *Bolboschoenus fluviatilis*: a. habit; b. lower portion of plant; c. inflorescence; d. spikelet; e. flower; f, g. scale, two views; h. achene; i–1. achenes, diagrammatic cross-sections, showing variations (Mason).

Fig. 300. *Bolboschoenus robustus*: a. habit; b. inflorescence; c. scale; d. achene; e. lenticular achene, diagrammatic cross-section (Mason).
Bolboschoenus maritimus subsp. *paludosus*: f. habit; g. inflorescence; h. achene (HCOT).

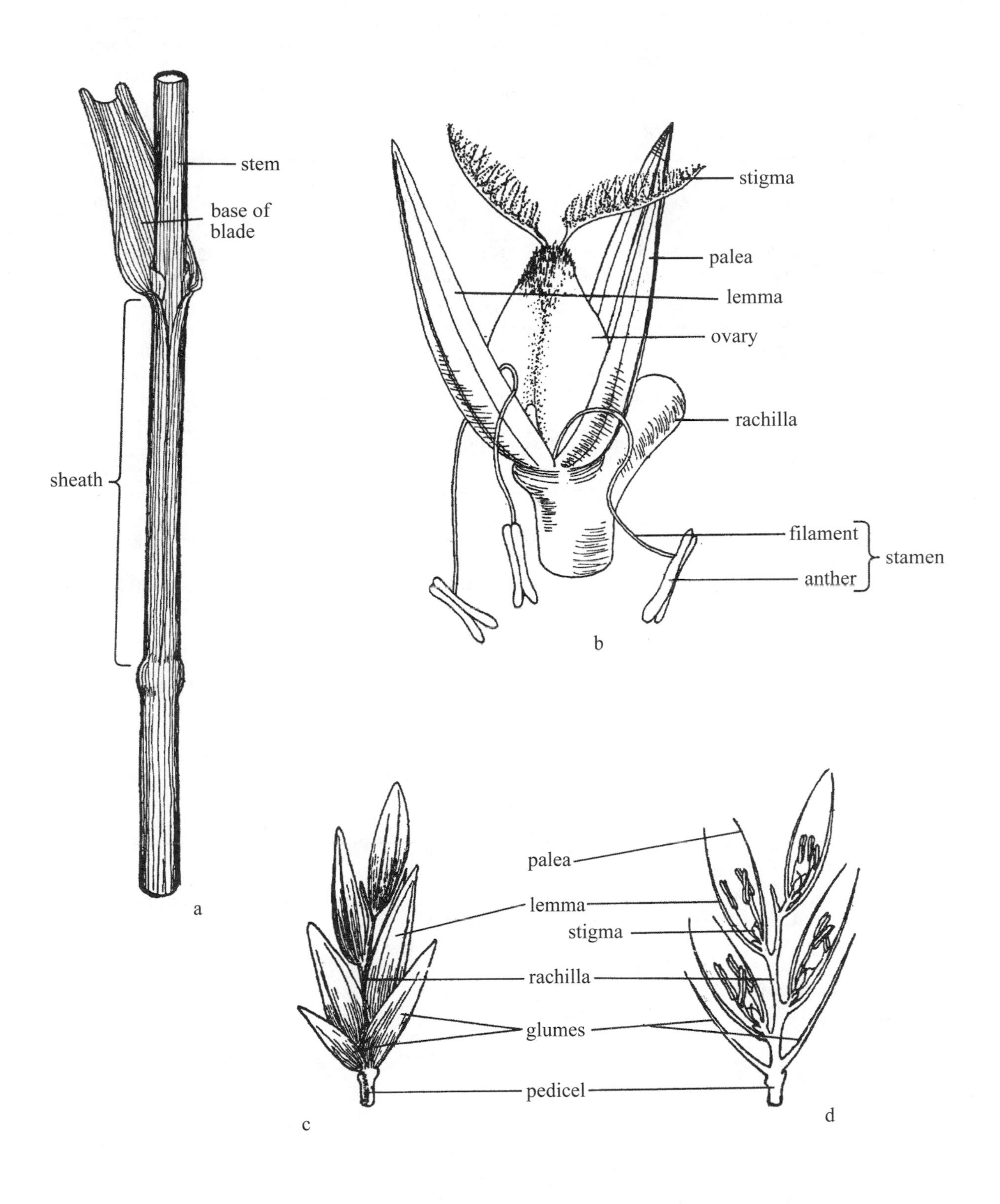

Fig. 301. Structure of Poaceae (grasses): a. section of stem with leaf base, generalized; b. floret, generalized; c. spikelet. side view, generalized; d. spikelet, diagram of longitudinal section (F).

1. Plants with woody, with much-branched stems (up to 8 m tall, frequently forming large colonies), often only vegetative, but when flowering (at intervals of 4–6 years), spikelets 4–7 cm long. . . . . 1. *Arundinaria*
1. Plants herbaceous (if appearing woody, then unbranched), flowering regularly, spikelets less than 4 cm long.
    2. Florets unisexual.
        3. Plants usually less than 0.6 m tall, dioecious; florets several per spikelet.
            4. Inflorescence of both pistillate and staminate plants a dense, spike-like panicle (fig. 303a); plants of salt marshes. . . . . 2. *Distichlis*
            4. Inflorescence of pistillate plants subglobose, inflorescence of staminate plants an open panicle; plants of non-saline sites (*E. reptans*). . . . . 10. *Eragrostis*
        3. Plants usually more than 1 m tall, monoecious; pistillate florets 1 per spikelet, staminate florets sometimes 2.
            5. Pistillate spikelets borne on upper branches, staminate spikelets borne on lower branches of inflorescence (fig. 304b). . . . . 3. *Zizania*
            5. Pistillate and staminate spikelets borne on same branches of inflorescence.
                6. Pistillate spikelets on upper portion of inflorescence branches, staminate spikelets on lower portion (fig. 305a). . . . . 4. *Zizaniopsis*
                6. Pistillate spikelets on lower portion of inflorescence branches, staminate spikelets on upper portion (fig. 306a). . . . . 5. *Tripsacum*
    2. Florets bisexual, or with 1 bisexual floret along with empty lemmas, or with 1 or 2 staminate florets.
        7. Spikelets with 2 to several bisexual florets.
            8. Spikelets sessile (fig. 303c,e), 2 at a node, 2-ranked on opposite sides of rachis, forming simple spikes; glumes and lemmas tipped with long awn. . . . . 6. *Elymus*
            8. Spikelets pedicellate (fig. 307b), not forming simple spikes (some species with spike-like inflorescences, but spikelets not sessile); glumes and lemmas awned or not.
                9. Plants large, coarse reeds, 1–6 m tall: inflorescence plume-like; rachilla bearded, with abundant long silky hairs as long as or longer than lemmas (fig. 307c). . . . . 7. *Phragmites*
                9. Plants less robust, seldom exceeding 1.5 m tall; inflorescence various, but not plume-like; rachilla lacking beard or, if bearded, then hairs short.
                    10. Lemmas densely villous on callus (fig. 308c); ovary pubescent at summit; plants of northern Great Plains. . . . . 8. *Scolochloa*
                    10. Lemmas not villous on callus; ovary glabrous.
                        11. Lemmas 3-nerved.
                            12. Glumes and lemmas rounded dorsally, not keeled; florets lacking cobwebby hairs at base.
                                13. Glumes and lemmas firm, yellowish-brown to purplish, truncate and irregularly cut at summit; plants of northern sites. . . . . 9. *Catabrosa*
                                13. Glumes and lemmas thin and dry, pale to red, acute to acuminate; plants widespread. . . . . 10. *Eragrostis*
                            12. Glumes and lemmas keeled; florets with a tuft of cobwebby hairs at base (except *P. annua*). . . . . 11. *Poa*
                        11. Lemmas 5- or more nerved (fig. 312b).
                            14. Leaf sheaths open; stems slender, weak, and decumbent or prostrate.
                                15. Lemmas conspicuously nerved; plants of freshwater habitats. . . . . 12. *Torreychloa*
                                15. Lemmas obscurely nerved; plants of saline habitats. . . . . 13. *Puccinellia*
                            14. Leaf sheaths closed; stems somewhat thick, erect. . . . . 14. *Glyceria*
        7. Spikelets with 1 bisexual floret (empty lemmas present, or 1 or 2 staminate florets in some species).
            16. Spikelets paired at each joint, 1 sessile, 1 pedicellate (fig. 315b) (in *Andropogon glomeratus* pedicellate spikelet reduced to a glume, or only villous pedicel present, but glume absent, fig. 321h).
                17. Spikelets dissimilar, sessile spikelet fertile, pedicellate spikelet reduced to a glume or absent (fig. 321h). . . . . 15. *Andropogon*
                17. Spikelets alike, sessile and pedicellate both of same size and fertile. . . . . 16. *Saccharum*
            16. Spikelets 1 at each joint, or if appearing paired, then arranged in 2 rows of closely overlapping spikelets on one side of rachis.
                18. Spikelets laterally compressed, with midrib or keel at edge.
                    19. Spikelets in 2 rows along one side of rachis, forming 1-sided spikes (figs. 316a, 319a).
                        20. Spikelets lanceolate to narrowly lanceolate; glumes unequal (fig. 316c); rhizomatous perennials, all but 1 species halophytic. . . . . 17. *Spartina*
                        20. Spikelets subglobose (fig. 319e); glumes subequal, tips crossing; tufted annuals, of freshwater sites. . . . . 18. *Beckmannia*
                    19. Spikelets not forming 1-sided spikes.

21. Inflorescence a panicle, spikelets with ciliate or hispidulous margins.
22. Florets several, overlapping in a single row (fig. 320d); glumes absent. . . . . 19. *Leersia*
22. Florets 3, upper 1 bisexual, lower 2 staminate, not appearing in a row; glumes present, as long as and enclosing florets. . . . . 20. *Anthoxanthum*
21. Inflorescence spike-like or paniculate, spikelets lacking fringed margins.
23. Spikelet with 2 sterile florets (empty lemmas) below bisexual floret. . . . . 21. *Phalaris*
23. Spikelet lacking sterile florets below bisexual floret.
24. Inflorescence spike-like, densely compact, cylindrical (fig. 321c–e); glumes strongly folded, conspicuously keeled. . . . . 22. *Alopecurus*
24. Inflorescence paniculate, open and loose, or if contracted, then not densely compact; glumes not strongly folded or conspicuously keeled.
25. Lemmas surrounded basally by tufts of long hairs from callus (fig. 323b); lemmas dorsally awned. . . . . 23. *Calamagrostis*
25. Lemmas lacking tufts of basal hairs; lemmas awnless (or gradually tapering to sharp tip).
26. Glumes much shorter than lemmas; panicle delicate, diffuse, with branches and pedicels capillary, or slender, densely contracted. . . . . 24. *Muhlenbergia*
26. Glumes as long as or longer than lemmas; panicle open, but not diffuse, branches and pedicels slender, but not capillary. . . . . 25. *Agrostis*
18. Spikelets dorsally compressed (fig. 562), with midrib or keel in middle of flattened side.
27. Spikelets in 2 or 4 rows restricted to one side of a flattened rachis (figs. 325d, 326g, 327a). . . . . 26. *Paspalum*
27. Spikelets not in rows restricted to one side of rachis.
28. Spikelets subtended by 1–12 slender bristles, these persisting on rachis after spikelets fall. . . . . 27. *Setaria*
28. Spikelets not subtended by bristles.
29. Spikelets gibbous because of swollen base of second glume (fig. 328b). . . . . 28. *Sacciolepis*
29. Spikelets not gibbous.
30. Spikelets spiny-hispid (fig. 329b), clustered on thick, spike-like lateral branches of panicle. . . . . 29. *Echinochloa*
30. Spikelets glabrous, or hairy, but then not hispid, borne in open panicles, or if inflorescence contracted and spike-like, then not with thick, lateral branches.
31. Basal leaves distinctly different from cauline leaves, forming a winter rosette. . . . . . . 30. *Dichanthelium*
31. Basal leaves similar to cauline leaves, not forming a winter rosette. . . . . 31. *Panicum*

1. *Arundinaria* (Cane, Canebrake Bamboo)

Perennials, arborescent; stems 0.5–8 m tall, becoming woody, much-branched, leafy; leaves with both outer and inner ligule, sheath summit with conspicuous bristles; inflorescence an open raceme or panicle; spikelets several-flowered, 3–7 cm long; reproductive stems dying after fruiting.

References: Clark and Triplett, 2007; McClure, 1973.

1. Plants 2–8 m tall; stem internodes usually grooved (away from branches); leaves along stem deciduous, sheaths 9–15 cm long. . . . . 1. *A. gigantea*
1. Plants usually sorter than 2.5 m tall; stem internodes usually round (terete); leaves along stem persistent or becoming deciduous, sheaths 11–18 cm long. . . . . 2. *A. tecta*

1. *A. gigantea* (Walter) Muhl. Giant Cane, Wild Cane, Switch Cane, Wild Bamboo Fig. 302
Riverbanks, swamps, sloughs and bayous, often forming extensive cane-brakes. L.I., N.Y., N.J., Del. and Md. w. to W.Va., s. Ohio, s. Ind., s. Ill., s. Mo., and e. Kans., s. to n. Fla. and e. Tex., centered on the Mississippi embayment. Species in this complex are extremely difficult to distinguish without considerable experience with this group. The stems of this native bamboo are used in making baskets and for fishing rods. (*A. gigantea* subsp. *macrosperma* (Michx.) McClure; *A. tecta* var. *distachya* Rupr.)

2. *A. tecta* (Walter) Muhl.
Swampy woods, wet to moist pine barrens, sandy margins of streams. Chiefly Atlantic Coast, se. N.Y., N.J., Pa. and Md. s. to Va., N.C., Ga. and Fla., w. to La. and e. Tex.; Tenn. This species tends to prefer more moist sites than *A. gigantea* (*A. gigantea* subsp. *tecta* (Walter) McClure)

Fig. 302. *Arundinaria gigantea*: a. habit; b. summit of leaf sheath, inner and outer surfaces showing ligule and auricles; c. floret, two views (Hitchcock).

2. *Distichlis* (Spike Grass, Salt Grass, Alkali Grass)

Perennial herbs, spreading readily by rhizomes or stolons, halophytic; stems spreading to ascending; leaves stiffly erect-spreading, conspicuously 2-ranked; stems usually hidden by overlapping leaf sheaths; plants dioecious (occasionally bisexual); inflorescence a spike-like panicle or raceme, the spikelets compressed.

Reference: Barkworth, 2003.

1. *D. spicata* (L.) Greene Spike Grass Fig. 303
Salt marshes and wet alkaline or saline soil. Coastal, N.S. and Que., w. to N.W.T. and B.C. s. to Fla., w. to Tex., N.M., Calif.; Mex. and Cuba. Growing in alkaline and saline sites of the Great Plains and deserts of the West, as well as along the West Coast; Cuba. In the East our plants are almost strictly coastal, especially in salt marshes. The species, with its conspicuously 2-ranked leaves, is distinctive even in the vegetative stage. Although western plants have been treated as a separate variety, Barkworth (2003) recognizes no infraspecific taxa. (*D. spicata* var. *stricta* (Torr.) Beetle)

3. *Zizania* (Wild-rice)

Annuals with short roots (easily pulled up); stems slender to thick, reaching 4 m tall; leaves flat, long, often floating when young; spikelets in much-branched panicles; flowers unisexual, spikelets of the upper inflorescence branches pistillate, spikelets of lower branches staminate, the latter red, greenish, or yellow; lemmas of pistillate spikelets long-awned. *Zizania* is sometimes confused with *Zizaniopsis*.

Both species of *Zizania* are important food for wildlife, especially wildfowl. *Zizania palustris* was (and locally still is) an important food source for native Americans, especially in Wisconsin and Minnesota, and is now commercially harvested for domestic markets.

References: Aiken, 1986; Aiken et al., 1988; Chambliss, 1940; Dore, 1969; Duvall and Biesboer, 1988; Fassett, 1924; Terrell, 2007; Terrell et al., 1997; Warwick and Aiken, 1986.

---

1. Pistillate lemmas thin and papery, dull, finely striate; rudimentary (aborted) spikelets very narrow, thread-like, 0.4–1 mm wide (fig. 304f); branches of pistillate inflorescence divergent at maturity.
  2. Plants 1–4 m tall; leaves (5)10–72 mm wide; ligules (6)10–20(25) mm long; awn of pistillate lemma 10–70(90) mm long. 1a. *Z. aquatica* var. *aquatica*
  2. Plants 0.3–1 m tall; leaves 3–12(20) mm wide; ligules ca. 3 mm long; awn of pistillate lemma 1–8 mm long; plants of St. Lawrence Seaway. 1b. *Z. aquatica* var. *brevis*
1. Pistillate lemmas firm and tough, lustrous, coarsely ribbed; rudimentary (aborted) spikelets 1.5–2.6 mm wide (fig. 304d); branches of pistillate inflorescence typically appressed at maturity (or with few somewhat spreading).
  3. Plants 0.7–2 m tall; leaves 3–15(21) mm wide, ligules 3–5(10) mm long; lower pistillate branches with 2–8 spikelets. 2a. *Z. palustris* var. *palustris*
  3. Plants 0.9–3 m tall; leaves (10)20–40 mm wide, ligules 10–15 mm long; lower pistillate branches with 9–30 spikelets. 2b. *Z. palustris* var. *interior*

---

1. *Z. aquatica* L. Southern Wild-rice

1a. *Z. aquatica* var. *aquatica* Fig. 304
Fresh or saline waters of river mouths, lakes and ponds; usually in shallow water. N.E. and s. Que. w. to s. Ont., Wisc., and se. Minn., s. to e. Ohio, c. Ind., c. Ill. and ne. Mo.; coastal plain s. to Fla, w. to s. La. (*Z. aquatica* var. *subbrevis* Boivin)

1b. *Z. aquatica* var. *brevis* Fassett
Tidal waters and tributaries of the St. Lawrence R., vicinity of Quebec City, Que.

2. *Z. palustris* L. Northern Wild-rice

2a. *Z. palustris* var. *palustris*
Quiet waters. N.S. and N.B. w. to Alta., s. to N.E., c. N.Y., nw. Pa., n. Ind., Iowa, Minn. and Man.; westward by deliberate plantings to Mont., Oreg., and B.C. (Terrell, 2007). Electrophoretic studies by Warwick and Aiken (1986) supported treatment of *Z. palustris* as distinct from *Z. aquatica* (Dore, 1969; Terrell et al., 1997), in contrast with Fassett's (1924) taxonomic treatment of the complex as consisting of only one species. (*Z. aquatica* var. *angustifolia* Hitchc.)

2b. *Z. palustris* var. *interior* (Fassett) Dore Fig. 304
Quiet waters and marshes. N.S., N.B. and s. Que. w. to s. Mich., Minn., and se. Man., s. to Ohio, Ill., Neb. and ne. Kans. (*Z. aquatica* var. *interior* Fassett; *Z. interior* (Fassett) Rydb.)

Fig. 303. *Distichlis spicata*: a. habit; b. floret (Hitchcock).
Elymus riparius: c. portion of inflorescence (Hitchcock).
Elymus virginicus: d. upper portion of plant (F); e. diagram of pair of spikelets with florets subtended by awned glumes (from node above) (*top*) and spikelets with florets subtended by awned glumes (from node below) (*bottom*) (Chase).

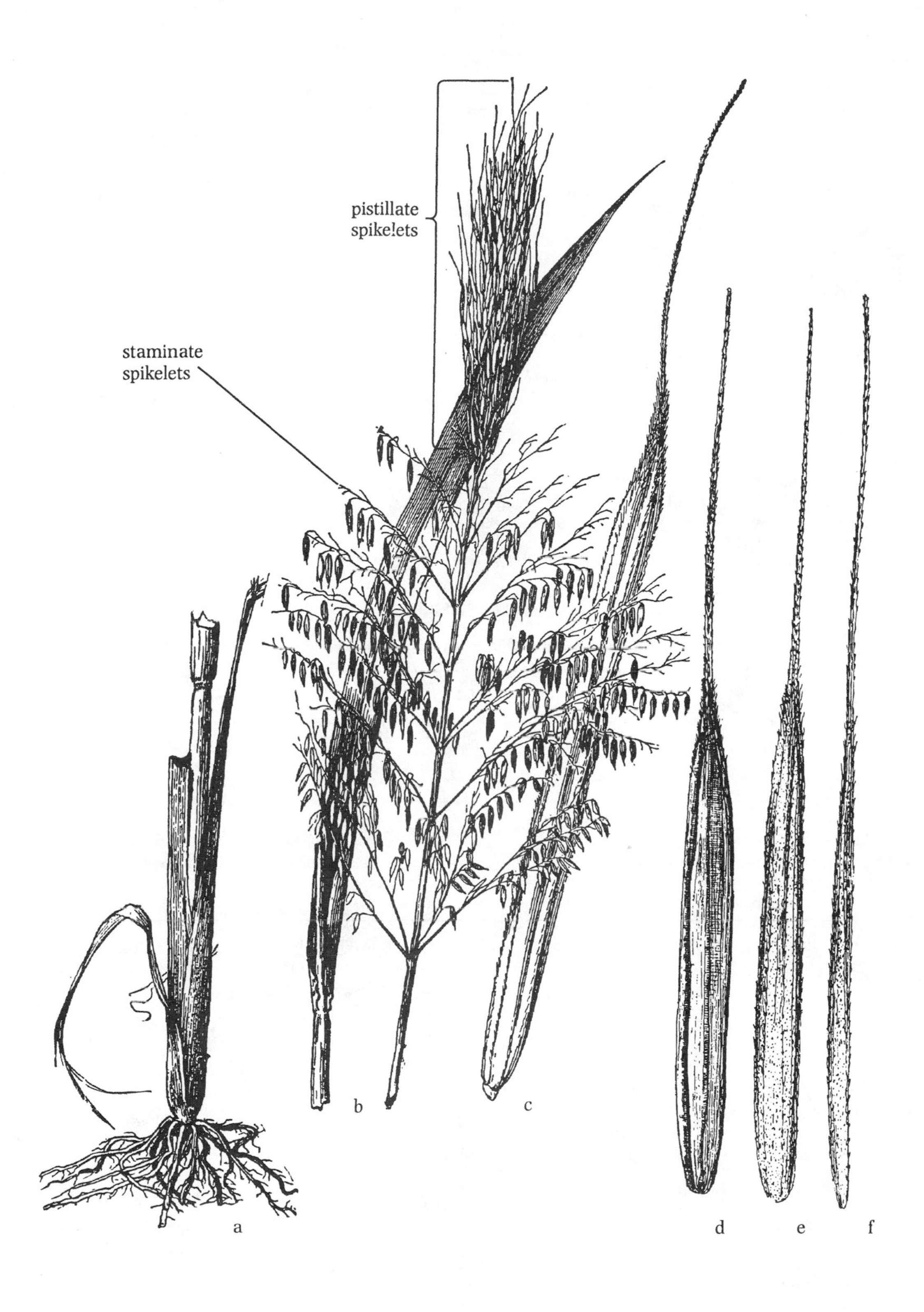

Fig. 304. *Zizania palustris* var. *interior*: a. basal portion of plant (Hitchcock); b. leaf and panicle (Hitchcock); c. pistillate spikelet, fertile, filled with grain (Hitchcock); d. pistillate spikelet, aborted (F). *Zizania aquatica* var. *aquatica*: e. pistillate spikelet, fertile, filled with grain; f. pistillate spikelet, aborted (F).

4. *Zizaniopsis* (Water-millet, False Wild-rice)

Perennial herbs; stems 1–3 m or more tall; leaves long, narrow, with rough margins; spikelets unisexual; pistillate and staminate spikelets on same branches, staminate spikelets on lower half of branch, pistillate spikelets on upper half.

REFERENCE: Terrell, 2007.

1. *Z. miliacea* (Michx.) Döll & Asch. Water-millet, Giant Cutgrass, Southern Wild-rice, False Wild-rice Fig. 305
Shallow water, swamps, marshes, ponds and stream margins; fresh or saline waters. Md. w. to Ky., c. Ill., se. Mo, Ark., and Okla., s. to Fla., La. and Tex.

5. *Tripsacum* (Gamma Grass)

Perennial herbs, with thick creeping rhizomes; stems erect, up to 5 m; leaves broad, flat; spikelets in terminal or axillary spikes, segmenting at the joints at maturity; flowers unisexual, staminate florets in upper portion of spike, pistillate florets in lower portion.

REFERENCE: Barkworth, 2007.

1. *T. dactyloides* (L.) L. Eastern Gamma Grass Fig. 306
Moist to wet grasslands and low ground, less frequently swales and shores. S. N.E. and N.Y. w. to se. Mich., Ill., s. Iowa, se. Neb. and Kans., s. to Fla., Tex., and n. Mex.; W.I.

6. *Elymus* (Wild-rye)

Perennial herbs; inflorescence a dense, bristly spike; spikelets 2 or 3 at each rachis node, each several-flowered, lemma long-awned.

REFERENCES: Barkworth et al., 2007; Church, 1967.

1. Glumes 0.5–2.3 mm wide, flat; inflorescence erect, the bases often sheathed. . . . . 1. *E. virginicus*
1. Glumes 0.4–0.8 mm wide, bristle-like; inflorescence arching. . . . . 2. *E. riparius*

1. *E. virginicus* L. Virginia Wild-rye Fig. 303
Alluvial woods, riverbanks, wet meadows, marshy shores, and margins of salt marshes, tidal waters, and coastal shores. Widespread in N.Am., Nfld. and Lab. w. to Alta., B.C., and ne. Wash., s. to n. Fla., Tex., e. Wyo., N.M. and Ariz. Several intergrading varieties have been described in this complex species (Barkworth et al., 2007), but appear to be geographically ill-defined and not included here.

2. *E. riparius* Wiegand Riverbank Wildrye Fig. 303
Riverbanks, alluvial woods, thickets and borders of streams. Me. and s. Que. w. to Ont., Wisc. and Minn. s. to w. N.C., n. G., n. Ala. and n. Ark.

E. canadensis, *E. villosa*, and *E. wiegandii* occur in dryish to wet woods and sometimes grow in alluvial soils.

7. *Phragmites* (Common Reed)

Perennial herbs, arising from rhizomes, colonial; stems 2–6 m tall; leaf blades broad, with tendency of leaves to be present on only one side of the stem; inflorescence large, plume-like; rachis bearded, with long silky hairs.

Phragmites is sometimes confused with *Zizania* (Wild-rice), but in the young, vegetative state, it can be distinguished by its thick rhizomes, which are deep in the substrate, making it difficult to pull up.

REFERENCES: Clayton, 1968; Haines, 2010, 2011; Saltonstall, 2002; Saltonstall et al., 2004; Saltonstall and Hauber, 2007.

1. Plants more robust, stems occurring in dense clones; leaf blades dark green to gray-green (sometimes yellowish green), with ligules 0.4–1 mm long (excluding fringe); middle and upper internodes of stem dull, ridged, greenish to tan (sheaths persisting); rhizomes flattened, more than 15 mm in width. . . . . 1. *P. australis*
1. Plants less robust, stems occurring in clones with stems sparsely set apart; leaf blades yellow-green, with ligules 1–1.7 mm long; middle and upper internodes of stem smooth, highly lustrous, reddish to purplish (exposed because sheaths readily deciduous; rhizomes terete, 15 mm or less in width. . . . . 2. *P. americanus*

Fig. 305. *Zizaniopsis miliacea*: a. habit; b. staminate spikelet; c. pistillate spikelet; d. ripe grain (Hitchcock).

Fig. 306. *Tripsacum dactyloides*: a. habit; b. pistillate spikelet; c. pair of staminate spikelets (Hitchcock).

1. *P. australis* (Cav.) Trin. ex Steud. Common Reed; The Phrag Fig. 307

Freshwater, brackish, or saline marshy areas, pond margins, ditches, and borders of salt marshes and tidal waters, especially disturbed sites. Extremely invasive; Nfld., and Que. w. to Man., Alta., s. N.W.T. and B.C., s. to S.C., Fla., Tex. and Calif. This plant is distinctive because of its large size and densely colonial habit. There is considerable confusion as to the nomenclature within *Phragmites*. In a study of *Phragmites* in North America Saltonstall (2002) recognized three genetic lineages, one introduced from Europe, *P. australis* subsp. *australis*, which is incredibly invasive and considered to be worldwide in distribution. The other two genetic entities are native: the more northern taxon has been recognized by Saltonstall and Hauber (2007, as *P. australis* subsp. *americanus*); the other native to the Atlantic coast of Florida, around the Gulf of Mexico westward to the Gulf of California and southward through Mexico, Central America to South America, referable to *P. australis* subsp. *berlandieri*. Clayton (1968) considered the European entity (*P. communis*) to be conspecific with the Australian *P. australis* (a cosmopolitan entity), hence the shift in nomenclature. But Saltonstall et al. (2004) suggest that the genetic relationship of the European plants and the Australian populations is poorly understood—the two entities possessing different chloroplast DNA haplotypes. Thus, it is not clear what is the correct nomenclature for our introduced, European invasive taxon in North America; it may well be that the name *P. communis* will be resurrected for this invasive plant (*P. communis* Trin.)

2. *P. americanus* (Saltonstall, P.M. Peterson & Soreng) A. Haines American Reed Fig. 307

Freshwater, brackish, or saline natural marshy areas, fens, swales, lakeshores, and borders of salt marshes and tidal waters. Based on a suite of characters that readily distinguish the native from the invasive European plants, Haines (2010) nomenclaturally elevated the subspecies (*P. australis* subsp. *americanus*) of Saltonstall et al. (2004) to the rank of species. Ironically, our native species is rare and in need of conservation, ranking from "special concern" to "endangered" status in various states. (*P. australis* subsp. *americanus* Saltonstall, P.M. Peterson & Soreng)

8. *Scolochloa* (Rivergrass)

Perennial herbs, stout, arising from rhizomes, up to 2 m tall; leaves flat; inflorescence terminal, an open panicle; spikelets several-flowered, with unequal glumes, the lower glume shorter than lowest lemma, the upper glume nearly as long as the uppermost floret.

Reference: Barkworth, 2007.

1. *S. festucacea* (Willd.) Link Whitetop Grass, Sprangle-top, Common Rivergrass Fig. 308

Shallow water and marshes. Man. w. to s. N.W.T. and Alask., s. to Minn., nw. Iowa, n. Neb., Kans., Wyo., nw. Mont., se. Oreg. and B.C.

9. *Catabrosa* (Brook Grass)

Perennial herbs, arising from rhizomes; stems creeping at base and rooting at nodes; leaves flat; inflorescence a pyramidal panicle; spikelets relatively short obtuse, florets loosely overlapping; lemmas 3-nerved, truncate at tip.

Reference: Barkworth, 2007.

1. *C. aquatica* (L.) Beauv. Brook Grass, Water Hair Grass, Water Whorlgrass Fig. 309

Wet meadows, stream margins, shores of lakes and ponds. Se. Lab., w. Nfld., and Gulf of St. Lawrence w. to Sask., B.C. and sw. Alask., s. to Wisc., Iowa, Neb., Colo., n. N.M., Ariz., Nev. and s. Oreg.

10. *Eragrostis* (Love Grass)

Perennial or annual (ours) herbs, small; inflorescence an open panicle, spikelets many-flowered. Similar to *Glyceria*, but lemmas only 3-nerved, and nerves not conspicuous; glumes and lemmas deciduous, but paleas persistent on rachilla.

Reference: Peterson, 2003.

1. Stems creeping, rooting at nodes and forming mats.
    2. Plants bisexual; stems and sheaths glabrous; lemmas glabrous; panicles open; anthers 2. . . . . 1. *E. hypnoides*
    2. Plants unisexual (dioecious); stems and sheaths pubescent; lemmas pubescent; pistillate panicles 1–3 cm long, about as wide as long; staminate panicles open; anthers 3. . . . . 2. *E. reptans*
1. Stems erect or ascending, not rooting at nodes.
    3. Spikelets linear, usually more than 5 mm long, with 5–15 florets. . . . . 3. *E. pectinacea*
    3. Spikelets oval, 2–3 mm long, with 2–5(7) florets. . . . . 4. *E. frankii*

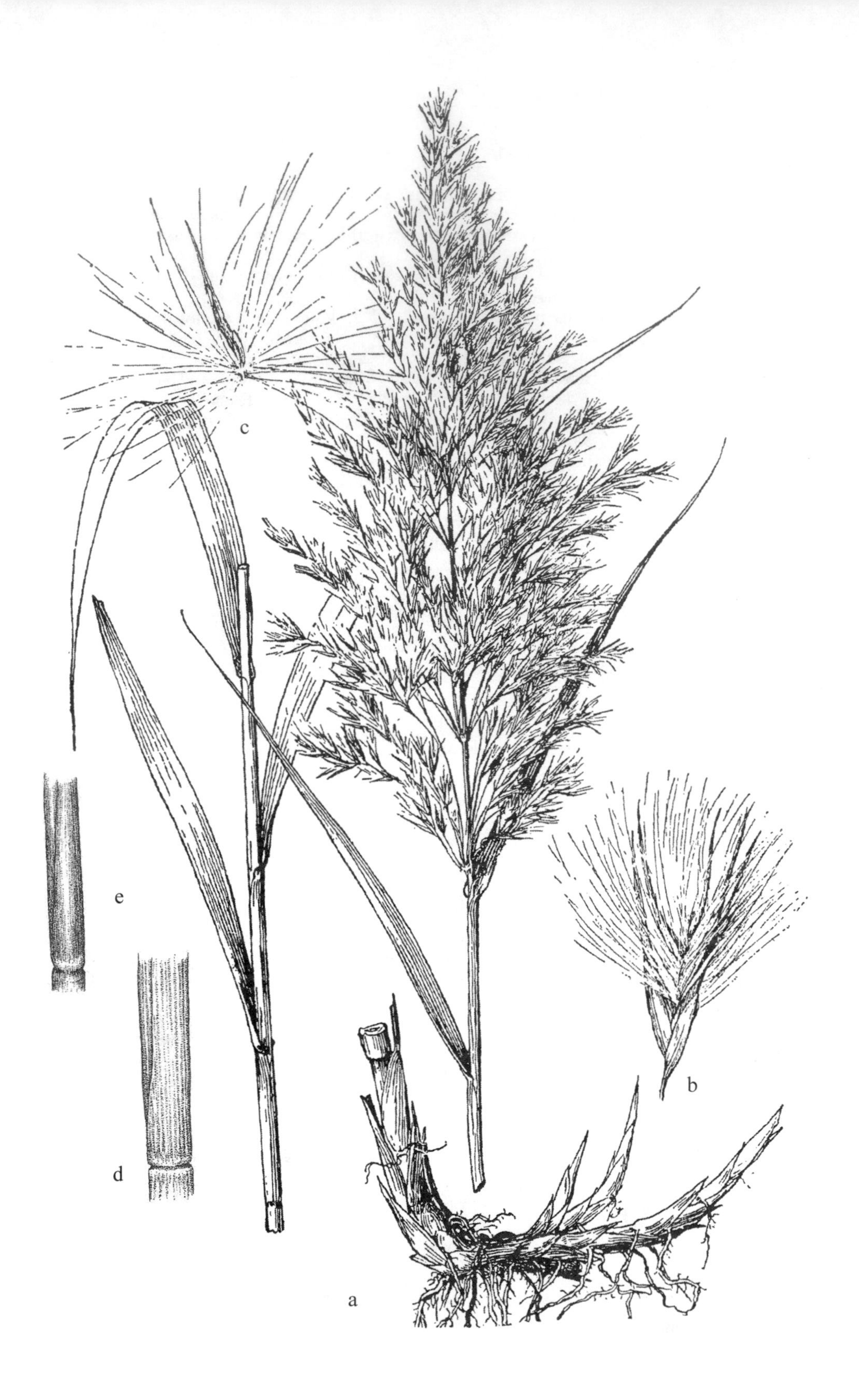

Fig. 307. *Phragmites australis*: a. habit; b. spikelet; c. floret (Hitchcock); d. rigid, dull, pale internode (Elizabeth Farnsworth, in *Flora Novae Angliae*, NPT).
Phragmites americanus: e. smooth lustrous, dark internode (Elizabeth Farnsworth, in *Flora Novae Angliae*, NPT).

Fig. 308. *Scolochloa festucacea*: a. habit; b. spikelet; c. floret (Hitchcock).

Fig. 309. *Catabrosa aquatica*: a. habit; b. spikelet; c. floret (Hitchcock).
Eragrostis frankii: d. panicle and floret (F).
Eragrostis pectinacea: e. panicle and floret (Hitchcock).

1. *E. hypnoides* (Lam.) Britton, Sterns & Poggenb. Creeping Love Grass, Teel Love Grass Fig. 310
Muddy shores and sandbars of streams and rivers. Me. and Que. w. to Sask. and Wash., s. to Fla., e. Tex., N.M. and Calif.; Mex., W.I. and S.Am.

2. *E. reptans* (Michx.) Nees Creeping Love Grass Fig. 310
River margins and shores, and frequently beds of streams and ponds after drought. W.Va., Ky. and s. Ill. w. to S.D., s. to Ga., n. Fla., La., Tex., N.M. and e. Mex; S.Am.

3. *E. pectinacea* (Michx.) Nees ex Steud. Small Love Grass Fig. 309
Sandy and gravelly shores, river sandbars, shallow water and mud of bottomlands and ditches; weedy in wet disturbed sites. Me. and sw. Que. w. to Ont., N.D., Mont. and B.C., s. to Fla., Ariz., Calif. and Mex.; most common in the eastern U.S. Our taxon is the widespread var. *pectinacea*.

4. *E. frankii* C. A. Mey. ex Steud. Sandbar Love Grass Fig. 309
River sandbars and muddy banks and seasonally wet soils. N.B., N.H., Vt. and sw. Que. w. to s. Ont., s. Mich. and s. Minn., and Neb., s. to Fla., Ark., Tex. and N.M. (*E. frankii* var. *brevipes* Fassett)

11. *Poa* (Bluegrass, Meadow Grass)

Perennial or annual herbs, arising from rhizomes or tufted; leaves with flat blades ending in boat-shaped apex; spikelets small, often only 2- or 3-flowered; upper glume shorter than lowest lemma; glumes and lemmas keeled and distinctly nerved, awnless, usually with a tuft of cobwebby hairs at base.

REFERENCE: Soreng, 2007.

1. Plants annual, 5–20 cm tall; lemmas lacking cobwebby tuft of hairs at base. . . . . 1. *P. annua*
1. Plants perennial, over 25 cm tall; lemmas with cobwebby tuft of hairs at base.
    2. Panicle many-flowered, branches usually 3–5 per node; plants somewhat robust, stems usually with about 6 leaves, the leaf sheaths closed 10–20% of the length. . . . . 2. *P. palustris*
    2. Panicle few-flowered, branches 2 per node; plants slender, stems usually with 2 or 3 leaves, the leaf sheaths closed 25–60% of the length. . . . . 3. *P. paludigena*

1. *P. annua* L. Annual Bluegrass
Disturbed sites, weedy, sometimes found in flooded sites or muddy shores. Greenl., Lab. and Nfld. w. to N.W.T., Alask, s. to Fla. and Calif.; intro. from Eur.

2. *P. palustris* L. Fowl Bluegrass Fig. 312
Meadows, shores and thickets. Lab. and Nfld. w. to sw. N.W.T. and Alask., s. to N.E., Va., w. N.C., Ohio, n. Ill., nw. Mo., n. N.M. and n. Calif. This species is often found only in vegetative condition, with floating leaves.

3. *P. paludigena* Fernald & Wiegand Bog Bluegrass Fig. 311
Bogs, fens, swamps, wet woods and springy places, especially calcareous sites. Rare; c. N.Y. w. to Mich., Wisc., and Minn., s. to Pa., Va., W.Va., n. Ind., and n. Ill. and Iowa.

Poa autumnalis, *P. sylvestris*, and *P. trivialis* tend to grow in drier sites, but may occur in alluvial soils, springy places, or seeps.

12. *Torreyochloa* (False Manna Grass)

Perennial herbs; stems slender, weak, and decumbent or procumbent; panicles contracted, spikelets small, awnless; lemmas dorsally rounded. This genus is similar to *Puccinellia*, which has obscure nerves on the lemmas, and *Glyceria* which differs in having closed sheaths.

REFERENCE: Davis, 2007.

1. Plants 10–60 cm tall; leaf blades 1–3.5 mm wide; spikelets 3–5 mm long; lemmas 2–2.8 mm long. . . . . 1. *T. fernaldii*
1. Plants 30–100 cm tall; leaf blades 4–10 mm wide; spikelets 5–7 mm long; lemmas 2.5–3.5 mm long. . . . . 2. *T. pallida*

1. *T. fernaldii* (Hitchc.) Church Fernald's False Manna Grass Fig. 312
Shallow water and wet, open places. Nfld. w. to Mich. and Minn., s. to N.E., L.I., N.Y. and Pa. (*Glyceria fernaldii* (Hitchc.) St. John; *Puccinellia fernaldii* (Hitchc.) E. G. Voss)

2. *T. pallida* (Torr.) Church Pale False Manna Grass Fig. 313
Shallow water, pond margins, sloughs and wet open places. Sw. N.S. and s. Me. w. to s. Ont. and Mich., s. to Va., Tenn. and se. Mo. (*Glyceria pallida* (Torr.) Trin.; *Puccinellia fernaldii* (Hitchc.) E. G. Voss)

Fig. 310. *Eragrostis hypnoides*: a. habit; b. floret (Hitchcock).
Eragrostis reptans: c. pistillate and staminate plants with pistillate and staminate florets (Hitchcock).

Fig. 311. *Poa paludigena*: a. habit (NYS Museum); b. spikelet (NYS Museum); c. panicle (Hitchcock); d. floret (Hitchcock).

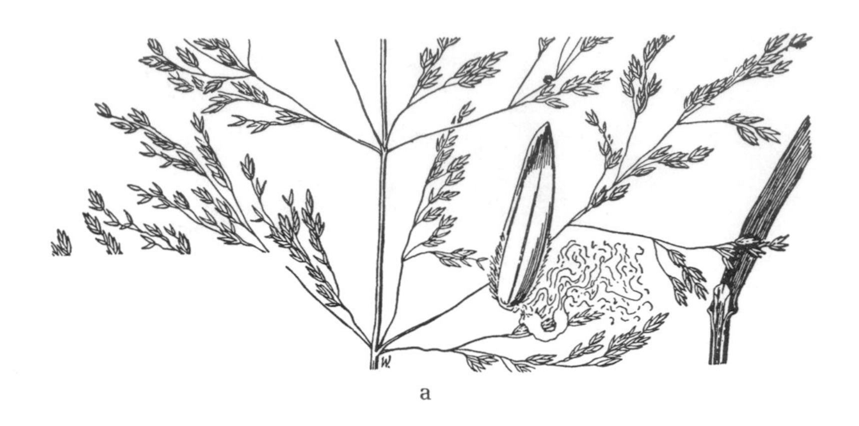

Fig. 312. *Poa palustris*: a. panicle and floret (with cobwebby tuft of hairs at base), and leaf sheath with ligule (Hitchcock).
Torreyochloa fernaldii: b. portion of plant with floret (rounded back of lemma with conspicuous nerves) (Hitchcock).
Puccinellia maritima: c. portion of plant with floret (Hitchcock).

Fig. 313. *Torreyochloa pallida*: a. panicle; b. floret (Hitchcock).
Glyceria obtusa: c. panicle; d. floret. two views (Hitchcock).
Glyceria striata: e. habit; f. spikelet; g. floret (Hitchcock).
Glyceria septentrionalis var. *arkansana*: h. portion of panicle and floret (lemma with rounded back and conspicuous nerves) (Hitchcock).

13. *Puccinellia* (Alkali Grass)

Perennial herbs; stems slender, weak, and decumbent or procumbent; panicles contracted, spikelets small, awnless; lemmas dorsally rounded. This genus is similar to *Torreyochloa*, but are plants of saline habitats, and to *Glyceria*, which differs in having closed sheaths.

Reference: Davis and Consaul, 2007.

---

1. Plants 15–100 cm tall; spikelets 5–12 mm long; lower glume 2–2.5(3) mm long; upper glume (2.5)3–4(4.5) mm long; anthers 1.5–2.6 mm long. . . . . 1. *P. maritima*
1. Plants 2–45 cm tall; spikelets 3–7 mm long; lower glume 1–2(2.5) mm long; upper glume 1.5–2.5(2.8) mm long; anthers 0.5–1.2 mm long. . . . . 2. *P. pumila*

---

1. *P. maritima* (Huds.) Parl. Seaside Alkali Grass Fig. 312
Saline shores and salt marshes. Coastal, St. P. et Miq. and Que. s. to N.E.

2. *P. pumila* (Vasey) Hitchc. Smooth Alkali Grass
Saline shores and salt marshes. Coastal, Greenl. and Baffin I., Nfld. and Lab. w. to Man.; Mass; Alask. and B.C. s. to Calif. (*Puccinellia langeana* subsp. *alaskana* (Schribn. & Merr.) T. J. Sørensen ex Hultén)

14. *Glyceria* (Manna Grass)

Perennial herbs, arising from rhizomes; stems thick, erect; leaf blades broad and flat, sheaths closed; spikelets several-flowered, awnless; lemmas dorsally rounded, conspicuously nerved. *Glyceria* is unique among the grasses in having closed sheaths.

References: Barkworth and Anderton, 2007; Church, 1949; Dore, 1947; Freckmann and Reed, 1979; Gutteridge, 1954.

---

1. Spikelet linear (fig. 314a), 10–40 mm long, nearly round in cross-section.
    2. Lemmas acute, conspicuously shorter than paleas (fig. 314c). . . . . 1. *G. acutifolia*
    2. Lemmas rounded (fig. 314d), only slightly shorter than paleas.
        3. Lemmas 5–6(8) mm long. . . . . 2. *G. fluitans*
        3. Lemmas 3–4(5.5) mm long.
            4. Lemmas glabrous between nerves, nerves slightly scabrous. . . . . 3. *G. borealis*
            4. Lemmas scabrous to somewhat hispid between nerves, nerves distinctly scabrous.
                5. Lemmas thick, obscurely nerved, scabrous on nerves, hairs (prickles) ca. 0.05 mm long; leaves 2–15 mm wide . . . . 4a. *G. septentrionalis* var. *septentrionalis*
                5. Lemmas thin, conspicuously nerved, hispidulous on nerves, hairs ca. 0.1 mm long; leaves 10–20 mm wide. . . . . 4b. *G. septentrionalis* var. *arkansana*
1. Spikelet ovate (fig. 313f), less than 10 mm long, more or less laterally compressed.
    6. Panicles contracted (fig. 313c); pedicels usually as long as or slightly longer than spikelets.
    6. Panicles thick, compact; spikelets 5–9 mm long. . . . . 5. *G. obtusa*
        7. Panicles long and slender; spikelets about 4 mm long. . . . . 6. *G. melicaria*
        7. Panicles open, with lax, spreading branches (fig. 313e); pedicels usually much longer than spikelets.
            8. Spikelets 3–5 mm wide; lemmas with veins visible, but not distinctly raised. . . . . 7. *G. canadensis*
            8. Spikelets 1–2.5 mm wide; lemmas with conspicuously raised veins.
                9. Upper (second) glume 0.7–1.4 mm long; spikelets 2–4.5 mm long; leaves 2–5(8) mm wide; panicles 10–20 cm long. . . . . 8. *G. striata*
                9. Upper (second) glume (1.8)2–4 mm long; spikelets 4–8(12) mm long; leaves 6–20 mm wide panicles 20–40 cm long.
                    10. Leaf sheaths smooth, leaf blades 6–12(15) mm wide; upper glume 1.8–2(2.5) mm long. . . . . 9. *G. grandis*
                    10. Leaf sheaths scaberulous; leaf blades 8–18(20) mm wide; upper glume 2.5–4 mm long. . . . . 10. *G. maxima*

---

1. *G. acutiflora* Torr. Creeping Manna Grass Fig. 314
Shallow water and wet soils. S. N.H. w. to Mich., s. to Va., W.Va., Tenn. and Mo.

2. *G. fluitans* (L.) R. Br. Floating Manna Grass Fig. 314
Shallow water. Widely scattered localities, Nfld., Gaspé Pen., Que., and N.S.; Nantucket I., Mass.; N.Y., n. N.J., se. Pa. and Md.; Tenn. and Ark.; Black Hills, S.D., Ida. and n. Calif.; intro. from Eur.

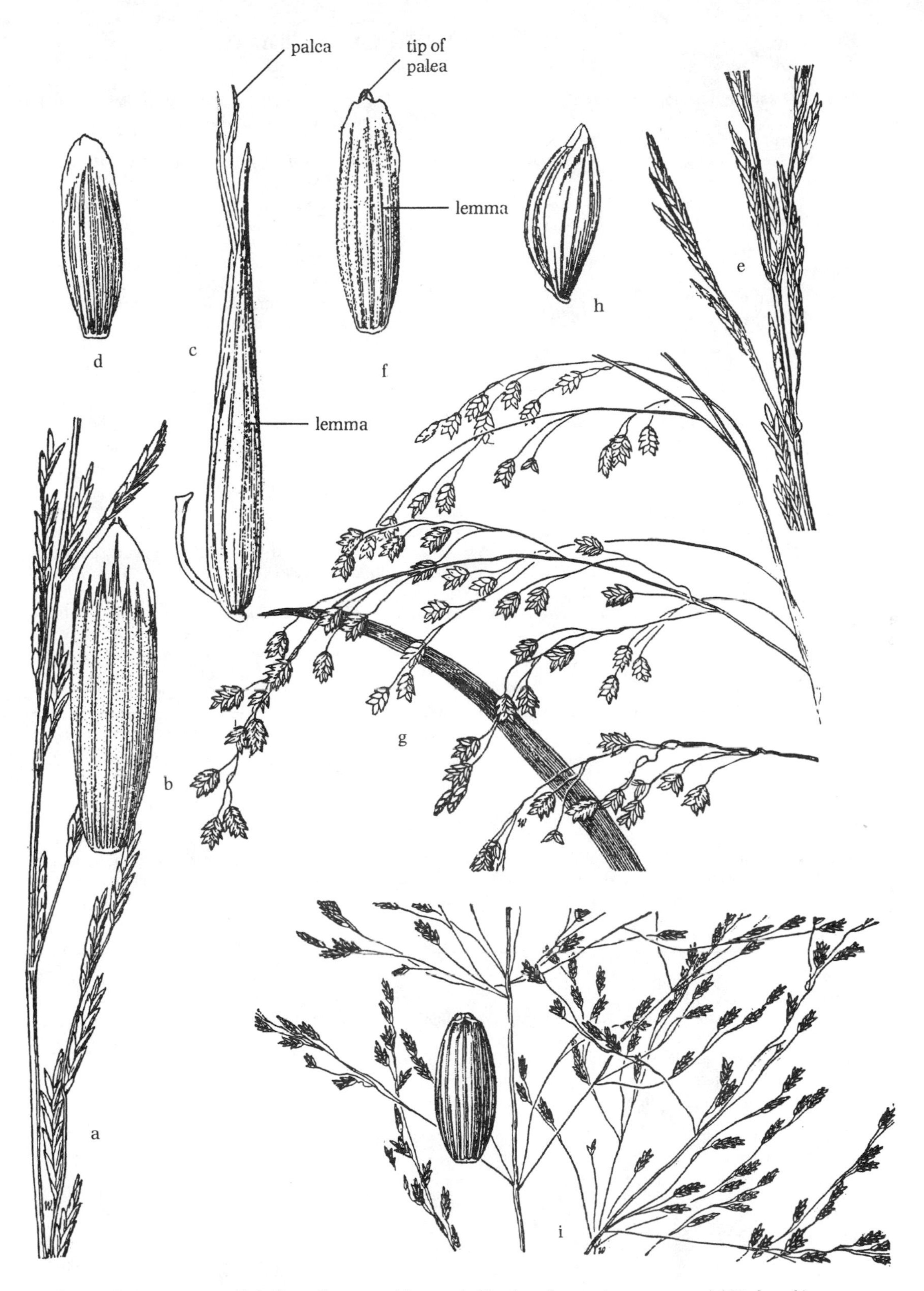

Fig. 314. *Glyceria fluitans*: a. panicle; b. floret (lemma with rounded back and conspicuous nerves) (Hitchcock).
Glyceria acutiflora: c. floret (Hitchcock).
Glyceria borealis: d. floret (lemma with rounded back and conspicuous nerves) (Hitchcock).
Glyceria septentrionalis: e. panicle; f. floret (Hitchcock).
Glyceria canadensis: g. panicle; h. floret. side view (Hitchcock).
Glyceria grandis: i. panicle and floret (Hitchcock).

3. *G. borealis* (Nash) Batchelder Small Floating Manna Grass Fig. 314
Shallow water and wet places. Nfld. w. to Man., N.W.T., and Alask., s. to N.E., Pa., n. Ohio, n. Ill., Iowa, S.D., mts. to Colo., Ariz. and Calif.

4. *G. septentrionalis* Hitchc. Manna Grass Fig. 314

4a. *G. septentrionalis* var. *septentrionalis* Eastern Manna Grass Fig. 314
Shallow water, marshy shores, swamps, woodland pools and wet places. Que. and N.E. w. to s. Ont., Mich. and Minn., s. to Ga., c. Fla. Ala., La. and e. Tex.

4b. *G. septentrionalis* var. *arkansana* (Fernald) Steyerm. & Kucera Arkansana Manna Grass Fig. 313
Wet bottomlands, swamps, sloughs and shallow water of pond and lake margins, especially the Mississippi floodplain. S. Ind. and s. Ill. w. to Okla., s. to w. Ky., Tenn., Miss., Ark., La. and Tex. (*G. arkansana* Fernald)

5. *G. obtusa* (Muhl.) Trin. Atlantic Manna Grass Fig. 313
Swales, open swamps, shores, and wet peaty and sandy soils. Usually coastal; N.S. s. to N.C.

6. *G. melicaria* (Michx.) F. T. Hubbard Melic Manna Grass
Wet woods, swamps, and wet peaty soils. N.B. w. to Ohio and Ont., s. to N.E., Md., and mts. of N.C., Ky. and Tenn.

7. *G. canadensis* (Michx.) Trin. Rattlesnake Manna Grass Fig. 314
Bogs, meadows, swales, shores and wet places. Nfld. w. to Minn., s. to N.E., Va., mts. of Tenn., ne. Ohio and n. Ill. A taller plant with somewhat smaller spikelets has been treated as var. *laxa* (Scribn.) Hitchc., but it intergrades with var. *canadensis* and may not deserve varietal status.

8. *G. striata* (Lam.) Hitchc. Fowl Manna Grass, Meadow Grass Fig. 313
Moist meadows, wet places and pond margins. Nfld. w. to N.W.T., Alask., s. to Fla., Tex., Ariz., Calif. and Mex. Barkworth and Anderton (2007) note that variation is continuous and therefore have not nomenclaturally recognize varieties; they indicate that western plants tend to be more robust and have been treated as var. *stricta* (Fernald) as distinct from the eastern plants. (*G. striata* subsp. *stricta* (Scribn.) Hultén)

9. *G. grandis* S. Wats. Reed Manna Grass, Reed Meadow Grass Fig. 314
Stream banks, marshes and wet places. Nfld. w. to Man., sw. N.W.T., and Alask., s. to Va., mts. of Tenn., ne. Ohio, n. Ill., Neb., mts. to Colo., N.M., Ariz. and Oreg. This taxon is sometimes treated as a subspecies of *G. maxima*. (*G. maxima* subsp. *grandis* (S. Watson) Hultén)

10. *G. maxima* (Hartm.) Holmb. Reed Grass, Tall Manna Grass
Marshes and wet ditches. Local, scattered populations, Nfld., Que. and Ont.; ne. Mass. and se. Conn.; e. Wisc.; B.C.; intro. from Eurasia. Canadian introductions are more robust than our own large native *G. grandis* and are said to be reddish on the lower portion of stems, making them distinctive in the vegetative state (Gutteridge, 1954). Freckmann and Reed (1979) noted that in Wisconsin this species has potential as a troublesome wetland weed and is presently tracked by the Wisconsin DNR (https://sewisc.org/images/managers/document_manager_folders/resources/Invasive%20Species%20Information/Glyceria%20maxima%20fact%20sheet.pdf, accessed August 2020.

15. *Andropogon* (Beardgrass)

Perennial herbs, usually tufted; stems solid, often tall, erect; inflorescences lateral and terminal, racemose or spike-like (densely clustered racemes in ours); spikelets paired, 1 sessile and bisexual, the other pedicellate and sterile (ours), reduced to a glume or often just a pedicel (fig. 557h), or pedicel bearing staminate floret, pedicel long-villous.

References: Campbell, 1983, 2003.

1. *A. glomeratus* (Walter) Britton, Sterns. & Poggenb. Bushy Bluestem Fig. 321
Peaty, boggy, or wet ground, swamps and marshes. Se. Mass. and R.I. w. to L.I., N.Y., N.J., Pa., W.Va., Ohio, Ky., Ill., Ark., s. to Fla. and La. A variable species, Campbell (1983, 2003) recognizes several varieties; range given is for our taxon, var. *glomeratus*. A variety of the southeast coastal plain, var. *glaucopsis* (Elliott) C.Mohr, grows in southern bogs, swamps, flatwoods, pond margins and ditches, and might be encountered in e. Md. or e. Va.; these are plants with leaves glabrous, smooth and glaucous, whereas ours have green, scabrous blades.

16. *Saccharum* (Plume Grass, Beardgrass)

Perennial herbs, tall, reed-like; leaves long, flat; inflorescence a terminal panicle; spikelets 1-flowered, paired at each joint, 1 sessile and 1 pedicellate; most species with spikelets subtended by a ring of silky hairs.

Reference: Webster, 2003.

1. Inflorescence strongly contracted; branches stiffly ascending (fig. 315a); spikelets not subtended by a ring of hairs. . . . . 1. *S. baldwinii*
1. Inflorescence with loosely ascending to spreading branches (fig. 315c); spikelets subtended by a ring of silky hairs (fig. 315b,d).
    2. Basal hairs long, longer than spikelets (fig. 315d), giving panicle a plume-like appearance. . . . . 2. *S. giganteum*
    2. Basal hairs shorter than spikelets (fig. 315b), panicle not plumose. . . . . 3. *S. brevibarbe*

1. *S. baldwinii* Spreng. Narrow Plume Grass
Fig. 315
Swales, marshes, swamp margins, ditches and wet places. Coastal plain, Va. s. to Fla., w. to e. Tex.; inland n. to Ark., Mo. and Tenn. (*Erianthus strictus* Baldw. ex Elliott)

2. *S. giganteum* (Walter) Pers. Sugarcane Plume Grass
Fig. 315
Swales, marshes, margins of swamps, wet lowlands and ditches. Coastal plain and lower Piedmont, N.Y. s. to Fla., w. to e. Tex.; inland n. to Ark., Tenn., and Ky. (*Erianthus giganteus* (Walter) Muhl.

3. *S. brevibarbe* (Michx.) Pers. Shortbeard Grass
Fig. 315
Sandy or peaty swales, swamp margins, ditches and wet openings. Del., Md. and Va. w. to Ky., s. Ill., Ark. and Okla., s. to n. Fla., La. and e. Tex. Webster (2003) noted that this a species restricted to the se. U.S., that barely gets into our range; he recognized two sympatric varieties, var. *brevibarbe* and var. *contortum* (Baldwin) R. D. Webster. (*Erianthus brevibarbis* Michx.)

17. *Spartina* (Cord Grass)

Perennial herbs; stems thick and erect, or weak, slender, and usually matted; spikelets 1-flowered, in 2 rows on one side of rachis, forming 1-sided spikes.

References: Barkworth, 2003; McDonnell and Crow, 1979; Mobberly, 1956.

1. Principal leaves strongly involute, 0.5–2 mm wide; stems 1–6 mm in diameter at base, typically weak and matted down. . . . . 1. *S. patens*
1. Principal leaves flat, 5–20 mm wide, becoming involute at apex; stems 5–25 mm in diameter at base, erect.
   2. Leaf margins smooth; keels of glumes smooth. . . . . 2. *S. alterniflora*
   2. Leaf margins scabrous (often involute); keels of glumes with short, stiff spine-like hairs (fig. 319b).
      3. First glume nearly as long as floret; second glume with conspicuous awn, 5–10 mm long; leaves 4–10(15) mm wide. . . . . 3. *S. pectinata*
      3. First glume about half the length of floret; second glume sharp-pointed, but not awned; leaves 10–25 mm wide. . . . . .4. *S. cynosuroides*

1. *S. patens* (Aiton) Muhl. Salt-meadow Grass, Salt-meadow Cord Grass Fig. 316
Salt marshes, saline flats, low dunes and tidal shores; often dominant in salt marshes. Coastal, se. Nfld. and e. Que., s. to Va. and Fla., w. to e. Tex. A purported hybrid *S. ×caespitosa* A. A. Eaton (*S. patens × S. pectinata*), forming tufts is occasionally encountered in upper, drier portions of salt marshes from N.S. and Me. s. to Del. and Md. (Barkworth, 2003; McDonnell and Crow, 1979).

2. *S. alterniflora* Loisel. Saltwater Cord Grass, Smooth Cord Grass Fig. 317
Salt marshes and tidal creeks. Coastal, Nfld. and e. Que. s. to Fla., w. to e. Tex. It has become established along the Pacific Coast; it has hybridized with *S. pectinata* in Massachusetts, with *S. foliosa* in California, and in Europe with *S. maritima* (Barkworth, 2003).

3. *S. pectinata* Saltwater Cord Grass, Prairie Cord grass
Fig. 319
Freshwater marshes, edges of salt marshes, wet prairies and gravels of shores and rivers. Nfld. w. to James Bay, Man. and sw. N.W.T., s. to w. N.C., Ark., Tex., N.M. and Oreg. This occurs along coastal sites but is also our common inland species of *Spartina*.

4. *S. cynosuroides* (L.) Roth Big Cord Grass, Salt Reed Grass Fig. 318
Salt marshes, tidal waters and freshwater coastal marshes. Coastal, Mass. w. to Fla, w. to e. Tex.

S. gracilis Trin. (Alkali Cord Grass) was discovered in nw. Minnesota in 1980 (Coffin and Pfannmuller, 1988). In western North America this species frequently occurs at margins of alkali lakes, small streams and river bottoms, but in Minnesota it occurs in prairie sites.

18. *Beckmannia* (Slough Grass)

Annuals, stout, tufted; stems erect; leaf blades flat; spikelets in 2 rows along one side of rachis, forming 1-sided spike-like branches of the panicles.

References: Hatch, 2007; Reeder, 1953.

1. *B. syzigachne* (Steud.) Fernald Slough Grass Fig. 319
Sloughs, marshes, ditches and wet ground. N.S., N.B. and Que. Man., sw. N.W.T. and Alask., s. to Me., N.Y., n. Ohio, n. Ill., Mo., Kan., N.M., Ariz. and n. Calif. Our plants are referable to subsp. *syzigachne*. (*B. syzigachne* var. *uniflora* (Scribn. ex. A. Gray) Boivin)

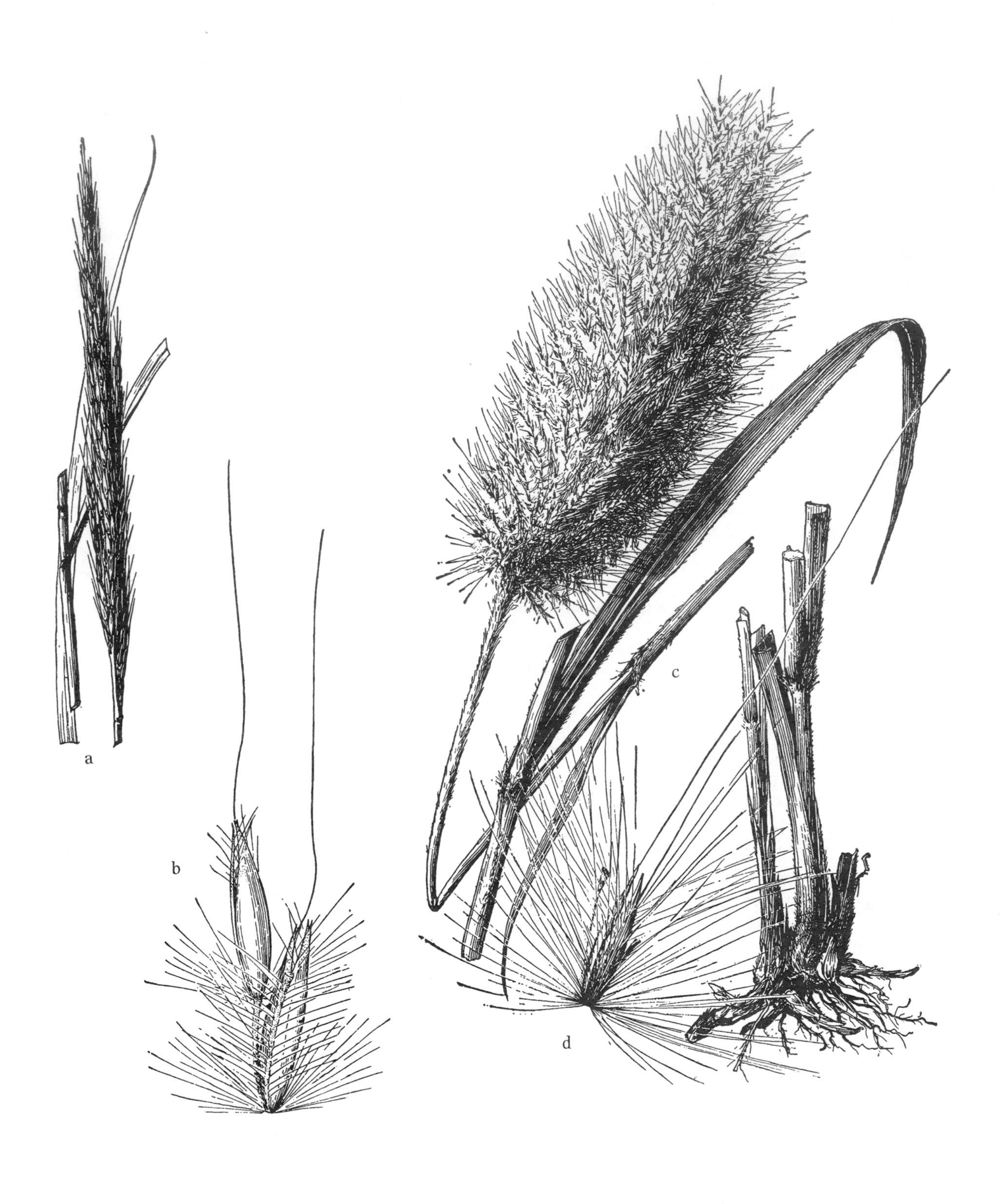

Fig. 315. *Saccharum baldwinii*: a. upper portion of plant (Hitchcock).
Saccharum brevibarbe: b. pair of spikelets, one sessile, one pedicellate (Hitchcock).
Saccharum giganteum: c. habit; d. spikelet, showing very long basal hairs (Hitchcock).

Fig. 316. *Spartina patens*: a. habit; b. ligule; c. spikelet (G&W).

Fig. 317. *Spartina alterniflora*: a. habit; b. ligule; c. spikelet (G&W).

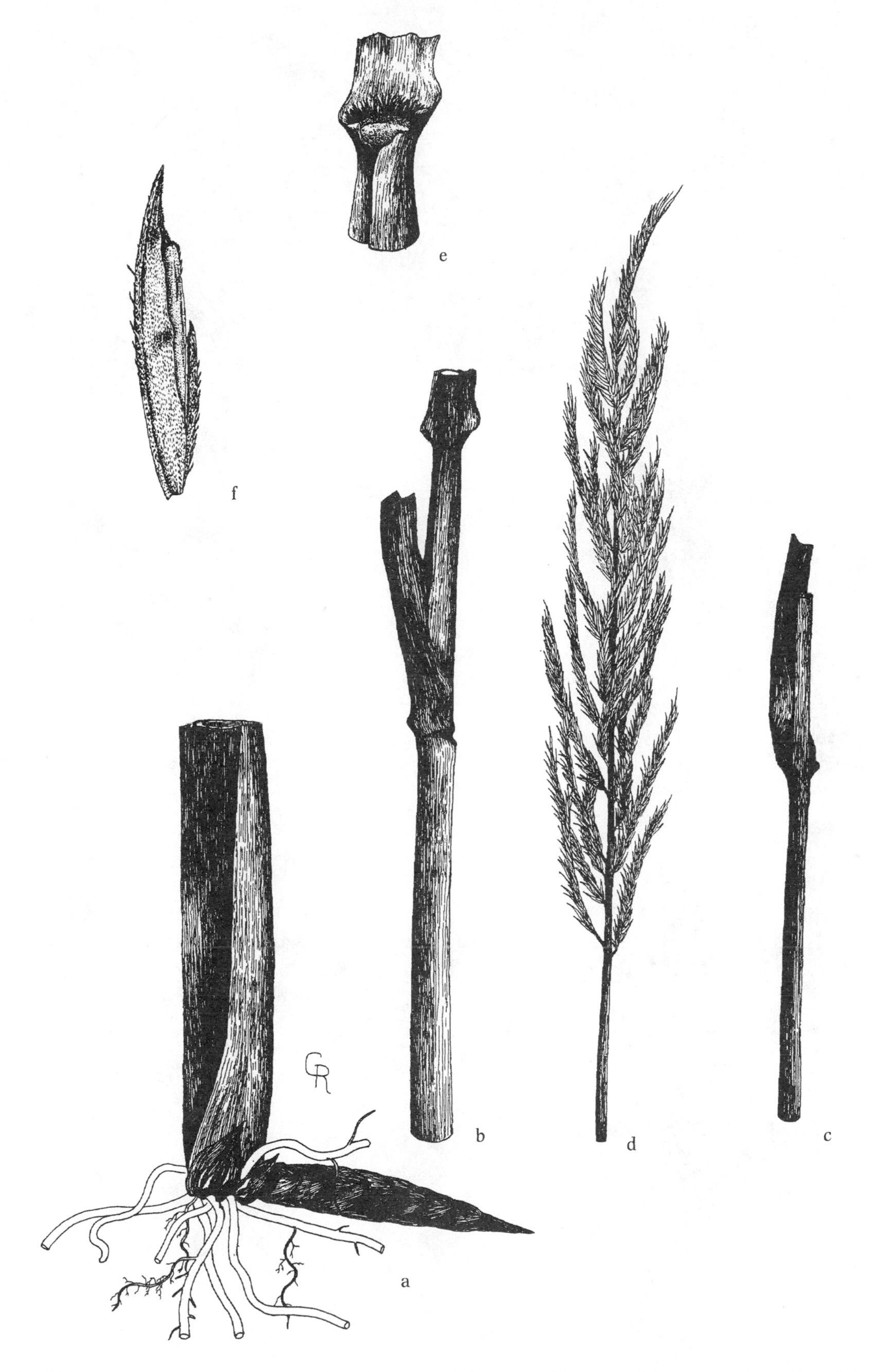

Fig. 318. *Spartina cynosuroides*: a. basal portion of plant; b. lower midsection of stem; c. upper section of stem; d. panicle; e. ligule; f. spikelet (G&W).

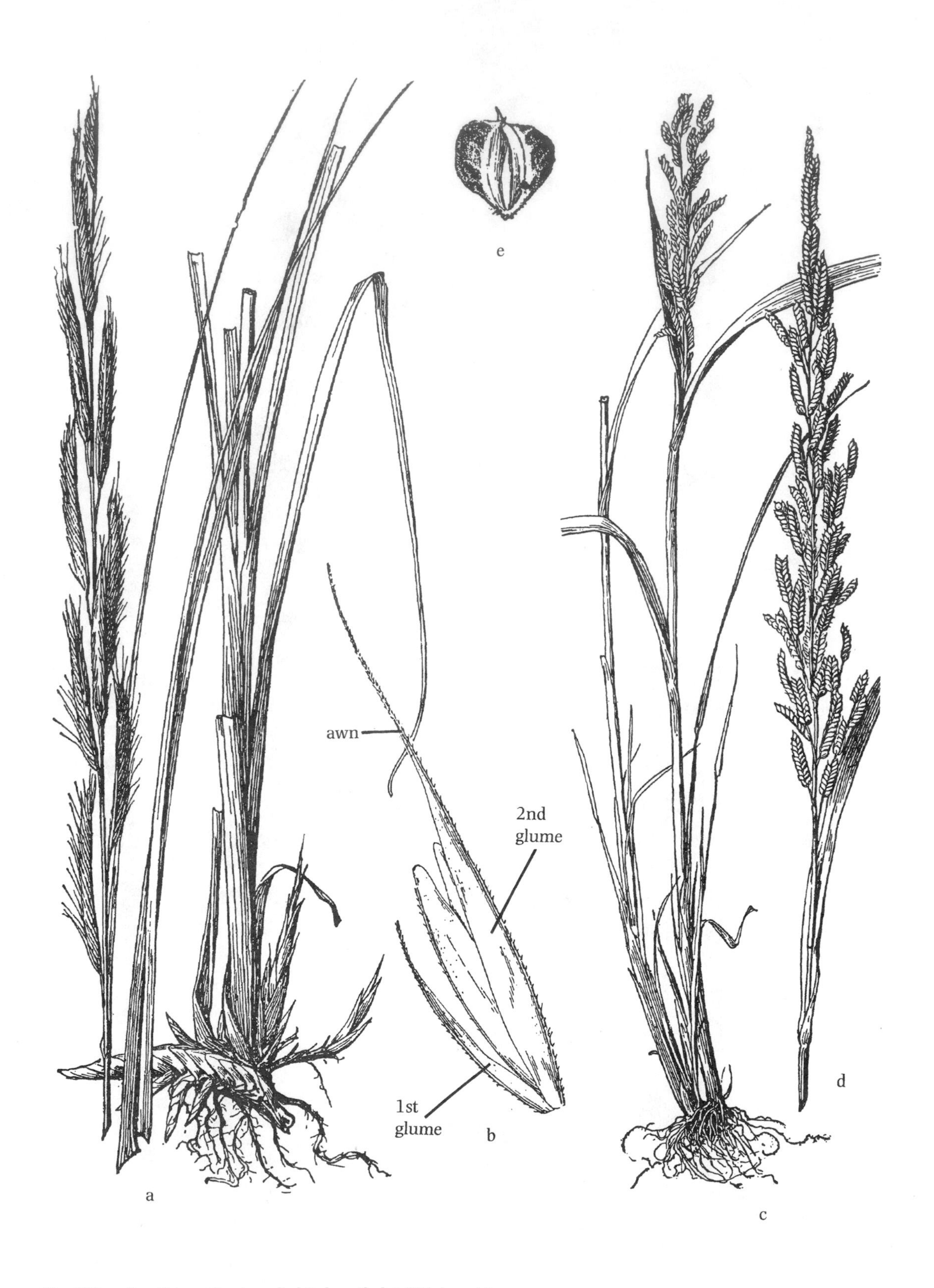

Fig. 319. *Spartina pectinata*: a. habit; b. spikelet (Hitchcock).
Beckmannia syzigachne: c. habit; d. upper portion of plant; e. spikelet (Hitchcock).

19. *Leersia* (Cutgrass)

Perennial herbs, arising from rhizomes; stems somewhat weak, often bearing flat leaves with cutting, scabrous margins; inflorescence a panicle; spikelets 1-flowered, bisexual, lemmas with bristly ciliate to hispidulous margins, florets overlapping in a single row.

REFERENCES: Pyrah, 1969, 2007.

---

1. Spikelets broadly oval to orbicular, 3–4 mm wide, closely overlapping (fig. 320d); principal leaf blades usually 10–15 mm wide. . . . . 1. *L. lenticularis*
1. Spikelets oblong, 1–2 mm wide, loosely overlapping (figs. 320a, 321a,b); principal leaf blades 2–10 mm wide.
   2. Panicle open, branches spreading and typically lacking spikelets on lower portion (fig. 320a); anthers 2 or 3.
      3. Lower branches of panicle whorled or appearing whorled; anthers 3; leaf sheaths conspicuously retrorse-scabrous; blade margins spinulose; rhizomes slender. . . . . 2. *L. oryzoides*
      3. Lower branches of panicle alternate; leaf sheaths glabrous or weakly scabrous; anthers 2; blade margins weakly scabrous; rhizomes thick. . . . . 3. *L. virginica*
   2. Panicle narrow, branches strongly ascending and bearing spikelets to base (fig. 321a); anthers 6. . . . . 4. *L. hexandra*

---

1. *L. lenticularis* Michx. Catchfly Grass Fig. 320
Swamps, marshes, low woods and wet ditches. Coastal plain, Md. and Va. s. to n. Fla., w. to La., and e. Tex.; major river drainages into Mississippi R. Valley n. to Ind., Ill., w. Wisc., se. Minn., Mo., e. Kan. and e. Okla.

2. *L. oryzoides* (L.) Sw. Rice Cutgrass Fig. 320
Swamps, marshes, shores, and wet places. N.S., N.B. and Que. w. to B.C., s. to Fla., Tex., N.M., Ariz. and Calif.; n. Mex. This species often forms zones around ponds and may be so abundant as to make walking difficult (the rough leaves snagging clothes and scraping skin). In late summer the plants often bear cleistogamous inflorescences hidden within the leaf sheaths (fig. 556c).

3. *L. virginica* Willd. White Grass Fig. 321
Moist to wet or marshy stream banks, swales, swamps and low woods. N.B., c. Me., and s. Que. w. to s. Ont., s. Minn., se. N.D. and Neb., s. to n. Fla. and e. Tex. (*L. virginica* var. *ovata* (Poir.) Fernald)

4. *L. hexandra* Sw. Southern Cutgrass Fig. 321
Shallow water, ditches and wet places. Coastal plain and piedmont, Va. s. to Fla., w. to e. Ark., e. Tex.; Tenn.; Trop. Am.

20. *Anthoxanthum* (Sweetgrass)

Perennial herbs, arising from rhizomes (ours) or tufted; stems erect, with short blades; inflorescence a panicle; spikelets bronze-colored, with 1 terminal bisexual floret and 2 staminate florets; glumes as long as and enclosing florets; lemmas hispidulous, especially along margins.

REFERENCE: Allred and Barkworth, 2007.

1. *Anthoxanthum nitens* (G. H. Weber) Y. Schouten & Veldkamp Northern Sweetgrass
Meadows, marshes, bogs, shores and moist places. Lab. and Nfld. w. to Alask., s. to N.E., N.J., Md., Va., Ohio, Ill., Iowa, S.D., N.M., Ariz. and Oreg. *Axthoxanthum odoratum* is very similar and typically occurs in uplands, but can occasionally be found in wetlands, whereas our plant is *A. nitens* and is typically in wetlands. (*Hierochloe hirta* (Schrank) Borbás)

21. *Phalaris* (Canary Grass)

Perennial herbs, arising from rhizomes (ours), or annuals; stems erect, with flat leaves; inflorescence a dense, cylindrical spike-like panicle; spikelets 1–3(4) florets, (ours) with 1 fertile terminal floret and 2 staminate florets; lemmas with 2 tufts of silky hairs at base.

REFERENCES: Anderson, 1961; Barkworth, 2007.

---

1. Plants perennial, rhizomatous; bisexual floret acute at apex. . . . . 1. *P. arundinacea*
1. Plants annual, lacking rhizomes; bisexual floret strongly acuminate to beaked at apex. . . . . 2. *P. carolinana*

---

1. *P. arundinacea* L. Reed Canary Grass Fig. 322
Meadows, swales, shores and wet ground. Nfld. w. to Man., sw. N.W.T. and Alask., s. to Va., w. N.C., Ky., Ill., Mo., Okla., N.M., Ariz. and ne. Calif. This is a common grass of wetlands, often forming large colonies. Although native, some populations represent introductions from Eurasia. A variegated form with white striped leaves is sometimes cultivated and known as "Ribbon Grass" or "Gardener's-Garters."

Fig. 320. *Leersia oryzoides*: a. upper portion of plant; b. base of plant; c. growth form with inflorescence nearly hidden by sheathing leaf (F).
Leersia lenticularis: d. portion of inflorescence and leaf; e. base of plant (F).

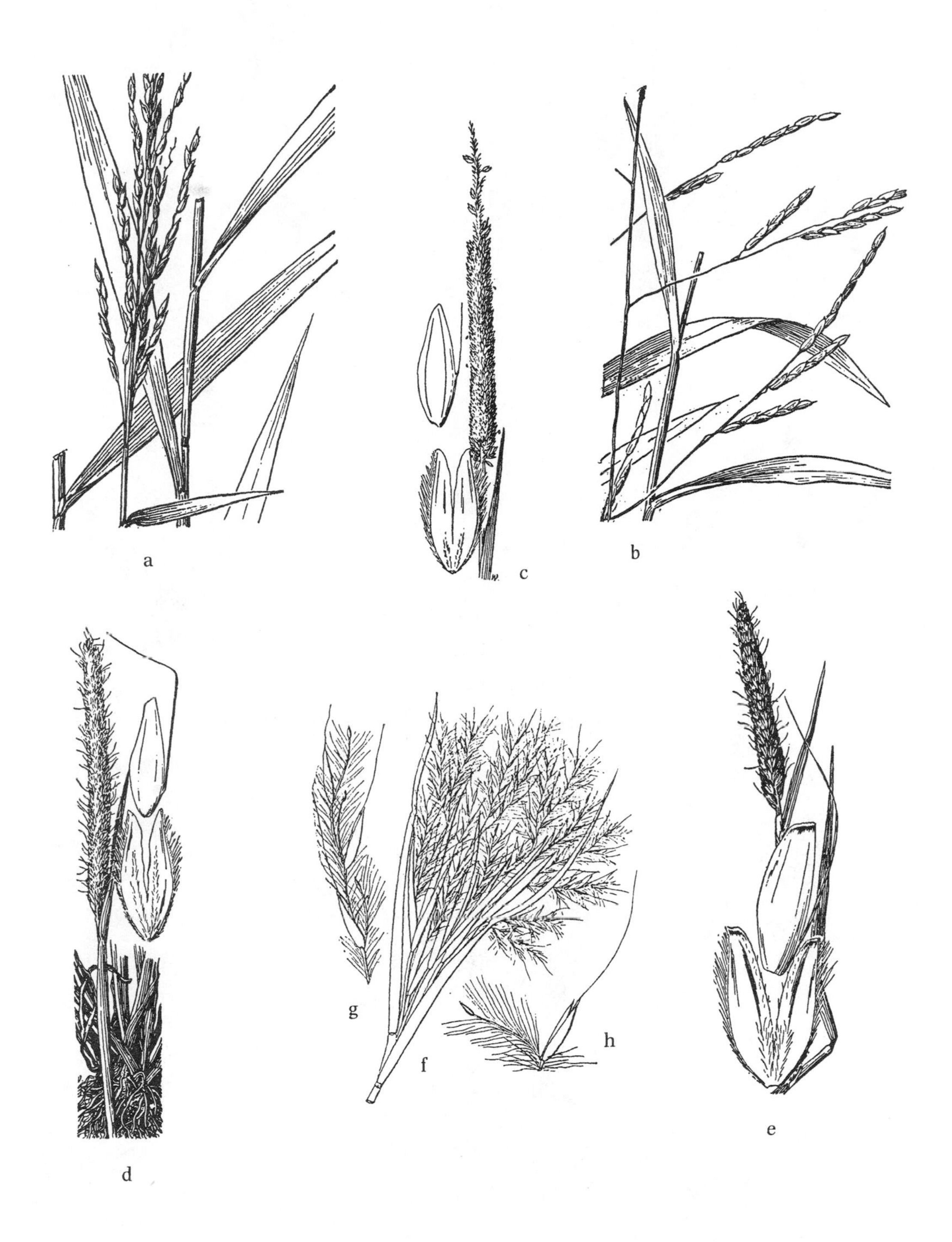

Fig. 321. *Leersia hexandra*: a. portion of plant (Hitchcock).
Leersia virginica: b. portion of plant (Hitchcock).
Alopecurus aequalis: c. inflorescence with glumes and floret (Hitchcock).
Alopecurus carolinianus: d. portion of plant with glumes and floret (Hitchcock).
Alopecurus geniculatus: e. portion of plant with glumes and floret (Hitchcock).
Andropogon glomeratus: f. upper portion of plant; g. branch of inflorescence; h. pair of spikelets, sessile spikelet fertile, pedicellate spikelet sterile at base (B&B).

Fig. 322. *Phalaris arundinacea*: a. habit; b. ligule; c. spikelet; d. florets; e. grain (Reed).

2. *P. caroliniana* Walter Carolina Canary Grass
Wet, marshy or swampy sites. Del. and Md. w. to Ohio, w. Ky., Mo., Kans. and se. Colo., s. to n. Fla., La., Tex., Ariz. and Calif.; sw. Oreg.

P. aquatica L., a native of the Mediterranean region, has been widely planted for its forage value; now well established in western N.Am., it has been reported from Virginia and mostly occurs in disturbed sites that seasonally flood.

22. *Alopecurus* (Foxtail)

Perennial or annual herbs; leaves flat; inflorescences spike-like, cylindrical panicles; spikelets 1-flowered, strongly folded, keeled; lemmas dorsally awned, giving bristly appearance.

Reference: Crins, 2007.

1. Awn inserted near middle of lemma, straight, barely (if at all) longer than glumes (fig. 321c); anthers 0.5–1.2 mm long. . . . . . . 1. *A. aequalis*
1. Awn inserted near base of lemma, geniculate, 2–4 mm longer than glumes (fig. 321d,e).
   2. Plants perennial, stems often decumbent, rooting at nodes; spikelets 2.5(3) mm long. . . . . .2. *A. geniculatus*
   2. Plants annual, tufted, stems erect; spikelets 2–2.5 mm long. . . . . 3. *A. carolinianus*

1. *A. aequalis* Sobol. Short-awn Foxtail Fig. 321
Shallow water, shores, stream margins, ditches, wet opening in woods and wet ground. Greenl. and Lab. w. to Man., w. N.W.T., and Alask., s. to N.E., Md., Ohio, sw. Ky., Ill., Mo., N.M., Ariz. and Calif. A growth form with lax, floating or creeping stems has been called var. *natans* (Wahlenb.) Fernald, but Crins (2007) indicates this form is not worthy of taxonomic recognition. This is our most widespread species of the genus.

2. *A. geniculatus* L. Water Foxtail Fig. 321
Shallow water and wet places. Se. Lab. and Nfld. w. to Sask. B.C. and Alask., s. to N.E., Va., Pa., Ohio, Ill., Ia., sporadic s. to Okla., Wyo., N.M., Ariz. and Calif. A native of Eurasia and western N.Am., plants of the Northeast represent introductions from Eurasia. (*A. pallescens* Piper & Beetle)

3. *A. carolinianus* Walter Annual Foxtail; Tufted Foxtail Fig. 321
Shores, ditches, low ground, and waste places. Mass., Conn., N.Y. and N.J. w. to s. Mich, Minn., N.D., Sask., and B.C., s. to Fla. and Tex., N.M., Ariz. and Calif.

23. *Calamagrostis* (Reed Bentgrass, Reedgrass, Bluejoint)

Perennial herbs, usually arising from rhizomes; stems simple or branched; inflorescence an open or a contracted, spike-like panicle; spikelets 1-flowered; first glume 1-nerved, second glume 3-nerved; lemma dorsally awned, awn often twisted and/or geniculate.

References: Greene, 1980, 1984, 1987; Marr et al., 2007.

1. Lemma awns geniculate (fig. 323d), twisted 1 or 2 times on drying.
   2. Collar at leaf sheath summit glabrous; awn attached near middle of lemma. . . . . .1. *C. pickeringii*
   2. Collar at leaf sheath summit usually bearing short, stiff lateral tufts of hairs; awn attached near base of lemma. . . . . 2b. *C. stricta* subsp. *inexpansa*
1. Lemma awns straight (sometimes weakly twisted).
   3. Panicle somewhat open, lax; callus hairs as long as lemma, not tufted. . . . . 3. *C. canadensis*
   3. Panicle contracted, branches stiff, erect; callus hairs half to as long as lemmas, in 2 lateral tufts.
      4. Spikelets 2–5 mm long; awn attached near base of lemma; grain glabrous.
         5. Leaf blades thin, 1.5–2.5(3) mm wide, involute, smooth to scabrous on upper surface; ligule 1–3.5 mm long; spikelets 2–2.5(3) mm long. . . . . 2a. *C. stricta* subsp. *stricta*
         5. Leaf blades thicker, firm, 2–6 mm wide, usually flat, strongly scabrous on upper surface; ligule 2–6 mm long; spikelets 3–4(5) mm long. . . . . 2b. *C. stricta* subsp. *inexpansa*
      4. Spikelets 5–7 mm long; awn attached near tip of lemma; grain pubescent. . . . . .4. *C. cinnoides*

1. *C. pickeringii* A. Gray Pickering's Reedgrass Fig. 323
Sphagnum bogs, wet meadows, wet alpine peats and margins of mountain streams and ponds. Nfld., N.S., and mts. of Me., N.H. and n. N.Y.; ne. Mass., L.I., N.Y. and N.J. (*C. pickeringii* var. *debilis* (Kearney) Fernald & Wiegand)

2. *C. stricta* (Timm) Koeler

2a. *C. stricta* subsp. *stricta* Slim-stem Reedgrass
Peaty and sandy soils, fens, marshes, shores of lakes and streams and wet tundra meadows. Greenl. and Lab. w. to Alask., s. to

Fig. 323. *Calamagrostis canadensis*: a. habit. with glumes and floret (Hitchcock).
Calamagrostis cinnoides: b. panicle with glumes and floret (Hitchcock).
Calamagrostis stricta subsp. *inexpansa*: c. panicle with glumes and floret (Hitchcock).
Calamagrostis pickeringii: d. panicle with glumes and floret (Hitchcock).

N.S., n. Me., n. N.H., e. N.Y., Mich., Ill., Minn., N.D., mts. to Colo., Ariz. and n. Calif. (*C. neglecta* var. *stricta* (Timm) Griseb.; *C. neglecta* var. *micrantha* (Kearney) Stebbins)

2b. *C. stricta* subsp. *inexpansa* (A. Gray) C. W. Greene Northern Reedgrass Fig. 323
Wet meadows, sphagnum bogs, and shores of lakes and streams. Lab. and Nfld. w. to Hudson Bay and Alask., s. to N.E., N.Y., Ont., Ind., n. Iowa, Neb., Colo. and Calif. Subspecies *inexpansa* has been recognized as consisting of apomictic plants derived from sexually reproducing subsp. *stricta*. Other variants previously recognized are not regarded as meriting nomenclatural recognition (Greene, 1984; Marr et al., 2007).

3. *C. canadensis* (Michx.) Beauv. Bluejoint Fig. 323
Wet meadows open swamps, thickets along streams, bog margins, wet prairies, forest openings and wet montane soils. Greenl. and lab. w. to N.W.T. and Alask. s. to N.E., N.J., mts. to N.C. and Ga, W.Va., n. Ohio, Ind., n. Mo., Kans., Neb., and mts. to Colo. and Calif. Greene (1980) recognized three varieties in eastern and inland portions of the range, with var. *canadensis* being more common, thus the range is given for var. *canadensis*. The other varieties are: var. *macouniana* (Vasey) Stebbens representing a smaller extreme, with less densely flowered panicles and smaller spikelets, and is best represented on the Great Plains; var. *langsdorfii* (Link) Inman is a far-northern taxon and tends to occur at high elevations in our range. *C. canadensis* is closely related to the European *C. purpurea* (Trinn.) Trinn. with which it could be considered conspecific (Marr et al., 2007).

4. *C. cinnoides* (Muhl.) W. P. C. Barton Fig. 323
Wet peaty or sandy soils, open swamps, and wet woods. N.S., s. Me. and N.H. sw. to s. N.Y., Pa., Ohio and Ky., s. to Va., S.C. n. Ga., Ala. and La.

24. *Muhlenbergia* (Muhly)

Perennial (ours) or annual herbs; stems slender (ours), often decumbent; inflorescence usually an open panicle with slender branches, sometimes densely contracted; spikelets chiefly 1-flowered, glumes acute; lemma membranous, acute.

REFERENCE: Peterson, 2003.

1. Inflorescence an open panicle, 2.5–6 cm wide; lemmas not pilose at base; plants tufted, not rhizomatous, stems arising from bases of old stems. . . . . 1. *M. uniflora*
1. Inflorescence slender, densely contracted, 0.3–1.8 cm wide; lemmas pilose at base; plants rhizomatous, stems arising from scaly rhizomes. . . . . 2. *M. glomerata*

1. *M. uniflora* (Muhl.) Fernald Bog Muhly Fig. 324
Bogs, swales, sandy shores and damp ground. Nfld. and Que. w. to Ont. and ne. Minn., s. to N.E., N.J., Md., Pa., n. Mich. and Wisc.; B.C. and sw. Oreg. This is a small delicate plant that could be mistaken at first glance for a species of *Panicum* or *Agrostis*.

2. *M. glomeratum* (Willd.) Trin. Spike Muhly, Marsh Wild-Timothy
Marshes, bogs, fens, tamarack swamps, sedge meadows, calcareous shores, stream banks and ditches. Nfld., Lab. and Que. w. to N.W.T. and Yuk., s. to N.C., Ky., Ill., Iowa, Neb., Colo., Utah, nw. Nev. and sw. Oreg.

Muhlenbergia capillaris, *M. expansa*, *M. frondosa*, *M. schreberi*, and *M. sylvatica* are occasionally found in wet sites and may occur in alluvial soils.

25. *Agrostis* (Bentgrass)

Perennial or annual herbs; stems slender, erect or creeping. Inflorescence a panicle, spikelets 1-flowered.

REFERENCE: Harvey, 2007.

1. Plants erect, underground rhizomes present, but stolons absent; flowering stems up to 120 cm; larger leaf blades mostly 3–7(10) mm wide; longest lower panicle branches 4–9 cm long. . . . . 1. *A. gigantea*
1. Plants weak, decumbent, rooting at nodes, creeping by stolons (often forming dense mats), rhizomes absent; flowering stems ascending, up to 60 cm tall; larger leaf blades 1.7–3(4) mm wide; longest lower panicle branches 2–6 cm long. . . . . 2. *A. stolonifera*

1. *A. gigantea* Roth Redtop Fig. 324
Swales, shores, wet meadows and damp thickets. Introduced from Eurasia and commonly planted as a pasture or lawn grass, this plant has escaped extensively in cooler regions, especially the Northeast. The panicles are broadly open during flowering, but later become much contracted and plants can be mistaken for *Phalaris arundinaceae*. (*A. stolonifera* var. *major* (Gaudin) Farw.; *A. alba* var. *major* Wimm. & Grab.)

Fig. 324. *Agrostis gigantea*: a. habit; b. spikelet (Hitchcock, as *A. alba*).
Muhlenbergia uniflora: c. habit with glumes and floret (Hitchcock).
Muhlenbergia glomerata: d. portion of inflorescence, scaly rhizomes, stem portion at node; e. spikelet with floret and glumes below (FNA).

2. *A. stolonifera* L. Creeping Bentgrass
Shores, shallow water, freshwater marshes, salt marshes, along streams and lake margins and damp fields. Se. Lab. and Nfld. w. to Alask., s. to Va., nw. Fla., La., Tex., N.M., Ariz. and Calif. (*A. alba* var. *palustris* (Huds.) Pers.; *A. palustris* Huds.)

26. *Paspalum* (Paspalum Grass)

Perennial and annual herbs; stems often stoloniferous; leaves flat; inflorescence with 1–many spike-like branches; spikelets in 2 or 4 rows on one side of flattened rachis; spikelets 2-flowered, orbicular, elliptic or ovate, lower spikelet sterile or staminate, upper floret lemma and palea hard, tightly enclosing grain.

Reference: Allen and Hall, 2003.

---

1. Rachis broadly winged, as wide as the 2 rows of spikelets.
   2. Leaf blades narrow, 2–5 mm wide; sheaths narrow, not appearing inflated; branches of inflorescence 2–4, branch axis flat, 1.8–3 mm wide; spikelets 1.7–2.1 mm long, 1.1–1.4 wide, glabrous . . . . . 1. *P. dissectum*
   2. Leaf blades broad, 3–25 mm wide; sheaths broad, appearing inflated-spongy (fig. 325a) (plants often forming floating mats); branches of inflorescence 20–50, branch axis folded (conduplicate), 0.7–1.5 mm wide; spikelets 1.1–1.9 mm long, 0.5–0.8 mm wide, pubescent . . . . . 2. *P. repens*
1. Rachis flat but not as wide as the 2 or 4 rows of spikelets, not distinctly winged.
   3. Plants annual; upper (fertile) floret dark brown and shiny at maturity . . . . . 3. *P. boscianum*
   3. Plants perennial; upper (fertile) floret pale brown or straw-colored, dull at maturity.
      4. Spikelets solitary on one side of rachis, forming 2 rows (fig. 326a) . . . . . 4. *P. laeve*
      4. Spikelets paired on one side of rachis, forming 4 rows (fig. 326g).
         5. Stems decumbent, rooting at nodes; spikelets obovate to elliptic, 2.8–3.6 mm wide . . . . . 5. *P. pubiflorum*
         5. Stems erect; spikelets orbicular to nearly orbicular, 2–2.8 mm wide . . . . . 6. *P. praecox*

---

1. *P. dissectum* (L.) L. Mudbank Paspalum Fig. 326
Shallow water and wet places, marshy shores, frequently abundant after water recedes. S. N.J., Del., e. Md. and e. Va. w. to w. Ky, s. Ill., s. Mo., and Kans., s. to c. Fla., La. and e. Tex.

2. *P. repens* P. J. Berg. Floating Paspalum; Water Paspalum Fig. 325
Ponds, lakes, swamps, quiet waters of streams, muddy stream banks and ditches. Md. and Va. w. to s. Ohio, s. Ind., Ill., Mo. and e. Kans., s. to Fla., La. and e. Tex.; Trop. Am. This is our most aquatic species of the genus, the plants often occurring submersed, with elongate, spongy stems and developing numerous dark, feathery roots at nodes; but it also forms extensive floating mats, especially in the tropics. Pohl (1980) regards *P. fluitans* (strictly tropical) as distinct from *P. repens* Berg., but Allen and Hall (2003) treat them as conspecific. (*P. fluitans* (Elliott) Kunth; *P. repens* var. *fluitans* (Elliott) Wipff & S. D. Jones)

3. *P. boscianum* Flüggé Bull Crown-grass Fig. 327
Shores, wet sands, peaty or mucky sites and low woods. Pa., Md. and Va. w. to w. Ky., and Ark., s. to s. Fla., La. and e. Tex.; Trop. Am.

4. *P. laeve* Michx. Field Paspalum Fig. 326
Wet sands, sandy peats or mucks, shores, thickets, savannas and disturbed sites. Mass. w. to Pa., sw. Mich., Ill., Mo., and Kans., s. to c. Fla., La. and se. Tex. (*P. laeve* var. *circulare* (Nash) Fernald; *P. laeve* var. *pilosum* Scribn.)

5. *P. pubiflorum* Rupr. ex Fourn. Hairy-seed Paspalum Fig. 327
Open low, wet places, stream banks, margins of ponds and lakes, wet meadows and disturbed places. Md. and Va. w. to s. Ohio, s. Ind., s. Ill., s. Mo., and Kans., s. to Fla., La. and Tex.; Mex. and Cuba.

6. *P. praecox* Walter Early Paspalum Fig. 326
Swamps, southern pitcher-plant bogs, wet pine savanna, and flatwoods. Coastal plain, se. Va. s. to c. Fla., w. to s. Ark, La. and se. Tex.; s. Ill. (*P. praecox* var. *curtisianum* (Steud.) Vasey)

Paspalum hispidum, *P. distichum*, *P. floridanum*, and *P. urvillei* may sometimes occur in wet sites in the southeastern portion of our range.

27. *Setaria* (Bristly Foxtail)

Perennial and annual herbs; leaves flat; inflorescence usually cylindrical, compact, and spike like; spikelets 2-flowered, lower floret sterile or staminate, upper floret bisexual spikelets subtended by an involucre of 1-numerous slender bristles, these persisting on rachis after spikelets fall off.

References: Rominger, 1962, 2003.

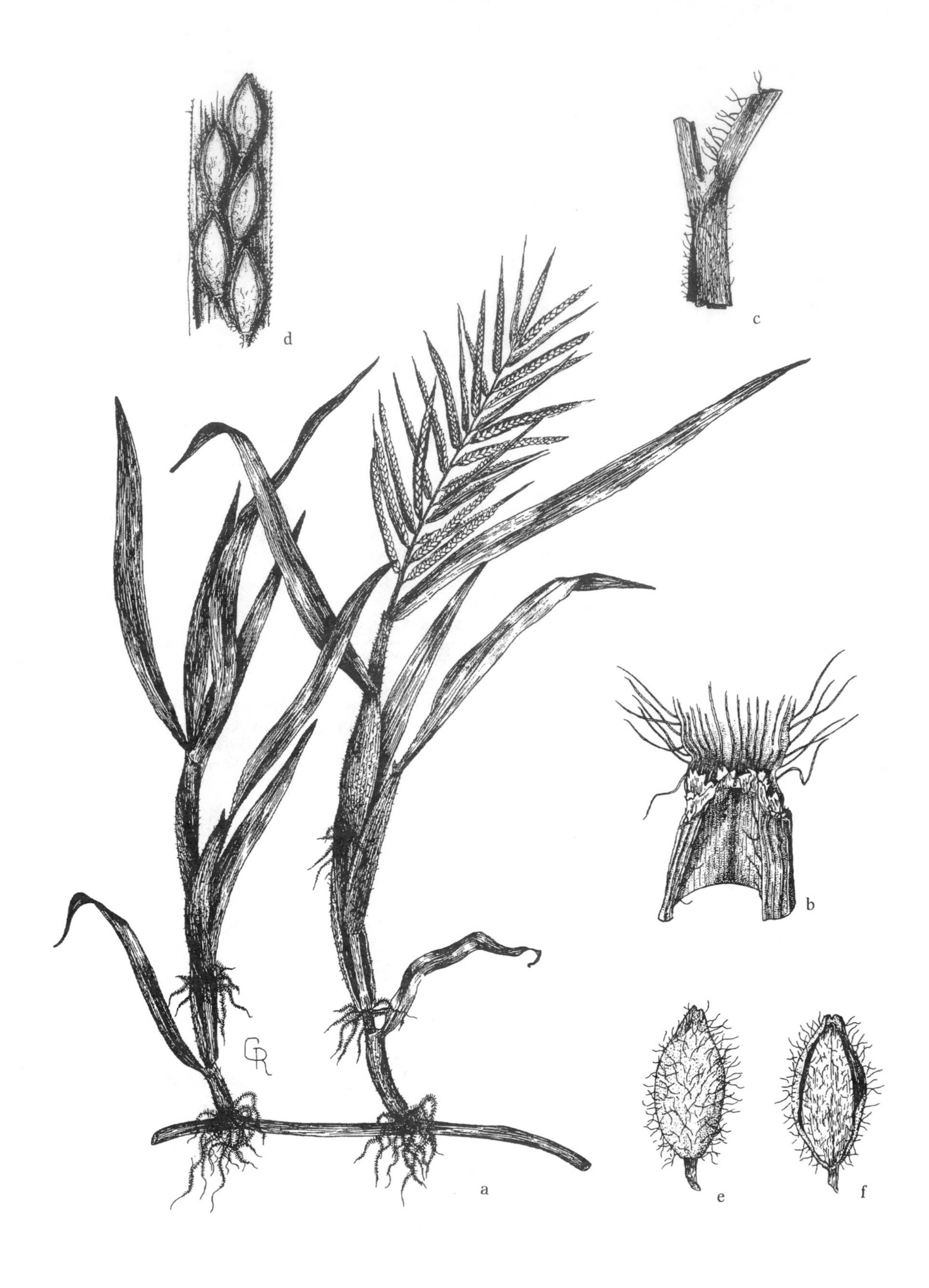

Fig. 325. *Paspalum repens*: a. habit; b. membranous ligule with ciliate base of blade; c. side view showing ciliate base of blade at junction with sheath; d. portion of raceme; e, f. spikelet, two views (G&W).

Fig. 326. *Paspalum laeve*: a. habit; b. spikelet, two views; c. upper floret (Hitchcock).
Paspalum dissectum: d. inflorescence; e. spikelet, two views; f. upper floret (Hitchcock).
Paspalum praecox: g. inflorescence; h. spikelet, two views; i. upper floret (Hitchcock).

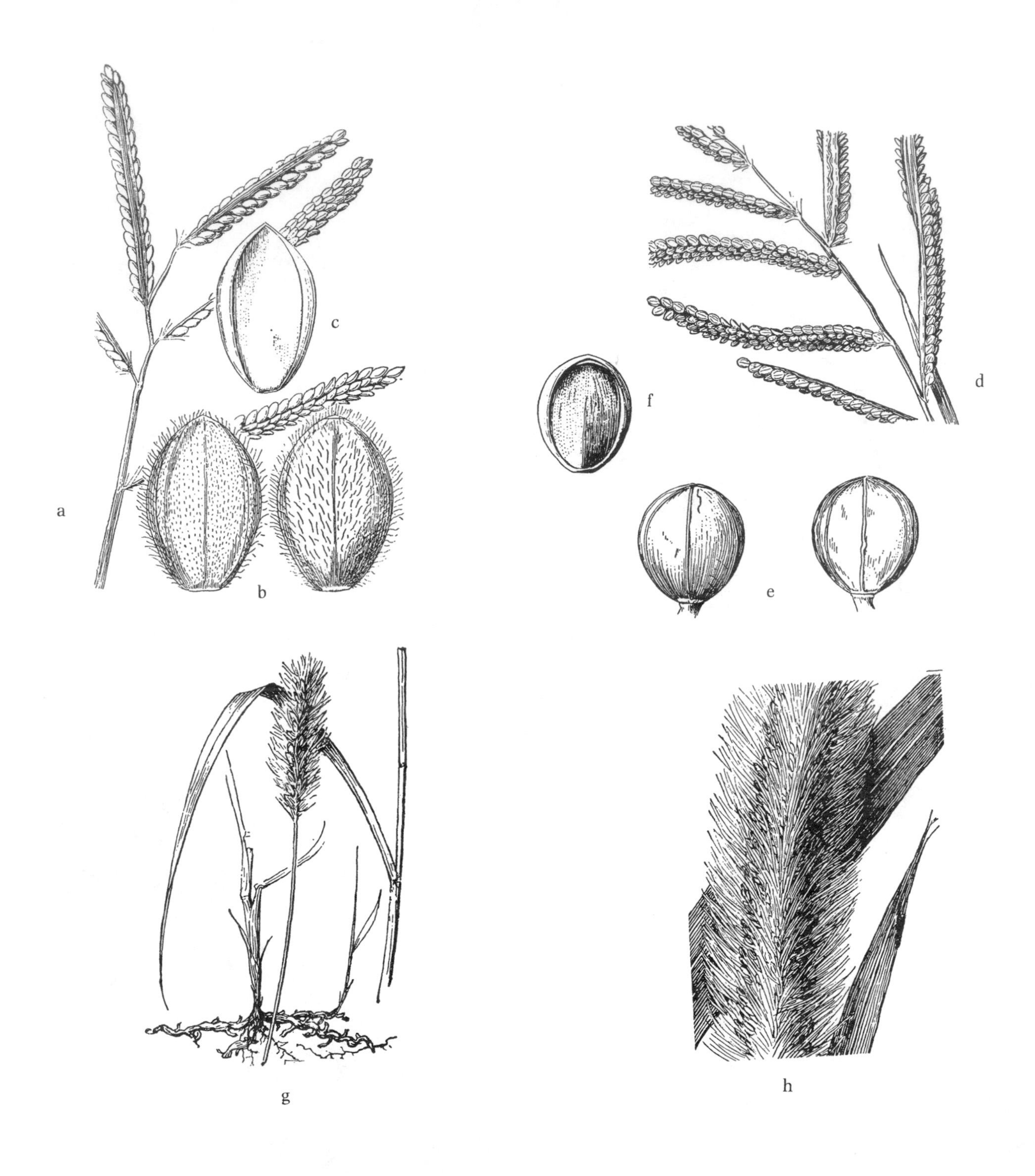

Fig. 327. *Paspalum pubiflorum*: a. inflorescence; b. spikelet. two views; c. upper floret (Hitchcock).
Paspalum boscianum: d. inflorescence; e. spikelet. two views; f. upper floret (Hitchcock).
Setaria parviflorum: g. habit (Hitchcock).
Setaria magna: h. portion of inflorescence with leaf (Hitchcock).

1. Plants perennial, with creeping, knotty rhizomes; stems slender, 0.4–1 m tall; spikelets subtended by involucre of 8–12 bristles. .... 1. *S. parviflora*
1. Plants annual, lacking rhizomes; stems thick (up to 2–3 cm in diameter at base), 3–4(6) m tall; spikelets subtended by involucre of 1–3 bristles. .... 2. *S. magna*

1. *S. parviflora* (Poir.) Kerguélen Knotroot Bristlegrass, Marsh Bristlegrass, Perennial Foxtail Fig. 327
Freshwater or saline marshes, shores, swales and seasonally wet sites. Se. N.E. w. to N.Y., Pa., Ohio, Ill., Ia., and Kans., s. to Fla., Tex., N.M., Ariz. and Calif.; Ida., Wash., Utah, s. Nev.; W.I. and S.Am. (*S. geniculata* (Lam.) Beauv.)

2. *S. magna* Griseb. Giant Foxtail, Giant Bristle-grass Fig. 327
Swamps, marshes, fresh and saline swales, shores and bayous. Coastal plain: N.J. s. to Fla., w. to Tex.; disjunct inland, Ark. and N.M.; Mex., Ber., W.I. and C. AM.

28. *Sacciolepis* (Cupscale Grass)

Perennial (ours) or annual herbs, tall: stems often decumbent; leaves flat, broad; inflorescence a contracted panicle; spikelets gibbous because of round swelling at base of second glume, 2-flowered, lower floret sterile or staminate, upper floret bisexual.

Reference: Wipff, 2003.

1. *S. striata* (L.) Nash American Cup-scale Fig. 328
Freshwater marshes, shores, swamps, ditches and wet places. Coastal plain, Me., Mass., s. N.J. and Del. s. to Fla., w. to e. Tex.; inland n. to Ark. Tenn., s. Mo. and Okla.; W.I.

29. *Echinochloa* (Wild Millet)

Perennial or annual herbs; leaves flat; inflorescence a panicle with spike-like lateral branches; spikelets plano-convex, spiny-hispid, glumes unawned to short awn-tipped, with 2(3) florets, upper floret fertile, the lemma unawned to awned, the lower floret sterile or staminate, the lemma with or without awn.

References: Fassett, 1949a; Gould et al., 1972; Michael, 2003; Wiegand, 1921.

1. Leaf sheaths with bristly hairs (fig. 328e); spikelets elliptic, about 3 times as long as wide; second (larger) glume awned. .... 1. *E. walteri*
1. Leaf sheaths glabrous; spikelets ovate, about twice as long as wide; second (larger) glume lacking awn.
   2. Lateral inflorescence branches simple (fig. 329d), 1–2 cm long; spikelets crowded into about 4 rows on one side of rachis; panicle pale green. .... 2. *E. colona*
   2. Lateral inflorescence branches compound (fig. 329a), more than 2 cm long; spikelets crowded and irregularly clustered; panicle dark green to purple.
      3. Smooth, lustrous portion of upper (fertile) lemma sharply differentiated from dark, dull, wrinkled tip, with a ring of minute setae present just below junction of body and tip; second glume and lower (sterile) lemma usually lacking stiff papillate-based hairs. .... 3. *E. crus-galli*
      3. Smooth, lustrous portion of upper (fertile) lemma gradually blending to the dull, wrinkled tip, lacking a ring of setae on upper portion of lemma at junction of body and tip; second glume and lower (sterile) lemma covered with stiff papillate-based hairs. .... 4. *E. muricata*

1. *E. walteri* (Pursh) Heller Coast Cock-spur Grass Fig. 328
Alkaline and saline marshes, swamps and shallow water of ponds and ditches. Que. and Ont. w. to s. Mich., Wisc., and Iowa, s. to Fla., e. Tex. and e. Mex.; Cuba. The spikes of this species often have a more bristly aspect than our other species.

2. *E. colona* (L.) Link Awnless Barnyard Grass; Jungle Rice Fig. 329
Moist to wet open places, shores, sand and gravel bars, ditches, cultivated fields and disturbed sites; weedy. Mass. and Vt.; N.J. and Va. w. to Ky., s. Ill., s. Mo. and e. Kans., s. to Fla., Tex., N.M., Ariz. and s. Calif.; sporadic in Mont., Wash. and Oreg; intro. from Eurasia.

3. *E. crus-galli* (L.) P. Beauv. Barnyard Grass Fig. 329
Moist to wet ground, marshes, shores, ditches cultivated fields and disturbed sites; weedy. Widely natzd. throughout the U.S.; intro. from Eurasia.

4. *E. muricata* (P. Beauv.) Fernald Rough Barnyard Grass
Low ground, ditches, and margins of swales and ponds. N.B. and Que. w. to Mich., Minn. and Alta., s. to Fla., Tex., N.M. and Calif. Two varieties have been recognized (Michael, 2003), both occurring with considerable overlap within the range of the species: var. *muricata* is the common variety in eastern N.Am., whereas var. *microstachya* (Wiegand) Shinners is the common taxon in western N.Am. This native species is very similar to the Eurasian *E. crus-galli*.

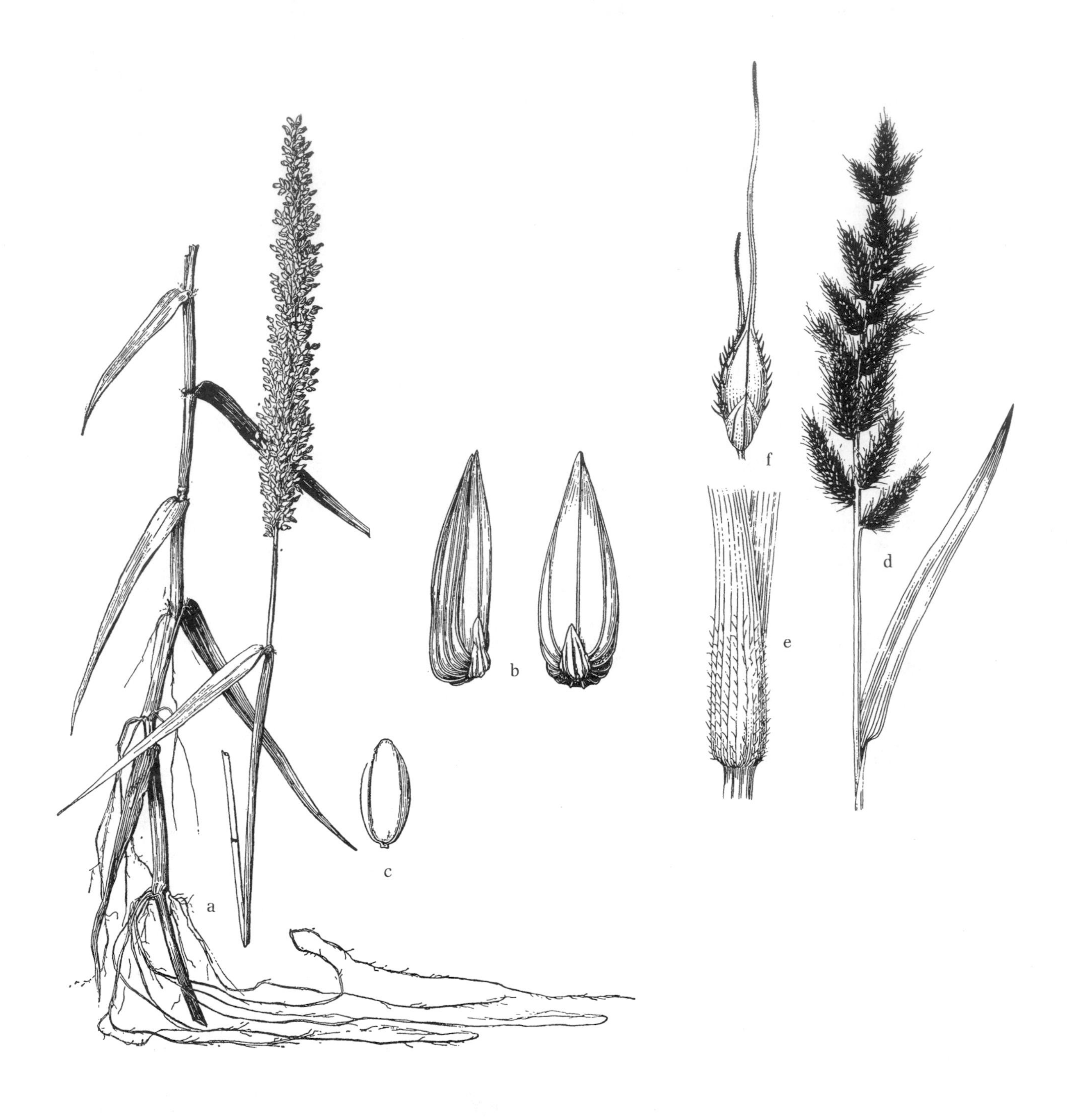

Fig. 328. *Sacciolepis striata*: a. habit; b. spikelet, two views; c. upper floret (Hitchcock). *Echinochloa walteri*: d. inflorescence; e. leaf sheath; f. spikelet (Beal).

Fig. 329. *Echinochloa crusgalli*: a. habit; b. spikelet, two views of lower, sterile floret; c. upper fertile floret (not visible in b.) (Hitchcock).
Echinochloa colona: d. inflorescence and portion of plant (Hitchcock).

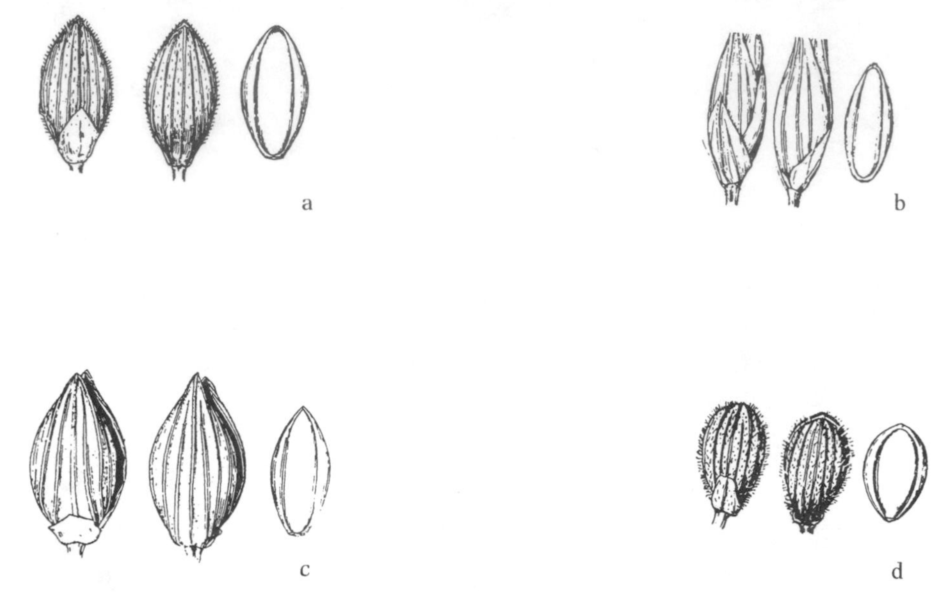

Fig. 330. *Dichanthelium boreale*: a. spikelet, two views (*left*) and upper floret (*at right*) (Hitchcock). *Panicum rigidulum*: b. spikelet, two views (*left*) and upper floret (*at right*) (Hitchcock). *Dichanthelium scabriusculum*: c. spikelet, two views (*left*) and upper floret (*at right*) (Hitchcock). *Dichanthelium acuminatum*: d. spikelet, two views (*left*) and upper floret (*at right*) (Hitchcock).

30. *Dichanthelium* (Panic Grass)

Dichanthelium is a large and complex genus once included within *Panicum*, but molecular evidence reinforces the morphologic evidence for treating it as a distinct genus (Barkworth, 2003). Flowering occurs early in the growing season, producing terminal panicles, whereas autumnal flowering produces lateral inflorescences. In many species the basal leaves are different from the stem leaves, forming a winter rosette (ours).

REFERENCE: Barkworth, 2003.

1. Ligular hairs conspicuous, 3–5 mm long. . . . . . 1. *D. acuminatum*
1. Ligular hairs obscure, less than 1 mm long, or lacking.
   2. Spikelets pubescent, elliptic, obtuse, 2–2.2 mm long. . . . . . 2. *D. boreale*
   2. Spikelets glabrous or obscurely puberulent, ovate, acute, 2.3–2.6 mm long. . . . . . 3. *D. scabriusculum*

1. *D. acuminatum* (Sw.) Gould & C. A. Clark Hairy Panic Grass Figs. 330, 331

Damp, sandy or gravelly shores to bogs, peaty or marshy ground, often calcareous or saline; often dry, sterile sites. Que. w. to s. Ont., Minn., Sask. and B.C., s. to Fla., Tex., N.M. and Calif.; Mex., C.Am., W.I., n. S.Am. Stephenson (1984) suggested that recognition of infraspecific taxa was unwarranted in this highly variable species, especially variation regarding pubescence vs. glabrous plants. Barkworth (2003) noted that this species is the most troublesome of the genus, and although she recognized nine subspecies for N.Am., she indicated that widespread introgression occurs both within the species as well as with other species. While we do not distinguish infraspecific taxa here, it is noteworthy that four subspecies of wet habitats recognized by Barkworth that occur within our range are treated by some authors as distinct species: subsp. *implicatum* (as *D. implicatum* (Schribn.) Kerguélen; subsp. *lanuginosum* (as *D. lanuginosum* (Elliott) Gould); subsp. *lindheimeri* (as D. *lindheimeri* (Nash) Gould); subsp. *spretum* (as *D. spretum* (Schult.) Freckmann. (*Panicum acuminatum* Sw.; *Panicum lanuginosum* Elliott; *P. spretum* Schult.)

2. *D. boreale* (Nash) Freckmann Northern Panic Grass Fig. 330

Shores, wet meadows, and damp open ground and woods. Nfld. and Que. w. to s. Ont. and Minn., s. to N.E., N.J., Va., N.C., Ga., Tenn. and Mo. (*Panicum boreale* Nash)

3. *D. scabriusculum* (Elliott) Gould & C. A. Clark Tall Swamp Panic Grass Fig. 330

Moist ground along ditches, streams and swamps, often in shallow water. Coastal plain, s. N.J. s. to n. Fla., w. to Ark. and e. Tex. (*Panicum scabriusculum* Elliott)

31. *Panicum* (Panic Grass)

Perennial and annual herbs; inflorescence typically an open panicle, sometimes contracted; spikelets 2-flowered, lower floret sterile, consisting of a lemma only, upper floret bisexual, with lemma and palea hard, cartilaginous, enclosing the grain.

Reference: Freckmann and Lelong, 2003.

1. Spikelets with wart-like tubercles on second glume and lower (sterile) lemma (fig. 331h). . . . . 1. *P. verrucosum*
1. Spikelets lacking tubercles.
   2. Plants perennial, rhizomatous or densely tufted, with basal leaves persisting from previous year.
      3. Panicle contracted, spike-like (fig. 332a); ligule membranous. . . . . 2. *P. hemitomon*
      3. Panicle open; ligule a fringe of cilia. . . . . 3. *P. rigidulum*
   2. Plants annual, lacking rhizomes; stems solitary or branched at base, lacking basal leaves from previous year.
      4. Spikelets with long, tapering tips (fig. 331c), 3–3.5 mm long; internodes of stem pilose. . . . . 4. *P. flexile*
      4. Spikelets with short-pointed tips (fig. 331a), 1.7–2 mm long; internodes of stem glabrous (except sometimes papillose-hispid below nodes). . . . . 5. *P. philadelphicum*

1. *P. verrucosum* Muhl. Warty Panic Grass Fig. 331
Moist or wet sands, peats and shores, often shady. Chiefly coastal plain, piedmont and Mississippi embayment, Mass. s. to Fla., w. to e. Tex.; inland n. to Ky., s. Ohio, nw. Ind. and sw. Mich., Ill., Okla., Ark. and se. Mo.

2. *P. hemitomon* Schultes Maidencane Fig. 332
Moist ground, riverbanks, ditches, margins of swamps, lakes and ponds, often in shallow water; sometimes weedy in low cultivated fields. Chiefly coastal plain and piedmont, s. N.J. s. to Fla., w. to Ala., La. and e. Tex.; Tenn.; S.Am.

3. *P. rigidulum* Bosc ex Nees Redtop Panicum Fig. 330
Swamps, marshy margins of rivers, ponds and lakes, floodplain forests, wet woods, wet pine savannahs and ditches. N.S., N.E. and Ont. w. to Mich. and Wisc., s. to Fla. and Tex.; B.C., Oreg. and Calif. Treated by some authors as *Coleataenia rigidula* (Bosc ex Nees) LeBlond; we have followed Freckmann and Lelong (2003).

4. *P. flexile* (Gatt.) Scribn. Wiry Witchgrass Fig. 331
Damp sandy or gravelly calcareous shores, fens, marshy places, meadows, ledges, calcareous barrens and open woods, especially on limestone. Sw. Que. and w. N.E. w. to Ont., Mich., Wisc. and N.D., s. to Fla. and e. Tex.; Utah.

5. *P. philadelphicum* Bernh. ex Trin. Philadelphia Witchgrass Fig. 331
Borders of rivers, creeks, marshes, and lakes, but more often dry, open ground. N.S., N.B., N.E. and sw. Que. w. to s. Mich. and s. Minn., s. to Ga., Ala., se. Kans. Okla. and Tex.

Haemodoraceae / Bloodwort Family

1. *Lachnanthes* (Red-root)

Perennial herbs, arising from slender rhizomes that exude a red juice; stems villous-tomentose; leaves sword-shaped, basal and scattered along stem; inflorescence a flat-topped compound cyme, the branches helicoid; perianth yellow, tomentose, outer parts shorter than inner parts (fig. 585d); stamens 3; fruit a globose capsule; seeds reddish-brown, flat, discoid.

Reference: Robertson, 2002.

1. *L. caroliniana* (Lam.) Dandy Red-root Fig. 333
Wet acidic soils, bogs, sandy and peaty shores, swamps, savannas, pinelands, hammocks, pocosins and cranberry bogs. Coastal plain, N.S.; se. Mass. and R.I. s. to N.J., Del., se. Va., S.C. and Fla., w. to La.; c. Tenn.; Cuba. (*L. tinctoria* (Walter) Elliott, *Gyrotheca tinctoria* (Walter) Salisb.) Sometimes becoming a troublesome weed in commercial coastal plain cranberry bogs.

Pontederiaceae / Pickerel-weed Family

References: Crow 2003a; Hellquist and Crow, 1982; Horn, 2002; Rosatti, 1987.

1. Plants free-floating; leaves with blades nearly circular to broadly elliptic, base widely cuneate, petioles usually inflated, often bulbous (fig. 334a). . . . . 1. *Eichhornia*
1. Plants rooted, emersed or submersed; leaves with blades ovate to lanceolate (fig. 335, 336) or reniform (fig. 338c), base cordate (fig. 336c) to truncate (fig. 336e,f), petioles not inflated or leaves linear and ribbon-like (fig. 337b).

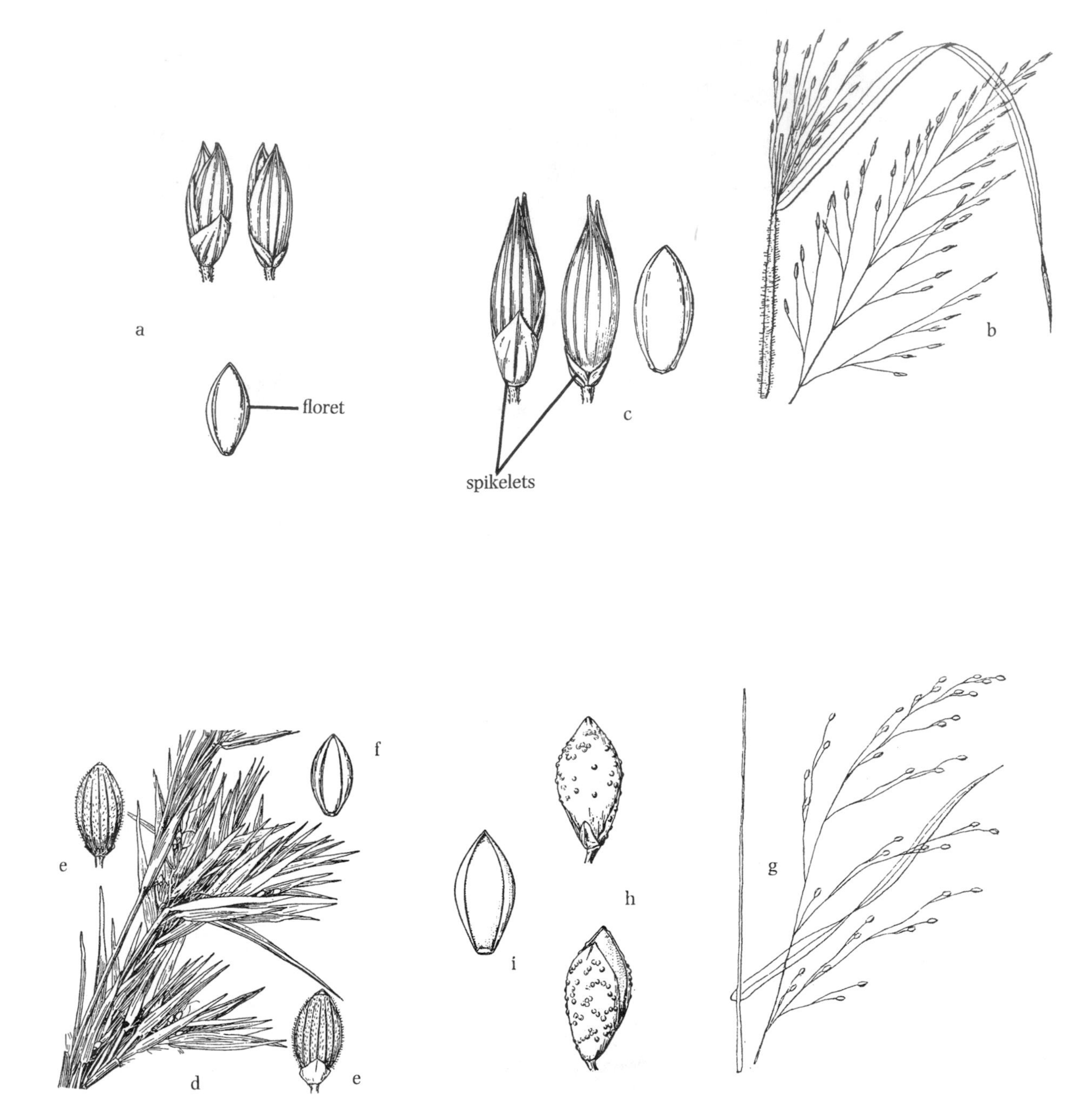

Fig. 331. *Panicum philadelphicum*: a. spikelet, showing glumes and lemmas, two views (*top*), with upper fertile floret (*shown below*) (Hitchcock).
Panicum flexile: b. inflorescence (B&B); c. spikelet, two views (*left*), with upper fertile floret (*at right*) (Hitchcock).
Dichanthelium acuminatum: d. inflorescence; e. spikelet, two views: f. upper fertile floret (Hitchcock).
Panicum verrucosum: g. habit (B&B); h. spikelet, two views (Hitchcock); i. upper fertile floret (Hitchcock).

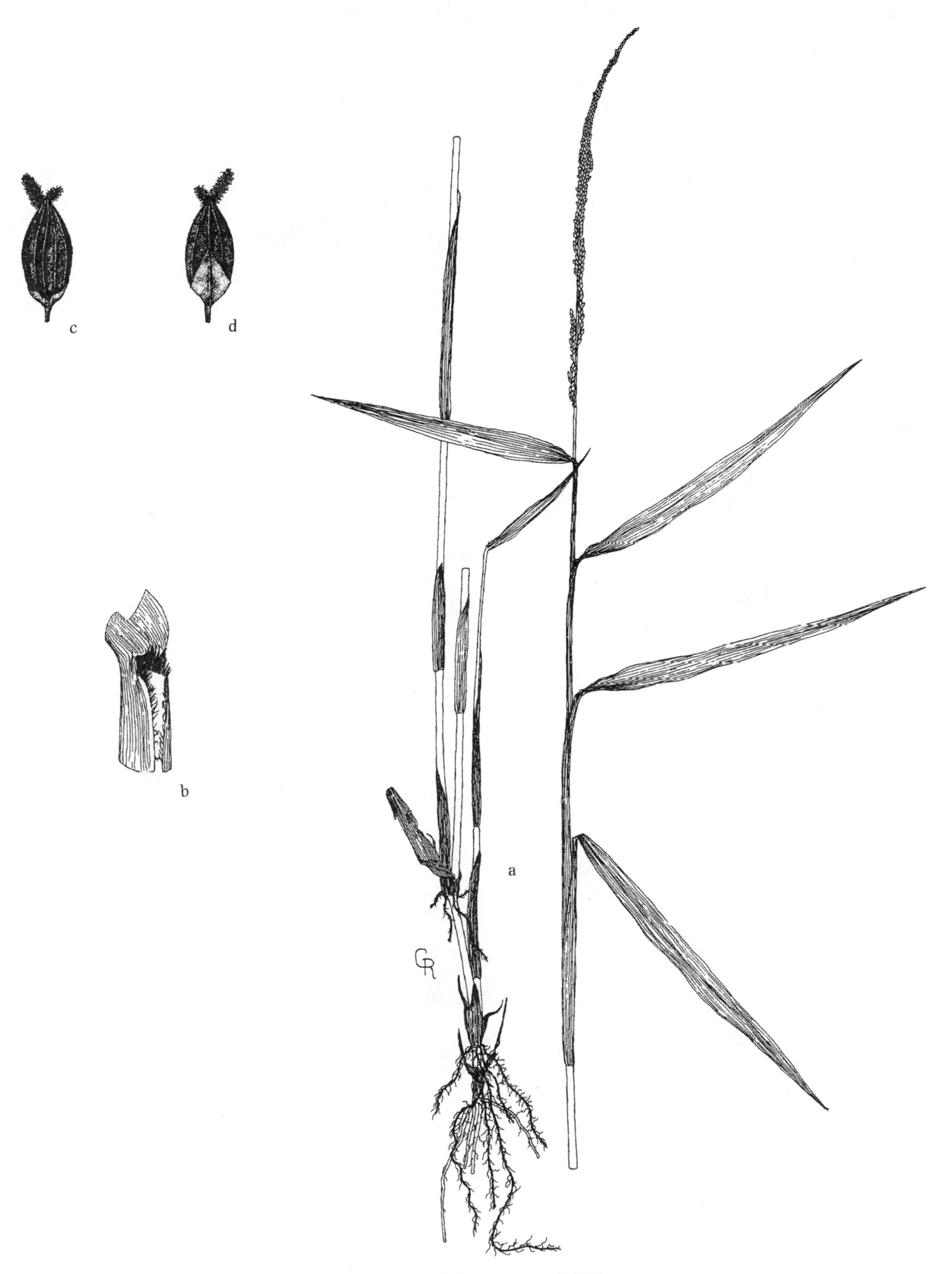

Fig. 332. *Panicum hemitomon*: a. habit; b. ligule; c, d. spikelet, two views (G&W).

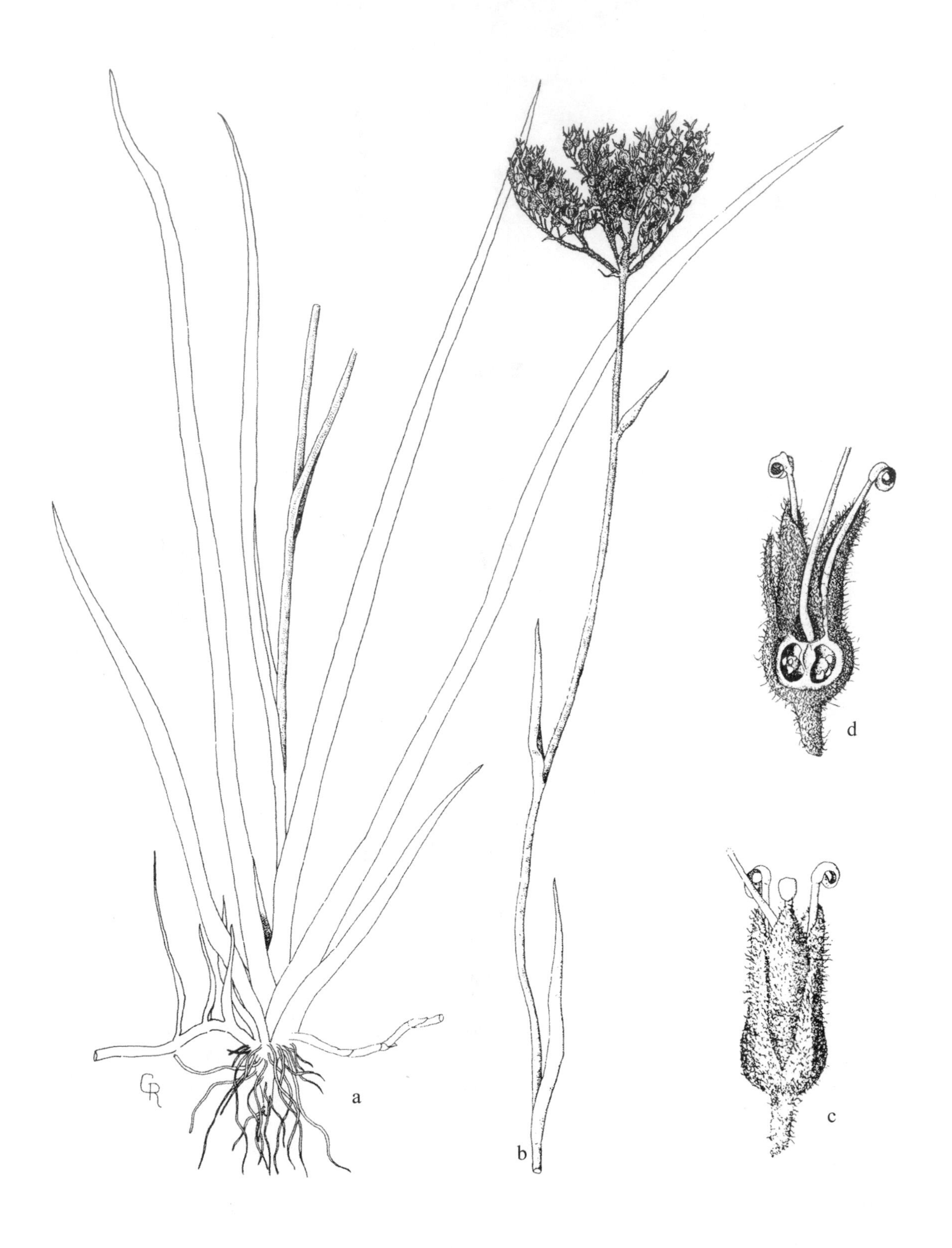

Fig. 333. *Lachnanthes caroliniana*: a. base of plant; b. upper portion of plant; c. flower; d. flower, longitudinal section (G&W).

2. Stamens 6; flowers 2-lipped, in a dense spike-like inflorescence, blue-violet; leaves cordate, hastate, or narrowly to broadly lanceolate; base of submersed linear leaves usually light lavender colored. . . . . 2. *Pontederia*
2. Stamens 3; flowers regular, solitary, yellow, blue, or rarely white; leaves reniform, lanceolate, or linear and ribbon-like; base of submersed linear-leaved green (on *H. dubia*). . . . . 3. *Heteranthera*

---

1. *Eichhornia* (Water-hyacinth)

Perennial and annual herbs; leaves subcircular, ovate to broadly elliptic, petioles often inflated, often bulbous; inflorescence spike-like; flowers violet-blue with yellow streaks, subtended by a spathe-like bract; fruit a capsule.

REFERENCE: Penfound and Earle, 1948.

1. *E. crassipes* (Mart.) Solms Water Hyacinth Fig. 334
Ponds, canals and ditches. Intro., se. Va. s. to Fla., w. to Mo., Tex. and Mex.; Wash. Oreg., Calif. Ariz. and Colo.; Trop. Am.; intro. widely in Old World tropics. In the southern United States, the species is locally abundant and sometimes a troublesome weed. This attractive free-floating aquatic plant native to Brazil is popular for aquatic gardens; occasionally escaped plants have been documented from Mass., Conn., N.Y., N.J. and s. Ont.; these do not persist in the northern states and Canada. Some government websites refer to this plant by an old name—*Pontederia crassipes* (USGS Website, Non-indigenous Aquatic Species, see https://nas.er.usgs.gov/taxgroup/Plants/default.aspx) (*Pontederia crassipes* Mart.)

2. *Pontederia* (Pickerel-weed)

Perennial herbs, arising from creeping rhizomes; leaves basal, erect, long-petioled, usually cordate; flowers violet-blue, in a dense spike subtended by a spathe-like bract; fruit an utricle, crested with 6-toothed ridges.

REFERENCES: Horn, 2000; Lowden, 1973.

1. *P. cordata* L. Pickerel-weed Figs. 335, 336
Margins of lakes, ponds, and streams, marshes, swamps and ditches. P.E.I. and N.S. w. to s. Que., s. Ont., Mich., Wisc., and Minn., s. to Fla., Ala., Miss., La. and Tex.; Cuba, Belize and S.Am. The seedlings consist of submersed rosettes of linear leaves light purple at base and often have 1 or 2 petiolate, slightly lanceolate blades. The range of variability in the leaves of this species is considerable. (*P. cordata* var. *lancifolia* (Muhl.) Torr.; *P. lanceolata* Nutt.)

3. *Heteranthera* (Mud-plantain)

Perennial or annual herbs; leaves petiolate, in a basal rosette or ribbon-like, sessile, along a submersed stem; flowers yellow, white, blue or pale purple, perianth united, tubular, emerging from sheathing leaf bases or nodes; stamens 3, unequal (1 longer than the 2 lateral), fruit a capsule.

REFERENCES: Horn, 1983, 1986, 1988, 2002; Thieret, 1971.

---

1. Leaf blades lax, long, linear, ribbon-like, usually submersed and floating, lacking petioles or stipule; flowers solitary, yellow. . . . . . 1. *H. dubia*
1. Leaf blades ovate to orbicular, reniform, oblong, oblanceolate or linear (but never ribbon-like), emersed or stranded, petioles and stipules present; flowers solitary or in inflorescences of 2–24 cm long, white, blue and pale purple.
   2. Inflorescences 1-flowered, flowering stems 2–24 cm long, floral tubes 11–49 mm long, actinomorphic.
      3. Leaf blade round to oblong, typically cordate at base or truncate, apex obtuse, uppermost portion of central perianth lobe flaring out with pair of lateral flanges (at yellow spot). . . . . 2. *H. rotundifolia*
      3. Leaf blade oblong to ovate, typically tapering to base or truncate, apex; uppermost portion of perianth lobe lacking lateral flanges. . . . . 3. *H. limosa*
   2. Inflorescences 2–24 flowered, flowering stems 1–9 cm long, floral tubes 3–12 mm long, zygomorphic.
      4. Spikes shorter than spathes; flowers white, uppermost lobe green or green and yellow basally; stamen filament hairs white; spathes subtending inflorescence equal to or slightly longer than the internode of the flowering stem. . . . . 4. *H. reniformis*
      4. Spikes longer than spathes, flowers pale purple or white, uppermost lobe dark purple basally with 1 or 2 yellow spots, stamen filament hairs purple; spathes subtending inflorescence 3 or more times longer than the internode of the flowering stem. . . . . 5. *H. multiflora*

Fig. 334. *Eichhornia crassipes*: a. habit, floating, with feathery roots; b. leaf with inflated petiole; c. flower; d. flower, longitudinal section; e. ovary, cross-section (Mason).

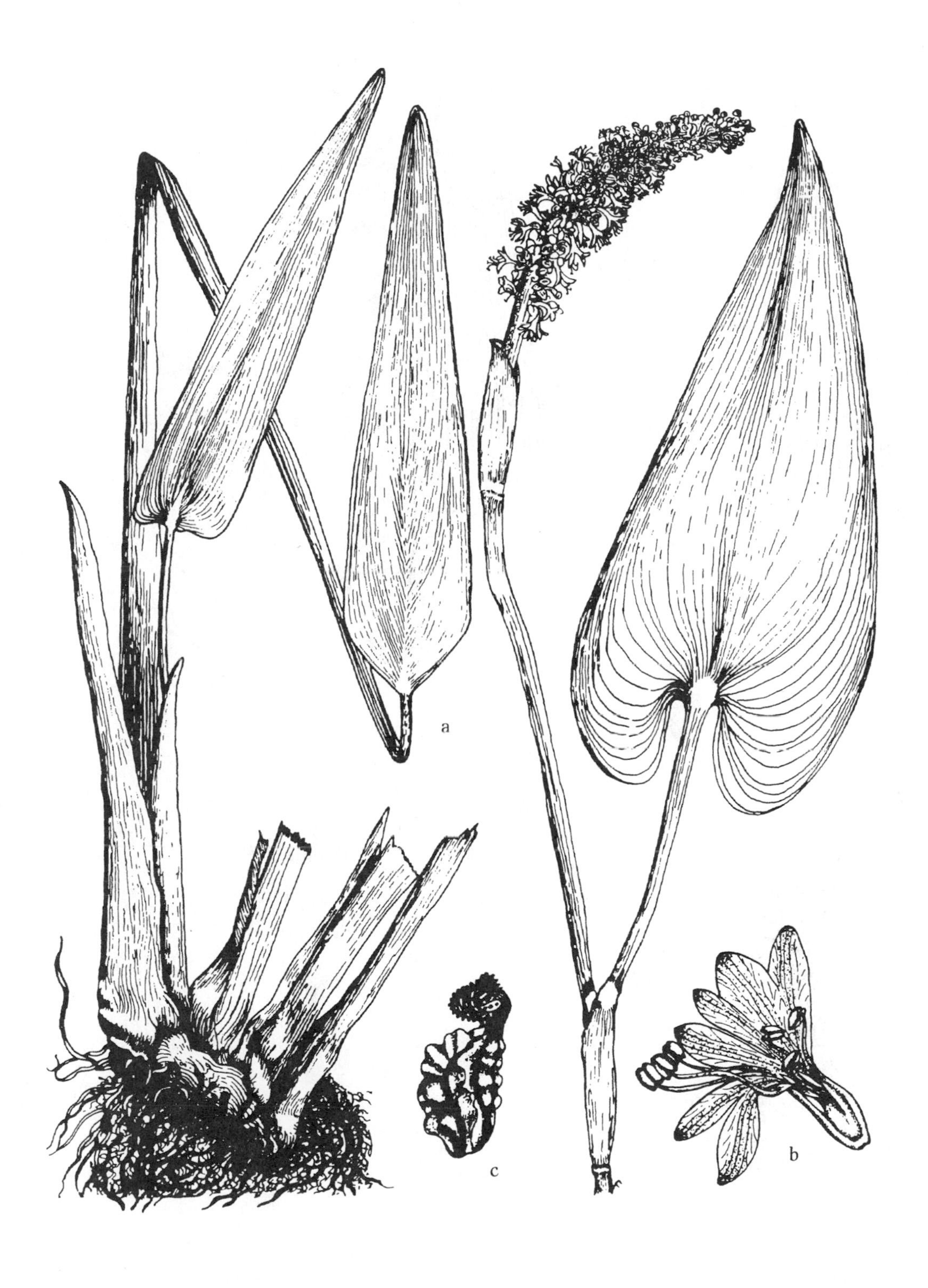

Fig. 335. *Pontederia cordata*: a. habit; b. flower; c. withering perianth, revolute-coiled, exposing shape of developing ovary (C&C).

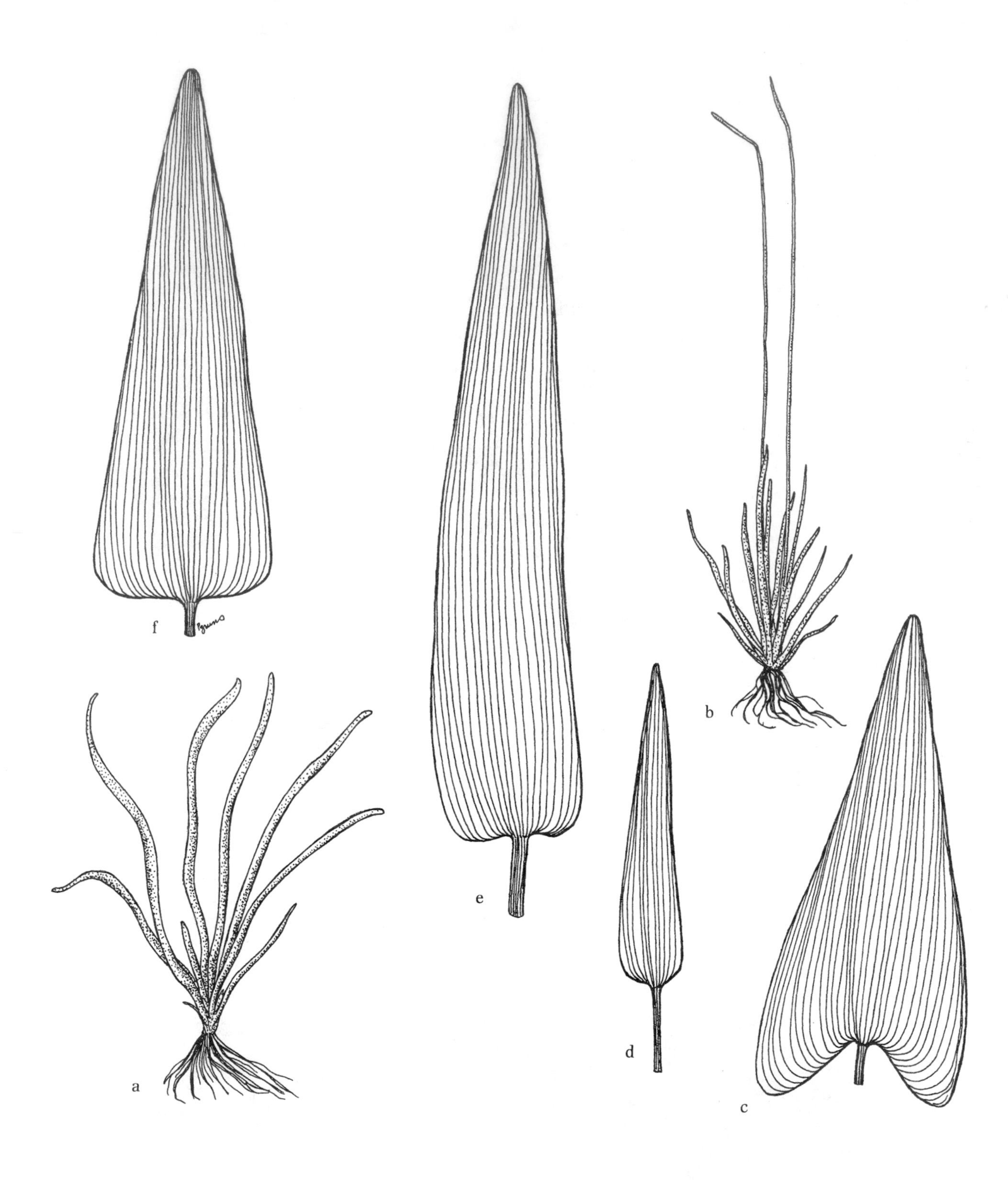

Fig. 336. *Pontederia cordata*: a. habit, submersed form; b. habit, submersed juvenile plant; c–f. leaf variations (NHAES).

1. *H. dubia* (Jacq.) MacMill. Water Star-grass Fig. 337
Alkaline and neutral waters or muddy shores of ponds, lakes and streams. N.B., Me., and sw. Que. w. to s. Ont., Minn., Ida., B.C. and Oreg. s. to Fla., Tex., Ariz., Calif. and Mex. This is often confused with a few of the linear-leaved species of *Potamogeton*, especially *P. zosteriformis*, but the presence of an obscure midvein in *H. dubia* is distinctive in contrast with the rather conspicuous midvein in *Potamogeton*. Plants stranded in the mud or floating in dense mats where the foliage reaches the surface tend to produce conspicuous yellow flowers; cleistogamous flowers often occur on submersed stem later in the season. (*Zosterella dubia* (Jacq.) Small)

2. *H. rotundifolia* (Kunth) Griseb. Round-leaf Mud-plantain
Shallow water ephemeral pools, ditches and ponds. Ky. and Mo. w. to S.D. and Colo. s. to Tex., N.M., Ariz. and Calif.; C.Am., S.Am. This species is closely related to *H. limosa*. (*H. limosa* var. *rotundifolia* Kunth)

3. *H. limosa* (Sw.) Willd. Blue Mud-plantain Fig. 338
Muddy shores and quiet waters of canals and ponds. Ky. and s. Ill. w. to S.D., Neb., and Colo., s. to w. Fla., Okla., Tex., N.M., Ariz., Calif. and Mex.; W.I., C.Am. and S.Am.

4. *H. reniformis* Ruiz & Pav. Kidney-leaf Mud-plantain Fig. 338
Muddy shores and shallow water. S. Conn. and c. N.Y. w. to se. Mo. and Ark., s. to Va., N.C., Ga. and w. Fla, La., sw. Tex. and Mex.; W.I., C.Am. and S.Am.

5. *H. multiflora* (Griseb.) C. N. Horn Bouquet Mud-plantain
Muddy shores and shallow water. N.J. and e. Pa. s. to ne. N.C.; sw. Ill. and Mo. w. to Neb., s. to sw. Tenn., Miss., La., Okla., s. Tex. and Mex.; S.Am.

Marantaceae / Arrowroot Family

1. *Thalia* (Thalia)

Perennial herbs, large, erect, arising from thick rhizomes; leaves large, alternate, basal, long-petioled; inflorescence a narrow to spreading panicle; flowers bisexual, in pairs, subtended by 2 bracts; sepals 3, small; petals 3, purple, 1 lanceolate, 2 obovate; ovary inferior; fruit a fleshy bluish-purple utricle.

Reference: Kennedy, 2000.

1. *T. dealbata* Fraser ex Roscoe Powdery Thalia, Powdery Alligator-flag Fig. 339
Swamps, pond margins, margins of streams and ditches. Coastal plain, S.C. and Ga., w. to Miss., La. and Tex.; Mississippi embayment n. to e. Okla., Ark., se. Mo. and s. Ill. Often grow as a handsome ornamental in tropical and temperate aquatic gardens.

Fig. 337. *Heteranthera dubia*: a. habit (NHAES); b. portion of filiform growth form (NHAES); c. flower (NHAES); d. habit, terrestrial form (F).

Fig. 338. *Heteranthera limosa*: a. habit; b. flower (C&C).
Heteranthera reniformis: c. habit (NHAES); d. flower (C&C).

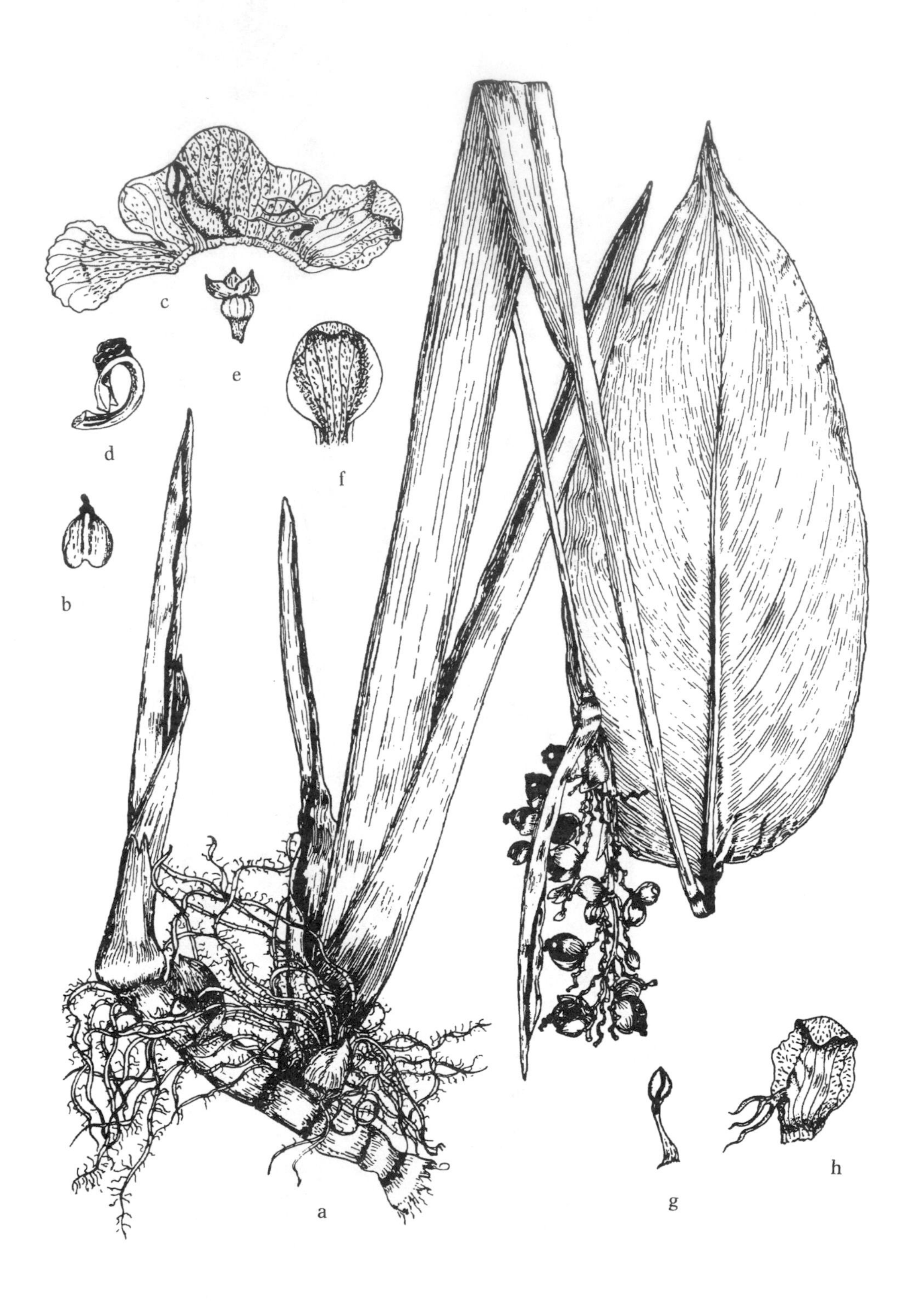

Fig. 339. *Thalia dealbata*: a. habit; b. individual bract with withered apex; c. portion of flower showing outer staminode, stamen, calloused staminode, and hooded staminode; d. pistil and stigmatic surface; e. ovary with bracts at summit; f. outer staminode; g. stamen separated from flower; h. hooded staminode (C&C).

Eudicots

Ceratophyllaceae / Hornwort Family

1. *Ceratophyllum* (Coontail, Hornwort)

Perennials, free-floating submersed aquatic, lacking roots; leaves whorled, sessile or with short petioles, dichotomously divided into narrow segments (seedling leaves often opposite, simple); plants monoecious; flowers unisexual, axillary, 1-several per node; fruit a 1-seeded, ovoid-oblong achene.

REFERENCES: Aboy, 1936; Fassett, 1953b; Hellquist and Crow, 1984; Jones, 1931; Les, 1980, 1985, 1986a, 1986b, 1988a, 1988b, 1988c, 1988d, 1989, 1991, 1997; Lowden, 1978; Muenscher, 1940; Wood, 1959.

1. Principal leaves firm, stiff, forked no more than 2 order (no pair of ultimate segment results from more than 2 consecutive forkings of leaf axis) (fig. 340b); leaf segments with distinct serrations (fig. 340b), marginal teeth with broad bases (fig. 340c): fruit smooth or slightly warty, with 2 basal spine, marginal spines absent (fig. 340d); first seedings simple. . . . . 1. *C. demersum*
1. Principal leaves delicate, somewhat flaccid, often forked 3 or 4 orders (ultimate segment pairs result from up to 3 consecutive forking of leaf axis) (fig. 340b); leaf segments lacking distinct serrations, marginal teeth when present, with narrow bases (fig. 340h), fruit warty; marginal spines present (fig. 340j); first seedling leaves divided at least once. . . . . .2. *C. echinatum*

1. *C. demersum* L. Fig. 340
Ponds, lakes, marshes, swamps, quiet streams and ditches. Weedy; N.S., P.E.I., Que. and Me. w. to N.W.T. and Alask., s. to Fla. and Calif., Mex.: W.I. and C.Am.

2. *C. echinatum* A. Gray Fig. 340
Ponds, lakes, marshes and swamps, especially soft waters. N.S., N.B., and s. Que. w. to Ont. and se. Man., s. to Fla. and e. Tex; B.C., w. Wash. and Oreg. Unlike the weedy *C. demersum*, this delicate species appears to be relatively uncommon, or at least overlooked; it's never weedy. Lowden (1978) regarded this taxon as conspecific with *C. muricatum* Cham.; Les (1985, 1991) presented evidence to support recognition of the two species as distinct.

Ranunculaceae / Crowfoot Family

Mostly herbaceous, perennial plants; leaves typically alternate (sometimes opposite), with sheathing leaf bases, simple and entire or often lobed, compound or dissected; flowers bisexual or unisexual, petals present or lacking, the sepals then typically petaloid; stamens numerous, pistils simple, few to many, borne on a receptacle; fruit a follicle, achene, or capsule.

REFERENCES: Whittemore and Parfitt, 1997

1. Perianth in 1 whorl or appearing to be so; petals if present, represented by small yellow gland-like structures between sepals and stamens (sometimes interpreted as staminodes).
    2. Sepals yellow or yellowish to white, persistent; fruit a follicle.
        3. Leaves simple, unlobed, merely toothed or crenate to sometimes entire; petals absent. . . . . 1. *Caltha*
        3. Leaves palmately divided into (3)5–7 segments; petals small, appearing as yellow glands between stamens and perianth. . . . . 2. *Trollius*
    2. Sepals greenish, early deciduous; fruit an achene.. . . . . 3. *Thalictrum*
1. Perianth clearly in 2 whorls; petals present.
    4. Flowers radially symmetrical; perianth white or yellow.
        5. Leaves simple, narrowly linear, forming a basal tuft; flowers with numerous pistils on an elongate, spike-like (tail-like) receptacle sepals spurred (fig. 343). . . . . 4. *Myosurus*

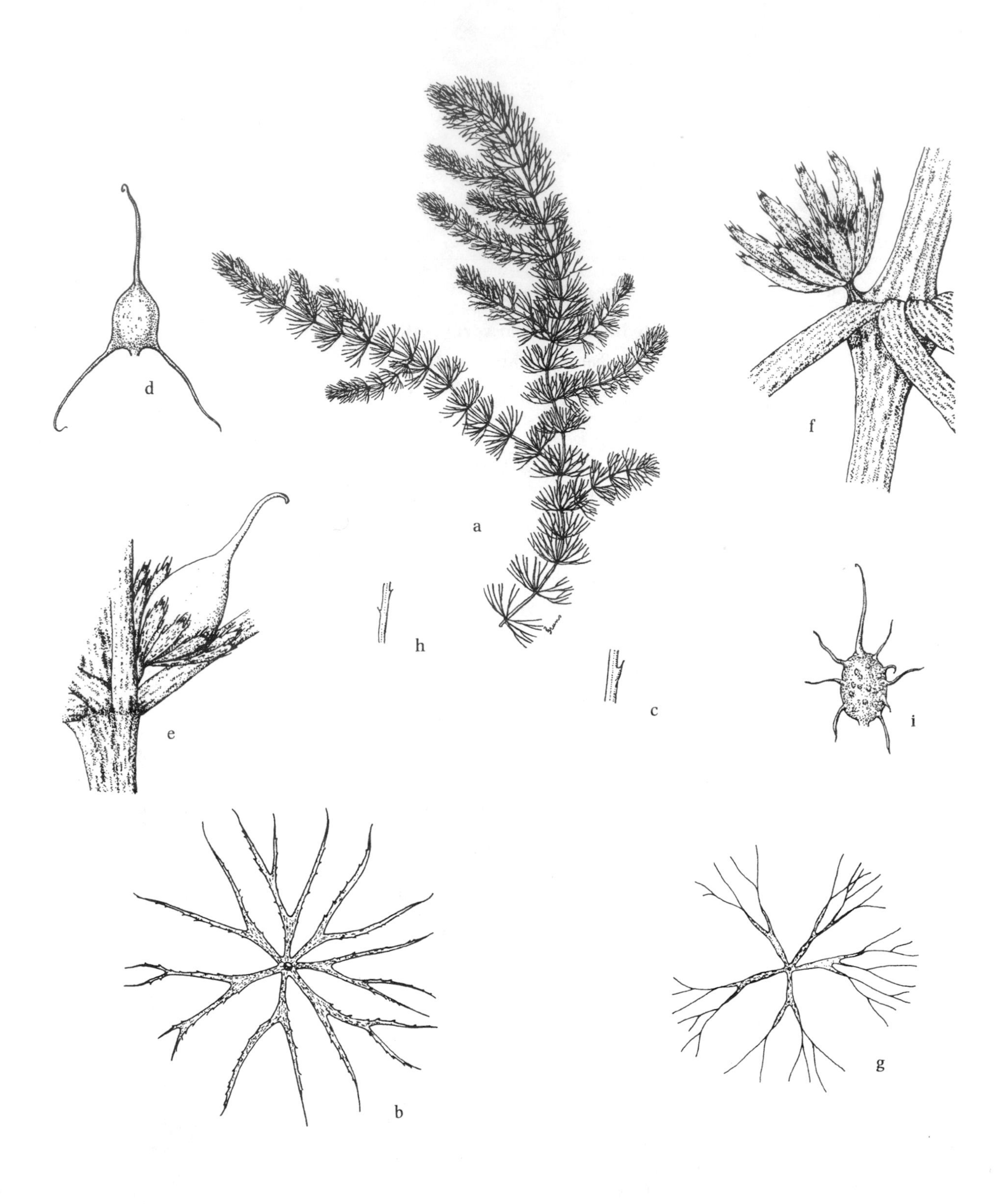

Fig. 340. *Ceratophyllum demersum*: a. habit (NHAES); b. leaf whorl (NYS Museum); c. leaf segment, showing broad-based marginal tooth (NHAES); d. fruit (NHAES).
Ceratophyllum echinatum: e. pistillate flower (NYS Museum); f. staminate flower (NYS Museum); g. leaf whorl (NYS Museum); h. leaf segment (NHAES); i. fruit (NHAES).

5. Leaves dissected, deeply divided, or if simple, then cauline or cauline and basal; flowers with few to many pistils on a short, nearly globose, receptacle; sepals lacking spurs. . . . . .5. *Ranunculus*
4. Flowers bilaterally symmetrical, uppermost sepal forming a distinct, petaloid hood (fig. 352b); perianth purple to lavender-blue. . . . . .6. *Aconitum*

1. *Caltha* (Marsh-marigold, Cowslip)

Perennial herbs; stems hollow; leaves basal and cauline; sepals yellow or white to pinkish, petaloid; petals absent; fruit a follicle.

REFERENCES: Ford, 1997; Smit, 1973; Walsh, 1986.

1. Flowers yellow, large, 15–40 mm wide; follicles (4)5–10(12), curved (arched); ascending (plants typically in large clumps, the older plants becoming decumbent, rooting at nodes and developing shoots). . . . . 1. *C. palustris*
1. Flowers white or pinkish, 10–12 mm wide; follicles numerous, ca. 30(20–55), straight; stems mostly floating or creeping, rooting at the nodes.. . . . . .2. *C. natans*

1. *C. palustris* L. Marsh-marigold, Cowslip Fig. 341
Swamps, wet meadows, and wet woods. Lab. and Nfld. w. to N.W.T. and Alask., s. to N.E., N.J., mts. to N.C. and Tenn., Ind., Iowa, e. Neb., Sask., Alta., B.C. Wash. and Oreg. A circumboreal species, our taxon is var. *palustris*.

2. *C. natans* Pall. Floating Marsh-marigold Fig. 341
Shallow water or creeping on mud. N.W.T. w. to Alask., s. to w. U.P. of Mich., n. Wisc., n. Minn., nw. Ont., and Alta.

2. *Trollius* (Globe-flower)

Perennial herbs; stems ascending; leaves alternate, palmately divided; sepals yellow to greenish-yellow, petaloid; petals represented by small yellow gland-like structures between sepals and stamens (sometimes interpreted as staminodes); pistils numerous; fruit a many-seeded follicle.

REFERENCE: Parfitt, 1997.

1. *T. laxus* Salisb. Spreading Globe-flower Fig. 341
Rich meadows and swamps. Very rare; w. Conn. to N.Y., s. to N.J., Pa., Del. and Ohio. *Trollius albiflorus* (A. Gray) Rydberg, previously treated as as *T. laxus* var. *albiflorus* A. Gray, is now regarded as a distinct, western species (Parfitt, 1997).

3. *Thalictrum* (Meadow-rue)

Perennial herbs; leaves ternately or otherwise decompound; inflorescence a raceme, panicle, or corymb; flowers unisexual, sometimes bisexual or plants polygamodioecious; sepals 4 or 5, green, early deciduous; petals absent; fruit an achene.

REFERENCES: Boivin, 1944; Keener, 1976b; Park and Festerling, 1997.

1. Plants bisexual, 4–20(35) cm tall; leaves chiefly in a basal rosette; inflorescence a raceme. . . . . .1. *T. alpinum*
1. Plants unisexual, 50–260 cm tall; leaves basal and cauline; inflorescence a panicle.
   2. Inflorescence more or less rounded; stigmas about half the length of achene body; anthers 0.8–1.5 mm long. . . . . 2. *T. pubescens*
   2. Inflorescence more or less pyramidal; stigmas about same length as achene body; anthers 1.5–3.5(4) mm long. . . . . 3. *T. dasycarpum*

1. *T. alpinum* L. Alpine Meadow-rue 342
Peaty coastal or alpine meadows and bogs, wet calcareous ledges and wet gravels. Greenl. and Lab. w. to Alask., s. to Nfld. and Gaspé Pen., Que.; mts. s. to Colo., Utah, Nev., and Calif.

2. *T. pubescens* Pursh Tall Meadow-rue, Muskrat-weed Fig. 342
Wet meadows, swamps, marshes, bogs, thickets, and rich woods. S. Lab. and Nfld. w. to Ont., s. to Ga. Tenn., Ala. and Miss. This species is regarded as a single, widespread, polymorphic species, with no infraspecific taxa worthy of recognition (Keener (1976b; Park and Festerling, 1997). (*T. polygamum* Muhl. ex Spreng.)

3. *T. dasycarpum* Fisch. & Avé-Lall. Purple Meadow-rue Fig. 342
Damp thickets, swamps, and wet meadows. Ont. w. to Alta., s. to Pa., Ohio, Ind., Ill., Ala., Miss., La., Tex., N.M., and Ariz. (*T. dasycarpum* var. *hypoglaucum* (Rydb.) Boivin.)

Fig. 341. *Caltha palustris*: a. habit; b. flower; c. fruit (NYS Museum).
Caltha natans: d. habit (*Flora Sibirica*).
Trollius laxus: e. upper portion of plant (Crow); f. fruit (NYS Museum).

Fig. 342. *Thalictrum pubescens*: a. habit, upper portion of plant; b. staminate flower; c. pistillate flower; d. fruit (NYS Museum).
Thalictrum alpinum: e. habit; f. fruit (Gleason).
Thalictrum dasycarpum: g. inflorescence; h. leaf; i. fruit; j. stamen (B&B).

4. *Myosurus* (Mousetail)

Annuals; leaves basal. forming a tuft, simple, narrowly linear; flowers solitary, on a scape; sepals 5, spurred; petals clawed, with nectary at claw; stamens numerous, giving flowers whitish appearance; pistils numerous; fruit an achene.

REFERENCES: Campbell, 1952; Sorrie, 1990; Whittemore, 1997.

1. *M. minimus* L. Tiny Mousetail Fig. 343
Wet and muddy places. S. Ont. and w. Ind. w. to Ill., sw. Minn., Mont., s. Man., Alta., and s. B.C., s. to Tenn., Ala., La., Tex. and Calif.; coastal plain and piedmont, se. Mass., se. Pa., and se. Va. s. to S.C. Whittemore (1997) recognized not infraspecific taxa.

5. *Ranunculus* (Crowfoot, Buttercup)

Perennial or annual herbs, submersed, amphibious, or terrestrial; basal leaves frequently present, often differing from cauline leaves; cauline leaves alternate, petioles sheathing stem; flowers usually solitary; sepals smaller than petals, not petaloid; petals yellow, sometimes white, with nectary at base; pistils numerous; fruit a head of achenes.

REFERENCES: Benson, 1948, 1954; Cook, 1963, 1966; Drew, 1936; Duncan, 1980; Keener, 1976a; Mitchell and Dean, 1982; Voss, 1985; Whittemore, 1997; Wiegleb et al., 2017.

---

1. Leaves dissected into thread-like or flat ribbon-like divisions (fig. 345b,c); leaves submersed, not floating (terrestrial forms develop in *R. flabellaris*, fig. 345d,e, and *R. gmelinii*, fig. 346d).
  2. Leaf segments thread-like; petals white, sometimes yellowish at base; achenes transversally wrinkled (fig. 344g).
    3. Leaves with distinct petiole above sheath before branching (fig. 344a,b), divisions typically limp, collapsing when removed from water. . . . . 1. *R. trichophyllus*
    3. Leaves branching immediately above sheath (fig. 344e,f) or with very short petiole, divisions somewhat firm, not collapsing when removed from water.
      4. Capillary segments of leaves pubescent, sheaths pubescent or glabrous; sepals, 3–4 mm long; petals 7–10 mm long, white with large yellow claw, the blades wide and overlapping at anthesis; achenes 30–80 (typically ca. 40); achene beak 0.2–0.5 mm long. . . . . 2. *R. subrigidus*
      4. Capillary segments of leaves glabrous, sheaths glabrous; sepals 2–3 mm long; petals 6–8 mm long, white with small yellow claw, the blades narrower, not overlapping at anthesis; achenes 7–25 (typically ca. 16); achene beak 0.7–1.1 mm long. . . . . 3. *R. longirostris*
  2. Leaf segments flat, sometimes very narrow; petals yellow; achenes not transversally wrinkled.
    5. Leaves finely dissected triternately, segments 1–2 mm wide; petals (6)7–15(17) mm long; body of achenes 2 mm long, with conspicuous corky wing-keel (fig. 345f). . . . . 4. *R. flabellaris*
    5. Leaves once or twice divided (or lobed), segments 15–25 mm wide; petals 4–7 mm long; body of achenes 1–1.5 mm long with corky thickenings beside inconspicuous keel. . . . . 5. *R. gmelinii*
1. Leaves simple, entire, lobed or divided, but not dissected into thread-like or ribbon-like divisions; leaves floating or aerial (plants aquatic, amphibious, or terrestrial).
  6. Leaves simple, unlobed.
    7. Stems branched little, if at all, from the base; achene body 1. 3–2 mm long, distinctly beaked; perennials.
      8. Leaves cordate-ovate to reniform; achenes longitudinally nerved or striate, 50–200 per head, head columnar. . . . . . 6. *R. cymbalaria*
      8. Leaves filiform to linear, lanceolate to ovate or elliptic, or obovate to oblanceolate; achenes not nerved or striate, fewer than 50 per head, head globose.
        9. Stems mostly erect, rooting only at lower nodes; principal cauline leaves 1–3 cm wide; sepals 5–7 mm long; beak of achene more than 1 mm long, horizontal. . . . . 7. *R. ambigens*
        9. Stems erect to prostrate, or creeping, then stoloniferous, rooting at nearly every node; principal leaves 0.5–1(1.3) cm wide, often in fascicles at rooted nodes; sepals 2–4 mm long; beak of achene less than 1 mm long, usually 0.3–0.5 mm long, erect.
          10. Stems erect to prostrate; sepals 3–4 mm long; petals 5–7 long, 3–4 mm wide. . . . . 8. *R. flammula*
          10. Stems creeping, rooting at every node, occasionally ascending; sepals 1–3 mm long; petals 3–5 mm long, 1–3 mm wide. . . . . 9. *R. reptans*
    7. Stems branching at base; achene body 0.6–1(1.5) mm long, minutely beaked; annuals.
      11. Petals usually 5(10), twice as long as sepals, 6–9 mm long. . . . . 10. *R. laxicaulis*
      11. Petals 1–3, minute, inconspicuous, about as long as sepals, 1–1.2 mm long. . . . . 11. *R. pusillus*

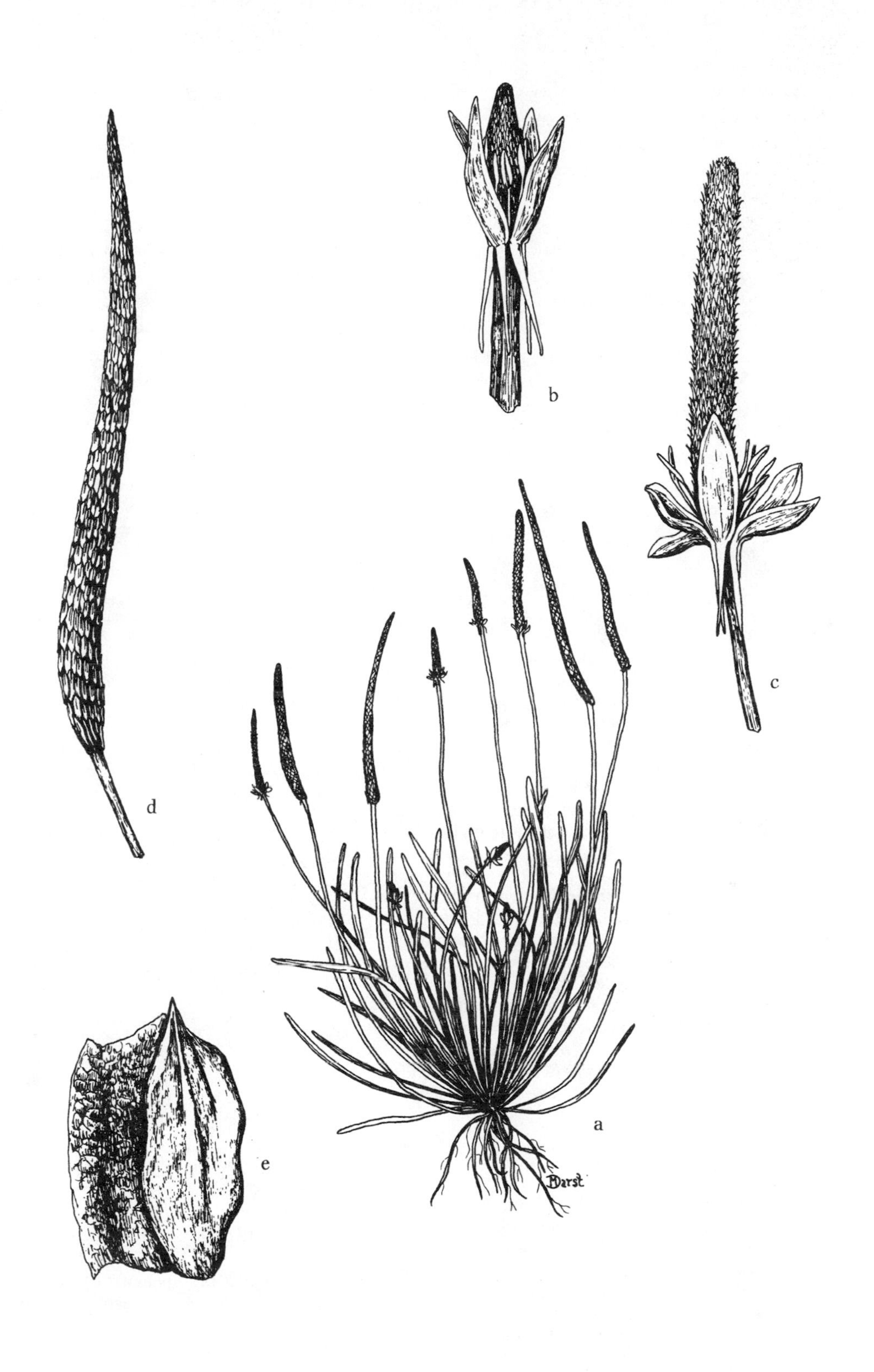

Fig. 343. *Myosurus minimus*: a. habit; b. flower, at anthesis; c. flower, post-anthesis; d. fruiting spike; e. achene (G&W).

6. Leaves lobed or divided.
  12. Petals white, sometimes yellowish at base; achenes wrinkled; floating leaves shallowly 3-lobed. 12. *R. hederaceus*
  12. Petals yellow; achenes smooth; floating leaves, if present, deeply cut or lobed.
    13. Basal leaves conspicuously lobed, but not compound (fig. 347d); achenes with style and beak minute (fig. 347e).
      14. Stems creeping, rooting at the nodes, submersed or amphibious, leaves arising from creeping stem; achenes smooth. . . . . 13. *R. hyperboreus*
      14. Stems not creeping, submersed plants with floating leaves (all basal), or amphibious plants with erect stems and leaves both basal and cauline; achenes roughened, faces ridged or with ring of fine depressions at margin of corky thickening.
        15. Basal leaves deeply cleft into 3 main divisions, segments usually shallowly lobed, ultimate segments broad; achene faces transversely ridged. . . . . 14a. *R. sceleratus* subsp. *sceleratus*
        15. Basal leaves deeply cleft into 3 main divisions, segments usually deeply lobed, ultimate segments narrow; achene faces with ring of fine depressions at margin of corky thickenings. . . . . 14b. *R. sceleratus* subsp. *multifidus*
    13. Basal leaves compound, with stalked leaflets (fig. 351a,d); achenes with conspicuous style and beak (fig. 351f,g).
      16. Petals 8–14 mm long, greatly exceeding sepals; head of achenes globose.
        17. Sepals clearly reflexed. . . . . 15a. *R. hispidus* var. *nitidus*
        17. Sepals spreading.
          18. Beak of achene long, usually longer than 1 mm slender, more or less straight; achenes broadly wing-keeled. . . . . 15b. *R. hispidus* var. *caricetorum*
          18. Beak of achene short, usually 1 mm long or less stout, strongly recurved; achenes not wing-keeled. . . . . 16. *R. repens*
      16. Petals 2–3 mm long, about half the length of sepals; head of achenes cylindrical. . . . . 17. *R. pensylvanicus*

---

1. *R. trichophyllus* Chaix ex Vill. White Water-crowfoot, Threadleaf Water-crowfoot Fig. 344
Shallow ponds, pools, slow streams and ditches. Lab. and Nfld. w. to Alask., s. to N.E., N.J., Va., n. Tenn., Iowa, S.D., N.M., Ariz. and Calif.; subcosmopolitan. Cook (1966) recognized only two infraspecific taxa: subsp. *trichophyllus*, the more widepread taxon which is characterized by having a more robust, spreading to erect submersed habit, with plants rooting at lower nodes; subsp. *eradicatus* (Laestad.) W. B. Drew, a more northern, arctic-alpine taxon, which is distinguished by its finer, prostrate submersed habit, with plants rooting at almost every node. But these two subspecies are not generally recognized today. The treatment in FNA (Whittemore, 1997) lumped the capillary-leaved Ranunculi (sect. *Batrachium*) as *R. aquatilis* L. Based on morphological and molecular data, this has now been shown to be inaccurate by Wiegleb et al. (2017); according to them *Ranunculus aquatilis* is not found in the western hemisphere, and those west of our coverage (Cordilleras of North and South America) that have been called *R. aquatilis*, but have laminar floating leaves, belong to *R. mongolicus* Serg. (*Batrachium trichophyllum* (Chaix ex. Vill.) Borsch; *R. aquatilis* var. *diffusus* With.; *R. trichophyllus* var. *calvescens* W. B. Drew; *R. trichophyllus* var. *eradicatus* (Laestad.) W. B. Drew)

2. *R. subrigidus* W. B. Drew White Water-crowfoot
Shallow ponds, lakes, pools, and streams. Nfld. and Que. w. to B.C., s. to Vt., nw. Mass., Ont., n. Mich., Iowa, Va., Tenn., Tex. and Mex. Although the FNA treatment of *Ranunculus* sect. *Batrachium* places *R. subrigidus* within *R. aquatilis* var. *diffusus* With. (Whittemore, 1997), the detailed study including morphological and molecular data done by Wiegleb et al. (2017) indicates that *R. aquatilis* does not occur in North America and resurrects *R. subrigidus* as a distinct species. (*R. circinatus* var. *subrigidus* (W. B. Drew) Benson; *R. aquatilis* var. *subrigidus* (W. B. Drew) Breitung)

3. *R. longirostris* Godr. White Water-crowfoot, Long-beak Buttercup Fig. 344
Submersed; calcareous ponds, lakes and slow streams; hard waters and brackish waters. Sw. Que. w. to s. Man., s. Sask., Mont., se. Ida., s. to w. N.E., Del., Pa., Tenn., Ala., Kans., n. Tex., N.M. and Utah. (*Batrachium longirostris* (Godr.) Schultz; *R. aquatilis* var. *longirostris* (Godr.) P. Lawson)

4. *R. flabellaris* Raf. Yellow Water-crowfoot Fig. 345
Amphibious; shallow water or muddy shores. N.B., Me. and s. Que. w. to Ont., s. Alta. and. B.C., s. to N.J., mts. to Va., N.C., Tenn. and Ala., Pa., Ohio, Ill., La., Okla., Tex., Ida., Utah, Nev., Wash., Oreg. and Calif. (*R. delphinifolius* Torr.)

5. *R. gmelinii* DC. Gmelin's Buttercup Fig. 346
Amphibious; shallow water of lakes and ponds, marshes, streams and muddy shores. Nfld. w. to N.W.T. and Alask., s. to N.S., n. Me., n. Mich., Iowa, e. Neb., Mont., mts. to N.M., Utah, Nev. and e. Oreg. (*R. gmelinii* var. *hookeri* (D. Don) Benson; *R. purshii* Richardson)

6. *R. cymbalaria* Pursh Seaside Crowfoot Fig. 346
Amphibious; saline or calcareous muds of streams and marshes. Greenl. and Lab. w. to Ont., Man., N.W.T., and Alask., s. to Nfld., N.E., N.J., w. N.Y., n. Ill., Ark., n. Tex., n. N.M., Wash., Oreg. and Calif.

7. *R. ambigens* S. Watson Water-plantain Spearwort Fig. 347
Swamps, marshes, muddy shores and ditches. S. Me. s. to N.C. Tenn., Ill., Ky., Ala. and La.

8. *R. flammula* L. Spearwort Fig. 349
Marshy sites especially coastal. Nfld., St. P. et. Miq., and N.S. Apparently this was introduced from Europe in early fishing ports of colonial British and French North America, but surprisingly has not spread.

9. *R. reptans* L. Creeping Spearwort Fig. 349
Marshy or wet sandy shores of lakes, streams and ditches. Greenl. and Lab. w. to N.W.T., and Alask., s. to N.E., N.J., Pa.,

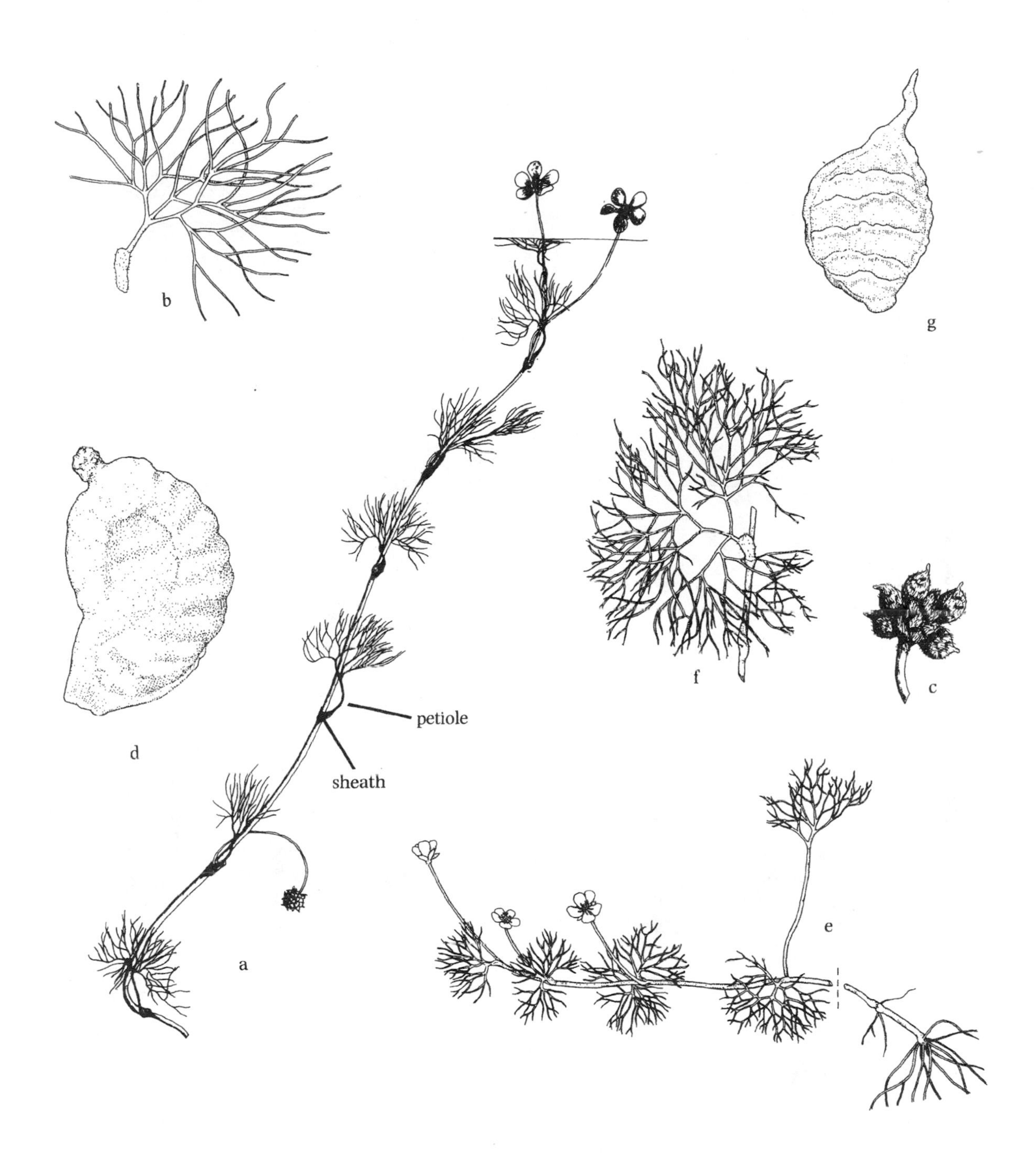

Fig. 344. *Ranunculus trichophyllus*: a. habit (F); b. leaf (NYS Museum); c. head of achenes (F); d. achene (NYS Museum).
Ranunculus longirostris: e. habit; f. leaf; g. achene (NYS Museum).

Fig. 345. *Ranunculus flabellaris*: a. habit (NYS Museum); b. leaf, aquatic form (NYS Museum); c. leaf, aquatic form (F); d. leaves, terrestrial form (F); e. leaf, terrestrial form (NYS Museum); f. achene (NYS Museum).

Fig. 346. *Ranunculus cymbalaria*: a. habit; b. achene (NYS Museum).
Ranunculus gmelinii: c. upper portion of plant; d. upper portion of terrestrial form; e. achene (F).

Fig. 347. *Ranunculus ambigens*: a. habit (F); b. flower (NYS Museum); c. achene (NYS Museum).
Ranunculus sceleratus: d. habit; e. achene (NYS Museum).

Fig. 348. *Ranunculus pusillus*: a. habit; b. achene (NYS Museum).
Ranunculus laxicaulis: c. habit (F).

Fig. 349. *Ranunculus flammula*: a. habit (Gleason).
Ranunculus reptans var. *reptans*: b. habit B&B); c. habit; d. achene (NYS Museum).
Ranunculus reptans var. *ovalis*: e. habit (NYS Museum).

Minn., Man., mts. to Mont., Wyo., and Colo. and Wash. This is a circumboreal species that tends to be rare and sometimes in need of conservation in Europe. Herein, *R. reptans* is recognized as distinct from the European *R. flammula*; however various American authors regard the two as conspecific. A terrestrial form with slightly wider blades has been regarded as *R. reptans* var. *ovalis* (J. M. Bigelow) Torrey & A. Gray. Whittemore (1997) noted that this wide ranging var. *ovalis* (his *R. flammula* var. *ovalis* (J. M. Bigelow) Benson) is essentially differentiated on the basis of leaf width from the filiform-leaved var. *R. reptans* var. *reptans* (0.4–1 mm vs. 2–5 mm wide in var. *ovalis*); the difference in leaf morphology is apparently the result of the plant growing emersed rather than submersed or becoming recently stranded, and that the two variants should perhaps be merged. Whittemore (1997) also noted that considerable intermediates between *R. reptans* and *R. flammula*, especially of the Pacific slope occur; further investigation is needed to resolve the relationships. The western populations may represent an entity sufficiently different from the eastern *R. reptans*, warranting taxonomic recognition, perhaps as an unnamed taxon within *R. flammula*. (*R. flammula* var. *filiformis* (Michx.) Hook.)

10. *R. laxicaulis* (Torr. & A. Gray) Darby Many-flowered Spearwort, Mississippi Buttercup Fig. 348
Shallow water or mud of marshes, pools, wet meadows, ditches and wet cultivated fields. Chiefly coastal plain, Del. s. to Ga. and Fla. w. to e. Okla. and e. Tex.; inland n. to Kan., Mo., Tenn., c. Ill. and s. Ind. (*R. flammula* var. *laxicaulis* Torr. & A. Gray) *R. texensis* Engelm.; *R. mississippiensis* Small)

11. *R. pusillus* Poir. Low Spearwort Fig. 348
Shallow water or mud of swamps, marshes, wet meadows, vernal pools, stream margins, ditches, and wet cultivated fields. Chiefly coastal plain and piedmont, se. N.Y. s. to Fla., w. to Tex.; inland n. to e. Mo., s. Ind. and Ohio; Calif. Our taxon is var. *pusillus*. (*R. tener* C. Mohr)

12. *R. hederaceus* L. Fig. 350
Pools and wet, sandy depressions. Se. Nfld.; coastal plain, e. Pa. and Md. s. to S.C. Wiegleb et al. (2017) regard this species as amphi-Atlantic in distribution; it occurs in western Eur., but they think there is a high likelihood that *R. hederaceus* was introduced into N.Am. rather than being native here. (*Batrachium hederaceum* (L.) S. E. Gray)

13. *R. hyperboreus* Rottb. High Northern Crowfoot Fig. 350
Shallow fresh or saline waters, pools, margins of ponds, wet shores and mudflats. Greenl. and Lab. w. to N.W.T. and Alask., s. to n. Nfld., Que., James Bay w. Alta., B.C. and Alask.; s. in the mountains to Mont., Wyo., Ida., Colo., Nev. and Utah.

14. *R. scleratus* L. Cursed Crowfoot Fig. 347

14a. *R. scleratus* subsp. *scleratus*
Swampy meadows, marshes, sloughs, ponds, margins of lakes and ditches. Chiefly eastern; Lab. and Nfld. w. to Alta., s. to n. Fla., La., and Tex.; intro in B.C., w. Wash. nw. Oreg. and Calif.

14b. *R. scleratus* subsp. *multifidus* (Nutt.) Hultén
Swampy meadows, boggy shores, pools, and springy places. Chiefly western and northern; Nfld., w. Que. and w. Ont. w. to N.W.T. and Alask., s. to n. Mich., n. Iowa, w. Minn., mts. to N.M. and Ariz. and n. Calif.

15. *R. hispidus* Michx. Hispid Buttercup

15a. *R. hispidus* var. *nitidus* (Elliott) T. Duncan Fig. 351
Swamps, wet woods, ditches and floodplains. N.Y. w. to s. Ont., Mich., Wisc. and. Minn. and S.D., s. to n. Fla., Ala. and e. Tex. (*R. nitidus* Elliott; *R. repens* var. *nitidus* (Elliott) Chapm.; *R. septentrionalis* Poir.; *R. carolinianus* DC.)

15b. *R. hispidus* var. *caricetorum* (Greene) T. Duncan Fig. 351
Swamps, marshes, ditches and floodplains. Nfld. and N.B. w. to Ont., e. Man. and e. N.D., s. to Va., Ky., and Mo. and e. Okla. (*R. caricetorum* Greene; *R. septentrionalis* var. *caricetorum* (Greene) Fernald)

16. *R. repens* L. Creeping Buttercup Fig. 350
Wet fields, meadows, lawns, roadsides, ditches, marshes and shores. Widely established weed in N.Am., especially in the East and West, sparingly in central states; intro. from Eur.

17. *R. pensylvanicus* L. f. Bristly Crowfoot Fig. 351
Marshes, wet meadows, ditches, and wet woods. Widely scattered range; s. Lab. and Nfld. w. to Man., Alta., s. N.W.T. and Alask., s. to N.J., Del., W.Va., Ohio, Ill., Iowa, and Neb., mts. to Colo., N.M. and Ariz., Ida. and Oreg.

6. *Aconitum* (Monkshood)

Perennial herbs, often tuberous; leaves palmately lobed or dissected; flowers borne in racemes or panicles, bilaterally symmetrical; sepals blue to purple or white to yellowish, uppermost sepal forming a hood; upper petals well-developed, spur-shaped, hidden by hood; lower petals minute or absent; pistils 3–5; fruit a follicle.

Reference: Brink and Woods, 1997.

1. *A. columbianum* Nutt. Columbian Monkshood Fig. 352
Crevices of rocks along stream margins, wet mossy talus, moist cliff bases, wet ravines in rich woods, and open seeps. N.Y., ne. Ohio, sw. Wisc. and Iowa; B.C. s. to Mont., Colo., N.M. Wash., Oreg. and Calif. Very rare in the East. Our populations had previously been described as *A. novaeboracense* A. Gray and had once had protection under the Federal Endangered Species Act; however, Brink and Woods (1997) regard these eastern populations as part of the exceedingly variable *A. columbianum* complex, and not worthy of formal taxonomic recognition. (*A. uncinatum* subsp. *noveboracense* (A. Gray) Hardin)

Fig. 350. *Ranunculus repens*: a. habit; b. achene (NYS Museum).
Ranunculus hederaceus: c. habit (Gleason).
Ranunculus hyperboreus: d. habit; e. cluster of achenes; f. achene (PB).

Fig. 351. *Ranunculus pensylvanicus*: a. habit (F); b. achene (NYS Museum); c. flower (NYS Museum). *Ranunculus hispidus* var. *caricetorum*: d. habit; e. flower; f. achene (NYS Museum). *Ranunculus hispidus* var. *nitidus*: g. achene (NYS Museum).

Fig. 352. *Aconitum columbianum*: a. habit; b. flower; c. follicle (NYS Museum, as *A. novaeboracense*).

Platanaceae / Plane-tree Family

1. *Platanus* (Sycamore)

Large trees, monoecious; bark creamy white, becoming brown and mottled, exfoliating with age; leaves simple, alternate, deciduous, palmately veined and irregularly lobed, margins coarsely toothed; leaf blades covered with a dense pubescence on young leaves, becoming glabrous except along veins; petiole base completely covering axillary bud; stipules large, leaf-like; flowers small, unisexual, borne in large numbers in long-stalked, spherical balls; fruiting heads usually 1(2), pendent on long peduncle, about 2.5–3 cm in diameter, achenes subtended by long tawny hairs.

REFERENCE: Kaul, 1997.

1. *Platanus occidentalis* L. Sycamore, Buttonwood, Plane-tree Fig. 353

River and stream margins, bottomlands and floodplains. S. Me. and s. N.H., w. to N.Y., s. Ont., s. Mich., s. Wisc., s. Iowa and e. Neb., s. to Fla., Ala., Miss., La. and e. Tex. This species is often extremely abundant along rivers and streams. Our native tree was cultivated widely as early as the 1750s in the area around Philadelphia (Kalm, 1771) and continues to be a frequently cultivated street tree in many towns and cities; it can be confused with the hybrid ornamental Sycamore or London Plane-tree (*Platanus* ×*acerifolia* (Aiton) Willd.)—a purported hybrid between *P. occidentalis* and *P. orientalis* L. of the Middle East, which would key out here, but does not apparently escape in North America. *Platanus* ×*acerifolia* is recognized by the lobes being somewhat longer and narrower, often longer than wide, and the fruiting heads typically 2, whereas *P. occidentalis* has a single carpellate head and the lobes of the leaf blade shorter than wide.

Nelumbonaceae / Water Chinquapin Family

1. *Nelumbo* (Lotus, Water Chinquapin, Sacred Bean)

Perennials, arising from rhizomes; leaves large, circular, peltate, elevated high above the water (fig. 52a) or floating; flowers large, showy, 12–25 cm wide, elevated above the water; perianth sulfur-yellow or pink; stamens numerous, readily deciduous; carpels numerous, separate, imbedded in a broad, flat-topped receptacle; fruit indehiscent nutlets, often remaining in cavities of the woody receptacle.

REFERENCES: Crow, 2015; Fairbrothers, 1958; Hall and Penfound, 1944; Islam et al., 2020; Moseley and Uhl, 1985; Peattie, 1928; Schneider and Buchanan, 1980; Sohmer, 1975, 1977; Sohmer and Sefton, 1978; Taylor, 1927; Ward, 1977; Wiersema, 1997.

1. Flowers pale yellow; fruits globose, 10–16 long, 8–13 mm wide (mostly less than 1.25 times longer than wide). . . . . *1. N. lutea*
1. Flowers pink (fading to white); fruits ovoid, 10–20 long, 7–13 mm wide (mostly more than 1.5 times longer than wide). . . . . *2. N. nucifera*

1. *N. lutea* Willd. Yellow Water Lotus, Water Chinquapin Fig. 354

Ponds, lakes, marshes, pools in marshes, and slow streams and backwaters. Se. Me. and Mass. w. to N.Y., s. Ont., s. Mich., s. Wisc., se. Minn., Iowa and e. Neb., s. to Fla., Ala., La., Okla. and Tex. According to Islam et al. (2020), based on microsatellite markers from 326 plants sampled from 19 populations across the range, *N. lutea* exhibited overall low genetic diversity geographically and low levels of gene flow among populations, with most populations showing recent demographic bottlenecks, thus prompting conservation concerns, suggesting wild *N. lutea* populations are vulnerable to extinction. (*N. pentapetala* (Walter) Fernald)

2. *N. nucifera* Gaertn. Indian Lotus, Sacred Lotus

Ponds, pools and shallow water of lake margins. E. Mass., se. N.Y., N.J., e. Pa. and Md.; widely scattered sites, w. W.Va. and s. Ohio, w. to Tenn. and Mo., s. to Fla. Ala., Miss. and La. Native to eastern and se. Asia, and n. Australia this escaped aquatic ornamental is locally and sporadically spreading from cultivation.

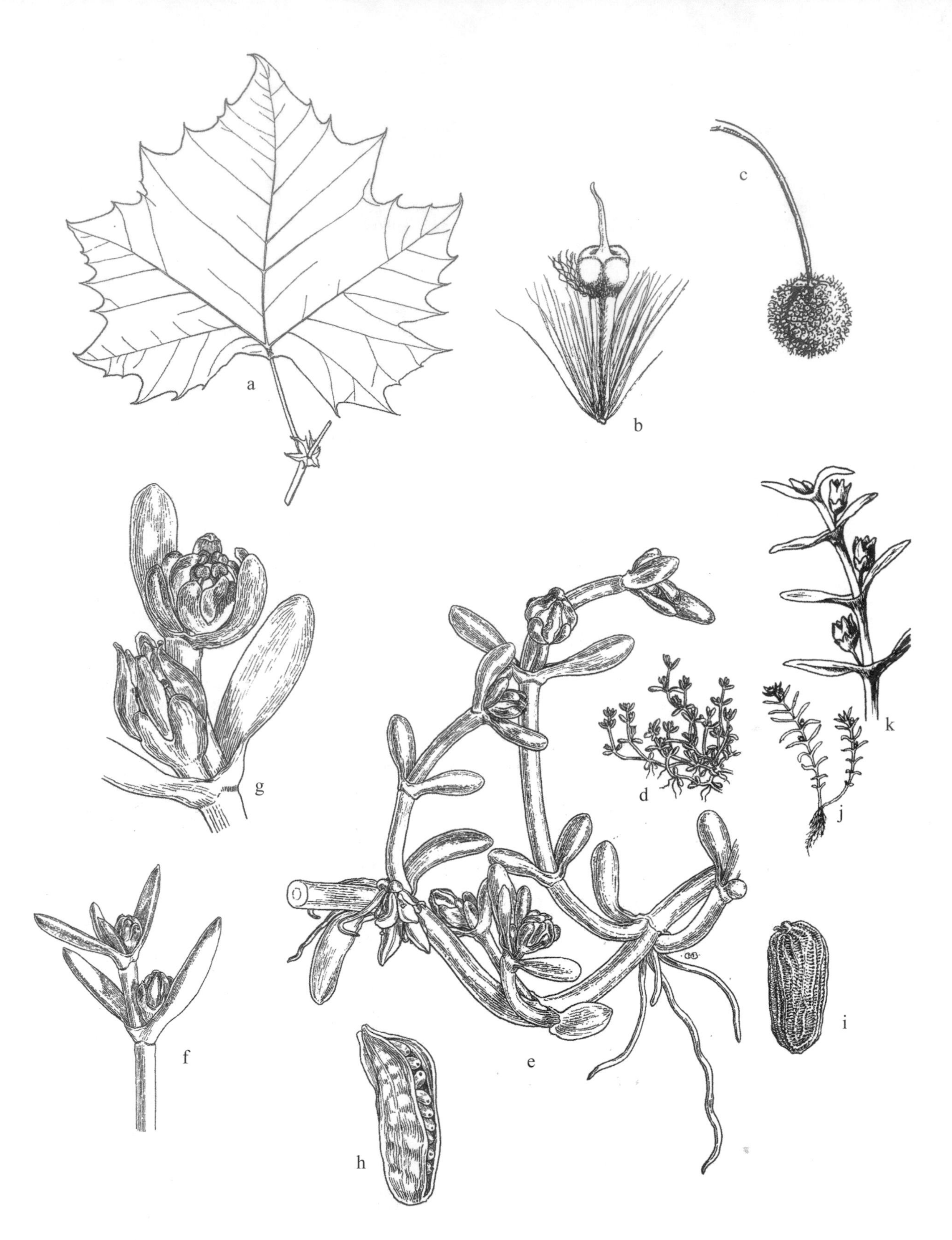

Fig. 353. *Platanus occidentalis*: a. leaf (G&W); b. achene; c. twig with fruit (C&C).
Crassula aquatica: d. habit; e. portion of plant showing different stages in anthesis; f. tip of plant showing flowers solitary in leaf axils; g. flowers showing different stages of anthesis; h. single follicle after dehiscence; i. seed (C&C); j. habit; k. upper portion of plant (F).

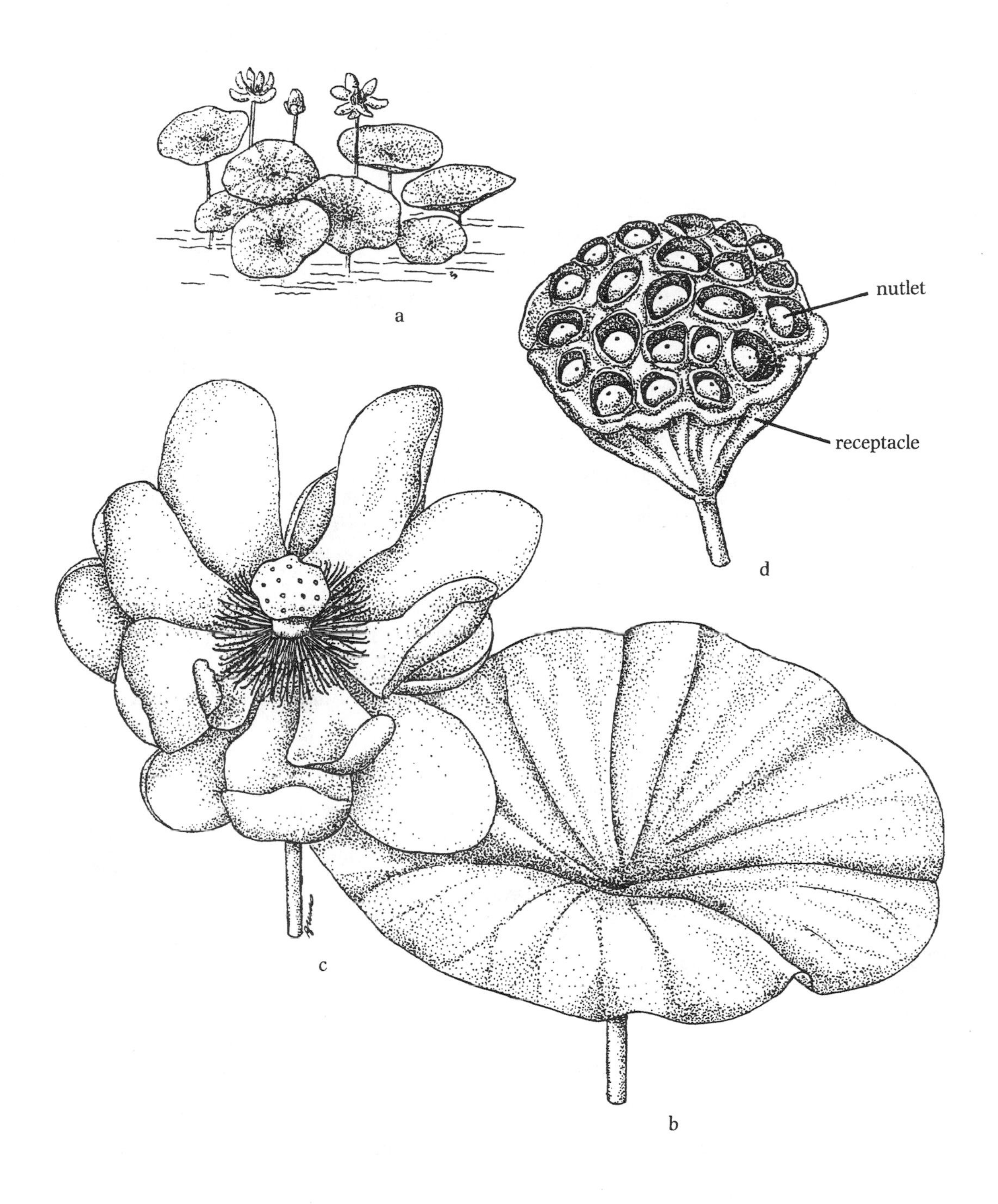

Fig. 354. *Nelumbo lutea*: a. habit; b. leaf; c. flower; d. woody receptacle with fruits (NHAES).

Rosids

Crassulaceae / Stonecrop Family

1. *Crassula* (Pigmyweed)

Annual, mat-forming succulents; leaves opposite, connate at base, entire; flowers 3- or 4-merous (ours), small, solitary, borne in leaf axils (ours); petals greenish-white (ours), borne on a hypanthium; pistils 3–4, separate or united at base; fruit a cluster of follicles.

References: Cody, 1954; Moran, 2009; Spongberg, 1978.

1. *C. aquatica* (L.) Schönland Pigmyweed Fig. 353
Brackish to fresh mud flats, peaty or muddy pond margins and shores. Lab., Nfld. and s. Que. w. to Ont. and Minn., s. to N.S. Me., s. N.H., s. Vt., Mass., R.I., se. N.Y., se. Pa. and e. Md.; Ga., Ala., La., s. Ark. e. Okla. and e. Tex.; Alask., Yuk. and N.W.T. s. to BC., Sask., Wash., Ore. and Ida., Calif, ne. Utah, e. Ariz. and w. N.M., Mex.; Eurasia. (*Tillaea aquatica* L.)

Penthoraceae / Penthorum Family

1. *Penthorum* (Ditch Stonecrop)

Perennial herbs, stoloniferous; leaves alternate, serrate; flowers 5-merous, borne in 2–4 spike-like branches; sepals green or reddish; petals inconspicuous, borne on a stipitate-glandular hypanthium; pistils 5, separate, united only at base; fruit a cluster of 5 follicles, united at base.

1. *P. sedoides* L. Ditch Stonecrop Fig. 362
Wet low ground, marshes and muddy shores. N.B. and sw. Que. w. to Ont., Man., Minn., e. N.D. and Neb., s. to nw. Fla. and Tex.; B.C., sw. Wash. and nw. Oreg.

Haloragaceae / Water-milfoil Family

1. Flowers 4-merous; fruit a schizocarp, splitting into four 1-seeded nutlets; emersed leaves reduced, bracteate, less than 1 cm long (except in *M. aquaticum*). . . . . 1. *Myriophyllum*
1. Flowers 3-merous; fruit an indehiscent, 3-angled nutlet (fig. 361c); emersed leaves leaf-like, more than 1 cm long. . . . . 2. *Proserpinaca*

1. *Myriophyllum* (Water-milfoil)

Submersed or amphibious perennial herbs (overwintering chiefly by winter buds); leaves cauline, whorled or alternate, usually finely pinnately dissected; flowers unisexual, solitary, sessile, in axils of leaves or bracts; fruit a schizocarp, splitting into 1-seeded nutlets.

References: Aiken, 1976, 1978, 1979, 1981; Aiken et al., 1979; Aiken and Cronquist, 1988; Aiken and McNeill, 1980; Aiken and Picard, 1980; Aiken and Walz, 1979; Ceska et al., 1986; Ceska and Ceska, 1986; Couch and Nelson, 1985, 1988; Crow and Hellquist, 1983; Fassett, 1939b; Grace and Wetzel, 1978; McAlpine et al., 2007; Moody and Les, 2007, 2010; Nelson and Couch, 1985a; Ritter and Crow, 1998; Scribailo and Alix, 2021.

1. Leaves scale-like, roundish (fig. 355d), or sometimes absent; vegetative stems short, quill-like (fig. 355d). . . . . 1. *M. tenellum*
1. Leaves pinnately divided, with filiform segments (fig. 358b,h); vegetative stems elongate (fig. 355a).

2. Leaves alternate, or both alternate and whorled on same stem.
   3. Leaves strictly alternate; fruits rounded dorsally, smooth or minutely papillate. . . . . 2. *M. humile*
   3. Leaves both alternate and whorled on same stem; fruits with prominent dorsal tuberculate ridges (fig. 356f).
      4. Flowers and fruit borne in axils of submersed leaves; winter buds formed in fall. . . . . 3. *M. farwellii*
      4. Flowers and fruit borne in axils of emersed bracteate leaves, forming an erect spike; winter buds absent. . . . . 4. *M. pinnatum*
2. Leaves whorled.
   5. Uppermost flowers alternate; leaves 3–12(22) mm long. . . . . 5. *M. alterniflorum*
   5. Uppermost flowers opposite; leaves mostly 8–45 mm long.
      6. Spikes with conspicuous, emersed, feathery leaves, similar to submersed leaves (fig. 360c). . . . . 6. *M. aquaticum*
      6. Spikes with reduced bracteate leaves (fig. 358c,d).
         7. Bracts usually more than twice as long as pistillate flowers.
            8. Bracts throughout inflorescence pectinate to pinnatifid (fig. 358i); staminate flowers with 8 stamens; winter buds well-formed, clavate, falling off early winter and readily dispersed (fig. 159g). . . . . 7. *M. verticillatum*
            8. Bracts of upper portion of inflorescence serrate (fig. 358c), somewhat pectinate at the water line (fig. 358d); staminate flowers with 4 stamens; winter buds lacking (new shoots forming near base late in in season, but not forming winter bud). . . . . 8. *M. heterophyllum*
         7. Bracts usually less than twice as long as pistillate flowers.
            9. Bracts of upper portion of inflorescence lanceolate, entire to denticulate, not glaucous; rhizomes not whitish; lowermost nodes of young shoots lacking entire leaves.
               10. Middle leaves with 14–21 segments on each side of rachis (fig. 359b); many of the uppermost leaves truncate at apex (fig. 359b); stem diameter below inflorescence greater, up to twice the diameter of lower stem; stem tips usually reddish; winter buds not formed. . . . . 9. *M. spicatum*
               10. Middle leaves with 6–11 segments on each side of rachis (fig. 359f); uppermost leaves rounded at apex (fig. 359f); stem diameter below inflorescence the same as diameter of flower stem; stem tips usually green; winter buds formed in fall. . . . . 10. *M. sibiricum*
            9. Bracts of upper portion of inflorescence deltoid, dentate, glaucous; rhizomes whitish; lowermost nodes of young shoots with 1-several pairs of entire leaves (fig. 360b). . . . . 11. *M. quitense*

---

1. *M. tenellum* Bigelow Slender Water-milfoil Fig. 355
Sandy or muddy bottoms of acid ponds and lakes. Nfld. w. to Ont., s. to N.S., N.E., L.I., N.Y., c. Pa., Va., N.C., Mich. and Minn. This species occurs chiefly in the sterile submersed form, flowering infrequently.

2. *M. humile* (Raf.) Morong Low Water-milfoil Fig. 356
Submersed or amphibious (producing a terrestrial form), in shallow acid waters of ponds, lakes and streams. N.S. w. sparingly to Ont. and Minn., Mainly coastal, s. to N.E., e. N.Y., e. Pa. and e. Md. Extremely variable, three ecological variants are recognized by some authors at the rank of forma. *Myriophyllum laxum* Shuttlew. ex Chapm., a coastal plain species of southeastern U.S. vegetatively similar to *M. humile*, but with typically whorled leaves and fruits strongly tuberculate, might be found to occur in Virginia (Scribailo and Alix, 2014).

3. *M. farwellii* Morong Farwell's Water-milfoil Fig. 356
Acid ponds and streams. Nfld. and N.S. w. to n. Mich. and c. Minn., s. to Me., N.H., Vt., w. Mass, c. N.Y. and Pa.; n. Ont.; Alask. and B.C. Sterile plants of *M. farwellii* can be easily confused with *M. humile* and *M. alterniflorum*. Positive identification requires fruit, or by the presence of winter buds (in *M. farwellii*).

4. *M. pinnatum* (Walter) Britton, Sterns & Poggenb. Cutleaf Water-milfoil Fig. 357
Peaty and muddy margins of ponds, lakes, streams, swamps and ditches. S. N.E. w. to W.Va., Ind., Ill., N.D., Iowa, Sask. and B.C. s. to Fla., sw. Okla. and Tex. (*M. scabratum* Michx.)

5. *M. alterniflorum* DC. Alternate-flowered Water-milfoil Fig. 355
Lakes, ponds and rivers. Greenl. and Nfld. w. to Sask., N.W.T. and Alask., s. to N.S., N.E., n. N.Y., n. Mich., Wisc. and n. Minn.; Eurasia.

6. *M. aquaticum* (Vell.) Verdc. Parrot-feather, Water-feather Fig. 360
Shallow water and margins of ponds, lakes, streams and ditches. S. Me., Conn., L.I. and c. N.Y. w. to Ohio, Mo., Kans., N.M. and Ariz., s. to Fla., La., Tex. and N.M.; B.C. n. Ida., Wash., Oreg., Ariz. and Calif. This South American native is a popular plant to grow in aquatic gardens and aquaria, but where it has become naturalized in portions of North America, Mexico and Central America, it is regarded as an invasive aquatic weed. (*M. brasiliense* Cambess.)

7. *M. verticillatum* L. Whorl-leaf Water-milfoil Fig. 358
Quiet waters of lakes, ponds, rivers and streams. Nfld. w. to Alask., s. to N.S., Conn., Del., Md., Ind., Neb., ne. Tex., Utah and Calif. Club-shaped winter buds are produced in late summer and fall, readily abscising and sinking to the sediments to overwinter; turion germination begins before ice-out in Michigan as early as March, responding to daylength increase (Weber and Noodén, 1974). Terrestrial forms often produced if stranded; these can be confused with the terrestrial forms of *M. spicatum* or *M. heterophyllum*.

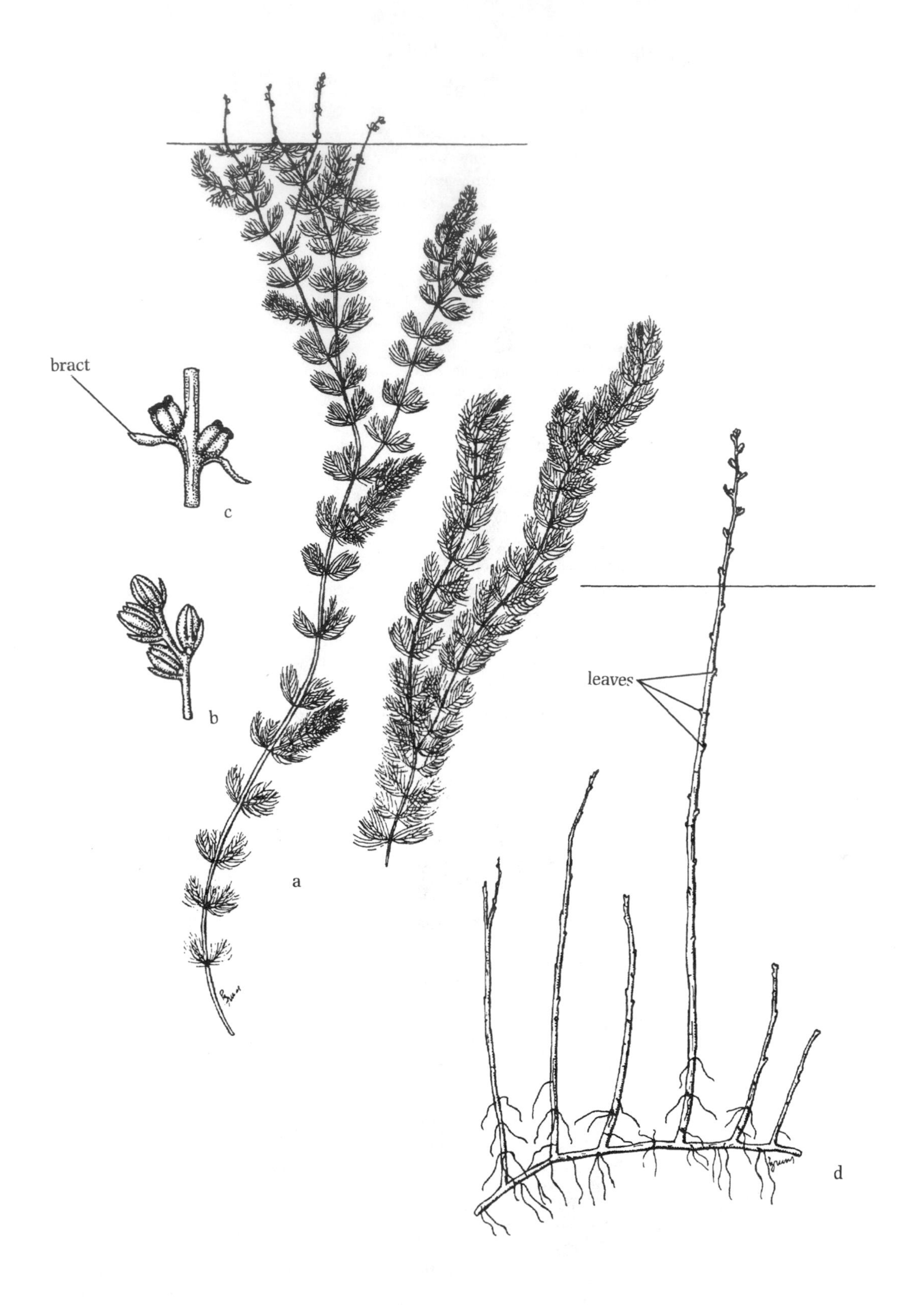

Fig. 355. *Myriophyllum alterniflorum*: a. habit; b. inflorescence; c. fruit (NHAES).
Myriophyllum tenellum: d. habit with scale-like leaves (NHAES).

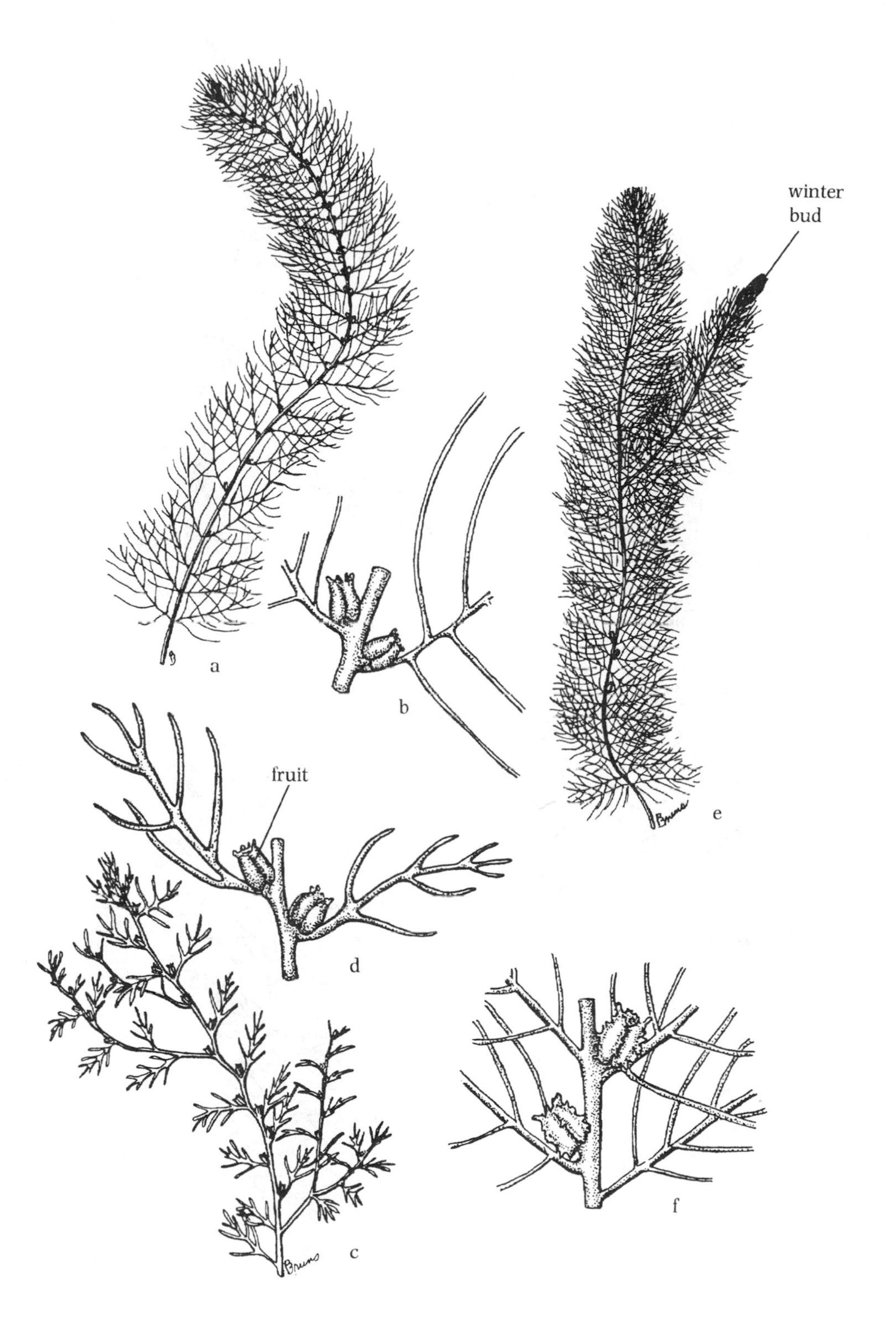

Fig. 356. *Myriophyllum humile*: a. habit, submersed plant; b. fruits on submersed plant; c. habit, terrestrial growth form; d. fruits on terrestrial plant (NHAES).
Myriophyllum farwellii: e. habit, submersed plant with winter bud; f. fruits (NHAES).

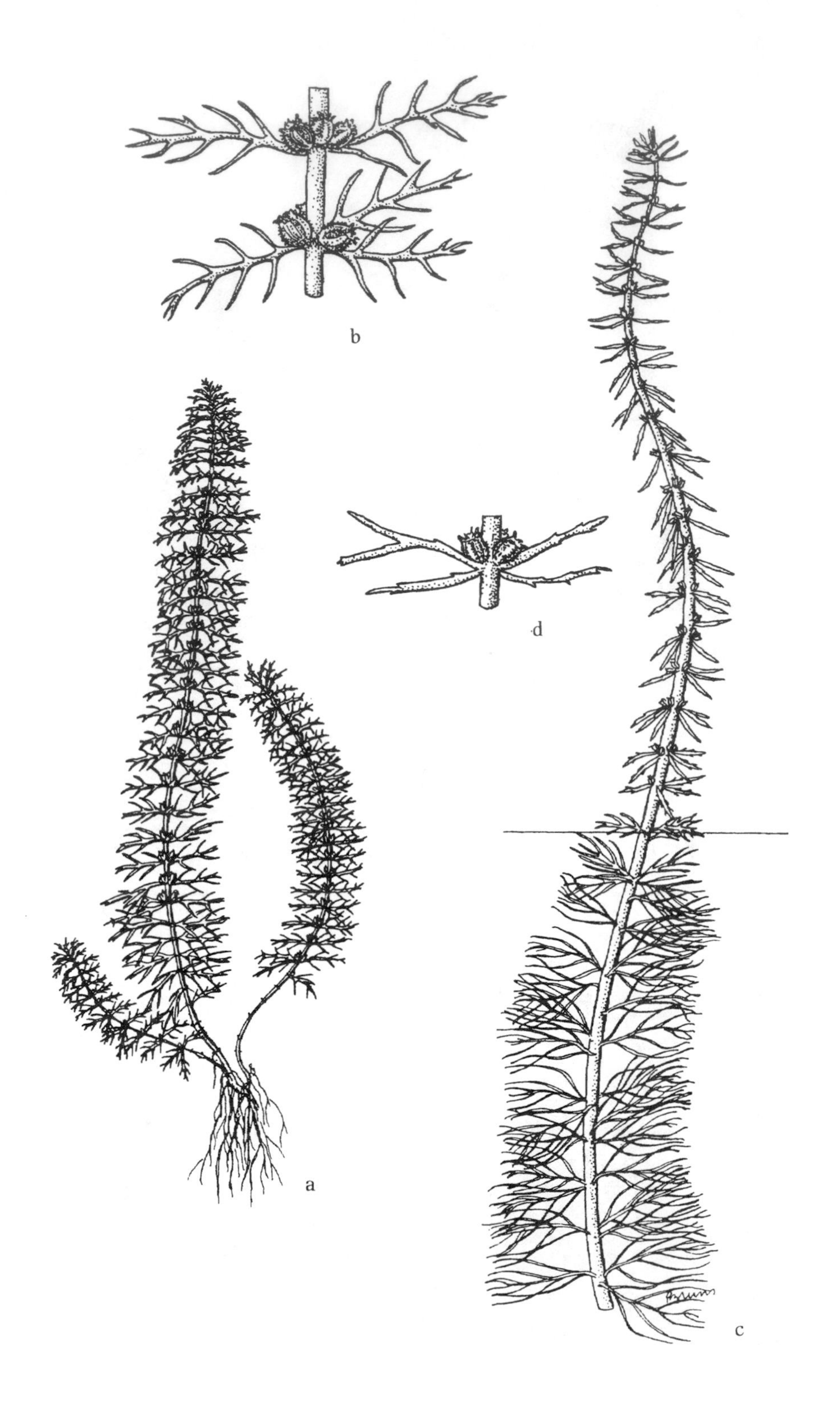

Fig. 357. *Myriophyllum pinnatum*: a. habit, terrestrial form; b. fruit on terrestrial form; c. habit, submersed form with emersed inflorescence; d. fruits on emersed portion of plant (NHAES).

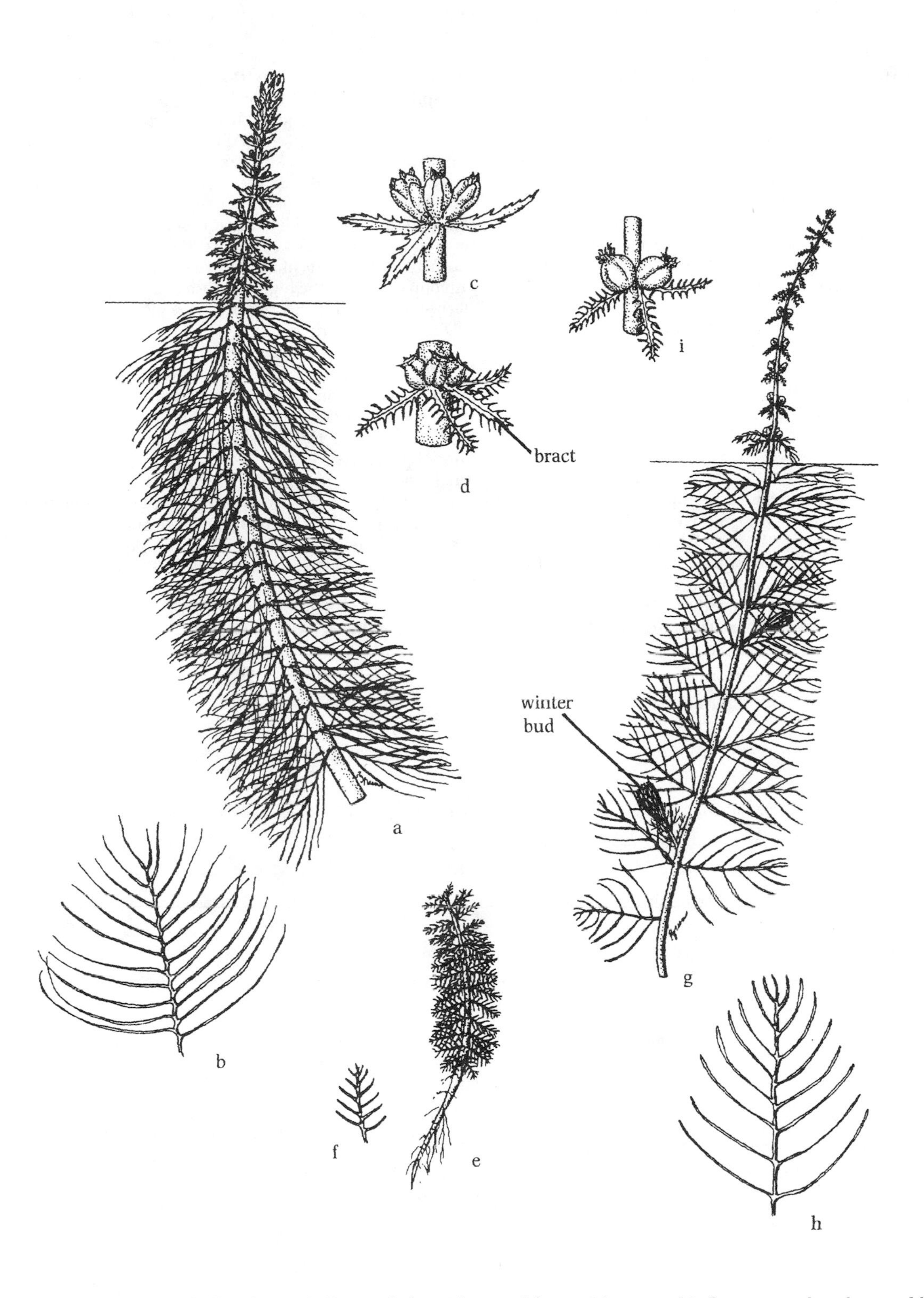

Fig. 358. *Myriophyllum heterophyllum*: a. habit. submersed form with emersed inflorescence; b. submersed leaf; c. flowers with bracts; d. fruits with bracts; e. habit, terrestrial form; f. leaf of terrestrial form (NHAES). *Myriophyllum verticillatum*: g. habit, submersed plant with winter buds and emersed inflorescence; h. submersed leaf; i. fruits with bracts (NHAES).

8. *M. heterophyllum* Michx. Variable Water-Milfoil Fig. 358
Ponds, lakes, and streams. Locally aggressive; N.B., sw. Que., s. Me., and c. N.H. w. to Ont., Mich., and N.D., s. to Fla., Okla., Tex. and N.M.; B.C. (intro.). A small terrestrial form (fig. 358e) often develops when stranded on shore as the water level drops; can be easily confused with terrestrial forms of *M. spicatum* or *M. verticillatum*. Moody and Les (2002) indicate that both native (non-aggressive) and non-native (invasive) populations occur in New England. Les and Mehrhoff (1999) speculated that highly invasive populations in the Northeast resulted from populations spreading northward from a southerly native range; this is confirmed by Thrum et al. (2011) based on molecular genetics, indicating multiple introductions from native populations both from the southern and midwestern portions of the native range. The picture is further complicated by the presence in New England of hybrid populations of *M. heterophyllum* × *M. pinnatum*, which always exhibit invasiveness (Moody and Les, 2002).

9. *M. spicatum* L. Eurasian Water-milfoil Fig. 359
Alkaline or occasionally acid waters of ponds, lakes and impoundments. Locally aggressive, N.B. s. Que. Ont. B.C. and Alask. throughout the U.S. (with few exceptions); natzd. from Eurasia, and strongly invasive. Vegetatively very similar to *M. sibiricum*, but our native species typically produces winter buds (turions) in the northern portion of North America (fig. 359e), whereas the non-native *M. spicatum* does not produce turions. While the number of leaf segments can be a useful vegetative character in differentiating *M. spicatum* (14–21 pairs) from *M. sibiricum* (6–11 pairs), overlap in segment numbers can occur; Moody and Les (2002) attribute this overlap as a consequence of hybridization between the two species, but it requires molecular fingerprinting to confirm identity.

10. *M. sibiricum* Komarov Northern Water-milfoil; Short-spike Water-milfoil Fig. 359
Alkaline waters of ponds, lakes and streams. Greenl. Nfld. and se. Lab. NB., Que. Ont. and Alask., s. to N.S., N.E., Md., W.Va., Ohio, Mich., nw. Kans. and Tex.; N.M., Ariz. and. Calif. Vegetatively, the stem tips of *M. spicatum* appear tassel-like and are usually red; vegetative tips of the native *M. sibiricum* are knob-shaped and generally lack reddish color, except during the time of winter bud formation (Aiken et al., 1979). In New England the southern populations of our native *M. sibiricum* have often been outcompeted with the highly invasive noxious weed *M. spicatum* (Eurasian Water-milfoil). (*M. exalbescens* Fernald; *M. spicatum* var. *exalbescens* (Fernald) Jeps.)

11. *M. quitense* Kunth Andean Water-milfoil Fig. 360
Lakes, especially on coves and windward sides of large bodies of water, and flowing rivers. P.E.I and N.B.; s. Mont., Wyo., n. Ida., B.C., Wash., Oreg., Calif., Utah, and Ariz. A native of high elevations of the Andes Mts. from Tierra del Fuego northward to Venezuela, with disjunct populations in western N.Am. and in the Canadian Maritimes, this species is now considered to be native in N.Am. rather than introduced (Ceska et al., 1986; Moody and Les, 2010; Ritter and Crow, 1998). McAlpine et al. (2007) have noted that it is well established in the southern St. John River system in New Brunswick; it is a matter of time until it will be found in the eastern portion of the United States. It appears as a stiff *M. sibiricum*, but it has extremely large broad bracts. (*M. elatinoides* Gaud.)

2. *Proserpinaca* (Mermaid-weed)

Submersed or amphibious herbs; stems weak, leaves alternate, pinnatifid to serrate; flowers sessile, axillary, solitary or 2–5; petals absent; fruit a 3-angled, 3-seeded nutlet.

REFERENCES: Alix and Scribailo, 2021; Burns, 1904; Crow and Hellquist, 1983; Davis, 1967; Fernald and Griscom, 1935; Kane and Albert, 1982; Wallenstein and Albert, 1963.

1. Leaves heteromorphic, those of emersed portion of plant serrate (fig. 361a) or serrate to deeply pinnatifid to pectinate (fig. 361a,b). .... 1. *P. palustris*
1. Leaves monomorphic, those of emersed portion of plant all pectinate (fig. 361d), never serrate.......... 2. *P. pectinata*

1. *P. palustris* L. Marsh Mermaid-weed Fig. 361
Shallow water of lakes, streams, ponds, marshes and ditches. Nfld. and N.S. w. to Minn., s. to Fla., La., e. Okla. and e. Tex. This heterophyllous, amphibious species exhibits considerable variability in vegetative morphology. Submersed leaves are divided and tend to grade toward pectinate to pinnatifid to serrate as the water level drops and the plants continue to grow. Serrate leaves (adult) are typically associated with flowering and fruiting. However, photoperiod appears to influence leaf morphology in emergent or stranded plants, and a reversion to a juvenile leaf form (pectinate) is related to short-day photoperiodicity. This suggests that plants that have long been referred to as *P. intermedia* actually fall within the normal range of variability of this species. Furthermore, variability in fruit size and shape does not support recognition of varieties. (*P. intermedia* Mackenz.)

2. *P. pectinata* Lam. Comb-leaf Mermaid-weed Fig. 361
Sandy margins of acid ponds, lakes, streams, ditches and damp clearings, occasionally submersed. Coastal plain, s. Lab., Nfld., N.S., Me. and c. N.H., s. to N.J., e. Pa., Del., Md., Va., N.C. and Fla., w. to se. Tex.; Ky., Tenn. and. w. Mich.

Saxifragaceae / Saxifrage Family

1. Plants decumbent, mat-forming; flowers 4-merous, petals absent. .... 1. *Chrysosplenium*
1. Plants erect; flowers 5-merous, petals present. .... 2. *Micranthes*

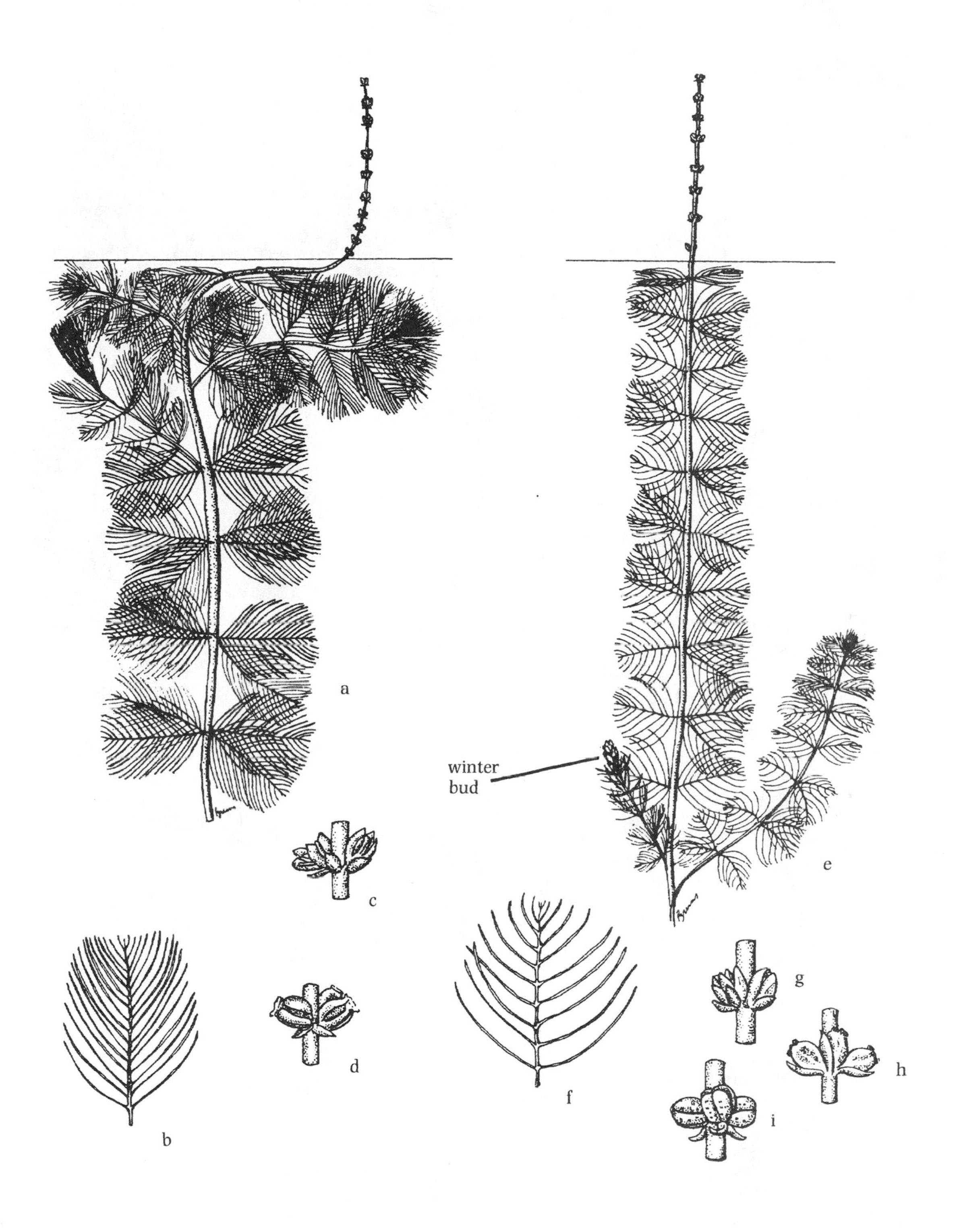

Fig. 359. *Myriophyllum spicatum*: a. habit, submersed form with emersed inflorescence; b. leaf; c. flowers; d. fruits (NHAES).
Myriophyllum sibiricum: e. habit, submersed form with emersed inflorescence; f. leaf; g. flowers; h. immature fruit; i. mature fruit (NHAES).

Fig. 360. *Myriophyllum quitense*: a. upper portion of plant; b. basal portion of plant (Ceska et al., 1986). *Myriophyllum aquaticum*: c. upper portion of plant with dense emersed leaves above and sparse submersed leaves below; d. emersed leaf (PB).

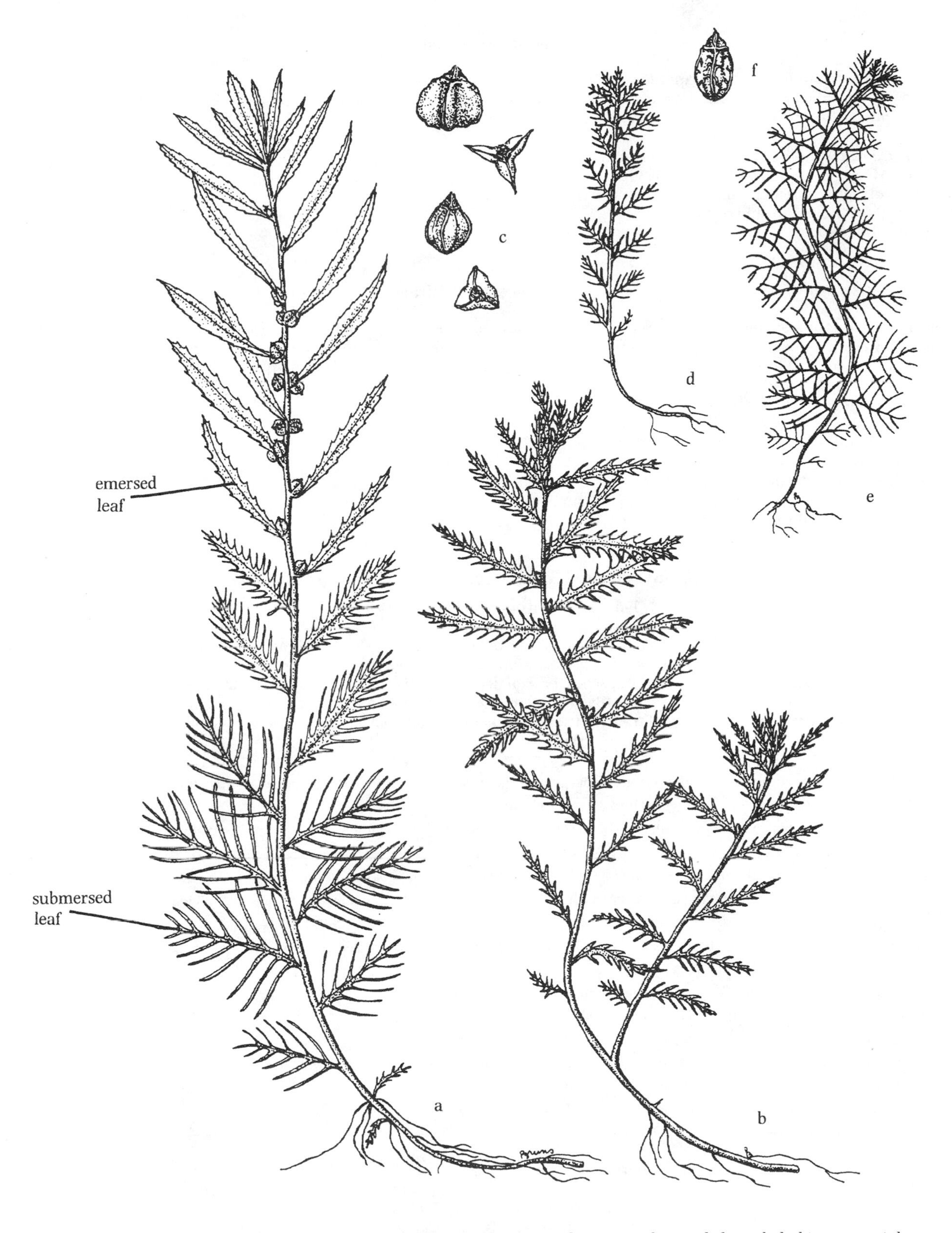

Fig. 361. *Proserpinaca palustris*: a. habit of plant, upper portion with terrestrial growth form; b. habit, terrestrial form with pinnatifid leaves; c. fruits, side and top views (NHAES).
Proserpinaca pectinata: d. habit, terrestrial growth form; e. habit, submersed plant; f. fruit (NHAES).

1. *Chrysosplenium* (Golden Saxifrage)

Perennial herbs, decumbent (ours, forming mats) or erect; leaves small, solitary, and slightly succulent; flowers small, terminal, solitary or in few-flowered cymes; sepals 4, petals absent; stamens usually 4 or 8, with orange-red anthers, inserted on a conspicuous disk (hypanthium); pistils 2, united at base, styles 2; fruit a 1-locular, 2-lobed capsule.

1. *C. americanum* Schwein. ex Hook. Water-mat, Water-carpet; American Golden Saxifrage Fig. 362
Shallow, cool, springy waters, often in the shade, Gaspé Pen., Que. and Ont. w. to e. Minn. and se. Man., s. to N.E., Md., Va., Ga., n. Ala., Ill., and Iowa.

2. *Micranthes* (Saxifrage)

Herbs, rhizomatous or stoloniferous; leaves basal, blades various, margins entire, crenate, serrate, or dentate; apex acute to obtuse, or rounded; flowering scape ascending, naked; inflorescences cymes or thryses, occasionally solitary; flowers mostly radial; sepals 5, green; petals 5, usually white or cream or various; hypanthium present; fruit a capsule, follicle-like or 2(3)-beaked.

Reference: Brouillet and Elvander, 2009

1. *M. pensylvanica* (L.) Haw. Swamp Saxifrage, Wild-beet Fig. 362
Wet meadows, swamp, thickets and wet prairies. S. Me. w. to Ont., Minn., Man. and Sask. s. to Va., nw. N.C., e. Tenn., W.Va., Ohio, nw. Ky., Ill. and Mo. (*Saxifraga pensylvanica* L.)

Celastraceae / Bittersweet or Staff-tree Family

1. *Parnassia* (Grass-of-parnassus, Bog-stars)

Perennial herbs; leaves basal; flowers solitary, borne on a scape, often with a single leaf-like bract; sepals and petals 5, white; stamens alternating with branched or unbranched, gland-tipped staminodes; fruit a 4-valved capsule.

Variously treated as a member of the Saxifragaceae, or as a distinct family, the Parnassiaceae, this monogeneric clade has been placed in the Celastraceae by APG IV (2016) and (Ball, 2016) on the basis of DNA sequence analyses that aligned Parnassiaceae with the Celastraceae as either a basal member of the Celastraceae (subfam. Parnassioideae) or at least as a sister family.

1. Leaves thin; calyx lobes herbaceous, not scarious at margin, ascending in flower, appressed in fruit; petals up to 3 times the length of calyx lobes; staminodes unlobed or divided into 3 to several gland-tipped filaments.
    2. Petals with 3 principal veins, shorter than or as long as sepals; scapes lacking leaf-like bract, or if present, along lower fifth of stem. . . . . . 1. *P. kotzebuei*
    2. Petals with 5–13 principal veins, longer than sepals; scapes with leaf-like bract near middle of stem.
        3. Basal leaf blades cordate at base; leaf-like bract cordate-clasping; petals 8–17(20) mm long; staminodes divided into 9–many gland-tipped filaments. . . . . . 2. *P. palustris*
        3. Basal leaf blades not cordate at base; leaf-like bract narrowed to base; petals 3.5–10 mm long; staminodes divided into 5–7(9) gland-tipped filaments. . . . . . 3. *P. parviflora*
1. Leaves leathery; calyx lobes leathery, somewhat scarious at margin, reflexed in fruit; petals 3 or more times the length of calyx lobes; staminodes 3-parted, with basal portion not dilated.
    4. Staminodes longer than the fertile stamens, 10–16 mm long; stamens 7–9 mm long. . . . . . 4. *P. grandifolia*
    4. Staminodes as long as or shorter than the fertile stamens, 4–9 mm long; stamens 7–11.5 mm long.
        5. Basal leaf blades reniform, wider than long; staminodes 5–9 mm long. . . . . . 5. *P. asarifolia*
        5. Basal leaf blades oval, longer than wide; staminodes 4–7 mm long. . . . . . 6. *P. glauca*

1. *P. kotzebuei* Cham. Kotzebuei's Grass-of-parnassus Fig. 363
Wet meadows, thickets, calcareous rocks and soils. Greenl. and Lab. w. to Alask., s. to e. Que., Ont., Man., mts. to Mont., Wyo., Colo, Nev. and Wash.

2. *P. palustris* L. Marsh Grass-of-parnassus Fig. 364
Wet calcareous soils and wet meadows. Lab. w. to Alask., s. to nw. Nfld., n. Mich., n. Minn., N.D., Mont., Wyo., Colo. N.M., Ariz. and Calif. An amphi-Atlantic species, the North American taxon is subsp. *neogaea* (Fernald) Hultén.

Fig. 362. *Chrysosplenium americanum*: a. habit; b. flower, side view; c. flower, top view (B&B).
Penthorum sedoides: d. upper portion of plant; e. fruit (F).
Micranthes pensylvanica: f. habit; g. flower; h. fruit (B&B).

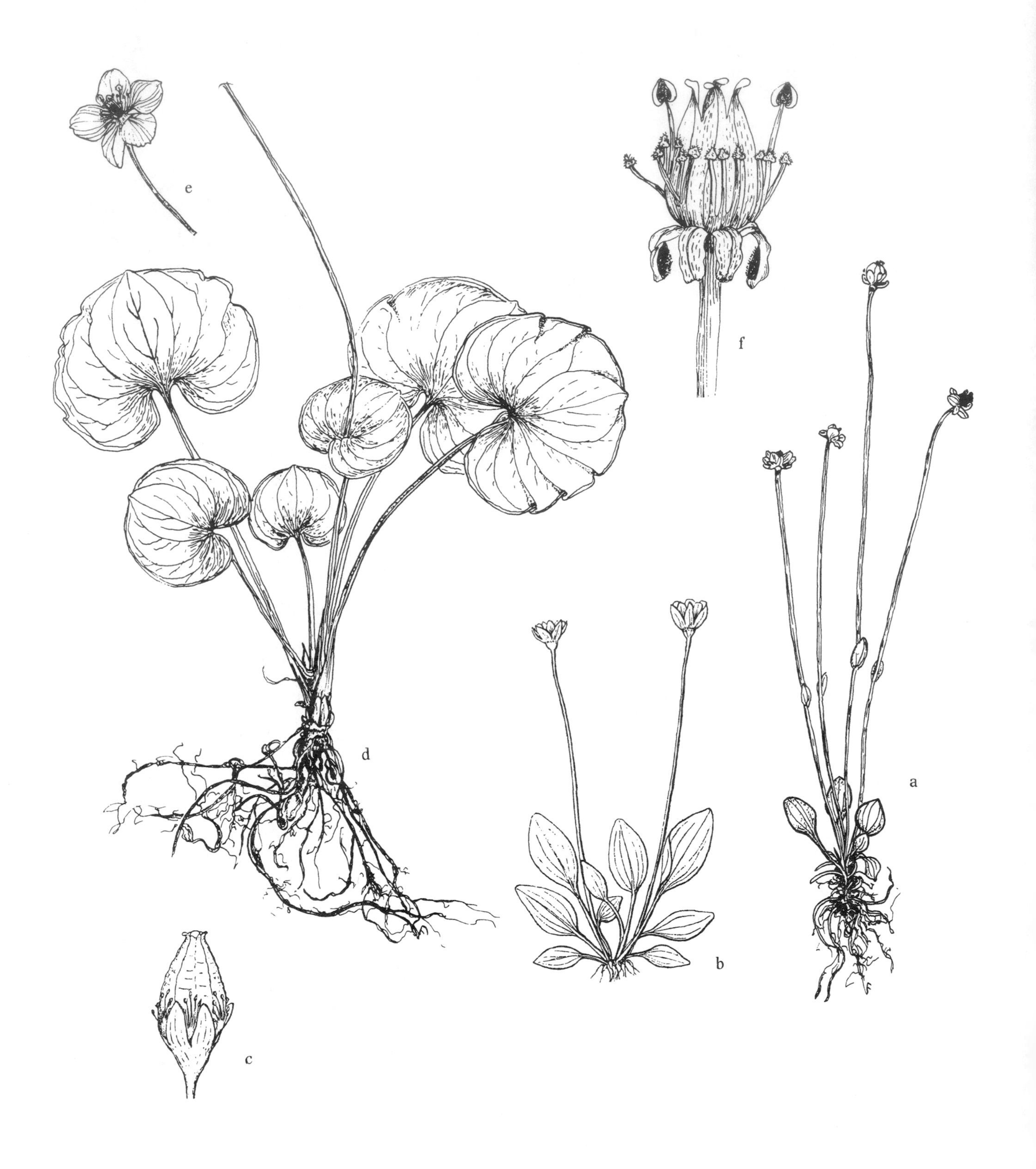

Fig. 363. *Parnassia parviflora*: a. habit (C&C).
Parnassia kotzebuei: b. habit; c. capsule (B&B).
Parnassia asarifolia: d. habit; e. flower; f. capsule (C&C).

Fig. 364. *Parnassia grandifolia*: a. lower portion of plant; b. flower; c. capsule (C&C).
Parnassia palustris: d. habit (USFS).
Parnassia glauca: e. habit (Gleason).

3. *P. parviflora* DC. Small-flowered Grass-of-parnassus Fig. 363
Wet calcareous soils, shores, meadows, bogs, and wet cliffs. Nfld. w. to Nunavut, N.W.T., and B.C., s. to C.B.I., P.E.I., Ont., n. Mich., N.D., S.D., Colo., Utah. Ariz. and Calif.

4. *P. grandifolia* DC. Large-leaf Grass-of-parnassus Fig. 364
Wet calcareous soils, shores, meadows, bogs and wet cliffs. Va., W.Va., s. Ky. and Tenn. w. to s. Mo. and Okla. s. to N.C., Ga., Fla., Ala. and e. Tex.

5. *P. asarifolia* Vent. Kidneyleaf Grass-of-parnassus Fig. 363
Bogs, wet woods, seeps and springs. Mts. of Va., W.Va., s. Ky., e. Tenn., s. to n. Ga. and ne. Ala., Ark. and e. Tex.

6. *P. glauca* Raf. Fen Grass-of-parnassus Fig. 364
Wet calcareous soils, meadows, fens and wet cliffs. Nfld. w. to Man. and Sask., s. to N.B., N.E., N.J., Pa., Ohio, n. Ill., Iowa, N.D. and S.D.

Hypericaceae / St. John's-wort Family

REFERENCES: Adams, 1973; Robson, 2015.

1. Flowers pink to light purple to flesh-colored or greenish (usually with purple veins); staminodal glands 3, alternating with clusters of stamens.. . . . . 1. *Triadenum*
1. Flowers yellow; staminodal glands absent. . . . . 2. *Hypericum*

1. *Triadenum* (Marsh St. John's-wort)

Perennial herbs; leaves simple, opposite, mostly with translucent dots (seen when held to light); flowers borne in terminal and axillary clusters; petals pink to flesh-colored or greenish; stamens 9, in 3 clusters with filaments connate toward base; staminodal glands 3, yellow, alternating with stamens; fruit a capsule, dehiscing by 3 valves.

REFERENCES: Gleason, 1947; Reznicek, 1985.

1. Leaves with translucent dots and dark-punctate on lower surface.
    2. Leaf base rounded to slightly cordate, sessile; stamen filaments connate only at base.
        3. Styles 2–3.5 mm long; sepals acute to acuminate, 4–7(8) mm long. . . . . 1. *T. virginicum*
        3. Styles 0.5–1.5 mm long; sepals obtuse to rounded, 2.5–5(5.2) mm long.. . . . . 2. *T. fraseri*
    2. Leaf base tapering to a short petiole; stamen filaments connate above the middle (fig. 366d). . . . . 3. *T. walteri*
1. Leaves lacking translucent dots and not dark-punctate on lower surface. . . . . 4. *T. tubulosum*

1. *T. virginicum* (L.) Raf. Virginia St. John's-wort Fig. 365
Bogs, marshes, swamps, shores and wet sands. N.S. and s. Que. w. to s. Ont. and s. Mich., s. to N.Y., W.Va., mts. to w. N.C. Tenn. and n. Ga., Ohio, Ky., n. Ind. and ne. Ill.; coastal plain, N.J. s. to Fla., w. to Ala., La., e. Okla. and e. Tex. (*Hypericum virginicum* L.)

2. *T. fraseri* (Spach) Gleason Fraser's St. John's-wort
Bogs, marshes, swamps and wet shores. S. Lab. and Nfld. w. to s. Man. and Sask., s. to se. Conn., n. N.J., Md., N.C., ne. Tenn., Ohio, n. Ind., n. Ill., Iowa and Neb.; intro. in B.C. and nw. Wash. In New England and Michigan *T. fraseri* tends to be a more northern species than *T. virginicum*, although in Michigan it sometimes overlaps in distribution and can even occur in the same bog (Crow, pers. obs.) (*T. virginicum* subsp. *fraseri* (Spach) Cooperrider; *Hypericum virginicum* var. *fraseri* (Spach) Fernald)

3. *T. walteri* (Gmel.) Gleason Greater St. John's-wort Fig. 366
Swamps, marshes and bayous. S. N.J., Del. and Md. w. to Ohio, s. Ind., s. Ill., se. Mo. and Okla., s. to Fla., Miss., La. and e. Tex. (*Hypericum tubulosum* var. *walteri* (Gmel.) Lott.; *T. petiolatum* (Walter) Britton)

4. *T. tubulosum* (Walter) Gleason Lesser St. John's-wort. Fig. 366
Swampy or marshy ground in woods, cypress and gum swamps. Md. and se. Va. w. to Ohio, Ind., s. Ill., se. Mo. and Okla.; s. to Fla., w. to La. and se. Tex. (*Hypericum tubulosum* Walter; *T. longifolium* Small)

2. *Hypericum* (St. John's-wort)

Perennial herbs, shrubs, or subshrubs, deciduous or evergreen; leaves simple, opposite, mostly with translucent dots, margins entire; flowers solitary or borne in terminal or axillary clusters; sepals 4 or 5; petals 4 or 5, yellow; stamens few to mostly numerous, separate (continuous and distinct) or often in clusters (connate at base); fruit a capsule.

REFERENCES: Adams, 1957; Adams and Robson, 1961; Gillett and Robson, 1981.

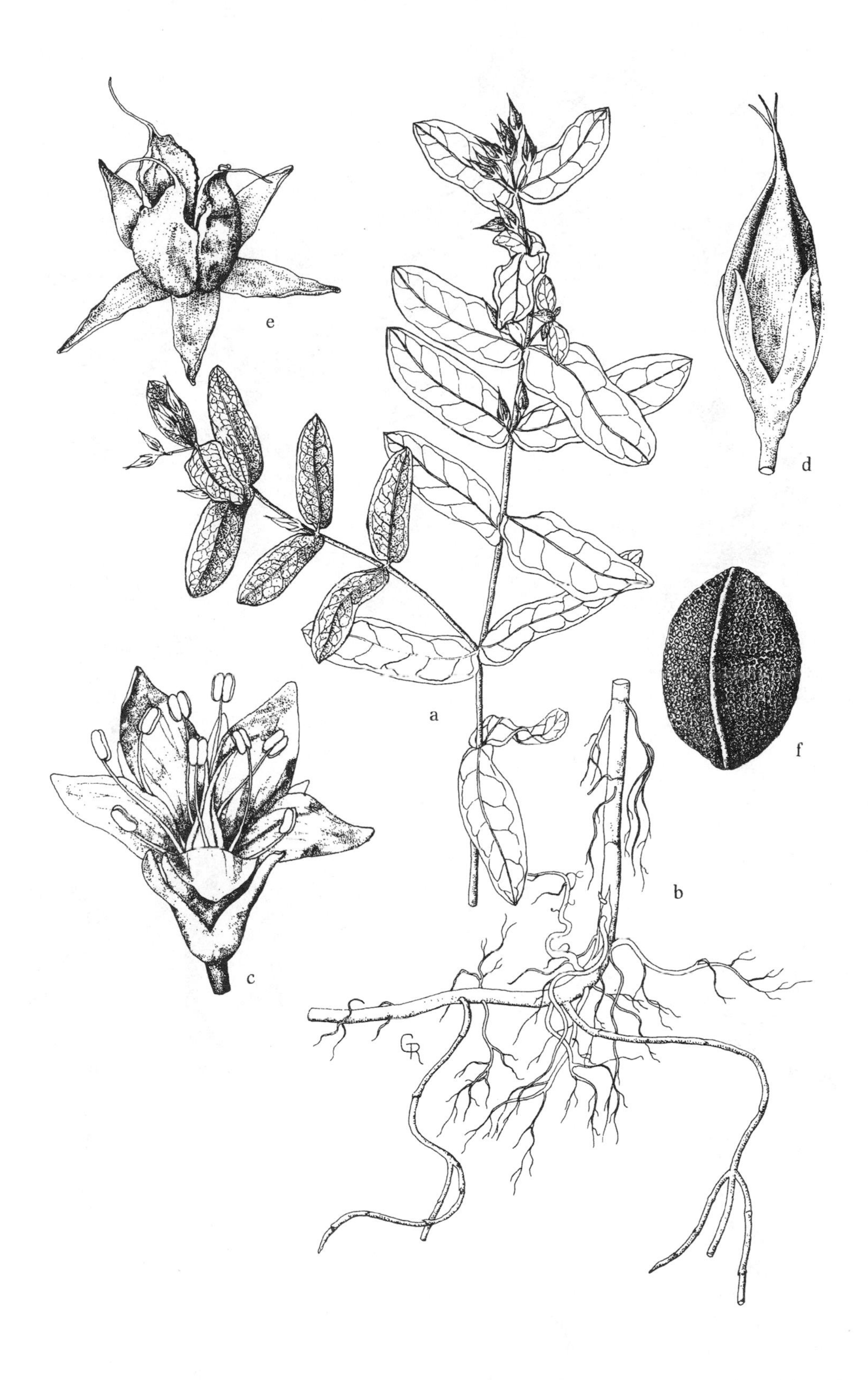

Fig. 365. *Triadenum virginicum*: a. upper portion of plant; b. base of plant; c. flower; d. capsule; e. dehisced capsule; f. seed (G&W).

Fig. 366. *Triadenum walteri*: a. upper portion of plant (C&C); b. flower (C&C); c. flower beginning to shed corolla (C&C); d. flower with corolla and anthers removed, showing united filaments (C&C); e. capsule with corolla still adhering to tip (C&C); f. section of capsule with young seeds (C&C); g. mature fruit (G&W); h. seed (G&W).
Triadenum tubulosum: i. upper portion of plant (B&B).

1. Plants woody, shrubs or subshrubs.
   2. Plants evergreen; leaves leathery; petals 4 (rarely 5), cross-shaped (cruciform); sepals 4, in unequal pairs.
      3. Leaves linear to oblanceolate, tapering to base, principal leaves 2–4(5) mm wide; styles 2: inner sepals much smaller than outer.
         4. Plants erect, unbranched from base; leaf blades usually narrowly elliptic to oblong, broadest near middle.. . . . . 1a. *H. hypericoides* subsp. *hypericoides*
         4. Plants low-growing, decumbent to prostrate, branching from the base (mat forming); leaf blades mostly oblanceolate, broadest beyond the middle. . . . . 1b. *H. hypericoides* subsp. *multicaule*
      3. Leaves broadly elliptic-oblong, subcordate to clasping at base, principal leaves 10–20 mm wide; styles 3(rarely 4); inner sepals only slightly smaller than outer. . . . . 2. *H. crux-andreae*
   2. Plants deciduous; leaves herbaceous; petals 5; sepals 5, equal in length.
      5. Flowers relatively few, in a loose cyme, 2–3.5 cm wide; styles 5; capsules 6.5–10 mm long. . . . . 3. *H. kalmianum*
      5. Flowers numerous, in a dense cyme 1.2–1.7 cm wide; styles 3; capsules 3.5–6.5 mm long. . . . . 4. *H. densiflorum*
1. Plants herbaceous.
   6. Plants submersed, stems flaccid (fig. 369a); leaves lacking translucent dots; plants usually sterile (fertile, erect plants having translucent dots usually on nearby shore).
      7. Leaves with 3 strong parallel veins arising at leaf base (fig. 369b). . . . . 5. *H. boreale*
      7. Leaves with 1 strong midvein and pinnate lateral veins (fig. 368d). . . . . 6. *H. ellipticum*
   6. Plants emersed or at water's edge, erect; leaves with translucent dots; plants usually fertile.
      8. Stems, petals and sepals with numerous black dots (especially on surfaces and/or margins). . . . . 7. *H. punctatum*
      8. Stems, petals, and sepals lacking black dots.
         9. Flowers 2.5–5 cm wide; styles 5; capsules 15–20 mm long. . . . . 8. *H. ascyron*
         9. Flowers less than 2 cm wide; styles 3; capsules 3–7.5 mm long.
            10. Styles separate to base, capsule not beaked (styles, if persisting, remaining separate at maturity); leaves with 3–7 strong veins arising from base (fig. 369e), or with a single midvein lacking lateral veins (fig. 370a) or sometimes having 2 weak lateral veins arising from base.
               11. Stamens fewer, 5–21; styles persisting on mature capsules up to 1 mm long, erect.
                  12. Sepals (4)4.5–6.5(7) mm long; capsules 4–8 mm long. . . . . 9. *H. majus*
                  12. Sepals 2.5–4.5 mm long; capsules 4–6 mm long.
                     13. Bracts of inflorescence leaf-like, elliptic, similar to cauline leaves (fig. 369b). . . . . 5. *H. boreale*
                     13. Bracts of inflorescence reduced, subulate to linear-setaceous.
                        14. Leaves linear to linear-oblong to linear-oblanceolate, tapering to base, 1–3-veined.
                           15. Capsules 4–6 mm long; sepals at maturity 4–6 mm long; leaves linear to linear-oblanceolate, 0.5–5.5 mm wide. . . . . 10. *H. canadense*
                           15. Capsules 3–4(5) mm long (often sterile, not filled out); sepals at maturity (2)2.5–4.5 mm long; leaves linear-oblong to oblanceolate, 2–6(10) mm wide.. . . . . 11. *H. ×dissimulatum*
                        14. Leaves broad, ovate to elliptic to orbicular, often cordate at base or clasping, 5–7-veined.
                           16. Flowers 4.–7 mm diam.; capsule narrowly conic-ellipsoid, 4–5 mm long; sepals lanceolate, widest below middle, 3–5 mm long; plants much branched.. . . . . 12. *H. gymnanthum*
                           16. Flowers 3–5 mm diam.; capsule narrowly ovoid to cylindric-ellipsoid, 2.5–3.5(4) mm long; sepals narrowly elliptic, 2.5–3.5 mm long; plants seldom branched below inflorescence (fig. 370f). . . . . 13. *H. mutilum*
               11. Stamens numerous (50–80, irregularly clustered), more than 20, styles on mature capsules 2–4 mm long, usually reflexed.. . . . . 14. *H. denticulatum*
            10. Styles united toward base, (appearing as one) persistent, forming a beak on mature capsule; leaves having a single strong midvein with several pinnate lateral veins.
               17. Leaves elliptic to elliptic-oblong, margins flat; sepals oblanceolate to oblong, bracts linear to lanceolate to oblong; bracts linear to lanceolate.. . . . . 6. *H. ellipticum*
               17. Leaves linear-lanceolate to lanceolate to narrowly oblong, margins slightly involute; sepals lanceolate to ovate; bracts minute, subulate.. . . . . 15. *H. adpressum*

1. *H. hypericoides* (L.) Crantz St. Andrew's Cross Fig. 367

1a. *H. hypericoides* subsp. *hypericoides*
Dry, open sandy woods to moist to wet shady, rich woods. S. N.J., Del. and Md. w. to Ky., Ill., Mo., s. to Fla., Okla. and Tex.

1b. *H. hypericoides* subsp. *multicaule* (Michx. ex Willd.) Crantz
Swampy ground, wet peats and floodplains. Nantucket I., Mass., L.I., N.Y., s. N.J., s. Pa., w. to s. Ohio, s. Ind., s. Ill., s. Mo. and se. Kans., s. to S.C., Ga., Ala., La., Okla., e. Tex. and e. Mex.; Ber., W.I., and C.Am. The more prostrate to decumbent subsp. *multicaule* (Michx.) is more likely to be encountered in our range than the erect shrub, subsp. *hypericoides*. (*Ascyrum hypericoides* (L.) Crantz; *Hypericum stragulum* W. P. Adams & N. Robson)

2. *H. crux-andreae* (L.) Crantz St. Peter's-wort Fig. 367
Dry to moist sandy woods, bogs, savannas, flatwoods, wet meadows, swales, cypress depressions, marshy and boggy shores, lake and pond margins, and sometimes upland sites. L.I., N.Y., s. N.J., se. Pa. and Del. w. to se. Ky., Tenn., Ark. and e. Okla., s. to Fla., Ala., La. and e. Tex. (*Ascyrum stans* Michx.; *H. stans* (Michx.) W. P. Adams & N. Robson)

3. *H. kalmianum* L. Kalm's St. John's-wort Fig. 367
Moist sands shores and wet pannes, mostly calcareous sites. W. Que. and w. N.Y. w. to Ont., Mich. and Wisc., s. to n. Ohio, n. Ind. and n. Ill.; especially well established near the shores of the Great Lakes.

4. *H. densiflorum* Pursh Bushy St. John's-wort Fig. 368
Wet, acid, sandy or peaty soils, meadows, swamps, wet thickets, stream margins and floodplains. Mass., L.I., N.Y., N.J. and Del. w. to sw. Pa., and W.Va., s. to S.C., Ala., La., Okla. and e. Tex.

5. *H. boreale* (Britton) E. P. Bicknell Northern St. John's-wort Fig. 369
Wet sandy or mucky soils, bogs, meadows, swamps, shallow water, shores and river margins. Nfld. and Lab. w. to Ont., Man. and Minn., s. to N.E., N.J., Va., N.C., W.Va., n. Ohio, n. Ind., n. Ill. and e. Iowa; B.C., w. Wash. and sw. Oreg. Submersed plants tend to be elongate, flexuous, and lack translucent dots; reminiscent of *Callitriche*, this growth form has been described as *H. boreale* forma *callitrichoides* Fassett (fig. 101a).

6. *H. ellipticum* Hook. Pale St. John's-wort Fig. 368
Sandy and gravelly shores, meadows, swamps, stream margins and gravel bars; sometimes submersed. Nfld. and Lab. w. to Ont., Minn. and N.D., s. to N.E., Md., Va., N.C., ne. Tenn., W.Va., Ind., Ill. and Iowa.

7. *H. punctatum* Lam. Spotted St. John's-wort Fig. 369
Moist or dry soils, fields, open woods and damp openings. Nfld., and s. Que. w. to Ont., Minn., se. Neb. and e. Kans., s. to Fla. and Tex.

8. *H. ascyron* L. Great St. John's-wort Fig. 368
Moist to wet thickets, meadows bordering rivers and open woods. Que. and N.E. w. to Ont., Mich. and Minn., s. to w. N.J., Md., Pa., Ohio, Ill., n. Mo., and ne. Kan. According to Robson (2016) the North American populations have typically been treated as *H. pyramidatum*, but he regards our taxon as subsp. *pyramidatum* (Aiton) N. Robson, one of three subspecies worldwide. (*H. pyramidatum* Aiton)

9. *H. majus* (A. Gray) Britton Large St. John's-wort Fig. 369
Wet meadows, swamps, lakeshores, wet prairies and wet roadsides. P.E.I., N.S., and N.B. w. to Ont., Sask. and B.C., s. to N.J., Pa., n. Ohio, Ill., Mo., Okla., Colo., Mont., Ida., Wash. and Oreg.

10. *H. canadense* L. Narrow-leaved St. John's-wort, Canadian St. John's-wort Fig. 370
Wet meadows, sandy or muddy shores and bogs. Nfld., Que. and Ont. w. to Minn., s. to n. Fla., se. Miss. and Iowa; w. Wash. and w. Oreg. (*H. canadense* var. *galiiforme* Fernald)

11. *H.* ×*dissimulatum* E. P. Bicknell
Bogs, wet sandy or peaty soils and shores. Nfld., N.B., N.S. w. to Que., Ont. and N.Y., s. to N.E., N.J. Pa., se. Va. and N.C. This hybrid appears to be a continuous series of hybrids between *H. canadense* and *H. muticum* or *H. boreale* (Robson, 2015).

12. *H. gymnanthum* Engelm. & A. Gray Small-flowered St. John's-wort Fig. 370
Wet sandy, peaty, or muddy shores, swales and moist low ground. L.I., N.Y., s. N.J. and Pa. w. to Ill., Mo. and Kans., s. to Fla., La. and e. Tex.

13. *H. mutilum* L. Dwarf St. John's-wort Fig. 370
Wet meadows, marshy sites, swamps, bogs, stream margins and low, open ground. Nfld. and Que. w. to Ont., Minn. and Neb., s. to Fla. and Tex.; Sask., B.C., Wash., Colo., Utah. and Calif. (*H. mutilum* var. *latisepalum* Fernald; *H. mutilum* var. *paviflorum* (Willd.) Fernald)

14. *H. denticulatum* Walter Coppery St. John's-wort
Bogs, marshes, shores, wet sandy soil, wet or dry woods and ditches. Coastal plain, N.Y., N.J. w. to Ohio and s. Ill., s. to n. Fla. Tenn. and Ala. (*H. denticulatum* var. *ovalifolium* (Britton) Blake)

15. *H. adpressum* W. P. C. Barton Creeping St. John's-wort Fig. 370
Marshes, wet meadows, swales and shores. Local, coastal plain, e. Mass. and R.I. s. to Ga.; inland to W.Va., Ind., Ill., Mo., Ky. and c. Tenn.

Podostemaceae / Riverweed Family

1. *Podostemum* (Riverweed, Threadfoot, Orchid-of-the-waterfall)

Submersed perennial herbs; roots somewhat flattened or ellipsoid, thalloid, photosynthetic, creeping across substrate (usually rocks) attached by fleshy disks; ascending stems bearing long, repeatedly forking leaves, segments linear to filiform or awl-shaped; flowers solitary, inconspicuous, axillary, enclosed in a spathella; tepals small, linear or awl-shaped, with acute apex; stamens usually 2; fruit a capsule, dehiscing by 2 valves, 1 soon dropping off, the other persisting.

References: Fassett, 1939b; Graham and Wood, 1975; Meijer, 1976; Philbrick, 1981, 1982; Philbrick and Bogle, 1988; Philbrick and Crow, 1983, 1992, 2015; Philbrick and Novelo, 2004; Royen, 1954.

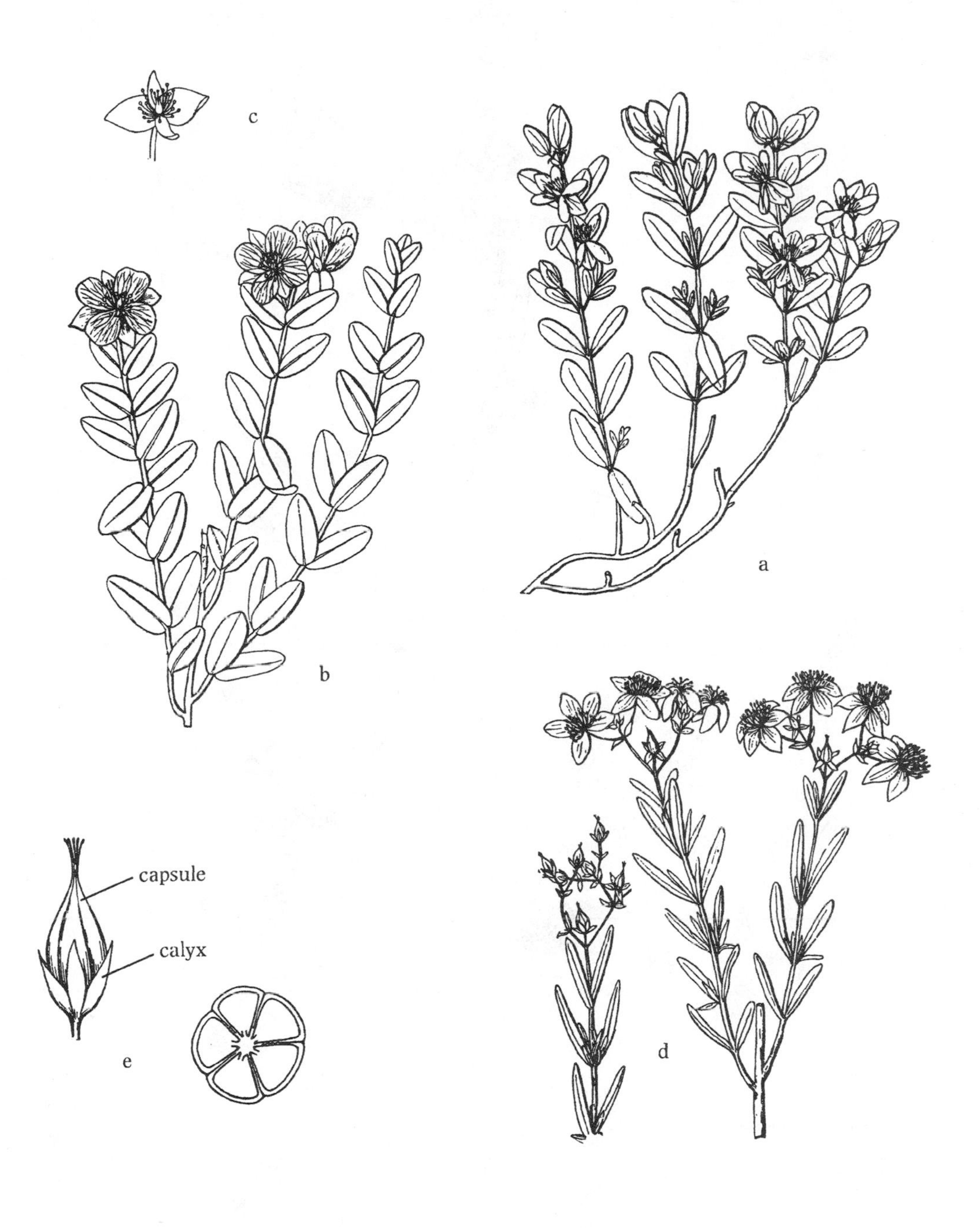

Fig. 367. *Hypericum hypericoides*: a. upper portion of plant (B&B).
Hypericum crux-andreae: b. upper portion of plant; c. flower (B&B).
Hypericum kalmianum: d. upper portion of plant; e. capsule with cross-section (B&B).

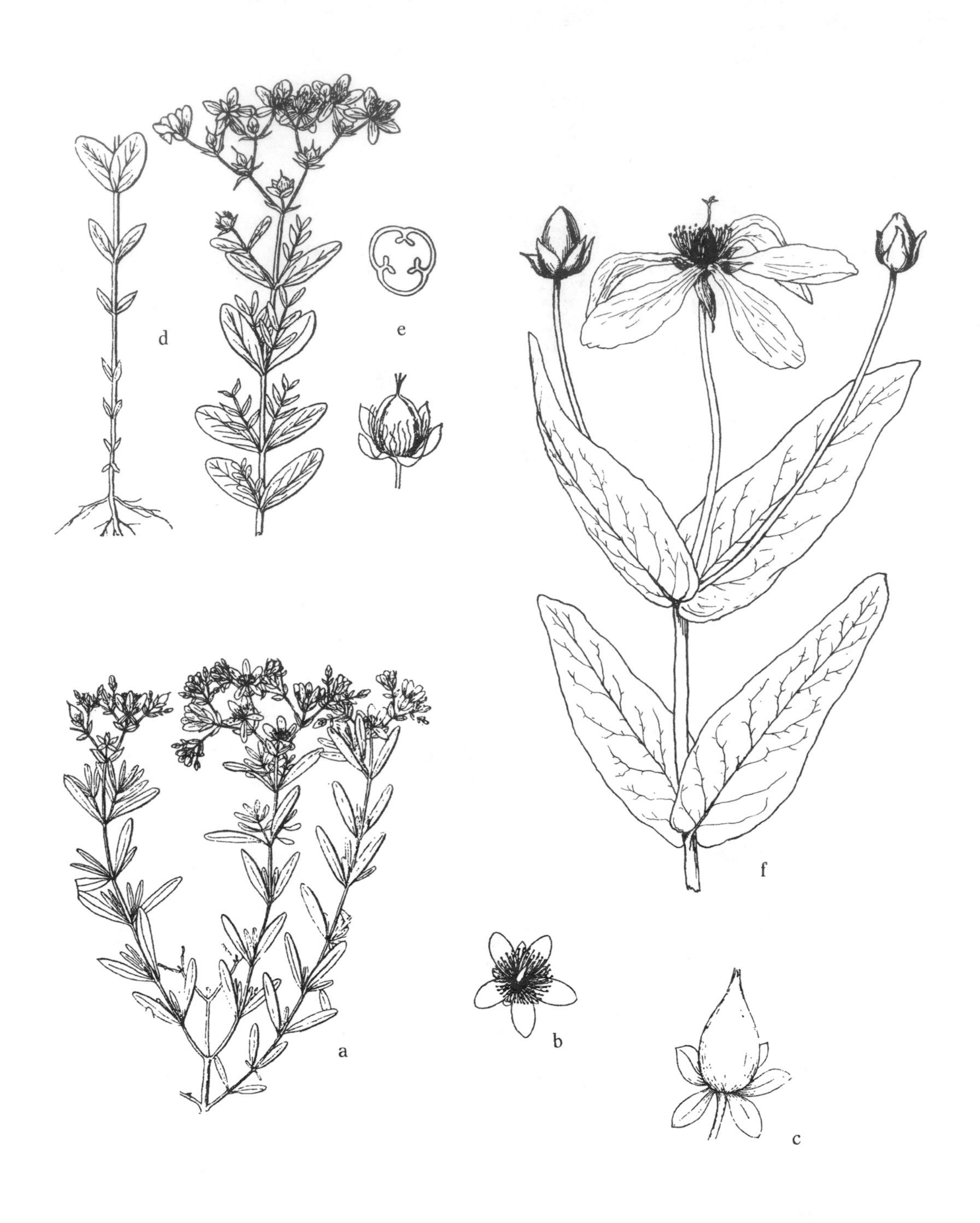

Fig. 368. *Hypericum densiflorum*: a. upper portion of plant; b. flower; c. capsule (B&B).
Hypericum ellipticum: d. habit; e. capsule with cross-section (B&B).
Hypericum ascyron: f. upper portion of plant (F).

Fig. 369. *Hypericum boreale*: a. habit. aquatic form (F); b. habit, terrestrial or emersed form (F); c. capsule (B&B).
Hypericum punctatum: d. upper portion of plant (F).
Hypericum majus: e. upper portion of plant (F).

Fig. 370. *Hypericum canadense*: a. upper portion of plant (F).
Hypericum gymnanthum: b. habit; c. capsule (B&B).
Hypericum adpressum: d. upper portion of plant; e. flower (B&B).
Hypericum mutilum: f. habit; g. upper portion of plant; h. flower (C&C).

1. *P. ceratophyllum* Michx. Riverweed, Orchid-of-the-waterfall Fig. 371
Fast-moving streams, especially rapids of piedmont and coastal plain. N.S., N.B., sw. Que., e. Ont., Me., N.H., Mass. and Conn. sw. to Pa., Ky., Ark. and e. Okla., s. to Ga., Ala. and e. La.; W.I., C.Am. This plant is easily overlooked because of its often-inaccessible habitat and its inconspicuous appearance, sometimes being mistaken for a moss or alga. Furthermore, there is considerable variability in growth form (Fassett, 1939b), ranging from the most slender, lax, and elongate forms (fig. 371a,c,d) to the most coarse and rigid morphological extremes of the species (fig. 371b).

Elatinaceae / Waterwort Family

1. Leaves usually less than 8 mm long; flowers solitary in leaf axils, sessile, (0)2–3-merous; plants glabrous. . . . . 1. *Elatine*
1. Leaves 10–30 mm long; flowers 1–3, in leaf axils, pedicellate (3–8 mm), 5-merous; plants usually glandular-puberulent. . . . . 2. *Bergia*

References: Tucker, 1986, 2016.

1. *Elatine* (Waterwort)

Annual or short-lived perennial, aquatic herbs; stems erect or creeping; leaves opposite; flowers inconspicuous, axillary, 2–4-merous; submersed plants with cleistogamous flowers; fruit a thin-walled capsule; seeds pitted (areolate). Seed surface morphology is important for identification.

The differentiation of species of Waterworts is based largely on mature seeds, which must be examined under the microscope. Superficially all the species are quite similar. In fact, members of the same species growing on mud versus in the water may differ more in appearance than different species growing within the same habitat. The general form of the plant is greatly influenced by submergence or emergence (mudflat stranded plants), thus the seeds remain diagnostically critical.

References: Fassett, 1939a; Fernald, 1917a, 1941; Razifard et al., 2016.

1. Petals 2; fruit 2-lobed; seeds with rounded pits, pits nearly uniform in size toward ends of seeds (fig. 372c), pit walls thick (rounded in *E. brachysperma*, but then fruits 3-lobed); leaves 0.7–5 mm long. . . . . 1. *E. minima*
1. Petals 3 (or lacking); fruit 3-lobed; seeds with angular, 6-sided pits, pits becoming smaller and rows irregular toward ends of seeds (fig. 373i), pit walls thin; leaves 2–15 mm long.
    2. Pedicels recurved in fruit, 0.5–2.5 mm. . . . . 2. *E. ambigua*
    2. Pedicels erect, 0–0.5 mm.
        3. Seed pits angular-hexangular, (13)16–25(30) per row; plants forming large mats, up to 200 cm wide.
            4. Leaves light green to green, blades linear lanceolate to narrowly oblong or obovate to broadly spatulate, 1.5–3.5 mm wide; seed pits 2–3 times as long as wide.
            4. Leaves linear lanceolate or narrowly oblong, 3–10 mm long, 1.5–3.5 mm wide; stems loosely matted. . . . . 3. *E. triandra*
                5. Leaves obovate to broadly spatulate, 3–8 mm long, 0.5–3.5 mm wide, rounded to shallowly notched at apex; stems densely matted. . . . . 4. *E. americana*
                5. Leaves reddish-green, blades oblong-lanceolate (0.5)1–2(2.2) mm wide; seed pits more or less as wide as long (sometimes up to 2 times as long as wide). . . . . 5. *E. rubella*
        3. Seed pits rounded, 9–15 per row; plants forming very small mats, usually 1–5 cm wide. . . . . 6. *E. brachysperma*

1. *E. minima* (Nutt.) Fisch. & C. A. Mey. Small Waterwort Fig. 372
Sandy pond and lake shores and sometimes tidal waters, chiefly acid waters. Nfld., Lab., N.S., and Me. w. to Ont., Mich., Wisc., and Minn. and Sask., s. to N.Y., Md., ne. Va., S.C. Tenn. and Ill.

2. *E. ambigua* Wight Asian Waterwort Fig. 373
Pools, marshes and rice fields. Mass. and Conn.; Va., and S.C.; Calif. Razifard et al. (2016) noted that this submersed Eurasian species is a popular aquarium plant and has been sold under the name *E. triandra*; apparently introduced in the East by way of aquarium disposal or fish stocking.

3. *E. triandra* Schkuhr Three-stamen Waterwort Figs. 372, 373
Lakes and muddy intermittent pools. Ont., Wisc. and Minn. w. to s. Sask., Alta. and N.W.T., s. to Ga., La., Tex., Colo., Utah, Calif. and Ariz.; local, Skowhegan, Me., w. Mass. and Brooklyn, N.Y., possibly intro.

Fig. 371. *Podostemum ceratophyllum*: a–d. habit; e, f. leaf apices, showing variations; g. base of leaf with stipules; h. fruit with spathe (F).

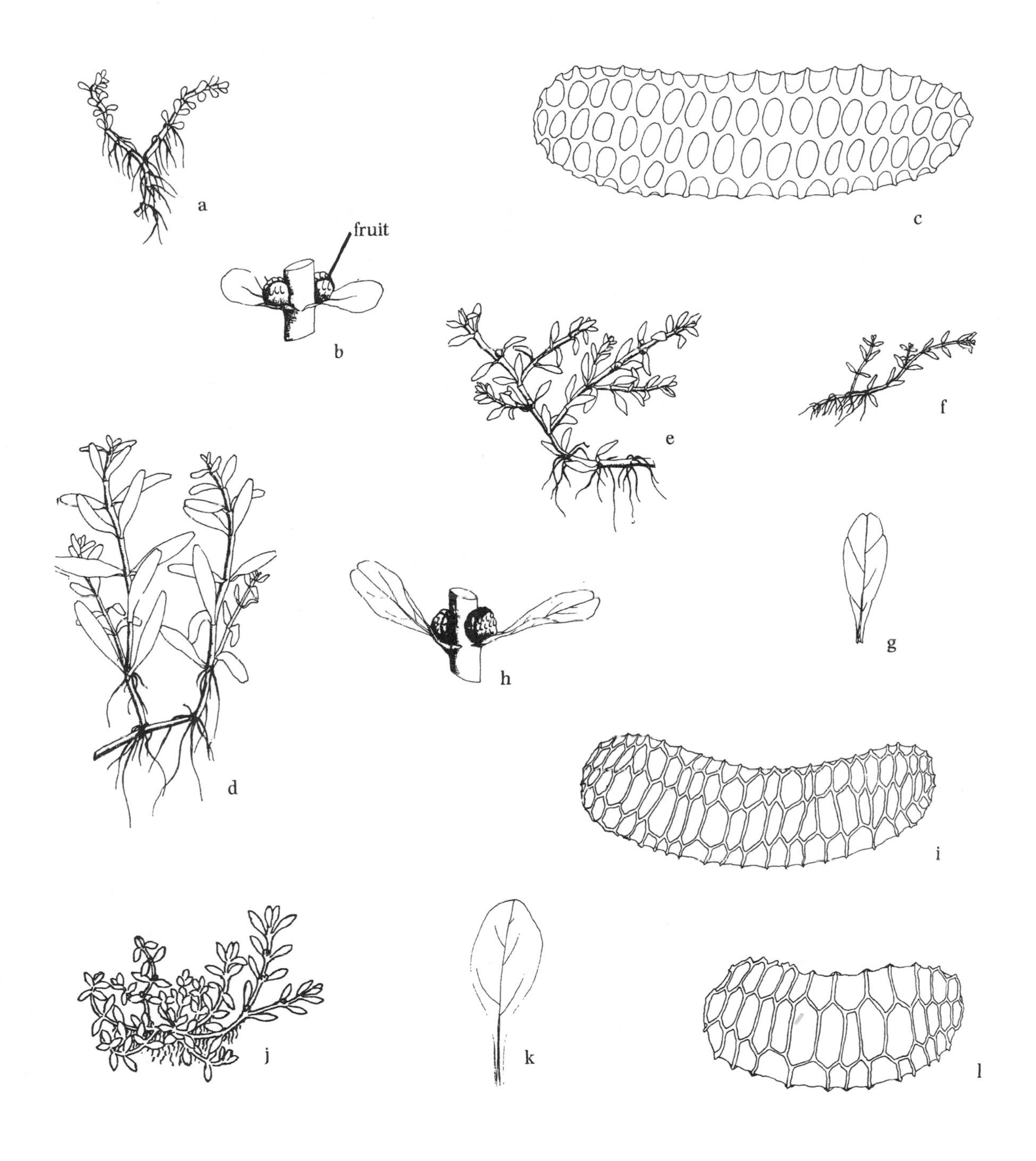

Fig. 372. *Elatine minima*: a. habit; b. fruits in leaf axils; c. seed (F).
Elatine triandra: d. habit, submersed form; e. habit, shallow-water form; f. habit, terrestrial form; g. leaf; h. fruits in leaf axils; i. seed (F).
Elatine americana: j. habit (B&B); k. leaf (F).
Elatine brachysperma: l. seed (F).

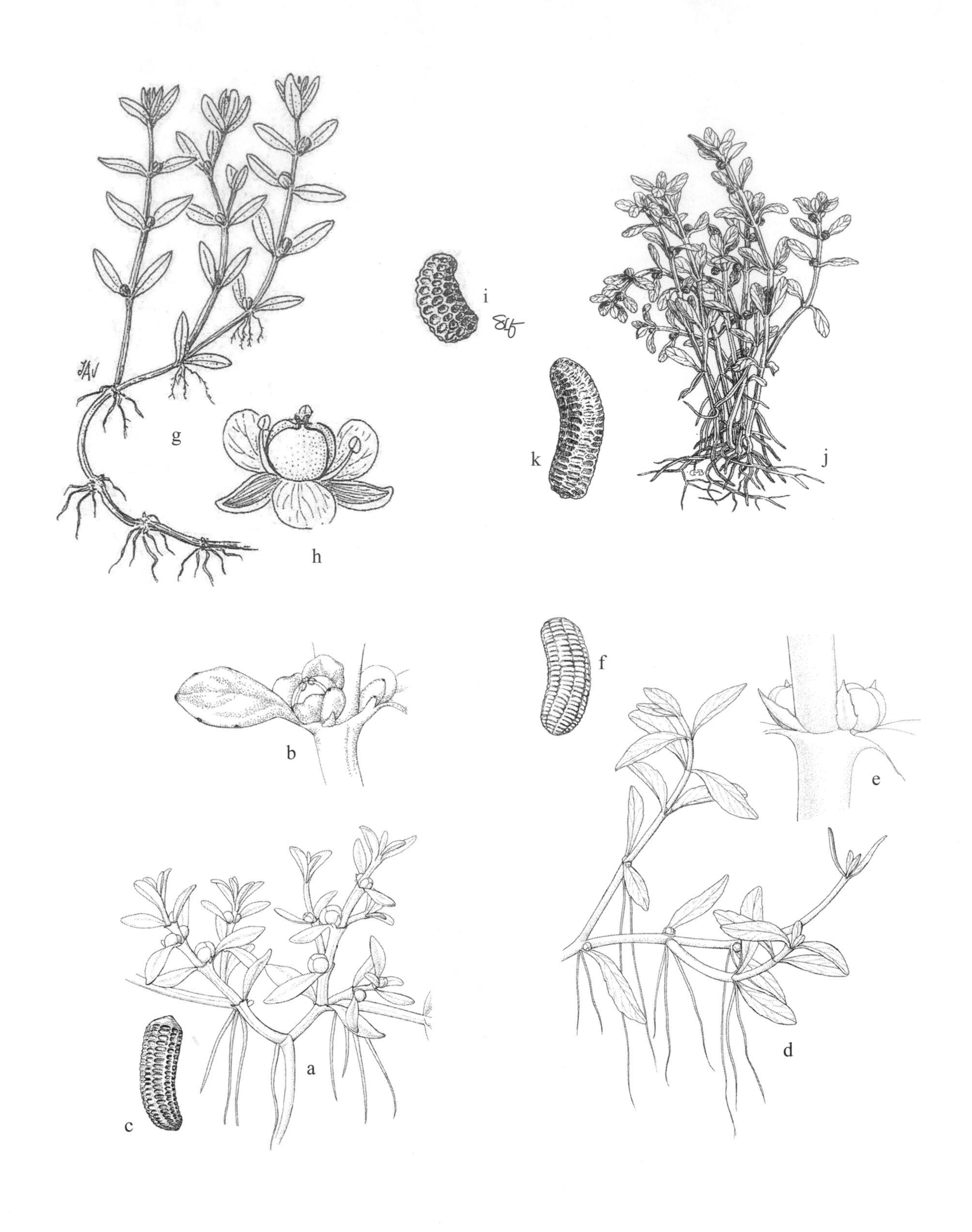

Fig. 373. *Elatine triandra*: a. habit; b. fruit in leaf axil; c. seed (FNA).
Elatine rubella: d. habit; e. fruit in leaf axil; f. seed (FNA).
Elatine brachysperma: g. habit; h. flower; i. seed (Jepson Flora).
Elatine ambigua: j. habit; k. seed (Mason).

4. *E. americana* (Pursh) Arn. American Waterwort Fig. 372
Muddy shores, especially tidal rivers and muddy margins of ponds and streams. St. Lawrence R. and Ottawa R., Que. and N.B. w. to Ont., Man. and N.W.T., s. to Del. and e. Va.; w. N.C., Mo., Okla., S.D. and Mont.; Calif. (*E. triandra* var. *americana* (Pursh) Fassett; *E. triandra* subsp. *americana* (Pursh) Á. Löve & D. Löve)

5. *E. rubella* Rydb. Southwestern Waterwort Fig. 373
Muddy shores, shallow vernal pools. N.D. and S.D. w. to Ida. and Oreg., s. to Neb., N.M., Ariz. and Calif. There is a discrepancy regarding reports from USDA PLANTS database including populations in Minn., Wisc. and Iowa as being in our range; these are not reported by Razifard et al. (2016); furthermore, they note that reports of *E. rubella* in Wash. and B.C. were based on misidentifications. It is possible the species might be expanding eastward in the Plains states; hence we have included the species here.

6. *E. brachysperma* A. Gray Shortseed Waterwort Figs. 372, 373
Shallow water and intermittent pools, Rare and local; Ohio, Ill., Ga. w. to Neb., Mont., and B.C. s. to Okla., Tex., N.M. Wyo., Nev., Ariz. and Calif. (*E. triandra* Schkuhr var. *brachysperma* (A. Gray) Fassett)

2. *Bergia* (Bergia)

Perennial herbs; stems prostrate to erect, sometimes becoming woody at base; leaves opposite, whitish, glandular-puberulent; flowers inconspicuous. 1–several, in axillary clusters, 5-merous; fruit a globose capsule; seeds shiny, obscurely pitted.

Reference: Tucker, 2016.

1. *B. texana* (Hook.) Seub. Texas Bergia Fig. 374
Swamps, marshes, muddy shores, intermittent pools, wet banks and ditches. S. Ill. and Mo. w. to s. S.D., se. Mont. and Wash., s. to Ark., Okla., La., Tex., Colo., N.M., Utah, Nev. and Calif.; Mex. (*Elatine texana* (Hook.) Torr. & A. Gray)

Violaceae / Violet Family

1. *Viola* (Violet)

Annual or perennial herbs; stemless, with leaves all basal (ours) or with basal and stem leaves; leaves simple, often cordate, sometimes deeply lobed to divided; petals blue to violet, yellow, or white, 5, separate, the lower one spurred; cleistogamous flowers produced somewhat later in season; fruit a 3-valved capsule.

References: Little and McKinney, 2015; Russell, 1965

---

1. Rhizomes slender, fibrous; plants stoloniferous (fig. 375b)
  2. Petals white, with purple veins.
    3. Leaves lanceolate to linear, more than 3 times as long as wide.................... 1. *V. lanceolata*
    3. Leaves oblong to ovate, less then 3 times as long as wide, or broadly cordate, blades usually wider than long.
      4. Leaves oblong to ovate, truncate to cuneate at base (fig. 375b).................... 2. *V. primulifolia*
      4. Leaves reniform to broadly cordate, usually wider than long, cordate at base.................... 3. *V. macloskeyi*
  2. Petals light blue to lavender.................... 4. *V. palustris*
1. Rhizomes thick; plants not stoloniferous.
  5. Leaves densely pubescent on both surfaces; all 5 petals villous; petioles villous at base.................... 5. *V. novae-angliae*
  5. Leaves nearly glabrous to only finely pubescent on upper surface; only 2 or 3 petals pubescent; petioles glabrous.
    6. Lower petal glabrous, lateral petals with knob-tipped hairs; cleistogamous flowers slender; earlier leaves cordate, green on lower surface.................... 6. *V. cucullata*
    6. Lower petal and lateral petals with cylindrical hairs; cleistogamous flowers ovoid; earlier leaves reniform, purplish on lower surface.................... 7. *V. nephrophylla*

---

1. *V. lanceolata* L. Lance-leaved Violet Fig. 375
Moist meadows, shores of ponds and lakes, stream banks and bogs. Nfld., Lab., P.E.I., and N.S. w. to Ont. and Minn., s. to nw. Fla., Neb., Okla. and c. Tex; B.C. s. Oreg.

2. *V. primulifolia* L. Primrose-leaved Violet Fig. 375
Wet meadows, boggy sites and borders of streams. N.E. w. to Pa., Ohio, Ind., and Minn., s. to Fla., e. Okla. and e. Tex. Our eastern taxon is var. *primulifolia*. (*V. primulifolia* subsp. *villosa* (D.C. Eaton) Russell; *V. primulif olia* var. *acuta* (Bigelow) Torr. & A. Gray)

3. *V. macloskeyi* Lloyd Smooth White Violet Fig. 375
Springy or wet woods, bogs, wet meadows and along cold streams, often in shallow water. Nfld., Lab. and Que. w. to N.W.T. and B.C., s. to N.E., n. N.J., Del., mts. to Tenn., S.C. and n. Ga., Ohio, Ind., e. Iowa, Minn., mts. to Colo., Mont., Wash., Ariz. and Calif. (*V. pallens* (Banks ex DC.) Brainerd)

4. *V. palustris* L. Alpine Marsh Violet
Subalpine and alpine brooks and wet peaty banks. Greenl., Nfld. and Lab. w. to Nunavut, N.W.T., and B.C., s. to Nfld. and Gaspé

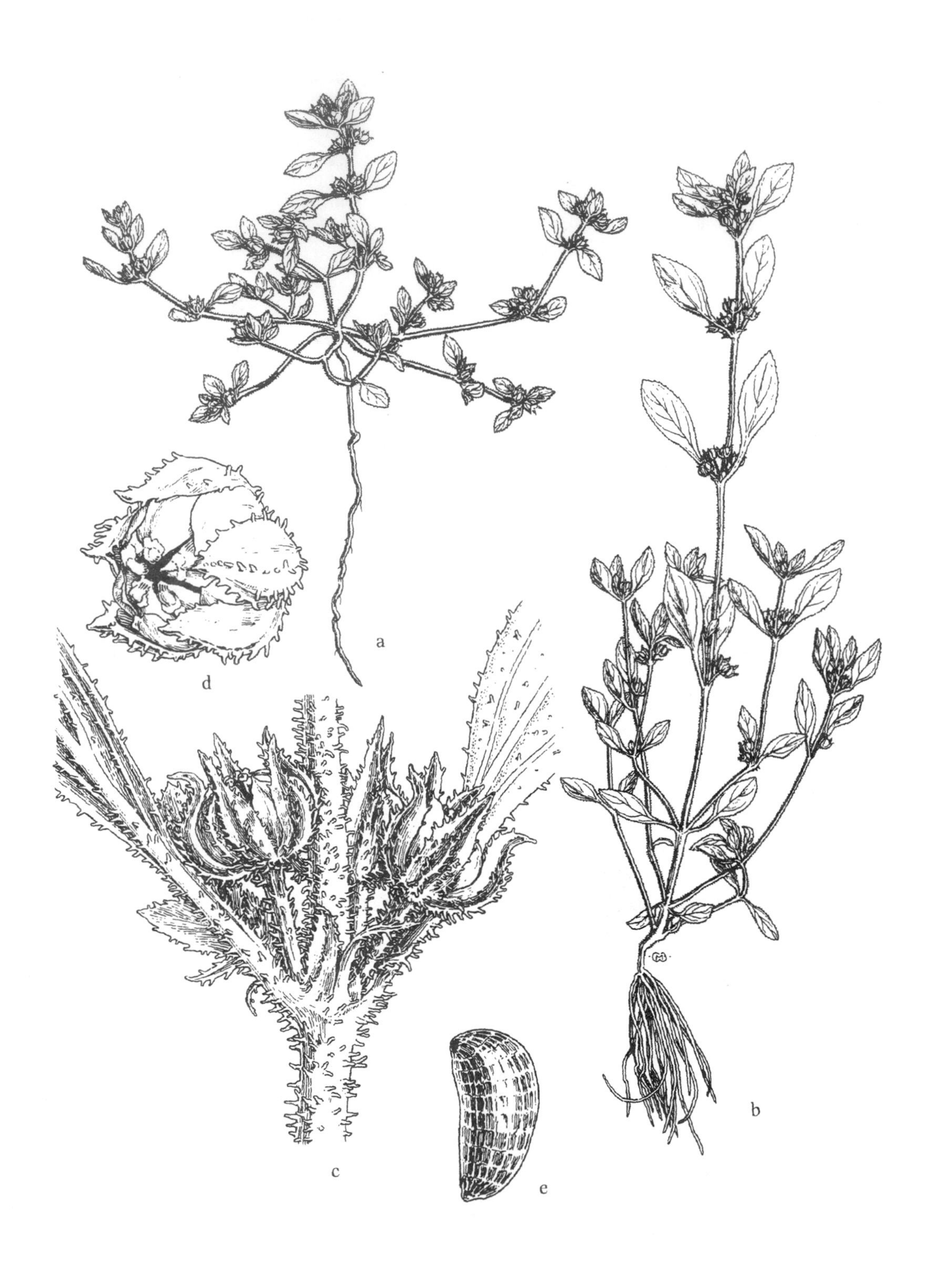

Fig. 374. *Bergia texana*: a. habit, prostrate form; b. habit, erect form; c. flowers in leaf axils; d. capsule with persistent calyx; e. seed (Mason).

Fig. 375. *Viola lanceolata*: a. habit (F).
Viola primulifolia: b. habit (Russell).
Viola macloskeyi: c. habit (Russell).

Pen., Que., mts. of Me. and N.H., Ont., Man., Colo., Utah, Mont., Ida. and Calif.

5. *V. novae-angliae* House New England Violet Fig. 376
Gravelly and sandy shores. Very rare; N.B. and Me.; w. Ont. w. to Man., s. to n. Mich., n. Wisc. and ne. Minn.

6. *V. cucullata* Aiton Blue Marsh Violet Fig. 376
Wet meadows, swamps, bogs, spring woods and stream margins. Nfld. and N.S. w. to Ont. and e. Minn., s. to N.C., n. Ga., Tenn., se. Mo., e. Ark. and Miss.

7. *V. nephrophylla* Greene Fig. 376
Gravelly riverbanks, rocky shores, wet meadows and bogs often calcareous sites. Nfld. w. to Nunavut, N.W.T., Yuk. and B.C., s. to N.E., n. N.Y., Pa., W.Va., Mich., Ind., Iowa, Neb., Ark., Okla., Tex., Colo., Ariz. and Calif.

Salicaceae / Willow Family

1. Leaves at least 3 times as long as wide; buds with a single bud scale; male plants with flowers in catkins, each with 2–9 stamens; female plants with pistillate flowers in catkins, each subtended by 1 or 2 glands. . . . . 1. *Salix*
1. Leaves usually less than twice as long as wide; buds with several overlapping bud scales; petioles strongly compressed laterally (leaves readily quivering in a light breeze) or if round in cross-section, often somewhat grooved on upper surface; male plants with staminate flowers in catkins, each with numerous stamens; female plants with pistillate flowers in catkins subtended by a cup-like disk. . . . . 2. *Populus*

1. *Salix* (Willow)

Trees, shrubs, or subshrubs, deciduous; leaves alternate, simple; plants dioecious, flowers unisexual, borne in erect to spreading or pendulous catkins; flowers lacking a perianth; pistillate flowers subtended by 1 or 2 glands; fruit a capsule; seeds with a tuft of hairs. Willows are difficult to determine, showing much variation in leaf width and pubescence. Furthermore, hybrids are common.

Frequently structures appearing like little pine cones can be found on willow twigs; these are insect galls.

References: Argus, 1973, 1980, 1986a, 1986b, 2010; Dorn, 1976.

1. Plants vegetative (often with flowers having fallen, possibly with fruits still remaining earlier in summer).
   2. Leaf margins with sharp teeth.
      3. Petiole coarsely glandular or glandular-lobed at base of blade.
         4. Leaf apices short-acuminate (fig. 377e); leaves whitish-pubescent on lower surface; stipules absent. . . . . 7. *S. serissima*
         4. Leaf apices attenuate (fig. 378a); leaves glabrous or rusty-pubescent on lower surface; stipules present. . . . . 2. *S. lucida*
      3. Petiole not coarsely glandular or glandular-lobed at base of blade, or glands, if present, obscure.
         5. Leaves at maturity green on lower surface.
            6. Leaf margins with 3–5 teeth per cm. . . . . 1. *S. interior*
            6. Leaf margins with more than 5 teeth per cm.
               7. Leaf apices acute to attenuate.
                  8. Bud apex sharp-pointed; bud scale margins free and overlapping; mature plants, trees. . . . . 4. *S. nigra*
                  8. Bud apex blunt; bud scale margins united; mature plants shrubs. . . . . 5. *S. eriocephala*
               7. Leaf apices blunt. . . . . 3. *S. myrtillifolia*
         5. Leaves at maturity white to whitish, on lower surface, pubescent or glaucous or subglaucous.
            9. Mature leaves glaucous to subglaucous on lower surface.
               10. Leaves narrowed toward apex from below middle, long-tapering, some with caudate apex.
                   11. Leaf apices acuminate; petioles 2–7 mm long; branchlets usually pubescent; stipules 5–15 mm long, broad and prominent. . . . . 8. *S. caroliniana*
                   11. Leaf apices acuminate to caudate; petioles 10–30 mm long; branchlets glabrous; stipules about 1 mm long, soon falling. . . . . 9. *S. amygdaloides*
               10. Leaves gradually narrowed toward apex from above middle, acute or acuminate, not long-tapering.
                   12. Leaf bases acute; stipules absent or minute. . . . . 11. *S. petiolaris*
                   12. Leaf bases rounded to subcordate; stipules present, conspicuous.
                       13. Leaves blackening when dried; expanding leaves not red-purple at apex. . . . . 12. *S. myricoides*
                       13. Leaves not blackening when dried; expanding leaves red-purple at apex. . . . . 5. *S. eriocephala*

Fig. 376. *Viola novae-angliae*: a. habit; b. flower (Crow).
Viola cucullata: c. habit (Russell).
Viola nephrophylla: d. habit (Russell).

9. Mature leaves pubescent or sericeous on lower surface.
    14. Branchlets flexible at base, difficult to break; stipules absent or minute and falling early.. .... 3. *S. petiolaris*
    14. Branchlets brittle at base; stipules present, 3–13 mm long, sometimes deciduous.
        15. Stipules cordate-ovate or nearly reniform; branchlets densely velutinous. .... 6. *S. cordata*
        15. Stipules lanceolate to nearly ovate, sometimes deciduous; branchlets puberulent to glabrous. ....13. *S. sericea*

2. Leaf margins entire, wavy or crenate.
    16. Leaves glabrous on lower surface.
        17. Stipule absent or minute; leaf margins slightly to obviously revolute.
            18. Leaves with sub-impressed veins on lower surface; leaf margins revolute; eaves narrow, linear-lanceolate to oblanceolate, attenuate at apex; large shrubs or small trees, 3–5 m tall. .... 14. *S. pellita*
            18. Leaves with slightly raised veins, finely reticulate, on upper surface; leaf margins only slightly revolute; leaves broader, obovate to oblanceolate, acute to obtuse at apex; slender, creeping, and stoloniferous shrubs, 0.3–1 m tall. .... 15. *S. pedicellaris*
        17. Stipule present; leaf margins not revolute...10. *S. discolor*
    16. Leaves pubescent, pilose, sericeous, or tomentose on lower surface.
        19. Leaves tomentose on lower surface.
            20. Leaves oblong to linear-lanceolate, 4–12 cm long; revolute leaf margin entire or wavy; mature shrubs 0.5–2 m tall; plants widespread. .... 16. *S. candida*
            20. Leaves elliptic to elliptic-ovate, 2–6 cm long, revolute leaf margins with appressed crenate teeth; mature shrubs 2–5 m tall; plants of Mingan I. (Gulf of St. Lawrence) and Nfld. ....17. *S.×cryptodonta*
        19. Leaves pubescent, pilose, or sericeous on lower surface.
            21. Stipules absent or minute; leaves whitened, sericeous or gray-pubescent on lower surface.
                22. Leaves green on upper surface, whitened and sericeous on lower surface; leaf margins revolute. .... 14. *S. pellita*
                22. Leaves somewhat gray-to white-pubescent on both surfaces; leaf margins not revolute. .... 18. *S. bebbiana*
            21. Stipules large; leaves rusty-pubescent on leaves lower surface...10. *S. discolor*

1. Plants in flower with either male or female catkins present (plants dioecious), often preceding leaf expansion.
    23. Catkins pistillate.
        24. Pistils or capsules villous, pilose, sericeous, or tomentose.
            25. Capsules densely tomentose, with tangled hairs.
                26. Catkin scales brown; pedicels 0.2–0.7 mm long; styles 0.3–1.9 mm long; capsules white-tomentose; plants widespread.. .... 16. *S. candida*
                26. Catkin scales reddish brown; pedicels 1–2.5 mm long; styles 0.8–1 mm long; capsule gray-tomentose; plants of Mingan I. (Gulf of St. Lawrence) and n. Nfld. .... 17. *S.×cryptodonta*
            25. Capsules sericeous, villous, or pilose.
                27. Capsules tapered to a beak (fig. 384j,k,l).
                    28. Catkin scale yellow or orange. .... 18. *S. bebbiana*
                    28. Catkin scale dark brown to nearly black.
                        29. Stipe shorter than scale (fig. 384l). ....10. *S. discolor*
                        29. Stipe longer than scale (fig. 384j). ....11. *S. petiolaris*
                27. Capsules not beaked (fig. 384g,h).
                    30. Capsules lanceolate; styles 0.6–1.5 mm long. .... 14. *S. pellita*
                    30. Capsule ovoid-oblong; styles 0.2–0.4 mm. long. ....13. *S. sericea*
        24. Pistils or capsules glabrous.
            31. Scale at base of each pistil light-colored, deciduous before capsule matures.
                32. Stipules lacking or if present, then bearing white glands.
                    33. Capsules 7–10(12) mm long; stipules absent. ....7. *S. serissima*
                    33. Capsules 3–6.5 mm long; stipules present (sometimes lacking in *S. amygdaloides*).
                        34. Stipe shorter than width of capsule; peduncle 0.5–1.0 mm long; leaves glossy on upper surface. .... 2. *S. lucida*
                        34. Stipe as long or longer than width of capsule; peduncle 1.5–4 mm long; leaves dull on upper surface.
                            35. Leaves lanceolate, narrow, apex acuminate; branchlets dark to light brown, usually pubescent. stipules usually conspicuous and persistent; styles 0.1–0.2 mm long. ....8. *S. caroliniana*
                            35. Leaves broadly to narrowly lanceolate, apex acuminate to caudate; branchlets yellowish tan grayish, glabrous; stipules small or lacking; styles 0.3–0.5 mm long. ....9. *S. amygdaloides*
                32. Stipules present, lacking white glands.

36. Capsules 7–10 mm long; leaves with 3–5 teeth per cm. .... 1. *S. interior*
36. Capsules leaves with more than 5 teeth per cm. .... 4. *S. nigra*
31. Scale at base or each pistil dark brown or reddish at tip, persistent.
37. Scales and rachis or catkins densely villous; leaf margins toothed.
38. New and old branchlets densely gray-tomentose, styles 0.5–1 mm long.......... 6. *S. cordata*
38. New and old branchlets yellow to brown, somewhat pubescent; styles 1–2.5(4) mm long.
39. Capsules 4.5–6.5 mm long; pedicels 1–1.5 mm long; styles 0.2–0.8(1) mm long.
40. Styles 0.5–0.8(10 mm long; leaves glossy on upper surface. .... 3. *S. myrtillifolia*
40. Styles 0.2–0.5 mm long; leaves dull on the upper surface.......... 5. *S. eriocephala*
39. Capsules 6–8(9) mm long; pedicels 1–2.5(4) mm long; styles 0.7–1.5 mm long. .... 12. *S. myricoides*
37. Scales and rachis or catkins with few or no hairs; leaf margins entire. .... 15. *S. pedicellaris*
23. Catkins staminate (plants dioecious).
41. Stamens 3–9 per flower (fig. 385a,f,g).
42. Catkins 1–2 cm long; stipules absent. ....7. *S. serissima*
42. Catkins 2–7 cm long; stipules present (may be absent is *S. amygdaloides*).
43. Catkins slender, loosely flowered.
44. Leaves glabrous on lower surface.
45. Petioles 2–7 mm long; leaves lanceolate to very narrowly so; leaf apices acuminate; branchlets dark to light brown, usually pubescent; stipules 5–15 mm wide, persistent.. .... 8. *S. caroliniana*
45. Petioles 10–30 mm long; leaves broadly to narrowly lanceolate; leaf apices acuminate to caudate; branchlets yellowish or tan, or grayish, glabrous; stipules about 1 mm wide or absent. .... 9. *S. amygdaloides*
44. Leaves non-glaucous on lower surface, rarely weakly glaucous. .... 4. *S. nigra*
43. Catkins thickish, densely flowered. .... 2. *S. lucida*
41. Stamens 2 per flower (fig. 385m,n).
46. Catkin scales yellowish.
47. Scales pilose. .... 18. *S. bebbiana*
47. Scales thinly pubescent. .... 1. *S. interior*
46. Catkin scales reddish-yellow, dark orange, brown, or black.
48. Scales villous, pilose, or pubescent-tomentose on outer surface (fig. 385a).
49. Young twigs densely tomentose.
50. Catkin scales reddish-yellow, villous at tip; plants of Mingan I. (Gulf of St. Lawrence) and Nfld. .... 17. *S.* ×*cryptodonta*
50. Catkin scales brown, pubescent-tomentose throughout; plants widespread. .... 16. *S. candida*
49. Young twigs glabrous to puberulent or velvety.
51. Catkins sessile. .... 10. *S. discolor*
51. Catkins subsessile or pedunculate.
52. Twigs velvety with shiny white hairs. .... 5. *S. eriocephala*
52. Twigs puberulent to glabrous.
53. Twigs grayish. .... 5. *S. cordata*
53. Twigs reddish, yellowish, greenish, or blackish.
54. Branchlets brittle at base, easily broken.
55. Catkin 20–39 mm long, 7–20 mm wide; nectary oblong or narrowly oblong, 0.6–1 mm, filaments separate; leaf margins entire to wavy-crenate. .... 14. *S. pellita*
55. Catkin 13.5–40 mm long, 4–9 mm wide; nectary ovate to oblong, 0.3–0.8 mm, filaments connate less than 1.5 their length or separate; leaf margins toothed. .... 13. *S. sericea*
54. Branchlets tenacious, not easily broken.
56. Catkins stout, 12–29 mm long, 6–17 mm wide, somewhat pendulous. .... 11. *S. petiolaris*
56. Catkins 19–44(51) mm long, 7–14(22) mm wide, erect.
57. Catkin scales dark brown, crinkly pilose; leaves not blackening when dried; expanding leaves red-purple at apices. .... 5. *S. eriocephala*
57. Catkin scales brownish-black, long-villous; leaves blackening when dried expanding leaves not red-purple at apices. .... 12. *S. myricoides*
48. Scales glabrate on outer surface or with a few hairs at tip.

58. Catkins 11.5–39 mm long, 5–14 mm wide; scales oblong-oblanceolate to oblanceolate, glabrate; young branchlets yellowish; leaves glossy on upper surface, dark green, margins toothed, emerging when flowering. .... 3. *S. myrtillifolia*

58. Catkins 11–21 mm long, 4–8 mm wide; scales oblong, glabrate, or occasionally with a few hairs at tip: young branchlets olive-green to dark brown; leaves dull on upper surface, margins entire, purplish when emerging. .... 15. *S. pedicellaris*

---

1. *S. interior* Rowlee Sandbar Willow Figs. 377, 384, 385
Sanbars, mudbars, beaches and alluvial soils. N.B. and s. Que, w., to Alask., s. to Md., Va., Tenn., Miss., e. Tex. Mont., Wyo., Colo., Alta. and B.C.; e. Mex. Argus (2010) treated the eastern Sandbar Willow as a species distinct from *S. exigua* of western N.Am. (*S. exigua* subsp. *interior* (Rowlee) Cronq.; *S. exigua* var. *exterior* (Fernald) C. F. Reed; *S. longifolia* Muhl.)

2. *S. lucida* Muhl. Shining Willow Figs. 378, 384
Swamps shores and low ground. Lab. and Nfld. w. to n. Man., N.W.T. and Alask., s. to Md., Va., W.Va., n. Ohio, n. Ill., Iowa and S.D.

3. *S. myrtillifolia* Anderss. Myrtle-leaved Willow Fig. 382
Calcareous swamps, shores, wet ground and alpine barrens. Lab., Nfld. and Nunavut, w. to N.W.T. and Alask., s. to Ont., Man., Sask. and B.C.; mts. of Wyo. and Colo.

4. *S. nigra* Marsh. Black Willow Figs. 378, 385
Pond and lake shores, stream banks, meadows and low woods. N.B. w. to Que., Ont., Minn. and Neb., s. to Ga., n. Fla., Tex. and e. Mex. (*S. nigra* var. *falcata* (Pursh) Torr.)

5. *S. eriocephala* Michx. Missouri River Willow Figs. 379, 385
Stream banks, shores, low thickets and ditches. Nfld. w. to Ont., Minn. and N.D., s. to Ky., Ga. n. Fla., Ala., Ark. and Kans. (*S. cordata* Muhl.; *S. rigida* Muhl.)

6. *S. cordata* Michx. Broad Heart-leaved Willow Figs. 383, 384
Sandy, gravelly and alluvial shores. Se. Lab. and Nfld. w. to n. Ont., s. to n. N.Y., Mich. and Wisc. s. to Ind. and Ill. (*S. cordata* var. *abrasa* Fernald.; *S. adenophylla* Hook.; *S. synticola* Fernald)

7. *S. serissima* (L. H. Bailey) Fernald Autumn Willow Figs. 377, 384, 385
Calcareous marshes, swamps and bogs. Lab. and Nfld. w. to Nunavut, N.W.T. and B.C., s. to w. Conn., n. N.J., Pa., n. Ohio, n. Ind., n. Ill., Wisc., Minn., S.D., Mont. and Colo. (*S. lucida* var. *serrisima* L. H. Bailey)

8. *S. caroliniana* Michx. Ward's Willow, Carolina Willow Figs. 377, 385
Rocky shores, riverbanks and low woods. N.J., Del., Md., and s. Pa. w. to s. Ohio, s. Ind., s. Ill., Mo., and e. Kan., s. to Fla., La. and Tex.; Cuba and C.Am. (*S. longipes* Shuttlew. ex Anderss.)

9. *S. amygdaloides* Anderss. Peach-leaved Willow Figs. 378, 384, 385
Shores, woods and alluvial swamps. Sw. Que., Ont. and Vt. w. to se. B.C., s. to w. Mass., N.Y., Pa., Ohio, Ill., Ky., Mo., Tex., Ariz. and Calif.

10. *S. discolor* Muhl. Large Pussy Willow Figs. 380, 384, 385
Swamps, thickets and shores. Lab. and Nfld. w. to N.W.T. and B.C., s. to Md., N.C., Ky., Mo., S.D. Colo., Ida. and Mont. (S. *discolor* var. *overi* C. R. Ball)

11. *S. petiolaris* Sm. Slender Willow Figs. 379, 384, 385
Meadows, swales, stream banks and lakeshores. N.S., P.E.I. and Que. w. to Man., N.W.T. and B.C., s. to n. Mass., w. Conn., N.Y., N.J., Pa., Ohio, n. Ill, ne. Iowa, Neb., Colo., Sask. and Wash. (*S. gracilis* Anderss.; *S. petiolaris* var. *gracilis* (Anderss.) Anderss.)

12. *S. myricoides* Muhl. Dune Willow, Blue-leaved Willow Figs. 379, 384
Gravelly shores, rich thickets, sand dunes and beaches; often calcareous soils. Nfld. and Lab., w. to n. Ont., s. to Me. and Que., Pa., Ohio., n. Ind., n. Ill. and e. Wisc.; especially along shores of Great Lakes. (*Salix glaucophylloides* Fernald)

13. *S. sericea* Marshall Silky Willow Figs. 380, 381, 383, 384, 385
Moist rocky shores, banks and low thickets. Sw. N.S., N.B. and s. Que., w. to Wisc., se. Minn. and e. Iowa, s. to N.E., S.C., ne. Ga., Tenn., n. Ala., se. Mo. and n. Ark. (*S. coactilis* Fernald)

14. *S. pellita* Anderss. ex Schneid. Figs. 383, 384
Stream banks, shores and swamps. Lab. and Nfld. w. to n. Ont. and s. Sask., s. to n. N.H., n. Vt., n. Mich., n. Wisc. and n. Minn.

15. *S. pedicellaris* Pursh Bog Willow Figs. 381, 384, 385
Acid bogs and swamps. Nfld., Lab. and Nunavut, w. to N.W.T. and Yuk., s. to N.E., N.J., Pa., Ohio, Ind., n. Ill., n. Iowa, Ida., N.D., Sask., Alta., Ida., Wash. and n. Oreg. (*S. pedicellaris* var. *hypoglauca* Fernald; *S. pedicellaris* var. *tenuescens* Fernald; *S. myrtilloides* var. *hypoglauca* (Fernald) C. R. Ball)

16. *S. candida* Flügge ex Willd. Hoary Willow, Sage-leaved Willow Figs. 381, 382, 384, 385
Calcareous bogs and thickets. Lab., Nfld. and Nunavut w. to N.W.T., Yuk. and Alask., s. to n. and w. N.E., n. N.J., Pa., n. Ohio, n. Ind., n. Ill., n. Iowa, S.D., Colo., Ida., ne. Wash. and B.C. (*S. candida* var. *denudata* Anderss.)

17. *S.* ×*cryptodonta* Fernald Fig. 383
Shores, thickets, stream banks and glades. Nfld. and Mingan I. (Gulf of St. Lawrence, Que.). Described by Fernald as a distinct species, Argus (2010) considered this to be a hybrid between *S. bebbiana* and *S. candida* and commonly occurring in Newfoundland; it has densely wooly juvenile leaves reminiscent of *S. candida*, with mature leaves becoming sparsely to moderately wooly beneath, and with strongly revolute margins; it resembles *S. bebbiana* in having long-beaked capsules and long stipes.

18. *S. bebbiana* Sarg. Long-beaked Willow Figs. 381, 384, 385
Wet to dry habitats. Lab. and Nfld. w. to Nunavut, N.W.T. and Alask., s. to N.E., N.Y., Md., Pa., n. Ohio, n. Ind., n. Ill., ne. Iowa, S.D., w. Neb., N.M., Ariz., Nev. and n. Calif.

Fig. 377. *Salix caroliniana*: a. leaf and stipules (F).
Salix interior: b–d. leaves and stipules (F).
Salix serissima: e. leaves; f. staminate branch; g. pistillate branch; h. capsule (Gleason).

Fig. 378. *Salix lucida*: a. leaf and stipules (F).
Salix nigra: b. leaves and stipules (F).
Salix amygdaloides: c. leaves (F).

Fig. 379. *Salix myricoides*: a. leaf and stipule (F).
Salix eriocephala: b–e. leaves and stipules (F).
Salix petiolaris: f. leaves (F).

Fig. 380. *Salix sericea*: a. leaves (F).
Salix discolor: b. leaves and stipules (F).

Fig. 381. *Salix pedicellaris*: a. leaves (F).
Salix sericea: b. leaves (F).
Salix bebbiana: c. leaves (F).
Salix candida: d. leaves (F).

Fig. 382. *Salix myrtillifolia*: a. staminate branch; b. pistillate branch (USDA).
Salix candida: c. leaves; d. staminate branch; e. pistillate branch (USDA).

Fig. 383. *Salix ×cryptodonta*: a. pistillate branch; b. capsule (Gleason).
Salix cordata: c. leaves; d. staminate branch; e. pistillate branch; f. capsule (Gleason).
Salix pellita: g. leaves; h. staminate branch; i. pistillate branch; j. capsule (Gleason).
Salix sericea: k. leaves; l. staminate branch; m. pistillate branch; n. capsule (Gleason).

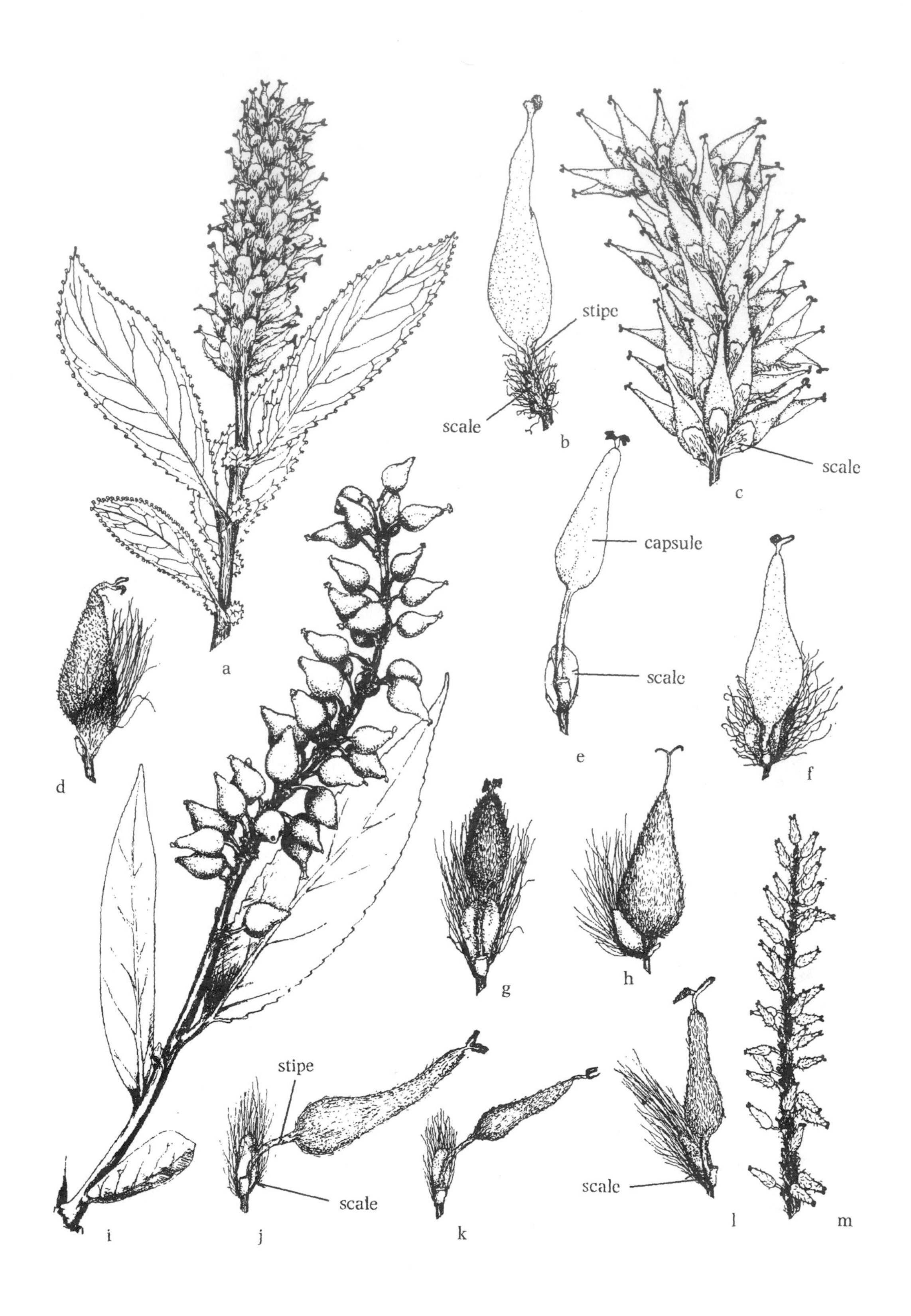

Fig. 384. *Salix* pistillate catkins and fruits (all F).
a. *Salix serissima*; b. *Salix myricoides* (as *S. glaucophylloides*); c. *Salix lucida*; d. *Salix candida*; e. *Salix pedicellaris*; f. *Salix eriocephala*; g. *Salix sericea*; h. *Salix pellita*; i. *Salix amygdaloides*; j. *Salix petiolaris*; k. *Salix bebbiana*; l. *Salix discolor*; m. *Salix interior* (as *S. longifolia*).

Fig. 385. Staminate catkins and flowers (all F).
a. *Salix nigra*; b. *Salix serissima*; c. *Salix interior* (as *S. longifolia*); d. *Salix sericea*; e. *Salix nigra*; f. *Salix amygdaloides*; g. *Salix caroliniana* (as *S. longipes*); h. *Salix bebbiana*; i. *Salix discolor*; j. *Salix eriocephala*; k. *Salix candida*; l. *Salix discolor*; m. *Salix petiolaris*; n. *Salix pedicellaris*; o. *Salix eriocephala.*

2. *Populus* (Cottonwood, Poplar)

Trees, deciduous; leaves alternate, simple; petioles strongly compressed laterally (leaves readily quivering in a light breeze) or if round in cross-section, often somewhat grooved on upper surface; flowers unisexual, borne in catkins; plants dioecious; flowers lacking a perianth; pistillate flowers subtended by a cup-like disk; fruit a capsule; seeds with a tuft of hairs.

REFERENCE: Eckenwalder, 2010.

1. Petioles strongly compressed (leaves readily quivering in a light breeze).
   2. Leaf blade apices short-acuminate; pedicels 1–13(17) mm when in fruit. . . . . 1a. *P. deltoides* subsp. *deltoides*
   2. Leaf bade apices long-acuminate, pedicel lengths 1–6(8) mm when in fruit. . . . . 1b. *P. deltoides* subsp. *monilifera*
1. Petioles round in cross-section, often somewhat grooved on upper surface.
   3. Leaves obtuse or rounded at apex; expanding blades densely cottony-pubescent, becoming glabrous at maturity, except for persistent patch of pubescence around veins at base; buds not sticky. . . . . 2. *P. heterophylla*
   3. Leaves acute to acuminate at apex; expanding blades glabrous or somewhat puberulent, becoming glabrous at maturity; buds very sticky. . . . . 3. *P. balsamifera*

1. *P. deltoides* W. Bartram ex Marshall Cottonwood Fig. 386

1a. *P. deltoides* subsp. *deltoides* Southern Cottonwood
Floodplains, swamp forests, riverbanks, shores and low, wet woods. S. Que. and N.E. w. to N.Y., s. Ont., Mich., Ill. and Iowa, s. to nw. Fla., La. and Tex.

1b. *P. deltoides* subsp. *monilifera* (Ait.) Eckenwalder
Plains Cottonwood
Stream and lake shores and dunes. N.Y., Mich. and Ont. w. to Alta., s. to Ohio, Ill., Mo., Okla., Tex. e. N.M., Mont. and B.C. Frequently cultivated.

2. *P. heterophylla* L. Swamp Cottonwood, Black Cottonwood
Fig. 386
Swamps, shores, and low, wet woods. Mass., R.I., Conn., and se. N.Y., s. on coastal plain to nw. Fla., w. to La.; Mississippi embayment n. to e. Ark., w. Tenn., se. Mo., w. Ky., s. Ill., Ind., Ohio and s. Mich.

3. *P. balsamifera* L. Balsam Poplar Fig. 386
Cool wet soils of stream margins and river gravels, shores, swamp forests and bog margins in boreal forests. Lab. and Nfld. and Nunavut w. to N.W.T. and Alask., s. to N.E., Pa., Del., Va., n. W.Va., ne. Ohio, Mich., n. Ind., n. Ill., n. Iowa, mts. to S.D., Colo., Mont. and n. B.C. *Populus* ×*jackii* Sarg. is a hybrid between *P. balsamifera* and *P. deltoides* and can be common in riparian and other wet habitats where the two species overlap (Eckenwalder, 2010)

Fabaceae (Leguminosae) / Bean Family

1. Plants viny; leaves with 2–9 pairs of leaflets, not sensitive to touch in fresh plants; fruit a legume (fig. 387b,d).
   2. Leaves terminated by a tendril (fig. 387a); stems narrowly winged; sprawling, non-climbing (although attached to low vegetation). . . . . 1. *Lathyrus*
   2. Leaves terminated by a leaflet, not a tendril (fig. 387c); stems round in cross-section, twining, climbing vine. . . . . 2. *Apios*
1. Plants large, erect (to 1–2 m tall) almost shrubby; leaves with more than 9 pairs of leaflets; sensitive to touch (leaflets folding) in fresh plants; fruit a loment fig. 387e). . . . . 3. *Aeschynomene*

1. *Lathyrus* (Vetchling, Wild Pea, Sweet Pea)

Perennial, herbaceous vines; stem often winged; leaves alternate, even-pinnately compound with 2–5 pairs (ours 2–3) of ovate to linear leaflets, the terminal leaflet modified as a tendril; flowers borne in a raceme; petals purple to violet; fruit a legume.

1. *L. palustris* L. (Vetchling) Fig. 387
Damp shores, thickets, wet meadows, fens, swales marshes, riverbanks and confer swamps. Lab. and Nfld. w. to Alask., s. to N.C., Tenn., Mo. and Oreg. Three varieties have been recognized: the stout, glabrous form has been described as var. *palustris*; the slender glabrous, wingless-stemmed plant as var. *myrtifolius* (Muhl. ex Willd.) A. Gray, and the slender pubescent wing-stemmed plant as var. *pilosus* (Cham.) Hultén.

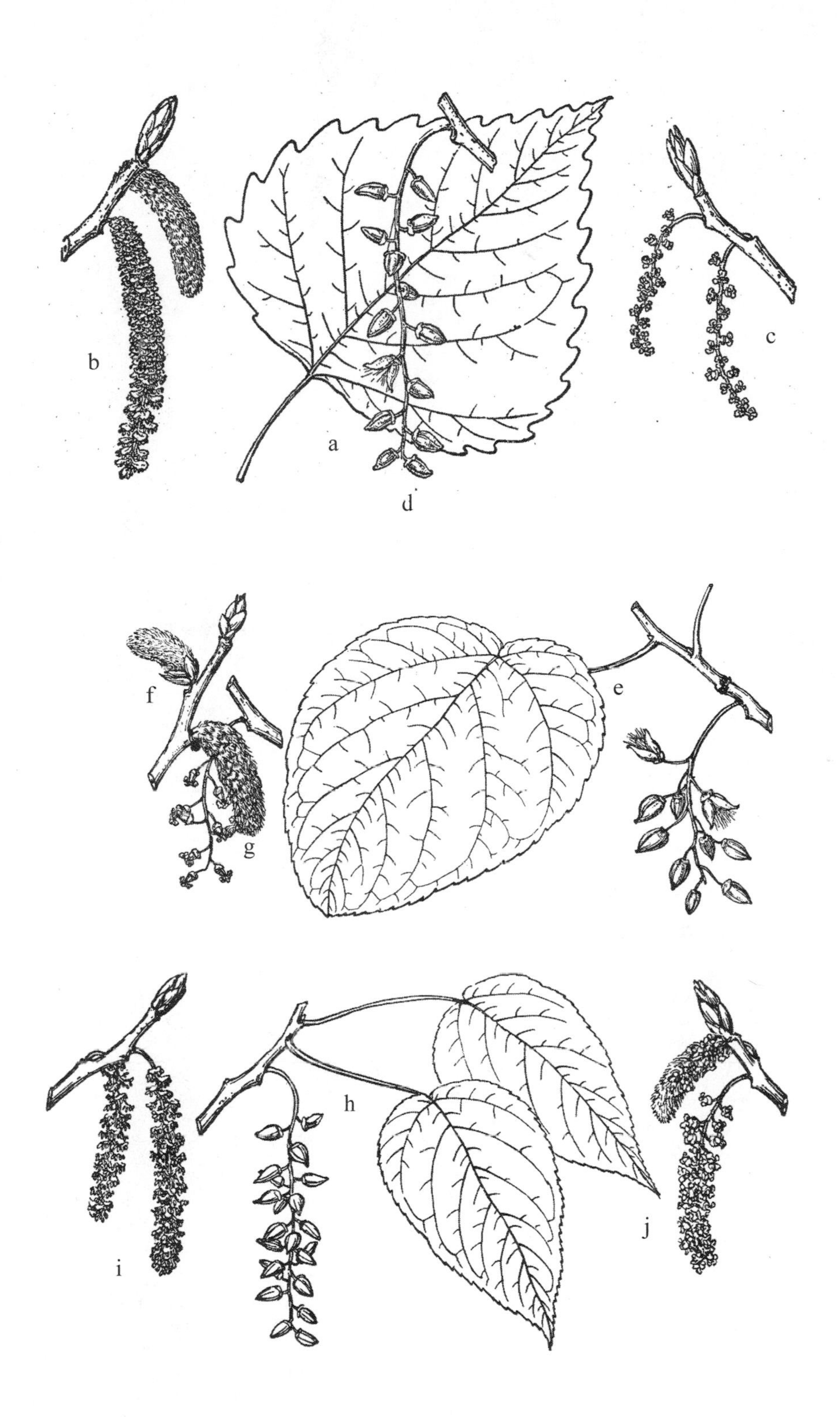

Fig. 386. *Populus deltoides*: a. leaf showing truncate leaf base; b. branch with male catkins; c. branch with female catkins; d. fruiting branch with one capsule dehisced (Sargent, 1905).
Populus heterophylla: e. portion of branch with leaf and female catkin with mature capsules (2 dehisced); f. male catkin; g. female catkin (Sargent, 1905).
Populus balsamifera: h. portion of branch with leaves and mature capsules; i. male catkins; j. female catkins (Sargent, 1905).

Fig. 387. *Lathyrus palustris*: a. upper portion of plant (B&T); b. legume (B&B).
Apios americana: c. habit (B&T); d. legume (B&B).
Aeschynomene virginica: e. upper portion of plant with loments; f. flower (B&B).

2. *Apios* (Groundnut, Wild Bean, Potato-bean)

Perennial herbaceous, twining vines; rhizomes developing a string of tuberous swellings (groundnuts); leaves odd-pinnately compound, with 3–7 leaflets; flowers borne in compact racemes; petals purple-brown; fruit a plump, linear-oblong legume.

1. *A. americana* Medik. Groundnut, Wild Bean, Potato-bean Fig. 387
Swamp borders, wet thickets, meadows and bottomlands. N.S. and N.B. w. to Ont., Minn. and e. N.D., s. to Fla., e. Tex.; Colo.

3. *Aeschynomene* (Sensitive Joint-vetch, Shy-leaves)

Annual (ours) or perennial herbs or shrubs; leaves even-pinnately compound, with many leaflets, often sensitive to light and touch (leaflets folding); flowers borne in few-flowered racemes; petals yellowish or reddish; fruit a loment.

1. *A. virginica* (L.) Britton, Sterns & Poggenb. Virginia Joint-vetch Fig. 387
Swales, ditches, seasonally wet sites, fresh or brackish tidal shores; sometimes in dense stands. Coastal, s. N.J. and Del. s. to e. Md., se. Va. and e. N.C.; Cuba; C.Am.

Rosaceae / Rose Family

References: Brouillet, 2014; Phipps, 2014; Robertson, 1974.

---

1. Plants woody.
 2. Leaves with a row of black short-linear glands along midvein on upper surface; pistils 1, ovary inferior, fused to hypanthium (fig. 388b); fruit a berry-like pome. . . . . 1. *Aronia*
 2. Leaves lacking midvein glands; pistils more than 1 (except *Sanguisorba*), ovaries superior, not fused to hypanthium; fruit a cluster of achenes (enclosed in a hip in *Rosa*) or follicles.
  3. Leaves simple; fruit dehiscent (follicles).
   4. Leaves lobed, palmately veined from base of blade; stipules narrowly ovate or stipule scars present; follicles 3–5, bright red to brownish; bark of older stems shreddy. . . . . 2. *Physocarpus*
   4. Leaves not lobed, pinnately veined from midrib; stipules absent; follicles 5, light brown; bark of older stems tight, not shreddy. . . . . 3. *Spiraea*
  3. Leaves compound, fruit indehiscent (achenes), borne on an open hypanthium (*Dasiphora*) or enclosed in hip (*Rosa*).
   5. Stems with prickles, flowers pink. . . . . 4. *Rosa*
   5. Stems lacking prickles, flowers yellow. . . . . 5. *Dasiphora*
1. Plants herbaceous, or slightly woody only at base.
 6. Leaves simple, shallowly lobed (fig. 392a); flowers white, 5-merous, fruit an aggregate of amber to reddish drupelets (fig. 392b). . . . . 6. *Rubus*
 6. Leaves compound, or if simple, then deeply lobed; flowers variously colored, 5-merous, except when white, then 4-merous; fruit dry, 1 or more achenes in a cluster.
  7. Flowers white, borne in dense spikes; sepals 4, petals absent, achenes solitary. . . . . 7. *Sanguisorba*
  7. Flowers purple, maroon, yellowish purple or yellow, solitary or borne in cymes; sepals 5, petals 5; achenes in clusters.
   8. Plants strongly erect; styles at maturity elongate and conspicuously geniculate, somewhat plumose. . . . . 8. *Geum*
   8. Plants prostrate or decumbent to weakly ascending; styles at maturity not elongate, geniculate or plumose.
    9. Flowers purple to maroon, plants herbaceous, with stems thick, becoming somewhat woody on lowermost portion, decumbent to ascending, without runners. . . . . 9. *Comarum*
    9. Flowers yellow; plants trailing prostrate herbs spreading by runners. . . . . 10. *Potentilla*

---

1. *Aronia* (Chokeberry)

Shrubs; leaves alternate, simple, margins glandular-serrate, midvein of upper surface bearing a row of glands; flowers borne in corymbose clusters; petals white; hypanthium adnate to ovary; fruit fleshy, a red, purple, or black pome.

References: Hardin, 1973; Pankhurst, 2014; Robertson et al., 1991; Uttal, 1984.

1. Fruit purple, purple-black, or black, 6–11 mm in diameter, hypanthium glabrous; plants usually glabrous or sparsely pubescent.
   2. Mature leaves glabrous on lower surface or rarely with a few scattered hairs, remaining green in autumn. . . . . 1. *A. melanocarpa*
   2. Mature leaves sparsely pubescent on lower surface, turning red in autumn. . . . . 2. *A. prunifolia*
1. Fruit red, 4–7 mm in diameter; plants usually densely pubescent. . . . . 3. *A. arbutifolia*

1. *A. melanocarpa* (Michx.) Elliott Black Chokeberry
Acid bogs, wet thickets, swamps, dry clearings, bluffs and cliffs. S. Lab. and Nfld. w. to Ont. and e. Minn., s. to N.J., Pa., mts. to S.C., n. Ga., and n. Ala., Fla., Tenn., Ark. and se. Mo. Robertson et al. (1991) reviewed generic delimitations of genera in the subfamily Maloideae (Rosaceae) and generally adopted a narrower circumscription of genera. Yet in the case of *Aronia*, they have merged this genus within their concept of the genus *Photina*, but according to (Pankhurst, 2014) DNA analyses have not supported that viewpoint (*Photina melanocarpa* (Michx.) Robertson & Phipps)

2. *A. prunifolia* (Marshall) Rehder Purple Chokeberry
Acid bogs, dune hollows, wet thickets and wet to dry clearings. Nfld. w. to Ont., Mich. and Wis., s. to Va., Ga., Ala. and Miss. *A. prunifolia* is morphologically intermediate between *A. melanocarpa* and *A. arbutifolia*, and Hardin (1973) has made a strong case to support the concept of a hybrid origin for *A. prunifolia* with these other two species as the parental taxa. His studies suggest that the hybrid taxon may be perpetuated by both facultative agamospermy as well as continued generation of new hybrids. Hardin (1973) placed *A. prunifolia* in synonymy with *A. melanocarpa*, however, we have followed Uttal's (1984) interpretation and recognize *A. prunifolia* as a species of hybrid origin, that forms independent, self-reproducing populations (Voss and Reznicek, 2012). (*A. arbutifolia* var. *atropurpurea* (Britton) Seymour)

3. *A. arbutifolia* (L.) Elliott Red Chokeberry Fig. 388
Acid bogs, savannas, swamps, thickets and moist ledges. Nfld. and N.S. w. to Que, and Ont. s. to N.Y., Pa., W.Va., Ky. and Ark., s. to Fla., Miss., La, Okla. and e. Tex.; most common along the coastal plain. (*A. arbutifolia* var. *glabra* Elliott)

2. *Physocarpus* (Ninebark)

Shrubs; stems with bark exfoliating in strips; leaves alternate, simple, often somewhat 3-lobed, palmately veined, stipules present; flowers small, borne in a corymb; petals white to slightly pinkish; hypanthium shallow; ovaries (1)3–5, superior; fruit a cluster of follicles.

1. *P. opulifolius* (L.) Maxim. Fig. 388
Stream banks, shores, wet pastures and thickets. Que. w. to Hudson Bay and Minn., s. to w. S.C., nw. Fla., Ala. Tenn. and Ark. This species is frequently planted as an ornamental and sometimes escapes from cultivation. (*P. opulifolius* var. *intermedius* (Rydb.) B. L. Rob.)

3. *Spiraea* (Spiraea)

Shrubs; leaves alternate, simple, stipules absent; flowers small, borne in a panicle or corymb; petals white, pink or rose; hypanthium shallow; ovaries usually 5, superior; fruit a follicle.

References: Kugel, 1958; Lis, 2014.

1. Leaves glabrous on both surfaces; petals white; follicles glabrous.
   2. Leaves more than 4 times as long as wide, finely serrate; inflorescence branches pubescent; flowers 5–8 mm across. . . . . 1a. *S. alba* var. *alba*
   2. Leaves less than 3 times as long as wide, coarsely serrate; inflorescence branches glabrous; flowers 3–5 mm across. . . . . 1b. *S. alba* var. *latifolia*
1. Leaves tomentose on lower surface; petals pink to rose; follicles pubescent. . . . . 2. *S. tomentosa*

Fig. 388. *Aronia arbutifolia*: a. branch with flowers; b. flower; c. branch with fruit; d. fruit (C&C). *Physocarpus opulifolius*: e. upper portion of branch with inflorescence; f. flower; g. fruit (B&B).

1. *S. alba* Du Roi Meadow-sweet Fig. 389

1a. *S. alba* var. *alba* Fig. 389
Meadows shorelines and low grounds W. Que, w. to Sask. and Alta., s. to Del., in mts. to w. N.C., ne. Tenn. and e. Ky., Ohio, Ill., Mo., N.D. and S.D. Kugel (1958) noted that this chiefly midwestern species and *S. latifolia* commonly hybridize to form intermediates where their ranges overlap, particularly in the northern Lower Peninsula of Michigan and northeastward into Ontario and western Quebec.

1b. *S. alba* var. *latifolia* (Aiton) Dippel Meadow-sweet Fig. 389
Moist low ground and uplands. Nfld. and Lab. w. to Mich. and Ont., s. to L.I., N.Y., and interior of Va. and N.C. Plants with larger flower but shorter panicles growing in alpine ravines and snowbank communities in the northern Appalachian mts. has been segregated as *S. alba* var. *septentrionalis* (Fernald) Seymour, but these populations are not clearly differentiated morphologically from var. *latifolia* and are not recognized by Lis (2014).

2. *S. tomentosa* L. Steeple-bush, Hardhack Fig. 389
Sandy shores, bogs, tamarack swamps, wet meadows, and wet low ground, usually acidic soils. N.S. and P.E.I. w. to s. Ont. and Minn., s. to S.C., Tenn., n. Ga., ne. Miss., La., Ark. and Kans.; Oreg. (*S. tomentosa* var. *rosea* (Raf.) Fernald)

4. *Rosa* (Rose)

Shrubs; stems with prickles; leaves alternate, pinnately compound, stipules usually adnate to petiole; flowers solitary or in corymbs; hypanthium globose (hip), enclosing numerous separate ovaries; fruit a cluster of achenes retained in the red hip.

REFERENCE: Lewis et al., 2014

---

1. Leaflets finely toothed, 12–25 teeth above middle, teeth along widest portion of leaflet 0.3–1.2 mm high (average ca. 0.5 mm); stipules 4 narrower, 2.5–5.5 mm wide.
   2. Flowers 2.5–5 cm across, petals 14–18 mm long; thorns just below stipules stout, usually somewhat recurved, internodal prickles absent; hip 2.5–3.5 mm long. . . . . 1. *R. palustris*
   2. Flowers 4–5 cm across, petals 19–23 mm long; thorns just below stipules very slender, scarcely differing from the numerous internodal prickles (reddish to purplish); hip 3–5 mm long. . . . . 2. *R. nitida*
1. Leaflets were coarsely toothed, 5–17 teeth above middle, teeth along widest portion of leaflet 0.6–2 mm high (average ca. 1 mm); stipules broad, 4–9 mm wide. . . . . 3. *R. virginiana*

---

1. *R. palustris* Marshall Swamp Rose Fig. 391
Swamps, bogs, marshes, shores of ponds and lakes and wet thickets. N.S. and N.B. w. to Wisc. and Iowa, s. to Fla., Miss. and Ark.

2. *R. nitida* Willd. Northeastern Rose Fig. 390
Bogs, wet shores and wet thickets. Nfld. s. Que. and s. Ont., s. to N.E., se. N.Y., e. N.J. and Ohio.

3. *R. virginiana* Mill. Virginia Rose Fig. 390
Swamps, shores and damp or dry thickets. Nfld. w. to s. Ont., s. to Va., N.C., Ga., Ala., Mo. and Ark.

5. *Dasiphora* (Shrubby Cinquefoil)

Perennial woody, herbs: Stems prostrate to upright; leaves deciduous or marcescent; stipules persistent adnate to petiole, blades bluntly ovate to obovate, margins entire; surfaces often densely strigose on veins and margins; inflorescences terminal; pedicels present; flowers; 4–7 mm in diameter, Sepals 5, petals 5 mostly yellow; fruits aggregated achenes, hypanthium persistent; sepals persistent.

REFERENCE: Ertter and Reveal, 2014.

1. *D. fruticosa* (L.) Rydb. Shrubby Cinquefoil, Golden Hardhack Fig. 391
Calcareous fens, wet meadows, barrens and ledges. Nfld. and s. Lab. w. to Nunavut, N.W.T. and Alask., s. to Pa., Ohio, n. Ind., n. Iowa, S.D., N.M. Ariz. and Calif. This is a widely cultivated shrub that has bountiful flowers; numerous horticultural cultivars have been produced including plants with pink or white flowered plants. (*P. fruticosa* subsp. *floribunda* (Pursh) Elkington (*P. fruticosea* L.; *P. fruticosa* var. *tenuifolia* Lehm.; *Pentaphylloides floribunda* (Pursh) Á. Löve)

Fig. 389. *Spiraea alba* var. *latifolia*: a. inflorescence (Ryan).
Spiraea tomentosa: b. inflorescence (OHIO).
Spiraea alba var *alba*: c. upper portion of branch (OHIO).

Fig. 390. *Rosa nitida*: a. flowering stems, showing internodal prickles; b. leaflet (Ryan). *Rosa virginiana*: c. fruiting stem, thorns at nodes; d. hips; e. leaflet (Ryan).

Fig. 391. *Rosa palustris*: a. flowering stem (OHIO).
Dasiphora fruticosa: b. flowering stem; c. vegetative stem (HCOT).
Comarum palustre: d. flowering stem; e. flower (F).

6. *Rubus* (Bramble)

Perennial; stems usually cane-like, or herbaceous; leaves alternate, mostly compound bisexual (ours dioecious), flowers 5-merous, petals white to pink or rosy-red; hypanthium shallow; ovaries numerous, superior; fruit an aggregate of drupelets.

1. Leaves simple, merely lobed and toothed, blades rotund to reniform. . . . . 1. *R. chamaemorus*
1. Leaves trifoliolate, leaflet broadly ovate to obovate or rhombic-ovate.
   2. Flowers solitary; petals deep rose-pink, 10–16(18) mm long. . . . . 2. *R. acaulis*
   2. Flowers 1–3(4); petals white or greenish-white (turning pinkish with age), 4–8 mm long. . . . . 3. *R. pubescens*

1. *R. chamaemorus* L. Bakeapple, Cloudberry Fig. 392
Coastal raised bogs and alpine bogs. Greenl. and Lab. w. to Alask., s. to coastal ne. Me. and mts. of w. Me. and n. N.H.; n. Ont., ne. Minn., Man. and B.C.; Montauk Point, L.I., N.Y. (population extirpated).

2. *R. acaulis* Michx. Dwarf Raspberry
Cold bogs, peaty meadows, fens and openings in conifer swamps. Baffin Is., Lab. and Nfld. w. to N.W.T. and Alask., s. to e. Que., Me., Ont., n. Mich., n. Minn., Man., Sask., mts. to Wyo., Colo., Wash. and Oreg.

3. *R. pubescens* Raf. Dwarf Raspberry
Bogs, White Cedar and Tamarack swamps and moist to wet woods. Lab. and Nfld. w. to Que., Ont., Man., Sask., N.W.T., Yuk. and B.C., s. to N.E., n. N.J., Pa., W.Va., n. Ohio, n. Ind., n. Ill., Iowa, Neb., mts. to Colo., Ida., Wash. and Oreg.

7. *Sanguisorba* (Burnet)

Perennial herbs; leaves pinnately compound; leaflets 7–17, serrate (ours) to pinnatifid; flowers numerous, borne in dense spikes or heads; sepals 4, petaloid; petals absent; hypanthium urn-shaped; ovary single, superior; fruit an achene.

1. *S. canadensis* L. American Burnet Fig. 392
Wet meadow, marshes, bogs, peaty soils, seeps and moist prairies. Lab., Nfld. and Que. w. to s. Mich., s. to N.J., Pa., Ohio and Ind., mts. to N.C., S.C., Ga., Ky. and Tenn.

8. *Geum* (Avens)

Perennial herbs; lower leaves mostly pinnate; cauline leaves 3-foliolate or 3-lobed; flowers few, solitary; hypanthium shallow; ovaries numerous, superior; fruit a globose cluster of achenes with plumose, geniculate styles.

1. *Geum rivale* L. Water Avens, Purple Avens Fig. 392
Wet meadows, bogs, wooded swamps, peaty slopes and shore. Greenl., Nfld. and Lab. w. to Nunavut and B.C., s. to N.E., n. N.J., Pa., W.Va., Ohio, n. Ill., Mo., S.D. Mont., Colo., N.M., Ida. and Wash. *Geum laciniatum* Murray and *G. macrophyllum* Willd. sometimes occur in wet sites.

9. *Comarum* (Marsh Cinquefoil)

Perennial herbs, rhizomatous and stoloniferous, Stems decumbent to ascending, horizontal stems often rooting at nodes; leaves often marcescent, cauline, alternate, pinnate to sub-palmate, leaflets (3)5–7, surfaces glabrous or finely appressed hairs; inflorescences, 1–10-flowered, terminal, hypanthium slightly cup-shaped, flowers, sepals 5, spreading to reflexed, petals 5 deep red to purple, rarely pink, carpels 30–100, glabrous; fruits aggregated achenes; hypanthium persistent, reflexed.

Reference: Ertter and Reveal, 2014.

1. *C. palustre* L. Marsh Cinquefoil, Marsh Five-finger Fig. 391
Marshes, wet meadows, bogs, and stream and pond margins, often floating or partly submerged. Greenl. and Lab. w. to Nunavut, N.W.T. and Alask., to n. N.J., Pa., Ohio, n. Ill., n. Iowa, N.D., Wyo., Colo, Utah and Calif. (*Potentilla palustris* (L.) Scop.; *P. palustris* var. *parviflora* (Raf.) Fernald & Long; *P. palustris* var. *villosa* (Pers.) Lehm.)

10. *Potentilla* (Cinquefoil)

Perennial herbs or sometimes shrubs; stems erect or creeping; leaves pinnately or palmately compound; flowers solitary or in cymes; calyx subtended by bractlets alternating with sepals; hypanthium shallow; ovaries numerous, superior; fruit a head of achenes, usually enclosed in persistent calyx.

References: Elven and Murray, 2014; Rousi, 1965.

Fig. 392. *Rubus chamaemorus*: a. habit; b. aggregation of druplets; c. stony endocarp (B&B).
Geum rivale: d. habit; e. achene (HCOT).
Sanguisorba canadensis: f. leaf and portion of stem; g. inflorescence; h. flower; i. achene enclosed in 4-angled calyx; j. achene (B&B).

1. Epicalyx bractlets as long or longer than the sepals, usually 2-fid or toothed; leaves silvery-sericeous on lower surface, densely hairy, with both curly cottony hairs and with long straight hairs on and between the veins; achenes with a dorsal groove; plants of freshwater, inland site. . . . . 1a. *P. anserina* subsp. *anserina*
2. Epicalyx bractlets shorter than sepals, rarely 2-fid or toothed; leaves dull white-pubescent on lower surface, densely hairy, with hairs mostly curly, long hairs absent or sparse on veins; achenes lacking a dorsal groove; plants of saline coastal sites, seldom inland.. . . . . 1b. *P. anserina* subsp. *pacifica*

1. *P. anserina* L Silverweed

1a. *P. anserina* subsp. *anserina* Silverweed Fig. 393
Wet to damp sandy and gravelly shores. Nfld. w. to Alask., s. to n. N.E., N.Y., Ind., Iowa, N.M., and Calif. The Silverweeds are placed in the genus *Argentina* by some authors. (*Argentina anserina* (L.) Rydb.)

1b. *P. anserina* subsp. *pacifica* (Howell) Rousi
Pacific Silverweed
Salt marshes and wet coastal sands and gravels. Coastal, Greenl. and Lab. to N.E. and L.I., N.Y.; Alask. s. to Calif. Elven and Murray (2014) recognize subsp. *groenlanica* Tratt., another taxon of coastal saline sites that occurs chiefly in the Arctic, having glabrous leaves (long hairs rarely along veins) which occurs as far south as Lab. and might be encountered in Nfld. (*Argentina egedii* (Wormsk.) Rydb.; *P. pacifica* Howell; *P. egedii* subsp. *pacifica* (Howell) L. A. Sergienko)

Rhamnaceae / Buckthorn Family

1. Leaves dark green to olive-green on both sides, not glossy, margins with small, serrate to crenulate teeth; flowers unisexual, sepals 5, petals 0; fruits with stones 3; low shrubs; buds with scales glabrous, but having ciliate margins. . . . . 1. *Rhamnus*
1. Leaves glossy on upper surface, dull green below, margins entire; flowers bisexual, sepals 5, petals 5, yellowish; drupes with stones 2(3); small trees; buds naked, lacking scales, pubescent. . . . . 2. *Frangula*

1. *Rhamnus* (Buckthorn)

Shrubs or small trees, deciduous; leaves alternate (sometimes subopposite), lanceolate-oblong to elliptic to lanceolate-ovate, margins serrate, veins pinnate, arching; flowers unisexual (rarely bisexual), petals 0 or 4(5) (ours absent); fruit a black drupe, stones 2–2(4) (ours 3).

Reference: Nesom and Sawyer, 2016.

1. *R. alnifolia* L'Her. Alder-leaved Buckhorn Fig. 394
Swamps, meadows and low woods; often calcareous soils. Nfld. w. to B.C., s. to n. and w. N.E., n. N.J., Pa., W.Va., Tenn., n. Ind., S.D., Neb., Utah, Wyo. and Calif.

Rhamnus cathartica L., another aggressive Eurasian woody plant that has widely invaded many woodlands and meadows, and can occasionally be found in swamps or occur at the margins of wetlands.

2. *Frangula* (Buckthorn)

Shrubs or small trees, deciduous; leaves alternate, pinnately veined (nearly straight, parallel), margins entire; flowers bisexual; sepals 5 (rarely 4), petals 5 (rarely 4), yellowish, hooded; stamens (4)5; fruit a black drupe, stones 2–3(4) (ours mostly 2).

Reference: Sawyer and Nesom, 2016.

1. *F. alnus* Mill. Glossy Buckthorn Fig. 394
Bogs, fens, tamarack and white cedar swamps, shores, river margins, ditches and thickets. S. Que., N.S. and N.B. w. to Sask., s. to N.E., N.J., Va., Tenn., Iowa, Neb., Colo. and Ida. This European plant is highly aggressive, invading many wetland sites and one of the few becoming established in bogs and fens. (*R. frangula* L.)

Ulmaceae / Elm Family

1. Flowers bisexual, fruit a samara; bark ridged and furrowed or scaly, not exposing reddish inner bark. . . . . 1. *Ulmus*
1. Flowers usually unisexual, with a few bisexual flowers usually on same inflorescence; fruits nutlike; bark scaly and flaky, exposing reddish inner bark. . . . . 2. *Planera*

Fig. 393. *Potentilla anserina* subsp. *anserina*: a. habit; b. stipules; c. young achene with basal part of style; d. mature achene (Mason).

Fig. 394. *Rhamnus alnifolia*: a. branch with fruits; b. leaf; c. branch with flowers; d. winter twig (Ryan); e. branch with flowers; f. flower; g. fruits (B&B).
Frangula alnus: h. branch with flowers; i. flower; j. fruit (B&B).

1. *Ulmus* (Elm)

Tree; bark ridged or furrowed; leaves simple, alternate, lateral veins pinnate, straight and parallel from midrib, margins serrate or doubly serrate; inflorescence axillary, racemes or fascicles, flowers emerging from terminal winter bud, pendent on long pedicels, bisexual; calyx united at least basally usually 4–5 lobed; corolla lacking; pistil 1, ovary superior, 2-carpellate; fruit an ovoid samara.

REFERENCE: Sherman-Broyles, 1997.

1. *Ulmus americana* L. American Elm Fig. 395
Floodplain forests, swamp forests, wet to moist rich deciduous woods, pastures and old fields. N.S. and P.E.I. w. to Ont., Man. and Sask., s. to Fla. and e. Tex.

2. *Planera* (Planer-tree, Water-elm)

Small tree or shrub, deciduous; leaves alternate, simple, ovate to lanceolate, base somewhat oblique, margins serrate; flowers bisexual and unisexual (plants polygamomonoecious), emerging from winter buds, in axillary clusters; fruit an ellipsoidal, leathery drupe.

REFERENCE: Barker, 1997.

1. *P. aquatica* (Walter) J. E. Gmel. Water-elm, Planer tree Fig. 396
Swamps, riverbanks, stream margins and floodplain forests. Coastal plain, ne. N.C. s. to La. and e. Tex.; Mississippi embayment n. to Ark., sc. Okla., sc. Mo., w. Tenn., s. Ill. and w. Ky.

Cannabaceae / Hemp Family

1. *Celtis* (Hackberry)

Medium to large tree, bark gray, smooth and frequently bearing corky-warty growths; leaves with principal lateral veins arising from the base, blades deltate to ovate, asymmetric at base, margins entire or serrate (ours); inflorescences cymes or fascicles; flowers usually unisexual, staminate and pistillate flowers on same plant; ovaries sessile ovoid; fruits fleshy drupes persisting in autumn.

REFERENCE: Barker, 1997.

1. *Celtis occidentalis* L. Hackberry Fig. 395
Rich moist soil along streams or in floodplains. Me. and Que. w. to Ont., Man. Wyo., Mont. s. to S.C., Ga, Miss., Okla. and nw. Tex.; Colo., N.M. and Utah.

Urticaceae / Nettle Family

REFERENCE: Bufford, 1997.

1. Flowers and fruits in axillary spikes; sepals united, enclosing achene (fig. 397f); stem firm, opaque. .... 1. *Boehmeria*
1. Flowers and fruits in axillary panicles; sepals separate, not enclosing achene (fig. 398b); stem watery-succulent, somewhat translucent. .... 2. *Pilea*

1. *Boehmeria* (False Nettle)

Perennial herbs (ours), shrubs, or small trees; leaves simple, opposite (ours) or alternate, stipulate; plants monoecious or dioecious; flowers in axillary spikes, greatly reduced; petals absent; sepals united, enclosing the ovary; fruit an achene, closely subtended by persistent calyx.

1. *B. cylindrica* (L.) Sw. Bog-hemp, False Nettle Fig. 397
Marshes, marshy shores, wet woods, bogs, swamps and stream margins. N.B., s. Que., and s. Ont. w. to s. Minn. and se. S.D., s. to Fla. and Tex.; N.M., Utah and Ariz.

Fig. 395. *Ulmus americana*: a. leafy branch; b. flowering branch; c. fruiting branch; d. branch with winter buds; e. fruit (Sargent).
Celtis occcidentalis: f. leafy branch with fruit (B&B).

Fig. 396. *Planera aquatica*: a. vegetative branch; b. leaves; c. flowering branch; d. fruit with fleshy projections and staminate flower below; e. staminate flower (C&C).

Fig. 397. *Boehmeria cylindrica*: a. portion of plant; b. staminate buds and cluster of pistillate flowers; c, e. clusters of pistillate flowers/fruits in different stages; d. staminate flower; f. fruit (C&C).

2. *Pilea* (Clearweed, Richweed, Coolwort)

Annual or perennial herbs; leaves opposite, stipulate; plants monoecious or dioecious; flowers in axillary panicles, greatly reduced; petals absent; sepals united basally, deeply lobed; fruit an achene, subtended by persistent calyx.

1. Leaves lustrous, translucent; petioles of larger leaves one third as long or as long as the blade length; leaf base cuneate (fig. 398a); achenes 0.7–1(1.1) mm wide, straw-colored to light green, often with purple markings, smooth. . . . . 1. *P. pumila*
1. Leaves somewhat dull, opaque; petioles of larger leaves one fifth to half the blade length; leaf base rounded; achenes 1.1–1.4 mm wide, black, roughened. . . . . 2. *P. fontana*

1. *P. pumila* (L.) A. Gray Clearweed Fig. 398
Wet woodlands, floodplains, swamps, margins of streams and wet openings. N.B. and Que. w. to Minn. and N.D., s. to n. Fla., La. and e. Tex.

2. *P. fontana* (Lunell) Rydb.
Swamps, wet shores and boggy or springy soils. Vt., N.Y. and Ont. w. to Minn. and N.D., s. to Mass., N.J., Del., Va., Ky., Ill., Iowa and e. Neb.; coastal plain, s. to Ga., n. Fla. and Ala.

Fagaceae / Beech Family

1. *Quercus* (Oak)

Trees and shrubs (monoecious), deciduous or evergreen; twigs with winter buds clustered at apex; leaves lobed or unlobed, thin or leathery, margin entire, toothed (often bristle-tipped); staminate flowers borne in pendant catkins (emerging with the leaves); pistillate flowers solitary or in few-flowered clusters at the base of the season's developing shoot, each subtended by a cupule of bracts; pistils with 3 styles; fruit a nut (acorn).

Reference: Nixon, 1997.

1. Leaves deeply lobed with U-shaped sinuses, teeth bristle-tipped (fig. 399a); blades lustrous green, glabrous on upper surface, lower surface pale or dull green, with pubescence in axils of veins; mature bark smooth or deeply furrowed, not scaly. . . . . 1. *Q. palustris*
1. Leaves shallowly lobed, lobes rounded, or if pointed, then lacking bristle-tips (fig. 399h,i); blades glabrous on upper surface, gray to pale green on lower surface, pubescent with stellate hairs; mature bark scaly or papery, rarely deeply furrowed.. . . . . .2. *Q. bicolor*

1. *Q. palustris* Münchh. Pin Oak Fig. 399
Floodplains, bottomlands, edges of periodically flooded sites and poorly drained upland clay soils. Mass., Ont. and N.Y. w. to s. Mich., e. Iowa and Mo., s. to N.C., Tenn. and Ark. Pin Oak is widely planted along streets throughout eastern U.S.

2. *Q. bicolor* Willd. Swamp White Oak Fig. 399
Swamp forests, moist slopes and poorly drained sites. Me., Que., Ont., s. Mich. and s. to Tenn., N.C., n. Ala. and Mo.

Betulaceae / Birch Family

1. Bracts of pistillate catkins thick, becoming hard and woody, long persistent; buds stalked (fig. 400e) (sessile in *A. viridis*). . . . . .1. *Alnus*
1. Bracts of pistillate catkins thin, deciduous at time of fruit dispersal; buds sessile. . . . . 2. *Betula*

1. *Alnus* (Alder)

Shrubs or trees, deciduous; winter terminal buds present, axillary buds usually stalked; leaves alternate, simple; flowers unisexual, borne in catkins; staminate bracts subtending 3(6) flowers; calyx usually 4-parted; stamens 2–4; pistillate bracts (woody) subtending 2 flowers; fruit a nutlet, usually thin-margined, occasionally winged.

References: Furlow, 1979, 1997.

1. Axillary buds sessile (fig. 400a) with 5 or more unequal bud scales; pistillate inflorescences emerging with leaves; fruit winged. . . . . 1. *A. viridis*
1. Axillary buds stalked (fig. 400e) with 2 or 3 equal bud scales; pistillate inflorescences emerging long before leaf expansion or late summer to early autumn (*A. maritimus*); fruit thin-margined, but not distinctly winged.

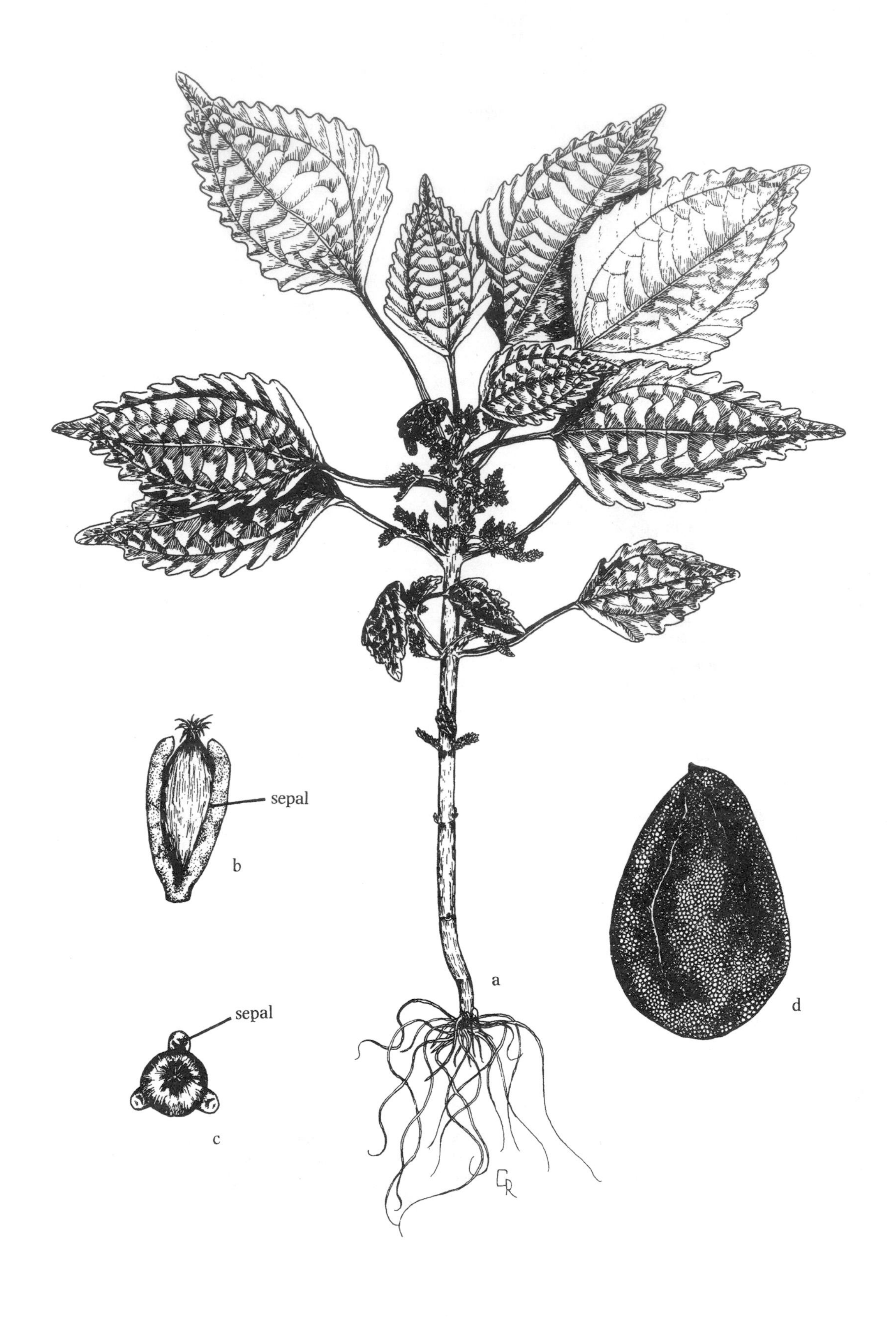

Fig. 398. *Pilea pumila*: a. habit; b. pistillate flower, side view; c. pistillate flower, top view; d. achene (G&W).

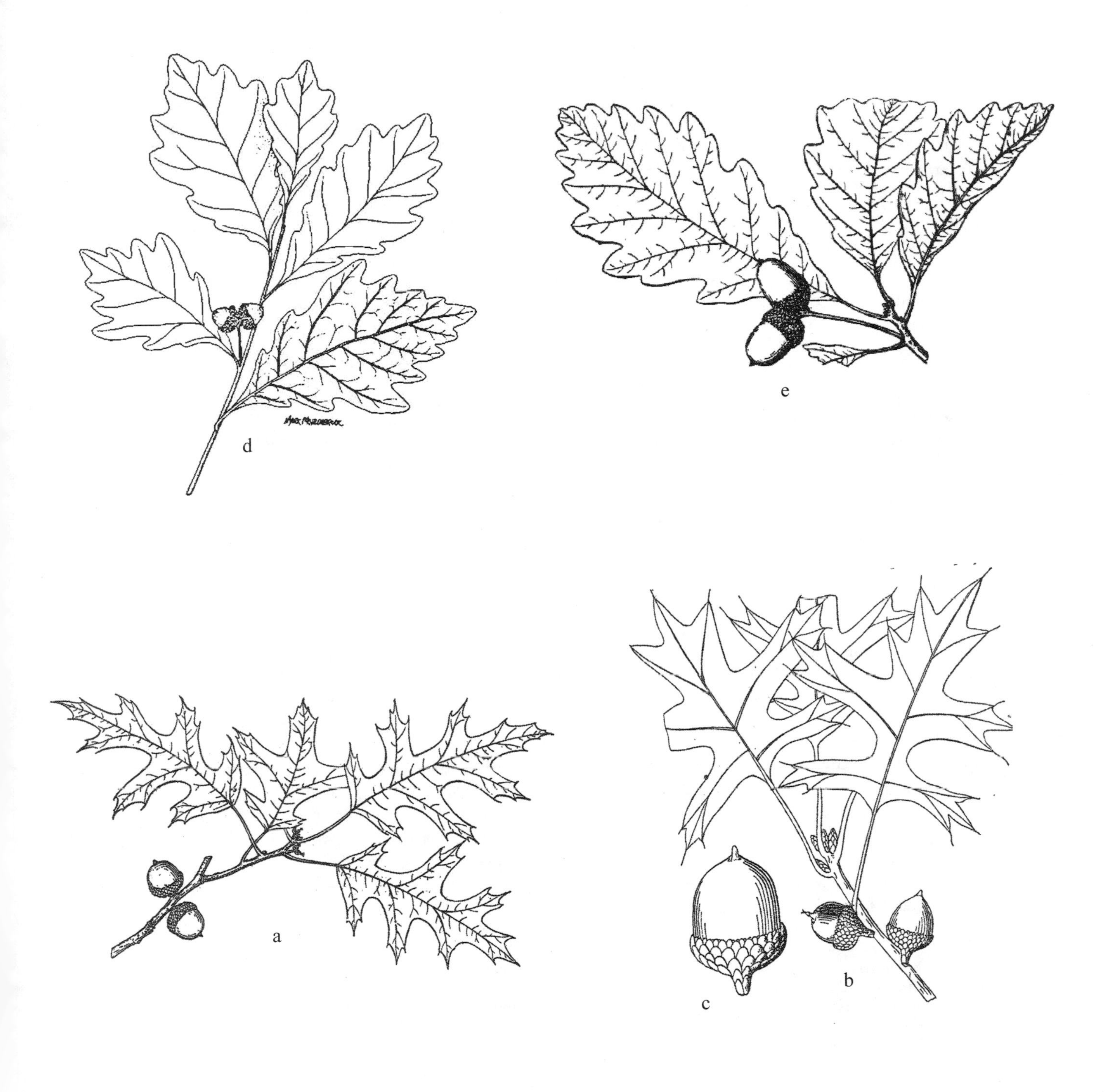

Fig. 399. *Quercus palustris*: a. leafy branch with acorn (Sargent, 1905); b. leafy branch with acorn; c. acorn (B&B). *Quercus bicolor*: d. leafy branch with acorns (USDA-NRCS); e. leafy branch with acorn (Sargent, 1905).

2. Leaves usually orbicular (fig. 400d); trees (to 20 m). . . . . 2. *A. glutinosa*
2. Leaves ovate, elliptic to obovate, but not orbicular; shrubs.
 3. Lateral veins of leaves extending into the marginal teeth; leaves with 9–12 lateral veins on each side; pistillate inflorescences in racemose clusters.
  4. Leaves narrowly ovate to elliptic; leaf margins doubly serrate, major teeth coarse, regular (secondary teeth larger). . . . . 3. *A. incana*
  4. Leaves broadly elliptic to obovate; leaf margins simply serrulate (lacking larger secondary teeth), occasionally somewhat serrate. . . . . 4. *A. serrulata*
 3. Lateral veins of leaves interconnecting with other veins near the margins, seldom extending into the marginal teeth leaves with 5–8 lateral veins on each side; inflorescences solitary. . . . . 5. *A. maritima*

---

1. *A. viridis* (Chaix) DC. Green Alder, Mountain Alder Fig. 400
Singly or forming dense thickets along streams, shores, coasts and in bogs. Greenl., Nfld. and Lab. w. to N.W.T. and Alask., s. to N.E., n. N.Y., Ont., n. Mich., n. Minn. and Sask.; disjunct to mts. of Pa. w. N.C. and e. Tenn. Our taxon is subsp. *crispa* (Aiton) Turrill. (*A. crispa* (Aiton) Pursh)

2. *A. glutinosa* (L.) Gaertn. Black Alder, European Alder Fig. 400
Riverbanks, shores and wet places. Occasional, N.S., Vt. and e. Mass. w. to s. Ont. and s. Mich. and c. Minn., s. to N.J., Del., n. Va., Ky., Tenn., Ala., Mo. and Kans. Escaped from cultivation and locally natzd.

3. *A. incana* (L.) Moench Speckled Alder, Tag Alder, Swamp Alder, Hazel Alder, Hoary Alder Fig. 400
Lakeshores, stream margins, borders of bogs and swamps, swales and wet fields. Lab. and Nfld. w. to n. Ont., Man. and Sask., s. to n. N.J., Pa., Md., Va., W.Va., Ohio, Ind., Ill., ne. Iowa and e. N.D. Our taxon is subsp. *rugosa* (DuRoi) R. T. Clausen. (*A. rugosa* (DuRoi) Spreng.)

4. *A. serrulata* (Aiton) Willd. Smooth Alder, Common Alder, Tag Alder, Red Alder Fig. 401
Wet soils, lakeshores, margins of streams, bogs, sloughs and swampy sites. N.S., N.B. and s. Que. sw. to N.Y., Ohio, Ky., Mo., and Okla., s. to n. Fla., Ala., La. and e. Tex.

5. *A. maritima* Muhl. ex Nutt. Seaside Alder, Brook Alder, Oklahoma Alder Fig. 401
Margins of ponds and streams. Delmarva Pen., Del. and Md.; disjunct to sc. Okla.

2. *Betula* (Birch)

Trees or shrubs, deciduous; winter buds sessile; leaves alternate, simple; flowers unisexual, borne in catkins; staminate bracts subtending 3 flowers of 2 stamens each; pistillate bracts usually 3-lobed, subtending 2 or 3 flowers; fruit a nutlet, usually with 2 thin wings.

Reference: Furlow, 1997.

---

1. Trees; leaves with 5–12 lateral veins on each side of midrib; trunks shaggy, bark irregularly exfoliating in thin layers. . . . . 1. *B. nigra*
1. Shrubs, erect, matted, or depressed; leaves with 2–3 lateral veins on each side of midrib; bark not exfoliating.
 2. Plants erect, 0.5–3 m tall; leaves 0.8–4(7) cm long; bracts of pistillate catkins 3-lobed (fig. 402b); nutlets winged (fig. 402g).
  3. Leaf blade 2.5–5(7) cm long; twigs not glandular-warty, heavily pubescent; lower leaf surface whitened. . . . . 2. *B. pumila*
  3. Leaf blades 1–2(4) cm long; twigs glandular-warty, pubescent to glabrous; lower leaf surface green. . . . . 3. *B. glandulosa*
 2. Plants with slender; creeping, subterranean stems, ascending branches 0.1–0.6 m long; leaves 0.5–1 cm long, bracts of pistillate catkins unlobed (fig. 403h); nutlets wingless (fig. 403g). . . . . 4. *B. michauxii*

---

1. *B. nigra* L. River Birch, Red Birch Fig. 402
Swamps, riverbanks and floodplains. S. N.H. and e. Mass. w. to N.Y., Ohio, Ind., Wisc. and Minn., s. to n. Fla. and s. Tex.

2. *B. pumila* L. Swamp Birch, Bog Birch, Dwarf Birch Fig. 402
Bogs, swamps, calcareous fens, muskegs, and lakeshores. Nfld. and Lab., w. to Ont., Man. Sask., N.W.T., Yuk., e. B.C. s. to N.J., Pa., Ohio, n. Ill., Wisc., Minn., sw. S.D., Iowa, Mont., Wash. and n. Calif.; mts s. to sw. Wyo. and Colo.

3. *B. glandulosa* Michx. Dwarf Birch, Resin Birch Fig. 403
Bogs, muskegs and wet alpine slopes. Greenl. and Lab. w. to Alask., s. to higher mts. of Me., n. N.H., and n. N.Y., n. Ont., Man., w. S.D., mts. to Colo., n. N.M., Utah, Ida., Wash., Oreg. and n. Calif.

4. *B. michauxii* Spach Newfoundland Dwarf Birch Fig. 403
On pool margins of sphagnum bogs and peaty meadows, heaths and wet peaty barrens. Lab. and n. Que. s. to Nfld. and N.S.

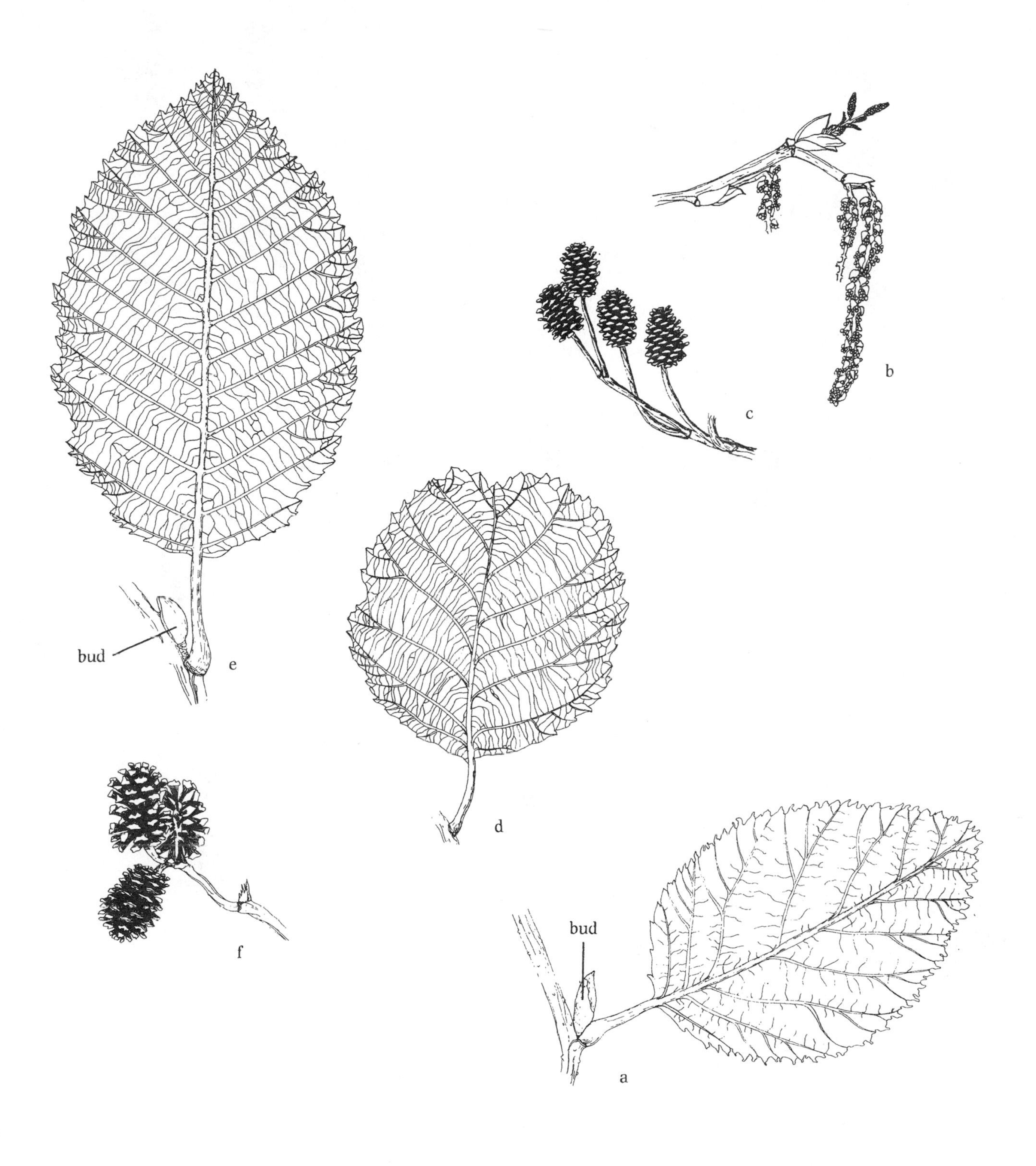

Fig. 400. *Alnus viridis*: a. leaf and bud; b. branch with staminate catkins; c. pistillate catkins of previous year (F).
Alnus glutinosa: d. leaf (F).
Alnus incana: e. leaf and bud; f. pistillate catkins of previous year (F).

Fig. 401. *Alnus serrulata*: a. branch with staminate catkins (K&G); b. leaves (C&C); c. staminate catkin (C&C); d. pistillate catkins (F); e. pistillate catkin (C&C).
Alnus maritima: f. branch with pistillate catkins; g. branch with staminate catkins (C&C).

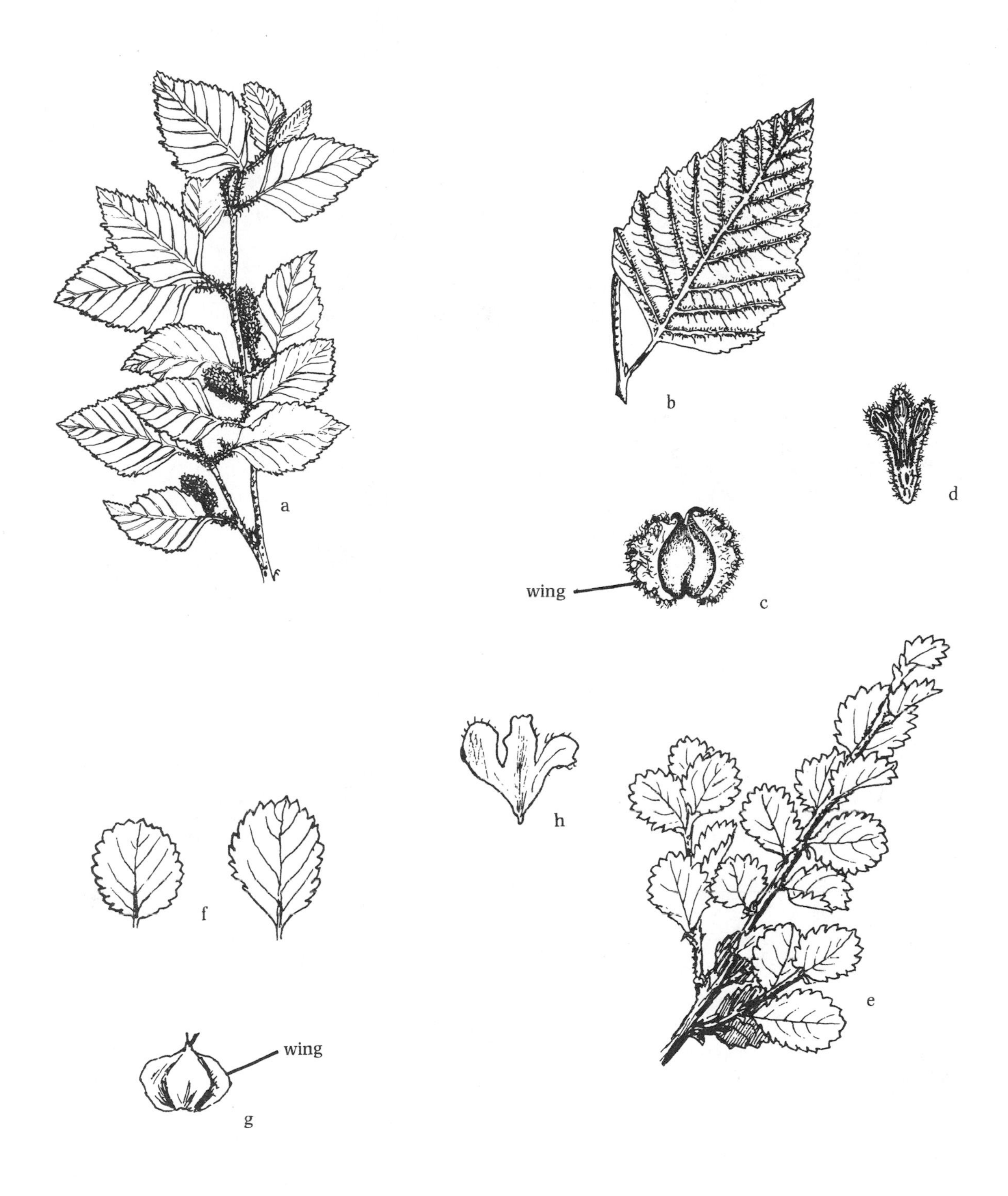

Fig. 402. *Betula nigra*: a. portion of branch with catkins; b. leaf; c. winged nutlet; d. bract of female catkin (C&C). *Betula pumila*: e. portion of branch; f. leaves; g. winged nutlet; h. bract of female catkin (Ryan).

Fig. 403. *Betula glandulosa*: a. portion of branch with male catkin (Crow); b. leaf (Ryan); c. winged nutlet (Ryan); d. bract of female catkin (Ryan).
Betula michauxii: e. portion of branch; f. leaves; g. nutlet; h. bract of female catkin (Ryan).

Myricaceae / Wax-myrtle Family

1. *Myrica* (Sweet Gale, Wax-myrtle, Bayberry)

Woody, shrubs (ours) or small trees; leaves evergreen or deciduous, alternate, resin-dotted and aromatic, stipules lacking; plants usually dioecious; staminate flowers in catkins, from axillary buds; pistillate flowers in short globose or cylindric catkins, from axillary buds; fruit a nutlet, in globose or ovoid clusters, often waxy or resin-dotted.

References: Bornstein, 1997; Elias, 1971.

---

1. Leaves with small resinous dots, but lacking punctate dots; catkins borne at summit of previous year's growth (emerging before leaves); pistillate bracts persistent around nutlets, forming wings, fruiting catkins bur-like; fruit a nutlet, resin-dotted, but not covered with white wax. . . . . 1. *M. gale*
1. Leaves with both small resinous dots and punctate dots; catkins borne scattered on previous year's growth, and often in leaf axils of previous year's evergreen leaves; pistillate bracts deciduous, fruiting catkins not bur-like; fruit a nut-like drupe covered with white wax.
   2. Leaves evergreen, narrow, oblanceolate, densely glandular on both surfaces; fruits 2–3.5(4) mm in diameter. . . . . 2. *M. cerifera*
   2. Leaves evergreen or deciduous, broader, ovate to elliptic, densely glandular on lower surface, but sparsely glandular or lacking glands on upper surface; fruits (2)3–5.5 mm in diameter.
      3. Leaves deciduous, margins distinctly revolute; twigs whitish-gray; inflorescences emerging with new leaves, borne on older wood, always below leafy tips; fruits 3.5–5.5 mm in diameter. . . . . 3. *M. pensylvanica*
      3. Leaves evergreen, margins not revolute; twigs black; inflorescences emerging before new leaves, borne frequently in leaf axils of previous year's evergreen leaves or on leafless twigs; fruits (2)3–4.5 mm in diameter. . . . . 4. *M. heterophylla*

---

1. *M. gale* L. Sweet Gale, Meadow-fern Fig. 404
Shallow water, swamps and shores. Lab. and Nfld. w. to N.W.T. and Alask., s. to N.Y., N.J., ne. Pa., Mich., n. Wisc., ne. Minn., Man., Sask., Alta., B.C., Wash and Oreg.; w. N.C.

2. *M. cerifera* L. Wax-myrtle, Southern Bayberry Fig. 404
Fresh or slightly saline shores, banks, swales, swamps, marshes, sand flats and pinelands. Coastal plain, s. N.J., Del., Md. and Va. s. to Fla., w. to e. Tex.; inland n. to se. Okla. and Ark.; Ber., W.I. (*Morella cerifera* (L.) Small)

3. *M. pensylvanica* Mirb. Northern Bayberry, Tallow Bayberry, Swamp Candleberry Fig. 404
Sand dunes and dry to wet sand flats, borders of salt marshes and shores. Chiefly coastal, s. Nfld., N.S., e. Que., w. to s. Ont., and s. Mich., s. to N.E., N.J., Del., se. Va., ne. N.C., Pa. and ne. Ohio. This species is a commonly planted landscape shrub and some populations represent escapees, especially in disturbed sites, whereas others in essentially undisturbed sites are likely to be native. (*Morella pensylvanica* (Mir.) Kartesz)

4. *M. heterophylla* Raf. Bayberry, Wax-myrtle
Bogs, pocosins, wet pine savannas and flatwoods. Coastal plain and piedmont, N.J., Pa., Va., and N.C., s. to c. Fla., w. to Ala., Miss. and Tex. Haines et al. (2011) treats the waxy-fruited group in the genus *Morella* and treats both *Myrica pensylvanica* and *Myrica heterophylla* as synonyms under *Morella caroliniensis* (Mill.) Small.

Onagraceae / Evening-primrose Family

Reference: Wagner et al., 2007.

---

1. Petals white, rose, pink, violet or purple; calyx segments deciduous after flowering; fruit several times longer than wide.
   2. Leaves mostly alternate, petals entire; stamens 8, mostly equal in length, in 1 whorl. . . . . 1. *Chamaenerion*
   2. Leaves opposite at base; petals notched; stamens 8, in unequal lengths, in 2 whorls. . . . . 2. *Epilobium*
1. Petals yellow or absent; calyx segments persistent in fruit up to about twice as long as wide. . . . . 3. *Ludwigia*

---

1. *Chamaenerion* (Fireweed)

Perennial, rhizomatous often forming colonies; stems unbranched, strigose or glabrous; leaves alternate, mostly fine-toothed; inflorescence a raceme; flowers strongly nodding (reflexed) in bud; flowers in long, showy terminal racemes; hypanthium (floral tube) not prolonged beyond ovary; sepals 4, spreading; petals 4, entire, clawed at base, stamens 8; floral tube absent; ovary inferior; fruit an elongate capsule, dehiscing by 4 valves; seeds with persistent tuft of white hairs.

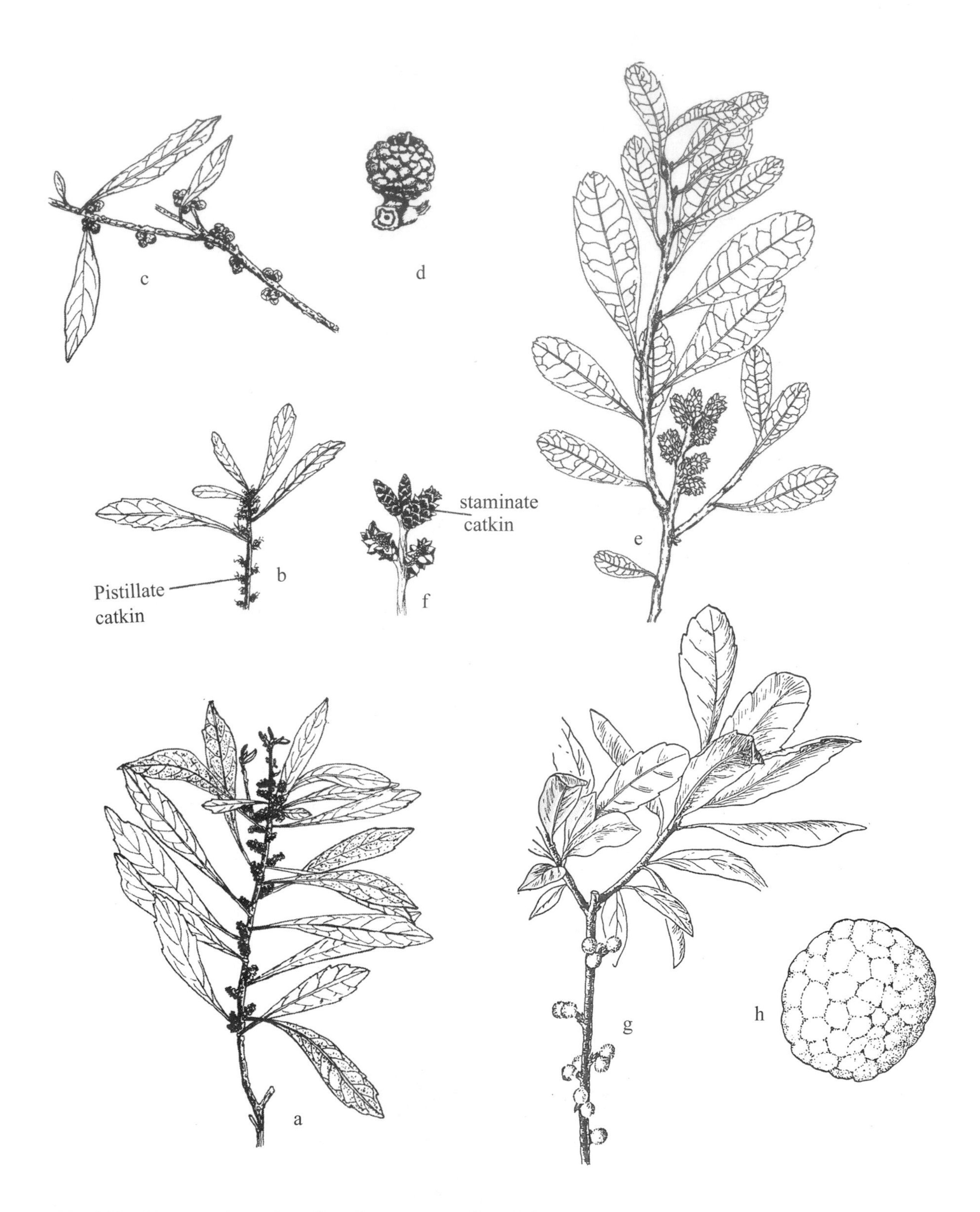

Fig. 404. *Myrica cerifera*: a. branch with staminate catkins; b. branch with pistillate catkins; c. fruiting branch; d. fruit (C&C).
Myrica gale: e. branch; f. branch with staminate catkins (F).
Myrica pensylvanica: g. branch with leaves and fruit on older portion of stem; h. fruit with waxy coating (Gleason).

According to Sennikov (2011) the generic name *Chamaenerion* Ség., published in 1754, is the correct name for the group of species with somewhat zygomorphic flowers and alternate leaves when it is segregated from the genus *Epilobium* (in the broader sense); the name *Chamerion* (Raf.) Raf. ex Holub, established in 1972 therefore becomes a heterotypic synonym.

1. *C. latifolium* (L.) Holub Dwarf Fireweed, River-beauty Fig. 405
River gravels, stream margins and damp slopes. Greenl. and Lab. w. to N.W.T. and Alask., s. to Nfld., Gaspé Pen., Que., Ont., Man., mts. to sw. S.D., Colo., n. Utah, n. Nev. and Calif.

2. *Epilobium* (Willow-herb)

Mostly perennial herbs; cauline leaves mostly alternate (or sometimes upper ones opposite); flowers 4-merous, single or few to many in terminal racemes; hypanthium (floral tube) present, extending beyond ovary, short; petals white, rose, pink, violet, or purple, notched at apex; stamens 4, ovary inferior; fruit an elongate, 4-valved capsule, seeds with tuft of hairs.

References: Stuckey, 1970; Ugent, 1962.

1. Petals (9)11–17 mm long; leaves strongly villous-hirsute on both surfaces. .... 1. *E. hirsutum*
1. Petals less than 10 mm long; leaves glabrous or pubescent to puberulent, but not villous-hirsute.
  2. Leaf margins revolute, entire to slightly wavy.
    3. Stems softly pubescent, with straight spreading hairs. .... 2. *E. strictum*
    3. Stems glabrous to minutely pubescent, with appressed or incurved hairs.
      4. Leaves densely pubescent with incurved hairs on upper surface. .... 3. *E. leptophyllum*
      4. Leaves glabrous or with few remote incurved hairs on upper surface. .... 4. *E. palustre*
  2. Leaf margins not revolute, toothed.
    5. Tuft of hairs (coma) at seed apex brown. .... 5 *E. coloratum*
    5. Tuft of hairs (coma) at seed apex white.
      6. Flowers 3–6 mm long; petals white or pale pink; fruiting pedicels 0.5–3 cm long; coma falling early. .... 6a. *E. ciliatum* subsp. *ciliatum*
      6. Flowers 4–9 mm long; petals pink-lilac; fruiting pedicels 0.2–1.5 cm long; coma persistent. .... 6b. *E. ciliatum* subsp. *glandulosum*

1. *E. hirsutum* L. Great Hairy Willow-herb Fig. 406
Meadows, marshes, damp stream banks and waste places. N.S., s. Que. and s. Ont. w. to Mich. and e. Wisc., s. to s. N.E., N.Y., Md., Ky. and n. Ill.; B.C., Wash. and n. Oreg.; natzd. from Eur., rapidly spreading.

2. *E. strictum* Muhl. ex Spreng. Downy Willow-herb
Bogs, swamps, meadows, marshy shores and margins of springs. Gaspé Pen., Que., w. to Ont. and Minn., s. to N.S., N.E., Md., n. Ohio, n. Ill. and Iowa.

3. *E. leptophyllum* Raf. Linear-leaved Willow-herb, Bog Willow-herb
Low ground, wet shores, stream margins, swamps, wet meadows and ditches. Nfld., Lab., and Que., w. to Man., N.W.T., B.C. and Alask., s. to N.S., N.E., N.C., Tenn., Okla., scattered in mts. to N.M., Utah and Calif.

4. *E. palustre* L. Marsh Willow-herb Fig. 406
Bogs, wet meadows, limestone barrens and swales. Greenl. and Lab. w. to N.W.T. and Alask., s.to ne. Mass., N.Y., n. N.J., ne. Pa., n. Mich., Minn., Kans., Colo., n. N.M., n. Utah, n. Nev. and n. Calif.

5. *E. coloratum* Bichler Purple-leaved Willow-herb Fig. 406
Marshes swamps, meadows and bogs. Nfld., Lab., s. Que and Ont., w. to S.D. s. to Ga., Ala., Tenn., Ark., Okla. and n. Tex.

6. *E. ciliatum* Raf. Fringed Willow-herb

6a. *E. ciliatum* subsp. *ciliatum* Fig. 406
Springy areas, wet rocks and marshes. Nfld. w. to B.C., S.C., Tenn., Iowa, Tex., N.M., Ariz. and Calif. (*E. americanum* Hausskn.; *E. adenocaulon* Hausskn.; *E. adenocaulon* var. *perplexans* Trel.)

6b. *E. ciliatum* subsp. *glandulosum* (Lehm.) Hoch & Raven Fig. 405
Damp thickets, marshes and coastal shores. S. Lab. and Nfld. w. to N.W.T. and Alask., s. to N.E., n. N.J., Ohio, n. Wisc., Minn., S.D., n. N.M., Ariz. and Calif. (*E. glandulosum* Lehm.)

Several species, such as *Epilobium alpinum*, *E. anagallidifolium*, *E. davuricum*, *E. hornemannii*, *E. leptocarpum*, and *E. saximontanum*, sometimes occur in wet sites, especially in wet peaty alpine habitats in the northern portion of our range.

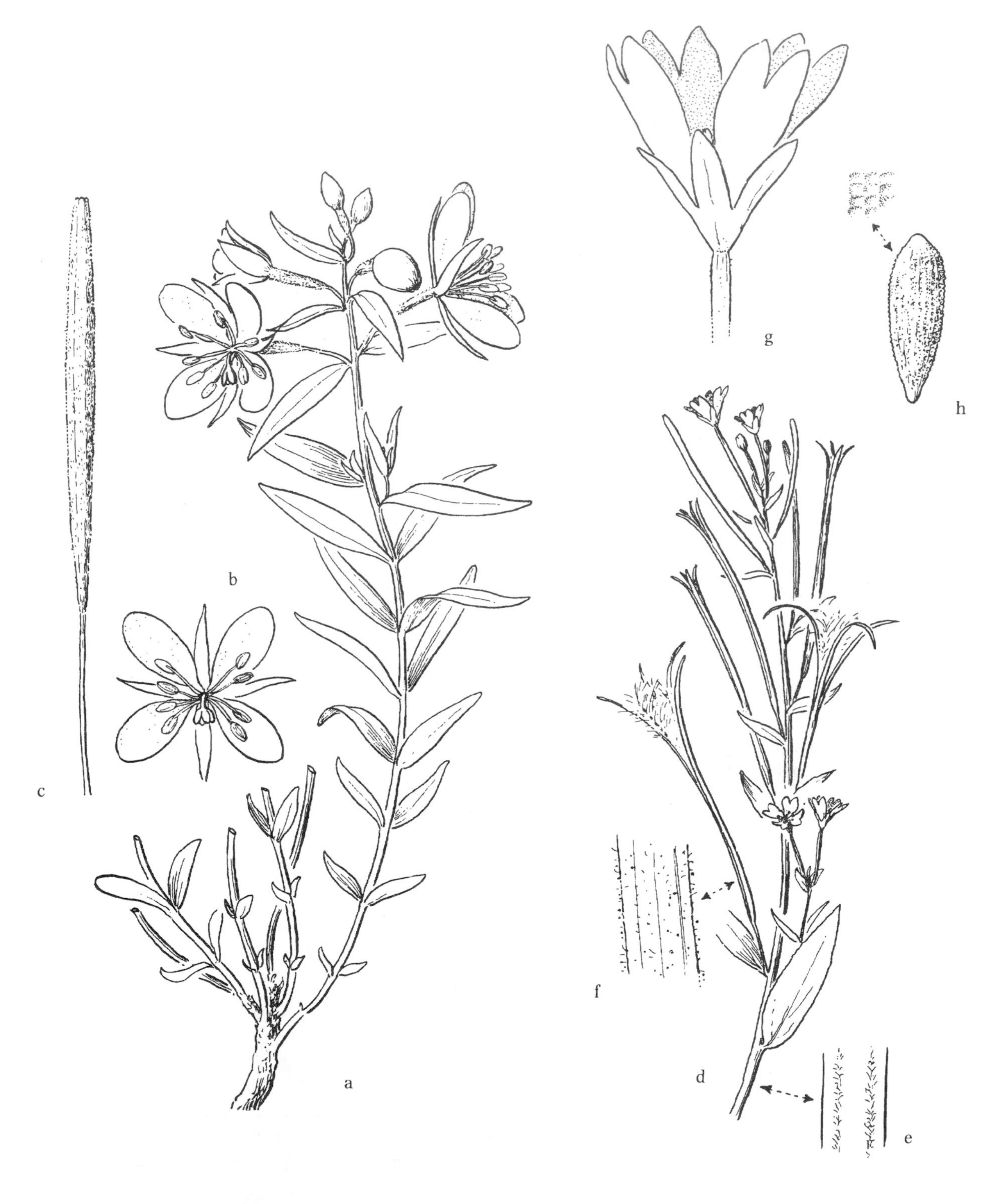

Fig. 405. *Chamaenerion latifolium*: a. upper portion of plant; b. flower; c. capsule (HCOT).
Epilobium ciliatum subsp. *glandulosum*: d. upper portion of plant; e. stem surface; f. capsule surface; g. flower; h. seed (HCOT).

Fig. 406. *Epilobium hirsutum*: a. upper portion of plant; b. flower (inferior ovary) (B&B).
Epilobium palustre: c. habit; d. capsule; e. seed (B&B).
Epilobium coloratum: f. upper portion of plant; g. capsule; h. seed (B&B).
Epilobium ciliatum subsp. *ciliatum*: i. upper portion of plant; j. flower; k. capsule; l. seed (B&B).

3. *Ludwigia* (False Loosestrife, Primrose-willow, Water-primrose, Seedbox)

Annual or perennial herbs, sometimes subshrubs; submersed roots and stems often spongy (aerenchyma tissue surrounding stem); leaves opposite or alternate, margins entire to slightly toothed; calyx segments 4 or 5(6), persistent; corolla 4 or 5(6), yellow (readily detaching, caducous), or absent; hypanthium present, but not extending beyond the inferior ovary; fruit a many-seeded capsule dehiscing terminally or longitudinally.

References: Monachino, 1945; Munz, 1942, 1944; Peng, 1988, 1989; Ramamoorthy and Raven, 1963; Zardini, 1987; Zardini et al., 1991.

---

1. Leaves alternate.
  2. Stamens 8–10; capsule 10–50 mm long; submersed roots and stems often spongy (aerenchyma tissue surrounding stem).
    3. Calyx segments 4; internodes conspicuously winged........ 1. *L. decurrens*
    3. Calyx segments 5(7); internodes not winged.
      4. Stems erect; pedicels 1–15 mm; petals 5–11 mm long........ 2. *L. leptocarpa*
      4. Stems mostly prostrate; pedicels 13–27(29); petals 15–29 mm.
        5. Bracteoles lanceolate to lance-ovate; sepals 8–19; petals 15–29 mm.
          6. Stems mostly glabrous; petals 15–20 mm; sepals 8–12(15) mm; pedicel 13–25(27) mm; seeds 0.8–1.0 mm long. .... 3. *L. grandiflora*
          6. Stems mostly pubescent; petals 18–29 mm; sepals 12–19 mm; pedicels 14–26 mm; seeds 1.2–1.5 mm long. .... 4. *L. hexapetala*
        5. Bracteoles deltoid; sepals 6–10 mm; petals 9–13 mm........ 5. *L. peploides*
  2. Stamens 4; capsule 2–10 mm long; submersed roots and stems not spongy.
    7. Capsule conspicuously longer than wide (fig. 411d).
      8. Petals 4; calyx 2–3 mm long; seeds 0.5 mm long........ 6. *L. linearis*
      8. Petals none; calyx 1–2 mm long; seeds 0.8 mm long........ 7. *L. glandulosa*
    7. Capsule about as wide as long, or only slightly longer (fig. 414c).
      9. Flowers and capsules distinctly pedicellate.
        10. Leaves tapering to base and apex; pedicel shorter than capsule........ 8. *L. alterniflora*
        10. Leaves rounded or obtuse at base and apex; pedicel as long as or longer than capsule.
          11. Plant glabrous to minutely puberulent; principal leaves 3–7 mm wide; calyx lobes reflexed; style 7–10 mm long. .... 9. *L. virgata*
          11. Plant villous-hirsute; principal leaves 5–12 mm wide; calyx lobes erect or spreading; style 3 mm or less........ 10. *L. hirtella*
      9. Flowers and capsules sessile or appearing sessile.
        12. Capsules pubescent; plants glabrous to strongly pubescent.
          13. Plants velvety-pilose; bracteoles nearly as long as floral tube........ 11. *L. pilosa*
          13. Plants glabrous or sparsely short-pubescent; bracteoles very short (fig. 416b)........ 12. *L. sphaerocarpa*
        12. Capsules glabrous; plants glabrous.
          14. Capsule 1–2 mm long; leaves obovate........ 13. *L. microcarpa*
          14. Capsule 2.4–7 mm long; leaves lanceolate, linear, linear-lanceolate, or linear-oblong.
            15. Capsule narrowly winged on angles, 2.4–4 mm long, as wide as long........ 14. *L. alata*
            15. Capsule with rounded edges, 4–7 mm long, slightly longer than wide........ 15. *L. polycarpa*
1. Leaves opposite.
  16. Flowers and capsules pedicellate; calyx lobes 3.5–6 mm long........ 16. *L. brevipes*
  16. Flowers and capsules sessile; calyx lobes 1–2.5 mm long.
    17. Capsule 2–4.5 mm long, with 4 longitudinal green bands; bracteoles absent, or if present, minute, borne at base of hypanthium........ 17. *L. palustris*
    17. Capsule 4–8 mm long, without longitudinal green bands; bracteoles borne on hypanthium at base of ovary (fig. 421c). .... 18. *L. repens*

---

1. *L. decurrens* Walter Wing-leaf Primrose-willow Fig. 407

Ditches, swamps and stream banks, often in shallow water. Se. Pa., Md., Va. and W.Va. w. to s. Ohio, s. Ind., s. Ill., s. Mo., Kans. and Okla., s. to Fla. and e. Tex.; Trop. Am. (*Jussiaea decurrens* (Walter) DC.)

2. *L. leptocarpa* (Nutt.) Hara Angle-stem Primrose-willow Fig. 408

Marshes, shores, and ditches. Coastal plain, N.C. s. to Fla., w. to Ark., e. Tex. and Mex.; Mississippi embayment and Ohio R. valley to se. Mo., s. Ill., n. Ky., s. Ind., s. Ohio and nw. W.Va.; Trop. Am. (*Jussiaea leptocarpa* Nutt.)

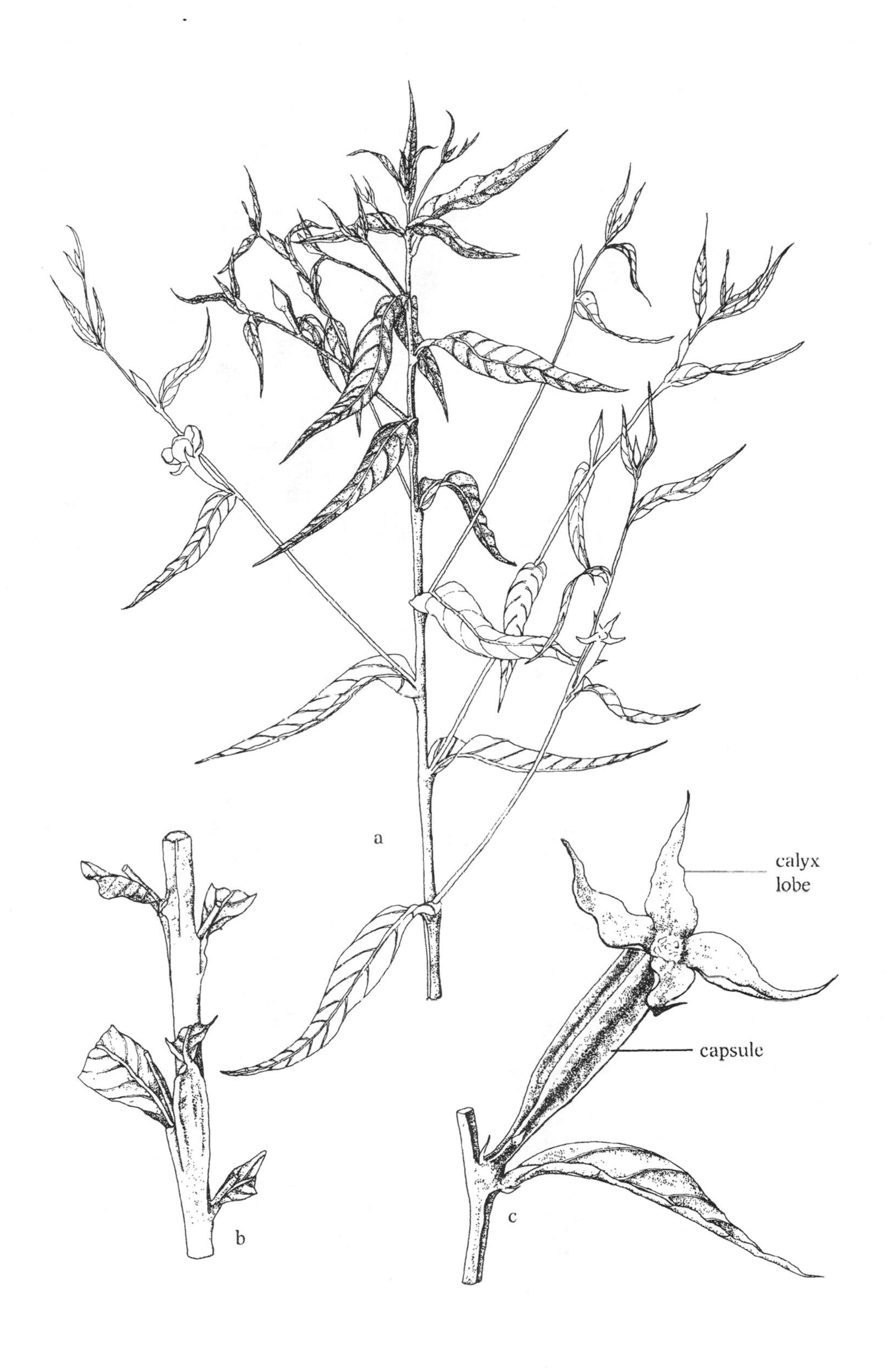

Fig. 407. *Ludwigia decurrens*: a. upper portion of plant; b. midsection of stem; c. node with fruit (G&W).

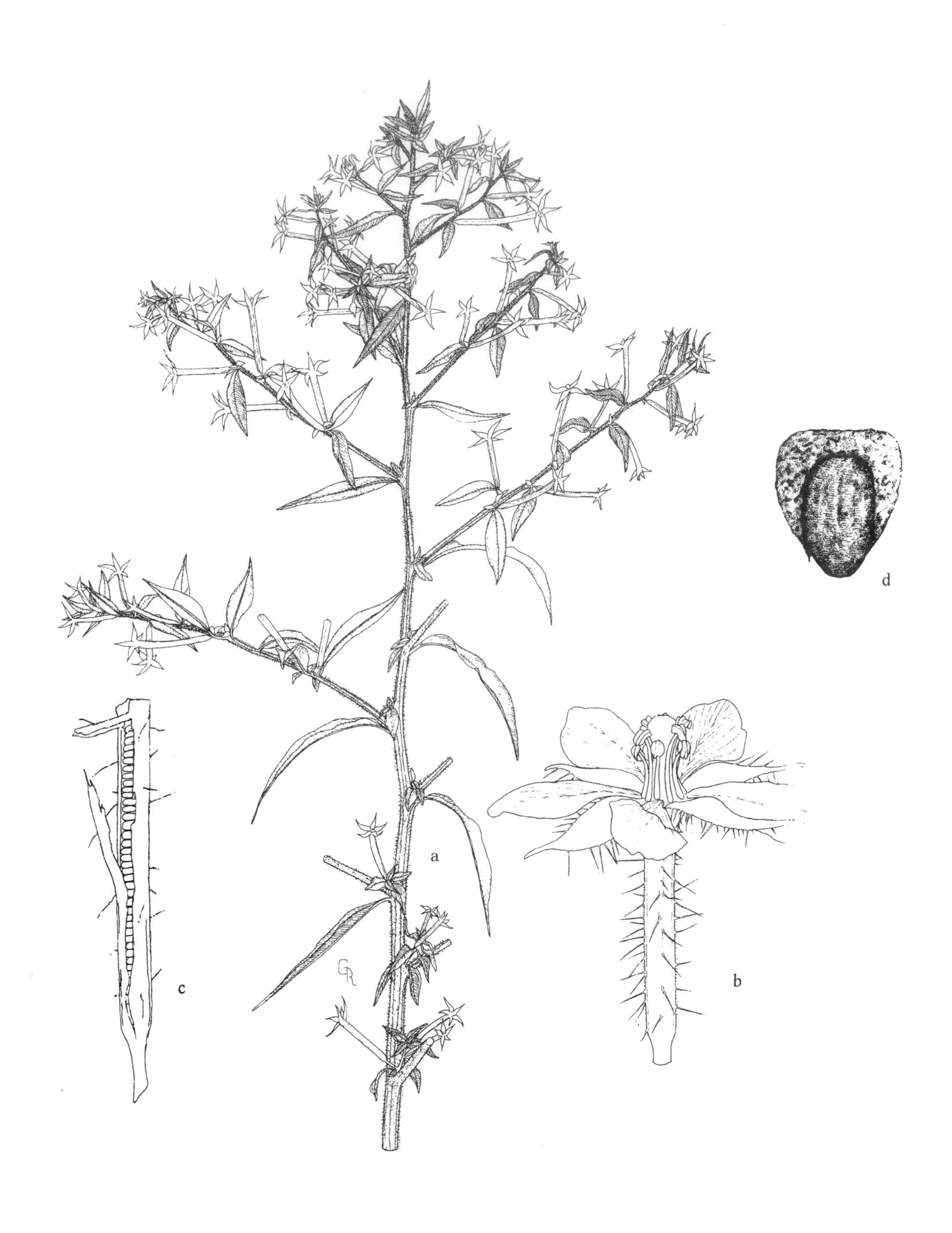

Fig. 408. *Ludwigia leptocarpa*: a. end of branch; b. flower; c. portion of capsule; d. seed with embracing segment of capsule (G&W).

3. *L. grandiflora* (Michx.) Greuter & Burdet Large-flower Primrose-willow
Marshes, pond and lake shores and ditches. s. N.Y., N.J., se. Pa. and n. Va. w. to w. Ky., sw. Mo. and Okla., s. to Fla. La. and Tex.; sw. Wash., w. Oreg. and Calif.; C.Am. and S.Am.; introd. in N.Am.

4. *L. hexapetala* (Hook & Arn.) Zardnii, H. Y. Gu & P. H. Raven Large-flower Primrose-willow Fig. 409
Marshes, swamps, pond and lake margins and ditches. N.C., Ky. and Ark., s. to Fla. and La.; Oreg. Only Ky. is within our range. (*L. grandiflora* subsp. *hexapetala* (Hook. & Arn.) G. L. Nesom & Kartesz)

5. *L. peploides* (Kunth) Raven Floating Primrose-willow Fig. 410
Ponds, streams, and ditches. L.I., N.Y., s. N.J. and Pa. w. to s. Ohio, Ill., Iowa, and Neb., s. to Fla. and Tex. Our taxon is subsp. *glabrescens* (Kuntze) P. H. Raven. (*Jussiaea repens* var. *peploides* (Kunth) Griseb.; *J. repens* var. *glabrescens* Kuntze)

6. *L. linearis* Walter Narrow-leaf Primrose-willow Fig. 411
Wet pine barrens, swamps, and ditches, often in shallow water. Chiefly coastal plain and piedmont, s. N.J., Del., Md. and Va., s. to Fla., w. to La and e. Tex.; inland to e. Okla., Ark. and c. Tenn.

7. *L. glandulosa* Walter Cylindric-fruit Primrose-willow Fig. 412
Swamps, wet woods, marshes, depressions and ditches. Coastal plain and piedmont, e. Md. and se. Va. s. to Fla., w. to La. and e. Tex.; Mississippi embayment n. to e. Okla., Ark., se. Mo., Tenn., s. Ill., w. Ky. and sw. Ind.; e. Kans. Our taxon is subsp. *glandulosa.*

8. *L. alternifolia* L. Seed-box, Rattle-box, Cylindric-fruit, Primrose-willow Fig. 413
Swamps, stream banks, marshy shores and ditches. Mass., Vt. and N.Y. w. to s. Ont., s. Mich., c. Wisc., Iowa, se. Neb., Kans. and Colo., s. to Fla. and e. Tex. (*L. alternifolia* var. *pubescens* Palmer & Steyerm.; *L. alternifolia* var. *linearifolia* Britton)

9. *L. virgata* Michx. Savanna Primrose-willow Fig. 414
Bogs, savannas and wet pinelands. Coastal plain, se. Va. s. to Fla. w. to s. Ala. and s. Miss.

10. *L. hirtella* Raf. Spindle-root Fig. 415
Pine barrens, sandy and peaty swamps and ditches. Coastal plain, s. N.J., Del., Md. and e. Va. s. to Fla., w. to e. Tex.; inland n. to se. Okla., Ark., Tenn. and Ky.

11. *L. pilosa* Walter Hairy Primrose-willow Fig. 415
Pools, bottomlands, wet pine barrens, canals and ditches. Coastal plain, se. Va. s. to Fla., w. to se. Tex.

12. *L. sphaerocarpa* Elliott Globefruit Primrose-willow Fig. 416
Sandy and peaty pond margins, shores, marshes, ditches and shallow water. Coastal plain, e. Mass., R.I., Conn., and se. N.Y., s. to Fla., w. to e. Tex.; disjunct n. to c. Tenn., ne. Ill., nw. Ind. and sw. Mich.

13. *L. microcarpa* Michx. Small-fruit Primrose-willow Fig. 417
Wet thickets, marly prairies, wet sands, peaty sands, shores and ditches. Coastal plain, N.C. s. to Fla., w. to La.; inland n. to Tenn., Ark. and se. Mo.

14. *L. alata* Elliott Winged Primrose-willow Fig. 418
Brackish and tidal marshes, freshwater swamps, ponds and ditches. Coastal plain, se. Va. s. to Fla., w. to Miss., e. La. (*L. simulata* Small; *L. lanceolata* Elliott.)

15. *L. polycarpa* Short & Peter Many-fruit Primrose-willow Fig. 419
Pond shores, pools savannas, brackish swales and ditches. Me., Vt., Mass., Conn., Pa., and Va.; s. Ont. w., to s. Mich., Wisc. and se. Minn., s. to w. W.Va., Ky., Ind., Ill., Mo., e. Kans., Neb. and Ark.; n. Ida.

16. *L. brevipes* (Long) Eames Long Beach Primrose-willow
Sands, shallow pools, marshes and swamps. Coastal plain; s. N.J. and Md. s. to S.C. and e. Ga.

17. *L. palustris* (L.) Elliott Water-purslane, Marsh Seedbox Fig. 420
Margins of ponds, rivers and ditches; often submerged. N.S., Que. and Ont., w. to Minn. and Neb.; s. to n. Fla., La., Tex.; B.C. Wash. and Ida. s. to Calif., Ariz., N.M. and Mex.; W.I. and S.Am. (*L. palustris* var. *americana* (DC.) Fernald & Griscom; *L. palustris* var. *nana* Fernald & Griscom)

18. *L. repens* J. R. Forst. Creeping Primrose-willow Fig. 421
Clear, cold water of springs, pools, and rivers; ditches and canals. Coastal plain; Va., and N.C. s. to Fla., w. to Tex.; inland n. to Tenn., Mo. Okla.; N.M., se. Nev., Ariz. and s. Calif. and c. Mex.; Ber. and W.I. (*L. natans* Elliott)

Lythraceae / Loosestrife Family

References: Graham, 1964, 1975.

---

1. Flowers in cymose clusters in leaf axils; stems woody at base, surrounded by loose, spongy tissue, becoming herbaceous above the water line, strongly arching, often touching the water and rooting at tips (forming dense populations, even floating mats)........ 1. *Decodon*
1. Flowers solitary in leaf axils or in axillary clusters, forming terminal spike-like; stems herbaceous throughout, lacking spongy tissue, not arching or rooting at tips.
   2. Plants emergent in shallow water or stranded on shores or mud; leaves all similar, not filiform or inflated at petioles.
      3. Floral tube cylindrical to bell-shaped about twice as long as wide (fig. 423b,e,g)....................................... 2. *Lythrum*
      3. Floral tube campanulate to globose, as wide as or wider than long (fig. 424c).

Fig. 409. *Ludwigia hexapetala*: a. habit; b. stem; c. flower (UF/IFAS, as *L. grandiflora*).

Fig. 410. *Ludwigia peploides*: a. habit; b. flower; c. capsule with cross-section; d. seeds with segments of capsule; e. seeds (Reed).

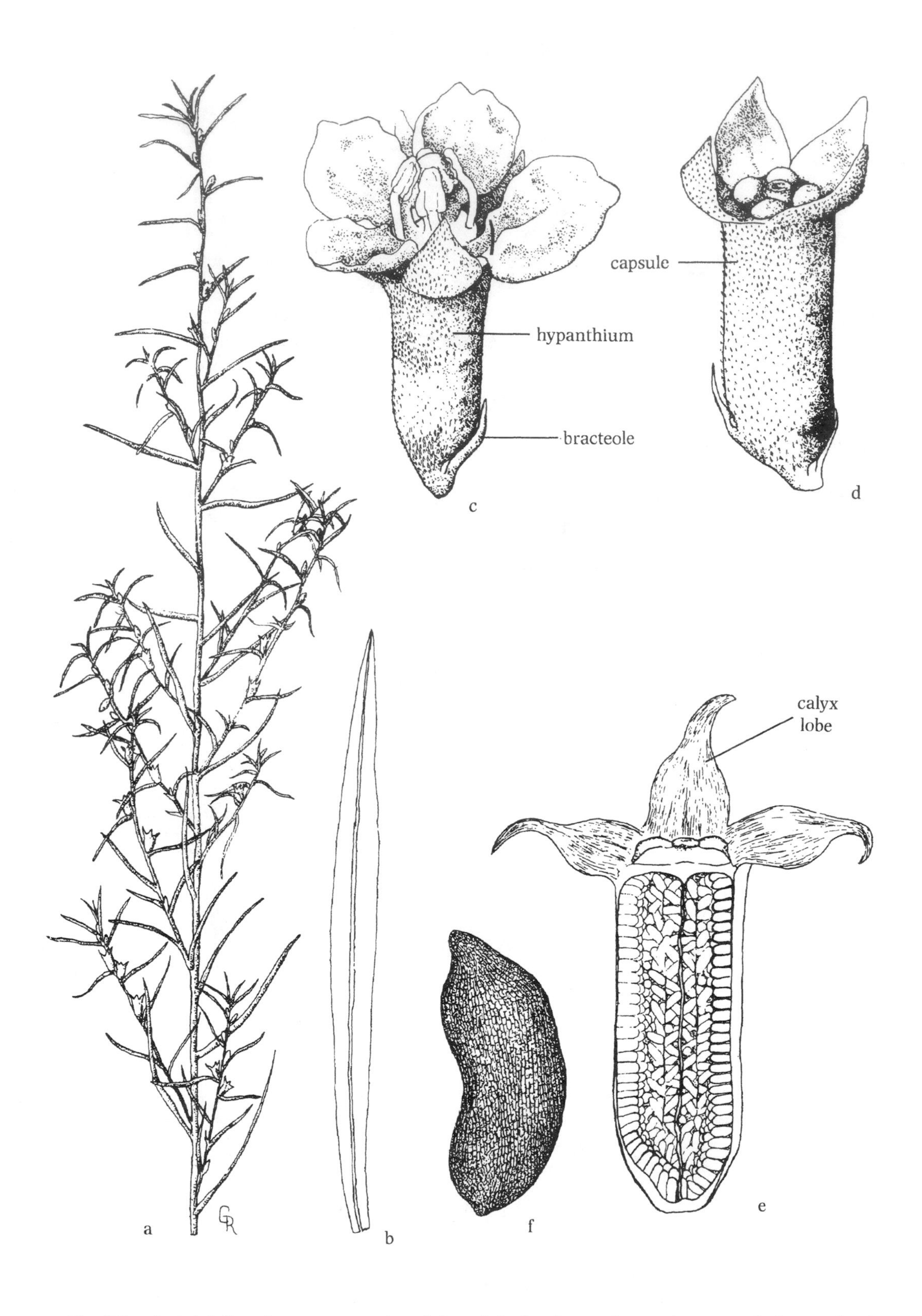

Fig. 411. *Ludwigia linearis*: a. upper portion of plant; b. leaf; c. flower; d. capsule; e. capsule, longitudinal section; f. seed (G&W).

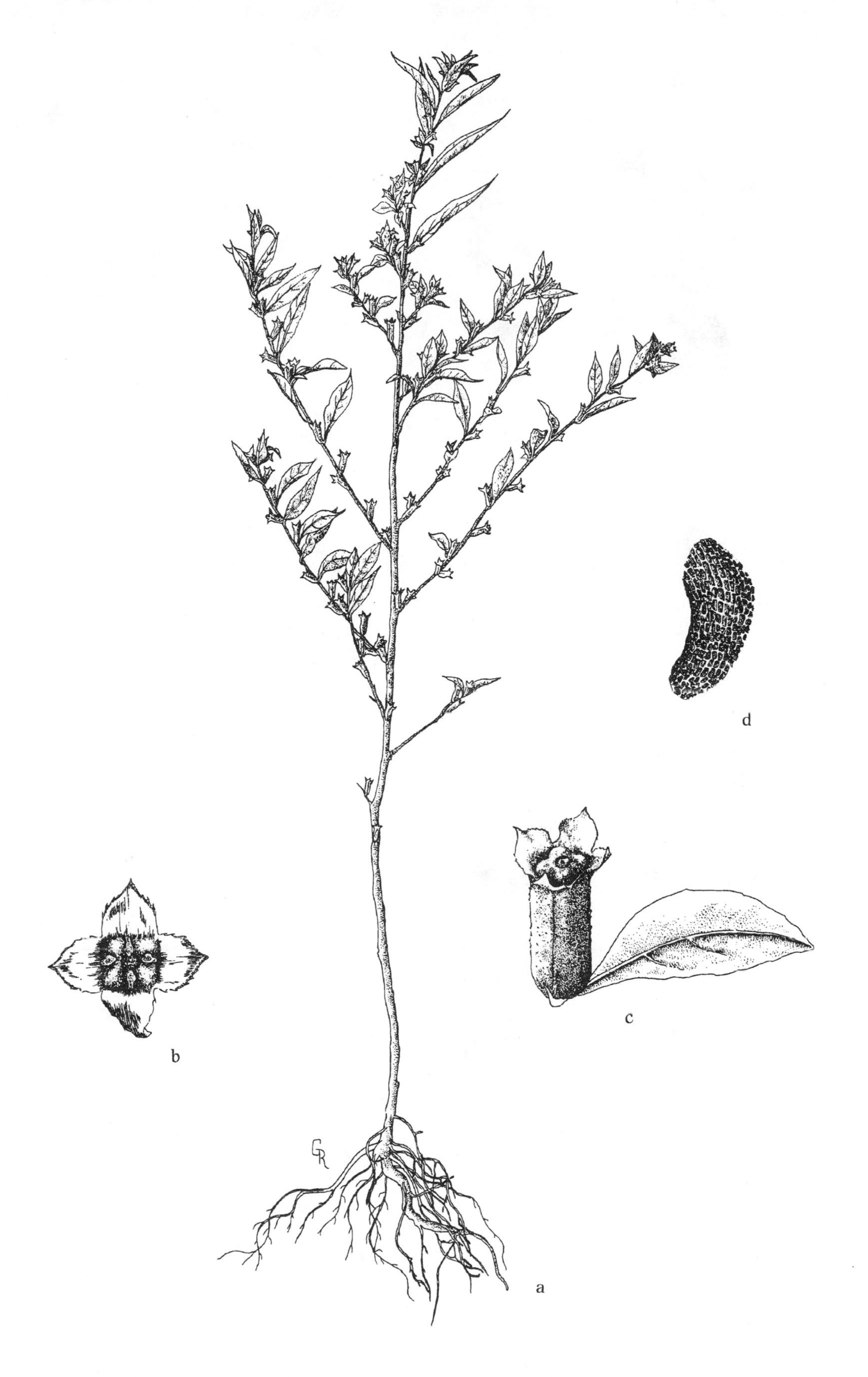

Fig. 412. *Ludwigia glandulosa*: a. habit; b. flower, top view; c. capsule; d. seed (G&W).

Fig. 413. *Ludwigia alternifolia*: a. base of plant; b. midsection of stem; c. tip of a branch; d. flower, top view; e. capsule, portion cut away and turned back (G&W).

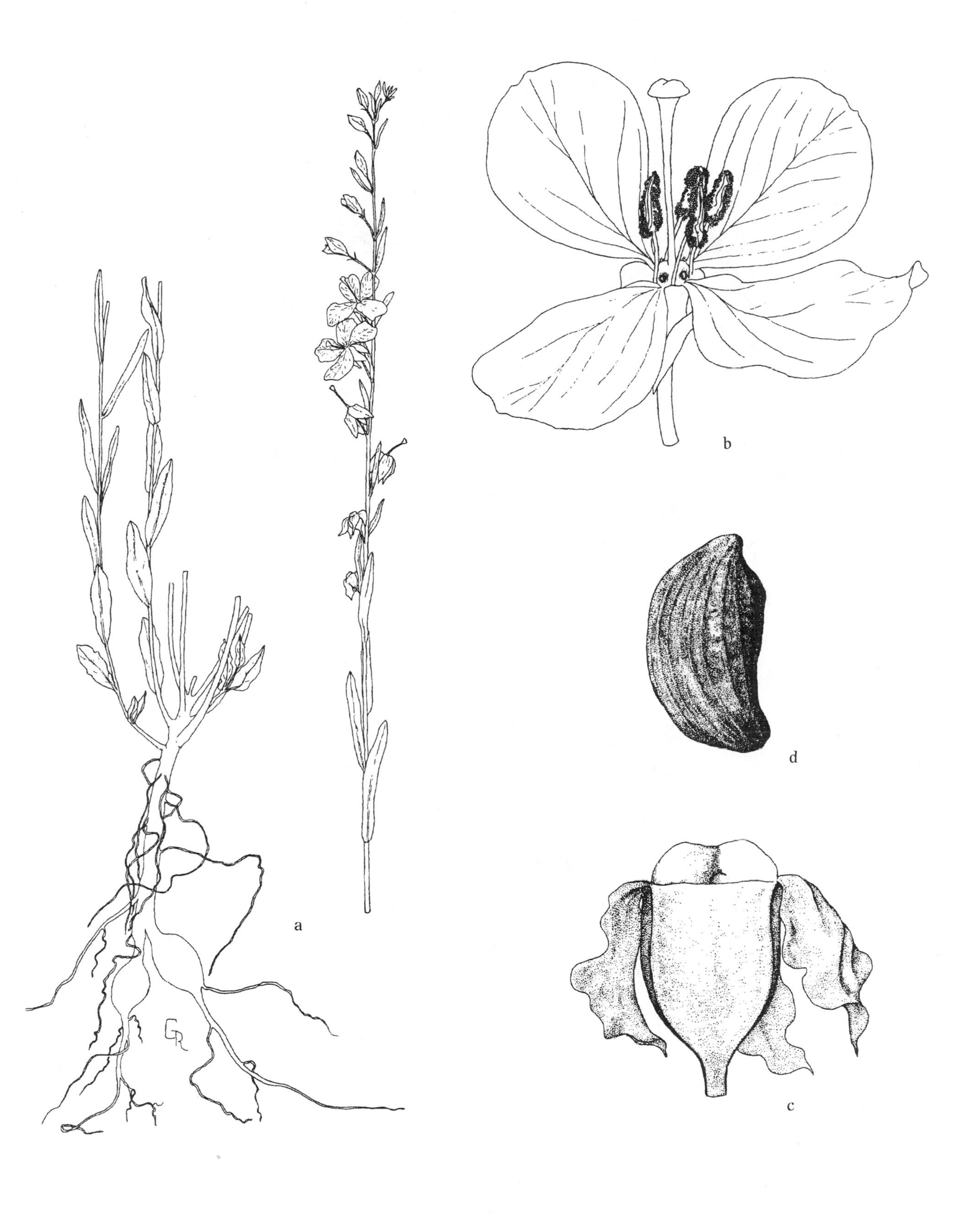

Fig. 414. *Ludwigia virgata*: a. habit; b. flower; c. capsule; d. seed (G&W).

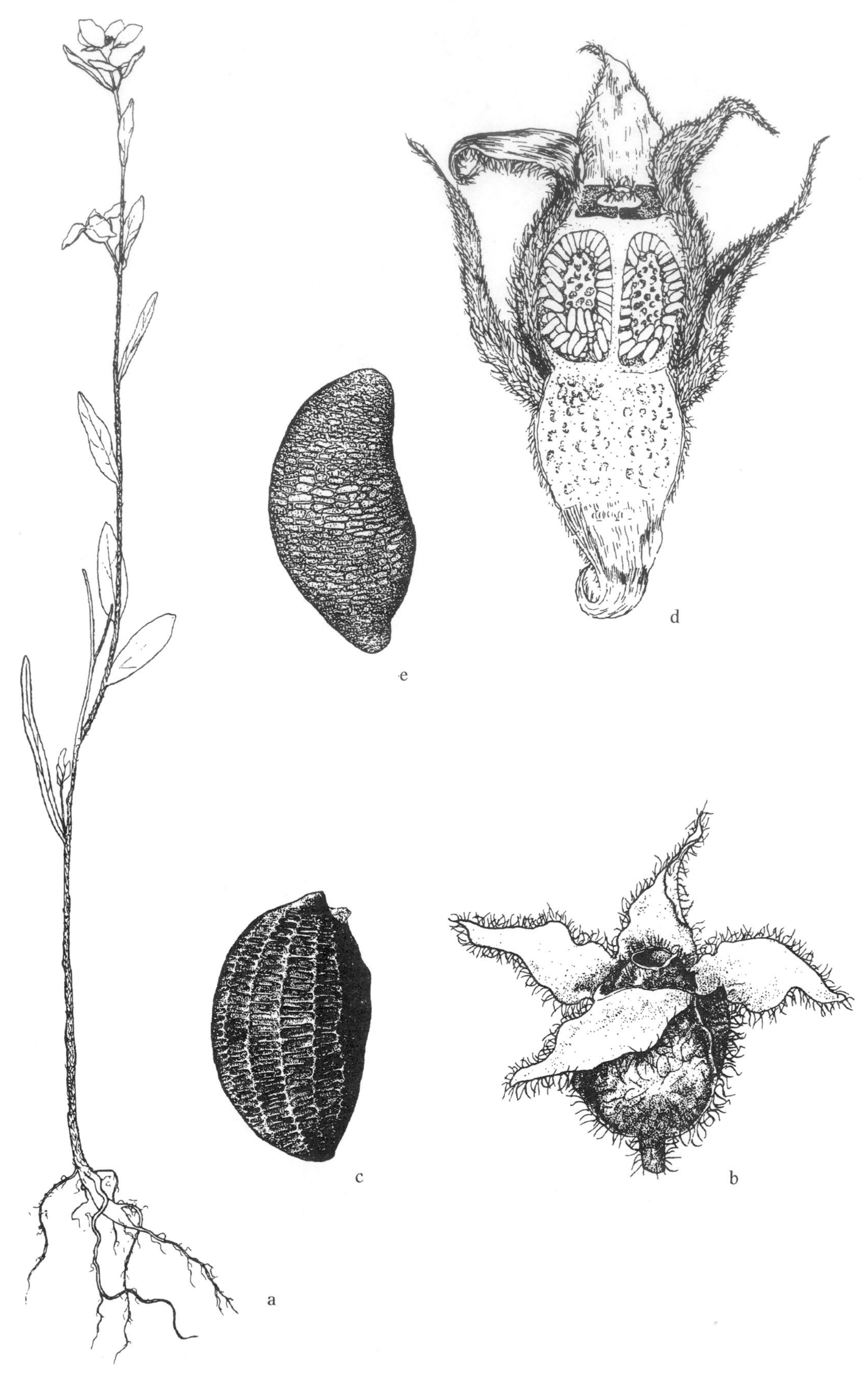

Fig. 415. *Ludwigia hirtella*: a. habit; b. fruit; c. seed (G&W).
Ludwigia pilosa: d. fruit, longitudinal section; e. seed (G&W).

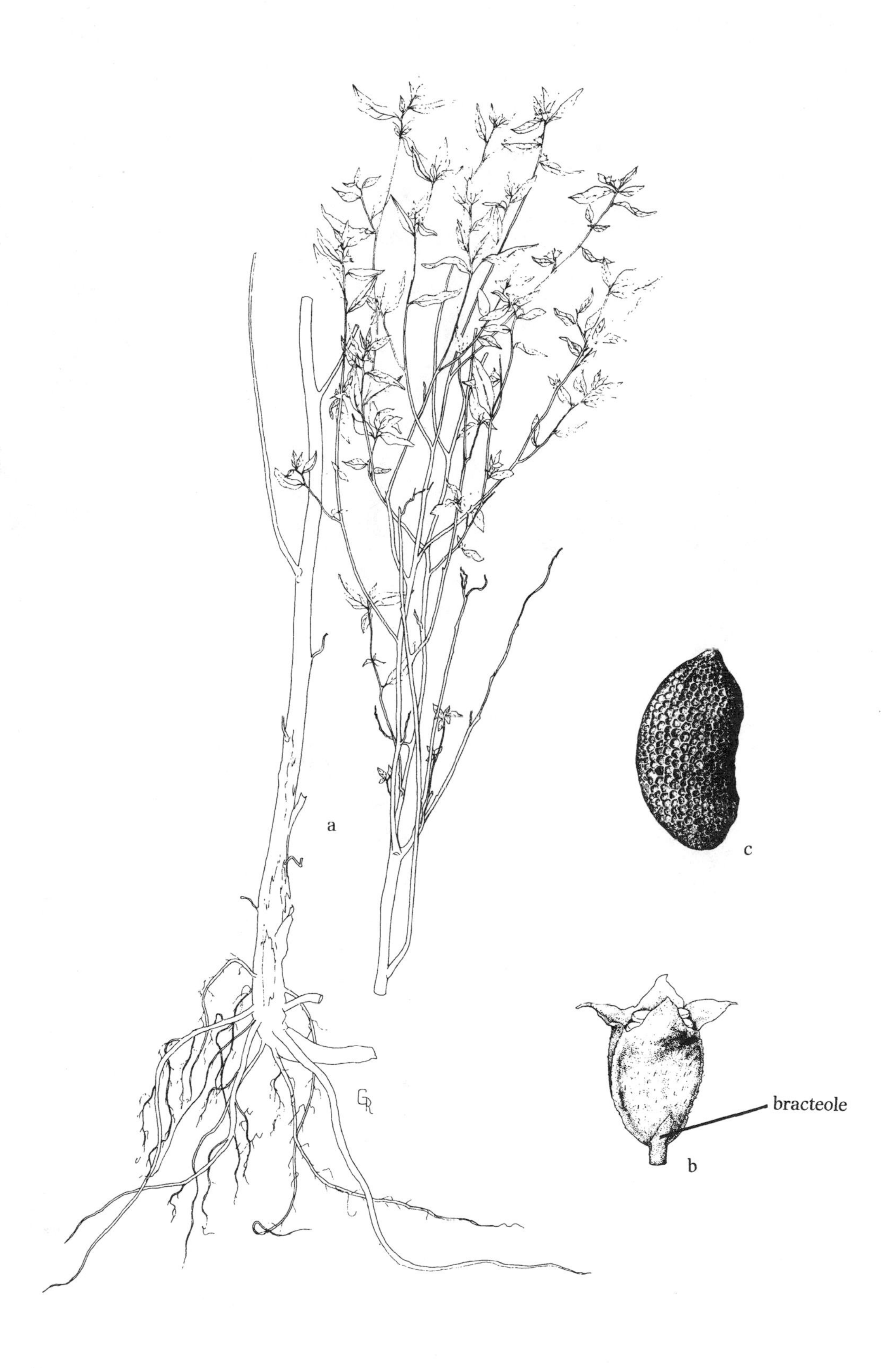

Fig. 416. *Ludwigia sphaerocarpa*: a. habit; b. flower subtended by bracteole; c. seed (G&W).

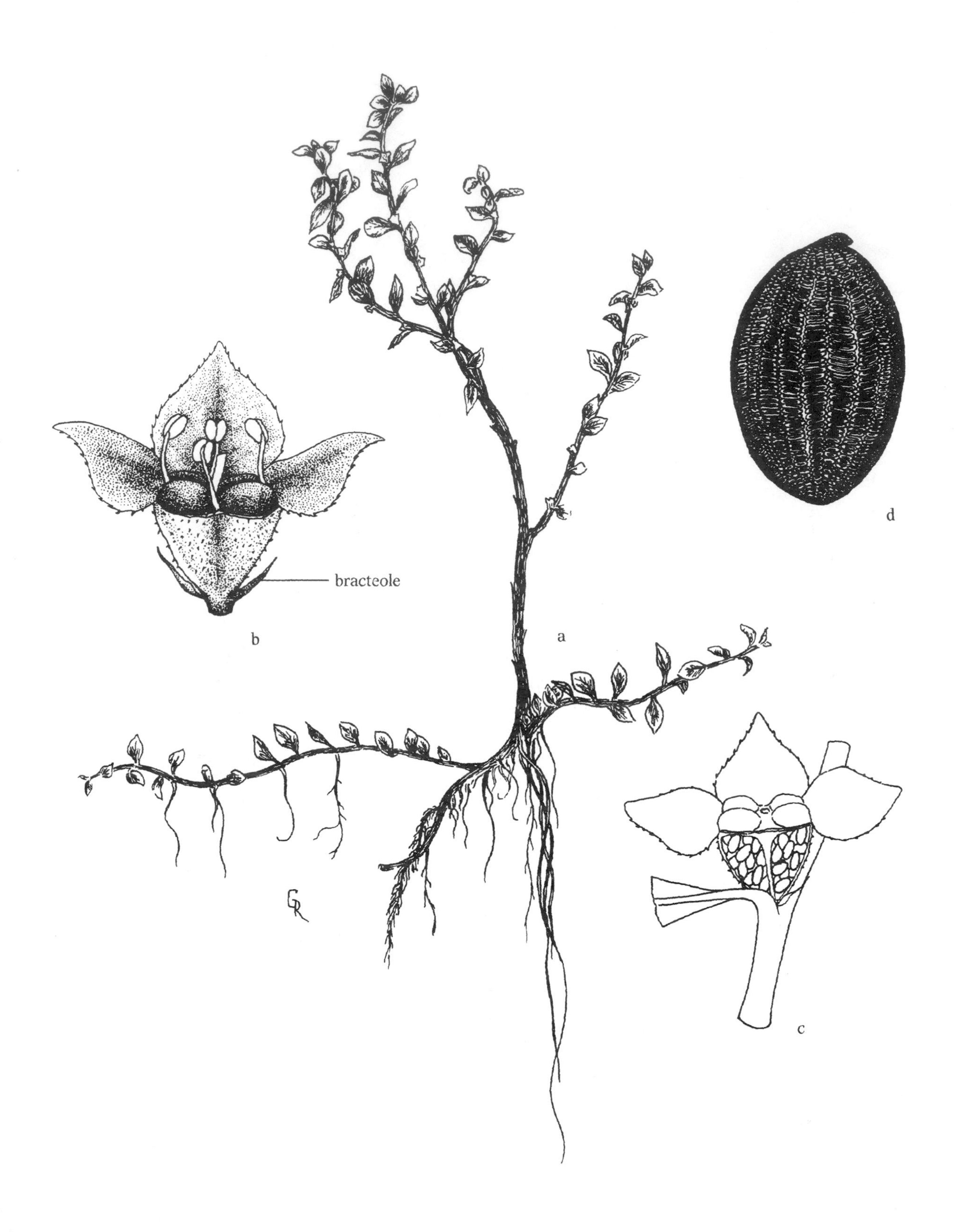

Fig. 417. *Ludwigia microcarpa*: a. habit; b. flower; c. capsule, side removed; d. seed (G&W).

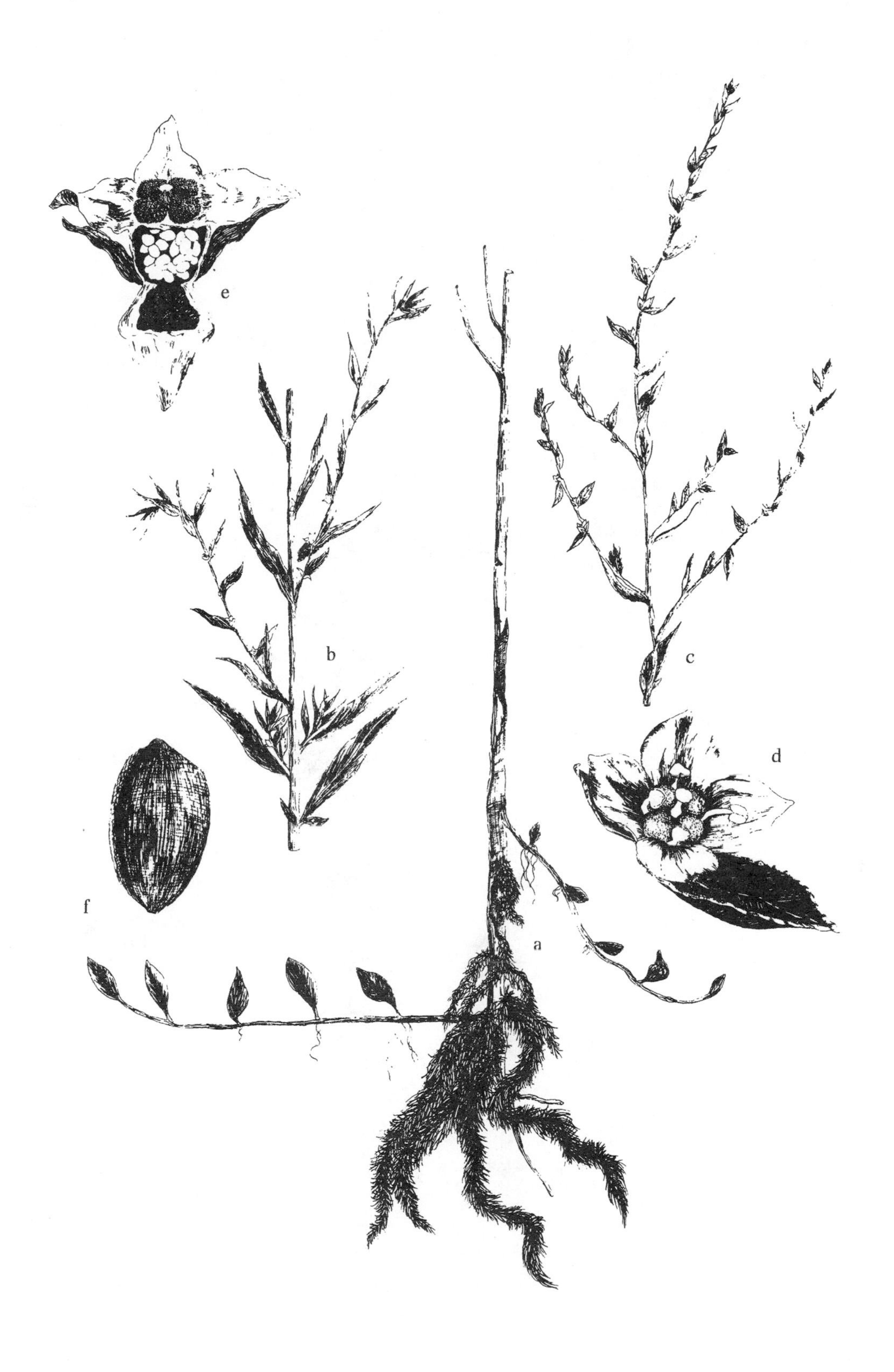

Fig. 418. *Ludwigia alata*: a. base of plant; b. midsection of stem; c. upper portion of plant; d. flower, top view; e. capsule, portion cut away and turned back; f. seed (G&W).

Fig. 419. *Ludwigia polycarpa*: a. habit; b. capsule (F).

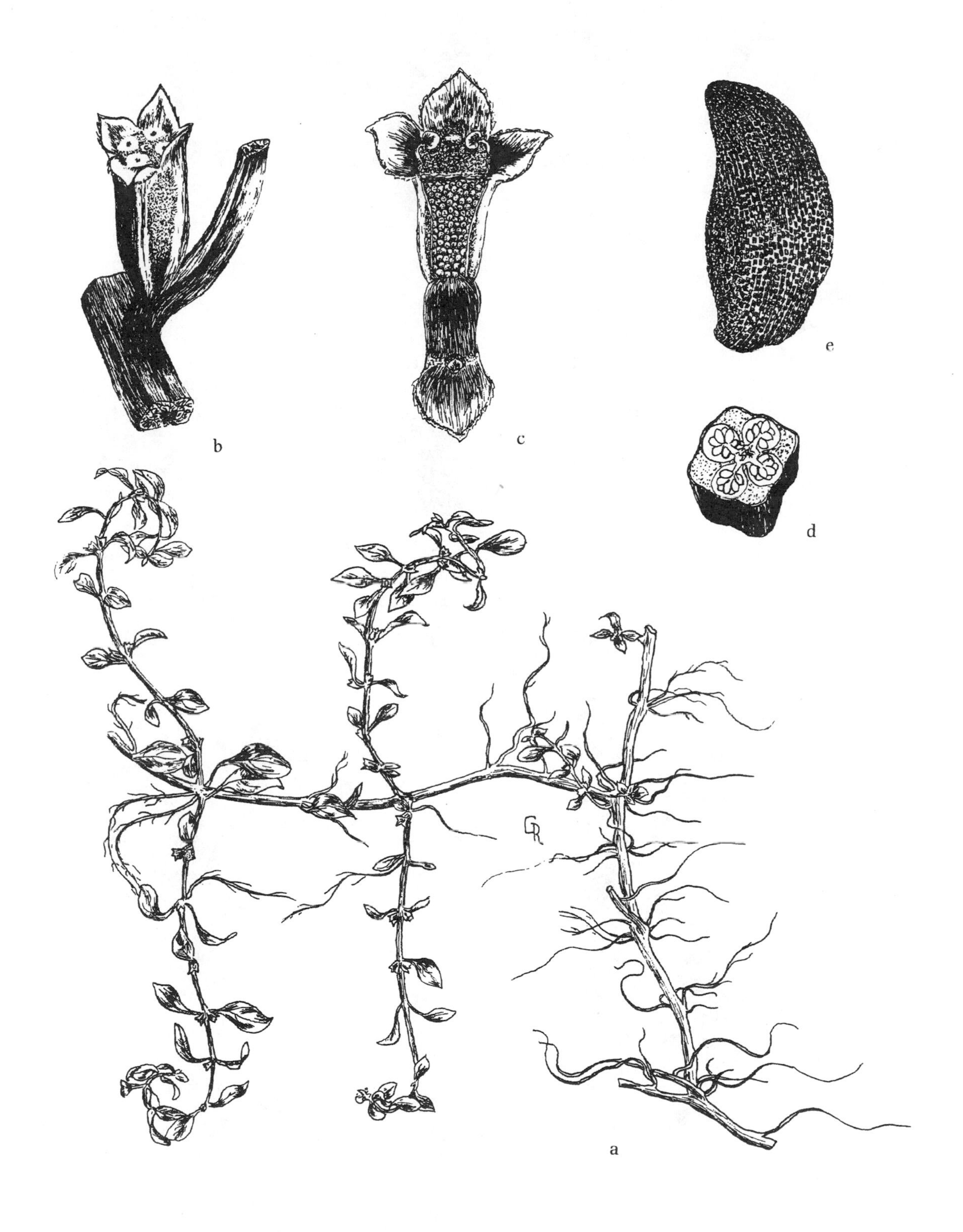

Fig. 420. *Ludwigia palustris*: a. habit; b. capsule; c. capsule, portion cut away and turned back; d. fruit, cross-section; e. seed (G&W).

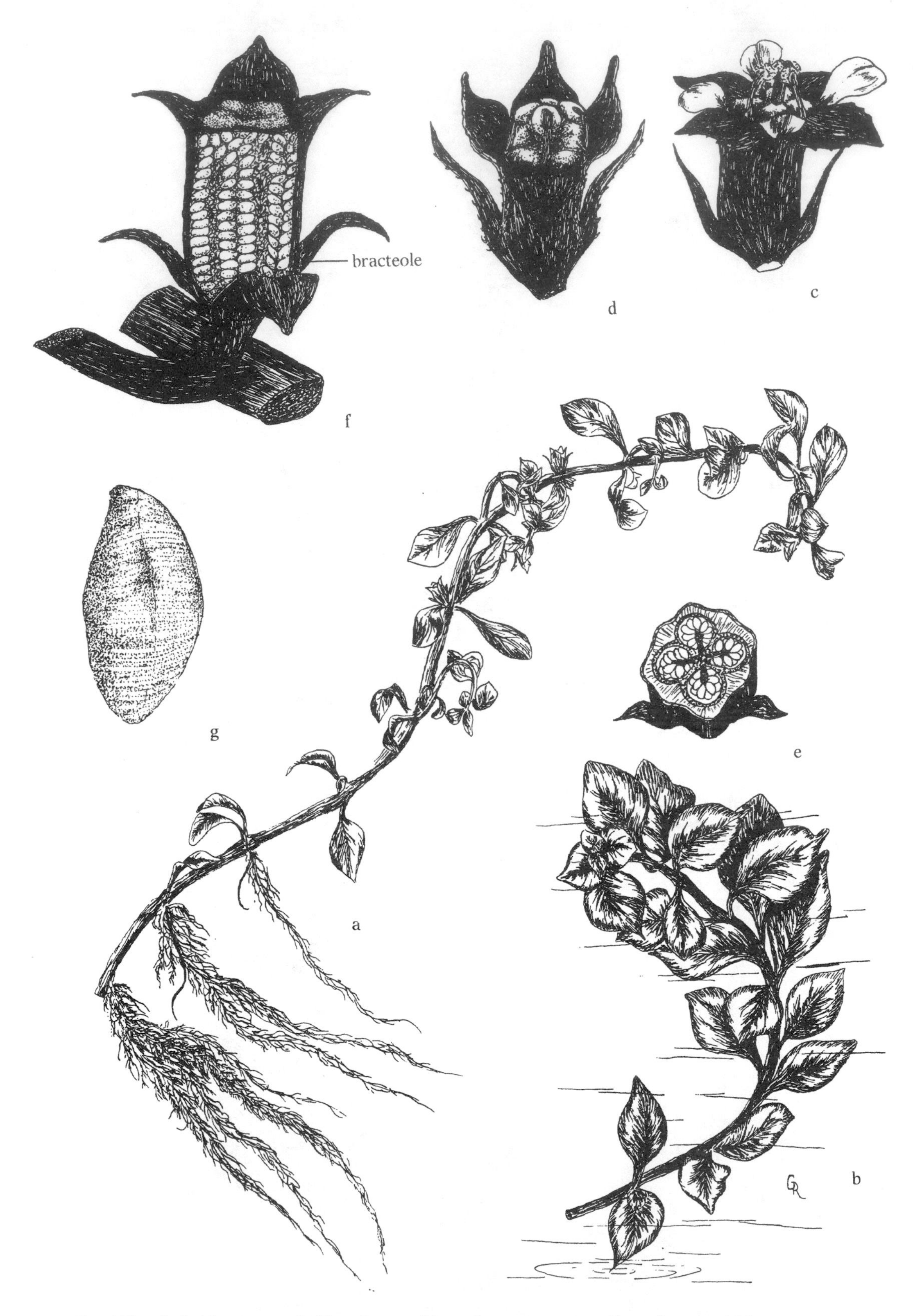

Fig. 421. *Ludwigia repens*: a. habit, submersed form; b. upper portion of branch, emersed form; c. flower; d. flower, lacking petals; e. capsule, cross-section; f. capsule, longitudinal section; g. seed (G&W).

4. Flowers/fruits 2-several in leaf axils. . . . . 3. *Ammannia*
4. Flowers/fruits solitary in leaf axils.
   5. Calyx with small appendages alternating with calyx lobes; leaves narrowed to a petiole; petals present, plants emersed, not flaccid. . . . . 4. *Rotala*
   5. Calyx lacking appendages; leaves sessile; petals absent; plants usually submersed, flaccid (fig. 425a). . . . . 5. *Didiplis*

2. Plants always strictly aquatic (typically forming floating mats); leaves dimorphic, floating leaves borne in a rosette of rhombic, coarsely toothed leaves with inflated petioles, submersed leaves opposite, finely pinnately dissected (having the appearance of roots, fig. 167); fruit a "woody" nut with 2 or 4 divergent sharp barbed spines. . . . . 6. *Trapa*

---

1. *Decodon* (Swamp Loosestrife, Water-willow)

Perennials; lower portion of stems woody, surrounded by spongy tissue (aerenchyma) at and below water level, herbaceous on middle and upper portions, arching, rooting at tips; leaves lanceolate, nearly sessile, opposite and whorled; flowers in axillary clusters; petals 4–5, magenta; tristylous (plant with 1 of 3 possible style lengths), with stamens 8 or 10 (of 2 lengths, opposite of the length of the style); fruit a globose, 3–5-locular capsule, surrounded by persistent floral tube.

1. *D. verticillatus* (L.) Elliott   Water-willow, Swamp Loosestrife   Fig. 422
Swamps, bogs, river margins, mucky pond margins and lakeshores. N.S. and Me. w. to s. Ont. and Minn., s. to Fla. and e. Tex. Plants vary from glabrous to softly pubescent, however, intermediate forms occur widely, and segregation at the varietal rank does not appear warranted. (*D. verticillatus* var. *laevigatus* Torr. & A. Gray)

2. *Lythrum* (Loosestrife)

Annual and perennial herbs; leaves opposite, or opposite on lower portion of stem and alternate on upper, sessile; flowers axillary, solitary or in clusters, mostly distylous or tristylous; calyx segments 5–7; petals magenta, pale purple, pink, or white; fruit a subcylindrical, 2-locular capsule.

Reference: Stuckey, 1980a.

---

1. Flowers large, in axillary clusters, forming terminal spike-like inflorescences; bracts and floral tubes densely pubescent. . . . . 1. *L. salicaria*
1. Flowers small, solitary, in axils of upper leaves; bracts and floral tubes glabrous.
   2. Plants annual; flowers borne the length of the plant.
      3. Hypanthium broadly bell-shaped, 1–2 mm, stem creeping or decumbent, rooting at nodes. . . . . 2. *L. portula*
      3. Hypanthium cylindric 4–6 mm, stems prostrate to ascending, rooting at base. . . . . 3. *L. hyssopifolia*
   2. Plants perennial; flowers borne on upper portion of plant.
      4. Main leaves of stem opposite, mostly shorter than the internodes, 1.4–4 mm wide; plants of coastal saline marshes. . . . . 4. *L. lineare*
      4. Main leaves of stem opposite on lower portion, alternate on upper, longer than the internodes, 5–12 mm wide; plants of freshwater marshes.
         5. Leaves ovate to oblong, with rounded or cordate bases. . . . . 5a. *L. alatum* var. *alatum*
         5. Leaves linear to lanceolate, with tapering bases. . . . . 5b. *L. alatum* var. *lanceolatum*

---

1. *L. salicaria* L.   Purple Loosestrife, Spike Loosestrife   Fig. 423
Meadows, marshes, floodplains, river margins and lakeshores. Nfld. w. to B.C., s. to Ga. Ala., Miss., Okla., n. Tex., N.M. and Calif. This introduced species has become extremely aggressive in portions of the Northeast and difficult to eradicate. A cultivated species, *L. virgatum* L., is similar to *L. salicaria* but is less robust, glabrous throughout, has leaves attenuate to the base, and has calyx lobes about as long as the appendages. Native to wetlands in Europe, *L. virgatum* has been reported as an escape from cultivation in the Boston, Mass., area and has the potential to spread.

2. *L. portula* (L.) D. A. Webb   Water-purslane   Fig. 423
Shallow water of lakes, ponds and shorelines. Mass. and Conn.; Ohio; B.C. s. to Wash., Oreg. Nev. and Calif.; intro. from Eur. This very small plant with small obovate leaves looks similar to *Callitriche* when first encountered.

3. *L. hyssopifolia* L.   Hyssop Loosestrife   Fig. 423
Marshes and damp sandy soils. Coastal, s. Me. s. to N.J. and Pa.; Ohio, Mich.; Pacific Coast, B.C., Ida. and Wash. s. to Calif. Coastal populations in N.E. are regarded as native, although some N.E. populations may be introduced; introd. elsewhere in N.Am. from Eur.

4. *L. lineare* L.   Wand Lythrum   Fig. 423
Brackish or saline marshes. Coastal plain, Conn. L.I., N.Y., s. to Fla., w. to La. and Tex.

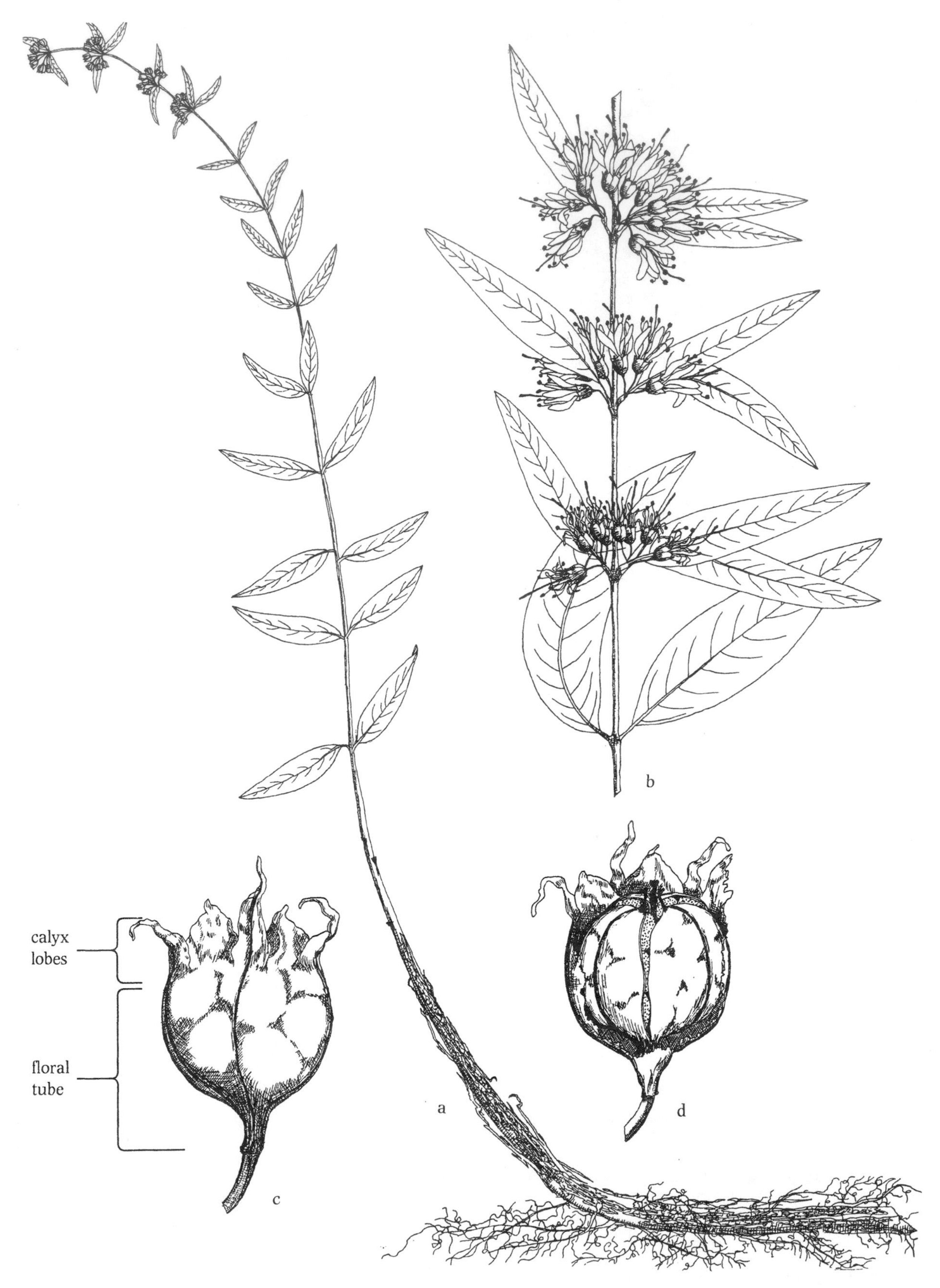

Fig. 422. *Decodon verticillatus*: a. habit; b. upper portion of plant; c. calyx and floral tube; d. fruit exposed by removal of portion of floral tube (F).

Fig. 423. *Lythrum salicaria*: a. upper portion of plant; b. flower (F).
Lythrum alatum var. *alatum*: c. upper portion of plant (F).
Lythrum hyssopifolia: d. upper portion of plant; e. flower (B&B).
Lythrum lineare: f. upper portion of plant; g. flower (B&B).
Lythrum portula: h. habit; i. fruit; j. flower (G. M. S. Easy in P&C).

5. *L. alatum* Pursh Winged Lythrum

5a. *L. alatum* var. *alatum* Fig. 423
Swamps, marshes, meadows, wet prairies and ditches. Me. w. to Ont. se. N.D., Mont. and, s. to Ga., Fla., n. Ala., n. Ark. and ne. Okla.

5b. *L. alatum* var. *lanceolatum* (Elliott) Torr. & A. Gray
Swamps, meadows, marshes, wet prairies and ditches. Se. Va. and sw. Ky. w. Okla., to Fla. and e. Tex.

3. *Ammannia* (Toothcup)

Annual; leaves narrow, opposite, usually auriculate; flowers axillary, solitary or in small clusters, falling early; perianth white, purple, or pink; fruit a 2–4-locular capsule.

REFERENCE: Graham, 1985.

1. Leaves all, or all but the lowest, auriculate at base; calyx teeth long, prominent.
   2. Capsules 4–6 mm in diameter, usually shorter than or as long as calyx teeth; peduncles absent; flowers pale lavender, 2.5–5 mm long, usually 1–3 per axil. . . . . 1. *A. robusta*
   2. Capsules 3.5–5 mm in diameter, as long as or longer than calyx teeth; peduncles 0–4(9) mm long; flowers rose-purple, 2–3.5 mm long, usually 3 or more per axil. . . . . 2. *A. coccinea*
1. Leaves of flower stem tapered at base, upper and middle leaves auriculate; calyx teeth short, obscure. . . . . 3. *A. latifolia*

1. *A. robusta* Heer & Regel Grand Redstem
Pond shores, marshes and ditches. S. N.J.; Ont. and Ohio w. to Ill., Iowa, N.D., Mont. and B.C., s. to w. Tenn., Ark., Tex., Colo., Utah, Calif. and Mex.

2. *A. coccinea* Rottb. Valley Redstem Fig. 424
Wet shores, marshes and ditches. S. N.J., n. Del., se. Pa. and ne. Md.; s. Ohio w. to Ill., Iowa, and N.D., s. to N.C., S.C., Fla. and Tex.; N.M., Ariz., Calif. and Mex., W.I., C.Am. and S.Am.

3. *A. latifolia* L. Pink Redstem Fig. 424
Tidal marshes and shores. Coastal and local, n. N.J. s. to Fla., w. to se. Tex.; Trop, Am.

4. *Rotala* (Toothcup)

Annuals; stems erect or decumbent, round to square in cross-section on upper portion; leaves opposite, linear-lanceolate, lanceolate, or oblanceolate, petioled; flowers solitary, sessile, subtended by small bracts; petals white or pink; floral tube 4-angled; fruit a capsule.

1. *R. ramosior* (L.) Koehne Toothcup, Lowland Ramosior Fig. 425
Damp shores, swamps, shallow pools and ditches. Mass. w. to Ont., Mich., Minn., Mont. and B.C., s. to Fla., Tex., N.M., Ariz., Calif. and Mex.; W.I., C.Am. and S.Am.

5. *Didiplis* (Water-purslane)

Annual; stems weak, erect to procumbent; leaves opposite, linear-lanceolate to linear-subulate, sessile, flaccid when submersed; flowers axillary, solitary; calyx greenish, lobes deltoid; petals absent; fruit a capsule.

1. *D. diandra* (DC.) Alph. Wood Fig. 425
Shallow water and margins of ponds, streams, lakes and temporary pools. Va. nw. to Ind., Wisc. and Minn., s. to Fla., Kans. and Tex.; Utah. (*Peplis diandra* Nutt. ex DC.)

6. *Trapa* (Water-chestnut)

Floating annual with slender roots; submersed leaves linear, opposite, falling early, replaced by fine, pinnately branched, leaf-like adventitious roots; floating leaves alternate, blades rhomboid, petioles inflated; flowers borne in axils of floating leaves; petals white or slightly pink; fruit a large "woody" nut with 4 or 2 sharp, barbed spines (a caltrop). Often treated as a monogeneric family, Trapaceae.

REFERENCES: Countryman, 1977; Crow and Hellquist, 1983; Dodd et al., 2019; Muenscher, 1934; Smith, 1955; Winne, 1935.

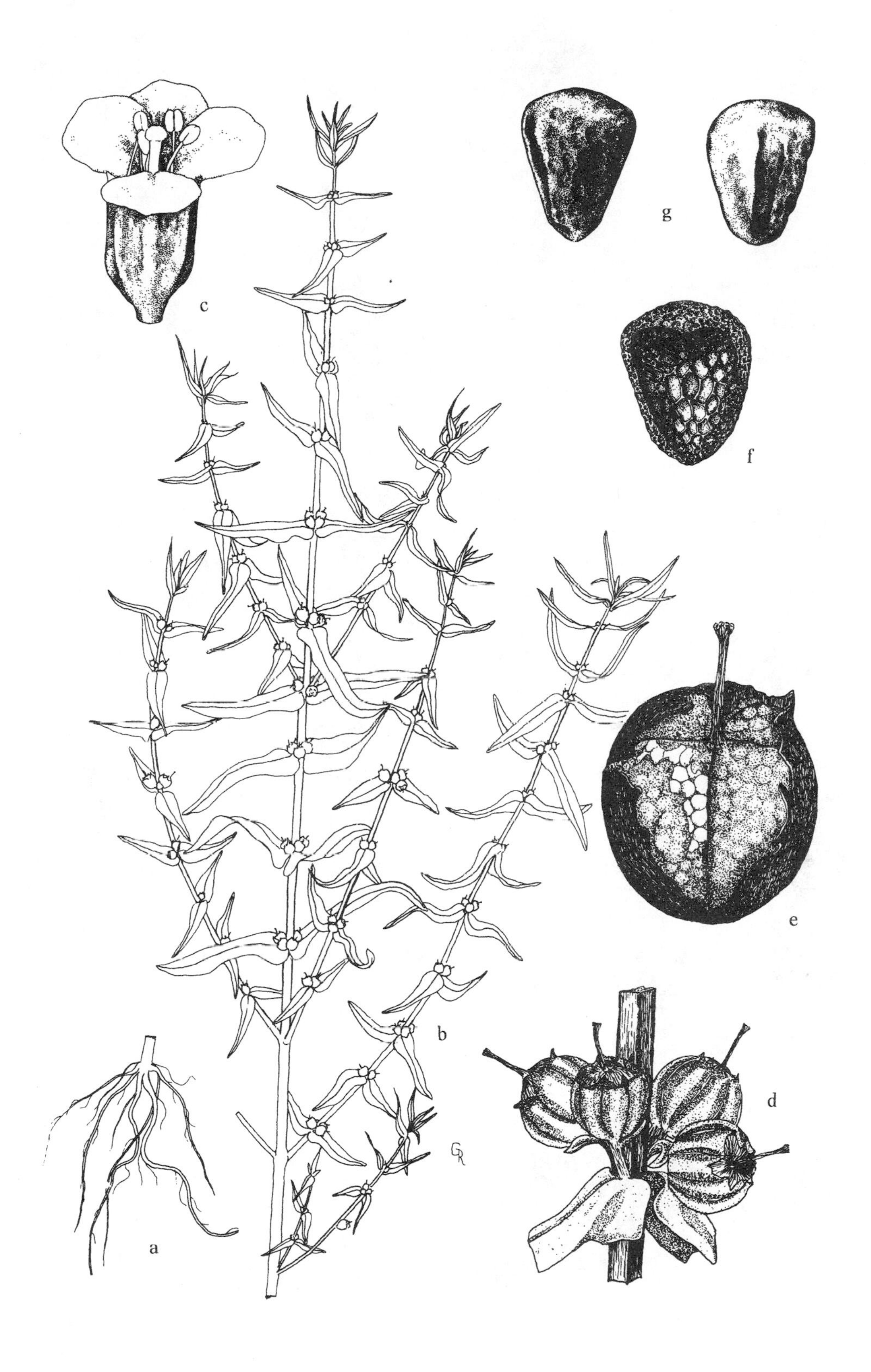

Fig. 424. *Ammannia coccinea*: a, b. habit; c. flower; d. cluster of fruits; e. fruit with wall partly disintegrated; f. seed (G&W).
Ammannia latifolia: g. seed, two views (G&W).

Fig. 425. *Didiplis diandra*: a. habit; b. fruit; c. seed (C&C).
Rotala ramosior: d. habit; e. flower; f. fruit; g. seed (G&W).

1. Plants producing 4-spined (rarely 2) woody nuts. . . . . 1. *T. natans*
2. Plants producing 2-spined woody nuts. . . . . 2. *T. bispinosa*

1. *T. natans* L. Water-chestnut Fig. 426
Shallow water of streams, rivers, lakes and ponds. Locally aggressive, S. Que. and Ont. N.H., Mass., w. Vt., and e. N.Y. s. to Pa., Md., Del. and n. Va.; natzd. from Eurasia. This species is especially abundant in the area of L. Champlain in Vt. and N.Y. and the Mohawk R. valley, N.Y. as well as the Potomac R. basin, often forming large floating mats.

2. *T. bispinosa* Roxb. Two-horned Water-chestnut Fig. 426
Ponds and streams in ne. Va. and e. Md., introd. from se. Asia, apparently as aquatic garden escapees and is beginning to spread, having first been discovered in 1995 and first observed in Md. in 2022. DNA analysis shows that it is a distinct, though cryptic, species (Dodd et al., 2019). This taxon is treated by some as *T. natans var. bispinosa* (Roxb.) Makino (https://powo.science.kew.org/taxon/urn:lsid:ipni.org:names:77098653-1, accessed 14 January 2023.)

Melastomataceae / Melastome Family

1. *Rhexia* (Meadow-beauty)

Perennial herbs; stems erect, hirsute and/or glandular-pubescent, usually 4-angled; leaves opposite, usually conspicuously 3-nerved, sessile or short-petiolate; flowers with a persistent urn-shaped hypanthium (floral tube); sepals 4, petals 4 and stamens 8, arising from hypanthium; filaments in 2 whorls of unequal length, downwardly curved (giving flower zygomorphic appearance); anthers elongate, dehiscing by terminal pores; ovary superior; fruit a 4-locular capsule; seeds typically curved with concentric lines of papillae, tubercles or protuberances.

References: James, 1956; Kral and Bostick, 1969; Wurdack and Kral, 1982.

1. Flowers sessile, the petals somewhat ascending at anthesis; anthers straight, 1–2(3) mm long; leaves 1–1.5 cm long. . . . . 1. *R. petiolaris*
1. Flowers pedicellate, petals spreading to somewhat reflexed at anthesis; anthers curved, more than 5–11 mm long; leaves 2–7 cm long.
    2. Faces of squared stem (mid-stem) all equal, flat to convex; stems sharp-angled or winged.
        3. Calyx lobes longer than neck of hypanthium, tapering to distinct awn-tip; hypanthium bristles non-glandular; stems glabrous at internodes, nodes pubescent. . . . . 2. *R. aristosa*
        3. Calyx lobes shorter than, or as long as, neck of hypanthium, acute to acuminate, but not awned; hypanthium bristles, if present, glandular-tipped, stems pubescent to hirsute throughout (occasionally glabrous in *R. virginica*).
            4. Roots tuberous; neck of hypanthium as long as body; stem angles conspicuously winged at mid-stem; seeds 0.65–0.8 mm long. . . . . 3. *R. virginica*
            4. Roots non-tuberous; neck of hypanthium typically than body; stem angles wingless (or narrowly winged) at mid-stem; seeds 0.5–0.6 mm long.
                5. Seeds with low, flattened protuberances on surface, in concentric parallel lines, surface appearing uniform (fig. 428c); plants of central U.S. . . . . 4b. *R. maritima* var. *interior*
                5. Seeds with papillose surface, in concentric parallel lines (fig. 430c); plants of Atlantic coastal plain. . . . . 4c. *R. maritima* var. *ventricosa*
    2. Faces of squared stem (mid-stem) unequal, opposite-paired, one pair wider, convex to rounded, dark green, the other pair narrower, flat to concave, pale green; stems not winged.
        6. Petals bearing scattered glandular hairs on outer edge of lower surface (easily seen before flower opens); mature petals 20–25(27) mm long, dull lavender; mature hypanthium glabrous, 10–15(20) mm long. . . . . 5. *R. nashii*
        6. Petals lacking glandular hairs; mature petals 12–15 mm long, white to pale lavender (rarely bright lavender); mature hypanthium bearing scattered hairs, 6–10 mm long. . . . . 4a. *R. mariana* var. *mariana*

1. *R. petiolata* Walter Fringed Meadow-beauty Fig. 427
Wet sands and peats of pine savannas, bogs, depressions and ditches. Coastal plain, Md. and se. Va. s. to Fla., w. to Ala. and se. Tex.

2. *R. aristosa* Britton Awn-petal Meadow-beauty Fig. 427
Damp peats and sands of pine savannas, bogs and ditches. Coastal plain, local, s. N.J., Del. and e. Va., s. to N.C., S.C., Ga. and s. Ala.

3. *R. virginica* L. Virginia Meadow-beauty, Handsome Harry Fig. 428
Wet sands, peats, gravelly shores, savannas and ditches. N.S. and Me. w. to s. Ont., sw. Mich., Wisc. and e. Iowa, s. to Fla., e. Okla. and e. Tex.

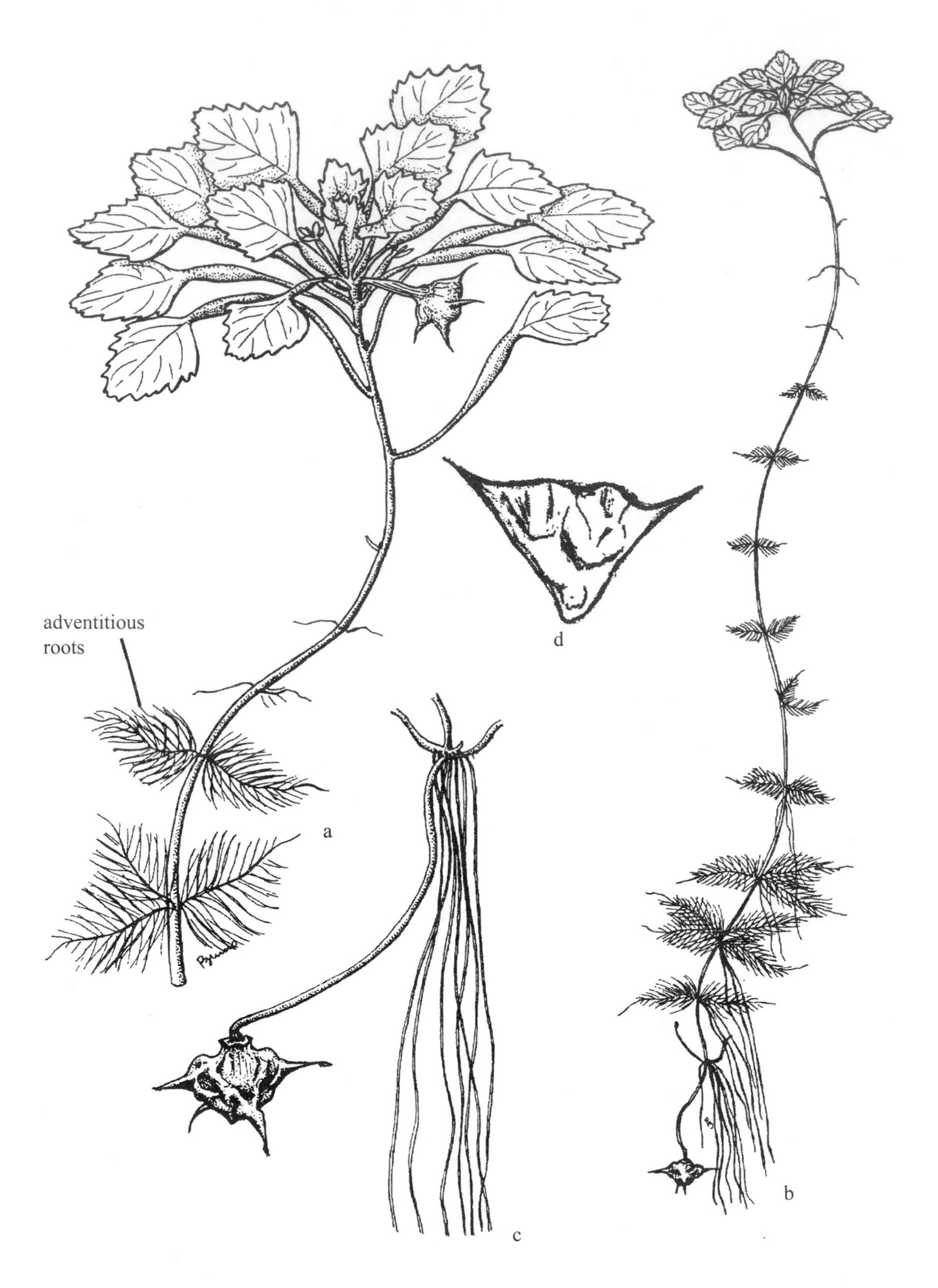

Fig. 426. *Trapa natans*: a. habit, upper portion of plant with floating leaves; b. habit; c. basal portion with nut (caltrop) attached (NHAES).
Trapa bispinosa: d. nut (caltrop) (CBH).

Fig. 427. *Rhexia petiolata*: a. habit; b. hypanthium with persisting calyx; c. seed (Kral and Bostick, 1969). *Rhexia aristosa*: d. habit; e. leaf; f. hypathium with persisting calyx; g. seed (Kral and Bostick, 1969).

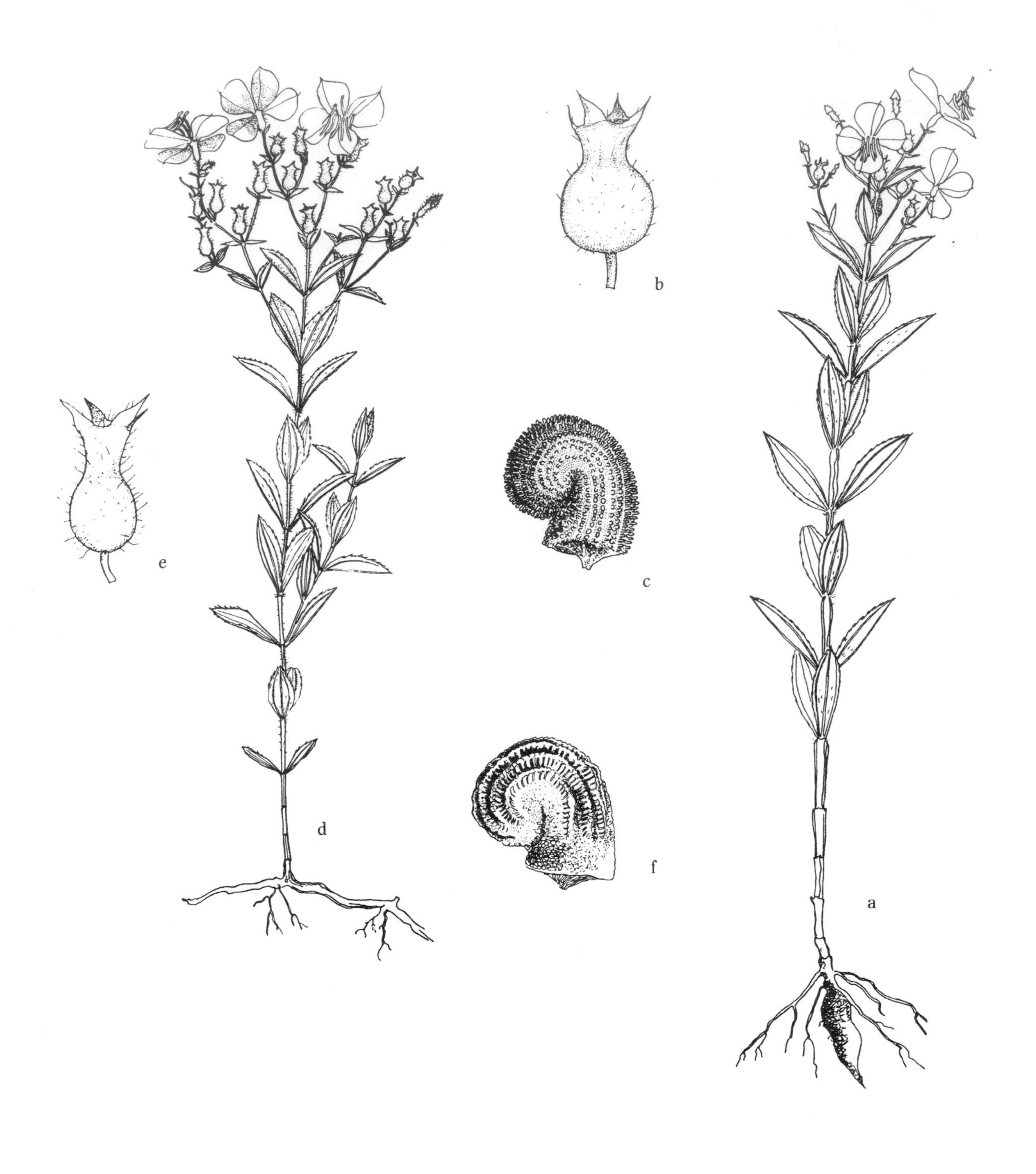

Fig. 428. *Rhexia virginica*: a. habit; b. hypanthium with persisting calyx; c. seed (Kral and Bostick, 1969). *Rhexia mariana* var. *interior*: d. habit; e. hypanthium with persisting calyx; f. seed (Kral and Bostick, 1969).

4. *R. mariana* L. Maryland Meadow-beauty

4a. *R. mariana* var. *mariana* Fig. 429
Damp sands and peats of savannas, bogs, pond shores and ditches. Coastal plain, se. Mass., s. to Va.; inland from Va. w. to sw. Mich., se. N.Y., s. N.J., se. Pa. w. to Ky., s. Ind., s. Ill., se. Mo. and se. Okla., s. to Fla. and e. Tex.

4b. *R. mariana* var. *interior* (Pennell) Kral & Bostick Fig. 428
Damp to wet soils, pond shores, wet prairies, swamps and ditches. S. Ind. w. to sw. Mo. and se. Kans., s. to Tenn., Ala., n. La., and e. Okla. and e. Tex. (*R. interior* Pennell)

4c. *R. mariana* var. *ventricosa* (Fernald & Griscom) Kral & Bostick Fig. 430
Pine barrens, savannas, bogs, shores and ditches. Coastal plain, s. N.J., Del., and Md. s. to S.C. (*R. ventricosa* Fernald & Griscom)

5. *R. nashii* Small Maid Marion Fig. 431
Wet sands and peats of swamps and bogs, flatwoods and ditches. Coastal plain, Md., e. Va. and N.C. s. to Fla., w. to Ala. and e. La. (*R. mariana* var. *purpurea* Michx.)

Anacardiaceae / Cashew Family

1. *Toxicodendron* (Poison Ivy, Poison Oak, Poison Sumac)

Shrubs (ours) or vines, with poisonous resins, deciduous; leaves palmately or pinnately (ours) compound; flowers unisexual, in a drooping panicle, mostly in axils of lower leaves; fruit a yellowish-white drupe.

References: Barkley, 1937; Gillis, 1971; McNair, 1925.

1. *T. vernix* (L.) Kuntze Poison Sumac, Swamp Sumac Fig. 432
Bogs and wooded swamps. N.S., Me. w. to s. Ont., Mich., Wisc. and se. Minn., s. to Fla. and e. Tex. (*Rhus vernix* L.)

Toxicodendron radicans (L.) Kuntze, Poison Ivy, is often found in damp woods, flood plains and occasionally in water at borders of ponds and streams. It occurs both as a climbing and as a sprawling vine.

Sapindaceae / Soapberry Family

1. *Acer* (Maple)

Trees, deciduous; leaves opposite, simple and palmately lobed (once compound); plants dioecious or polygamodioecious, flowers unisexual or unisexual and bisexual (bisexual flowers sometimes functionally unisexual), small, in clusters, emerging before leaves; fruits winged (samaroid schizocarp, splitting into 2 1-seeded mericarps).

1. Leaves simple, lobed.
   2. Leaves deeply lobed half to a third the distance to the base (fig. 432d); petals absent; ovary and young fruit white villous; fruits (mericarps) 4–8 cm long. . . . . 1. *A. saccharinum*
   2. Leaves usually lobed less than half the distance to the base (fig. 432a); petals present, red; ovary and young fruit glabrous; fruits (mericarps) 1.5–5 cm long. . . . . 2. *A. rubrum*
1. Leaves compound (fig. 432c). . . . . 3. *A. negundo*

1. *A. saccharinum* L. Silver Maple, Soft Maple, White Maple, River Maple Fig. 432
Riverbanks, floodplains and bottomlands. N.B., Me., and s. Que., w. to Ont., N.D. and Sask., s. to Fla., Tex. and e. N.M.; Wash. and n. Calif.

2. *A. rubrum* L. Red Maple, Swamp Maple, Soft Maple, Scarlet Maple Fig. 432
Swamps, bogs, marshes, floodplains, wet woodlands and uplands. Nfld. and s. Que. w. to Ont., Minn., N.D. and se. Man., s. to Fla. and e. Tex.

3. *A. negundo* L. Box-elder, Ash-leaved Maple Fig. 432
Floodplains, riverbanks and lowlands. N.S. and Que., w. to N.W.T. and B.C. s. to Fla., e. Tex., N.M., Ariz. and Calif.; Mex. and Guat.

Malvaceae / Mallow Family

1. Fruit a 5-angled schizocarp, shorter than wide, splitting into five 1-seeded segments. . . . . 1. *Kosteletzkya*
1. Fruit a ovoid to spheroid capsule, with several seeds per locule. . . . . 2. *Hibiscus*

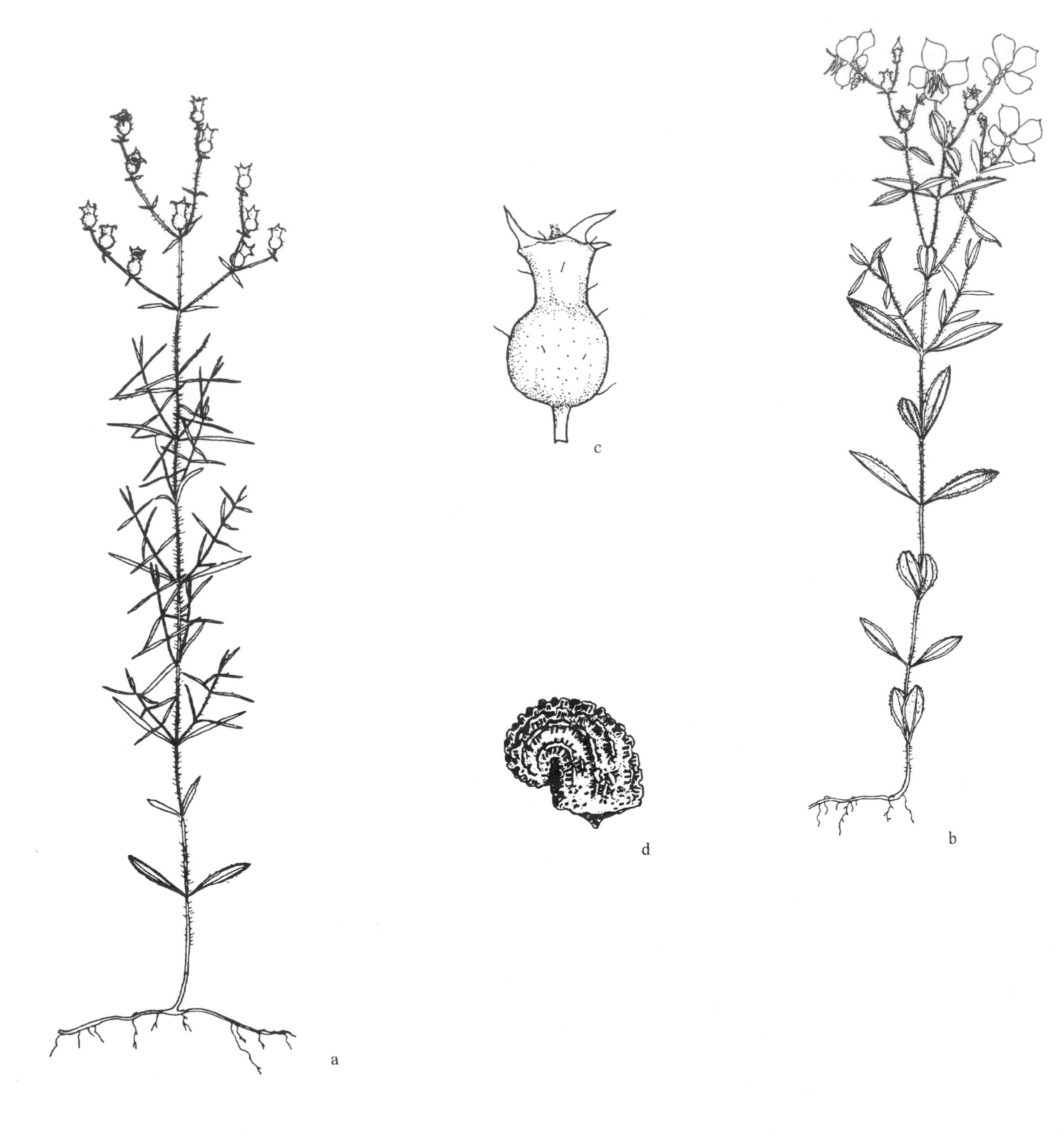

Fig. 429. *Rhexia mariana* var. *mariana*: a. habit, narrow-leaved form; b. habit, broad-leaved form; c. hypanthium with persisting calyx; d. seed (Kral and Bostick, 1969).

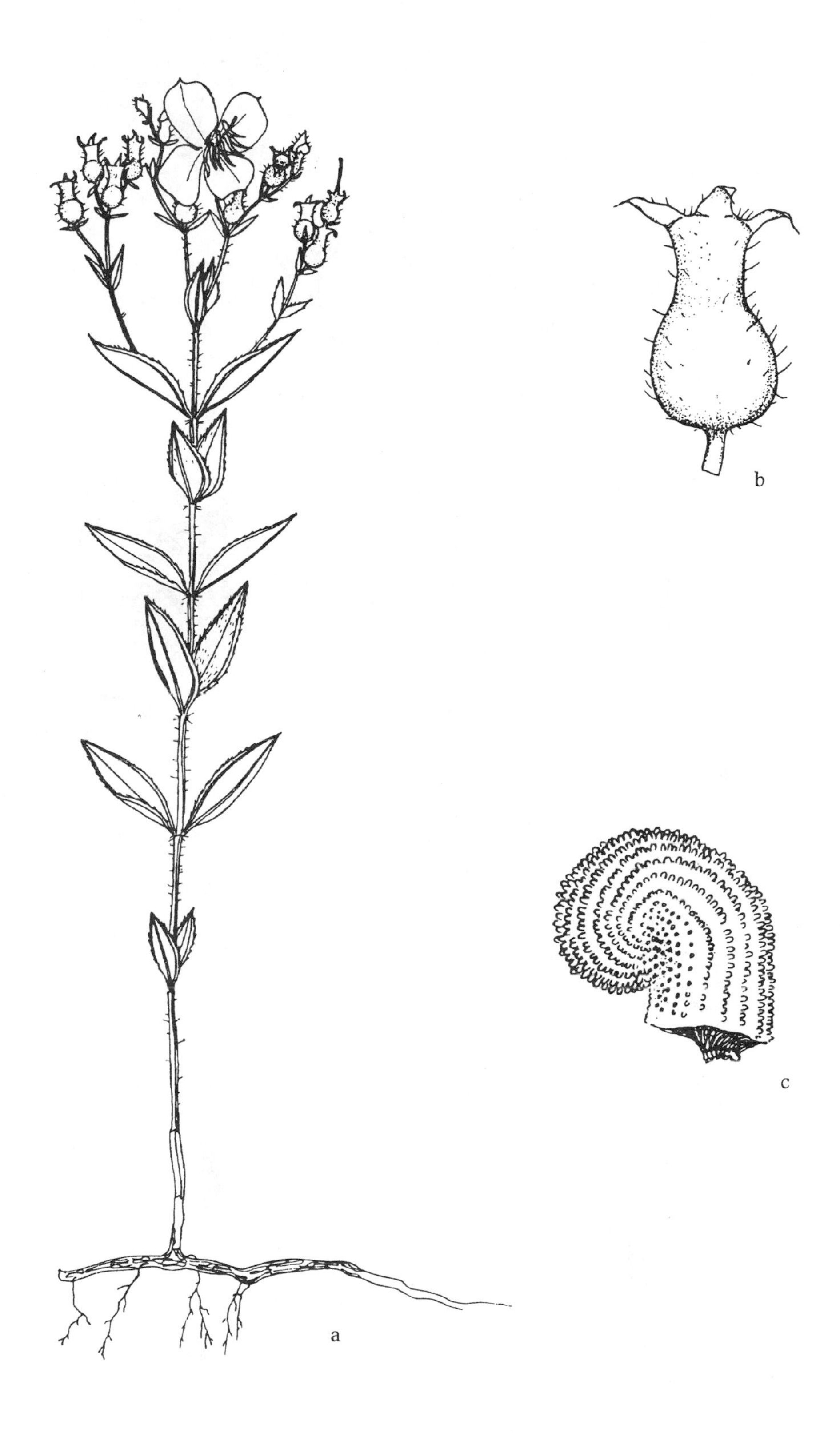

Fig. 430. *Rhexia mariana* var. *ventricosa*: a. habit; b. hypanthium with persisting calyx; c. seed (Kral and Bostick, 1969).

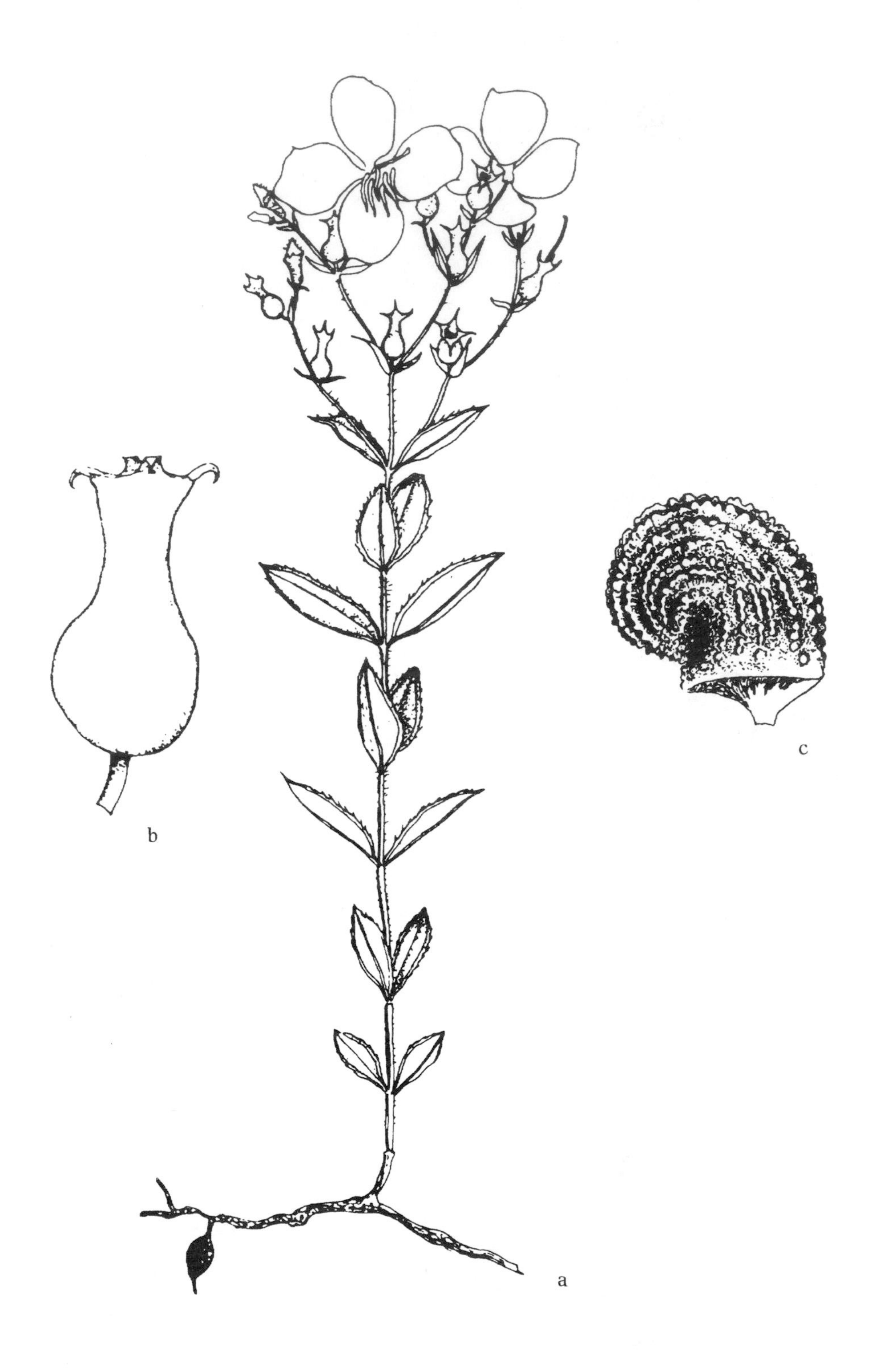

Fig. 431. *Rhexia nashii*: a. habit; b. mature hypanthium with persisting calyx; c. seed (Kral and Bostick, 1969).

Fig. 432. *Acer rubrum*: a. leaves and fruits; b. flowers emerging from buds (WVA).
Acer negundo: c. leaves and fruits (WVA).
Acer saccharinum: d. leaves; e. fruits (WVA).
Toxicodendron vernix: f. leaf; g. fruiting stem; h. winter twig (OHIO).

1. *Kosteletzkya* (Seashore Mallow)

Perennial herbs; stems erect, often branched, plants up to 2.5 m tall; plants pubescent, hairs simple and stellate; leaves alternate, simple or lobed, serrate; flowers showy; sepals united, subtended by a whorl of bracts (epicalyx); petals separate, pink or white; stamens monadelphous; fruit a 5-angled schizocarp, splitting into five 1-seeded segments.

REFERENCE: Blanchard, 2015.

1. *K. pentacarpos* (L.) Ledeb. Seashore Marsh-mallow Fig. 433

Salt marshes, freshwater marshes, marshy shores, sloughs, swamp borders, wet open woods and ditches. Outer coastal plain, L.I. N.Y., s. to Ga. and Fla., w. to e. Tex.; Ber. and Cuba. Blanchard (2015) noted that the North American populations (typically treated as *K. virginica*) cannot be distinguished from the Eurasian *K. pentacarpos*, thus nomenclaturally that name must be applied to our plants. (*K. virginica* (L.) Presl. ex A. Gray)

2. *Hibiscus* (Rose-mallow)

Annual or perennial herbs (ours) or shrubs; stems thick, plants up to 2 m tall; leaves alternate, simple or lobed, serrate; flowers showy; sepals united, subtended by a whorl of bracts (epicalyx); petals pale rose, creamy, or white, often deep purplish toward center of corolla, separate; stamens monadelphous; fruit a spheroid capsule, with several seeds per locule.

REFERENCE: Blanchard, 2015.

---

1. Stems and leaves glabrous or nearly so; leaves on lower portion of stem often hastate. . . . . 1. *H. laevis*
1. Stems and leaves pubescent; especially younger leaves; lower leaf surface usually velvety, with dense stellate hairs; leaves on the lower portion of the stem unlobed or only shallowly lobed, not hastate.
   2. Leaf blade upper surfaces mostly glabrous; capsules glabrous. . . . . 2a. *H. moscheutos* subsp. *moscheutos*
   2. Leaf blade upper surfaces mostly stellate hairy; capsules hairy. . . . . 2b. *H. moscheutos* subsp. *lasiocarpos*

---

1. *H. laevis* All. Halberd-leaved Rose-mallow, Halberd-leaved Marsh-mallow Fig. 435

Marshes, shallow water, riverbanks, marshy river margins and ditches. Se. Pa. and Md. w. to s. Ont., se. Mich., sw. Wisc., s. Minn. and e. Neb., s. to Ga., nw. Fla., La. and e. Tex. (*H. militaris* Cav.)

2. *H. moscheutos* L. Swamp Rose-mallow

2a. *H. moscheutos* subsp. *moscheutos* Figs. 433, 436

Brackish and freshwater marshes, floodplain pools, beaver ponds and ditches. se. N.H., e. Mass., and N.Y., w. to s. Ont., s. Mich., se. Wis., Mo. and e. Kans., s. to Fla., La. and e. Tex. (*H. moscheutos* subsp. *incanus* (J. C. Wendl.) H. E. Ahles: *H. incanus* J. C. Wendl.: *H. moscheutos* subsp. *palustris* (L.) R. T. Clausen)

2b. *H. moscheutos* subsp. *lasiocarpos* (Cav.) O. J. Blanchard Wooly Rose-mallow Figs. 434, 436

Freshwater marshes, pond and stream shores and ditches. Ill., Ky., Tenn. and Ala. w. to Kans., Okla., Tex. and N.M.; Calif. and Mex. This subspecies tends to be primarily west of the Mississippi River and is treated by some authors as a distinct species. (*H. lasiocarpos* Cav.; *H. moscheutos* var. *lasiocarpos* (Cav.) B. L. Turner).

Brassicaceae (Cruciferae) / Mustard Family

The Mustard Family is a large and diverse group, yet is easily recognized by the following characteristics: leaves basal, basal and cauline, or cauline only, the blades often simple, but frequently crenate, dentate, lobed, cleft, or even dissected, less frequently compound; flowers cross-shaped (cruciform), with 4 sepals, 4 petals, and 6 stamens (4 long and 2 outer ones shorter); fruit a silique (long and slender) or a silicle (short and broad), consisting of 2 carpels, the valves of which fall away at maturity leaving a membranous partition. Although the family is distinctive, differentiation between genera or species is sometimes unclear because of the great range of variability often exhibited. The leaves, even in a single species, may present a confusing series of variations. However, the fruits are quite characteristic for each genus and are important for identification.

REFERENCE: Al-Shehbaz, 2010a.

---

1. Leaves all basal, quill-like (fig. 437a), simple, margins entire. . . . . 1. *Subularia*
1. Leaves, at least some, cauline, blades flat, variously divided or simple, but margins not entire.
   2. Petals yellow or absent. . . . . 2. *Rorippa*
   2. Petals white, pinkish, or purplish.
      3. Submersed leaves highly dissected into capillary divisions (fig. 440a); fruits ovoid, 1-locular. . . . . 2. *Rorippa*
      3. Submersed leaves simple, divided into flat segments, or absent: fruits elongate, 2-locular.

Fig. 433. *Kosteletzkya pentacarpos*: a. flowering branch; b. capsule, partly dehisced; c. seed (G&W). *Hibiscus moscheutos* subsp. *moscheutos*: d. upper portion of plant (Beal).

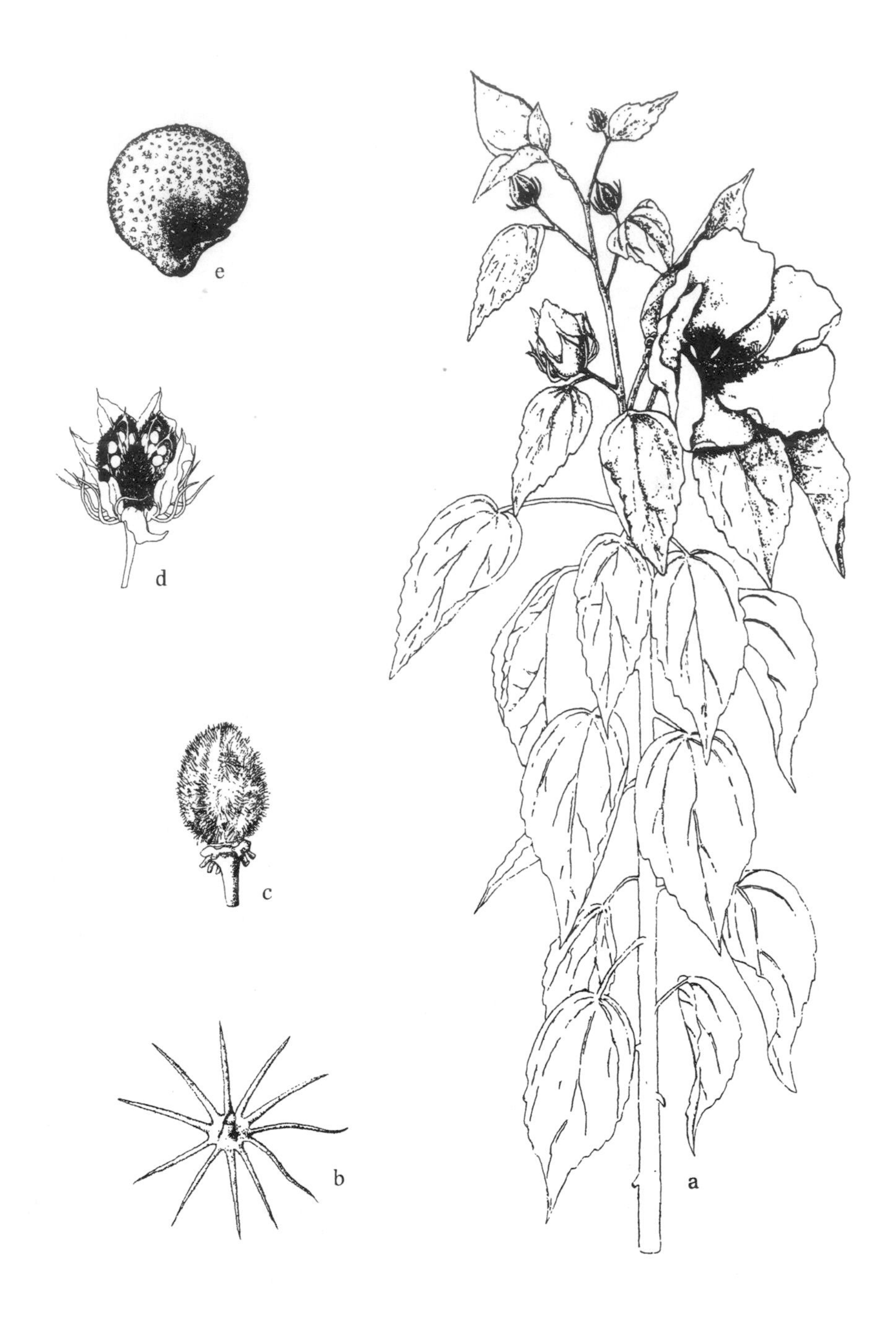

Fig. 434. *Hibiscus moscheutos* subsp. *moscheutos*: a. upper portion of plant; b. epicalyx, bottom view; c. capsule with calyx and epicalyx removed; d. capsule, dehisced, with calyx and epicalyx present; e. seed (G&W, as *H. moscheutos* subsp. *incana*).

Fig. 435. *Hibiscus laevis*: a. flowering branch; b. fruiting branch with persisting calyx and epicalyx of flowers; c. anther; d. seed (C&C, as *H. militaris*).

Fig. 436. *Hibiscus moscheutos* subsp. *moscheutos*: a. leaves; b. flower (F, as *H. palustris*).
Hibiscus moscheutos subsp. *lasiocarpos*: c. expanding flower bud with bracts (epicalyx) (F, as *H. lasiocarpos*).

4. Fruits flattened; stems not rooting at nodes. . . . . 3. *Cardamine*
4. Fruits round in cross-section; stems rooting at nodes. . . . . 4. *Nasturtium*

1. *Subularia* (Awlwort)

Annual, usually submersed, flowering when emersed; leaves basal, in a rosette, subulate; scape naked, few-flowered; flowers minute, 4-merous; petals white; stamens 4 long, 2 short; fruit an ovoid or globose silicle.

REFERENCE: Mulligan and Calder, 1964.

1. *S. aquatica* L. Fig. 437
Shallow water of sandy, gravelly, or muddy lakes and ponds and slow streams. Greenl. and Lab. w. to N.W.T. and Alask., s. to n. N.E., n. N.Y., Ont., ne. Minn., Sask. and mts. to Wyo., Colo., Utah and Calif. A circumpolar species, our taxon is subsp. *americana* Mulligan and Calder, differing little from the Eurasian subsp. *aquatica*.

2. *Rorippa* (Marsh Cress, Yellow Cress)

Annual, perennial, or sometimes biennial herbs; basal rosette often present and well-developed; stems 1-several, erect or submersed; leaves sessile to short-petiolate, simple or divided, or finely dissected; inflorescence a raceme; flowers 4-merous; petals yellow or white (sometimes whitish on drying) or absent; stamens 4 long, 2 short (rarely fewer); fruit a silique.

REFERENCES: Al-Shehbaz, 2010b; Al-Shehbaz and Bates, 1987; Butters and Abbe, 1940; Fernald, 1940; La Rue, 1943; Les, 1994; Stuckey, 1966a, 1966b, 1972.

1. Petals white; submersed leaves finely dissected, transitioning to pinnatifid to lobed to toothed when emergent. . . . . 1. *R. aquatica*
1. Petals yellow; submersed leaves, if present, simple or divided.
 2. Petals absent, stamens usually 4(3–6); pedicels 0.5–1.5 mm long. . . . . 2. *R. sessilifolia*
 2. Petals present; stamens 6; pedicels more than 2 mm long.
  3. Plants annual or biennial, from taproots (lacking rhizomes), petals shorter than or as long as sepals.
   4. Leaves glabrous on lower surface; stems glabrous (rarely sparsely pubescent on lower portion). . . . . 3a. *R. palustris* subsp. *palustris*
   4. Leaves hirsute on lower surface; stems densely hirsute throughout. . . . . 3b. *R. palustris* subsp. *hispida*
  3. Plants perennial, rhizomatous or with creeping stems; petals longer than sepals.
   5. Plants prostrate or decumbent; basal leaves present, but not forming rosettes; upper leaves auriculate. . . . . 4. *R. sinuata*
   5. Plants erect; basal rosettes present, at least on younger plants; upper leaves not auriculate.
    6. Petals (2.7)3.2–5.3(6); fruits 10–25 mm long. . . . . 5. *R. sylvestris*
    6. Petals 2.5–3.5; fruits 3–6 mm long. . . . . 6. *R. amphibia*

1. *R. aquatica* (Eaton) Palmer & Steyerm. Lake Cress Fig. 440
Quiet waters of lakes and ponds, sluggish streams, muddy shores and springy places. S. Que. and Vt. w. to Ont., Mich., Wisc. and Minn., s. to nw. Fla., e. Okla. and e. Tex. (*Armoracia aquatica* (Eaton) Wiegand; *A. lacustris* (A. Gray) Al-Shehbaz & V. Bates; *Neobeckia aquatica* (Eaton) Greene)

2. *R. sessiliflora* (Nutt.) Hitchc. Stalk-less Yellow Cress Fig. 437
Shallow water, wet sites and river gravels. Mass.; Md. and Va. w. to Ohio, Ind., Wisc., se. Minn. and Neb., s. to nw. Fla., La. and Tex. *Nasturtium sessiliflorum* Nutt.; *N. limnosum* Nutt.)

3. *R. palustris* (L.) Besser Bog Yellow Cress Fig. 438
Stuckey (1972) noted that *Rorippa islandica* (Oeder ex Murray) Borbás, a European species, does not occur in North America and that such references to *R. islandica* should be referred to *R. palustris*.

3a. *R. palustris* subsp. *palustris* Fig. 438, 439
Wet shores, creek banks and damp openings. Nfld., Lab., N.B. and w. Que. w. to N.W.T. and Alask., s. throughout the U.S.; Mex., C.Am., S.Am. (intro.); Eurasia. Al-Shehbaz (2010b) noted that many poorly defined varieties were described based on stem height and fruit length, but are not worthy of taxonomic recognition. (*R. palustris* var. *fernaldiana* (Butters & Ahles) Stuckey; *R. palustris* subsp. *glabra* (O. E. Schultz) Stuckey)

3b. *R. palustris* subsp. *hispida* (Desv.) Jonsell Fig. 439
Shores, creek banks, sandy beaches and wet openings (often naturally disturbed sites). Lab. and Nfld. w. to N.W.T. and Alask., s. to Md., Va., Ky., Ill., Iowa, Kans., mts. to N.M., Nev. and n. Calif. (*Nasturtium hispidum* DC.; *R. hispida* (Desv.) Britton; *R. palustris* var. *hispida* (Desv.) Rydb.)

4. *R. sinuata* (Nutt.) Hitchc. Spreading Yellow Cress Fig. 437
Sandy or rocky shores, moist fields and ditches. Md.; Ont.; Ill. nw. to Minn., N.D. s. Sask., B.C. and Wash. s. to w. Ky., Ark., n. Tex., N.M., Ariz. and Calif.

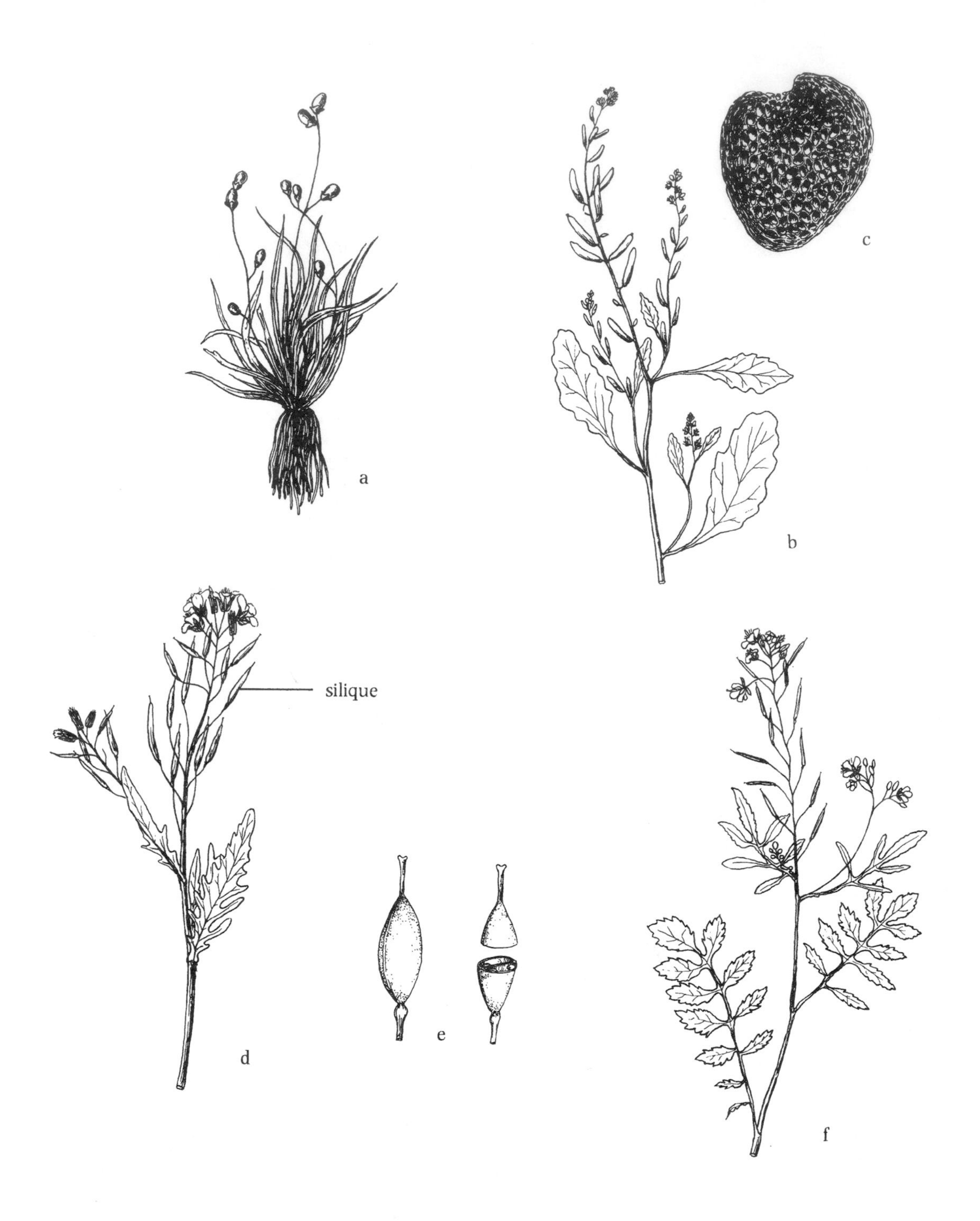

Fig. 437. *Subularia aquatica*: a. habit (F).
Rorippa sessiliflora: b. habit; c. seed (F).
Rorippa sinuata: d. habit (F).
Rorippa amphibia: e. fruit with cross-section (F).
Rorippa sylvestris: f. habit (F).

Fig. 438. *Rorippa palustris*: a–d. emersed leaf variations; e, f. submersed leaf variations (F).

Fig. 439. *Rorippa palustris* subsp. *hispida*: a. habit; b. leaf (F). *Rorippa palustris* subsp. *palustris*: c. habit, upper portion; d. section of stem with inflorescence in leaf axil (F).

Fig. 440. *Rorippa aquatica*: a. basal portion of stem, submersed portion; b. midsection of stem; c. upper midsection of stem; d. flowering stem; e. flower; f. fruit (G&W).

5. *R. sylvestris* (L.) Besser Creeping Yellow Cress Fig. 437
Wet meadows, shores and roadsides; weedy. Nfld. and Que., w. to Alta and B.C. and Alask., s. to N.C., Ala., La., Kans., Colo., Utah, Ariz. and Calif.; intro. from Eur.

6. *R. amphibia* (L.) Besser Great Water Cress Fig. 437
Swamps and shallow water. Very local; Que., sc. Me., and c. Mass. and sw. Conn.; intro. from Eur.

3. *Cardamine* (Bitter Cress)

Annual, biennial, or perennial herbs, often with a basal rosette; stems somewhat short, seldom branched; leaves simple or pinnately divided; flowers 4-merous; petals white, pink, lilac or purplish; fruit a silique, usually linear.

References: Al-Shehbaz et al., 2010; Fernald, 1920; Stuckey, 1962.

---

1. Stem leaves simple, sometimes with 1 or 2 tiny lateral lobes on the petiole (figs. 442i, 443a).
    2. Stems erect, bulbous at base (fig. 441a).
        3. Petals white; stems glabrous; lowermost leaves petiolate, becoming sessile on middle and upper portions of stem.......... 1. *C. bulbosa*
        3. Petals purplish to pink; upper stems sparsely to densely hirsute; leaves all sessile.................................... 2. *C. douglassii*
    2. Stems prostrate to weakly ascending (fig. 442i), often rooting on lower portion, not bulbous at base.
        4. Flowers conspicuous, petals present; pedicels greatly elongating after flowering, 10–20 mm long at maturity..........3. *C. rotundifolia*
        4. Flowers inconspicuous, petals absent; pedicels barely elongating after flowering, 1–3(6) mm long at maturity. ....4. *C. longii*
1. Stem leaves pinnately divided or compound.
    5. Petioles auriculate at base; basal leaves simple. .... 5. *C. clematis*
    5. Petioles not auriculate at base; basal leaves divided or compound (or withered away).
        6. Flowers small, inconspicuous, petals 2–3.5(4) mm long...6. *C. pensylvanica*
        6. Flowers large, conspicuous, petals (6)8–15(18) mm long.. ....7. *C. pratensis*

---

1. *C. bulbosa* (Muhl.) Britton, Sterns & Poggenb. Bulbous Cress, Spring Cress Fig. 441
Wet meadows, springs, brooks and wet woods. N.H., Vt., and sw. Que. w. to Minn., Man. and e. S.D., s. to n. Fla. and e. Tex.

2. *C. douglassii* Britton Purple Cress, Spring Cress Fig. 442
Calcareous springy sites, wet woods, floodplain and bottomlands. W. Mass. and w. Conn. w. to s. Ont., Minn. and Wisc., s. to Va., S.C., Tenn., Ala., and Mo. (*C. bulbosa* var. *purpurea* (Torr.) Britton, Sterns & Poggenb.)

3. *C. rotundifolia* Michx. Mountain Water Cress Fig. 443
Springy sites, margins of brooks, low woods, and swampy places. Chiefly montane; N.Y. w. to Ohio, s. to Va., N.C., nw. Ga. and Tenn.

4. *C. longii* Fernald Long's Bitter Cress Fig. 442
Tidal flats, muds, and banks, swampy areas and along shores of tidal rivers, usually shaded; rarely inland. Rare and local, mostly coastal; Me., Mass., R.I., Conn., s. N.Y. and Chesapeake Bay, Md. and Va. and e. N.C.

5. *C. clematitis* Shuttlew. ex A. Gray Fig. 442
Springy sites and rocky streams. Mts. of sw. Va. s. to w. N.C., w. S.C., and e. Tenn.

6. *C. pensylvanica* Muhl. ex Willd. Pennsylvania Bitter Cress Figs. 441, 443
Streams, swamps, wet woods, clearings and ditches. S. Lab. and Nfld. w. to Man., sw. N.W.T., B.C. and Alask., s. to Fla., Ala., Ark., Okla., Tex., N.M. and n. Calif. (*C. pensylvanica* var. *brittoniana* Farw.)

7. *C. pratensis* L. Cuckoo Flower Fig. 441
Bogs, swamps and springs, chiefly calcareous. Lab., Nfld. and Que., w. to N.W.T. and Alask., s. to n. N.J., Md., W.Va., n. Ohio, n. Ind. n. Ill., Wisc., n. Minn., Sask., Alta. and B.C. (*C. palustris* (Wimm. & Grab.) Peterm.)

4. *Nasturtium* (Water Cress)

Perennial, mat-forming herbs (often floating); stems prostrate to decumbent, rooting at nodes; leaves pinnately compound; flowers 4-merous; petals white; fruit a linear silique.

References: Al-Shehbaz, 2010b; Al-Shehbaz and Price, 1998; Green, 1962.

---

1. Fruits short, relatively broad 2–2.5(3) mm long; seeds appearing to lie in 2 rows (fig. 444g); seed coat with large reticulations, ca. 25–50(60) areolae per side. ....1. *N. officinale*
1. Fruits long, more slender, 1–27 mm long; seeds appearing to lie in 1 row (fig. 445d); seed coat with small reticulations ca. (75)100–150(175) areolae.. 2. *N. microphyllum*

Fig. 441. *Cardamine bulbosa*: a. habit (F).
Cardamine pensylvanica: b. habit (F).
Cardamine pratensis: c. habit (F).

Fig. 442. *Cardamine douglassii*: a. lower portion of plant; b. upper portion of plant with flowers; c. upper portion of plant with fruit; d. fruit (B&B).
Cardamine clematitis: e. habit; f. flower; g. fruit; h. seed (B&B).
Cardamine longii: i. habit, procumbent; j. fruit (Crow).

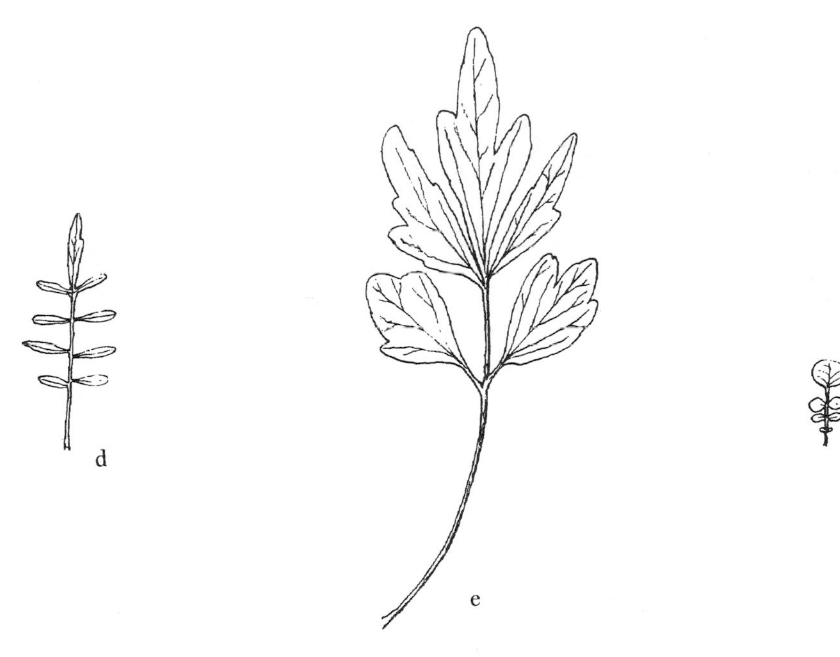

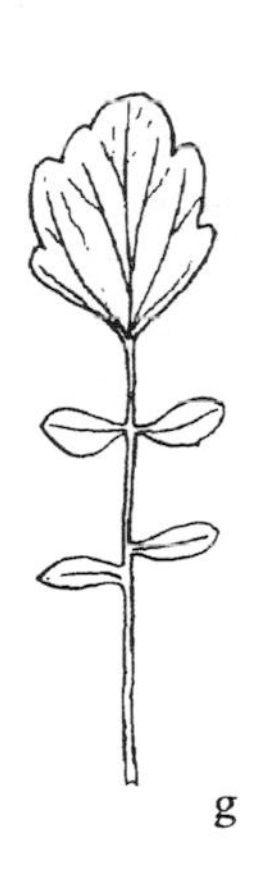

Fig. 443. *Cardamine rotundifolia*: a. habit; b. flower; c. fruit (B&B).
Cardamine pensylvanica: d–g. leaf variations (F).

1. *N. officinale* R. Br. Water Cress Fig. 444
Cold brooks, springs, seeps and muddy shores of sluggish streams. Widespread in temperate N.Am., C.Am. and S.Am.; intro. from Eur. (cultivated for green salads). (*Rorippa nasturtium-aquaticum* (L.) Hayek)

2. *N. microphyllum* Boenn. ex Reichenb. Water Cress Fig. 445
Springs and cool streams, ponds, swales, wet meadows and seeps. Nfld. and Lab., w. to Man., s. to N.E., to N.Y., Mich., Ind. and Ky.; local, B.C., Ida., Oreg., Wyo. and N.M.; intro. from Eur. The rare hybrid, *N. ×sterilis* Airy Shaw, is a sterile hybrid resulting from a cross between *N. officinale* and *N. microphyllum* and can be recognized by poorly formed seeds. (*Rorippa microphylla* (Boenn. ex Reichenb.) Hyl. ex Á. Löve & D. Löve; *N. officinale* var. *microphyllum* (Reichenb.) Thell.; *N. uniseriatum* Howard & Manton)

Droseraceae / Sundew Family

1. Free-floating aquatic plants, lacking roots; leaves whorled along the stem, with inflated petioles and blades with folded edges (appearing half-moon shaped), forming active snap traps. . . . . 1. *Aldrovanda*
1. Rooted terrestrial plants; leaves all in a basal rosette, margins and surface bearing stalked gland-tipped hairs, forming passive sticky traps. . . . . 2. *Drosera*

1. *Aldrovanda* (Waterwheel Plant)

Aquatic free-floating herbs; stem simple or sparsely branched; leaves in whorls; petiole swollen, flattened expanding upwards, dividing into apical bristles of somewhat semicircular upturned halves with sensitive hairs and digestive glands, blades snapping closed, trapping organisms; flowers solitary, axillary, white, sepals, stamens and petals 5; fruit a 5-valved capsule.

REFERENCES: Adamec, 2018; Westermeier et al., 2018.

1. *A. vesciculosa* L. Waterwheel Fig. 446
Floating in acidic waters among sedges and Ericaceous plants in bogs and among shoreline emergent vegetation. Introd. into n. N.J., s. N.Y. and s. N.H. This is a monotypic species native to Europe, Asia, Africa and Australia; ours have been introduced from Japan via insectivorous plant enthusiasts.

2. *Drosera* (Sundew)

Annual, biennial, or perennial (ours) herbs; insectivorous; leaves in a basal rosette, spreading to erect, upper surfaces and margins bearing gland-tipped hairs, exuding sticky drops of fluid which glisten in the sun, attracting and trapping insects, sessile glands secreting digestive enzymes; flowers small, borne in a scorpioid cyme; petals white or pink; fruit a capsule.

REFERENCES: McPherson and Schnell, 2012; Mellichamp, 2015; Schnell, 2002; Shinners, 1962; Wood, 1960; Wynne, 1944.

1. Leaves with expanded blades, 3–12 mm wide, suborbicular to obovate, spatulate, or linear; plants not bulbous at base.
   2. Leaf blade suborbicular to slightly wider than long; seeds elongate-fusiform, with longitudinal striations (but not pitted). . . . . 1. *D. rotundifolia*
   2. Leaf blade spatulate to obovate or linear; seeds variously ornamented, if elongate-fusiform, then with elongate pits (giving a somewhat striate appearance).
      3. Leaf blades linear (10–30 mm long, 1.5–3 mm wide); seeds rhomboid. . . . . 2. *D. linearis*
      3. Leaf blades spatulate to obovate (13–50 mm long, 3–12 mm wide); seeds various, but not rhomboid.
         4. Scapes glandular-pubescent nearly to base; blades tapering to broad petioles, often not clearly differentiated from blades, stipules absent. . . . . 3. *D. brevifolia*
         4. Scapes glabrous; blades clearly differentiated from petioles, stipules present (adnate or free).
            5. Leaf blades spatulate to elongate-spatulate; petioles glabrous; flowers white.
               6. Leaf blades spatulate, 8–20 mm long; stipules free; seeds oblong, 0.7–1 mm long, papillate. . . . . 4. *D. intermedia*
               6. Leaf blades elongate-spatulate, 15–35(50) mm long; stipules adnate to petioles, except at tip; seeds elongate-fusiform, 1–1.5 mm long, with elongate pits giving striate appearance. . . . . 5. *D. anglica*
            5. Leaf blades broadly spatulate to orbiculate; petioles bearing long, non-glandular hairs; flowers pink. . . . . 6. *D. capillaris*
1. Leaves long-filiform, blades 1.5–1 mm wide, barely, if at all, differentiated from the flattened petioles (coiled fiddiehead-like when emerging) (fig. 447c); plants bulbous at base. . . . . 7. *D. filiformis*

Fig. 444. *Nasturtium officinale*: a. upper portion of plant (F); b. robust form, emersed and trailing in shallow water (F); c. submersed form, deep water (F); d. submersed form, shallow water (F); e. terrestrial form (F); f. flower, 1 petal removed (F); g. fruit, seeds in 2 rows (G&W); h. seed (G&W).

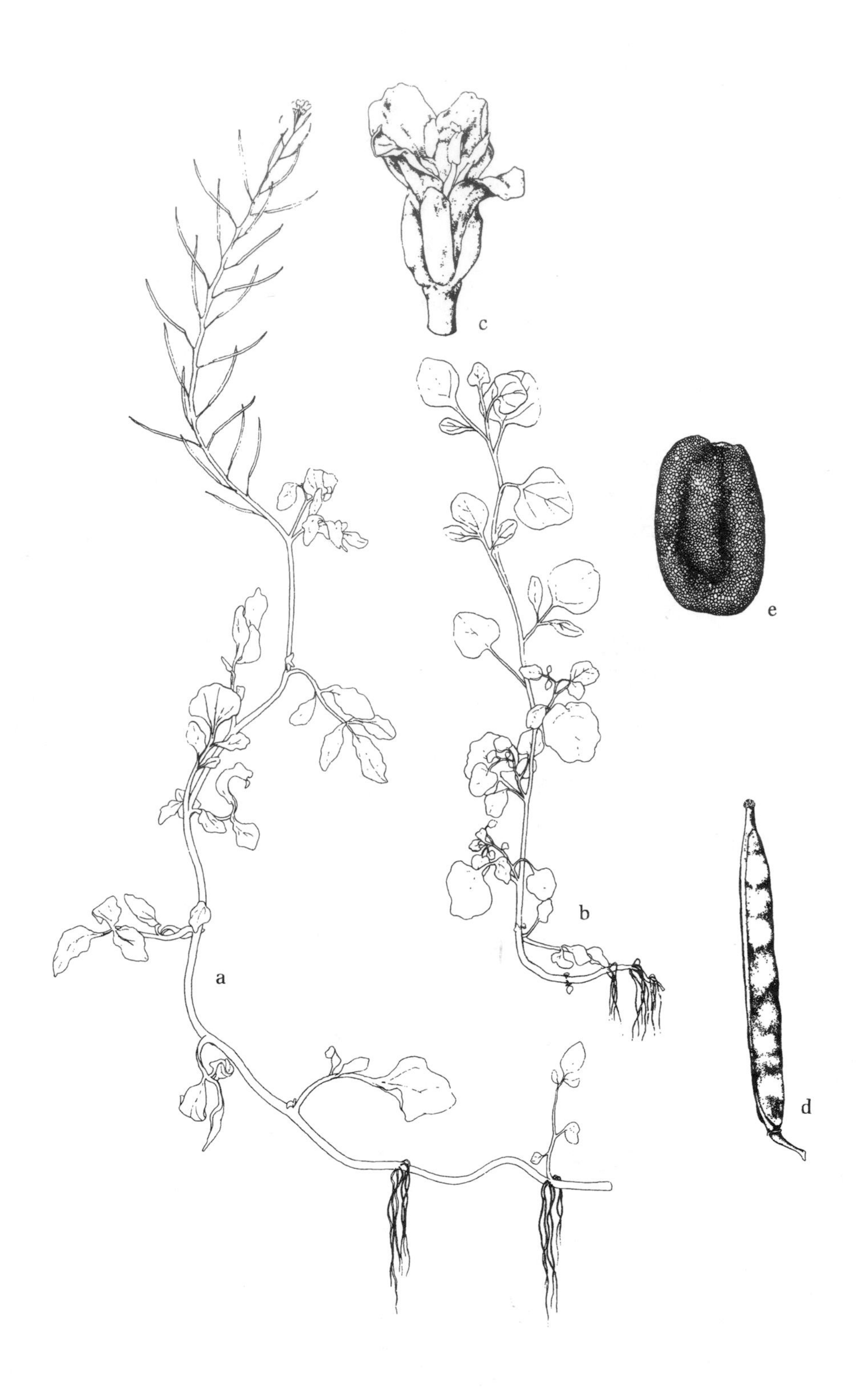

Fig. 445. *Nasturtium microphyllum*: a. habit. flowering and fruiting; b. habit, vegetative; c. flower; d. fruit, seeds in single row; e. seed (G&W).

Fig. 446. *Aldrovanda vesiculosa*: a. habit, submersed; b. whorl of leaves; c. leaf (FNTA).
Drosera rotundifolia: d. habit (*Fl. China*).
Drosera intermedia: e. habit; f. flower; g. capsule (FNA).

1. *D. rotundifolia* L. Round-leaved Sundew Fig. 446
Bogs, wet peaty or sandy shores, sandy swamps, often on hummocks. Greenl., Nfld. and Lab. w. to N.W.T. and Alask., s. to N.E., N.J., chiefly in mts. to N.C., Ga., and Miss. Ill., Ky., Tenn. Minn., Iowa, Mont., Colo., Oreg. and n. Calif. (*D. rotundifolia* var. *comosa* Fernald)

2. *D. linearis* Goldie Linear-leaved Sundew Fig. 447
Calcareous bogs and wet pannes, marly shores and wet limy sites. W. Nfld., Lab. and Que. w. s. Man., Sask., and sw. N.W.T. and B.C. s. to n. Me., Bruce Pen., Ont., n. Mich., Wisc., n. Minn., Mont. and B.C.

3. *D. brevifolia* Pursh Shortleaf Sundew, Dwarf Sundew Fig. 447
Wet peats and sands, savannas, wet pinelands and wet ditches. Coastal plain, e. Va. s. to Fla., w. to La. and e. Tex., Mississippi embayment n. to Ark., e. Okla. and e. Kans.; local, Ky. and Tenn.

4. *D. intermedia* Hayne Spatulate-leaved Sundew Fig. 446
Bogs, wet acid sands and peats and borders of swamps, often in shallow pools. Nfld. and Lab w. to Ont. and ne. Minn., s. to N.E., N.J., mts. to Tenn. and S.C., Ohio, and n. Ill.; coastal plain, N.J. s. to Fla., w. to Miss. and e. Tex.

5. *D. anglica* Huds. Great Sundew, English Sundew Fig. 447
Calcareous bogs and wet pannes, marly shores and wet limy sites. Local, s. Lab. and w. Nfld. w. to N.W.T. and Alask., s. to n. Me., Mich., n. Wisc., n. Minn., Sask., Mont., Wyo., Colo, Ida. Wash., Oreg. and n. Calif.; Eurasia, Hawaii. According to Mellichamp (2015) this species a fertile tetraploid apparently derived of hybrid origin, with *D. rotundifolia* and *D. linearis* as parental species; furthermore a sterile hybrid may be encountered wherever these two species grow together. Schnell (2002) noted that *D. angelica* typically has wider flowers (8–10 mm vs. 2–7 mm) than the sterile hybrid, as well as thicker scapes (1.5–2 mm vs. 1–1.2 mm).

6. *D. capillaris* Poir. Pink Sundew
Wet peats and sands, savannas, wet pinelands and wet sandy ditches. Coastal plain, Del. and se. Va. s. to Fla., w. to e. Tex.; inland to Tenn. and Ark.; Trop. Am.

7. *D. filiformis* Raf. Dew-thread, Thread-leaf Sundew Fig. 447
Damp sands, and sandy or gravelly pond margins. Coastal plain, N.S. e. Mass. s. to s. N.J., N.C. and nw. Fla.

Polygonaceae / Smartweed Family

References: Graham and Wood, 1965; Mitchell and Dean, 1978.

1. Perianth parts 4 or 5, in similar whorls, not enlarging in fruit flowers fascicled in spike-like inflorescences or in axillary pedunculate clusters.
   2. Flowers in axillary, nearly sessile fascicles, (2–4) flowered; flowers usually closed, anthers whitish-yellow; leaf blades linear, elliptic to obovate, plants of coastal saline and inland saline waters (ours). . . . . 1. *Polygonum*
   2. Flowers fascicled in spike-like inflorescences (terminal, axillary or both); flowers open, anthers yellow, pink, or red; leaf blades lanceolate, ovate, hastate or sagittate; plants of freshwater locations. . . . . 2. *Persicaria*
1. Perianth parts 6, in 2 dissimilar whorls, the inner 3 becoming enlarged, forming conspicuous wing-like valves surrounding fruit (fig. 459b); flowers in whorls (fig. 461d). . . . . 3. *Rumex*

1. *Polygonum* (Knotweed)

Annual (rarely perennial), stems mainly prostrate or decumbent; stipular sheaths (ocrea) 4–12-veined; petiole articulated to proximal part of ocrea; flowers mostly on axillary inflorescence 1–7(10)-flowered, subtended by sheathing bracts (ocreolae); flowers usually closed, perianth of sepals, united at base (petals absent), outer perianth lobes equaling or somewhat larger than inner, anthers whitish yellow; fruits included or exserted from persistent perianth.

Reference: Costa et al., 2005.

1. Leaves gray-green, strongly glaucous; stipular sheaths conspicuous, upper portion silvery to hyaline, persistent; achenes brown to blackish, 3–4 mm long, exserted from perianth (fig. 448k). . . . . 1. *P. glaucum*
1. Leaves bluish-green to yellow-green, sometimes weakly glaucous; stipular sheaths not conspicuous, upper portion brownish or hyaline, not persistent, soon becoming lacerate; achenes often dimorphic, brown, 2–3.5 mm long, included in perianth, or pale yellow-brown to greenish, 4–6.5 mm long, exserted from or included within perianth (fig. 448g).
   2. Leaves strongly dimorphic, yellowish-green fresh or dried; pedicels 2.5–6 mm long, exserted from stipular sheath (fig. 448b). . . . . 2a. *P. ramosissimum* subsp. *ramosissimum*
   2. Leaves usually alike, not strongly dimorphic, bluish-green fresh, turning dark brown or black after drying; pedicels 1–2 mm long, shorter than and included within stipular sheaths (fig. 448e). . . . . 2b. *P. ramosissimum* subsp. *prolificum*

Fig. 447. *Drosera linearis*: a. habit (Crow).
Drosera anglica: b. habit (Crow).
Drosera filiformis: c. habit; d. flower (Gleason).
Drosera brevifolia: e. habit; f. leaf (B&W).

1. *P. glaucum* Nutt. Seaside Knotweed, Seabeach Knotweed Fig. 448

Seashores, margins of saline ponds and salt marshes, dune hollows, wet pannes and borders of tidal streams. Coastal, Mass. s. to Ga.

2. *P. ramosissimum* Michx.

2a. *P. ramosissimum* subsp. *ramosissimum* Bushy Knotweed Fig. 448

Sandy and gravelly soils, shores of ponds and lakes, salt marshes and sometimes disturbed sites. Chiefly midwestern, N.S., N.B., N.E. and s. Que. w. to Sask., N.W.T., B.C. and e. Wash., s. to Ga., Tenn., Tex., N.M., Ariz. and Calif.

2b. *P. ramosissimum* subsp. *prolificum* (Small) Castea & Tardif Bushy Knotweed Fig. 448

Wet saline places, salt marshes and coastal shores, wet prairies and waste places. P.E.I. and N.E. w. to Ont., Sask., and B.C. s. to N.C., Tenn., Ala., Tex. N.M., Utah and Calif. (*P. ramosissimum* var. *prolificum* Small; *P. prolificum* (Small) B. L. Rob.).

2. *Persicaria* (Smartweed, Knotweed)

Annual or perennial herbs, terrestrial, aquatic, or amphibious; leaves alternate, petiolate or sessile, stipules modified, forming a tubular or 2-lobed sheath (ocrea) surrounding stem, nodes usually swollen; flowers fascicled in relatively short spike-like inflorescences or in clusters on terminal or axillary peduncles, subtended by sheathing bracts (ocreolae); sometimes peduncles barely longer than ocreolae; perianth of sepals (petals absent), united at base, tepal lobes 4 or 5(6); fruit an achene, enclosed in persistent perianth.

References: Brooks and Mertens, 1972; Fassett, 1949b; Fernald, 1917b; Hinds and Freeman, 2005; Mertens, 1965; Mitchell, 1968, 1971, 1976; Park, 1988; Savage and Mertens, 1968; Wolf and McNeill, 1986.

---

1. Stems bearing recurved barbs (fig. 449a,e).
   2. Leaves sagittate, strongly barbed on midvein of lower surface (fig. 449b); flowers in globose clusters; peduncles glabrous; achene strongly 3-angled. . . . . 1. *P. sagittata*
   2. Leaves hastate, not barbed on midvein; flowers in few-flowered racemes; peduncles hispid (fig. 449f); achene biconvex. . . . . 2. *P. arifolia*
1. Stems lacking barbs.
   3. Flowers borne in much elongated spike-like, chiefly terminal, inflorescences with flowers widely separated from each other; styles elongated, persisting in fruit, hooked at the tip (fig. 450d). . . . . 3. *P. virginiana*
   3. Flowers fascicled in spike-like inflorescences, terminal or terminal and axillary, flowers often close together (inflorescences interrupted in *P. punctata* and *P. robustior*); styles deciduous in fruit.
      4. Plants perennial, with rhizomes or stolons, prostrate or, if ascending, then rooting at the lower nodes.
         5. Inflorescences usually solitary (occasionally 2), terminal only. . . . . 4. *P. amphibia*
         5. Inflorescences usually several per stem, terminal and/or axillary.
            6. Stipular sheaths entire, lacking bristles (fig. 452b) (older sheaths may become fractured and shattered, with persistent nerves appearing as bristle). . . . . 5. *P. glabra*
            6. Stipular sheaths with fringe of bristles (fig. 453f).
               7. Perianth lobes with numerous, small glandular dots (fig. 452e) (sometimes obscure on fresh material).
                  8. Sheathing bracts (ocreolae) subtending flowers within inflorescence lacking cilia (fig. 452d); plants robust, leaves large, 2–4.5 cm wide; inflorescences interrupted, flowers somewhat sparse (tiny leaves sometimes occurring in inflorescences). . . . . 6. *P. robustior*
                  8. Sheathing bracts (ocreolae) subtending flowers within inflorescence ciliate (fig. 453); plants slender, leaves smaller 0.5–2.5 cm wide; inflorescence essentially uninterrupted, flowers numerous. . . . . 7. *P. punctata*
               7. Perianth lobes lacking glandular dots (but sometimes in *P. hydropiperoides* with dots restricted to perianth tube or within perianth).
                  9. Perianth lobes white to greenish-white to creamy; stipular bristles 5–12(15) mm long. . . . . 8. *P. setacea*
                  9. Perianth lobes pink to purplish; sometimes greenish-white; stipular bristles (1)2–4 mm long. . . . . 9. *P. hydropiperoides*
      4. Plants annual with taproots and fibrous roots, lacking rhizomes or stolons.
         10. Stipular bracts of stems with fringe of bristles.
             11. Perianth lobes with glandular dots.
                 12. Achenes dull, minutely roughened; plants with strong peppery taste. . . . . 10. *P. hydropiper*
                 12. Achenes shiny, smooth; plants lacking peppery taste. . . . . 7. *P. punctata*
             11. Perianth lobes lacking glandular dots.
                 13. Peduncles and stems strongly glandular-pubescent (fig. 456a,b). . . . . 11. *P. careyi*
                 13. Peduncles glabrous, stems glabrous to sparsely pubescent, but not glandular.

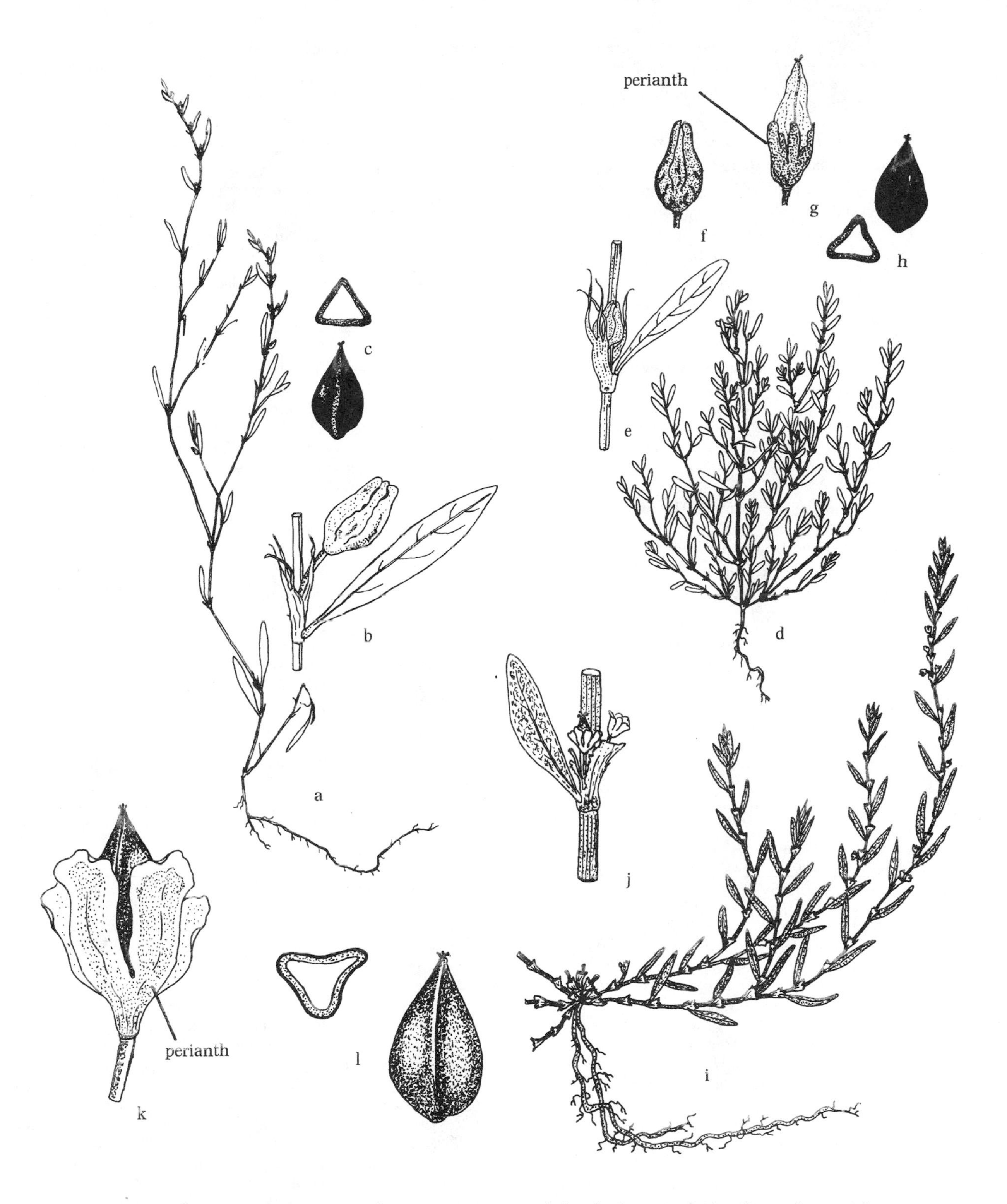

Fig. 448. *Polygonum ramosissimum* subsp. *ramosissimum*: a. habit; b. flower in leaf axil; c. achene with cross-section (NYS Museum).
Polygonum ramosissimum subsp. *prolificum*: d. habit; e. flower in leaf axil; f. achene enclosed by perianth; g. achene exserted from perianth; h. achene with cross-section (NYS Museum).
Polygonum glaucum: i. habit; j. flowers in leaf axil; k. achene slightly exserted from perianth; l. achene with cross-section (NYS Museum).

14. Bristles of inflorescence ocreolae subtending flowers 0.2–1.3(2) mm long.
    15. Perianth at anthesis 1.8–2 mm long; inflorescence 4–6.5 mm wide; leaf lacking. . . . . 12. *P. puritanorum*
    15. Perianth at anthesis 2–3 mm long; inflorescence 7–11 mm wide; leaf with distinct leaf blotch. . . . . 13. *P. maculosa*
14. Bristles of inflorescence ocreolae subtending flowers (0.5)1–4(6) mm long. . . . . 14. *P. longiseta*

10. Sheathing bracts of stems entire or with very short bristles to 1 mm long.
    16. Outer perianth lobes with anchor-shaped veins, perianth lobes 4(5); inflorescence usually nodding. . . . . 15. *P. lapathifolia*
    16. Outer petals lacking anchor-shaped veins, perianth lobes 5; inflorescence mostly erect.
        17. Flowers heterostylous; achenes usually with a central hump; ocreolae subtending flowers without bristles, or if present, to 0.8 mm. . . . . 16. *P. bicornis*
        17. Flowers homostylous; achenes with a central hump (other side slightly concave); ocreolae subtending flowers with or without bristles, if present, only to 0.5 mm. . . . . 17. *P. pensylvanica*

---

1. *P. sagittata* (L.) H. Gross Tearthumb, Arrow-vine Fig. 449
Wet meadows, marshes, wet woods and thickets, stream banks, ditches and margins of tidal creeks. Nfld. w. to Sask, s. to n. Fla. and e. Tex.; Oreg. A more slender ecological variant of fresh and tidal marshes of southeastern Virginia has been described as var. *gracilentum* Fernald. (*Polygonum sagittatum* L.)

2. *P. arifolia* (L.) Haraldson Tearthumb, Halberd-leaved Tearthumb Fig. 449
Wet soils, marshes, wet meadows and tidal marshes. P.E.I. and N.B. w. to s. Ont. and Minn., s. to Ga. and Mo.; Wash. Plants of the se. U.S. tend to have larger achenes and are sometimes segregated as var. *pubescens* (Keller) Fernald. (*Polygonum arifolium* L.)

3. *P. virginiana* (L.) Gaertner Jumpseed Fig. 450
Swamp woods, alluvial woods, stream banks and moist thickets. P.E.I., sw. Que., Ont. and w. N.H. w. to s. Ont. and Minn., s. to Fla., Tex. and Mex. (*Polygonum virginicum* L.)

4. *P. amphibia* (L.) S. F. Gray Water Smartweed Fig. 451
Amphibious; shallow water and shores of lakes and ponds, quiet streams, swamps, marshes, prairies and dry pond or lake beds. Nfld., Que. and N.S. w. to Alta, Yuk. and Alask., B.C. and Wash. s. to S.C., Tenn., Miss., Tex., and Calif. This is an extremely variable species with aquatic and emergent growth forms as well as stranded and terrestrial forms. According to Mitchell (1968, 1976) and Mitchell and Dean (1978) much of the phenotypic variability observed in this species is apparently strongly clinal. Mitchell recognized two varieties native to North America, var. *emersa* and var. *stipulacea*, and noted that intermediates which frequently occur between these varieties are usually amphibious and resemble var. *stipulacea*, but with somewhat longer inflorescences. However, Hinds and Freeman (2005) point out that formal recognition of varieties is highly questionable when Eurasian populations are considered. Therefore, we have refrained from recognizing infraspecific taxa. (*Polygonum amphibium* var. *stipulaceum* N. Coleman; *Persicaria amphibia* var. *emersa* (Michx.) J. C. Hickman)

5. *P. glabra* (Willd.) M. Gómez Smooth Smartweed, Denseflower Knotweed Fig. 452
Typically in water of swampy woods, wet thickets, marshy shores and margins of pools. Chiefly coastal plain, s. N.J. s. to Fla., w. to Tex.; Mississippi embayment n. to se. Mo. and Tenn.; W.I., C.Am. and S.Am. Although previously treated as *P. densiflorum*, Hinds and Freeman (2005) note that differences between our North American plants and those of Eurasian *P. glabra* are very minor and have merged the two taxa. (*Persicaria densiflora* (Meisn.) Moldenke; *Polygonum densiflorum* Meisn.)

6. *P. robustior* (Small) E. P. Bicknell Stout Smartweed, Water Smartweed Fig. 452
Shallow water of ponds and lakes, stream margins, swampy places and wet soils. Chiefly coastal plain, N.S., and Que. w. to Ont. and Mich., s. to Va., Ky. and Mo.; Fla. and Tex. (*Polygonum punctatum* var. *majus* (Meisn.) Fassett; *Polygonum robustius* (Small) Fernald)

7. *P. punctata* (Elliott) Small. Dotted Smartweed, Water Smartweed Fig. 453
Shallow water of lakes and ponds, shores, marshes, swamps, flood plains, tidal shores and wet soils. N.S. and Que. w. to Sask., and B.C., s. to Fla., Tex. and Calif. and Mex; C.Am. and S.Am. This wide-ranging species is extremely variable, and numerous varieties have been described from the New World. Occurrence of intermediates, however, is widespread, and taxonomic recognition of some of these variants is questionable. (*Polygonum punctatum* Elliott.; *Polygonum punctatum* var. *confertiflorum* (Meisn.) Fassett)

8. *P. setacea* (Baldwin) Small Bog Smartweed Small Smartweed Fig. 453
Pond and lake margins, swamps, wet woods and clearings. S. N.E. and N.Y. s. to Fla., Ill., Mo., Kans. and Tex.; Wash. (*Polygonum setaceum* Baldwin)

9. *P. hydropiperoides* (Michx.) Small Mild Water-pepper, Water Smartweed, Swamp Smartweed Fig. 454
Shallow water, shores, swales, swamp margins, wet sandy and peaty shores, ditches, thickets and bottomlands. N.S. and N.S. w. to s. Ont., Minn. and B.C., s. to Fla., Tex., Calif.; inland n. to Okla., s. Mo. and s. Ill.; Mex.; S.Am. Two varieties were recognized in the previous edition: var. *hydropiperoides* (fig. 454c,d), with plants having pink to white perianth and achenes enclosed by the perianth and var. *opelousanum* (fig. 454a,b), with plants having greenish perianth, leaves with pale flat plate-like glands on the lower surface of the leaves, and achenes typically exserted from perianth. While these are sometimes recognized at the species level, variation in this complex is considerable, thus Hinds and Freeman (2005) recognize no infraspecific taxa. (*Persicaria opelousana* (Riddell ex Small) Small; *Polygonum hydropiperoides* Michx.; *Polygonum opelousanum* Riddell ex Small; *Polygonum hydropiperoides* var. *adenocalyx* (Stanford) Gleason)

10. *P. hydropiper* (L.) Opiz Water-pepper, Common Smartweed; Marshpepper Knotweed Fig. 455
Wet meadows and pastures, bottomlands, marshy shores and wet disturbed sites. Widespread in N.Am.; intro. weed from Eurasia. (*Polygonum hydropiper* L.)

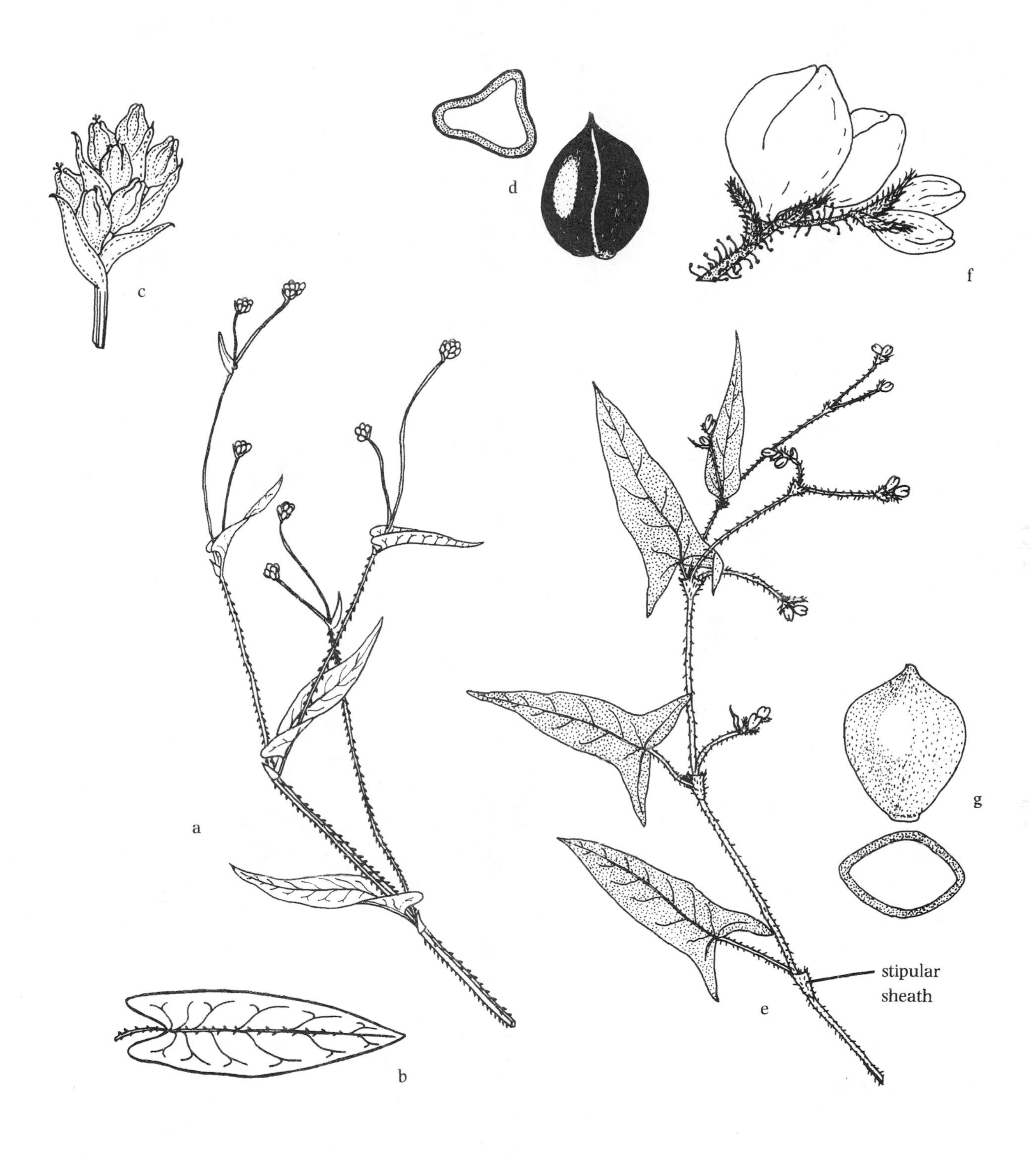

Fig. 449. *Persicaria sagittata*: a. habit; b. leaf. lower surface; c. inflorescence; d. achene with cross-section (NYS Museum).
Persicaria arifolia: e. habit; f. inflorescence; g. achene with cross-section (NYS Museum).

Fig. 450. *Persicaria virginiana*: a. habit; b. stipular sheath; c. flower with developing achene; d. achene; e. achene, cross-section (G&W).

Fig. 451. *Persicaria amphibia*: a. habit, aquatic form; b. stipular sheath; c. flower cluster with bilobed bracts; d. flowers; e. achene; f. habit, terrestrial form; g. rhizomes; h. inflorescence; i. achene, two views (Reed).

Fig. 452. *Persicaria glabra*: a. upper portion of plant (F); b. stipular sheath (G&W).
Persicaria robustior: c. upper portion of plant; d. inflorescence; e. perianth enclosing achene; f. achene (NYS Museum).
Persicaria puritanorum: g. upper portion of plant (CBH).

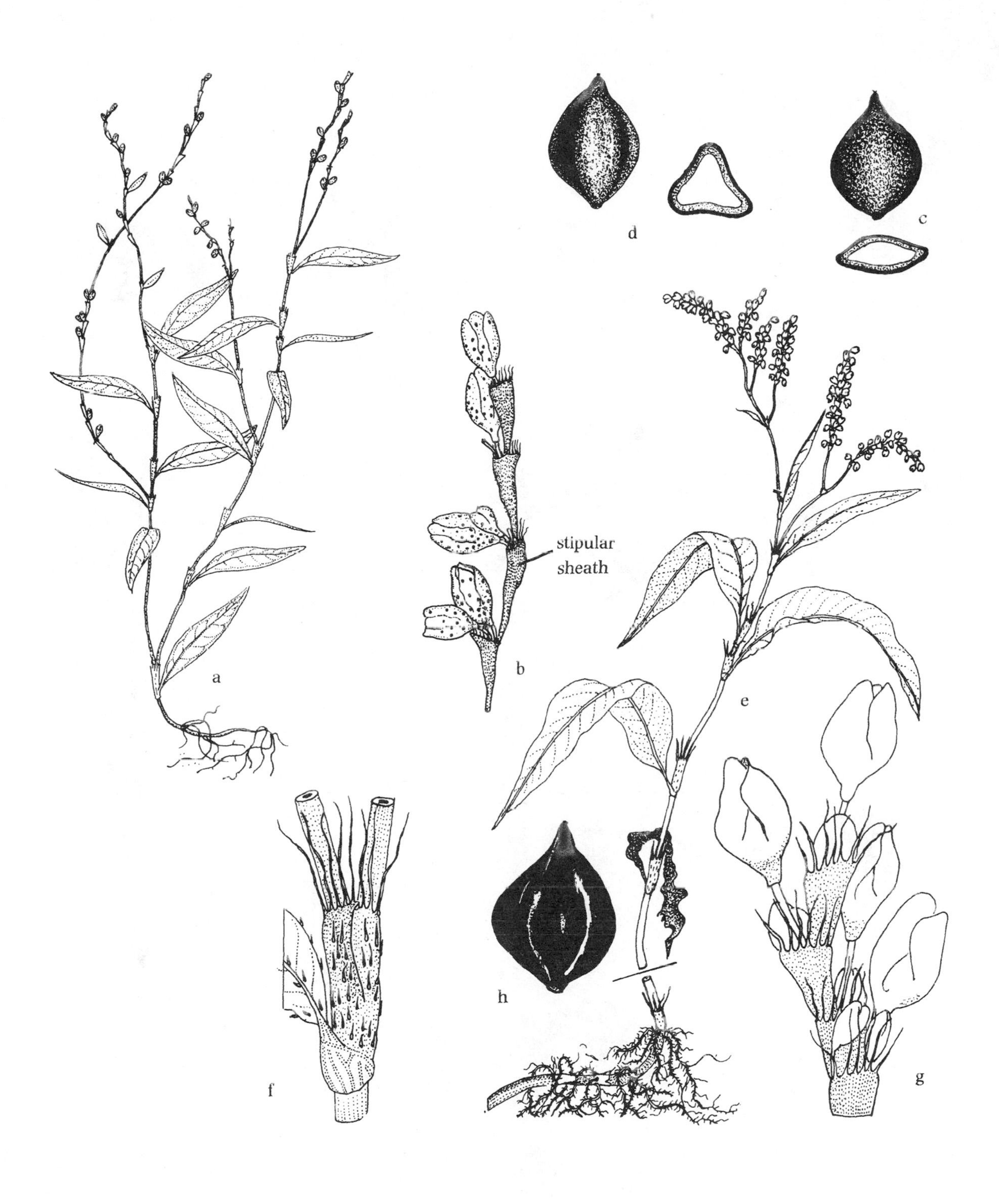

Fig. 453. *Persicaria punctata*: a. habit; b. inflorescence; c. achene; with cross-section; d. achene with cross-section (NYS Museum).
Persicaria setacea: e. habit; f. section of stem with stipular sheath; g. inflorescence; h. achene (NYS Museum).

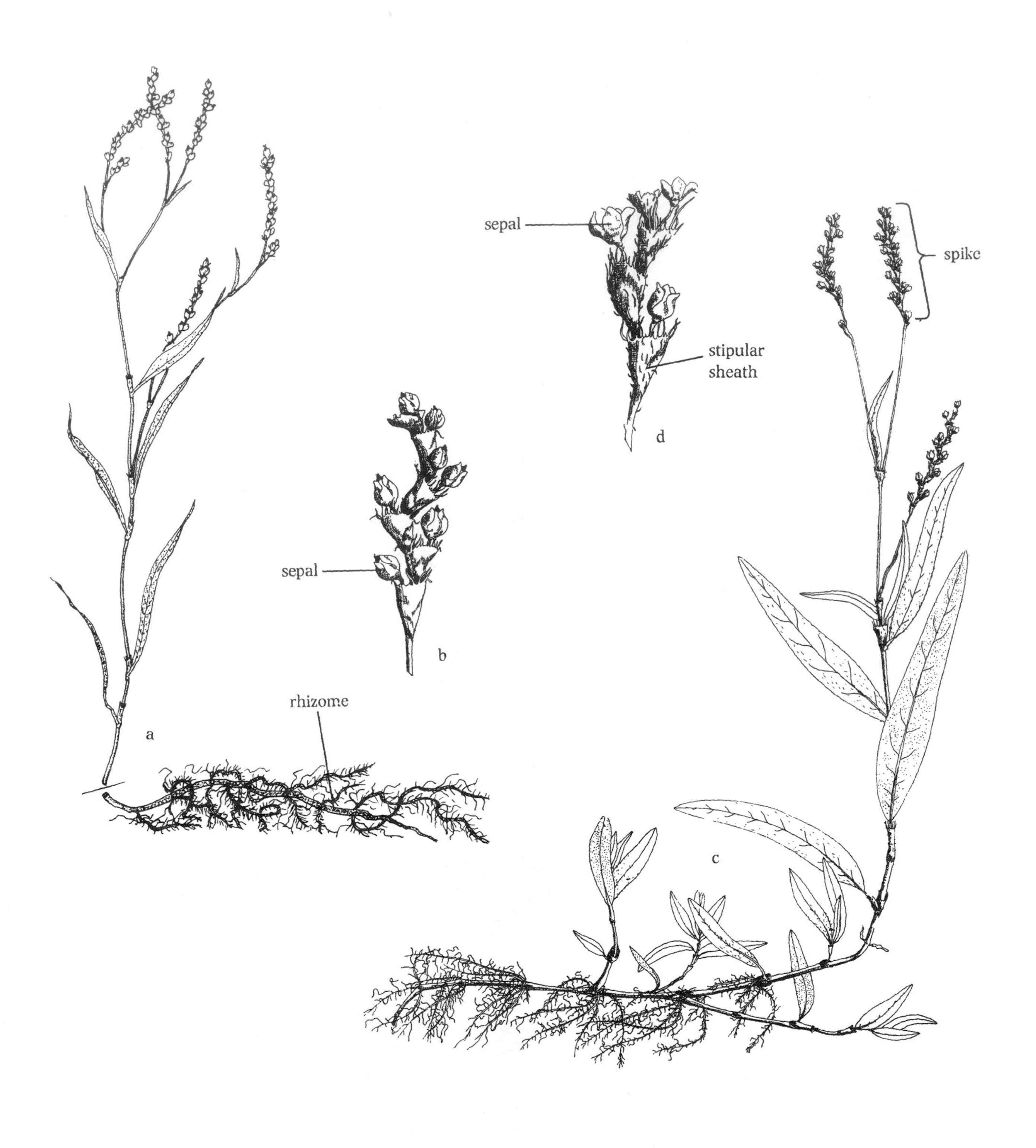

Fig. 454. *Persicaria hydropiperoides*: a. habit (NYS Museum); b. inflorescence (F); c. habit; d. inflorescence (F).

Fig. 455. *Persicaria hydropiper*: a. habit; b. stipular sheath (ochrea), 2 views; c. inflorescence; d. flower; e. achenes, with cross-sections, showing variations (Reed).

11. *P. careyi* (Olney) Greene Carey's Smartweed, Pinkweed Fig. 456
Swamps, thickets, meadows, wet fields, riverbeds and wet disturbed sites. Me. w. to Ont., Wisc. and Minn., s. to Del. and n. Ind. (*Polygonum careyi* Olney)

12. *P. puritanorum* (Fernald) Soják Water Smartweed Fig. 452
Sandy and gravelly coastal shores. Endemic, Me., Mass. and R.I.

13. *P. maculosa* A. Gray Spotted Lady's Thumb, Lady's Thumb, Heart's-ease, Heartweed, Smartweed. Fig. 457
Damp disturbed sites, ditches, stream banks, and shores; weedy. Introd. from Eurasia, throughout N.Am. and Greenl. (*Polygonum persicaria* L.)

14. *P. longiseta* (Bruijn) Kitag. Oriental Lady's Thumb, Smartweed Fig. 456
Shores, damp ground, roadsides and disturbed ground. N.B., and Que. w. to Mich., Wisc., Minn. and Neb., s. to Fla., La., Tex. and N.M; Alta. and B.C.; intro. from e. Asia. (*Polygonum longisetum* Bruijn; *Polygonum cespitosum* Blume; *Persicaria caespitosa* var. *longiseta* (Bruijn) C. E. Reed)

15. *P. lapathifolia* (L.) S. F. Gray Pale Smartweed, Dock-leaved Smartweed, Willow-weed, Curlytop Knotweed Fig. 458
Weedy; damp soils and cultivated fields, swampy thickets, shallow water and shores. Widespread in N.Am.; intro. from Eur. into Can. and Alask., native throughout U.S. A quite variable species and although several infraspecific taxa have been described, these are no longer taxonomically recognized. (*Polygonum lapathifolium* L.)

16. *P. bicornis* (Raf.) Nieuwl. Pink Smartweed
Moist, disturbed areas, permanent and ephemeral wetlands, ditches, shorelines of ponds and lakes. Ill. w. to S.D. and Wyo. s. to Mo, La., Tex. and N.M. (*Persicaria longistyla* (Small) Small; *Polygonum bicorne* Raf.; *P. longistylum* Small)

17. *P. pensylvanica* (L.) Small Pennsylvanica Smartweed Pinkweed, Smartweed Fig. 458
Weedy; thickets, marshy shores, swales, coastal gravels, ditches and wet disturbed sites. Nfld. and Que. w. to Man., Mont., Oreg. s. to Fla., Tex. and Calif.; Alask. This species is extremely variable and, while several infraspecific taxa had been described, they are no longer recognized as taxonomically distinct. (*Polygonum pensylvanicum* L.)

3. *Rumex* (Dock, Sorrel)

Annual, biennial, or perennial herbs; stems mostly erect; leaves alternate, simple, margins finely toothed, crinkled, or entire; stipule modified, forming a tubular sheath (ocrea) surrounding stem; flowers fascicled, borne in terminal or axillary panicles, fascicles subtended by sheathing, usually scarious, bracts (ocreolae); sepals 6, the inner 3 becoming enlarged, forming wing-like valves surrounding fruit; petals absent; fruit an achene, enclosed by persistent inner sepals.

References: Mitchell, 1978; Mosyakin, 2005; Rechinger, 1937; Sarkar, 1958.

---

1. Margins of mature sepal valves with long bristles (fig. 459d), or conspicuous, spine-like teeth (fig. 459b).
    2. Plants perennial, stems firm, somewhat woody; lower cauline leaves narrowly to broadly ovate, red-veined, leaf base cordate; sepal valves with spine-like teeth; fruiting sepals with 1 tubercle; plants of non-saline soils.. . . . . 1. *R. obtusifolius*
    2. Plants annual, stems herbaceous; lower cauline leaves linear-lanceolate, usually green-veined, leaf base usually truncate to attenuate; sepal valves with long bristles; fruiting sepals with 3 tubercles; plants of saline or brackish soils.
        3. Tubercles of sepals plump, ellipsoidal to ovoid (fig. 459e); marginal spines about as long as width of valves. . . . . 2. *R. persicarioides*
        3. Tubercles of sepals narrower, not plump, linear to lanceolate (fig. 459d); marginal spines longer than width of valve.. . . . 3. *R. fueginus*
1. Margins of mature sepal valves entire or merely weakly toothed (fig. 462d).
    4. Stems with little or no lateral branching; leaves usually dark green to reddish.
        5. Pedicels jointed near middle; leaves wavy and convoluted at margins. . . . . 4. *R. crispus*
        5. Pedicels jointed below middle; leaves not strongly wavy or convolute. . . . . 5. *R. orbiculatus*
    4. Stems with well-developed lateral branching; leaves with a pale sheen.
        6. Fruiting pedicels strongly reflexed from their immediate bases, but straight (fig. 461e), 3–5 times as long as the valves. . . . . 6. *R. verticillatus*
        6. Fruiting pedicels mostly curved (fig. 462b,d), 0.5–2 times as long as the valves.
            7. Tubercles broad, conspicuous, more than half the width of the valves (fig. 462b). . . . . 7. *R. pallidus*
            7. Tubercles narrower, much less than half the width of the valves (fig. 462d).
                8. Valves triangular; tubercles consistently 3; leaf blades linear-lanceolate, 5–6 times as long as wide.. . . . . 8. *R. triangulivalvis*
                8. Valves ovate-rounded; tubercles 1 or 2(3); leaf blades ovate-lanceolate, elliptic-lanceolate or lanceolate, 2.5–4 times as wide. . . . . 9. *R. altisimus*

---

1. *R. obtusifolius* L. Bitter Dock, Red-veined Dock, Blunt-leaved Dock Fig. 459
Moist soils of disturbed sites and pastures. Greenl., Nfld., Que., Ont., B.C. and Alask., s. throughout the U.S.

2. *R. persicarioides* L. Golden Dock Fig. 459
Sandy beaches and margins of coastal marshes. Coastal, Que., P.E.I., N.B. and N.S., ne. Mass. and L.I., N.Y.; B.C. s. to. Calif. (*R. maritimus* var. *persicarioides* (L.) R. S. Mitchell)

Fig. 456. *Persicaria careyi*: a. habit (NYS Museum); b. glandular pubescence of stem (F); c. flower (NYS Museum); d. achene.
Persicaria longiseta: e. habit; f. inflorescence; g. achene (NYS Museum, as *Polygonum cespitosum*).

Fig. 457. *Persicaria maculosa*: a. habit; b. inflorescence; c. stipular sheath; d. achene, two views (Reed, as *Polygonum persicaria*).

Fig. 458. *Persicaria lapathifolia*: a. habit; b. inflorescence; c. flower with anchor-shaped veins; d. achene. with cross-section (NYS Museum).
Persicaria pensylvanica: e. habit; f. inflorescence; g. achene (Reed).

Fig. 459. *Rumex obtusifolius*: a. upper portion of plant with basal leaf; b. fruit (NYS Museum).
Rumex fueginus: c. upper portion of plant; d. fruit (NYS Museum).
Rumex persicarioides: e. fruit (NYS Museum).

3. *R. fueginus* Phil. American Golden Dock, Fuegian Dock Fig. 459
Saline, brackish, or alkaline marshes and shores. N.S., P.E.I. and Que. w. to B.C. and Alask., s. to R.I. and N.Y., Del., Md., Ky. and Tex., N.M., Ariz., Calif. and Baja Calif.; S.Am. Originally described from Tierra del Fuego, Chile, but is native and more widely distributed in N.Am. It is similar to *R. maritimus* and specimens are frequently misidentified as such. (*R. maritimus* var. *fueginus* (Phil.) Dusén)

4. *R. crispus* L. Yellow Dock, Curly Dock, Sour Dock Fig. 460
Ditches, disturbed sites and low ground. Weedy; widespread in N.Am.; intro. from Eur.

5. *R. orbiculatus* A. Gray Great Water Dock Fig. 461
Swamps, meadows, shallow water and shores. N.E. w. to N.Y., Mich., Wisc., Minn. and N.D., s. to N.J., Pa., Ohio, Ind., Ill., Iowa, N.D. and Neb. This species is referred to in some works as *R. britannica* L., an unfortunate name for a species native to N.Am. and not known from Britain. There is actually considerable confusion over the correct name. Asa Gray (1867) recognized both *R. britannica* and his newly named *R. orbiculatus* in his 5th edition of *Gray's Manual*. Linnaeus' (1753) concept of *R. britannica* was based on a fragment of a specimen obtained from Gronovius, giving the source as "Habitat in Virginia" and citing Gronovius' *Flora Virginica* (based on a manuscript sent to him by John Clayton and published on behalf of him). Our *R. orbiculatus* does not occur in Virginia—not south of Pa. and N.J., as was noted by Fernald (1945) when he wrestled with determining the correct name. Since Linnaeus (1753) cited Gronovius' description from *Flora Virginica* the specimen in the Linnaean Herbarium would serve as dubious material to designate as a lectotype to typify the name. Furthermore, Gray studied the material of the Clayton specimen from the Gronovius herbarium and noted that it differs from that in the Linnaean Herbarium and more readily is identifiable as another species, *Rumex obtusifolius* L., which does occur as an adventive in Virginia (Fernald, 1945). This was further complicated by the use of the name *R. britannica* by Michaux and by Pursh, which may have referred to *R. altissimus* Alph. Wood. Given the confusion, it seems more prudent to apply the name *Rumex orbiculatus* A. Gray for this taxon.

6. *R. verticillatus* L. Swamp Dock, Water Dock Fig. 461
Swamps, marshes, and margins of ponds, lakes and streams. S. Que. w. to s. Ont., Minn. and Neb. s. to Fla. and e. Tex.

7. *R. pallidus* Bigelow White Dock, Seabeach Dock Fig. 462
Salt marshes and sandy or rocky coasts. Coastal, Nfld. and lower St. Lawrence R., Que. P.E.I., N.S., s. Me., N.H., Vt. and Mass.

8. *R. triangulivalvis* (Danser) Rech. f. Willow-leaved Dock Fig. 462
Moist, solid, disturbed sites, sometimes saline. Nfld. w. to Man., N.W.T., Yuk. and B.C., s. to N.E., N.Y., N.J., Ky., Mo., Tex., N.M., Calif., and Mex.

9. *R. altissimus* Alph. Wood Pale Dock, Tall Dock Fig. 463
Swamps, stream margins and alluvial soils, wet meadows and ditches. N.E. w. to Minn., Wyo. and Colo., s. to Fla., Tex. and Ariz.

Plumbaginaceae / Leadwort Family

1. *Limonium* (Sea-lavender, Marsh-rosemary)

Mostly perennial herbs; basal rosette from crown of thick, woody rootstock; leaves simple, elliptic to obovate to oblanceolate; flowers 5-merous radially symmetrical; calyx united, white, persistent; corolla lavender, white or yellow, united at base; stamens 5, adnate to base of petals; pistil 1, placentation basal; fruit thin-walled, a 1-seeded utricle.

References: Luteyn, 1976; Smith, 2005.

1. *L. carolinianum* (Walter) Britton Sea-lavender, Marsh-rosemary, Canker-root, Statice, Seaside-thrift Fig. 463
Salt marshes, saline shores, interdunal swales, saline ditches and mangrove swamps. Coastal, Lab., Nfld. and St. Lawrence R., Que., s. to N.E., se. N.Y., N.J., Del., Md., Va. and s. Fla., w. to Tex. and ne. Mex; Ber. Although several taxa have been described, Luteyn (1976) noted rather continuous variation thought North America and thus treated *Limonium* in eastern North America as a single polymorphic species; establishment by seed is rare and dispersal is primarily through rhizomes. (*L. nashii* Small; *L. carolinianum* var. *nashii* (Small) Boivin)

Amaranthaceae / Amaranth, Goosefoot Family

Herbs, rarely subshrubs, annuals or perennials; leaves alternate or opposite, usually petiolate; blade margins entire, crisped, or erose; inflorescences cymules, arranged in spikes, panicles, heads, glomerules, clusters or racemes; flowers bisexual or unisexual, generally small to minute, tepals mostly (1)–5 or absent, distinct or basally united; stamens usually as many as tepals; ovary typically superior, with a single ovule; fruits dry, utricles or small nuts.

Reference: Robertson and Clemants, 2003.

1. Stems succulent, jointed; leaves minute, scale-like, opposite. . . . . 1. *Salicornia*
1. Stems not succulent or jointed; leaves flat or linear, not reduced, alternate or opposite.
    2. Leaves flat, linear, linear-lanceolate or hastate, not distinctly succulent (slightly fleshy in *Atriplex*).

Fig. 460. *Rumex crispus*: a. habit; b. fruit surrounded by persistent sepals; c. fruit, showing 3 sepal valves; d. achene (Reed).

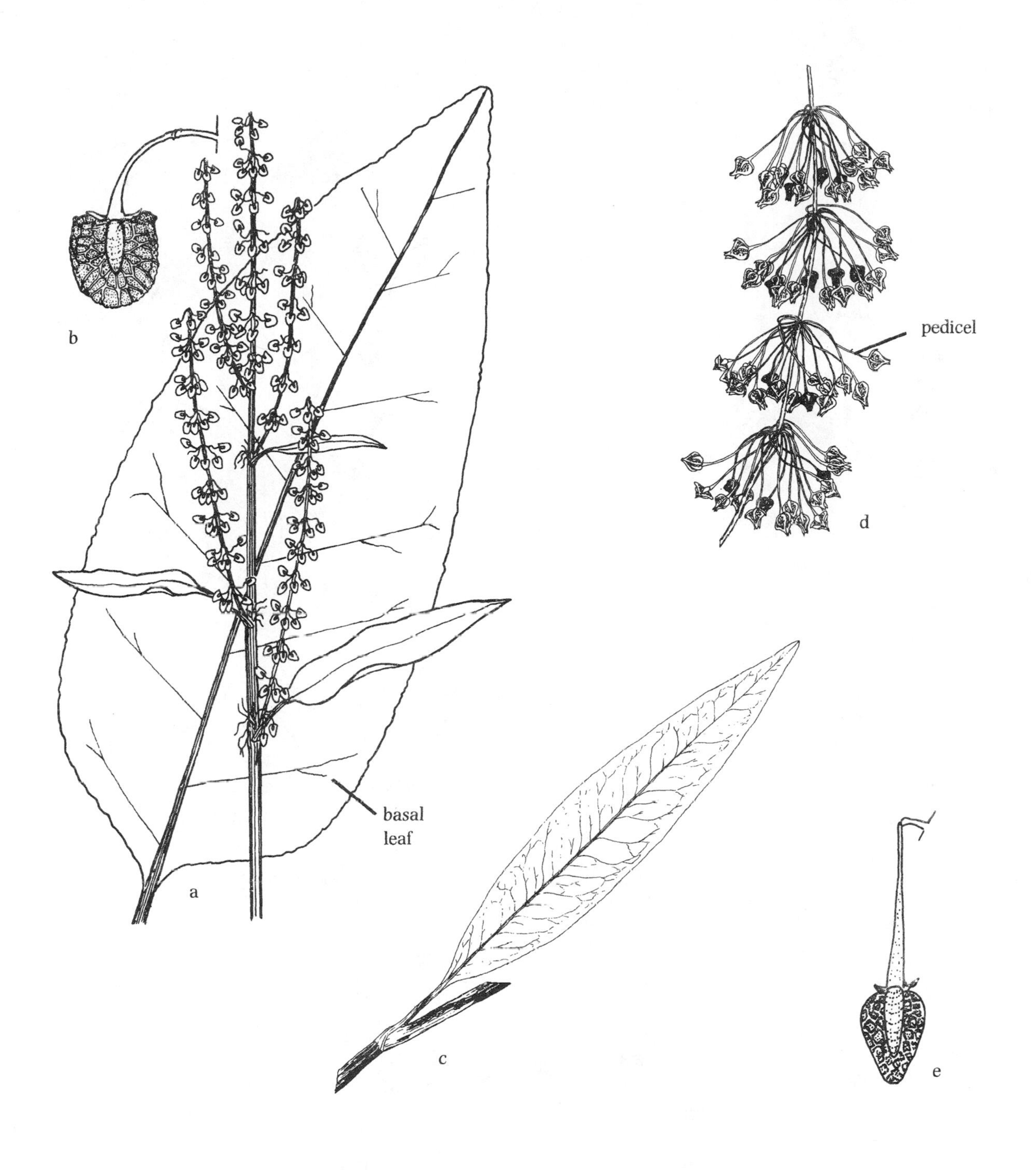

Fig. 461. *Rumex orbiculatus*: a. upper portion of plant with basal leaf; b. fruit (NYS Museum). *Rumex verticillatus*: c. leaf (F); d. inflorescence (F); e. fruit (NYS Museum).

Fig. 462. *Rumex pallidus*: a. habit; b. fruit (NYS Museum).
Rumex triangulivalvis: c. habit; d. fruit (NYS Museum).

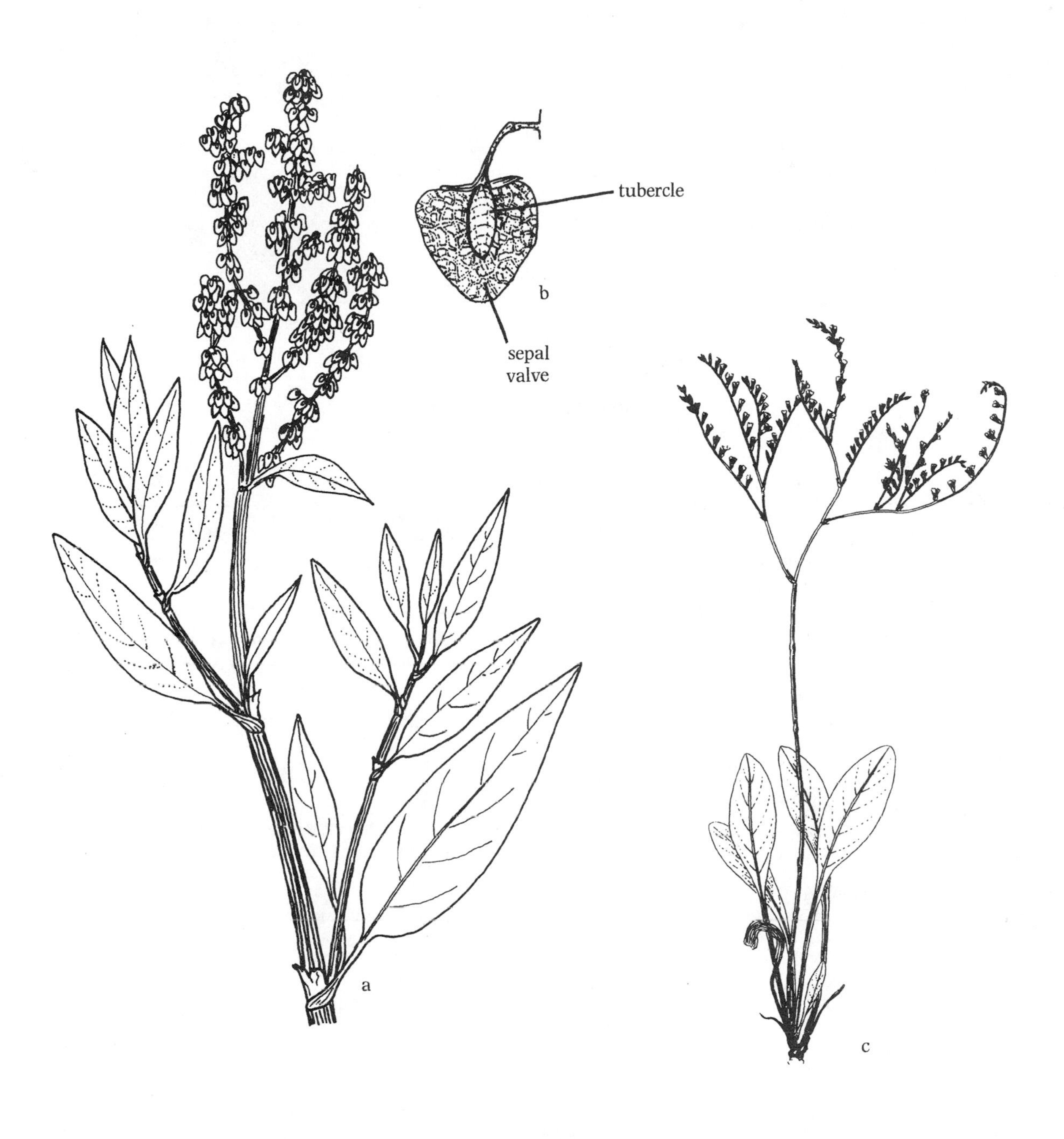

Fig. 463. *Rumex altissimus*: a. habit; b. fruit (NYS Museum).
Limonium carolinianum: c. habit (Beal).

3. Plants annual; leaves alternate, petiolate, flowers in axillary clusters or glomerules, very small, green non-showy, lacking perianth.
    4. Plants mostly sprawling, decumbent or procumbent or weakly ascending; leaves mostly hastate, or linear to ovate-lanceolate (then thickish, slightly succulent), surface mealy; inflorescences on shorter spikes, 2–9 cm, mostly axillary. .... 2. *Atriplex*
    4. Plants erect to strongly ascending (1–3 m), rarely prostrate; leaves linear, linear-lanceolate, or ovate to obovate on lower stem and oblong to elliptic to narrowly lanceolate above, surface not mealy; inflorescences very long narrow spikes, terminal .... 3. *Amaranthus*
3. Plants perennial; leaves opposite, sessile; flowers in axillary globose heads, larger, conspicuous, to 6 mm long, tepals white, papery. .... 4. *Alternanthera*
2. Leaves nearly round in cross-section, clearly succulent, never hastate. .... 5. *Suaeda*

---

1. *Salicornia* (Glasswort, Saltwort, Samphire)

Annual or perennial herbs; stems succulent, jointed, branched or simple; leaves minute, opposite, scale-like; inflorescence spike-like; flowers inconspicuous, appearing embedded in succulent axis; fruit membranous, a 1-seeded utricle.

REFERENCE: Ball, 2003.

---

1. Plants annual, with taproot; stems herbaceous at base; middle flower higher than lateral flowers at joints (fig. 464b).
    2. Scale-like leaves conspicuous, mucronate; spike 4.5–6 mm wide. .... 1. *S. bigelovii*
    2. Scale-like leaves obscure, blunt; spike 1.5–5 mm wide.
        3. Fertile segments of spike longer than wide.; plants chiefly coastal.
            4. Anthers 0.3–0.5 mm, always exserted, dehiscing in open. .... 2. *S. depressa*
            4. Anthers (0.1)0.2–0.3 mm, typically not exserted, mostly dehiscing within flower. .... 3. *S. maritima*
        3. Fertile segments of spike as wide as or wider than long; plants of Great Plains. .... 4. *S. rubra*
1. Plants perennial, rhizomatous; stems becoming woody at base; middle and lateral flowers at nearly the same level at joints (fig. 465e). .... 5. *S. pacifica*

---

1. *S. bigelovii* Torr. Dwarf Saltwort Fig. 464
Salt marshes. Coastal, s. Me. s. to S.C. and Fla., Tex.; s. Calif.; Yucatan Pen., Mex.; W.I.

2. *S. depressa* Standl. Samphire, Chicken Claws, Pigeon-foot Fig. 465
Salt marshes and sea strands. Atlantic Coast and James Bay, Que., N.S. and N.B., s. to N.E., N.J., Va., and Ga.; inland in saline sites in N.Y.; Pacific Coast, Alask. and B.C. s. to Calif. This widespread taxon of coastal North America was previously believed to be *S. europaea*, which is an Old World species restricted to Europe.

3. *S. maritima* S. L. Wolff & Jefferies
Upper salt marsh and tidal channels; rarely salt springs inland. Coastal; s. Nfld., s. Que., Ont. (St. Lawrence Seaway), N.S., P.E.I. and N.B.

4. *S. rubra* A. Nelson Saltwort Fig. 464
Seasonally wet depressions and pools, saline or alkaline soils, and inland salt marshes. S. Mich., w. Minn. and Man. w. to n. B.C., s. to Kan., Wyo., Nev. and Wash.

5. *S. pacifica* Standl. Perennial Saltwort, Woody Glasswort, Lead-grass Fig. 465
Sandy sea strands and borders of salt marshes. Mainly coastal, Nfld., N.S., P.E.I., N.B., Me., Que. and Ont., s. to Fla.; Calif., and Mex. The perennial species are segregated by some authors (Ball, 2003) as the genus *Sarcocornia*, thus the name would be *Sarcocornia pacifica* (Standley) A. J. Scott; we have taken the broader view, retaining it in *Salicornia*. The name *Salicornia virginica* L. has long been misapplied to our eastern perennial species, but examination of Linnaeus's type specimen clearly shows it to be an annual while ours is clearly perennial (Ball, 2003; http://linnean-online.org/41/, accessed 13 September 2021).

2. *Atriplex* (Orach)

Annual (ours) or perennial herbs and shrubs; leaves usually alternate, usually with mealy or scurfy surface; flowers unisexual; staminate flowers borne in a terminal spike-like inflorescence; pistillate flowers borne in axillary or spike-like clusters (glomerules), enclosed between 2 leafy bracts; perianth minute or lacking; fruit a 1-seeded utricle; seeds homomorphic or dimorphic (ours, large brown type and small black type), with curved embryo forming a peripheral ring.

REFERENCES: Bassett et al., 1983; Taschereau, 1972; Welsh, 2003.

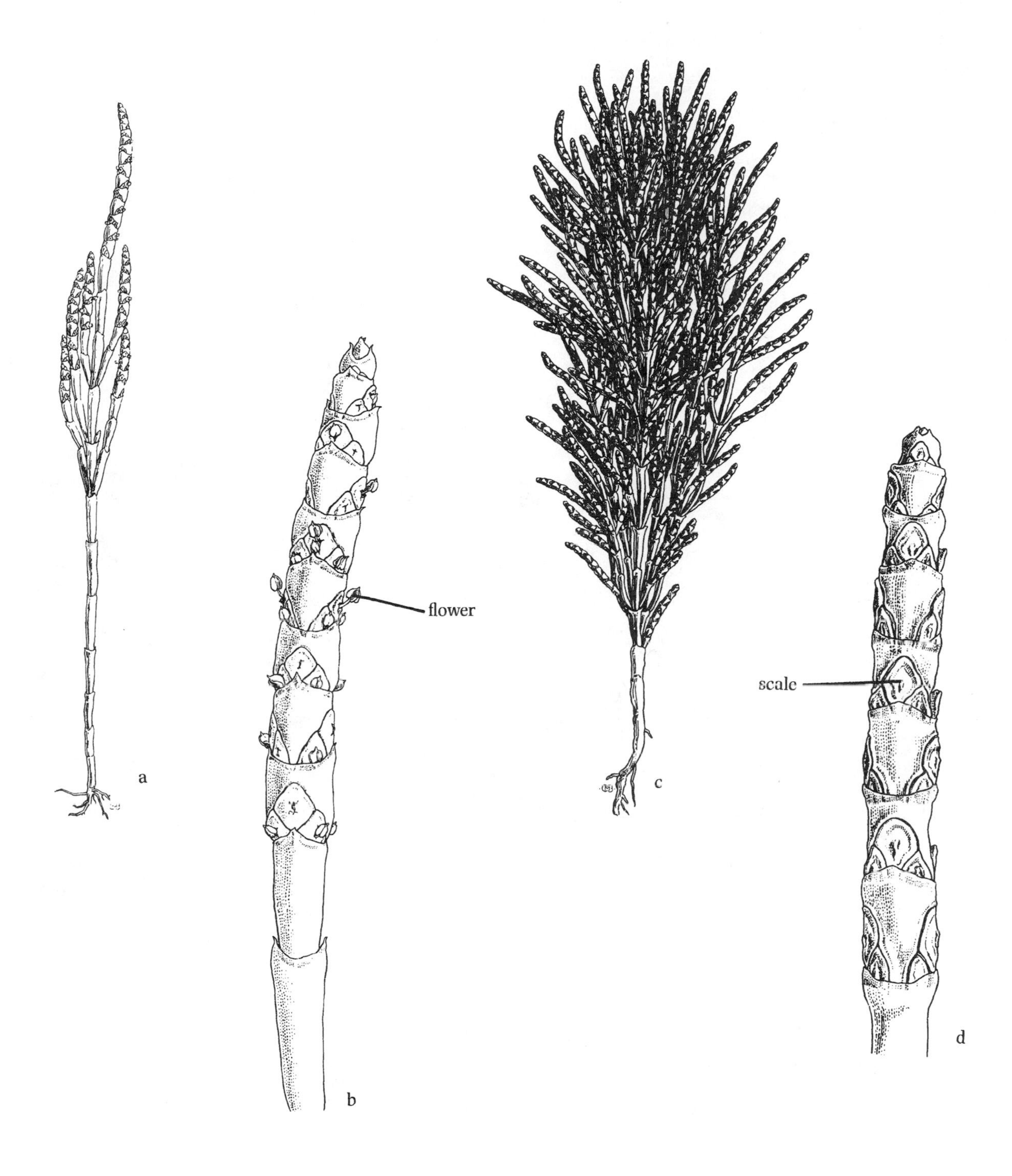

Fig. 464. *Salicornia bigelovii*: a. habit; b. spike with triads of flowers (Mason).
Salicornia rubra: c. habit; d. spike with blunt scales (Mason).

Fig. 465. *Salicornia depressa*: a. habit; b. portion of branch with blunt scales (Gleason, as *S. europaea*). *Salicornia pacifica*: c. habit; d. stem with budding branches; e. flowering stem (G&W, as *Salicornia virginica*).

1. Leaves mostly triangular or lance-hastate with lobes pointed upward (antrorse), with thin texture; inflorescence with leafy bracts nearly to tip; seeds not distinctly dimorphic, brown only, or brown and some black. . . . . 1. *A. glabriuscula*
1. Leaves linear to ovate-lanceolate or deltoid-hastate, with lobes pointed outward, with thick texture (slightly succulent); inflorescences with leafy bracts only near base; seeds distinctly dimorphic, small, black, glossy, as well as some larger, dull brown.
   2. Principal lower and upper leaves deltoid-hastate, broad-based, with lateral-pointing basal lobes. . . . . 2. *A. prostrata*
   2. Principal lower leaves ovate-lanceolate to linear-lanceolate, lacking basal lobes, and upper leaves broad-based, with forward-curved or lateral-pointing basal lobes. . . . . 3. *A. dioica*

1. *A. glabriuscula* Edmundston Northeast Salt Bush
Salt marshes, sea beaches and exposed cobbles. Nfld., Lab. and N.S. w. to n. Man. and n. Alta, s. to N.E. coast and Pa. Our taxon is var. *acadiensis* (Taschereau) S. L. Welsh. Welsh (2003) recognized two other varieties, var. *glabriuscula* and var. *franktonii*, which occur in our range, but they occur on coastal wet sandy beaches and rocky cobbles, ledges and boulders, often with fucoid algae; all 3 varieties are treated as distinct species by Traschereau (1972) as well as in some floras (Haines, 2011).

2. *A. prostrata* Boucher ex DC. Orach, Halberd-leaved Atriplex Fig. 466
Saline soils, muds, sands and cobbles of sea beaches, salt marshes, and waste places. Nfld. w. to Alta. and Wash., s. to S.C., Ohio, Ill., Mo., Neb., w. N.M., n. Ariz. and Calif. (*A. patula* var. *hastata* (L.) A. Gray; *A. triangularis* Willd.)

3. *A. dioica* Raf. Spike Saltbush Fig. 466
Saline soils of upper salt marshes, bays, tidal rivers and inland saline habitats. Coastal, Nfld. and Maritime Provinces s. to N.E. and N.J.; James Bay; w. Ont. and Man. w. to Yuk. and B.C. and scattered localities in western states. According to Taschereau (1972) this species was not previously recognized as occurring in eastern North America, with eastern plants treated in earlier manuals as *A. patula* var. *hastata* (L.) A. Gray, a name now placed in synonymy with *A. prostrata*. Welsch (2003) notes that Rafinesque's name, *Atriplex dioica*, was published a mere 6 months prior to Nuttall's publication of the basionym, *Chenopodium subspicatum*, later transferred to *Atriplex* by Rydberg, as *A. subspicata* (Nutt.) Rydb.; thus the name *A. dioica* Raf. has nomenclatural priority. (*A. subspicata* (Nutt.) Rydb.)

Atriplex littoralis L., a species very similar to *A. dioica*, may become established on upper margins of salt marshes; it is more typically a plant of drier sites such as beaches and old ballast dumps near early port cities.

3. *Amaranthus* (Amaranth, Pigweed)

Chiefly annuals; leaves alternate; inflorescences axillary or terminal or both; flowers minute, bisexual or unisexual (ours dioecious); petals absent; stamen filaments separate, anthers 4-celled; fruit a 1-seeded utricle.

References: Mosyakin and Robertson, 2003; Sauer, 1955.

1. Pistillate flowers with tepals absent or 1–2, usually less that 2 mm; fruit indehiscent, firm, angled or ribbed; bracts shorter than fruit; seeds 2–3.5 mm in diameter; staminate flowers with bracts usually less than 1 mm long; plants of saline sites. . . . . 1. *A. cannabinus*
1. Pistillate flowers with tepals 5, usually longer than 2 mm; fruit dehiscent or indehiscent, thin, not angled or ribbed; bracts as long as or longer than fruit; seeds about 1 mm in diameter; staminate flowers with bracts 1–2 mm long; plants of freshwater sites. . . . . 2. *A. tuberculatus*

1. *A. cannabinus* (L.) J. D. Sauer Water-hemp, Salt-marsh Water-hemp Fig. 467
Salt marshes and tidal shores. Coastal, s. Me. s. to Ga., Ala., and n. Fla. (*Acnida cannabina* L.)

2. *A. tuberculatus* (Moq.) J. D. Sauer Water-hemp
Margins of ponds, lakes. and rivers, marshes, boggy sites and wet disturbed ground. P.E.I., Vt. and s. Que. w. to s. Ont., Wisc., Minn., Man. and N.D., s. to Conn., N.Y., Ohio, Tenn., Fla., La., Tex., Mo., Neb. Kans., Colo. and e. N.M.; Wash., Ida., Nev. and Calif. (*Acnida altissima* (Riddell) Moq. ex Standl.)

4. *Alternanthera* (Chaff-flowers)

Perennial herbs (ours) or subshrubs; leaves opposite; flowers bisexual, minute, subtended by bracts and bractlets; calyx usually 5-parted, lobes unequal; petals absent; stamens with filaments united to form a tube, anthers 2-celled; fruit a 1-seeded utricle.

References: Clemants, 2003; Penfound, 1940a.

Fig. 466. *Atriplex prostrata*: a. habit; b. bracteole variations; c. seed variations (Bassett et al., 1983).
Atriplex dioica: d. habit; e. bracteole variations; f. seed variations (Bassett et al., 1983).

Fig. 467. *Amaranthus cannabinus*: a. habit, upper portion of plant (Beal).
Alternanthera philoxeroides: b. habit; c. flower; d. fruit with persistent perianth; e. achenes; f. seeds (Reed).

1. *A. philoxeroides* (Mart.) Griseb. Alligator Weed Fig. 467
Ponds, lakes, streams, canals and ditches. Coastal plain, Va. s. to Fla., w. to Tex.; Tenn., s. Ill. and Ark.; Calif.; C.Am. and S.Am. This prostrate, stoloniferous, mat-forming plant, introduced from South America, has become a noxious aquatic weed in North America, where it appears to never produce seeds and reproduces solely by vegetative means. (*Achyranthes philoxeroides* (Mart. (Standl.)

5. *Suaeda* (Sea-blite)

Annual or perennial herbs; stems succulent; leaves alternate, linear, nearly round in cross-section, succulent; flowers usually bisexual, sessile; calyx succulent, 5-parted, fused at base; fruit a 1-seeded utricle; seed with embryo spirally coiled.

Characters of these succulent plants are visible when plants are fresh, but may be difficult to see in dried specimens.

References: Bassett and Crompton, 1978; Ferren and Schenk, 2003; Hopkins and Blackwell, 1977; McNeill et al., 1977.

---

1. Calyx lobes unequal, 1 or 2 sharply pointed at apex, lateral appendages present, strongly hood-shaped. . . . . 1. *S. calceoliformis*
1. Calyx lobes more or less equal, all rounded or rounded and keeled, lacking lateral appendages.
   2. Calyx lobes rounded and keeled; branches erect; flowers borne in panicles or almost spike-like inflorescences. . . . . 2. *S. linearis*
   2. Calyx lobes rounded, but not usually keeled; branches decumbent; flowers borne in few-flowered glomerules along stem.
      3. Plants prostrate to erect; leaves pale green, glaucous, blades 10–30(50) mm long; seeds 1.5–2 mm in diameter, reddish-brown to black. . . . . 3a. *S. maritima* subsp. *maritima*
      3. Plants prostrate, forming mats up to 0.5 m across; leaves dark green, not glaucous, blades 3–15 mm long; seeds 1–1.5 mm in diameter, black. . . . . 3b. *S. maritima* subsp. *richii*

---

1. *S. calceoliformis* (Hook.) Moq. Matted Sea-blite
Salt marshes, sea strands and highly saline or alkaline soils. Se. Nfld. and e. Que. s. to N.S., N.E., and N.J.; James Bay, Man. and w. Minn. nw. to Sask., B.C., N.W.T., Yuk. and s. Alask., s. to w. Mo., Kan., Tex., N.M., Ariz. and Calif. Although this is chiefly a western species, Bassett and Crompton (1978) agree with Nuttall's earlier assessment that the eastern *S. americana* should be treated as conspecific with the western taxon, previously known under the name *S. depressa*, a name McNeill and colleagues (1977) have shown to be incorrectly applied to this taxon. (*S. americana* (Pers.) Fernald)

2. *S. linearis* (Elliott) Moq. Annual Seep-weed Fig. 468
Salt marshes and sandy seashores. Me. s. to Fla., w. to Tex. and Mex.; W.I.

3. *S. maritima* (L.) Dumort.

3a. *S. maritima* subsp. *maritima* White Sea-blite, Herbaceous Seep-weed Fig. 468
Salt marshes, saline mud flat and sea strands. Nfld., N.S. and Que. s. to N.E., N.J., and Va.; Man.; intro. from Eurasia.

3b. *S. maritima* subsp. *richii* (Fernald) Bassett & C.W. Crompton Rich's Seep-weed Fig. 468
Salt marshes and sea strand; Se. Nfld. s. to N.S., Me., N.H., R.I. and ne. Mass. This native taxon is rare and of conservation concern in Me., N.H. and Mass. (*Suaeda richii*)

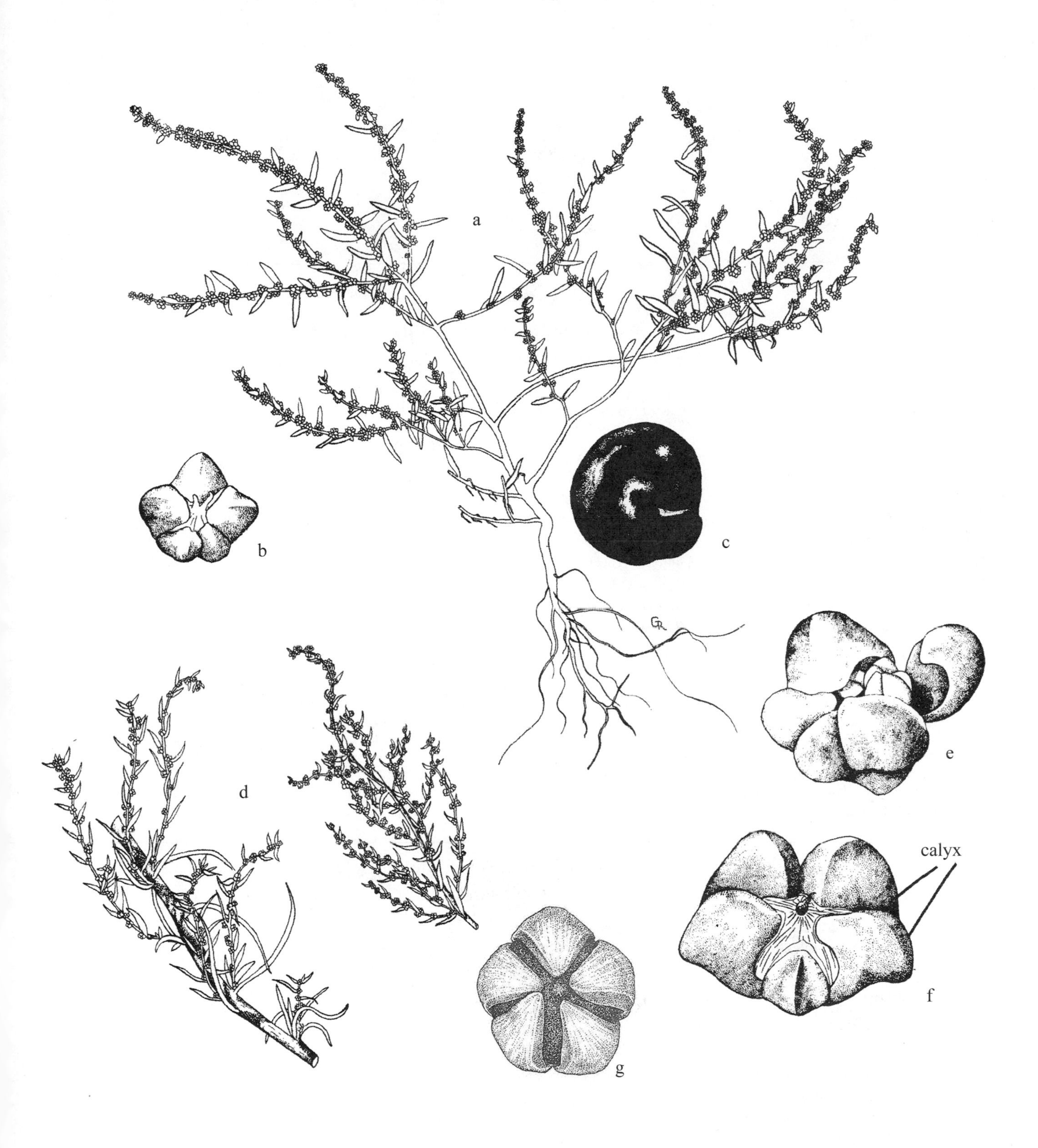

Fig. 468. *Suaeda maritima* subsp. *maritima*: a. habit; b. flower; c. seed (G&W).
Suaeda linearis: d. upper and middle branch portion; e. flowers; f. fruit enclosed by calyx (G&W).
Suaeda maritima subsp. *richii*: g. fruit enclosed by calyx (NPT).

Asterids

Nyssaceae / Sour Gum Family

1. *Nyssa* (Sour Gum, Tupelo)

Trees, deciduous; twigs with diaphragmed pith (longitudinal section); leaves alternate, simple, pinnately veined; flowers bisexual or unisexual, inconspicuous, axillary, pedunculate, solitary or borne in 2s or in racemes; ovary inferior, fruit a 1-seeded ovoid or ellipsoidal drupe, with persisting sepals at tip.

REFERENCES: Eyde, 1963, 1966; Tucker, 2016.

---

1. Leaf petiole 30–60 mm long, blade usually 8–11(18) cm long, margin with a few coarse serrations at base; pistillate flowers and fruits solitary, ovaries pubescent; fruits 24–27 mm long, punctate.......1. *N. aquatica*
1. Leaf petiole 5–15(30) mm long, blade usually 3.7–12(16) cm long, margins entire; pistillate flowers and fruits in clusters of 3–8, ovaries glabrous; fruits 7–14 mm long.
    2. Leaf blades herbaceous, (2)3.5–6.5(9.1) cm wide, about twice as long as wide, ovate to broadly elliptic.......2. *N. sylvatica*
    2. Leaf blades somewhat leathery, 1.5–3.5 cm wide, about 3 times as long as wide, linear-elliptic to oblanceolate.......3. *N. biflora*

---

1. *N. aquatica* L. Cotton-gum, Tupelo, Water Tupelo Fig. 469
Swamps and ponds. Coastal plain, Va. s. to n. Fla., w. to e. Tex.; Mississippi embayment n. to se. Mo., s. Ill., w. Ky. and Tenn.

2. *N. sylvatica* Marshall Black Gum, Sour Gum Fig. 469
Upland and wet woodlands, marshes and lakeshores. C. Me. and N.H. w. to s. Ont., Mich., se. Wisc., se. Mo. and e. Kans., s. to Fla., e. Tex. and Mex.

3. *N. biflora* Walter Black Gum, Swamp Tupelo Fig. 469
Swamps, pond and lake margins, wet savannas and floodplains. S. N.J., Del., Md. and Va. w. to Ky., s. Ill. and s. Mo., s. to Fla., e. Tex. (*N. sylvatica* var. *biflora* (Walter) Sarg.)

Cornaceae / Dogwood Family

1. *Cornus* (Dogwood)

Shrubs (ours) or small trees, deciduous; leaves opposite, simple, margins entire, veins running parallel curved toward apex; flowers small, bisexual, borne in cymes; fruit a globose to ellipsoid drupe.

Flora Nova Angliae (Haines et al., 2011) treats the species of *Cornus* in our key as belonging to the segregate genus *Swida*.

REFERENCES: Farwell, 1931; Murrell and Poindexter, 2016; Rickett, 1944; Wilson, 1965.

---

1. Leaves with secondary veins (4)5–9 per side; stems gray or pinkish-gray.
    2. Leaves scabrous on upper surface, pilose on lower surface; fruit bluish-white.......1. *C. drummondii*
    2. Leaves with sparse appressed hairs on both lower surfaces (becoming glabrate late summer); fruit blue.......2. *C. foemina*
1. Leaves with secondary veins 3–4(5) per side; stems reddish (bright red or blood red in late summer and winter).
    3. Pith of twigs white; leaves glaucous and finely pubescent on lower surface, fruit white.......3. *C. sericea*
    3. Pith of twigs brown; leaves with minutely appressed pubescence on lower surface; fruit blue.......4. *C. amomum*

Fig. 469. *Nyssa aquatica*: a. leaf and fruits (F).
Nyssa sylvatica: b. leaves and fruits; c. leaves (F).
Nyssa biflora: d. leaves (F).

1. *C. drummondii* C. A. Meyer Fig. 470
Shores, thickets, floodplains and bluffs. N.Y., w. to s. Ont., s. Mich., Wisc., S.D. and Neb., s. to Tenn., Ala., Miss., La. and e. Tex. (*C. asperifolia* var. *drummondii* (C.A. Meyer) J. M. Coulter & W. H. Evans; *Swida drummondii* (C. A. Meyer) Soják)

2. *C. foemina* Mill. Stiff Dogwood, Gray Dogwood Fig. 471
Swamps, wet woods, riverbanks and thickets, S. N.J. and s. Del., Md. and Va. w. to Ky., s. Ind., s. Ill. and se. Mo., s. to Fla., Okla. and e. Tex. Murrell and Poindexter (2016) recognize *C. racemosa* as a distinct species whereas other authors include it within *C. foemina* as subsp. *racemosa* (Lam) J. S. Wilson, a taxon that tends to be in drier sites, but can occur in wetlands and occurs throughout our range. (*C. stricta* Lam.; *Swida foemina* (Mill.) Small)

3. *C. sericea* L. Red-stemmed Dogwood, Red-osier Dogwood Fig. 472
Shores, thickets and bog borders. Nfld. and s. Lab. w. to Alask., s. to N.J., Va., W.Va., Ind., Ill., n. Iowa, Neb., N.M., Ariz., Calif. and Mex. (*C. stolonifera* Michx; *C. stolonifera* var. *baileyi* (J. M. Coult. & W.H. Evans) Dresche; *Swida sericea* (L.) Holub)

4. *C. amomum* Mill. Red Willow Fig. 470
Shores, thickets and swamps. Me, w. to Mich., Ill. and Iowa, s. to Fla., Miss. and Mo. (*Swida amomum* (Mill.) Small)

Balsaminaceae / Touch-me-not Family

1. *Impatiens* (Touch-me-not, Jewelweed, Balsam, Snapweed)

Annuals; stems succulent, hollow; leaves ovate, oval, or lanceolate, coarsely toothed; flowers bisexual, in axillary clusters; corolla united, yellow, orange, or red, spurred; fruit a succulent, turgid, 5-valved capsule with explosive dehiscence.

---

1. Petals pale yellow; spur 5 mm long, bent at right angle to corolla tube (fig. 473a)........ 1. *I. pallida*
1. Petals orange to reddish; spur 6–12 mm long, reflexed, parallel to corolla tube (fig. 473b)........ 2. *I. capensis*

---

1. *I. pallida* Nutt. Pale Touch-me-not, Jewelweed, Snapweed Fig. 473
Calcareous wet woods, meadows and springy areas. Nfld. w. to Ont. and Minn. s. to Ga., Tenn., Ala., Mo., e. Kans., n. Ark. and ne. Okla.

2. *I. capensis* Meerb. Spotted Touch-me-not, Jewelweed, Snapweed Fig. 473
Wet woods, meadows, swamps and springy areas. Nfld. w. to N.W.T. and Alask., s. to Fla., Ala., e. Tex.; Colo., Ida. and w. Oreg. (*I. biflora* Walter)

Polemoniaceae / Polemonium Family

1. *Polemonium* (Greek Valerian, Jacobs's Ladder)

Perennial herbs; leaves alternate, pinnately compound; flowers borne in terminal paniculate or thrysoid clusters blue to purple, united, campanulate; stamens epipetalous, exerted beyond corolla; fruit a capsule.

1. *P. vanbruntiae* Britton Appalachian Jacob's Ladder, Bog Jacob's Ladder Fig. 474
Swamps, stream banks, sphagnum bogs and mossy glades. Very local; N.B., s. Que., e. Me., nw. Vt., and c. N.Y. s. to n. N.J., Pa., Md. and W.Va.

Primulaceae / Primrose Family

Annual or perennial herbs (ours); leaves simple, opposite or alternate, lacking stipules; flowers bisexual, with radial symmetry; sepals and petals united at base forming a tube, usually 5-merous; stamens 5; pistil 1, with superior ovary or partly inferior; placentation free-central; fruit a capsule.

The Primulaceae have sometimes been split up into 4 very closely related families, the Myrsinaceae, Primulaceae, Theophrastaceae, and Maesaceae. Angiosperm Phylogeny Group treats these at the rank of subfamily, within the Primulaceae.

---

1. Plants submersed; leaves deeply pinnatifid to finely dissected; inflorescence a whorl of inflated peduncles, floating (*H. inflata*, fig. 475a) or not inflated (*H. palustris*, fig. 475b)........ 1. *Hottonia*
1. Plants emersed or terrestrial; leaves simple, not pinnatifid or dissected; peduncles of inflorescence not inflated.
    2. Basal rosette present; cauline leaves alternate; ovary inferior to half-inferior........ 2. *Samolus*
    2. Basal rosette absent; cauline leaves opposite or whorled; ovary superior.
        3. Flowers pedicellate, yellow; corolla present........ 3. *Lysimachia*
        3. Flowers sessile, red, pink, or white; corolla absent (*L. maritima*)........ 3. *Lysimachia*

Fig. 470. *Cornus drummondii*: a. branch with fruits (OHIO).
Cornus amomum: b. branch with flowers; c. flower; d. stony endocarp; e. branch with flowers (Gleason).

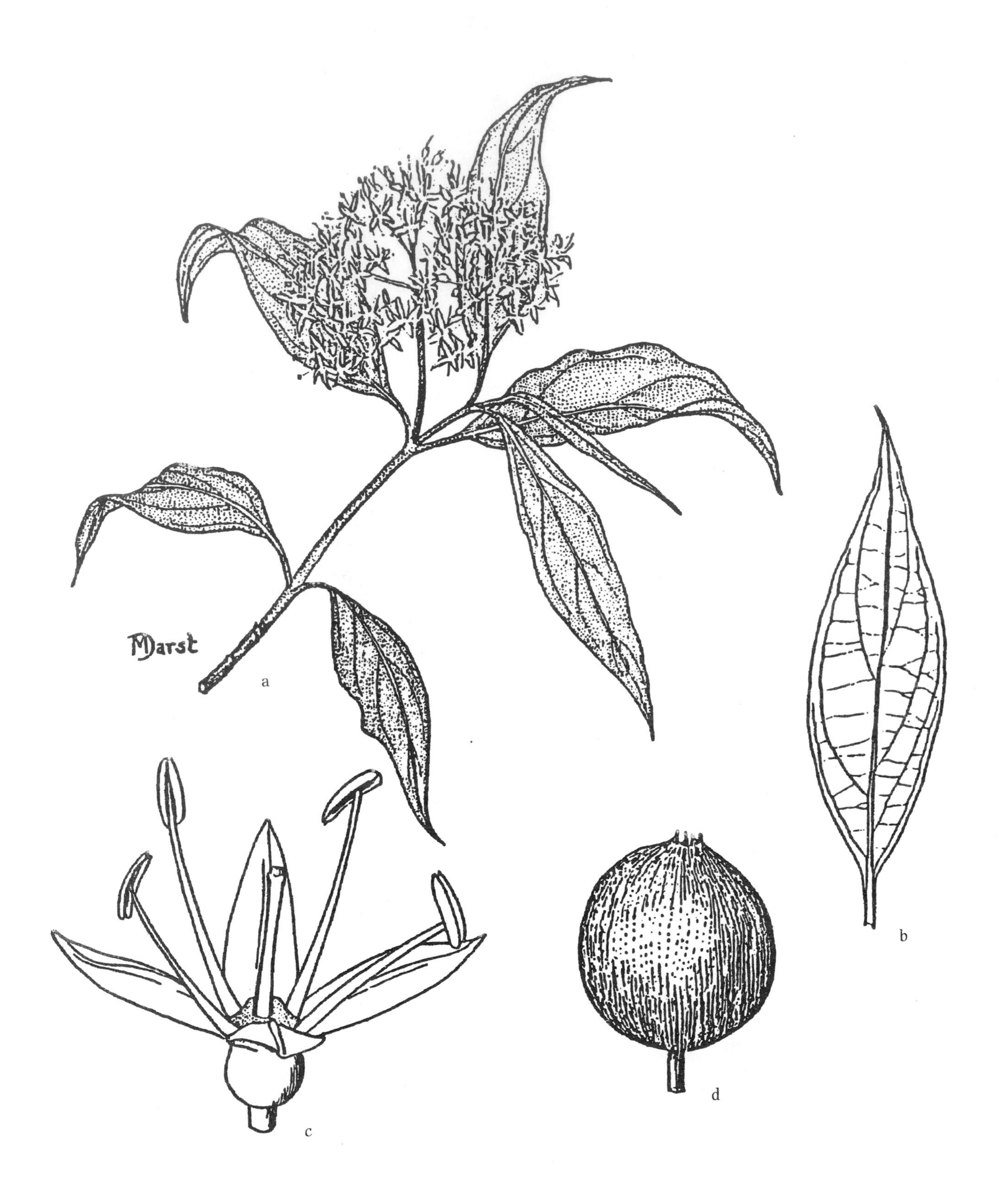

Fig. 471. *Cornus foemina*: a. branch with flowers; b. leaf; c. flower; d. fruit (G).

Fig. 472. *Cornus sericea*: a. branch with flowers (C&C); b. flower (C&C); c. branch with fruits (C&C); d. branch with fruits (F); e. fruit (C&C).

Fig. 473. *Impatiens pallida*: a. upper portion of plant (WVA).
Impatiens capensis: b. upper portion of plant (WVA).

Fig. 474. *Polemonium vanbruntiae*: a. habit (USDA); b. habit (WVA).
Lysimachia thrysifolia: c. upper portion of plant (F).
Samolus parviflorus: d. habit. (F).

1. *Hottonia* (Featherfoil)

Annual or perennial aquatic herbs; submersed stems flaccid, inflated, or non-inflated; leaves submersed, deeply pinnatifid to finely dissected, mostly crowded near surface of water; inflorescence a whorl of inflated peduncles (*H. inflata*), or a single ascending peduncle, non-inflated (*H. palustris*); flowers with radial symmetry whorled, corolla white or light lavender, united at base; fruit a capsule.

REFERENCES: Cholewa, 2009; Prankerd, 1911.

---

1. Stems just below water surface and inflorescence stems (peduncles) at surface inflated, forming a floating whorl; flowers tiny, inconspicuous, white, corolla shorter than calyx (green and foliar in appearance), stamens not exerted; plants annual, not producing stolons, growing in shallow water, submersed stems reaching 0.3–0.5 m long reaching the surface to flower. . . . . 1. *H. inflata*
1. Stems and inflorescence stems not inflated; flowers showy, light lavender or white, with a yellow throat, corolla longer than calyx, stamens exerted; plants perennial, stoloniferous, growing in deeper water, rooted in mud, submersed stems often ca. 1 m long, reaching surface to flower. . . . . 2. *H. palustris*

---

1. *H. inflata* Elliott Featherfoil Fig. 475
Quiet waters of lakes, ponds, swamps, canals and ditches. Chiefly coastal plain, s. Me. s. to Ga. w. to La., se. Okla. and e. Tex.; Mississippi embayment n. to Ark., Mo., w. Tenn., w. Ky., s. Ill., s. Ind., s. and ne. Ohio and w. W.Va. This species is amphibious, growing in shallow waters with fluctuating water levels, and forming a terrestrial growth form of succulent pectinated tufts of leaves along the stranded stems. Populations may have an abundance of plants in a given year, and then disappear for several to many years, and then return with robustness.

2. *H. palustris* L. Water Violet Fig. 475
Quiet waters of ponds and swamps. Reported from a small private pond in Me, but clearly has become established in a bay of Lake Winnipesaukee, N.H., exhibiting invasive qualities (Nichols et al., 2022). This species is sold as a popular plant for aquaria and outdoor aquatic gardens and can be purchased from various sources via the Internet; it will likely escape and spread.

2. *Samolus* (Water Pimpernel)

Perennial, biennial, or rarely annual herbs; leaves simple; basal leaves in a rosette; cauline leaves alternate; flowers 5-merous; corolla white (ours) or pink, united; ovary inferior to half-inferior; fruit a capsule.

1. *S. parviflorus* Raf. Seaside Brookweed Fig. 474
Shallow water, shores, swamps, wet soils, ditches and borders of salt marshes. N.B. and P.E.I. w. to Que., Ont., Mich., Wisc., Ill., Mo., and Kans. s. to Fla. and Tex.; sw. Wash., w. Oreg., Nev., Utah, N.M., Ariz., Calif. and Mex.; W.I., C.Am. and S.Am. (*S. floribundus* Kunth; *S. valerandi* subsp. *parviflorus* (Raf.) Hultén).

3. *Lysimachia* (Loosestrife)

Perennial herbs; stems round or often square in cross-section, some succulent; leaves opposite, sometimes appearing whorled, or alternate, simple, margins entire; flowers axillary or in racemes; calyx red to pink, white or green; corolla yellow or absent, united; stamens usually monadelphous; fruit an ovoid or subglobose capsule.

REFERENCES: Cholewa, 2009; Cooperrider and Brockett, 1974, 1976; Iltis and Shaughnessey, 1960; Ray, 1956.

---

1. Flowers white to red, with one whorl of perianth, corolla absent, calyx petaloid; leaves succulent, sessile. . . . . 1. *L. maritima*
1. Flowers yellow, with calyx and corolla both present; leaves not succulent, petiolate or tapering to petiole-like base.
    2. Stems erect or ascending, never prostrate (decumbent or trailing in *L. radicans*); leaves ovate to lanceolate to linear, petioles ciliate or pubescent; plants not evergreen.
        3. Flowers in short, dense, spike-like axillary racemes (fig. 474d); petals pale yellow, usually 6-merous (4–9); corolla lobes linear, shorter than filaments of stamens. . . . . 2. *L. thyrsiflora*
        3. Flowers solitary, in leaf axils or in axillary or elongate terminal racemes; petals yellow, 5-merous; corolla lobes lanceolate to elliptic to broadly oval, longer than filaments of stamens.
            4. Flowers solitary, or few-flowered racemes, in leaf axils (fig. 478a); leaves not punctate.

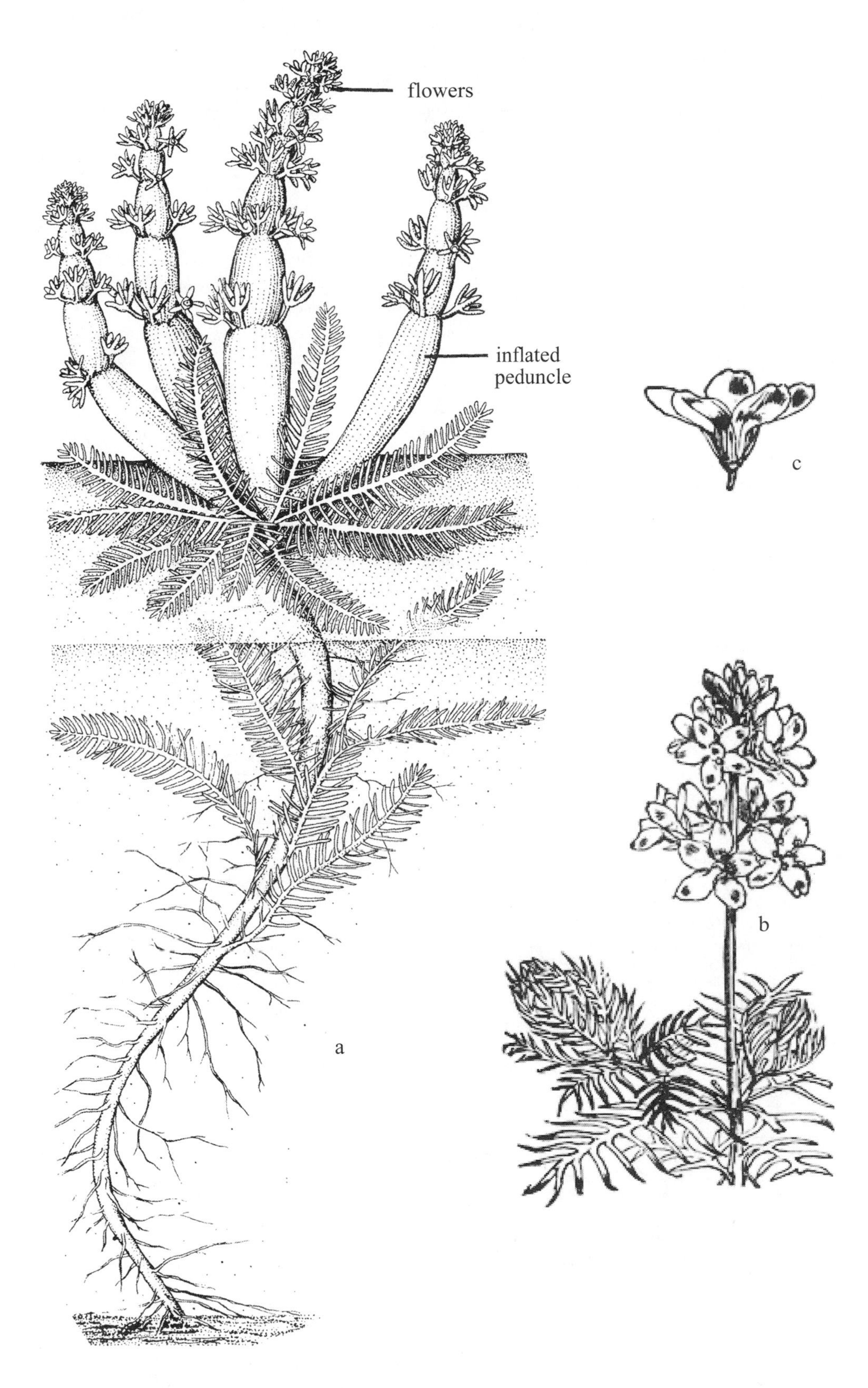

Fig. 475. *Hottonia inflata*: a. habit (NYS Museum).
Hottonia palustris: b. habit; c. flower (Fitch and Smith, 1880).

5. Leaves firm, somewhat leathery, margins strongly revolute, lateral veins obscure. . . . . 3. *L. quadriflora*
5. Leaves thin, margins not revolute, lateral veins conspicuous on lower and middle portions of stem, becoming sessile on upper portion; petioles without cilia or with cilia only on portion of the petiole.
   6. Stems weakly erect, reclining or decumbent creeping, often (sometimes rooting at nodes); flowers small, corolla lobes 2–6 mm long. . . . . 4. *L. radicans*
   6. Stems erect, not rooting at nodes; flowers large, corolla lobes 4–12 mm long.
      7. Leaf blades ovate to ovate-lanceolate; broadly rounded to subcordate at base; leaves all distinctly petiolate; petioles conspicuously long-ciliate, the entire length (fig. 478b). . . . . 5. *L. ciliata*
      7. Leaf blades narrowly lanceolate, elliptic or linear, tapering to base or somewhat rounded; leaves petiolate.
         8. Calyx lobes nerveless or weakly nerved; leaves dimorphic, the lowermost leaves distinctly petiolate (the petioles as long as the blades), the upper blades more narrow, usually sessile or narrowed to a subpetiolar base; capsules 2–5 mm long; plants weakly erect to reclining, up to 0.7 m tall. . . . . 6. *L. lanceolata*
         8. Calyx lobes with 3–5 distinct nerves; leaves monomorphic; capsules 4–6.5 mm long; plants robust, stem rigid, erect, up to 1 m tall. . . . . 7. *L. hybrida*
4. Flowers in terminal racemes (fig. 477b), or lower flowers appearing axillary, grading into a terminal raceme (fig. 479d); leaves punctate.
   9. Flowers in a distinct inflorescence, all subtended by bracts. . . . . 8. *L. terrestris*
   9. Flowers appearing axillary on lower portion of stem, grading upward into a raceme; leaves becoming smaller, grading into bracts (fig. 479d). . . . . 9. *L. ×producta*
2. Stems prostrate, creeping, rooting at the nodes (often forming mats); leaves nearly circular, petioles glabrous; plants evergreen. . . . . 10. *L. nummularia*

---

1. *L. maritima* (L.) Galasso, Banfi & Soldano Sea Milkwort Fig. 476

Salt marshes, saline shores, and wet or damp alkaline soils. Nfld. and St. Lawrence R., Que., w. to James Bay, N.W.T., and Alask., s. along Atlantic Coast to N.E., N.J., Md., and ne. Va.; Minn. w. to Wash., s, to N.D., Colo., N.M. and Calif. (*Glaux maritima* L.; *G. maritima* var. *obtusifolia* Fernald)

2. *L. thyrsiflora* L. Swamp Loosestrife, Tufted Loosestrife Fig. 474

Cold swamps, bogs, fens, marshes, pond margins and river bottoms. Lab., Nfld., N.S., and P.E.I. w. to Que., Ont., Sask. and Alask., s. to n. N.J., Md., W.Va., Ohio, Ill., n. Mo., Neb., Kans., n. Colo., n. Utah and n. Calif. (*Naumbergia thyrsiflora* (L.) Reichenb.)

3. *L. quadriflora* Sims Linear-leaf Loosestife, Four-flower Loosestrife Fig. 477

Calcareous bogs, swales, wet prairies, swamps, shores, riverbanks and ditches. Mass. and w. N.Y. w. to Ont., Mich. and se. Man., s. to Md., mts. to w. Va. and nw. Ga. and Ala., Ohio, Ky., Ill., Mo. and n. Ark. and Okla. (*Steironema revolutum* (Nutt.) Raf.; *S. quadriflorum* (Sims) Hitchc.)

4. *L. radicans* Hook. Trailing Loosestrife Fig. 479

Swamps, wet woods and stream banks. Mississippi embayment, s. Ill., w. Ky., w. Tenn. and se. Mo. s. to Ala., La., Okla. and e. Tex.; Va. (*Steironema radicans* (Hook.) A. Gray)

5. *L. ciliata* L. Fringed Loosestrife Figs. 477, 478

Damp woods, wet thickets, stream banks, meadows and shores. N.S. and Que. w. to Man., Alta., and s. B.C., s. to n. Fla., Ark., Kan., Okla., N.M., n. Utah and n. Oreg. (*Steironema ciliatum* (L.) Baudo)

6. *L. lanceolata* Walter Lance-leaf Loosestrife Fig. 477

Swamps, thickets, pond margins, sloughs, meadows, wet prairies and ditches. Me. and sw. Que. w. to Mich., Wisc. and Iowa, s. to N.E., Del., S.C., n. Fla., La., Tenn., n. Ark., e. Okla. and e. Tex.

7. *L. hybrida* Michx. Mississippi Loosestrife, Lowland Loosestrife Fig. 478

Marshes, wet meadows, swamps, stream banks and wet depressions. N.B. and Que. w. to Alta., s. to Fla. and Okla.; sc. Wash., Oreg., nw. N.M. and ne. Ariz. This species has long been treated as a variety of *L. lanceolata*. (*L. lanceolata* var. *hybrida* (Michx.) A. Gray)

8. *L. terrestris* (L.) Britton, Sterns & Poggenb. Yellow Loosestrife, Swamp Loosestrife Figs. 477, 478

Swamps, thickets, shores, bogs and river bottoms. Nfld. and Lab. w. to Minn. and Man., s. to N.J., S.C., Ga., Tenn., Ill., Iowa; Okla.; intro. on Vancouver I., B.C., in w. Wash. and Oreg. Often sterile plants may be found with bulblets in the leaf axils (fig. 142c). (*L. terrestris* var. *ovata* (E. L. Rand & Redfield) Fernald)

9. *L. ×producta* (A. Gray) Fernald Fig. 479

Open woods, damp thickets, shores, swamp margins and sandy fields. Me. and sw. Que. w. to Ont., Mich. and Wisc., s. to N.J., Va., N.C., S.C and Tenn. A fertile hybrid of *L. quadriflora* × *L. terrestris* occurring as isolated populations or with one or both parents present. (*L. foliosa* Small)

10. *L. nummularia* L. Moneywort, Creeping Jenny Fig. 476

Moist roadsides, fields, shores, floodplains and stream banks, often in water; N.S. and N.B. w. to Ont. and Minn., s. to Ga., La., Neb., Kans. and Colo.; B.C. s to Calif.; intro. from Eur., escaped from cultivation.

Fig. 476. *Lysimachia nummularia*: a. habit, prostrate (WVA); b. habit, prostrate; c. flower, showing glands and red spots; d. flower, longitudinal section (Mason).
Lysimachia maritima: e. habit, unbranched plant; f. flowers at anthesis; g. young fruit (Mason).

Fig. 477. *Lysimachia quadriflora*: a. habit (F).
Lysimachia terrestris: b. upper portion of plant (F).
Lysimachia ciliata: c. upper portion of plant (F).
Lysimachia lanceolata: d. habit (F).

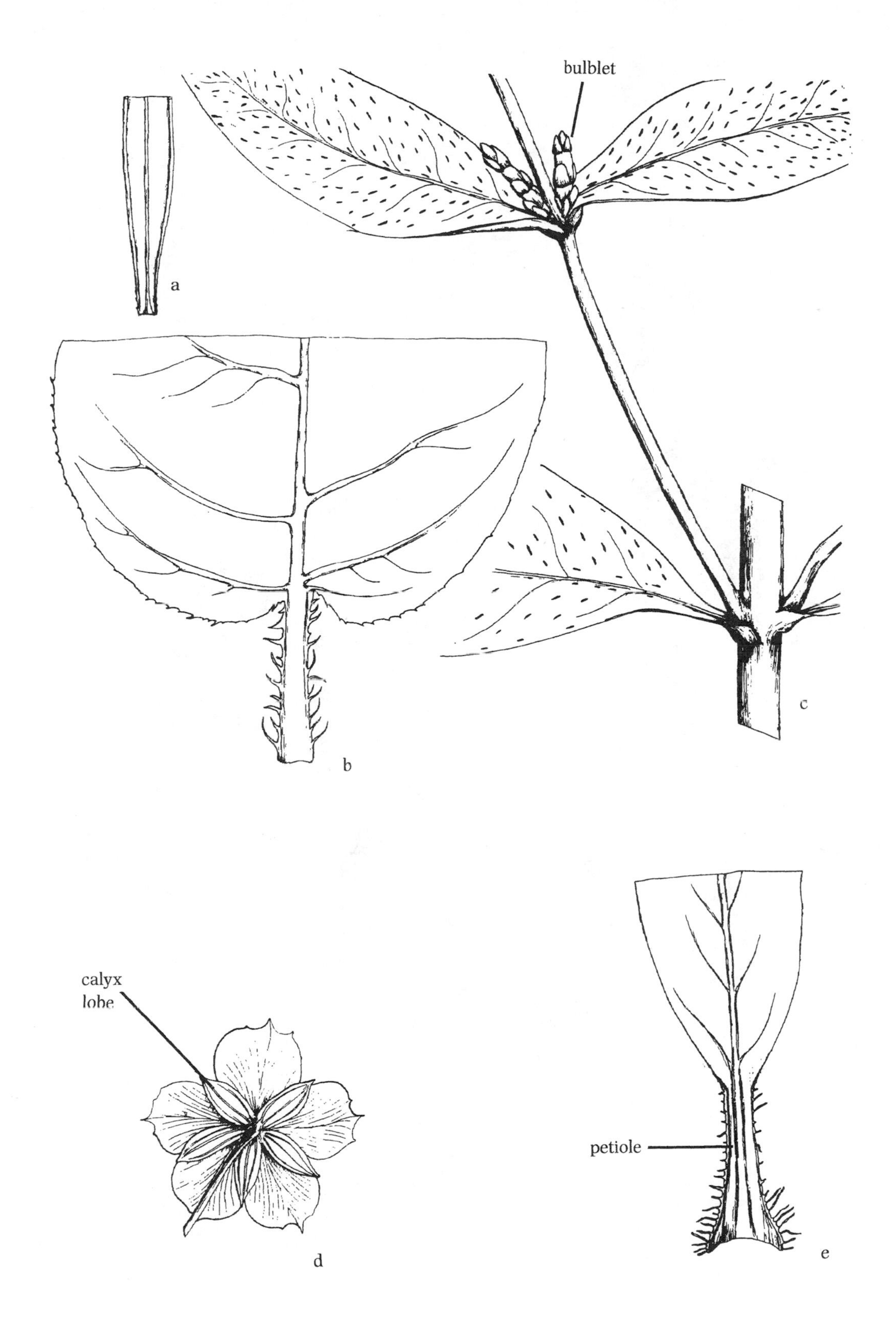

Fig. 478. *Lysimachia quadriflora*: a. leaf base with inrolled margins (F).
Lysimachia ciliata: b. leaf base with ciliate petiole (F).
Lysimachia terrestris: c. stem, leaves, and bulblets (F).
Lysimachia hybrida: d. flower, bottom view; e. leaf base with ciliate petiole (F).

Fig. 479. *Lysimachia radicans*: a. habit; b. flower; c. capsule (C&C).
Lysimachia ×producta: d. upper portion of plant (WVA).

Sarraceniaceae / Pitcher-plant Family

1. *Sarracenia* (Pitcher-plant)

Perennial herbs, insectivorous; leaves pitcher-shaped or trumpet-shaped, with a wing on one side and a hood-like expansion at summit; flowers nodding; sepals persistent; petals purple (ours), maroon, red, or yellow, readily dropping; stamens numerous, shedding pollen into the upside-down umbrella-shaped style; fruit a many-seeded capsule.

References: McDaniel, 1971; Mellichamp and Case, 2009; Schnell, 2002.

---

1. Pitchers with white semi-translucent areolae on upper portions and hoods. . . . . 1. *S. leucophylla*
1. Pitchers lacking white areolae on upper portions and hoods.
    2. Leaves broadly pitcher-shaped, spreading to weakly ascending, wing broad (fig. 480h; petals purple or reddish-purple. . . . . 2. *S. purpurea*
    2. Leaves slender, trumpet-shaped, straight, ascending (fig. 480g), wing scarcely developed; petals yellow. . . . . 3. *S. flava*

---

1. *S. leucophylla* Raf. White-top Pitcher Plant Fig. 480
Bogs, wet pine savannas, cypress depressions, and freshwater tidal thickets. Ga., Fla., Ala. and Miss; intro into se. Va. and Skagit Co., Wash.

2. *S. purpurea* L. Pitcher-plant Fig. 480
Bogs, peaty barrens, wet pine savannas and flatwoods. Nfld. w. to Man., Sask., n. Alta., sw. N.W.T., and ne. B.C., s. to N.J., Md., Ohio, Ind., n. Ill., and Minn.; chiefly coastal plain, se. Va. s. to Ga.; local, mts. of Tenn.

3. *S. flava* L. Trumpets, Trumpet-leaf Fig. 480
Wet pinelands, bogs, cypress swamps, and wet, peaty thickets. Chiefly coastal plain, se. Va. s. to n. Fla., w. to Ala. and se. Miss.; intro. to se. N.Y. and nw. Wash.

Clethraceae / White-alder Family

1. *Clethra* (White-alder)

Shrubs (ours) or small trees, deciduous; leaves alternate, margins serrate; flowers borne in crowded, terminal or axillary racemes; petals white; fruit a 3-valved capsule, enclosed in persistent calyx.

Reference: Tucker and Jones, 2009.

1. *C. alnifolia* L. Sweet Pepperbush Fig. 481
Swamps, thickets, shores, riverbanks, wet pine savannas, flatwoods, *Chamaecyparis* swamps, and shrub bogs; sometimes dry sandy woods and rocky acidic ridge tops. Chiefly coastal plain and piedmont, N.S., Me. and s. N.H., s. to se. N.Y., N.J., nw. and e. Pa., Va., N.C. and n. Fla., w. to Ala., La. and se. Tex.

Ericaceae / Heath Family

---

1. Plants erect, low to tall shrubs; fruit a capsule, or if fleshy (berry or drupe), then blue to black.
    2. Leaves densely white to rusty-tomentose on lower surface; petals separate to base (*R. groenlandicum*). . . . . 6. *Rhododendron*
    2. Leaves glabrous or pubescent, but not tomentose on lower surface; petals united.
        3. Ovary superior; fruit a capsule; flowers various, urceolate, rotate, or funnel-form to campanulate.
            4. Leaves opposite or whorled; flowers rotate, the stamens bent back with anthers held with tension in pouches in corolla, springing inward when touched. . . . . 1. *Kalmia*
            4. Leaves alternate; flowers urceolate or funnelform, the stamens not bent and anthers not held in pouches.
                5. Flowers urceolate (figs. 483e, 484b,f); capsules loculicidal.
                    6. Flowers and fruits solitary, axillary; leaves progressively smaller toward tip of stem, bearing rusty-peltate. . . . . 2. *Chamaedaphne*
                    6. Flowers and fruits in inflorescences; leaves not progressively smaller toward tip of stem, lacking peltate scales.
                        7. Leaves evergreen, leathery.
                            8. Shrubs low, somewhat creeping; leaves linear, with strongly revolute margins, white-glaucous on lower surface inflorescence terminal. . . . . 3. *Andromeda*
                            8. Shrubs erect; leaves broader, narrowly to broadly elliptic, obovate, or slightly ovate, margins not revolute, glaucous on lower surface; inflorescence axillary. . . . . 4. *Lyonia*
                        7. Leaves deciduous, not leathery.

Fig. 480. *Sarracenia leucophylla*: a. habit; b. upper portion of leaf showing opening to pitcher-trap and hood; c. flower hanging "upside down" with style disk and showing numerous stamens (petals removed); d. sepals; e. bracts; f. flower with petals present, hanging between lobes of style (J. C. Putnam Hancock, in Chafin et al., 2007).
Sarracenia flava: g. habit (Beal).
Sarracenia purpurea: h. habit (G&W).

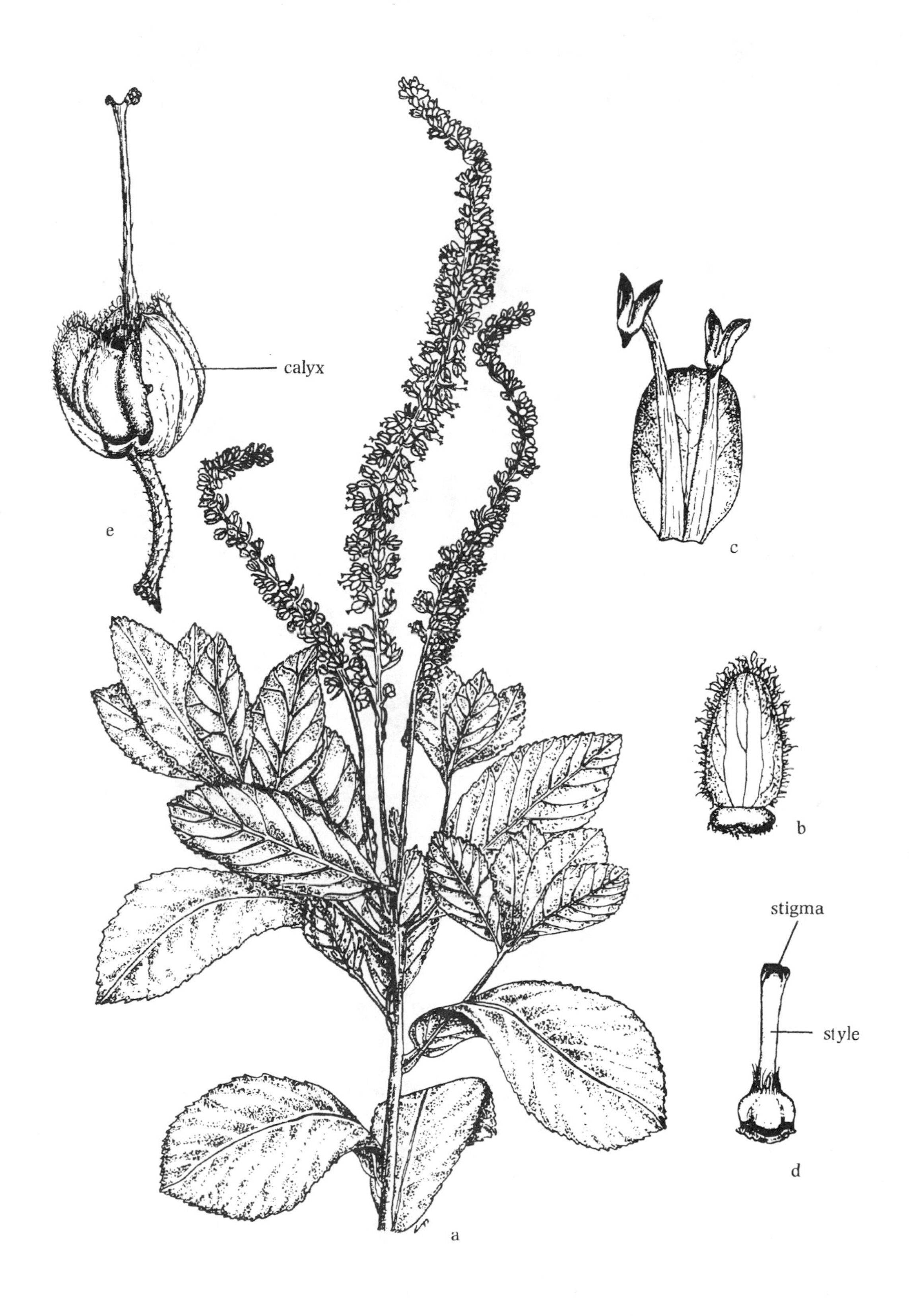

Fig. 481. *Clethra alnifolia*: a. upper portion of plant; b. inner surface of sepal; c. inner surface of petal with stamens; d. young style and stigmas; e. capsule with part of calyx removed (C&C).

9. Flowers in axillary fascicles, racemes of fascicles, or racemes of racemes; capsules with thickened, lighter-colored sutures.......... 4. *Lyonia*
9. Flowers in simple racemes; capsules lacking thickened sutures.......... 5. *Eubotrys*
5. Flowers funnelform to somewhat campanulate, sometimes strongly irregular; capsules septicidal.......... 6. *Rhododendron*
3. Ovary inferior; fruit fleshy, a berry or berry-like drupe (with persisting calyx at apex); flowers urceolate.
10. Leaves with numerous amber, glandular dots or stipitate glands on lower surface; fruit a drupe with 10 seed-like nutlets. .... 7. *Gaylussacia*
10. Leaves lacking glands; fruit a many-seeded berry.......... 8. *Vaccinium*
1. Plants low, trailing, creeping subshrubs, with weakly ascending branches (figs. 489a, 490a); fruit fleshy, red or black.
11. Leaves alternate, flat with revolute margins; flowers bisexual, petiolate; petals 4, pink, strongly reflexed; fruit a many-seeded pale to red-speckled berry, ripening to red late in season (cranberries remaining firm until frost); seeds numerous, ca. 1 mm. .... 8. *Vaccinium*
11. Leaves short-needle-like, spirally arranged (appearing whorled); flowers unisexual (on separate plants), with occasional perfect flowers, sessile; petals 3, white, tiny, inconspicuous, dropping off early (staminate flowers with 3 pink to purple stamens); fruit a black fleshy drupe; seeds (1-seeded stones) 6–9, 1.5–3 mm.......... 9. *Empetrum*

1. *Kalmia* (Laurel)

Shrubs, mostly evergreen; leaves alternate, opposite, or whorled, mostly leathery, margins entire, often revolute; flowers borne in racemose to umbelliform corymbs; corolla reddish-purple, pink, pale pink, or white, united, rotate to campanulate; stamens 10, strongly bent over with anthers held in pouches in the corolla, springing inward when touched; ovary superior; fruit a septicidal capsule.

References: Liu et al., 2009; Southall and Hardin, 1974.

1. Leaves linear to narrowly lanceolate, margins revolute (fig. 482d); inflorescence terminal.......... 1. *K. polifolia*
1. Leaves narrowly to broadly elliptic, margins not revolute; inflorescence lateral.
2. Leaves glabrous on lower surface or only sparsely puberulent (especially young leaves); calyx glandular-pubescent.......... 2a. *K. angustifolia* var. *angustifolia*
2. Leaves densely pubescent on lower surface; calyx short pubescent, but not glandular.......... 2b. *K. angustifolia* var. *carolina*

1. *K. polifolia* Wang. Pale Laurel, Bog Laurel, Swamp Laurel
Fig. 482
Bogs, muskegs and peaty soils. Lab. w. to N.W.T. and Sask., s. to N.E., n. N.J., Pa., Mich., Wisc. and Minn.

2. *K. angustifolia* L. Sheep laurel, Lambkill, Dwarf laurel
Fig. 482

2a. *K. angustifolia* var. *angustifolia*
Wet to moist conifer forests, bogs, margins of peatlands, and moist acid soils of oak and aspen forests. Nfld. and Lab. w. to Ont. and Mich., s. to N.E., Pa, Va. and N.C.

2b. *K. angustifolia* var. *carolina* (Small) Fernald
Sheep Laurel, Lambkill, Wicky
Damp to wet thickets, swamp margins, wet peats, savannas, bottomland woods, montane bogs and rocky montane woods. Coastal plain, se. Va. s. to S.C.; mts. of sw. Va. s. to ne. Tenn., w. N.C., and ne. Ga. Some authors treat this subspecies as a distinct species, *K. carolina* Small.

2. *Chamaedaphne* (Leatherleaf)

Shrubs, evergreen; stems low, much-branched; leaves alternate, leathery, margins obscurely toothed, both surfaces covered with brown dot-like scales, leaves progressively smaller toward tip of stem; flowers solitary, in axils of reduced leaves; corolla white, urceolate; ovary superior; fruit a subglobose, loculicidal capsule.

1. *C. calyculata* (L.) Moench Fig. 483
Acid bogs, muskegs, peaty swales and peaty shores. Lab. and Nfld. w. to N.W.T. and Alask., s. to N.E., N.J., local in coastal plain and mts. to N.C. and Ga., Ohio, n. Ind., n. Ill., Wisc., Minn., n. Iowa, Sask., Alta. and n. B.C. (*C. calyculata* var. *latifolia* (Aiton) Fernald; *Cassandra calyculata* (L.) D. Don)

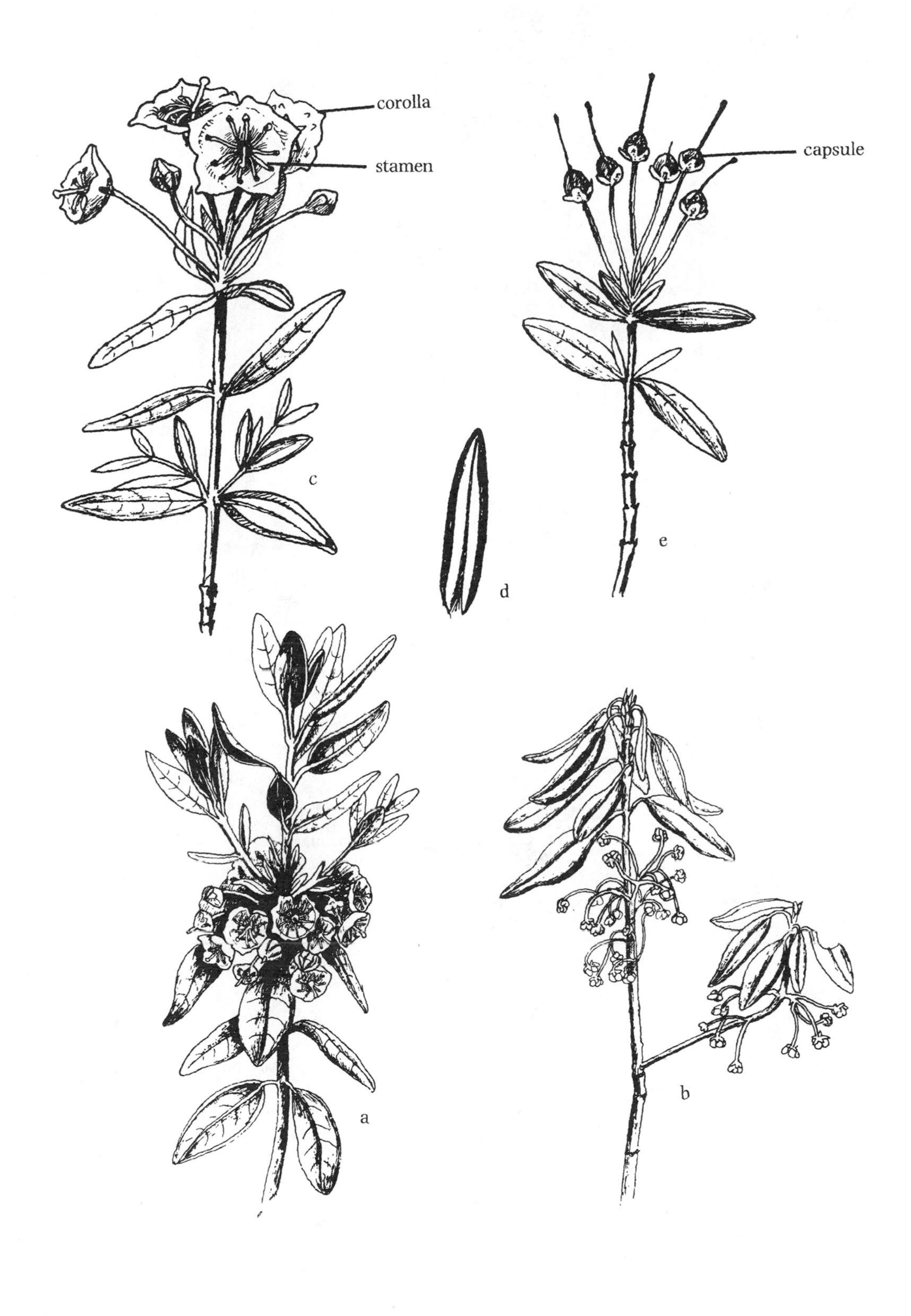

Fig. 482. *Kalmia angustifolia*: a. flowering branch; b. fruiting branch (Ryan).
Kalmia polifolia: c. flowering branch, flowers with stamens bent, anthers in corolla pockets; d. leaf; e. fruiting branch (Ryan).

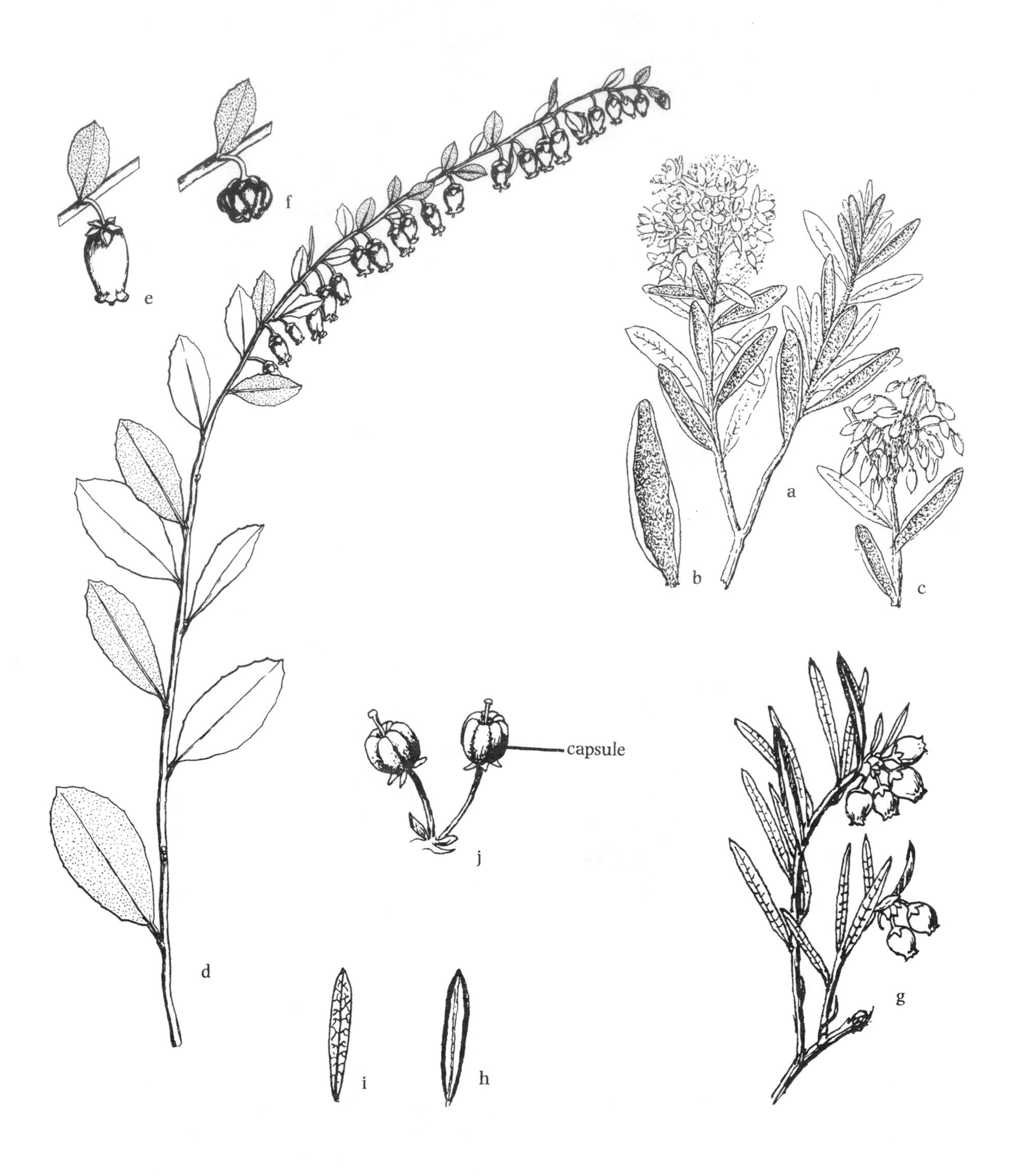

Fig. 483. *Rhododendron groenlandicum*: a. flowering branch; b. leaf with revolute margins; c. fruiting branch (B&B).
Chamaedaphne calyculata: d. flowering branch; e. flower; f. fruit (F).
Andromeda polifolia: g. flowering branch; h. leaf, lower surface; i. leaf, upper surface; j. fruit (Ryan, as *A. glaucophylla*).

3. *Andromeda* (Bog-rosemary)

Shrubs, evergreen; stems low, much branched; leaves alternate, leathery, linear with revolute margins, glaucous on lower surface; flowers borne in terminal umbelliform clusters; corolla pink to white, urceolate; ovary superior; fruit a subglobose, loculicidal capsule.

REFERENCE: Fabijan, 2009.

1. *A. polifolia* L. Bog Rosemary Fig. 483
Acid bogs and peaty shores. Greenl. and Lab. w. to Man. and Sask., s. to N.E., n. N.J., Pa., W.Va., n. Ind., Wisc. and Minn. Fabijan (2009) treats our taxon as var. *latifolia* Aiton, which has long been referred to as *A. glaucophylla* in N.Am., and some authors maintain this as a distinct N.Am. species (*A. polifolia* subsp. *glaucophylla* (Link) Hultén; *A. glaucophylla* Link)

4. *Lyonia* (Maleberry, Stagger-bush)

Shrubs (ours) or small trees, evergreen or deciduous; leaves alternate, margins entire to serrate; flowers borne in racemes, panicles, or fascicles; corolla white to red, urceolate to cylindrical; ovary superior; fruit a loculicidal capsule with thickened, light-colored sutures.

REFERENCES: Judd, 1981, 2009.

1. Leaf margins serrulate; flowers in subpaniculate inflorescences (racemes of clusters) growing from stem tip of previous year; corolla urceolate (subglobose) . . . . 1. *L. ligustrina*
1. Leaf margins entire; flowers in axillary fascicles, subtended by leaves or by leaf scars of stem of previous year; corolla cylindrical (elongate-urceolate).
   2. Leaves evergreen, leathery; flower clusters subtended by leaves; corolla pink (sometimes red or white), 2–5 mm wide, swollen at base. . . . . 2. *L. lucida*
   2. Leaves deciduous, thin; flower clusters subtended by leaf scars; corolla white (rarely pink), 4.5–9 mm wide, not swollen at base. . . . . 3. *L. mariana*

1. *L. ligustrina* (L.) DC. Maleberry, He-huckleberry, Male-blueberry Fig. 484
Acid swamps, bogs, pond and river margins, moist to dry acid woods and thickets. Me. sw. to N.Y., Pa., and s. Ohio, Ky., Tenn., s. to S.C., Ga. and Ala. Our taxon is var. *ligustrina*; Judd (1981, 2009) also recognizes a coastal plain taxon, var. *foliosiflora* (Michx.) Fernald with conspicuous, leaf-like bracts, which reaches its northern limit in southeastern Virginia.

2. *L. lucida* (Lam.) K. Koch Fetterbush Lyonia, Hurrah-bush
Swamps, especially bald cypress swamps, pond margins, shrub bogs, thickets, pine savannas and flatwoods. Coastal plain and piedmont; se. Va. s. to Fla., Ala. and La.; Cuba.

3. *L. mariana* (L.) D. Don Piedmont Stagger-bush, Wicks Fig. 485
Pine savannas, flatwoods, swamps, pond margins, shrub bogs, thickets and woodland margins. Coastal plain and piedmont; R.I., sw. Conn. and se. N.Y., s. to Va., Ga. and Fla.; se. Mo. s. to Ark., se. Okla., n. La. and e. Tex.

5. *Eubotrys* (Hobblebush, Dog-laurel, Fetterbush)

Shrubs; stems erect, branches spreading; leaves deciduous, blade oblong to oblanceolate or obovate, margins spinulose-serrulate, surfaces glabrous, often unicellular hairy on major veins on lower surface; inflorescences axillary, fascicled or solitary racemes, 8–25-flowered; sepals 5, connate basally; corolla cylindric-urceolate, corolla lobes 5, white to pale pink; stamens 8(10); fruits a dry depressed-globose capsule. *Eubotrys* is sometimes included within the genus *Leucothoe*, but morphological and molecular work indicates an affinity with *Chamaedaphne* (Tucker 2009).

REFERENCE: Tucker, 2009.

1. *E. racemosa* (L.) Nutt. Swamp Dog-hobble, Fetterbush Fig. 484
Swamps, bogs, cypress-gum ponds and depressions, thickets along shores, wet sandy woods and pine flatwoods. Coastal plain and piedmont; e. Mass, s. to Fla. w. to se. La.; c. Tenn. (*Leucothoe racemosa* (L.) A. Gray; *L. racemosa* var. *projecta* Fernald; *Eubotrys elongata* (Small) Small)

Fig. 484. *Eubotrys racemosa*: a. flowering branch; b. flower; c. fruiting branch; d. fruit (C&C).
Lyonia ligustrina: e. flowering branch; f. flower; g. fruiting branch; h. new fruit; i. branch with fruit of previous season; j. remains of old fruit (C&C).

Fig. 485. *Rhododendron maximum*: a. flowering branch (WVA).
Rhododendron periclymenoides: b. flowering branch (WVA).
Lyonia mariana: c. flowering branch; d. flower; e. fruiting branch (B&B).

6. *Rhododendron* (Rhododendron, Rosebay, Azalea)

Shrubs, evergreen or deciduous; leaves alternate, leathery or thin, with or without an indumentum on lower surface; flowers showy, born in umbelliform clusters; corolla campanulate to funnelform, lobes usually broadly spreading, symmetry regular to irregular; ovary superior; fruit a septicidal capsule.

Reference: Judd and Kron, 2009.

---

1. Leaves evergreen, thick, leathery, flowers white.
    2. Plants less than 1 m tall, creeping shrubs, prostrate to ascending; leaves 2–5 cm long; leaf lower surface rusty to white tomentose; flowers with corolla of separate petals. . . . . 1. *R. groenlandicum*
    2. Plants greater than 1 m tall, erect shrubs, leaves mostly 9–20 cm long, leaf lower surface glabrous or sparsely pubescent; flowers with corolla united. . . . . 2. *R. maximum*
1. Leaves deciduous, thin, flowers rosy-purple, pink to pinkish-white or white.
    3. Flowers strongly irregular; corolla tube short, not funnelform; stamens 10; capsules glaucous-puberulent. . . . . 3. *R. canadense*
    3. Flowers nearly regular; corolla tube elongate, funnelform; stamens 5; capsules stipitate-glandular, hirsute, or both, or densely pubescent, but never glaucous.
        4. Flowers emerging early before or with the leaves, flowers pink or pinkish-white.
            5. Pedicels, calyces, and capsules stipitate-glandular; plants strongly colonial by subterranean runners, short usually 0.2–0.6 m tall. . . . . 4. *R. atlanticum*
            5. Pedicels, calyces, and capsules lacking stipitate glands; plants not strongly colonial, tall, up to 5 m.
                6. Leaves glabrous on lower surface, except for short appressed hairs along midrib; corolla lobes about as long as tube. . . . . 5. *R. periclymenoides*
                6. Leaves sparsely to densely gray-pubescent on lower surface; corolla lobes clearly shorter than tube. . . . . 6. *R. canescens*
        4. Flowers emerging after the leaves, flowers white. . . . . 7. *R. viscosum*

---

1. *R. groenlandicum* (Oeder) Kron & Judd Labrador Tea Fig. 483
Bogs, wet peaty shores and peaty subalpine-alpine sites. Greenl., Nfld. and Lab. w. to Nunavut, N.W.T. and Alask., s. to N.E., n. N.J., n. Pa., Ohio, Mich., Minn., Sask., Alta., Ida., B.C., w. Wash. and nw. Oreg. Long treated in the segregate genus *Ledum*, molecular data along with morphological and phylogenetic analysis place this unique plant within the genus *Rhododendron*. (*Ledum groenlandicum* Oeder)

2. *R. maximum* L. Great Laurel, Rosebay Fig. 485
Stream banks, pond margins, swamps, wet woods and moist uplands. N.E. w. to N.Y. and Ohio, s. to Va. Ky. and Tenn., chiefly in mts. and piedmont to n. Ga. and n. Ala.

3. *R. canadense* (L.) Torr. Rhodora Fig. 486
Bogs, wet thickets, acid barrens, rocky slopes and mountain tops. Nfld. and Que. s. to N.E., n. N.J. and ne. Pa.

4. *R. atlanticum* (Ashe) Rehder Dwarf Azalea Fig. 486
Borders of shrub-tree bogs, pine savannas, flatwoods and open bottomlands. Coastal plain; s. N.J., se. Pa., and Del. s. to Ga.

5. *R. periclymenoides* (Michx.) Shinners Pinxter-flower, Wild Azalea Fig. 485
Moist or dry-woods, damp thickets, stream margins, swamps and bogs. Me., N.H. and Mass. w. to N.Y. and n. and s. Ohio, s. to Ga., Ky., s. Ill., Tenn. and Ala. (*R. nudiflorum* (L.) Torrey)

6. *R. canescens* (Michx.) Sweet Hoary Azalea, Wild Azalea Fig. 487
Wet woods, stream banks, springy sites, swamps, margins of shrub-tree bogs, pine flatwoods and open dry woods. S. Pa., and Md.; N.C., w. to sw. Ky., Tenn., Ill. and Ark. s. to Fla., Ala., Miss., La., embayment n. to w. Tenn., Ark. and se. Okla. and e. Tex.

7. *R. viscosum* (L.) Torrey Swamp-honeysuckle, Clammy Azalea Fig. 486
Swamps, lakeshores, thickets, wet woods, bogs, and wet pine flatwoods. Sw. Me. w. to s. to Pa., S.C., e. Tenn., Fla., Miss., Ark., Okla. and Tex.

7. *Gaylussacia* (Huckleberry)

Shrubs, deciduous (ours) or evergreen; leaves alternate, with numerous glandular dots or stipitate glands; flowers borne in axillary racemes; corolla white to pink-tinged or greenish, urceolate; ovary inferior; fruit a drupe with 10 seed-like nutlets (endocarps).

Reference: Sorrie et al., 2009.

---

1. Bracts of raceme large, leaf-like, persistent; sepals glandular-ciliate; ovary and fruit glandular-pubescent (fig. 488b). . . . . 1. *G. bigeloviana*
1. Bracts of raceme small, soon falling off; sepals not ciliate; ovary and fruit glabrous. . . . . 2. *G. baccata*

Fig. 486. *Rhododendron canadense*: a. flowering branch (Ryan).
Rhododendron atlanticum: b. flowering branch (Gleason).
Rhododendron viscosum: c. flowering branch (Gleason).
Vaccinium corymbosum: d. flowering branch; e. fruiting branch (WVA).

Fig. 487. *Rhododendron canescens*: a. flowering branch; b. flower (C&C).

1. *G. bigeloviana* (Fernald) Sorrie & Weakley Bog Huckleberry Fig. 488
Sphagnum bogs and wet peats, wet sandy soil, pinelands and dry barrens. Coastal plain, Nfld., N.S. e. N.B., Que. and s. to Del. and Md.; disjunct to sw. N.C. and nw. S.C. Sorrie et al. (2009) separate this taxon from the closely related *G. dumosa* (Andrews) Torrey & A. Gray, a smaller shrub, with smaller flowers, and a more southerly distribution.

2. *G. baccata* (Wang.) K. Koch Black Huckleberry Fig. 488
Dry or moist woods, thickets and margins of bogs and swamps. Nfld. and Que. w. to Man., s. to n. Ga., Ark. and Miss.

8. *Vaccinium* (Blueberry, Cranberry)

Shrubs or subshrubs, evergreen or deciduous; stems much branched, tall and erect, low and spreading, or trailing; leaves alternate, margins serrate or entire; flowers solitary or borne in racemes; corolla white to pinkish, urceolate, campanulate, or united basally with strongly reflexed lobes; ovary inferior (sepals persisting at apex of fruit); fruit a many-seeded berry.

References: Vander Kloet, 1980, 1983, 2009.

---

1. Plants tall (up to 0.5 m), erect shrubs; flowers not nodding, corolla cylindric-urceolate; fruits blue to blue-black and glaucous or dull black. . . . . 1. *V. corymbosum*
1. Plants low, trailing subshrubs (viny); flowers nodding on long pedicels, corolla basally united, lobes strongly reflexed (fig. 489a); fruits pale to red speckled, becoming red late in season.
    2. Pedicels with a pair of leaf-like bracts, 2–4 mm long, 1–2 mm wide, borne just above middle; leaves narrowly elliptic, margins not revolute or weakly so. . . . . 2. *V. macrocarpon*
    2. Pedicels with a pair (sometimes 0–5) of scale-like bracts, 1.5 mm or less long, less than 1 mm wide, borne near top; leaves ovate, margins often strongly revolute. . . . . 3. *V. oxycoccos*

---

1. *V. corymbosum* L. Highbush Blueberry Fig. 486
Swamps, bogs, wet thickets, low woods and sometimes dry uplands. S. N.S., N.B., Que. and Ont. w. to Mich., Wisc. and n. Ill., s. to Fla., Ky., Ark., se. Okla., and e. Tex. Vander Kloet (1980, 2009) argues that while at least 25 morphological variants have been at one time or another recognized at the species rank, the highbush blueberry group is best treated as a single, polymorphic species. (*V. atroccocum* (A. Gray) Heller)

2. *V. macrocarpon* Aiton Cranberry Fig. 489
Acid bogs, swamps, wet peaty or sandy shores and interdunal swales. Nfld. w. to Ont. and Minn. s. to N.J., Del., Va., mts. to Tenn. and N.C., Ohio, n. Ind. and n. Ill. (*Oxycoccus macrocarpus* (Aiton) Pers.)

3. *V. oxycoccos* L. Small Cranberry Fig. 489
Sphagnum bogs, fens, muskegs and arctic-alpine tundra. Greenl. and Lab. w. to N.W.T. and Alask., s. to N.E., n. N.J., mts. and coastal plain, Va., W.Va., Ohio, n. Ind., ne. Ill., Wisc., Minn., Sask., n. Ida., w. Wash., w. Oreg. and n. Calif. (*V. microcarpum* (Turcz.) Hook. f.; *O. microcarpus* Turcz ex Rupr.)

9. *Empetrum* (Crowberry)

Low, spreading, much-branched shrubs; leaves stiff, evergreen, linear to elliptic, needle-like, with revolute margins; flowers bisexual or more typically unisexual (dioecious plants), axillary; sepals 3, small; petals 3, tiny, readily fall off; fruit a black (ours), purple, or red drupe. *Empetrum* has long been treated as distinct at the family level, Empetraceae, but the Angiosperm Phylogeny Group IV (APG IV, 2017) includes the group as a tribe within the Ericaceae, subfamily Ericoideae.

References: Fernald and Wiegand, 1913; Li et al., 2002; Löve, 1960; Moore et al., 1970; Murray et al., 2009; Soper and Voss, 1964.

1. *E. nigrum* L. Black Crowberry Fig. 490
Bogs, heaths, sea cliffs and coastal headlands, moist sandy bluffs and arctic-alpine barrens. Greenl. and Lab. w. to Alask., s. to N.S., coastal Me., alpine sites of n. N.E. and n. N.Y., n. Mich., ne. Minn., Man., Alta., B.C. and mts. to n. Calif.

Rubiaceae / Madder Family

---

1. Plants woody, shrubs. . . . . 1. *Cephalanthus*
1. Plants herbaceous.
    2. Leaves whorled. . . . . 2. *Galium*
    2. Leaves opposite. . . . . 3. *Diodia*

Fig. 488. *Gaylussacia bigeloviana*: a. flowering branch; b. fruiting branch; c. leaf (Ryan, as *G. dumosa*). *Gaylussacia baccata*: d. flowering branch; e. leaf (Ryan).

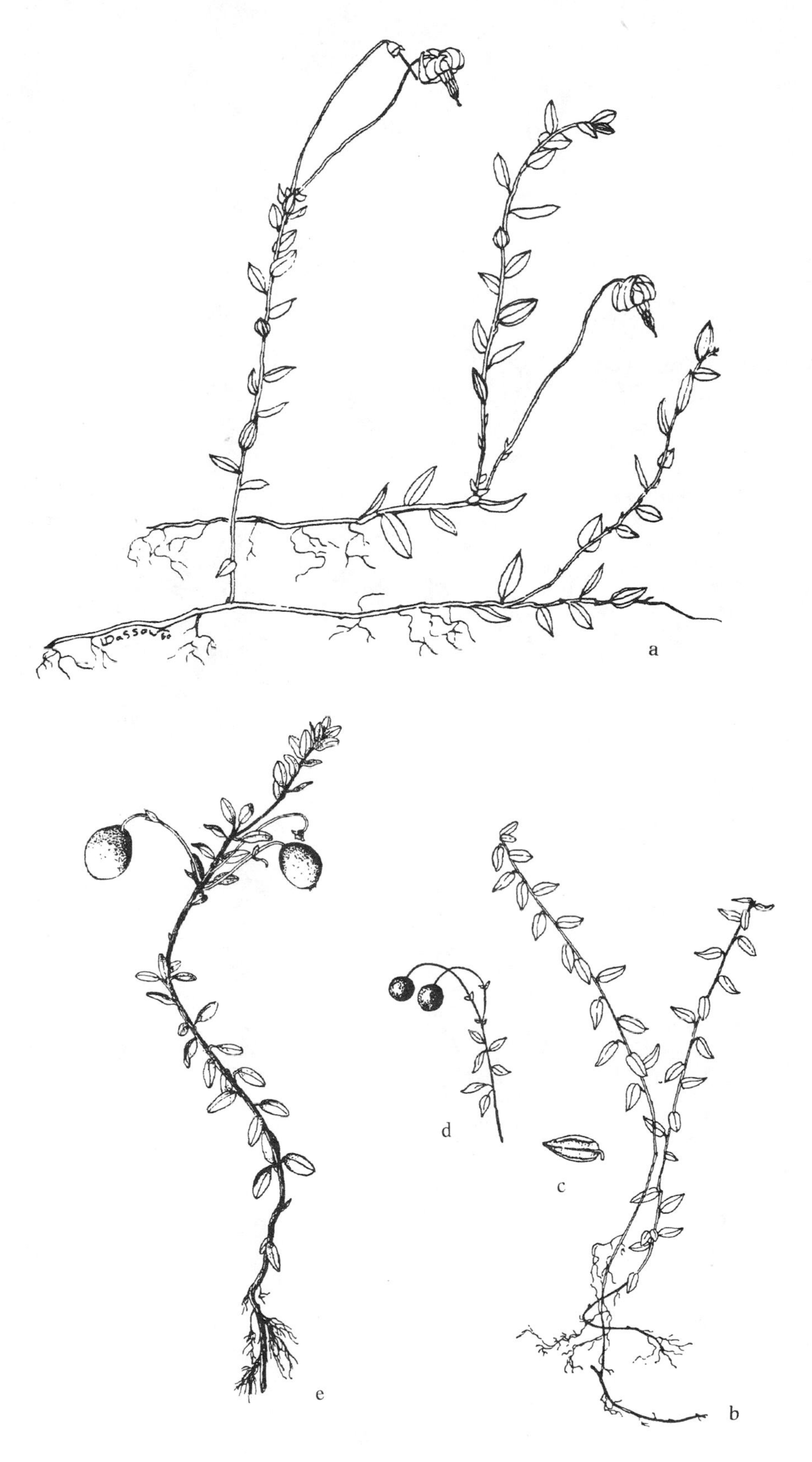

Fig. 489. *Vaccinium oxycoccos*: a. flowering habit (USFS); b. habit; c. leaf, lower surface; d. tip of fruiting branch (OHIO).
Vaccinium macrocarpon: e. fruiting habit (OHIO).

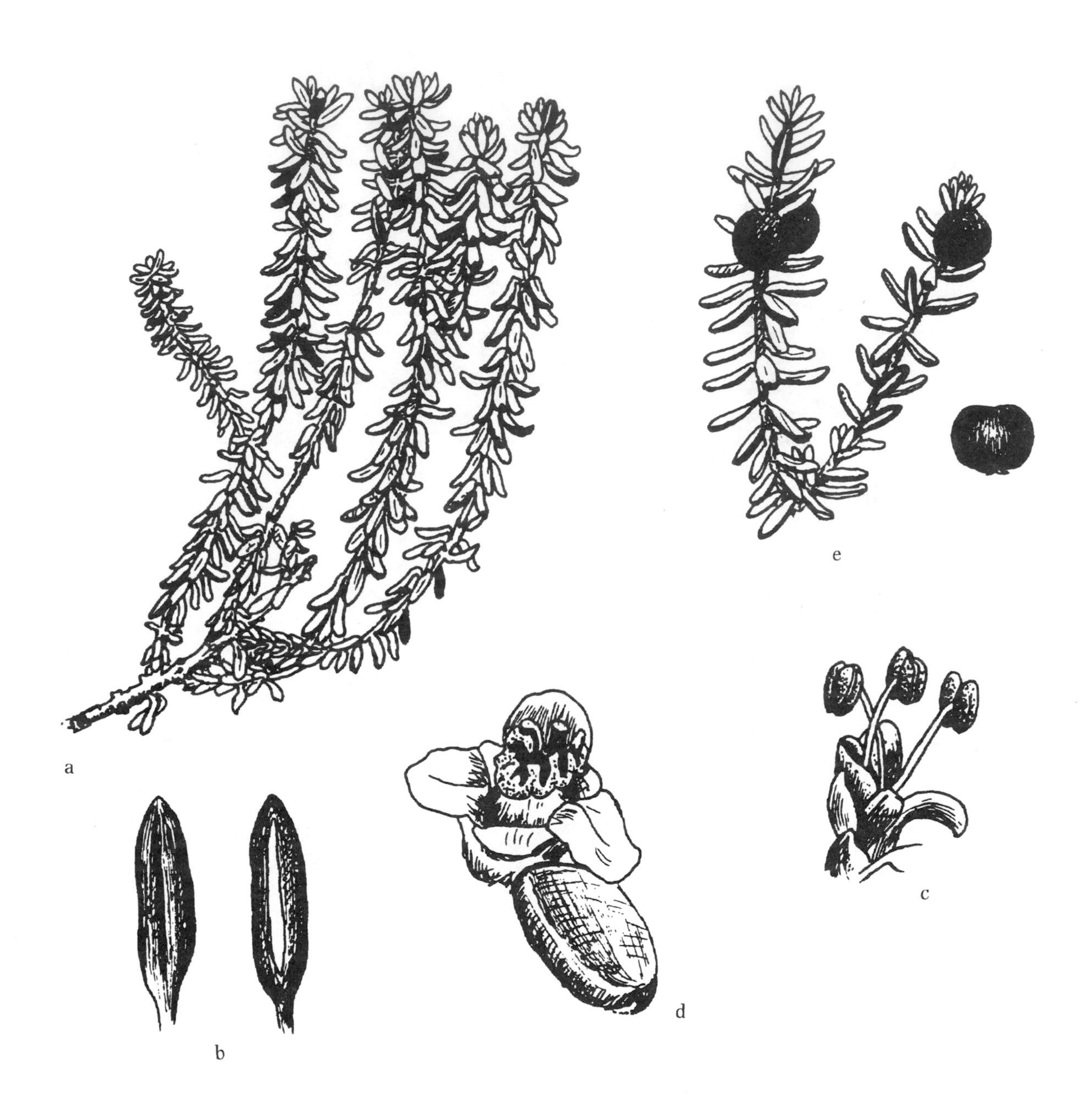

Fig. 490. *Empetrum nigrum*: a. habit; b. leaves; c. staminate flower; d. pistillate flower; e. branch tip with fruit (Ryan).

1. *Cephalanthus* (Buttonbush)

Shrubs; leaves deciduous, opposite and/or in whorls of 3 or 4, petiolate; blades pinnately veined, oval, oblong-oval, elliptic, or ovate; flowers small, borne in dense, globose, long-pedunculate heads, axillary or terminal; corolla white, united, funnelform; stamens long, with filaments conspicuously exerted (giving globose inflorescence a "spiky" appearance); fruit splitting into 2 or 4 nutlets.

1. *C. occidentalis* L. Common Buttonbush Fig. 491
Swamps, bogs, lakeshores and stream margins. N.S., N.B. and sw. Que. w. to s. Ont., Mich., Wisc., e. Minn. and e. Neb., s. to Fla., and Tex.; Ariz. and Calif.; Mex. and W.I. Our taxon is var. *occidentalis*. (*C. occidentalis* var. *pubescens* Raf.)

2. *Galium* (Bedstraw)

Annual or perennial herbs, low, often prostrate, trailing, or matted; leaves linear, sessile, or short-petiolate, whorled; flowers tiny, borne in single or branched, terminal or axillary cymes; calyx lacking; corolla white or yellow, united, lobes 3–4(5); ovary inferior; fruit bristly or glabrous, separating into 2 seed-like segments.

References: Puff, 1976, 1977; Urban and Iltis, 1957.

---

1. Leaves bristle-tipped, usually 6 per whorl. . . . . 1. *G. asprellum*
1. Leaves blunt at apex, usually 4 or fewer per whorl, or if 5 or 6, then not bristle-tipped.
   2. Corolla lobes 4, corolla 2–4 mm across.
      3. Flowers and fruits several, (usually 10–25) in much-branched cymes (fig. 491e). . . . . 2. *G. palustre*
      3. Flowers and fruits fewer, (usually 1–6) in once- or twice-branched cymes.
         4. Leaves usually more or less recurved, (5)–12(16) mm long, less than 2.5 mm wide, margins revolute; fruiting pedicels 1–2.5(4) mm long; fruits 2–3(3.5) mm long, both carpels usually developing. . . . . 3. *G. labradoricum*
         4. Leaves ascending or spreading, (8)10–25(30) mm long, 2.5–6(7) mm wide, margins flat; fruiting pedicels (3)5–12(15) mm long; fruits 4–5 mm long, often only 1 carpel developing. . . . . 4. *G. obtusum*
   2. Corolla lobes 3 (with occasional individual 4-lobed flowers), corolla not more than 2 mm across.
      5. Fruiting pedicels curved, reflexed (fig. 491); leaves linear or linear-oblanceolate or oblanceolate (mostly 4 leaves in a whorl of main axis).
         6. Flowers and fruits in clusters of (2)3, with pedicels nearly equal in length: corolla 1.2–2.5 mm in diameter; fruits (2)2.2–3.5(4) mm long; leaves (7)8–20(25) mm long. . . . . 5. *G. trifidum*
         6. Flowers and fruits usually solitary, or if 2, then pedicels of unequal length (1 flower/fruit subsessile); corolla 0.8–1.8 mm in diameter; fruits 0.8–1.2 mm long; leaves 2.5–7(10) mm long. . . . . 6. *G. brevipes*
      5. Fruiting pedicels straight, more or less stiff, strongly divergent; leaves broadly oblanceolate to oblong-spatulate (in 4, 5, or 6 leaves in a whorl of main axis). . . . . 7. *G. tinctorium*

---

1. *G. asprellum* Michx. Rough Bedstraw
Wet woods and thickets, stream margins, swamps and meadows. Nfld. w. to w. Ont. and Minn., s. to N.E., N.Y., mts. to w. NC. and Tenn., Ohio, Ill., Iowa and Mo.

2. *G. palustre* L. Common Marsh Bedstraw Fig. 491
Wet meadows and banks. Nfld. w. to Ont. and Man., s. to N.E., N.Y., se. Pa., W.Va., Mich. and Ill.; Alask., Mont., Wash. and Oreg.

3. *G. labradoricum* (Wiegand) Wiegand Northern Bog Bedstraw
Bogs, *Thuja* swamps, mossy thickets and sedge meadows. Lab., Nfld. and Nunavut w. to N.W.T. and B.C., s. to n. Me., N.H., w. Mass., w. Conn., nw. N.J., N.Y., n. Ind., n. Ill., n. Iowa, N.D., Sask. and Alta. (*G. tinctorium* var. *labradoricum* Wiegand)

4. *G. obtusum* Bigel. Bluntleaf Bedstraw
Neutral to alkaline sites, swampy thickets, alluvial swamps, low woods, wet prairies and wet meadows which tend to dry out. N.S. and sw. Que. w. to s. Ont., Mich., Wisc. and Minn., s. to Ga., Fla., Ala., La. and e. Tex. The widespread taxon of our range is subsp. *obtusum*. Puff (1977) also recognizes subsp. *filifolium* (Wiegand) Puff, a taxon of the piedmont and coastal plain, which extends from Fla. northward into Va., Md., Del. and N.J., occurring in coastal saline sites at swampy areas at the mouths of rivers; it differs primarily in having leaves filiform-linear to very narrowly oblanceolate (1–2 mm wide) and smaller fruits; however, the two taxa intergrade morphologically.

5. *G. trifidum* L. Threepetal Bedstraw Fig. 491
Swamps, wet shores, thickets and banks. Greenl., S. Lab. and Nfld. w. to N.W.T. and Alask., s. to N.E., Pa., n. Ohio, n. Ill., Minn., ne. Neb., Tex., w. to. Wash., Oreg. and Calif. Our widespread taxon is subsp. *trifidum*. A coastal, succulent, entirely glabrous form, occurring along brackish and saline shores from Newfoundland and the St. Lawrence River to Massachusetts has been described as subsp. *halophilum* (Fernald & Wiegand) Puff. (*G. brandegeei* A. Gray)

6. *G. brevipes* Fernald & Wiegand Limestone Swamp Bedstraw
Chiefly calcareous sites, mud flats, riverbanks, sandy shores, mossy swales and rarely acid boggy soils. Gaspé Pen., Que. w. to

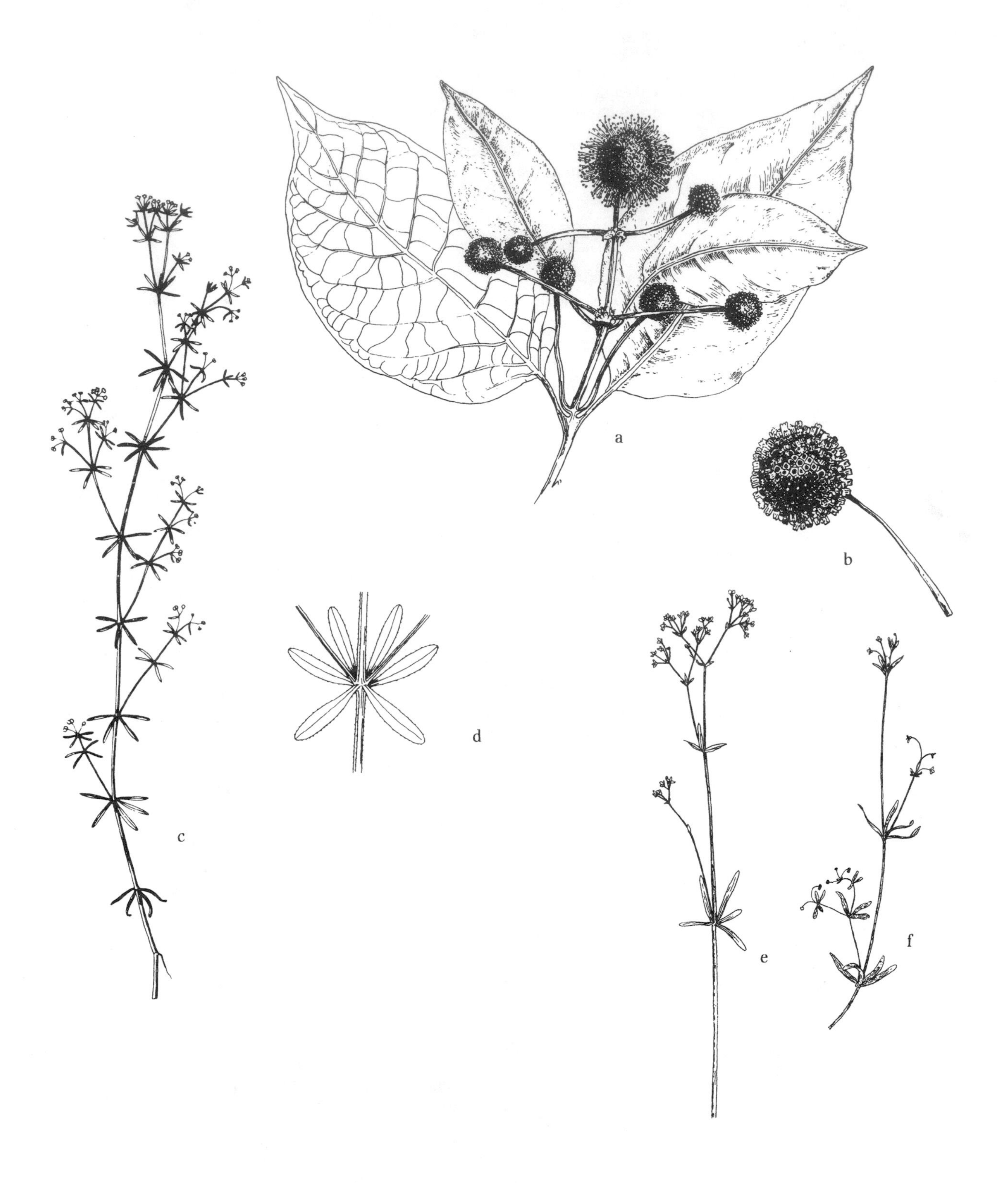

Fig. 491. *Cephalanthus occidentalis*: a. branch tip; b. head of fruits (C&C).
Galium tinctorium: c. upper portion of plant; d. leaf whorl (Beal).
Galium palustre: e. upper portion of plant (F).
Galium trifidum: f. upper portion of plant (F).

Ont., Mich. se. Man., and e. Sask., s. to Me., n. N.H., w. Mass., N.Y., n. Ind., Wisc., Minn. and ne. S.D. Although Gleason (1952) treats G. *brevipes* as conspecific with G. *brandegeei*, Puff (1976) placed G. *brandegeei* as a synonym within his concept of G. *trifidum* subsp. *subbiflorum*, a western taxon. (G. *trifidum* subsp. *brevipes* (Fernald & Wiegand) Á Löve & D. Löve.)

7. *G. tinctorium* L. Stiff marsh Bedstraw Fig. 491
Swamps, marshes, swampy meadows, wet depressions and ditches. Nfld. and Lab. w. to Ont. and Minn., s. to S.C., Fla., Neb. and Tex. Our widespread taxon is subsp. *tinctorium*. (*G. claytoni* Michx.; *G. trifidum* subsp. *tinctorium* (L.) Hara; *G. tinctorium* var. *floridanum* Wieg; *G. tinctorium* subsp. *floridanum* (Wiegand) Puff).

3. *Diodia* (Buttonweed)

Perennial herbs; stems prostrate, decumbent, or weakly ascending; leaves opposite, sessile, entire, with membranous stipules and elongate bristles; flowers small, sessile, axillary, usually solitary; corolla white, united; ovary inferior, 2-locular; fruit separating into 2 segments.

1. *D. virginiana* L. Virginia Buttonweed Fig. 492
Wet shores, low grounds. swales. wet prairies and ditches. N.J. w. to s. Ill. and Mo. and Kans., s. to Fla., e. Okla. and Tex. (*D. virginiana* var. *attenuata* Fernald)

Loganiaceae / Logania Family

1. *Mitreola* (Miterwort, Hornpod)

Annuals; leaves opposite, simple, elliptic to lanceolate-oblong; flowers small, actinomorphic, 5-merous, borne in a 1-sided terminal cyme; calyx united, corolla united, forming a tube, white or pink; stamens 5, epipetalous; ovary superior; fruit a capsule.

---

1. Leaves petiolate or subpetiolate, sometimes sessile on upper portion of stem, blade narrow, lance-elliptic to oblanceolate to almost linear, (2)3–8 cm long; corolla twice as long as calyx. . . . . 1. *M. petiolata*
1. Leaves sessile, blade ovate to suborbicular or elliptic, 2 cm or less long; corolla slightly longer than. . . . . 2. *M. sessilifolia*

---

1. *M. petiolata* (Walter ex J. F. Gmel.) Torr. & A. Gray Lax Hornpod Fig. 493
Wet sandy shores, floodplains, boggy areas and ditches. Se. Va. w. to Tenn., se. Mo., Ark. and se. Okla., s. to Fla. and c. Tex.; Trop. Am. (*Cynoctonum mitreola* (L.) Britton)

2. *M. sessilifolia* (J. F. Gmel.) G. Don Swamp Hornpod Fig. 493
Southern bogs, pine savannas, damp shores and low woods. Coastal plain, se. Va. s. to Fla., w. to Ark., Okla., e. Tex. and N.M. (*Cynoctonum sessilifolium* J. F. Gmel.)

Gentianaceae / Gentian Family

---

1. Leaves scale-like, 5 mm long or less (fig. 494); corolla lobes overlapping in bud. . . . . 1. *Bartonia*
1. Leaves scale-like, greater than 5 mm long; corolla lobes convolute in bud.
    2. Corolla pink or white, corolla lobes longer than corolla tube, style strongly bent to one side at anthesis. . . . . 2. *Sabatia*
    2. Corolla blue; corolla lobes shorter than corolla tube; styles straight at anthesis.
        3. Plants annual or biennial; flowers usually 4-merous, corolla not plicate, lobes somewhat spreading seeds spherical. . . . . 3. *Gentianopsis*
        3. Plants perennial; flowers 5-merous; corolla plicate (fig. 498b,d,f), often with appendages between the incurved lobes (fig. 498d); seeds flattened and winged. . . . . 4. *Gentiana*

---

1. *Bartonia* (Screw-stem)

Annual or biennial herbs; stems slender, seldom branched; leaves opposite, subopposite or alternate; tiny, scale-like, awl-shaped; flowers borne in panicles or racemes; corolla white, creamy-white, greenish, of greenish-yellow, united; fruit a capsule.

References: Gillett, 1959; Henson, 1985; Reznicek and Whiting, 1976.

---

1. Stem leaves mostly opposite; petals oblong, abruptly rounded at tip with a sharp mucro at apex. . . . . 1. *B. virginica*
1. Stem leaves mostly alternate; petals lance-oblong, with acute tip to acuminate apex.

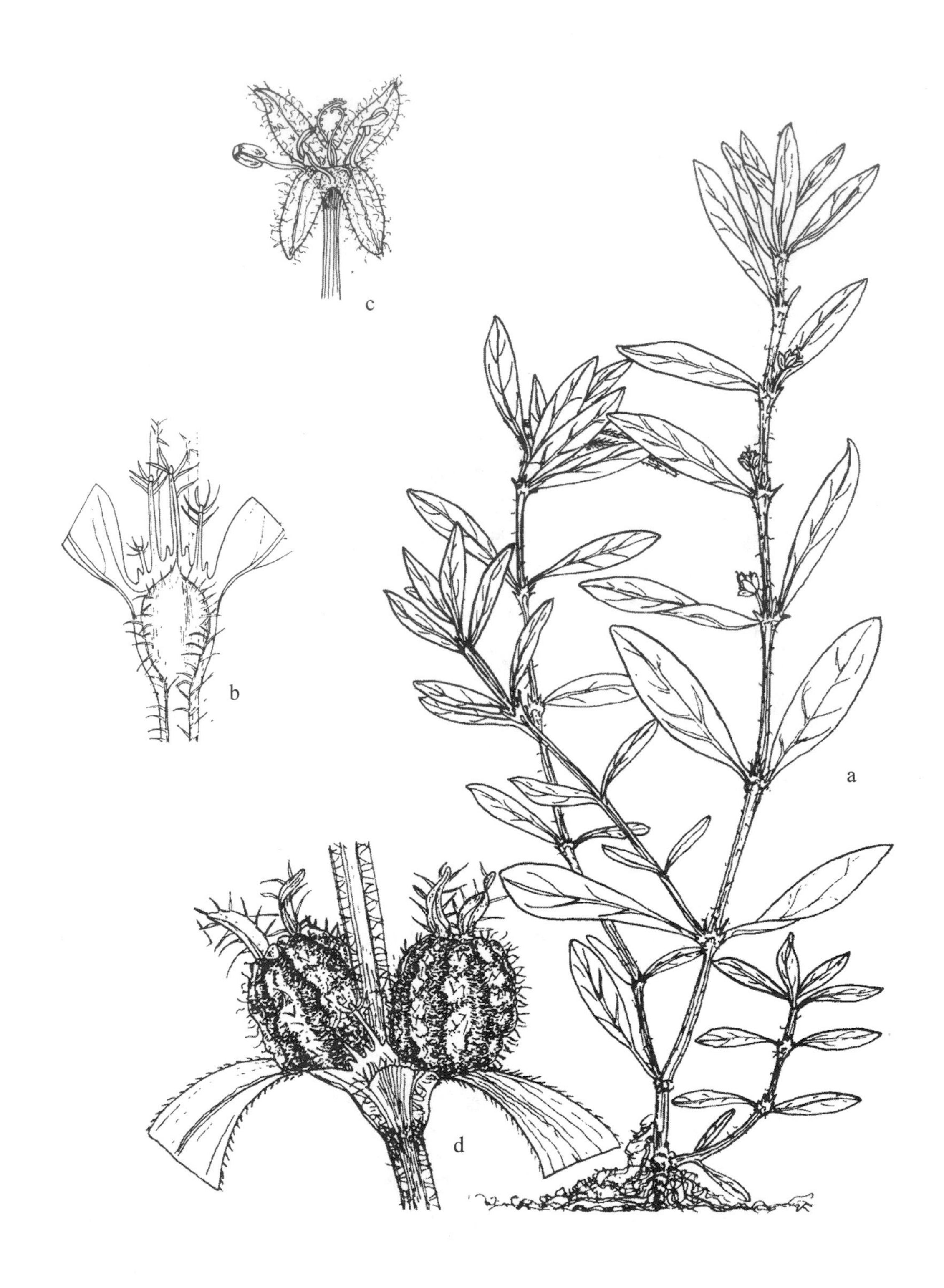

Fig. 492. *Diodia virginiana*: a. habit; b. leaf base with stipules; c. flower; d. fruit (C&C).

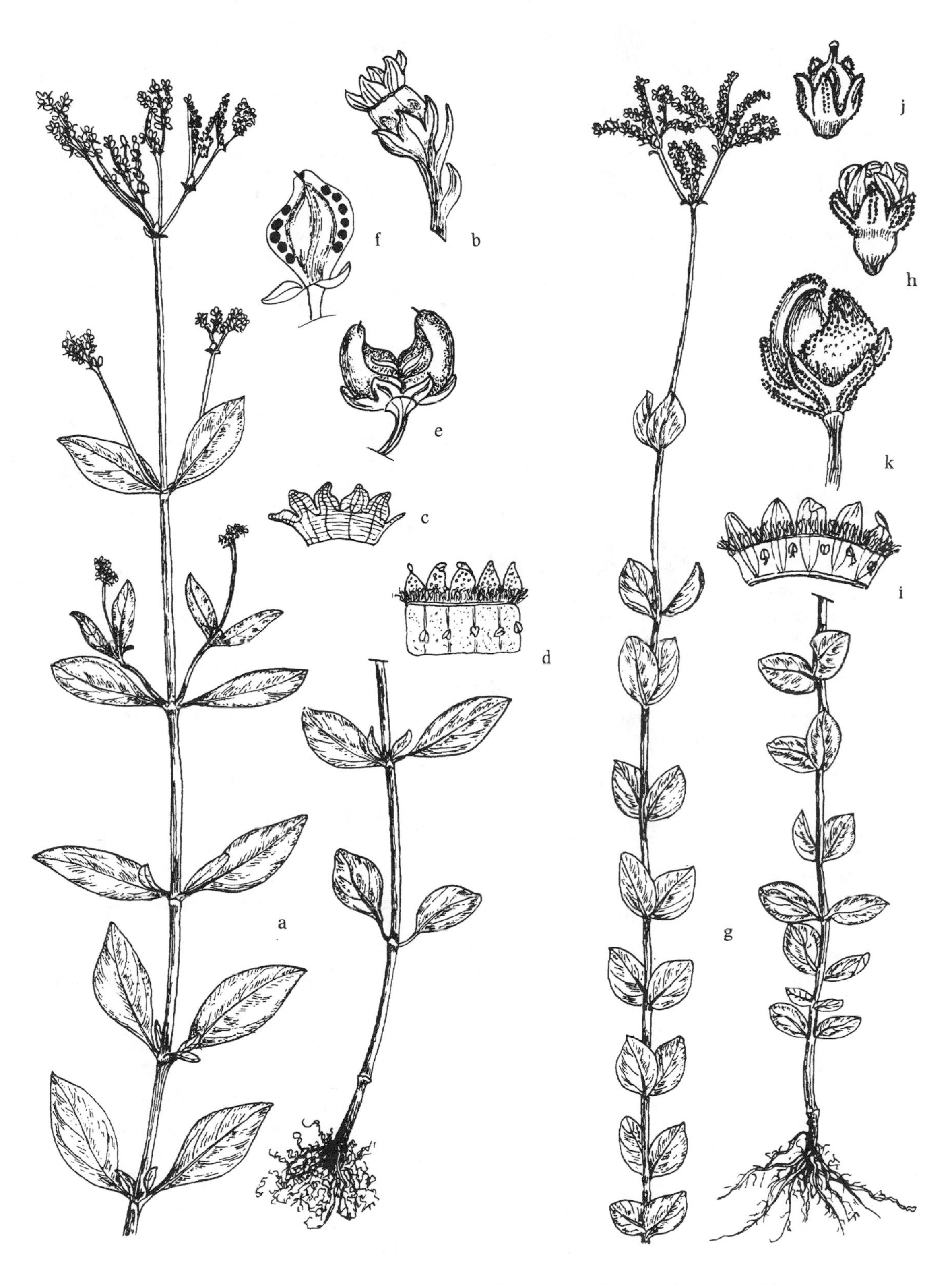

Fig. 493. *Mitreola petiolata*: a. habit; b. flower; c. calyx, spread out; d. corolla, spread out; e. 2-lobed capsule; f. dehiscent capsule (C&C, as *Cynoctonum mitreola*).
Mitreola sessilifolia: g. habit; h. flower; i. corolla, spread out; j. young capsule; k. 2-lobed capsule (C&C, as *Cynoctonum sessilifolia*).

2. Stems slender, lax (frequently twining), green; anthers usually yellow, 0.3–0.5 mm long. . . . . .2. *B. paniculata*
2. Stems somewhat thick, ascending, purple or purple-tinged; anthers usually purple, 0.5–0.6 mm long. 3. *B. iodandra*

---

1. *B. virginica* (L.) Britton, Sterns & Poggenb. Yellow Screw-stem Fig. 494
Wet acid soil, bogs, wet meadows and marshes. Nfld. and N.S. w. to sw. Que., n. N.Y., s. Ont., Mich., n. Ill. and Minn., s. to Fla. and La.

2. *B. paniculata* (Michx.) Muhl. Twining Screw-stem Fig. 494
Wet peats and sands, swamps and marshes. Coastal plain, Nfld. and N.S. s. to Fla., w. to e. Tex.; inland n. to Ky. and Ark.; Mich. and s. Ont.

3. *B. iodandra* B. L. Rob. Purple Screw-stem
Bogs, fens, swamps, meadows and fields. N.S., N.B., Me., Mass., R.I., Conn. and s. Ont. This northern species is often treated as a subspecies of *B. paniculata*, but has some characters somewhat intermediate between *B. paniculata* and *B. virginica* (Haines, 2011); Gillette (1959) has suggested a hybrid origin for the species.

2. *Sabatia* (Rose-gentian, Marsh-pink)

Annual or perennial herbs; leaves below the branches opposite and decussate, margins entire; calyx united at base; corolla rose-pink, pink, rose, purple, or white, united, tubular at base, lobes spreading; style bent to one side; fruit a capsule.

Reference: Wilbur, 1955.

---

1. Calyx and corolla lobes usually 5 (up to 7 in *S. calycina*)
   2. Corolla white; branches of inflorescence mainly opposite. . . . . 1. *S. difformis*
   2. Corolla pink (rarely white); branches of inflorescence mainly alternate.
      3. Leaves thin; calyx lobes leaf-like, oblanceolate, 2–5 mm wide. . . . . .2. *S. calycina*
      3. Leaves firm; calyx lobes liner, less than 2 mm wide.
         4. Plants annual, stem solitary; leaves tapering to base and apex; calyx lobes linear-lanceolate, up to three-fourths the length of the corolla.. . . . . 3. *S. stellaris*
         4. Plants perennial, cespitose, stems 2-nerved; leaves sessile, rounded; calyx lobes setaceous, three-fourths the length of to slightly longer than corolla.. . . . . 4. *S. campanulata*
1. Calyx and corolla lobes (7)9–13.
   5. Primary leaves acute at apex; calyx lobes firm, 0.5–1.5 mm wide; calyx tube nerveless. . . . . .5. *S. kennedyana*
   5. Primary leaves mostly blunt at apex; calyx lobes herbaceous, 1–3 mm wide; calyx tube nerved or corrugated. . . . . .6. *S. dodecandra*

---

1. *S. difformis* (L.) Druce Lance-leaf Rose Gentian
Bogs, sandy swamps, peaty shores, wet pine barrens and ditches. Coastal plain, s. N.J. s. to Fla., w. to Ala. and Miss.

2. *S. calycina* (Lam.) Heller Coastal Rose Gentian
River swamps, wet woodlands and ditches. Coastal plain, se. Va. s. to Fla., w. to Ark. and e. Tex.; Cuba.

3. *S. stellaris* Pursh Sea-pink, Marsh-pink Fig. 495
Saline and brackish marshes, marly prairies, interdunal swales and wet banks. Coastal, s. Mass., and N.Y., s. to Fla., w. to La.; c. Mex., Cuba and Bahamas.

4. *S. campanulata* (L.) Torr. Slender Rose Gentian Fig. 495
Damp sands and peats, pine savannas and ditches. Coastal plain, Mass. and N.Y. s. to Fla., w. to La., Ark. and e. Tex.; local, s. Appalachians, Tenn., Ky.; Ind. and s. Ill. (*S. campanulata* var. *gracilis* (Michx.) Fernald; *S. gracilis* (Michx.) Salisb.)

5. *S. kennedyana* Fernald Plymouth Gentian, Large Sabatia Fig. 495
Sandy and peaty margins of lakes, ponds and streams, often in shallow water. Rare and local. Coastal plain, s. N.S., se. Mass., R.I., Va., se. N.C. and ne. S.C. (*S. dodecandra* var. *kennedyana* (Fernald) Ahles)

6. *S. dodecandra* (L.) Britton, Sterns & Poggenb. Large Marsh-pink, Sea-pink Fig. 496
Saline, brackish or occasionally freshwater marshes. Coastal, Conn. and N.Y. s. to Fla., w. to Ala. and Tex.

3. *Gentianopsis* (Fringed Gentian)

Annual or biennial herbs; leaves opposite, sessile, entire; flowers solitary, 4-merous; corolla blue, rarely white, united, lobes fringed or toothed at tip; fruit a 2-valved capsule.

References: Gillett, 1957; Iltis, 1965; Mason and Iltis, 1965.

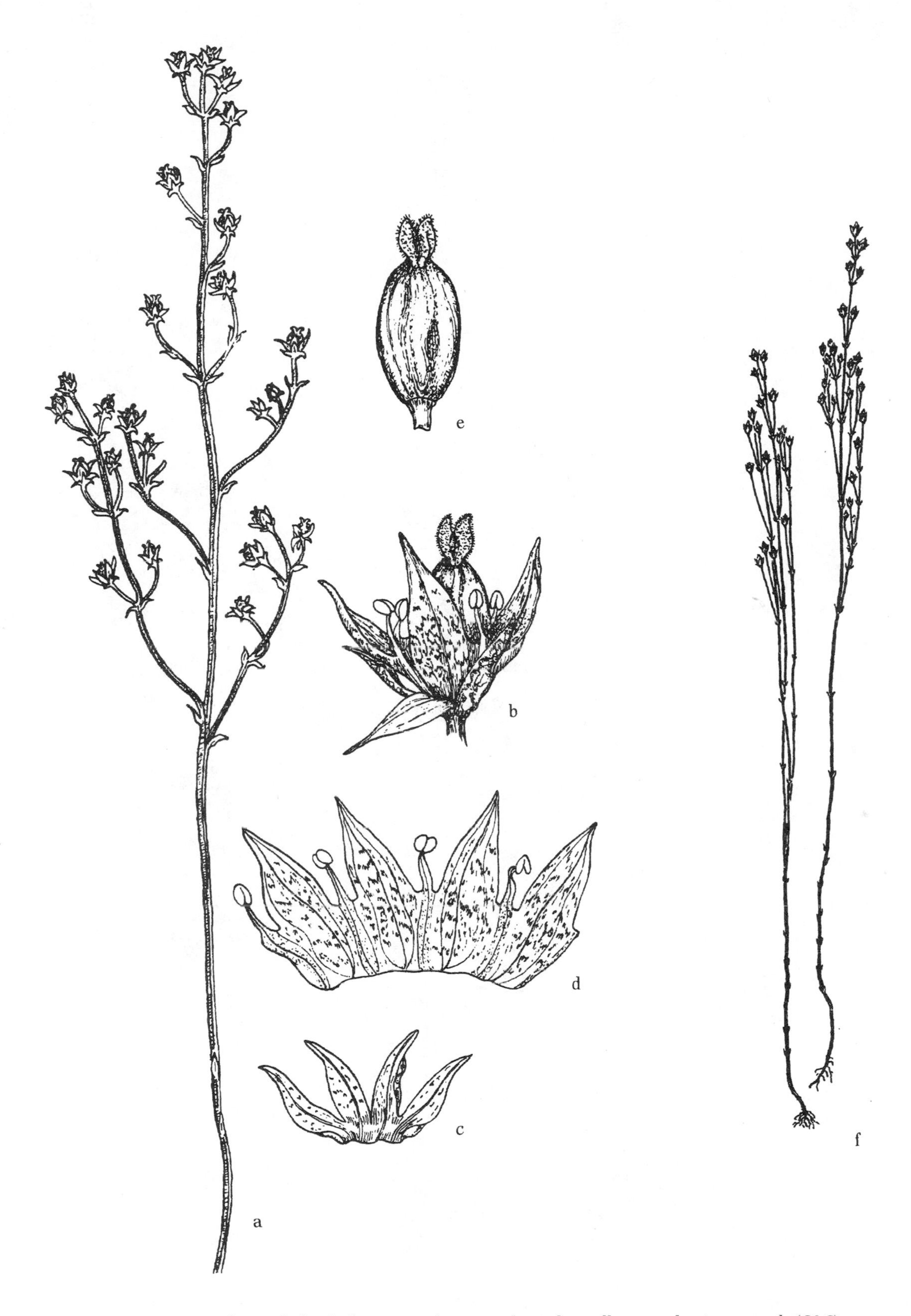

Fig. 494. *Bartonia paniculata*: a. habit; b. flower; c. calyx, spread out; d. corolla, spread out; e. capsule (C&C). *Bartonia virginica*: f. habit (WVA).

Fig. 495. *Sabatia campanulata*: a. upper portion of stem; b. flower (Gleason).
Sabatia stellaris: c. flower and section of stem (Beal).
Sabatia kennedyana: d. upper portion of stem; e. capsule with persistent calyx (Crow).

Fig. 496. *Sabatia dodecandra*: habit (C&C).

1. Leaves on upper portion of stem ovate-lanceolate; corolla lobes fringed across top and partly along sides (fig. 497a)............. 1. *G. crinita*
1. Leaves on upper portion of stem linear to narrowly lanceolate; corolla lobes toothed across top and fringed along sides in some (497b).
   2. Calyx lobes with papillose-scabrous keels; corolla lobes fringed along sides.. 2. *C. virgata*
   2. Calyx lobes with smooth of slightly granulose scabrous keels; corolla lobes not fringed along sides........................ 3. *G. macounii*

1. *G. crinita* (Froel.) Ma Greater Fringed Gentian Fig. 497
Meadows, brook and pond shores, wet thickets and ditches. Me. and sw. Que. w. to s. Man., s. to Pa., N.C., Tenn., Ind., S.D. and Iowa, s. locally in mts. to Ga. (*Gentiana*)

2. *G. virgata* (Raf.) Holub Lesser Fringed Gentian Fig. 497
Wet marly shores, sandy swamps, wet prairies and wet calcareous rocks. W. N.Y., s. Ont., Man. and Sask., s. to Ohio, n. Ill., Iowa and S.D. A second taxon, subsp. *victorinii* (Fernald) Lammars, has been recognized from the tidal estuary of the St. Lawrence R., Que., originally described by Fernald as a distinct species. (*Gentiana procera* Holm; *Gentianella crinita* subsp. *procera* (Holm) J. M. Gillett)

3. *G. macounii* (T. Holms) Iltis Macoun's Fringed Gentian
Brackish, slightly saline and freshwater marshes and shores. Gaspé Pen. and St. Lawrence R., Que.; Ungava, Hudson and James Bay; Man. nw. to N.W.T., s. to Ont., nw. Minn., N.D., S.D., Mont., Alta. and B.C. (*Gentiana macounii* Holm; *Gentiana tonsa* (Lunell) Vict.; *Gentianella crinita* subsp. *macounii* (Holm) J. Gillett; *G. procera* subsp. *macounii* (Holm) Iltis)

4. *Gentiana* (Gentian)

Perennial herbs; leaves opposite, sessile, entire; flowers solitary or in clusters, 5-merous; corolla blue, blue-violet, or violet-purple, rarely white, united, lobes incurved, spreading, with membranous teeth, appendages, or plaits (folds) between the lobes; fruit a 2-valved capsule.

REFERENCES: Mason and Iltis, 1965; Pringle, 1967.

1. Calyx lobes and leaves ciliate.
   2. Corolla closed at anthesis (fig. 498a); plants as wide and long as or wider and longer than corolla lobes.
      3. Corolla lobes truncate and minutely apiculate (fig. 498b); plants wider and longer than corolla lobes. .... 1. *G andrewsii*
      3. Corolla lobes rounded; plaits as wide as long as corolla lobes (fig. 497d)... 2. *G. clausa*
   2. Corolla open at anthesis (fig. 497e); plants narrower and shorter than corolla lobes.
      4. Corolla lobes extending beyond plants by 1 mm or less; stems glabrous... 3. *G. saponaria*
      4. Corolla lobes extending beyond plaits by 3–5 mm; stems minutely puberulent...................................... 4. *G. catesbaei*
1. Calyx lobes ad leaves not ciliate.
   5. Leaves at middle to upper portion of stem cordate or rounded at base; corolla lobes acute, extending beyond plaits by 3–5 mm. .... 5. *G. rubricaulis*
   5. Leaves all narrowed to base; corolla lobes rounded or obtuse, extending beyond plaits by 1–2 mm......................... 6. *G. linearis*

1. *G. andrewsii* Griseb. Closed Gentian, Bottle Gentian Fig. 498
Damp shores, meadows, prairies and woodlands. Sw. Que. and Ont. w. to Sask., s. to Ky., Iowa and Colo.

2. *G. clausa* Raf. Closed Gentian, Bottle Gentian, Blind Gentian Fig. 497
Damp woods, thickets, stream banks and meadows. S. Que. and N.E. w. to Pa. and ne. Ohio, s. to N.J., Md. and W.Va., s. in the mts. and piedmont to N.C. and e. Tenn.

3. *G. saponaria* L. Soapwort Gentian, Harvest Bells Fig. 497
Swamps. shores, bogs, wet woodlands, and thickets. N.Y. w. to Mich. s. to Fla., Ill. Ark., Okla. and e. Tex.

4. *G. catesbaei* Walter Catesby's Gentian, Sampson's Snakewort
Damp sands, peats, sphagnous bogs, wet woodlands, pinelands, swamps and ditches. Coastal plain, Del. s. to Fla., w. to Ala. (*G. catesbaei* var. *nummulariaefolia* Fernald; *G. elliottii* Chapm.)

5. *G. rubricaulis* Schwein. Closed Gentian, Red-stemmed Gentian Fig. 498
Damp sands, shores, meadows and woodlands. N.B. and n. Me., w. to s. Ont., n. Mich., Wisc., Minn. and Man. (*G. linearis* var. *latifolia* A. Gray; *G. linearis* var. *lanceolata* A. Gray; *G. grayi* Kusnezow)

6. *G. linearis* Froel. Narrow-leaved Gentian Fig. 498
Wooded lake and stream shores, swampy woods, meadows and bogs. Lab. w. to w. Ont., s. to Mass., Md., W.Va., Va., Tenn. and Mich. (*G. saponaria* var. *linearis* (Froel.) Griseb.)

Fig. 497. *Gentianopsis crinita*: a. upper portion of plant (Mason and Iltis, 1965).
Gentianopsis virgata: b. upper portion of plant (Mason and Iltis, 1965, as *G. procera* subsp. *procera*).
Gentiana clausa: c. upper portion of plant; d. corolla lobes and plaits (WVA).
Gentiana saponaria: e. upper portion of plant; f. corolla lobes and plaits (WVA).

Fig. 498. *Gentiana andrewsii*: a. flower clusters; b. corolla lobes and plaits (Mason and Iltis, 1965). *Gentiana rubricaulis*: c. flower clusters; d. corolla lobes and plaits (Mason and Iltis, 1965). *Gentiana linearis*: e. upper portion of stem; f. corolla lobes and plaits (WVA).

Apocynaceae / Dogbane Family

1. *Asclepias* (Milkweed)

Perennial herbs, most with a milky sap; leaves simple, opposite or whorled, rarely alternate; flowers solitary or in racemose cymes or umbels; calyx united at base; corolla reflexed, deeply 5-lobed, with corona (hood) and an incurved horn, or corolla lacking a corona, pistils 1, ovary 2-carpellate, deeply divided, the stigma surrounded by fused anthers forming a gynostegium; pollen adhering in masses (pollinia), or not adhering; fruit a pair of follicles (often only 1 developing), seeds with a silky tuft of hairs; seeds with or without a coma. The milkweeds (*Asclepias*) have long been treated in a distinct family, Asclepiadaceae, but now merged with the Apocynaceae (APG IV, 2016).

---

1. Flowers flame-red to orange. . . . . 1. *A. lanceolata*
1. Flowers purplish-red to lavender.
   2. Leaves sessile, with veins transverse, running toward margins (fig. 500). . . . . 2. *A. rubra*
   2. Leaves mostly petiolate, with veins ascending toward apex (fig. 501)
      3. Stems and lower surfaces of leaves glabrous or sparsely pilose. . . . . 3a. *A. incarnata* subsp. *incarnata*
      3. Stems and lower surfaces of leaves densely pubescent. . . . . 3b. *A. incarnata* subsp. *pulchra*

---

1. *A. lanceolata* Walter Few-flowered Milkweed Fig. 499
Brackish to freshwater marshes, swamps, bogs, wet pinelands and ditches. Coastal plain, s. N.Y., s. N.J. s, to Fla. and Tex.; Tenn.

2. *A. rubra* L. Red Milkweed Fig. 500
Swamps, bogs, wet pinelands and wet marshes. Coastal plain, s. N.Y., s. N.J. and e. Pa. s. to Fla., w. to Ark. and e. Tex. (*A. laurifolia* Michx.)

3. *A. incarnata* L. Swamp Milkweed Fig. 501

3a. *A. incarnata* subsp. *incarnata*
Swamps, wet thickets, wet prairies and shores. N.S. and Que. w. to Man., Mont. and Ida., s. to Fla. Tex., N.M. and Nev.

3b. *A. incarnata* subsp. *pulchra* (Ehrh. ex Willd.) Woodson
Swamps, thickets and slopes. Me. sw. to N.Y. and W.Va., Ky. and Tenn., s. to Ga. and Fla.; Tex. (*A. incarnata* var. *neoscotica* Fernald)

Hydroleaceae / Hydrolea or False Fiddleleaf Family

1. *Hydrolea* (Hydrolea)

Annual or perennial; usually spiny herbs; leaves alternate, entire; flowers blue, inflorescence corymbose or cymose; corolla rotate-campanulate, exceeding or equaling the calyx; stamens included or exserted; capsules ovoid or globose, many seeded; seeds minute, striate or rugose.

REFERENCE: Davenport, 1988.

---

1. Flowers in axillary fascicles; styles about 5 mm long, same length as ovary.
   2. Stem and calyx glabrous or minutely puberulent. . . . . 1. *H. uniflora*
   2. Stem and calyx sparsely pubescent, with long, jointed hairs. . . . . 3. *H. quadrivalvis*
1. Flowers in terminal panicles; styles about 10 mm long, longer than ovary. . . . . 3. *H. ovata*

---

1. *H. uniflora* Raf. One-flower False Fiddleleaf Fig. 502
Shallow water, marshy shores, swampy woods and ditches. Chiefly Mississippi embayment, w. Ky., s. Ind., s. Ill., se. Mo. and e. Okla., s. to Ala., La. and e. Tex.

2. *H. quadrivalvis* Walter Water-pod Fig. 504
Swampy woods, streams, marshy shores and ditches. Chiefly coastal plain, Md. and se. Va., s. to Fla., w. to La.; inland to Tenn. (*Nama quadrivalvis* (Walter) Kuntze)

3. *H. ovata* Nutt. ex Choisy Ovate False Fiddlehead Fig. 503
Swamps, streams, ponds, bayous and ditches. W. Ky. and se. Mo. w. to e. Okla., s. to sw. Ga. and e. Tex.

Solanaceae / Nightshade Family

1. *Solanum* (Nightshade)

Annual or perennial (ours) shrubs, vines (ours), or herbs; stems often woody at base (ours); leaves simple, entire or with lobed bases; corolla color various (ours purple), united, rotate or broadly campanulate; fruit a 2-locular berry.

Fig. 499. *Asclepias lanceolata*: a. habit; b. flower; c. portion of corona with incurved horn (C&C).

Fig. 500. *Asclepias rubra*: a. habit; b. flower (C&C).

Fig. 501. *Asclepias incarnata*: a. habit; b. basal portion of plant; c. flower (C&C).

Fig. 502. *Hydrolea uniflora*: a. habit; b. corolla. spread open; c. stamen; d. ovary (3 styles atypical); e. capsule with 2 styles (C&C).

Fig. 503. *Hydrolea ovata*: a. upper portion of plant; b. basal portion of plant; c. corolla, spread open; d. stamen; e. ovary with calyx (C&C).

Fig. 504. *Hydrolea quadrivalvis*: a. habit (Beal).
Solanum dulcamara: b. upper portion of plant (F).

1. *S. dulcamara* L. Bittersweet, Nightshade Fig. 504
Moist thickets, stream banks, swales, swamps, flood plain forests, bog moats, fencerows, disturbed depressions and low ground and upland habitats. Natzd. from Eurasia throughout much of Can. and U.S.; occasional in the Northwest. Plants with pubescent branches and leaves have been segregated as *S. dulcamara* var. *villosissimum* Desv.

Boraginaceae / Borage Family

1. Ovary distinctly 4-lobed, style arising from base between lobes (fig. 505b,e)....1. *Myosotis*
1. Ovary shallowly lobed or appearing unlobed, style terminal (fig. 506g). .... 2. *Heliotropium*

1. *Myosotis* (Forget-me-not, Scorpion-grass)

Annual, biennial or perennial (ours); herbs; leaves alternate, sessile, entire; flowers borne in scorpoid cymes; corolla blue, pink, or white, changing to blue with a yellow center, united; ovary deeply 4-lobed; fruit enclosed by calyx, separating into four 1-seeded nutlets.

1. Stem angled; flowers 6–10 mm broad; calyx lobes shorter than calyx tube (fig. 505b); style longer than nutlets. ....1. *M. scorpioides*
1. Stem rounded; flowers 2–5 mm broad; calyx lobes as long as calyx tuber; style shorter than nutlets.... 2. *M. laxa*

1. *M. scorpiodes* L. True Forget-me-not Fig. 505
Wet stream banks, marshy shores and wet meadows. Nfld. w. to Ont., Man., s. to N.E. and mainly in the mts. to Ga., Tenn., Mo. and La.; w. S.D., Mont., Ida. and B.C. s. in mts. to Colo., N.M., Ariz. and Calif.; natzd. from Eur. (*M. palustris* (L.) Hill)

2. *M. laxa* Lehm. Bay forget-me-not Fig. 505
Shallow water and shores, marshes and bogs. Nfld. w. to Ont. and Minn., s. in mts. to N.C. Tenn., n. Ga. and n. Ala.; Alta. and B.C. s. to Neb., Wyo. n. Utah, Ida., nw. Nev. and n. Calif.

2. *Heliotropium* (Heliotrope)

Annual (ours) or perennial herbs; plants of saline sites may be succulent; leaves linear, ovate, lanceolate, or elliptic, long-petioled to sessile; flowers solitary or borne in a spike-like scorpioid cyme; corolla blue or white, united; ovary shallowly 4-lobed or appearing unlobed; fruit separating into four 1-seeded nutlets.

1. *H. curassavicum* L. Seaside Heliotrope Fig. 506
Saline and brackish shores, saline marshes, mangrove swamps and sandy or muddy flats. Chiefly coastal, se. Me., N.Y., N.J., Del., Md. and e. Va. s. to Ga., and Fla., w. to Tex., N.M. and Ariz.; Mississippi embayment n. to Ark., Mo., and s. Ill.; Neb. Our taxon is var. *curassavicum*; the well-marked var. *obovatum* DC. has larger flowers and broader leaves, ranging west of the Mississippi R.; Man. w. to Sask., s. to. Ill., Mo., Okla., s. N.M., Ariz. and s. Calif.; Mex., W.I., C.Am., and S.Am.

Oleaceae / Olive Family

1. Leaves pinnately compound; fruit a samara (fig. 507b,d,f).... 1. *Fraxinus*
1. Leaves simple; fruit an elongate dark blue to black waxy drupe (fig. 507g,h). .... 2. *Forestiera*

1. *Fraxinus* (Ash)

Trees; twigs stout; leaves opposite, pinnately compound; flowers small, unisexual (plants dioecious), in dense clusters, short racemes or panicles; calyx 4-cleft, toothed or entire, or absent; petals 4 or absent; stamens usually 2; fruit a pendant cluster of samaras.

1. Samaras thickened over mature seed, little winged at base, if at all; calyx present; leaves and young twigs pubescent or glabrous.
   2. Leaflets distinctly petiolulate; wing of samaras mostly 7–10 mm wide, sometimes notched at tip (fig. 507f). .... 1. *F. profunda*
   2. Leaflets short-petiolulate to subsessile; samaras mostly 4.5–6.5(8.5) mm wide, not notched at tip. ....2. *F. pennsylvanica*
1. Samaras distinctly flattened (including seed), winged to the base; calyx absent or minute; leaves and twigs glabrous.

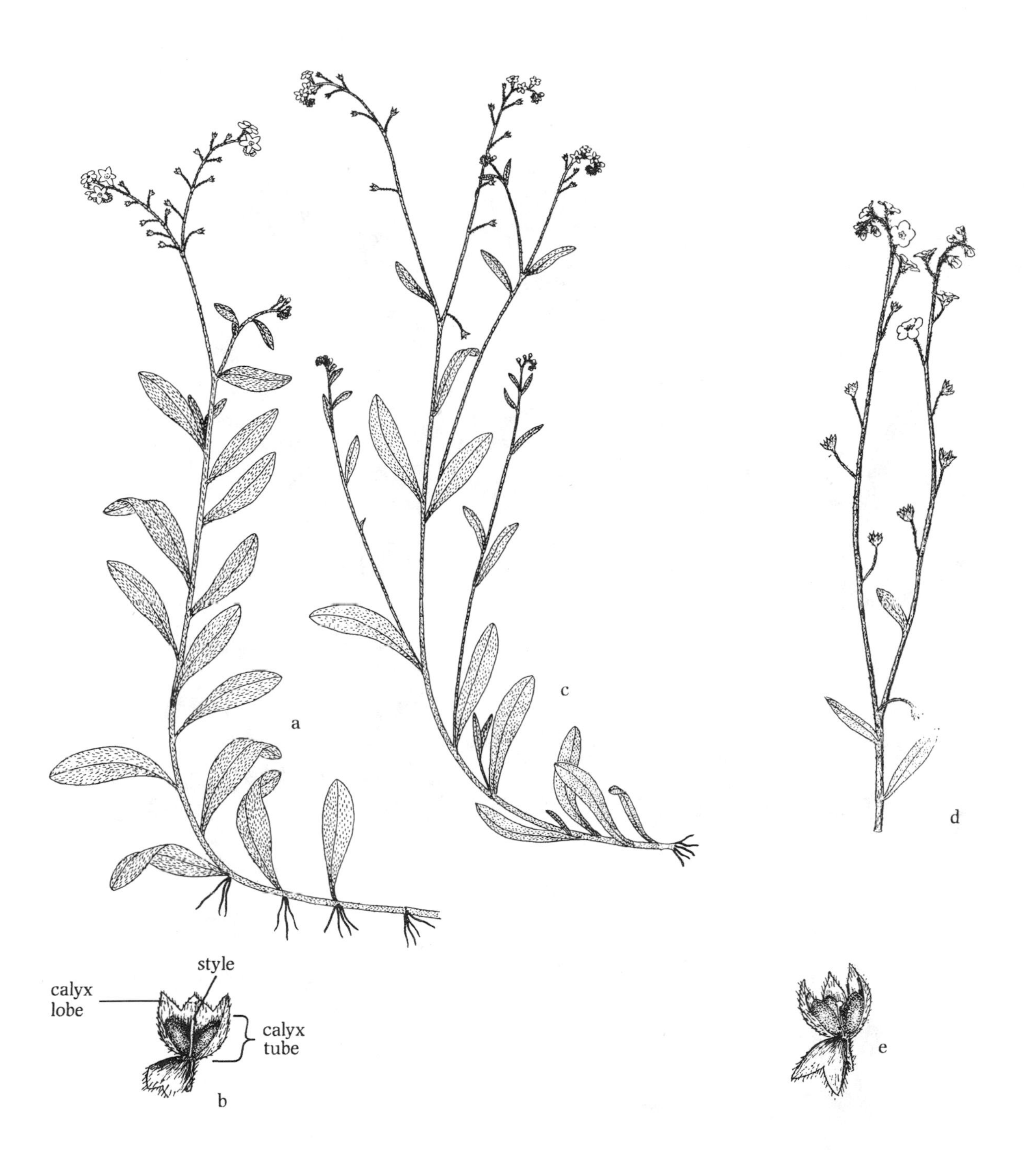

Fig. 505. *Myosotis scorpioides*: a. habit (Beal); b. calyx and fruit (F).
Myosotis laxa: c. habit (Beal); d. upper portion of plant (F); e. calyx and fruit (F).

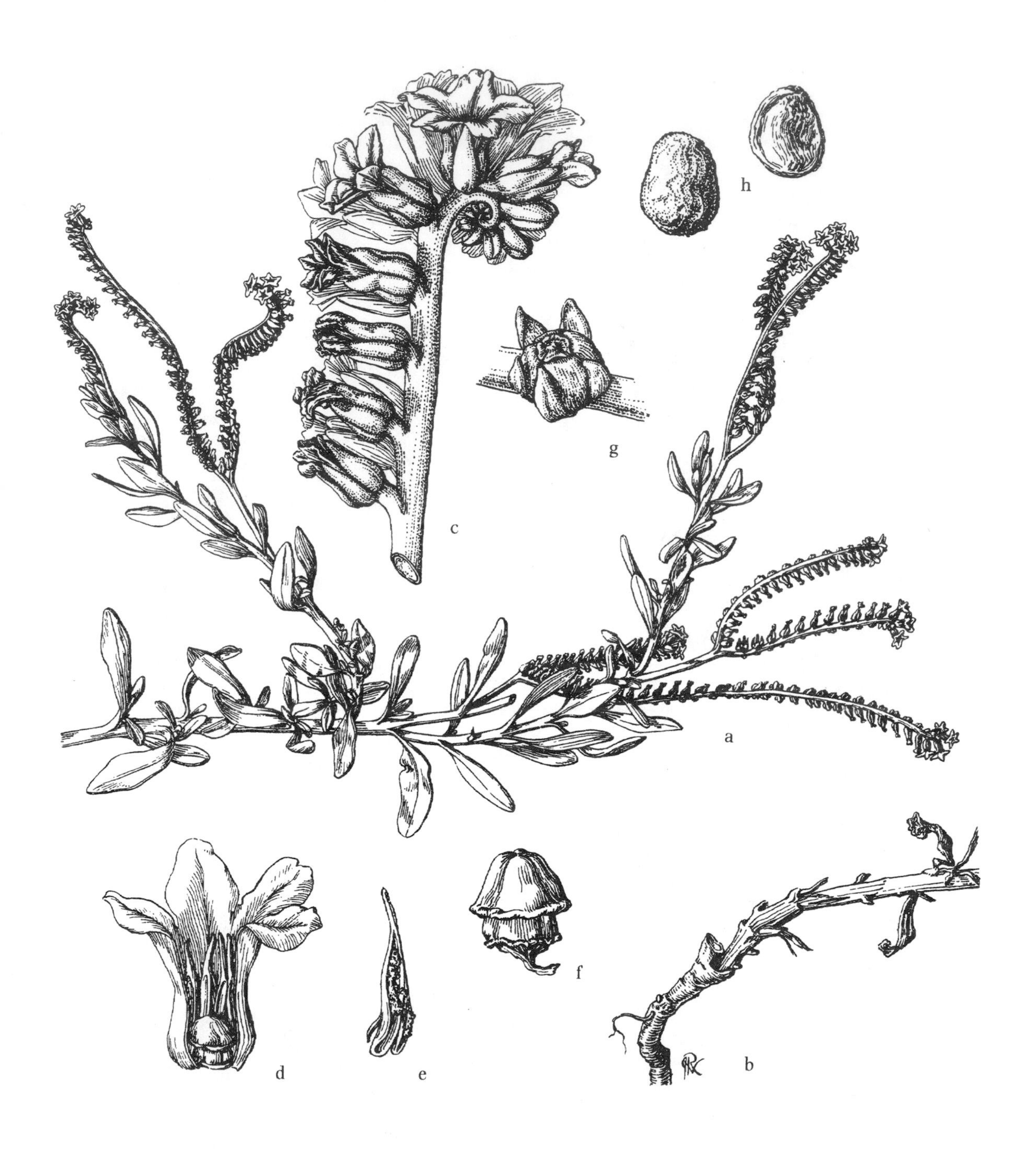

Fig. 506. *Heliotropium curassavicum*: a. branch, showing semi-prostrate habit; b. lower portion of stem and thick root; c. upper portion of inflorescence; d. flower, longitudinal section; e. anther; f. pistil; g. ovary with calyx; h. nutlet, two views (Mason).

3. Leaflets sessile (fig. 260a) often conspicuously serrate; calyx absent at base of samara; plants chiefly northern. . . . . 3. *F. nigra*
3. Leaflets petiolulate, margin entire; calyx minute at base of samara; plants southern, chiefly coastal plain. . . . . 4. *F. caroliniana*

---

1. *F. profunda* (Bush) Bush Red Ash, Pumpkin Ash Fig. 507
Flooded swamps and bottomlands. Chiefly coastal plain, e. N.Y., N.J., Md. and e. Va., s. to n. Fla., w. to La.; inland, Ont. and s. Mich., s. to sw. Ohio, Ind., s. Ill., Tenn., se. Mo., and Ark. (*F. tomentosa* Michx. f.; *F. michauxii* Britton)

2. *F. pennsylvanica* Marshall Green Ash, Red Ash
Moist to wet woods, lake margins, stream banks, floodplains and swamps along rivers. Que. w. to Ont., Man., Sask. and Mont., s. to Fla. Ala., and Tex.; Wyo., Colo., and Utah; intro. in Wash., Oreg., Ida., Ariz. and N.M. Plants with glabrous twigs and leaflets are called Green Ash; those with pubescent young twigs and leaflets are called Red Ash and have been described as var. *subintegerrima* (Vahl) Fernald, but are not regarded as taxonomically distinct. Glabrous plants might be confused with our common upland tree, *Fraxinus americana*, which has leaves whitish beneath, and a terminal bud that is rounded at apex, unlike *F. pennsylvanica* with pale green leaves beneath and a terminal bud that is pointed at apex.

3. *F. nigra* Marshall Black Ash Fig. 507
Swamps, floodplains and shores, sometimes a dominant tree. Nfld. w. to Man., s. to Del., Md., W.Va., Ky., Ohio, Ill. and Iowa. Some areas have been greatly impacted by infestation of the Emerald Ash Borer Beetle (*Agrilus planipennis*), with dead trees of many Black Ash swamps being replaced with *Acer saccharium* (Silver Maple).

4. *F. caroliniana* Mill. Carolina Ash, Water Ash, Pop Ash Fig. 507
Flooded swamps, bottomlands and wet shores. Coastal plain, sc. Va. s. to Fla., w. to Ark. and Tex. (*F. caroliniana* var. *oblanceolata* (M. A. Curtis) Fernald & Schub.; *F. pauciflora* Nutt.)

2. *Forestiera* (Swamp-privet)

Woody shrubs or small trees; leaves opposite, simple, margins often indistinctly crenate, usually entire toward base; flowers unisexual (plants dioecious); staminate flowers in dense, nearly sessile, lateral fascicles; pistillate flowers in short panicles; fruit a small, black 1-seeded drupe.

1. *F. acuminata* (Michx.) Poir. Swamp-privet Fig. 507
Riverbanks, swamps, wet woodlands and sand and gravel bars. S.C. w. to Tenn., s. Ind., c. Ill., s. Mo. and se. Kans., s. to Fla. and Tex.; especially Mississippi River Valley.

Plantaginaceae / Plantain Family

---

1. Leaves basal.
 2. Flowers numerous, flowers numerous, in spikes; fruit dehiscent, 2-many-seeded. . . . . 1. *Plantago*
 2. Flowers 1 or 2, on short peduncles or sessile, fruit indehiscent, 1-seeded. . . . . 2. *Littorella*
1. Leaves cauline to mostly cauline.
 3. Leaves 3–12 whorled. . . . . 3. *Hippuris*
 3. Leaves opposite.
  4. Plants submersed (*C. terrestris, emersed*), stems mostly limp, submersed leaves notched; fruit heart-shaped. . . . . 4. *Callitriche*
  4. Plants emersed (submersed form also in *Gratiola*), stems stiff to prostrate, submersed leaves uncommon, not notched; fruit not heart-shaped.
   5. Calyx 4-parted or 4-lobed. . . . . 5. *Veronica*
   5. Calyx 5-parted or 5-lobed.
    6. Leaves pinnately lobed (fig. 519b). . . . . 6. *Leucospora*
    6. Leaves not lobed.
     7. Flowers in a spike, terminal or terminal and lateral. . . . . 7. *Chelone*
     7. Flowers solitary, in leaf axils.
      8. Pedicels with a pair of small bractlets at base (in leaf axils). . . . . 8. *Mecardonia*
      8. Pedicels lacking bractlets, or if present, then at apex (just below calyx).
       9. Pedicels with pair of bracts below calyx (523e): leaf margins toothed (slightly to entire in G. aurea). . . . . 9. *Gratiola*
       9. Pedicels lacking bracts, leaf margins entire. . . . . 10. *Bacopa*

Fig. 507. *Fraxinus nigra*: a. leaf; b. fruit (F).
Fraxinus caroliniana: c. leaf; d. fruit (F).
Fraxinus profunda: e. upper portion of leaf; f. fruit (F).
Forestiera acuminata: g. fruiting branch; h. fruits (G&W).

1. *Plantago* (Plantain, Ribwort)

Annual or perennial herbs: leaves alternate on a short stem, appearing basal; flowers small, bisexual, sessile or subsessile in the axils of bracts, borne in a spike: petals scarious; fruit a circumscissile capsule.

REFERENCE: Shipunov, 2019.

1. Plants smaller; leaves linear, 1–2 cm × 1–1.5 cm (sometimes lanceolate or oblanceolate along Gulf of St. Lawrence), succulent. .... 1. *P. maritima*
1. Plants large; leaves broadly ovate to cordate, 10–30 cm × 8–20 cm, not succulent.......... 2. *P. cordata*

1. *P. maritima* L. Seaside Plantain, Goose Tongue Plantain Fig. 508
Salt marshes, coastal ledges, cliffs and sands; inland salt marshes, alkaline and saline flats, roadsides. Coastal, Greenl., Nfld. and Lab. s. to N.E. and N.J.; Hudson Bay; Man., Sask. and Alta., Alask. s. to B.C., Wash., Oreg and Calif. A wide ranging, very variable species, Shipunov (2019) recognizes no infraspecific taxa. (*P. juncoides* Lam.)

2. *P. cordata* Lam. Heartleaf Plantain Fig. 508
Shallow water in marshes, wet woodlands and freshwater tidal shores. E. N.Y. and Ont. w. to Wisc., Ark. and Mo., s. to N.C., Ga., Fla. and La.

2. *Littorella* (Littorella, Shoreweed)

Perennial herbs, arising from rhizomes; leaves linear, straight or arching, basal; flowers small, unisexual; petals scarious; staminate flowers solitary at end of short peduncles; pistillate flowers sessile, basal; fruit an achene, enclosed by a persistent calyx.

REFERENCE: Shipunov, 2019.

1. *L. americana* Fernald American Shore-weed Fig. 508
Shores and margins of lakes and ponds or submersed. Nfld. and Lab. w. to Ont. s. to Me., n. N.H., n. Vt., n. N.Y., n. Mich., n. Wisc. and n. Minn. Molecular data supports recognition of the N.Am. populations as a species distinct from the closely related European *L. uniflora* Ascherson (Shipunov, 2019. (*L. uniflora* var. *americana* (Fernald) Gleason)

3. *Hippuris* (Mare's-tail)

Emersed or submersed perennials, arising from a rhizome; leaves whorled, entire, linear-attenuate to oblanceolate, elliptic, or oblong-ovate; flowers sessile, bisexual or unisexual, axillary in middle and upper leaves; fruit a small, ovoid nut.

REFERENCES: Crow and Hellquist, 1983; Elven et al., 2019; McCully and Dale, 1961a, 1961b.

1. Leaves 6–12 in whorl. linear to linear-attenuate; plants of freshwater.......... 1. *H. vulgaris*
2. Leaves 4(6) in whorl, oblanceolate, elliptic, or oblong-ovate; plants of brackish or saline water.......... 2. *H. tetraphylla*

1. *H. vulgaris* L. Common Mare's-tail Figs. 509, 510
Damp shores and shallow water of ponds, lakes, streams and ditches. Greenl. and Lab. w. to Alask., s. to Nfld., N.S., n. N.E., N.Y., n. Ind., n. Ill., Minn., Neb., and N.M., Ariz. and Calif.; s. S.Am.; Eurasia. Leaves of submersed plants markedly different from emergent leaves, being thin, longer and flaccid.

2. *H. tetraphylla* L. f. Four-leaf Mare's-tail Fig. 510
Coastal tidal marshes. Lab. and Gaspé Pen., Que. w. to James Bay, Hudson Bay, Nunavut, N.W.T., Alask., and B.C. (*H. maritima* Hellen.; *H. lanceolata* Retz.)

4. *Callitriche* (Water-starwort, Water-chickweed)

Annual or perennial, submersed, stranded, or terrestrial herbs; leaves opposite, simple, entire, submersed, floating, or emersed, highly variable; flowers tiny, axillary, unisexual; plants monoecious; perianth absent; fruit a schizocarp, splitting into four 1-seeded mericarps. Mature fruit critical for identification.

REFERENCES: Fassett, 1951; Fernald, 1932a; Lansdown, 2009, 2019; Philbrick, 1984, 1989; Philbrick et al., 1996; Philbrick and Anderson, 1992; Svenson, 1932b.

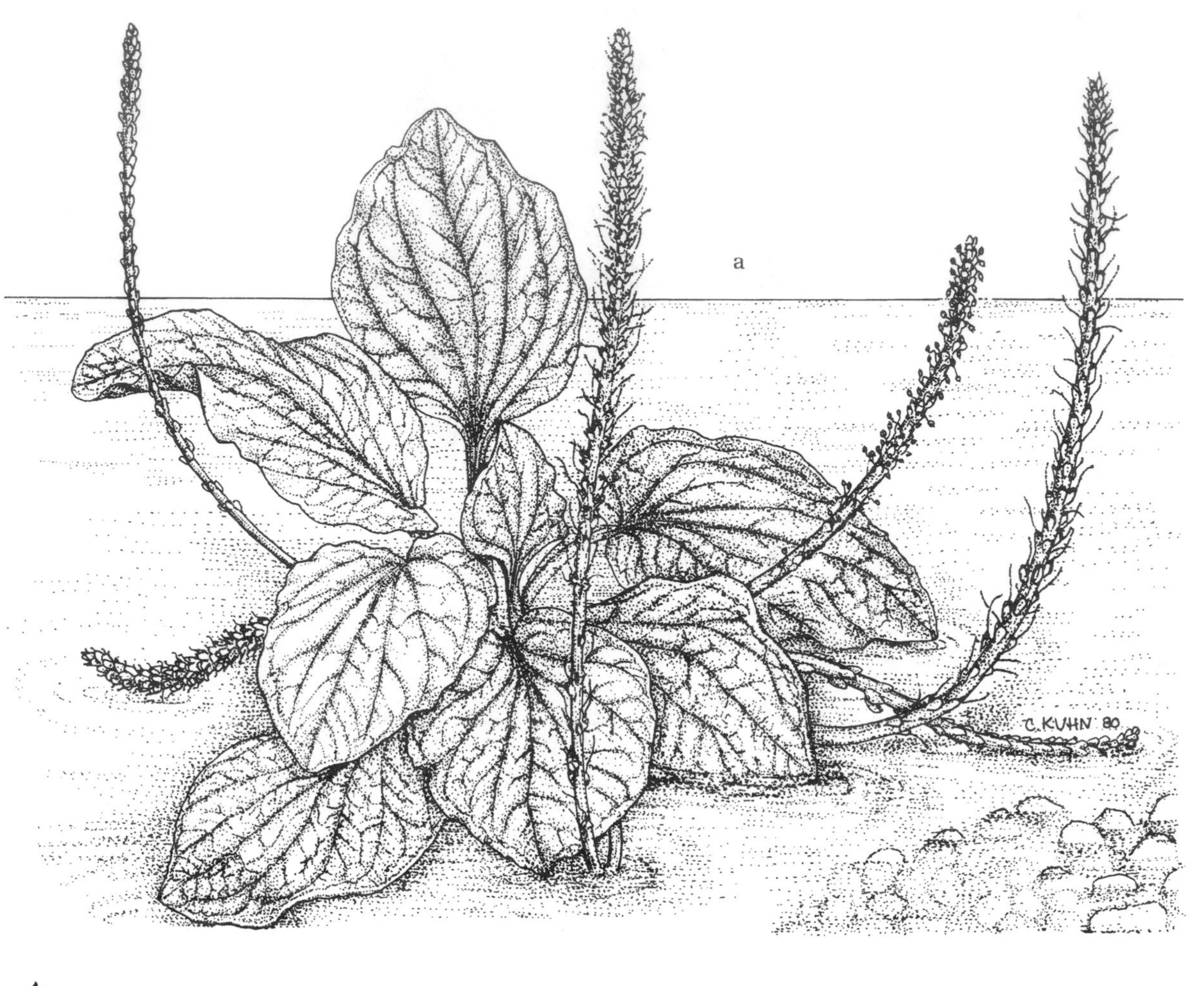

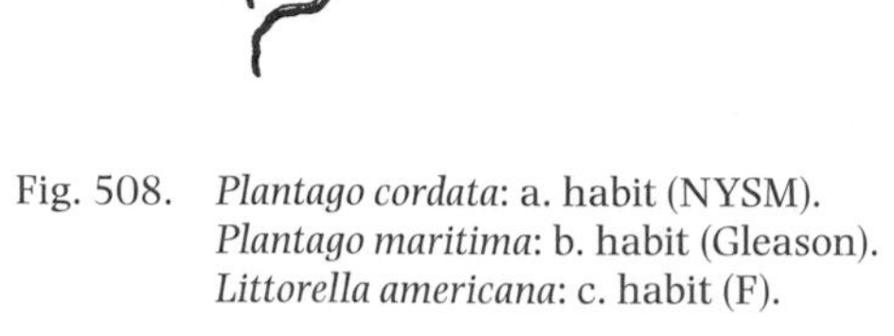
Fig. 508. *Plantago cordata*: a. habit (NYSM).
Plantago maritima: b. habit (Gleason).
Littorella americana: c. habit (F).

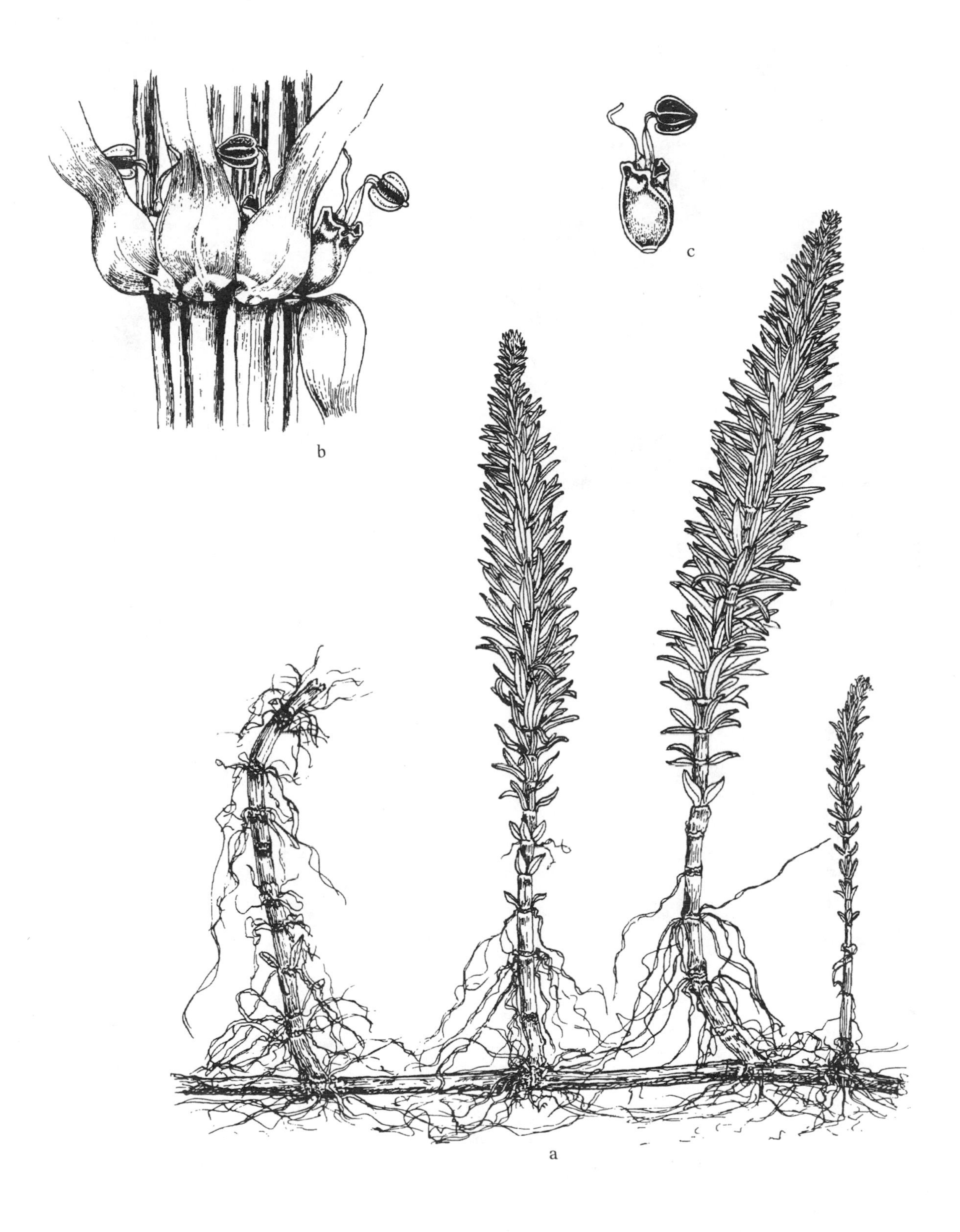

Fig. 509. *Hippuris vulgaris*: a. habit, vegetative; b. node with flowers in leaf axils; c. flower (C&C).

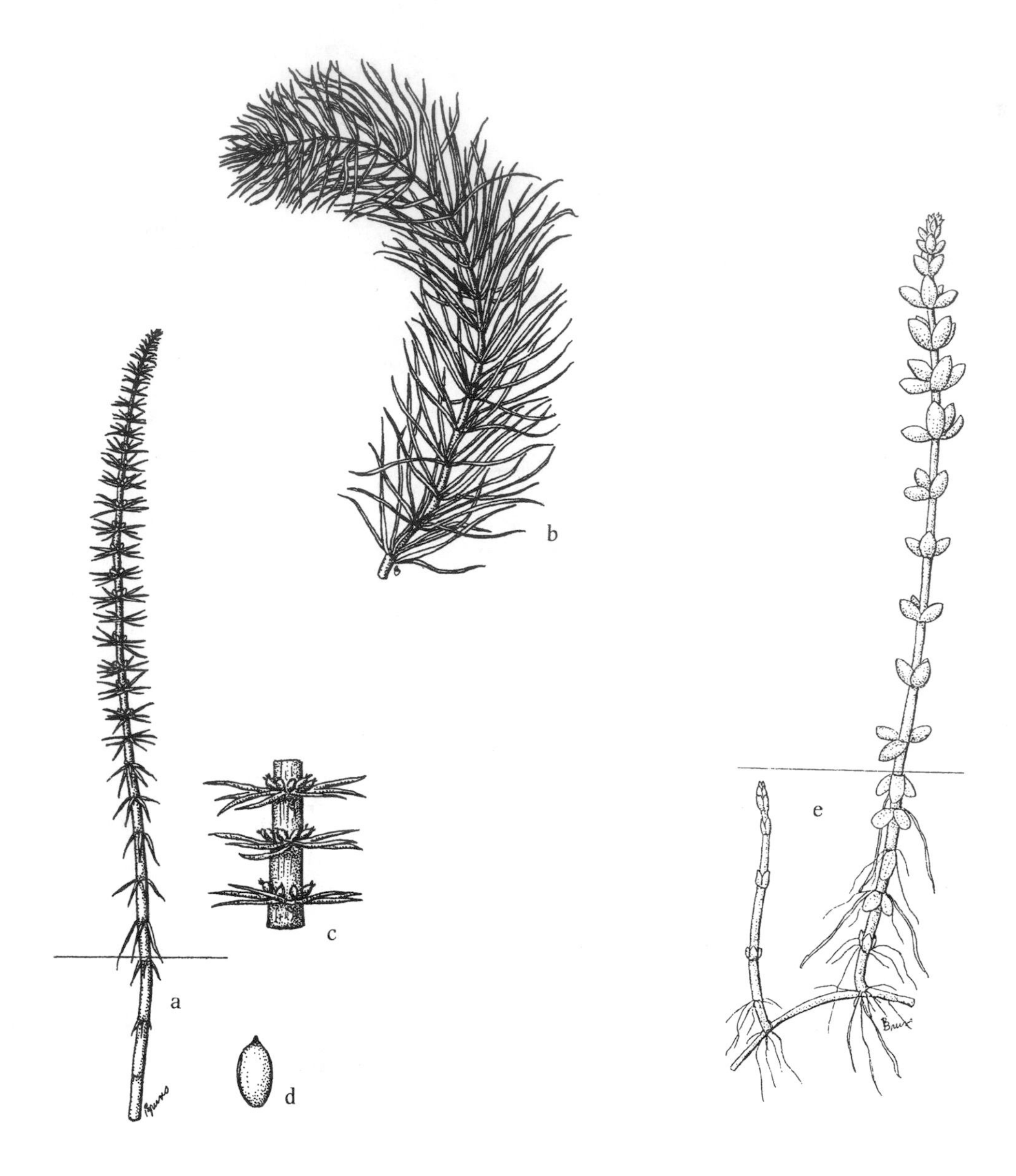

Fig. 510. *Hippuris vulgaris*: a. habit, emersed portion fertile; b. habit, submersed form, vegetative; c. section of emersed stem with fruits in leaf axils; d. fruit (NHAES).
Hippuris tetraphylla: e. habit (PB).

1. Flowers and young fruits with 2 thin, often whitish, bracts at base (fig. 511d); leaves often dimorphic, with floating rosette of spatulate leaves and submersed linear leaves.
   2. Fruit margin distinctly winged, wing nearly 0.1 mm wide, extending evenly from summit to base of fruit; fruit globose...... 1. *C. stagnalis*
   2. Fruit margin not winged or with a wing less than 0.05 mm wide, extending three-fourths the distance or less from summit to base, wing becoming narrower toward base; fruit ellipsoidal, obovoid, or nearly heart-shaped.
      3. Fruit longer than wide by 0.2 mm or more, ellipsoidal. .... 2. *C. palustris*
      3. Fruit as wide as long, obovoid or nearly heart-shaped........ 3. *C. heterophylla*
1. Flowers and fruits lacking bracts at base; leaves all similar, never forming a floating rosette.
   4. Plants submersed, perennial; leaves linear, widest toward base.
      5. Fruit wings narrower than seed width........ 4. *C. stenoptera*
      5. Fruit wings as wide or wider than seed width........ 5. *C. hermaphroditica*
   4. Plants terrestrial, in seasonally wet areas, annual: leaves obovate to oblanceolate or spatulate, narrowed toward base. ..... 6. *C. terrestris*

1. *C. stagnālis* Scop. Pond Water-starwort Fig. 511
Ponds, slow-moving water, and cool springs. E. Que., N.H., and Mass. s. to Pa. and Md.; Wisc., B.C., Wash., Oreg. and Calif.; natzd. from Eur.

2. *C. palustris* L. Vernal Water-starwort Fig. 512
Ponds, lakes, sphagnum bogs, and slow-moving streams; also stranded on wet soil. Greenl. and Lab. w. to Alask., s. to N.E., w. Va., W.Va., Mich., Ill., Neb., Tex., N.M., Calif. and Mex. After examining Fernald's type material of *C. anceps*, Philbrick (1989) concluded that the taxon should be regarded as merely an ecological variant of *C. palustris*. Lansdown (2009) also noted that live populations were needed to clarify the status of *C. anceps*. (*C. anceps* Fernald; *C. verna* L.)

3. *C. heterophylla* Pursh Two-headed Water-starwort Fig. 513
Ponds, lakes, streams and ditches; sometimes stranded on wet soil. Nfld. and Lab. w. to Wash., s. to Fla., Tex., Calif., and Mex.

4. *C. stenoptera* Lansdown Narrow-wing, Water-starwort Fig. 515
Ponds, and lakes, Nfld. and Ont. w. to Alask. and B. C., s. to Minn., Mont., Ida. Oreg., Neb., N.M. and Ariz.

5. *C. hermaphroditica* L. Northern Water-starwort Fig. 514
Quiet waters of lakes and streams, occasionally in brackish water. Greenl. and Lab. w. to Alask., s. to Que., Vt., n. N.Y., Mich., Wisc., Minn., Sask., Colo., N.M., Utah, Ariz. and Calif. (*C. autumnalis* L.)

6. *C. terrestris* Raf. Terrestrial Water-starwort Fig. 515
Damp soils, muddy shores, pathways and fields. N.B., Me. and N.Y. w. to Ohio, Ind., and Wisc., s. to Fla., Mo., Neb. and Tex; Mex. (*C. austini* Engelm.; *C. deflexa* A. Braun)

5. *Veronica* (Speedwell)

Annual or Perennial herbs; foliage leaves opposite; bracteal leaves and bracts mostly alternate; flowers borne in a raceme; corolla blue violet, white or occasionally pink; united, weakly irregular, the lower lobe narrower than other 3; fruit a capsule.

References: Albach, 2019; Les and Stuckey, 1985.

1. Leaves short-petiolate.
   2. Leaves widest near base, pointed at apex (fig. 516a).......... 1. *V. americana*
   2. Leaves widest near middle, rounded at apex (fig. 516e).......... 2. *V. beccabunga*
1. Leaves sessile.
   3. Leaf bases rounded, somewhat clasping (fig. 517a), usually more than l cm wide; capsule about as wide as long, not strongly 2-lobed.
      4. Racemes (20)30–60-flowered; pedicels 3–7(10) mm long (curved upward in fruit); capsules globose, (2.5)3–3.5(4) × 2.5–3.5(4), rounded or barely notched at apex (Fig. 517b). .... 3. *V. anagallis-aquatica*
      4. Racemes 5–25-lowered; pedicels (3)5–10 mm (straight in fruit); capsules nearly globose, 2–3(3.5) × 4–5 mm, distinctly notched at apex. .... 4. *V. catenata*
   3. Leaf bases tapering, not clasping, 1 cm or less wide; capsule much wider than long, strongly 2-lobed (fig. 517f)........... 5. *V. scutellata*

1. *V. americana* (Raf.) Schwein. ex Benth. American Brooklime, American Speedwell Fig. 516
Shallow water of springs, trickles, streamlets, stream banks and swamps. Nfld. w. to Alask., s. to w. N.C., e. Tenn., Mo., Okla., Tex. Ariz., and Calif.

2. *V. beccabunga* L. European Brooklime, European Speedwell, Fig. 516
Brooks, muddy shores and stream banks. Local, Que. (especially St. Lawrence R.) w. to Mich., s. to N.Y., N.J., W.Va., Ohio, Wisc. and Ill., La.; Nev. and Calif.; intro. from Eur.

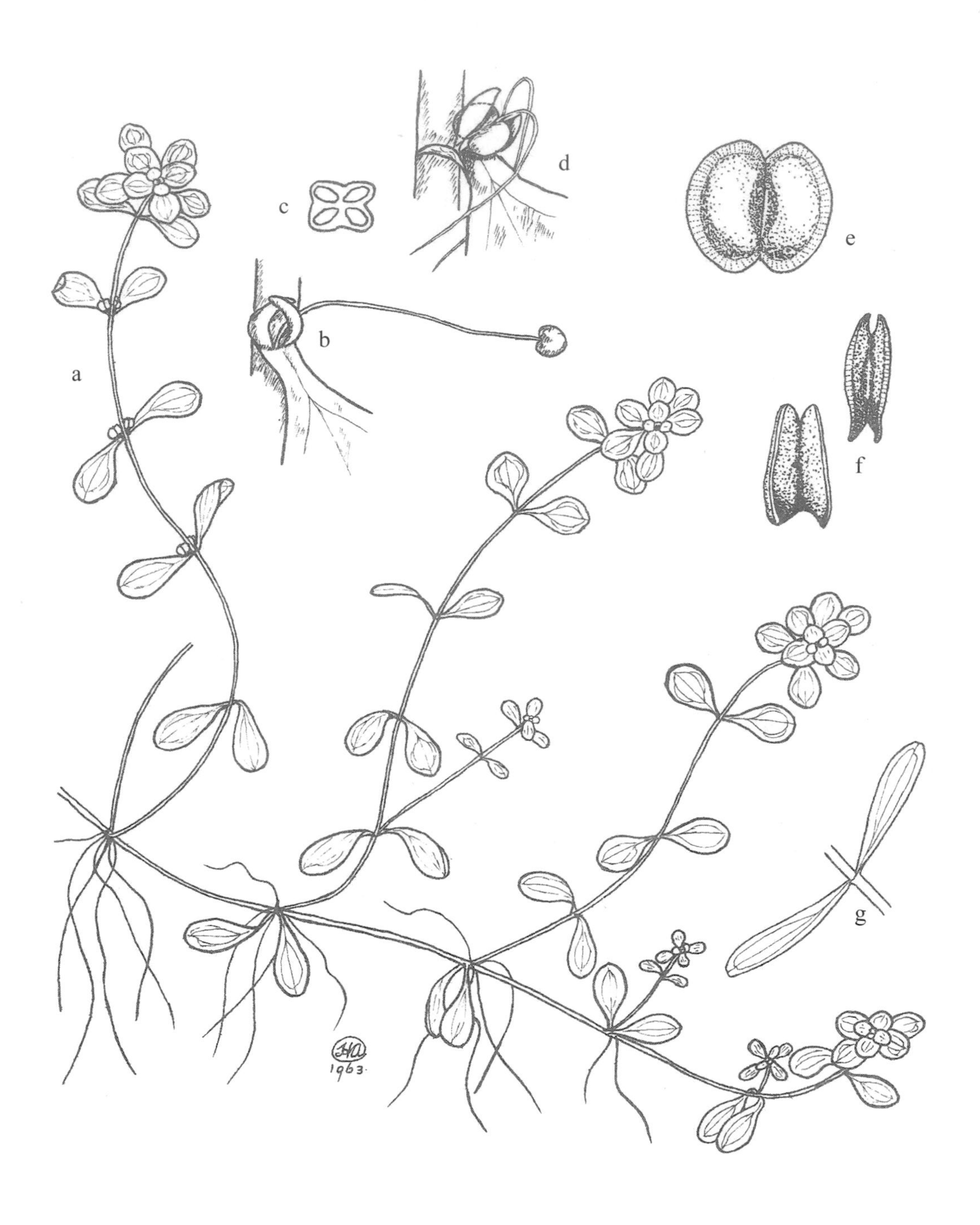

Fig. 511. *Callitriche stagnalis*: a. habit showing slender, branching stems rooting at nodes, opposite leaves with leaves 3–7-nerved; apical rosettes well-defined; fruits axillary; b. male flower in leaf axil with two falcate bracteoles and a single stamen; c. female flower showing 4-lobed ovary (diagrammatic); d. female flower in leaf axil, showing the two bracteoles, and two long, curved styles; e. fruit, face view, showing wing and shallow facial groove; f. fruit, side view, showing deep commissural groove; g. elongated submersed leaves (APOA).

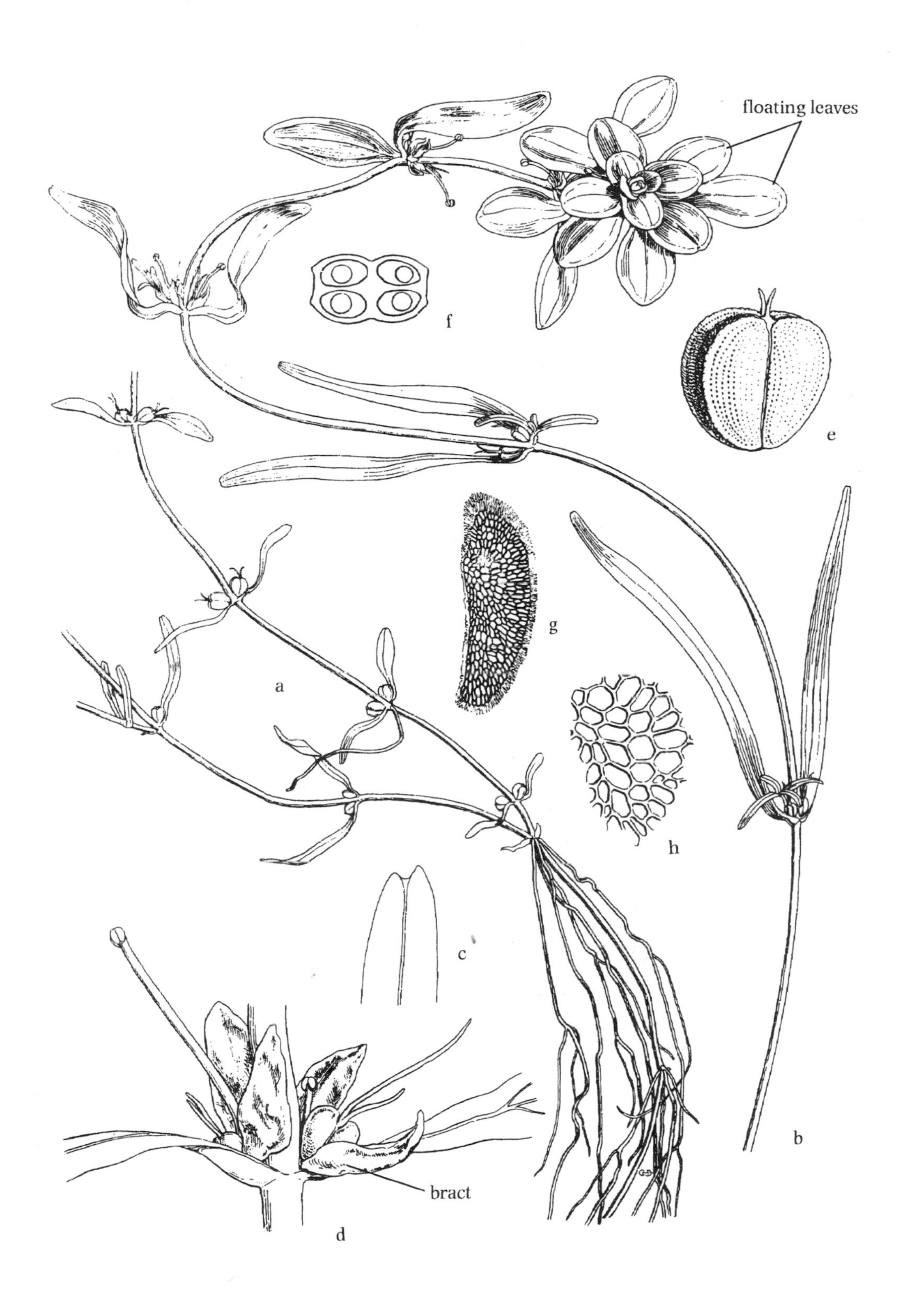

Fig. 512. *Callitriche palustris*: a. habit, lower submersed portion of plant; b. habit, upper portion of plant with floating rosette at apex; c. notched tip of submersed leaf; d. axillary flowers with bracts; e. fruit; f. fruit, cross-section (diagrammatic); g. mature carpel with wing; h. reticulate pattern on carpel surface (Mason).

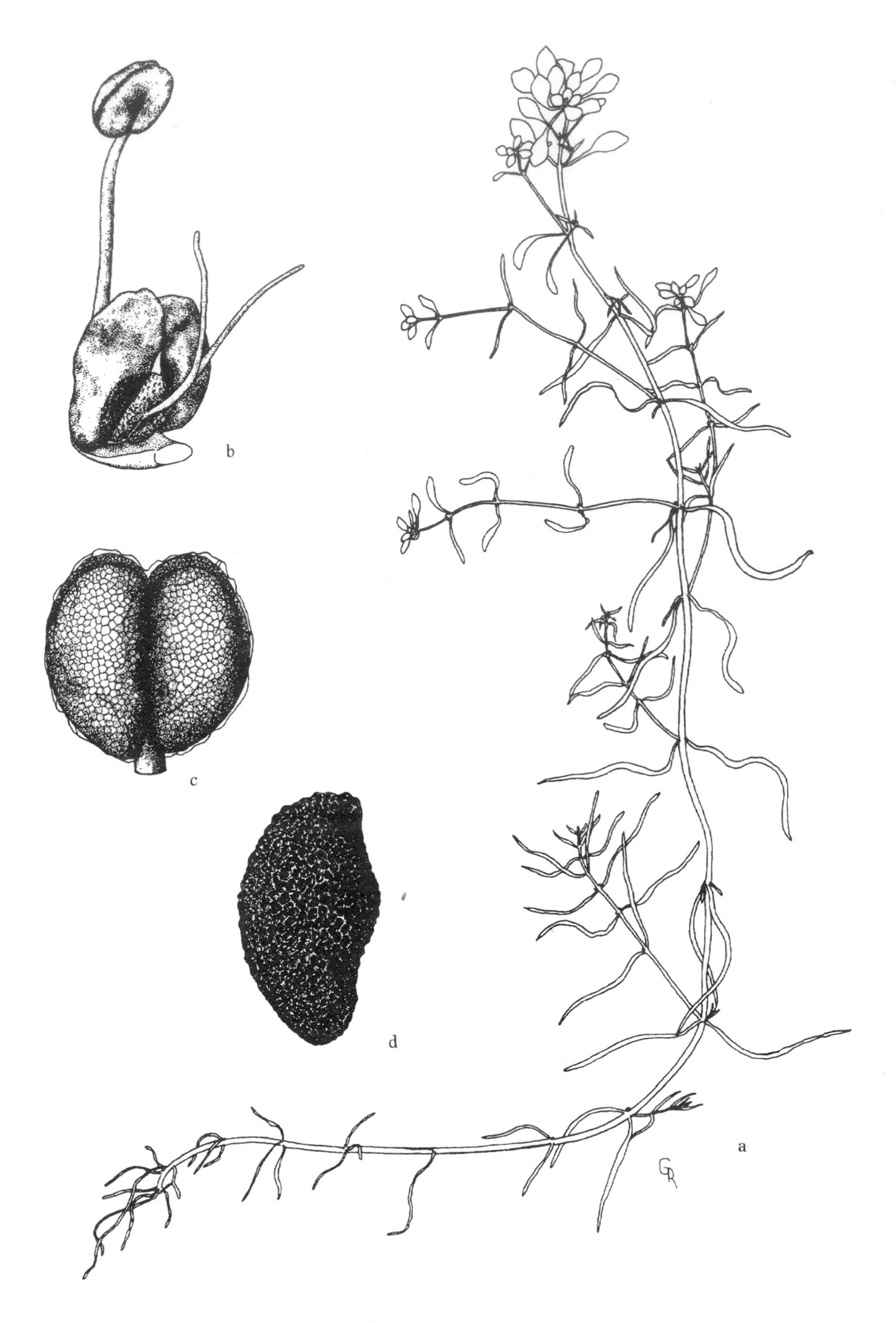

Fig. 513. *Callitriche heterophylla*: a. habit; b. flower; c. fruit; d. seed (G&W).

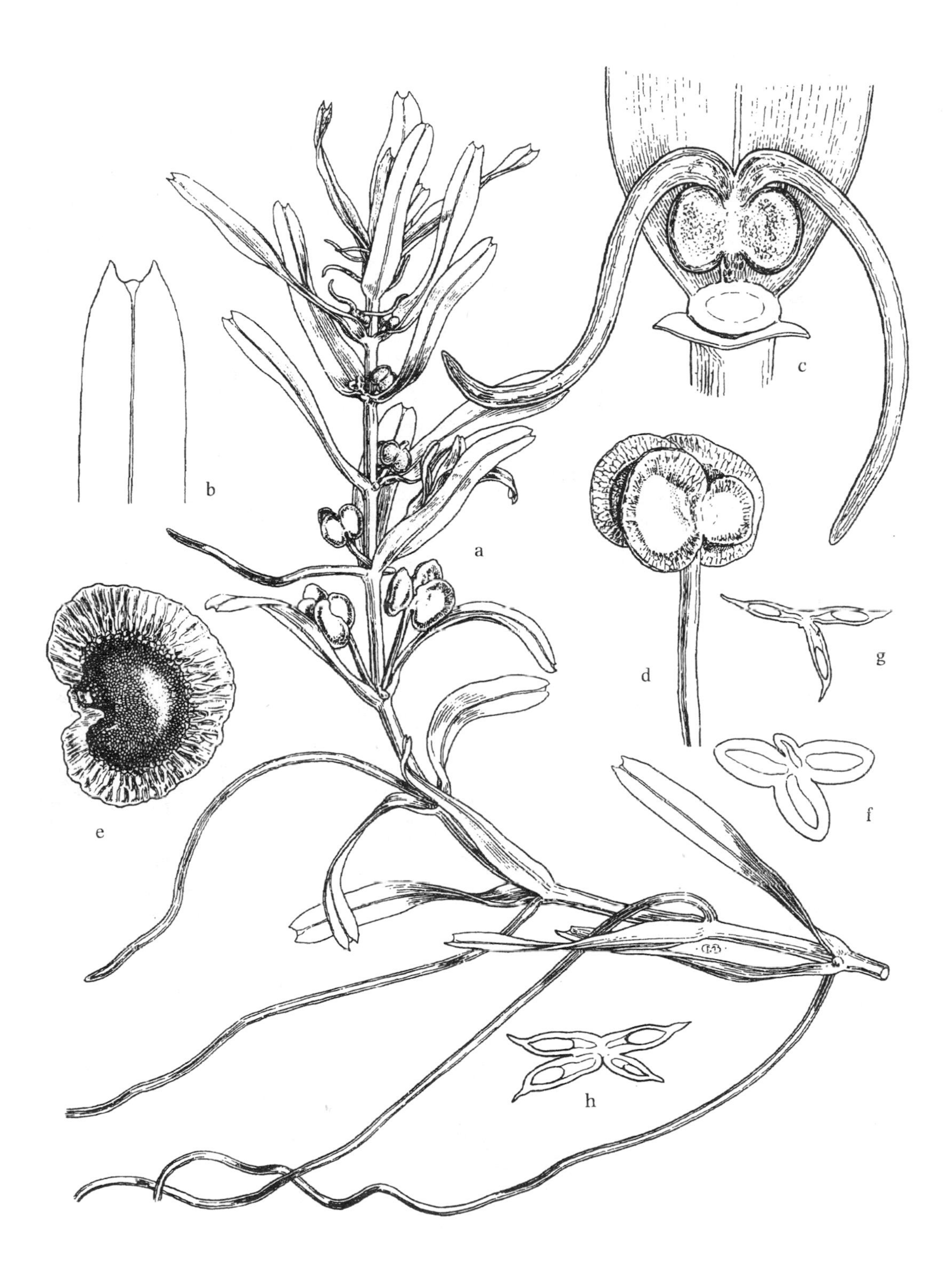

Fig. 514. *Callitriche hermaphroditica*: a. upper portion of submersed plant; b. notched leaf apex; c. young flower; d. mature fruit; e. mature carpel with wing; f. young fruit, cross-section; g, h. mature fruits, cross-sections, showing variation (diagrammatic) (Mason).

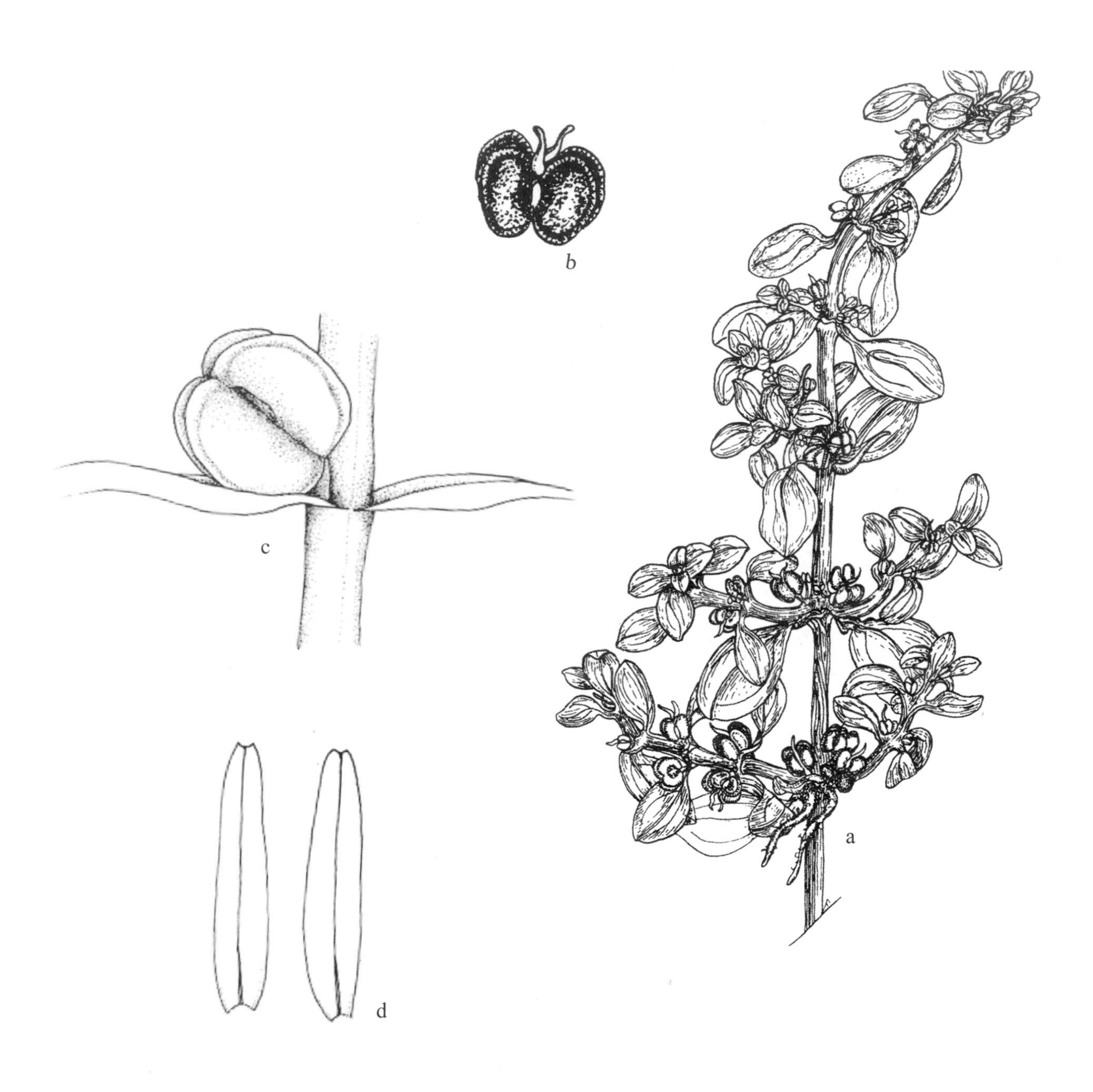

Fig. 515. *Callitriche terrestris*: a. habit; b. mature fruit (C&C).
Callitriche stenoptera: c. fruit; d. leaves (Lansdown, 2009).

Fig. 516. *Veronica americana*: a. habit (Mason); b. upper portion of plant (F); c. flower (Mason); d. capsule (Mason). *Veronica beccabunga*: e. leaf (F).

3. *V. anagallis-aquatica* L. Water Speedwell, Brook-pimpernel. Figs. 517, 518
Springs, trickles, streamlets, stream banks, swamps, flowing water and wet meadows; chiefly calcareous sites. Que. w. to B.C., s. to N.C. Ga. and Fla., w. to Ark. And Tex., w. to Calif; Eur. (*V. catenata* Pennell).

4. *V. catenata* Pennell Fig. 517
Lakeshores, stream channels, ditches, wet places; rarely running water. Que. and N.E. w. to B. C., s. to Va., Tenn., Mo., Tex., N.M., and Calif. This taxon is closely related to *V. anagallis-aquatica* and is sometimes treated as conspecific.

5. *V. scutellata* L. Skullcap Speedwell Fig. 517
Shores, swamps and meadows; often found in submersed form. Nfld. and Lab. w. to Alask., s. to N.C., Tenn., La. Iowa, Neb., Colo., Nev. and Calif. (*V. scutellata* var. *villosa* Schumacher)

6. *Leucospora* (Leucospora)

Spreading annual; leaves opposite, pinnately lobed; flowers small. Solitary, axillary; corolla white, tinged pink, or lavender, bilaterally symmetrical, 2-lipped; fruit a septicidal capsule.

1. *L. multifida* (Michx.) Nutt. Narrowleaf Pale-seed Fig. 519
Shores of lakes, ponds, and streams, mud flats and ditches. N.Y., s. Ont. and Mich. w. to Iowa and Neb., s. to Fla., w. to Tex.

7. *Chelone* (Turtlehead, Snakehead)

Perennial herbs; stems erect; leaves opposite, serrate; flowers borne in short, spike-like racemes; calyx united at base, 5-lobed; corolla white, pink, greenish-white, or purple, united, bilaterally symmetrical; 2-lipped, upper lip keeled in center, notched at apex, lower lip 3-lobed, center lobe smallest, throat of corolla inflated; fruit a septicidal capsule.

References: Crosswhite, 1965; Nelson, 2019.

---

1. Leaves sessile, broadly rounded to subcordate; staminode apices purple. . . . . 1. *C. cuthbertii*
1. Leaves petiolate or tapering to winged or obscure petioles; staminode apices white or green.
   2. Leaves with short petioles or subsessile; corolla white, greenish-yellow, or rose, occasionally purple-tinged toward apex; bracts at flower bases not fringed; staminode apices green. . . . . 2. *C. glabra*
   2. Leaves with distinct petioles; corolla purple; bracts at flower bases fringed; staminode apices white.
      3. Petioles 3–20 mm long; principal leaves 1–3.5(5) cm wide, cuneate at base; inflorescence bracts 4–10(17) mm long. . . . . . 3. *C. obliqua*
      3. Petioles (2)10–40 mm long; principal leaves (2)3–5(8) cm wide, rounded to truncate at base; inflorescence bracts 2–7 mm long.. . . . . 4. *C. lyonii*

---

1. *C. cuthbertii* Small, Cuthbert's Turtlehead
Bogs, swamps, wet thickets, wet meadows, stream banks and springy stream-heads. Se. Va., mts. of N.C., w. S.C., n. Ga.

2. *C. glabra* L. White Turtlehead Fig. 520
Bogs, swales, low thickets, wet woodlands, shores and ditches. Nfld. w. to e. Man., s. to Ga., Ala., Mo. and Ark. This species is extremely variable and while varieties have been described, Nelson (2019) has determined that taxonomic recognition is not warranted.

3. *C. obliqua* L. Red Turtlehead Fig. 520
Alluvial swamps, stream banks, cypress swamps, and wet woods. Rare, Mass.; Md. w. to se. Mich. and s. Minn., s. to Fla., Ala., Miss., and Ark. A larger-flowered midwestern plant (especially upper Mississippi drainage, Mich. and Minn. s. to Mo. and Ark.) has been treated as var. *speciosa* Pennell & Wherry, whereas the smaller-flowered var. *obliqua* occurs in se. U.S., and in our range only in Va. and Md.; var. *erwiniae* (Pennell & Wherry) Pennell is very rare and endemic to the Blue Ridge mt. region of se. Ky., w. N.C. and w. S.C.

4. *C. lyonii* Pursh Pink Turtlehead
Stream banks, shores, thickets and coniferous forests at higher elevations. Va., W.Va., N.C., S.C. and Tenn.; escaped from cultivation locally in Me., Mass., R.I., Conn. and N.Y.

8. *Mecardonia* (Mercardonia)

Perennial herbs; leaves opposite, sessile or short-petiolate, somewhat glandular-punctate, margins toothed toward apex; flowers solitary, axillary; corolla white (ours) or yellow, united forming a corolla tube, bilaterally symmetrical, 2-lipped, upper lip 2-lobed, lower lip 3-lobed: fruit a loculicidal capsule.

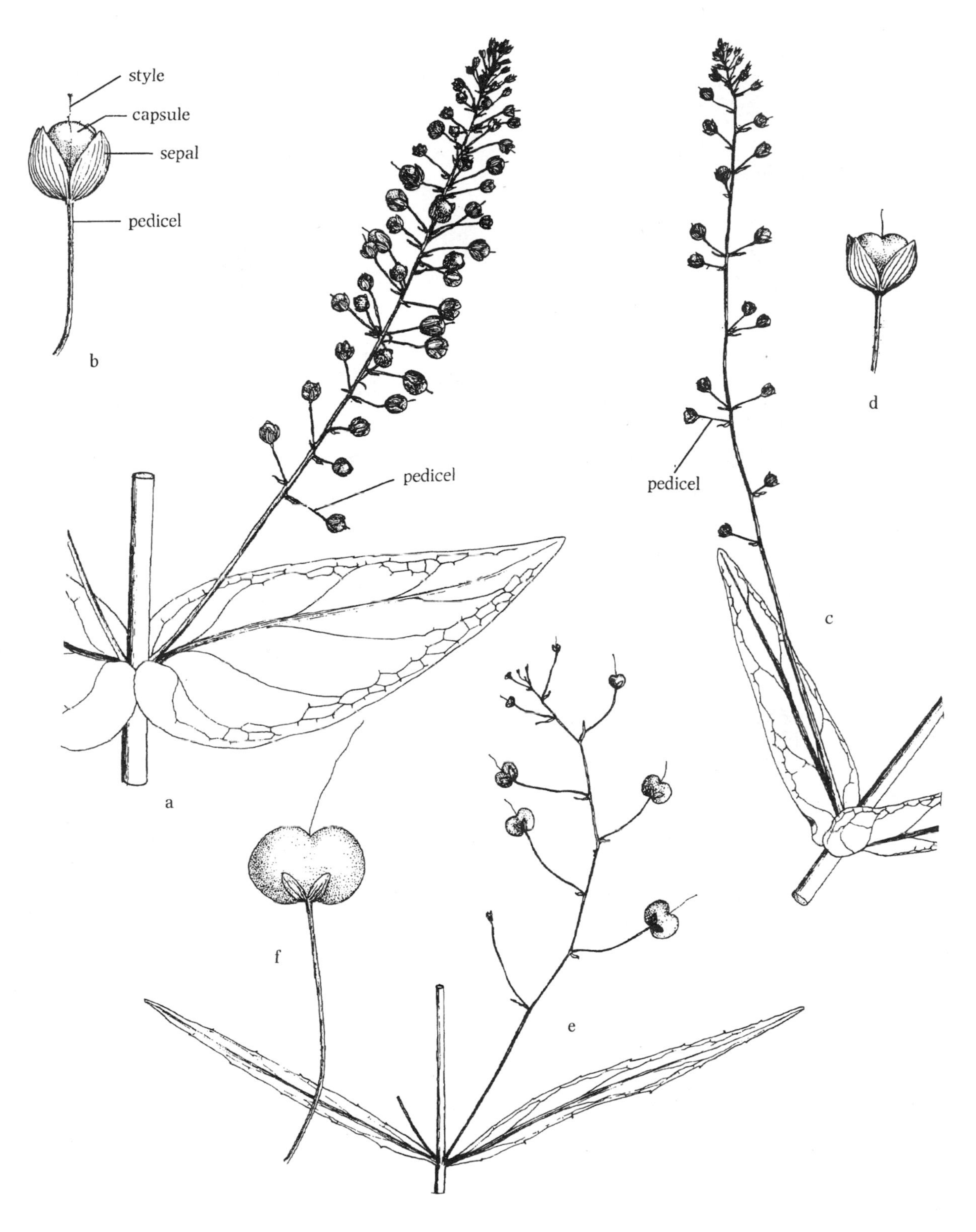

Fig. 517. *Veronica anagallis-aquatica*: a. leaves and inflorescence; b. fruit (F).
Veronica catenata: c. leaves and inflorescence; d. fruit (F).
Veronica scutellata: e. leaves and inflorescence; f. fruit (F).

Fig. 518. *Veronica anagallis-aquatica*: a. habit; b. flower; c. corolla, showing epipetalous stamens; d. capsule; e, f. seeds, two views (Mason).

Fig. 519. *Leucospora multifida*: a. habit (C&C); b. upper portion of plant (F); c. flower (C&C); d. placenta, after shedding seeds, removed from capsule (C&C); e. seed (C&C).

Fig. 520. *Chelone glabra*: a. upper portion of plant; b. leaf; c. leaf (F).
Chelone obliqua: d. upper portion of plant; e. flower (B&T).

1. *M. acuminata* (Walter) Small Axilflower Fig. 521
Damp fields, marshy shores, pine savannas, prairies, and ditches. Coastal plain, Del. and Md. s. to Fla., w. to e. Tex.; inland n. to s. Mo. and Tenn. (*Bacopa acuminata* (Walt.) Robins.)

9. *Gratiola* (Hedge-hyssop)

Annual or perennial, somewhat succulent herbs; leaves opposite, sessile or short-petiolate, toothed or entire, often glandular punctate; flowers sessile or pedicellate, axillary; corolla white, yellow, or lavender, united in a corolla tube, bilaterally symmetrical, upper lip entire or 2-lobed, lower lip 3-lobed; fruit a capsule, dehiscing by 4 valves.

References: Freeman, 2019; Spooner, 1984.

---

1. Plants villous-hirsute (fig. 522a); flower and fruits subsessile; corolla as long as or slightly longer than calyx, 5–9 mm long....... 1. *G. pilosa*
1. Plants glabrous or pubescent, but not villous-hirsute; flowers and fruits distinctly pedicellate; corolla much longer than calyx, 8–15 mm long.
    2. Leaves widest at base, more or less clasping; capsules 1–3 mm long; perennials with creeping rhizomes.
        3. Corolla golden-yellow (rarely white); leaves entire or slightly toothed. .... 2. *G. lutea*
        3. Corolla pale-lavender to white, often with brown or purple lines; leaves dentate to serrate.
            4. Leaf teeth 1–3 per side above middle; leaf blades obscurely 1-veined. .... 3. *G. ramosa*
            4. Leaf teeth more than 3 per side above middle; leaf blades 3–5-veined. .... 4. *G. viscidula*
    2. Leaves narrowed toward base, not clasping; capsules 3–7 mm long; annuals with a taproot.
        5. Pedicels thick, usually 1–4 mm long; capsule (3)4–7 mm long. .... 5. *G. virginiana*
        5. Pedicels slender, 10–2.5 mm long; capsule 3–5 mm long. .... 6. *G. neglecta*

---

1. *G. pilosa* Michx. Shaggy Hedge-hyssop Fig. 522
Bogs, pine savannas, flatwoods, and dry to wet habitats. S. N.J. s. to Fla., w. to. Tex.; inland n. to Tenn., Ky., Ark. and c. Okla. (*Sophronanthe pilosa* (Michx.) Small)

2. *G. lutea* Raf. Golden-pert, Golden Hedge-hyssop Fig. 522
Sandy, gravelly, and peaty shores often submersed. Coastal plain, Nfld. s. to Fla. and Ala.; inland, N.Y. and Pa. w. to Mich., Ill. and N.D. The submersed form with small, pointed leaves is very common (fig. 275i). (*G. aurea* Muhl. ex Pursh)

3. *G. ramosa* Walter Branched Hedge-hyssop Fig. 523
Damp sandy pinelands, pond margins and depressions. Coastal plain, Del. and Md.; N.C. s. to Fla., w. to sw. La.

4. *G. viscidula* Pennell Short's Hedge-hyssop Fig. 524
Swales, bogs, marshes, pond and lake margins, stream banks, seeps, wet meadows and wet woodlands. Del., Md., and Va. w. to W.Va., s. Ohio, Ky. and s. to. Ga., ne. Fla, and Tenn.

5. *G. virginiana* L. Roundfruit Hedge-hyssop Fig. 523
Brooks, ponds, ditches and tidal habitats. N.J. and Md. w. to W.Va., Ohio, Ill., Mich. and Kans., s. to Fla. and c. Tex. The estuarine plants have been described as var. *aestuariorum* Pennell, but may not be worthy of taxonomic recognition.

6. *G. neglecta* Torr. Clammy Hedge-hyssop Fig. 524
Wet, muddy habitats, swamps, marshy shores and ditches. N.S. and Que. w. to B.C., s. to Ga., Ala., Tex., and Calif. Estuarine plants along the St. Lawrence River, Que., have been described as var. *glaberrima* Fernald, and are distinguished by their white flowers and somewhat clasping, glabrous leaves; considering the wide distribution and broad ecological amplitude of this species, recognition of varieties may not be warranted (Freeman, 2019).

10. *Bacopa* (Water-hyssop)

Usually slightly succulent, perennial herbs; leaves opposite, sessile or subsessile, often glandular-punctate; flowers solitary (rarely paired) in leaf axils; corolla variously colored, united, nearly regular, tubular-campanulate, with (3)5 subequal lobes; fruit a capsule.

Reference: Ahedor, 2019.

---

1. Leaf blades 1-veined, oblanceolate to narrowly wedge-shaped-obovate, tapering to narrow base, not clasping (fig. 525a,e); pedicels longer than subtending leaves. .... 1. *B. monnieri*
1. Leaf blades with (3)5–9 or more palmate veins, subcircular to ovate, or broadly obovate, usually with somewhat clasping bases (fig. 525f,j); pedicels usually shorter than subtending leaves.
    2. Leaves ovate, punctate; corollas violet-blue with violet-blue throats; bruised fresh plants lemon-scented. .... 2. *B. caroliniana*

Fig. 521. *Mecardonia acuminata*: a. habit; b. flower; c. capsule with calyx; d. seed, pitted-reticulate (G&W).

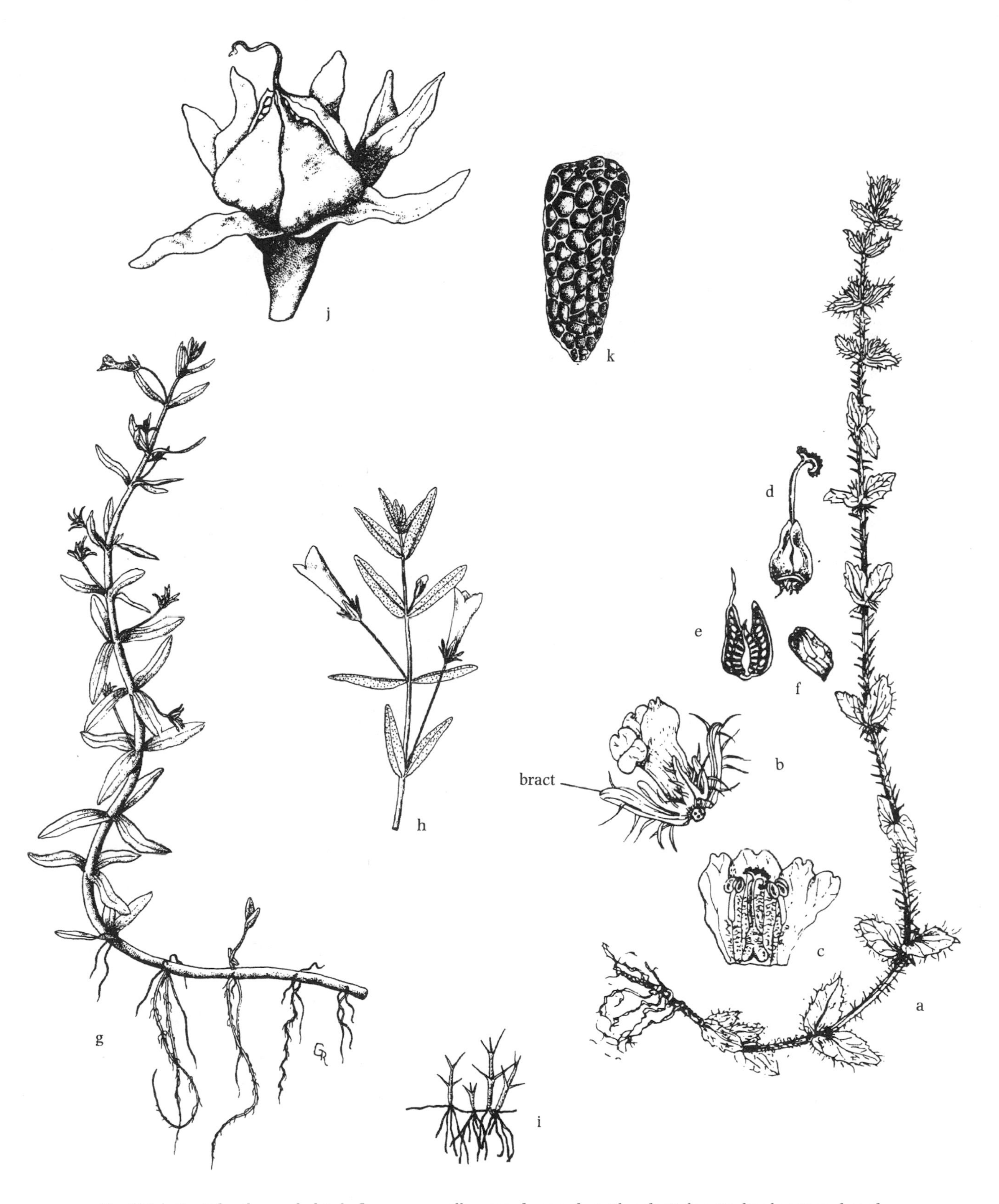

Fig. 522. *Gratiola pilosa*: a. habit; b. flower; c. corolla, spread open; d. pistil; e. fruit, longitudinal section; f. seed, pitted-reticulate (C&C).
Gratiola lutea: g. habit (G&W); h. upper portion of plant (F); i. habit, sterile submersed form (F); j. capsule with subtending bracts and bracteoles (G&W); k. seed, pitted-reticulate (G&W).

Fig. 523. *Gratiola ramosa*: a. habit (G&W).
Gratiola virginiana: b. habit (C&C); c. upper portion of plant (F); d. leaf (P); e. flower (C&C).

Fig. 524. *Gratiola neglecta*: a. upper portion of plant (C&C); b. upper portion of plant (F); c. capsule with persistent perianth (C&C).
Gratiola viscidula: d. upper portion of plant (F).
Bacopa rotundifolia: e. habit (C&C); f. leaves and flower (F); g. flower (C&C).

2. Leaves subcircular, ovate, elliptic, or broadly obovate, not punctate; corollas white with yellow or white throats; bruised plants not lemon-scented.
   3. Leaves 5–13-veined; pedicels 2–4 times as long as calyx, 5–20 mm long, corolla white with yellow throat, 5–10 mm long. .... 3. *B. rotundifolia*
   3. Leaves obscurely 3–5-veined; pedicels approximately same length as calyx, 3–6 mm long, corolla white with white throats, 2–5 mm long........ 4. *B. innominata*

---

1. *B. monnieri* (L.) Pennell Herb of Grace Fig. 525
Sand flats, fresh or brackish shores, interdunal swales, pools, and ditches. Coastal, Md. and se. Va. s. to Fla., w. to La., Okla., Tex.; Ariz. and s. Calif.; Trop. Am.

2. *B. caroliniana* (Walter) B. L. Rob. Blue Water-hyssop Fig. 525
Pond and lake margins, pineland streams, canals, and drainage ditches. Coastal plain, Md. and se. Va. s. to Fla., w. to La. and e. Tex.

3. *B. rotundifolia* (Michx.) Wettst. Ditch Water-hyssop Fig. 524
Ponds, pools, lakeshores, and ditches. Md., Va., and N.C. w. to Ky., Ind., Ill., s. Wis., w. Minn., N.D., Mont., Alta., and Ida., s. to Ala., Miss., La., Tex., N.M. Ariz. and c. Calif. (*B. simulans* Fernald)

4. *B. innominata* (G. Maza) Alain Tropical Water-hyssop Fig. 526
Alluvial deposits along rivers and streams, wet ditched, muddy banks and shores. Chiefly coastal plain, se. Md. s. to S.C., Ga. and Fla.; W.I., n. S.Am. (*B. stragula* Fernald)

Scrophulariaceae / Figwort Family

1. *Limosella* (Mudwort)

Small annual, stoloniferous, forming mats; leaves basal, succulent; flowers small, solitary at tips of recurved peduncles; corolla white or purple, united, rotate; fruit a septicidal capsule.

---

1. Leaves flat, with a lanceolate to elliptic blade, 10–20 in a rosette (fig. 527d); corolla about 2 mm wide, dull white to purplish, lobes acute. .... 1. *L. aquatica*
1. Leaves round in cross-section, linear, bladeless, 5–10 in a rosette (fig. 527a); corolla about 3 mm wide, white, tinged with lavender, lobes rounded........ 2. *L. australis*

---

1. *L. aquatica* L. Water Mudwort Fig. 527
Fresh to brackish muddy and sandy shores. Coastal or local, Greenl., Lab. and Nfld. w. to Alask., s. to Ont., Minn., w. Neb., N.M. and Ariz.; S.Am.

2. *L. australis* R. Br. Atlantic Mudwort Fig. 527
Brackish muddy and sandy shores. Coastal, Nfld. s. to N.E., N.J., ne. Md., se. Va. and ne. N.C.; S.Am., S. Afr. (*Limosella subulata* Ives; *L. aquatilis* var. *tenuifolia* (J. P. Wolff) Hook. f.)

Linderniaceae / False Pimpernel Family

---

1. Calyx lobes 5; flowers small, but more than 2 mm long; corolla pale lavender, light blue, or white, persisting; stamens 4 (2 fertile, 2 sterile)........ 1. *Lindernia*
1. Calyx lobes 4; flowers tiny, less than 2 mm long; corolla white, readily falling off; stamens. .... 2. *Micranthemum*

---

1. *Lindernia* (False Pimpernel)

Annual or perennial herbs; leaves opposite, simple, entire or dentate; flowers solitary, axillary; corolla pale lavender, light blue, or white, united, rotate; fruit a septicidal capsule.

References: Cooperrider and McCready, 1975; Lewis, 2019.

1. *L. dubia* (L.) Pennell Yellow-seed false Pimpernell Fig. 528
Sandy shores, damp open sites, ponds, wet meadows, tidal shores and fresh tidal shores. N.S., N.B. and s. Que. w. to Ont., Minn., e. N.D., e. S.D., and Neb., s. to Fla., La. and e. Tex.; B.C., Wash. Oreg. and Ida. s. in mts. to Colo., N.M., Ariz. and Calif. Several varieties have been described, but Lewis (2019) noted that extreme morphological variability is expressed in *L. dubia* and therefore recognized no infraspecific taxa. (*L. dubia* var. *inundata* (Pennell) Pennell; *L. dubia* var. *riparia* (Raf.) Fernald.; *L. anagallidea* (Michx.) Pennell)

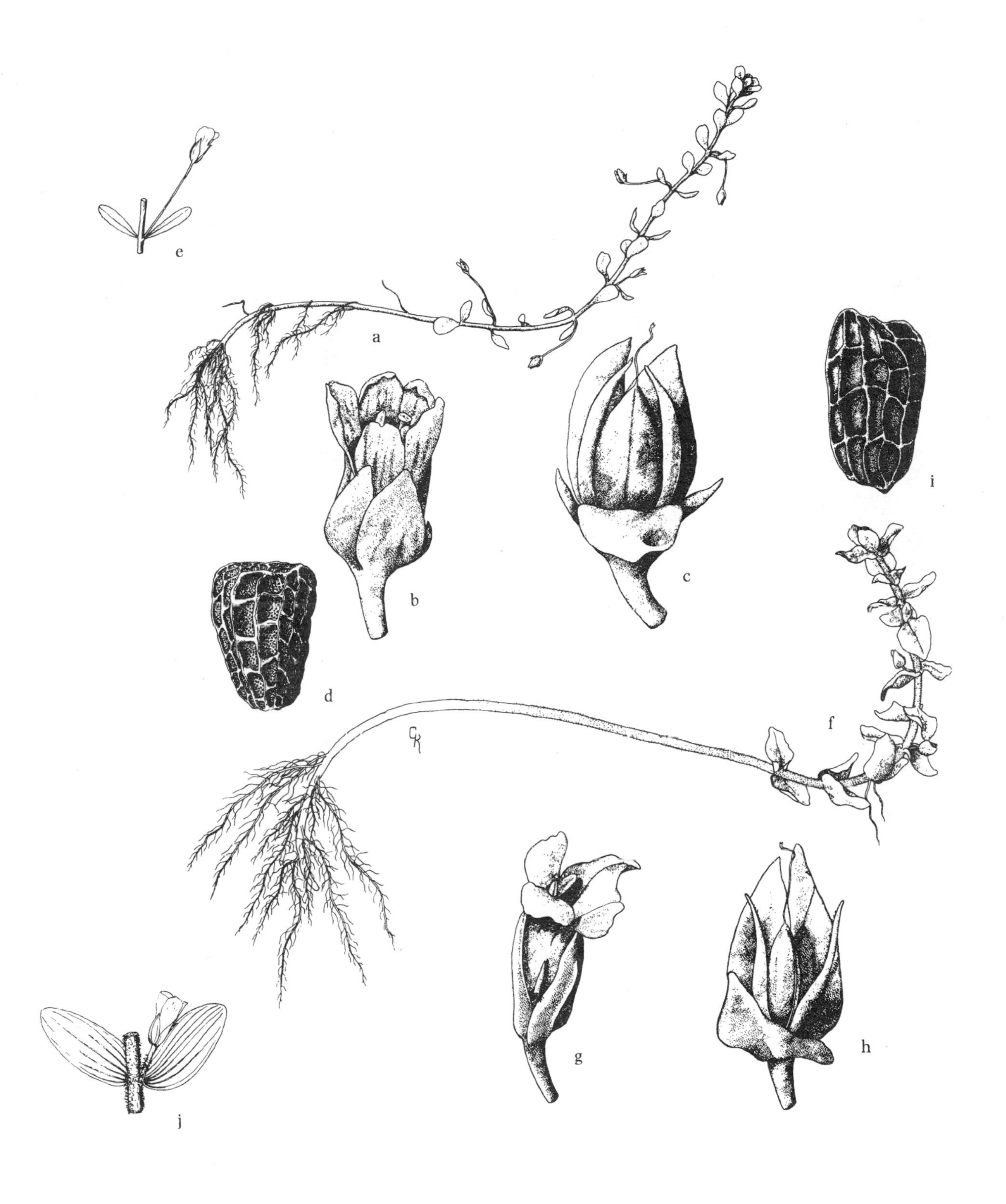

Fig. 525. *Bacopa monnieri*: a. habit (G&W); b. flower (G&W); c. capsule (G&W); d. seed, pitted-reticulate (G&W); e. leaves and flower (F).
Bacopa caroliniana: f. habit (G&W); g. flower (G&W); h. capsule (G&W); i. seed, pitted-reticulate (G&W); j. leaves and flower (F).

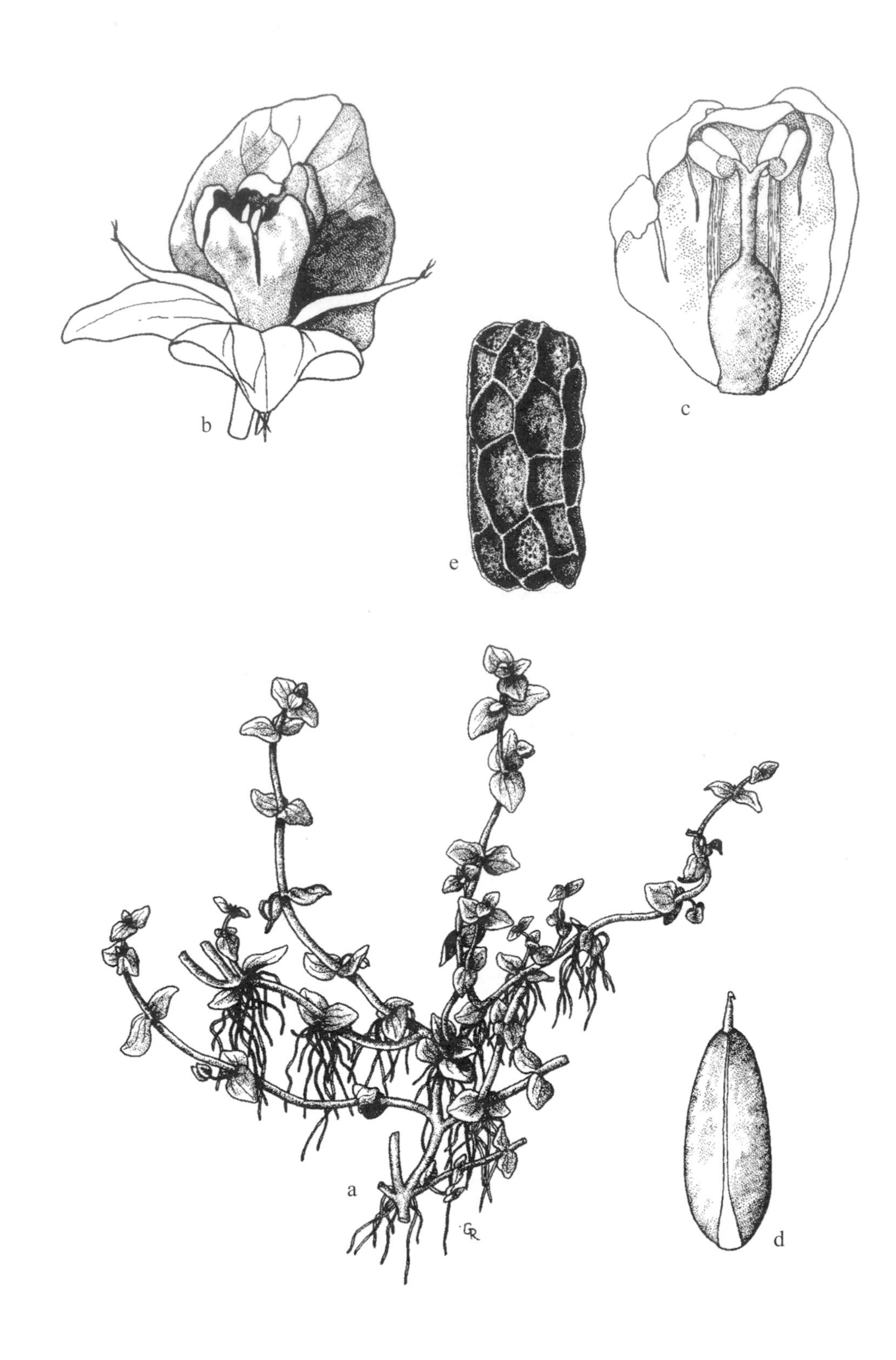

Fig. 526. *Bacopa innominata*: a. habit; b. flower; c. corolla, split open, showing two stamens and the pistil; d. capsule; e. seed, pitted-reticulate (G&W).

Fig. 527. *Limosella australis*: a. habit; b. corolla, spread open; c. seed, pitted-reticulate (Mason). *Limosella aquatica*: d. habit; e. corolla, spread open; f. seed, pitted-reticulate (Mason).

Fig. 528. *Lindernia dubia*: a. habit (C&C); b. leaves and fruit (F); c. habit (Mason); d. leaves and fruit (F); e. capsule (Mason); f. seed, pitted-reticulate (Mason).

2. *Micranthemum* (Mudflower)

Annual; stems prostrate or weakly ascending, leaves opposite or whorled, sessile, entire, rounded or spatulate; flowers tiny, solitary, in leaf axils; flowers bilaterally symmetrical 4-merous, corolla white or purple united forming a tube; fruit a 1-locular capsule, dehiscence irregular.

Reference: Keener, 2019

1. Leaves elliptic to oblanceolate, 3–5 mm long; flowers strongly bilaterally symmetrical; corolla longer than calyx, forming a prominent lip. . . . . 1. *M. micranthemoides*
1. Leaves orbicular, ovate or obovate, 2–9 mm long; flowers weakly bilaterally symmetrical; corolla shorter than or equal to calyx, not forming a prominent lip. . . . . 2. *M. umbrosum*

1. *M. micranthemoides* (Nutt.) Wettst. Nuttall's Mudflower Fig. 529
Fresh tidal muds. Local, se. N.Y., N.J., se. Pa., Del., Md. and e. Va. Possibly extinct in wild (last collected in 1941), but may yet occur in aquarium plant trade (Keener, 2019).

2. *M. umbrosum* (Walter) Blake Shade Mudflower Fig. 529
Stream shores, low woods, pools, swamps and ditches. Coastal plain and piedmont, se. Va. s. to Fla., w. to La., s. Ark. and e. Tex.; W.I., C.Am. and S.Am.

Acanthaceae / Acanthus Family

1. *Justicia* (Water-willow)

Perennials, arising from rhizomes; leaves opposite, petiolate, margins entire; flowers solitary or borne in spikes or panicles; corolla white, pink, or purple, united, upper lip 2-lobed, lower lip 3-lobed; fruit a capsule.

Reference: Penfound, 1940b

1. Leaves lanceolate-linear, elongate-lanceolate, or oblanceolate, 8–20 cm long; flowers in a congested, capitate spike, terminating a stiff peduncle. . . . . 1. *J. americana*
1. Leaves lanceolate to elliptic to oblanceolate, 3–10 cm long; flowers borne singly or paired on a spike, not congested (fig. 531a), peduncle slender and flexuous. . . . . 2. *J. ovata*

1. *J. americana* (L.) Vahl Water-willow Fig. 530
Shores and shallow water of streams, rivers, lakes and ditches. Sw. Que. and nw. Vt. w. to Ont., Mo., and Kans., s. to Fla., Okla. and Tex. (*J. americana* var. *subcoriacea* Fernald; *Dianthera americana* L.)

2. *J. ovata* (Walter) Lindau Looseflower Water-willow Fig. 531
Wooded swamps, cypress swamps, bottomlands, depressions and ditches. Coastal plain, se. Va. s. to Fla., w. to e. Tex.; inland n. to w. Ky., se. Mo., Ark. and se. Okla. Two varieties may be distinguished: var. *ovata*, a taxon of the Atlantic coastal plain, and var. *lanceolata* (Chapm.) R. W. Long, occupying the Gulf coastal plain and Mississippi embayment. (*J. lanceolata* (Chapm.) Small; *J. humilis* Michx.; *Dianthera ovata* Walter)

Lentibulariaceae / Bladderwort Family

1. Plants aquatic or semi-terrestrial, leaves not in rosettes, blades dissected or laminar, adaxial surfaces glabrous; flowers usually 2–20-flowered racemes (sometimes 1 open at a time), sometimes solitary, subtended by a single bract, sometimes also a pair of bracteoles; calyx 2-lipped. . . . . 1. *Utricularia*
1. Plants terrestrial, leaves in rosettes, blades not dissected, margins involute, adaxial surface glandular hairy; flowers solitary, borne on a scape, bracts none, calyx 5-lobed, not 2-lipped. . . . . 2. *Pinguicula*

1. *Utricularia* (Bladderwort)

Perennial aquatic and semi-aquatic herbs (ours), many producing turions (winter buds), or tropical epiphytes; plants consisting of stem systems, roots absent, the aquatic species often free-floating or anchored by rhizoids or stolons; lateral leaves (leaf-like segments)

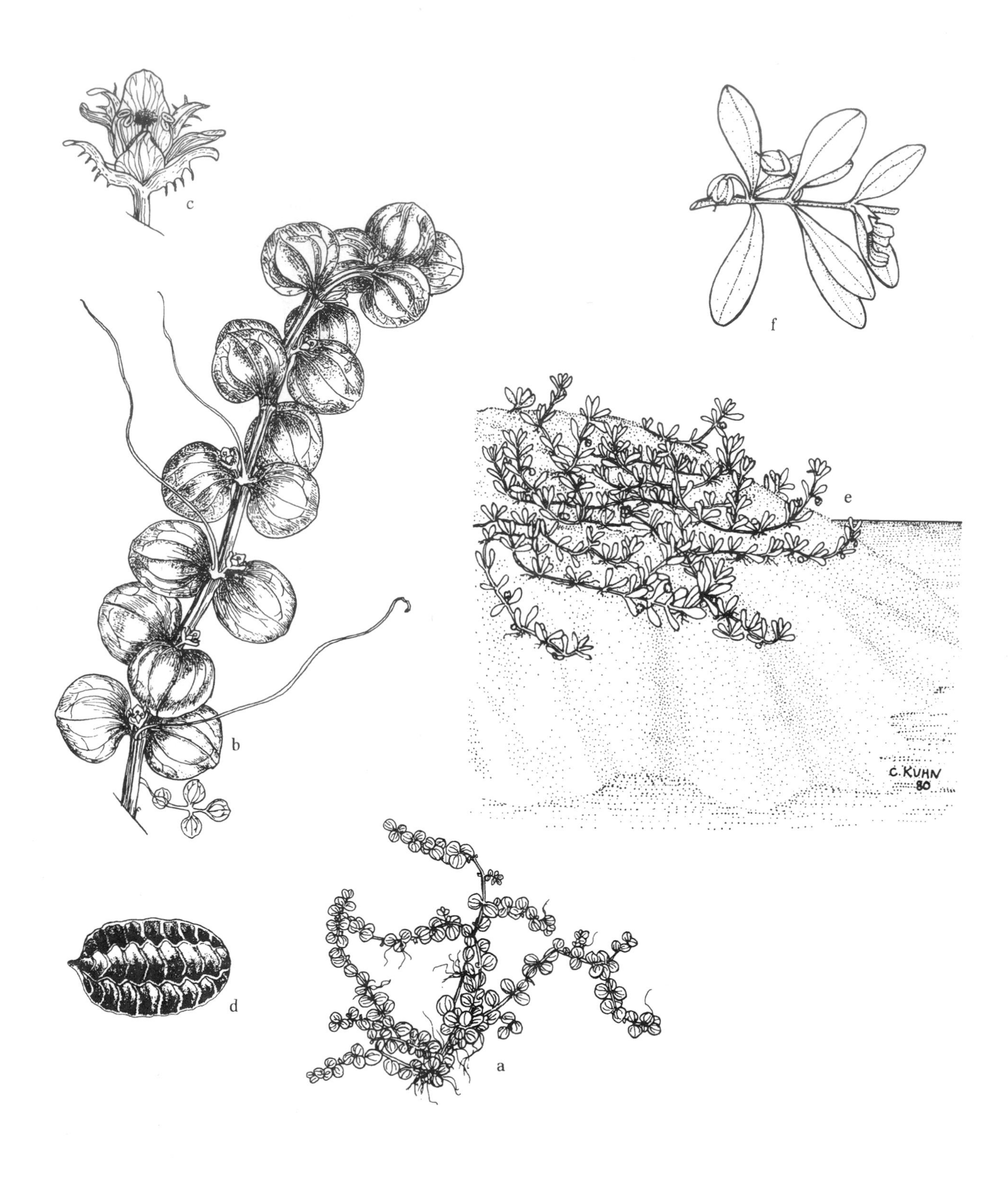

Fig. 529. *Micranthemum umbrosum*: a. habit; b. portion of plant; c. flower; d. seed, pitted-reticulate (C&C). *Micranthemum micranthemoides*: e. habit; f. section of stem with leaves and young fruit (NYSM).

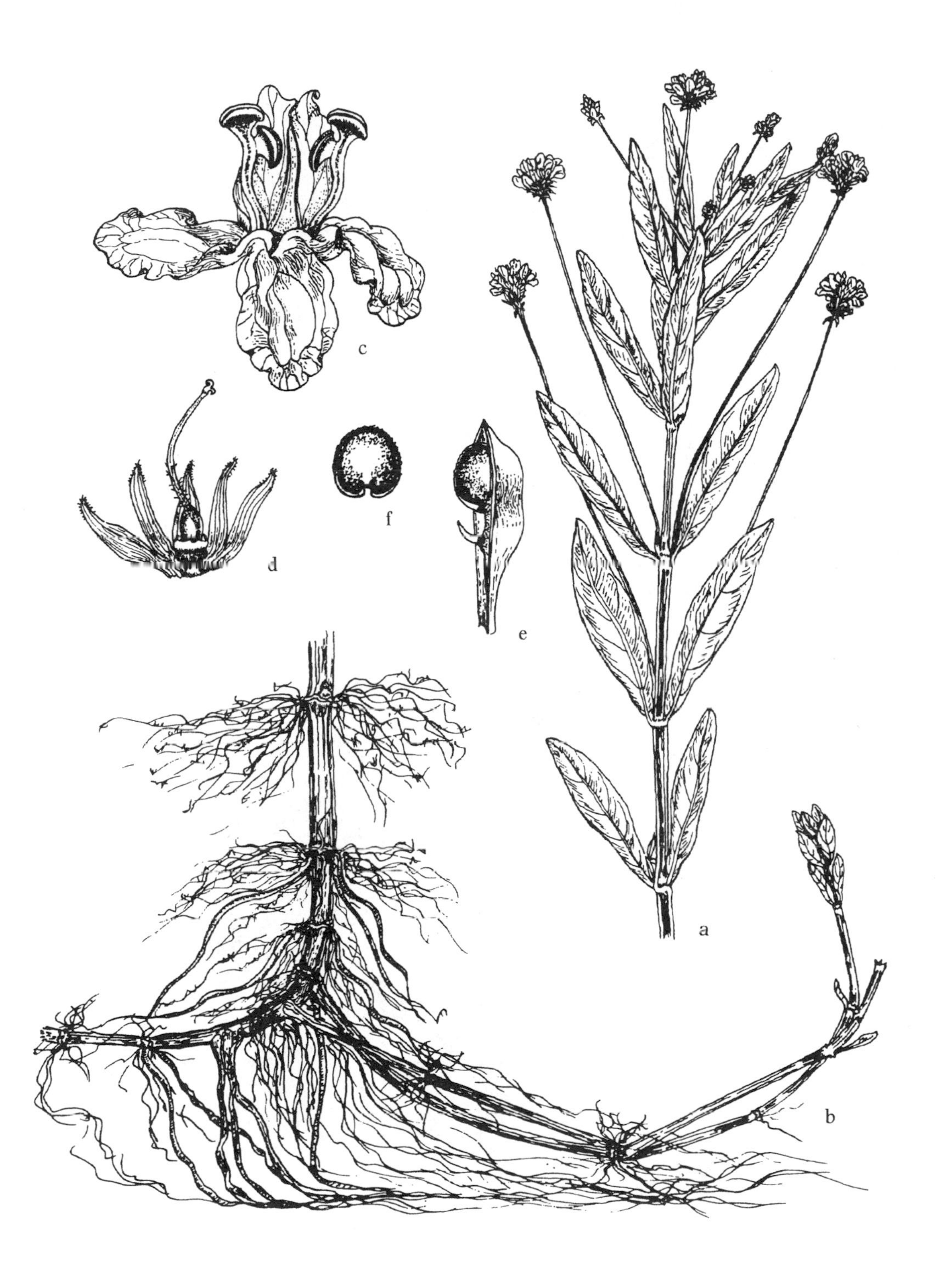

Fig. 530. *Justicia americana*: a. upper portion of plant; b. basal portion of plant; c. flower; d. calyx with ovary and style; e. capsule, 1 valve removed; f. seed (C&C).

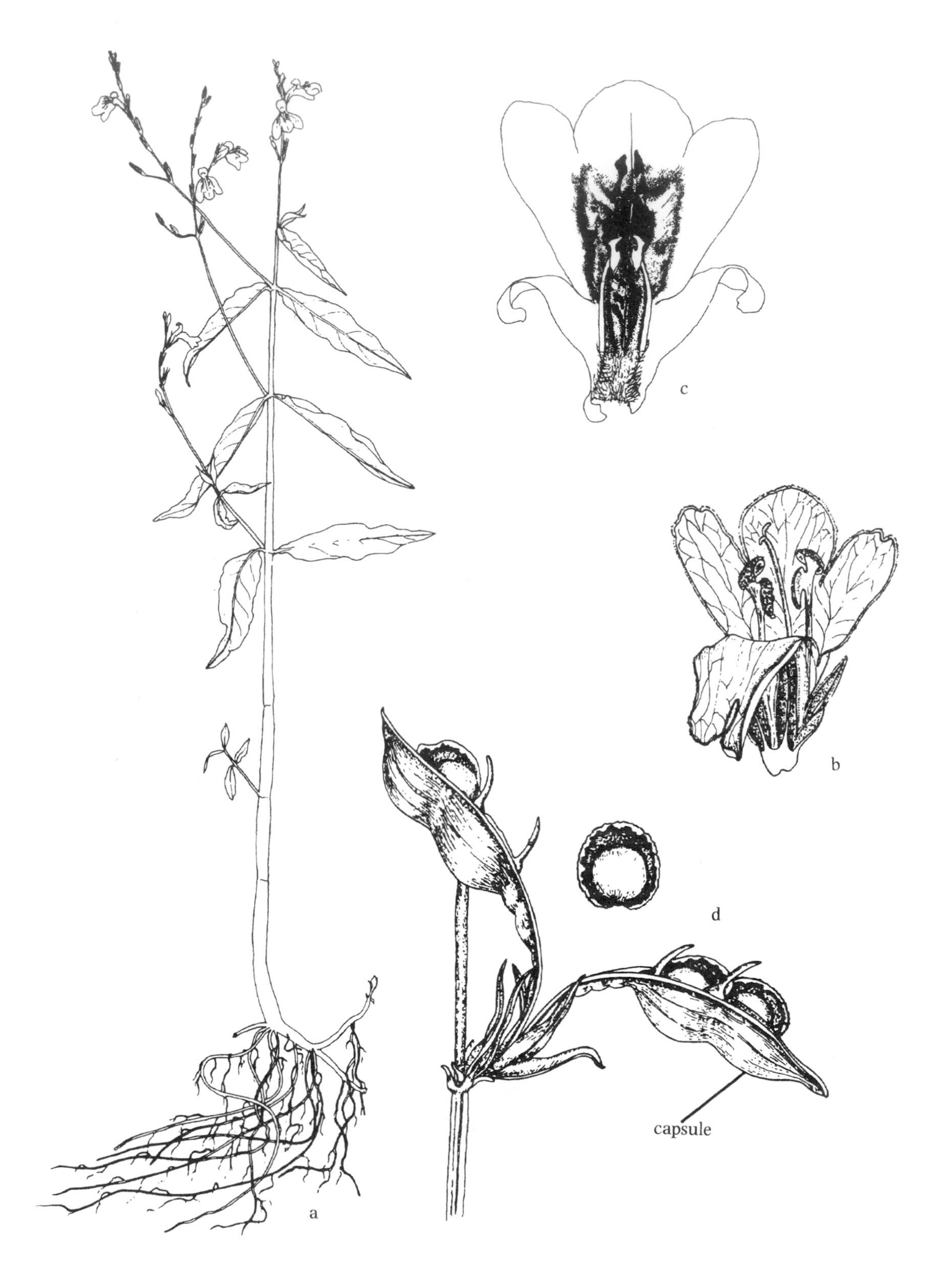

Fig. 531. *Justicia ovata*: a. habit (G&W); b. flower (C&C); c. flower, longitudinal section, with epipetalous stamens (G&W); d. capsule (dehisced) and seeds (C&C).

green, usually bearing bladders (carnivorous traps), bladders sometimes borne on separate, non-photosynthetic segments; flowers borne on scapes elevated above water surface, solitary or in racemes; corolla yellow to greenish-yellow, purple or white, united, bilaterally symmetrical, 2-lipped, spurred; fruit a 2-valved capsule.

Aquatic species of *Utricularia* in cold temperate regions perennate by turions ("hibernacula") that are usually quite distinctive by species (figs. 534c,f, 536c). Semi-terrestrial and tropical epiphytic species usually develop tubers by which they survive dry conditions. Bladders (carnivorous suction-traps in *Utricularia*, figs. 534b,j, 536g), bear external "trigger" appendages associated with trap mouths (fig. 536g), and internal trap glands, comprised of abundant specialized quadrifid trichomes lining the trap (associated with setting trap, secreting enzymes and absorbing nutrients), as well as bifid trichomes (fewer), forming a ring adjacent to trap opening (ring of unbranched trichomes in *U. resupinata*).

REFERENCES: Beal and Quay, 1968; Crow, 2004, 2007, 2015b, 2022, in press; Crow and Hellquist, 1985; Gates, 1939; Hellquist, 1974; Kondo et al., 1978; Lloyd, 1933, 1935; Reinert and Godfrey, 1962; Rossbach, 1939; Tans, 1987; Taylor, 1964, 1989; Uttal, 1956.

---

1. Plants aquatic, submersed, unattached (rootless) or creeping over substrate (sometimes stranded on mud, appearing anchored); leaves divided into filiform segments.
   2. Leaves whorled; bladders borne only at tips of leaf segments (fig. 535a,b,); corollas purple or pink (rarely white), lateral lobes of lower lip saccate. . . . . 1. *U. purpurea*
   2. Leaves not whorled; bladders scattered; corollas yellow or white, lateral lobes of lower lip not saccate.
      3. Scapes subtended by inflated, floating branches (fig. 533a,g).
         4. Racemes (3)9–14(18)-flowered, pedicels 1–3.5 cm; corollas bright yellow, spur distinctly notched at apex; pedicels strongly recurved in fruit; inflated branches 3–8 cm; individual float branches: margins tapering to scape; main stem (stolon) below float 2–3 mm diam. (at 5 cm below floats), leaves on submersed stolons divided into unequal primary segments, much-branched, bushy, ultimate segments filiform; (fig. 533h), bladders of 2 sizes, 1.5–2(3) mm and 0.7–1 mm. . . . . 2. *U. inflata*
         4. Racemes (1)3–4(5)-flowered, pedicels 0.2–1.8 cm; corollas dull yellow, spur rounded at apex (rarely with slight notch); pedicels ascending in fruit; inflated branches 1–4 cm; individual float branches: margins parallel most of length; main stem (stolon) below float to 0.7 mm diam. (at 5 cm below floats), leaves on submersed stolons divided into 2 equal primary segments, less densely branched, ultimate segments threadlike, not bushy; bladders all of similar size, 1.5–2 mm. . . . . . 3. *U. radiata*
      3. Scapes not subtended by inflated, floating branches.
         5. Stems dimorphic, with whitish, bladder-bearing stolons (often buried in substrate), and green stolons bearing dissected leaves, often lacking bladders or with 1–few bladders.
            6. Elongate green leafy branches appearing cylindric in outline (foxtail-like) (fig. 532a), leaf segments fine, terete, narrower beyond each dichotomy, ultimate divisions threadlike, midveins not visible; bracts basifixed, clasping pedicel; corolla lips nearly equal length. . . . . 4. *U. striata* (in part)
            6. Elongate green leafy branches flat (leaves in one plane), leaf segments flat, more or less same width throughout, with midvein visible in ultimate divisions; bracts conspicuously auriculate, not clasping at base; corolla lower lip length 2 times upper lip.
               7. Corolla bright yellow, spur slightly shorter than lower lip, slender cylindric, narrowed/constricted at base, more-or-less appressed to lower lip; bladders restricted to non-green stolons, (1.5)2.5–5.5 mm; ultimate green leaf divisions toothed, lateral teeth (5)9–12(20), strongly setulose (bristles sometimes in fascicles of 2–4), apex more-or-less obtuse; turions (winter buds), when present, oblong, 7–11(15) mm long, strongly setulose with silvery-white bristles on margins of scale-like leaves. . . . . 5. *U. intermedia*
               7. Corolla light or pale yellow, spur nearly lacking, to half as long as lower lip, short-conical to pyramidal, broad at base, oriented at 45° to 90° angle to lower lip; bladders typically on both white subterranean stolons and on green leaves, 0.3–2.2(3.8) mm; ultimate green leaf divisions lacking teeth or with 1–7(9) lateral setulose teeth, always acute at apex; turions, when present, globose-ovoid, (1.5–)2–3(3.5), weakly setulose (appearing naked, green).
                  8. Flowers, dull yellow, spur nearly lacking, broadly conical, 1.5–3.2 mm long; bracts and scales purplish; lateral margins of lower lip strongly curved downward; palate obscure, elongate, with slightly raised lateral margins; lateral margins of ultimate leaf segments entire, not toothed.. . . . . 6. *U. minor* (in part)
                  8. Flowers light yellow, spur about half as long as the lower lip, 2.2–3.5(–5) mm long, short-conical to pyramidal, oriented at approximately a 45–90° angle from the lower lip (fig. 534); bracts and scales green; lateral margins of lower lip nearly flat to slightly curved upward or slightly deflexed; palate conspicuous, rounded; lateral margins of ultimate leaf segments sparsely toothed, weakly setulose. . . . . 7. *U. ochroleuca*

5. Stems of one type, not dimorphic, with green vegetative branches having leaves bearing bladders (or in *U. olivacea* green bladder-bearing stolons, but lacking leaves).
    9. Corollas white-translucent, (1.3)2–2.3(3.5) mm, stolons extremely delicate, coiled at tips, leaves absent (plants extremely diminutive, often entangled in other submersed vegetation and easily overlooked). 8. *U. olivacea*
    9. Corollas yellow, more than (3.5–)5 mm; stolons variable, but leaves present (plants, if small, not extremely diminutive).
        10. Corollas pale yellow (purplish tinged or striped near base); lower corolla lip strongly recurved; bracts auriculate, purplish; leaf segments flat, with central vein. . . . . 6. *U. minor* (in part)
        10. Corollas bright yellow; lower corolla lip spreading; bracts not auriculate, green (sometimes auriculate in *U. vulgaris*, but then green); leaf segments filiform or if slightly flattened, then without visible central vein.
            11. Vegetative stems less than 30 cm; leaves divided 1–4(8) times, ultimate segments hair-like; flowers usually (1)2–3(6); lower corolla lip 2-lobed, smaller than upper lip; plants, if flowering usually entangled among floating vegetation at surface, or stranded with flowers exposed above substrate and vegetative portion usually buried in sand/mud. . . . . 9. *U. gibba* (in part)
            11. Vegetative stems 30–100 cm; leaves divided 6 or more times, ultimate segments not hair-like; flowers 2–15; lower corolla lip 3-lobed, larger than upper; submersed and free-floating near surface, not typically tangled among other vegetation.
                12. Racemes 2–5(8)-flowered; corolla with lower lip distinctly 3-lobed, not red-streaked (upper and lower lip oriented upward giving saddle-like appearance similar to *U. gibba*); apetalous cleistogamous flowers 1–2, 1.5–2.5 mm diam., at base of scape (sometimes only cleistogamous flowers present); vegetative branches 1–3.5 cm across; leaf segments with margins entire or sparsely setulose. 10. *U. geminiscapa*
                12. Racemes (3)6–15-flowered; corolla with lower lip very obscurely 3-lobed, streaked with red on palate and spur; cleistogamous flowers lacking; vegetative branches 3–12 cm across; leaf segments with margins conspicuously setulose (with or without teeth).. . . . . 11. *U. vulgaris*

1. Plants semi-aquatic, in wetlands or anchored on wet shores (appearing semi-terrestrial), typically only flowers and scapes visible with vegetative part in substrate (leaves very slender, terete, linear to thin, obovate to oblanceolate) (figs. 538a, 539a, 540a, 541a,b).
    13. Corolla light purple to rose-pink (rarely white); inflorescences 1-flowered (flower tipped backward, appearing upside-down); each flower subtended by, with a single pair of opposite, fused bracts at base of pedicel (fig. 538e); distinctly septate (sometimes septa only 1–2, toward apex) (fig. 538a). . . . . 12. *U. resupinata*
    13. Corolla yellow; inflorescences with 2 or more flowers (sometimes 1 open at a time), subtended by a single bract or 1 bract and 2 bracteoles; leaf blades not septate.
        14. Corolla pale yellow to greenish yellow, with conspicuous veins, red streaks at base of lower lip, continuing into spur; spur usually with shallow notch at apex, sometimes obscurely 3-fid; stolons radiating from base of inflorescence, whitish, bearing rhizoids. . . . . 4. *U. striata* (in part)
        14. Corolla bright yellow, without conspicuous veins; spur not notched at apex (sometimes slightly denticulate in *U. subulata*); base of inflorescence not bearing radiating stolons.
            15. Upper corolla lip 3-lobed, slightly larger than lower lip. . . . . 9. *U. gibba* (in part)
            15. Upper corolla lip entire to 2-lobed, smaller than lower lip.
                16. Scape bracts peltate (fig. 539b,c); flowers each subtended by a bract, bracteoles none; spur length slightly shorter than to slightly longer than lower corolla lip, usually reddish; palate bilobed. . . . . 13. *U. subulata*
                16. Scape bracts basifixed (fig. 540j); flowers each subtended by a bract and 2 bracteoles; spur longer than lower lip, never reddish; palate a conical hump (not bilobed).
                    17. Inflorescences congested, (1)3–5(9)-flowered, usually clustered distally; corollas 15–25(30) mm, spur 7–12(14) mm; scapes green to yellowish green, stout, 0.5–1.5 mm diameter near base. . . . . 14. *U. cornuta*
                    17. Inflorescences elongate, several to many-flowered, usually widely distributed along scape; corolla 9–15 mm, spur 4–6(7) mm; scapes greenish purple to purple, slender, wiry, 0.4–1 mm diameter near base.. . . . . 15. *U. juncea*

---

1. *U. purpurea* Walter Eastern Purple Bladderwort Fig. 535
Acidic waters of lakes, ponds, sloughs, swamps, sluggish streams and ditches. Nfld., N.S. and N.B. w. to N.Y. s. Ont., Mich., n. Ind., n. Ill., Wisc., and Minn., s. chiefly on the coastal plain to Fla., w. to La. and e. Tex.; Mex., W.I. and C.Am.

2. *U. inflata* Walter Floating Bladderwort, Large Swollen Bladderwort Fig. 533
Ponds, lakes, swamps, sloughs, canals and drainage ditches. Chiefly coastal plain, se. Me., e. Mass., se. N.Y. e. Pa. and N.J. s. to S.C., Ga., Fla., w. to La., Tenn., Ark., e. Okla. and e. Tex.; B.C. and nw. Wash. (intro.).

3. *U. radiata* Small Little Floating Bladderwort, Small Swollen Bladderwort Fig. 533
Ponds, lakes, swamps and drainage ditches. Chiefly coastal plain, N.S., N.B. and N.E. s. to N.J., Va., N.C., and Fla., w. to Ala., La., and e. Tex.; inland n. to e. Okla., Ark. w. Tenn.; nw. Ind., and sw. Mich. (*U. inflata* var. *minor* Chapm.)

4. *U. striata* Leconte ex Torr. Striped Bladderwort Fig. 532
Ponds, pools, swamps, wet peats and sandy shores. Coastal plain, se. Mass., R.I., and c. Conn. s. to N.J., e. Pa., Del., Va., S.C., Ga. and Fla., w. to e. Okla. and e. Tex.; Calif. (intro.). Plants becoming stranded following a drop in the water table can be confused with the terrestrial growth form of the Atlantic and Gulf coastal plain phase of *U. gibba*, the terrestrial phase of *U. striata* having several whitish subterranean stolons at the base of the inflorescence bearing tiny rhizoids near the summit of the stolons. A single site in Butte Co., California, has been documented, apparently a weed associated with rice cultivation in that area. According to Taylor (1989) the name *U. fibrosa* Walter has been applied widely to this taxon, but his careful interpretation of Walter's original descriptions has led Taylor to conclude that *U. fibrosa* is conspecific with *U. biflora* Lam.; Taylor has placed both these names in synonymy under *U. gibba* L.

5. *U. intermedia* Hayne Flatleaf Bladderwort Fig. 532
Shallow pools and ponds, peaty soils and wet sands. Greenl., Lab. and Nfld. w. to Nunavut, N.W.T. and Alask., s. to N.E., N.Y., N.J., n. Del., Md., ne. Pa., Ohio, n. Ind., n. Ill., nw. Iowa, N.D., Wyo., Utah., ne. Nev. and Calif.

6. *U. minor* L. Lesser Bladderwort Fig. 534
Acidic waters of shallow pools, wet meadows, bogs, marshes and shores. Greenl. and Lab. w. to N.W.T. and Alask., s. to N.E., n. N.J., n. Ohio, n. Ill., nw. Iowa, ne. S.D., Neb., Colo., and Calif.; Eurasia. Populations are often encountered only in the vegetative state.

7. *U. ochroleuca* R. W. Hartman Yellowish-white Bladderwort Fig. 534
Bogs, boggy meadows, marshes, often shallow water, tending to remain vegetative if in deeper water of streams and lakes. Rare, widely scattered localities, Greenl. and Lab. w. to Nunavut, N.W.T. and Alask., s. to N.S., Que., n. Mich., w. Ohio, Ont., Man., Mont., Wyo., Alta., B.C., Wash., Oreg. and n. Calif.; Eurasia. More populations have been located in Yellowstone National Park than any other locality (Hellquist et al., 2014; Routledge et al., 2020). Flowers of *U. ochroleuca* vary in regard to streaking (presence of reddish streaks/veins, conspicuous to sometimes faint) or absence of streaks on the palate and spur. The spur is especially distinctive with its short-conical/pyramidal shape and 45–90°-angle orientation (in contrast to *U. intermedia* with its spur appressed to and only slightly shorter than the lower lip, and constriction near base of spur). Turions of *U. ochroleuca* appear naked, whereas the densely packed turion leaves of *U. intermedia* are conspicuously setulose, appearing whitish on herbarium specimens.

U. ochroleuca is regarded as of hybrid origin with purported *U. intermedia* and *U. minor* parentage, and appears to be a vegetative apomict, persisting and dispersing via turions (P. Taylor, 1989). Thor (1988) employed morphology of the quadrifid trichomes lining the inner surface of bladder traps as taxonomic characters. Plachno and Adamec (2007) found that measurement of the angle between the two shorter arms to be statistically most diagnostic. Crow (2015b, 2022) found them to be reliable in distinguishing *U. ochroleuca* (with mean angle 128.9°; range 111–146°) from *U. intermedia* (with mean angle 28.6°; range 16–42 or arms closed and no angle), and from *U. minor* (arms usually strongly reflexed), but not reliably distinct from depauperate, sterile specimens of *U. vulgaris* subsp. *macrorhiza* (with mean angle 133.9°; range 114–154°). A few populations from North America that appear to fit Thor's (1988) concept of the very similar *U. stygia* G. Thor, described from Europe, having quadrifids that are intermediate between *U. intermedia* and *U. ochroleuca* (the most problematic populations occurring in northern California), but considering the variability observed, it appears more practical to treat the *U. ochroleuca* complex in the broader sense. (*U. occidentalis* A. Gray; *U. stygia* G. Thor)

8. *U. olivacea* C. Wright ex Griseb. Piedmont Bladderwort, Pygmy Bladderwort Fig. 535
Ponds and lakes. Very rare, coastal plain, s. N.J.; Va., s. to N.C., S.C., Ga., Fla., w. to Ala. and Miss.; W.I., C.Am. and S.Am. *U. olivacea* is undoubtedly overlooked, partly because of the diminutive nature of the vegetative plant body and tiny flowers, and partly because flowering tends to occur when the plants are stranded on the wet substrate when water recedes or when the plants become greatly entangled with floating vegetation.

9. *U. gibba* L. Humped Bladderwort Figs. 535, 537
Shallow water of lakes, ponds, pools, swamps, bogs, marshes, sandy shores, ditches and sluggish streams. N.S., N.B., and s. Que. w. to s. Ont., Mich. and Minn., s. to Fla., La., and Tex.; B.C., Wash., Oreg. and Ida., s. to Calif. and Mex.; W.I., C.Am., and S.Am. Plants tend to remain vegetative when in deeper water, chiefly flowering when the water level drops to a few cm deep, or plants become stranded on mud, or becomes tangled in floating vegetation. Plants within this complex having somewhat larger flowers and with spurs relatively longer and more slender than "true" *U. gibba* have traditionally been recognized as *U. biflora* Lam. However, Taylor (1989) observed that in his study of the group many specimens of intermediate flower size could not clearly be placed in one species or the other, and has taken the view that the name *U. biflora* is best treated as a synonym under the variable and wide-ranging (pantropical) *U. gibba*. (*U. biflora* Lam.; *U. fibrosa* Walter; *U. pumila* Walter)

10. *U. geminiscapa* Benj. Hidden-fruit Bladderwort Fig. 534
Quiet waters of lakes, ponds, bogs and sluggish streams. Nfld., N.S. and N.B. w. to e. Ont., Mich., Wisc., and Iowa, s. to N.E., N.J., Pa., Del., Va., n. W.Va., n. Ohio, and n. Ind. (*U. clandestina* Nutt.)

11. *U. vulgaris* L. Common Bladderwort Fig. 536
Quiet waters of lakes, ponds, pools, bogs, swamps and sluggish streams. Widespread in N.Am.; Lab. and Nunavut w. to N.W.T. and Alask., s. to N.E., Va., Fla., Tex., N.M., Ariz., Calif. and Mex. The North American taxon is treated herein as subsp. *macrorhiza* (Laconte) R. T. Clausen. This taxon has been widely treated as distinct at the species level from the European *Utricularia vulgaris*, but the differences, largely associated with the spur, are very minor. Taylor (1989) noted that in the European taxon the spur is 2.5–6(8) mm, shorter than the lower lip, with a broad conical base, and tapering to a narrowly cylindric or narrowly conical, blunt to somewhat acute apex, and typically straight (sometimes somewhat concave or convex), and glands are present only on the internal dorsal surface of the spur. In contrast, subsp. *macrorhiza* has a spur as long as the lower lip, 4–7(9) mm, basally more narrowly conical, with the cylindrical distal portion clearly curved upward, and with an acute apex; internal glands are present on both dorsal and ventral surfaces. Crow (2022) found that in both taxa the internal spur glands are usually not visible on herbarium specimens, and spurs must be dissected in fresh material to

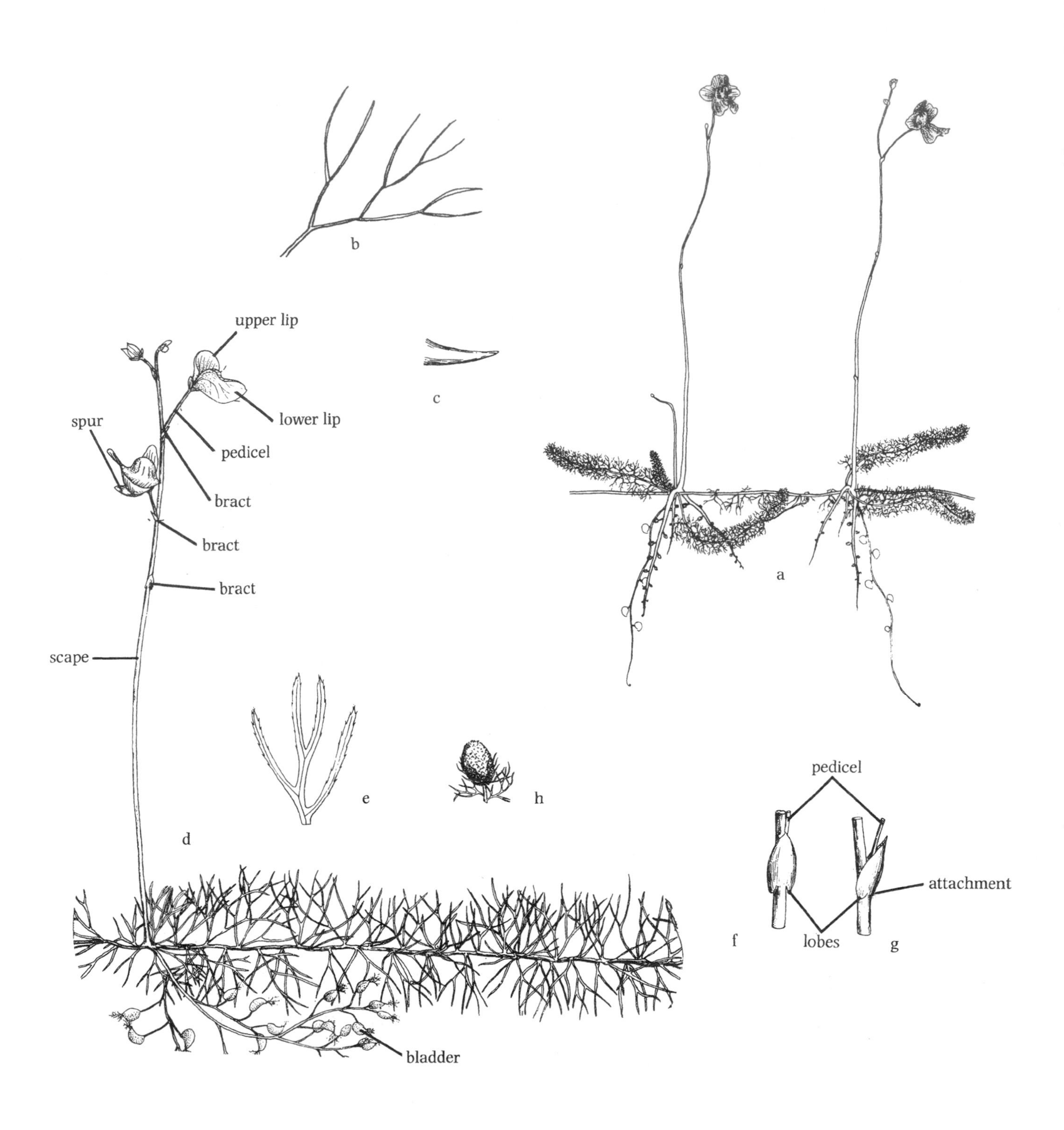

Fig. 532. *Utricularia striata*: a. habit (G&W); b. filiform leaf segments (NHAES); c. spur of corolla (F). *Utricularia intermedia*: d. habit; e. flattened leaf segment; f. bract, view of outer surface; g. bract, side view; h. winter bud (turion) (F).

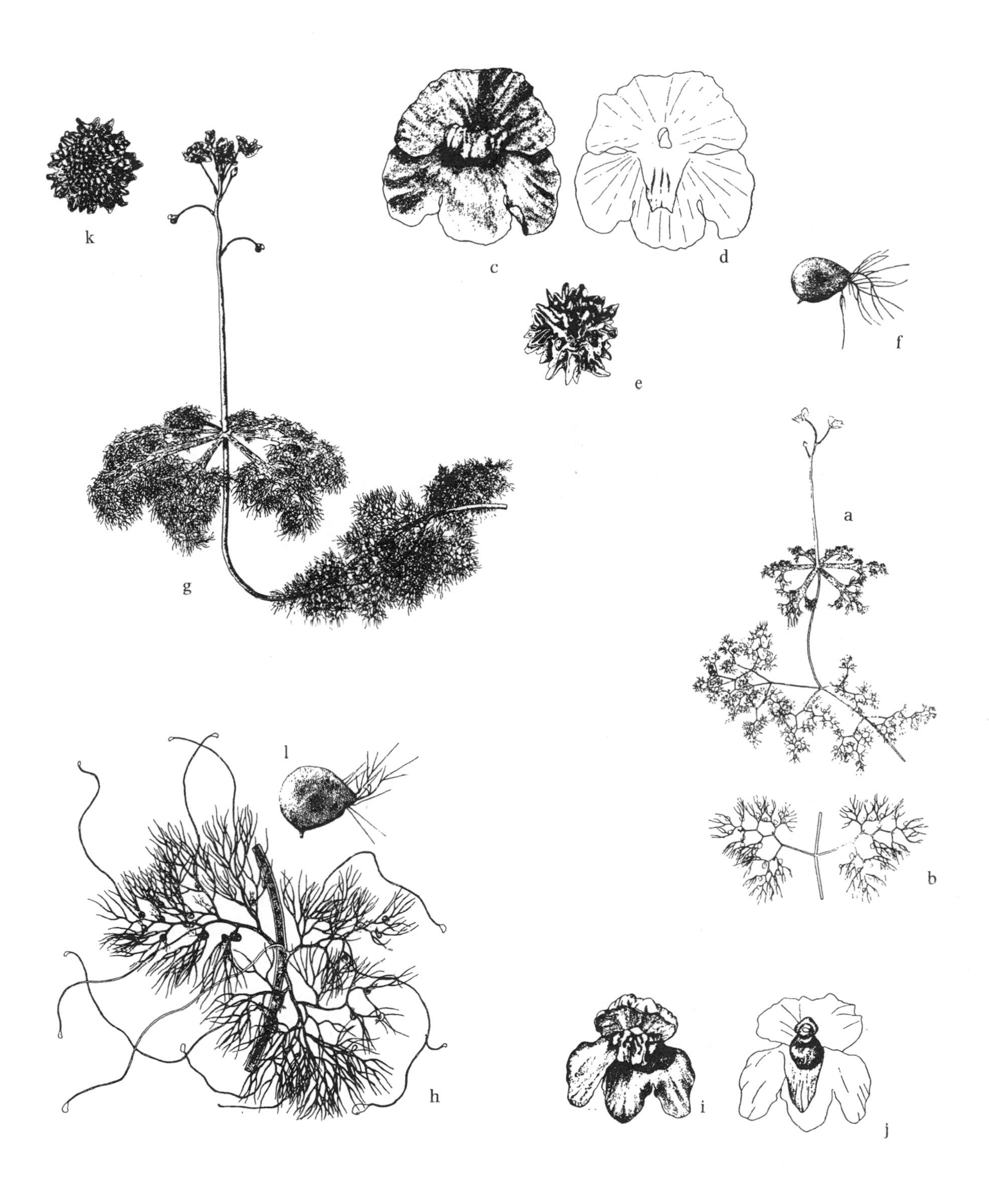

Fig. 533. *Utricularia radiata*: a. habit, portion of plant; b. threadlike leaf segments; c, d. corolla, two views; e. seed; f. bladder (trap) (Reinert and Godfrey, 1962).
Utricularia inflata: g. habit, portion of plant; h. threadlike leaf segments; i, j. corolla, two views; k. seed; bladder (trap) (Reinert and Godfrey, 1962).

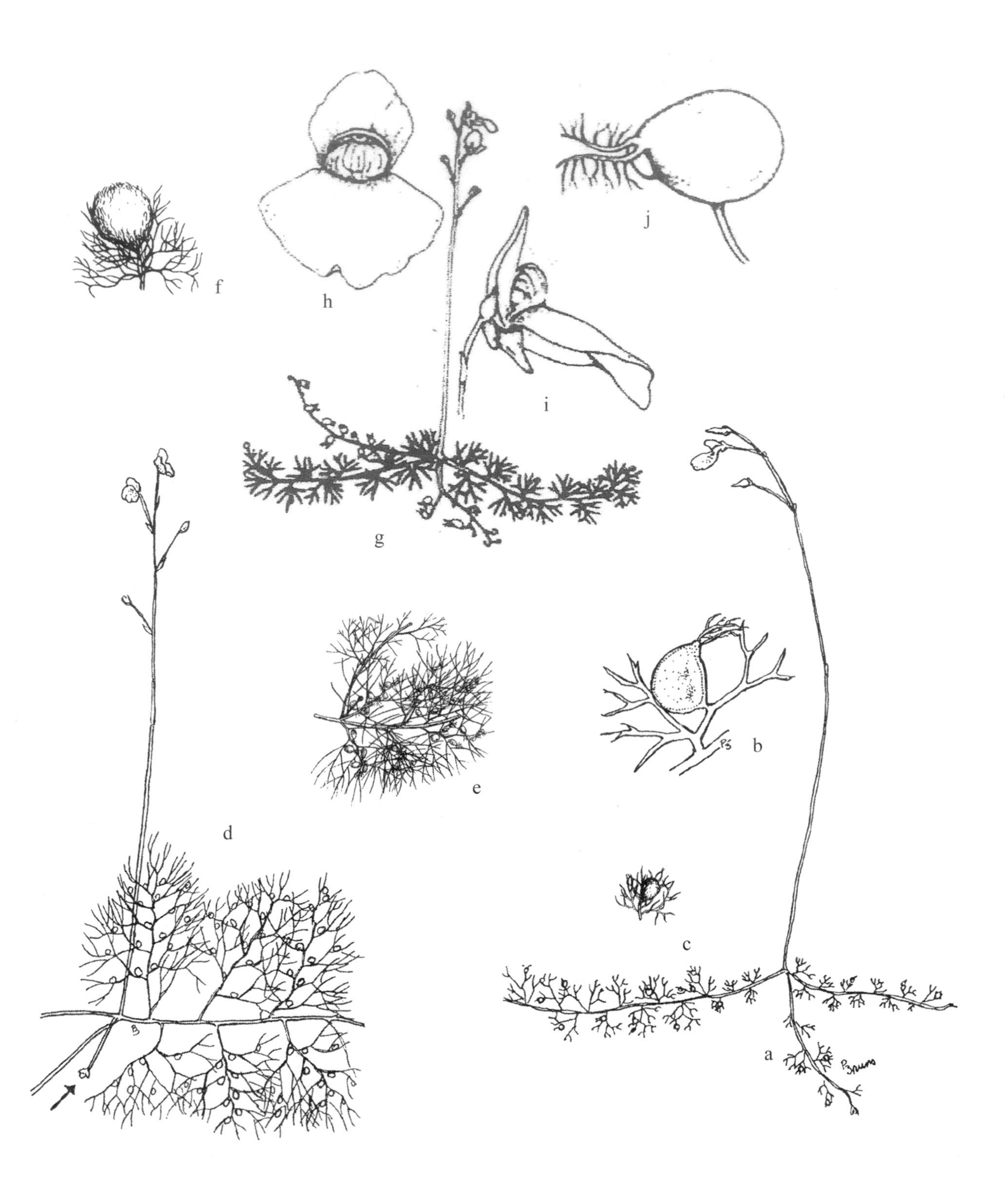

Fig. 534. *Utricularia minor*: a. habit (NHAES); b. bladder trap on flattened leaf segments (NHAES); c. winter bud (turion) (F).
Utricularia geminiscapa: d. habit, section of plant with cleistogamous flower, at arrow (NHAES); e. filiform leaf segments (F); f. winter bud (turion) (F).
Utricularia ochroleuca: g. habit, flattened leaf segments; h. flower, front view; i. lateral view showing spur at ca. 45° angle; j. bladder (trap) (FNA).

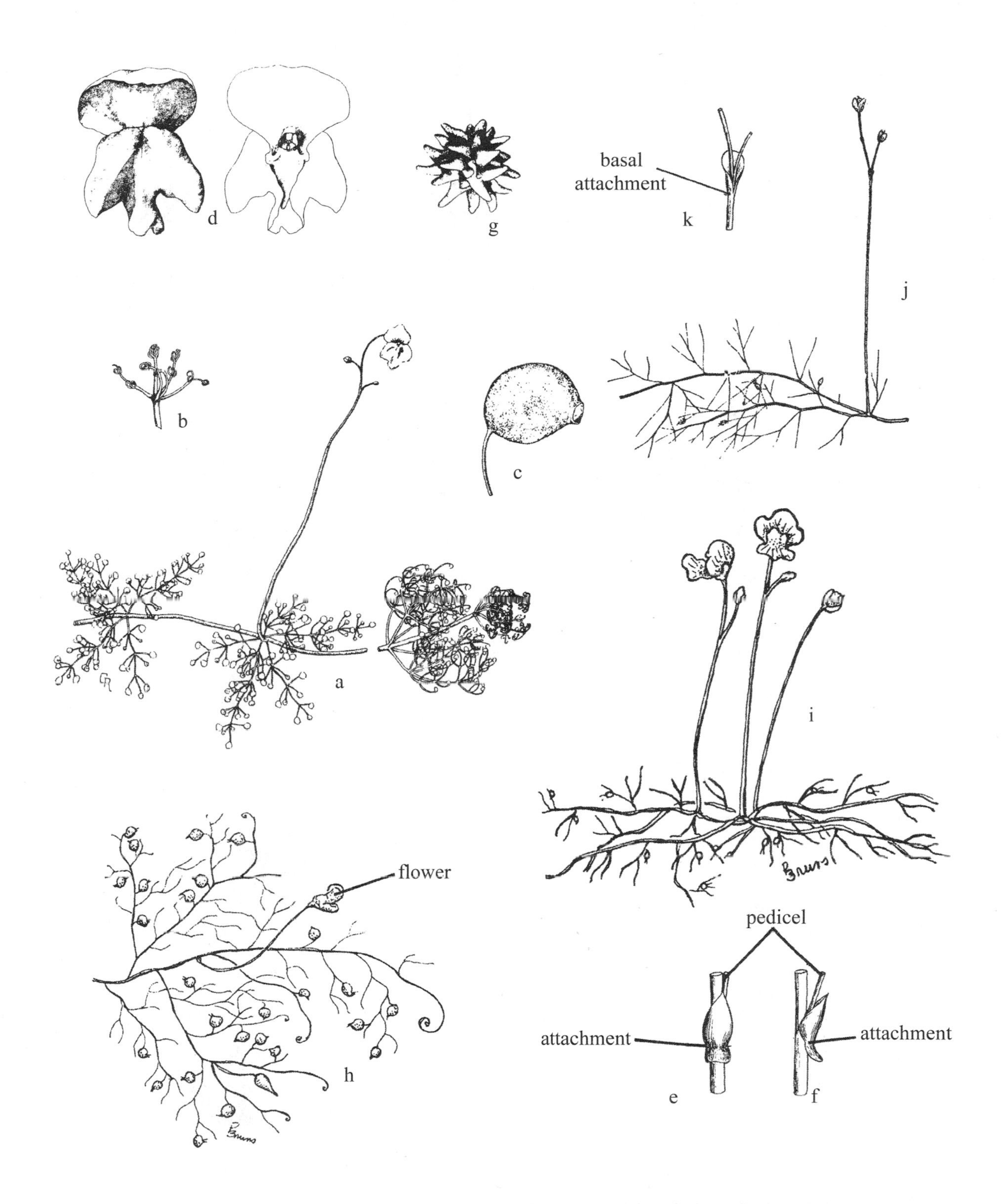

Fig. 535. *Utricularia purpurea*: a. habit, whorled leaf segments (G&W); b. winter bud (loosely formed) (F); c. bladder (trap) (G&W); d. corolla, two views (G&W); e. bract, view of outer surface (F); f. bract, side view (F); g. seed. *Utricularia olivacea*: h. habit, portion of plant (PB). *Utricularia gibba*: i. habit (NHAES, as *U. biflora*); j. habit with fruit; k. portion of scape with basally attached bract (F).

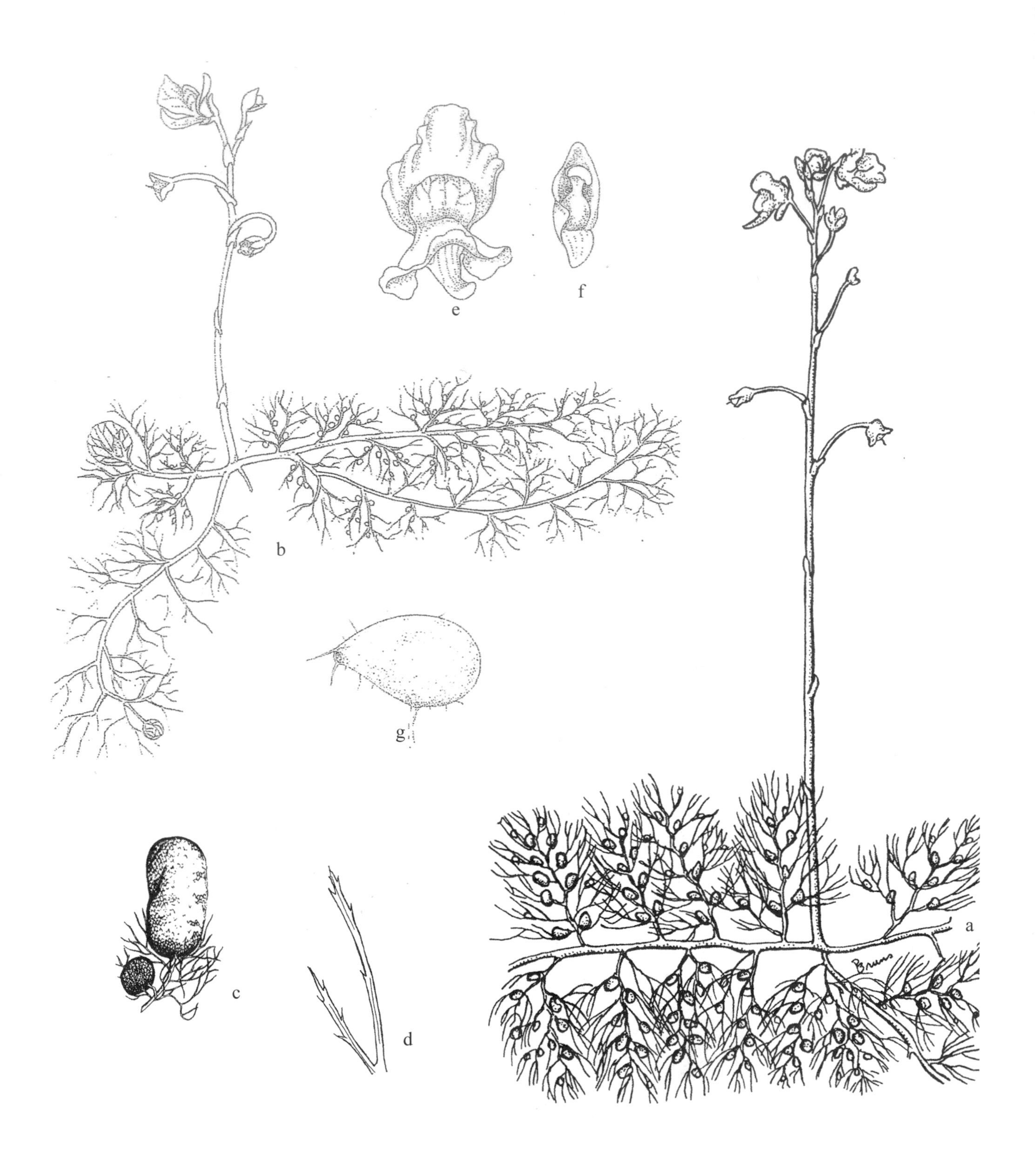

Fig. 536. *Utricularia vulgaris* subsp. *macrorhiza*: a. habit, portion of plant with bladders (NHAES); b. habit with two winter buds at branch tips (FNA); c. winter bud (turion) (F); d. leaf segment with setulose margins (NHAES); e. flower (FNA); f. young fruit enclosed by persisting calyx (FNA); g. bladder (trap) (FNA).

Fig. 537. *Utricularia gibba*: a. habit (G&W); b. rhizoid (G&W); c. flower spur (F); d. flower, front view (G&W); e. flower, diagrammatic view of back showing spur (G&W); f. capsule with persisting calyx; g, h. seed, two views (G&W).

be seen. When fruiting, both subspecies have recurved pedicels, while the sepals are somewhat to strongly divergent. The European subsp. *vulgaris* is vegetatively less robust, but Taylor acknowledged that the two cannot be distinguished vegetatively. While subsp. *macrorhiza* has been known to extend into northeast Asia, Taylor noted that it was not clear to what extent either subsp. *macrorhiza* or subsp. *vulgaris* occurred in western Siberia. However, Crow (2022) reported that during field work in the region west of Novosibirsk and in the Altai Mountains, it was found that the Siberian material was indistinguishable from our American taxon, and more recent reexamination of that Siberian material by him confirmed that those specimens belong to subsp. *macrorhiza*. Hence, subsp. *macrorhiza* appears to be geographically more widely distributed than subsp. *vulgaris*. (*U. vulgaris* var. *americana* Gray; *U. macrorhiza* Leconte)

12. *U. resupinata* B. D. Greene ex Bigelow
Lavender Bladderwort Fig. 538
Margins of pools, ponds, and lakes, and wet ditches. N.S. and Que. w. to Mich., n. Wisc., ne. Minn., and w. Ont., s. to N.E., N.Y., nw. Pa., and n. Ind.; coastal plain, N.J., Del., and e. Va., s. to Fla. and s. Ala.; mts. of Tenn.

13. *U. subulata* L. Zigzag Bladderwort Fig. 539
Shallow water, bogs, wet peats and sandy margins of ponds, wet pine savannas, flatwoods, cypress pond borders and peaty ditches. Chiefly coastal plain, sw. N.S., se. Mass. and R.I., s. to e. Pa., Va., S.C., Fla., w. to Tex.; inland n. to Tenn., Ark., e. Okla., Mo., nw. Ind. and sw. Mich.; W.I., C.Am., and S.Am.

14. *U. cornuta* Michx. Horned Bladderwort Fig. 540
Bogs, wet peaty and sandy shores, marly wet pannes and boggy ditches. Lab., Nfld. and Que. w. to Ont., Sask., and ne. Alta., s. to Pa., Ohio, n. Ind., ne. Ill., Wisc. and n. Minn.; scattered localities, especially coastal plain, N.J. s. to Fla., w. to La., Ark. and e. Tex.; Tenn. and Ky.; Bahamas and Cuba.

15. *U. juncea* Vahl Southern Bladderwort Fig. 541
Shallow water, bogs, sandy-peaty pond margins and shores, wet pine savannas, flatwoods, seepages and peaty ditches. Coastal plain, se. N.Y., and s. N.J. s. to Fla., w. to Miss., La., e. Okla. and e. Tex.; e. Mex., W.I., C.Am. and n. S.Am. (*U. virgatula* Barnh.)

2. *Pinguicula* (Butterwort)

Perennial herbs; leaves in basal rosettes, broad, entire, margins entire, strongly involute (often increasingly in-rolled toward apex), sticky, insectivorous, with both short-stalked and sessile glands; flowers solitary, borne on scapes; calyx 2-lipped, corolla violet (ours), yellow, white, to light blue, blue, light lavender or violet, united, bilaterally symmetrical, 2-lipped, spurred; fruit a capsule.

References: Cieslak et al., 2005; Crow, 2022, in press; Kondo and Shimai, 2006.

1. Rosettes 2.3–9 cm across; corolla not purple-veined (except within corolla tube on white blotch), lips lobed; scapes never villous, with short glandular hairs entire length or sparsely glandular above, glabrous below, or sometimes completely glabrous; corollas (10)14–36 mm (including spur); spur slender, (1)3–9(11) mm. . . . . 1. *P. vulgaris*
1. Rosettes 0.8–2(3) cm across; corolla purple-veined from tube into spur, lips not lobed; scapes usually white-villous proximally, densely glandular-pubescent on upper portion; corollas 6–10 mm (including spur); spur conical, (1.5)2.5–5(6) mm. . . . . 2. *P. villosa*

1. *P. vulgaris* L. Fig. 542
Wet rocks, bogs, calcareous wet sandy shores, seeps and meadows; chiefly calcareous sites. Greenl. and Lab. w. to Alask., s. to Nfld., n. N.E., N.Y., n. Mich., n. Wisc., ne. Minn., Man., Sask., nw. Mont., Wash., Oreg. and n. Calif.; Japan and Russian Far East. Plants often form clusters by producing plantlets from short stolons arising from stems below the rosettes. In western North America, *P. vulgaris* tends to be distributed less commonly in coastal areas, whereas *P. macroceras* tends to occur at mid-elevations in the mountains and along coasts, ranging from northern California to the Aleutian Islands, and extending to the Russian Far East and northern Japan. Larger-flowered plants of *P. vulgaris* appearing more like *P. macroceras* occasionally occur in areas such as Greenland, Newfoundland and southern Labrador, along James Bay, the Northwest Territories, and Yukon. The two species appear to overlap in the Canadian Rockies and western Montana where *P. macroceras* tends to have lower corolla lobes that may be somewhat obovate but that often do not overlap. Some intermediate specimens occur in eastern Oregon and eastern Washington. Frequently specimens occur that are difficult to clearly place in one taxon or the other.

2. *P. villosa* L. Hairy butterwort Fig. 542
Sphagnum bogs, muskegs, hummocks along streams and pools, arctic/timberline and turfy alpine tundra. Nfld. w. to Que., Nunavut, N.W.T., Sask., Yuk. and Alask., s. to Ont., Sask., Alta. and B.C.; Eurasia. Plants small, often hidden in deep sphagnum moss; over-wintering by small hibernacula. Populations of *P. villosa* with white corollas occasionally occur.

Verbenaceae / Vervain Family

1. Flowers borne in elongate spikes terminating branches; corolla 5-lobed, only slightly irregular; calyx 5-lobed; fruit splitting into 4 nutlets. . . . . 1. *Verbena*
1. Flowers borne in short, dense, many-flowered capitate spikes on long axillary peduncles; corolla 4-lobed, 2-lipped (the upper lobe notched, the lower 3-lobed); calyx 2–4-toothed or lobed; fruit splitting into 2 nutlets. . . . . 2. *Phyla*

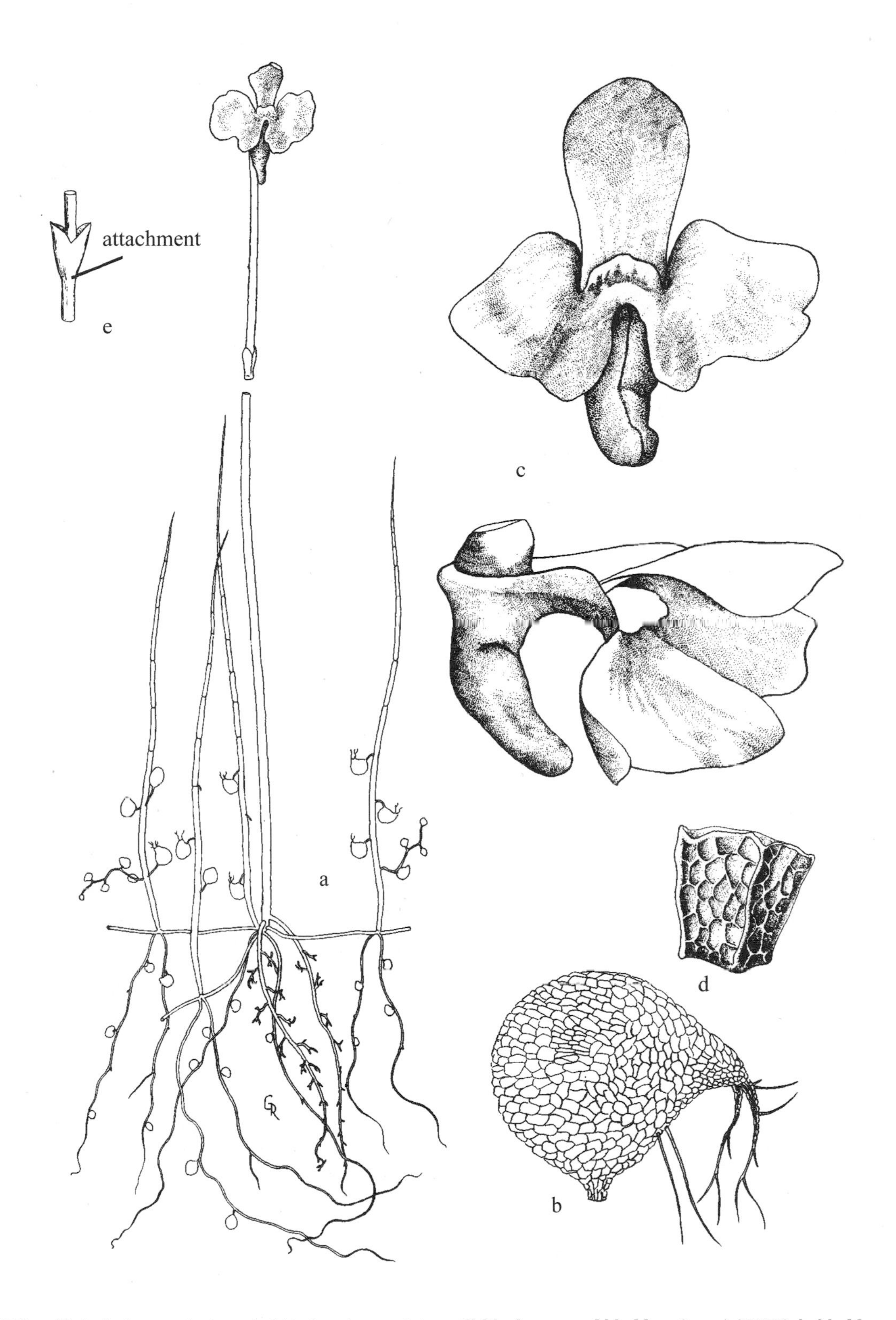

Fig. 538. *Utricularia resupinata*: a. habit, showing septate quill-like leaves and bladders (traps) (G&W); b. bladder (trap) (G&W); c. flower, two views (G&W); d. seed, pitted-reticulate (G&W); e. pair of fused bracts with basal attachment (F).

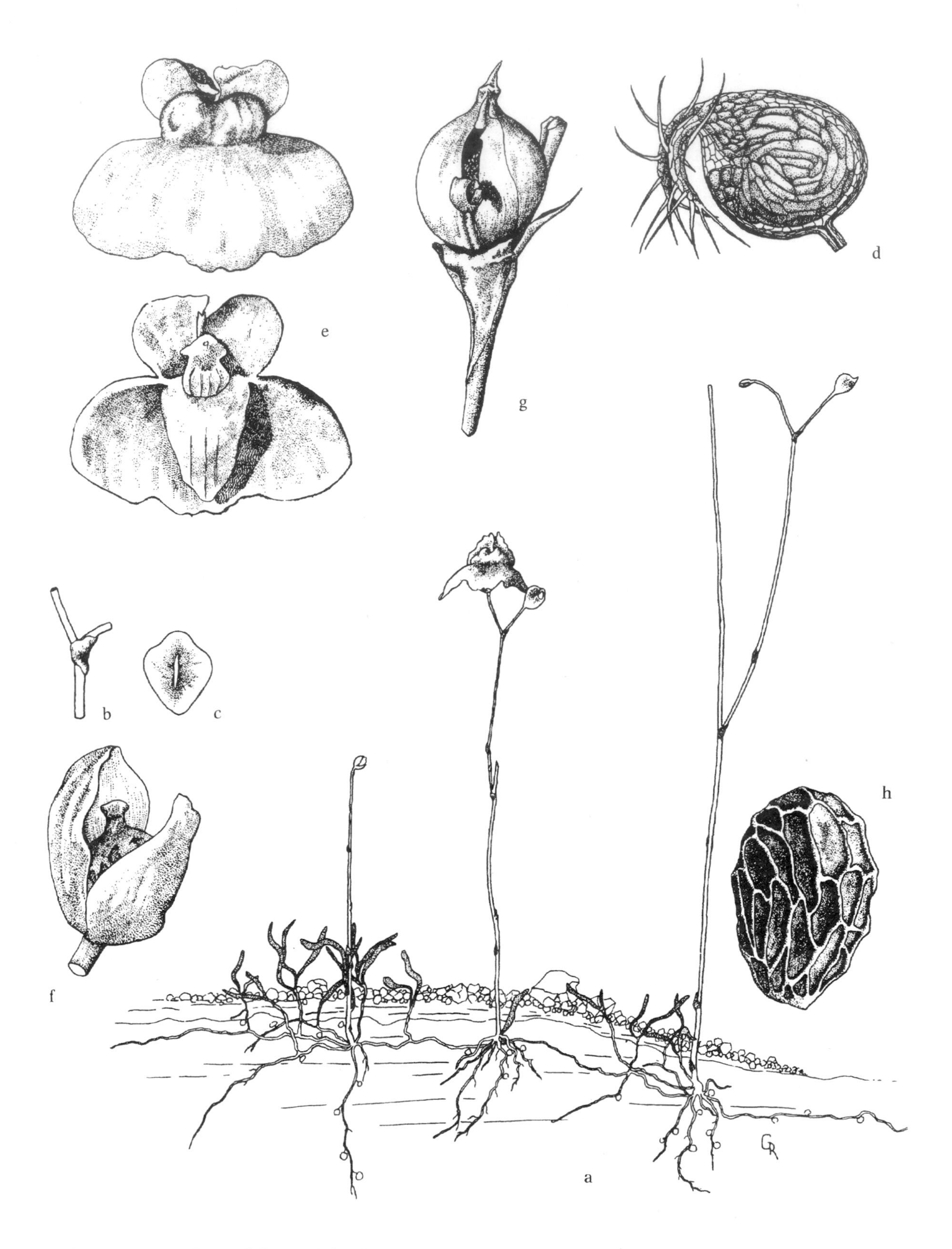

Fig. 539. *Utricularia subulata*: a. habit; b. bract, peltate attachment, side view; c. bract, view of outer surface, showing peltate attachment; d. bladder; e. corolla, two views; f. capsule with calyx; g. capsule, split open; h. seed, pitted-reticulate (G&W).

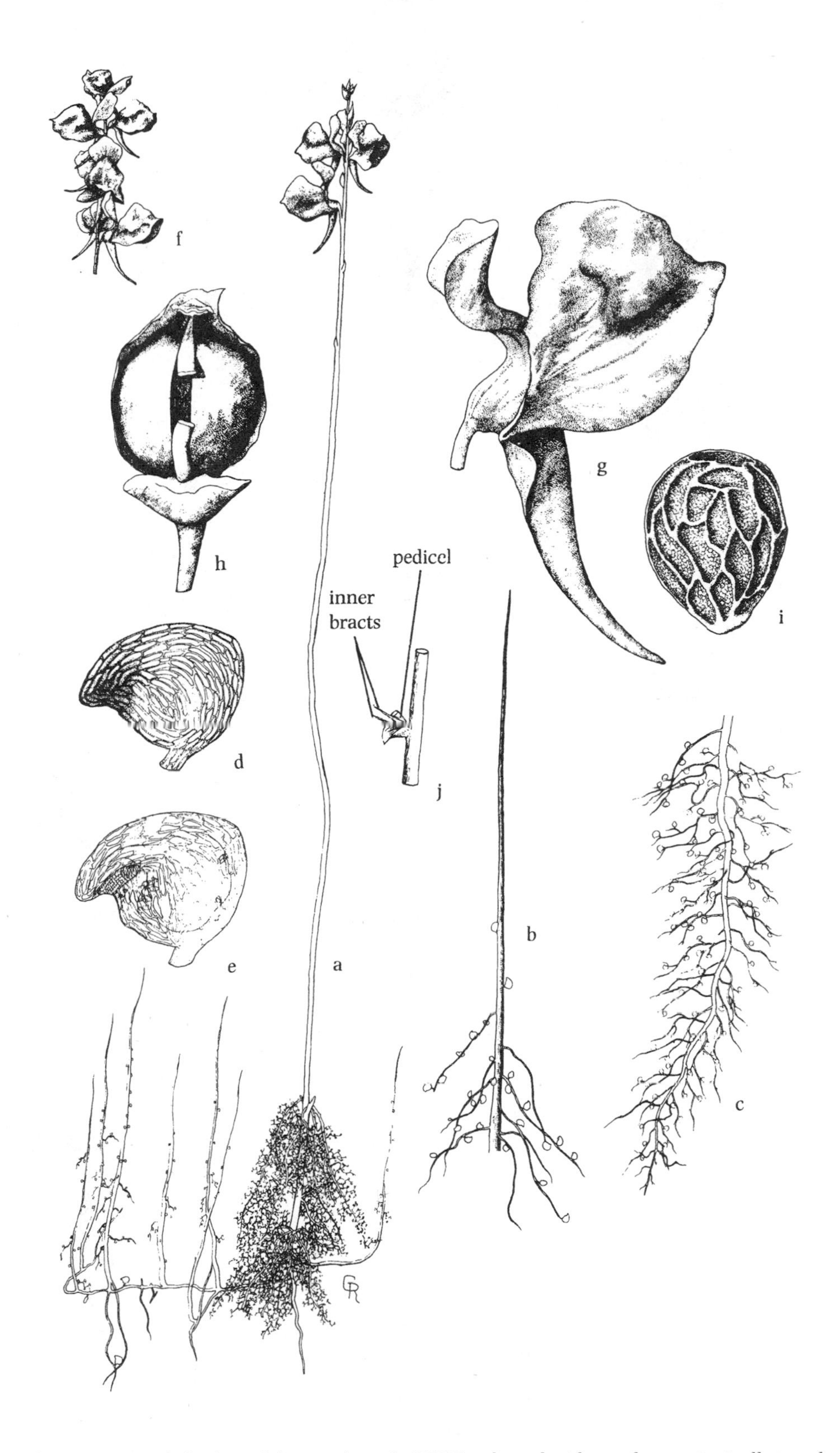

Fig. 540. *Utricularia cornuta*: a. habit (G&W); b. erect branch (G&W); c. branch at base of scape, typically in substrate (G&W); d, e. bladder variations (G&W); f. inflorescence (G&W); g. flower (G&W); h. capsule, partly dehisced (G&W); i. seed (G&W); j. bract (F).

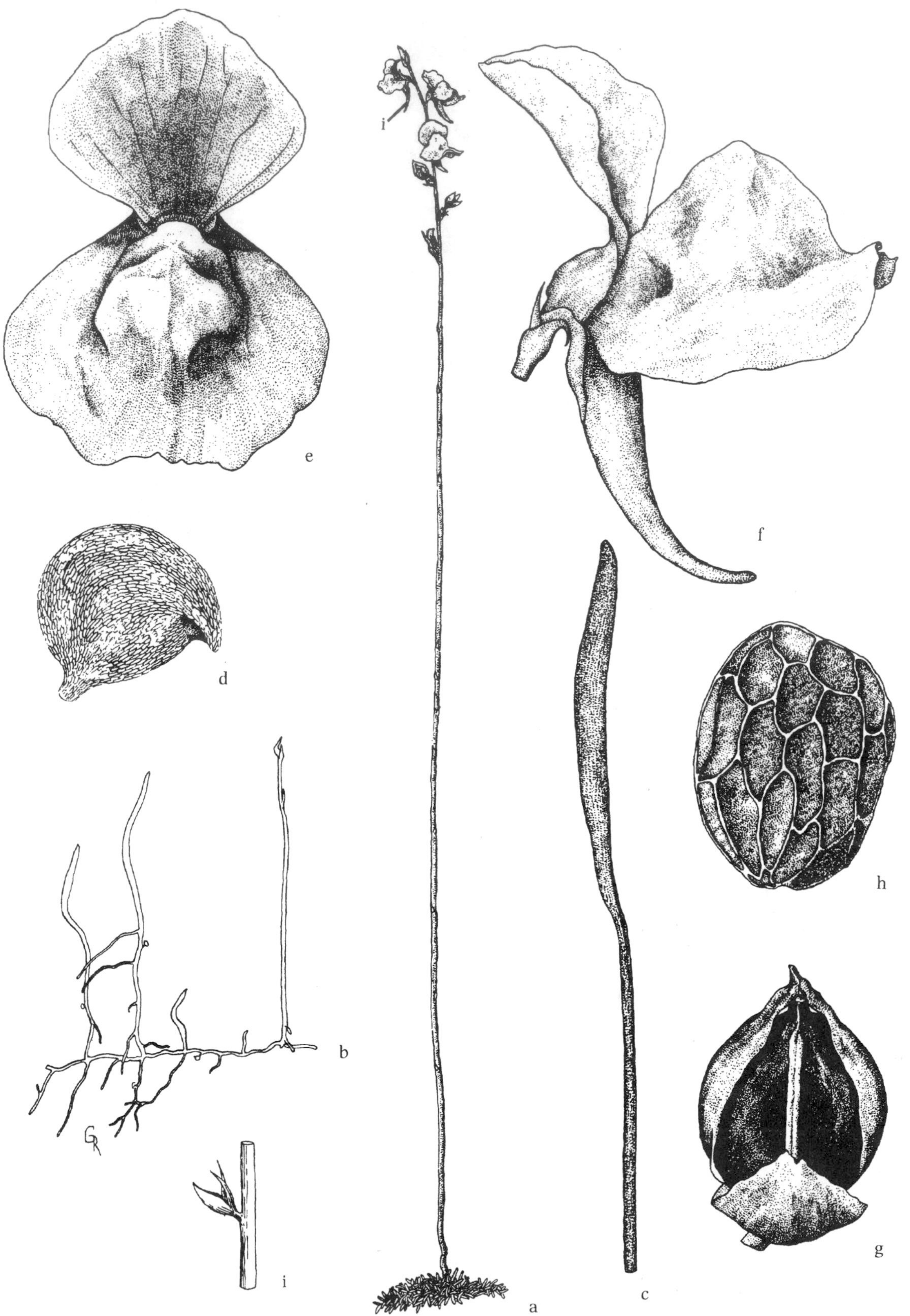

Fig. 541. *Utricularia juncea*: a. habit with inflorescence on long wiry scape, vegetative portion in substrate (G&W); b. portion of vegetative plant (G&W); c. phyllode-like leaf, much enlarged, usually present only under very favorable conditions (G&W); d. bladder (G&W); e, f. flower, two views (G&W); g. capsule (sepal pulled back) (G&W); h. seed (G&W); i. scape with bract and two bracteoles (F).

Fig. 542. *Pinguicula vulgaris*: a. habit (USFS).
Pinguicula villosa: b. habit; flower (Ill. Fl. B.C.).

1. *Verbena* (Vervain)

Annual or perennial (ours) herbs; stems square; leaves simple, opposite, scabrous (ours); flowers in loose to congested terminal spikes; calyx 5-lobed; corolla 5-lobed, appearing only slightly irregular; fruit a schizocarp enclosed by persistent calyx, splitting into 4 nutlets.

1. Spikes thick, fruits overlapping. . . . . 1. *V. hastata*
1. Spikes slender to filiform, fruit not overlapping. . . . . 2. *V. scabra*

1. *V. hastata* L. Blue Vervain, Swamp Vervain Fig. 543
Samps, thickets, shores, meadows and wet woodlands. N.S. w. to s. B.C., s. throughout U.S.

2. *V. scabra* Vahl. Fig. 543
Swamps, shores, wet woodlands and wet thickets. Coastal plain, se. Va. s. to Fla., w. to Ala., La., Tex., s. N.M., s. Ariz. and s. Calif.

2. *Phyla* (Fog-fruit, Frog-fruit)

Perennial herbs; stems square, creeping or procumbent; leaves simple, opposite; flowers small, borne in dense, many-flowered capitate spikes on axillary peduncles; corolla pink, blue, or white, united, weakly 2-lipped, the upper lip notched, the lower 3-lobed; fruit a schizocarp, splitting into 2 nutlets.

1. Leaves tapering to acuminate apex, margins with (5)7–11 teeth per side. . . . . 1. *P. lanceolata*
1. Leaves blunt at apex, margins with (1)3–4(7) teeth per side. . . . . 2. *P. nodiflora*

1. *P. lanceolata* (Michx.) Greene Northern Frog-fruit, Lance-leaf Fog-fruit Fig. 544
Brackish or freshwater marshes, sloughs, shores, wet woodlands and ditches. S. Ont. w. to s. Mich., Minn. and S.D., s. Ga., nw. Fla., Ala., La. and Tex.; Colo., N.M., Ariz., Nev., Calif. and n. Mex. (*P. lanceolata* var. *recognita* (Fernald & Griscom) Soper; *Lippia lanceolata* Michx.)

2. *P. nodiflora* (L.) Michx. Common Frog-fruit, Turkey Tangle Fog-fruit Fig. 545
Low ground, wet sands, mud flats, beaches, depressions and ditches. Chiefly coastal plain; N.J., Pa. and se. Va. s. to Fla., w. to Tex.; Mississippi embayment n. to c. Ky., se. Mo., Ark., and Okla.; Colo., sw. Utah, se. Nev., N.M., Ariz. and Calif.; Trop. Am. (*Lippia nodiflora* (L.) Michx.)

Lamiaceae (Labiatae) / Mint Family

The mints constitute a large family readily recognized by the following characteristics: stems square; leaves opposite; flowers bilaterally symmetrical, corolla 2-lipped, united in a corolla tube; stamens 2 or 4; ovary deeply 4-lobed; fruit separating into four 1-seeded nutlets. The members of this family are often difficult to identify.

1. Calyx tube with a protuberance on upper side. . . . . 1. *Scutellaria*
1. Calyx tube lacking a protuberance.
    2. Corolla appearing 1-lipped, with 5 lobes (one much larger than the others). . . . . 2. *Teucrium*
    2. Corolla distinctly 2-lipped or weakly 2-lipped (almost regular in *Mentha*).
        3. Flowers in terminal inflorescences.
            4. Flowers 1 per bract of inflorescence 1 or 2 per node, not whorled. . . . . 3. *Physostegia*
            4. Flowers 2-many per bract of inflorescence (sometimes leaf-like) whorls with 4 or more flowers.
                5. Corolla strongly 2-lipped; stamens with elongate filaments ascending (somewhat hidden) under upper corolla lip; plants lacking strong minty odor. . . . . 4. *Stachys*
                5. Corolla weakly bilateral, 1 upper lip with lower lip 3-lobed); stamens with short filaments, not exerted from corolla; plants with strong minty odor. . . . . 5. *Mentha*
        3. Flowers axillary (usually dense clusters).
            6. Flowers sessile; stamens 2, staminodes 2; plants lacking minty odor. . . . . 6. *Lycopus*
            6. Flowers pedicellate: stamens 4, staminodes absent; plants with minty odor.
                7. Flowers many per node: leaf margins conspicuously toothed. . . . . 5. *Mentha*
                7. Flowers few, 2–8 per node; leaf margins entire or with 2–4 small teeth. . . . . 7. *Clinopodium*

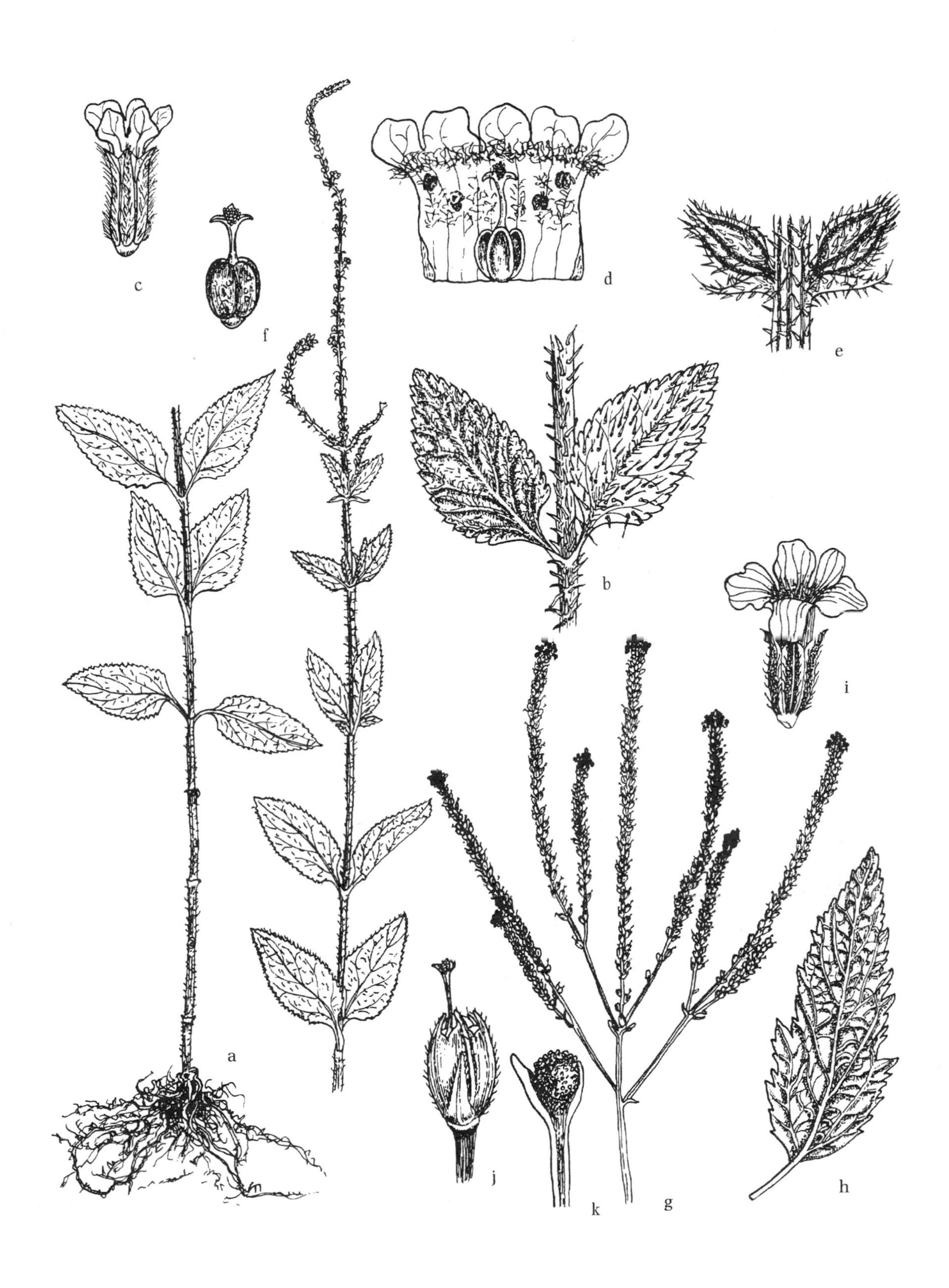

Fig. 543. *Verbena scabra*: a. habit; b. leaves; c. flower; d. flower, spread open; e. fruit enclosed in calyx; f. 4-lobed fruit with style and stigma, before splitting into 4 nutlets (C&C).
Verbena hastata: g. inflorescence; h. leaf; i. flower; j. fruit enclosed in calyx; k. style and stigma (C&C).

Fig. 544. *Phyla lanceolata*: a. habit (C&C); b. upper portion of plant, narrow-leaved form (F); c. pair of leaves, broad-leaved form (F).

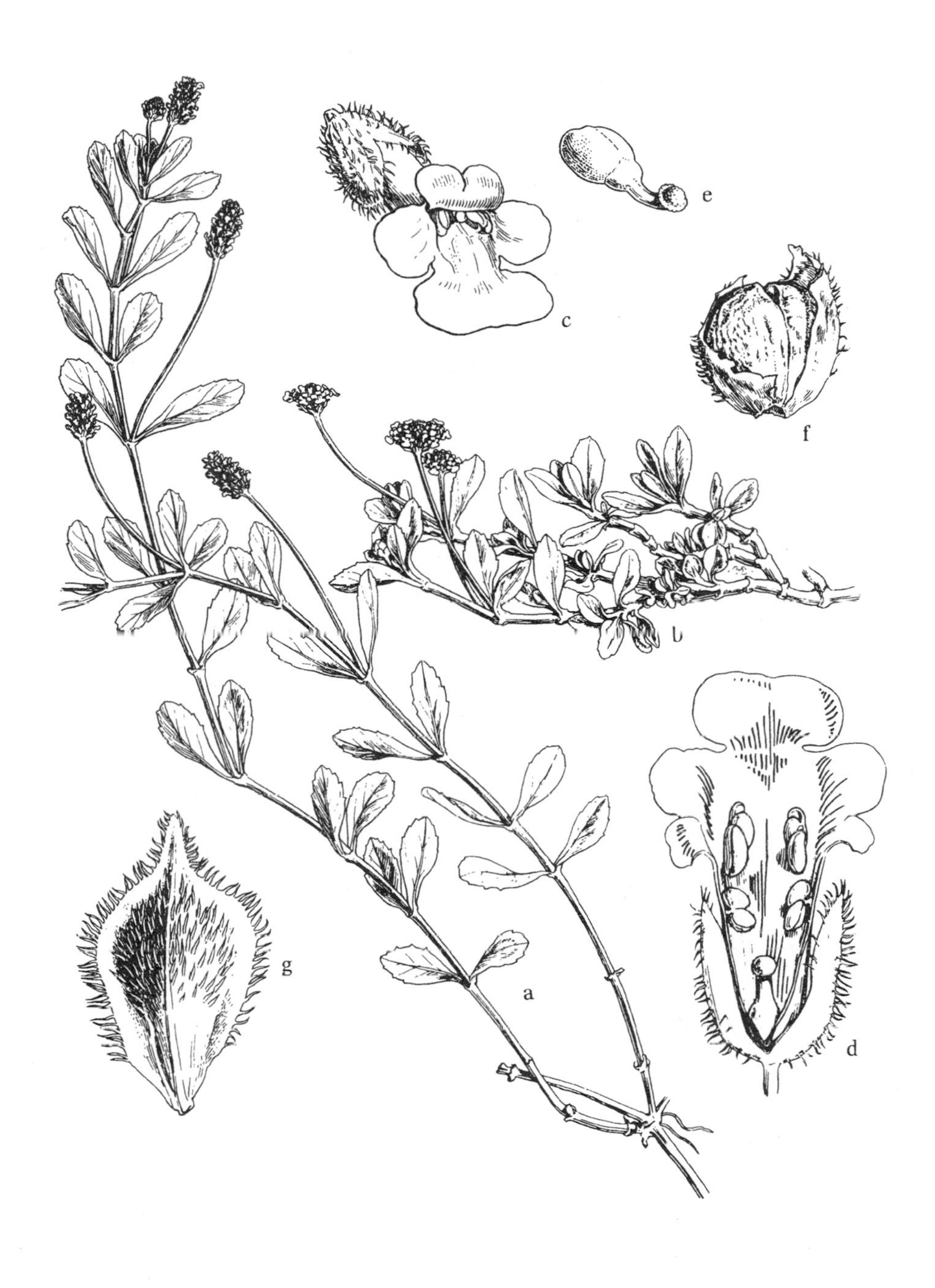

Fig. 545. *Phyla nodiflora*: a. habit, decumbent plant; b. habit, prostrate plant; c. bilabiate flower; d. flower, longitudinal section; e. pistil; f. nutlets with calyx; g. bract (Mason).

1. *Scutellaria* (Skullcap)

Perennial herbs, not aromatic; stems 4-angled; leaves opposite, simple; flowers solitary in leaf axils or borne in axillary racemes; corolla united, 2-lipped, blue (rarely white or pink); calyx 2-lipped, calyx tube with a projection on upper side; ovary deeply 4-lobed; fruit separating into four 1-seeded nutlets.

1. Leaves short-petioled, petiole 1–4 mm long.. .... 1. *S. galericulata*
1. Leaves long-petioled, petiole 3–20 mm long.
    2. Flowers 5–8 mm long, on-sided on terminal and axillary racemes. .... 2. *S. lateriflora*
    2. Flowers 10–15 mm long, solitary, axillary. .... 3. *S.* ×*churchilliana*

1. *S. galericulata* L. Common Skullcap, Marsh Skullcap Fig. 546
Shores, meadows, swamps and thickets. Nfld. w. to Alask., s. to Mo., Kans., Tex. N.M. and Calif. (*S. epilobiifolia* A. Hamilton; *S. galericulata* var. *epilobiifolia* (A. Hamilton) Jordal)

2. *S. lateriflora* L. Blue Skullcap Fig. 546
Swamps, meadows, swampy woods, shore and thickets. Nfld. w. to s. B.C., s. throughout the U.S. to n. Fla. and s. Calif.

3. *S.* ×*churchilliana* Fernald
Shores and thickets. N.S., Que., N.B., Me., Vt. and Mich. This species is considered a hybrid of *S. galericulata* × *S. lateriflora*, having characteristics intermediate between the two species; its occurrence within its range is sporadic.

2. *Teucrium* (Germander)

Perennial herbs, arising from rhizomes; stems 4-angled; leaves opposite simple, serrate, upper surfaces pubescent; flowers borne in a narrow terminal raceme; corolla lavender, pink, or creamy, united, appearing 1-lipped, with 4 upper lobes nearly equal, lower lobe much larger; ovary deeply 4-lobed; fruit separating into four 1-seeded nutlets.

1. *T. canadense* L. Canada Germander Fig. 547
Shores, swamps wet woodland, prairies, meadows and thickets. Nfld., P.E.I., N.S. and s. Que. w. to Minn. and B.C., s. to Fla., Tex., N.M., Calif. and Mex.; Cuba.

3. *Physostegia* (False Dragonhead, Obedient Plant)

Perennial herbs; stems 4-angled; leaves opposite, sessile or lower leaves petiolate, simple, serrate, dentate, or wavy; flowers showy, borne in a spike-like inflorescence; corolla pink, rose, or rose-lavender, extending beyond calyx, united, 2-lipped, upper lip erect, nearly entire, lower lip 3-parted; ovary deeply 4-lobed; fruit separating into four 1-seeded triangular nutlets.

References: Cantino, 1981, 1982.

1. Leaves clasping stem (at least some).
    2. Larger stem leaves usually acute to tapering at apex; leaves serrate; raceme axis densely pubescent. .... 1. *P. angustifolia*
    2. Larger stem leaves obtuse at apex; leaves entire, wavy, or bluntly toothed; raceme axis minutely puberulent.
        3. Calyx tube (1)2–4 mm long; flowers usually 9–19 mm long; uppermost stem leaves usually widest at base, clasping stem; base of stem often conspicuously swollen. .... 2. *P. intermedia*
        3. Calyx tube 3.5–6 mm long; flowers 14–30 mm long, often longer than 20 mm; uppermost stem leaves usually widest or below middle of blade (but not at base), not clasping stem; base of stem seldom conspicuously swollen. .... 3. *P. leptophylla*
1. Leaves sessile or petiolate, not clasping stem.
    4. Larger leaves usually sharply serrate.
        5. Raceme axis densely pubescent. .... 1. *P. angustifolia*
        5. Raceme axis minutely puberulent. .... 4. *P. virginiana*
    4. Larger leaves obscurely crenate to shallowly dentate or bluntly serrate.. .... 3. *P. leptophylla*

1. *P. angustifolia* Fernald Narrowleaf False Dragonhead Fig. 548
Swales, thickets and ditches. Ill., Mo. and e. Kans. s. to sw. Ga., Ala. and c. Tex. This species typically has stiff, sessile, clasping leaves.

2. *P. intermedia* (Nutt.) Engelm. & A. Gray Slender False Dragonhead
Swamps, marshes, wet meadows, bottomlands and ditches. Ky, Ill. and se. Mo. w. to se. Okla., s. to La. and e. Tex.

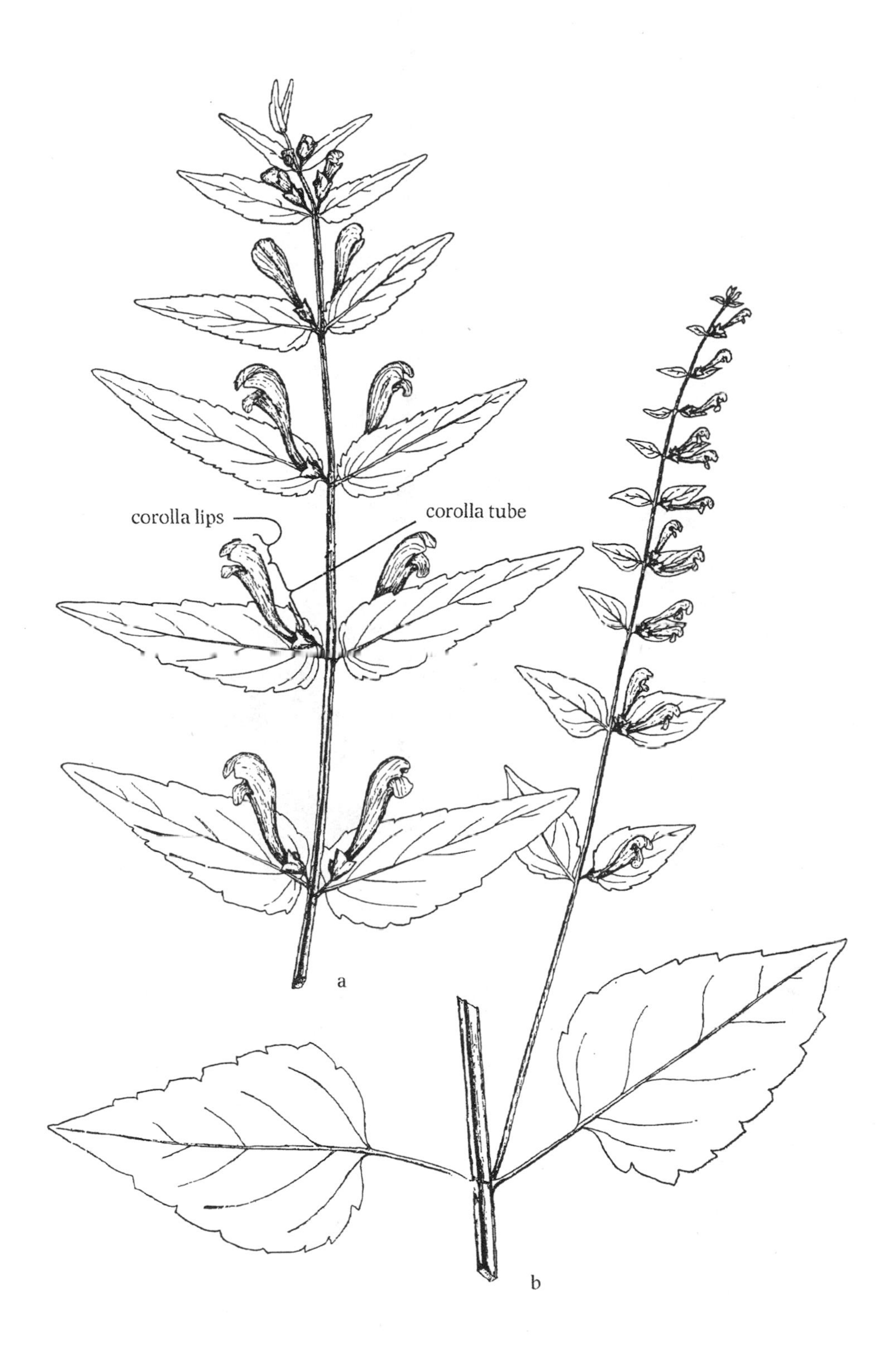

Fig. 546. *Scutellaria galericulata*: a. upper portion of plant (F).
Scutellaria lateriflora: b. leaves and axillary inflorescence (F).

Fig. 547. *Teucrium canadense*: a. upper portion of plant; b. lower portion of plant; c. section of stem with leaf; d. leaf; e. upper portion of inflorescence; f. lower portion of inflorescence; g. flower; h. fruit with calyx, in leaf axils; i. 4 nutlets (matured from 4-lobed ovary) (C&C).

Fig. 548. *Physostegia angustifolia*: habit (Lundell).

3. *P. leptophylla* Small Slenderleaf False Dragonhead Fig. 549
Flooded river swamps, often in deep shade, stream margins, freshwater and brackish marshes and shallow water. Coastal plain, se. Va. s. to Fla. (*P. aboriginorum* Fernald)

4. *P. virginiana* (L.) Benth. Obedient Plant Fig. 550
Riverbanks, thickets, lakeshores, swales and ditches. S. Que. w. to Minn., Man., and e. Sask., s. to Md., N.C., Fla., La., Tex. and se. N.M. This is often grown as an ornamental; garden escapes are frequent, especially in the Northeast.

4. *Stachys* (Hedge-nettle)

Annual or perennial herbs; stems 4-angled; leaves opposite, simple; flowers borne in whorls in whorls in axils of reduced leaves; calyx with bell-shaped tube and 4 or 5 long, deltoid, sharp teeth; corolla purple, united, 2-lipped, upper lip entire, lower 3-part; ovary deeply 4-lobed; fruit separating into four 1-seeded nutlets.

References: Mulligan and Munro, 1989; Nelson, 1981; Nelson and Fairey, 1979.

---

1. Leaves with petioles 10–20 mm long. . . . . 1. *S. tenuifolia*
1. Leaves sessile, or with petioles 5 mm or less long.
   2. Stems glabrous, or if pubescent, only on angles.
      3. Calyx tube and lobes hirsute to hirsute-ciliate (fig. 552f). . . . . 2. *S. hispida*
      3. Calyx tube sparsely villous to glabrous; calyx lobes glabrous.
         4. Leaves entire or with a few low, inconspicuous teeth (fig. 552c); stems glabrous or rarely sparsely pubescent on angles. . . . . 3. *S. hyssopifolia*
         4. Leaves sharply serrate; stems glabrous on sides, bristly on angles. . . . . 4. *S. aspera*
   2. Stems glandular-pubescent to villous or hirsute on sides and angles.
      5. Leaves ovate to oblong-elliptic; calyx teeth half the length of calyx tube. . . . . 5. *S. eplingii*
      5. Leaves lanceolate-narrowly ovate; calyx teeth three-fourths the length of calyx tube up to the same length.
         6. Leaf blades with sides parallel below middle (fig. 552h), usually less than 2.5 cm wide; sides of stems with shorter hairs than those on angles. . . . . 6. *S. palustris*
         6. Leaf blades narrowed below middle (fig. 552j), 2 cm or more wide; sides of stems with bristly, reflexed hairs, like those on angles. . . . . 7. *S. pilosa*

---

1. *S. tenuifolia* Willd. Smooth Hedge-hyssop Fig. 551, 552
Marshes, meadows, swamps, and wet woodlands, S. Que. w. to Minn., s. to S.C., n. Ga., and e. Tex.

2. *S. hispida* Pursh Hairy Hedge-nettle Fig. 552
Marshes, swamps and wet woodlands. N.B., Mass. and Vt. w. to Man., s. to Md., Ky. and Ark. (*S. palustris* var. *hispida* (Pursh) Boivin; *S. tenuifolia* var. *hispida* (Pursh) Fernald)

3. *S. hyssopifolia* Michx. Fig. 552
Sandy, gravelly, or peaty shores and bogs. Coastal plain, Me., N.H. and N.Y. se. Mass. s. to S.C. and Fla., mts. of Va. and N.C., s. Mich. and nw. Ind., Ill., Mo., Ky. and Fla.

4. *S. aspera* Michx. Hyssopleaf Hedge-nettle Fig. 552
Swamps, marshes, damp sands and prairies. N.Y., N.J., se. Pa. and Md. w. to Ohio, Ind., Ill. and Iowa, s. to se. Ga., Fla. Ala., La. and e. Mo.; adv. in c. Me. (S. *ambigua* (A. Gray) Britton; *S. hyssopifolia* var. *ambigua* A. Gray)

5. *S. eplingii* J. B. Nelson Epling's Hedge-nettle
Meadows, bogs and forests. Va., and W.Va., w. to Ky., Ark. and Okla., s. to nw. S.C., Ga. and Ala.

6. *S. palustris* L. Woundwort, Marsh Hedge-nettle Fig. 552
Wet meadows and ditches. Nfld. and Que., w. to s. Ont., Man., B.C., Yuk. and Alask., Minn., s. to N.E., N.Y., N.J., Md., Ohio, Mich., and Ill.; natzd. from Eurasia.

7. *S. pilosa* Nutt. Hairy Hedge-nettle Fig. 552
Shores, meadows and fields. Que. and Ont. w. to Man., N.W.T., Yuk. and Alask., s. to N.C., Ky., Ark., Okla., N.M., Ariz. and Calif. Two varieties occur throughout much of its range: var. *pilosa*, with soft hairs on stem angles, and var. *arenicola* (Britton) G. Mulligan & D. Munro, having stiff, typically pustulose-based hairs (notably thicker than hairs of the stem faces), which tends to be more eastern in distribution (*S. palustris* var. *homotricha* Fernald).

5. *Mentha* (Mint)

Perennial herbs, stems, 4-angled, leaves opposite, simple, punctate-glandular on lower surface, with strong minty odor; flowers small, borne in axillary clusters or spike-like inflorescences, calyx tube 5-lobed lobes subequal; corolla pale lavender or white, united, 2-lipped, upper lip single, entire or notched, lower lip 3-lobed, ovary deeply 4-lobed; fruit separating into four 1-seeded nutlets.

Fig. 549. *Physostegia leptophylla*: a. habit; b. flower; c. fruit with calyx spread open (G&W).

Fig. 550. *Physostegia virginiana*: a, b. upper portion of plants; c, d. 5-lobed calyces; e. flower (F).

Fig. 551. *Stachys tenuifolia*: a. upper portion of plant; b. basal portion of plant; c. flower; d. flower, longitudinal section; e. 4-lobed fruit prior to splitting, hidden by persistent calyx (axillary); f. nutlet (C&C).

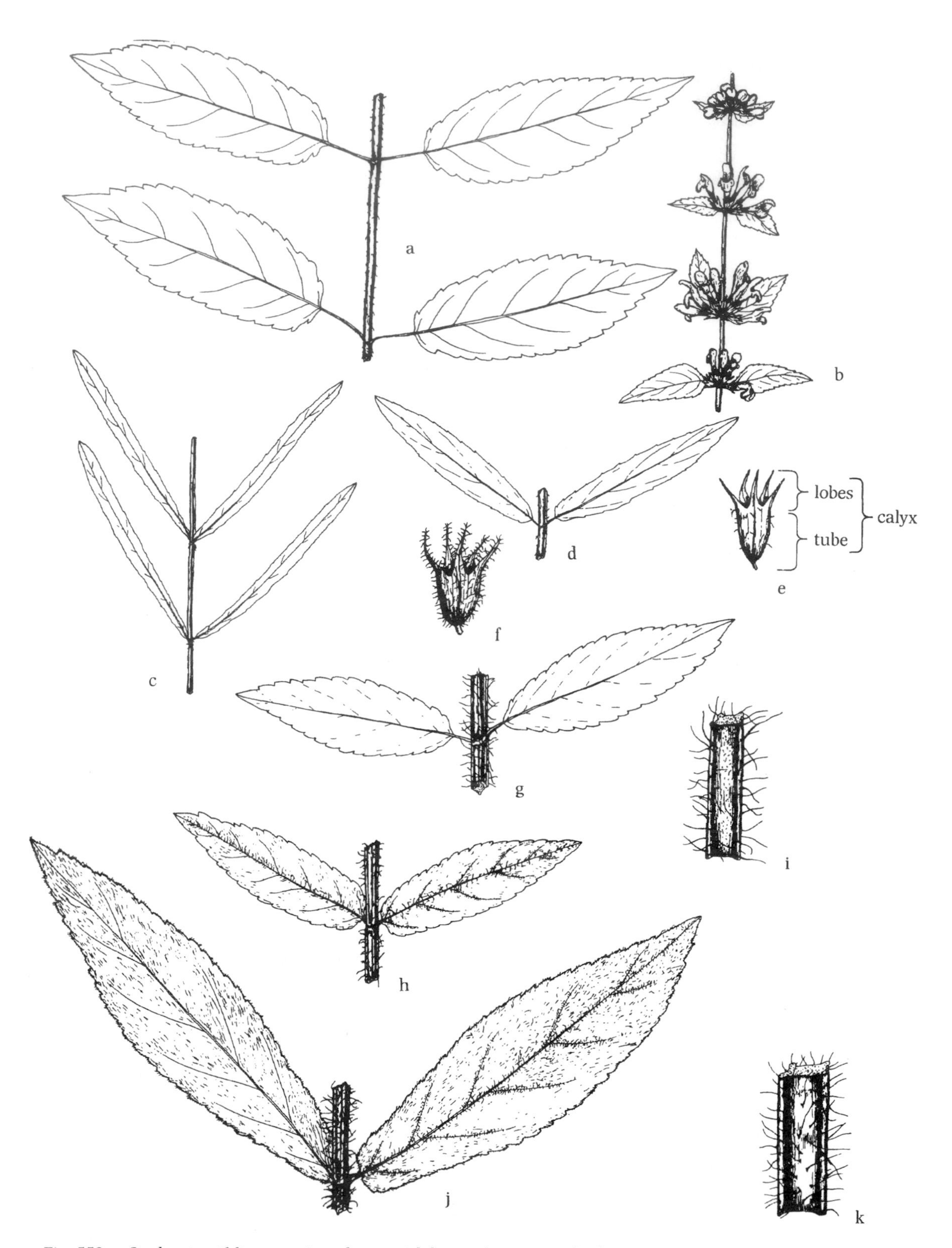

Fig. 552. *Stachys tenuifolia*: a. section of stem with leaves; b. portion of inflorescence (F).
Stachys hyssopifolia: c. section of stem with leaves (F).
Stachys aspera: d. section of stem with leaves; e. calyx (F).
Stachys hispida: f. calyx; g. section of stem with leaves (F).
Stachys palustris: h. section of stem with leaves; i. section of stem (F).
Stachys pilosa: j. section of stem with leaves; k. section of stem (square) (F, as *S. palustris* var. *homotricha*).

1. Flowers clustered in leaf axils, uppermost leaves often lacking flowers.
   2. Upper leaves conspicuously reduced, 2–3 times longer than floral cluster radius. . . . . 1. *M. ×gracilis*
   2. Upper leaves slightly reduced or same size as others, many times longer than floral cluster radius (fig. 553b). . . . . 2. *M. canadensis*
1. Flowers in crowded or interrupted terminal spikes.
   3. Leaves more than twice as long as wide; stem and leaves glabrous, or nearly so. . . . . 3. *M. ×piperita*
   3. Leaves usually less than twice as long as wide; stem and leaves usually pubescent.
      4. Leaves sessile or with short petioles 2–3 mm long; leaf margins regularly toothed and wavy. . . . . 4. *M. spicata*
      4. Leaves with petioles much longer than 3 mm long; leaf margins regularly toothed, not wavy. . . . . 5. *M. aquatica*

1. *M. ×gracilis* Sole Ginger-mint, Scotch Spearmint
Wet meadow, shores and waste places. Nfld. w. to Mich., Ont., Wisc., Minn. and Neb.; s. to Va., S.C., n. Ky Ill., Ark. and Kans.; B.C. and Sask. Regarded as hybrid between *M. arvensis* and *M. spicata*. Intro. from Eur. This plant is cultivated as the flavoring for spearmint gum. (*M. cardiaca* Gerarde ex Baker)

2. *M. canadensis* L. Wild Mint Fig. 553
Damp soil, shores and swamps. Lab., Nfld. and Que. w. to Alask., s. throughout the U.S to Fla., La., Tex., Ariz., Calif. and Mex. (*M. arvensis* subsp. *borealis* (Michx.) Taylor & MacBryde; *M. arvensis* var. *canadensis* (L.) Fernald)

3. *M. ×piperita* L. Peppermint Fig. 554
Brooksides, wet meadows, pond and lakeshores, thickets and ditches. Natzd. from Eur., escaped from cultivation throughout N.Am. This species is considered a hybrid of *M. aquatica* × *M. spicata*.

4. *M. spicata* L. Spearmint Fig. 554
Wet areas, bogs, seeps and pond shores. Natzd. from Eur., especially from cultivation throughout N.Am. (*M. crispa* L.)

5. *M. aquatica* L. Water Mint
Wet areas. Local, N.S. w. to Ont., Mich. and Wisc. s. to Del., Ga., Mo. and Ark.; B.C. s. to Calif. and Utah; Natzd. from Eur.

6. *Lycopus* (Water-horehound, Bugleweed)

Perennial herbs, stoloniferous, some bearing tubers; stems 4-angled; leaves opposite, simple, usually toothed or pinnatifid; flowers small, borne in compact axillary clusters; calyx tube with 4 or 5 lobes; corolla white, united, 2-lipped, 4- or 5-lobed; stamens 2, staminodes 2 or more; ovary deeply 4-lobed; fruit separating into four 1-seeded nutlets.

References: Andrus and Stuckey, 1981; Henderson, 1962; Hermann, 1936; Stuckey, 1969; Stuckey and Phillips, 1970; Webber and Ball, 1980.

1. Calyx lobes acute, but not sharp-tipped, usually shorter than mature nutlets (fig. 555b,c).
   2. Plant base slender, with mostly non-tuberiferous stolons; filaments shorter than corolla tube, anthers included; fruit clusters dense, 8–15 mm wide; calyx lobes mostly obscured by mature nutlets (fig. 555b). . . . . 1. *L. virginicus*
   2. Plant base tuberous, stolons terminated by a tuber; filaments usually longer than corolla tube, anthers exserted (fig. 557b); fruit clusters not dense, 4–9 mm wide; calyx lobes not obscured by nutlets. . . . . 2. *L. uniflorus*
1. Calyx lobes sharp-tipped, much longer than mature nutlets (fig. 555a,d).
   3. Lower and middle leaves petioled; stems not tuberous at base.
      4. Leaves strigose on upper surface; nutlets 1.5–2 mm long, 1–1.3 mm wide. . . . . 3. *L. europaeus*
      4. Leaves glabrous on upper surface; nutlets 1–1.6 mm long, 0.6–1.0 mm wide.
         5. Lower and middle leaves deeply lobed (fig. 558a,c), sharply dentate or incised at base; calyx lobes tipped with a rigid spine (fig. 555a); nutlets smooth-margined at summit.
            6. Stem with blunt angles; leaves usually incised toward base; nutlets with corky ridge; plants of freshwater inland. . . . . 4. *L. americanus*
            6. Stem with thin-winged angles; leaves rarely incised, merely sharp-dentate; nutlets lacking corky ridge; plants of freshwater estuarine shores. . . . . 5. *L. laurentianus*
         5. Lower and middle leaves serrate; calyx lobes acute, not spine-tipped; nutlets wavy at summit (fig. 559b,c). . . . . 6. *L. rubellus*
   3. Lower and middle leaves sessile; stems tuberous at base.
      7. Stem pubescent; leaves scabrous, larger leaves with 6–12 teeth per side; flower clusters subtended by conspicuous bracts (fig. 555d). . . . . 7. *L. asper*
      7. Stem glabrous or minutely puberulent; leaves smooth, larger leaves with 3–7 teeth per side; flower clusters subtended by inconspicuous bracts (fig. 555e). . . . . 8. *L. amplectens*

Fig. 553. *Mentha canadensis*: a. habit; b. inflorescence; c. flower; d. flower, spread open showing 4-lobed ovary and epipetalous stamens; e. calyx, open to show nutlets; f, g. nutlet, two views; h. stem, cross-section (Mason, as *M. arvensis*).

Fig. 554. *Mentha spicata*: a. upper portion of plant; b. leaf apex; c. flower; d. corolla, spread open, stamens epipetalous; e. pistil (C&C).
Mentha ×piperita: f. upper portion of plant; g. flower; h. calyx, spread open; i. corolla, spread open (C&C).

1. *L. virginicus* L. Virginia Water-horehound
Figs. 555, 556
Swamps, wet woodlands, thickets and shores. N.E. w. to Que., Ont. Minn. and s. Mo., s. to Fla. and e. Tex.

2. *L. uniflorus* Michx. Northern Bugleweed Figs. 555, 557
Sandy shores, marshes, bogs and wet woodlands. Nfld. w. to N.W.T., Alask. and B.C., s. to N.C., Ga., n. Ky., Ill., Iowa, Ark. and Okla.; Mont. and Oreg. s. to n. Calif. (*L. virginicus* var. *pauciflorus* Benth.)

3. *L. europaeus* L. Gypsy-wort
Swales, springheads, shores, waste areas and roadsides. N.S., St. Lawrence R., Que., L. Ontario, Ont. and N.Y., L. Erie, Ohio and Mich. and Bruce Pen., L. Huron, Ont.; chiefly coastal and Great Lakes shores, e. Mass. s. to Va. and N.C.; B.C.; intro. from Eur., local, especially near seaports.

4. *L. americanus* Muhl. ex Bart. American Water-horehound
Figs. 555, 558
Wet sandy or peaty swamps, wet woodlands, swales and ditches. Nfld. w. to B.C., s. to n. Fla. and Tex.; more abundant in n. portion of range. (*L. americanus* var. *longii* Benner; *L. americanus* var. *scabrifolius* Fernald)

5. *L. laurentianus* Roll.-Germ. St. Lawrence Water-horehound
Freshwater estuaries. St. Lawrence Seaway, Que. and N.B.

6. *L. rubellus* Moench Taper-leaf water-horehound
Figs. 559, 560
Depressions, marshy shores, wet woodlands and seeps. Me. w. to Ill. and Mo., s. to Fla. and e. Tex.; Oreg. (*L. angustifolius* Elliott)

7. *L. asper* Greene Western Water-horehound, Rough Bugleweed Figs. 555, 562
Marshes, sandy and peaty shores and meadows. Que. and Ont. w. to Mich., Sask., Alta., B.C. and Alask., s. to N.E., N.Y., n. Ohio, n. Ill., Mo., Kans., Tex., N.M., Ariz. and Calif.; natzd in e. N.Am. (*L. lucidus* subsp. *americanus* (A. Gray) Hultén)

8. *L. amplectens* Raf. Clasping Water-horehound
Figs. 555, 561
Marshes, sandy and peaty shores and ditches. Coastal plain; se. Mass. s. to Fla., w. to s. Ala. and Miss.; Ind. (*L. sessilifolius* A. Gray)

7. *Clinopodium* (Wild basil)

Annual or perennial herbs; aromatic, stoloniferous; stems 4-angled; leaves opposite, sessile or petiolate, simple; inflorescences various; corolla purple or white, united distinctly 2-lipped, upper lip erect, lower lip 3-parted, spreading; ovary 4-lobed; stamens 4, fruit separating into four 1-seeded nutlets.

Reference: Cantino and Wagstaff, 1998.

1. *C. arkansanum* (Nutt.) House Limestone Calamint
Fig. 562
Damp calcareous soils, banks, shores, seeps and meadows. Ont. w. to Minn., s. to w. N.Y., Tenn., Ark., Okla., Tex. and N.M. (*Satureja arkansana* (Nutt.) Briq.; *S. glabella* (Michx.) Briq.; *S. glabra* (Nutt.) Fernald; *Calamintha arkansana* (Nutt.) Shinners)

Phrymaceae / Lopseed Family

1. Plants tiny, with leaves paired, directly arising from slender creeping stem, leaf margins entire; calyx 3-lobed. . . . . 1. *Glossostigma*
1. Plants robust, with leaves opposite, arising along and erect or procumbent stem; leaf margins toothed; calyx 5-lobed.
    2. Plants typically erect; corolla blue to light violet or pinkish (rarely white); pedicel of flowers shorter than calyx; leaf venation with secondary veins joining with secondary veins immediately above. . . . . 2. *Mimulus*
    2. Plants typically prostrate, suberect or decumbent; corolla yellow (ours); leaf venation with lower (basal) secondary veins arching upward. . . . . 3. *Erythranthe*

1. *Glossostigma*

Reference: Les and Capers, 2019.

1. *G. cleistanthum* W. R. Barker Fig. 563
Freshwater tidal marshes or flats and river margins and lakes. Mass., R.I., Conn., N.J., e. Pa. and Del. Native to New Zealand and Australia, this tiny plant has recently been introduced in New England, identity confirmed by molecular analysis (Les et al., 2006).

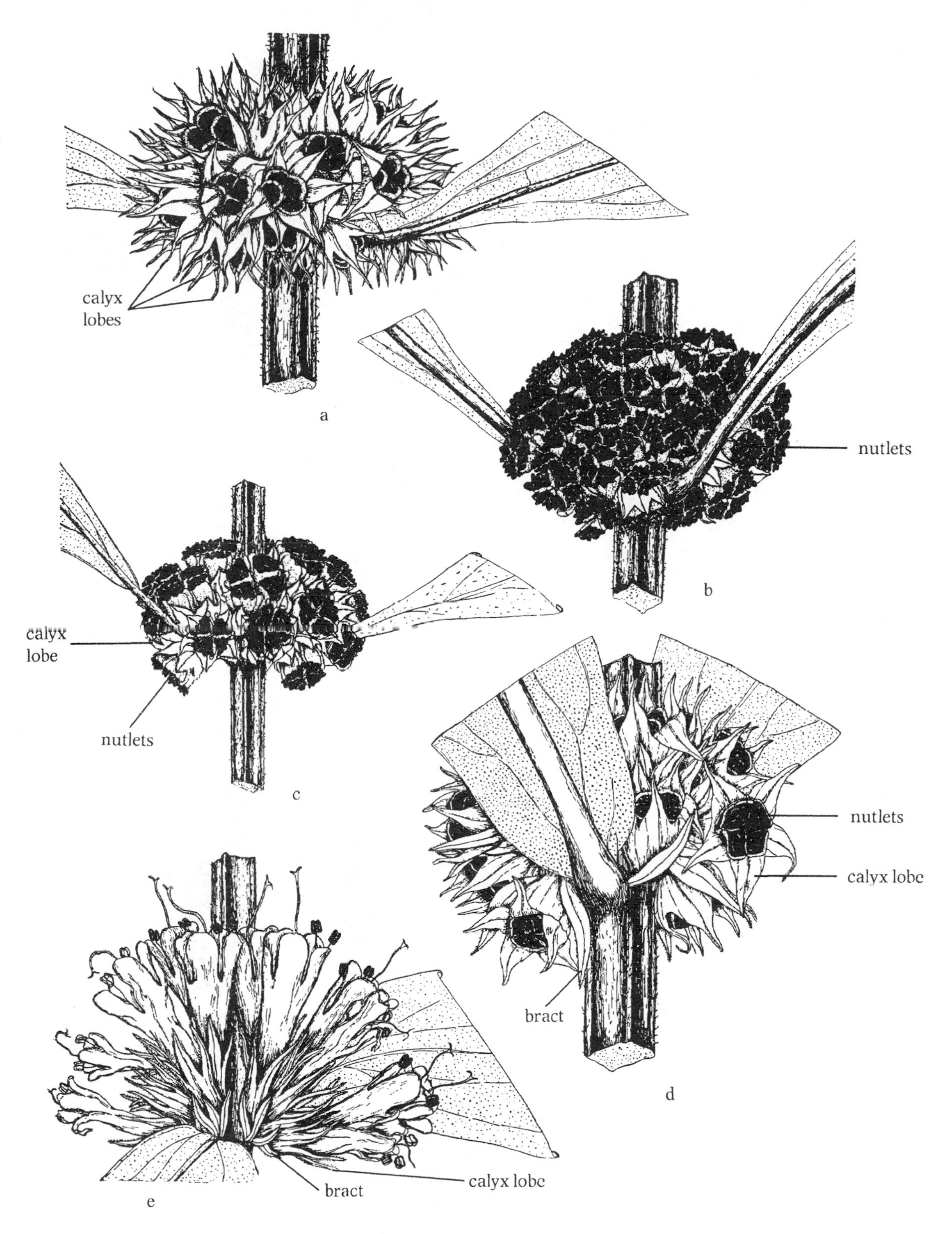

Fig. 555. *Lycopus americanus*: a. axillary fruit clusters (F).
Lycopus virginicus: b. axillary fruit clusters (F).
Lycopus uniflorus: c. axillary fruit clusters (F).
Lycopus asper: d. axillary fruit clusters (F).
Lycopus amplectens: e. axillary flower clusters (F).

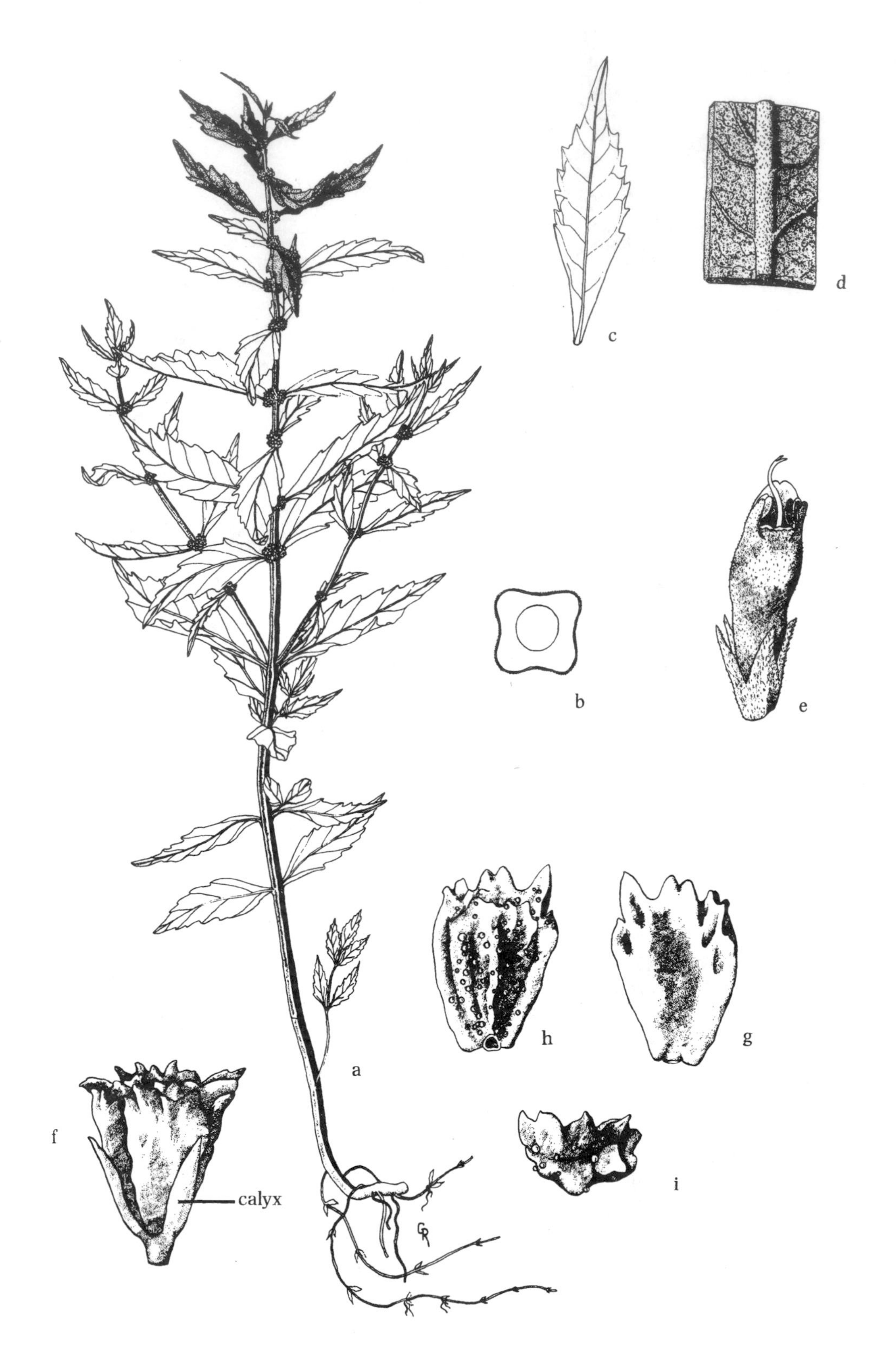

Fig. 556. *Lycopus virginicus*: a. habit; b. square stem (cross-section, diagrammatic); c. leaf from midsection of stem; d. portion of lower leaf surface; e. flower; f. nutlets with persistent calyx; g. outer face of nutlet; h. inner face of nutlet; i. nutlet, top view (G&W).

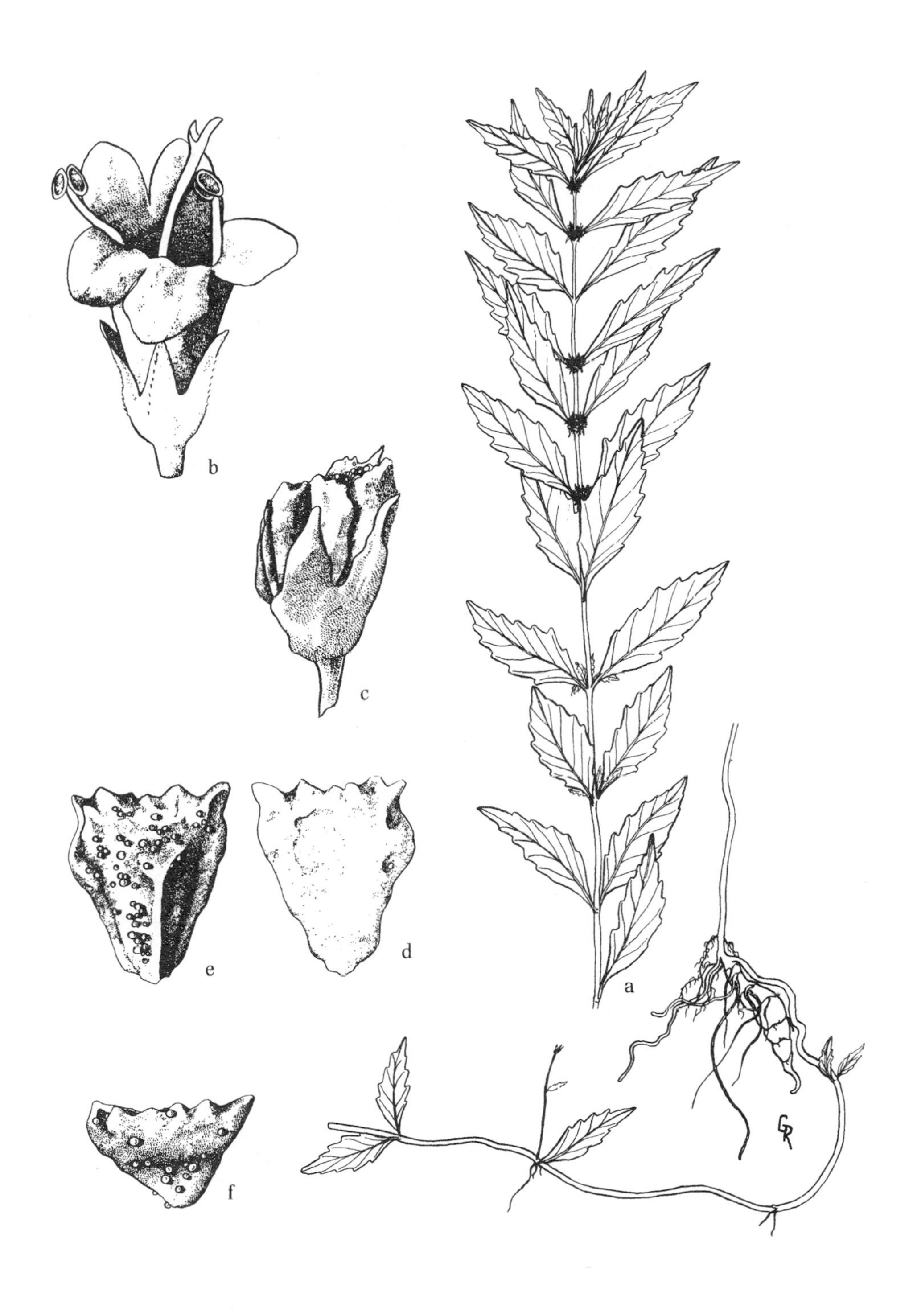

Fig. 557. *Lycopus uniflorus*: a. habit (showing tuber at base); b. flower; c. calyx with 4 enclosed nutlets; d. outer face of nutlet; e. inner face of nutlet; f. nutlet, top view (G&W).

Fig. 558. *Lycopus americanus*: a. habit; b. square stem (cross-section, diagrammatic); c. leaf variations on a single plant; d. calyx with 3 subtending bracts; e. outer face of nutlet; f. inner face of nutlet; g. nutlet, top view (G&W).

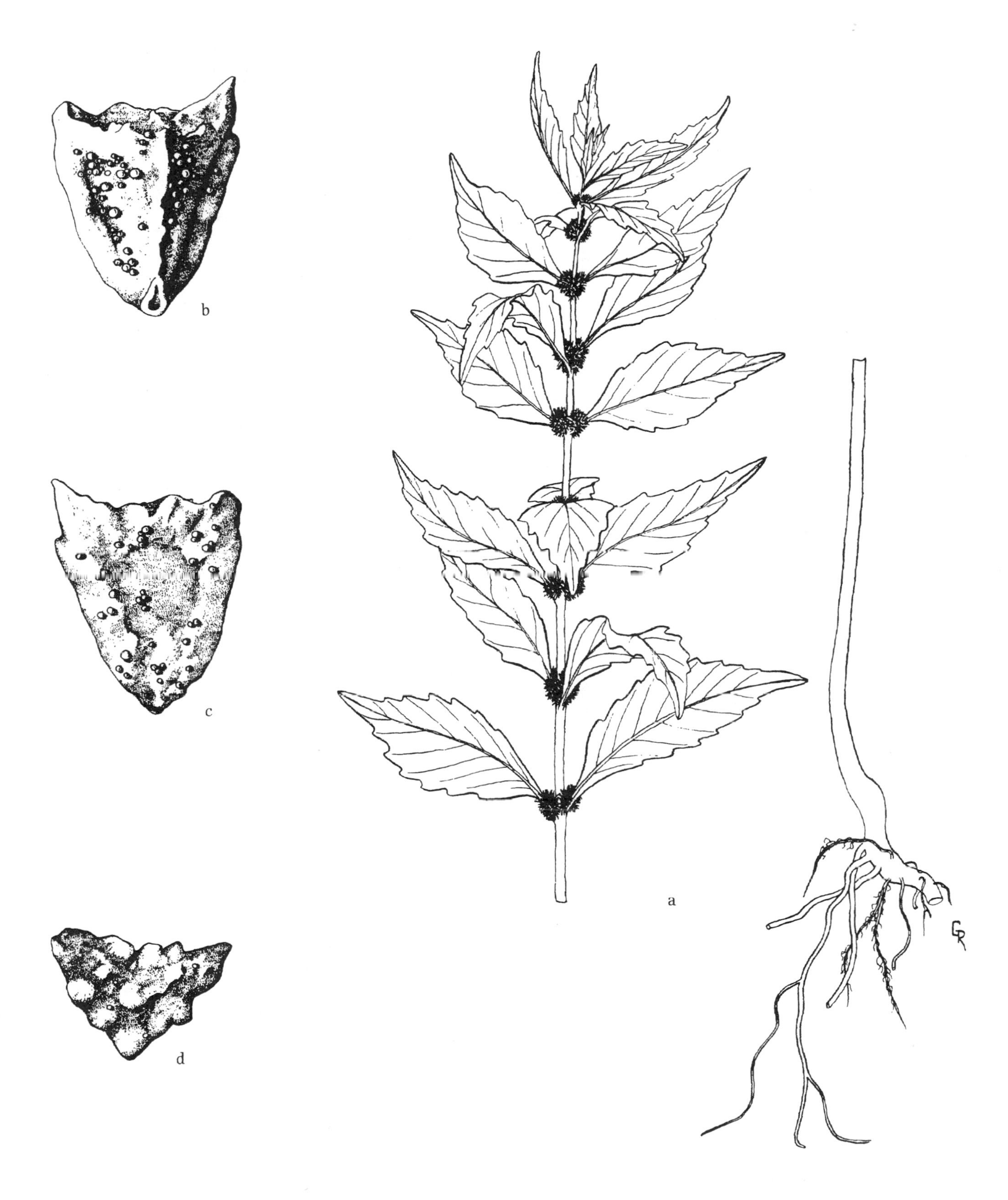

Fig. 559. *Lycopus rubellus*: a. habit; b. inner face of nutlet; c. outer face of nutlet; d. nutlet, top view (G&W).

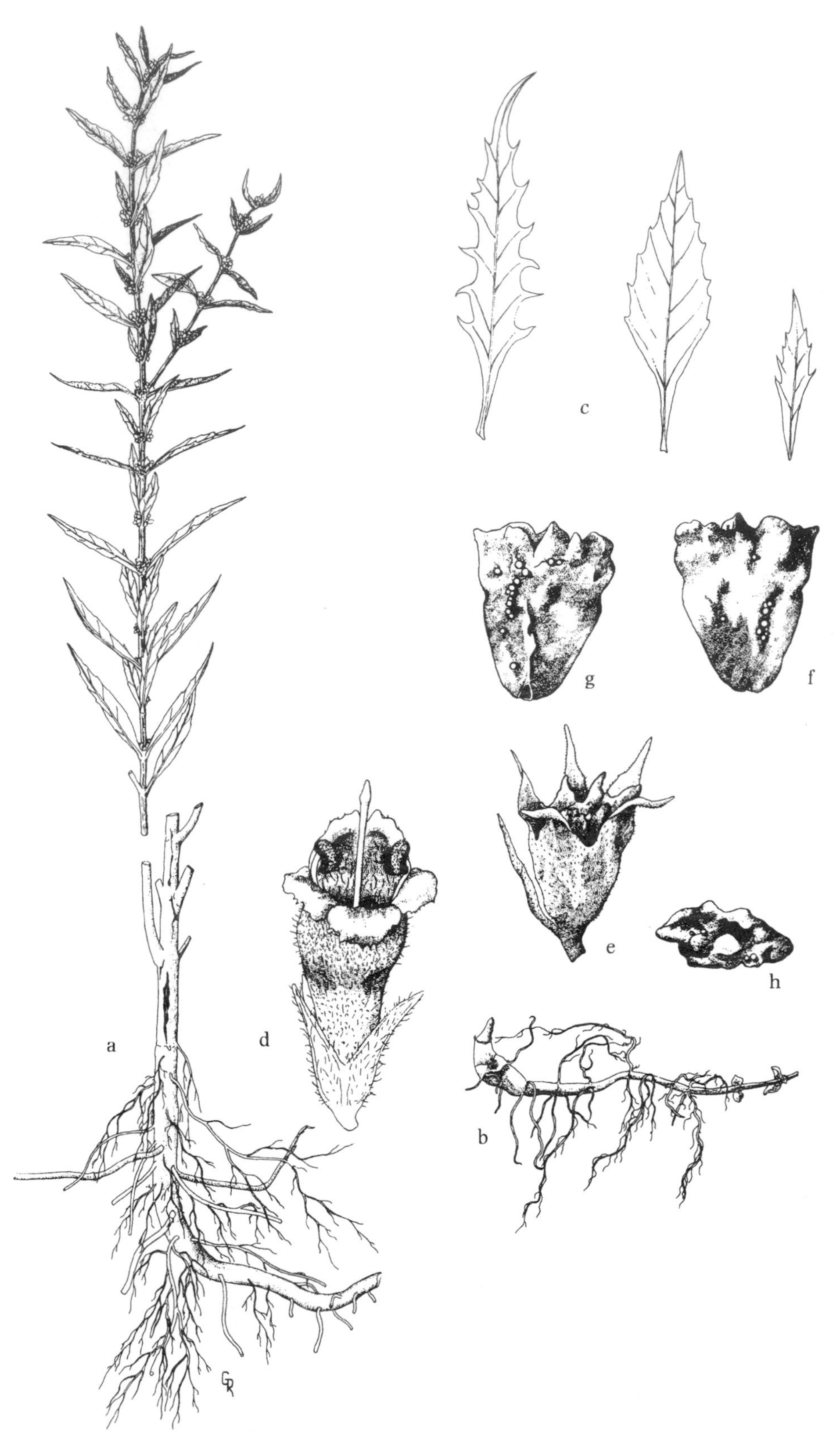

Fig. 560. *Lycopus rubellus*: a. habit; b. fusiform tuber at end of rhizome; c. (*left to right*) lower, middle, and upper stem leaves; d. flower; e. calyx with 4 enclosed nutlets and subtending bract; f. outer face of single nutlet; g. inner face of single nutlet; h. nutlet, top view (G&W).

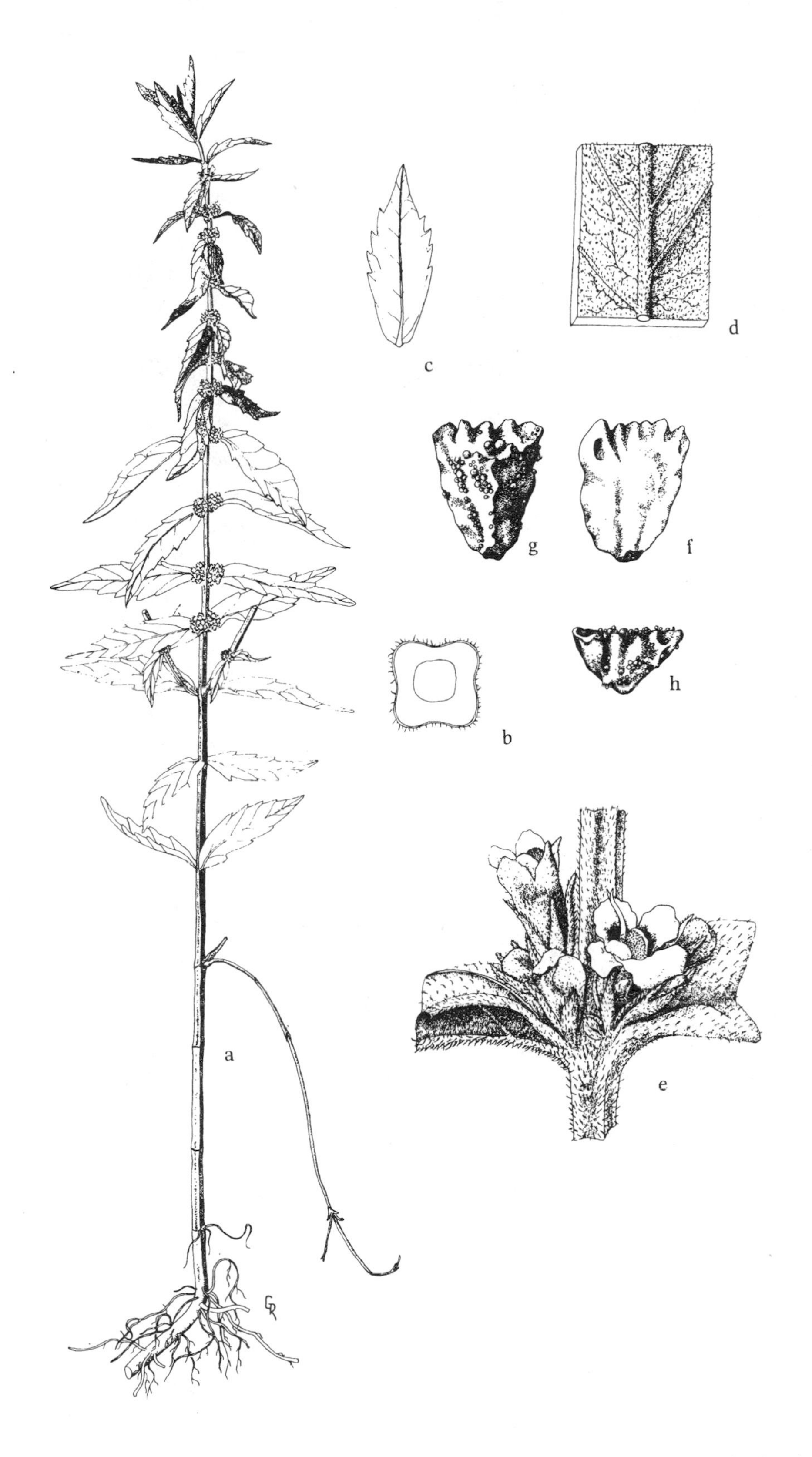

Fig. 561. *Lycopus amplectens*: a. habit; b. stem, cross-section; c. leaf; d. portion of lower leaf surface; e. node with flower cluster; f. outer face of nutlet; g. inner face of nutlet; h. nutlet, top view (G&W).

Fig. 562. *Lycopus asper*: a. habit (C&C); b. nutlet, top view (F); c. inner face of nutlet (F). *Clinopodium arkansana*: d. habit; e. flower (C&C).

Fig. 563. *Mimulus ringens*: a. habit (Crow); b. upper portion of stem (F); c. flower with palate nearly closing corolla tube (Crow).
Glossostigma cleistanthum: d. habit, showing cleistogamous flowers; e. stem with sessile capsule from cleistogamous flower; f. stalked capsule from chasmogamous flower (FNA).

2. *Mimulus* (Monkey-flower)

References: Nesom and Fraga, 2019; Posto and Prather, 2003; Windler et al., 1976.

1. Leaves sessile; stems rounded (not winged on angles); pedicel of flowers longer than calyx. . . . . 1. *M. ringens*
1. Leaves petiolate; stems narrowly winged on angles (fig. 563a); pedicel of flowers shorter than calyx. . . . . 2. *M. alatus*

1. *M. ringens* L. Square-stem Monkey-flower Fig. 563
Shores, meadows, wet woodlands, and estuaries. P.E.I. and N.S. w. to Man. and Alta., s. to Ga., La., Tex., Colo., Oreg. and Calif. Nesom (2019) recognizes *M. ringens* var. *colpophilus* Fernald; (Greene) Grant) as a diminutive plant of freshwater tidal shores, mud flats and adjacent wet meadows from Que., Me., and Vt.

2. *M. alatus* Aiton Winged Monkey-flower Fig. 564
Swamps, wet woods, floodplains, marshy shores and ditches. Mass. and Conn. w. to s. Ont., s. Mich., and e. Neb., s. to Fla. and e. Tex.

3. *Erythranthes* (Stalked Monkey-flower)

References: Nesom, 2014; Nesom and Fraga, 2019.

1. Stems with long viscid white hairs; calyx not inflated at maturity. . . . . 1. *E. moschata*
1. Stems glabrous or slightly pubescent, not viscid; calyx inflated at maturity.
    2. Flowers 20–40 mm long; corolla throat nearly closed by 2 hairy ridges; pedicel much longer than leaves subtending (fig. 564e); stems suberect or decumbent, rooting only at lower nodes. . . . . 2. *E. guttata*
    2. Flowers 8–18(22) mm long; corolla throat open; pedicel shorter or somewhat longer than (but less than twice as long as) leaves subtending it (fig. 564f); stems prostrate, rooting freely at nodes.
        3. Flowers 8–12 mm long; leaf margins entire to irregularly denticulate, not wavy or only slightly wavy; stems prostrate throughout. . . . . 3. *E. geyeri*
        3. Flowers 12–18(22) mm long; leaf margin distinctly dentate, wavy (fig. 564g); stems ascending toward tip. . . . . 4. *E. michiganensis*

1. *E. moschata* (Douglas ex Lindl.) G. L. Nesom Muskflower Fig. 564
Stream margins, springy places and ditches. Nfld. w. to Ont., s. to Mass., N.Y. and W.Va.; Mont. w. to s. B.C., s. to Colo., Utah and Calif.; locally natzd., indigenous eastward only in Nfld., M.I., Que., and Mich. (*Mimulus moschtus* Douglas ex Lindl.)

2. *E. guttata* (DC.) G. L. Nesom Common or Seep Monkeyflower Fig. 564
Brooks and meadows. Natzd. and local, N.B., Conn., N.Y., Pa., nw. Mich. and Ont.; sw. Sask., N.W.T., and Alask. s. to N.M., Ariz. and Mex. (*Mimulus guttatus* DC.)

3. *E. geyeri* (Torrey) G. L. Nesom Geyer's Monkeyflower Fig. 564
Wet shores, springs, and brooks, often floating along shores. Que. w. to Ont., s. Man., Sask., and Alta., s. to Pa., Mich., Ill., Mo., Tex., N.M., Ariz. and Mex. (*Mimulus glabratus* var. *fremontii* (Benth) Grant; *M. glabratus* var. *jamesii* (Torr. & A. Gray) A. Gray)

4. *E. michiganensis* (Pennell) G. L. Nesom Fig. 564
Wet shores, springs, wooded swamps and springy beaches. N. Michigan, especially around n. end of Lake Michigan. Based on molecular data this federally listed endangered species and narrow Michigan endemic has been elevated to species status (Posto and Prather, 2003). Its total range, so far as known, is in the Straits of Mackinac and Grand Traverse regions of Michigan (about 20 sites altogether in Benzie, Leelanau, Emmet, Cheboygan, and Mackinac Cos., and Beaver Island). (*M. glabratus* var. *michiganensis* (Pennell) Fassett; *Mimulus michiganensis* Posto and Prather)

Orobanchaceae / Broom-rape Family

1. Leaves broader, pinnately lobed; corollas strongly asymmetric, the upper corolla lobe forming a hood; stamens 4. . . . . 1. *Pedicularis*
1. Leaves narrow, unlobed; corollas campanulate; stamens 2. . . . . 2. *Agalinis*

Fig. 564. *Mimulus alatus*: a. section of stem with leaves and fruiting calyces, with outline of stem cross-section above (F).
Erythranthe moschata: b. habit; c. flower; d. flower, top view (HCOT, as *Mimulus moschatus*).
Erythranthe guttata: e. upper portion of plant (F, as *Mimulus guttatus*).
Erythranthe geyeri: f. portion of plant (F, as *Mimulus glabratus* var. *fremontii*).
Erythranthe michiganensis: g. leaf (F, as *Mimulus michiganensis*).

1. *Pedicularis* (Lousewort)

Annual, biennial, or perennial herbs; hemiparasitic; stems erect; leaves alternate or opposite, pinnately or bipinnately lobed; flowers borne in terminal spikes or racemes; corolla yellow, pink, red, or purple; united, bilaterally symmetrical; ovary superior; fruit a loculicidal capsule.

---

1. Flowers yellow; lip of corolla curved (fig. 566b), tapering to a short beak; plants of areas south and west of Maritime Provinces. .... 1. *P. lanceolata*
1. Flowers rose-purple; upper lip of corolla hood-like (fig. 566d); plants of Maritime Provinces and e. Que.. .... 2. *P. palustris*

---

1. *P. lanceolata* Michx. Fig. 566
Wet meadows, bogs, marshy shores and wet woodlands, often in calcareous sites. Mass. w. to Man., s. to Va., mts. to N.C., Tenn., and n. Ga., Ohio, Mo., n. Ark. and Neb.

2. *P. palustris* L. Swamp Lousewort; European Purple Lousewort Fig. 566
Marshes and wet meadows. Nfld., M.I., e. Que., and N.S.; amphi-Atlantic.

2. *Agalinis* (Gerardia)

Annual or, rarely, perennial herbs; hemiparasitic; stems erect, branching; leaves opposite or upper leaves alternate on flowering stems, rarely all but the lowest leaves alternate or subopposite, sessile, entire; calyx 5-lobed; corolla pink, lavender, or rarely white, united, bilaterally symmetrical, more or less 2-lipped, lobes 5, usually subequal; fruit a loculicidal capsule.

---

1. Plants perennial, with a slender, creeping rhizome; stems somewhat spongy, with aerenchyma; corolla slightly succulent. ...... 1. *A. linifolia*
1. Plants annual, with fibrous roots; stems lacking aerenchyma; corolla thin, not succulent.
   2. Pedicels 2–6 times as long as calyx. .... 2. *A. tenuifolia*
   2. Pedicels shorter than, as long as, or somewhat longer than calyx, but less than twice its length.
      3. Calyx lobes usually longer than calyx tube (3–5 mm long).. .... 3. *A. neoscotia*
      3. Calyx lobes usually shorter than calyx tube.
         4. Leaves weakly succulent; calyx lobes rounded; plants of coastal salt marshes. .... 4. *A. maritima*
         4. Leaves not succulent; calyx lobes acute or acuminate; plants of inland sites, seldom of salt marshes.
            5. Stem angled at base, smooth or slightly scabrous; calyx lobes 0.5–5.5 mm long.
               6. Calyx lobes usually less than half the length of calyx tube, 0.5–2 mm long; (18)20–38 mm long. .... 5. *A. purpurea*
               6. Calyx lobes over half the length of calyx tube, 1.5–3.5 mm long; corolla 10–20(23) mm long. ....6. *A. paupercula*
            5. Stem almost round in cross-section at base, extremely scabrous; calyx lobes 0.6–1.6 mm long. .... 7. *A. fasciculata*

---

1. *A. linifolia* (Nutt.) Britton Flax-leaf Foxglove
Pond margins, pinelands, marshy shores, marshes and ditches. Coastal plain, local in Md. and Del. s. to e. Va., e. N.C. and e. S.C. s. to Fla., w. to e. La.

2. *A. tenuifolia* (Vahl) Raf. Slender-leaf False Foxglove
Low woods, prairies, peaty areas, stream banks and dry habitats. N.S., Me. and sw. Que. w. to Ont., Minn., Man., N.D., e. Wyo., N.M. and Colo., s. to Ala., Miss., Ark. and e. Tex. Several varieties have been recognized in this variable species.

3. *A. neoscotia* (Greene) Fernald Middletown False Foxglove
Damp sand, boggy and peat sites and edges of salt marshes. Sable I., N.S., w. N.S. and ne. Me.

4. *A. maritima* (Raf.) Raf. Seaside Gerardia Fig. 565
Salt marshes. Coastal, N.S. and se. Me. s. to se. N.Y., N.J., Del., Va., N.C. and Fla., w. to e. Tex. and ne. Mex; W.I. The taller, more southern taxon with narrower, more acute leaves is sometimes separated as var. *grandiflora* (Benth.) Shinners reaches n. to our range in se. Va.

5. *A. purpurea* (L.) Pennell Purple Gerardia Fig. 566
Damp acid soils, clearings, savannas, marshy shores and bogs. S. Me., Mass., and N.Y., w. to Ont., Mich., Minn. and e. Neb., s. to Fla. and e. Tex.

6. *A. paupercula* (A. Gray) Britton Small-flower False Foxglove
Damp ground, shores, bogs and sandy shores. N.S., N.B. and Ont., w. to Minn. and Man., s. to N.Y., N.J., Pa., n. Ohio., n. Ind., Ill., and e. Iowa; sw. Va. Northern plants with a smaller corolla are here treated as a separate species but are sometimes included in *A. purpurea* as var. *parviflora* (Benth.) B. Boivin.

7. *A. fasiculata* (Elliott) Raf. Beach False-foxglove
Moist to wet habitats, swales, dune hollow, and tidal marshes. Coastal Plain, s. Md. s. to Fla., w. to e. Tex.; inland n. to Ark., sw. Mo., Okla. and se. Kans.

Fig. 565. *Agalinis maritima*: a. habit; b. flower; c. flower, longitudinal section; d. capsule; e. seed (pitted-reticulate) (G&W).

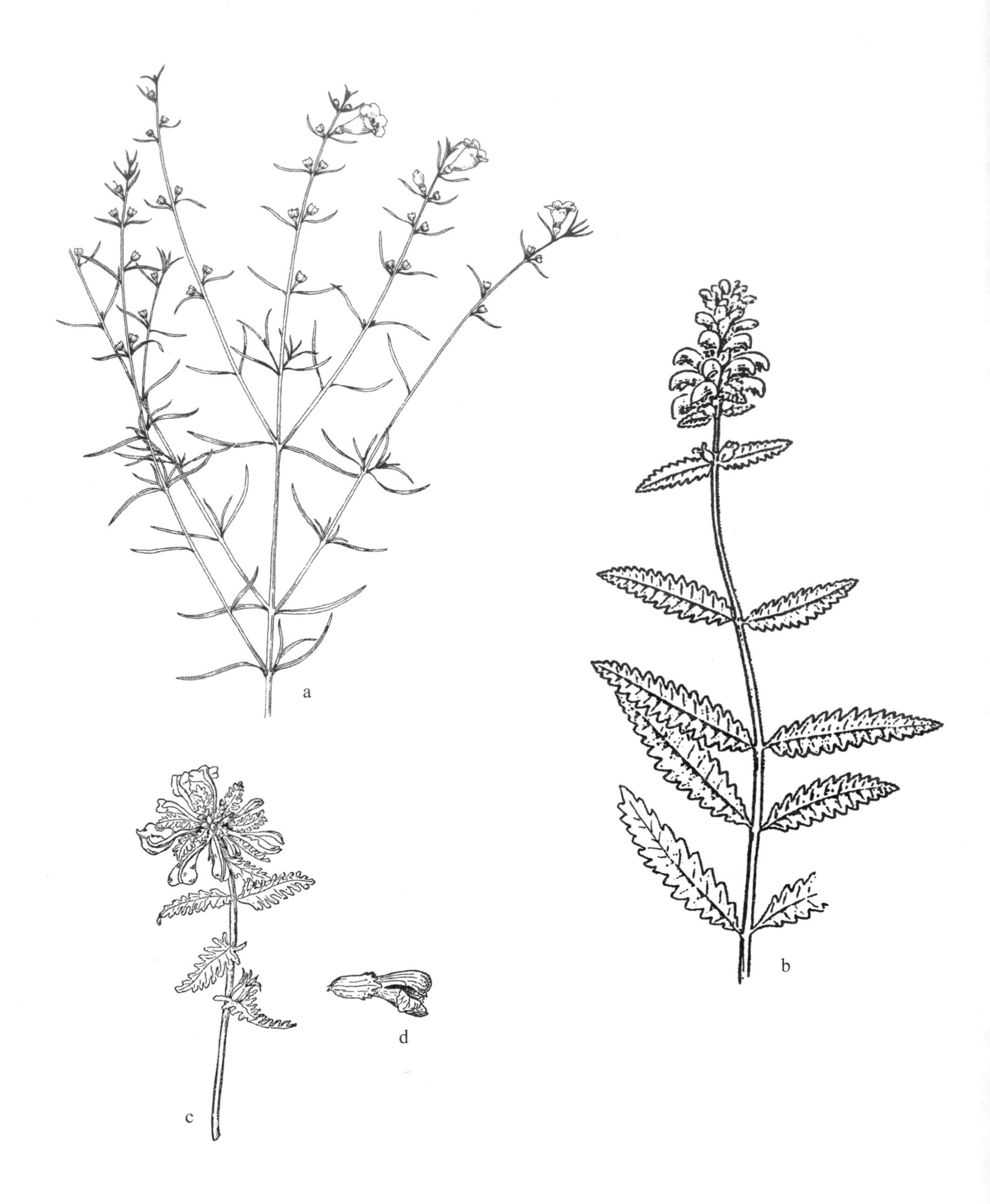

Fig. 566. *Agalinis purpurea*: a. upper portion of plant (B&T).
Pedicularis lanceolata: b. upper portion of plant (WVA).
Pedicularis palustris: c. upper portion of plant; d. flower (Gleason).

Aquifoliaceae / Holly Family

1. *Ilex* (Holly)

Shrubs or small trees, deciduous or evergreen; leaves alternate, simple; dioecious, margins entire, serrate, dentate or sinuate-spinose; plants dioecious, flowers unisexual, axillary, white, greenish or yellow, solitary or 2-several in a cyme; fruits red to purple black or yellow (rarely albino) drupe.

---

1. Leaf margins crenate to serrate, at least at apex, apex somewhat obtuse (fig. 567) to acute (fig. 567a) to acuminate (fig. 567d); flowers usually in cluster, pedicels less than 1 cm long; petals white, slightly united at base, oblong or obovate.
   2. Leaves deciduous, lacking punctate glands on lower surface; fruits red (rarely yellowish).
   2. Leaves dull on upper surface, margins coarsely toothed; calyx segment ciliate or pubescent, obtuse. . . . . 1. *I. verticillata*
      3. Leaves shiny on upper surface, margins finely toothed; calyx segments sparsely ciliate or entire acute. . . . . 2. *I. laevigata*
      3. Leaves evergreen (leathery), with punctate glands on lower surface; fruits black (rarely albino). . . . . 3. *I. glabra*
1. Leaf margins entire, apex mucronate (fig. 567h); flowers solitary; pedicels more than 1 cm long; petals yellowish or greenish, not united at base. . . . . 4. *I. mucronata*

---

1. *I. verticillata* (L.) A. Gray   Winterberry, Black-alder   Fig. 567
Swamps, marshy thickets and margins of ponds and bog. Nfld. and s. Que. w. to e. Minn., s. to n. Fla., Miss., e. La. and Tex. This is a highly variable taxon.

2. *I. laevigata* (Pursh) A. Gray   Smooth Winterberry   Fig. 567
Wooded swamps and bogs. S. Me. and N.H. w. to N.Y., s. to Md., N.C., n. S.C. and Ga.

3. *I. glabra* (L.) A. Gray   Inkberry, Gallberry   Fig. 567
Swamps, bogs, wet woods, flatwoods and pine savannas. Chiefly coastal plain, N.S., e. Me. and se. N.H. s. to Fla., w. to Ark., La. and e. Tex.

4. *I. mucronata* (L.) Powell, Savolainen & Andrews
Mountain Holly   Fig. 567
Damp woods, bog, swamps, thickets and high elevations. Nfld. and Que. w. to Ont. and e. Minn., s. to Md., W.Va., n. Ohio, n. Ind., n. Ill. and Wisc. (*Nemopanthus mucronatus* (L.) Trel.)

Campanulaceae / Bellflower Family

REFERENCE: Rosatti, 1986.

---

1. Flowers radially symmetrical, carpels 3 (stigmas 3-branched); capsules opening by lateral pores.
   2. Stems weak, retrorsely scabrous on angles (clinging to other vegetation); leaves up to 0.8 cm wide; flowers solitary, stalked, white to bluish-white, bell-shaped. . . . . 1. *Campanula*
   2. Stems sturdy, erect, glabrous; leaves 2–6 cm wide; flowers in terminal spikes (sometimes in small clusters or solitary), sessile or subsessile, rotate (nearly flat), light blue. . . . . 2. *Campanulastrum*
1. Plants bilaterally symmetrical (2-lipped); stamens with anthers and filaments united into a tube surrounding the style; carpels 2 (stigmas 2-lobed); capsules opening by valves at apex. . . . . 3. *Lobelia*

---

1. *Campanula* (Bellflower)

Perennial (ours) and annual herbs; leaves alternate, serrate, petiolate or sessile: stems scabrous (ours) or smooth; flowers solitary or borne in panicles; corolla blue, purplish-blue, or white, united, radially symmetrical, campanulate to rotate; stamens 5, borne on base of corolla tube; ovary inferior; fruit a capsule opening by apical pores.

REFERENCE: Shetler, 1963.

1. *C. aparinoides* Pursh   Marsh Bellflower   Fig. 568
Wet meadows, marshes, bogs, swales, shores, stream margins and wet thickets. N.S., N.B., Que. w. to Ont., Minn., Man. and Sask., s. to N.E., Va., mts to w. N.C., n. Ga., Tenn., and s. Mo., Neb. and Colo; Wash. The larger-flowered form is sometimes segregated as *C. uliginosa* Rydb., but Shetler (1963) indicated that, although it probably merits recognition as a separate race, it is hardly a good species. (*C. aparinoides* var. *grandiflora* Holz.)

Fig. 567. *Ilex verticillata*: a. branch with flowers; b. flower; c. branch with fruits, winter condition; d. leaf (Ryan).
Ilex glabra: e. branch with evergreen leaves and fruits; f. fruit (USDA-NRCS).
Ilex laevigata: f. branch with flowers; g. flower (B&B).
Ilex mucronata: h. flower; i. branch with fruits; j. branch, winter condition; k. leaves (Ryan).

Fig. 568. *Campanula aparinoides*: a. habit; b. fruit; c. section of scabrous stem (B&B); d, e. upper portion of plant (F). *Campanulastrum americanum*: f. inflorescence (B&B); g. inflorescence (WVA).

2. *Campanulastrum* (Bellflower)

Annual, with erect stems, glabrous; flowers and median leaves with short, margined petioles, upper sessile; margins serrate, apices acuminate; flowers in just below apex, spicate racemes, terminating branches; flowers 2–2.5 cm across; corolla light blue; tubular basally, style well exerted beyond corolla; capsule obconic, dehiscing by apical pores.

1. *C. americana* (L.) Small Tall Bellflower, American Bellflower Fig. 568
Wet rich woods and shores. N.Y. and s. Ont. w. to s. Mich., Wisc., s. Minn. and Neb., s. to Va., N.C., Ga. nw. Fla., Ala., Ark. and e. Okla. (*Campanula americana* L.)

3. *Lobelia* (Lobelia)

Annual or perennial herbs (ours) or shrubs, often with an acrid, poisonous, milky or colored sap; stems upright, leaves alternate and simple; flowers borne in racemes; corolla red, purple, blue, or white, united, bilaterally symmetrical, upper lip with 2 nearly erect lobes, lower lip 3-lobed; anthers united around style (fig. 292g), filaments upper portion united in a tube surrounding style (partly monadelphous), free at base; ovary inferior, 2-carpellate (stigma 2-lobed); fruit a capsule.

The only strictly aquatic species is *L. dortmanna* which, given its habit as a submersed rosette of leaves, with only its inflorescence being emergent, is very different in appearance from all others.

References: Bowden, 1959; McVaugh, 1936; Spaulding and Barger, 2016.

---

1. Leaves hollow (a pair of tubes in cross-section) (fig. 569b), all in a basal rosette (fig. 569a); cauline leaves absent; plants submersed. .... 1. *L. dortmanna*
1. Leaves flat, strictly cauline or cauline with basal rosettes; plants emergent or terrestrial.
  2. Flowers 1.5–4.5 cm long; leaves wider, mostly 5–40(60) mm wide (except *L. glandulosa*, then with pubescent pedicels).
    3. Corolla brilliant red (rarely pink or albino), 3–4.5 cm long.......... 2. *L. cardinalis*
    3. Corolla blue, often with a white spot at throat (rarely albino), 1.8–3.3 cm long.
      4. Calyx lobes bristly ciliate, with conspicuous lobes or auricles at base; filament tube 12–15 mm long (fig. 570c)...... 3. *L. siphilitica*
      4. Calyx with auricles lacking or very small; filament tube 6–11 mm long.
        5. Calyx lobes entire, lacking prominent glandular teeth; pedicels arching to one side in fruit.......... 4. *L. elongata*
        5. Calyx lobes with prominent glandular teeth; pedicels erect to ascending in fruit.
          6. Pedicels glabrous; lower lip of corolla glabrous; cauline leaves elliptic or short-ovate.......... 5. *L. georgiana*
          6. Pedicels pubescent; lower lip of corolla hirsute at base; cauline leaves narrowly lanceolate to linear.......... 6. *L. glandulosa*
  2. Flowers 0.7–1.6 cm long; leaves linear, filiform or narrowly lanceolate, 1–5 mm wide.
    7. Pedicels with a pair of minute bracts near middle (fig. 570d); plants mostly northern (Pa., Ohio, and Ill. northward)....... 7. *L. kalmii*
    7. Pedicels lacking bracts or with a pair of minute bracts nearly hidden at base; plants mostly of coastal plain (N.J. and L.I., N.Y. southward).
      8. Leaves filiform, ca. 0.5 mm wide (often deciduous); calyx glabrous; plants rhizomatous, stems often spongy-thickened at base.......... 8. *L. boykinii*
      8. Leaves 1–4 mm wide; calyx pubescent or glabrous; plants not rhizomatous, stems not spongy-thickened at base.
        9. Corolla blue, with white center at throat, lower lip glabrous; pedicels and calyx glabrous or with scattered straight hairs.......... 9. *L. nuttallii*
        9. Corolla blue, with no white center, lower lip pubescent within near throat of corolla tube; pedicels and calyx typically strongly antrorsely scabrid.......... 10. *L. canbyi*

---

1. *L. dortmanna* L. Water Lobelia Fig. 569
Sandy or gravelly shores, submersed with inflorescences typically emersed, in acid waters. Nfld. w. to Minn., N.W.T. and Alask., s. to N.E., n. N.J., Md., ne. Pa., n. Mich., n. Wisc., ne. Minn., Man., Alta., B.C., Wash. and Ore.

2. *L. cardinalis* L. Cardinal Flower Fig. 569
Damp shores, meadows, swamps, stream banks or submersed at stream margins. N.B. and N.E. w. to Wisc. and se. Minn., Neb., Colo., Utah and s. Calif. s. to Fla., Tex., N.M. and Ariz.; C.Am., n. S.Am. (*L. splendens* Willd.)

3. *L. siphilitica* L. Great Lobelia, Big Blue Lobelia Fig. 570
Rich low woods, swamps, stream banks and wet meadows. N.E. and N.Y. w. to Ont., Mich., Minn., Man., N.D. and Colo., s. to w. N.C., n. Ga., Ala., La., e. Okla. and ne. Tex.

4. *L. elongata* Small Long-leaf Lobelia
Fresh to brackish tidal marshes, swamps and low wet ground. Coastal: Del. and Md. s. to Ga. and La.

Fig. 569. *Lobelia dortmanna*: a. habit; b. single hollow leaf (cross-section showing 2 tubes) (F).
Lobelia cardinalis: c. basal portion of plant; d. inflorescence; e. leaves from medial portion of stem; f. flower; g. anthers fused forming a tube and exserted portion of style with 2-lobed stigma; h. capsule within persistent calyx segments; i. capsule, cross-section; j. seed (G&W).

Fig. 570. *Lobelia siphilitica*: a. upper portion of plant (WVA); b. section of stem with leaves (F); c. flower with portion of stamen exposed to show fused filaments forming a tube and fused stamens (with stigma exerted from anther tube) (F).
Lobelia kalmii: d. habit (F).
Lobelia nuttallii: e. habit (Beal).

5. *L. georgiana* McVaugh Southern Lobelia
Meadows, peaty soils, wet woods and stream banks. Va. w. to Tenn., s. to n. Fla. and w. to Ala. and La. (*L. amoena* var. *glandulifera* A. Gray)

6. *L. glandulosa* Walter Glade Lobelia Fig. 571
Damp pinelands, swamps, stream banks and ditches. Coastal plain; Md. and se. Va. s. to Fla., w. to sw. Ala. and n. Miss.

7. *L. kalmii* L. Ontario Lobelia Fig. 570
Calcareous fens, marly shores, wet ledges, and meadows. Nfld. w. to N.W.T. and B.C., s. to n. and w. N.E., N.J., Pa., Ohio, n. Ill., n. Iowa, S.D., n. Mont., Wash. and Oreg.

8. *L. boykinii* Torr. & A. Gray ex A. DC. Boykin's Lobelia
Cypress ponds, wet pinelands, depressions, ponds and ditches. Coastal plain; s. N.J. and s. Del.; N.C., S.C. s. to Fla., w. to Ala. and Miss.

9. *L. nuttallii* Roem. & Schult. Nuttall's Lobelia Fig. 570
Sandy or grassy swamps, meadows, low woods, ditches and sometimes brackish marshes. Coastal plain and piedmont; se. N.Y., s. N.J., se. Pa. and Del. s. to Ga., nw. Fla., w. to Ala. and se. La.; inland n. to se. Ky., e. Tenn., and Okla.

10. *L. canbyi* A. Gray Canby's Lobelia
Sandy swamps, wet pinelands and swales. Coastal plain; s. N.J., Del. and Md. s. to S.C.; mts of n. Ga. and c. Tenn.

Menyanthaceae / Buckbean Family

1. Leaves compound, trifoliolate, emersed; flowers borne in a raceme. . . . . 1. *Menyanthes*
1. Leaves simple, ovate, floating; flowers borne in an umbel (usually 1 open at a time, at water surface). . . . . 2. *Nymphoides*

1. *Menyanthes* (Buckbean)

Perennial herbs; leaves alternate, nearly basal, trifoliolate; flowers 5-merous; corolla white, united, lobes bearded on upper surface, appearing fringed; fruit a capsule.

Reference. Mason and Iltis, 1965.

1. *M. trifoliata* L. Bogbean, Buckbean Fig. 572
Bogs, marshes, pond margins and shallow water. Greenl., Lab. and Nfld. w. to Nunavut, N.W.T., and Alask., s. to N.C., Ohio, Ill., Mo., N.M. Ariz. and Calif.; circumboreal.

2. *Nymphoides* (Floating-heart)

Perennial herbs; basal leaves small, submersed; floating leaves larger, cordate-ovate, subcircular or reniform (similar to *Nymphaea*) on long petioles; flowers 5-merous; corolla white or yellow, united; fruit a capsule.

References: Clark, 1938; House, 1937; Stuckey, 1974; Tippery and Les, 2011.

1. Corolla white; stems with clusters of submersed, tuberous roots, borne just below floating leaf near water surface (fig. 573b,c).
    2. Floating leaves 1.5–5 cm wide; calyx lobes 2–3 mm long, green; stem below inflorescence green (not-red-punctate) capsule 4–5 mm long. . . . . 1. *N. cordata*
    2. Floating leaves 4–15 cm wide; calyx lobs 4–5 mm long, red-punctate; stem below inflorescence red-punctate, capsule 10–14 mm long. . . . . 2. *N. aquatica*
1. Corolla yellow; stems lacking clusters of tuberous roots borne below floating leaf near the water surface. . . . . 3. *N. peltata*

1. *N. cordata* (Elliott) Fernald Little Floating-heart Fig. 573
Ponds, lakes and quiet streams, Nfld. w. to se. Ont. s. to Fla. and La.

2. *N. aquatica* (J. F. Gmel.) Kuntze Big Floating-heart, Banana Plant Fig. 574
Ponds, lakes and quiet streams. Coastal plain; s. N.J. s. to Fla. w. to e. Tex.

3. *N. peltata* (J. F. Gmel.) Kuntze Yellow Floating-heart Fig. 575
Ponds, impoundments, quiet waters, and canals. Natzd.; intro. ornamental from Eur. and e. Asia; in Que., Ont. and N.E. and N.Y. w. to Ohio, Ill. and Mo., s. to Ky., Tenn., Miss. Mo., Ark. Okla., ne. Tex. Ariz. and Calif.: Wash.

Fig. 571. *Lobelia glandulosa*: a. habit; b. leaf from middle portion of stem; c. fruit enclosed within calyx; d. seed (G&W).

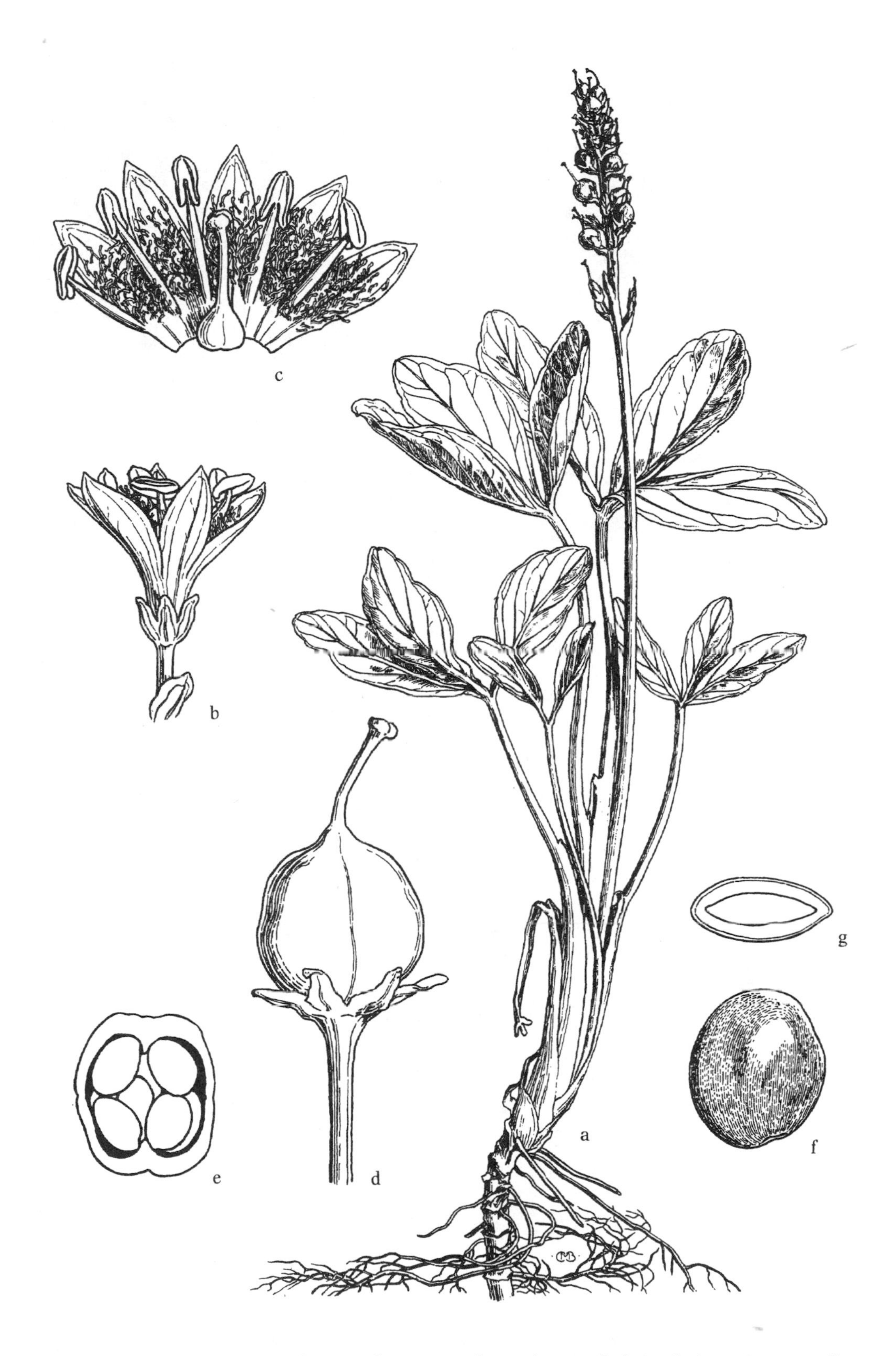

Fig. 572. *Menyanthes trifoliata*: a. habit; b. flower; c. flower, spread open showing fimbriate hairs on inner corolla surface; d. capsule; e. capsule, cross-section; f. seed; g. seed, cross-section (Mason).

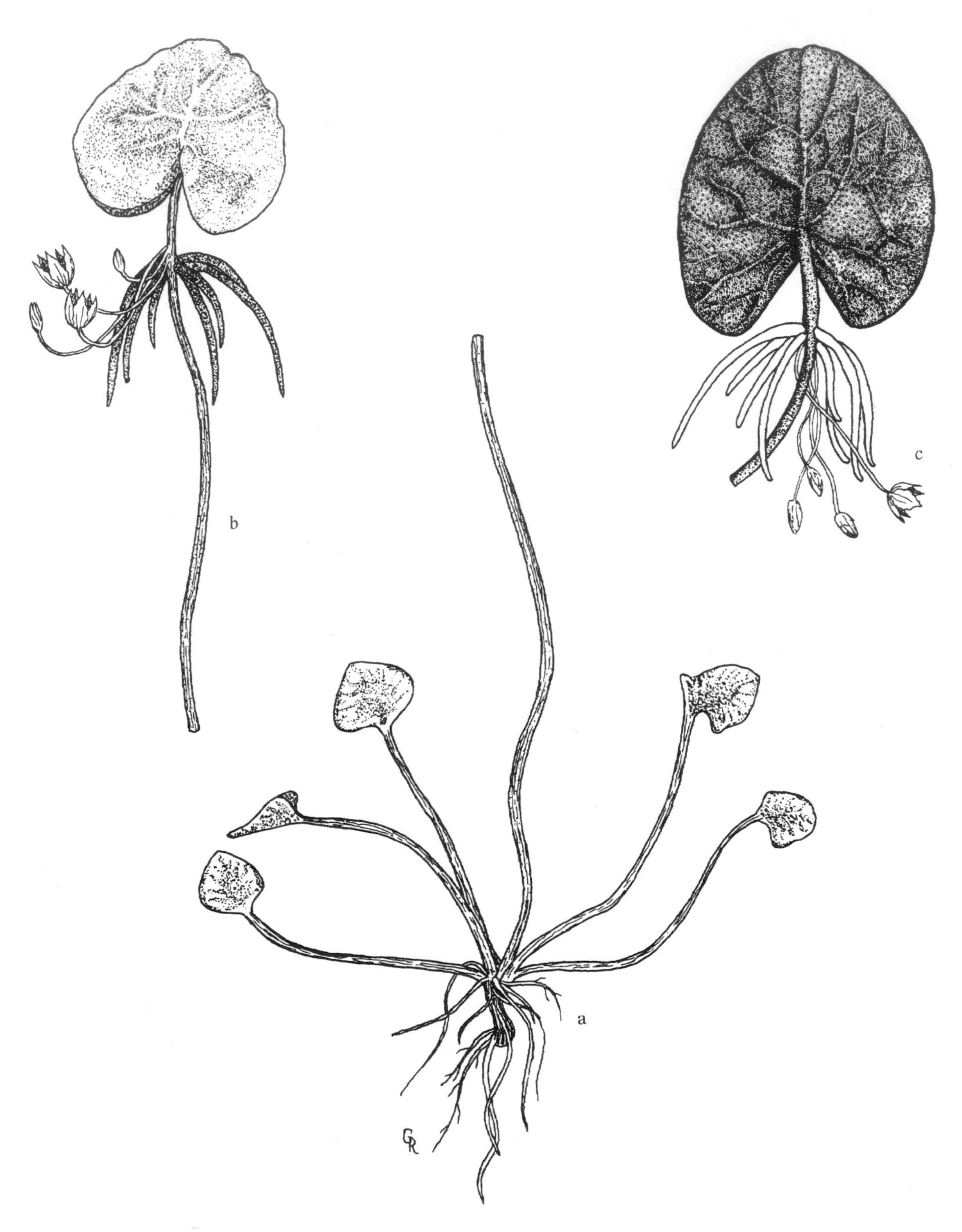

Fig. 573. *Nymphoides cordata*: a. base of seedling with stem, submersed; b. floating leaf with flowers and cluster of tuberous roots at base of petiole and pedicels; c. lower surface of floating leaf with flowers and tuberous roots (G&W).

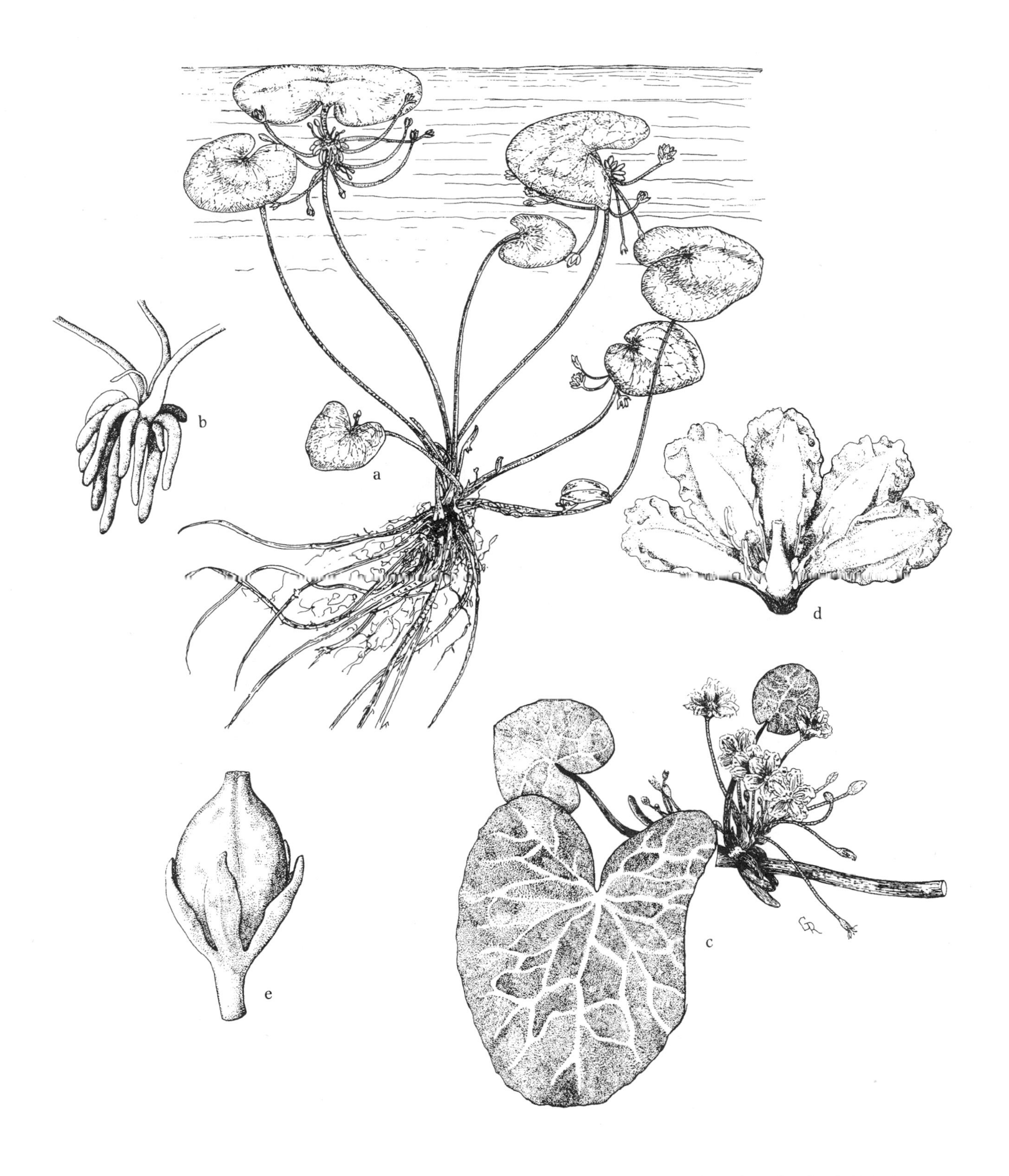

Fig. 574. *Nymphoides aquatica*: a. habit (C&C); b. cluster of tuberous roots at base of petiole and pedicels (G&W); c. floating leaves with flowers and tuberous roots (G&W); d. flower, corolla spread open (G&W); e. capsule (G&W).

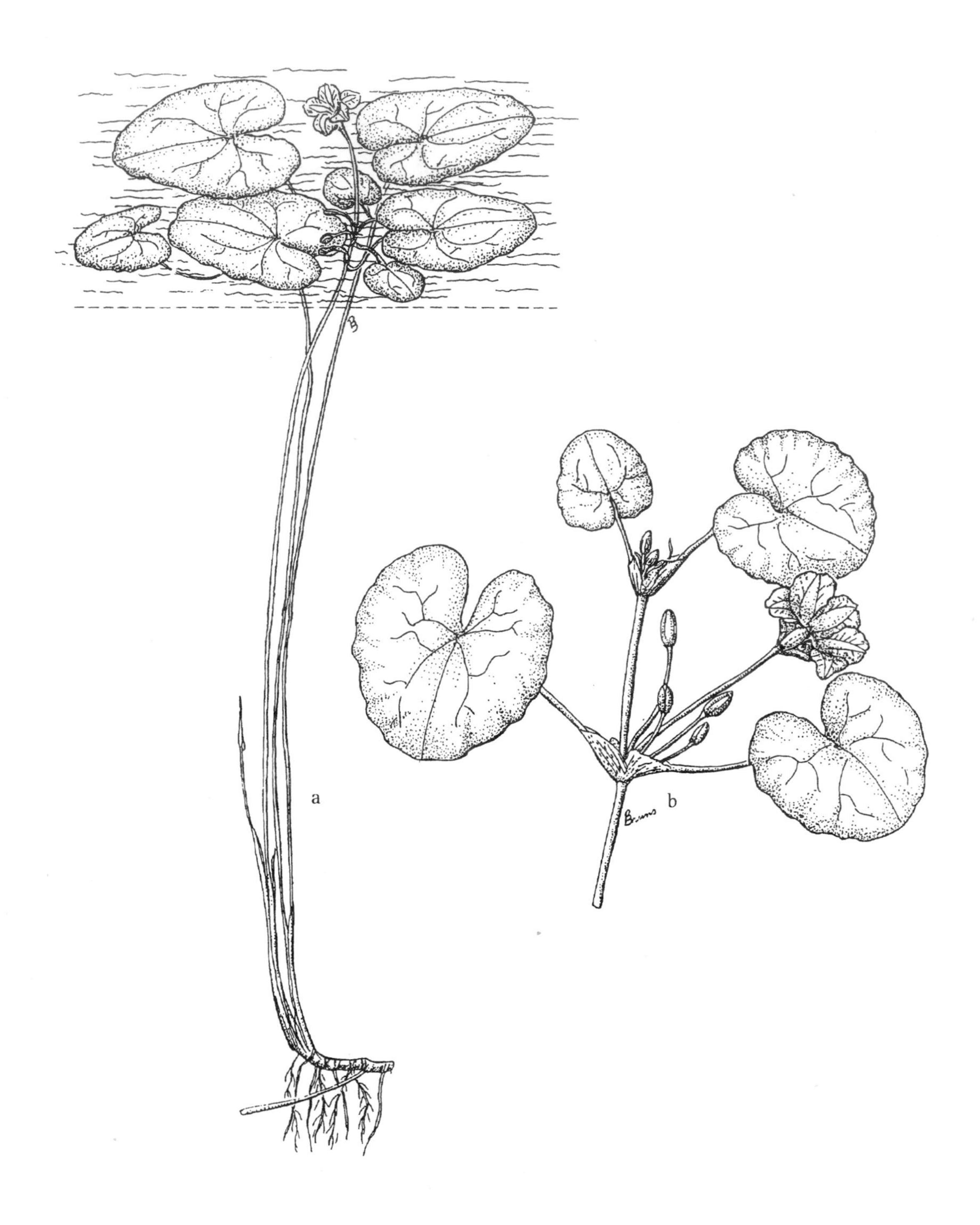

Fig. 575. *Nymphoides peltata*: a. habit; b. upper portion of plant (PB).

Asteraceae (Compositae) / Aster Family

The Aster Family is the largest and most advanced family of the Eudicots, numbering more than 25,000 species; it is also taxonomically difficult. Although chiefly a terrestrial group, a number of taxa are wetland plants, but only a few, such as *Bidens beckii*, a submersed species, and *Sclerolepis uniflora*, an amphibious plant, are truly aquatic. The family is readily characterized by: flowers aggregated in heads (each simulating a single flower), each head subtended by 1 or more series of involucral bracts; flowers of two types, tubular flowers (disk flowers) in the central portion of head, ligulate flowers (ray flowers) at margin of head; the head may contain both types of flowers or consist of only disk flowers (ray flowers being absent), or consist of only ray flowers (although none of ours entirely ligulate); stamens borne on the corolla, the filaments free, but the anthers fused (dehiscing inward); ovary inferior; fruit a head of achenes, often with persistent pappus of bristles, capillary bristles, scales, or awns, or the pappus absent.

---

1. Flower heads consisting of ray and disk flowers (fig. 576).
  2. Ray flowers yellow or orange, occasionally purplish at base.
    3. Stems and leaves flexuous, supported by water; leaves highly dissected, filiform, opposite, but often appearing whorled (*B. beckii*, fig. 576). . . . . . . 1. *Bidens*
    3. Stems and leaves upright; leaves simple or divided, but never dissected, opposite or alternate.
      4. Plants herbaceous, not shrubby.
        5. Leaves opposite. . . . . . . 1. *Bidens*
        5. Leaves alternate.
          6. Pappus, at least in part of capillary bristles (fig. 592d,h).
            7. Involucral bracts of equal length, in only 1 series; ray flowers arranged in peripheral circle. . . . . . . 2. *Packera*
            7. Involucral bracts of unequal length, usually in 2 rows or more series; ray flowers or irregular arrangement around periphery of head (fig. 593h,l).
              8. Heads corymbose (figs. 592e,j, 593a,e,j).
                9. Leaves glandular-punctate; lower cauline leaves falling off early . . . . 3. *Euthamia*
                9. Leaves not glandular-punctate; lower cauline leaves persistent. . . . . . . 4. *Solidago*
              8. Heads paniculate; main branches often spreading or arcing (fig. 594c). . . . . . . 4. *Solidago*
          6. Pappus of translucent scales (fig. 596b). . . . . . . 5. *Helenium*
      4. Plants woody (at least at base), shrubby. . . . . . . 6. *Borrichia*
  2. Ray flowers purple, violet, pink or white.
    10. Leaves alternate.
      11. Leaf apices not apiculate; pappus of long capillary bristles.
        12. Inflorescence umbelliform, ligulate flowers white. . . . . . . 7. *Doellingeria*
        12. Inflorescence various, but if corymbiform, then ligulate flowers not white.
          13. Phyllaries blunt to round-tipped, in several series. . . . . . . 8. *Eurybia*
          13. Phyllaries acute to acuminate.
            14. Flower buds nodding, phyllaries in 1–2 series, hyaline with green midrib. . . . . . . 9. *Oclemena*
            14. Flower heads erect; phyllaries in 1–2 or several series; phyllaries green throughout or with prominent, often rhomboid-green zone. . . . . . . 10. *Symphyotrichum*
      11. Leaf apices apiculate; pappus of several small bristles (not capillary) and 2 somewhat longer awns (sometimes absent) . . . . 11. *Boltonia*
    10. Leaves opposite. . . . . . . 12. *Eclipta*
1. Flower heads consisting of only disk flowers (figs. 607b, 619b).
  15. Leaves linear to linear-subulate, tipped by a tiny callus, strictly whorled in 3s, 4s, or 5s (Fig. 604a); plants submersed and flexuous (remaining vegetative), when water level drops, stranded stems giving rise to erect, emersed to terrestrial fertile stems. . . . . . . 13. *Sclerolepis*
  15. Leaves of various shapes (but not linear, or if linear to linear lanceolate then with sheathing leaf base, usually accompanied by lobed or toothed leaves—*Cotula*), not tipped with by a callus, alternate, opposite, or whorled (*Eutrochium*); plants emersed to terrestrial, erect or viny and twining (*Mikania*).
    16. Receptacle with bristles or chaff subtending flowers.
      17. Leaf margins spiny; pappus of soft, white, plumose bundles of bristles. . . . . . . 14. *Cirsium*
      17. Leaf margins not spiny; pappus of awns, or pappus absent.
        18. Plants herbaceous, not succulent; flowers yellow; pappus of awns present. . . . . . . 1. *Bidens*
        18. Plants woody, shrubs, somewhat succulent on upper portion; flowers greenish-white; pappus absent. . . . . . . 15. *Iva*

16. Receptacle lacking bristles or chaff.
  19. Leaves alternate.
    20. Leaves clasping at base, irregularly lobed, toothed or entire; plants lacking a fetid odor. .... 16. *Cotula*
    20. Leaves not clasping, regularly toothed; plants with a fetid odor. .... 17. *Pulchea*
  19. Leaves opposite or whorled.
    21. Plants twining vines.. .... 18. *Mikania*
    21. Plants erect, non-twining.
      22. Receptacle conic; flowers blue.. .... 19. *Conoclinum*
      22. Receptacle flat; flowers purple, white or pink.
        23. Flowers purple or pink.
          24. Leaves mostly whorled.. .... 20. *Eutrochium*
          24. Leaves opposite. .... 21. *Fleishmannia*
        23. Flowers white. .... 22. *Eupatorium*

---

1. *Bidens* (Beggar-ticks, Bur-marigold, Stick-tight)

Annuals, biennial, or perennial herbs, leaves opposite, simple to pinnately or ternately compound or, in *B. beckii*, highly dissected; flower heads yellow, rarely white or pink; ray flowers, if present, sterile; receptacular bracts (chaff) subtending flowers; disk flowers bisexual, fertile; achenes flattened pappus usually with 2–4 teeth or awns. Beggar-ticks, with their bright yellow flowers, are sometimes very conspicuous in late summer. In the fall many members of this genus force attention on themselves by their barbed fruits fastening in great number to one's clothes.

Although most species of *Bidens* can be recognized in flowering condition if one is familiar with them, in some cases, mature achenes are necessary for certain identification. Like many annuals, *Bidens* is quite variable in habit, and almost any species growing along a lakeshore may range in size from dwarf, unbranched individuals of only a few centimeters tall to copiously branched plants up to a meter or more tall.

References: Roberts, 1982, 1985; Sherff, 1937, 1965; Strother and Weedon, 2006.

---

1. Plants submersed, flexuous (vegetative), the leaves highly dissected, reaching the surface, the flowering stems emergent, with emergent leaves transitioning to simple (pectinate to cleft or entire) (fig. 576). .... 1. *B. beckii*
1. Plants emergent in shallow water to somewhat terrestrial; the leaves simple, cleft or compound never dissected.
  2. Primary leaves simple (fig. 577) or deeply 3(7)-cleft (not compound), with terminal lobe on winged stalk (fig. 582).
    3. Leaves sessile (except occasionally the lowest).
      4. Mature disk 1.3–2.8 cm in diameter; fruiting heads often nodding; leaves conspicuously toothed; achenes obscurely striate.
        5. Ray flowers 1.5–3 cm long; achenes flat, wingless, deep brown to purple; stems firm, glabrous; receptacular bracts with upper portions red. .... 2. *B. laevis*
        5. Ray flowers usually absent, or up to 1.7 cm long; achenes with wing-like margins, olive-green; stems soft, often hispid; receptacular bracts with upper portions yellow. .... 3. *B. cernua*
      4. Mature disk rarely up to 1.5 cm in diameter; fruiting heads erect; leaves subentire or slightly dentate achenes conspicuously striate. .... 4. *B. hyperborea*
    3. Leaves petiolate, petioles 1–4 cm long.
      6. Heads consisting of ray and disk flowers, ray flowers 1–2 cm long. .... 5. *B. mitis*
      6. Heads consisting of disk flowers only, or if ray flowers present, then less than 0.5 cm long.
        7. Achenes strongly 4-angled, midribs prominent, sometimes almost wing-like. .... 8. *B. connata*
        7. Achenes flat or flattish, midribs slender or obscure.
          8. Heads narrow, usually 8–30-flowered.
            9. Achenes slightly pubescent or glabrous; awns usually less than half the length of achene body. .... 6. *B. eatonii*
            9. Achenes extremely pubescent; awns usually more than half the length of achene body. .... 7. *B. bidentoides*
          8. Heads broad, usually 30–150-flowered.
            10. Central achenes 1–2 mm wide; primary leaves slender-petiole.. .... 9. *B. tripartita*
            10. Central achenes 2.4–3 mm wide; primary leaves tapering to wing-petioled. .... 10. *B. heterodoxa*
  2. Primary leaves deeply pinnately lobed or pinnately or ternately compound (figs. 588, 590).
    11. Heads showy; both disk and ray flowers present, ray flowers 1 cm long or longer, conspicuously longer than outer involucral bracts (figs. 587a, 591a).

12. Achenes 2.5–4.5(5) mm long, margins smooth or scarcely ciliate. . . . . 5. *B. mitis*
12. Achenes 4–10 mm long, margins distinctly ciliate.
  13. Achenes 1.5–2(2.5) times as long as wide, with a thin hispid-ciliate margin.
    14. Outer involucral bracts 8–10, shorter than inner, glabrous or short pubescent. . . . . 11. *B. aristosa*
    14. Outer involucral bracts 12–21, mostly longer than inner, hispid-ciliate. . . . . 12. *B. polylepis*
  13. Achenes 2.5–4 times as long as wide, with a slightly strigose-ciliate margin. . . . . 13. *B. trichosperma*
11. Heads not showy; only disk flowers present, or if ray flowers present, then less than 5 mm long, shorter than outer involucral bracts (fig. 590b).
  15. Outer involucral 5–16(21), ciliate at least toward base.
    16. Disks 10–15 mm in diameter, 10 mm high, outer involucral bracts 5–10.. . . . . 14. *B. frondosa*
    16. Disks up to 20 mm in diameter, 1 mm high; outer involucral bracts 10–20. . . . . 15. *B. vulgata*
  15. Outer involucral bracts 3–5, not ciliate. . . . . 16. *B. discoidea*

---

1. *B. beckii* Torr. ex Spreng. Water-marigold Fig. 576
Ponds, lakes and slow streams, C.B.I. and N.B. w. to Sask., s. to N.J., Md., Pa., n. Ohio, Ind., Wisc. and e. Mo.; s. B.C. Wash. Mont., Ida and Oreg. This species has been observed fruiting as far north as northern Minnesota and Manitoba. Overwinters by turions. (*Megalodonta beckii* (Torr. ex Spreng.) Greene)

2. *B. laevis* (L.) Britton, Sterns & Poggenb. Smooth Beggar-ticks Fig. 577
Marshes, sluggish streams stream banks, wet meadows and ditches. Chiefly coastal plain; Me. and N.H. s. to Fla., w. Calif.; local inland; Ohio, Ill. Mo, Colo., Nev.; Trop. Am.

3. *B. cernua* L. Nodding Beggar-ticks Fig. 578
Marshes, sloughs, springs, shores and shallow water. N.S., P.E.I., and Que. w. to B.C. and Alask., s. to Ga. Tex. and Calif. (*B. cernua* var. *integra* Wiegand; (*B. cernua* var. *minima* (Huds.) Pursh); *B. cernua* var. *elliptica* Wiegand.; *B. cernua* var. *oligodonta* Fernald & St. John)

4. *B. hyperborea* Greene Estuary Beggar-ticks Fig. 579
Muds of freshwater and tidal estuaries. Local, St. Lawrence R. system, Que., w. to James Bay, s. to N.B., N.S., Me., Mass., s. N.Y. and n. N.J. This is a highly variable species with several varieties having been described from numerous estuarine systems, but these ecological variants are no longer given taxonomic recognition. (*B. hyperborea* var. *colpophila* (Fernald & St. John) Fernald; *B. hyperborea* var. *cathancensis* Fernald; *B. hyperborea* var. *arcuans* Fernald; *B. hyperborea* var. *gaspensis* Fernald.)

5. *B. mitis* (Michx.) Sherff Small-fruit Beggar-ticks Fig. 591
Marshes, shallow water of ponds and lakes, swampy woodlands, ditches and brackish swamps. Coastal plain, s. N.J., Del. and Md. s. to Fla., w. to e. Tex.; inland Tenn., Ark. and Mo.

6. *B. eatonii* Fernald Eaton's Beggar-ticks Fig. 580
Tidal shores and brackish marshes. Coastal, N.B., St. Lawrence R., Que., Kennebec R., Me., Merrimack R., Mass., Connecticut R. and Quinnipiac R., Conn. and Hudson R., N.Y. This species is extremely variable with several doubtful varieties described, including four named from the same estuary; these are no longer recognized. Examination of type specimens and other specimens collected at the type locality of *B. eatonii* along the tidal flats of the Merrimack River, Massachusetts, by Caldwell and Crow (1992) revealed that there is variability in the size of flower heads and that the heads are not consistently few-flowered. Strother and Weedon (2006) suggest that *B. eatonii* as well as *B. heterodoxa* might better be included within *B. tripartita*.

7. *B. bidentoides* (Nutt.) Britton Delmarva Beggar-ticks Fig. 581
Fresh to brackish tidal shores. Se. N.Y., s. N.J., se. Pa., Del. and ne. Md. (*B. mariana* Blake; *B. bidentoides* var. *mariana* (Blake) Sherff)

8. *B. connata* Muhl. ex Willd. Purplestem Beggar-ticks Figs. 582, 583
Wet sandy or peaty shores, swamps and bogs. Nfld., N.S. and s. Que. w. to Minn., N.D. and Mont., s. to Va. and Ga., Tenn., Mo. and Kans. Numerous varieties have been described.

9. *B. tripartita* L. Threelobe Beggar-ticks Fig. 584
Swampy thickets and waste places. Que. w. to Alask., s. throughout the U.S., worldwide. Our plants have been long treated as *B. comosa*, which tend to differ from European plants by having smooth achene bodies that lack tubercles, but is now lumped in with *B. tripartita*. Some authors include both *B. comosa* and *B. connata* within *B. tripartita*. (*B. comosa* (A. Gray) Wiegand)

10. *B. heterodoxa* Fernald & H. St. John Connecticut Beggar-ticks Fig. 583
Freshwater, brackish or saline marshes. Local, M.I., P.E.I., N.B. Que. and Conn. Extremely variable. Strother and Weedon (2006) suggest that *B. heterodoxa* as well as *B. eatonii* might better be included within *B. tripartita*. (*B. tripartita* var. *heterodoxa* Fernald)

11. *B. aristosa* (Michx.) Britton Tickseed-sunflower, Bearded Beggar-ticks Fig. 585
Low grounds, meadows, marshes, fields and ditches. S. Me., s. N.J. and Mass., to s. Mich., s. Wisc. and s. Minn., s. to S.C., Ala., La., Okla. and e. Tex.; Colo. Populations from the Great Plains are native; natzd. eastward. (*B. aristosa* var. *fritcheyi* Fernald; *B. aristosa* var. *mutica* A. Gray)

12. *B. polylepis* Blake Bearded Beggar-tick Fig. 586
Low prairies, marshes, swales, shores, meadows and disturbed sites. N.J., Pa. w. to s. Ont., s. Mich., Ill. Iowa, Kans. and Colo., s. to S.C., Tenn., Ala., La., Tex. and N.M. (*B. polylepis* var. *retrorsa* Sherff; *B. involucrata* (Nutt.) Britton)

13. *B. trichosperma* Michx. Tickseed-sunflower, Crowned Beggar-ticks Fig. 587
Bogs, fens, tamarack swamps, swales, prairies, hardwood swamps and coastal marshes. S. Que., Mass. and N.Y. w. to s. Ont., Mich., Minn. and s. S.D., s. to Fla., Ala., La., Ark. and Neb. (*B. coronata* (L.) Britton)

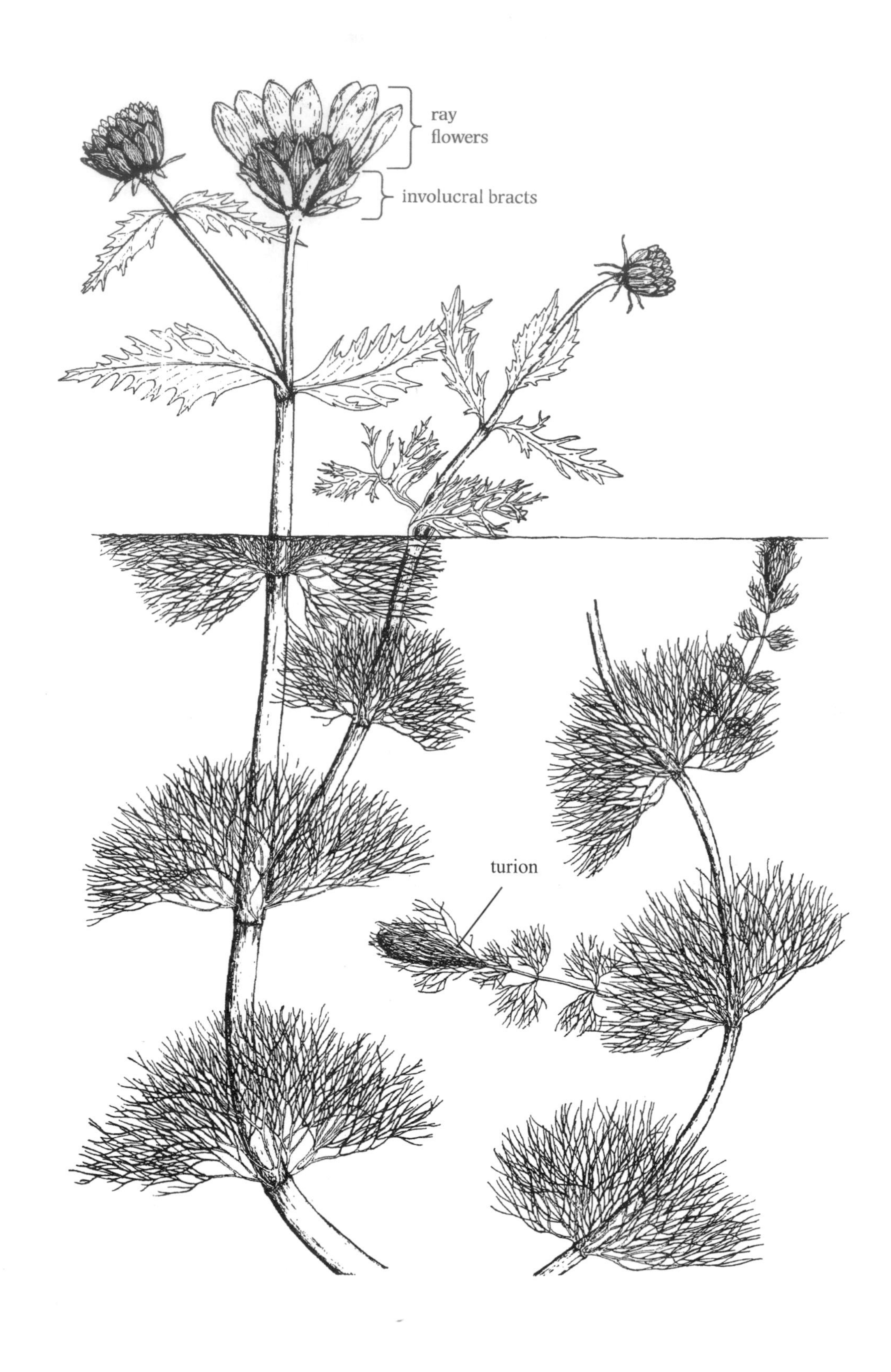

Fig. 576. *Bidens beckii*: submersed and emersed portions of plant (F).

Fig. 577. *Bidens laevis*: a. upper portion of plant; b. achene (F).

Fig. 578. *Bidens cernua*: a. upper portion of plant; b. habit, small growth form; c. leaf; d. achene (F).

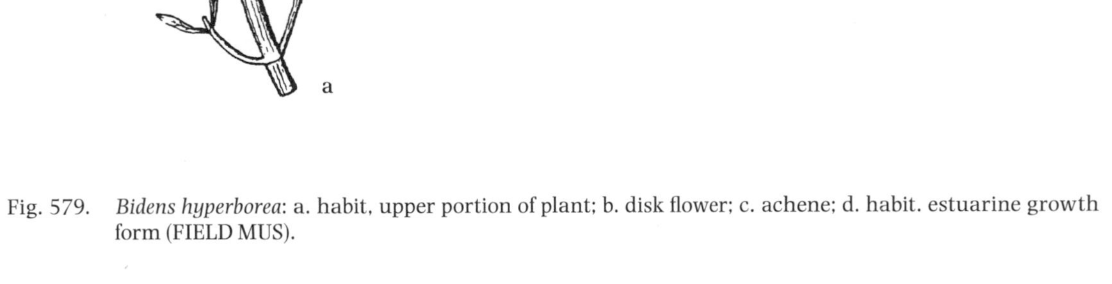

Fig. 579. *Bidens hyperborea*: a. habit, upper portion of plant; b. disk flower; c. achene; d. habit, estuarine growth form (FIELD MUS).

Fig. 580. *Bidens eatonii*: a. upper portion of plant; b. achene (FIELD MUS).

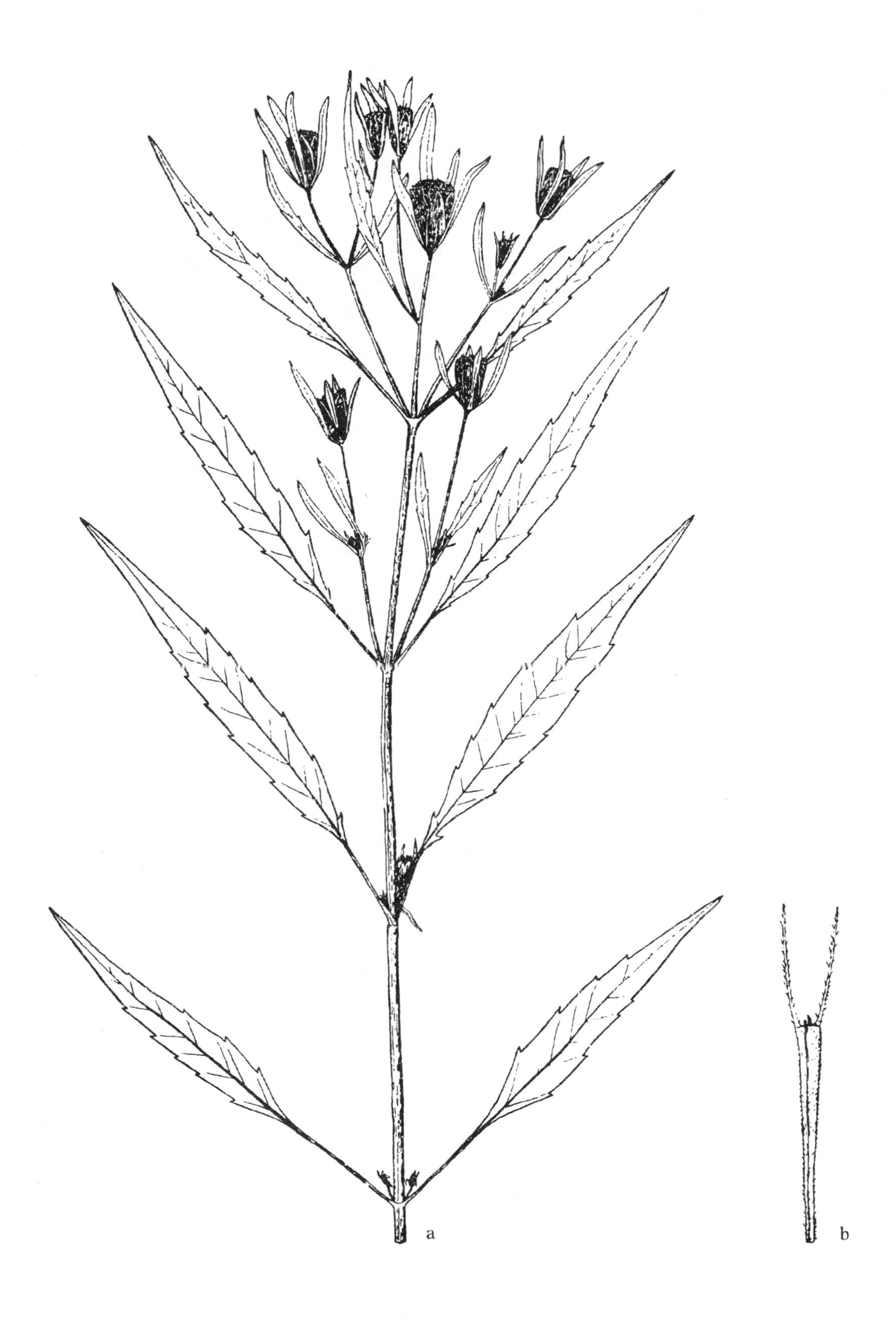

Fig. 581. *Bidens bidentoides*: a. upper portion of plant; b. achene (FIELD MUS).

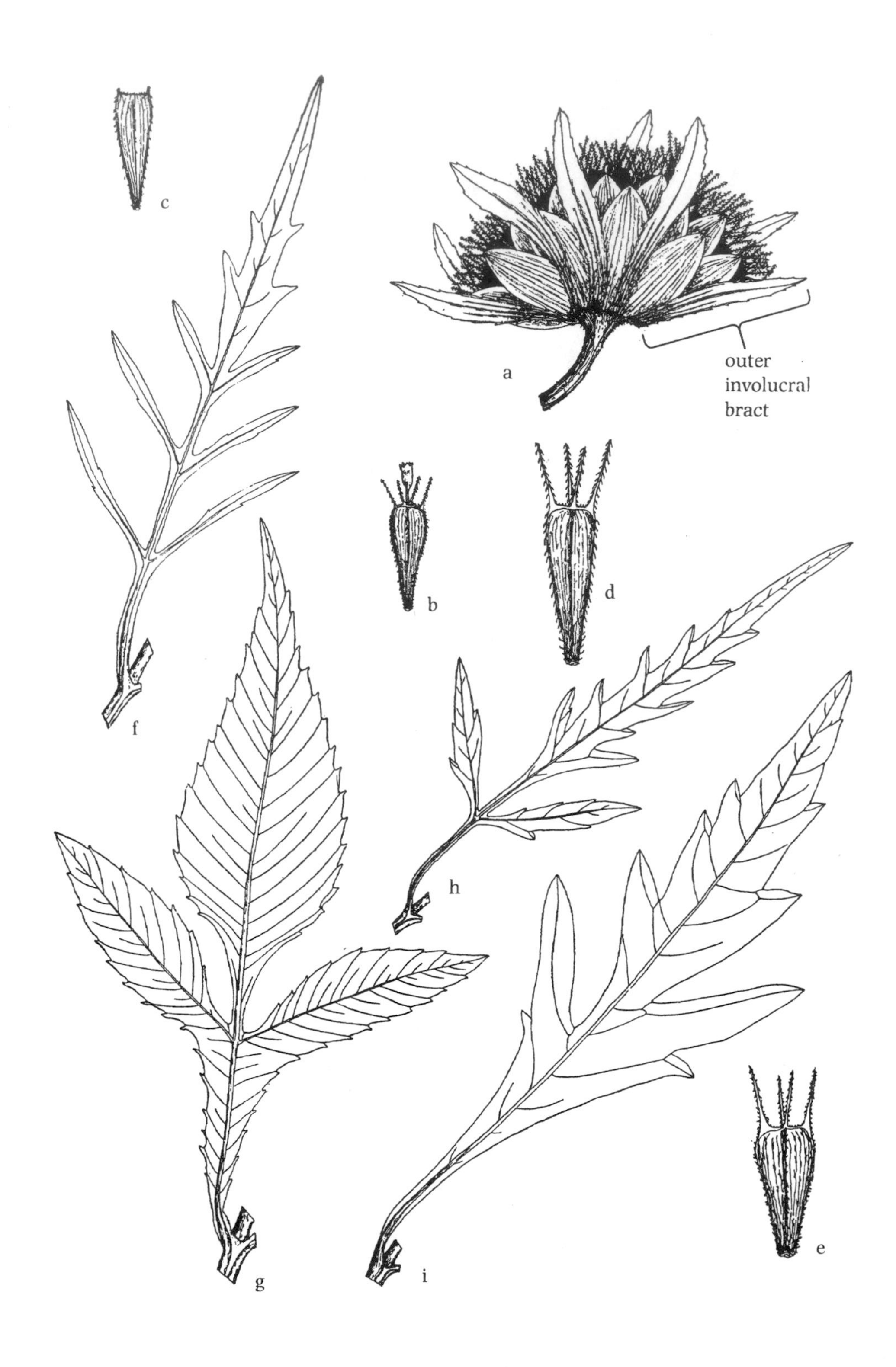

Fig. 582. *Bidens connata*: a. flower head; b. young achene with corolla persisting; c–e. achene variations; f–i. leaf variations (F).

Fig. 583. *Bidens connata*: a. upper portion of plant; b. leaf; c. flower head; d. achene (F). *Bidens heterodoxa*: e, f. achene variation (F).

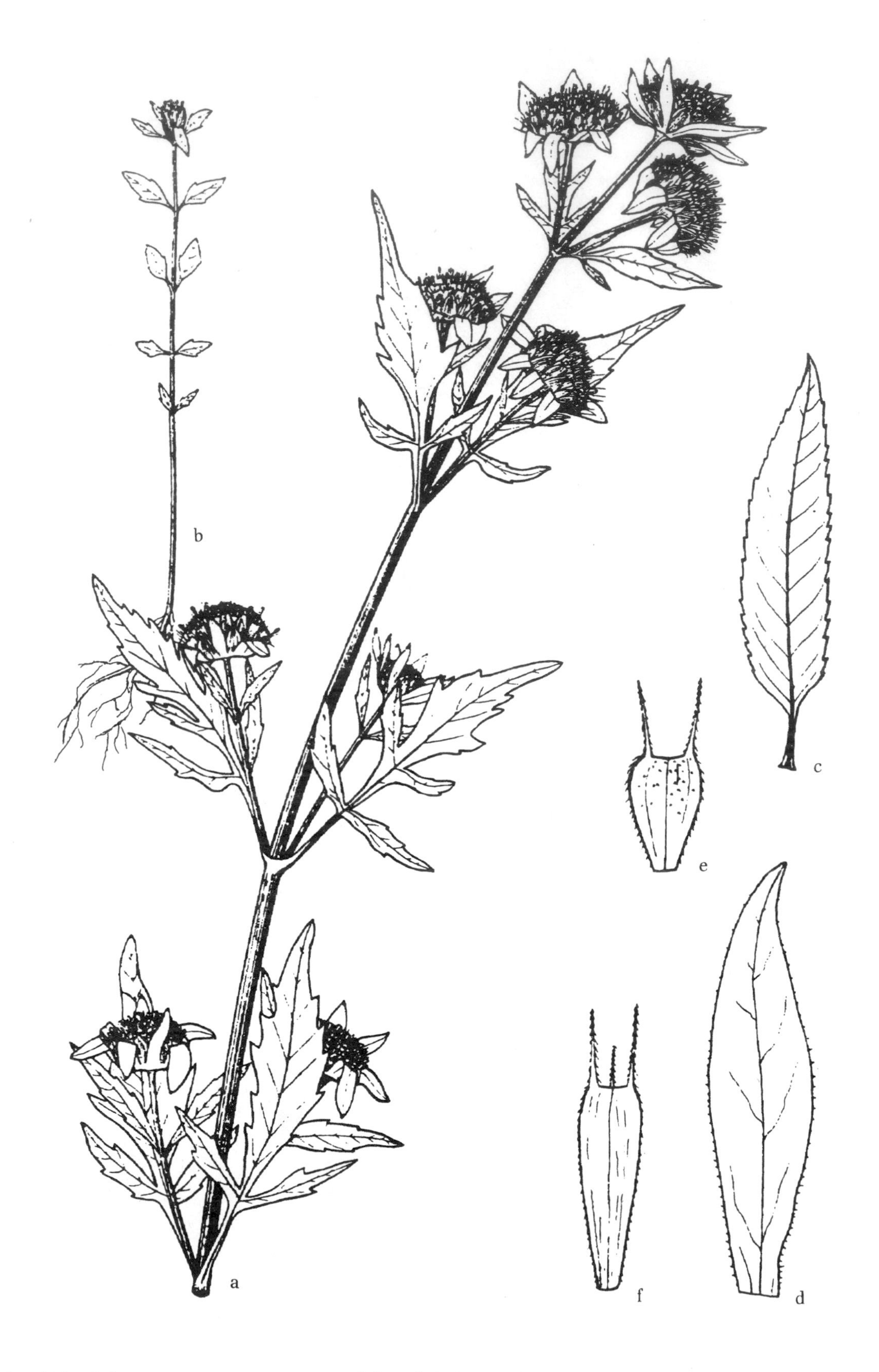

Fig. 584. *Bidens tripartita*: a. upper portion of plant; b. habit, small plant; c, d. leaf variations; e, f. achene variations (FIELD MUS).

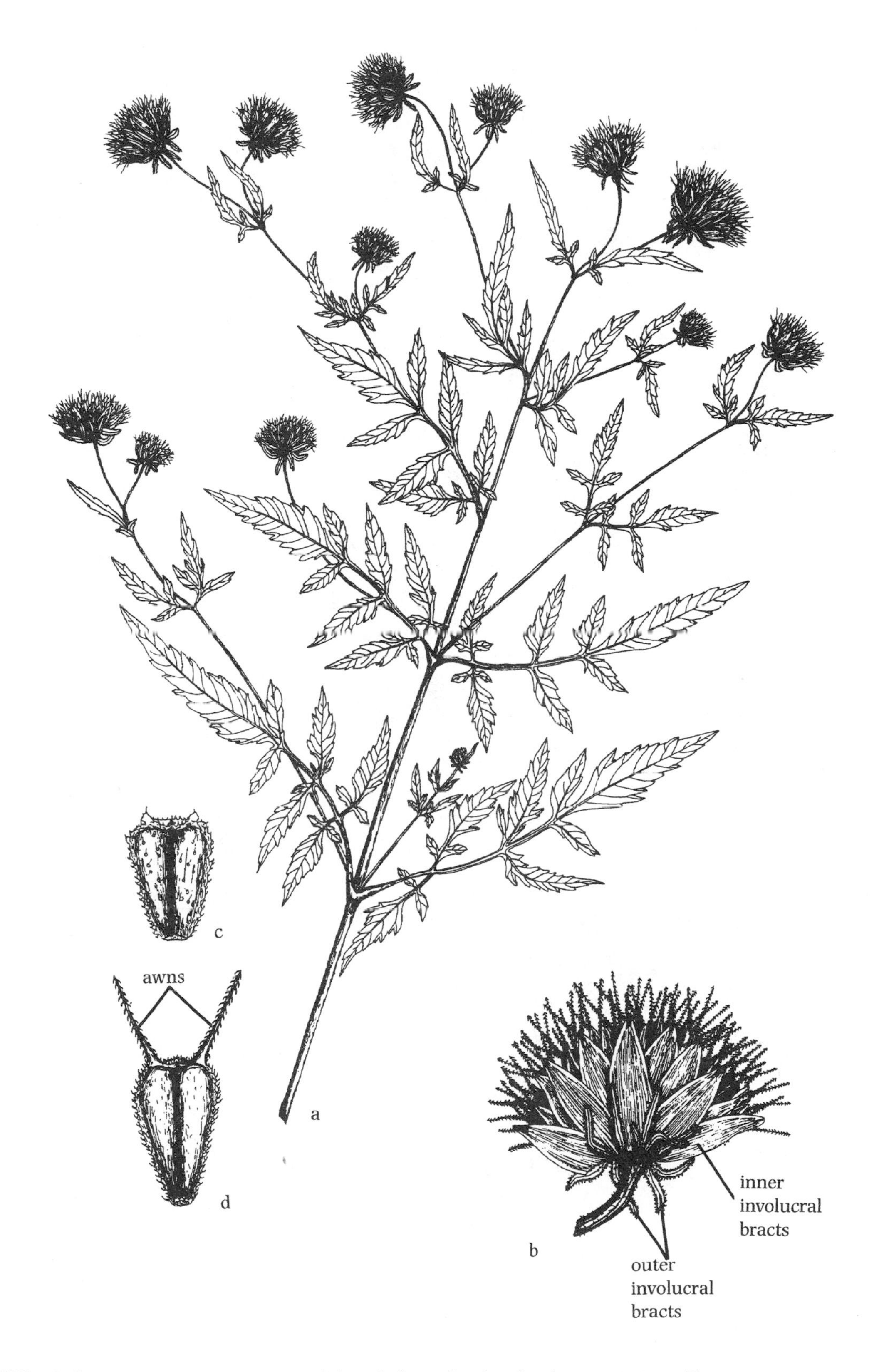

Fig. 585. *Bidens aristosa*: a. upper portion of plant; b. flower head; c, d. achene variations (F).

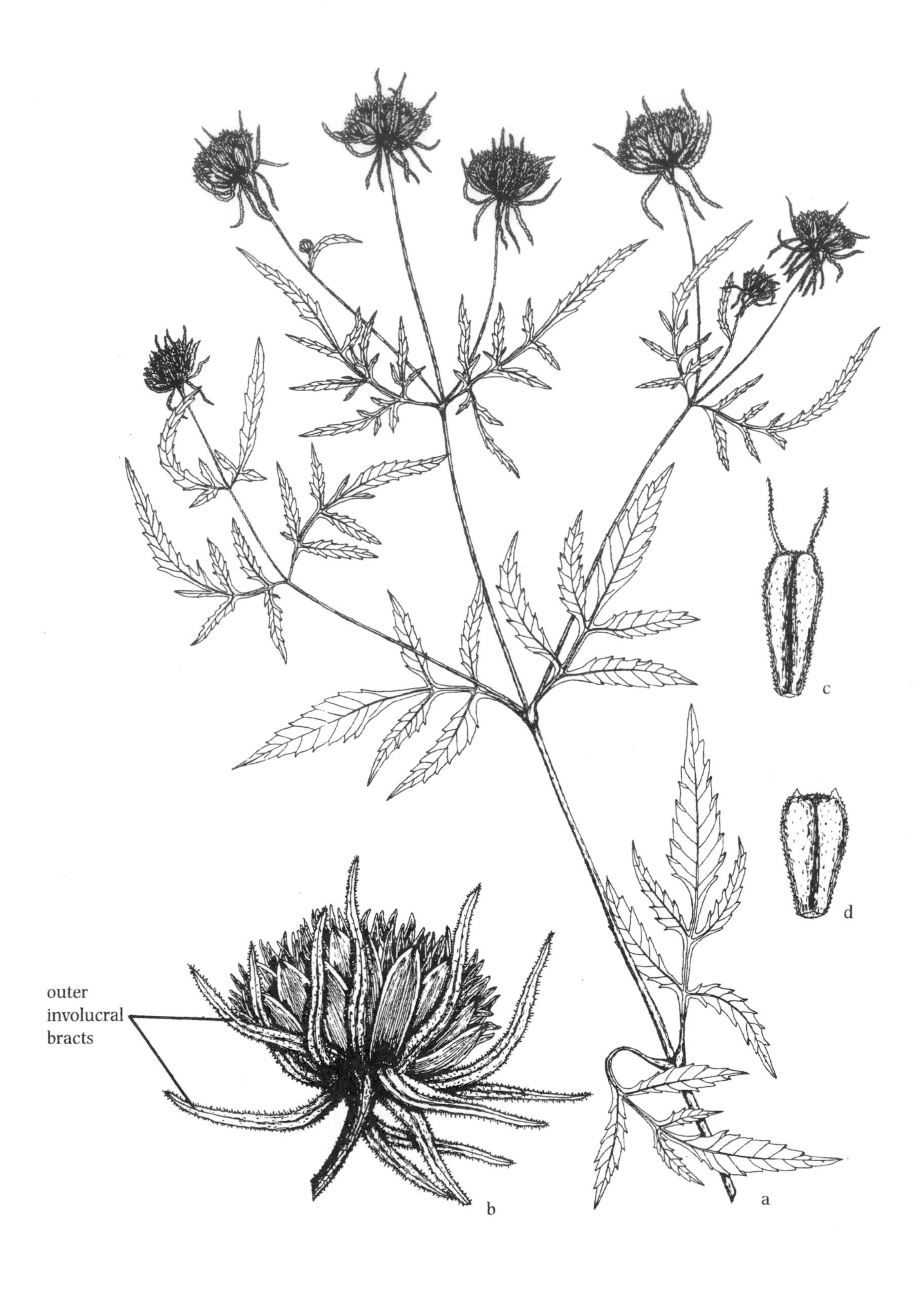

Fig. 586. *Bidens polylepis*: a. upper portion of plant; b. flower head; c, d. achene variations (F).

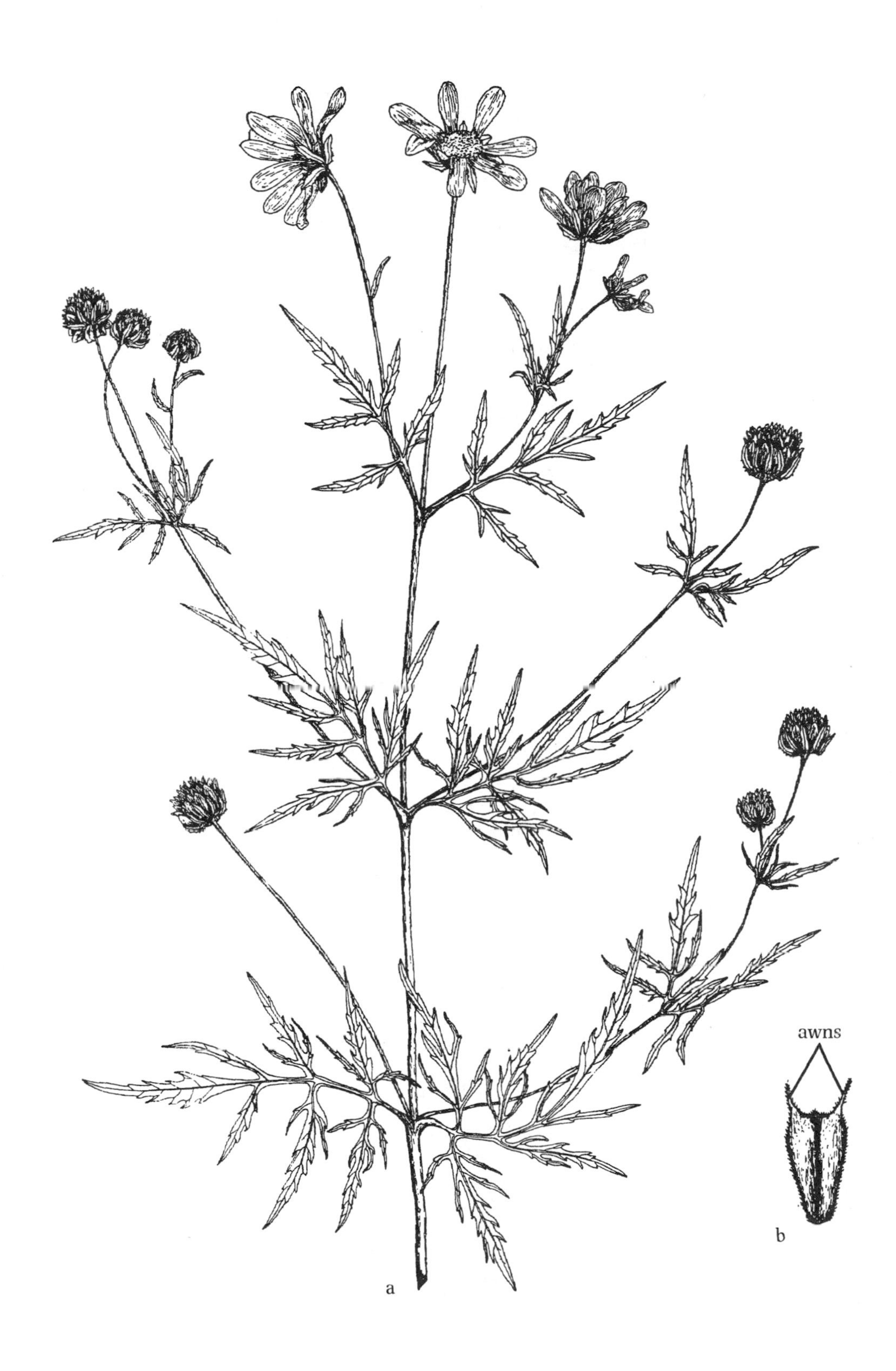

Fig. 587. *Bidens trichosperma*: a. upper portion of plant; b. achene (F).

14. *B. frondosa* L. Devil's Beggar-ticks Fig. 588
Wet sites, wet thickets, stream banks, marshy shores, ditches and waste places. Nfld. w. to B.C. and Alask., s. to Fla. Tex., Calif. and Mex. (*B. frondosa* var. *anomala* Porter ex Fernald; *B. frondosa* var. *stenodonta* Fernald & H. St. John)

15. *B. vulgata* Greene Beggar-ticks, Stick-tight Fig. 589
Wet sites, thickets, low gound and ditches. N.S., N.B. and s. Que. w. to Alta., B.C. and Wash., s. to Va., mts. of N.C., S.C. and Ga., Tenn., La., Mo., Kans., Okla., Wyo., N.M., Ida., Nev. and Calif. (*B. vulgata* var. *schizantha* Lunell)

16. *B discoidea* (Torr. & Gray) Britton Small Beggar-ticks Fig. 590
Peaty or Sandy swamps, shores, debris in pools and floodplains. N.S., N.B., Que. and s. Ont. w. to ne. Minn., s. to Fla., se. Okla. and se. Tex.

2. *Packera* (Ragwort, Groundsel, Squaw-weed)

Annual or perennial (ours) herbs; leaves alternate, unlobed to dissected; heads in a corymbose, cymose, or umbelliform inflorescences; ray flowers, pistillate, 0 or (1)5–13, and disk flowers, bisexual, 20–80 or more; flowers pale yellow to deep orange-red; achenes prominently ribbed.

REFERENCES: Barkley, 1978; Trock, 2006.

1. Basal and lower cauline leaves with cordate, subcordate, or truncate base (fig. 591g).
   2. Ray flowers 8–10 mm, golden; basal leaf margins crenate to sometimes coarsely serrate; basal leaf blades cordate to reniform, 1.75–3.5 times as long as wide. . . . . 1. *P. aurea*
   2. Ray flowers 4–7 mm, yellow; basal leaf margins sharply to finely serrate; basal leaf blades narrowly ovate to oblong-lanceolate, 0.8–1.5 times as long as wide. . . . . 2. *P. schweinitziana*
1. Basal and lower cauline leaves tapering to a petiolate base, or truncate to cuneate, but not cordate (fig. 591f).
   3. Heads with both disk and ray flowers, the ray flowers pale yellow (rarely lacking ray flowers). . . . . 3. *P. paupercula*
   3. Heads with disk flowers only or if ray flowers present, then very short, less than 5 mm, deep yellow. . . . . 4. *P. indecora*

1. *P. aurea* (L.) Á. Löve & D. Löve Golden Ragwort, Squaw-weed Fig. 591
Meadows, boggy swales, low woods, thickets and swamps. Nfld. and Lab., w. to Que., w. to Man. and. Minn., s. to n. Ga., nw. Fla., S.D., Ark. and e. Okla. (*Senecio aureus* L.; *S. aureus* var. *intercursus* Fernald; *S. aureus* var. *gracilis* (Pursh) Hook.; *S. aureus* var. *semicordatus* (Mackenz. & Bush) Greenm.; *S. aureus* var. *aquilonius* Fernald)

2. *P. schweinitziana* (Nutt.) W.A. Weber & Á. Löve New England Groundsel
Peaty meadows, swales, swampy woods, thickets and fields. Que. s. to P.E.I., N.S., n. N.E. and ne. N.Y.; mts. of N.C. and Tenn. (*Senecio schweinitziana* Nutt.; *S. robbinsii* Oakes ex Rusby)

3. *P. paupercula* (Michx.) Á. Löve & D. Löve Balsam Groundsel Fig. 591
Calcareous damp sites, among wet rocks, bogs and wet meadows. Lab. and Nfld. w. to Alask., s. to Va. and N.C., nw. Ga., Tenn., Ill., Iowa, Neb., Colo., Utah, Wash. and Ida. (*Senecio pauperculus* Michx.; *S. pauperculus* var. *balsamitae* (Muhl. ex Willd.) Fernald; *S. pauperculus* var. *neoscoticus* Fernald; *pauperculus* var. *praelongus* (Greenm.) House; *S. gaspensis* Greenm.; *S. crawfordii* Britton)

4. *P. indecora* (Greene) Á. Löve & D. Löve Elegant Groundsel
Rich thickets, swales, stream banks and bogs. Nfld. and Lab. and e. Que. w. to Nunavut, N.W.T. and Alask., s. to n. Mich., ne. Minn., Wyo., Ida. and n. Calif. (*Senecio indecorus* Greene)

3. *Euthamia* (Flat-topped Goldenrod)

Perennial herbs; stems erect, branching in upper portion of plant; leaves sessile, linear to narrowly lanceolate, surfaces punctate; flower heads yellow, small, in corymbs; ray flowers more numerous than disk flowers; achenes many-ribbed, nearly round in cross-section, pappus a single series of capillary bristles.

REFERENCES: Haines; 2006; Sieren, 1970, 1981.

1. Leaves 1-nerved, or sometimes with a faint vein on either side of the main vein.
   2. Involucre about 4.5–6 mm long; plants of se. U.S. . . . . 1. *E. leptocephala*
   2. Involucre about 3–4.5 mm long; plants of coastal plain. . . . . 2. *E. caroliniana*
1. Leaves 3-nerved, usually with 1 or 2 pairs of additional obscure lateral nerves. . . . . 3. *E. graminifolia*

Fig. 588. *Bidens frondosa*: a. upper portion of plant; b. leaf; c. flower head; d, e. achene variations (F).

Fig. 589. *Bidens vulgata*: a. upper portion of plant; b. flower head; c. achene (F).

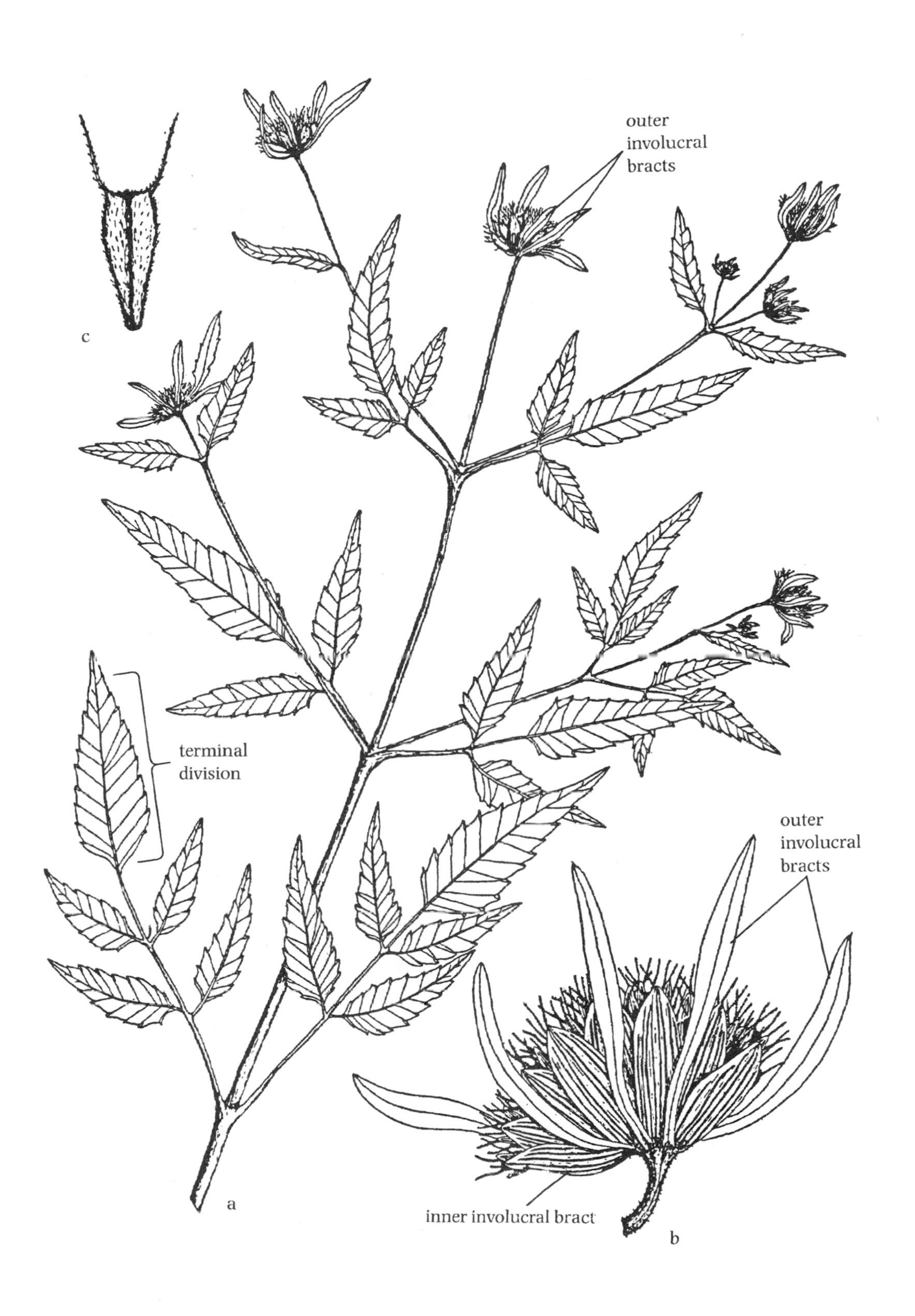

Fig. 590. *Bidens discoidea*: a. upper portion of plant; b. flower head; c. achene (F).

Fig. 591. *Bidens mitis*: a. upper portion of plant; b–e. leaf variations (FIELD MUS).
Packera paupercula: f. habit (Gleason).
Packera aurea: g. habit, including basal leaves (WVA).

1. *E. leptocephala* (Torr. & A. Gray) Greene Bushy Goldentop
Damp sandy soil, glades, meadows and swales. N.C. and Tenn. w. to se. Mo. and e. Okla., s. to Miss., La. and e. Tex. (*Solidago leptocephala* Torr. & A. Gray)

2. *E. caroliniana* (L.) Greene ex Porter & Britton Slender-leaved Goldenrod Fig. 592
Sandy damp shores, wet marshes, brackish and saline habitats, often in shallow water. Coastal plain; N.S. s. to Fla., w. to La.; s. Mich. and n. Ind. (*E. galetorum* Greene; *E. tenuifolia* (Pursh) Greene)

3. *E. graminifolia* (L.) Nutt. ex Cass. Narrow-leaved Goldenrod Fig. 592
Damp to dry shores, thickets and marshes. Nfld. and Que. w. to N.W.T. and B.C., s. to N.C., S.C., Miss., La., Okla. and Tex., mts. to Colo. and N.M., Ida. and Oreg. (*Solidago graminifolia* var. *nuttallii* (Greene) W. Stone; *S. graminifolia* var. *major* (Michx.) Moldenke)

4. *Solidago* (Goldenrod)

Perennial herbs; stems erect, unbranched below inflorescence; basal and lower cauline leaves petiolate, sub-petiolate, or sessile; upper cauline leaves sessile; flower heads borne in corymbs or panicles, cream-colored or yellow (ours); achenes many-ribbed, nearly round in cross-section, pappus a single series of numerous equal or unequal capillary bristles.

References: Semple and Cook, 2006; Laureto and Pringle, 2010.

---

1. Inflorescence corymbose (fig. 593e,j)
   2. Rays 1.5–3 mm long, involucre ca. 3.5–5.5(6.5) mm long.
      3. Peduncles 6.3–8.5 mm long; leaf blades flat, with one longitudinal vein, occasionally with lateral veins. . . . . 1. *S. ohioensis*
      3. Peduncles 2.8–4 mm long; leaf blades mostly folded inwards in a V-shaped, with 3 or more veins arising at base. . . . . 2. *S. riddellii*
   2. Rays 3–4.5(7) mm long, involucre ca. 5–9 mm long.
      4. Peduncles 5–6.2 mm long; larger involucres 5–7(8) mm long; rays 0.5–0.6 mm wide; basal leaves entire; plants mostly 30–60 cm tall. . . . . 3. *S. houghtonii*
      4. Peduncles 6–12 mm long; large involucres 7–9 mm long; rays 1.2–2 mm wide; basal leaves sparsely serrulate, plants mostly 50–80 cm tall. . . . . 4. *S. vossii*
1. Inflorescence paniculate (fig. 594).
   5. Leaves somewhat succulent (or leathery); plants of maritime habitats.
      6. Involucres 4–7 mm, rays 12–17; disk flowers 17–22. . . . . 5a. *S. sempervirens* subsp. *sempervirens*
      6. Involucres 3–4 mm, rays 7–11; disk flowers 10–16.. . . . . 5b. *S. sempervirens* subsp. *mexicana*
   5. Leaves not succulent; plants mainly of inland habitats, occasionally maritime.
      7. Basal leaves and lower cauline flowers much longer than middle cauline leaves.
         8. Leaves usually glabrous on upper surface.
            9. Plants rhizomatous; inflorescence glabrous; achenes pubescent. . . . . 6. *S. stricta*
            9. Plants not rhizomatous inflorescence pubescent; achenes glabrous. . . . . 7. *S. uliginosa*
         8. Leaves scabrous on upper surface. . . . . 8. *S. patula*
      7. Basal leaves and lower cauline leaves usually shorter than middle cauline leaves.
         10. Stem glabrous below inflorescence; leaves gradually narrowed to base, often with short petiole. . . . . 9. *S. latissimifolia*
         10. Stem somewhat pubescent below inflorescence; leaves sessile, bases broad or rounded.. . . . . 10. *S. fistulosa*

---

1. *S. ohioensis* Riddell Ohio Goldenrod Fig. 593
Swamps, beaches, wet prairies and calcareous bogs. W. N.Y. and e. Ont. w. to Ohio, Mich., Ind., n. Ill. and se. Wisc. (*Aster ohioensis* (Riddell) Kuntze; *Oligoneuron ohioense* (Riddell) G. N. Jones)

2. *S. riddellii* Frank Riddell's Goldenrod Fig. 593
Wet prairies, swamps, wet meadows and ditches. S. Ont. w. to Man. and Minn., s. to Ohio, Ill. and Mo.

3. *S. houghtonii* Torr. & A. Gray Houghton's Goldenrod Fig. 593
Damp mossy ground, marly bogs, swamps, alvar (stone pavements) and wet beaches. Ont., Mich., and N.Y. A similar amphiploid hybrid with the putative parent species *S. ohioensis* and *S. uliginosa* is found in western New York. This differs from *S. houghtonii* in that it lacks a basal leaf.

4. *S. vossii* J. S. Pringle & Laureto Voss's Goldenrod Fig. 594
Moist swales in Jack Pine plains. This is a rare, very narrow endemic species presently only known from a small area from Crawford and Kalkaska Counties, Michigan, commemorates revered life-long botanist and student of the flora of Michigan, Edward G. Voss. It is a fertile octaploid, shown through molecular techniques to be of hybrid origin sharing phylogenetic ancestry with *S. houghtonii*, a federally protected hexaploid species (Laureto and Pringle, 2010).

5. *S. sempervirens* L. Seaside Goldenrod Fig. 594

5a. *S. sempervirens* subsp. *sempervirens*
Saline or brackish seashores, dunes, salt marshes and tidal shores, rarely freshwater sites; recently becoming established somewhat inland along highway ditches. Coastal; Nfld. s. to Va.; inland Ont., Mich., Ohio and Ill., saline areas and along heavily salted highways.

Fig. 592. *Euthamia caroliniana*: a. upper portion of plant; b. leaf; c. flower head; d. fruiting head with pappus at summit of achenes; e. upper portion of plant; f. leaf; g. flower head; h. fruiting head with pappus at summit of achenes (Gleason).
Euthamia graminifolia: i. upper portion of plant; j. leaf; k. fruiting head (WVA).

Fig. 593. *Solidago riddellii*: a. upper portion of plant; b. basal leaf; c. flower head; d. achene (Gleason).
Solidago ohioensis: e. upper portion of plant; f. cauline leaves and section of stem; g. basal leaf; h. flower head; i. achene (Gleason).
Solidago houghtonii: j. upper portion of plant; k. midsection of plant; l. flower head (Gleason).

Fig. 594. *Solidago sempervirens*: a. upper portion of plant; b. flower head (Gleason).
Solidago patula: c. upper portion of plant (WVA).
Solidago uliginosa: d. upper portion of plant (WVA).
Solidago vossii: e. habit; f. head with three flowers showing; g. mature achene with corolla and pappus attached at summit; h. phyllary with chlorophyllous zone dark (P. J. Laureto, in Laureto and Pringle, 2010).

5b. *S. sempervirens* subsp. *mexicana* (L.) Semple
Brackish marshes, coastal; Mass. and R.I. s. to Fla., w. to Tex.; Mex., W.I. and C.Am. (*S. mexicana* L.)

6. *S. stricta* Aiton Wand Goldenrod
Bogs, wet pine savannas, floodplains, damp sands and coastal marshes. N.J. and Del. s. to Fla., w. to Tex. and Mex.; W.I. Our taxon, subsp. *gracillima* (Torr. & A. Gray) Semple, has plants with entire basal leaves with elongate proximal branches and occurs mainly along the inner coastal plain. The more southernly taxon from the outer coastal plain is subsp. *stricta* with basal leaves sparsely to obviously serrate, without elongate proximal branches.

7. *S. uliginosa* Nutt. Bog Goldenrod Fig. 594
Acid swamps, calcareous fens, wet sands, peaty barrens, and wet to dryish thickets. Nfld. w. to Ont., Man., Minn. and s. Wisc., s. to Del., Md., and mts. to N.C., Tenn., Ga. and Ala.

8. *S. patula* Muhl. Rough-leaved Goldenrod Fig. 594
Bogs, swamps, wet woods and meadows. Vt. w. to Wisc. and Iowa s. to Ga. and La. A more slender southern plant, subsp. *strictula* (Torr. & A. Gray) Semple, occurs from N.C. to Tex., with only s. Mo. in our range.

9. *S. latissimifolia* Mill. Elliott's goldenrod
Swamps, wet thickets, bogs and wet woodlands. Coastal plain; N.S.; e. Mass. s. to Fla and Ala. Appearing in other manuals as *S. elliottii* Torr. & A. Gray, Utall and Porter (1988) have found *S. latissimifolia* to be an older name for this taxon, thus having nomenclatural priority. (*S. elliottii* var. *ascendens* Fernald; *S. edisoniana* Mack.; *S. mirabilis* Small)

10. *S. fistulosa* Mill. Pine Barren Goldenrod
Low ground, wet woodlands and depressions. Coastal plain; s. N.J. s. to Fla., w. to Tex.

5. *Helenium* (Sneezeweed)

Annual or perennial (ours) herbs; stems erect, winged (ours) or unwinged; leaves simple, alternate, sessile to sub-petiolate; flower heads terminal, solitary or numerous, in corymbs; ray flowers yellow, purplish at base in some; achenes ribbed, obconic, pappus of 5–8 thin chaffy scales.

REFERENCE: Bierner, 2006.

1. Disk flowers yellow. . . . . 1. *H. autumnale*
1. Disk flowers brown or purplish. . . . . 2. *H. flexuosum*

1. *H. autumnale* L. Common Sneezeweed Fig. 595
Thickets, meadows, flood plains and shores, Que. and w. N.E. w. to N.W.T. and B.C. s. to Fla.; Ariz. and Calif. (*H. latifolium* Mill.; *H. autumnale* var. *canaliculatum* Lam.; *H. autumnale* var. *parviflorum* (Nutt.) Fernald; *H. parviflorum* Nutt.; *H. virginicum* Blake)

2. *H. flexuosum* Raf. Purplehead Sneezeweed Fig. 596
Pine savannas, flatwoods, damp plains, meadows and ditches, N.E. w. to Ont., Mich. and Mo., s. to Fla., se. Okla. and e. Tex. (*H. nudiflorum* Nutt.)

6. *Borrichia* (Sea-oxeye)

Woody shrubs; leaves opposite, sub-petiolate or gradually tapering to petioles; flower heads consisting of both ray and disk flowers, receptacle flat; ray flowers yellow; disk flowers brownish yellow; achenes 3- or 4-angled, pappus a short-toothed crown (fig. 596e)

REFERENCE: Semple, 2006.

1. *B. frutescens* (L.) DC. Bushy Seaside Tansy Fig. 596
Saline or brackish marshes, mud flats, shores and mangrove swamps. Coastal; Md, se. Va. s. to Fla. w. to Tex. and e. Mex.; W.I., Ber.

7. *Doellingeria* (Tall Flat-topped Aster)

Perennials; from rhizomes or crowns; stems erect; leaves basal and cauline, alternate or sessile, lanceolate, margins entire or serrate; inflorescence corymbiform, often flat-topped; phyllaries in several series, green with green midrib; ray flowers pistillate, ligules white, disk flowers bisexual; pappus of fine capillary bristles, outer short and inner long.

REFERENCE: Semple and Chmielewski, 2006.

1. *D. umbellata* (Miller) Nees
Moist soils, clearings, thickets and forest near streams. Nfld. and Lab. w. to Ont. and Sask., s. to Ga., Ala., Ill., Iowa, and Neb. (*Aster umbellatus* Mill.)

Fig. 595. *Helenium autumnale*: a. upper portion of plant; b. node and small section of stem, showing decurrent petiolar wings; c. flower head with disc and ray flowers; d. fruiting head; e. achene with pappus scales at summit (G&W).

Fig. 596. *Helenium flexuosum*: a. habit; b. achene with pappus scales at summit (C&C).
Borrichia frutescens: c, d. upper portion of plant, showing leaf variations; e. achene (C&C).

8. *Eurybia* (Aster)

Perennials; rhizomes often become woody; stems ascending to erect; leaves basal and cauline, alternate, sessile or petiolate margins entire or serrate; heads radiate, corymbiform; involucres campanulate; phyllaries 20–140 (in 3–7 series); receptacles flat to slightly convex; ray flowers pistillate; corollas white to purple; disk flowers bisexual, corollas yellow becoming purple at maturity; pappus bristles unequal, soft to stiff.

Reference: Brouillet, 2006.

1. *E. radula* (Sol. ex Aiton) G. L. Nesom
Bogs, swamps, stream margins and low woods. Nfld., Que. and Ont. s. to N.E., Del., Pa., Md., w. Va. and W.Va. (*Aster radula* Sol. ex Aiton)

9. *Oclemena* (Aster)

Perennials; roots fibrous or thick and fleshy; stems erect, simple, minutely villous; leaves cauline, alternate, sessile or short-petiolate; heads radiate in loose corymbiform arrays or single; involucres cylindric to campanulate; phyllaries 5–50 in 3–4 series; receptacles plano-convex pitted or smooth; ray flowers mainly 7–25, pistillate, fertile, corollas white or pink; disk flowers mainly 14–35, bisexual, fertile, corollas pale or pinkish yellow, reddening at maturity; pappus bristles persistent, tan and barbellate.

Reference: Brouillet, 2006.

---

1. Stems 1–2 mm in diameter; leaves numerous (30)40–75(100), crowded (especially toward base of stem); largest leaves 5–12 mm wide, margins revolute, entire. . . . . 1. *O. nemoralis*
1. Stems 1.5–4 in diameter; leaves fewer, 20–40, more or less evenly spaced; largest leaves 7–24 mm wide, margins flat, often toothed. . . . . 2. *O. ×blakei*

---

1. *O. nemoralis* (Aiton) Greene Bog Aster Fig. 597
Bogs, wet peats, ponds and lakeshores. Nfld. w. to Ont., s. to N.E., N.J., n. Pa., N.Y. and n. Mich. (*Aster nemoralis* Aiton)

2. *O. ×blakei* (Porter) G. L. Nesom
Damp thickets, low woods, shores and rarely bogs. Sw. Nfld. and C.B.I. w. to sw. Que. and n. N.Y., s. to N.E. and N.J. Pike (1970) presented strong evidence for the possible hybrid origin of *O. blakei* with *O. nemoralis*, a plant of bogs and wet peaty soils, as one parent, and *A. acuminatus*, a plant of dry to moist woods and clearings, as the other parent; it's been confirmed with isozyme studies, with hybrid populations including backcrosses with parents (Brouillet, 2006). (*Aster nemoralis* var. *blakei* Porter)

10. *Symphyotrichum* (Aster)

Annual, tap-rooted, or perennial, rhizomatous or stems arising from a caudex; leaves basal and cauline, the stem leaves sessile or petiolate; inflorescence paniculate or racemose to corymbose, often with reduced bracteal leaves, subtending branchlets; phyllaries in 3–7 series; ray flowers, fertile, white, bluish-white, blue, violet or pink; disk flowers bisexual, fertile yellow or white becoming purplish to reddish or pinkish at maturity; pappus of capillary filiform bristles, elongate.

References: Brouillet et al., 2006; Nesom, 1994, 1997.

---

1. Plants annual; ray flowers short, obscure, about as long as involucral bracts (fig. 598c). . . . . 1. *S. subulatum*
1. Plants perennial ray flowers long, conspicuous, longer than involucral bracts (fig. 599).
   2. Leaves thick, succulent, margins not scabrous; plants of salt marshes and saline shores. . . . . 2. *S. tenuifolium*
   2. Leaves not succulent; plants of various wet habitats, but not salt marshes.
      3. Cauline leaves weakly to strongly auriculate-clasping at base.
         4. Involucral bracts densely glandular; ray flowers usually deep purple or rose. . . . . 3. *S. novae-angliae*
         4. Involucral bracts not glandular; ray flowers light violet, violet-blue, blue to pale blue, occasionally white.
            5. Leaves narrow, usually 1.5–5 (rarely 9) mm wide, margins entire to subentire, leaves weakly auriculate. . . . . 4. *S. boreale*
            5. Leaves wider, usually 5–70 mm wide, margins serrate or dentate, occasionally entire, leaves strongly auriculate (fig. 597f).
               6. Stems mostly flexuous, petioles and leaf bases strongly dilated; leaf margins serrate. 5. *S. prenanthoides*
               6. Stems straight; petiole and leaf bases not dilated, leaf margins mostly serrate of cuneate-serrate.

Fig. 597. *Symphyotrichum novae-angliae*: a. upper portion of plant (WVA).
Oclemena nemoralis: b. flowering portion of plant; c. ray flower (pistillate with pappus bristles); d. disk flower (bisexual with pappus bristles) (B&B).
Symphyotrichum elliottii: e. branch and inflorescence (Beal).
Symphyotrichum puniceum: f. flowering portion of plant (USDA-NRCS); g. section of stem with leaves (F).

7. Leaf upper surfaces scabrous or glabrate, occasionally pilose on the midveins of lower surface; involucres mainly 8–12 mm long; ray flowers 20–50. . . . . 6. *S. puniceum*
7. Leaf upper surfaces glabrous, occasionally hirsute on midveins of lower surface; involucres mainly 6–9 mm long; ray flowers 15–35. . . . . 7. *S. novi-belgii*
3. Cauline leaves tapered to rounded at base, not clasping.
8. Involucre 7–15 mm long, bracts long-attenuate, only slightly overlapping; ray flowers usually pink, occasionally lilac or purple. . . . . 8. *S. elliottii*
8. Involucre 2.5–11 mm long, bracts rounded to acute, but not attenuate, distinctly overlapping.
9. Ray flowers violet, blue-violet or purple (rarely white); involucre 5–7(8) mm long. . . . . 9. *S. praealtum*
9. Ray flowers white, rarely pink, purple, or purple tinged; involucre 2.5–6.6 mm long.
10. Leaves pubescent along midrib on lower surface; ray flowers 9–14. . . . . 10. *S. lateriflorum*
10. Leaves glabrous along midrib on lower side; ray flowers 15–40.
11. Heads few per plant; largest leaves 2–10 cm long or longer. . . . . 11 *S. tradescantii*
11. Heads numerous per plant; largest leaves 10 cm or longer. . . . . 12. *S. lanceolatum*

---

1. *S. subulatum* (Michx.) G. L. Nesom Annual Saltmarsh Aster Fig. 598
Brackish, saline or freshwater tidal marshes, thickets, lowland fields and ditches. Coastal; ne. N.B., s. Me. and s. Ont., s. to Fla., Miss. and Tex.; inland w. to c. N.Y. s. Mich., Ohio, Ill., Nebr. and Ark. Our taxon is var. *subulatum* (*A. exilis* Ell.; *Aster subulatus* Michx.)

2. *S. tenuifolium* (L.) G. L. Nesom Perennial Salt Marsh Aster Fig. 599
Saline and brackish marshes and shores. Coastal; Me. and N.H. s. to Fla. w. to se. Tex. Our taxon is var. *tenuifolium*. (*Aster tenuifolius* L.)

3. *S. novae-angliae* (L.) G. L. Nesom New England Aster Fig. 597
Damp shores, thickets, meadows, stream banks, woodland borders and ditches; often in drier habitats. N.S. and N.B. w. to Ont., Mich., Minn., Man., N.D., Mont. and B.C., s. to Va., Ga., Miss., Okla., N.M., Utah and n. Calif. This is one of our most showy asters. (*Aster novae-angliae* L.)

4. *S. boreale* (Torr. and A. Gray) Á. Löve & D. Löve Rush Aster, Northern Bog Aster Fig. 600
Bogs, swamps and wet shores. Nfld., Lab., Anticosti I., P.E.I. and Gaspé Pen., Que., w. to Alask., s. to N.B., n. and w. N.E., n. N.J., n. Pa., Va., Ohio, n. Ind., Wisc., Neb., Colo., Wyo., Ida. and Wash. (*Aster borealis* Prov.; *A. junciformis* Rydb.; *A. laxifolius* var. *borealis* Torr. & A. Gray)

5. *S. prenanthoides* (Muhl. ex Willd.) G. L. Nesom Crookedstem Aster
Moist or swampy grounds, woods thickets, meadows, seeps and stream banks. Mass. w. to Ont., Mich., Minn. s. to N.C., Ky., Ill. and Iowa (*Aster prenanthoides* Muhl. ex Willd.)

6. *S. puniceum* (L.) Á. Löve & D. Löve Purple-stemmed Aster Fig. 597
Shores, bogs, marshes, wet meadows, stream banks and thickets; occasionally in brackish or saline habitats. Nfld. w. to s. Man. and B.C., s. to L.I., N.Y., and mts. and piedmont to Ga., Ala., Miss., e. Tex., se. Mo., Iowa and S.D. (*Aster puniceus* L.)

7. *S. novi-belgii* (L.) G. L. Nesom New York Aster
Wet thickets, meadows, marshes, borders of swamps, salt marshes and pine savannas. Chiefly near coast and along coastal plain; Nfld. and s. Que. w. to N.Y., s. to N.E., N.J., Va. and S.C. (*Aster novi-belgii* L.)

8. *S. elliottii* (Torr. & A. Gray) G. L. Nesom Marsh American Aster Fig. 597
Swamps, wet thickets and fresh to brackish shores. Coastal plain; se. Va. s. to Fla., w. to La. (*Aster elliotti* Torr. & A. Gray)

9. *S. praealtum* (Poir.) G. L. Nesom Willow Aster Fig. 600
Meadows, prairies, low woods, fields, swales and ditches. Me. and Ont. w. to Mich., Wisc., Iowa and S.D., s. to Tenn., Fla., Miss., La. Tex. and N.M. (*Aster praealtus* Poir.)

10. *S. lateriflorum* (L.) Á. Löve & D. Löve Calico Aster, Starved Aster Fig. 600
Clearings, thickets, dry to moist fields, shores, wet woodlands and swamps. N.S. N.B., M.I. and Me. w. to Ont., Man. and Minn., s. to Va., Fla. and e. Tex. (*Aster lateriflorus* (L.) Britton; *A. vimineus* Lam.)

11. *S. tradescanti* (L.) G. L. Nesom Shore Aster Fig. 601
Damp shores and stream banks. Nfld. Lab, and Que. s. to n. N.E., N.Y and N.J. (*Aster tradescanti* L.)

12. *S. lanceolatum* (Willd.) G. L. Nesom Panicled Aster Fig. 601
Thickets, meadows, banks, swales and shores. Nfld. w. to Sask, N.W.T. and B.C., s. to Ga., Fla., Ark., Okla., Tex. and Calif. This species is quite variable, with several varieties having been recognized (Semple and Chmielewski, 1987; Brouillet et. al, 2006). (*Aster lanceolatus*)

11. *Boltonia* (Boltonia)

Perennial herbs; leaves alternate, sessile, blades narrow, elongate, one-nerved; inflorescence little-branched, with few heads, to much-branched, with many heads, somewhat paniculate; ray flowers slender, white, pink, purple, or blue; disk flowers yellow; achenes flattened, somewhat winged, pappus of several minute bristles and/or scales.

REFERENCE: Karaman-Castro and Urbatsh, 2006.

Fig. 598. *Symphyotrichum subulatum*: a, b. habit; c. portion of inflorescence, showing involucral bracts in several series; d. ray flower; e. disk flower; f. mature achene with pappus (Mason, as *Aster exilis*, a synonym under *A. subulatus* var. *ligulatus*, the western phase of our species).

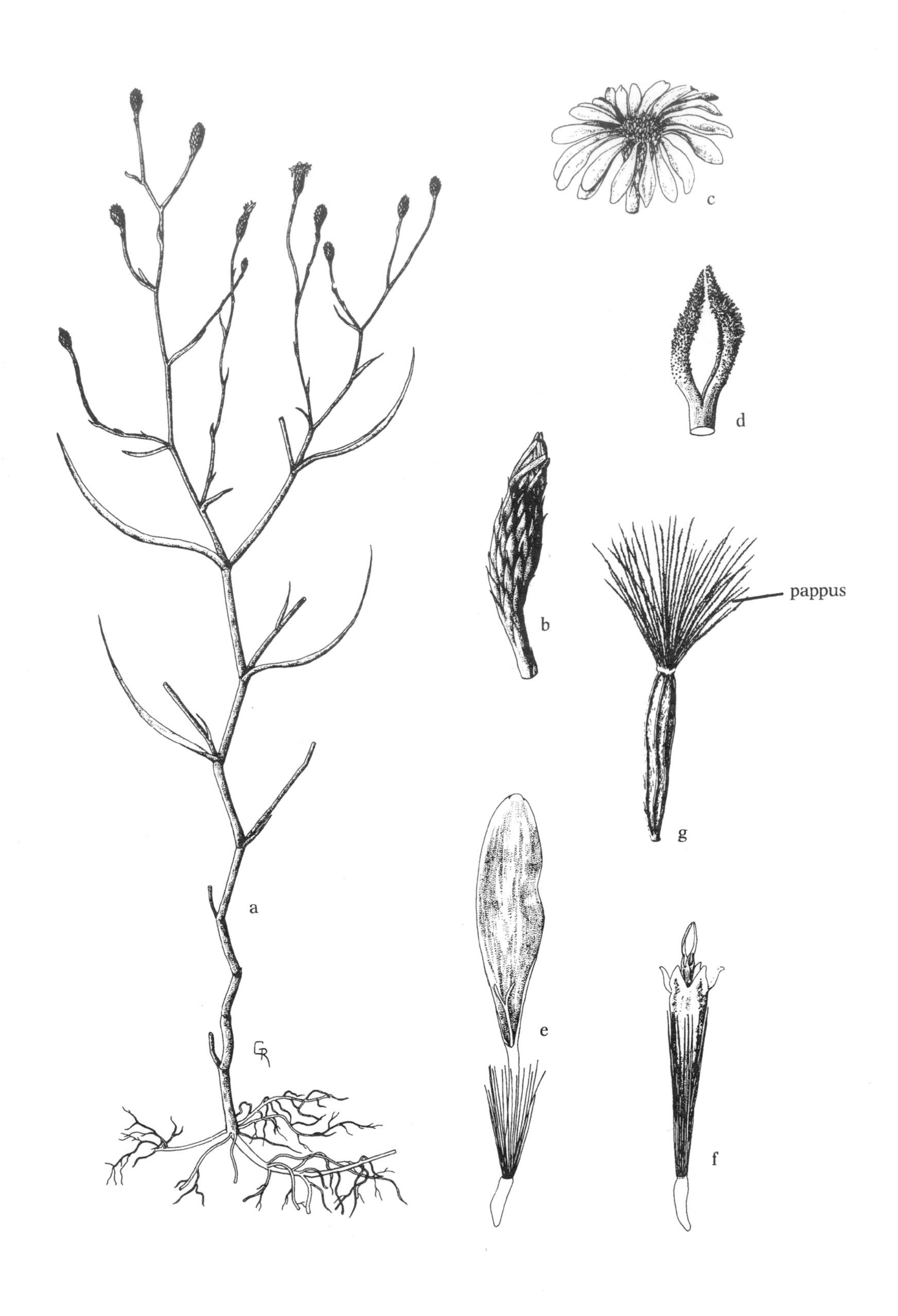

Fig. 599. *Symphyotrichum tenuifolium*: a. habit; b. bud; c. flower head; d. style branches; e. ray flower; f. disk flower; g. achene with capillary pappus (G&W).

Fig. 600. *Symphyotrichum boreale*: a. leaf; b. flower head (F).
Symphyotrichum praealtum: c. leaf and branch; d. leaf base; e. corolla of disk flower (F).
Symphyotrichum lateriflorum: f. upper portion of plant (WVA); g. portion of plant (F); h. flower head (F).

Fig. 601. *Symphyotrichum tradescantii*: a. flowering portion of plant; b. single head of achenes (USDA-NRCS); c. disk flower with pappus of capillary bristles at summit of ovary; d. ray flower with pappus of capillary bristles at summit of ovary (B&B); e. single leaf (F).
Symphyotrichum lanceolatum: f. upper portion of plant with flower heads; g. disk flower with pappus of capillary bristles at summit of ovary; h. ray flower with pappus of capillary bristles at summit of ovary; i. bract subtending flower head (B&B); j. stem with leaves and flower head (F).

1. *B. asteroides* (L.) L'Her White Doll's-daisy Fig. 602
Gravelly shores, bottomlands, moist thickets, tidal marshes and upland habitats. N.E., N.Y., and n. N.J. w. to n. Ohio, Wisc., Minn., and N.D., s. to Ga., La., Okla., and e. Tex.; Man. and Sask. This is a highly variable species with numerous named varieties.

Boltonia decurrens (Torr. & A. Gray) Alph. Wood, Clasping-leaf Doll's-daisy, is a plant with conspicuously decurrent leaves and sometimes treated as a variety of *B. asteroides*; it occurs locally in alluvial soils along the Illinois River, s. Ill., and locally in moist ground and alluvial soils, occasionally bordering sloughs and on prairie pond margins in se. Mo.

12. *Eclipta* (Eclipta)

Annual (ours) or perennial herbs; leaves opposite, simple, entire, or toothed; flower head 1–3, borne in leaf axils; ray flowers white (rarely yellow), obscure; disk flowers white, numerous; achenes 3- or 4-angled, often warty, pappus absent or obscure.

Reference: Strother, 2006

1. *E. prostrata* L. False Daisy Fig. 603
Alluvial soils, waste places, shallow water and ditches. Mass. and s. Ont. w. to Ind., Wisc., Minn. and S.D., s. to Fla., Tex., N.M., Ariz. and Calif.; Trop. Am. (*E. alba* var. *prostrata* (L.) Hassk.)

13. *Sclerolepis* (Bog-button)

Perennial herbs; submersed stems flexuous, emersed stems erect; leaves linear to linear-subulate, sessile, in whorls of 3–6; flower heads solitary, terminal, pink, lavender, or rarely white; achenes 5-angled, pappus of 5 short straw-colored scales.

References: Lamont, 2006; Mehrhoff, 1983.

1. *S. uniflora* (Walter) Britton, Sterns & Poggenb. Pink Bog-button Fig. 604
Sandy or peaty shores, depressions, bogs, swamps and ditches; often in shallow water and ponds with a fluctuation of water table. C. N.H., Mass. and R.I., coastal plain, s. N.J. s. to Fla., w. to Ala. This plant remains entirely vegetative in the submersed stage, but as the water level drops the flexuous horizontal stems become stranded and rooted at the nodes along the shore, then developing numerous fertile uprights shoots of a terrestrial growth form, each producing a solitary, terminal head of flowers.

14. *Cirsium* (Thistle)

Biennial (ours) or perennial herbs, very spiny; leaves alternate, usually lobed to pinnatifid, with numerous spines or spine-like teeth; heads consisting of disk flowers only, purple (ours), white, or yellow; receptacle with dense, soft bristles; achenes obscurely angled or compressed, pappus of many plumose, soft, capillary bristles.

Reference: Keil, 2006.

1. Stem spiny-winged; involucre 1–1.5 cm long; fruiting pappus about 0.9–1.1 cm long....1. *C. palustre*
1. Stem not spiny; involucre 1.7–3 cm long; fruiting pappus usually 1.2–2 cm long.
    2. Leaves pale green on lower surface, thinly tomentose and more or less webby; involucral bracts not spine-tipped....2. *C. muticum*
    2. Leaves gray on lower surface, densely tomentose; outer and middle involucral bracts spine-tipped.... 3. *C. virginianum*

1. *C. palustre* (L.) Scop. Marsh Thistle Fig. 605
Damp woods, thickets and wet roadside ditches. Nfld.; local; N.S., St. P. et. Miq. and Ont. w. to n. Mich., s. to N.H., Mass. and N.Y.: B.C.; intro. from Eurasia.

2. *C. muticum* Michx. Swamp Thistle Fig. 605
Low woods, swamps, meadows and ditches. Nfld., St. P. et Miq. w. to Sask. and N.D. s. to Md., mts. to N.C., Tenn., Fla., Mo., Okla., La. and e. Tex. (*Carduus muticus* (Michx.) Pers.)

3. *C. virginianum* (L.) Michx. Virginia Thistle
Bogs, wet pine woods, swales, and clearings. Coastal plain; s. N.J., Del., and Va., s. to Ga. and ne. Fla., Del. s. to Ga.

Fig. 602. *Boltonia asteroides*: a. habit (Beal); b. upper portion of plant (F).

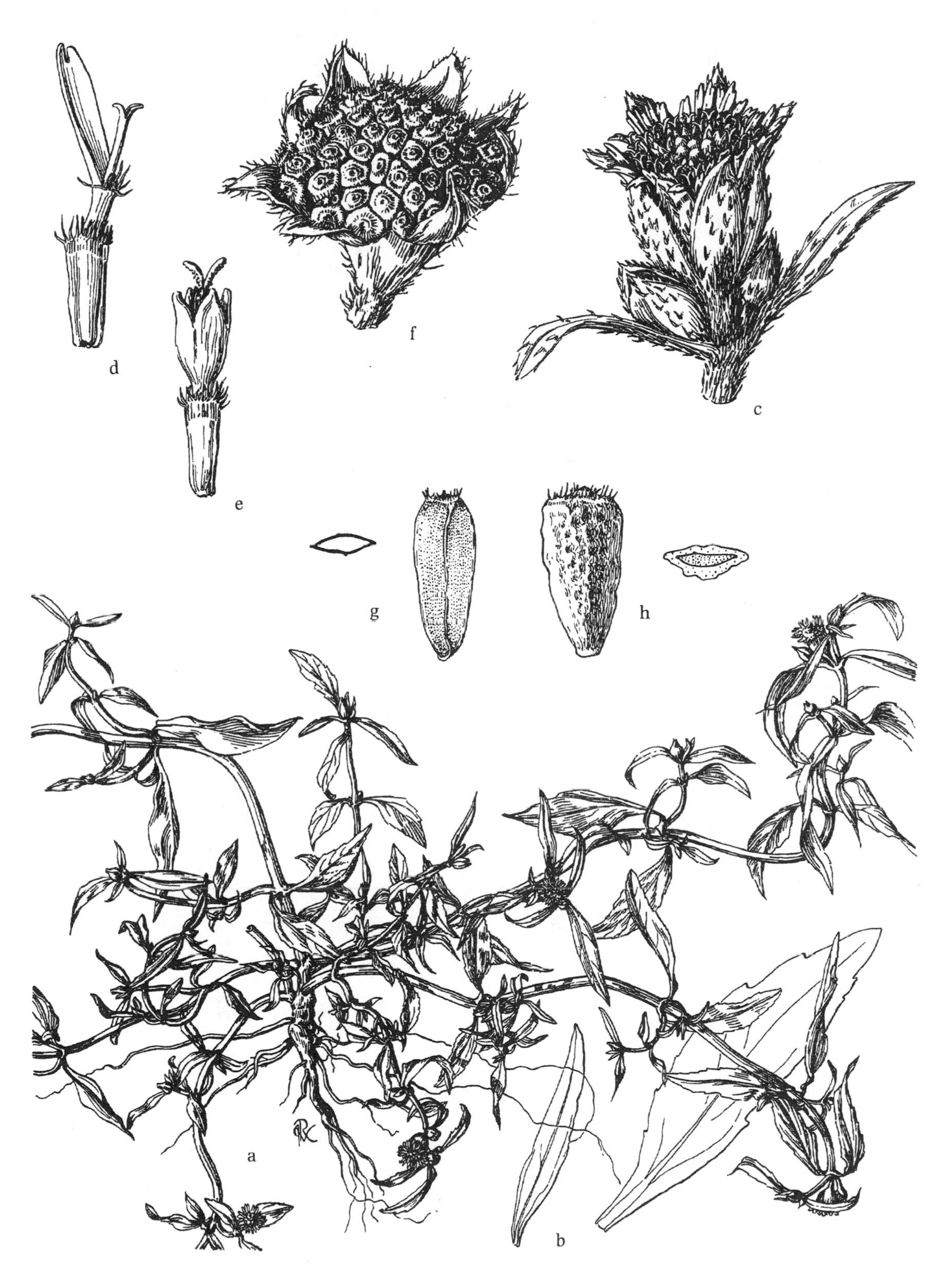

Fig. 603. *Eclipta prostrata*: a. habit; b. leaves; c. flower head; d. ray flower; e. disk flower; f. fruiting head; g. achene, thin-walled type, with cross-section; h. achene, thick-walled tuberculate type, with cross-section (Mason).

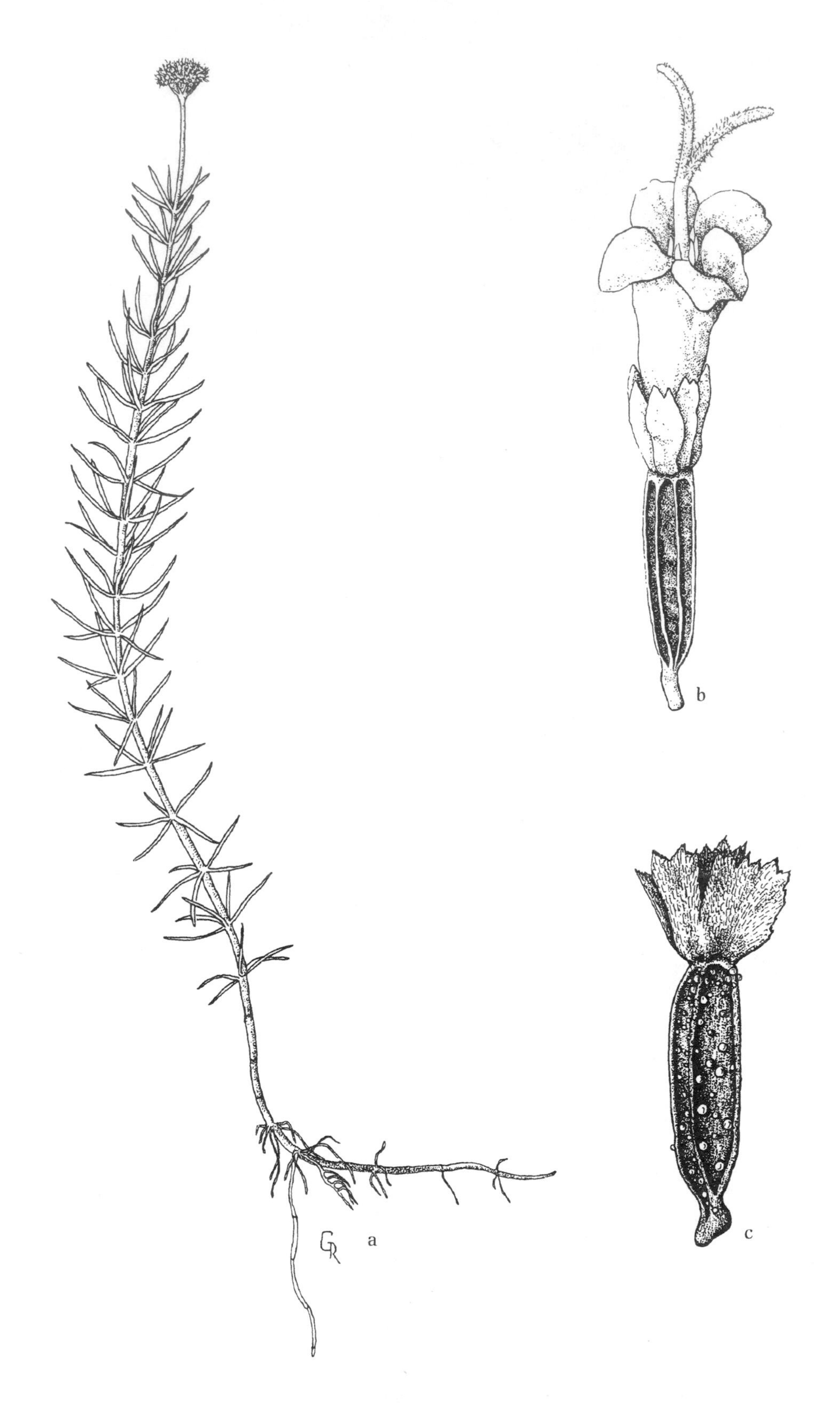

Fig. 604. *Sclerolepis uniflora*: a. habit. terrestrial form; b. disk flower; c. achene (G&W).

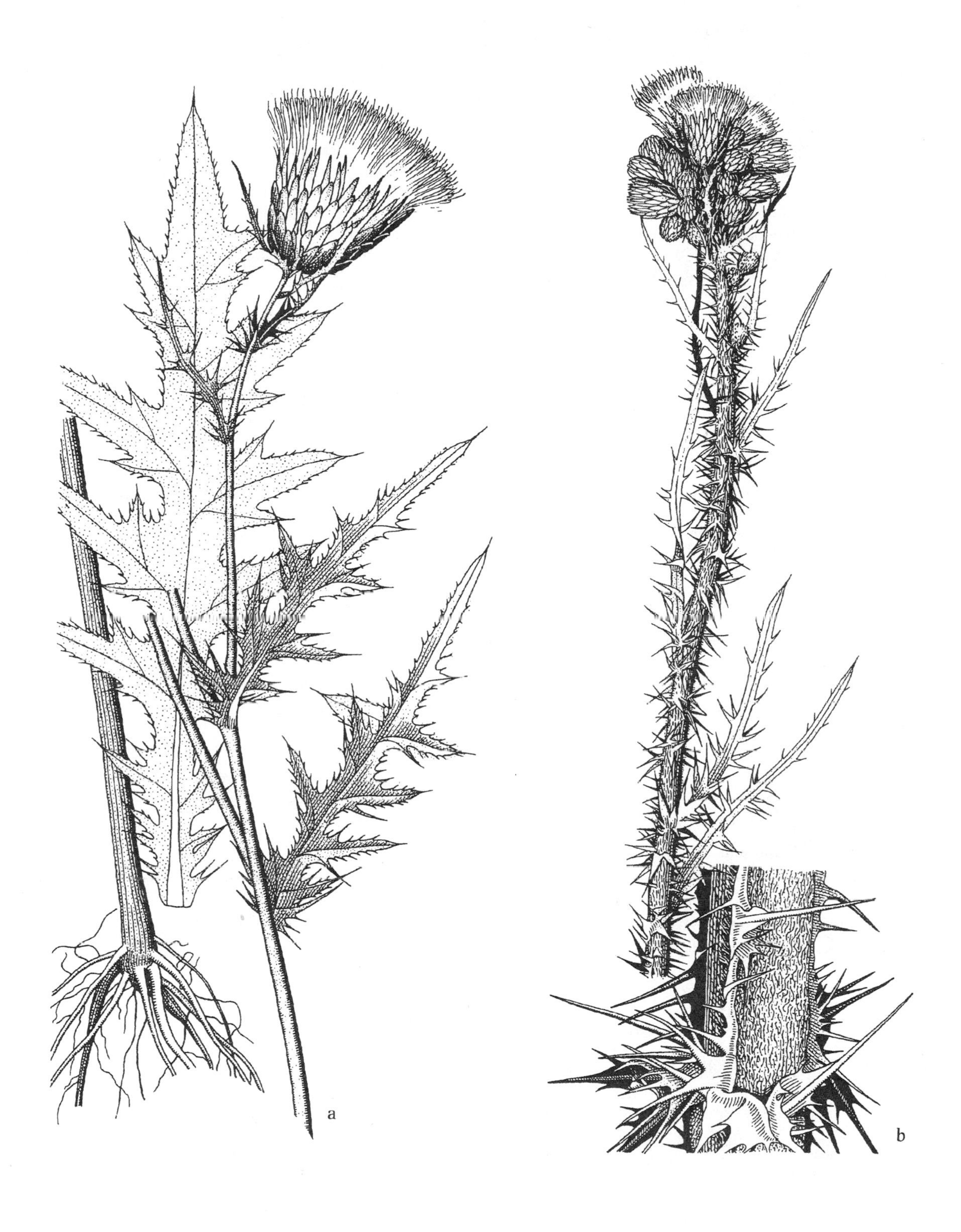

Fig. 605. *Cirsium muticum*: a. habit (Gleason).
Cirsium palustre: b. habit (Gleason).

15. *Iva* (Marsh-elder, Sumpweed)

Annual or perennial herbs, some shrubby (ours); leaves generally all opposite, except upper ones reduced, simple; inflorescence spiciform (ours) or racemiform ray flowers absent; disk flowers several; achenes biconvex or obovate, pappus absent.

REFERENCES: Jackson, 1960; Strother, 2006.

1. *I. frutescens* L. Maritime Marsh-elder, Jesuit's Bark Fig. 606
Saline marshes, swamps and mud flats and beaches. Coastal; s. N.S. and s. Me. s. to Fla., w. to Tex. The more northern taxon, subsp. *oraria* (Bartlett) R. C. Jackson, occurs from Virginia northward and is characterized by having elliptic, oval, or broadly lanceolate leaves, 2–5 cm wide; and somewhat larger achenes, 2.6–3.3 mm long. The more southern taxon, subsp. *fructescens*, occurs southward from Virginia and has lanceolate leaves, 1–3 cm wide, and smaller achenes, 2–2.6 mm long. (*I. fructescens* var. *oraria* (Bartlett) Fernald & Griscom)

16. *Cotula* (Brass-buttons)

Annuals or perennial (ours) herbs, prostrate, decumbent to erect, rooting at nodes; stems spongy; leaves sessile, clasping at base, irregularly toothed or lobed or entire; flower heads yellow to pale yellow, solitary; discoid flowers with outer ones pistillate, lacking corolla, 12–40+ in single series; inner discoid flowers bisexual (but functionally staminate), corolla present, yellow; ray flowers absent (or if present, then very few, with corolla white); fruit an achene, obovoid to oblong, flattened, winged, pappus lacking, outer fruits of head stalked, with inner wing face papillate, inner fruits lacking wings.

REFERENCE: Watson, 2006.

1. *C. coronopifolia* L. Brass-buttons Fig. 606
Saline marshes, freshwater marshes, mud flats and still or slow-flowing waters along streams and stream banks; sometimes becoming dense and emersed in water up to ca. 0.5 m deep. Que., N.S., P.E.I. and N.B.; Mass.; Md.; Alaska; B.C., Wash., Oreg., Calif., w. Nev. and w. Ariz.; Mex. (Baja California); S.Am.; intro. from southern Africa. Few populations in ne. N.Am.; more robustly established in w. N.Am.

17. *Pluchea* (Marsh-fleabane, Stinkweed)

Annual or perennial herbs, with a strong fetid or camphoric odor; leaves alternate, simple, petiolate or sessile; flower heads in paniculate or corymbose inflorescences; ray flowers absent; disk flowers lavender-pink, lavender-purple, rose-purple, or white; achenes grooved, pappus a single series of barbed bristles.

REFERENCE: Nesom, 2006.

---

1. Leaves sessile or somewhat clasping; flowers usually creamy white, sometimes yellowish or pale pink. . . . . 1. *P. foetida*
1. Leaves petiolate; flowers pink to rose-purplish, or purple.
   2. Involucral bracts conspicuously glandular on outer surface, outer bracts short-pubescent and short-ciliate; upper leaves distinctly petiolate. . . . . 2. *P. camphorata*
   2. Involucral bracts sparsely and obscurely glandular on outer surface, outer and middle bracts short-pubescent, but lacking cilia; upper leaves sessile or short-petiolate, tapering at base.. . . . . 3. *P. odorata*

---

1. *P. foetida* (L.) DC. Stinking Fleabane Fig. 606
Swamps, wooded and open wet areas, marshy shores and ditches. Coastal plain; s. N.J. s. to Fla., w. to se. Tex. and Mex.; inland n. to Okla., Ark. and se. Mo.; W.I. (*P. tenuifolia* Small)

2. *P. camphorata* (L.) DC. Camphorweed, Stinkweed
Freshwater to brackish marshes, riverbanks, swamps, wet meadows, swales, shores and ditches. Del. w. to Ohio, Ill., s. Mo., and e. Kans., s. to Fla. and Tex.

3. *P. odorata* (L.) Cass. Sweet-scent, Saltmarsh Fleabane, Shrubby Camphorweed Fig. 607
Saline, brackish, or freshwater marshes, swales and mud flats. Coastal plain; s. Me. and Mass. s. to Fla., w. to Tex., and Mex.; inland n. to Ark. Okla., Kans., N.M., Utah, Nev., Ariz. and Calif.; W.I., C.Am. and S.Am. Cronquist (1980) treated the coastal populations from Maryland northward as var. *succulenta* (Fernald) Cronq., distinguished by shorter stature and larger heads. (*P. purpurascens* var. *succulenta* Fernald)

18. *Mikania* (Climbing Hempweed)

Twining and climbing perennial vines; leaves opposite, petiolate; inflorescence corymbose, axillary; ray flowers absent; disk flowers 4 per head, white, pink, or blue; achenes 5-angled, pappus of numerous capillary bristles.

REFERENCE: Holmes, 2006.

Fig. 606. *Cotula coronopifolia*: a. habit; b. achene of female flower, outer face; c. achene of female flower, showing papillate inner face and broad membranous margins; d. bisexual flower, showing tubular corolla above the ovary, the papillate inner face of the ovary visible; e. achene of bisexual flower, showing smooth outer face and narrow margins (APOA).
Iva fructescens: f. upper portion of plant (Beal).
Pluchea foetida: g. upper portion of plant (Beal).

Fig. 607. *Pluchea odorata*: a. upper portion of plant; b. flower head; c. outer involucral bracts; d, e. inner view of involucral bracts; f. central disk flower (bisexual); g. tubular disk flower, slit open; h. outer pistillate flower (C&C).

1. *M. scandens* (L.) Willd. Fig. 608
Thickets, swamps, marshes, stream banks and wet woodlands. S. Me. w. to Mich., s. Ill. and se. Mo., s. to Fla., se. Okla. and se. Tex.; Trop. Am.

19. *Conoclinium* (Mistflower)

Perennials; usually rhizomatous; stems erect to decumbent; leaves cauline, opposite, petiolate, blades mostly 3-nerved, margins dentate or lobed; heads discoid inflorescences tightly corymbiform; involucres hemispheric 3–6 mm diameter; phyllaries persistent ca. 25 in 2–3 series obscurely 2–3-nerved; receptacles conic; flowers 35–70 or more; corolla blue to purple or violet, rarely white; pappus of barbed bristles, persistent; achenes ribbed, glabrous or sparsely gland-dotted.

Reference: Patterson and Nesom, 2006.

1. *C. coelestinum* (L.) DC. Blue Mistflower, Blue Boneset, Ageratum Fig. 609
Wet areas, stream banks, floodplains, damp thickets and ditches. N.J. and sc. Pa. w. to W.Va., s. Ont., Mich., Ill., Neb. and Kans., s. to Fla. and Tex.; Cuba. (*Eupatorium coelestinum* L.)

20. *Eutrochium* (Joe-pye Weed)

Perennials; stems, erect, unbranched; leaves mostly cauline, usually whorled (3–7 per node), petiolate, margins serrate, teeth typically gland tipped, upper blade surface usually gland-dotted and hirsute, puberulent, pubescent, scabrous, lower surface mostly puberulent to scabrous-hirsute; heads discoid in compound corymbose inflorescence; involucres cylindric, 2.5–7 mm in diameter; phyllaries persistent 10–22 in several series, typically pale pink to purple; flowers 4–22; corollas purplish or pinkish, rarely white; pappus of 1 whorl of barbed bristles, persistent, cream to pinkish-purple.

Reference: Lamont, 2006.

1. Leaves in whorls of 3 or 4; blades 3-nerved, lowest pair of lateral veins longer than other lateral veins (fig. 610b). . . . . 1. *E. dubium*
1. Leaves in whorls of 4–7; blades pinnately veined, all lateral veins approximately same length.
    2. Stem speckled, or sometimes becoming purplish, not glaucous: flowers 8–20 per head, purple; inflorescence flat-topped. . . 2. *E. maculatum*
    2. Stems more or less purplish throughout, strongly glaucous: flowers 3–8 per head, purple-pink; inflorescence dome-shaped to convex. . . . . 3. *E. fistulosum*

1. *E. dubium* (Willd. ex Poir.) E. E. Lamont Coastal Plain Joe-pye Weed Fig. 610
Swamps, moist thickets, shores, marshes and ditches. N.S. e. N.Y., s. along the coastal plain and piedmont to S.C. (*Eupatorium dubium* Willd. ex. Poir.; *Eupatoriadelphis dubius* (Willd. ex Poir.) King & H. E. Robins.)

2. *E. maculatum* (L.) E. E. Lamont Spotted Joe-pye Weed Fig. 610
Shores, damp meadows, thickets, and ditches. Nfld. and n. Ont. w. to B.C., s. to Pa., mts. to N.C. and Tenn., n. Ind., Iowa, Neb., N.M., and Wash. Three varieties are recognized. (*Eupatorium maculatum* L.; *Eupatoriadelphis maculatus* (L.) King & H. E. Robins.)

3. *E. fistulousum* (Barratt) E. E. Lamont Joe-pye Weed, Trumpet-weed Fig. 611
Damp thickets, alluvial woodland, stream banks, meadows and thickets. S Me. w. to N.Y., Mich., Ind., Iowa and Okla., s. to Fla. and e. Tex. (*Eupatorium fistulosum*) Barratt; *Eupatoriadelphis fistulosus* (Barratt) King & H. E. Robins.)

21. *Fleischmannia* (Thoroughwort)

Annual or perennials; stems erect, simple or little branched; leaves opposite, petiolate, blades 3-nerved, margins more or less crenate to serrate, surfaces glabrous. Heads discoid in loose corymbiform arrays; involucres obconic to hemispheric, 2–4 mm in diameter; phyllaries persistent 20–30 in 2 to 4 series; receptacles flat to slightly convex; flowers 15–25; corolla bluish, pinkish, purplish or white; pappus of barbed bristles in a single whorl, persistent (or fragile, dropping off).

Reference: Nesom, 2006.

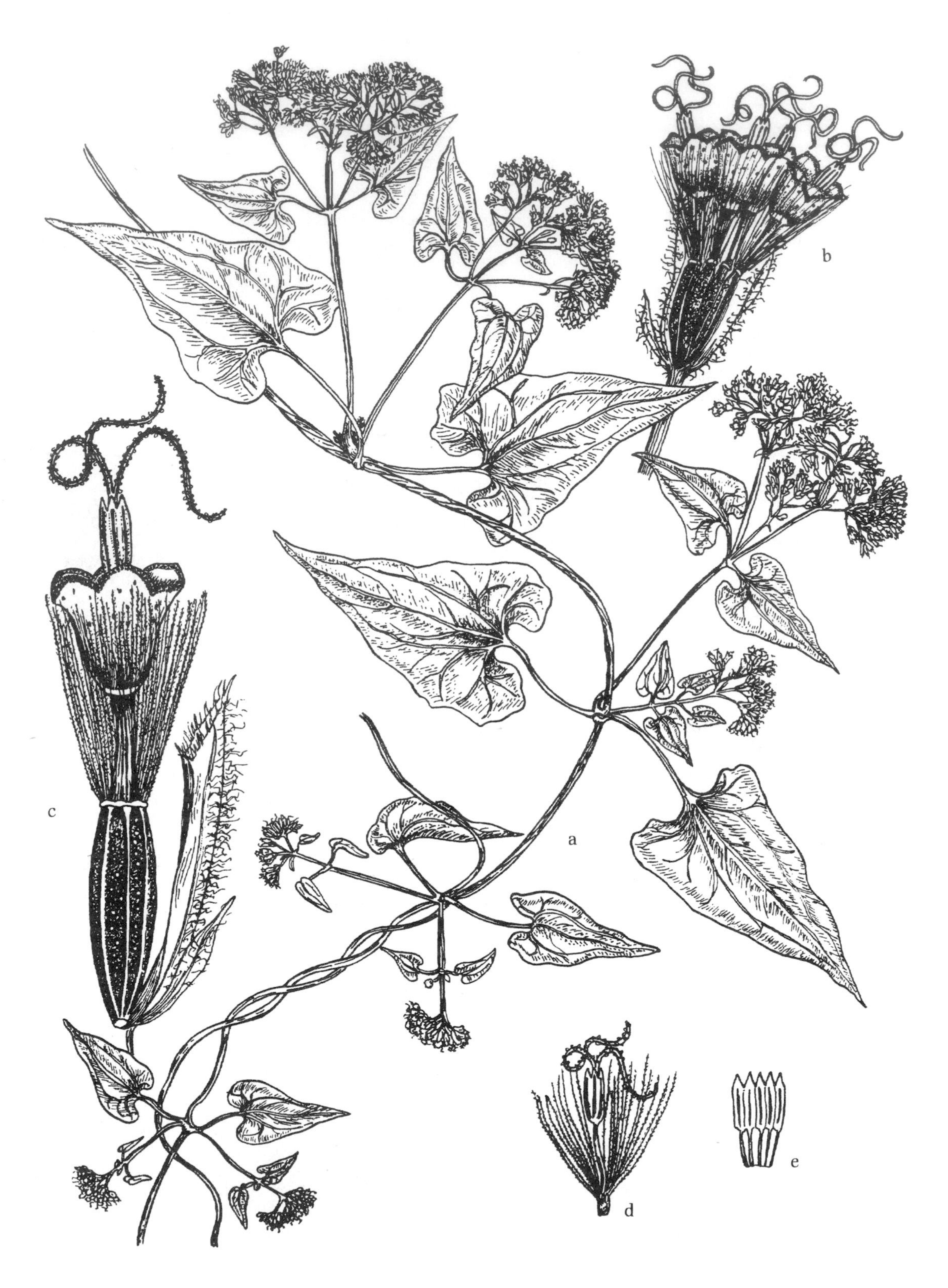

Fig. 608. *Mikania scandens*: a. portion of viny plant; b. flower head; c. disk flower; d. flower with corolla removed, showing anthers united in a ring surrounding style; e. anther ring, spread open to show internal surface (C&C).

Fig. 609. *Conoclinium coelestinum*: a. habit; b. flower head; c. receptacle with 1 disk flower; d. anther ring spread open to show internal surface (C&C).

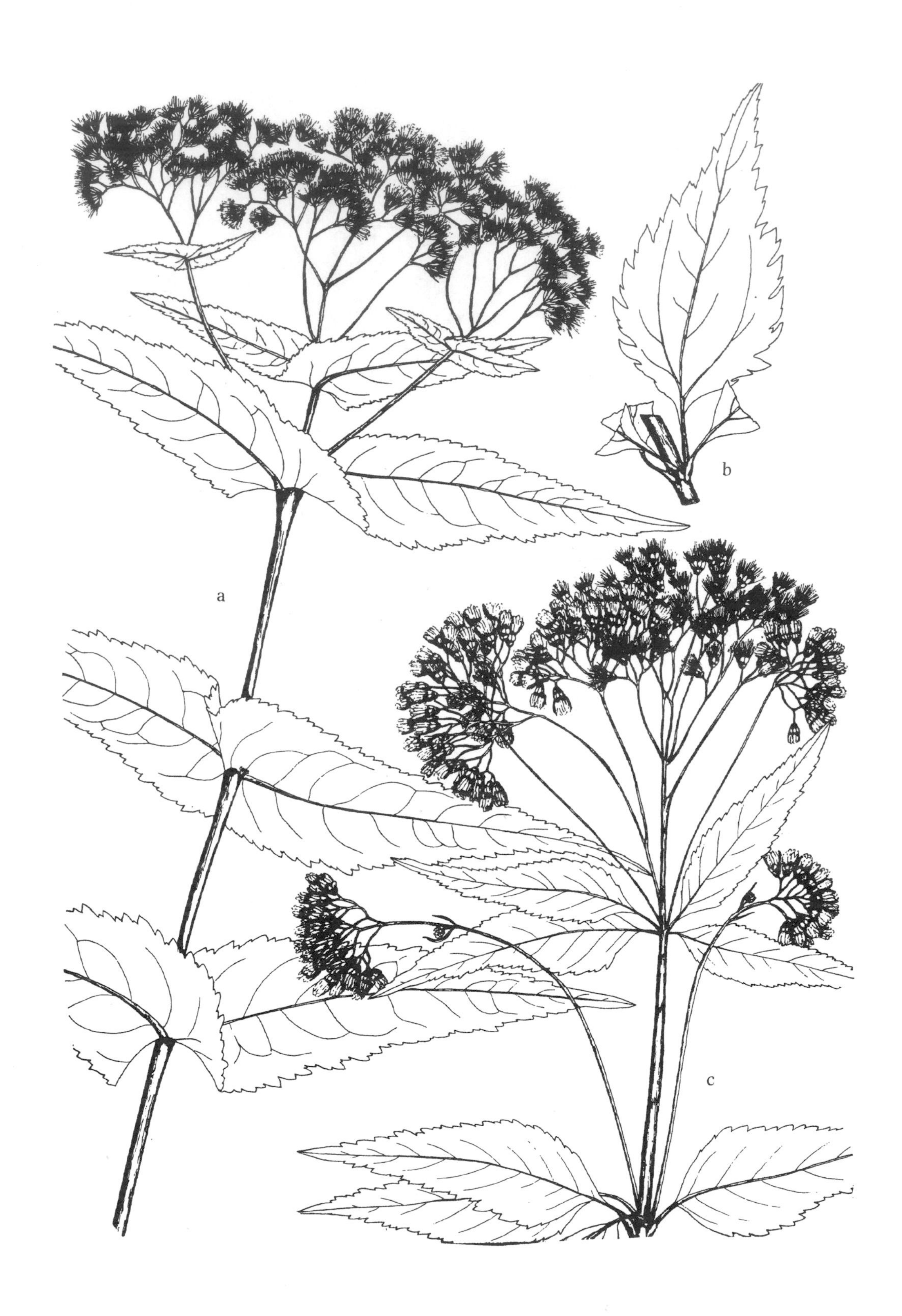

Fig. 610. *Eupatorium perfoliatum*: a. upper portion of plant, the leaves connate-perfoliate (F).
Eutrochium dubium: b. section of stem with leaf showing 3 strong veins, the lowest lateral vein 2 longer than other lateral veins (F).
Eutrochium maculatum: c. upper portion of plant (F).

Fig. 611. *Eupatorium novae-angliae*: a. upper portion of plant; b. flower head; c. achene; d. involucral bract (Crow, as *Eupatorium leucolepus* var. *novae-angliae*).
Fleishmannia incarnata: e. upper portion of plant; f. flower head (WVA).
Eutrochium fistulosum: g. upper portion of plant; h. flower head (WVA).

1. *F. incarnata* (Walter) King & H. E. Robins.
Pink Thoroughwort Fig. 611
Wooded swamps, rich woods and ditches. Se. Va. w. to W.Va., s. Ohio, s. Ind., Ky., s. Ill., se. Mo., and Okla., s. to Fla., e. Tex. and Mex. (*Eupatorium incarnatum* Walter)

22. *Eupatorium* (Thoroughwort)

Perennials; stems arising from rhizomes or caudices, erect; leaves mostly cauline, usually opposite, petiolate or sessile; blades 3-nerved, margins entire or toothed, surfaces usually gland dotted, glabrous or puberulent to pubescent, scabrous, or setulose; heads discoid in corymbose or paniculiform inflorescences; involucres obconic to ellipsoid 1–3(5) mm in diameter; phyllaries persistent, in 2–3(4) series; flowers (3)5(15+); corollas usually white, rarely pink; pappus of barbed bristles in a single whorl, persistent.

REFERENCES: LeBlond and Weakley, 2007; Siripun and Schilling, 2006.

1. Flowers (3)5(7) per head.
    2. Principal leaves dissected into many filiform divisions (fig. 612c), annual. . . . . 1. *E. capillifolium*
    2. Principal leaves entire, toothed, or lobed; perennial.
        3. Involucral bracts acuminate to mucronate.
            4. Leaves strongly folded along midribs, the blades curved, 4–10 mm wide; widespread in. . . . . 2. *E. leucolepis*
            4. Leaves weakly folded along midribs, the blades not curved, 10–15 mm wide; confined to e. Mass. . . . . 3. *E. novae-angliae*
        3. Involucral bracts broadly rounded to acute, not acuminate.
            5. Leaves broad (10)15–65 mm aide, cuneate to nearly truncate, rounded or subcordate at base. . . . . 4. *E. rotundifolium*
            5. Leaves narrower, 3–20(30) mm wide, tapering to narrow base.
                6. Plants with a conspicuous, tuberous, short rhizome. . . . . 5. *E. mohrii*
                6. Plants with a short caudex or caudex-like rhizome (not tuberous). . . . . 6. *E. semiserratum*
1. Flowers usually more than 9 per head.
    7. Leaves sessile or connate-perfoliate.
        8. Leaf bases connate-perfoliate; stem villous. . . . . 7. *E. perfoliatum*
        8. Leaf bases sessile; stem puberulent or minutely pilose. . . . . 8. *E. resinosum*
    7. Leaves petiolate. . . . . 9. *E. serotinum*

1. *E. capillifolium* (Lam.) Small Dog-fennel
Fig. 612
Moist to wet thickets, pond and lake borders and upland locations. Chiefly coastal plain and piedmont; N.J. s. to Fla., w. to Tenn., Mo., Okla., s. Ark., La., and e. Tex.

2. *E. leucolepis* (DC) Torr. & A. Gray Justiceweed
Pine barrens, wet meadows and shores of lakes, ponds and rivers. Coastal; N.J. s. to Fla. and w. to Tex.; Okla.

3. *E. novae-angliae* (Fernald) V. I. Sullivan ex A. Haines & Sorrie New England Justiceweed Fig. 611
Bogs, wet sands, wet meadows and pond margins. Coastal plain, se. Mass. and R.I. This narrowly endemic rare plant is believed to be of hybrid origin between *E. perfoliatum* and the recently described southern species, *E. paludicola* E. E. Shill. & LeBlond, endemic to the Atlantic coastal plain of the Carolinas, but apparently of no genetic connection to the previously purported relative, *E. leucolepis*. Thus *E. novae-angliae* is treated as a distinct species, and these populations remain of conservation concern (Siripun and Shilling, 2006; LeBlond et al., 2007). (*E. leucolepus* var. *novae-angliae* Fernald)

4. *E. rotundifolium* L. Roundleaf Thoroughwort
Bogs, damp to wet sands and thickets: occasionally in dry locations. S. Me., s. N.H. and Mass., w. to s. Pa., s. Ohio, s. Ind., and Mo., s. to Fla., Miss., Okla. and e. Tex.

5. *E. mohrii* Greene Mohr's Thoroughwort
Pond margins, shores, moist ground and ditches. Coastal plain; se. Va. s. to Fla., w. to La. and Tex. (*E. recurvans* Small)

6. *E. semiserratum* DC. Smallflower Thoroughwort Fig. 613
Wet to dry open woods, clearings, shores and swamps. Md. and Va. w. to Ky., e. Mo. and Okla., s. to n. Fla Ala., and e. Tex. (*E. cuneifolium* var. *semiserratum* (DC.) Fernald & Griscom)

7. *E. perfoliatum* L. Thoroughwort, Common Boneset Fig. 610
Low woods, thickets, wet shores and ditches. N.S. and Que. w. to se. Man., s. to Fla., Okla. and Tex. Plants treated as var. *colpophilum* Fernald & Griscom, with puberulent to glabrate leaves, occur along fresh to brackish tidal shores from Que. to Me. and appear to be ecological response to fluctuation of water level; Siripun and Shilling (2006) do not give this tidal variant taxonomic recognition.

8. *E. resinosum* Torr. Pine Barren Thoroughwort
Pine barren bogs and swamps. Coastal plain; N.J., N.C. and S.C.

9. *E. serotinum* Michx. Late-flowering Thoroughwort
Fig. 613
Wet woods, bottomlands, damp to dryish thickets, borders of freshwater to brackish marshes, meadows and pastures. Mass. Conn. and s. N.Y. w. to s. Mich., s. Wisc., Minn. and e. Neb., s. to Fla., La. and Tex.

Fig. 612. *Eupatorium capillifolium*: a. basal and upper portions of plant; b. portion of panicle of heads; c. section of stem with leaves; d. disk flower; e. achenes. one with pappus removed (Reed).

Fig. 613. *Eupatorium semiserratum*: a. upper portion of plant; b. leaf; c. flower head; d. disk flower (G&W). *Eupatorium serotinum*: e. upper portion of plant; f. flower head (WVA).

Araliaceae / Aralia Family

1. *Hydrocotyle* (Water Pennywort, Navelwort, Marsh-pennywort)

Perennial herbs; stems creeping; leaves solitary, circular, peltate, or deeply notched; flowers borne in simple or compound.

The genus has long been treated as a member of the Apiaceae, but Angiosperm Phylogeny Group IV (APG IV, 2017) includes *Hydrocotyle* within the Araliaceae, as subfamily Hydrocotyloideae.

---

1. Leaves peltate (fig. 614c,e).
    2. Flowers in a simple, subglobose umbel (fig. 614e); fruit notched at base and summit (fig. 614f)........................... 1. *H. umbellata*
    2. Flowers in simple or forking, interrupted, spike-like whorls (fig. 614c); fruit not notched at base and summit (fig. 614d).. 2. *H. verticillata*
1. Leaves deeply notched at base (fig. 614a).
    3. Leaves shallowly lobed (fig. 614a); inflorescence appearing sessile.................................. 3. *H. americana*
    3. Leaves deeply lobed (fig. 614b); inflorescence.................................. 4. *H. ranunculoides*

---

1. *H. umbellata* L. Many-flowered Marsh-pennywort, Water Pennywort Fig. 614

Shores or shallow water of ponds, swamps and ditches. N.S. and e. Mass. w. to N.Y., ne. Ohio, Mich. and Minn., s. to Fla. and Tex.; s. Oreg. s. to Calif. and Mex.; C.Am. and S.Am.

2. *H. verticillata* L. Whorled Marsh-pennywort, Water Pennywort Fig. 614

Flooded shores, swamps, wet woodlands, floodplains and low ground. Coastal plain and slightly inland, se. Mass. s. to Fla., w. to Tex.; Mo., Okla., Utah, Nev. and Calif.; W.I., C.Am. and S.Am. Two varieties have been recognized: var. *verticillata*, with flowers and fruits on stalks 2 mm or less long with a rarely forking inflorescence; var. *featherstoniana* (Jennings) Mathias, with sparsely and minutely hirsute petioles. (*H. australis* J. M. Coult. & Rose; *H. canbeyi* J. M. Coult. & Rose; *H. triradiata* (A. Rich.) Fernald)

3. *H. americana* L. American Marsh-pennywort, Water Pennywort Fig. 614

Meadows, damp woods, bogs, slopes and seeping ledges. Nfld. w. to Minn., s. to Md., w. N.C., w. S.C., c. Ky., Ind. and Ark.

4. *H. ranunculoides* L. f. Floating Marsh-pennywort, Water Pennywort Fig. 614

Shallow water, shores, sloughs and ditches. N.Y., Pa., Del. and W.Va. w. to Ark., Kans., Okla., and Ariz., s. to Fla. and Tex.; Pacific Coast, B.C. s. to Calif. and Ariz.; C.Am. and S.Am.

Apiaceae (Umbelliferae) / Parsley Family

This is a large and well-marked family, distinguished by leaves usually compound to deeply dissected, with typically sheathing leaf bases; flowers borne in simple or, more often, compound umbels; 5-merous; sepals tiny, sometimes absent; petals small, separate, ovary inferior, carpels 2, style swollen at base (forming a stylopodium, superficially confused with the ovary); fruit a schizocarp, splitting into two 1-seeded segments (mericarps), with each segment ribbed. The species are often difficult to identify; fruits are particularly important for diagnostic purposes.

Reference: Mathias and Constance, 1944–1945.

---

1. Leaves all simple.
    2. Flowers in dense, sessile, compact bracteate heads (fig. 615a).................................. 1. *Eryngium*
    2. Flowers in simple (fig. 616b) or compound umbels (fig. 619b).
        3. Stems horizontal, rhizomatous, leaves and inflorescences arising from nodes of the rhizome.
            4. Leaves phyllodial, septate, without expanded blades (fig. 616b,c).................................. 2. *Lilaeopsis*
            4. Leaves with distinct blades, the petioles not septate (fig. 616d,e).................................. 3. *Centella*
        3. Stems erect, or if prostrate then not rhizomatous.
            6. Leaves with distinct blades.
                7. Leaf blades lanceolate to ovate, toothed; estuarine growth form in tidal rivers.................................. 4. *Sium*
                7. Leaf blades linear to narrowly oblanceolate, entire; plants of inland southern bogs and pineland.................. 5. *Oxypolis*
            6. Leaves phyllodial, without expanded blades (fig. 621b).................................. 6. *Ptilimnium*
1. Leaves compound (at least some).
    8. Leaf segments filiform (figs. 621, 622), up to 1 mm wide.................................. 6. *Ptilimnium*
    8. Leaf segments (1)1.5 mm or more wide (segments on some upper leaves of plant occasionally narrower).
        9. Leaves once compound, pinnate or palmate.

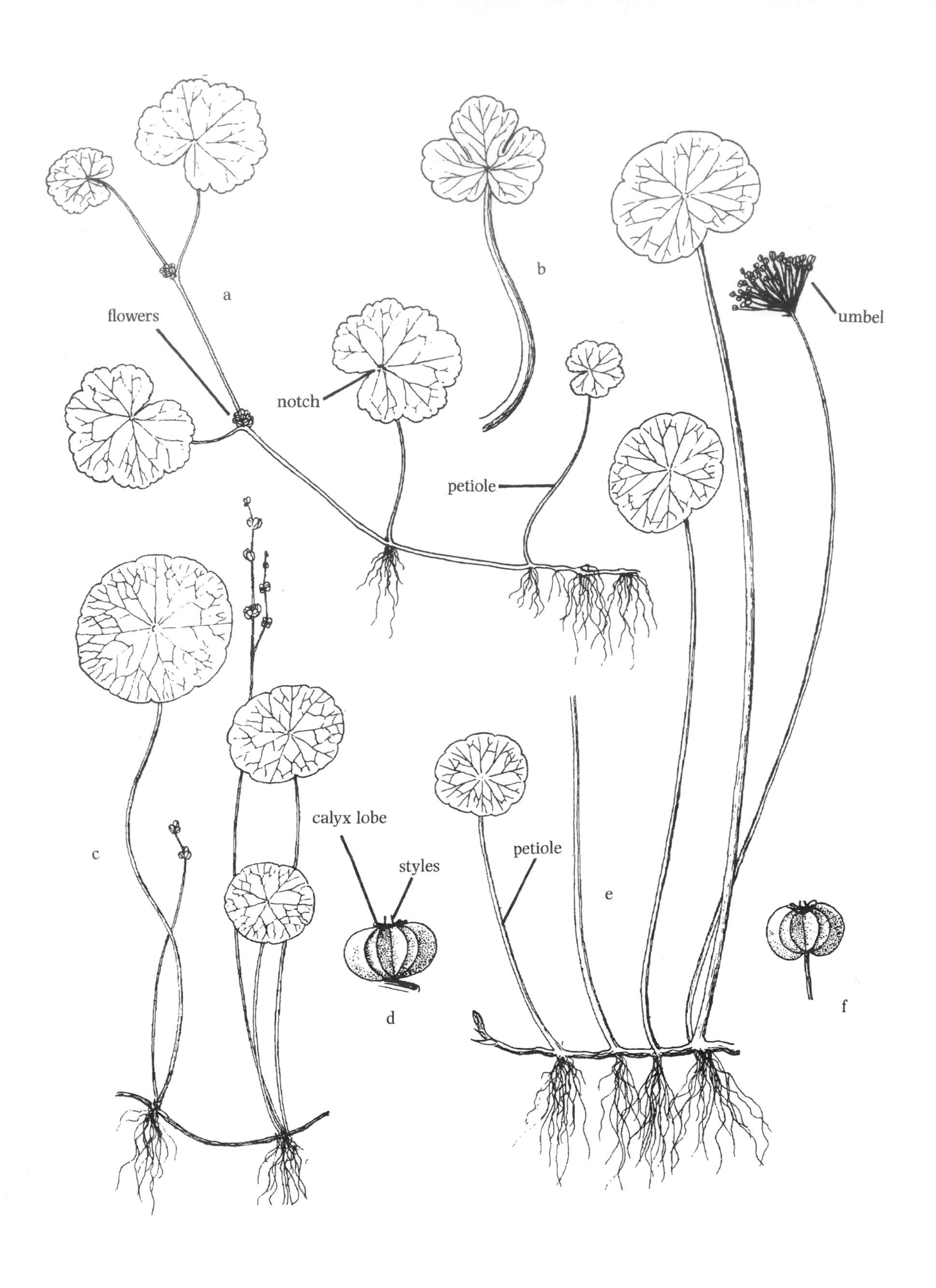

Fig. 614. *Hydrocotyle americana*: a. habit (F).
Hydrocotyle ranunculoides: b. leaf (F).
Hydrocotyle verticillata: c. habit; d. fruit (F).
Hydrocotyle umbellata: e. habit; f. fruit (F).

10. Leaflets entire, linear to narrowly elliptic.
  11. Leaf segments not septate; fruit flattened, winged dorsally (fig. 619). . . . . 5. *Oxypolis*
  11. Leaf segments septate; fruit almost round in cross-section, wingless.
    12. Main leaves with 3(6) palmately arranged leaflets or segments; fruit beaked (fig. 623b). . . . . 7. *Cynosciadium*
    12. Main leaves with typically more than 3(2–9) pinnately arranged leaflets or segments; fruit beakless. . . . . 8. *Limnosciadium*
10. Leaflets serrate, lobed or incised, lanceolate to circular.
  13. Leaflets with 5 or fewer coarse teeth per margin. . . . . 5. *Oxypolis*
  13. Leaflets with numerous teeth (fig. 618) or when incised, not doubly compound (fig. 625).
    14. Plants stoloniferous (stem rooting at nodes), basal rosette of leaves absent; leaflets of upper leaves irregularly incised; fruit ribs obscured by corky pericarp. . . . . 9. *Berula*
    14. Plants not stoloniferous present (stem rooted at base), basal rosette of leaves present; leaflets of upper leaves regularly serrate; fruit ribs conspicuous. . . . . 4. *Sium*
9. Leaves twice or 3 times compound.
  15. Flowers white.
    16. Leaf sheaths 5–10 cm long, conspicuously expanded or inflated.. . . . . 10. *Angelica*
    16. Leaf sheaths of most leaves seldom more than 3 cm long, scarcely expanded or inflated.
      17. Sap watery; leaf segments coarsely toothed (a few on a plant sometimes entire), not pinnatifid. . . . . 11. *Cicuta*
      17. Sap milky; leaf segments pinnatifid, lobes entire to finely serrulate. . . . . 12. *Peucedanum*
  15. Flowers yellow . . . . 13. *Zizia*

---

1. *Eryngium* (Eryngo)

Biennial or perennial herbs; stems creeping or erect; leaves basal and/or cauline, alternate, membranous or leathery, often toothed; flowers borne in single, axillary heads, or in terminal heads; petals blue (ours) or white; fruit a subglobose to obconic or short-oblong schizocarp.

---

1. Plants prostrate, stoloniferous; flower heads borne singly in leaf axils; bracts small, entire. . . . . 1. *E. prostratum*
1. Plants erect, not stoloniferous; flower heads borne in terminal cymes; bracts large, spiny-toothed. . . . . 2. *E. aquaticum*

---

1. *E. prostratum* Nutt. ex DC. Creeping Eryngo Fig. 616
Damp sands, shores, floodplains, depressions and ditches. Se. Va. w. to Ky., Ill., se. Mo., and e. Okla., s. to Fla. and e. Tex. (*E. prostratum* var. *disjunctum* Fernald)

2. *E. aquaticum* L. Rattlesnake-master Fig. 615
Wet soils or shallow water of freshwater to brackish marshes, streams, ponds, bogs and ditches. Coastal plain, N.Y., N.J. s. to Fla. and Miss. intro. to Ont. and B.C. (*E. aquaticum* var. *ravenelii* (A. Gray) Mathias & Constance; *E. ravenelii* A. Gray; *E. virginianum* Lam.; *E. floridanum* J. M.Coult. & Rose)

2. *Lilaeopsis* (Grasswort)

Perennial herbs; stems rhizomatous, horizontal; leaves phyllodial, septate; flowers borne in umbels; petals white; fruit an ovate to almost circular schizocarp with thick, corky lateral ribs.

---

1. Phyllodes 1–7 cm long; peduncles as long as or longer than phyllodes (fig. 616b). . . . . 1. *L. chinensis*
1. Phyllodes 10–30 cm long; peduncles shorter than phyllodes (fig. 616c). . . . . 2. *L. carolinensis*

---

1. *L. chinensis* (L.) Kuntze Eastern Grasswort Fig. 616
Brackish marshes and tidal shores. Coastal plain, N.S. s. to Fla., w. to La. and Tex. (*L. lineata* (Michx.) Greene)

2. *L. carolinensis* J. M. Coult. & Rose Carolina Grasswort Fig. 616
Shallow water and muddy margins of pools, ponds, marshes and ditches. Coastal plain, Va. s. to Fla., w. to La. and Ark.; S.Am. (*L. attenuata* (Hook. & Arn.) Fernald)

Fig. 615. *Eryngium aquaticum*: a. habit; b. leaf with sheathing leaf base; c. base of leaf sheath; d. flower with inferior ovary, subtending bractlet; e–g. bractlet variations (G&W).

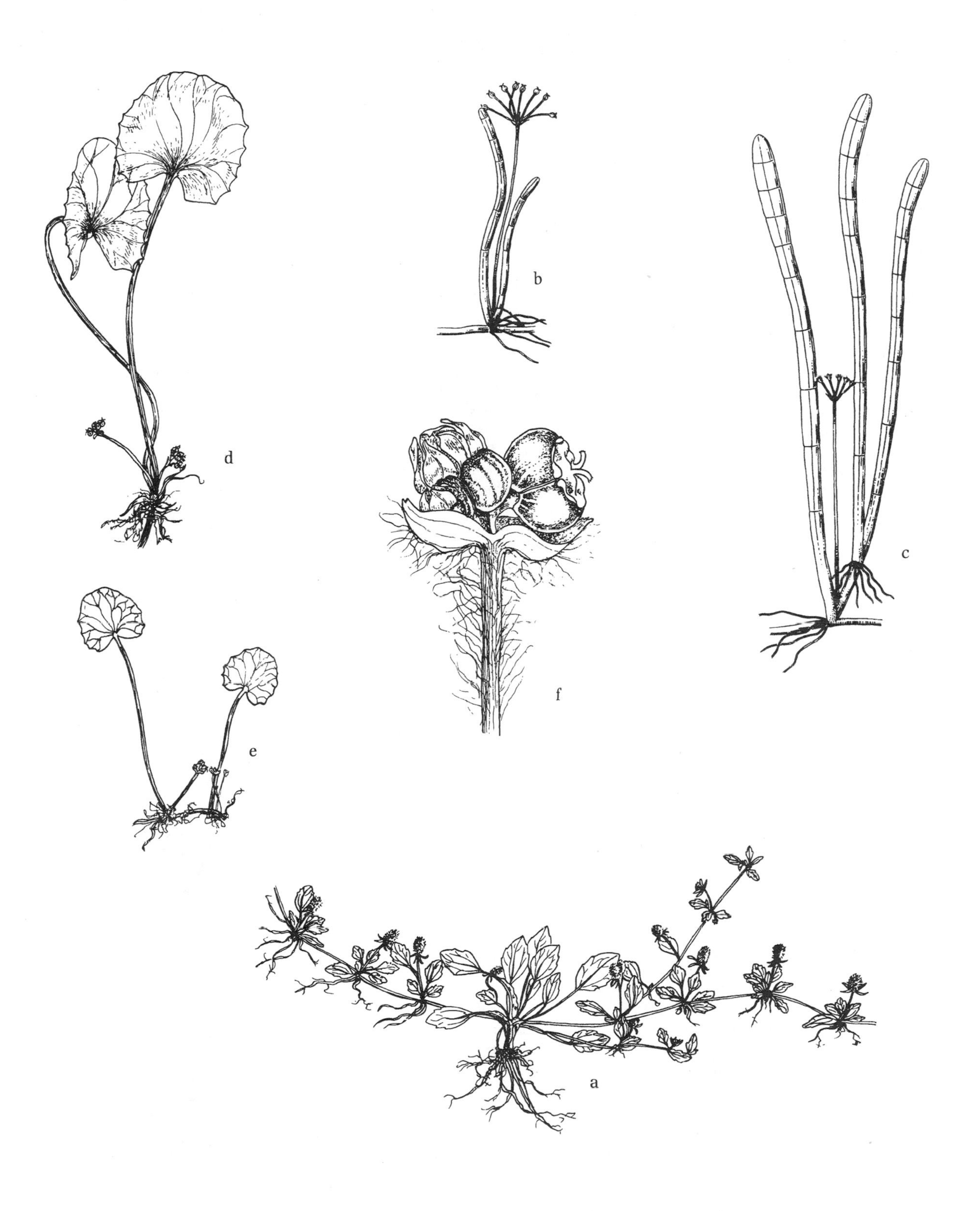

Fig. 616. *Eryngium prostratum*: a. habit, prostrate branching from basal rosette (C&C).
Lilaeopsis chinensis: b. habit (Beal).
Lilaeopsis carolinensis: c. habit (Beal).
Centella erecta: d, e. habit; f. inflorescence (C&C, as *C. asiatica*).

3. *Centella* (Centella)

Perennial herbs; leaves arising from slender, creeping stem, simple, ovate to oblong, with cordate to truncate bases; flowers 1–4 per umbel; petals white to greenish, often pink-tinged; fruit a schizocarp, flattened laterally, conspicuously veined.

1. *C. erecta* (L. f.) Fernald Erect Centella Fig. 616
Wet or dried sands and clays along margins of streams, ponds, lakes and ditches. Coastal plain, s. N.J., se. Md. and Del. s. to Fla., w. to Miss., La., Ark., e. Tex. and Mex.; WI. and C.Am.

4. *Sium* (Water-parsnip)

Perennial herbs; leaves compound, odd-pinnate, rarely simple, serrate, long-petioled; flowers borne in twice-compound umbels; petals white; fruit an ovoid schizocarp, slightly compressed laterally, with prominently corky-winged ribs.

1. *S. suave* Walter Water-Parsnip, Hemlock Water-parsnip Figs. 617, 618
Submersed and emersed, in ponds, lakes, streams, meadows, swamps, tidal river shores and floodplain forests. Nfld. w. to N.W.T., B.C. and Alask., s. throughout the U.S. Submersed juvenile leaves with very slender leaflets (Fig. 617). This is a highly variable species, especially responding to inundation in tidal rivers and to fluctuating seasonal changes in water depth. A number of forms have been described for these variants, but are not worthy of taxonomic recognition.

5. *Oxypolis* (Hog-fennel)

Perennial herbs; leaves pinnately or ternately compound (ours), rarely reduced to slender septate phyllodes; flowers borne in compound umbels; petals white; fruit a schizocarp, dorsally flattened, with corky wings.

1. Leaves pinnately compound or divided (fig. 619) .... 1. *O. rigidior*
1. Leaves ternately divided (fig. 620a) or with the uppermost simple .... 2. *O. ternata*

1. *O. rigidior* (L.) Raf. Cowbane, Water-dropwort Fig. 619
Bogs, swamps, wet woodlands, marshes and ditches. N.Y. w. to Ont. and Minn. s. to Fla. and Tex.

2. *O. ternata* (Nutt.) Heller Piedmont Cowbane Fig. 620
Damp pinelands, bogs and savannas. Coastal plain and piedmont, se. Va. and N.C. s. to n. Fla.; e. Tex.

6. *Ptilimnium* (Mock Bishop-weed)

Annual herbs; leaves phyllodial. hollow, and round in cross-section, or compound, petiolate; flowers borne in compound axillary and terminal umbels; petals white; anthers rose to purple; fruit a schizocarp, ovoid, laterally flattened, with dorsal ribs.

References: Easterly, 1957; Kral, 1981.

1. Leaves compound, dissected into filiform segments.
    2. Leaf segments crowded, ultimate segments appearing whorled along leaf rachis; styles 1.5–3 mm long .... 1. *P. costatum*
    2. Leaf segments not crowded or appearing whorled, the divisions 2 or 3 per node on leaf rachis; styles 0.2–1.5 mm long.
        3. Uppermost umbels with 2–16 primary rays; petals 0.5–1 mm wide; fruits 1.5–3 mm long .... 2. *P. capillaceum*
        3. Uppermost umbels with 12–30 primary rays; petals 1–2 mm wide; fruits 1–1.5 long. .... 3. *P. nuttallii*
1. Leaves phyllodial (appearing as a sheathing petiole, lacking a blade) .... 4. *P. nodosum*

1. *P. costatum* (Elliott) Raf. Ribbed Mock Bishop-weed Fig. 621
Swamps, wet barrens, pond margins and stream banks. N.C. w. to Tenn., Ill., and Mo., s. to Ga. and e. Tex.

2. *P. capillaceum* (Michx.) Raf. Herb-William Fig. 622
Brackish and freshwater marshes, swamps, pond and lake shores and ditches. Coastal plain, se. Mass. and N.Y. s. to Fla., w. to Tex.; inland n. to w. Ky., s. Ill., Kans. and Mo.; S.D.

Fig. 617. *Sium suave*: a. portion of branch; b. section of stem; c. leaf from lower portion of stem; d. submersed leaf; e. section of petiole; f. flower; g. fruit (G&W).

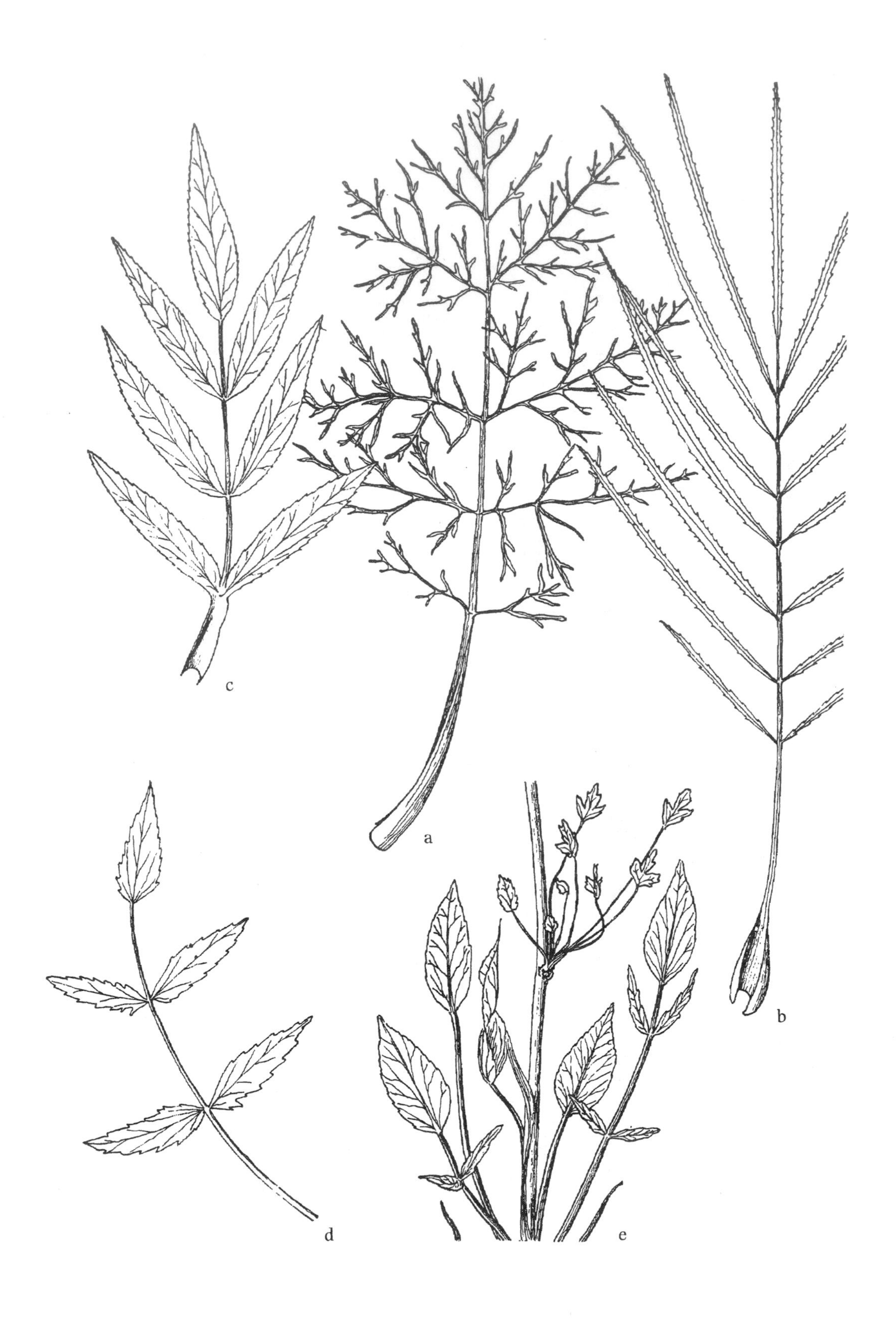

Fig. 618. *Sium suave*: a, b. submersed leaf forms; c, d. emersed leaf forms; e. habit, growth form of plants of tidal shores and ponds/lakes with fluctuation water level (F).

Fig. 619. *Oxypolis rigidior*: a. base of plant; b. inflorescence; c, d. leaf variations; e. flower; f. fruit (C&C).

Fig. 620. *Oxypolis ternata*: a. habit (G&W).
Ptilimnium nodosum: b. habit; c. stem with node and sheathing leaf base (Kral, 1981).

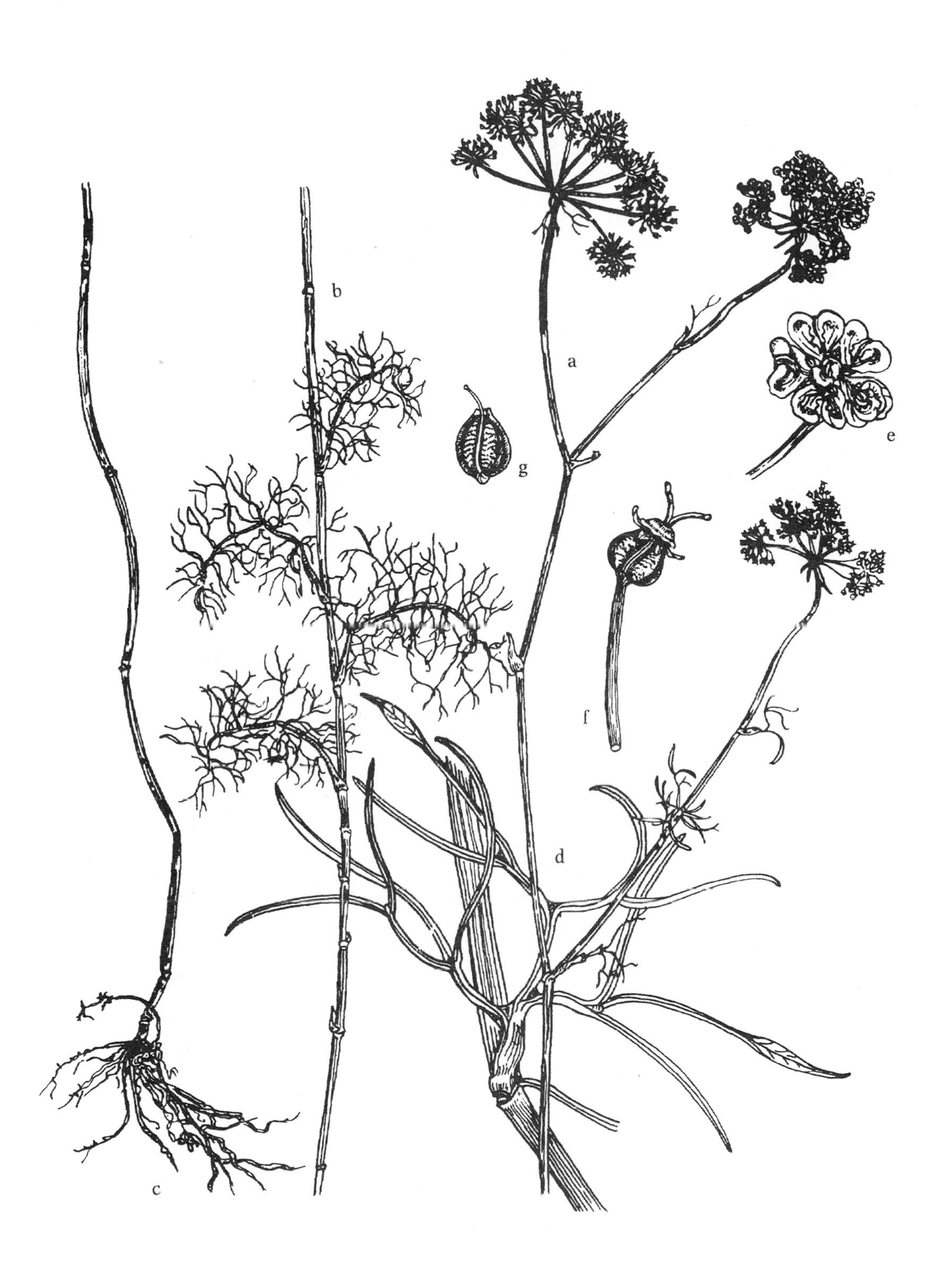

Fig. 621. *Ptilimnium costatum*: a. inflorescence; b. midsection of stem; c. base of plant; d. highly dissected leaf on stem; e. flower; f. fruit (schizocarp) showing pair of mericarps not yet split apart; g. fruit, showing single mericarp (C&C).

Fig. 622. *Ptilimnium capillaceum*: a. upper and basal portions of plant; b. leaf from midsection of stem; c. upper leaf; d. flower; e. fruit (schizcarp) (C&C).

3. *P. nuttallii* (DC.) Britton Lace-flower
Damp prairies, wet crevices in rocks, shores, wet woodlands and ditches. W. Ky. and Ill. w. to Kans., s. to Ala. and La. and e. Tex.

4. *P. nodosum* (Rose) Mathias Piedmont Mock Bishop-weed Fig. 620
Riverbanks. Piedmont, w. Md., and Va., s. to Va., Ga. and Ala.; Ark. Field studies by Kral (1981) indicated that *P. fluviatile* is conspecific with the rare *P. nodosum*, and that the extremes expressed in the range of variability are ecologically induced. The name *P. nodosum* has nomenclatural priority. (*P. fluviatile* (Rose) Mathias)

7. *Cynosciadium* (Cynosciadium)

Annual herbs; basal leaves linear to lanceolate, septate, cauline leaves palmately divided; flowers borne in compound umbels; petals white; fruit a schizocarp, long, ovoid, dorsal ribs narrow, lateral ribs prominently corky-winged.

1. *C. digitatum* DC. Finger Dog-shade Fig. 623
Wet sites, lowland woods and ditches. Okla., se. Mo. and s. Ill., s. to Ala., Miss., La. and e. Tex.

8. *Limnosciadium* (Dog-shade)

Annual herbs; basal leaves linear-lanceolate, septate, simple or pinnate; main cauline leaves pinnately divided; flowers borne in compound umbels; petals white; fruit an oblong to oval schizocarp, lateral ribs corky-winged.

1. *L. pinnatum* (DC.) Mathias & Constance Tansy Dog-shade Fig. 624
Pond shores, low wet open places and ditches. Ill., sw. Mo. and se. Kans., s. to Miss., La. and Tex. (*Cynosciadium pinnatum* DC.)

9. *Berula* (Water-parsnip)

Perennial herbs; leaves pinnate, narrowly oblong; leaflets oblong, subentire, serrate or lobed; submersed leaves compound; cauline leaves reduced; flowers borne in compound umbels; petals white; fruit a schizocarp, ellipsoidal to globose, compressed laterally, ribs obscure in corky pericarp.

1. *B. erecta* (Huds.) Coville Cutleaf Water-parsnip Fig. 625
Swamps, marshes and streams. N.Y. and Ont. w. to N.D., Mont. and B.C., s. to Ill., Okla., N.M. and Calif. (*B. erecta* var. *incisa* (Torr.) Cronq.; *B. incisa* (Torr.) G. Jones; *B. pusilla* (Nutt.) Fernald)

10. *Angelica* (Angelica)

Stout perennial herbs; leaves large, sessile, compound, 2 or 3 times ternately divided; flowers borne in large umbels; petals white; fruit a schizocarp, oblong-ellipsoidal, with thin, flat, broad lateral wings.

1. *A. atropurpurea* L. Alexanders Fig. 626
Thickets, bottomlands, swamps and wet open ground. S. Lab. and Nfld. w. to Ont., Wisc. and Minn., s. to Md., W.Va., mts. of N.C., Ky, and Tenn., Ill. and Iowa. (*A. atropurpurea* var. *occidentalis* Fassett)

11. *Cicuta* (Water-hemlock)

Perennial or biennial herbs; leaves 1–3 times pinnately compound; leaflets serrate; flowers borne in axillary or terminal compound umbels; petals white; fruit a schizocarp, ovoid, ellipsoidal, or subglobose, slightly flattened laterally, with flattish corky ribs. These plants are extremely poisonous.

---

1. Bulblets present, clustered in leaf axils (fig. 628d); leaflets linear. . . . . 1. *C. bulbifera*
1. Bulblets absent; leaflets linear-lanceolate to ovate-oblong.
    2. Leaflets lanceolate to ovate-oblong; fruits ellipsoidal, ovoid, or subglobose, with conspicuous pale, subequal lateral and intermediate ribs, with dark alternating grooves (fig. 629e).

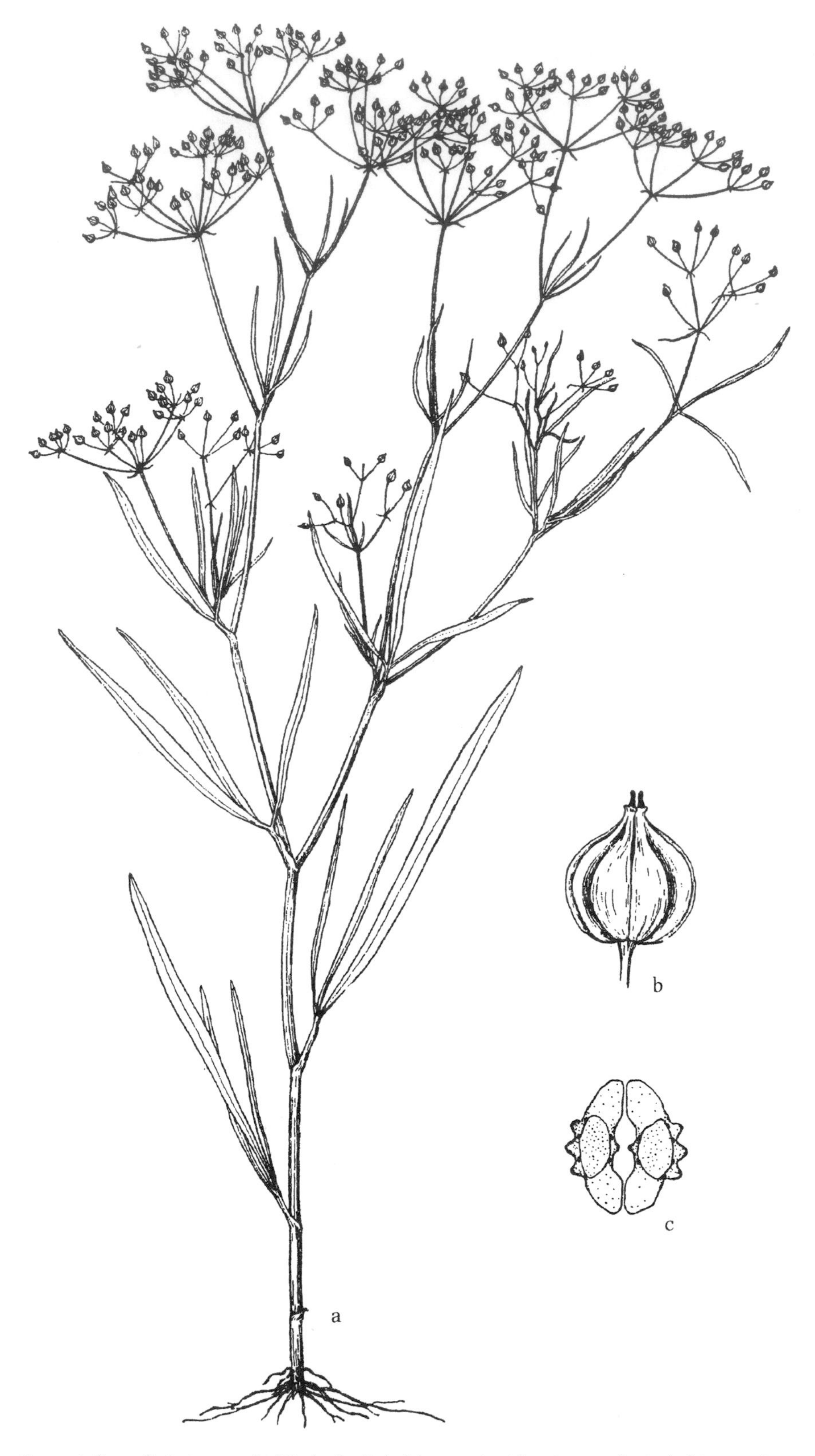

Fig. 623. *Cynosciadium digitatum*: a. habit; b. fruit (schizocarp), side view; c. fruit (schizocarp), cross-section (Lundell).

Fig. 624. *Limnosciadium pinnatum*: a. habit; b. fruit (schizocarp), side view; c. fruit (schizocarp), cross-section (Lundell).

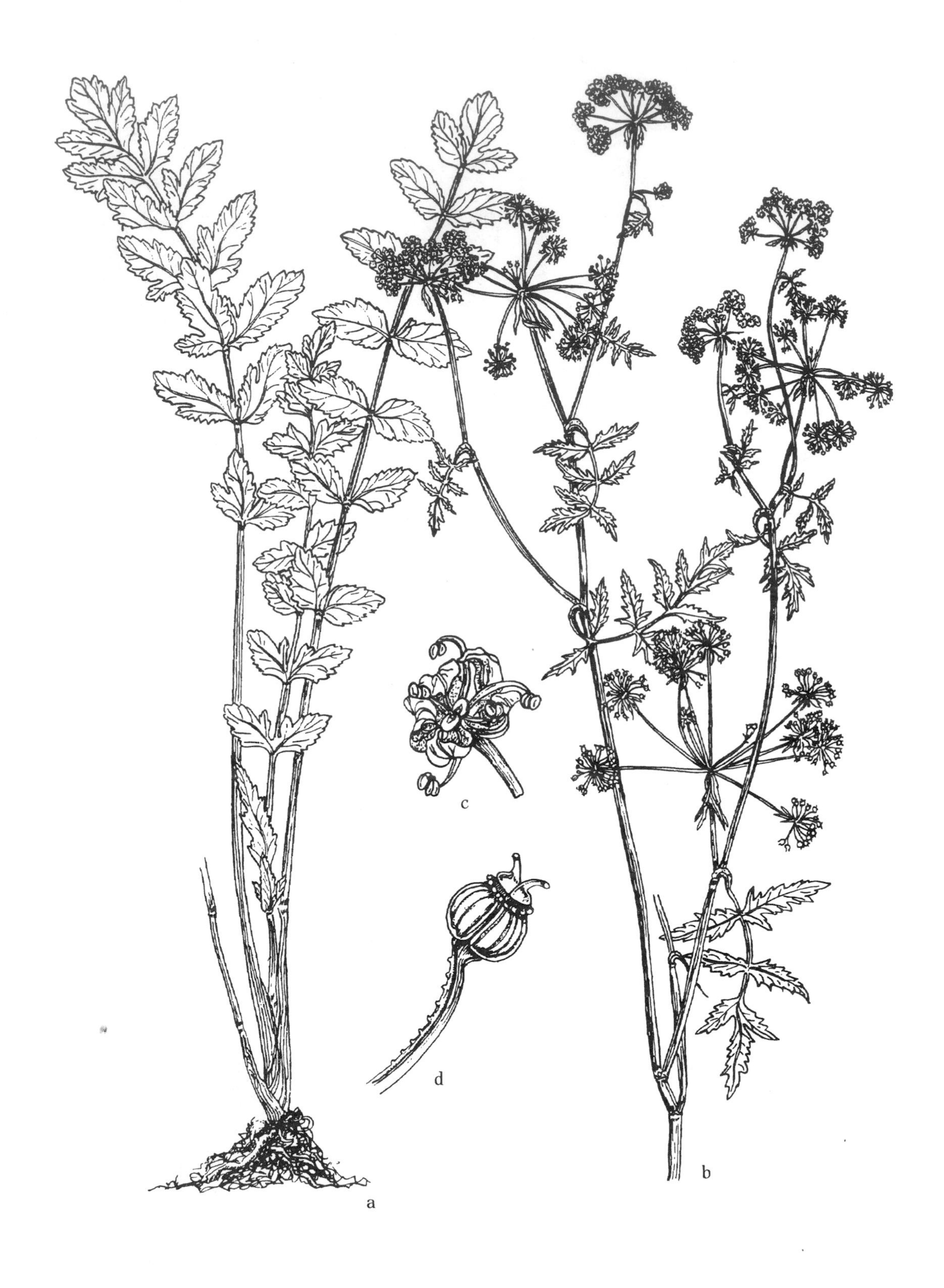

Fig. 625. *Berula erecta*: a. basal leaves; b. upper portion of plant; c. flower; d. fruit (schizocarp) (C&C).

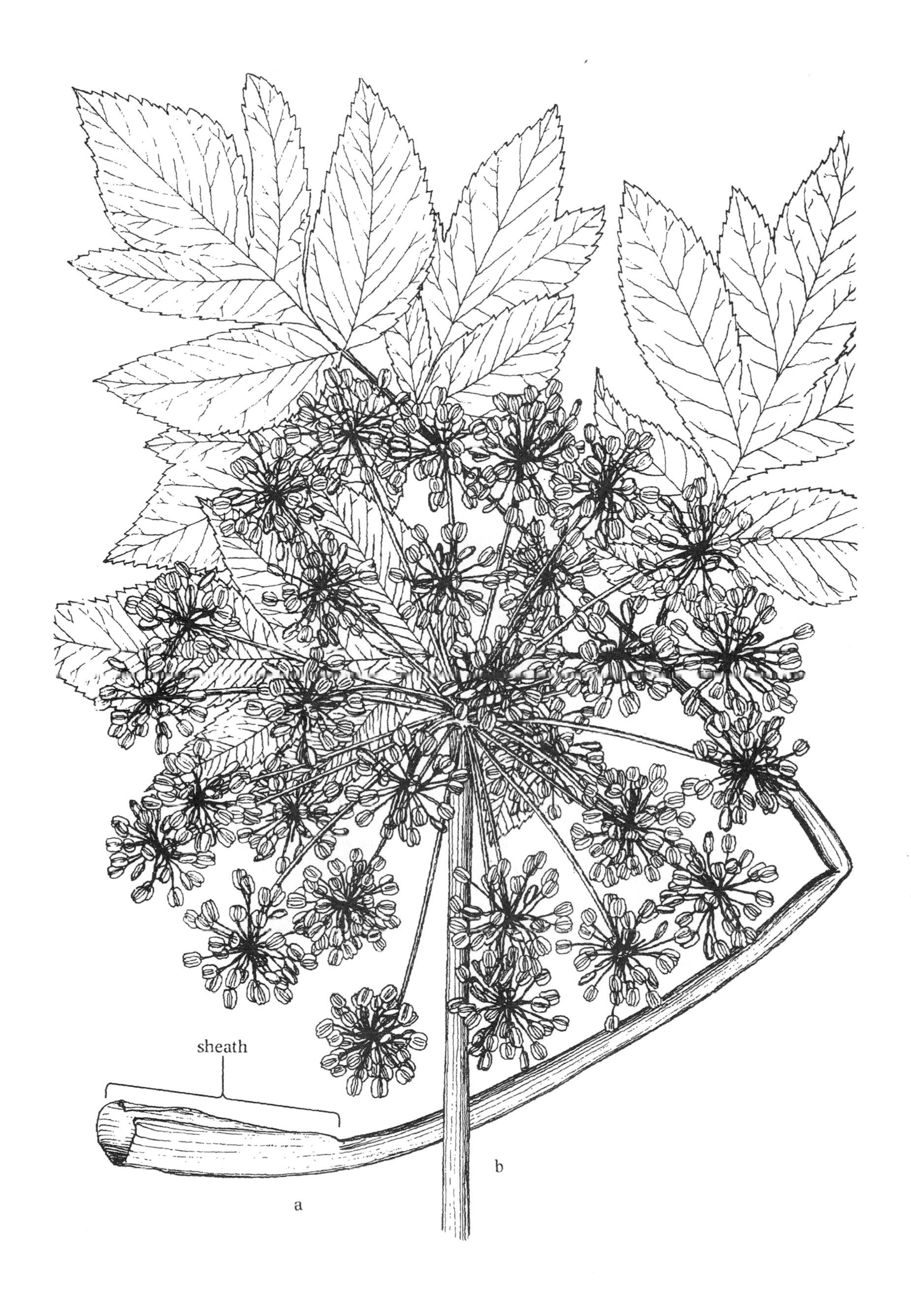

Fig. 626. *Angelica atropurpurea*: a. lower leaf with conspicuous leaf sheath; b. umbellate infructescence (F).

3. Larger leaflets 3–5.5 cm wide; fruit 2–3 mm long, subglobose or kidney-shaped-globose, ribs separate at constriction of fruit. . . . . 2. *C. mexicana*
3. Larger leaflets 0.6–3(4) cm wide; fruit 3–4 mm long, ellipsoidal or ovoid, rarely subglobose, ribs convergent at constriction of fruit. . . . . 3a. *C. maculata* var. *maculata*
2. Leaflets linear-lanceolate; fruits reniform to cordate-ovoid, with conspicuous lateral ribs, but obscure dorsal and intermediate ribs. . . . . 3b. *C. maculata* var. *victorinii*

1. *C. bulbifera* L. Bulb-bearing Water-hemlock Fig. 628
Swamps, pond and lake shores and wet thickets. Nfld. w. to B.C., s. to N.E., Va. and N.C., Tenn., Ill., Iowa, Kans., Wyo. and Oreg.

2. *C. mexicana* J. M. Coutl. & Rose Mexican Water-hemlock Fig. 629
Stream margins, thickets and marshy shores. Se. Va. s. to Fla., w. to Tex. and e. Mex.

3. *C. maculata* L. Water Hemlock Figs. 627, 628

3a. *C. maculata* var. *maculata* Water-hemlock, Spotted Cowbane, Musquash-root, Beaver-poison Figs. 627, 628
Marshes, swamps, meadows, thickets and ditches. N.S., P.E.I. and Que. w. to e. Man., s. to Md., w. N.C., Tenn., Mo., Okla. and Tex. Two additional varieties have been described but are not recognized by all authors; *Cicuta maculata* var. *angustifolia* Hook. is predominately west and north, while var. *bolanderi* (S. Wats.) G. Mulligan is more southern and western. All varieties of *C. maculata* are extremely poisonous. (*C. maculata* var. *curtissii* (J. M. Coult. & Rose) Fernald.; *C. curtissii* J. M. Coult. & Rose)

3b. *C. maculata* var. *victorinii* (Fernald) Boivin
Freshwater tidal flats. St. Lawrence R., Que. (*C. maculata* var. *victorinii* (Fernald) Boivin) (*C. victorinii* Fernald)

12. *Peucedanum* (Hog's Fennel)

Perennial, biennial (ours), or annual herbs; leaves once to several times pinnately (ours) or ternately compound; flowers borne in compound umbels; petals white (ours), yellow, or pinkish; fruit a schizocarp, globose, ovoid (ours), or oblong, strongly flattened dorsally.

1. *P. palustre* (L.) Moench Milk Parsley, Marsh Hog's Fennel
Marshes, shores and wet places. S. Me. and e. Mass.; intro. from Eur. Plants were first reported for North America in 1959 and thought to be a potential invasive (Harris, 1959; Richardson, 1971; Svenson and Pyle, 1979), but after more than 50 years, it has remained in few scattered localities.

13. *Zizia* (Golden Alexanders)

Perennial herbs, glabrous; stem erect, to about 0.6–1 m tall, basal leaves simple or compound (ours), stem leaves ternately compound, leaflets finally serrate; umbels compound, central flower sessile, others pedicellate; flowers yellow; sepals minute, triangular; petals obovate, abruptly narrowed at apex; stylopodium lacking; fruit laterally flattened with narrow winged ribs.

1. *Zizia aurea* (L.) W. D. J. Koch Golden Alexanders Fig. 628
Wet meadows, fens, sedge meadows, swamp forests, tamarack swamps, floodplains, moist thickets, roadside ditches. N.S., N.B. and Que. w. to Ont., Mich., Minn., Man., N.D., S.D. and se. Mont., s. to n. Fla., Ala., e. Okla. and e. Tex.

Caprifoliaceae / Honeysuckle Family

1. Plants woody, shrubs (ours); leaves simple. . . . . 1. *Lonicera*
1. Plant herbaceous; leaves pinnately compound. . . . . 2. *Valeriana*

1. *Lonicera* (Honeysuckle)

Woody twinning vines or shrubs (ours); leaves opposite, simple, deciduous; flowers fragrant, borne in axillary or terminal clusters; corolla funnelform to tubular, slightly irregular; ovary inferior, subtended by bracts and bracteoles; fruit a berry.

1. Peduncles 2–7(15) cm long, shorter than to about the same length as flowers; flowers nearly actinomorphic; bracts longer than the ovaries; berries blue; leaf margins with brownish hairs or cilia. . . . . 1. *L. villosa*
1. Peduncles 1.1–3.5(4) cm long, usually longer than flowers; flowers strongly 2-lipped, zygomorphic; bracts minute (often early deciduous); berries red; leaf margins not ciliate. . . . . 2. *L. oblongifolia*

Fig. 627. *Cicuta maculata*: a. upper portion of plant; b. basal portion of plant and basal leaf; c. flower; d. petal; e. fruit (schizocarp); f. fruit, showing single mericarp after schizocarp split (C&C).

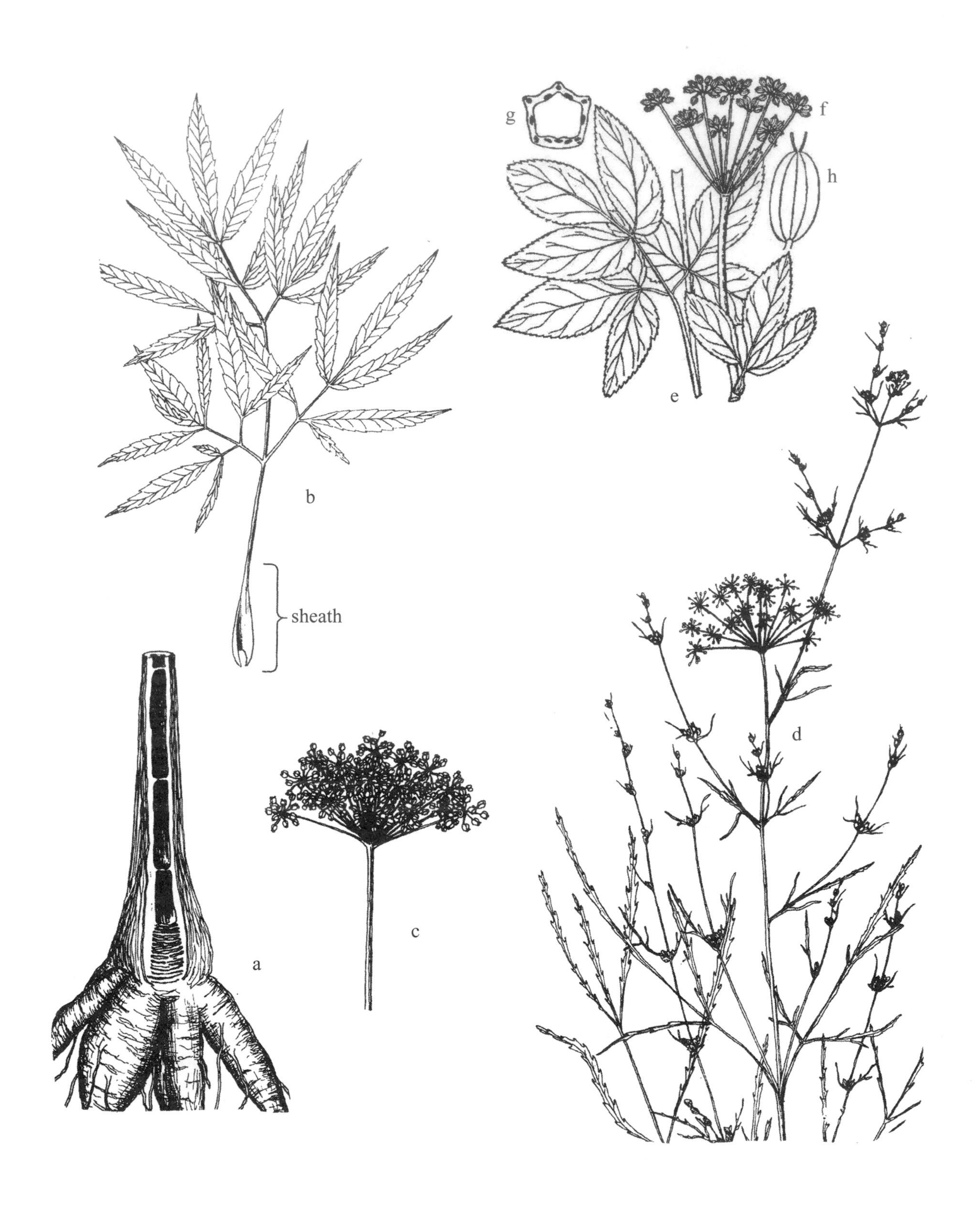

Fig. 628. *Cicuta maculata*: a. basal portion of plant, showing roots and longitudinal stem longitudinal section; b. leaf with sheathing leaf base; c. inflorescence (F).
Cicuta bulbifera: d. upper portion of plant (F).
Zizia aurea: e. compound leaf; f. inflorescence; g. stem, cross-section; h. fruit (schizocarp) (B&B).

Fig. 629. *Cicuta mexicana*: a. basal portion of plant; b. autumnal tuberous rhizome; c. compound leaf sheathing midsection of stem, with axillary branch; d. small branch with inflorescence; e. fruit (schizocarp) (G&W).

1. *L. villosa* (Michx.) Schult. Mountain Fly Honeysuckle Fig. 630
Peaty barrens, bogs, fens, coniferous swamps, boggy meadows, and wet thickets. Lab. and Nfld. and Lab. w. to Ont., Nunavut, Sask., s. to N.E., Pa., ne. Ohio, Mich., Wisc. and n. Minn. (*L. villosa* var. *calvescens* (Fernald) Wiegand; *L. villosa* var. *fulleri* Fernald; *L. villosa* var. (Eaton) Fernald; *L. villosa* var. *tonsa* Fernald)

2. *L. oblongifolia* (Goldie) Hook. Swamp Fly Honeysuckle Fig. 630
Bogs, fens, coniferous swamps, thickets, wet woods, and stream margins; especially calcareous sites. N.S., N.B., se. Que. w. to Man. and Sask., s. to n. Me., Vt., N.Y., n. Pa., ne. Ohio, Mich., Wisc. and n. Minn.

2. *Valeriana* (Valerian)

Perennial herbs; leaves opposite, simple or pinnate; flowers small, bisexual (ours) or unisexual, regular or slightly irregular, borne in terminal panicles or capitate cymes; fruit an achene.

The Angiosperm Phylogeny Group (APG IV, 2016; Lee et al., 2021) now treats the segregate family Valerianaceae as a tribe of ca. 375 species (Valerianoideae) within a reconstituted Caprifoliaceae.

References: Bell and Donoghue, 2005; Lee et al., 2021.

1. *V. uliginosa* (Torr. & A. Gray) Rydb. Marsh Valerian Fig. 631
Calcareous swamps, fens, and wet woods. NB. and Que. w. to w. Ont., s. to n. N.E. and Mass., s. N.Y., ne. Ohio, Mich., n. Ind., n. Ill. and e. Wisc. (*V. sitchense* subsp. *uliginosa* (Torr. & A. Gray) F. G. Meyer)

Viburnaceae / Arrow-wood Family

According to Wilson (2016) a proposal submitted to the General Committee for Botanical Nomenclature to "super-conserve" Adoxaceae over Viburnaceae was rejected, thus Viburnaceae is now the correct name for this family as circumscribed by APG IV (2016).

1. Leaves pinnately compound. . . . . 1. *Sambucus*
1. Leaves simple. . . . . 2. *Viburnum*

1. *Sambucus* (Elderberry)

Shrubs with soft stems, pith thick, soft; herbage with rank odor when crushed; leaves opposite, pinnately compound, deciduous; flowers small, borne in many-flowered cymes, flat-topped (ours) or conical; corolla white, united at base; stamens 5; ovary inferior; fruit a red or purplish-black (ours), berry-like drupe with 3 small seed-like pits.

1. *S. canadensis* L. American Black Elderberry, Elderberry Fig. 630
Wet or damp soil, swamps, and ditches, banks of channels and ditches, wet disturbed sites. N.S. and Que. w. to Man., N.D. and Mont. s. to Fla. and Tex. Ariz. and Calif.; Mex and W.I.

2. *Viburnum* (Viburnum, Arrow-wood)

Shrubs or trees; leaves simple, opposite, deciduous; flowers small, borne in flat-topped cymes; corolla white or cream-colored, united; fruit a 1-seeded drupe.

1. Leaves unlobed, blades pinnately veined arising laterally from the midvein; fruits blue-black.
    2. Leaves entire or finely toothed: lateral veins anastomosing before reaching margin.
        3. Leaves minutely pilose on lower surface, especially on margins, margins wavy to irregularly crenulate or appearing entire (rarely dentate); winter buds reddish-brown to grayish-brown (late summer). . . . . 1a. *V. nudum* var. *nudum*
        3. Leaves glabrous, punctate, or scurfy on lower surface, margins dentate, crenate, or crenulate; winter buds yellow-brown or golden (late summer). . . . . 1b. *V. nudum* var. *cassinoides*
    2. Leaves coarsely toothed; lateral veins extending into teeth.
        4. Branchlets, petioles, and often leaf surfaces pubescent: fruit stone with deep furrow-like ventral groove. . . . . 2. *V. dentatum*

Fig. 630. *Lonicera oblongifolia*: a. flowering branch; b. flower (Gleason).
Lonicera villosa: c. flowering branch with paired flowers; d. fruiting branch; e. leaf; f. branch, winter condition (Ryan).
Sambucus canadensis: g. inflorescence and compound leaves; h. infructescence (WVA).

Fig. 631. *Valeriana uliginosa*: a. upper portion of plant, leaves compound; b. middle portion of plant; c. flower (Crow).

4. Branchlets, petioles, and leaf surfaces glabrous or leaves pubescent only on veins of lower surface; fruit stone with shallow, broad, trough-like ventral groove. . . . . 3. *V. recognetum*

1. Leaves lobed, veins palmately veined, with 3(5) veins arising from the base; petioles with conspicuous glands just below the junction with the blades; fruits red.

5. Petiolar glands sessile, with the surface concave (indented), (0.8)0.9–1.5(2) mm long; blades glabrous on upper surface middle lobe about as long as wide.. . . . . 4. *V. opulus*

5. Petiolar glands typically stalked and flat-topped 4–0.8 mm long; blades with sparse hairs over the upper surface, middle lobe longer than wide. . . . . 5. *V. trilobum*

---

1. *V. nudum* L. Withe-rod, Wild-raisin

1a. *V. nudum* var. *nudum* Possum-haw Fig. 632

Swamps, floodplains, wet pinelands, and bogs. R.I., Conn., se. N.Y., N.J. and e. Pa. s. to Fla, w. to W.Va., Ky., Tenn., and Ark., and e. Tex. (*V. nudum* var. *angustifolium* Torr. & A. Gray)

1b. *V. nudum* var. *cassinoides* (L.) Torr. & A. Gray Wild-raisin, Withe-rod Fig. 633

Thickets, swamps, and clearings. Nfld. w. to Ont., Ohio, n. Ind., Mich., Wisc. and ne. Ill., s. to N.J., Pa. and Md., mts. to e. Tenn., n. Ga. and ne. Ala. Some authors treat this as a distinct species, *V. cassinoides* L.

2. *V. dentatum* L. Southern Arrow-wood Fig. 633

Moist or dry sandy thickets, stream banks, bogs, and well-drained soil. New England w. to s. N.Y., Pa., Ohio, Ill., and Iowa, s. to n. Fla., La. and e. Tex.

3. *V. recognetum* Fernald Arrow-wood Fig. 633

Moist or dry sandy thickets, stream banks, bogs, and well-drained soil. N.S., N.B. and s. Que. w. to s. Ont., Mich., and Wisc., s. to Fla., se. Mo., Ark. and e. Tex. This species is treated by some authors as the glabrate phase of *V. dentatum*, as var. *lucidum* Aiton.

4. *V. opulus* L. European Cranberrybush; Highbush-cranberry

Ditch banks, wet thickets, edges of forests, naturalized from cultivation. N.S., N.B. and N.E. w. to s. Ont., Mich., Wisc. and Iowa, s. to Virg., Tenn., and Mo.; Mont. and Wash.; intro. from Eur.

5. *V. trilobum* Marshall American Cranberrybush; American Highbush-cranberry

Swamps, shores, fens, thickets along rivers and streams, wet roadsides and ditches. Nfld. and Que. w. to Ont., Sask. and B.C., s. to N.E., N.J., Pa., W.Va., ne. Ky., n. Ohio., n. Ind., Ill., Iowa, e. Neb., w. S.D., ne. Wyo., s. Mont., Ida. and Wash.; N.M. This species is more likely to be found in wet areas, than the introduced European taxon, but both can be found in the same areas. The two are sometimes treated as conspecific; our native species is then referred to as *V. opulus* subsp. *trilobum* (Marshall) R. T. Clausen. (*V. opulus* var. *americanum* Aiton)

Fig. 632. *Viburnum nudum* var. *nudum*: a. branch with flowers and young leaves; b. flowers; c. calyx and inferior ovary, subtended by 2 bracts; d. branch with fruit and mature leaves (C&C).

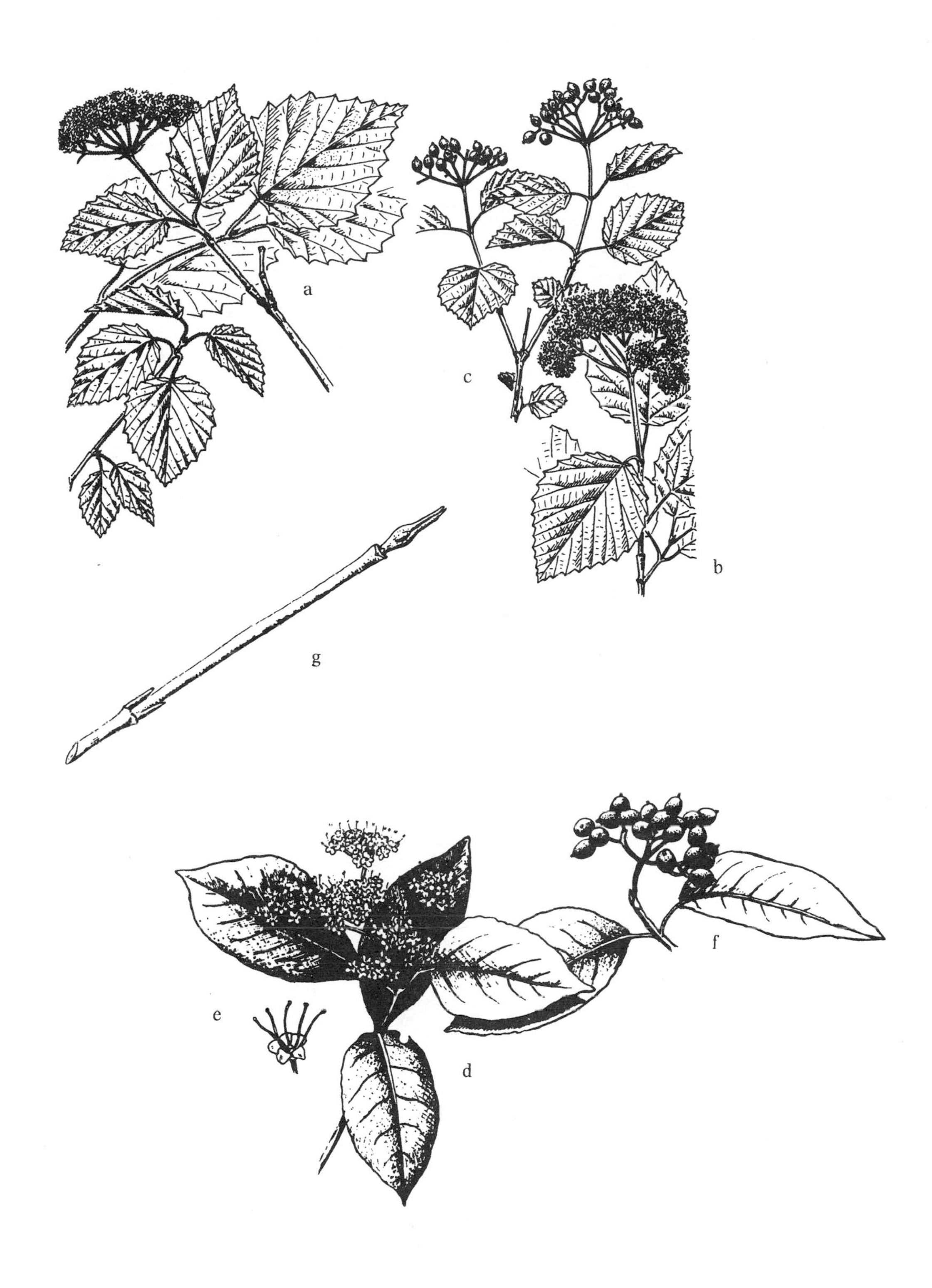

Fig. 633. *Viburnum dentatum*: a. flowering branch (WVA).
Viburnum recognetum: b. flowering branch; c. fruiting branch (WVA).
Viburnum nudum var. *cassinoides*: d. flowering branch; e. flower; f. fruiting branch; g. winter twig (Ryan, as *V. cassinoides*).

Glossary of Plant Terms

Abaxial. Pertaining to the side of an organ away from the axis, such as the lower surface of a leaf; cf. *adaxial*.

Abscise. To separate from.

Acaulescent. Stemless or apparently so, or with the stem subterranean.

Achene. A hard, dry, indehiscent, one-seeded, one-locular fruit.

Actinomorphic. Descriptive of a flower or set of flower parts which can be cut through the center into equal and similar parts along two or more planes; radially symmetrical; regular.

Acuminate. Long-tapering to a pointed apex, the sides of the angle somewhat concave.

Acute. Sharply angled, the sides of the angle essentially straight, the angle less than 90 degrees.

Adaxial. Pertaining to the side of an organ toward the axis, such as the upper surface of a leaf; cf. *abaxial*.

Adnate. Pertaining to the fusion of one structure to another, such as the stipules to the petioles in some species of *Potamogeton*.

Adventitious. Pertaining to organs arising from an irregular or unusual position, such as roots from the stem, or leaves or buds from a root.

Aerenchyma. Type of plant tissue consisting of thin-walled cells with numerous, large intercellular spaces.

Agamospermy. Pertaining to the development of seeds without fertilization; a type of apomixis.

Aggregate. Crowded into a dense cluster but remaining separate.

Alternate. Leaves or other parts borne singly at nodes along the stem (axis).

Ament. A catkin; a lax, usually pendulous, spike-like inflorescence, usually bearing apetalous, unisexual flowers.

Anastomosing. Pertaining to veins or ridges branching and remerging so as to form a network.

Androecium. A collective term for the stamens.

Androgynous. Pertaining to an inflorescence having both staminate and pistillate flowers, the staminate flowers in the apical, or upper, portion; cf. *gynecandrous*.

Annual. A plant which completes its life history within one year.

Anther. The pollen-bearing part of the stamen.

Anthesis. Flowering, or strictly the time of expansion of a flower when pollination may take place; often used to designate the flowering period.

Antrorse. Having hairs or processes directed upward or forward; cf. *retrorse*.

Apetalous. Without petals.

Apex. The tip or distal end.

Aphyllous. Lacking leaves.

Apical. Pertaining to the apex.

Apiculate. Tapering to an abrupt short, pointed tip.

Apocarpous. Gynoecium with separate carpels.

Apomictic. Typically, setting seed without fertilization; production of viable offspring without fertilization (including reproduction via vegetative propagules).

Appressed. Lying close and pressed against; adpressed.

Aquatic. Living in or on water for all or most of the life span.

Arborescent. Approaching the size and habit of a tree.

Arcuate. Arching; bowed; moderately curved.

Areolate. Marked out into small spaces.

Aril. A fleshy outgrowth from the hilum or funiculus covering the outer portion of the seed coat; often appears as a fleshy seed.

Arillate. Pertaining to a seed having an aril.

Aristate. Bearing a stiff bristle-like awn or seta; tapered to a very narrow, much elongate apex.

Armed. Provided with any kind of strong or sharp defense structures such as thorns, spines, prickles, barbs.

Articulate. Jointed; provided with joints or places where separation may naturally take place.

Ascending. Rising upward; produced somewhat obliquely or indirectly upward.

Attenuate. Gradually narrowed to a point at apex and/or base.

Auricle. An ear-shaped part of appendage, such as the projections at the base of some leaves or petals.

Auriculate. Having ear-like lobes; having auricles.

Awl-shaped. Narrow and sharp-pointed, gradually tapering from the base to a slender, stiff point; subulate.

Awn. A stiff, bristle-like appendage, often at the end of a structure.

Axil. The angle between an organ and the axis which bears it, such as between the leaf and the stem.

Axillary. Located in the axil.

Axis. The main line of development of a plant or organ, such as the main stem.

Barbed. Having rigid points or short bristles, usually reflexed.

Beaked. Ending in a firm, prolonged slender tip, such as beaks of fruits or pistils; rostrate.

Bearded. Bearing long stiff hairs or bristles; pertaining to a tuft or zone of pubescence.

Berry. A fleshy fruit with few to many seeds, but lacking stones; cf. *drupe*.

Biconvex. Lenticular; having both surfaces rounded (lens shaped).

Biennial. A plant which requires two years to complete its life cycle, the first year growing only vegetatively, often producing only a basal rosette, then flowering, fruiting, and dying the second year.

Bifid. Two-cleft; split into two segments, such as at the apex of some leaves or petals.

Bilabiate. Two-lipped.

Bilaterally symmetrical. Descriptive of a flower or set of flower parts which can be cut through the center into two equal and similar parts along only one plane; zygomorphic; irregular.

Bipinnate. Twice, or doubly, pinnately compound.

Bipinnatifid. Twice pinnatifid; cf. *pinnatifid*.

Bisexual. Having both stamens (male parts) and pistil (female parts) present and functional in one flower, as in bisexual flowers.

Biternate. Pertaining to leaves which are twice divided into threes (three main divisions themselves divided into three parts: nine leaflets); cf. *triternate*.

Bladder. A thin, sac-like structure; as in the bladder traps of *Utricularia* (Bladderwort).

Blade. The expanded, usually flattened, portion of a leaf, bract, petal, or other structure.

Bloom. A whitish, usually waxy covering on the surface; cf. *glaucous*.

Bract. A reduced leaf or leaf-like structure, particularly subtending a flower or an inflorescence.

Bracteate. Bearing bracts.

Bracteole. The diminutive of bract; bractlet.

Bractlet. A very small bract, or sometimes applied to bracts in a secondary position; bracteole.

Branchlet. Ultimate division or divisions of a branch.

Bristle. A stiff hair or trichome.

Bud. An embryonic axis with its appendages, such as an unexpanded flower or a bud in the axil of a leaf.

Bulb. A short underground stem surrounded by succulent leaves or scales.

Bulblet. A small bulb-like structure (vegetative propagule) often borne in the axil of a leaf, inflorescence, or other unusual area.

Bulbose. Having bulbs; the structure of a bulb.

Bulbous. Bulb-like.

Callous. Having the texture of a callus.

Callus. A thick, leathery, or hardened protuberance; in grasses, the tough swelling at the base, or point of insertion, of the lemma or palea.

Calyx. Collective term for sepals; the outer, usually green, whorl of floral parts; outer series of perianth parts, frequently enclosing the rest of the flower in the bud stage; sometimes colored or petal-like, or sometimes greatly reduced or lacking.

Campanulate. Bell-shaped.

Canescent. Gray-pubescent; surface covered with a gray or whitish pubescence.

Capillary. Thread-like or hair-like, very slender.

Capitate. Aggregated into a dense, compact cluster; formed like a head.

Capitulum. A dense inflorescence, consisting of an aggregation of usually sessile flowers; a head.

Capsule. A dry dehiscent fruit resulting from the maturation of a compound ovary (of one or more fused carpels).

Carpel. A segment or unit of a compound pistil or the single unit of a simple pistil; the ovule-bearing units of an ovary.

Cartilaginous. Firm and tough, but flexible.

Catkin. An ament; a lax, usually pendulous, spike-like inflorescence, usually bearing apetalous, unisexual flowers.

Caudate. Having a tail-like appendage; as at leaf apices.

Caudex. A persistent base of stem, often woody, from which new aerial herbaceous stems arise each year.

Cauline. Pertaining to the stem; as opposed to basal.

Cespitose. Growing in tufts; in dense clumps; having several to many stems in a close basal tuft.

Chaff. A thin, dry, membranous scale or bract, particularly the bracts subtending flowers in flower heads of the Asteraceae.

Chaffy. Having or resembling chaff.

Channeled. Having a deep longitudinal groove; furrowed.

Chartaceous. Thin but stiff; papery texture; usually not green.

Chlorophyllous. Containing chlorophyll, thus typically giving a green or greenish appearance; as opposed to achlorophyllous, lacking chlorophyll.

Ciliate. Having hairs extending from the margins or edges.

Ciliolate. The diminutive of ciliate.

Clasping. Partly or completely surrounding the stem.

Clavate. Club-shaped, gradually thickened toward the top.

Cleft. Deeply cut; having divisions to or nearly to the middle, such as a palmately or pinnately cleft leaf.

Cleistogamous. Self-fertilized in bud without the flower opening.

Collar. Outer side of grass leaf at the junction of sheath and blade.

Column. In the Orchidaceae, a coalescence of stamens and style into a single, somewhat elongate body.

Commissure. The face or plane along which two carpels join, such as the face by which one mericarp joins another in a schizocarp.

Comose. Bearing a tuft or tufts of soft hairs at one end.

Compound. Composed of several parts united as a common whole, such as a compound pistil, or a compound leaf composed of two or more distinct leaflets.

Compressed. Flattened laterally.

Concave. Pertaining to a surface or margin rounded inwardly; cf. *convex*.

Cone. A structure consisting of seed-bearing scales arranged on a usually elongate axis, such as a pine cone.

Congeneric. Belonging to the same genus.

Conglomerate. Densely clustered.

Conic. Cone-shaped.

Connate. United; especially applied to like structures, such as the bases of two opposite leaves.

Conspecific. Belonging to the same species.

Contiguous. Touching, but not fused.

Contorted. Twisted; bent or twisted on itself.

Convex. Pertaining to a surface or margin that is rounded outwardly; cf. *concave*.

Convolute. Rolled or twisted together in bud stage; rolled longitudinally in winter buds of some *Potamogeton* species.

Cordate. Heart-shaped; with a sinus and rounded lobes at the base; often applied to the basal portion rather than the entire organ.

Coriaceous. Leathery in texture, such as the texture of many evergreen leaves.

Corm. A thick, solid, bulb-like stem surrounded by papery scales, usually underground.

Corolla. A collective term for petals; the inner whorl of a perianth, usually colored or otherwise differentiated; frequently the showy part of the flower, but sometimes reduced or lacking; if petals are united, the corolla may be divided into a corolla tube and lobes or lips, and is then said to be sympetalous or gamopetalous.

Corona. A crown or inner petal-like appendage between the petals and stamens, such as in *Asclepias*.

Corrugated. Having a surface that is wrinkled or in folds.

Corymb. A flat-topped, racemose type of inflorescence in which the lower pedicels are successively elongate, the flowering sequence indeterminate, with the marginal flowers opening first.

Corymbose. Pertaining to flowers or fruits arranged in or resembling a corymb.

Creeper. A trailing shoot, usually rooting at most of the nodes.

Creeping. Pertaining to a shoot running along at or near the surface of the ground and often rooting at the nodes; repent.

Crenate. Having a margin with low, rounded lobes or teeth.

Crenulate. Finely crenate; having very small, rounded lobes or teeth.

Crested. Having a ridge-like projection on the surface or across the top.

Crisped. Having a wavy surface toward the margin.

Cruciform. Cross-shaped, such as flowers of Brassicaceae.

Cucullate. Hooded or hood-shaped.

Culm. The aerial stem of a grass, sedge, or rush.

Cuneate. Wedge-shaped; narrowly triangular, tapering toward the point of attachment.

Cuspidate. Tipped with a short, tooth-like rigid point.

Cyme. A broad, flat-topped inflorescence in which the branches arise from the same point, and the flowering sequence is determinate, with the central flowers opening first.

Deciduous. Not persistent, falling at the end of one season of growth; not evergreen.

Decompound. More than once compound.

Decumbent. Growing reclining on the ground, but with the apical portion ascending.

Decurrent. Extending downward from the point of insertion, usually applied to leaves apparently prolonged downward along the stem, often as wings.

Decussate. Pertaining to organs, especially leaves, borne oppositely on the stem, each pair alternating at right angles with the pairs above and below.

Dehiscent. Opening and shedding contents; applied to stamens and fruits as opening by valves, slits, pores, etc.

Deltoid. Pertaining to shapes that are triangular in outline.

Dentate. Toothed, with teeth directed outward, not forward.

Denticulate. Minutely or finely dentate.

Depauperate. Stunted, much reduced and imperfect in structure and development.

Depressed. Somewhat flattened from above; pressed down.

Determinate. A defined number or growth; an inflorescence in which the terminal or central flower opens first, thus the axis prolongation is arrested.

Diadelphous. Having stamens which are united in two groups by their filaments; as in Fabaceae.

Diaphragmed. Having cross-partitions; as in the pitch of *Nyssa*.

Dichotomous. Forking in pairs.

Diffuse. Loosely branching or spreading, descriptive of open growth.

Digitate. Having finger-like divisions; a compound with members arising from one point.

Dimorphic. Occurring in two forms, such as two types of leaves or two types of seeds.

Dioecious. Flowers that are unisexual, with staminate and pistillate flowers on separate plants.

Diploid. Having two full sets of chromosomes in each cell.

Discoid. Having only disk flowers, as in some Asteraceae; disk-shaped.

Disk. A development of the receptacle at or around the base of the pistil; in the Asteraceae the central portion of the flowering head, bearing tubular flowers.

Disk flower. Applied to the tubular flowers in the central portion of the flowering heads on many Asteraceae (distinguished from the ray flowers).

Dissected. Divided or cut into many small, slender segments, such as dissected leaves.

Distal. At or toward the end farthest from the point of attachment; cf. *proximal*.

Distichous. Two-ranked, on opposite sides of an axis and in the same plane.

Distylous. Pertaining to flowers of a given species having two different style lengths and anthers borne at corresponding levels; as in the Lythraceae; cf. *tristylous*.

Divergent. Spreading broadly; inclining away from each other.

Divided. Lobed or separated to the base; often referring to the leaf blade when it is cut into distinct divisions to or almost to the midvein.

Dorsal. Pertaining to or located on the back, or upper (adaxial), surface; cf. *ventral*.

Dorsiventral. Flattened with definite upper or front (adaxial) and lower, or back (abaxial), surfaces.

Downy. Pubescent, especially pertaining to a surface covered with fine, soft hairs.

Drupe. A fleshy, indehiscent fruit with the seed enclosed in a stony endocarp or inner layer of fruit wall (stone); stone fruit.

Druplet. One small drupe in an aggregate fruit made up of drupes; as in *Rubus* (blackberries, raspberries).

Elator. Band-like structure which is sensitive to changes in humidity, and is an aid in dispersal of the spores, as in *Equisetum*.

Ellipsoid. A solid with an elliptic outline.

Elliptic. Oval or oblong in outline, with the ends equally narrowed or rounded.

Emarginate. Having a shallow notch at the apex.

Embryo. The rudimentary plant within the seed.

Emergent. Extending out of the water; emersed.

Emersed. Extending out of the water; emergent.

Endocarp. The inner layer of a fruit wall, sometimes becoming hard or bony, as in a drupe.

Entire. Having a smooth margin without any teeth, lobes, or marginal glands (may or may not be hairy, or ciliate).

Epicalyx. A series of bracts closely subtending and resembling sepals or a calyx.

Epigynous. Pertaining to a flower in which the calyx, corolla, and stamens are borne at the summit of the ovary, with the ovary fused to an expanded receptacle or floral tube, the ovary thus said to be inferior.

Epipetalous. Descriptive of stamens when they are inserted on the corolla.

Epiphyte. A plant which grows upon another plant, but without parasitizing it.

Equitant. Descriptive of leaves overlapping in 1 plane (two-ranked); as in *Iris* and *Xyris*.

Erect. In an upright position; ascending.

Erose. Having an uneven margin, appearing gnawed, or having a margin with very small teeth of irregular shape and size.

Estuarine. Growing in estuaries.

Evergreen. Remaining green through the dormant season and functional the following growing season.

Exfoliate. To shed or come off in thin scale-like or flake-like layers.

Exocarp. The outermost layer of the fruit wall.

Exserted. Extending beyond; sticking out; such as stamens from the perianth.

Falcate. Curved and flat, like a sickle.

Farinose. Surface covered with a mealy powder.

Fascicle. A bundle or cluster, such as the leaves of *Pinus* or the stamens of *Hypericum*.

Fasciculate. Congested in close clusters or bundles.

Feather-veined. Having veins all branching from the sides of a midrib.

Fetid. Having a disagreeable odor.

Filament. The portion of the stamen bearing the anther.

Filiform. Thread-like, long, and very slender.

Fimbriate. Fringed; cut into regular segments giving a fringed appearance.

Fimbrillose. Abounding with fringe.

Flabellate. Fan-shaped or broadly wedge-shaped.

Flaccid. Limp, lax, and weak; floppy.

Flexuous. Having a wavy appearance; especially applied to submersed aquatics where plants are not rigid, thus they readily bend in flowing water.

Floccose. Surface covered with irregular tufts of soft wooly or cottony hairs which usually rub off readily.

Floral tube. Tubular structure in the flower derived by the fusion of the floral envelopes and the androecium, on which are borne the calyx, corolla, and stamens; sometimes referred to as a hypanthium.

Floret. A small flower, usually one of a dense cluster; in grasses, a unit consisting of the pistil and stamens surrounded by the lemma and palea.

Flower. An axis bearing stamens or pistils or both, typically subtended by a perianth, although the perianth can be lacking.

Foliaceous. Leaf-like in appearance or texture.

-foliolate. Suffix pertaining to the number of leaflets in a compound leaf; cf. *trifoliolate*.

Follicle. A dry dehiscent fruit splitting along only one suture; the product of a simple pistil.

Forked. Divided into nearly equal segments.

Foveolate. Pitted, especially diminutive, often describing surface of seeds, pollen, or spores.

Frond. In the Lemnaceae, the green thallus-like stem which functions as foliage; the blade of the leaf in ferns.

Fruit. The seed-bearing organ; the matured pistil.

Funnelform. Having the shape of a funnel, the tube gradually widening into a limb; usually applied to the shape of the corolla.

Furrowed. Having longitudinal channels or grooves; channeled.

Fusiform. Tapering at both ends; spindle-shaped.

Gemma. An asexual propagule, sometimes appearing as a vegetative bud.

Geniculate. Bent abruptly.

Gibbous. Swollen on one side, such as on a calyx or corolla tube.

Glabrate. Becoming glabrous with age.

Glabrous. Smooth, lacking pubescence; not hairy.

Gland. A secreting structure, often a protuberance or appendage, frequently the surface sticky, waxy, or glistening.

Glandular-pubescent. Having hairs, or trichomes, terminated by pinhead-like glands, often sticky.

Glanduliferous. Bearing glands.

Glaucous. Covered with a whitish waxy bloom or powdery coating.
Globose. Shaped like a sphere or globe.
Glochidia. Barbed or hooked hairs or bristles.
Glomerate. Densely or compactly clustered; often referring to flowers occurring in dense clusters.
Glomerulate. Having small, dense clusters of flowers.
Glomerule. A small cluster of flowers, usually consisting of a densely compacted cyme.
Glume. One of a pair of bracts lowermost on a grass spikelet.
Glutinous. Covered with a sticky exudate; viscid.
Grain. The seed-like fruit of a grass; the small hard structure in the fruiting structure of *Rumex*.
Granulose. Composed of or appearing covered with minute grains.
Gynecandrous. Pertaining to staminate and pistillate flowers in the same inflorescence, the pistillate flowers in the apical, or upper, portion; especially used in the Cyperaceae (*Carex*); cf. *androgynous*.
Gynoecium. Collective term for the pistil or pistils of a flower.
Gynostegium. A complex structure in the flowers of *Asclepias* whereby the upper portion of the gynoecium is united with the androecium.
Habit. The general growth form of a plant, such as herbaceous or woody.
Habitat. The type of environmental conditions under which a plant grows.
Halophyte. A plant of saline soil.
Hastate. Arrow-shaped, with basal lobes that spread or extend outward, sometimes nearly at right angles; halberd-shaped.
Head. An inflorescence consisting of a dense cluster of sessile or nearly sessile flowers on a very short axis; a capitulum; as in Asteraceae.
Helicoid. Spiraled or coiled like a helix.
Herb. A plant with no persistent woody stem above ground.
Herbaceous. Having the structure or texture of an herb, not woody; dying back to ground level each year; sometimes applied to bracts which are not stiff and cartilaginous.
Heteromorphic. Having different growth forms; also said of 2 or more forms of flowers.
Heterophyllous. Having different types of leaves.
Heterosporous. Producing spores of two different sizes, the larger (megaspore) developing into a female gametophyte, the smaller (microspore) developing into a male gametophyte; cf. *homosporous*.
Hibernaculum. Winter bud; turion; a specialized, highly compacted structure allowing plant to persist over the unfavorable season (such as winter or dry season).
Hip. Pertaining to a fruiting structure which consists of a fleshy, expanded vase-like hypanthium enclosing the achenes; used in *Rosa*.
Hirsute. Surface covered with long, rather coarse or stiff hairs.
Hirtellous. Minutely or softly hirsute.
Hispid. Surface covered with stiffish hairs that are sometimes bristly or spike-like.
Hispidulous. Minutely hispid; with short, stiff hairs.
Homomorphic. Having one growth form.
Homosporous. Producing spores which are all of one size; cf. *heterosporous*.
Hyaline. Having a thin, membranous, transparent or translucent texture.
Hybrid. A plant resulting from a cross between parents that are genetically distinct; in taxonomy, the offspring of two different species or infraspecific taxa.
Hydrophyte. An aquatic plant; life form of plants with resting buds in water, typically, at most, only the inflorescences being emergent.
Hypanthium. A cup-like receptacle derived from the fusion of floral envelopes and androecium, and on which are borne the calyx, corolla, stamens (especially Rosaceae); sometimes applied to a floral tube derived from the fusion of floral envelopes and androecium (as in Onagraceae).
Hypogynium. The perianth-like structure subtending the ovary in the Cyperaceae, especially *Scleria*.
Hypogynous. Pertaining to a flower having the calyx, corolla, and stamens attached below the ovary, the ovary is thus said to be superior.
Imbricate. Overlapping, vertically or spirally, with the upper portion covering the base of the next higher.
Immersed. Growing wholly under water; submersed.
Imperfect. Pertaining to flowers having functional stamens or pistils, but not both; unisexual.
Incised. Having deeply cleft margins.
Indehiscent. Pertaining to fruits that remain closed, not opening (dehiscing) at maturity.
Indusium. A thin, epidermal outgrowth covering the clusters of sporangia (sori) on ferns.
Inferior ovary. The calyx, corolla, and stamens arise from the summit of the ovary because of the fusion of tissue to the ovary wall; cf. *superior ovary*.
Inflated. Bladdery.
Inflorescence. An aggregation of flowers occurring together in a particular arrangement.
Infructescence. An inflorescence when the fruits develop.
Inserted. Attached to or growing out of.
Internode. The portion of a stem between the nodes or points where leaves are borne.
Introduced. Brought from another region; adventive (not native).
Involucral. Pertaining to an involucre.
Involucre. A cluster of bracts in an inflorescence subtending or surrounding a flower cluster, or head, or a single flower.
Involute. Having margins that are rolled inward (toward the upper, or adaxial, surface); cf. *revolute*.
Irregular. Descriptive of a flower or set of flower parts which can be cut through the center into two equal and similar parts along only one plane; bilaterally symmetrical; zygomorphic.

Jointed. Having one or more constrictions marking a point of articulation; as in *Equisetum*.

Keel. A central ridge, like the keel of a boat; as in some *Sparganium* leaves.

Labellum. Lip; the distinctive lower (by twist of the pedicel/ovary) petal of the Orchidaceae.

Labiate. Lipped, as applied to an unequally divided calyx or corolla.

Lacerate. Irregularly cleft.

Lacunae. Air spaces within a tissue, especially in the leaves, stems, and roots of submersed aquatic plants.

Lamina. The blade or expanded part of a leaf, bract, petal, etc.

Lanceolate. Shaped like a lance, several times longer than wide, broadest toward the base and narrowed to the apex.

Lateral. Pertaining to the sides.

Latex. The milky juice or sap of some plants, such as *Asclepias*.

Lax. Loose or somewhat limp; not rigid or tight.

Leaflet. A division of a compound leaf.

Legume. Fruit in Fabaceae produced from a simple pistil, with many seeds and splitting along two sutures.

Lemma. The lower of the two bracts surrounding the pistil and stamens in a flower in the grasses.

Lenticel. Corky spots on young bark of stems and roots, involved in gaseous exchange; corresponding to epidermal stomata.

Lenticular. Having both surfaces lens-shaped; biconvex.

Ligule. The collar-like appendage, membranous or composed of hairs, at the junction of the blade and the sheath in the leaf of grasses and a few *Carex*.

Linear. Long and narrow, with parallel or nearly parallel margins.

Lip. The upper or lower part of an unequally divided or bilabiate calyx or corolla; the labellum of Orchidaceae.

Lobe. Any segment of an organ, especially if rounded (e.g., lobes of a leaf or of the corolla).

Locule. The cavity of an ovary or anther.

Loculicidal. Pertaining to the dehiscence of a capsule along the back of the locule, not along the septum.

Loment. A specialized legume which is constricted between the seeds and breaks apart into one-seeded segments rather than splitting lengthwise along the sutures.

Massulae. A group or mass of cohering pollen grains, as in the Orchidaceae, or microspores, as in *Azolla*.

Median. Pertaining to the middle.

Megasporangium. A structure in which megaspores are produced.

Megaspore. The female or larger type of spore visible in *Isoetes*, *Marsilea*, *Azolla*, and *Salvinia* giving rise to the female gametophyte.

Membranous. Thin, rather soft, and more or less translucent and pliable; membranaceus.

Mericarp. A segment of a schizocarp, such as in the Apiaceae.

-merous. Suffix pertaining to parts or their number, typically referring to the perianth only; as a 5-merous (pentamerous) flower having corolla parts in fives.

Microsporangium. A structure in which microspores are produced.

Microspore. The smaller type of spore visible in *Isoetes*, *Marsilea*, *Azolla*, and *Salvinia* giving rise to the male gametophyte.

Microstrobilus. A strobilus, or cone-like structure, bearing microsporangia or pollen sacs; the male cones in Gymnosperms; androstrobilus.

Midrib. The main or central vein of a leaf or leaflet or similar structure.

Monadelphous. Having stamens which are united (fused) in a single tube or column by their filaments, the anthers remaining distinct.

Monoecious. Flowers that are unisexual, with staminate and pistillate flowers on the same plant.

Mucilaginous. Slimy; slimy material exuded by plant organs.

Mucronate. Having a short, sharp, abrupt tip on a rounded or truncate apex.

Naturalized. Having a foreign origin, but established and reproducing without cultivation as though native.

Nectary. Any area or appendage where nectar is secreted.

Nerve. A simple or unbranched vein or rib.

Netted. Reticulate; veins connected to form a network.

Node. The place along a stem at which one or more leaves or bracts are borne.

Nodulose. Having little knobs or knots.

Nut. A hard, indehiscent one-seeded fruit.

Nutlet. A small, hard, indehiscent one-seeded fruit, or 1-seeded segments from ovary breaking into parts (as in the 4-lobed ovaries of Lamiaceae, Boraginaceae).

Obconic. Conic or cone-shaped, with the point of attachment at the apex.

Obcordate. Heart-shaped, but inverted, with the point at the base.

Oblanceolate. Lanceolate, but broadest above the middle, long-tapered to the base.

Oblique. Unequal-sided or slanting.

Oblong. Longer than broad, with the sides nearly parallel.

Obovate. Ovate, but inverted, broadest above the middle; cf. *ovate*.

Obovoid. Ovoid, but inverted, with the broader end toward the apex; cf. *ovoid*.

Obtuse. With the sides or margins meeting at more than a 90-degree angle.

Ocrea. A stipular sheath at the nodes in the Polygonaceae.

Ocreolae. The smaller or secondary stipular sheaths, as in the inflorescences in the Polygonaceae.

Olivaceous. Olive-green.

Opaque. Not translucent.

Opposite. Pertaining to leaves or bracts occurring two at a node on opposite sides of the stem; in flower parts when one part occurs in front of another.

Orbicular. Circular in outline.

Oval. Broadly elliptic.

Ovary. The part of the pistil in which the seeds develop.

Ovate. Egg-shaped in outline, with the broader end toward the base.

Ovoid. A solid with an ovate outline.

Ovuliferous. Bearing ovules; in ovuliferous cones in Gymnosperms, the seed-bearing structures.

Palate. A rounded projection or prominence of the lower lip of a bilabiate corolla, causing the throat of the corolla to be closed or nearly so.

Palea. The tiny, often very thin, upper bract of a grass floret, enclosed by the lemma.

Palmate. Radiately lobed or divided.

Paniculate. Pertaining to flowers or fruits arranged in or resembling a panicle.

Panicle. A loose, irregularly branched inflorescence, the flowering sequence indeterminate, with the lower flowers opening first.

Papilla. A minute nipple-shaped projection.

Papillate. Having one or more papillae; papillose.

Papillose. Bearing minute nipple-shaped projections; papillate.

Pappus. The bristle-like, capillary, scaly, or chaffy structure (modified calyx) at the summit of the achene in the Asteraceae.

Papule. A relatively large papilla, or nipple-like projection.

Parallel-veined. Descriptive of an organ in which the veins run toward the apex in approximately parallel lines (especially leaves of Monocots).

Parted. Cleft to below the middle.

Pectinate. Descriptive of an organ which is cleft into divisions such as to resemble a comb; pinnatifid, with narrow, closely set segments.

Pedicel. The stalk of an individual flower or fruit.

Pedicellate. Borne on a pedicel.

Peduncle. The primary stalk of an entire inflorescence, or the stalk of a flower when borne singly.

Pedunculate. Borne on a peduncle.

Pellucid-punctate. Having translucent dots (best seen when holding to light).

Peltate. Pertaining to a flat structure that is attached by its lower surface, often toward the center, not at its margin; as in leaves of *Brasenia* and *Nelumbo*.

Pendulous. More or less hanging.

Perennate. To overwinter; to survive from year to year.

Perennial. Living three or more years; herbaceous perennials usually dying back each winter.

Perfect. Pertaining to a single flower having both functional stamens and a pistil or pistils, with a calyx and corolla either present or absent; bisexual.

Perfoliate. Pertaining to leaves or bracts that apparently have the stem passing through them; opposite or whorled leaves that are united in a collar-like structure around the stem.

Perianth. General term for a calyx and corolla together, or for either one if the other is absent.

Perigynium. A special sac-like structure which encloses the pistil and ultimately the fruit in *Carex*.

Perigynous. Pertaining to a flower having a calyx, corolla, and stamens borne on a hypanthium, but without these tissues being fused to the ovary, the ovary thus superior.

Persistent. Pertaining to an organ that remains attached after completing its normal biological function.

Petal. A unit of the corolla.

Petaloid. Having the appearance of petals, such as petaloid sepals.

Petiolate. Having a petiole.

Petiole. The stalk of a leaf.

Phyllary. An involucral bract in the Asteraceae.

Phyllodial. Pertaining to a more or less flattened, expanded, bladeless petiole that serves as a leaf blade; often linear or sometimes resembling a ribbon-like leaf.

Pilose. Having long, soft hairs.

Pinna. Leaflet or primary division of a pinnately compound leaf.

Pinnate. Compound, with leaflets arranged along the two sides of an axis.

Pinnate-pinnatifid. A pinnately compound leaf with the leaflets then cleft pinnately, such that the sinuses do not quite reach the midvein.

Pinnatifid. Pinnately cleft, the sinuses not quite reaching the midvein.

Pinnule. The secondary unit of a twice pinnately compound leaf.

Pistil. The seed-bearing organ of the flower, consisting of the ovary, style, and stigma, that may be simple or compound (with two or more carpels fused).

Pistillate. Pertaining to a flower that has a pistil or pistils, but not stamens; unisexual, with pistils.

Pith. The spongy central portion of a stem.

Placentation. Pertaining to the arrangement of the placenta, the interior area of the ovary bearing the ovules.

Plaited. Plicate; with folds (plaits), usually lengthwise.

Plano-convex. Having one surface rounded and the other surface flat.

Plicate. Plaited; with folds, usually lengthwise.

Plumose. Having fine, elongate hairs arranged like the plume of a feather.

Podogyne. A stalk bearing the ovary or fruit, sometimes resembling a pedicel, but arising as an elongation at the base of the ovary, within the calyx; gynophore; as in *Ruppia* and *Podostemum*.

Pollinium. A mass of waxy pollen or of coherent pollen grains, as in *Asclepias* and Orchidaceae.

Polygamodioecious. Chiefly dioecious, but bearing some bisexual flowers.

Polygamomonoecious. Chiefly monoecious, but bearing some bisexual flowers.

Polymorphic. Exhibiting more than one form; variable as to habit.

Pome. A fleshy, indehiscent fruit developed from an inferior ovary, with the seeds enclosed within a cartilaginous ovary wall, the fleshy part developed from the hypanthium.

Prickle. A small and more or less slender sharp outgrowth from the epidermis; internodal thorn.

Procumbent. Trailing or lying on the ground, but not typically rooting at the nodes.

Prophyllum. The bracteole at the base of an individual flower, as in *Juncus*.

Prostrate. Lying flat upon the ground.

Proximal. At or toward the base or the point of attachment; cf. *distal*.

Puberulent. Covered with a pubescence of very short hairs, not densely spaced.

Pubescent. Often a general term referring to a surface being covered with hairs, in contrast with glabrous; a type of pubescence with short, soft, downy hairs.

Punctate. Having depressed dots scattered over the surface.

Raceme. A simple inflorescence of pedicelled flowers along a common axis, the flowering sequence indeterminate, with the lower flowers opening first.

Racemose. Pertaining to flowers or fruits arranged in or resembling a raceme.

Rachilla. A secondary axis; specifically, in the Cyperaceae and Poaceae, the floral axis, in contrast with the axis of the spike or spikelet.

Rachis. The central axis of an inflorescence or of a compound leaf.

Radially symmetrical. Descriptive of a flower or set of flower parts which can be cut through the center into equal and similar parts along two or more planes; actinomorphic; regular.

Radiate. Spreading from a common center; in the Asteraceae, bearing ray, or ligulate, flowers.

Ray. The branch of an umbelliform inflorescence.

Ray flower. The strap-like, or ligulate, flower in a head of flowers of many Asteraceae.

Receptacle. The more or less expanded portion of an axis which bears the parts of the flower, or the cluster of flowers of a head.

Receptacular. Pertaining to the receptacle; as in receptacular bracts subtending the cluster of flowers in a head in the Asteraceae.

Recurved. Curved downward or backward.

Reflexed. Sharply bent downward or backward from the point of attachment.

Regular. Descriptive of a flower or set of flower parts which can be cut through the center into equal and similar parts along two or more planes; actinomorphic; radially symmetrical.

Reniform. Kidney-shaped.

Repent. Creeping; prostrate and rooting at the nodes.

Reticulate. Forming a network.

Retrorse. Having hairs or processes bent backward (away from the apex) or downward; cf. *antrorse*.

Revolute. Pertaining to margins that are rolled backward (toward the lower, or abaxial, surface); cf. *involute*.

Rhizoid. A simple, root-like structure.

Rhizome. Any prostrate or subterranean stem.

Rhomboid. Having a rhombic outline; an equilateral oblique-angled shape; four-sided, with a somewhat diamond outline.

Rib. Prominent veins or ridges on structures such as some fruits, seeds, bracts, perigynia, or stems.

Riparian. Growing by rivers or streams.

Rootstock. An underground stem; rhizome.

Rosette. A cluster of leaves or other organs radiating from an apparent common point of attachment.

Rostellum. A small beak; a slender extension between the stigmatic cavity and the anther in the Orchidaceae.

Rostrate. Having a beak.

Rotate. Pertaining to a sympetalous corolla having a short tube and flaring into a broad, flattened, and circular limb.

Rufescent. Tinged with red or becoming reddish or reddish-brown.

Rugose. Having a wrinkled surface, with sunken veins.

Runner. A very slender stolon; a horizontal stem with long internodes trailing along the surface of the ground.

Saccate. Sac- or pouch-shaped.

Sagittate. Shaped like an arrowhead, with the basal lobes directed downward or backward.

Salverform. Descriptive of a corolla having a slender tube that abruptly expands into a flat limb.

Samara. An indehiscent winged fruit.

Scaberulous. Descriptive of a surface that is slightly rough to the touch.

Scabrous. Descriptive of a surface that is rough to the touch.

Scale. A small, thin, flattened structure, usually a modified leaf, arising from a stem or other organ.

Scape. A leafless or nearly leafless stem bearing flowers.

Scapose. Bearing or resembling a scape.

Scarious. Thin, dry, and membranous, not green.

Schizocarp. A fruit which splits into one-seeded segments (mericarps), such as in the Apiaceae.

Scorpioid. An inflorescence coiled in bud like the tail of a scorpion, as in *Drosera* and *Myosotis*.

Scurfy. Pertaining to a surface covered with scale-like or bran-like particles.

Secund. Having a part or element directed to one side only.

Seed. A ripened ovule.

Segment. A subdivision or lobe of a deeply cleft or divided leaf or similar organ.

Sepal. A unit of the calyx.

Septate. Divided by partitions, as in roots of *Sagittaria* and *Eriocaulon* or stems of some *Juncus*.

Septicidal. Pertaining to the dehiscence of a capsule along the septum; cf. *loculicidal*.

Septum. Any kind of partition.

Sericeous. Silky; covered with soft, silky hairs which usually point in one direction.

Serrate. Having sharp teeth pointing forward.

Serrulate. Having very small marginal teeth; minutely serrate.

Sessile. Joined directly by the base, lacking a stalk, pedicel, or petiole.

Seta. A bristle or bristle-shaped structure.

Setaceous. Bearing bristles.

Setulose. Bearing minute bristles.

Sheath. The basal part of a leaf wrapped about the stem; a collar-like outgrowth at a node.

Sigmoid. Doubly curved or S-shaped.

Silicle. A short 2-celled fruit, usually wider than long, in the Brassicaceae in which at dehiscence the two valves fall away, exposing a persistent septum.

Silique. An elongate, 2-celled fruit in the Brassicaceae in which at dehiscence the two valves fall away, exposing a persistent septum.

Silky. Covered with soft, straight hairs, usually pointing in one direction; sericeous.

Simple. Consisting of a single, or unbranched, unit, sometimes cleft, but not divided into distinct parts; not compound.

Sinuate. Having a strongly wavy margin in the plane of the blade.

Sinus. The cleft or angle between two lobes.

Smooth. Glabrous, or lacking hairs; without roughness.

Sori. Clusters of sporangia in ferns.

Spadix. A spike with a fleshy axis, as in Araceae.

Spathe. A large, sheathing bract, often open on one side, enclosing an inflorescence.

Spatulate. Oblong, with the basal portion narrow, widening above, and rounded at the apex.

Spicate. Pertaining to flowers or fruits arranged in or resembling a spike.

Spicule. A small, slender, sharp-pointed structure, usually on a margin or surface.

Spiculose. Having a margin or surface covered with fine points.

Spike. An elongate inflorescence with flowers sessile or appearing sessile.

Spikelet. A secondary spike in some sedges; the segment of the inflorescence of grasses enclosed by the glumes.

Spine. A firm, slender, sharp-pointed structure; a sharp woody or rigid outgrowth from the stem, usually a modified stem.

Spinulose. Having small spines over the surface or along the margin.

Sporangium. A structure in which spores are produced and contained.

Spore. A microscopic, simple reproductive body, the product of meiosis, which is capable of developing into a new individual, particularly in Pteridophytes and lower plants.

Sporocarp. A specialized organ enclosing sporangia and spores; as in *Marsilea*, *Azolla*, and *Salvinia*.

Sporophyll. A leaf bearing sporangia and spores; sometimes aggregated into a strobilus, as in Lycopodiaceae.

Spur. A more or less elongate tubular or sac-like extension of some part of a flower, usually nectariferous.

Stamen. The pollen-bearing organ of the flower.

Staminate. Pertaining to a flower that has stamens but no pistil; unisexual, with stamens.

Staminode. A sterile stamen, or any structure resembling stamens in the staminal part of the flower; sometimes staminodes are petaloid.

Stellate. Star-shaped.

Sterile. Lacking functional reproductive organs.

Stigma. The portion of a pistil which is receptive to pollen for effective fertilization.

Stipe. A little stalk; a leaf stalk of a fern frond.

Stipitate. Having a stipe.

Stipulate. Having stipules.

Stipule. Appendages at the base of a leaf, sometimes fused to the petiole or leaf base.

Stolon. A stem which trails on the ground, often rooting at nodes; a runner; term often used to describe slender rhizomes near the surface of the ground.

Stoloniferous. Readily producing stolons.

Stramineous. Straw-colored.

Striate. Marked with fine longitudinal lines or streaks.

Strigose. Having a surface covered with appressed sharp, straight, and stiff hairs.

Strobilus. A dense aggregation of structures bearing the sporangia (often somewhat elongate or cone-like), as in *Equisetum*, Lycopodiaceae, and the female cones and pollen-bearing structures (microsporangia) of conifers.

Style. The usually elongate upper portion of a pistil between the ovary and the stigma; often indicative of the number of carpels in a compound pistil.

Stylopodium. The swollen, disk-like expansion at the base of the style, as in Apiaceae.

Sub-. A prefix meaning under, or almost, or not quite.

Subulate. Awl-shaped, narrow and tapering from base to apex, often somewhat stiff.

Succulent. Fleshy; soft and thickened in texture.

Superior ovary. Occurs in a flower whose calyx, corolla, and stamens arise from the receptacle below the ovary; cf. *inferior ovary*.

Supra-axillary. Borne at a distance above the axil, as in globose heads of pistillate flowers in some *Sparganium*.

Suture. A line of opening or dehiscence.

Sympetalous. Having the petals of the corolla united, at least at the base; gamopetalous.

Syncarpous. Pertaining to a gynoecium with carpels united in one pistil; a compound pistil.

Synsepalous. Having the sepals of the calyx united (fused); gamosepalous.

Taproot. The primary descending root; a direct continuation of the radicle of the embryo.

Taxon. A general term applied to any hierarchical taxonomic level or group; taxa (plural).

Tendril. A slender, twisting or clasping outgrowth by which the plant climbs or clings for support; may be a modified stem, leaf, leaflet, or stipule.

Tepal. A collective term applied to the units of the perianth when the calyx and corolla are of similar form and not readily differentiated.
Terete. Round in outline; circular in cross-section.
Terminal. At the end; apical; distal.
Ternate. Borne in threes.
Tetrad. A group of four structures forming a unit, as pollen or spores in tetrads.
Tetraploid. Having four full sets of chromosomes in each cell.
Thalloid. Resembling a thallus.
Thallus. A somewhat flattened vegetative body without differentiation into stem and leaf.
Thorn. A modified stem, leaf, or stipule, stiffened and terminating in a sharp point; cf. *spine*; an internodal thorn is the same as a prickle.
Throat. The opening into a sympetalous corolla or perianth; the place where the corolla tube and limb join.
Thyrse. A contracted cylindrical or ovoid and usually compact panicle.
Thyrsoid. Pertaining to flowers or fruits arranged in or resembling a thyrse.
Tomentose. Pertaining to a densely pubescent surface with matted, soft wooly hairs; densely wooly.
Trailing. Prostrate but not usually rooting.
Translucent. Thin and membranous, such that light can be transmitted through the tissue; partly transparent.
Trichome. Any hair-like outgrowth of the epidermis.
Trifoliate. Having three leaves.
Trifoliolate. Pertaining to a compound leaf having three leaflets.
Trigonous. Three-angled.
Tripinnate. Three times pinnately compound.
Triploid. Having three full sets of chromosomes in each cell.
Tristylous. Pertaining to flowers of a given species having three different style lengths and anthers borne at corresponding levels; in the Lythraceae and Pontederiaceae, flowers with a long style having mid-length and short stamens, flowers with a mid-length style have long and short stamens, and flowers with short styles have long and mid-length stamens.
Triternate. Pertaining to leaves which are three times divided into threes; typically 27 leaflets; cf. *biternate*.
Truncate. Ending abruptly, appearing as if cut square across the end; pertaining to a base or apex that is nearly straight across.
Tuber. A thickened, short, subterranean stem; sometimes a short thickening on a root.
Tubercle. A swollen appendage, especially on the fruits of some Cyperaceae; a small, rounded protruding body.
Tuberculate. Bearing tubercles.
Tuberous. Bearing or producing tubers.
Tubular. Descriptive of the corolla forming a slender tube.
Turion. A winter bud; hibernaculum.
Tussock. A dense tuft of grass or a grass-like plant.
Umbel. An inflorescence in which the peduncles and pedicels radiate from the same level, often flat-topped; in the Apiaceae the umbels are usually compound.
Umbelliform. Pertaining to an inflorescence resembling an umbel.
Undulate. Having a wavy surface or margin.
Unisexual. Pertaining to flowers that are either staminate or pistillate; pertaining to plants bearing either staminate or pistillate flowers, but not both.
Urceolate. Urn-shaped; hollow and cylindrical or ovoid with a constriction at or below the mouth.
Utricle. A small, dry, bladdery 1-seeded fruit; any small bladder-like body.
Valvate. Opening by valves; in buds, meeting by the edges without overlapping.
Valve. Each of the units resulting from the splitting of a capsule.
Vascular. Pertaining to the presence of conductive tissues, xylem and phloem, in a plant.
Veins. Threads of fibrovascular tissue in leaves and other organs.
Velum. The membranous indusium or covering of megaspores or microspores at the base of individual leaves in *Isoetes*.
Velutinous. Velvety; covered with fine, soft, short, spreading pubescence.
Venation. The arrangement of the veins.
Ventral. Pertaining to the lower, or abaxial, surface; cf. *dorsal*.
Verticil. A whorl.
Verticillate. Having structures such as leaves, flowers, or fruits arranged in whorls.
Vestigial. Rudimentary.
Villous. Covered with long, silky, straight hairs.
Viscid. Sticky; glutinous.
Viscidium. The sticky pad detaching with the pollinium in pollen transfer, as in the Orchidaceae.
Weak. Not strong, such as some stems at the base.
Weedy. Troublesome or aggressive, intruding where not wanted.
Whorled. Arranged in a circle at one level on the stem.
Wing. Any thin or membranous expansion bordering or surrounding an organ.
Winter bud. A small dense vegetative shoot which forms at stem tips and survives the winter underwater, renewing growth in spring; turion; hibernaculum; as in *Utricularia*, *Myriophyllum*, and *Potamogeton*.
Zygomorphic. Descriptive of a flower or set of flower parts which can be cut through the center into two equal and similar parts along only one plane; bilaterally symmetrical; irregular.

Glossary of Habitat Terms

This glossary provides definitions of habitat terms that may not be in everyday usage. Definitions of habitats such as ponds, lakes, streams, brooks, temporary pools, and many others which are readily understood by everyone are not included here. It must be realized that some terms have regional meanings. Thus, what is called a pond in Newfoundland is certainly large enough to be called a lake anywhere else. Likewise, the term "bog" in the South describes a somewhat different habitat from that in the North, yet there are strong enough similarities (in acidity, poor drainage, brown waters, and some of the same species) that the term "bog" is readily understood by a northerner visiting a southern bog.

Acid water. Water characterized by a pH value below 7, but more often below a pH of 6.

Alkaline water. Water that is high in calcium carbonate content. This water is also typically correlated with a pH above 7.

Alluvial plain. Flat, often broad, river margins with an accumulation of silt caused by seasonal flooding. Also called a floodplain.

Alluvial tidal freshwater shore. Flat shore with an accumulation of silty soil due to diurnal tidal flooding, caused by oceanic tides backing up the flow of freshwater, thus upstream from the area of penetration of saltwater.

Alpine bog. A bog or depression filled with peat found near or above the timber line at high elevations, or somewhat lower elevations at high latitudes.

Bayou. A slow-moving creek or stream or backwater of a river, often near the mouth of a major river, especially frequent near the coast; primarily in the Gulf coastal plain region, but also in backwaters of larger rivers near mouth of large lakes (as in bayous of the Grand River just before flowing into Lake Michigan).

Bog. In the North, an acid peat-accumulating wetland that has no significant inlets or outlets and supports various acidophilic plants. Sphagnum moss is typically present, and the habitat is nutrient poor. There is frequently a floating mat of vegetation on the margin of an open bog pond. In the South, bogs are open or semi-open areas with a peaty substrate, characterized by grass and sedge flora.

Bottomland. Low land found adjacent to a river or stream that experiences periodic flooding.

Brackish water. Non-coastal water with a moderate to high concentration of dissolved inorganic salts; salty.

Calcareous fen. Low marshy ground with peaty substrate covered wholly or partly with standing water over a limestone bedrock, and characterized by lime-loving plants. See *fen*.

Canal. A waterway originally dug for navigation, irrigation, or transporting water.

Canebrake. A dense thicket of Cane or Canebrake Bamboo, *Arundinaria gigantea* or *A. tecta*; found in the South; often extensive.

Cedar swamp. A swamp dominated by either Northern White Cedar, *Thuja occidentalis*, or by Atlantic Coast Cedar, *Chamaecyparis thyoides*. See *swamp*.

Cobble. An area characterized by numerous small to medium-sized stones.

Conifer swamp. A swamp dominated by Tamarack, *Larix laricina*, Northern White Cedar, *Thuja occidentalis*, and/or Black Spruce, *Picea mariana*. See *swamp*.

Cranberry bog. A bog where cranberries are in great abundance; also an artificial bog managed as an agricultural area for the cultivation of cranberries, *Vaccinium macrocarpon*.

Cypress swamp. A swamp with much standing water for the growing season, and characterized by stands of the deciduous conifer *Taxodium*.

Ditch. A manmade channel that is often seasonally wet.

Estuary. Shallow to deep saline tidal waters and adjacent saline wetlands where river systems meet the ocean. Salinity levels are lower than oceanic levels because of freshwater run-off from the land and incoming rivers and streams.

Eutrophic water. Water that is high in nutrients, particularly nitrates and phosphates.

Fen. A peat-accumulating wetland that receives some of its nutrients from the substrate and from runoff waters. It usually supports marsh-like vegetation and shrubs, and often has a near neutral pH of 7 to slightly alkaline pH.

Flats. Usually applied to broad, flat areas, especially behind dunes or along broad, sandy lakeshores that are occasionally flooded, and where the water table is close to the surface.

Floodplain. Flat, often broad, river margins with an accumulation of silt caused by seasonal flooding. See *alluvial plain*.

Fresh tidal flat. A mud flat that is subjected to fluctuation of water due to tides but is upstream of the area penetrated by tidal saltwater.

Freshwater tidal marsh. A marshy area along a tidal river flooded twice a day by a backup up freshwater prevented by tidal waters from running to the ocean, but upstream of the area penetrated by tidal saltwater; often characterized by graminoid vegetation.

Glade. A large open area in a woods or forest, often wet and marshy and dominated by sedges and grasses.

Impoundment. A manmade body of water. This is often subjected to great water level fluctuations because of drawdown and spring recharge.

Mangrove swamp. A tropical and subtropical forested wetland characterized by the presence of tides and saltwater, typically developing at the mouths of river systems.

Marl. Calcium carbonate (lime) that is precipitated by certain aquatic plants such as *Chara* and *Potamogeton* in the process of photosynthesis.

Marly sand. Damp to wet sand rich in marl.

Marly wet panne. Barren marl flat that is covered with a deposit of marl because of flooding and subsequent evaporation.

Marsh. A frequently or continuously inundated tract of land characterized by emersed, chiefly herbaceous wetland vegetation.

Moderately alkaline water. Here referring to water with a bicarbonate range of approximately 20–50 mg HCO_3^-/liter^{-1} or with a carbonate range of approximately 15–40 mg/liter $CaCO_3$.

Mossy talus. Wet, moss-covered area at the base of a cliff that is associated with springs or constant seeps. See *talus slope*.

Muck. Wet, heavy soil or well-composted peat.

Muskeg. A widely used term for extensive peatlands of Canada and Alaska with the characteristics of northern bogs; especially common in boreal and subarctic regions.

Panne (Pan). A highly saline depression or pool in a salt marsh. Also used to describe a wet depression associated with sand dunes or sandy shores.

Peat. Soil produced by incomplete decay of sedges and grasses or sphagnum moss or other vegetation.

Peat barren. A large flat area dominated by a peaty substrate, appearing "barren" because of the absence of forest.

Peaty slope. Peatland that occupies shallow depressions on gentle gradients that are often associated with springs, seeps of small brooks, or cold maritime coasts.

Pine barren bog. A highly acidic boggy area with a sandy/peaty substrate along the Atlantic and Gulf coastal plains consisting of pine flatlands dominated by such species as Pitch Pine, *Pinus rigida*, and Atlantic White Cedar, *Chamaecyparis thyoides*.

Pineland. Land dominated by pines, especially along the southern coastal plain.

Pine savanna. An area of waterlogged substrates or inundated with shallow surface water during seasons of heavy precipitation. This area may become quite dry during extended periods of droughts.

Pocosin. A broad, flat, evergreen shrub bog along the southern coastal plain that is generally water-logged, acidic, and nutrient poor, with a sandy or peaty substrate.

Prairie. An area of the Midwest and Plains states dominated by grasses growing on a thick, rich, loamy soil. Depressions on rolling prairie land tend to be wet, at least early in growing season.

Quagmire. Very soft, muddy or boggy ground.

Raised bog. A large, generally flat area of peat, accumulated to a level above the original basin, that receives its water only from precipitation and capillary action; characteristic of coastal plains of Newfoundland, the Maritimes, and northeastern Maine in N.AM.; Ireland, Scotland, Scandinavia, Russia, and Tierra del Fuego.

Rill. A tiny stream, trickle or rivulet.

Rivulet. A tiny stream; streamlet, trickle.

Saline marsh. A coastal marsh influenced by saltwater, with waterlogged soil during the growing season, and often covered with up to 1 meter of water; coastal and Great Plains.

Saline water. Salty water.

Salt marsh. A coastal marsh flooded by tidal saltwater, the uppermost portions with waterlogged soil, but flooded only by monthly spring tides.

Seep (Seepage). An area of slow leaking or oozing water.

Slough. A marshy, swampy area, very shallow lake system, or sluggish channel with deep mud or accumulating peat; especially of the prairie regions.

Sphagnum bog. A bog with a deep, often floating mat of sphagnum moss. See *bog*.

Spring. A flow of water from the ground; frequently serves as a source of a stream.

Swale. A natural, open, treeless hollow, meadow, or depression which is at least seasonally wet.

Swamp. A wooded area with waterlogged soil, covered by surface water much of the year. The term is often used loosely as a synonym for "marsh"; thus the term "wooded swamp" is sometimes employed to clarify the forested or wooded nature of the area.

Talus slope. Coarse rock debris formed at the base of a cliff or a steep slope, caused chiefly by gravitational rockslides.

Thicket. An area of very dense growth of shrubs, usually found along streams or on margins of lakes and ponds.

Vernal pool. An area of standing water for a period in the spring, but drying each summer.

Waste area. Any area that is greatly disturbed by human activities and is no longer actively used.

Wet meadow. Low grassland often waterlogged and/or with the water table near the surface, but without standing water most of the year.

Wet montane soil. Wet soil at high elevation.

References

Aboy, H. E. 1936. A study of the anatomy and morphology of *Ceratophyllum demersum*. M.S. thesis, Cornell Univ., Ithaca, N.Y.

Ackerman, J. D. 1986. Mechanistic implications for pollination in the marine angiosperm *Zostera marina*. Aquatic Bot. 24: 343–353.

Adamec, L. 2018. Biological flora of Central Europe: *Aldrovanda vesiculosa* L. Persp. Plant Ecol, Evol. and Syst. 35: 8–21.

Adams, F. S. 1969. Winter bud production and function in *Brasenia schreberi*. Rhodora 71: 417–433.

Adams, P. 1973. Clusiaceae of the southeastern United States. J. Elisha Mitchell Sci. Soc. 89: 62–71.

Adams, P., and R. K. Godfrey. 1961. Observations on the *Sagittaria subulata* complex. Rhodora 63: 247–266.

Adams, W. P. 1957. A revision of the genus *Ascyrum*. Rhodora 59: 73–95.

Adams, W. P., and N. K. Robson. 1961. A re-evaluation of the generic status of *Ascyrum* and *Crookea* (Guttiferae). Rhodora 63: 10–16.

Ahedor, A. R. 2019. *Bacopa*. Pp. 260–263, in: Flora of North America Editorial Committee, eds. 1993+. Flora of North America North of Mexico. 21+ vols. New York and Oxford. Vol. 17.

Aiken, S. G. 1976. Turion formation in watermilfoil, *Myriophyllum farwellii*. Michigan Bot. 15: 99–102.

Aiken, S. G. 1978. Pollen morphology in the genus *Myriophyllum* (Haloragaceae). Canad. J. Bot. 56: 976–982.

Aiken, S. G. 1979. North American species of *Myriophyllum* (Haloragaceae). Ph.D. diss., Univ. of Minnesota, St. Paul, Minn.

Aiken, S. G. 1981. A conspectus of *Myriophyllum* in North America. Brittonia 33: 57–69.

Aiken, S. G. 1986. The distinct morphology and germination of the grains of two species of Wild Rice (*Zizania*, Poaceae). Canad. Field-Nat 100: 237–240.

Aiken, S. G., and A. Cronquist. 1988. Lectotypification of *Myriophyllum sibiricum* Komarov (Haloragaceae). Taxon 37: 958–966.

Aiken, S. G., P. F. Lee, D. Punter, and J. M. Stewart. 1988. Wild rice in Canada. Agric. Canad. Publ. 1830.

Aiken, S. G., and J. McNeill. 1980. The discovery of *Myriophyllum exalbescens* Fernald (Haloragaceae) in Europe and the typification of *M. spicatum* L. and *M. verticillatum* L. J. Linn. Soc. Bot. 80: 213–222.

Aiken, S. G., P. R. Newroth, and I. Wile. 1979. Biology of Canadian weeds. 34. *Myriophyllum spicatum* I. Canad. J. Pl. Sci. 59: 201–215.

Aiken, S. G., and R. R. Picard. 1980. The influence of substrate on the growth and morphology of *Myriophyllwn exalbescens* and *Myriophyllum spicatum*. Canad. J. Bot. 58: 1111–1118.

Aiken, S. G., and K. F. Walz. 1979. Turions of *Myriophyllum exalbescens*. Aquatic Bot. 6: 357–363.

Al-Shehbaz, I. A. 2010a. Brassicaceae. Pp. 224–246, in: Flora of North America Editorial Committee, eds. 1993+. Flora of North America North of Mexico. 21+ vols. New York and Oxford. Vol. 7.

Al-Shehbaz, I. A. 2010b. *Nasturtium*, *Rorippa*. Pp. 489–492, 493–501, in: Flora of North America Editorial Committee, eds. 1993+. Flora of North America North of Mexico. 21+ vols. New York and Oxford. Vol. 7.

Al-Shehbaz, I. A., and V. Bates. 1987. *Armoracia lacustris* (Brassicaceae), the correct name for the North American Lake Cress. J. Arnold Arbor. 68: 357–359.

Al-Shehbaz, I. A., K. Marhold, and J. Lihová. 2010. *Cardamine*. Pp. 464–484, in: Flora of North America Editorial Committee, eds. 1993+. Flora of North America North of Mexico. 21+ vols. New York and Oxford. Vol. 7.

Albach, D. C. 2019. *Veronica*. Pp. 305–322, in: Flora of North America Editorial Committee, eds. 1993+. Flora of North America North of Mexico. 21+ vols. New York and Oxford. Vol. 17.

Alix, M. S., and R. W. Scribailo. 2021. *Proserpinaca*. Pp. 29–31, in: Flora of North America Editorial Committee, eds. 1993+. Flora of North America North of Mexico. 21+ vols. New York and Oxford. Vol. 10.

Allen, C. M., and D. W. Hall. 2003. *Paspalum*. Pp. 566–599, in: Flora of North America Editorial Committee, eds. 1993+. Flora of North America North of Mexico. 21+ vols. New York and Oxford. Vol. 25.

Allred, K. W., and M. E. Barkworth. 2007. *Anthoxanthum*. Pp. 758–764, in: Flora of North America Editorial Committee, eds. 1993+. Flora of North America North of Mexico. 21+ vols. New York and Oxford. Vol. 24.

Alston, A. H. G. 1955. The heterophyllous *Selaginella* of continental North America. Bull. Brit. Mus. (Nat. Hist.), Bot. 1: 219–274.

Anderson, D. E. 1961. Taxonomy and distribution of the genus *Phalaris*. Iowa State J. Sci. 36: 1–96.

Anderson, E. 1928. The problem of species in Northern Blue Flags, *Iris versicolor* L. and *Iris virginica* L. Ann. Missouri Bot. Gard. 15: 241–332.

Anderson, E. 1936. The species problem in *Iris*. Ann. Missouri Bot. Gard. 23: 457–509.

Anderson, L. C., C. D. Zeis, and S. F. Alam. 1974. Phytogeography and possible origins of *Butomus* in North America. Bull. Torrey Bot. Club 101: 292–296.

Andrus, M., and R. L. Stuckey. 1981. Introgressive hybridization and habitat separation in *Lycopus americanus* and

L. *europaeus* at the southwestern shores of Lake Erie. Michigan Bot. 20: 127–135.

APG IV. 2016. An update of the Angiosperm Phylogeny Group classification for the orders and families of flowering plants: APG IV. Bot. J. Linnean Soc. 181: 1–20. Also: Stevens, P. F. (2001 onwards). Angiosperm Phylogeny Website. Version 14, July 2017 [and more or less continuously updated since]. http://www.mobot.org/MOBOT/research/APweb/.

Arber, A. 1920. Water Plants. A Study of Aquatic Angiosperms. Reprinted 1972 with preface by W. T. Stearn. J. Cramer, Germany.

Argus, G. W. 1973. The genus *Salix* in Alaska and the Yukon. Natl. Mus. Canad. Nat. Sci. Publ. Bot. No. 2.

Argus, G. W. 1980. Typification and identity of *Salix eriocephala* Michx. (Salicaceae). Brittonia 32: 170–177.

Argus, G. W. 1986a. The genus *Salix* (Salicaceae) in the southeastern United States. Syst. Bot. Monogr. 9: 1–170.

Argus, G. W. 1986b. Studies of the *Salix lucida* and *Salix reticulata* complexes in North America. Canad. J. Bot. 64: 541–551.

Argus, G. W. 2010. *Salix*. Pp. 23–162, in: Flora of North America Editorial Committee, eds. 1993+. Flora of North America North of Mexico. 21+ vols. New York and Oxford. Vol. 7.

Ashton, H. I. 1977. Aquatic Plants of Australia. Melbourne Univ. Press, Melbourne, Australia.

Aurand, D. 1982. Nuisance Aquatic and Aquatic Plant Management Programs in the United States. Vol. 2. Southeastern Region. U.S. Env. Protect. Agency and the Mitre Corp., McLean, Va.

Ball, P. W. 2002. *Carex* sect. *Limosae*, *Carex* sect. *bicolores*. Pp. 416–418, 424–426, in: Flora of North America Editorial Committee, eds. 1993+. Flora of North America North of Mexico. 21+ vols. New York and Oxford. Vol. 23.

Ball, P. W. 2003. *Salicornia*, *Sarcocornia*. Pp. 382–384, 384–387 in: Flora of North America Editorial Committee, eds. 1993+. Flora of North America North of Mexico. 21+ vols. New York and Oxford. Vol. 4.

Ball, P. W., and D. E. Wujek. 2002. *Eriophorum*. Pp. 21–27, in: Flora of North America Editorial Committee, eds. 1993+. Flora of North America North of Mexico. 21+ vols. New York and Oxford. Vol. 23.

Ballard, H. E. 1990. Hybrids among three caulescent violets, with special reference to Michigan. Michigan Bot. 29: 43–54.

Barker, W. T. 1997. *Planera*, *Celtis*. Pp. 376, 376–379, in: Flora of North America Editorial Committee, eds. 1993+. Flora of North America North of Mexico. 21+ vols. New York and Oxford. Vol. 3.

Barkley, F. A. 1937. A monographic study of *Rhus* and its immediate allies in North and Central America, including West Indies. Ann. Missouri Bot. Gard. 24: 265–498.

Barkley, T. M. 1978. *Senecio*. North American Flora II, 10: 50–139.

Barkley, T. M., R. E. Brooks, and E. K. Schofield, eds. 1986. Flora of the Great Plains. Univ. Press of Kansas, Lawrence, Kans.

Barkworth, M. E. 2003. *Distichlis*, *Spartina*, *Dichanthelium*, *Tripsacum*. Pp. 25, 240–251, 406–450, 693–696, in: Flora of North America Editorial Committee, eds. 1993+. Flora of North America North of Mexico. 21+ vols. New York and Oxford. Vol. 25.

Barkworth, M. E. 2007. Poaceae: Key to Tribes, *Scolochloa*, *Catabrosa*, *Phalaris*. Pp. 7–10, 732–734, 610–611, 764–773, in: Flora of North America Editorial Committee, eds. 1993+. Flora of North America North of Mexico. 21+ vols. New York and Oxford. Vol. 24.

Barkworth, M. E., and L. K. Anderton. 2007. *Glyceria*. Pp. 68–88, in: Flora of North America Editorial Committee, eds. 1993+. Flora of North America North of Mexico. 21+ vols. New York and Oxford. Vol. 24.

Barkworth, M. E., J. J. N. Campbell, and B. Salomon. 2007. *Elymus*. Pp. 288–343, in: Flora of North America Editorial Committee, eds. 1993+. Flora of North America North of Mexico. 21+ vols. New York and Oxford. Vol. 24.

Bassett, I. J., and C. W. Crompton. 1978. The genus *Suaeda* (Chenopodiaceae) in Canada. Canad. J. Bot. 56: 581–591.

Bassett, I. J., C. W. Crompton, J. McNeill, and P. M. Taschereau. 1983. The Genus *Atriplex* (Chenopodiaceae) in Canada. Agric. Canad. Monogr. No. 31.

Bates, V. M., and E. T. Browne. 1981. *Azolla filiculoides* new to the southeastern United States. Amer. Fern J. 71: 33–34.

Bauters, K., I. Larridon, M. Reynders, W. Huygh, P. Asselman, A. Vrijdaghs, A. M. Muasya, and P. Goetghebeur. 2014. A new classification for *Lipocarpha* and *Volkiella* as infrageneric taxa of *Cyperus* (Cypereae, Cyperoideae, Cyperaceae): Insights from species tree reconstruction supplemented with morphological and floral developmental data. Phytotaxa 166: 1–32.

Beal, E. O. 1956. Taxonomic revision of the genus *Nuphar* Sm. of North America and Europe. J. Elisha Mitchell Sci. Soc. 72: 317–346.

Beal, E. O. 1960a. *Sparganium* (Sparganiaceae) in the southeastern United States. Brittonia 12: 176–181.

Beal, E. O. 1960b. The Alismataceae of the Carolinas. J. Elisha Mitchell Sci. Soc. 76: 68–79.

Beal, E. O. 1977. A Manual of Marsh and Aquatic Vascular Plants of North Carolina with Habitat Data. North Carolina Agric. Exp. Sta. Tech. Bull. 247.

Beal, E. O., and T. L. Quay. 1968. A review of *Utricularia olivacea* Wright ex Grisebach (Lentibulariaceae). J. Elisha Mitchell Sci. Soc. 84: 462–466.

Beal, E. O., and R. M. Southall. 1977. Taxonomic significance of experimental selection by vernalization in *Nuphar* (Nymphaeaceae). Syst. Bot. 2: 49–60.

Beal, E. O., and J. W. Thieret. 1986. Aquatic and Wetland Plants of Kentucky. Kentucky Nat. Preserves Comm., Sci. and Tech. Ser. No. 5.

Beal, E. O., J. W. Wooten, and R. B. Kaul. 1982. Review of the *Sagittaria engelmanniana* complex (Alismataceae) with environmental correlations. Syst. Bot. 7: 417–482.

Beetle, A. A. 1947. *Scirpus*. North American Flora 18: 481–504.

Bell, C. D., and M. J. Donoghue. 2005. Phylogeny and biogeography of Valerianaceae (Dipsacales) with special reference to the South American valerians. Organisms, Diversity, and Evolution 5: 147–159.

Benson, L. 1948. A treatise on the North American Ranunculi. Amer. Midl. Nat. 40: 1–261.

Benson, L. 1954. Supplement to a treatise on the North American Ranunculi. Amer. Midi. Nat. 52: 328–369.

Bierner, M. W. 2006. *Helenium*. Pp. 426–435, in: Flora of North America Editorial Committee, eds. 1993+. Flora of North America North of Mexico. 21+ vols. New York and Oxford. Vol. 21.

Björkquist, I. 1968. Studies in *Alisma* L. II. Chromosome studies, crossing experiments and taxonomy. Opera Bot. 19: 1–138.

Blake, S. F. 1912. The forms of *Peltandra virginica*. Rhodora 14: 102–106.

Blanchard, Jr., O. J. 2015. *Hibiscus, Kosteletzkya*. Pp. 252–267, 272–274, in: Flora of North America Editorial Committee, eds. 1993+. Flora of North America North of Mexico. 21+ vols. New York and Oxford. Vol. 6.

Bogin, C. 1955. Revision of the genus *Sagittaria* (Alismataceae). Mem. New York Bot. Gard. 9. 179–233.

Boivin, B. 1944. American Thalictra and their Old World allies. Rhodora 46: 337–377, 391–445, 453–487.

Bornstein, A. J. 1997. Myricaceae. Pp. 429–435, in: Flora of North America Editorial Committee, eds. 1993+. Flora of North America North of Mexico. 21+ vols. New York and Oxford. Vol. 3.

Boufford, D. E. 1997. Urticaceae. Pp. 400–413, in: Flora of North America Editorial Committee, eds. 1993+. Flora of North America North of Mexico. 21+ vols. New York and Oxford. Vol. 3.

Bowden, W. M. 1959. Cytotaxonomy of *Lobelia* L. section *Lobelia*. I. Three diverse species and seven small-flowered species. Canad. J. Genet. Cytol. 1: 49–64.

Bowden, W. M. 1960. Cytotaxonomy of *Lobelia* L. section *Lobelia*. II. Four narrow-leaved species and five medium-flowered species. Canad. J. Genet. Cytol. 2: 11–27.

Brackley, F. E. 1985. The orchids of New Hampshire. Rhodora 87: 1–117.

Braun, E. L. 1961. The Woody Plants of Ohio. Ohio State Univ. Press, Columbus, Ohio.

Braun, E. L. 1967. The Vascular Flora of Ohio. Vol. 1, Monocotyledonae. Ohio State Univ. Press, Columbus, Ohio.

Brayshaw, T. C. 1985. Pondweeds and Bur-reeds, and their relatives: Aquatic families of Monocotyledons in British Columbia. Occas. Pap. British Columbia Prov. Mus. 26: 1–167.

Brink, D. E., and J. A. Woods. 1997. *Aconitum*. Pp. 191–195, in: Flora of North America Editorial Committee, eds. 1993+. Flora of North America North of Mexico. 21+ vols. New York and Oxford. Vol. 3.

Britton, D. M., and J. P. Goltz. 1990. *Isoetes prototypus*, a new diploid species from eastern Canada. Canad. J. Bot. 69: 277–281.

Britton, N. L., and A. Brown. 1896–1898. An Illustrated Flora of the Northern United States, Canada, and the British Possessions. Charles Scribner's Sons, New York. [Vol. 1, 1896; Vol. 2, 1897; Vol. 3, 1898].

Brooks, G. M., and T. R. Mertens. 1972. A biosystematic study of *Polygonum ramosissimum* and *Polygonum tenue*. Proc. Indiana Acad. Sci. 81: 277–283.

Brooks, R. E., and S. E. Clemants. 2000. *Juncus*. Pp. 211–255, in: Flora of North America Editorial Committee, eds. 1993+. Flora of North America North of Mexico. 21+ vols. New York and Oxford. Vol. 22.

Brooks, R. E., and C. Kuhn. 1986. Seed morphology under SEM and light microscopy in Kansas *Juncus* (Juncaceae). Brittonia 38: 201–209.

Brouillet, L. 2006. *Oclemena, Eurybia*. Pp. 78–81, 365–382, in: Flora of North America Editorial Committee, eds. 1993+. Flora of North America North of Mexico. 21+ vols. New York and Oxford. Vol. 20.

Brouillet, L. 2014. Key to subfamilies and tribes of Rosaceae. Pp. 21–32, in: Flora of North America Editorial Committee, eds. 1993+. Flora of North America North of Mexico. 21+ vols. New York and Oxford. Vol. 9.

Brouillet, L., J. C. Semple, G. A. Allen, K. L. Chambers, and S. D. Sundberg. 2006. *Symphyotrichum*. Pp. 465–539, in: Flora of North America Editorial Committee, eds. 1993+. Flora of North America North of Mexico. 21+ vols. New York and Oxford. Vol. 20.

Browning, J. K., D. Gordon-Gray, and S. G. Smith. 1995. Achene structure and taxonomy of North American *Bolboschoenus* (Cyperaceae). Brittonia 47: 433–445.

Bruce, J. G., W. H. Wagner, Jr., and J. M. Beitel. 1991. Two new species of bog clubmosses, *Lycopodiella* (Lycopodiaceae), from southwestern Michigan. Michigan Bot. 30: 3–10.

Bruederle, L. P., and D. E. Fairbrothers. 1986. Allozyme variation in populations of the *Carex crinita* complex (Cyperaceae). Syst. Bot. 11: 583–594.

Brunton, D. F. 1986. Status of the Mosquito Fern, *Azolla mexicana* (Salviniaceae), in Canada. Canad. Field-Nat. 100: 409–413.

Brunton, D. F. 2019. A practical technique for preserving specimens of duckmeal, *Wolffia* (Araceae). Canad. Field-Nat. 133: 139–143.

Brunton, D. F., and D. M. Britton. 1991. *Isoetes ×hickeyi* (Isoetaceae: Pteridophyta) in Canada. Fern Gaz. 14: 17–23.

Brunton, D. F., and D. M. Britton. 1996. Taxonomy and distribution of *Isoetes valida*. Amer. Fern J. 86: 16–25.

Buck, W. R. 1977. A new species of *Selaginella* in the *S. apoda* complex. Canad. J. Bot. 55: 366–367.

Buddell, G. F., II, and J. W. Thieret. 1997. Saururaceae. Pp. 37–38, in: Flora of North America Editorial Committee, eds. 1993+. Flora of North America North of Mexico. 21+ vols. New York and Oxford. Vol. 3.

Buell, M. F. 1935. *Acorus calamus* in America. Rhodora 37: 367–369.

Burns, G. P. 1904. Heterophylly in *Proserpinaca palustris*. Ann. Bot. 18: 579–589.

Butters, F. K., and E. C. Abbe. 1940. The American varieties of *Rorippa islandica*. Rhodora 42: 25–32, pl. 588.

Caldwell, F. A., and G. E. Crow. 1992. A floristic and vegetation analysis of a freshwater tidal marsh on the Merrimack River, West Newbury, Massachusetts. Rhodora 94: 63–97.

Campbell, C. S. 1983. Systematics of *Andropogon virginicus* complex (Gramineae). J. Arnold Arbor. 64: 171–254.

Campbell, C. S. 1985. The subfamilies and tribes of Gramineae (Poaceae) in the southeastern United States. J. Arnold Arbor. 66: 123–199.

Campbell, C. S. 2003. *Andropogon*. Pp. 649–664, in: Flora of North America Editorial Committee, eds. 1993+. Flora of North America North of Mexico. 21+ vols. New York and Oxford. Vol. 25.

Campbell, G. R. 1952. The genus *Myosurus* L. (Ranunculaceae) in North America. El Aliso 2: 389–403.

Cantino, P. D. 1981. The *Physostegia purpurea-leptophylla-denticulata* problem: Taxonomic and nomenclatural clarification. Rhodora 83: 581–593.

Cantino, P. D. 1982. A monograph of the genus *Physostegia* (Labiatae). Contr. Gray Herb. 211: 1–105.

Cantino, P. D., and S. J. Wagstaff. 1998. A reexamination of North American *Satureja* s. l. (Lamiaceae) in light of molecular evidence. Brittonia 50: 53–70.

Case, F. W., Jr. 1987. Orchids of the Western Great Lakes Region. Rev. ed. Cranbrook Inst. Sci., Bloomfield Hills, Mich.

Case, F. W., Jr., and P. M. Catling. 1983. The genus *Spiranthes* in Michigan. Michigan Bot. 22: 79–92.

Catling, P. M. 1981. Taxonomy of autumn-flowering *Spiranthes* species of southern Nova Scotia. Canad. J. Bot. 59: 1253–1270.

Catling, P. M. 1982. Breeding systems of northeastern North American *Spiranthes* taxa (Orchidaceae). Canad. J. Bot. 60: 3017–3039.

Catling, P. M., and I. Dobson. 1985. The biology of Canadian weeds. 69. *Potamogeton crispus* L. Canad. J. Pl. Sci. 65: 655–668.

Catling, P. M., and W. G. Dore. 1982. Status and identification of *Hydrocharis morsus-ranae* and *Limnobium spongia* (Hydrocharitaceae) in northeastern North America. Rhodora 84: 523–545.

Catling, P. M., and Z. S. Porebski. 1995. The spread and current distribution of European Frogbit, *Hydrocharis morsus-ranae* L., in North America. Canad. Field-Nat. 109: 236–241.

Cayouette, J. 2004. A taxonomic review of the *Eriophorum russeolum—E. scheuchzeri* complex (Cyperaceae) in North America. Sida 21: 791–814.

Ceska, A., and O. Ceska. 1986. Notes on *Myriophyllum* (Haloragaceae) in the Far East: The identity of *Myriophyllum sibiricum* Komarov. Taxon 35: 95–100.

Ceska, O., A. Ceska, and P. D. Warrington. 1986. *Myriophyllum quitense* and *Myriophyllum ussuriense* (Haloragaceae) in British Columbia, Canada. Brittonia 38: 73–81.

Chafin, L. G., J. C. Putnam Hancock, H. O. Nourse, and C. Nourse. 2007. Field Guide to the Rare Plants of Georgia. State Botanical Garden of Georgia and The Georgia Plant Conservation Alliance, distributed by Univ. of Georgia Press, Athens, Ga.

Chambliss, C. E. 1940. The botany and history of *Zizania aquatica* L. (Wild Rice). J. Washington Acad. Sci. 30: 185–205. (Reprinted in Ann. Rep. Smithsonian Inst., 1940.)

Chase, A. 1964. First Book of Grasses. Smithsonian Inst. Press, Washington, D.C.

Chaw, S-M., A. Zharkikh, H-M. Sunng, T-C. Lau, and W. H. Li. 1997. Molecular phylogeny of extant Gymnosperms and Seed Plant evolution: Analysis of nuclear 18S rRNA sequences. Molecular Biology and Evolution 14: 56–68.

Cholewa, A. F. 2009. *Hottonia*, *Lysimachia*. Pp. 259, 308–318, in: Flora of North America Editorial Committee, eds. 1993+. Flora of North America North of Mexico. 21+ vols. New York and Oxford. Vol. 8.

Church, G. L. 1949. A cytotaxonomic study of *Glyceria* and *Puccinellia*. Amer. J. Bot. 36: 155–165.

Church, G. L. 1967. Taxonomic and generic relationships of eastern North American species of *Elymus* with setaceous glumes. Rhodora 69: 121–162.

Cieslak, T., J. S. Polepalli, A. White, K. Muller, T. Borsch, W. Barthlott, J. Steiger, A. Marchant, and L. Legendre. 2005. Phylogenetic analysis of *Pinguicula* (Lentibulariaceae): Chloroplast DNA sequences and morphology support several geographically distinct radiations. Amer. J. Bot. 92: 1723–1736.

Clark, H. L., and J. W. Thieret. 1968. The duckweeds of Minnesota. Michigan Bot. 7: 67–76.

Clark, L. G., and E. A. Kellogg. 2007. Poaceae, introduction. Pp. 3–6, in: Flora of North America Editorial Committee, eds. 1993+. Flora of North America North of Mexico. 21+ vols. New York and Oxford. Vol. 24.

Clark, L. G., and J. K. Triplett. 2007. *Arundinaria*. Pp. 17–20, in: Flora of North America Editorial Committee, eds. 1993+. Flora of North America North of Mexico. 21+ vols. New York and Oxford. Vol. 24.

Clark, O. M. 1938. Spread of *Nymphoides peltatum* in Lake Messina. Proc. Oklahoma Acad. Sci. 18: 21–22.

Clausen, R. T. 1936. Studies in the genus *Najas* in northern United States. Rhodora 38: 334–345.

Clausen, R. T. 1937. A new species of *Najas* from the Hudson River. Rhodora 39: 57–60.

Clausen, R. T. 1952. Suggestion for the assignment of *Torreyochloa* to *Puccinellia*. Rhodora 54: 42–45.

Clayton, W. D. 1968. The correct name of the Common Reed. Taxon 17: 168–169.

Clemants, S. E. 1990. Juncaceae (Rush Family) of New York State. Contr. to a Flora of New York State VII. New York State Mus. Bull. No. 475.

Clemants, S. E. 2003. *Alternanthera.* Pp. 447–451, in: Flora of North America Editorial Committee, eds. 1993+. Flora of North America North of Mexico. 21+ vols. New York and Oxford. Vol. 4.

Cochrane, T. S. 2002. *Carex* sect. *Heleoglochin*, *Carex* sect. *Physoglochin*, *Carex* sect. *Leucoglochin*, *Carex* sect. *Leptocephalae*. Pp. 278–281, 299–301, 530–531, 565–566, in: Flora of North America Editorial Committee, eds. 1993+. Flora of North America North of Mexico. 21+ vols. New York and Oxford. Vol. 23.

Cochrane, T. S., and R. F. C. Naczi. 2002. *Carex* sect. *Granulares*. Pp. 440–442, in: Flora of North America Editorial Committee, eds. 1993+. Flora of North America North of Mexico. 21+ vols. New York and Oxford. Vol. 23.

Cody, W. J. 1954. A history of *Tillaea aquatica* (Crassulaceae) in Canada and Alaska. Rhodora 56: 96–101.

Cody, W. J. 1961. *Iris pseudacorus* L. escaped from cultivation in Canada. Canad. Field-Nat. 75: 139–142.

Coffin, B., and L. Pfannmuller, eds. 1988. Minnesota's Endangered Flora and Fauna. Univ. of Minnesota Press, Minneapolis, Minn.

Conard, H. S. 1905. The Water-lilies, a Monograph of the Genus *Nymphaea*. Publ. Carnegie Inst. Washington, D.C. No. 4.

Cook, C. D. K. 1963. Studies on *Ranunculus* subgenus *Batrachium*. II. General morphological considerations in the taxonomy of the subgenus. Watsonia 5: 294–303.

Cook, C. D. K. 1966. A monographic study of *Ranunculus* subgenus *Batrachium* (DC.) A. Gray. Mitt. Bot. Staatss. München 6: 47–237.

Cook, C. D. K. 1985. *Sparganium*: Some old names and their types. Bot. Jahrb. Syst. 107: 269–276.

Cook, C. D. K. 1988. Wind pollination in aquatic angiosperms. Ann. Missouri Bot. Gard. 75: 768–777.

Cook, C. D. K., B. J. Gut, E. M. Rix, J. Schneller, and M. Seitz. 1974. Water Plants of the World. Dr. W. Junk b.v., The Hague.

Cook, C. D. K., and R. Lüönd. 1982a. A revision of the genus *Hydrocharis* (Hydrocharitaceae). Aquatic Bot. 14: 177–204.

Cook, C. D. K., and R. Lüönd. 1982b. A revision of the genus *Hydrilla* (Hydrocharitaceae). Aquatic Bot. 13: 485–504.

Cook, C. D. K., and M. S. Nicholls. 1986. A monographic study of the genus *Sparganium* (Sparganiaceae). Part 1. Subgenus *Xanthosparganium* Holmberg. Bot. Helv. 96: 213–267.

Cook, C. D. K., and M. S. Nicholls. 1987. A monographic study of the genus *Sparganium* (*Sparganiaceae*). Part 2. Subgenus *Sparganium*. Bot. Helv. 97: 1–44.

Cook, C. D. K., and K. Urmi-König. 1983a. A revision of the genus *Limnobium* including *Hydromystria* (Hydrocharitaceae). Aquatic Bot. 17: 1–27.

Cook, C. D. K., and K. Urmi-König. 1983b. A revision of the genus *Stratiotes*. Aquatic Bot. 16: 213–249.

Cook, C. D. K., and K. Urmi-König. 1985. A revision of the genus *Elodea* (Hydrocharitaceae). Aquatic Bot. 21: 111–156.

Cooperrider, T. S., and B. L. Brockett. 1974. The nature and status of *Lysimachia* ×*producta* (Primulaceae). Brittonia 26: 119–128.

Cooperrider, T. S., and B. L. Brockett. 1976. The nature and status of *Lysimachia* ×*producta* (Primulaceae)—II. Brittonia 28: 76–80.

Cooperrider, T. S., and G. A. McCready. 1975. On separating Ohio specimens of *Lindernia dubia* and *L. anagallidea* (Scrophulariaceae). Castanea 40: 191–197.

Cope, T. A., and C. A. Stace. 1978. The *Juncus bufonius* L. aggregate in western Europe. Watsonia 12: 113–128.

Core, E. L. 1941. *Butomus umbellatus* in America. Ohio J. Sci. 41: 79–85.

Correll, D. S. 1950. Native Orchids of North America. Stanford Univ. Press, Stanford, Calif.

Correll, D. S., and H. B. Correll. 1972. Aquatic and Wetland Plants of Southwestern United States. U.S. Env. Protect. Agency, U.S. Gov. Print. Off., Washington, D.C.

Costa, M., F. J. Tardiff, and H. R. Hinds. 2005. *Polygonum*. Pp. 547–571, in: Flora of North America Editorial Committee, eds. 1993+. Flora of North America North of Mexico. 21+ vols. New York and Oxford. Vol. 5.

Couch, R., and E. Nelson. 1985. *Myriophyllum spicatum* in North America. Proc. 1st Internatl. Symp. on Watermilfoil (*Myriophyllum spicatum*) and related Haloragaceae species 1: 8–18.

Couch, R., and E. Nelson. 1988. *Myriophyllum quitense* (Haloragaceae) in the United States. Brittonia 40: 85–88.

Countryman, W. D. 1977. Water chestnut (*Trapa natans* L.) in Lake Champlain. Proc. Lake Champlain Basin Env. Conf. (Miner Center, Chazy, N.Y.) 4: 3–10.

Cowie, I. D., P. S. Short, and M. Osterkamp Madsen. 2000. Floodplain Flora: A Flora of the Coastal Floodplains of the Northern Territory, Australia. Flora of Australia Supplementary Series Number 10.

Cranfill, R. B. 1993. *Woodwardia*. Pp. 226–227, in: Flora of North America Editorial Committee, eds. 1993+. Flora of North America North of Mexico. 21+ vols. New York and Oxford. Vol. 2.

Crins, W. J. 2002. *Trichophorum*, *Carex* sect. *Ceratocystis*. Pp. 28–31, 523–527, in: Flora of North America Editorial Committee, eds. 1993+. Flora of North America North of Mexico. 21+ vols. New York and Oxford. Vol. 23.

Crins, W. J. 2007. *Alopecurus*. Pp. 780–789, in: Flora of North America Editorial Committee, eds. 1993+. Flora of North America North of Mexico. 21+ vols. New York and Oxford. Vol. 24.

Crins, W. J., and P. W. Ball. 1989a. Taxonomy of the *Carex flava* complex (Cyperaceae) in North America and northern Eurasia. I. Numerical taxonomy and character analysis. Canad. J. Bot. 67: 1032–1047.

Crins, W. J., and P. W. Ball. 1989b. Taxonomy of the *Carex flava* complex (Cyperaceae) in North America and northern Eurasia. II. Taxonomic treatment. Canad. J. Bot. 67: 1048–1065.

Cronquist, A. 1980. Vascular Flora of the Southeastern United States. Vol. 1. Asteraceae. Univ. of North Carolina Press, Chapel Hill, N.C.

Crosswhite, F. S. 1965. Variation in *Chelone glabra* in Wisconsin (Scrophulariaceae). Michigan Bot. 4: 62–66.

Crow, G. E. 1982. New England's Rare, Threatened, and Endangered Plants. U.S. Gov. Print. Off., Washington, D.C.

Crow, G. E. 1993. Species diversity in aquatic angiosperms: Latitudinal patterns. Aquat. Bot. 44: 229–258.

Crow, G. E. 2002. Plantas acuáticas del Parque Nacional Palo Verde y Valle Tempisque, Costa Rica / Aquatic plants of Palo Verde National Park and the Tempisque Valley, Costa Rica. (A bilingual field guide). Instituto Nacional Biodiversidad (INBio), Santo Domingo de Heredia, Costa Rica.

Crow, G. E. 2003a. Pontederiaceae. Pp. 822–828, in: B. E. Hammel, M. H. Grayum, C. Herrera, and N. Zamora, eds. Manual de Plantas de Costa Rica. Vol. III: Monocotiledóneas. Monogr. Syst. Bot. Missouri Bot. Gard. 93: 1–884.

Crow, G. E. 2003b. Potamogetonaceae. Pp. 829–832, in: B. E. Hammel, M. H. Grayum, C. Herrera, and N. Zamora, eds. Manual de Plantas de Costa Rica. Vol. III: Monocotiledóneas. Monogr. Syst. Bot. Missouri Bot. Gard. 93: 1–884.

Crow, G. E. 2004. Lentibulariaceae. Pp. 211–212, in: N. P. Smith, S. A. Mori, A. Henderson, D. W. Stevenson, and S. V. Heald, eds. Families of Neotropical Flowering Plants. Princeton Univ. Press, Princeton, N.J.

Crow, G. E. 2007. Lentibulariaceae. Pp. 189–197, in: B. E. Hammel, M. H. Grayum, C. Herrera, and N. Zamora, eds. Manual dePlantasdeCostaRica. Vol. VI:Dicotiledóneas(Haloragaceae-Piperaceae). Monogr. Syst. Bot. Missouri Bot. Gard. 111: 1–933.

Crow, G. E. 2015a. *Nelumbo lutea* Willd. (Nelumbonaceae): Occurrence of a rare plant in western Michigan confirmed. Michigan Bot. 54: 124–130.

Crow, G. E. 2015b. The taxonomic value of internal bladder-trap quadrifids in recognizing and identifying *Utricularia ochroleuca* (Lentibulariaceae). Botanical Electronic News (BEN) No. 487, February 28, 2015. http://www.ou.edu/cas/botany-micro/ben/ben487.html.

Crow, G. E. 2020. Cabombaceae. Pp. 248–250, in: B. E. Hammel, M. H. Grayum, C. Herrera, and N. Zamora, eds. Manual de Plantas de Costa Rica. Vol. IV, Parte 2: Dicotiledóneas (Balanophoraceae-Clethraceae). Monogr. Syst. Bot. Missouri Bot. Gard. 138: 1–524.

Crow, G. E. 2022. Lentibulariaceae. Flora of North America North of Mexico, Provisional Publication. Flora of North America Association. http://floranorthamerica.org/File:Lentibulariaceae04n.pdf, accessed October 28, 2022.

Crow, G. E. In press. Lentibulariaceae. In: Flora of North America Editorial Committee, eds. 1993+. Flora of North America North of Mexico. 21+ vols. New York and Oxford. Vol. 18.

Crow, G. E., and C. B. Hellquist. 1981. Aquatic Vascular Plants of New England. Part 2. Typhaceae and Sparganiaceae. New Hampshire Agric. Exp. Sta. Bull. 517.

Crow, G. E., and C. B. Hellquist. 1982. Aquatic Vascular Plants of New England. Part 4. Juncaginaceae, Scheuchzeriaceae, Butomaceae, Hydrocharitaceae. New Hampshire Agric. Exp. Sta. Bull. 520.

Crow, G. E., and C. B. Hellquist. 1983. Aquatic Vascular Plants of New England. Part 6. Trapaceae, Haloragaceae, Hippuridaceae. New Hampshire Agric. Exp. Sta. Bull. 524.

Crow, G. E., and C. B. Hellquist. 1985. Aquatic Vascular. Plants of New England. Part 8. Lentibulariaceae. New Hampshire Agric. Exp. Sta. Bull. 528.

Czerepanov, S. K. 1981. Plantae Vasculares URSS. Komarov Botanical Inst. Leningrad. (In Russian).

Daubs, E. H. 1962. The occurrence of *Spirodela oligorrhiza* in the United States. Rhodora 64: 83–85.

Daubs, E. H. 1965. A monograph of Lemnaceae. Illinois Biol. Monogr. 34: 1–118.

Davenport, L. J. 1988. A monograph of *Hydrolea* (Hydrophyllaceae). Rhodora 90: 169–208.

Davis, G. J. 1967. *Proserpinaca*: Photoperiodic and chemical differentiation of leaf development and flowering. Pl. Physiol. 42: 667–669.

Davis, J. I. 2007. *Torreyochloa*. Pp. 607–609, in: Flora of North America Editorial Committee, eds. 1993+. Flora of North America North of Mexico. 21+ vols. New York and Oxford. Vol. 24.

Davis, J. I., and L. L. Consaul. 2007. *Puccinellia*. Pp. 459–477, in: Flora of North America Editorial Committee, eds. 1993+. Flora of North America North of Mexico. 21+ vols. New York and Oxford. Vol. 24.

Dawson, J. E. 1979. A biosystematic study of *Rumex* Section *Rumex* in Canada and the United States. Ph.D. thesis. Carleton Univ., Ottawa, Ont.

Delahoussaye, A. J., and J. W. Thieret. 1967. *Cyperus* subgenus *Kyllinga* (Cyperaceae) in the continental United States. Sida 3: 128–136.

Dennis, W. M., and D. H. Webb. 1981. The distribution of *Pilularia americana* A. Br. (Marsileaceae) in North America, north of Mexico. Sida 9: 19–24.

DePoe, C. E., and E. O. Beal. 1969. Origin and maintenance of clinal variations in *Nuphar* (Nymphaeaceae). Brittonia 21: 15–28.

Dodd, L., N. Rybicki, R. Thum, Y. Kadono, and K. Ingram. 2019. Genetic and Morphological Differences of Water Chestnut (Myrtales: Lythraceae: *Trapa*) Populations in the Northeastern United States, Japan, and South Africa. US Army Corps of Engineers Engineer Research and Development Center, Vicksburg, MS. Available at https://apps.dtic.mil/dtic/tr/fulltext/u2/1070329.pdf.

Dore, W. G. 1947. *Glyceria maxima* in Canada. Canad. Field-Nat. 61: 174.

Dore, W. G. 1954. Frog-bit (*Hydrocharis morsus-ranae L.*) in the Ottawa River. Canad. Field-Nat. 68: 180–181.

Dore, W. G. 1957. *Wolffia* in Canada. Canad. Field-Nat. 71: 10–16.

Dore, W. G. 1968. Progress of the European Frog-bit in Canada. Canad. Field-Nat. 82: 76–82.

Dore, W. G. 1969. Wild Rice. Canad. Dept. Agric. Res. Publ. 1393: 1–84.

Dore, W. G., and J. McNeill. 1980. Grasses of Ontario. Agric. Canad. Monogr. 26.

Dorn, R. D. 1976. A synopsis of American *Salix*. Canad. J. Bot. 54: 2769–2789.

Douglas, G. W., D. V. Meidinger, and J. Pojar. 1999. Illustrated Flora of British Columbia. Volume 3: Dicotyledons (Diapensiaceae through Onagraceae). British Columbia Ministry Environment, Lands and Parks and B.C. Ministry of Forests, Victoria, B.C.

Dressler, R. L. 1981. The Orchids. Natural History and Classification. Harvard Univ. Press, Cambridge, Mass.

Drew, W. B. 1936. North American representatives of *Ranunculus* sect. *Batrachium*. Rhodora 38: 1–47.

Duncan, T. 1980. A taxonomic study of the *Ranunculus hispidus* Michaux complex in the Western Hemisphere. Univ. California Publ. Bot. 77: 1–125.

Duncan, W. H., and M. B. Duncan. 1987. The Smithsonian Guide to Seaside Plants of the Gulf and Atlantic Coasts. Smithsonian Inst. Press, Washington, D.C.

Duvall, M. R., and D. D. Biesboer. 1988. Nonreciprocal hybridization failure in crosses between annual wild-rice species (*Zizania palustris* × *Z. aquatica*: Poaceae). Syst. Bot. 13: 229–334.

Easterly, N. W. 1957. A morphological study of *Ptilimnium*. Brittonia 9: 136–145.

Eckenwalder, J. E. 2010. *Populus*. Pp. 5–22, in: Flora of North America Editorial Committee, eds. 1993+. Flora of North America North of Mexico. 21+ vols. New York and Oxford. Vol. 7.

Eleuterius, L. N. 1975. The life history of the Salt Marsh Rush, *Juncus roemerianus*. Bull. Torrey Bot. Club 102: 135–140.

Elias, T. S. 1971. The genera of Myricaceae in the southeastern United States. J. Arnold Arb. 52: 305–318.

Elven, R., and D. F. Murray. 2014. *Potentilla* sect. *Anserina*. Pp. 126–130, in: Flora of North America Editorial Committee, eds. 1993+. Flora of North America North of Mexico. 21+ vols. New York and Oxford. Vol. 9.

Elven, R., D. F. Murray, and H. Solstad. 2019. *Hippuris*. Pp. 55–56, in: Flora of North America Editorial Committee, eds. 1993+. Flora of North America North of Mexico. 21+ vols. New York and Oxford. Vol. 17.

Ertter, B., and J. L. Reveal. 2014. *Dasiphora*, *Comara*. Pp. 295–297, 300–301, in: Flora of North America Editorial Committee, eds. 1993+. Flora of North America North of Mexico. 21+ vols. New York and Oxford. Vol. 9.

Eyde, R. H. 1963. Morphological and paleobotanical studies of the Nyssaceae. I. A survey of the modern species and their fruits. J. Arnold Arbor. 44: 1–59.

Eyde, R. H. 1966. The Nyssaceae in the southeastern United States. J. Arnold Arbor. 47: 117–125.

Fabijan, D. M. 2009. *Andromeda*. Pp. 503–505, in: Flora of North America Editorial Committee, eds. 1993+. Flora of North America North of Mexico. 21+ vols. New York and Oxford. Vol. 8.

Fairbrothers, D. E. 1958. A naturalized stand of Indian Lotus (*Nelumbo nucifera*) in New Jersey. Bull. Torrey Bot. Club 85: 70.

Fairey, J. E., III. 1967. The genus *Scleria* in the southeastern United States. Castanea 32: 37–71.

Farwell, O. A. 1931. Concerning some species of *Cornus* of Philip Miller. Rhodora 33: 68–72.

Fassett, N. C. 1924. A study of the genus *Zizania*. Rhodora 26: 153–160.

Fassett, N. C. 1939a. Notes from the herbarium of the University of Wisconsin—XVII. *Elatine* and other aquatics. Rhodora 41: 367–377.

Fassett, N. C. 1939b. Notes from the herbarium of the University of Wisconsin—XVIII. Rhodora 41: 524–529.

Fassett, N. C. 1949a. Some notes on *Echinochloa*. Rhodora 51: 1–3.

Fassett, N. C. 1949b. The variations of *Polygonum punctatum*. Brittonia 6: 369–393.

Fassett, N. C. 1951. *Callitriche* in the New World. Rhodora 53: 137–155, 161–182, 185–194, 210–222.

Fassett, N. C. 1953a. A monograph of *Cabomba*. Castanea 18: 116–128.

Fassett, N. C. 1953b. North American *Ceratophyllum*. Comun. Inst. Trop. Invest. Ci. Univ. El Salvador 2: 25–45.

Fassett, N. C. 1957. A Manual of Aquatic Plants. With revision appendix by E. C. Ogden. Univ. of Wisconsin Press, Madison, Wisc. (Originally published by McGraw-Hill, 1940.)

Fedchenko, B. A. 1934. *Butomaceae*. Pp. 228–230, in V. L. Komarov, ed. Flora of the U.S.S.R. Vol. 1. (English translation, IPST Press, Jerusalem.)

Fernald, M. L. 1903. Some variations in *Triglochin maritima*. Rhodora 5: 174–175.

Fernald, M. L. 1917a. The genus *Elatine* in eastern North America. Rhodora 19: 10–15.

Fernald, M. L. 1917b. The variations of *Polygonum pensylvanicum*. Rhodora 19: 70–73.

Fernald, M. L. 1918. The diagnostic character of *Vallisneria americana*. Rhodora 20: 108–110.

Fernald, M. L. 1920. Some variations of *Cardamine pratensis* in America. Rhodora 22: 11–14.

Fernald, M. L. 1922. Notes on *Sparganium*. Rhodora 24: 26–34.

Fernald, M. L. 1923. Notes on the distribution of *Najas* in northeastern America. Rhodora 25: 105–109.

Fernald, M. L. 1932a. *Callitriche stagnalis* on the lower St. Lawrence. Rhodora 34: 39.

Fernald, M. L. 1932b. The linear-leaved North American species of *Potamogeton* section *Axillares*. Mem. Amer. Acad. Arts 17: 1–183.

Fernald, M. L. 1940. The eastern American varieties of *Rorippa islandica*. Rhodora 42: 267–274.

Fernald, M. L. 1941. *Elatine americana* and *E. triandra*. Rhodora 43: 208–211.

Fernald, M. L. 1945. Botanical specialties of the Seward Forest and adjacent areas of southeastern Virginia. Rhodora 47: 93–142, 149–182, 191–204. (Reprinted as Contr. Gray Herbarium, Harvard University—No. CLVI [with Rhodora page numbers]).

Fernald, M. L. 1950. Gray's Manual of Botany. 8th ed. American Book Co., New York.

Fernald, M. L., and L. Griscom. 1935. *Proserpinaca palustris* varieties. Rhodora 37: 177–178.

Fernald, M. L., and K. M. Wiegand. 1913. The genus *Empetrum* in North America. Rhodora 15: 211–217.

Fernald, M. L., and K. M. Wiegand. 1914. The genus *Ruppia* in eastern North America. Rhodora 16: 119–127.

Ferren, W. R., Jr., and H. J. Schenk. 2003. *Suaeda*. Pp. 390–398, in: Flora of North America Editorial Committee, eds. 1993+. Flora of North America North of Mexico. 21+ vols. New York and Oxford. Vol. 4.

Fitch, W. H., and G. W. Smith. 1880. Illustrations of the British flora: A series of wood engravings, with dissections of British Plants. London: L. Reeve & Co. (available at https://www.biodiversitylibrary.org/page/22965846).

Ford, B. A. 1997. *Caltha*. Pp. 187–189, in: Flora of North America Editorial Committee, eds. 1993+. Flora of North America North of Mexico. 21+ vols. New York and Oxford. Vol. 3.

Ford, B. A., and P. W. Ball. 1988. A reevaluation of the *Triglochin maritimum* L. complex (Juncaginaceae) in eastern and central North America and Europe. Rhodora 90: 313–337.

Freckmann, R. W., and M. G. Lelong. 2003. *Panicum*. Pp. 450–488, in: Flora of North America Editorial Committee, eds. 1993+. Flora of North America North of Mexico. 21+ vols. New York and Oxford. Vol. 25.

Freckmann, R. W., and D. M. Reed. 1979. *Glyceria maxima*, a new, potentially troublesome wetland weed. Bull. Bot. Club Wisconsin 11: 30–35.

Freeman, C. C. 2019. *Gratiola*. Pp. 264–269, in: Flora of North America Editorial Committee, eds. 1993+. Flora of North America North of Mexico. 21+ vols. New York and Oxford. Vol. 17.

Frieland, S. 1941. The American species of *Hemicarpha*. Amer. J. Bot. 28: 855–861.

Fuller, A. M. 1933. Studies on the flora of Wisconsin. Part I. The orchids; Orchidaceae. Bull. Pub. Mus. Milwaukee 14: 1–284.

Furlow, J. J. 1979. The systematics of the American species of *Alnus* (Betulaceae). Rhodora 81: 1–21, 151–248.

Furlow, J. J. 1997. Betulaceae. Pp. 507–538, in: Flora of North America Editorial Committee, eds. 1993+. Flora of North America North of Mexico. 21+ vols. New York and Oxford. Vol. 3.

Gaiser, L. O. 1949. Further distribution of *Butomus umbellatus* in the Great Lakes region. Rhodora 51: 387–390.

Gale, S. 1944. *Rhynchospora*, section *Eurhynchospora*, in Canada, the United States and the West Indies. Rhodora 46: 89–134, 159–197, 207–249, 255–278.

Gaskin, J. F., J. Andreas, B. J. Grewell, P. Haefliger, and N. E. Harms. 2021. Diversity and origins of *Butomus umbellatus* (flowering rush) invasion in North America. Aquat. Bot. 173: 1–8.

Gates, F. C. 1939. Conditions for the flowering of *Utricularia resupinata*. Lilloa 5: 159–162.

Gaudet, J. J. 1965. The effect of various environmental factors on the leaf form of the aquatic fern *Marsilea vestita*. Physiol. Pl. 18: 674–686.

Gil, H., K. Lee, Y. Ha, C. Jang, and D. K. Kim. 2019. *Sparganium glomeratum* (Typhaceae): A new record from South Korea. Korean J. Pl. Taxon. 49: 374–279.

Gillett, J. M. 1957. A revision of the North American species of *Gentianella* Moench. Ann. Missouri Bot. Gard. 44: 195–269.

Gillett, J. M. 1959. A revision of *Bartonia* and *Obolaria* (Gentianaceae). Rhodora 61: 43–62.

Gillett, J. M., and N. K. B. Robson. 1981. The St. John's-worts of Canada. Natl. Mus. Canad. Nat. Sci. Publ. Bot. No. 11.

Gillis, W. T. 1971. The systematics and ecology of poison-ivy and the poison-oaks (*Toxicodendron*, Anacardiaceae). Rhodora 73: 72–159, 161–237, 370–443, 465–540.

Gleason, H. A. 1947. *Triadenum*. Phytologia 2: 288–291.

Gleason, H. A. 1952. The New Britton and Brown Illustrated Flora of the Northeastern United States and Adjacent Canada. 3 vols. Lancaster, Press, Lancaster, Pa.

Gleason, H. A., and A. Cronquist. 1991. Manual of Vascular Plants of Northeastern United States and Adjacent Canada. 2nd ed. New York Botanical Garden, Bronx, N.Y.

Gmelin, J. G. 1769. Flora Sibirica: Sive historia plantarum Siberiae. Vol. 4. St. Petersburg (illustration of *Caltha natans*, http://www.plantillustrations.org/illustration.php?id_illustration=213965).

Godfrey, R. K. 1988. Trees, Shrubs, and Woody Vines of Northern Florida and Adjacent Georgia and Alabama. Univ. of Georgia Press, Athens, Ga.

Godfrey, R. K., G. W. Reinert, and R. D. Houk. 1961. Observations on microsporocarpic material of *Azolla caroliniana*. Amer. Fern J. 51: 89–92.

Godfrey, R. K., and J. W. Wooten. 1979. Aquatic and Wetland Plants of Southeastern United States. Monocotyledons. Univ. of Georgia Press, Athens, Ga.

Godfrey, R. K., and J. W. Wooten. 1981. Aquatic and Wetland Plants of Southeastern United States. Dicotyledons. Univ. of Georgia Press, Athens, Ga.

Goldman, D. H., L. K. Magrath, and P. M. Catling. 2002. *Calopogon*. Pp. 597–602, in: Flora of North America Editorial Committee, eds. 1993+. Flora of North America North of Mexico. 21+ vols. New York and Oxford. Vol. 26.

Gould, E. W., M. A. Ali, and D. E. Fairbrothers. 1972. A revision of *Echinochloa* in the United States. Amer. Midl. Nat. 87: 36–59.

Gould, E. W., and C. A. Clark. 1978. *Dichanthelium* (Poaceae) in the United States and Canada. Ann. Missouri Bot. Gard. 65: 1088–1132.

Gould, F. W. 1975. The Grasses of Texas. Texas A&M Univ. Press, College Station, Tex.

Gould, F. W., and R. B. Shaw. 1983. Grass Systematics. 2nd ed. Texas A&M Univ. Press, College Station, Tex.

Govaerts, R., and D. A. Simpson. 2007. World Checklist of Cyperaceae. Sedges: 1–765. The Board of Trustees of the Royal Botanic Gardens, Kew.

Grace, J. B., and R. J. Wetzel. 1978. The production biology of Eurasia Watermilfoil (*Myriophyllum spicatum* L.). A review. J. Aquatic Pl. Managem. 16: 1–11.

Graham, S. A. 1964. The genera of the Lythraceae in the southeastern United States. J. Arnold Arbor. 45: 235–250.

Graham, S. A. 1975. Taxonomy of the Lythraceae in the southeastern United States. Sida 6: 80–103.

Graham, S. A. 1979. The origin of *Ammannia* ×*coccinea* Rottboell. Taxon 28: 169–178.

Graham, S. A. 1985. A revision of *Ammannia* (Lythraceae) in the Western Hemisphere. J. Arnold Arbor. 66: 395–420.

Graham, S. A., and C. Wood. 1965. The genera of Polygonaceae in the southeastern United States. J. Arnold Arbor. 46: 91–121.

Graham, S. A., and C. E. Wood, Jr. 1975. The Podostemaceae of the southeastern United States. J. Arnold Arbor. 56: 456–465.

Gray, A. 1867. A Manual of Botany of the Northern United States. 5th ed. New York, N.Y.

Grayum, M. H. 1987. A summary of evidence and arguments supporting removal of *Acorus* from the Araceae. Taxon 36: 723–729.

Grear, J. W., Jr. 1966. Cytogeography of *Orontium aquaticum*. Rhodora 68: 25–34.

Green, P. S. 1962. Watercress in the New World. Rhodora 64: 32–43.

Greene, C. W. 1980. The systematics of *Calamagrostis* (Gramineae) in eastern North America. Ph.D. diss., Harvard Univ., Cambridge, Mass.

Greene, C. W. 1984. Sexual and apomictic reproduction in Calamagrostis (Gramineae) from eastern North America. Amer. J. Bot. 71: 285–293.

Greene, C. W. 1987. *Calamagrostis pickeringii* in Maine. Rhodora 89: 333–336.

Gutteridge, R. L. 1954. *Glyceria maxima* on the Mississippi River, Ontario, 1953. Canad. Field-Nat. 68: 133–135.

Hagstrom, J. O. 1916. Critical researches on the Potamogetons. Kungl. Svenska Vetenskapskad. Handl. 55: 1–281.

Haines, A. 2006. *Euthamia*. Pp. 97–100, in: Flora of North America Editorial Committee, eds. 1993+. Flora of North America North of Mexico. 21+ vols. New York and Oxford. Vol. 20.

Haines, A. 2010. New combinations in the New England tracheophyte flora. Botanical Notes 13: 1–8, Addendum.

Haines, A. 2011. Flora Novae Angliae: A Manual for the Identification of Native and Naturalized Higher Vascular Plants of New England. Yale Univ. Press, New Haven, Conn.

Hall, T. F. 1940. The biology of *Saururus cernuus* L. Amer. Midi. Nat. 24: 253–260.

Hall, T. F., and W. T. Penfound. 1944. The biology of the American Lotus, *Nelumbo lutea* (Willd.) Pers. Amer. Midi. Nat. 31: 744–758.

Hämet-Ahti, L. 1980. *Juncus alpinoarticulatus*: The legitimate name for Juncus *alpinus*. Ann. Bot. Fenn. 17: 341–342.

Hämet-Ahti, L. 1986. North American races of *Juncus alpinoarticulatus* (Juncaceae). Ann. Bot. Fenn. 23: 277–281.

Hardin, J. W. 1973. The enigmatic chokeberries *Aronia* (Rosaceae). Bull. Torrey Bot. Club 100: 178–184.

Harms, V. L. 1973. Taxonomic studies of North American *Sparganium*. I. *S. hyperboreum* and *S. minimum*. Canad. J. Bot. 51: 1629–1641.

Harms, V. L., D. E Hooper, and L. Baker. 1986. *Plantago maritima* and *Carex mackenziei* new for Saskatchewan: Additional rare inland stations for two seacoast salt marsh species. Rhodora 88: 315–323.

Harms, V. L., and G. E. Ledingham. 1986. The Narrow-leaved Cat-tail, *T.* ×*glauca*, newly reported from Saskatchewan. Canad. Field-Nat. 100: 107–110.

Harris, S. K. 1959. *Peucedanum palustre*, an interesting addition to the flora of Essex County, Massachusetts. Rhodora 61: 181.

Hartog, C. den, and E. van der Plas. 1970. A synopsis of the Lemnaceae. Blumea 18: 355–368.

Harvey, M. J. 2007. *Agrostis*. Pp. 633–662, in: Flora of North America Editorial Committee, eds. 1993+. Flora of North America North of Mexico. 21+ vols. New York and Oxford. Vol. 24.

Hatch, S. L. 2007. *Beckmannia*. Pp. 484–486, in: Flora of North America Editorial Committee, eds. 1993+. Flora of North America North of Mexico. 21+ vols. New York and Oxford. Vol. 24.

Hauke, R. L. 1963. A monograph of the genus *Equisetum* subgenus *Hippochaete*. Beih. Nova Hedwigia 8: 1–123.

Hauke, R. L. 1965. An analysis of a variable population of *Equisetum arvense* and *E.* ×*litorale*. Amer. Fern J. 55: 123–135.

Hauke, R. L. 1978. A taxonomic monograph of *Equisetum* subgenus *Equisetum*. Nova Hedwigia 30: 385–455.

Hauke, R. L. 1993. Equisetaceae. Pp. 76–84, in: Flora of North America Editorial Committee, eds. 1993+. Flora of North America North of Mexico. 21+ vols. New York and Oxford. Vol. 2.

Haynes, R. R. 1974. A revision of North American *Potamogeton* subsection *Pusilli* (Potamogetonaceae). Rhodora 76: 564–649.

Haynes, R. R. 1978. The Potamogetonaceae in the southeastern United States. J. Arnold Arbor. 59: 170–191.

Haynes, R. R. 1979. Revision of North and Central America *Najas* (Najadaceae). Sida 8: 34–56.

Haynes, R. R. 1980. Aquatic and marsh plants of Alabama. I.—Alismatidae. Castanea 45: 31–51.

Haynes, R. R. 1985. A revision of the clasping-leaved *Potamogeton* (Potamogetonaceae). Sida 11: 173–188.

Haynes, R. R. 1988. Reproductive biology of selected aquatic plants. Ann. Missouri Bot. Gard. 75: 805–810.

Haynes, R. R. 2000. Butomaceae, Ruppiaceae, Zosteraceae. Pp. 3–4, 75–76, 90–94, in: Flora of North America Editorial Committee, eds. 1993+. Flora of North America North of Mexico. 21+ vols. New York and Oxford. Vol. 22.

Haynes, R. R., and C. B. Hellquist. 1996. New combinations in North American Alismatidae. Novon 6: 370–371.

Haynes, R. R., and C. B. Hellquist. 2000. Juncaginaceae, Potamogetonaceae. Pp. 43–46, 47–74, in: Flora of North America Editorial Committee, eds. 1993+. Flora of North America North of Mexico. 21+ vols. New York and Oxford. Vol. 22.

Haynes, R. R., and L. B. Holm-Nielsen. 1987. The Zannichelliaceae in the southeastern United States. J. Arnold Arbor. 68: 259–268.

Haynes, R. R., and L. B. Holm-Nielsen. 2003. Potamogetonaceae. Flora Neotropica 85: iii–iv, 1–52.

Haynes, R. R., and D. C. Williams. 1975. Evidence for the hybrid origin of *Potamogeton longiligulatus* (Potamogetonaceae). Michigan Bot. 14: 94–100.

Hellquist, C. B. 1974. A white-flowered form of *Utricularia purpurea* from New Hampshire. Rhodora 76: 805.

Hellquist, C. B. 1980. Correlation of alkalinity and the distribution of *Potamogeton* in New England. Rhodora 82: 331–344.

Hellquist, C. B. 1984. Observations of *Potamogeton hillii* Morong in North America. Rhodora 86: 101–111.

Hellquist, C. B. 1993. Taxonomic considerations in aquatic vegetation assessments. Lake and Reserv. Managem. 7: 175–183.

Hellquist, C. B., and G. E. Crow. 1980. Aquatic Vascular Plants of New England. Part 1. Zosteraceae, Potamogetonaceae, Zannichelliaceae, Najadaceae. New Hampshire Agric. Exp. Sta. Bull. 515.

Hellquist, C. B., and G. E. Crow. 1981. Aquatic Vascular Plants of New England. Part 3. Alismataceae. New Hampshire Agric. Exp. Sta. Bull. 518.

Hellquist, C. B., and G. E. Crow. 1982. Aquatic Vascular Plants of New England. Part 5. Araceae, Lemnaceae, Xyridaceae, Eriocaulaceae, and Pontederiaceae. New Hampshire Agric. Exp. Sta. Bull. 523.

Hellquist, C. B., and G. E. Crow. 1984. Aquatic Vascular Plants of New England. Part 7. Cabombaceae, Nymphaeaceae, Nelumbonaceae, and Ceratophyllaceae. New Hampshire Agric. Exp. Sta. Bull. 527.

Hellquist, C. B., and G. E. Crow. 1986. *Potamogeton ×haynesii* (Potamogetonaceae), a new species from northeastern North America. Brittonia 38: 415–419.

Hellquist, C. B., and R. L. Hilton. 1983. A new species of *Potamogeton* (Potamogetonaceae) from northeastern United States. Syst. Bot. 8: 86–92.

Hellquist, C. B., C. T. Philbrick, and R. L. Hilton. 1988. The taxonomic status of *Potamogeton lateralis* Morong (Potamogetonaceae). Rhodora 90: 15–19.

Hellquist, C. E., C. B. Hellquist, J. J. Whipple. 2014. New records for rare and under-collected aquatic vascular plants of Yellowstone National Park. Madroño 61: 159–176.

Henderson, N. C. 1962. A taxonomic revision of the genus *Lycopus* (Labiatae). Amer. Midl. Nat. 68: 95–138.

Henderson, N. C. 2000. *Iris*. Pp. 371–395, in: Flora of North America Editorial Committee, eds. 1993+. Flora of North America North of Mexico. 21+ vols. New York and Oxford. Vol. 26.

Hendricks, A. J. 1957. A revision of the genus *Alisma* (Dill.) L. Amer. Midl. Nat. 58: 470–493.

Henry, L. K., W. E. Buker, and D. L. Pearth. 1975. Western Pennsylvania orchids. Castanea 40: 93–168.

Henson, D. 1985. *Bartonia paniculata*, new to Michigan. Michigan Bot. 24: 19–20.

Hermann, F. J. 1936. Diagnostic characters in *Lycopus*. Rhodora 38: 373–375.

Hermann, F. J. 1970. Manual of the Carices of the Rocky Mountains and Colorado Basin. U.S.D.A. Forest Serv. Agriculture Handbook No. 374

Hermann, F. J. 1975. Manual of the Rushes (*Juncus* spp.) of the Rocky Mountains and Colorado Basin. U.S.D.A. Forest Serv. Gen. Tech. Rep. RM–18.

Heslop-Harrison, Y. 1955. *Nuphar* Sm. J. Ecol. 43: 342–364.

Hess, W. J. 1986. *Wolffia papulifera* Thompson (Lemnaceae), new to Michigan. Sida 11: 407–411.

Hickey, J. R. 1986. *Isoetes* megaspore surface morphology: Nomenclature, variation and systematic importance. Amer. Fern J. 76: 1–16.

Hillman, W. S. 1961. The Lemnaceae, or duckweeds: A review of the descriptive and experimental literature. Bot. Rev. 27: 221–287.

Hillman, W. S., and D. D. Culley, Jr. 1978. The uses of duckweeds. Amer. Sci. 66: 442–451.

Hinds, H. R., and C. C. Freeman. 2005. *Persicaria*. Pp. 574–594, in: Flora of North America Editorial Committee, eds. 1993+. Flora of North America North of Mexico. 21+ vols. New York and Oxford. Vol. 5.

Hitchcock, A. S. 1950. Manual of the Grasses of the United States. 2nd ed., revised by A. Chase. U.S.D.A. Misc. Publ. 200.

Hitchcock, C. L., A. Cronquist, M. Ownbey, and J. W. Thompson. 1955–1969. Vascular Plants of the Pacific Northwest. 5 parts. Univ. of Washington Press, Seattle, Wash.

Holmes, W. C. 2006. *Mikania*. Pp. 545–547, in: Flora of North America Editorial Committee, eds. 1993+. Flora of North America North of Mexico. 21+ vols. New York and Oxford. Vol. 21.

Holmgren, N. H. 1998. Illustrated Companion to Gleason and Cronquist's Manual: Illustrations of the Vascular Plants of Northeastern United States and Adjacent Canada. The New York Botanical Garden, Bronx, N.Y.

Holub, J. 1977. *Stuckenia* Börner 1912–the correct name for *Coleogeton* (Potamogetonaceae). Preslia, Praha 69: 361–366.

Hopkins, C. O., and W. H. Blackwell, Jr. 1977. Synopsis of *Suaeda* (Chenopodiaceae) in North America. Sida 7: 147–173.

Horn, C. N. 1983. The annual growth cycle of *Heteranthera dubia* in Ohio. Michigan Bot. 23: 29–34.

Horn, C. N. 1986. Typifications and a new combination in *Heteranthera* (Pontederiaceae). Phytologia 59: 290.

Horn, C. N. 1988. Developmental heterophylly in the genus *Heteranthera* (Pontederiaceae). Aquatic Bot. 197–209.

Horn, C. N. 2002. Pontederiaceae. Pp. 37–46, in: Flora of North America Editorial Committee, eds. 1993+. Flora of North America North of Mexico. 21+ vols. New York and Oxford. Vol. 26.

Hotchkiss, N., and H. L. Dozier. 1949. Taxonomy and distribution of North America cattails. Amer. Midi. Nat. 41: 237–254.

House, H. D. 1937. A new plant joins the Hudson River flora. Torreya 37: 80–82.

Hultén, E. 1950. Flora of Alaska and Yukon. X. Lunds Univ. Arsskrift N. E, Sweden, Avd. 2, Bd. 46.

Hunt, G. S., and R. W Lutz. 1959. Seed production by Curly-leaved Pondweed and its significance to waterfowl. J. Wildl. Managem. 23: 405–408.

Iltis, H. H. 1965. The genus *Gentianopsis*: Transfers and phytogeographic comments. Sida 2: 129–154.

Iltis, H. H., and W. M. Shaughnessey. 1960. Preliminary reports on the flora of Wisconsin. No. 43. Primulaceae—Primrose family. Trans. Wisconsin Acad. Sci., Arts, and Letters 49: 113–135.

Islam, M. R., Y. Zhang, Z. Z. Li, H. Liu, J. M. Chen, and X. Y. Yang. 2020. Genetic diversity, population structure, and historical gene flow of *Nelumbo lutea* in USA using microsatellite markers. Aquat. Bot. 160: 1–7.

Jackson, R. C. 1960. A revision of the genus *Iva* L. Univ. Kansas Sci. Bull. 41: 793–875.

Jacobs, D. L. 1947. An ecological life history of *Spirodela polyrhiza*. Ecol. Monogr. 17: 437–469.

James, C. W. 1956. A revision of *Rhexia* (Melastomataceae). Brittonia 8: 201–230.

Jermy, A. C. 1990. Selaginellaceae. Pp. 39–45, in: K. Kubitzki et al., eds. 1990+. The Families and Genera of Vascular Plants. Berlin, Germany.

Johnson, D. M. 1986. Systematics of the New World species of *Marsilea*. Syst. Bot. Monogr. 11: 1–87.

Johnson, D. M. 1988. Proposal to conserve *Marsilea* L. (Pteridophyta: Marsileaceae) with *Marsilea quadrifolia* Linnaeus, *typ. conserv.* Taxon 37: 483–486.

Johnson, D. M. 1993. Marsileaceae. Pp. 331–335, in: Flora of North America Editorial Committee, eds. 1993+. Flora of North America North of Mexico. 21+ vols. New York and Oxford. Vol. 2.

Jones, E. N. 1931. The morphology and biology of *Ceratophyllum demersum*. Stud. Bot. Univ. Iowa 13: 11–46.

Judd, W. S. 1981. A monograph of *Lyonia* (Ericaceae). J. Arnold Arbor. 62: 63–128.

Judd, W. S. 2009. *Lyonia*. Pp. 500–503, in: Flora of North America Editorial Committee, eds. 1993+. Flora of North America North of Mexico. 21+ vols. New York and Oxford. Vol. 8.

Judd, W. S., and K. A. Kron. 2009. *Rhododendron*. Pp. 455–473, in: Flora of North America Editorial Committee, eds. 1993+. Flora of North America North of Mexico. 21+ vols. New York and Oxford. Vol. 8.

Kalm, P. 1771. The America of 1750: Peter Kalm's travels in North America. The English version of 1770 rev. from the original Swedish and edited by Adolph B. Benson, with a translation of new material from Kalm's diary notes. Dover Publ., New York.

Kane, M. E., and L. S. Albert. 1982. Environmental and growth regulator effects on heterophylly and growth of *Proserpinaca intermedia* (Haloragaceae). Aquatic Bot. 13: 73–85.

Kaplan, Z. 2008. A taxonomic revision of *Stuckenia* (Potamogetonaceae) in Asia, with note on the diversity and variation of the genus on a worldwide scale. Folia Geobot. 43: 159–234.

Kaplan, Z., and J. Fehrer. 2009. New combinations revealed by molecular analysis: The unknown side of North American pondweed diversity (*Potamogeton*). Syst. Bot. 34: 625–642.

Kaplan, Z., J. Fehrer, V. Bambasová, and C. B. Hellquist. 2018. The endangered Florida pondweed (*Potamogeton floridanus*) is a hybrid: Why we need to understand biodiversity thoroughly. PloS ONE 13: e0195241. https://doi.org/10.1371/journal.pone.0195241.

Kaplan, Z., and K. Marhold. 2012. Multivariate morphometric analysis of the *Potamogeton compressus* group (Potamogetonaceae). Bot. J. Linn. Soc. 170: 112–130.

Kaplan Z., and J. Štěpánek. 2003. Genetic variation within and between populations of *Potamogeton pusillus* agg. Plant Syst. Evol. 239: 95–112.

Karaman-Castro, V., and L. E. Urbatsch. 2006. *Boltonia*. Pp. 353–357, in: Flora of North America Editorial Committee, eds. 1993+. Flora of North America North of Mexico. 21+ vols. New York and Oxford. Vol. 20.

Kartesz, J. T. 1994. A Synonymized Checklist of the Vascular Flora of the United States, Canada and Greenland. 2nd ed. Vol. I—Checklist. Vol. 2—Thesaurus. Timber Press, Portland, Ore.

Kaul, R. B. 1997. Platanaceae. Pp. 358–361, in: Flora of North America Editorial Committee, eds. 1993+. Flora of North America North of Mexico. 21+ vols. New York and Oxford. Vol. 3.

Kaul, R. B. 2000. *Sparganium*. Pp. 270–277, in: Flora of North America Editorial Committee, eds. 1993+. Flora of North America North of Mexico. 21+ vols. New York and Oxford. Vol. 22.

Keener, B. R. 2019. *Micranthemum*. Pp. 358–359, in: Flora of North America Editorial Committee, eds. 1993+. Flora of North America North of Mexico. 21+ vols. New York and Oxford. Vol. 17.

Keener, C. S. 1976a. Studies in the Ranunculaceae of the southeastern United States. V. *Ranunculus* L. Sida 6: 266–283.

Keener, C. S. 1976b. Studies in the Ranunculaceae of the southeastern United States. II. *Thalictrum* L. Rhodora 78: 457–472.

Keil, D. J. 2006. *Cirsium*. Pp. 95–164, in: Flora of North America Editorial Committee, eds. 1993+. Flora of North America North of Mexico. 21+ vols. New York and Oxford. Vol. 19.

Kennedy, H. 2000. Marantaceae. Pp. 315–319, in: Flora of North America Editorial Committee, eds. 1993+. Flora of North America North of Mexico. 21+ vols. New York and Oxford. Vol. 22.

Kerguélen, M., G. Bosc., and J. Lambinon. 1987. Données taxonomiques, nomenclaturales et chorologiques pour une révision de la flore de France. Lejeunia, n.s., 120: 1–263.

Kessler, J. W. 1988. A treatment of *Scleria* (Cyperaceae) for North America north of Mexico. Sida 12: 391–407.

Key, J. S. 1982. Field Guide to Missouri Ferns. Missouri Dept. Conserv., Jefferson City, Mo.

Kim, C., and H.-K. Choi. 2011. Molecular systematics and character evolution of *Typha* (Typhaceae) inferred from nuclear and plastid DNA sequence data. Taxon 60: 1417–1428.

Knapp, W. M., and R. F. C. Naczi. 2008. Taxonomy, morphology, and geographic distribution of *Juncus longii* (Juncaceae). Syst. Bot. 33: 685–694.

Kondo, K., M. Segawa, and K. Nehira. 1978. Anatomical studies on seeds and seedlings of some *Urtricularia* (Lentibulariaceae). Brittonia 30: 89–95.

Kondo, K., and H. Shimai. 2006. Phylogenetic analysis of the northern *Pinguicula* (Lentibulariaceae) based on internal transcribed spacer (ITS) sequence. Acta Phytotax. Geobot. 57: 155–164.

Kott, L. S. 1981. *Isoetes acadiensis*, a new species from eastern North America. Canad. J. Bot. 59: 2592–2594.

Kott, L. S., and D. M. Britton. 1980. Chromosome numbers for *Isoetes* in northeastern North America. Canad. J. Bot. 58: 980–984.

Kott, L. S., and D. M. Britton. 1983. Spore morphology and taxonomy of *Isoetes* in northeastern North America. Canad. J. Bot. 61: 3140–3163.

Kott, L. S., and D. M. Britton. 1985. Role of morphological characters of leaves and the sporangial region in the taxonomy of *Isoetes* in northeastern North America. Amer. Fern J. 75: 44–55.

Koyama, T. 1962. The genus *Scirpus* Linn., some North American aphylloid species. Canad. J. Bot. 40: 913–937.

Koyama, T. 1963. The genus *Scirpus* Linn., critical species of the section *Pterolepis*. Canad. J. Bot. 40: 913–937.

Kozlowski, G., R. A. Jones, and F. L. Nicholls. 2008. Biological Flora of Central Europe: *Baldellia ranunculoides* (Alismataceae). Persp. Plant Ecol., Evol. and Syst. 10: 109–142.

Kral, R. 1966a. Eriocaulaceae of continental North America north of Mexico. Sida 2: 285–332.

Kral, R. 1966b. *Xyris* (Xyridaceae) of the continental United States and Canada. Sida 2: 177–260.

Kral, R. 1971. A treatment of *Abildgaardia, Bulbostylis* and *Fimbristylis* (Cyperaceae) for North America. Sida 4: 57–227.

Kral, R. 1978. A synopsis of *Fuirena* (Cyperaceae) for the Americas north of South America. Sida 7: 309–354.

Kral, R. 1981. Notes on some "quill"-leaved Umbellifers. Sida 9: 124–134.

Kral, R. 1983. The Xyridaceae in the southeastern United States. J. Arnold Arbor. 64: 421–429.

Kral, R. 1988. The genus *Xyris* (Xyridaceae) in Venezuela and contiguous northern South America. Ann. Missouri Bot. Gard. 75: 522–722.

Kral, R. 2000. Xyridaceae, Eriocaulaceae. Pp. 154–167, 198–210, in: Flora of North America Editorial Committee, eds. 1993+. Flora of North America North of Mexico. 21+ vols. New York and Oxford. Vol. 22.

Kral, R. 2002. *Fimbristylis, Rhynchospora*. Pp. 121–131, 200–239, in: Flora of North America Editorial Committee, eds. 1993+. Flora of North America North of Mexico. 21+ vols. New York and Oxford. Vol. 23.

Kral, R., and P. E. Bostick. 1969. The genus *Rhexia* (Melastomataceae). Sida 3: 387–440.

Kramer, K. U. 1993. Systematics of the Pteridophytes. In: H. D. Behnke, U. Lüttge, K. Esser, J. W. Kadereit, and M. Runge, eds. Progress in Botany/Fortschritte der Botanik, vol. 54. Springer, Berlin, Heidelberg.

Kugel, A. R. 1958. Variation in the *Spiraea alba-latifolia* complex. Ph.D. diss., Univ. of Michigan, Ann Arbor, Mich.

Kurz, H., and R. K. Godfrey. 1962. Trees of Northern Florida. Univ. of Florida Press, Gainesville, Fla.

La Frankie, J. V., Jr. 1986a. Morphology and taxonomy of the New World species of *Maianthemum* (Liliaceae). J. Arnold Arbor. 67: 371–439.

La Frankie, J. V., Jr. 1986b. Transfer of the species of *Smilacina* Desf. to *Maianthemum* Wigg. (Liliaceae). Taxon 35: 584–589.

Lakela, O. 1941. *Sparganium glomeratum* in Minnesota. Rhodora 43: 83–85.

Lamont, E. E. 2006. *Eutrochium, Sclerolepis*. Pp. 474–478, 488–490, in: Flora of North America Editorial Committee, eds. 1993+. Flora of North America North of Mexico. 21+ vols. New York and Oxford. Vol. 21.

Landolt, E. 1980. Key to determination of taxa within the family of Lemnaceae. Veroff. Geobot. Inst. ETH Stiftung Rübel, Zurich 70: 13–21.

Landolt, E. 1981. Distribution pattern of the family Lemnaceae in North Carolina. Veroff. Geobot. Inst. ETH Stiftung Rübel, Zurich 77: 112–148.

Landolt, E. 1986. The family of Lemnaceae—a monographic study. (Vol. 1.) Veroff. Geobot. Inst. ETH Stiftung Rübel, Zurich 71: 1566.

Landolt, E., 2000. Lemnaceae. Pp. 143–153, in: Flora of North America Editorial Committee, eds. 1993+. Flora of North America North of Mexico. 21+ vols. New York and Oxford. Vol. 22.

Landolt, E., and R. Kandeler. 1987. The family of Lemnaceae—A monographic study. Vol. 2. Veröff. Geobot. Inst. E. T. H. Stiftung Rübel Zürich 95.

Lansdown, R. V. 2009. Nomenclatural Notes on *Callitriche* (Callitrichaceae) in North America. Novon 19: 364–369.

Lansdown, R. V. 2019. *Callitriche*. Pp. 49–54, in: Flora of North America Editorial Committee, eds. 1993+. Flora of North America North of Mexico. 21+ vols. New York and Oxford. Vol. 17.

Larridon, I., K. Bauters, M. Reynders, W. Huygh, and P. Goetghebeur. 2014. Taxonomic changes in C_4 *Cyperus* (Cypereae, Cyperoideae, Cyperaceae): Combining the sedge genera *Ascolepis*, *Kyllinga* and *Pycreus* into *Cyperus s.l.* Phytotaxa 166: 33–48.

La Rue, C. D. 1943. Regeneration in *Radicula aquatica*. Pap. Michigan Acad. Sci. 28: 51–61.

Laureto, P. J., and J. S. Pringle. 2010 [2011]. *Solidago vossii* (Asteraceae): A new species of Goldenrod from Northern Michigan. Michigan Bot. 49: 105–117.

LeBlond, R. J., E. E. Schilling, R. D. Porcher, B. A Sorrie, J. F. Townsend, P. D. McMillan, and A. S. Weakley. 2007. *Eupatorium paludicola* (Asteraceae): Anew species from the coastal plain of North and South Carolina. Rhodora 109: 137–144.

LeBlond, R. J., and A. S. Weakley. 2002. *Schizaea pusilla* in North Carolina. Rhodora 104: 86–91.

Lee, A. K., J. S. Gilman, M. Srivastav, A. D. Lerner, M. J. Donoghue, and W. L. Clement. 2021. Reconstructing Dipsacales phylogeny using Angiosperms 353: Issues and insights. Amer. J. Bot. 108: 1122–1142.

Lee, D. W. 1975. Population variation and introgression in North American *Typha*. Taxon. 24: 633–641.

Lellinger, D. B. 1985. A Field Manual of the Ferns and Fern-allies of the United States and Canada. Smithsonian Inst. Press, Washington, D.C.

Lelong, M. G. 1984. New combinations for *Panicum* subgenus *Panicum* and subgenus *Dichanthelium* (Poaceae) of the southeastern United States. Brittonia 36: 262–273.

Les, D. H. 1980. Contributions to the Biology and Taxonomy of *Ceratophyllum* in the Eastern United States. M.S. thesis, Eastern Michigan Univ., Ypsilanti, Mich.

Les, D. H. 1983. Taxonomic implications of aneuploidy and polyploidy in *Potamogeton* (Potamogetonaceae). Rhodora 85: 301–323.

Les, D. H. 1985. The taxonomic significance of plumule morphology in *Ceratophyllum* (Ceratophyllaceae). Syst. Bot. 10: 338–346.

Les, D. H. 1986a. The phytogeography of *Ceratophyllum demersum* and *C. echinatum* (Ceratophyllaceae) in glaciated North America. Canad. J. Bot. 64: 498–509.

Les, D. H. 1986b. The evolution of achene morphology in *Ceratophyllum* (Ceratophyllaceae). I. Fruit spine variation and relationships of *C. demersum*, *C. submersum*, and *C. apiculatum*. Syst. Bot. 11: 549–558.

Les, D. H. 1988a. The evolution of achene morphology in *Ceratophyllum* (Ceratophyllaceae). II. Fruit variation and systematics of the "spiny-margined" group. Syst. Bot. 13: 73–86.

Les, D. H. 1988b. The evolution of achene morphology in *Ceratophyllum* (Ceratophyllaceae). III. Relationships of the "facially-spined" group. Syst. Bot. 13: 509–518.

Les, D. H. 1988c. The origin and affinities of the Ceratophyllaceae. Taxon 37: 326–345.

Les, D. H. 1988d. Breeding systems, population structure, and evolution in hydrophilous angiosperms. Ann. Missouri Bot. Gard. 75: 819–835.

Les, D. H. 1989. The evolution of achene morphology in *Ceratophyllum* (Ceratophyllaceae). IV: Summary of proposed relationships and evolutionary trends. Syst. Bot. 14: 254–262.

Les, D. H. 1991. Genetic diversity in the monoecious hydrophile *Ceratophyllum* (Ceratophyllaceae) Amer. J. Bot. 78: 1070–1082.

Les, D. H. 1994. Molecular systematics and taxonomy of lake cress (*Neobeckia aquatica*; Brassicaceae), an imperiled aquatic mustard. Aquat. Bot. 49: 149–165.

Les, D. 1997. Ceratophyllaceae. Pp. 81–84, in: Flora of North America Editorial Committee, eds. 1993+. Flora of North America North of Mexico. 21+ vols. New York and Oxford. Vol. 3.

Les, D. H. 2020. Aquatic Monocotyledons of North America: Ecology, Life History, Systematics. CRC Press, Boca Raton, Fl.

Les, D. H., and R. S. Capers. 2019. *Glossostigma*. Pp. 369–370, in: Flora of North America Editorial Committee, eds. 1993+. Flora of North America North of Mexico. 21+ vols. New York and Oxford. Vol. 17.

Les, D. H., R. S. Capers, and N. P. Tippery. 2006. Introduction of *Glossostigma* (Phrymaceae) to North America: A taxonomic and ecological overview. Amer. J. Bot. 93: 927–939.

Les, D. H., M. H. Cleland, and C. T. Philbrick. 1995. Taxonomic realignments in the Potamogetonaceae: Evidence from molecular data. Abstract. Amer. J. Bot. 82: 144.

Les, D. H., M. A. Cleland, and M. Waycott. 1997b. Phylogenetic studies in Alismatidae, II: Evolution of marine angiosperms (seagrasses) and hydrophily. Syst. Bot. 22: 443–463.

Les, D. H., and R. R. Haynes. 1996. *Coleogeton* (Potamogetonaceae), a new genus of pondweeds. Novon 6: 389–391.

Les, D. H., S. W. L. Jacobs, N. P. Tippery, L. Chen., M. L. Moody, and M. Wilstermann-Hildebrand. 2008. Systematics of *Vallisneria* (Hydrocharitaceae). Syst. Bot. 23: 49–65.

Les, D. H., and L. J. Mehrhoff. 1999. Introduction of nonindigenous aquatic plants in southern New England: A historical perspective. Biological Invasions. 1: 281–300.

Les, D. H., L. J. Mehrhoff, M. A. Cleland, and J. D. Gabel. 1997. *Hydrilla verticillata* (Hydrocharitaceae) in Connecticut. J. Aquat. Pl. Manag. 35: 10–14.

Les, D. H., M. L. Moody, and C. L. Soros. 2006. A reappraisal of phylogenetic relationships in the monocotyledon family Hydrocharitaceae (Alismatidae). Aliso 22: 211–230.

Les, D. H., N. M. Murray, and N. P. Tippery. 2009. Systematics of two imperiled pondweeds (*Potamogeton vaseyi*, *P. gemmiparus*) and taxonomic ramifications for Subsection *Pusilli* (Potamogetonaceae). Syst. Bot. 34: 643–651.

Les, D. H., E. L. Peredo, U. M. King, L. K. Benoit, N. P. Tippery, C. J. Ball, and R. K. Shannon. 2015. Through thick and thin: Cryptic sympatric speciation in the submersed genus *Najas* (Hydrocharitaceae). Molec. Phylogen. and Evol. 82: 15–30.

Les, D. H., C. T. Philbrick, and R. A. Novelo. 1997a. The phylogenetic position of river-weeds (Podostemaceae): Insights from *rbc*L sequence data. Aquatic Bot. 57: 5–27.

Les, D. H., S. P. Sheldon, and N. P. Tippery. 2010. Hybridization in hydrophiles: Natural interspecific hybrids in *Najas* (Hydrocharitaceae). Syst. Bot. 35: 716–744.

Les, D. H., and D. J. Sheridan. 1990a. Hagstrom's concept of phylogenetic relationships in *Potamogeton* L. (Potamogetonaceae). Taxon 39: 41–58.

Les, D. H., and D. J. Sheridan. 1990b. Biochemical heterophily and flavonoid evolution in North American *Potamogeton*. Amer. J. Bot. 77: 458–465.

Les, D. H., and R. L. Stuckey. 1985. The introduction and spread of *Veronica beccabunga* (Scrophulariaceae) in eastern North America. Rhodora 87: 503–515.

Lewis, D. Q. 2002. Burmanniaceae. Pp. 486–488, in: Flora of North America Editorial Committee, eds. 1993+. Flora of North America North of Mexico. 21+ vols. New York and Oxford. Vol. 26.

Lewis, D. Q. 2019. *Lindernia*. 353–357, in: Flora of North America Editorial Committee, eds. 1993+. Flora of North America North of Mexico. 21+ vols. New York and Oxford. Vol. 17.

Lewis, W. H., B. Ertter, and A. Bruneau. 2014. *Rosa*. Pp. 75–119, in: Flora of North America Editorial Committee, eds. 1993+. Flora of North America North of Mexico. 21+ vols. New York and Oxford. Vol. 9.

Li, H. L. 1955. Classification and phylogeny of Nymphaeaceae and allied families. Amer. Midl. Nat. 54: 33–41.

Li, J., J. Alexander III, T. Ward, P. Del Tredici, and R. Nicholson. 2002. Phylogenetic relationships of Empetraceae inferred from sequences of chloroplast gen *matK* and nuclear ribosomal DNA ITS region. Molec. Phylogen. Evol. 25: 306–315.

Lichvar, R. W., N. C. Melvin, M. L. Butterwick, and W. N. Kirchner. 2012. National Wetland Plant List indicator rating definitions. U.S. Army Corps of Engineers, Engineer Research and Development Center, Cold Regions Research and Engineering Laboratory ERDC/CRREL TR-12-1. Vicksburg, Miss.

Lis, R. A. 2014. *Spiraea*. Pp. 398–411, in: Flora of North America Editorial Committee, eds. 1993+. Flora of North America North of Mexico. 21+ vols. New York and Oxford. Vol. 9.

Little, R. J., and L. E. McKinney. 2015. Violaceae. Pp. 106–164, in: Flora of North America Editorial Committee, eds. 1993+. Flora of North America North of Mexico. 21+ vols. New York and Oxford. Vol. 6.

Liu, S., K. E. Denford, J. E. Ebinger, J. G. Packer, and G. C. Tucker. 2009. *Kalmia*. Pp. 480–485, in: Flora of North America Editorial Committee, eds. 1993+. Flora of North America North of Mexico. 21+ vols. New York and Oxford. Vol. 8.

Lloyd, F. E. 1933. The structure and behavior of *Utricularia purpurea*. Canad. J. Res. 8: 234–252.

Lloyd, F. E. 1935. *Utricularia*. Biol. Review 10: 72–100.

Lloyd, F. E. 1942. The Carnivorous Plants. Chronica Botanica, Waltham, Mass.

Löve, Á., and D. Löve. 1957. Drug content and polyploidy in *Acorus*. Proc. Genet. Soc. Canad. 2: 14–17.

Löve, Á., and D. Löve. 1958. Biosystematics of *Triglochin maritimum* agg. Naturaliste Canad. 85: 156–165.

Löve, D. 1960. The red-fruited Crowberries in North America. Rhodora 62: 265–292.

Löve, D., and H. Lieth. 1961. *Triglochin gaspense*, a new species of Arrow Grass. Canad. J. Bot. 39: 1261–1272.

Lowden, R. M. 1973. Revision of the genus *Pontederia* L. Rhodora 75: 426–487.

Lowden, R. M. 1978. Studies on the submerged genus *Ceratophyllum* L. in the neotropics. Aquatic Bot. 4: 127–142.

Lowden, R. M. 1982. An approach to the taxonomy of *Vallisneria* (Hydrocharitaceae). Aquatic Bot. 13: 269–298.

Lowden, R. M. 1992. Floral variation and taxonomy of *Limnobium* L. C. Richard (Hydrocharitaceae). Rhodora 94: 111–134.

Lu, Y., J-H. Ran, D-M. Guo, Z-Y. Yang, and X-Q. Wang. 2014. Phylogeny and Divergence Times of Gymnosperms inferred from single-copy nuclear genes. PLoS ONE 9(9): e107679.

Luer, C. A. 1975. The Native Orchids of the United States and Canada Excluding Florida. New York Botanical Garden, Bronx, N.Y.

Lui, K., F. L. Thompson, and C. G. Eckert. 2005. Causes and consequences of extreme variation in reproductive strategy and vegetative growth among invasive populations of a clonal aquatic plant, *Butomus umbellatus* L. (Butomaceae). J. Botany 79: 294–299.

Lumpkin, T. A. 1993. Azollaceae. Pp. 338–342, in: Flora of North America Editorial Committee, eds. 1993+. Flora of North America North of Mexico. 21+ vols. New York and Oxford. Vol. 2.

Lundell, C. L. 1961–1969. Flora of Texas. 3 vols. Texas Research Foundation, Renner, Tex.

Luteyn, J. L. 1976. Revision of *Limonium* (Plumbaginaceae) in eastern North America. Brittonia 28: 303–317.

Mackenzie, K. K. 1931–1935. Cyperaceae. North American Flora 18: 1–478.

Mackenzie, K. K. 1940. North American Cariceae. 2 vols. New York Botanical Garden, Bronx, N.Y.

Magrath, L. K., and R. A. Coleman, 2002. *Listera*. Pp. 586–592, in: Flora of North America Editorial Committee, eds. 1993+. Flora of North America North of Mexico. 21+ vols. New York and Oxford. Vol. 26.

Malme, G. O. K. 1937. Xyridaceae. North American Flora 19: 3–15.

Marie-Victorin, F. 1931. L'Anacharis canadensis. Histoire et solution d'un imbroglio taxonomique. Contr. Lab. Bot. Univ. Montréal 18: 1–43.

Marie-Victorin, F. 1943. Les Vallisneries americaines. Contr. Inst. Bot. Univ. Montréal 46: 1–38.

Marr, K. L., R. J. Hebda, and C. W. Greene. 2007. *Calamagrostis*. Pp. 707–732, in: Flora of North America Editorial Committee, eds. 1993+. Flora of North America North of Mexico. 21+ vols. New York and Oxford. Vol. 24.

Mason, C. T., and H. H. Iltis. 1965. Preliminary reports on the flora of Wisconsin. No. 53. Gentianaceae and Menyanthaceae—Gentian and Buckbean families. Trans. Wisconsin Acad. Sci., Arts and Letters 54: 295–329.

Mason, H. L. 1957. A Flora of the Marshes of California. Univ. of California Press, Berkeley, Calif.

Mastrogiuseppe, J. 2002. *Dulichium*. Pp. 198, in: Flora of North America Editorial Committee, eds. 1993+. Flora of North America North of Mexico. 21+ vols. New York and Oxford. Vol. 23.

Mastrogiuseppe, J., P. E. Rothrock, A. C. Dibble, and A. A. Reznicek. 2002. *Carex* sect. *Ovales*. Pp. 332–378 in: Flora of North America Editorial Committee, eds. 1993+. Flora of North America North of Mexico. 21+ vols. New York and Oxford. Vol. 23.

Mathias, M. E., and L. Constance. 1944–1945. Umbelliferae. North American Flora 28B: 43–295.

McAlpine, D. F., G. Bishop, O. Ceska, M. L. Moody, and A. Ceska. 2007. Andean watermilfoil, *Myriophyllum quitense* (Haloragaceae), in the Saint John River Estuary System, New Brunswick, Canada. Rhodora 109: 101–107.

McClure, E. A. 1973. Genera of Bamboos native to the New World (Gramineae: Bambusoideae). Smithsonian Contr. Bot. No. 9.

McCully, M. E., and H. M. Dale. 1961a. Heterophylly in *Hippuris*, a problem in identification. Canad. J. Bot. 39: 1099–1116.

McCully, M. E., and H. M. Dale. 1961b. Variations in leaf numbers in *Hippuris*. Canad. J. Bot. 39: 611–625.

McDaniel, S. 1971. The genus *Sarracenia* (Sarraceniaceae). Bull. Tall Timbers Res. Sta. No. 9.

McDonnell, M. J., and G. E. Crow. 1979. The typification and taxonomic status of *Spartina caespitosa* A. A. Eaton. Rhodora 81: 123–129.

McNair, J. B. 1925. The taxonomy of Poison Ivy with a note on the origin of the generic name. Publ. Field Mus. Nat. Hist., Bot. Ser., Vol. 4, No. 3.

McNeill, J., I. J. Bassett, and C. W. Crompton. 1977. *Suaeda calceoliformis*, the correct name for *Suaeda depressa* Auct. Rhodora 79: 133–138.

McPherson, S., and D. Schnell. 2012. Carnivorous plants of the United States and Canada. Natural History Productions, Poole, Dorset, England.

McVaugh, R. 1936. Studies in the taxonomy and distribution of the eastern North American species of *Lobelia*. Rhodora 38: 241–263, 276–298, 305–329, 346–362.

Mehrhoff, L. J. 1983. *Sclerolepis uniflora* (Compositae) in New England. Rhodora 85: 433–438.

Meijer, W. 1976. A note of *Podostemum ceratophyllum* Michx. as an indicator of clean streams in and around the Appalachian Mountains. Castanea 41: 319–324.

Mellichamp, T. L. 2015. Droseraceae. Pp. 418–425, in: Flora of North America Editorial Committee, eds. 1993+. Flora of North America North of Mexico. 21+ vols. New York and Oxford. Vol. 6.

Mellichamp, T. L., and F. W. Case. 2009. Sarraceniaceae. Pp. 350–363, in: Flora of North America Editorial Committee, eds. 1993+. Flora of North America North of Mexico. 21+ vols. New York and Oxford. Vol. 8.

Meriläinen, J. 1968. *Najas minor* All. in North America. Rhodora 70: 161–175.

Mertens, T. R. 1965. Taxonomy of *Polygonum* section *Polygonum (Avicularia)* in North America. Madroño 18: 85–92.

Michael, P. W. 2003. *Echinochloa*. Pp. 390–403, in: Flora of North America Editorial Committee, eds. 1993+. Flora of North America North of Mexico. 21+ vols. New York and Oxford. Vol. 25.

Mickel, J. T. 1979. How to Know the Ferns and Fern Allies. Wm. C. Brown Co., Dubuque, Iowa.

Miller, M., and E. O. Beal. 1972. *Scirpus validus* and *S. acutus*—a question of distinctness. J. Minnesota Acad. Sci. 38: 21–23.

Mitchell, D. S., and P. A. Thomas. 1972. Ecology of waterweeds in the neotropics, and ecological survey of the aquatic weeds *Eichhornia crassipes* and *Salvinia* species and their natural enemies in the neotropics. Tech. Pap. Hydrology 12: 1–50.

Mitchell, R. S. 1968. Variation in the *Polygonum amphibium* complex and its taxonomic significance. Univ. California Publ. Bot. 45: 1–65.

Mitchell, R. S. 1971. Comparative leaf structure of aquatic *Polygonum* species. Amer. J. Bot. 58: 342–360.

Mitchell, R. S. 1976. Submergence experiments on nine species of semi-aquatic *Polygonum*. Amer. J. Bot. 63: 1158–1165.

Mitchell, R. S. 1978. *Rumex maritimus* L. versus *R. persicarioides* L. (Polygonaceae) in the Western Hemisphere. Brittonia 30: 293–296.

Mitchell, R. S., and E. O. Beal. 1979. Magnoliaceae through Ceratophyllaceae of New York State. Contr. to a Flora of New York State II. New York State Mus. Bull. No. 435.

Mitchell, R. S., and J. K. Dean. 1978. Polygonaceae (Buckwheat Family) of New York State. Contr. to a Flora of New York State I. New York State Mus. Bull. No. 431.

Mitchell, R. S., and J. K. Dean. 1982. Ranunculaceae (Crowfoot Family) of New York State. Contr. to a Flora of New York State IV. New York State Mus. Bull. No. 446.

Mitchell, R. S., and C. J. Sheviak. 1982. Rare Plants of New York State. New York State Mus. Bull. No. 445.

Mobberly, D. G. 1956. The taxonomy and distribution of the genus *Spartina*. Iowa State J. Sci. 30: 471–574.

Moldenke, H. N. 1937. Eriocaulaceae. North American Flora 19: 1750.

Monachino, J. 1945. *Jussiaea uruguayensis* in Staten Island, New York. Rhodora 47: 237–239.

Monson, P. H. 1960. Variation in *Nymphaea*, the White Waterlily, in the Itasca State Park region. Proc. Minnesota Acad. Sci. 25–26: 26–39.

Montgomery, J. D., and W. H. Wagner, Jr. 1993. *Dryopteris*. Pp. 280–288, in: Flora of North America Editorial Committee, eds. 1993+. Flora of North America North of Mexico. 21+ vols. New York and Oxford. Vol. 2.

Moody, M. L., and D. H. Les. 2007. Phylogenetic systematics and character evolution in the angiosperm family Haloragaceae. Amer. J. Bot. 94: 2005–2025.

Moody, M. L., and D. H. Les. 2010. Systematics of the aquatic angiosperm genus *Myriophyllum* (Haloragaceae). Syst. Bot. 35: 121–139.

Moore, A. W. 1969. *Azolla*: Biology and agronomic significance. Bot. Review (London) 35: 17–34.

Moore, D. M., J. B. Harborne, and C. A. Williams. 1970. Chemotaxonomy, variation and geographical distribution of the Empetraceae. J. Linn. Soc. Bot. 63: 277–293.

Moore, E. 1915. The Potamogetons in relation to pond culture. Bull. Bur. Fisheries 33: 251–291.

Moran, R. V. 2009. *Crassula*. Pp. 150–155, in: Flora of North America Editorial Committee, eds. 1993+. Flora of North America North of Mexico. 21+ vols. New York and Oxford. Vol. 8.

Morris, F., and E. E. Eames. 1929. Our Wild Orchids. Charles Scribner's Sons, New York.

Moseley, M. F., E. L. Schneider, and P. S. Williamson. 1993. Phylogenetic interpretations from selected floral vasculature characters in the Nymphaeaceae sensu lato. Aquatic Bot. 44: 325–342.

Moseley, M. F., and N. W. Uhl. 1985. Morphological studies of the Nymphaeaceae sensu lato. XV. The anatomy of the flower of *Nelumbo*. Bot. Jahrb. Syst. 106: 61–98.

Moseley, M. F., P. S. Williamson, and H. Kosakai. 1984. Morphological studies of the Nymphaeaceae sensu lato. XIII. Contributions to the vegetative and floral structure of *Cabomba*. Amer. J. Bot. 71: 902–924.

Mosyakin, S. L. 2005. *Rumex*. Pp. 489–533, in: Flora of North America Editorial Committee, eds. 1993+. Flora of North America North of Mexico. 21+ vols. New York and Oxford. Vol. 5.

Mosyakin, S. L., and K. R. Robertson. 2003. *Amaranthus*. Pp. 410–435, in: Flora of North America Editorial Committee, eds. 1993+. Flora of North America North of Mexico. 21+ vols. New York and Oxford. Vol. 4.

Muasya, A. M., and D. A. Simpson. 2002. A monograph of the genus *Isolepis* R. Br. (Cyperaceae). Kew Bull. 57: 257–362.

Muenscher, W. C. 1934. Aquatic vegetation of the Mohawk watershed. Pp. 228–249, in: A Biological Survey of the Mohawk-Hudson Watershed. New York State Dept. Conserv., Suppl. Ann. Rep. 1934, Biol. Surv. No. 9.

Muenscher, W. C. 1936. The germination of seeds of *Potamogeton*. Ann. Bot. (London) 50: 805–821.

Muenscher, W. C. 1940. Fruits and seedlings of *Ceratophyllum*. Amer. J. Bot. 27: 231–233.

Muenscher, W. C. 1944. Aquatic Plants of the United States. Comstock Publ. Co., Ithaca, N.Y.

Mulligan, G. A., and J. T. Calder. 1964. The genus *Subularia* (Cruciferae). Rhodora 66: 127–135.

Mulligan, G. A., and D. B. Munro. 1989. Taxonomy of species of North American *Stachys* (Labiatae) found north of Mexico. Naturaliste Canadien 116: 35–51.

Munz, P. A. 1942. Studies in Onagraceae. XII. A revision of the New World species of *Jussiaea*. Darwiniana 4: 179–284.

Munz, P. A. 1944. Studies in Onagraceae. XIII. The American species of *Ludwigia*. Bull. Torrey Bot. Club 71: 152–165.

Murray, D. F. 2002. *Carex* sect. *Racemosae*, *Carex* sect. *Capituligerae*. Pp. 401–414, 569–570 in: Flora of North America Editorial Committee, eds. 1993+. Flora of North America North of Mexico. 21+ vols. New York and Oxford. Vol. 23.

Murray, D. F., V. Mirré, and R. Elven. 2009. *Empetrum*. Pp. 486–489, in: Flora of North America Editorial Committee, eds. 1993+. Flora of North America North of Mexico. 21+ vols. New York and Oxford. Vol. 8.

Murrell, Z. E., and D. B. Poindexter. 2016. *Cornus*. Pp. 443–457, in: Flora of North America Editorial Committee, eds. 1993+. Flora of North America North of Mexico. 21+ vols. New York and Oxford. Vol. 12.

Naczi, R. F. C., R. J. Driskill, E. L. Pennell, N. E. Seyfried, A. O. Tucker, and N. H. Dill. 1986. New records of some rare DelMarVa sedges. Bartonia 52: 49–57.

Nauman, C. E. 1993. Salviniaceae. Pp. 336–337, in: Flora of North America Editorial Committee, eds. 1993+. Flora of North America North of Mexico. 21+ vols. New York and Oxford. Vol. 2.

Nelson, A. D. 2019. *Chelone*. Pp. 57–60 in: Flora of North America Editorial Committee, eds. 1993+. Flora of North America North of Mexico. 21+ vols. New York and Oxford. Vol. 17.

Nelson, E. N., and R. W. Couch. 1985a. History of the introduction and distribution of *Myriophyllum aquaticum* in North America. Proc. 1st Internat. Symp. on Watermilfoil (*Myriophyllum spicatum*) and Related Haloragaceae Species 1: 19–26.

Nelson, E. N., and R. W. Couch. 1985b. Aquatic Plants of Oklahoma. I. Submersed, Floating-leaved, and Selected Emergent Macrophytes. Oral Roberts Univ., Tulsa, Okla.

Nelson, J. B. 1981. *Stachys* (Labiatae) in southeastern United States. Sida 9: 104–123.

Nelson, J. B., and J. E. Fairey III. 1979. Misapplication of the name *Stachys nuttallii* (Lamiaceae) to a new southeastern species. Brittonia 31: 491–494.

Nesom, G. L. 2006. *Pluchea*, *Fleishmannia*. Pp. 478–484, 540–541, in: Flora of North America Editorial Committee, eds. 1993+. Flora of North America North of Mexico. 21+ vols. New York and Oxford. Vol. 21.

Nesom, G. L. 2014. Taxonomy of *Erythranthe* sect. *Erythranthe* (Phrymaceae). Phytoneuron 31: 1–41.

Nesom, G. L. 2019. *Mimulus*. Pp. 366–369, in: Flora of North America Editorial Committee, eds. 1993+. Flora of North America North of Mexico. 21+ vols. New York and Oxford. Vol. 17.

Nesom, G. L., and N. S. Fraga. 2019. *Erythranthe*. Pp. 372–425, in: Flora of North America Editorial Committee, eds. 1993+. Flora of North America North of Mexico. 21+ vols. New York and Oxford. Vol. 17.

Nesom, G. L., and J. O. Sawyer. 2016. *Rhamnus*. Pp. 45–52, in: Flora of North America Editorial Committee, eds. 1993+. Flora of North America North of Mexico. 21+ vols. New York and Oxford. Vol. 12.

Netherland, M. D. 1997. Turion ecology of *Hydrilla*. J. Aquat. Plant Manage. 35: 1–10.

Nichols, W. F., A. P. Smagula, A. Haines, and D. McGrady. 2022. *Hottonia palustris* (Primulaceae): A newly documented non-native aquatic plant species in New Hampshire, U.S.A. Rhodora 123: 221–228.

Nienaber, M. A. 2000. Scheuchzeriaceae. Pp. 41–42, in: Flora of North America Editorial Committee, eds. 1993+. Flora of North America North of Mexico. 21+ vols. New York and Oxford. Vol. 22.

Nixon, K. C. 1997. Fagaceae. Pp. 436–506, in: Flora of North America Editorial Committee, eds. 1993+. Flora of North America North of Mexico. 21+ vols. New York and Oxford. Vol. 3.

Oeder, G. C. 1767. Flora Danica. Vol. 2, Fasc. 6, Plate 337. Botanic Garden in Copenhagen, Copenhagen, Denmark.

Ogden, E. C. 1943. The broadleaved species of *Potamogeton* of North America north of Mexico. Rhodora 45: 57–105, 119–163, 171–214.

Ogden, E. C. 1981. Field Guide to Northeastern Ferns. New York State Mus. Bull. No. 444.

Ørgaard, M. 1991. The genus *Cabomba* (Cabombaceae)—A taxonomic study. Nordic J. Bot. 11: 179–203.

Osborn, J. M., and E. L. Schneider. 1988. Morphological studies of the Nymphaeaceae sensu lato. XVI. The floral biology of *Brasenia schreberi*. Ann. Missouri Bot. Gard. 75: 778–794.

Pace, M. C., and Cameron, K. M. 2017. The systematics of the *Spiranthes cernua* species complex (Orchidaceae): Untangling the Gordian Knot. Systematic Botany, 42: 1–30.

Packer, J. G. 2002. *Tofieldia* and *Triantha*. Pp. 60–64, in: Flora of North America Editorial Committee, eds. 1993+. Flora of North America North of Mexico. 21+ vols. New York and Oxford. Vol. 26.

Padgett, D. J. 1997. A biosystematic monograph of the genus *Nuphar* Sm. (Nymphaeaceae). Ph.D. diss., Univ. of New Hampshire, Durham, N.H.

Padgett, D. L. 2007. A monograph of *Nuphar* (Nymphaeaceae). Rhodora 109: 1–95.

Padgett, D. L., J. J. Carboni, and D. J. Schepis. 2010. The dietary composition of *Chrysemys picta picta* (eastern painted turtles) with special reference to the seeds of aquatic macrophytes. Northeastern Nat. 17: 305–312.

Padgett, D. J., and G. E. Crow. 1993a. Some unwelcome additions to the flora of New Hampshire. Rhodora 95: 348–351.

Padgett, D. J., and G. E. Crow. 1993b. A comparison of floristic composition and species richness within and between created and natural wetlands of southeastern New Hampshire. Pp. 171–186, in: F. J. Webb, Jr., ed. Proceedings of the Twentieth Annual Conference on Wetlands and Restoration and Creation. Hillsborough Community College. Tampa, Fla.

Padgett, D. J., and G. E. Crow. 1994a. Foreign plant stock: Concerns for wetland mitigation. Rest. Managem. Notes 12: 168–171.

Padgett, D. J., and G. E. Crow. 1994b. A vegetation and floristic analysis of a created wetland in southeastern New Hampshire. Rhodora 96: 1–29.

Padgett, D. J., D. H. Les, and G. E. Crow. 1998. Evidence for the hybrid origin of *Nuphar* ×*rubrodisca* (Nymphaeaceae). Amer. J. Bot. 85: 1468–1476.

Padgett, D. J., D. H. Les, and G. E. Crow. 1999. Phylogenetic relationships in *Nuphar* (Nymphaeaceae): Evidence from morphology, chloroplast DNA, and nuclear ribosomal DNA. Amer. J. Bot. 86: 1316–1324.

Padgett, D. J., J. E. Mendell, and C. R. Carbonell. 2021. An established *Nuphar advena* (Nymphaeaceae) in Massachusetts, U.S.A. Rhodora 123: 360–362.

Palmer, D. D. 2018. Michigan Ferns and Lycophytes: A Guide to Species of the Great Lakes Region. Univ. of Michigan Press, Ann Arbor, Mich.

Pankhurst, R. J. 2014. *Aronia*. Pp. 445–446, in: Flora of North America Editorial Committee, eds. 1993+. Flora of North America North of Mexico. 21+ vols. New York and Oxford. Vol. 9.

Parfitt, B. D. 1997. *Trollius*. Pp. 189–190, in: Flora of North America Editorial Committee, eds. 1993+. Flora of North America North of Mexico. 21+ vols. New York and Oxford. Vol. 3.

Park, C. W. 1988. Taxonomy of *Polygonum* section *Echinocaulon*. Mem. New York Bot. Gard. 47: 1–82.

Park, M. M., and D. Festerling, Jr. 1997. *Thalictrum*. Pp. 258–271, in: Flora of North America Editorial Committee, eds. 1993+. Flora of North America North of Mexico. 21+ vols. New York and Oxford. Vol. 3.

Paton, A. 1990. A global taxonomic investigation of *Scutellaria* (Labiatae). Kew Bull. 45: 399–450.

Patterson, T. F., and G. L. Nesom. 2006. *Conoclinium*. Pp. 478–480, in: Flora of North America Editorial Committee, eds. 1993+. Flora of North America North of Mexico. 21+ vols. New York and Oxford. Vol. 21.

Peattie, D. C. 1928. The celestial lotus: Sacred to the Oriental; enchanting to the Occidental. Nature Mag. 11: 294–297.

Penfound, W. T. 1940a. The biology of *Achyranthes philoxeroides* (Mart.) Standley. Amer. Midl. Nat. 24: 248–252.

Penfound, W. T. 1940b. The biology of *Dianthera americana* L. Amer. Midl. Nat. 24: 242–247.

Penfound, W. T., and T. T. Earle. 1948. The biology of the Water Hyacinth. Ecol. Monogr. 18: 447–472.

Peng, C.-I. 1988. The biosystematics of *Ludwigia* sect. *Microcarpium* (Onagraceae). Ann. Missouri Bot. Gard. 75: 970–1009.

Peng, C.-I. 1989. The systematics and evolution of *Ludwigia* sect. *Microcarpium* (Onagraceae). Ann. Missouri Bot. Gard. 76: 221–302.

Pennell, F. W. 1935. The Scrophulariaceae of eastern temperate North America. Acad. Nat. Sci. Philadelphia Monogr. 1: 1–650.

Peterson, P. M. 2003. *Eragrostis, Muhlenbergia*. Pp. 65–105, 145–200, in: Flora of North America Editorial Committee, eds. 1993+. Flora of North America North of Mexico. 21+ vols. New York and Oxford. Vol. 25.

Pfeiffer, N. 1922. Monograph on the Isoetaceae. Trans. Sci. St. Louis 9: 79–223.

Philbrick, C. T. 1981. Some notes regarding pollination in a New Hampshire population of *Podostemum ceratophyllum* Michx. (Podostemaceae). Rhodora 83: 319–321.

Philbrick, C. T. 1982. New locations for *Podostemum ceratophyllum* Michx. (Podostemaceae) in New Hampshire and Maine, with some comments on a unique floral form. Rhodora 84: 301–303.

Philbrick, C. T. 1983. Aspects of floral biology in three species of *Potamogeton* (Pondweeds). Michigan Bot. 23: 35–38.

Philbrick, C. T. 1984. Pollen tube growth within vegetative tissues of *Callitriche* (Callitrichaceae). Amer. J. Bot. 71: 882–886.

Philbrick, C. T. 1988. Evolution of underwater outcrossing from aerial pollination systems: A hypothesis. Ann. Missouri Bot. Gard. 75: 836–841.

Philbrick, C. T. 1989. Systematic studies in North American Callitrichaceae. Ph.D. diss., Univ. of Connecticut, Storrs, Conn.

Philbrick, C. T., R. A. Aakjar, Jr., and R. L. Stuckey. 1998. Invasion and spread of *Callitriche stagnalis* (Callitrichaceae) in North America. Rhodora 100: 25–38.

Philbrick, C. T., and G. J. Anderson. 1992. Pollination biology in the Callitrichaceae. Syst. Bot. 17: 282–292.

Philbrick, C. T., and A. L. Bogle. 1988. A survey of floral variation in five populations of *Podostemum ceratophyllum* Michx. (Podostemaceae). Rhodora 90: 113–121.

Philbrick, C. T., and G. E. Crow. 1983. Distribution of *Podostemum ceratophyllum* Michx. Rhodora 85: 325–341.

Philbrick, C. T., and G. E. Crow. 1992. Isozyme variation and population structure of *Podostemum ceratophyllum* Michx. (Podostemaceae) north of the glacial boundary. Aquat. Bot. 43: 311–325.

Philbrick, C. T., and D. H. Les. 1996. Evolution of aquatic angiosperm reproductive systems. Bioscience 46: 813–826.

Philbrick, C. T., and A. Novelo R. 2004. Monograph of *Podostemum* (Podostmaceae). Syst. Bot. Monogr. 70: 1–106.

Phipps, J. B. 2014. Rosaceae. Pp. 18–20, in: Flora of North America Editorial Committee, eds. 1993+. Flora of North America North of Mexico. 21+ vols. New York and Oxford. Vol. 9.

Pike, R. B. 1970. Evidence for the hybrid status of *Aster* ×*blakei* (Porter) House. Rhodora 72: 401–436.

Plachno, B. J., and L. Adamec. 2007. Differentiation of *Utricularia ochroleuca* and *U. stygia* populations in Třeboň Basin, Czech Republic, on the basis of quadrifid glands. Carniv. Pl. Newslett. 36: 87–95.

Pohl, R. W. 1980. Flora Costaricensis. Family no. 15, Gramineae. Fieldiana, Bot. Ser., n.s., 4: 1–608.

Posto, A. L., and L. A. Prather. 2003. The evolutionary and taxonomic implications of RAPD data on the genetic relationships of *Mimulus michiganensis* (comb. et stat. nov.: Scrophulariaceae). Syst. Bot. 28: 172–178.

Prankerd, T. L. 1911. On the structure and biology of the genus *Hottonia*. Ann. Bot. 25: 253–267.

Preston, C. D., and J. M. Coft. 1997. Aquatic Plants in Britain and Ireland. Illustrated by G. M. S. Easy. The Environment Agency, The Institute of Terrestrial Ecology and The Joint Nature Conservation Committee. Brill, Leiden and Boston.

Pridgeon, A. M., P. Cribb, M. W. Chase, and F. N. Rasmussen, eds. 2005. Genera Orchidacearum. Volume 4. Epidendroideae (Part 1). Oxford Univ. Press, Oxford.

Pringle, J. S. 1967. Taxonomy of *Gentiana* section *Pneumonanthae*, in eastern North America. Brittonia 19: 1–32.

Pteridophyte Phylogeny Group. 2016. A community-derived classification for extant lycophytes and ferns. J. Syst. and Evol. 54: 563–603. Also available at: https://onlinelibrary.wiley.com/doi/epdf/10.1111/jse.12229.

Puff, C. 1976. The *Galium trifidum* group (*Galium* sect. *Aparinoides*, Rubiaceae). Canad. J. Bot. 54: 1911–1925.

Puff, C. 1977. The *Galium obtusum* group (*Galium* sect. *Aparinoides*, Rubiaceae). Bull. Torrey Bot. Club 104: 202–208.

Pyrah, G. L. 1969. Taxonomic and distributional studies in *Leersia* (Gramineae). Iowa State J. Sci. 44: 215–270.

Pyrah, G. L. 2007. *Leersia*. Pp. 42–45, in: Flora of North America Editorial Committee, eds. 1993+. Flora of North America North of Mexico. 21+ vols. New York and Oxford. Vol. 24.

Ramamoorthy, T. P., and E. M. Zardini. 1987. The systematics and evolution of *Ludwigia* sect. *Myrtocarpus* sensu lato (Onagraceae). Missouri Bot. Gard. Monogr. Syst. Bot. 19: 1–120.

Raven, P. H. 1963. The Old World species of *Ludwigia* (including *Jussiaea*), with a synopsis of the genus (Onagraceae). Reinwardtia 6: 327–427.

Ray, J. D. 1956. The genus *Lysimachia* in the New World. Illinois Biol. Monogr. 24: 1–160.

Raymond, M. 1951. Two new *Eriophorum* hybrids from northeastern North America. Sv. Bot. Tidsk. 15: 523–531.

Raymond, M. 1958. Additional notes on some S.E. Asiatic *Scirpus*. Nat. Canad. 86: 225–242.

Razifard, H., A. J. Rosman, G. C. Tucker, and D. H. Les. 2017. Systematics of the cosmopolitan aquatic genus *Elatine*. Syst. Bot. 42: 73–86.

Razifard, H., G. C. Tucker, and D. H. Les. 2016. *Elatine*. Pp. 349–353, in: Flora of North America Editorial Committee, eds. 1993+. Flora of North America North of Mexico. 21+ vols. New York and Oxford. Vol. 12.

Rechinger, K. H., Jr. 1937. The North American species of *Rumex*. Field Mus. Nat. Hist., Bot. Ser. 17: 1–151.

Reed, C. F. 1965. *Isoetes* in southeastern United States. Phytologia 12: 369–400.

Reed, C. F. 1970. Selected Weeds of the United States. U.S.D.A. Agric. Res. Ser., Agric. Handb. 336.

Reed, P. B., Jr. 1988. National List of Plant Species That Occur in Wetlands: Northeast (Region 1). U.S. Fish Wildl. Serv. Biol. Rep. 88(26.1).

Reeder, J. R. 1953. Affinities of the grass genus *Beckmannia* Host. Bull. Torrey Bot. Club 80: 187–196.

Reinert, G. W., and R. K. Godfrey. 1962. Reappraisal of *Utricularia inflata* and *U. radiata* (Lentibulariaceae). Amer. J. Bot. 49: 213–220.

Reinking, M. 1981. *Juncus×stuckeyi* (Juncaceae), a natural hybrid from northern Ohio. Brittonia 33: 170–178.

Reveal, J. L. 1970. *Sparganum simplex* Huds., a superfluous name. Taxon 19: 796–797.

Reveal, J. L. 1990. The neotypification of *Lemna minuta* Humb., Bonpl. & Kunth, an earlier name for *Lemna minuscula* Herter (Lemnaceae). Taxon 39: 328–330.

Reznicek, A. A. 1985. *Triadenum virginicum* (L.) Raf. (Marsh St. John's-wort) in Ontario. Plant Press 3: 124–125.

Reznicek, A. A. 1990. Evolution in sedges (*Carex*, Cyperaceae). Canad. J. Bot. 68: 1409–1432.

Reznicek, A. A. 2002. *Carex* sect. *Stellulatae*, *Carex* sect. *Lupulinae*, *Carex* sect. *Rostrales*. Pp. 326–331, 511–514, 514–517, in: Flora of North America Editorial Committee, eds. 1993+. Flora of North America North of Mexico. 21+ vols. New York and Oxford. Vol. 23.

Reznicek, A. A., and P. W. Ball. 1974. The taxonomy of *Carex* series *Lupulinae* in Canada. Canad. J. Bot. 52: 2387–2399.

Reznicek, A. A., and P. W. Ball. 1980. The taxonomy of *Carex* section *Stellulate* in North America north of Mexico. Contr. Univ. Michigan Herb. 14: 153–203.

Reznicek, A. A., and R. S. W. Bobbette. 1976. The taxonomy of *Potamogeton* subsection *Hybridi* in North America. Rhodora 78: 650–673.

Reznicek, A. A., and P. M. Catling. 1986. *Carex striata*, the correct name for *C. walteriana* (Cyperaceae). Rhodora 88: 405–406.

Reznicek, A. A., and P. M. Catling. 2002. *Carex* sect. *Chordorrhizae*, *Carex* sect. *Holarrhenae*, *Carex* sect. *Davisae*, *Carex* sect. *Paludosae*, *Carex* sect. *Carex*. Pp. 298–299, 301–302, 302–306, 491–498, 498–501, in: Flora of North America Editorial Committee, eds. 1993+. Flora of North America North of Mexico. 21+ vols. New York and Oxford. Vol. 23.

Reznicek, A. A., J. E. Fairey III, and A. T. Whittemore. 2002. *Scleria*. Pp. 242–251, in: Flora of North America Editorial Committee, eds. 1993+. Flora of North America North of Mexico. 21+ vols. New York and Oxford. Vol. 23.

Reznicek, A. A., and B. A. Ford. 2002. *Carex* sect. *Vesicariae*. Pp. 501–511, in: Flora of North America Editorial Committee, eds. 1993+. Flora of North America North of Mexico. 21+ vols. New York and Oxford. Vol. 23.

Reznicek, A. A., and R. E. Whiting. 1976. *Bartonia* (Gentianaceae) in Ontario. Canad. Field-Nat. 90: 67–69.

Rhoades, R. W. 1962. The aquatic form of *Alisma subcordatum* Raf. Rhodora 64: 227–229.

Richardson, F. D. 1980. Ecology of *Ruppia maritima* L. in New Hampshire (U.S.A.) tidal marshes. Rhodora 82: 403–439.

Richardson, F. D. 1983. Variation, adaptation and reproductive biology in *Ruppia maritima* L. populations from New Hampshire coastal and estuarine tidal marshes. Ph.D. diss., Univ. of New Hampshire, Durham, N.H.

Richardson, L. E. 1971. *Peucedanum palustre* again in Massachusetts. Rhodora 73: 460–461.

Rickett, H. W. 1944. *Cornus stolonifera* and *Cornus occidentalis*. Brittonia 5: 149–159.

Ritter, N. P., and G. E. Crow. 1998. *Myriophyllum quitense* Kunth (Haloragaceae) in Bolivia: A terrestrial growth-form with bisexual flowers. Aquatic Bot. 60: 389–395.

Roberts, M. L. 1985. The cytology, biology and systematics of *Megalodonta beckii* (Compositae). Aquatic Bot. 21: 99–110.

Roberts, M. L. 1972. *Butomus umbellatus* in the Mississippi watershed. Castanea 37: 83–85.

Roberts, M. L. 1982. The systematics of North American *Bidens* section *Bidens*. Ph.D. diss., Ohio State Univ., Columbus, Ohio.

Roberts, M. L., R. L. Stuckey, and R. S. Mitchell. 1981. *Hydrocharis morsus-ranae* (Hydrocharitaceae): New to the United States. Rhodora 83: 147–148.

Robertson, K. R. 1974. The genera of Rosaceae in the southeastern United States. J. Arnold Arbor. 55: 303–332, 344–401, 611–662.

Robertson, K. R. 2002. Haemodoraceae, *Lophiola*. Pp. 47–49, 48–49, in: Flora of North America Editorial Committee, eds. 1993+. Flora of North America North of Mexico. 21+ vols. New York and Oxford. Vol. 26.

Robertson, K. R., and S. E. Clemants. 2003. Amaranthaceae. Pp. 405–456, in: Flora of North America Editorial Committee, eds. 1993+. Flora of North America North of Mexico. 21+ vols. New York and Oxford. Vol. 4.

Robertson, K. R., J. B. Phipps, J. R. Rohrer, and P. G. Smith. 1991. A synopsis of genera in Maloideae (Rosaceae). Syst. Bot. 16: 376–394.

Robson, N. K. B. 2015. Hypericaceae. Pp. 71–105, in: Flora of North America Editorial Committee, eds. 1993+. Flora of

North America North of Mexico. 21+ vols. New York and Oxford. Vol. 6.

Robuck, O. W. 1985. The Common Plants of the Muskegs of Southeast Alaska. U.S. Forest Serv., Pacific Northwest Forest and Range Exp. Sta. Misc. Publ.

Romero-González, G. A., G. C. Fernández-Concha, R. L. Dresser, L. K. Magrath, and G. W. Argus. 2002. Orchidaceae. Pp. 490–499, in: Flora of North America Editorial Committee, eds. 1993+. Flora of North America North of Mexico. 21+ vols. New York and Oxford. Vol. 26.

Rominger, J. M. 1962. Taxonomy of *Setaria* (Gramineae) in North America. Illinois Biol. Monogr. 29: 1–132.

Rominger, J. M. 2003. *Setaria*. Pp. 539–558, in: Flora of North America Editorial Committee, eds. 1993+. Flora of North America North of Mexico. 21+ vols. New York and Oxford. Vol. 25.

Rosatti, T. J. 1987. The genera of Pontederiaceae in the southeastern United States. J. Arnold Arbor. 68: 35–71.

Rossbach, G. B. 1939. Aquatic Utricularias. Rhodora 41: 113–128.

Rothrock, P. E. 2021. Sedges of Indiana and the adjacent states. Vol. II: The *Carex* species. Indiana Academy of Science, Indianapolis, Ind.

Rothrock, P. E., and A. A. Reznicek. 2002. *Carex* sect. *Paniceae*. Pp. 426–431, in: Flora of North America Editorial Committee, eds. 1993+. Flora of North America North of Mexico. 21+ vols. New York and Oxford. Vol. 23.

Rousi, A. 1965. Biosystematic studies on the species aggregate *Potentilla anserina* L. Ann. Bot. Fenn. 2: 47–112.

Routledge, R., A. Graeff, and G. E. Crow. 2020. The discovery of *Utricularia ochroleuca* (Lentibulariaceae), yellowish-white bladderwort, in Michigan. The Great Lakes Botanist 59: 239–245.

Royen, P. van. 1954. The Podostemaceae of the New World. Part III. Act. Bot. Neerl. 3: 215–263.

Russell, N. H. 1965. Violets (*Viola*) of central and eastern United States: An introductory survey. Sida 2: 1–113.

Ryan, G. A. 1978. Native Trees and Shrubs of Newfoundland and Labrador. Parks Div., Dept. Tourism, Gov. Newfoundland and Labrador, St. John's, Nfld.

Saltonstall, K. 2002. Cryptic invasion by a non-native genotype of the common reed, *Phragmites australis*, into North America. Proc. Natl. Acad. U.S.A. 99: 2445–2449.

Saltonstall, K., and D. Hauber. 2007. Notes on *Phragmites australis* (Poaceae: Arundinoideae) in North America. J. Bot. Res. Inst. Texas 1: 385–388.

Saltonstall, K., P. M. Peterson, and R. J. Soreng. 2004. Recognition of *Phragmites australis* subsp. *americanus* (Poaceae: Arundinoideae) in North America. Sida 21: 683–692.

Sargent, C. S. 1890–1902. The Silva of North America. 14 vols. Houghton, Mifflin and Co., Boston, Mass.

Sargent, C. S. 1905. Manual of the Trees of North America (Exclusive of Mexico). Houghton Mifflin Co., Boston and New York.

Sarkar, N. M. 1958. Cytotaxonomic studies on *Rumex* section *Axillares*. Canad. J. Bot. 36: 947–996.

Sauer, J. 1955. Revision of the dioecious Amaranths. Madroño 13: 5–46.

Savage, A. D., and T. R. Mertens. 1968. A taxonomic study of genus *Polygonum*, sect. *Polygonum* (*Avicularia*), in Indiana and Wisconsin. Proc. Indiana Acad. 77: 357–369.

Sawyer, J. O., and G. L. Nesom. 2016. *Frangula*. Pp. 52–59, in: Flora of North America Editorial Committee, eds. 1993+. Flora of North America North of Mexico. 21+ vols. New York and Oxford. Vol. 12.

Schloesser, D. W. 1986. A Field Guide to Valuable Underwater Aquatic Plants. Great Lakes Fishery Lab., U.S. Fish Wildl. Serv., Ann Arbor, Mich.

Schneider, E. L., and J. D. Buchanan. 1980. Morphological studies of the Nymphaeaceae. XI. The floral biology of *Nelumbo pentapetala*. Amer. J. Bot. 67: 182–193.

Schneider, E. L., and T. Chaney. 1981. The floral biology of *Nymphaea odorata* (Nymphaeaceae). Southw. Nat. 26: 159–165.

Schneider, E. L., and J. M. Jeter. 1982. Morphological studies of the Nymphaeaceae. XII. The floral biology of *Cabomba caroliniana*. Amer. J. Bot. 69: 1410–1419.

Schneider, E. L., and L. A. Moore. 1977. Morphological studies of the Nymphaeaceae. VII. The floral biology of *Nuphar lutea* ssp. *macrophylla*. Brittonia 29: 88–99.

Schneider, H., A. R. Smith, and K. M. Pryer. 2009. Is morphology really at odds with molecules in estimating fern phylogeny? Syst. Bot. 34: 455–475.

Schnell, D. 2002. Carnivorous Plants of the United States and Canada. 2nd ed. Timber Press, Portland, Ore.

Schrenk, W. J. 1978. North American Platantheras: Evolution in the making. Amer. Orchid Soc. Bull. 47: 429–437.

Schuyler, A. E. 1962a. A new species of *Scirpus* in the northeastern United States. Rhodora 64: 43–49.

Schuyler, A. E. 1962b. Sporadic culm formation in *Scirpus longii*. Bartonia 32: 1–5.

Schuyler, A. E. 1964a. Notes on five species of *Scirpus* in eastern North America. Bartonia 33: 1–6.

Schuyler, A. E. 1964b. A biosystematic study of the *Scirpus cyperinus* complex. Proc. Acad. Nat. Sci. Philadelphia 115: 283–311.

Schuyler, A. E. 1966. The taxonomic delineation of *Scirpus lineatus* and *Scirpus pendulus*. Notul. Nat. Acad. Nat. Sci. Philadelphia 390: 1–3.

Schuyler, A. E. 1967. A taxonomic revision of the North American leafy species of *Scirpus*. Proc. Acad. Nat. Sci. Philadelphia 119: 295–323.

Schuyler, A. E. 1968. A new status for an eastern North American *Scirpus*. Rhodora 69: 198–202.

Schuyler, A. E. 1974a. Typification and application of the names *Scirpus americanus* Pers., *S. olneyi* Gray, and *S. pungens* Vahl. Rhodora 76: 51–52.

Schuyler, A. E. 1974b. *Scirpus cylindricus*: An ecologically restricted eastern North American tuberous bulrush. Bartonia 43: 29–37.

Scribailo, R. W., and M. S. Alix. 2021. Haloragaceae, *Myriophyllum*. Pp. 12–13, 14–29, in: Flora of North America Editorial Committee, eds. 1993+. Flora of North America North of Mexico. 21+ vols. New York and Oxford. Vol. 10.

Sculthorpe, C. D. 1967. The Biology of Aquatic Vascular Plants. Edward Arnold, London.

Semple, J. C. 2006. *Borrichia*. Pp. 129–130, in: Flora of North America Editorial Committee, eds. 1993+. Flora of North America North of Mexico. 21+ vols. New York and Oxford. Vol. 21.

Semple, J. C., and J. G. Chmielewski. 2006. *Doellingeria*. Pp. 43–46, in: Flora of North America Editorial Committee, eds. 1993+. Flora of North America North of Mexico. 21+ vols. New York and Oxford. Vol. 20.

Semple, J. C., and R. E. Cook. 2006. *Solidago*. Pp. 107–166, in: Flora of North America Editorial Committee, eds. 1993+. Flora of North America North of Mexico. 21+ vols. New York and Oxford. Vol. 20.

Sennikov, A. N. 2011. *Chamerion* or *Chamaenerion* (Onagraceae)? The old story in new words. Taxon 60: 1485–1488.

Setchell, W. A. 1929. Morphological and phenological notes on *Zostera marina* L. Univ. California Publ. Bot. 14: 389–452.

Setchell, W. A. 1933. A preliminary survey of the species of *Zostera*. Proc. Natl. Acad. U.S.A. 19: 810–817.

Shaffer-Fehre, M. 1991. The position of *Najas* within the Alismatidae (Monocotyledones) in the light of new evidence from seed coat structures in the Hydrocharitoideae (Hydrocharitaceae). Bot. J. Linn. Soc. 107: 189–209.

Shehbaz, I. A., and R. A. Price. 1998. Delimitation of the genus *Nasturtium* (Brassicaceae). Novon 8: 124–126.

Sherff, E. E. 1937. The genus *Bidens*. Part 1. Field Mus. Nat. Hist., Bot. Ser. 16: 1–346.

Sherff, E. E. 1965. Notes on varieties of *Bidens connata* and a hybrid with *B. cernua*. Rhodora 67: 59–62.

Sherman-Broyles, S. L. 1997. *Ulmus*. Pp. 369–375, in: Flora of North America Editorial Committee, eds. 1993+. Flora of North America North of Mexico. 21+ vols. New York and Oxford. Vol. 3.

Sherman-Broyles, S. L., W. T. Barker, and L. M. Schulz. 1997. *Celtis*. Pp. 376–379, in: Flora of North America Editorial Committee, eds. 1993+. Flora of North America North of Mexico. 21+ vols. New York and Oxford. Vol. 3.

Shetler, S. G. 1963. A checklist and key to the species of *Campanula* native or commonly naturalized in North America. Rhodora 65: 319–337.

Sheviak, C. J. 1974. An Introduction to the Ecology of the Illinois Orchidaceae. Illinois State Mus. Sci. Pap. 45.

Sheviak, C. J. 1982. Biosystematic Study of the *Spiranthes cernua* Complex. New York State Mus. Bull. No. 448.

Sheviak, C. J. 2002. *Cypripedium*, *Platanthera*. Pp. 499–507, 551–571, in: Flora of North America Editorial Committee, eds. 1993+. Flora of North America North of Mexico. 21+ vols. New York and Oxford. Vol. 26.

Sheviak, C. J., and P. M. Brown, 2002. *Spiranthes*. Pp. 530–545, in: Flora of North America Editorial Committee, eds. 1993+. Flora of North America North of Mexico. 21+ vols. New York and Oxford. Vol. 26.

Shinners, L. H. 1962. *Drosera* (Droseraceae) in the southeastern United States: An interim report. Sida 1: 53–59.

Shipunov, A. 2019. *Littorella*, *Plantago*. Pp. 280–281, 281–293, in: Flora of North America Editorial Committee, eds. 1993+. Flora of North America North of Mexico. 21+ vols. New York and Oxford. Vol. 17.

Sieren, D. J. 1970. A taxonomic revision of the genus *Euthamia* (Compositae). Ph.D. diss., Univ. of Illinois, Urbana-Champaign, Ill.

Sieren, D. J. 1981. The taxonomy of the genus *Euthamia*. Rhodora 83: 551–579.

Siripun, K. C., and E. E. Schilling. 2007. *Eupatorium*. Pp. 462–474, in: Flora of North America Editorial Committee, eds. 1993+. Flora of North America North of Mexico. 21+ vols. New York and Oxford. Vol. 21.

Smit, P. G. 1973. A revision of *Caltha* (Ranunculaceae). Blumea 21: 119–150.

Smith, A. R. 1993. Thelypteridaceae. Pp. 206–222, in: Flora of North America Editorial Committee, eds. 1993+. Flora of North America North of Mexico. 21+ vols. New York and Oxford. Vol. 2.

Smith, A. R. 2005. *Limonium*. Pp. 606–610, in: Flora of North America Editorial Committee, eds. 1993+. Flora of North America North of Mexico. 21+ vols. New York and Oxford. Vol. 5.

Smith, A. R., K. M. Pryer, E. Schuettpelz, P. Kokrall, H. Schneider, and P. G. Wolf. 2006. A classification for extant ferns. Taxon 55: 705–731.

Smith, J. G. 1895. Revision of the North American species of *Sagittaria* and *Lophotocarpus*. Ann. Rep. Missouri Bot. Gard. 6: 27–64.

Smith, R. H. 1955. Experimental control of Water Chestnut (*Trapa natans*) in New York State. New York Fish and Game J. 2: 173–193.

Smith, S. G. 1967. Experimental and natural hybrids in North American *Typha* (Typhaceae). Amer. Midl. Nat. 78: 257–287.

Smith, S. G. 1969. Natural hybridization in the *Scirpus lacustris* complex in north central United States. Pp. 175–200, in J. E. Gunckel, ed., Current Topics in Plant Science. Academic Press, New York.

Smith, S. G. 1986. The cattails (*Typha*): Interspecific ecological differences and problems of identification. Lake and Reserv. Managem. 2: 13–16.

Smith, S. G. 1987. *Typha*: Its taxonomy and the ecological significance of hybrids. Arch. Hydrobiol. 27: 129–138.

Smith, S. G. 1995. New combinations in North American *Schoenoplectus*, *Bolboschoenus*, *Isolepis*, and *Trichophorum* (Cyperaceae). Novon 5: 97–102.

Smith, S. G. 2000. Typhaceae. Pp. 278–285, in: Flora of North America Editorial Committee, eds. 1993+. Flora of

North America North of Mexico. 21+ vols. New York and Oxford. Vol. 22.

Smith, S. G. 2002. *Bolboschoenus*, *Schoenoplectus*, *Isolepis*. Pp. 37–44, 44–60, 137–140, in: Flora of North America Editorial Committee, eds. 1993+. Flora of North America North of Mexico. 21+ vols. New York and Oxford. Vol. 23.

Smith, S. G. 2012. *Sparganium eurycarpum* var. *greenei*, in Jepson Flora Project, eds. *Jepson eFlora*, https://ucjeps.berkeley.edu/eflora/eflora_display.php?tid=66594, accessed on August 21, 2020.

Smith, S. G., J. J. Bruhl, M. S. González-Elizondo, and F. J. Menapace. 2002. *Eleocharis*. Pp. 60–120, in: Flora of North America Editorial Committee, eds. 1993+. Flora of North America North of Mexico. 21+ vols. New York and Oxford. Vol. 23.

Smith, S. G., and T. Gregor. 2014. North American distribution of *Eleocharis mamillata* (Cyperaceae) and confusion with *E. macrostachya* and *E. palustris*. Rhodora 116: 163–186.

Snogerup, S. 1963. Studies in the genus *Juncus*. III. Observations on the diversity of chromosome numbers. Bot. Not. 116: 142–156.

Snogerup, S. 1980. *Juncus*. Pp. 102–111, in Totin et al. Flora Europaea, Vol. 1. Cambridge Univ. Press, Cambridge.

Snogerup, S., P. F. Zika, and J. Kirschner. 2002. Taxonomic and nomenclatural notes on *Juncus*. Preslia, Praha 74: 247–266.

Snyder, D. B. 1986. Rare New Jersey plant species rediscovered. Bartonia 52: 44–48.

Snyder, E., A. Francis, and S. J. Darbyshire. 2016. Biology of invasive alien plants in Canada. 13. *Stratiotes aloides* L. Canad. J. Plant. Sci. 96: 225–242.

Soderstrom, T. R., K. W. Hilu, C. S. Campbell, and M. E. Barkworth, eds. 1987. Grass Systematics and Evolution. Smithsonian Inst. Press, Washington, D.C.

Sohmer, S. H. 1975. The name of the American *Nelumbo*. Taxon 24: 491–493.

Sohmer, S. H. 1977. Aspects of the biology of *Nelumbo pentapetala* (Walter) Fernald, the American Lotus of the upper Mississippi. Trans. Wisconsin Acad. Sci., Arts and Letters 65: 258–273.

Sohmer, S. H., and D. F. Sefton. 1978. The reproductive biology of *Nelumbo pentapetala* (Nelumbonaceae) on the upper Mississippi River. II. The insects associated with the transfer of pollen. Brittonia 30: 355–364.

Soper, J. H., and E. G. Voss. 1964. Black Crowberry in the Lake Superior region. Michigan Bot. 3: 35–38.

Soreng, R. J. 2007. *Poa*. Pp. 486–601, in: Flora of North America Editorial Committee, eds. 1993+. Flora of North America North of Mexico. 21+ vols. New York and Oxford. Vol. 24.

Sorrie, B. A. 1990. *Myosurus minimus* (Ranunculaceae) in New England with notes on floral morphology. Rhodora 92: 103–104.

Sorrie, B. A., A. S. Weakley, and G. C. Tucker. 2009. *Gaylussacia*. Pp. 530–535, in: Flora of North America Editorial Committee, eds. 1993+. Flora of North America North of Mexico. 21+ vols. New York and Oxford. Vol. 8.

Southall, R. M., and J. W. Hardin. 1974. A taxonomic revision of *Kalmia* (Ericaceae). J. Elisha Mitchell Sci. Soc. 90: 1–23.

Spaulding, D. D., and T. W. Barger. 2016. Keys, distribution, and taxonomic notes for the Lobelias (*Lobelia*, Campanulaceae) of Alabama and adjacent states. Phytoneuron 76: 1–60.

Spongberg, S. A. 1978. The genera of Crassulaceae in the southeastern United States. J. Arnold Arbor. 59: 197–248.

Spooner, D. M. 1984. Infraspecific variation in *Gratiola viscidula* Pennell (Scrophulariaceae). Rhodora 86: 79–87.

Standley, L. A. 1983. A clarification of the status of *Carex crinita* and *C. gynandra* (Cyperaceae). Rhodora 85: 229–241.

Standley, L. A. 1987a. Anatomical and chromosomal studies of *Carex* section *Phacocystis* in eastern North America. Bot. Gaz. (Crawfordsville) 148: 507–518.

Standley, L. A. 1987b. Taxonomy of the *Carex lenticularis* complex in eastern North America. Canad. J. Bot. 65: 673–686.

Standley, L. A. 2002. *Carex* sect. *Vulpinae*, *Carex* sect. *Multiflorae*, *Carex* sect. *Glaucescentes*. Pp. 273–278, 281–285, 421–422, in: Flora of North America Editorial Committee, eds. 1993+. Flora of North America North of Mexico. 21+ vols. New York and Oxford. Vol. 23.

Standley, L. A., J. Cayouette, and L. Bruederle. 2002. *Carex* sect. *Phacocystis*. Pp. 379–401, in: Flora of North America Editorial Committee, eds. 1993+. Flora of North America North of Mexico. 21+ vols. New York and Oxford. Vol. 23.

Standley, P. C. 1917. Amaranthaceae. North American Flora 21: 95–169.

Staniforth, R. J., and K. A. Frego. 1980. Flowering Rush (*Butomus umbellatus*) in the Canadian prairies. Canad. Field-Nat. 94: 333–336.

Stephenson, S. N. 1984. The genus *Dichanthelium* (Poaceae) in Michigan. Michigan Bot. 23: 107–119.

Steward, A. N., L. R. Dennis, and H. M. Gilkey. 1960. Aquatic Plants of the Pacific Northwest with Vegetative Keys. Oregon State Monogr., Stud. Bot. No. 11.

St. John, H. 1920. The genus *Elodea* in New England. Rhodora 22: 17–29.

St. John, H. 1965. Monograph of the genus *Elodea*. Part 4. Summary. Rhodora 67: 1–35, 155–180.

Stolze, R. G. 1987. *Schizaea pusilla* discovered in Peru. Amer. Fern J. 77: 64–65.

Stoutamire, W. P. 1974. Relationships of the Purple-fringed Orchis *Platanthera psycodes* and *P. grandiflora*. Brittonia 26: 42–58.

Strausbaugh, P. H., and E. L. Core. 1952–1964. Flora of West Virginia. West Virginia Univ., Morgantown, W.Va.

Strother, J. L. 2006. *Iva*, *Eclipta*. Pp. 25–28, 128–129, in: Flora of North America Editorial Committee, eds. 1993+. Flora of North America North of Mexico. 21+ vols. New York and Oxford. Vol. 21.

Strother, J. L., and R. R. Weedon. 2006. *Bidens*. Pp. 205–218, in: Flora of North America Editorial Committee, eds. 1993+.

Flora of North America North of Mexico. 21+ vols. New York and Oxford. Vol. 21.

Stuckey, R. L. 1962. Characteristics and distribution of the spring cresses, *Cardamine bulbosa* and *C. douglasii*, in Michigan. Michigan Bot. 1: 27–34.

Stuckey, R. L. 1966a. Differences in habitats and associates of varieties of *Rorippa islandica* in the Douglas Lake region of Michigan. Michigan Bot. 5: 99–108.

Stuckey, R. L. 1966b. The distribution of *Rorippa sylvestris* (Cruciferae) in North America. Sida 2: 361–376.

Stuckey, R. L. 1968. Distributional history of *Butomus umbellatus* (Flowering Rush) in the western Lake Erie and Lake St. Clair region. Michigan Bot. 7: 134–142.

Stuckey, R. L. 1969. The introduction and spread of *Lycopus asper* (Western Water Horehound) in the western Lake Erie and Lake St. Clair region. Michigan Bot. 8: 111–120.

Stuckey, R. L. 1970. Distributional history of *Epilobium hirsutum* (Great Hairy Willow-herb) in North America. Rhodora 72: 164–181.

Stuckey, R. L. 1972. Taxonomy and distribution of the genus *Rorippa* (Cruciferae) in North America. Sida 4: 279–430.

Stuckey, R. L. 1974. The introduction and distribution of *Nymphoides peltatum* (Menyanthaceae) in North America. Bartonia 42: 14–23.

Stuckey, R. L. 1979. Distributional history of *Potamogeton crispus* (Curly Pondweed) in North America. Bartonia 46: 22–42.

Stuckey, R. L. 1980a. Distributional history of *Lythrum salicaria* (Purple Loosestrife) in North America. Bartonia 47: 3–20.

Stuckey, R. L. 1980b. The migration and establishment of *Juncus gerardii* (Juncaceae) in the interior of North America. Sida 8: 334–347.

Stuckey, R. L. 1981. Distributional history of *Juncus compressus* (Juncaceae) in North America. Canad. Field-Nat. 95: 167–171.

Stuckey, R. L. 1985. Distributional history of *Najas marina* (Spiny Naiad) in North America. Bartonia 51: 2–16.

Stuckey, R. L. 1993. Phytogeographical outline of aquatic and wetland angiosperms in continental eastern North America. Aquatic Bot. 44: 259–301.

Stuckey, R. L., and W. L. Phillips. 1970. Distributional history of *Lycopus europaeus* (European Water-horehound) in North America. Rhodora 72: 351–367.

Sulman, J. D., B. T. Drew, C. Drummond, E. Hayasaka, and K. J. Sytsma. 2013. Systematics, biogeography, and character evolution of *Sparganium* (Typhaceae): Diversification of a widespread, aquatic lineage. Amer. J. Bot. 100: 2023–2039.

Svenson, H. K. 1932a. Monographic studies in the genus *Eleocharis*—II. Rhodora 34: 192–203, 215–227.

Svenson, H. K. 1932b. *Callitriche stagnalis* in eastern United States. Rhodora 34: 37–38.

Svenson, H. K. 1944. The New World species of *Azolla*. Amer. Fern J. 34: 69–84.

Svenson, H. K. 1957. Cyperaceae. Tribe 2, Scirpeae. North American Flora 18: 505–556.

Svenson, H. K., and R. W. Pyle. 1979. The Flora of Cape Cod. An Annotated List of the Ferns and Flowering Plants of Barnstable County, Massachusetts. Cape Cod Museum of Natural History, Brewster, Mass.

Tans, W. 1987. Lentibulariaceae: The Bladderwort family in Wisconsin. Michigan Bot. 26: 52–62.

Taschereau, P. M. 1972. Taxonomy and distribution of *Atriplex* species in Nova Scotia. Canad. J. Bot. 50: 1571–1594.

Taylor, H. J. 1927. The history and distribution of yellow *Nelumbo*, Water Chinquapin or American Lotus. Proc. Iowa Acad. Sci. 34: 119–124.

Taylor, P. 1964. The genus *Utricularia* L. (Lentibulariaceae) in Africa (south of the Sahara) and Madagascar. Kew Bull. 18: 1–245.

Taylor, P. 1989. The genus *Utricularia*—a taxonomic monograph. Kew Bull. Add. Ser. 14: 1–724.

Taylor, W. C., N. T. Luebke, D. M. Britton, R. J. Hickey, and D. F. Brunton. 1993. Isoetaceae. Pp. 64–75, in: Flora of North America Editorial Committee, eds. 1993+. Flora of North America North of Mexico. 21+ vols. New York and Oxford. Vol. 2.

Taylor, W. C., N. T. Luebke, and M. B. Smith. 1985. Speciation and hybridisation in North American quillworts. Proc. Roy. Soc. Edinburgh 86B: 259–263.

Terrell, E. E. 2007. *Zizania*, *Zizaniopsis*. Pp. 47–51, 52–53, in: Flora of North America Editorial Committee, eds. 1993+. Flora of North America North of Mexico. 21+ vols. New York and Oxford. Vol. 24.

Terrell, E. E., P. M. Peterson, J. L. Reveal, and M. R. Duvall. 1997. Taxonomy of North American species of *Ziziana* (Poaceae). Sida 17: 533–549.

Thieret, J. W. 1971. Observations on some aquatic plants in northern Minnesota. Michigan Bot. 10: 117–124.

Thomas, W. W. 1984. The systematics of *Rhynchospora* section *Dichromena*. Mem. New York Bot. Gard. 37: 1–116.

Thompson, F. L., and C. G. Eckert. 2004. Trade-offs between sexual and clonal reproduction in an aquatic plant: Experimental manipulations vs. phenotypic correlations. J. Evol. Biol. 17: 581–592.

Thompson, S. A. 2000. Acoraceae, Araceae. Pp. 124–127, 128–142, in: Flora of North America Editorial Committee, eds. 1993+. Flora of North America North of Mexico. 21+ vols. New York and Oxford. Vol. 22.

Thor, G. 1988. The genus *Utricularia* in the Nordic countries with special emphasis on *U. stygia and U. ochroleuca*. Nord. J. Bot. 8: 219–395.

Thorne, R. F. 1992. Classification and geography of the flowering plants. Bot. Rev. 58: 225–348.

Tiner, R. W., Jr. 1987. A Field Guide to Coastal Wetland Plants of the Northeastern United States. Univ. of Massachusetts Press, Amherst, Mass.

Tippery, N. P., and D. L. Les. 2011. Phylogenetic relationships and morphological evolution of *Nymphoides* (Menyanthaceae). 36: 1101–1113.

Toivonen, H. 2002. *Carex* sect. *Dispermae*, *Carex* sect. *Glareosae*. Pp. 298, 311–321, in: Flora of North America Editorial Committee, eds. 1993+. Flora of North America North of Mexico. 21+ vols. New York and Oxford. Vol. 23.

Trock, D. K. 2006. *Packera*. Pp. 570–602, in: Flora of North America Editorial Committee, eds. 1993+. Flora of North America North of Mexico. 21+ vols. New York and Oxford. Vol. 20.

Trudeau, P. N. 1982. Nuisance Aquatic Plants and Aquatic Plant Management Programs in the United States. Vol. 3. Northeastern and North Central Region. U.S. Env. Protect. Agency and the Mitre Corp., McLean, Va.

Tucker, G. C. 1983. The taxonomy of *Cyperus* (Cyperaceae) in Costa Rica and Panama. Syst. Bot. Monogr. 2: 1–85.

Tucker, G. C. 1984. A revision of the genus *Kyllinga* Rottb. (Cyperaceae) in Mexico and Central America. Rhodora 86: 507–538.

Tucker, G. C. 1985. *Cyperus flavicomus*, the correct name for *Cyperus albomarginatus*. Rhodora 87: 539–541.

Tucker, G. C. 1986. The genera of Elatinaceae in the southeastern United States. J. Arnold Arbor. 67: 471–483.

Tucker, G. C. 1987. The genera of Cyperaceae in the southeastern United States. J. Arnold Arbor. 68: 361–445.

Tucker, G. C. 1992. *Scirpus* (Cyperaceae) in Connecticut. Newslett. Connecticut Bot. Soc. 20: 3–10.

Tucker, G. C. 2002. *Kyllinga*, *Lipocharpha*, *Cladium*. Pp. 193–194, 195–197, 240–242, in: Flora of North America Editorial Committee, eds. 1993+. Flora of North America North of Mexico. 21+ vols. New York and Oxford. Vol. 23.

Tucker, G. C. 2009. *Eubotrys*. Pp. 510–512, in: Flora of North America Editorial Committee, eds. 1993+. Flora of North America North of Mexico. 21+ vols. New York and Oxford. Vol. 8.

Tucker, G. C. 2016. Elatinaceae, *Bergia*, Nyssaceae. Pp. 438, 349, 458–461, in: Flora of North America Editorial Committee, eds. 1993+. Flora of North America North of Mexico. 21+ vols. New York and Oxford. Vol. 12.

Tucker, G. C., and S. C. Jones. 2009. Clethraceae. Pp. 364–366, in: Flora of North America Editorial Committee, eds. 1993+. Flora of North America North of Mexico. 21+ vols. New York and Oxford. Vol. 8.

Tucker, G. C., B. G. Marcks, and J. R. Carter. 2002. *Cyperus*. Pp. 141–191, in: Flora of North America Editorial Committee, eds. 1993+. Flora of North America North of Mexico. 21+ vols. New York and Oxford. Vol. 23.

Tucker, G. C., and N. G. Miller. 1990. Achene microstructure in *Eriophorum* (Cyperaceae): Taxonomic implications and paleobotanical applications. Bull. Torrey Bot. Club 117: 266–283.

Ugent, D. 1962. Preliminary reports on the flora of Wisconsin. No. 47. The orders Thymelaeales, Myrtales, and Cactales. Trans. Wisconsin Acad. Sci., Arts and Letters 51: 83–134.

Urban, E. K., and H. H. Iltis. 1957. Preliminary reports on the flora of Wisconsin. No. 38. Rubiaceae—Madder family. Trans. Wisconsin Acad. Sci., Arts and Letters 46: 91–104.

Urbanska-Worythiewicz, K. 1975. Cytological variation within *Lemna* L. Aquatic Bot. 1: 377–394.

U.S. Army Corps of Engineers. 2016. National Wetland Plant List. Website (http://wetland-plants.usace.army.mil/nwpl_static/v33/home/home.html), accessed August 2020.

USDA, NRCS. 2022. The PLANTS Database. Website. National Plant Data Team, Greensboro, N.C. http://plants.usda.gov, accessed regularly, 2019–2022.

Uttal, L. J. 1956. Notes on *Utricularia biflora* and *U. fibrosa*. Rhodora 58: 41–43.

Uttal, L. J. 1984. Nomenclatural changes, lectotypification, and comments in *Aronia* Medikus (Rosaceae). Sida 10: 199–202.

Uttal, L. J., and D. M. Porter. 1988. The correct name for Elliot's Goldenrod. Rhodora 90: 157–168.

Valdespino, I. A. 1993. Selaginellaceae. Pp. 38–63, in: Flora of North America Editorial Committee, eds. 1993+. Flora of North America North of Mexico. 21+ vols. New York and Oxford. Vol. 2.

Vander Kloet, S. P. 1980. The taxonomy of the Highbush Blueberry, *Vaccinium corymbosum*. Canad. J. Bot. 58: 1187–1201.

Vander Kloet, S. P. 1983. The taxonomy of *Vaccinium* sect. *Oxycoccus*. Rhodora 85: 1–43.

Vander Kloet, S. P. 2009. *Vaccinium*. Pp. 515–530, in: Flora of North America Editorial Committee, eds. 1993+. Flora of North America North of Mexico. 21+ vols. New York and Oxford. Vol. 8.

Väre, H. 2007. Typification of names published by the Finnish botanist Fredrik Nylander. Ann. Bot. Fennici 44: 465–480.

Viereck, L. A., and E. L. Little, Jr. 1972. Alaska Trees and Shrubs. U.S. Gov. Print. Off., Washington, D.C.

Voss, E. G. 1966. Nomenclatural notes on Monocots. Rhodora 68: 435–463.

Voss, E. G. 1967. A vegetative key to the genera of submersed and floating aquatic vascular plants of Michigan. Michigan Bot. 6: 35–50.

Voss, E. G. 1972. Michigan Flora. Part I. Gymnosperms and Monocots. Cranbrook Inst. Sci. Bull. 55. Bloomfield Hills, Mich.

Voss, E. G. 1985. Michigan Flora. Part II Dicots (Saururaceae-Cornaceae). Cranbrook Inst. Sci. Bull. 59. Bloomfield Hills, Mich.

Voss, E. G. 1996. Michigan Flora. Part III. Dicots (Pyrolaceae-Compositae). Cranbrook Inst. Sci. Bull. 61. Bloomfield Hills, Mich.

Voss, E. G., and A. A. Reznicek. 2012. Field Manual of Michigan Flora. Univ. of Michigan Press, Ann Arbor, Mich.

Wagner, W. H., Jr. 1993. Schizaeaceae. Pp. 112–113, in: Flora of North America Editorial Committee, eds. 1993+. Flora of North America North of Mexico. 21+ vols. New York and Oxford. Vol. 2.

Wagner, W. H., Jr., and J. M. Beitel. 1992. Generic classification of modern North American Lycopodiaceae. Ann. Missouri Bot. Gard. 79: 676–686.

Wagner, W. H., Jr., and J. M. Beitel. 1993. Lycopodiaceae. Pp. 18–37, in: Flora of North America Editorial Committee, eds. 1993+. Flora of North America North of Mexico. 21+ vols. New York and Oxford. Vol. 2.

Wagner, W. L., P. C. Hoch, and P. H. Raven. 2007. A revised classification of the Onagraceae. Syst. Bot. Monogr. 83: 1–240.

Wallenstein, A., and L. S. Albert. 1963. Plant morphology: Its control in *Proserpinaca* by photoperiod, temperature and gibberellic acid. Science 140: 998–1000.

Walsh, S. 1986. Plants at their geographical range limits in Quetico Provincial Park, northwestern Ontario. Plant Press 4: 76–78.

Wang, X-Q., and J-H. Ran. 2014. Evolution and biogeography of gymnosperms. Molecular Phylogenetics and Evolution 75: 24–40.

Ward, D. B. 1977. *Nelumbo lutea*, the correct name for the American Lotus. Taxon 26: 227–234.

Warwick, S. I., and S. G. Aiken. 1986. Electrophoretic evidence for the recognition of two species in annual Wild Rice (*Zizania*, Poaceae). Syst. Bot. 11: 464–473.

Waterway, M. J. 2002. *Carex* sect. *Hymenochlaenae*. Pp. 461–475, in: Flora of North America Editorial Committee, eds. 1993+. Flora of North America North of Mexico. 21+ vols. New York and Oxford. Vol. 23.

Watson, F. D., and J. E. Eckenwalder. 1993. Cupressaceae. Pp. 399–422, in: Flora of North America Editorial Committee, eds. 1993+. Flora of North America North of Mexico. 21+ vols. New York and Oxford. Vol. 2.

Watson, L. E. 2006. *Cotula*. Pp. 543–544, in: Flora of North America Editorial Committee, eds. 1993+. Flora of North America North of Mexico. 21+ vols. New York and Oxford. Vol. 19.

Webber, J. W., and P. W. Ball. 1980. Introgression in Canadian populations of *Lycopus americanus* Muhl. and *L. europaeus* L. (Labiatae). Rhodora 82: 281–304.

Weber, J. A., and L. D. Noodén. 1974. Turion formation and germination in *Myriophyllum verticillatum*: Phenology and its interpretation. Michigan Bot. 13: 151–158.

Webster, R. D. 1988. Genera of the North American Paniceae (Poaceae: Panicoideae). Syst. Bot. 13: 576–609.

Webster, R. D. 2003. *Saccharum*. Pp. 609–616, in: Flora of North America Editorial Committee, eds. 1993+. Flora of North America North of Mexico. 21+ vols. New York and Oxford. Vol. 25.

Wehrmeister, J. R. 1978. An ecological life history of the pondweed *Potamogeton crispus* L. in North America. M.S. thesis, Ohio State Univ., Columbus, Ohio. (Reprinted, CLEAR Tech. Rep. No. 99.)

Wehrmeister, J. R., and R. L. Stuckey. 1992. Life history of *Potamogeton crispus*. Michigan Bot. 31: 3–16.

Welsh, S. L. 2003. *Atriplex*. Pp. 322–381, in: Flora of North America Editorial Committee, eds. 1993+. Flora of North America North of Mexico. 21+ vols. New York and Oxford. Vol. 4.

Wentz, W. A., and R. L. Stuckey. 1971. The changing distribution of the genus *Najas* (Najadaceae) in Ohio. Ohio J. Sci. 71: 292–302.

Westermeier, A. S., R. Sachse, S. Poppinga, P. Vögele, L. Adamec, T. Speck, and M. Bischoff. 2018. How the carnivorous waterwheel plant (*Aldrovanda vesiculosa*) snaps. Proc. R. Soc. B. 285: 20180012 (available at http://dx.doi.org/10.1098/rspb.2018.0012).

Whiting, R. E., and P. M. Catling. 1986. Orchids of Ontario. CanaColl Foundation, Ottawa.

Whittemore, A. T. 1997. *Ranunculus*, *Myosurus*. Pp. 88–135, 135–138, in: Flora of North America Editorial Committee, eds. 1993+. Flora of North America North of Mexico. 21+ vols. New York and Oxford. Vol. 3.

Whittemore, A. T., and B. D. Parfitt. 1997. Ranunculaceae. Pp. 85–87, in: Flora of North America Editorial Committee, eds. 1993+. Flora of North America North of Mexico. 21+ vols. New York and Oxford. Vol. 3.

Whittemore, A. T., and A. E. Schuyler. 2002. *Scirpus*. Pp. 8–21, in: Flora of North America Editorial Committee, eds. 1993+. Flora of North America North of Mexico. 21+ vols. New York and Oxford. Vol. 23.

Wiegand, K. M. 1921. The genus *Echinochloa* in North America. Rhodora 23: 49–65.

Wiegleb, G., A. A. Bobrov, and J. Zalewska-Gałosz. 2017. A taxonomic account of *Ranunculus* sect. *Batrachium* (Ranunculaceae). Phytotaxa 319: 1–55.

Wiegleb, G., and Z. Kaplan. 1988. An account of the species of *Potamogeton* L. (Potamogetonaceae). Folia Geobot. 33: 241–316.

Wiersema, J. H. 1988. Reproductive biology of *Nymphaea* (Nymphaeaceae). Ann. Missouri Bot. Gard. 75: 795–804.

Wiersema, J. H. 1996. *Nymphaea tetragona* and *Nymphaea leibergii* (Nymphaeaceae): Two species of diminutive water-lilies in North America. Brittonia 48: 520–531.

Wiersema, J. H. 1997. Nelumbonaceae, Cabombaceae. Pp. 64–65, 78–80, in: Flora of North America Editorial Committee, eds. 1993+. Flora of North America North of Mexico. 21+ vols. New York and Oxford. Vol. 3.

Wiersema, J. H., and R. R. Haynes. 1983. Aquatic and marsh plants of Alabama. III. Magnoliidae. Castanea 48: 99–108.

Wiersema, J. H., and C. B. Hellquist. 1997. Nymphaeaceae. Pp. 66–77, in: Flora of North America Editorial Committee, eds. 1993+. Flora of North America North of Mexico. 21+ vols. New York and Oxford. Vol. 3.

Wilbur, R. L. 1955. A revision of the North American genus *Sabatia* (Gentianaceae). Rhodora 57: 1–33, 43–71. 78–104.

Williams, G. R. 1970. Investigations in the white water-lilies (*Nymphaea*) of Michigan. Michigan Bot. 9: 72–86.

Wilson, J. S. 1965. Variation of the three taxonomic complexes of the genus *Cornus* in eastern United States. Trans. Kansas Acad. Sci. 67: 747–817.

Wilson, K. A. 1960. Genera of Arales in the southeast United States. J. Arnold Arbor. 41: 47–72.

Wilson, K. L. 2016. Report of the General Committee: 15. Taxon 65: 1150–1151.

Windler, D. R., B. E. Wofford, and M. W. Bierner. 1976. Evidence of natural hybridization between *Mimulus ringens* and *Mimulus alatus* (Scrophulariaceae). Rhodora 78: 641–649.

Winne, W. T. 1935. A Study of the Water Chestnut, *Trapa natans*, with a view of its control in the Mohawk River. M.S. thesis, Cornell Univ., Ithaca, N.Y.

Wipff, J. K. 2003. *Sacciolepis*. Pp. 404–405, in: Flora of North America Editorial Committee, eds. 1993+. Flora of North America North of Mexico. 21+ vols. New York and Oxford. Vol. 25.

Wohler, J. R., I. M. Wohler, and R. T. Hartman. 1965. The occurrence of *Spirodela oligorrhiza* in western Pennsylvania. Castanea 30: 230–231.

Wolf, S. J., and J. McNeill. 1986. Synopsis and achene morphology of *Polygonum* section *Polygonum* (Polygonaceae) in Canada. Rhodora 88: 457–479.

Wood, C. E. 1959. The genera of the Nymphaeaceae and Ceratophyllaceae in the southeastern United States. J. Arnold Arbor. 40: 94–112.

Wood, C. E. 1960. The genera of Sarraceniaceae and Droseraceae of the southeastern United States. J. Arnold Arbor. 41: 160–163.

Woods, K., K. W. Hilu, J. H. Wiersema, and T. Borsch. 2005. Pattern of variation and systematics of *Nymphaea odorata*: I. Evidence from morphology and Inter-Simple Sequence Repeats (ISSRs). Syst. Bot. 30: 471–480.

Wooten, J. W. 1970. Experimental investigations of the *Sagittaria graminea* complex: Transplant studies and genecology. J. Ecol. 58: 233–242.

Wooten, J. W. 1971. The monoecious and dioecious conditions in *Sagittaria latifolia* L. (Alismataceae). Evolution 25: 549–553.

Wooten, J. W. 1973a. Edaphic factors in species and ecotype differentiation of *Sagittaria*. J. Ecol. 61: 151–156.

Wooten, J. W. 1973b. Taxonomy of seven species of *Sagittaria* from eastern North America. Brittonia 25: 64–74.

Wurdack, J. J., and R. Kral. 1982. The genera of Melastomataceae in the southeastern United States. J. Arnold Arbor. 63: 429–439.

Wylie, R. B. 1917. The pollination of *Vallisneria spiralis*. Bot. Gaz. (Crawfordsville) 63: 135–145.

Wynne, F. E. 1944. *Drosera* in eastern North America. Bull. Torrey Bot. Club 71: 166–174.

Yeo, R. R. 1964. Life history of Common Cattail. Weeds 12: 284–288.

Yeo, R. R. 1965. Life history of Sago Pondweed. Weeds 13: 314–321.

Yeo, R. R., R. H. Falk, and J. R. Thurston. 1984. The morphology of hydrilla (*Hydrilla verticillata* (L. f.) Royle). J. Aquatic Pl. Managem. 22: 1–16.

Zardini, E. M., H. Gu, and P. H. Raven. 1991. On the separation of two species within the *Ludwigia uruguayensis* complex (Onagraceae). Syst. Bot. 16: 242–244.

Zenkert, C. A. 1960. The Old-World flowering rush: An attractive and aggressive immigrant. Science on the March, Mag. Buffalo Mus. Sci. 40: 71–75.

Zhengyi, W., and P. Raven. 2022. Flora of China Illustrations, Vol. 8. Brassicaceae through Saxifragaceae. Missouri Botanical Garden Press, St. Louis, Mo.

Zika, P. F. 2013. A synopsis of the *Juncus hesperius* group (Juncaceae, Juncotypus) and their hybrids in western North America. Brittonia 128–141.

Zika, P. F. 2015. *Juncus*, in Jepson Flora Project, eds. Jepson eFlora, Revision 3. https://ucjeps.berkeley.edu/eflora/eflora_display.php?tid=9148, accessed on August 20, 2020.

Zomlefer, W. B. 1997. The genera of Nartheciaceae in the southeastern United States. Harvard Pap. Bot. 2: 195–211.

Index of Common Names

Index of Scientific Names

Scientific names of species and genera in normal font are accepted names for this manual. Family names are in all capital letters. Names in italics are synonyms. Page numbers in bold represent illustrations.